ALGEBRA'S COMMON GRAPHS

P9-EMG-583

Identity Function

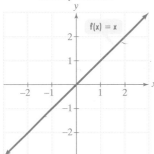

$f(x) = x$

Standard Quadratic Function

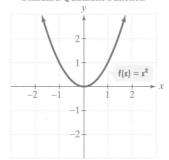

$f(x) = x^2$

Standard Cubic Function

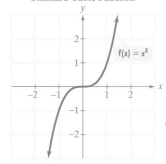

$f(x) = x^3$

Absolute Value Function

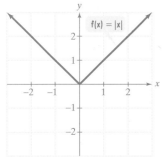

$f(x) = |x|$

Square Root Function

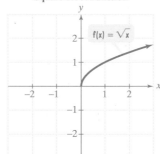

$f(x) = \sqrt{x}$

Greatest Integer Function

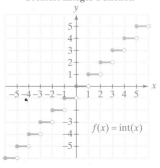

$f(x) = \text{int}(x)$

TRANSFORMATIONS

In each case, c represents a positive real number.

Function		Draw the graph of f and:
Vertical translations	$\begin{cases} y = f(x) + c \\ y = f(x) - c \end{cases}$	Shift f upward c units. Shift f downward c units.
Horizontal translations	$\begin{cases} y = f(x - c) \\ y = f(x + c) \end{cases}$	Shift f to the right c units. Shift f to the left c units.
Reflections	$\begin{cases} y = -f(x) \\ y = f(-x) \end{cases}$	Reflect f about the x-axis. Reflect f about the y-axis.
Stretching or Shrinking	$\begin{cases} y = cf(x); c > 1 \\ y = cf(x); 0 < x < 1 \end{cases}$	Stretch f, multiplying each of its y-values by c. Shrink f, multiplying each of its y-values by c.

DISTANCE AND MIDPOINT FORMULAS

1. The distance from (x_1, y_1) to (x_2, y_2) is
$$\sqrt{(x_2 - x_1)^2 + (y_2 - y_1)^2}.$$

2. The midpoint of the line segment with endpoints (x_1, y_1) and (x_2, y_2) is
$$\left(\frac{x_1 + x_2}{2}, \frac{y_1 + y_2}{2} \right).$$

QUADRATIC FORMULA

The solutions to $ax^2 + bx + c = 0$ with $a \neq 0$ are
$$x = \frac{-b \pm \sqrt{b^2 - 4ac}}{2a}.$$

FUNCTIONS

1. Linear Function: $f(x) = mx + b$
Graph is a line with slope m and y-intercept b.

2. Quadratic Function: $f(x) = ax^2 + bx + c, a \neq 0$

Graph is a parabola with vertex at $x = -\dfrac{b}{2a}$.

Quadratic Function: $f(x) = a(x - h)^2 + k$
In this form, the parabola's vertex is (h, k).

3. nth-Degree Polynomial Function: $f(x) =$
$a_n x^n + a_{n-1} x^{n-1} + a_{n-2} x^{n-2} + \cdots + a_1 x + a_0, a_n \neq 0$
For n odd and $a_n > 0$, graph falls to the left and rises to the right.
For n odd and $a_n < 0$, graph rises to the left and falls to the right.
For n even and $a_n > 0$, graph rises to the left and to the right.
For n even and $a_n < 0$, graph falls to the left and to the right.

4. Rational Function: $f(x) = \dfrac{p(x)}{q(x)}$, $p(x)$ and $q(x)$ are polynomials, $q(x) \neq 0$

5. Exponential Function: $f(x) = b^x, b > 0, b \neq 1$
Graphs:

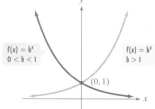

6. Logarithmic Function: $f(x) = \log_b x, b > 0, b \neq 1$
$y = \log_b x$ is equivalent to $x = b^y$.
Graph:

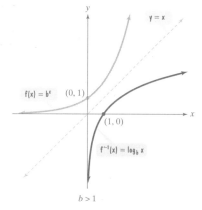

PROPERTIES OF LOGARITHMS

1. $\log_b(MN) = \log_b M + \log_b N$

2. $\log_b\left(\dfrac{M}{N}\right) = \log_b M - \log_b N$

3. $\log_b M^p = p \log_b M$

4. $\log_b M = \dfrac{\log_a M}{\log_a b} = \dfrac{\ln M}{\ln b} = \dfrac{\log M}{\log b}$

5. $\log_b b^x = x; \quad \ln e^x = x$

6. $b^{\log_b x} = x; \quad e^{\ln x} = x$

INVERSE OF A 2 × 2 MATRIX

If $A = \begin{bmatrix} a & b \\ c & d \end{bmatrix}$, then $A^{-1} = \dfrac{1}{ad - bc} \begin{bmatrix} d & -b \\ -c & a \end{bmatrix}$, where $ad - bc \neq 0$.

CRAMER'S RULE

If

$$a_{11}x_1 + a_{12}x_2 + a_{13}x_3 + \cdots + a_{1n}x_n = b_1$$
$$a_{21}x_1 + a_{22}x_2 + a_{23}x_3 + \cdots + a_{2n}x_n = b_2$$
$$a_{31}x_1 + a_{32}x_2 + a_{33}x_3 + \cdots + a_{3n}x_n = b_3$$
$$\vdots$$
$$a_{n1}x_1 + a_{n2}x_2 + a_{n3}x_3 + \cdots + a_{nn}x_n = b_n$$

then $x_i = \dfrac{D_i}{D}, D \neq 0$.

D: determinant of the system's coefficients

D_i: determinant in which coefficients of x_i are replaced by $b_1, b_2, b_3, \ldots, b_n$.

CONIC SECTIONS

Circle

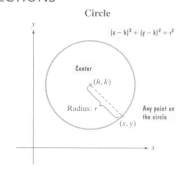

Ellipse

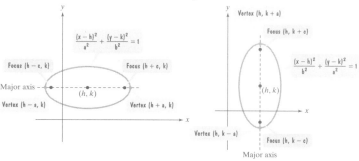

Hyperbola

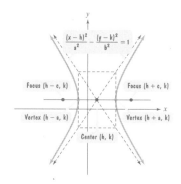
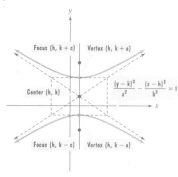

(continued on inside back cover)

Precalculus Essentials

Precalculus Essentials

Robert Blitzer
Miami-Dade Community College

PEARSON
Prentice Hall

Upper Saddle River, NJ 07458

Library of Congress Cataloging-in-Publication Data

Blitzer, Robert.
   Precalculus essentials / Robert Blitzer
     p. cm.
   Includes index.
   ISBN 0-13-109044-5
     1. Algebra.   I. Title.

QA154.3.B587 2004
512—dc22   2003060680

Senior Acquisitions Editor: *Eric Frank*
Project Manager: *Dawn Murrin*
Editor-in-Chief: *Sally Yagan*
Vice President/Director of Production and Manufacturing: *David W. Riccardi*
Executive Managing Editor: *Kathleen Schiaparelli*
Senior Managing Editor: *Linda Mihatov Behrens*
Production Management: *Elm Street Publishing Services, Inc./Barbara Mack*
Production Assistant: *Nancy Bauer*
Assistant Managing Editor, Math Media Production: *John Matthews*
Media Production Editor: *Donna Crilly*
Assistant Manufacturing Manager/Buyer: *Michael Bell*
Manufacturing Manager: *Trudy Pisciotti*
Senior Marketing Manager: *Halee Dinsey*
Marketing Assistant: *Rachel Beckman*
Editorial Assistant/Supplements Editor: *Tina Magrabi*
Art Director: *Jonathan Boylan*
Cover Designers: *Maureen Eide/Geoffrey Cassar*
Art Editor: *Thomas Benfatti*
Creative Director: *Carole Anson*
Director of Creative Services: *Paul Belfanti*
Director, Image Resource Center: *Melinda Reo*
Manager, Rights and Permissions: *Zina Arabia*
Interior Image Specialist: *Beth Boyd-Brenzel*
Cover Image Specialist: *Karen Sanatar*
Image Permission Coordinator/Photo Researcher: *Elaine Soares*
Cover Art: *Jalapeño pepper, photographed by John E. Kelly, copyright Getty Images*
Art Studio: *Artworks:*
  Managing Editor, AV Production & Management: *Patty Burns*
  Production Manager: *Ronda Whitson*
  Manager, Production Technologies: *Matt Haas*
  Project Coordinator: *Jessica Einsig*
  Illustrators: *Kathryn Anderson, Mark Landis*
  Quality Assurance: *Pamela Taylor, Timothy Nguyen*
Formatting Manager: *Jim Sullivan*
Assistant Manager, Formatting: *Allyson Graesser*
  Formatters: *Clara Bartunek, Julita Nazario, Judith R. Wilkens, Vicki L. Croghan*

© 2004 Pearson Education, Inc.
Pearson Prentice Hall
Pearson Education, Inc.
Upper Saddle River, New Jersey 07458

All rights reserved. No part of this book may be reproduced, in any form
or by any other means, without permission in writing from the publisher.

Preason Prentice Hall® is a trademark of Pearson Education, Inc.

Printed in the United States of America

10 9 8 7 6 5 4 3 2 1

ISBN 0-13-109044-5

Pearson Education LTD., *London*
Pearson Education Australia PTY, Limited, *Sydney*
Pearson Education Singapore, Pte. Ltd
Pearson Education North Asia Ltd, *Hong Kong*
Pearson Education Canada, Ltd., *Toronto*
Pearson Educación de Mexico, S.A. de C.V.
Pearson Education -- Japan, *Tokyo*
Pearson Education Malaysia, Pte. Ltd

Contents

Chapter 2

Chapter 3

Chapter 4

Chapter 5

Chapter 6

Shaded chapters available in Blitzer, Precalculus, 2nd ed.

Chapter 7

Chapter 8

Chapter 9

Chapter 10

Chapter 11

Appendix

Preface

Precalculus Essentials is designed and written to help students make the transition from intermediate algebra into calculus. The book has three fundamental goals:

1. To help students acquire a solid foundation in algebra and trigonometry, preparing them for other courses such as calculus, business calculus, and finite mathematics.
2. To show students how algebra and trigonometry can model and solve authentic real-world problems.
3. To enable students to develop problem-solving skills, while fostering critical thinking, within an interesting setting.

One major obstacle in the way of achieving these goals is the fact that very few students actually read their textbook. This has been a regular source of frustration for my colleagues and me in the classroom. Anectodal evidence gathered over years highlights two basic reasons that students do not take advantage of their textbook:

- "I'll never use this information."
- "I can't follow the explanations."

As a result, I've written every page of this book with the intent of overcoming these two objections. See the book's Walkthrough, beginning on page xvi, for the ideas and tools I've used to do so.

A Note from the Publisher on Essentials

By publishing this concise version of Robert Blitzer's *Precalculus*, 2nd ed., we provide a lighter, less expensive alternative to the traditional textbook on the subject. It is important to note that *Precalculus Essentials* differs from its predecessor *only in terms of length*, i.e., Chapters 7–11 have been eliminated. The omitted chapters and their sections are highlighted for your reference in the Table of Contents. *Precalculus Essentials* is ideal for courses that do not cover the entire scope of the traditional book. If the complete version of this textbook would better suit your needs, simply contact your Prentice Hall representative for assistance.

A Note on Technology

Technology, and specifically the use of a graphing utility, is covered thoroughly, although its coverage by an instructor is optional. If you require the use of a graphing utility in the course, you will find support for this approach, particularly in the wide selection of clearly designated technology exercises in each exercise set. If you wish to minimize or eliminate the discussion or use of a graphing utility, the book is written to enable you to do so. Regardless of the role technology plays in your course, the technology boxes with TI-83 screens that appear throughout the book should allow your students to understand what graphing utilities can do, enabling them to visualize, verify, or explore what they have already graphed or manipulated by hand. The book's technology coverage is intended to reinforce, but never replace, algebraic solutions.

A Note on Precalculus Students

Precalculus is not simply a condensed version of my **Algebra and Trigonometry** book. Precalculus students are different from algebra and trigonometry students, and this text reflects those differences. Here are a few specific examples:

- Functions, the core of any precalculus course, are introduced in Chapter 1. Linear equations, quadratic equations, and linear inequalities are reviewed in the prerequisites chapter. By contrast, my **Algebra and Trigonometry** book covers equations and inequalities in Chapter 1 before turning to functions in Chapter 2.

- Modeling from verbal conditions is covered within the context of functions rather than linear equations. Section 1.9 (Modeling with Functions) is not included in **Algebra and Trigonometry**.

- Chapter 2 presents polynomial and rational inequalities within the context of polynomial and rational functions. Information about these functions' graphs is used to understand the solution of the inequalities.

- Chapter 2 introduces an optimization strategy for solving word problems using quadratic functions. Students will be able to use this strategy in calculus when solving similar problems with the derivative.

- Chapter 4 develops trigonometry from the perspective of the unit circle.

- Liberal arts applications are often replaced by more scientific applications. For example, Newton's Law of Cooling is developed in Chapter 3.

Supplements

Student Supplements	Instructor Supplements
Student Solutions Manual Fully worked solutions to odd-numbered exercises. 0-13-142235-9	**Instructor's Solutions Manual** Fully worked solutions to all exercises in the text. 0-13-101923-6
CD Lecture Series Four CD-ROMs contain 20 minutes of lectures and tutorials per textbook section; objectives are reviewed and key examples from the textbook are worked out. These are available for students to purchase alone, or in a package with their book. 0-13-140130-0	**Instructor's Edition with Instructor's Resource CD** • Provides answers to all exercises in the back of the text. • Includes Instructor's Resource CD containing TestGen, Instructor's Solutions Manual, Additional Chapter Projects, and Test Item File. ISM and TIF are passcode-protected. 0-13-109045-3
VHS Lecture Series Same content as CD Lecture Series in VHS format. Instructors can order the VHS videos and make them available to students in the library or media lab. 0-13-101926-0	**Test Item File** A printed test bank derived from TestGen. 0-13-101927-9
PH Tutor Center Provides students with help when they need it most—while they're doing their homework. PH math tutors (trained college instructors) provide help via a toll-free phone number, fax, and email. 0-13-064604-0	**TestGen** Test-generating software enabling instructors to create tests from the text section objectives. Many questions are algorithmically generated, allowing for unlimited versions of any test. Instructors may also edit problems or create their own. 0-13-101925-2

MathPak 5.0

- Features MathPro 5.0, an online, customizable tutorial and assessment software package, integrated with the text at the learning objective level. The easy-to-use gradebook enables instructors to track and evaluate student performance on tutorial work, quizzes, and tests. An optional Diagnostics module allows students to identify weaknesses in prerequisite material, and to have a customized set of tutorials provided to them for additional practice on those identified weaknesses.

- Includes access to a website containing the Student Solutions Manual, Online Graphing Calculator Help, PowerPoint slides used in the lecture videos, and quizzes and tests allowing students to assess their skills and comprehension of the material.

Student Version: 0-13-140796-1
Instructor Version: 0-13-101931-7

PH Grade Assist

This online homework and assessment program enables instructors to create customized homework and tests by choosing problems from the text or algorithmic versions of those problems or creating their own problems. PHGA supports multiple question types including free response. The built-in parser is sophisticated, grading student responses while recognizing algebraic, numeric, and unit equivalents. The gradebook also allows instructors to easily track student performance.

Student Version: 0-13-102000-5
Instructor Version: 0-13-102002-1

Companion Website

Free website to all text users provides quizzes, chapter tests, PowerPoint slides available for download, and Online Graphing Calculator Help.

URL: www.prenhall.com/blitzer

Acknowledgments

I wish to express my appreciation to all of the reviewers of my precalculus series for their helpful criticisms and suggestions, frequently transmitted with wit, humor, and intelligence. In particular, I would like to thank the following for reviewing **College Algebra**, **Algebra and Trigonometry**, and **Precalculus.**

Reviewers of Current and Previous Editions

Timothy Beaver, *Isothermal Community College*
Bill Burgin, *Gaston College*
Jimmy Chang, *St. Petersburg College*
Donna Densmore, *Bossier Parish Community College*
Disa Enegren, *Rose State College*
Nancy Fisher, *University of Alabama*
Jeremy Haefner, *University of Colorado*
Joyce Hague, *University of Wisconsin at River Falls*
Mary Leesburg, *Manatee Community College*
Mike Hall, *Univeristy of Mississippi*
Christopher N. Hay-Jahans, *University of South Dakota*
Alexander Levichev, *Boston University*
Zongzhu Lin, *Kansas State University*
Benjamin Marlin, *Northwestern Oklahoma State University*
Marilyn Massey, *Collin County Community College*
David Platt, *Front Range Community College*
Janice Rech, *University of Nebraska at Omaha*
Judith Salmon, *Fitchburg State College*
Cynthia Schultz, *Illinois Valley Community College*
Chris Stump, *Bethel College*
Pamela Trim, *Southwest Tennessee Community College*
Chris Turner, *Arkansas State University*
Philip Van Veldhuizen, *University of Nevada at Reno*
Tracy Wienckowski, *Univesity of Buffalo*

Kayoko Yates Barnhill, *Clark College*
Lloyd Best, *Pacific Union College*
Diana Colt, *University of Minnesota-Duluth*
Yvelyne Germain-McCarthy, *University of New Orleans*
Cynthia Glickman, *Community College of Southern Nevada*
Sudhir Kumar Goel, *Valdosta State University*
Donald Gordon, *Manatee Community College*
David L. Gross, *University of Connecticut*
Joel K. Haack, *University of Northern Iowa*
Christine Heinecke Lehmann, *Purdue University North Central*
Celeste Hernandez, *Richland College*
Winfield A. Ihlow, *SUNY College at Oswego*
Nancy Raye Johnson, *Manatee Community College*
James Miller, *West Virginia University*
Debra A. Pharo, *Northwestern Michigan College*
Gloria Phoenix, *North Carolina Agricultural and Technical State University*
Juha Pohjanpelto, *Oregon State University*
Richard E. Van Lommel, *California State University-Sacramento*
Dan Van Peursem, *University of South Dakota*
David White, *The Victoria College*

Additional acknowledgments are extended to Pat Foard, for the Herculean task of preparing the solutions manuals; Teri Lovelace and the team at LaurelTech for preparing the answer section and serving as accuracy checker; Jim Sullivan, Allyson Graesser, and the Prentice Hall formatting team, for the countless hours they spent paging the book; and Cathy Schultz of Elm Street Publishing Services, whose talents as supervisor of production kept every aspect of this complex project moving through its many stages.

I would like to thank my editor at Prentice Hall, Eric Frank, and associate editor, Dawn Murrin, who guided and coordinated the book from manuscript through production. Thanks to the wonderful team of designers, including Jonathan Boylan and Maureen Eide, for the beautiful covers and interior design. Finally, thanks to Halee Dinsey and Patrice Jones, for your innovative marketing efforts, to Sally Yagan for your continuing support, and to the entire Prentice Hall sales force for your confidence and enthusiasm about the book.

To the Student

I've written this book so that you can learn about the power of algebra and trigonometry and how it relates directly to your life outside the classroom. All concepts are carefully explained, important definitions and procedures are set off in boxes, and worked-out examples that present solutions in a step-by-step manner appear in every section. Each example is followed by a similar matched problem, called a Check Point, for you to try so that you can actively participate in the learning process as you read the book. (Answers to all Check Points appear in the back of the book.) Study Tips offer hints and suggestions and often point out common errors to avoid. A great deal of attention has been given to applying algebra and trigonometry to your life to make your learning experience both interesting and relevant.

As you begin your studies, I would like to offer some specific suggestions for using this book and for being successful in this course:

1. **Attend all lectures.** No book is intended to be a substitute for valuable insights and interactions that occur in the classroom. In addition to arriving for lectures on time and being prepared, you will find it useful to read the section before it is covered in the lecture. This will give you a clear idea of the new material that will be discussed.

2. **Read the book.** Read each section with pen (or pencil) in hand. Move through the illustrative examples with great care. These worked-out examples provide a model for doing exercises in the exercise sets. As you proceed through the reading, do not give up if you do not understand every single word. Things will become clearer as you read on and see how various procedures are applied to specific worked-out examples.

3. **Work problems every day and check your answers.** The way to learn mathematics is by doing mathematics, which means working the Check Points and assigned exercises in the exercise sets. The more exercises you work, the better you will understand the material.

4. **Prepare for chapter exams.** After completing a chapter, study the summary, work the exercises in the Chapter Review, and work the exercises in the Chapter Test. Answers to all these exercises are given in the back of the book.

5. **Use the supplements available with this book.** A solutions manual containing worked-out solutions to the book's odd-numbered exercises, all review exercises, and all Check Points; a dynamic web page; and videotapes and CD-ROMs created for every section of the book are among the supplements created to help you tap into the power of mathematics. Ask your instructor or bookstore which supplements are available and where you can find them.

I wrote this book in beautiful and pristine Point Reyes National Seashore, north of San Francisco. It was my hope to convey the beauty of mathematics using nature as a source of inspiration and creativity. Enjoy the pages that follow as you empower yourself with the algebra and trigonometry needed to succeed in college, your career, and your life.

Regards,

Bob

Robert Blitzer

About the Author

Bob Blitzer is a native of Manhattan and received a Bachelor of Arts degree with dual majors in mathematics and psychology (minor: English literature) from the City College of New York. His unusual combination of academic interests led him toward a Master of Arts in mathematics from the University of Miami and a doctorate in behavioral sciences from Nova University. Bob is most energized by teaching mathematics and has taught a variety of mathematics courses at Miami-Dade Community College for nearly 30 years. He has received numerous teaching awards, including Innovator of the Year from the League for Innovations in the Community College, and was among the first group of recipients at Miami-Dade Community College for an endowed chair based on excellence in the classroom. In addition to *Precalculus*, Bob has written *Introductory Algebra for College Students*, *Intermediate Algebra for College Students*, *Introductory and Intermediate Algebra for College Students*, *Algebra for College Students*, *Thinking Mathematically*, *College Algebra*, and *Algebra and Trigonometry*, all published by Prentice Hall.

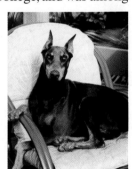

Finally, Bob loves to spend time with his pal, Harley, pictured to the right. He's so cute (Harley, not Bob) that we couldn't resist including him.

Why Blitzer's Precalculus Essentials?

This text was written to address students' most commonly cited reasons for not using their texts:

- "I'll never use this information ("When will I use this?")"
- "I can't follow the explanations."

"When Will I Use This?"

This text integrates dynamic applications that connect mathematics to the entire spectrum of students' interests.

The interesting and diverse applications . . .

- represent a wide range of disciplines.

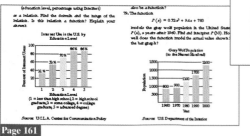

- feature unique and interesting data that show students that mathematics can be applied in many settings.

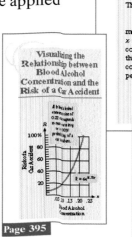

- are driven by real and sourced data, illustrating the power of algebra and trigonometry to model contemporary issues and problems.

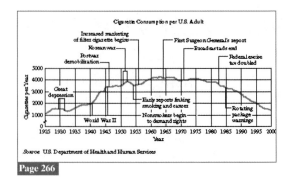

Page 266

Unique Chapter and Section Opening Vignettes

- Each chapter and section begins with a vignette highlighting an everyday scenario, posing a question about it, and exploring how the chapter section subject can be applied to answer the question.
- These are revisited in the course of the chapter or section in an example, discussion, or exercise.

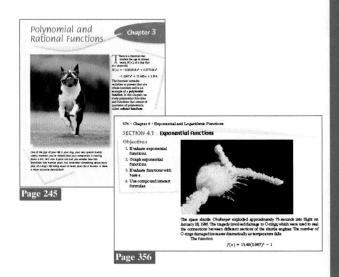

Page 245

Page 356

Enrichment Essay (and other interesting asides)

- Enrichment Essays provide historical, interdisciplinary, and otherwise interesting connections to the math under study, showing students that math is an interesting and dynamic discipline.

Page 283

"I Can't Follow the Explanations."

Clear & Friendly Writing Style

- Blitzer's language is clear, direct, and simple. He breaks down concepts in a conversational style, providing analogies and drawing connections to students' experiences whenever possible.

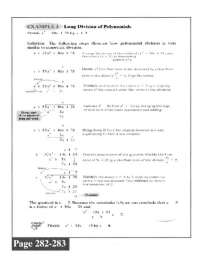

> Yogi Berra, catcher and renowned hitter for the New York Yankees (1946–1963), said it best: "Prediction is very hard, especially when it's about the future." At the start of the twenty-first century, we are plagued by questions about the environment. Will we run out of gas? How hot will it get? Will there be neighborhoods where the air is pristine? Can we make garbage disappear? Will there be any wilderness left? Which wild animals will become extinct? These concerns have led to the growth of the environmental industry in the United States.
>
> **EXAMPLE 10 The Growth of the Environmental Industry**

Page 396

Detailed Illustrations, Examples, and Check Points

Examples:

- are abundant, because students learn by example.
- are thoroughly annotated to the right of the algebraic steps. These annotations are in a conversational style, providing the voice of an instructor in the book, explaining key steps and ideas as the problem is solved.
- offer students the opportunity to stop and test their understanding of the example by working a similar exercise immediately following that example, called a **Check Point.**
- The answers to the **Check Points** are provided in the answer section.

Page 282-283

Exercise Sets that Precisely Parallel Examples (pages 193-196)

- An extensive collection of exercises is included at the end of each section.
- Exercises are organized by level within six category types: **Practice Exercises, Application Exercises, Writing in Mathematics, Technology Exercises, Critical Thinking Exercises,** and **Group Exercises.**
- The order of the practice exercises is exactly the same as the order of the section's illustrative examples. This parallel order enables students to refer to the titled examples and their detailed explanations to achieve success working the practice exercises.

Explanatory Voice Balloons

Voice balloons are used in a variety of ways to demystify mathematics. They:

- translate mathematical ideas into plain English.
- help clarify problem-solving procedures.
- present alternative ways of understanding concepts.
- connect complex problems to the basic concepts students have already learned.

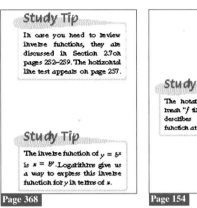

Page 489

Study Tips

Study Tip boxes:

- appear in abundance throughout the book.
- offer suggestions for problem solving.
- point out common mistakes.
- provide informal tips and suggestions.

Page 368

Page 154

Clearly Stated Section Objectives

Learning objectives:

- are clearly stated at the beginning of each section.
- help students recognize and focus on the most important ideas.
- appear in the margin at their point of use.
- form the foundation for the algorithms in MathPro (tutorial software) and in TestGen (test generator software).

Page 305

Chapter Review Grids

- summarize definitions and concepts for every section of the chapter.
- refer students to the examples that illustrate these key concepts.

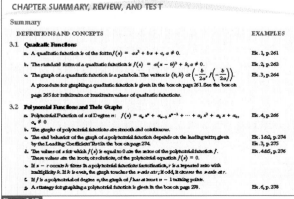

Page 347

xix

Applications Index

Prerequisites: Fundamental Concepts of Algebra

Chapter P

This chapter reviews fundamental concepts of algebra that are prerequisites for the study of college algebra. Algebra, like all of mathematics, provides the tools to help you recognize, classify, and explore the hidden patterns of your world, revealing its underlying structure. Throughout the new millennium, literacy in algebra will be a prerequisite for functioning in a meaningful way personally, professionally, and as a citizen.

Listening to the radio on the way to work, you hear candidates in the upcoming election discussing the problem of the country's 5.6 trillion dollar deficit. It seems like this is a real problem, but then you realize that you don't really know what that number means. How can you look at this deficit in the proper perspective? If the national debt were evenly divided among all citizens of the country, how much would each citizen have to pay? Does the deficit seem like such a significant problem now?

SECTION P.1 *Real Numbers and Algebraic Expressions*

Objectives

1. Recognize subsets of the real numbers.
2. Use inequality symbols.
3. Evaluate absolute value.
4. Use absolute value to express distance.
5. Evaluate algebraic expressions.
6. Identify properties of the real numbers.
7. Simplify algebraic expressions.

The U.N. Building is designed with three golden rectangles.

The United Nations Building in New York was designed to represent its mission of promoting world harmony. Viewed from the front, the building looks like three rectangles stacked upon each other. In each rectangle, the ratio of the width to height is $\sqrt{5} + 1$ to 2, approximately 1.618 to 1. The ancient Greeks believed that such a rectangle, called a **golden rectangle**, was the most visually pleasing of all rectangles.

The ratio 1.618 to 1 is approximate because $\sqrt{5}$ is an irrational number, a special kind of real number. Irrational? Real? Let's make sense of all this by describing the kinds of numbers you will encounter in this course.

1 Recognize subsets of the real numbers.

The Set of Real Numbers

Before we describe the set of real numbers, let's be sure you are familiar with some basic ideas about sets. A **set** is a collection of objects whose contents can be clearly determined. The objects in a set are called the **elements** of the set. For example, the set of numbers used for counting can be represented by

$$\{1, 2, 3, 4, 5, \ldots \}.$$

The braces, { }, indicate that we are representing a set. This form of representing a set uses commas to separate the elements of the set. The set of numbers used for counting is called the set of **natural numbers**. The three dots after the 5 indicate that there is no final element and that the listing goes on forever.

The sets that make up the real numbers are summarized in Table P.1. We refer to these sets as **subsets** of the real numbers, meaning that all elements in each subset are also elements in the set of real numbers.

Notice the use of the symbol $\approx$ in the examples of irrational numbers. The symbol means "is approximately equal to." Thus,

$$\sqrt{2} \approx 1.414214.$$

We can verify that this is only an approximation by multiplying 1.414214 by itself. The product is very close to, but not exactly, 2:

$$1.414214 \times 1.414214 = 2.00000123796.$$

Technology

A calculator with a square root key gives a decimal approximation for $\sqrt{2}$, not the exact value.

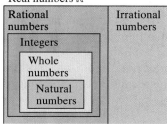

Real numbers ℝ

This diagram shows that every real number is rational or irrational.

Table P.1 Important Subsets of the Real Numbers

Name	Description	Examples
Natural numbers ℕ	$\{1, 2, 3, 4, 5, \ldots\}$ These numbers are used for counting.	$2, 3, 5, 17$
Whole numbers 𝕎	$\{0, 1, 2, 3, 4, 5, \ldots\}$ The set of whole numbers is formed by adding 0 to the set of natural numbers.	$0, 2, 3, 5, 17$
Integers ℤ	$\{\ldots, -5, -4, -3, -2, -1, 0, 1, 2, 3, 4, 5, \ldots\}$ The set of integers is formed by adding negatives of the natural numbers to the set of whole numbers.	$-17, -5, -3, -2, 0,$ $2, 3, 5, 17$
Rational numbers ℚ	The set of rational numbers is the set of all numbers which can be expressed in the form $\frac{a}{b}$, where a and b are integers and b is not equal to 0, written $b \neq 0$. Rational numbers can be expressed as terminating or repeating decimals.	$-17 = \frac{-17}{1}, -5 = \frac{-5}{1}, -3, -2,$ $0, 2, 3, 5, 17,$ $\frac{2}{5} = 0.4,$ $\frac{-2}{3} = -0.6666\cdots = -0.\overline{6}$
Irrational numbers 𝕀	This is the set of all numbers whose decimal representations are neither terminating nor repeating. Irrational numbers cannot be expressed as a quotient of integers.	$\sqrt{2} \approx 1.414214$ $-\sqrt{3} \approx -1.73205$ $\pi \approx 3.142$ $-\frac{\pi}{2} \approx -1.571$

Study Tip

Not all square roots are irrational numbers. For example, $\sqrt{25} = 5$ because $5 \times 5 = 25$. Thus, $\sqrt{25}$ is a natural number, a whole number, an integer, and a rational number $\left(\sqrt{25} = \frac{5}{1}\right)$.

The set of **real numbers** is formed by combining the rational numbers and the irrational numbers. Thus, every real number is either rational or irrational.

The Real Number Line

The **real number line** is a graph used to represent the set of real numbers. An arbitrary point, called the **origin**, is labeled 0; units to the right of the origin are **positive** and units to the left of the origin are **negative**. The real number line is shown in Figure P.1.

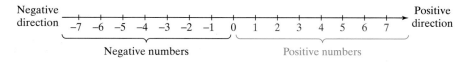

Figure P.1 The real number line

Real numbers are **graphed** on a number line by placing a dot at the correct location for each number. The integers are easiest to locate. In Figure P.2, we've graphed the integers -3, 0, and 4.

Figure P.2 Graphing -3, 0, and 4 on a number line

Every real number corresponds to a point on the number line and every point on the number line corresponds to a real number. We say there is a **one-to-one correspondence** between all the real numbers and all points on a real number line. If you draw a point on the real number line corresponding to a real number, you are **plotting** the real number. In Figure P.2, we are plotting the real numbers -3, 0, and 4.

2 Use inequality symbols.

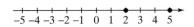

Figure P.3

Ordering the Real Numbers

On the real number line, the real numbers increase from left to right. The lesser of two real numbers is the one farther to the left on a number line. The greater of two real numbers is the one farther to the right on a number line.

Look at the number line in Figure P.3. The integers 2 and 5 are plotted. Observe that 2 is to the left of 5 on the number line. This means that 2 is less than 5:

$2 < 5$: 2 is less than 5 because 2 is to the *left* of 5 on the number line.

In Figure P.3, we can also observe that 5 is to the right of 2 on the number line. This means that 5 is greater than 2:

$5 > 2$: 5 is greater than 2 because 5 is to the *right* of 2 on the number line.

The symbols $<$ and $>$ are called **inequality symbols**. They may be combined with an equal sign, as shown in the following table:

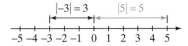

Study Tip

The symbols $<$ and $>$ always point to the lesser of the two real numbers when the inequality is true.

$2 < 5$ *The symbol points to 2, the lesser number.*

$5 > 2$ *The symbol points to 2, the lesser number.*

Symbols	Meaning	Example	Explanation
$a \le b$	a is less than or equal to b.	$3 \le 7$	Because $3 < 7$
		$7 \le 7$	Because $7 = 7$
$b \ge a$	b is greater than or equal to a.	$7 \ge 3$	Because $7 > 3$
		$-5 \ge -5$	Because $-5 = -5$

3 Evaluate absolute value.

Figure P.4 Absolute value as the distance from 0

Absolute Value

The **absolute value** of a real number a, denoted by $|a|$, is the distance from 0 to a on the number line. This distance is always taken to be nonnegative. For example, the real number line in Figure P.4 shows that

$$|-3| = 3 \quad \text{and} \quad |5| = 5.$$

The absolute value of -3 is 3 because -3 is 3 units from 0 on the number line. The absolute value of 5 is 5 because 5 is 5 units from 0 on the number line. The absolute value of a positive real number or 0 is the number itself. The absolute value of a negative real number, such as -3, is the number without the negative sign.

We can define the absolute value of the real number x without referring to a number line. The algebraic definition of the absolute value of x is given as follows:

Definition of Absolute Value

$$|x| = \begin{cases} x & \text{if } x \ge 0 \\ -x & \text{if } x < 0 \end{cases}$$

If x is nonnegative (that is $x \ge 0$), the absolute value of x is the number itself. For example,

$$|5| = 5 \qquad |\pi| = \pi \qquad \left|\frac{1}{3}\right| = \frac{1}{3} \qquad |0| = 0.$$

Zero is the only number whose absolute value is 0.

If x is a negative number (that is, $x < 0$), the absolute value of x is the opposite of x. This makes the absolute value positive. For example,

$$\left|-3\right| = -(-3) = 3 \qquad \left|-\pi\right| = -(-\pi) = \pi \qquad \left|-\frac{1}{3}\right| = -\left(-\frac{1}{3}\right) = \frac{1}{3}.$$

> This middle step is usually omitted.

EXAMPLE 1 Evaluating Absolute Value

Rewrite each expression without absolute value bars:

a. $\left|\sqrt{3} - 1\right|$ **b.** $\left|2 - \pi\right|$ **c.** $\dfrac{|x|}{x}$ if $x < 0$.

Solution

a. Because $\sqrt{3} \approx 1.7$, the expression inside the absolute value bars, $\sqrt{3} - 1$, is positive. The absolute value of a positive number is the number itself. Thus,

$$\left|\sqrt{3} - 1\right| = \sqrt{3} - 1.$$

b. Because $\pi \approx 3.14$, the number inside the absolute value bars, $2 - \pi$, is negative. The absolute value of x when $x < 0$ is $-x$. Thus,

$$\left|2 - \pi\right| = -(2 - \pi) = \pi - 2.$$

c. If $x < 0$, then $|x| = -x$. Thus,

$$\frac{|x|}{x} = \frac{-x}{x} = -1.$$

Study Tip

After working each Check Point, check your answer in the answer section before continuing your reading.

Check Point 1 Rewrite each expression without absolute value bars:

a. $\left|1 - \sqrt{2}\right|$ **b.** $\left|\pi - 3\right|$ **c.** $\dfrac{|x|}{x}$ if $x > 0$.

Listed below are several basic properties of absolute value. Each of these properties can be derived from the definition of absolute value.

Discovery

Verify the triangle inequality if $a = 4$ and $b = 5$. Verify the triangle inequality if $a = 4$ and $b = -5$.
When does equality occur in the triangle inequality and when does inequality occur? Verify your observation with additional number pairs.

Properties of Absolute Value

For all real numbers a and b,

1. $|a| \geq 0$ **2.** $|-a| = |a|$ **3.** $a \leq |a|$

4. $|ab| = |a||b|$ **5.** $\left|\dfrac{a}{b}\right| = \dfrac{|a|}{|b|}, b \neq 0$

6. $|a + b| \leq |a| + |b|$ (called the triangle inequality)

---

4 Use absolute value to express distance.

Distance between Points on a Real Number Line

Absolute value is used to find the distance between two points on a real number line. If a and b are any real numbers, the **distance between a and b** is the absolute value of their difference. For example, the distance between 4 and 10 is 6. Using absolute value, we find this distance in one of two ways:

$$\left|10 - 4\right| = |6| = 6 \quad \text{or} \quad \left|4 - 10\right| = |-6| = 6.$$

> The distance between 4 and 10 on the real number line is 6.

Notice that we obtain the same distance regardless of the order in which we subtract.

> **Distance between Two Points on the Real Number Line**
> If a and b are any two points on a real number line, then the distance between a and b is given by
> $$|a - b| \quad \text{or} \quad |b - a|.$$

EXAMPLE 2 Distance between Two Points on a Number Line

Find the distance between -5 and 3 on the real number line.

Solution Because the distance between a and b is given by $|a - b|$, the distance between -5 and 3 is

$$|-5 - 3| = |-8| = 8.$$

$a = -5 \qquad b = 3$

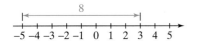

Figure P.5 The distance between -5 and 3 is 8.

Figure P.5 verifies that there are 8 units between -5 and 3 on the real number line. We obtain the same distance if we reverse the order of the subtraction:

$$|3 - (-5)| = |8| = 8.$$

Check Point 2 Find the distance between -4 and 5 on the real number line.

Algebraic Expressions

Algebra uses letters, such as x and y, to represent real numbers. Such letters are called **variables**. For example, imagine that you are basking in the sun on the beach. We can let x represent the number of minutes that you can stay in the sun without burning with no sunscreen. With a number 6 sunscreen, exposure time without burning is six times as long, or 6 times x. This can be written $6 \cdot x$, but it is usually expressed as $6x$. Placing a number and a letter next to one another indicates multiplication.

Notice that $6x$ combines the number 6 and the variable x using the operation of multiplication. A combination of variables and numbers using the operations of addition, subtraction, multiplication, or division, as well as powers or roots (which are discussed later in this chapter), is called an **algebraic expression**. Here are some examples of algebraic expressions:

$$x + 6, \quad x - 6, \quad 6x, \quad \frac{x}{6}, \quad 3x + 5, \quad \sqrt{x} + 7.$$

5 Evaluate algebraic expressions.

Evaluating Algebraic Expressions

Evaluating an algebraic expression means to find the value of the expression for a given value of the variable. For example, we can evaluate $6x$ (from the sunscreen example) when $x = 15$. We substitute 15 for x. We obtain $6 \cdot 15$, or 90. This means if you can stay in the sun for 15 minutes without burning when you don't put on any lotion, then with a number 6 lotion, you can "cook" for 90 minutes without burning.

Many algebraic expressions involve more than one operation. Evaluating an algebraic expression without a calculator involves carefully applying the following order of operations agreement:

The Order of Operations Agreement

1. Perform operations within the innermost parentheses and work outward. If the algebraic expression involves a fraction, treat the numerator and the denominator as if they were each enclosed in parentheses.
2. Evaluate all exponential expressions.
3. Perform multiplications and divisions as they occur, working from left to right.
4. Perform additions and subtractions as they occur, working from left to right.

EXAMPLE 3   Evaluating an Algebraic Expression

Evaluate:

a. $(25t^2 + 125t) \div (t^2 + 1)$ for $t = 3$.

b. $\dfrac{-b + \sqrt{b^2 - 4ac}}{2a}$ for $a = 2$, $b = 9$, and $c = -5$.

Solution

Expression	Value of the Variable(s)	Substitute	Evaluating the Expression
a. $(25t^2 + 125t) \div (t^2 + 1)$	$t = 3$	$(25 \cdot 3^2 + 125 \cdot 3) \div (3^2 + 1)$	$= (25 \cdot 9 + 125 \cdot 3) \div (9 + 1)$ $= (225 + 375) \div (9 + 1)$ $= 600 \div 10$ $= 60$
b. $\dfrac{-b + \sqrt{b^2 - 4ac}}{2a}$	$a = 2$ $b = 9$ $c = -5$	$\dfrac{-9 + \sqrt{9^2 - 4(2)(-5)}}{2(2)}$	$= \dfrac{-9 + \sqrt{81 - (-40)}}{4}$ $= \dfrac{-9 + \sqrt{121}}{4}$ $= \dfrac{-9 + 11}{4} = \dfrac{2}{4} = \dfrac{1}{2}$

Check Point 3   Evaluate:

a. $(3t^2 + 8t) \div (t^2 - 2)$ for $t = 2$.

b. $\dfrac{-b + \sqrt{b^2 - 4ac}}{2a}$ for $a = 1$, $b = -6$, and $c = 9$.

6 Identify properties of the real numbers.

Properties of Real Numbers and Algebraic Expressions

When you use your calculator to add two real numbers, you can enter them in any order. The fact that two real numbers can be added in any order is called the **commutative property of addition**. You probably use this property, as well as other properties of real numbers listed in Table P.2 on the next page, without giving it much thought. The properties of the real numbers are especially useful when working with algebraic expressions. For each property listed in Table P.2, a, b, and c represent real numbers, variables, or algebraic expressions.

The Associative Property and the English Language

In the English language, phrases can take on different meanings depending on the way the words are associated with commas.

Here are two examples:

- *Woman, without her man, is nothing.*
  Woman, without her, man is nothing.
- *What's the latest dope?*
  What's the latest, dope?

Table P.2   Properties of the Real Numbers

Name	Meaning	Examples
Commutative Property of Addition	Two real numbers can be added in any order. $a + b = b + a$	• $13 + 7 = 7 + 13$ • $13x + 7 = 7 + 13x$
Commutative Property of Multiplication	Two real numbers can be multiplied in any order. $ab = ba$	• $\sqrt{2} \cdot \sqrt{5} = \sqrt{5} \cdot \sqrt{2}$ • $x \cdot 6 = 6x$
Associative Property of Addition	If three real numbers are added, it makes no difference which two are added $(a + b) + c = a + (b + c)$	• $3 + (8 + x) = (3 + 8) + x$ $\qquad\qquad = 11 + x$
Associative Property of Multiplication	If three real numbers are multiplied, it makes no difference which two are multiplied first. $(a \cdot b) \cdot c = a \cdot (b \cdot c)$	• $-2(3x) = (-2 \cdot 3)x = -6x$
Distributive Property of Multiplication over Addition	Multiplication distributes over addition. $\overset{\frown}{a \cdot (b + c)} = a \cdot b + a \cdot c$ $= 15x + 35$	• $\overset{\frown}{7(4 + \sqrt{3})} = 7 \cdot 4 + 7 \cdot \sqrt{3}$ $\qquad\qquad = 28 + 7\sqrt{3}$ • $\overset{\frown}{5(3x + 7)} = 5 \cdot 3x + 5 \cdot 7$
Identity Property of Addition	Zero can be deleted from a sum. $a + 0 = a$ $0 + a = a$	• $\sqrt{3} + 0 = \sqrt{3}$ • $0 + 6x = 6x$
Identity Property of Multiplication	One can be deleted from a product. $a \cdot 1 = a$ $1 \cdot a = a$	• $1 \cdot \pi = \pi$ • $13x \cdot 1 = 13x$
Inverse Property of Addition	The sum of a real number and its additive inverse gives 0, the additive identity. $a + (-a) = 0$ $(-a) + a = 0$	• $\sqrt{5} + (-\sqrt{5}) = 0$ • $-\pi + \pi = 0$ • $6x + (-6x) = 0$ • $(-4y) + 4y = 0$
Inverse Property of Multiplication	The product of a nonzero real number and its multiplicative inverse gives 1, the multiplicative identity. $a \cdot \dfrac{1}{a} = 1, \quad a \neq 0$	• $7 \cdot \dfrac{1}{7} = 1$ • $\left(\dfrac{1}{x - 3}\right)(x - 3) = 1, x \neq 3$

Commutative Words and Sentences

The commutative property states that a change in order produces no change in the answer. The words and sentences listed here suggest a characteristic of the commutative property; they read the same from left to right and from right to left!

dad	Draw, o coward!	Revolting is error. Resign it, lover.
repaper	Dennis sinned.	Naomi, did I moan?
never odd or even	Ma is a nun, as I am.	Al lets Della call Ed Stella.

The properties in Table P.2 apply to the operations of addition and multiplication. Subtraction and division are defined in terms of addition and multiplication.

> **Definitions of Subtraction and Division**
> Let a and b represent real numbers.
>
> **Subtraction:** $a - b = a + (-b)$
> We call $-b$ the **additive inverse** or **opposite** of b.
>
> **Division:** $a \div b = a \cdot \frac{1}{b}$, where $b \neq 0$
> We call $\frac{1}{b}$ the **multiplicative inverse** or **reciprocal** of b. The quotient of a and b, $a \div b$, can be written in the form $\frac{a}{b}$, where a is the **numerator** and b the **denominator** of the fraction.

Because subtraction is defined in terms of adding an inverse, the distributive property can be applied to subtraction:

$$a(b - c) = ab - ac$$

$$(b - c)a = ba - ca.$$

For example,

$$4(2x - 5) = 4 \cdot 2x - 4 \cdot 5 = 8x - 20.$$

7 Simplify algebraic expressions.

Simplifying Algebraic Expressions

The **terms** of an algebraic expression are those parts that are separated by addition. For example, consider the algebraic expression

$$7x - 9y - 3,$$

which can be expressed as

$$7x + (-9y) + (-3).$$

This expression contains three terms, namely $7x$, $-9y$, and -3.

The numerical part of a term is called its **numerical coefficient**. In the term $7x$, the 7 is the numerical coefficient. In the term $-9y$, the -9 is the numerical coefficient.

A term that consists of just a number is called a **constant term**. The constant term of $7x - 9y - 3$ is -3.

A term indicates a product. The expressions that are multiplied to form the term are called its **factors**. **Like terms** have the same variable factors with the same exponents on the variables. For example, $7x$ and $3x$ are like terms because they have the same variable factor, x. The distributive property (in reverse) can be used to add these terms:

$$7x + 3x = (7 + 3)x = 10x.$$

Study Tip

To add like terms, add their numerical coefficients. Use this result as the numerical coefficient of the terms' common variable(s).

An algebraic expression is **simplified** when parentheses have been removed and like terms have been combined.

EXAMPLE 4   Simplifying an Algebraic Expression

Simplify: $6(2x - 4y) + 10(4x + 3y)$.

Solution

$$6(2x - 4y) + 10(4x + 3y)$$
$$= 6 \cdot 2x - 6 \cdot 4y + 10 \cdot 4x + 10 \cdot 3y \quad \text{Use the distributive property to remove the parentheses.}$$
$$= 12x - 24y + 40x + 30y \quad \text{Multiply.}$$
$$= (12x + 40x) + (30y - 24y) \quad \text{Group like terms.}$$
$$= 52x + 6y \quad \text{Combine like terms.}$$

Check Point 4 Simplify: $7(4x - 3y) + 2(5x + y)$.

Properties of Negatives

The distributive property can be extended to cover more than two terms within parentheses. For example,

This sign represents subtraction.

This sign tells us that the number being distributed is negative.

$$-3(4x - 2y + 6) = -3 \cdot 4x - (-3) \cdot 2y + (-3)(6)$$
$$= -12x - (-6y) + (-18)$$
$$= -12x + 6y - 18.$$

The voice balloons illustrate that negative signs can appear side by side. They can represent the operation of subtraction or the fact that a real number is negative. Here is a list of properties of negatives and how they are applied to algebraic expressions:

Properties of Negatives

Let a and b represent real numbers, variables, or algebraic expressions.

Property	Examples
1. $(-1)a = -a$	$(-1)4xy = -4xy$
2. $-(-a) = a$	$-(-6y) = 6y$
3. $(-a)b = -ab$	$(-7)4xy = -7 \cdot 4xy = -28xy$
4. $a(-b) = -ab$	$5x(-3y) = -5x \cdot 3y = -15xy$
5. $-(a + b) = -a - b$	$-(7x + 6y) = -7x - 6y$
6. $-(a - b) = -a + b$	$-(3x - 7y) = -3x + 7y$
$= b - a$	$= 7y - 3x$

Do you notice that properties 5 and 6 in the box are related? In general, expressions within parentheses that are preceded by a negative can be simplified by dropping the parentheses and changing the sign of every term inside the parentheses.

For example,

$$-(3x - 2y + 5z - 6) = -3x + 2y - 5z + 6.$$

EXERCISE SET P.1

 ## Practice Exercises

*In Exercises 1–4, list all numbers from the given set that are **a.** natural numbers, **b.** whole numbers, **c.** integers, **d.** rational numbers, **e.** irrational numbers.*

1. $\{-9, -\frac{4}{5}, 0, 0.25, \sqrt{3}, 9.2, \sqrt{100}\}$

2. $\{-7, -0.\overline{6}, 0, \sqrt{49}, \sqrt{50}\}$

3. $\{-11, -\frac{5}{6}, 0, 0.75, \sqrt{5}, \pi, \sqrt{64}\}$

4. $\{-5, -0.\overline{3}, 0, \sqrt{2}, \sqrt{4}\}$

5. Give an example of a whole number that is not a natural number.

6. Give an example of a rational number that is not an integer.

7. Give an example of a number that is an integer, a whole number, and a natural number.

8. Give an example of a number that is a rational number, an integer, and a real number.

Determine whether each statement in Exercises 9–14 is true or false.

9. $-13 \le -2$

10. $-6 > 2$

11. $4 \ge -7$

12. $-13 < -5$

13. $-\pi \ge -\pi$

14. $-3 > -13$

In Exercises 15–24, rewrite each expression without absolute value bars.

15. $|300|$

16. $|-203|$

17. $|12 - \pi|$

18. $|7 - \pi|$

19. $|\sqrt{2} - 5|$

20. $|\sqrt{5} - 13|$

21. $\dfrac{-3}{|-3|}$

22. $\dfrac{-7}{|-7|}$

23. $\|-3| - |-7\|$

24. $\|-5| - |-13\|$

In Exercises 25–30, evaluate each algebraic expression for $x = 2$ and $y = -5$.

25. $|x + y|$

26. $|x - y|$

27. $|x| + |y|$

28. $|x| - |y|$

29. $\dfrac{y}{|y|}$

30. $\dfrac{|x|}{x} + \dfrac{|y|}{y}$

In Exercises 31–38, express the distance between the given numbers using absolute value. Then find the distance by evaluating the absolute value expression.

31. 2 and 17

32. 4 and 15

33. -2 and 5

34. -6 and 8

35. -19 and-4

36. -26 and -3

37. -3.6 and-1.4

38. -5.4 and -1.2

In Exercises 39–48, evaluate each algebraic expression for the given value of the variable or variables.

39. $5x + 7; x = 4$

40. $9x + 6; x = 5$

41. $4(x + 3) - 11; x = -5$

42. $6(x + 5) - 13; x = -7$

43. $\dfrac{5}{9}(F - 32); F = 77$

44. $\dfrac{5}{9}(F - 32); F = 50$

45. $\dfrac{1 - (x - 2)^2}{1 + (x - 2)^2}; x = -1$

46. $\dfrac{2x^2 - 1}{2x^3 + 1}; x = -\dfrac{1}{2}$

47. $\dfrac{-b + \sqrt{b^2 - 4ac}}{2a}; a = 4, b = -20, c = 25$

48. $\dfrac{-b + \sqrt{b^2 - 4ac}}{2a}; a = 3, b = 5, c = -2$

In Exercises 49–58, state the name of the property illustrated.

49. $6 + (-4) = (-4) + 6$

50. $11 \cdot (7 + 4) = 11 \cdot 7 + 11 \cdot 4$

51. $6 + (2 + 7) = (6 + 2) + 7$

52. $6 \cdot (2 \cdot 3) = 6 \cdot (3 \cdot 2)$

53. $(2 + 3) + (4 + 5) = (4 + 5) + (2 + 3)$

54. $7 \cdot (11 \cdot 8) = (11 \cdot 8) \cdot 7$

55. $2(-8 + 6) = -16 + 12$

56. $-8(3 + 11) = -24 + (-88)$

57. $\dfrac{1}{(x + 3)}(x + 3) = 1, x \ne -3$

58. $(x + 4) + [-(x + 4)] = 0$

In Exercises 59–68, simplify each algebraic expression.

59. $5(3x + 4) - 4$

60. $2(5x + 4) - 3$

61. $5(3x - 2) + 12x$

62. $2(5x - 1) + 14x$

63. $7(3y - 5) + 2(4y + 3)$

64. $4(2y - 6) + 3(5y + 10)$

65. $5(3y - 2) - (7y + 2)$ **66.** $4(5y - 3) - (6y + 3)$

67. $7 - 4[3 - (4y - 5)]$ **68.** $6 - 5[8 - (2y - 4)]$

In Exercises 69–74, write each algebraic expression without parentheses.

69. $-(-14x)$ **70.** $-(-17y)$

71. $-(2x - 3y - 6)$ **72.** $-(5x - 13y - 1)$

73. $\frac{1}{3}(3x) + [(4y) + (-4y)]$ **74.** $\frac{1}{2}(2y) + [(-7x) + 7x]$

Application Exercises

75. Are first putting on your left shoe and then putting on your right shoe commutative?

76. Are first getting undressed and then taking a shower commutative?

77. Give an example of two things that you do that are not commutative.

78. Give an example of two things that you do that are commutative.

79. The algebraic expression $81 - 0.6x$ approximates the percentage of American adults who smoked cigarettes x years after 1900. Evaluate the expression for $x = 100$. Describe what the answer means in practical terms.

80. The algebraic expression $1527x + 31{,}290$ approximates average yearly earnings for elementary and secondary teachers in the United States x years after 1990. Evaluate the algebraic expression for $x = 10$. Describe what the answer means in practical terms.

81. The optimum heart rate is the rate that a person should achieve during exercise for the exercise to be most beneficial. The algebraic expression

$$0.6(220 - a)$$

describes a person's optimum heart rate, in beats per minute, where a represents the age of the person.

a. Use the distributive property to rewrite the algebraic expression.

b. Use each form of the algebraic expression to determine the optimum heart rate for a 20-year-old runner.

Writing in Mathematics

Writing about mathematics will help you learn mathematics. For all writing exercises in this book, use complete sentences to respond to the question. Some writing exercises can be answered in a sentence; others require a paragraph or two. You can decide how much you need to write as long as your writing clearly and directly answers the question in the exercise. Standard references such as a dictionary and a thesaurus should be helpful.

82. How do the whole numbers differ from the natural numbers?

83. Can a real number be both rational and irrational? Explain your answer.

84. If you are given two real numbers, explain how to determine which one is the lesser.

85. How can $\dfrac{|x|}{x}$ be equal to 1 or -1?

86. What is an algebraic expression? Give an example with your explanation.

87. Why is $3(x + 7) - 4x$ not simplified? What must be done to simplify the expression?

88. You can transpose the letters in the word "conversation" to form the phrase "voices rant on." From "total abstainers" we can form "sit not at ale bars." What two algebraic properties do each of these transpositions (called anagrams) remind you of? Explain your answer.

Critical Thinking Exercises

89. Which one of the following statements is true?

a. Every rational number is an integer.

b. Some whole numbers are not integers.

c. Some rational numbers are not positive.

d. Irrational numbers cannot be negative.

90. Which of the following is true?

a. The term x has no numerical coefficient.

b. $5 + 3(x - 4) = 8(x - 4) = 8x - 32$

c. $-x - x = -x + (-x) = 0$

d. $x - 0.02(x + 200) = 0.98x - 4$

In Exercises 91–93, insert either $<$ or $>$ in the box between the numbers to make the statement true.

91. $\sqrt{2} \,\square\, 1.5$ **92.** $-\pi \,\square\, -3.5$

93. $-\dfrac{3.14}{2} \,\square\, -\dfrac{\pi}{2}$

94. A business that manufactures small alarm clocks has a weekly fixed cost of $5000. The average cost per clock for the business to manufacture x clocks is described by

$$\frac{0.5x + 5000}{x}.$$

a. Find the average cost when $x = 100$, 1000, and 10,000.

b. Like all other businesses, the alarm clock manufacturer must make a profit. To do this, each clock must be sold for at least 50¢ more than what it costs to manufacture. Due to competition from a larger company, the clocks can be sold for $1.50 each and no more. Our small manufacturer can only produce 2000 clocks weekly. Does this business have much of a future? Explain.

SECTION P.2 *Exponents and Scientific Notation*

Objectives

1. Understand and use integer exponents.
2. Use properties of exponents.
3. Simplify exponential expressions.
4. Use scientific notation.

1 Understand and use integer exponents.

Powers of Ten

$$10 = 10^1$$
$$100 = 10^2$$
$$1000 = 10^3$$
$$10,000 = 10^4$$
$$100,000 = 10^5$$
$$1,000,000 = 10^6 \quad \text{million}$$
$$10,000,000 = 10^7$$
$$100,000,000 = 10^8$$
$$1,000,000,000 = 10^9 \quad \text{billion}$$

Technology

You can use a calculator to evaluate exponential expressions. For example, to evaluate 5^3, press the following keys:

Many Scientific Calculators

5 $\boxed{y^x}$ 3 $\boxed{=}$

Many Graphing Calculators

5 $\boxed{\wedge}$ 3 $\boxed{\text{ENTER}}$.

Although calculators have special keys to evaluate powers of ten and squaring bases, you can always use one of the sequences shown here.

Although people do a great deal of talking, the total output since the beginning of gabble to the present day, including all baby talk, love songs, and congressional debates, only amounts to about 10 million billion words. This can be expressed as 16 factors of 10, or 10^{16} words.

Exponents such as 2, 3, 4, and so on are used to indicate repeated multiplication. For example,

$$10^2 = 10 \cdot 10 = 100,$$
$$10^3 = 10 \cdot 10 \cdot 10 = 1000, \quad 10^4 = 10 \cdot 10 \cdot 10 \cdot 10 = 10,000.$$

The 10 that is repeated when multiplying is called the **base**. The small numbers above and to the right of the base are called **exponents** or **powers**. The exponent tells the number of times the base is to be used when multiplying. In 10^3, the base is 10 and the exponent is 3.

The formal algebraic definition of a natural number exponent summarizes our discussion:

Definition of a Natural Number Exponent

If b is a real number and n is a natural number,

$$b^n = \underbrace{b \cdot b \cdot b \cdot \cdots \cdot b}_{b \text{ appears as a factor } n \text{ times.}}$$

where the **Base** is b and the **Exponent** is n.

b^n is read "the nth power of b" or "b to the nth power." Thus, the nth power of b is defined as the product of n factors of b.

Furthermore, $b^1 = b$.

Study Tip

-3^4 is read "the opposite of 3 to the fourth power." By contrast, $(-3)^4$ is read "negative 3 to the fourth power."

EXAMPLE 1 Evaluating an Exponential Expression

Evaluate: $(-2)^3 \cdot 3^2$.

Solution

$$(-2)^3 \cdot 3^2 = (-2)(-2)(-2) \cdot 3 \cdot 3 = -8 \cdot 9 = -72$$

This is $(-2)^3$, read "−2 cubed."

This is 3^2, read "3 squared."

Check Point 1 Evaluate: $(-4)^3 \cdot 2^2$.

2 Use properties of exponents.

Properties of Exponents

The major properties of exponents are summarized in the box that follows.

Study Tip

When a negative integer appears as an exponent, switch the position of the base (from numerator to denominator or from denominator to numerator) and make the exponent positive.

Properties of Exponents

Property	Examples
The Negative Exponent Rule If b is any real number other than 0 and n is a natural number, then $$b^{-n} = \frac{1}{b^n}.$$	$\bullet\ 5^{-3} = \dfrac{1}{5^3} = \dfrac{1}{125}$ $\bullet\ \dfrac{1}{4^{-2}} = \dfrac{1}{\frac{1}{4^2}} = 4^2 = 16$
The Zero Exponent Rule If b is any real number other than 0, $$b^0 = 1.$$	$\bullet\ 7^0 = 1$ $\bullet\ (-5)^0 = 1$ $\bullet\ -5^0 = -1$ Only 5 is raised to the zero power.
The Product Rule If b is a real number or algebraic expression, and m and n are integers, $$b^m \cdot b^n = b^{m+n}.$$	$\bullet\ 2^2 \cdot 2^3 = 2^{2+3} = 2^5 = 32$ $\bullet\ x^{-3} \cdot x^7 = x^{-3+7} = x^4$

When multiplying exponential expressions with the same base, add the exponents. Use this sum as the exponent of the common base.

| **The Power Rule** If b is a real number or algebraic expression, and m and n are integers, $$(b^m)^n = b^{mn}.$$ | $\bullet\ (2^2)^3 = 2^{2 \cdot 3} = 2^6 = 64$ $\bullet\ (x^{-3})^4 = x^{-3 \cdot 4} = x^{-12} = \dfrac{1}{x^{12}}$ |

When an exponential expression is raised to a power, multiply the exponents. Place the product of the exponents on the base and remove the parentheses.

Study Tip

$\dfrac{4^3}{4^5}$ and $\dfrac{4^5}{4^3}$ represent different numbers:

$$\frac{4^3}{4^5} = 4^{3-5} = 4^{-2} = \frac{1}{4^2} = \frac{1}{16}$$

$$\frac{4^5}{4^3} = 4^{5-3} = 4^2 = 16.$$

The Quotient Rule

If b is a nonzero real number or algebraic expression, and m and n are integers,

$$\frac{b^m}{b^n} = b^{m-n}.$$

- $\dfrac{2^8}{2^4} = 2^{8-4} = 2^4 = 16$

- $\dfrac{x^3}{x^7} = x^{3-7} = x^{-4} = \dfrac{1}{x^4}$

When dividing exponential expressions with the same nonzero base, subtract the exponent in the denominator from the exponent in the numerator. Use this difference as the exponent of the common base.

Products Raised to Powers

If a and b are real number or algebraic expressions, and n is an integer,
$$(ab)^n = a^n b^n.$$

- $(-2y)^4 = (-2)^4 y^4 = 16y^4$
- $(-2xy)^3 = (-2)^3 x^3 y^3 = -8x^3 y^3$

When a product is raised to a power, raise each factor to the power.

Quotients Raised to Powers

If a and b are real numbers, $b \neq 0$, or algebraic expressions, and n is an integer,

$$\left(\frac{a}{b}\right)^n = \frac{a^n}{b^n}.$$

- $\left(\dfrac{2}{5}\right)^4 = \dfrac{2^4}{5^4} = \dfrac{16}{625}$

- $\left(-\dfrac{3}{x}\right)^3 = \dfrac{(-3)^3}{x^3} = -\dfrac{27}{x^3}$

When a quotient is raised to a power, raise the numerator to that power and divide by the denominator to that power.

3 Simplify exponential expressions.

Simplifying Exponential Expressions

Properties of exponents are used to simplify exponential expressions. An exponential expression is **simplified** when

- No parentheses appear.
- No expressions with powers are raised to powers.
- Each base occurs only once.
- No negative exponents appear.

EXAMPLE 2   Simplifying Exponential Expressions

Simplify:
  a. $(-3x^4 y^5)^3$ **b.** $(-7xy^4)(-2x^5 y^6)$ **c.** $\dfrac{-35x^2 y^4}{5x^6 y^{-8}}$ **d.** $\left(\dfrac{4x^2}{y}\right)^{-3}$.

Solution

  a. $(-3x^4 y^5)^3 = (-3)^3 (x^4)^3 (y^5)^3$ Raise each factor inside the parentheses to the third power.

$\qquad\qquad = (-3)^3 x^{4 \cdot 3} y^{5 \cdot 3}$ Multiply powers to powers.

$\qquad\qquad = -27x^{12} y^{15}$ $(-3)^3 = (-3)(-3)(-3) = -27$

  b. $(-7xy^4)(-2x^5 y^6) = (-7)(-2)xx^5 y^4 y^6$ Group factors with the same base.

$\qquad\qquad = 14x^{1+5} y^{4+6}$ When multiplying expressions with the same base, add the exponents.

$\qquad\qquad = 14x^6 y^{10}$ Simplify.

Visualizing Powers of 3

The triangles contain $3, 3^2, 3^3,$ and 3^4 circles.

c. $\dfrac{-35x^2y^4}{5x^6y^{-8}} = \left(\dfrac{-35}{5}\right)\left(\dfrac{x^2}{x^6}\right)\left(\dfrac{y^4}{y^{-8}}\right)$ Group factors with the same base.

$= -7x^{2-6}y^{4-(-8)}$ When dividing an expression with the same base, subtract the exponents.

$= -7x^{-4}y^{12}$ Simplify. Notice that $4 - (-8) = 4 + 8 = 12.$

$= \dfrac{-7y^{12}}{x^4}$ Move x^{-4}, the factor with the negative exponent, from the numerator to the denominator and make the exponent positive.

d. $\left(\dfrac{4x^2}{y}\right)^{-3} = \dfrac{4^{-3}(x^2)^{-3}}{y^{-3}}$ Raise each factor inside the parentheses to the -3 power.

$= \dfrac{4^{-3}x^{-6}}{y^{-3}}$ Multiply powers to powers.

$= \dfrac{y^3}{4^3x^6}$ Move factors with negative exponents from the numerator to the denominator (or vice versa) and change the sign of the exponent.

$= \dfrac{y^3}{64x^6}$ $4^3 = 4 \cdot 4 \cdot 4 = 64$

Check Point 2

Simplify:

a. $(2x^3y^6)^4$ **b.** $(-6x^2y^5)(3xy^3)$ **c.** $\dfrac{100x^{12}y^2}{20x^{16}y^{-4}}$ **d.** $\left(\dfrac{5x}{y^4}\right)^{-2}.$

Study Tip

Try to avoid the following common errors that can occur when simplifying exponential expressions.

Correct	Incorrect	Description of Error
$b^3 \cdot b^4 = b^7$	$b^3 \cdot b^4 = b^{12}$	The exponents should be added, not multiplied.
$3^2 \cdot 3^4 = 3^6$	$3^2 \cdot 3^4 = 9^6$	The common base should be retained, not multiplied.
$\dfrac{5^{16}}{5^4} = 5^{12}$	$\dfrac{5^{16}}{5^4} = 5^4$	The exponents should be subtracted, not divided.
$(4a)^3 = 64a^3$	$(4a)^3 = 4a^3$	Both factors should be cubed.
$b^{-n} = \dfrac{1}{b^n}$	$b^{-n} = -\dfrac{1}{b^n}$	Only the exponent should change sign.
$(a + b)^{-1} = \dfrac{1}{a + b}$	$(a + b)^{-1} = \dfrac{1}{a} + \dfrac{1}{b}$	The exponent applies to the entire expression $a + b$.

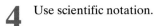 Use scientific notation.

Scientific Notation

The national debt of the United States is about \$5.6 trillion. A stack of \$1 bills equaling the national debt would rise to twice the distance from the Earth to the moon. Because a trillion is 10^{12}, the national debt can be expressed as

$$5.6 \times 10^{12}.$$

The number 5.6×10^{12} is written in a form called *scientific notation*. A number in **scientific notation** is expressed as a number greater than or equal to 1 and less than 10 multiplied by some power of 10. It is customary to use the multiplication symbol, $\times$, rather than a dot, to indicate multiplication in scientific notation. Here are two examples of numbers in scientific notation:

- Each day, 2.6×10^7 pounds of dust from the atmosphere settle on Earth.
- The diameter of a hydrogen atom is 1.06×10^{-8} centimeter.

We can use the exponent on the 10 to change a number in scientific notation to decimal notation. If the exponent is *positive*, move the decimal point in the number to the *right* the same number of places as the exponent. If the exponent is *negative*, move the decimal point in the number to the *left* the same number of places as the exponent.

EXAMPLE 3   **Converting from Scientific to Decimal Notation**

Write each number in decimal notation:

     a. 2.6×10^7 **b.** 1.016×10^{-8}.

Solution

a. We express 2.6×10^7 in decimal notation by moving the decimal point in 2.6 seven places to the right. We need to add six zeros.

$$2.6 \times 10^7 = 26{,}000{,}000$$

b. We express 1.016×10^{-8} in decimal notation by moving the decimal point in 1.016 eight places to the left. We need to add seven zeros to the right of the decimal point.

$$1.016 \times 10^{-8} = 0.00000001016$$

Check Point 3   Write each number in decimal notation:

     a. 7.4×10^9 **b.** 3.017×10^{-6}.

To convert from decimal notation to scientific notation, we reverse the procedure of Example 3.

- Move the decimal point in the given number to obtain a number greater than or equal to 1 and less than 10.
- The number of places the decimal point moves gives the exponent on 10; the exponent is positive if the given number is greater than 10 and negative if the given number is between 0 and 1.

EXAMPLE 4   **Converting from Decimal Notation to Scientific Notation**

Write each number in scientific notation:

     a. 4,600,000   **b.** 0.00023.

Solution

a. $4{,}600{,}000 = 4.6 \times 10^?$

Decimal point moves 6 places. 4.6×10^6

b. $0.00023 = 2.3 \times 10^{-?}$

Decimal point moves 4 places. 2.3×10^{-4}

Technology

You can use your calculator's $\boxed{\text{EE}}$ (enter exponent) or $\boxed{\text{EXP}}$ key to convert from decimal to scientific notation. Here is how it's done for 0.00023:

Many Scientific Calculators

Keystrokes	Display
.00023 $\boxed{\text{EE}}$ $\boxed{=}$	$2.3 - 04$

Many Graphing Calculators

Use the mode setting for scientific notation.

Keystrokes	Display
.00023 $\boxed{\text{ENTER}}$	$2.3\text{E} - 4$

Check Point 4 Write each number in scientific notation:

a. 7,410,000,000   **b.** 0.000000092.

Computations with Scientific Notation

The product and quotient rules for exponents can be used to multiply or divide numbers that are expressed in scientific notation. For example, here's how to find the product of 3.4×10^9 and 2×10^{-5}.

$$(3.4 \times 10^9)(2 \times 10^{-5}) = (3.4 \times 2) \times (10^9 \times 10^{-5})$$
$$= 6.8 \times 10^{9+(-5)}$$
$$= 6.8 \times 10^4 \quad \text{or} \quad 68,000$$

In our next example, we use the quotient of two numbers in scientific notation to help put a number into perspective. The number is our national debt. The United States began accumulating large deficits in the 1980s. To finance the deficit, the government had borrowed \$5.6 trillion as of the end of 2000. The graph in Figure P.6 shows the national debt increasing over time.

Technology

$(3.4 \times 10^9)(2 \times 10^{-5})$

On a Calculator:

Many Scientific Calculators

3.4 EE 9 ✕ 2 EE 5 +/- =

Display

6.8   04

Many Graphing Calculators

3.4 EE 9 ✕ 2 EE (−) 5 ENTER

Display (in scientific notation mode)

6.8E 4

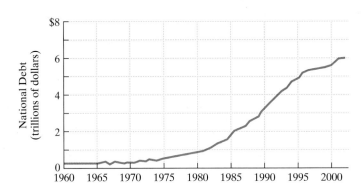

Figure P.6 The national debt

Source: Office of Management and Budget

EXAMPLE 5 The National Debt

As of the end of 2000, the national debt was \$5.6 trillion, or 5.6×10^{12} dollars. At that time, the U.S. population was approximately 280,000,000 (280 million), or 2.8×10^8. If the national debt were evenly divided among every individual in the United States, how much would each citizen have to pay?

Solution The amount each citizen must pay is the total debt, 5.6×10^{12} dollars, divided by the number of citizens, 2.8×10^8.

$$\frac{5.6 \times 10^{12}}{2.8 \times 10^8} = \left(\frac{5.6}{2.8}\right) \times \left(\frac{10^{12}}{10^8}\right)$$
$$= 2 \times 10^{12-8}$$
$$= 2 \times 10^4$$
$$= 20,000$$

Every U.S. citizen would have to pay about \$20,000 to the federal government to pay off the national debt. A family of three would owe \$60,000.

Technology

Here is the keystroke sequence for solving Example 5 using a calculator:

5.6 EE 12 ÷ 2.8 EE 8.

The quotient is displayed by pressing = on a scientific calculator and ENTER on a graphing calculator. The answer can be displayed in scientific or decimal notation. Consult your manual.

Check Point 5

In 2000, Americans spent 3.6×10^9 dollars on full-fat ice cream. At that time, the U.S. population was approximately 280 million, or 2.8×10^8. If ice cream spending is evenly divided, how much did each American spend?

An Application: Black Holes in Space

The concept of a black hole, a region in space where matter appears to vanish, intrigues scientists and nonscientists alike. Scientists theorize that when massive stars run out of nuclear fuel, they begin to collapse under the force of their own gravity. As the star collapses, its density increases. In turn, the force of gravity increases so tremendously that even light cannot escape from the star. Consequently, it appears black.

A mathematical formula, called the Schwarzchild formula, describes the critical value to which the radius of a massive body must be reduced for it to become a black hole. This formula forms the basis of our next example.

EXAMPLE 6   An Application of Scientific Notation

Use the Schwarzchild formula

$$R_s = \frac{2GM}{c^2}$$

where

$R_s = $ Radius of the star, in meters, that would cause it to become a black hole

$M = $ Mass of the star, in kilograms

$G = $ A constant, called the gravitational constant

$\quad = 6.7 \times 10^{-11} \dfrac{\text{m}^3}{\text{kg} \cdot \text{s}^2}$

$c = $ Speed of light

$\quad = 3 \times 10^8$ meters per second

to determine to what length the radius of the sun must be reduced for it to become a black hole. The sun's mass is approximately 2×10^{30} kilograms.

Solution

$$R_s = \frac{2GM}{c^2}$$

$$= \frac{2 \times 6.7 \times 10^{-11} \times 2 \times 10^{30}}{(3 \times 10^8)^2} \qquad \textit{Substitute the given values.}$$

$$= \frac{(2 \times 6.7 \times 2) \times (10^{-11} \times 10^{30})}{(3 \times 10^8)^2} \qquad \textit{Rearrange factors in the numerator.}$$

$$= \frac{26.8 \times 10^{-11+30}}{3^2 \times (10^8)^2} \qquad \textit{Add exponents in the numerator. Raise each factor in the denominator to the power.}$$

$$= \frac{26.8 \times 10^{19}}{9 \times 10^{16}}$$ *Multiply powers to powers:* $(10^8)^2 = 10^{8 \cdot 2} = 10^{16}$.

$$= \frac{26.8}{9} \times 10^{19-16}$$ *When dividing expressions with the same base, subtract the exponents.*

$$\approx 2.978 \times 10^3$$

$$= 2978$$

Although the sun is not massive enough to become a black hole (its radius is approximately 700,000 kilometers), the Schwarzchild model theoretically indicates that if the sun's radius were reduced to approximately 2978 meters, that is, about $\frac{1}{235,000}$ its present size, it would become a black hole.

Check Point 6

Pouiseville's law states that the speed of blood, S, in centimeters per second, located r centimeters from the central axis of an artery is

$$S = (1.76 \times 10^5)[(1.44 \times 10^{-2}) - r^2].$$

Find the speed of blood at the central axis of this artery.

EXERCISE SET P.2

Practice Exercises

Evaluate each exponential expression in Exercises 1–22.

1. $5^2 \cdot 2$

2. $6^2 \cdot 2$

3. $(-2)^6$

4. $(-2)^4$

5. -2^6

6. -2^4

7. $(-3)^0$

8. $(-9)^0$

9. -3^0

10. -9^0

11. 4^{-3}

12. 2^{-6}

13. $2^2 \cdot 2^3$

14. $3^3 \cdot 3^2$

15. $(2^2)^3$

16. $(3^3)^2$

17. $\dfrac{2^8}{2^4}$

18. $\dfrac{3^8}{3^4}$

19. $3^{-3} \cdot 3$

20. $2^{-3} \cdot 2$

21. $\dfrac{2^3}{2^7}$

22. $\dfrac{3^4}{3^7}$

Simplify each exponential expression in Exercises 23–64.

23. $x^{-2}y$

24. xy^{-3}

25. x^0y^5

26. x^7y^0

27. $x^3 \cdot x^7$

28. $x^{11} \cdot x^5$

29. $x^{-5} \cdot x^{10}$

30. $x^{-6} \cdot x^{12}$

31. $(x^3)^7$

32. $(x^{11})^5$

33. $(x^{-5})^3$

34. $(x^{-6})^4$

35. $\dfrac{x^{14}}{x^7}$

36. $\dfrac{x^{30}}{x^{10}}$

37. $\dfrac{x^{14}}{x^{-7}}$

38. $\dfrac{x^{30}}{x^{-10}}$

39. $(8x^3)^2$

40. $(6x^4)^2$

41. $\left(-\dfrac{4}{x}\right)^3$

42. $\left(-\dfrac{6}{y}\right)^3$

43. $(-3x^2y^5)^2$

44. $(-3x^4y^6)^3$

45. $(3x^4)(2x^7)$

46. $(11x^5)(9x^{12})$

47. $(-9x^3y)(-2x^6y^4)$

48. $(-5x^4y)(-6x^7y^{11})$

49. $\dfrac{8x^{20}}{2x^4}$

50. $\dfrac{20x^{24}}{10x^6}$

51. $\dfrac{25a^{13}b^4}{-5a^2b^3}$

52. $\dfrac{35a^{14}b^6}{-7a^7b^3}$

53. $\dfrac{14b^7}{7b^{14}}$

54. $\dfrac{20b^{10}}{10b^{20}}$

55. $(4x^3)^{-2}$

56. $(10x^2)^{-3}$

57. $\dfrac{24x^3y^5}{32x^7y^{-9}}$

58. $\dfrac{10x^4y^9}{30x^{12}y^{-3}}$

59. $\left(\dfrac{5x^3}{y}\right)^{-2}$

60. $\left(\dfrac{3x^4}{y}\right)^{-3}$

61. $\left(\dfrac{-15a^4b^2}{5a^{10}b^{-3}}\right)^3$

62. $\left(\dfrac{-30a^{14}b^8}{10a^{17}b^{-2}}\right)^3$

63. $\left(\dfrac{3a^{-5}b^2}{12a^3b^{-4}}\right)^0$

64. $\left(\dfrac{4a^{-5}b^3}{12a^3b^{-5}}\right)^0$

In Exercises 65–72, write each number in decimal notation.

65. 4.7×10^3 **66.** 9.12×10^5

67. 4×10^6 **68.** 7×10^6

69. 7.86×10^{-4} **70.** 4.63×10^{-5}

71. 3.18×10^{-6} **72.** 5.84×10^{-7}

In Exercises 73–80, write each number in scientific notation.

73. 3600   **74.** 2700

75. 220,000,000   **76.** 370,000,000,000

77. 0.027   **78.** 0.014

79. 0.000763   **80.** 0.000972

In Exercises 81– 88, perform the indicated operation and express the answer in decimal notation.

81. $(2 \times 10^3)(3 \times 10^2)$ **82.** $(5 \times 10^2)(4 \times 10^4)$

83. $(4.1 \times 10^2)(3 \times 10^{-4})$ **84.** $(1.2 \times 10^3)(2 \times 10^{-5})$

85. $\dfrac{12 \times 10^6}{4 \times 10^2}$ **86.** $\dfrac{20 \times 10^{26}}{10 \times 10^{15}}$

87. $\dfrac{6.3 \times 10^3}{3 \times 10^5}$ **88.** $\dfrac{9.6 \times 10^2}{3 \times 10^{-3}}$

In Exercises 89–92, rewrite the expression so that each number is in scientific notation. Then use scientific notation to perform the operation(s). Express the answer in scientific notation.

89. $\dfrac{480{,}000{,}000{,}000}{0.00012}$ **90.** $\dfrac{282{,}000{,}000{,}000}{0.00141}$

91. $\dfrac{0.00072 \times 0.003}{0.00024}$ **92.** $\dfrac{66{,}000 \times 0.001}{0.003 \times 0.002}$

Application Exercises

Use 10^{12} for one trillion and 2.8×10^8 for the U.S. population in 2000 to solve Exercises 93–95.

93. In 2000, the government collected approximately $1.9 trillion in taxes. What was the per capita tax burden, or the amount that each U.S. citizen paid in taxes? Round to the nearest hundred dollars.

94. In 2000, U.S. personal income was $8 trillion. What was the per capita income, or the income per U.S. citizen? Round to the nearest hundred dollars.

95. In the United States, we spend an average of $4000 per person each year on health care—the highest in the world. What do we spend each year on health care nationwide? Express the answer in scientific notation.

96. Approximately 2×10^4 people run in the New York City Marathon each year. Each runner runs a distance of 26 miles. Write the total distance covered by all the runners (assuming that each person completes the marathon) in scientific notation.

97. The mass of one oxygen molecule is 5.3×10^{-23} gram. Find the mass of 20,000 molecules of oxygen. Express the answer in scientific notation.

98. The mass of one hydrogen atom is 1.67×10^{-24} gram. Find the mass of 80,000 hydrogen atoms. Express the answer in scientific notation.

Writing in Mathematics

99. Describe what it means to raise a number to a power. In your description, include a discussion of the difference between -5^2 and $(-5)^2$.

100. Explain the product rule for exponents. Use $2^3 \cdot 2^5$ in your explanation.

101. Explain the power rule for exponents. Use $(3^2)^4$ in your explanation.

102. Explain the quotient rule for exponents. Use $\dfrac{5^8}{5^2}$ in your explanation.

103. Why is $(-3x^2)(2x^{-5})$ not simplified? What must be done to simplify the expression?

104. How do you know if a number is written in scientific notation?

105. Explain how to convert from scientific to decimal notation and give an example.

106. Explain how to convert from decimal to scientific notation and give an example.

Critical Thinking Exercises

107. Which one of the following is true?

   a. $4^{-2} < 4^{-3}$ **b.** $5^{-2} > 2^{-5}$

   c. $(-2)^4 = 2^{-4}$ **d.** $5^2 \cdot 5^{-2} > 2^5 \cdot 2^{-5}$

108. The mad Dr. Frankenstein has gathered enough bits and pieces (so to speak) for $2^{-1} + 2^{-2}$ of his creature-to-be. Write a fraction that represents the amount of his creature that must still be obtained.

109. If $b^A = MN, b^C = M$, and $b^D = N$, what is the relationship among A, C, and D?

Group Exercise

110. Putting Numbers into Perspective. A large number can be put into perspective by comparing it with another number. For example, we put the $5.6 trillion national debt into perspective by comparing it to the number of U.S. citizens. The total distance covered by all the runners in the New York City Marathon (Exercise 96) can be put into perspective by comparing this distance with, say, the distance from New York to San Francisco.

   For this project, each group member should consult an almanac, a newspaper, or the World Wide Web to find a number greater than one million. Explain to other members of the group the context in which the large number is used. Express the number in scientific notation. Then put the number into perspective by comparing it with another number.

SECTION P.3 *Radicals and Rational Exponents*

Objectives

1. Evaluate square roots.
2. Use the product rule to simplify square roots.
3. .Use the quotient rule to simplify square roots.
4. Add and subtract square roots.
5. Rationalize denominators.
6. Evaluate and perform operations with higher roots.
7. Understand and use rational exponents.

What is the maximum speed at which a racing cyclist can turn a corner without tipping over? The answer, in miles per hour, is given by the algebraic expression $4\sqrt{x}$, where x is the radius of the corner, in feet. Algebraic expressions containing roots describe phenomena as diverse as a wild animal's territorial area, evaporation on a lake's surface, and Albert Einstein's bizarre concept of how an astronaut moving close to the speed of light would barely age relative to friends watching from Earth. No description of your world can be complete without roots and radicals. In this section, we review the basics of radical expressions and the use of rational exponents to indicate radicals.

1 Evaluate square roots.

Square Roots

The **principal square root** of a nonnegative real number b, written $\sqrt{b}$, is that number whose square equals b. For example,

$$\sqrt{100} = 10 \text{ because } 10^2 = 100 \quad \text{and} \quad \sqrt{0} = 0 \text{ because } 0^2 = 0.$$

Observe that the principal square root of a positive number is positive and the principal square root of 0 is 0.

The symbol $\sqrt{}$ that we use to denote the principal square root is called a **radical sign**. The number under the radical sign is called the **radicand**. Together we refer to the radical sign and its radicand as a **radical**.

The following definition summarizes our discussion:

> **Definition of the Principal Square Root**
> If a is a nonnegative real number, the nonnegative number b such that $b^2 = a$, denoted by $b = \sqrt{a}$, is the **principal square root** of a.

In the real number system, negative numbers do not have square roots. For example, $\sqrt{-9}$ is not a real number because there is no real number whose square is -9.

If a number is nonnegative ($a \geq 0$), then $(\sqrt{a})^2 = a$. For example,

$$(\sqrt{2})^2 = 2, \quad (\sqrt{3})^2 = 3, \quad (\sqrt{4})^2 = 4, \quad \text{and} \quad (\sqrt{5})^2 = 5.$$

A number that is the square of a rational number is called a **perfect square**. For example,

64 is a perfect square because $64 = 8^2$.

$\frac{1}{9}$ is a perfect square because $\frac{1}{9} = \left(\frac{1}{3}\right)^2$.

The following rule can be used to find square roots of perfect squares:

> **Square Roots of Perfect Squares**
> $$\sqrt{a^2} = |a|$$

For example, $\sqrt{6^2} = 6$ and $\sqrt{(-6)^2} = |-6| = 6$.

② Use the product rule to simplify square roots.

The Product Rule for Square Roots

A square root is **simplified** when its radicand has no factors other than 1 that are perfect squares. For example, $\sqrt{500}$ is not simplified because it can be expressed as $\sqrt{100 \cdot 5}$ and $\sqrt{100}$ is a perfect square. The **product rule for square roots** can be used to simplify $\sqrt{500}$.

> **The Product Rule for Square Roots**
> If a and b represent nonnegative real numbers, then
> $$\sqrt{ab} = \sqrt{a}\sqrt{b} \quad \text{and} \quad \sqrt{a}\sqrt{b} = \sqrt{ab}.$$
> The square root of a product is the product of the square roots.

Example 1 shows how the product rule is used to remove from the square root any perfect squares that occur as factors.

EXAMPLE 1   Using the Product Rule to Simplify Square Roots

Simplify: **a.** $\sqrt{500}$ **b.** $\sqrt{6x} \cdot \sqrt{3x}$.

Solution

a. $\sqrt{500} = \sqrt{100 \cdot 5}$ 100 is the largest perfect square factor of 500.
$\phantom{\sqrt{500}} = \sqrt{100}\sqrt{5}$ $\sqrt{ab} = \sqrt{a}\sqrt{b}$
$\phantom{\sqrt{500}} = 10\sqrt{5}$ $\sqrt{100} = 10$

b. We can simplify $\sqrt{6x} \cdot \sqrt{3x}$ using the power rule only if $6x$ and $3x$ represent nonnegative real numbers. Thus, $x \geq 0$.

$\sqrt{6x} \cdot \sqrt{3x} = \sqrt{6x \cdot 3x}$ $\sqrt{a}\sqrt{b} = \sqrt{ab}$
$\phantom{\sqrt{6x} \cdot \sqrt{3x}} = \sqrt{18x^2}$ Multiply.
$\phantom{\sqrt{6x} \cdot \sqrt{3x}} = \sqrt{9x^2 \cdot 2}$ 9 is the largest perfect square factor of 18.
$\phantom{\sqrt{6x} \cdot \sqrt{3x}} = \sqrt{9x^2}\sqrt{2}$ $\sqrt{ab} = \sqrt{a}\sqrt{b}$
$\phantom{\sqrt{6x} \cdot \sqrt{3x}} = \sqrt{9}\sqrt{x^2}\sqrt{2}$ Split $\sqrt{9x^2}$ into two square roots.
$\phantom{\sqrt{6x} \cdot \sqrt{3x}} = 3x\sqrt{2}$ $\sqrt{9} = 3$ (because $3^2 = 9$) and $\sqrt{x^2} = x$ because $x \geq 0$.

Check Point 1 Simplify:
  a. $\sqrt{3^2}$ **b.** $\sqrt{5x} \cdot \sqrt{10x}$.

3 Use the quotient rule to simplify square roots.

The Quotient Rule for Square Roots

Another property for square roots involves division.

> **The Quotient Rule for Square Roots**
>
> If a and b represent nonnegative real numbers and $b \neq 0$, then
>
> $$\frac{\sqrt{a}}{\sqrt{b}} = \sqrt{\frac{a}{b}} \quad \text{and} \quad \sqrt{\frac{a}{b}} = \frac{\sqrt{a}}{\sqrt{b}}.$$
>
> The square root of a quotient is the quotient of the square roots.

EXAMPLE 2 Using the Quotient Rule to Simplify Square Roots

Simplify: **a.** $\sqrt{\dfrac{100}{9}}$ **b.** $\dfrac{\sqrt{48x^3}}{\sqrt{6x}}$.

Solution

a. $\sqrt{\dfrac{100}{9}} = \dfrac{\sqrt{100}}{\sqrt{9}} = \dfrac{10}{3}$

b. We can simplify the quotient of $\sqrt{48x^3}$ and $\sqrt{6x}$ using the quotient rule only if $48x^3$ and $6x$ represent nonnegative real numbers. Thus, $x \geq 0$.

$$\frac{\sqrt{48x^3}}{\sqrt{6x}} = \sqrt{\frac{48x^3}{6x}} = \sqrt{8x^2} = \sqrt{4x^2}\sqrt{2} = \sqrt{4}\sqrt{x^2}\sqrt{2} = 2x\sqrt{2}$$

$\sqrt{x^2} = x$ because $x \geq 0$.

Check Point 2 Simplify: **a.** $\sqrt{\dfrac{25}{16}}$ **b.** $\dfrac{\sqrt{150x^3}}{\sqrt{2x}}$.

4 Add and subtract square roots.

Adding and Subtracting Square Roots

Two or more square roots can be combined provided that they have the same radicand. Such radicals are called **like radicals**. For example,

$$7\sqrt{11} + 6\sqrt{11} = (7 + 6)\sqrt{11} = 13\sqrt{11}.$$

EXAMPLE 3 Adding and Subtracting Like Radicals

Add or subtract as indicated:
  a. $7\sqrt{2} + 5\sqrt{2}$ **b.** $\sqrt{5x} - 7\sqrt{5x}$.

A Radical Idea: Time is Relative

What does travel in space have to do with radicals? Imagine that in the future we will be able to travel at velocities approaching the speed of light (approximately 186,000 miles per second).

According to Einstein's theory of relativity, time would pass more quickly on Earth than it would in the moving spaceship. The expression

$$R_f\sqrt{1 - \left(\frac{v}{c}\right)^2}$$

gives the aging rate of an astronaut relative to the aging rate of a friend on Earth, R_f. In the expression, v is the astronaut's speed and c is the speed of light. As the astronaut's speed approaches the speed of light, we can substitute c for v:

$$R_f\sqrt{1 - \left(\frac{v}{c}\right)^2} \quad \text{Let } v = c.$$
$$= R_f\sqrt{1 - 1^2}$$
$$= R_f\sqrt{0} = 0$$

Close to the speed of light, the astronaut's aging rate relative to a friend on Earth is nearly 0. What does this mean? As we age here on Earth, the space traveler would barely get older. The space traveler would return to a futuristic world in which friends and loved ones would be long dead.

Solution

a. $7\sqrt{2} + 5\sqrt{2} = (7 + 5)\sqrt{2}$ Apply the distributive property.
$$= 12\sqrt{2} \quad \text{Simplify.}$$

b. $\sqrt{5x} - 7\sqrt{5x} = 1\sqrt{5x} - 7\sqrt{5x}$ Write $\sqrt{5x}$ as $1\sqrt{5x}$.
$$= (1 - 7)\sqrt{5x} \quad \text{Apply the distributive property.}$$
$$= -6\sqrt{5x} \quad \text{Simplify.}$$

Check Point 3   Add or subtract as indicated:

a. $8\sqrt{13} + 9\sqrt{13}$ **b.** $\sqrt{17x} - 20\sqrt{17x}$.

In some cases, radicals can be combined once they have been simplified. For example, to add $\sqrt{2}$ and $\sqrt{8}$, we can write $\sqrt{8}$ as $\sqrt{4 \cdot 2}$ because 4 is a perfect square factor of 8.

$$\sqrt{2} + \sqrt{8} = \sqrt{2} + \sqrt{4 \cdot 2} = 1\sqrt{2} + 2\sqrt{2} = (1 + 2)\sqrt{2} = 3\sqrt{2}$$

EXAMPLE 4   Combining Radicals That First Require Simplification

Add or subtract as indicated:
a. $7\sqrt{3} + \sqrt{12}$ **b.** $4\sqrt{50x} - 6\sqrt{32x}$.

Solution

a. $7\sqrt{3} + \sqrt{12}$
$$= 7\sqrt{3} + \sqrt{4 \cdot 3} \quad \text{Split 12 into two factors such that one is a perfect square.}$$
$$= 7\sqrt{3} + 2\sqrt{3} \quad \sqrt{4 \cdot 3} = \sqrt{4}\sqrt{3} = 2\sqrt{3}$$
$$= (7 + 2)\sqrt{3} \quad \text{Apply the distributive property. You will find that this step is usually done mentally.}$$
$$= 9\sqrt{3} \quad \text{Simplify.}$$

b. $4\sqrt{50x} - 6\sqrt{32x}$
$$= 4\sqrt{25 \cdot 2x} - 6\sqrt{16 \cdot 2x} \quad \text{25 is the largest perfect square factor of 50 and 16 is the largest perfect square factor of 32.}$$
$$= 4 \cdot 5\sqrt{2x} - 6 \cdot 4\sqrt{2x} \quad \sqrt{25 \cdot 2} = \sqrt{25}\sqrt{2} = 5\sqrt{2} \text{ and } \sqrt{16 \cdot 2} = \sqrt{16}\sqrt{2} = 4\sqrt{2}.$$
$$= 20\sqrt{2x} - 24\sqrt{2x} \quad \text{Multiply.}$$
$$= (20 - 24)\sqrt{2x} \quad \text{Apply the distributive property.}$$
$$= -4\sqrt{2x} \quad \text{Simplify.}$$

Check Point 4   Add or subtract as indicated:

a. $5\sqrt{27} + \sqrt{12}$ **b.** $6\sqrt{18x} - 4\sqrt{8x}$.

5 Rationalize denominators.

Rationalizing Denominators

You can use a calculator to compare the approximate values for $\dfrac{1}{\sqrt{3}}$ and $\dfrac{\sqrt{3}}{3}$. The two approximations are the same. This is not a coincidence:

$$\frac{1}{\sqrt{3}} = \frac{1}{\sqrt{3}} \cdot \boxed{\frac{\sqrt{3}}{\sqrt{3}}} = \frac{\sqrt{3}}{\sqrt{9}} = \frac{\sqrt{3}}{3}.$$

> Any number divided by itself is 1. Multiplication by 1 does not change the value of $\dfrac{1}{\sqrt{3}}$.

This process involves rewriting a radical expression as an equivalent expression in which the denominator no longer contains a radical. The process is called **rationalizing the denominator**. If the denominator contains the square root of a natural number that is not a perfect square, **multiply the numerator and denominator by the smallest number that produces the square root of a perfect square in the denominator**.

EXAMPLE 5   Rationalizing Denominators

Rationalize the denominator: **a.** $\dfrac{15}{\sqrt{6}}$ **b.** $\dfrac{12}{\sqrt{8}}$.

Solution

a. If we multiply numerator and denominator by $\sqrt{6}$, the denominator becomes $\sqrt{6} \cdot \sqrt{6} = \sqrt{36} = 6$. Therefore, we multiply by 1, choosing $\dfrac{\sqrt{6}}{\sqrt{6}}$ for 1.

$$\frac{15}{\sqrt{6}} = \frac{15}{\sqrt{6}} \cdot \frac{\sqrt{6}}{\sqrt{6}} = \frac{15\sqrt{6}}{\sqrt{36}} = \frac{15\sqrt{6}}{6} = \frac{5\sqrt{6}}{2}$$

Multiply by 1.   Simplify: $\dfrac{15}{6} = \dfrac{15 \div 3}{6 \div 3} = \dfrac{5}{2}$.

b. The *smallest* number that will produce a perfect square in the denominator of $\dfrac{12}{\sqrt{8}}$ is $\sqrt{2}$, because $\sqrt{8} \cdot \sqrt{2} = \sqrt{16} = 4$. We multiply by 1, choosing $\dfrac{\sqrt{2}}{\sqrt{2}}$ for 1.

$$\frac{12}{\sqrt{8}} = \frac{12}{\sqrt{8}} \cdot \frac{\sqrt{2}}{\sqrt{2}} = \frac{12\sqrt{2}}{\sqrt{16}} = \frac{12\sqrt{2}}{4} = 3\sqrt{2}$$

Check Point 5   Rationalize the denominator: **a.** $\dfrac{5}{\sqrt{3}}$ **b.** $\dfrac{6}{\sqrt{12}}$.

How can we rationalize a denominator if the denominator contains two terms? In general,

$$(\sqrt{a} + \sqrt{b})(\sqrt{a} - \sqrt{b}) = (\sqrt{a})^2 - (\sqrt{b})^2 = a - b.$$

Notice that the product does not contain a radical. Here are some specific examples.

The Denominator Contains:	Multiply by:	The New Denominator Contains:
$7 + \sqrt{5}$	$7 - \sqrt{5}$	$7^2 - (\sqrt{5})^2 = 49 - 5 = 44$
$\sqrt{3} - 6$	$\sqrt{3} + 6$	$(\sqrt{3})^2 - 6^2 = 3 - 36 = -33$
$\sqrt{7} + \sqrt{3}$	$\sqrt{7} - \sqrt{3}$	$(\sqrt{7})^2 - (\sqrt{3})^2 = 7 - 3 = 4$

EXAMPLE 6 **Rationalizing a Denominator Containing Two Terms**

Rationalize the denominator: $\dfrac{7}{5 + \sqrt{3}}$.

Solution If we multiply the numerator and denominator by $5 - \sqrt{3}$, the denominator will not contain a radical. Therefore, we multiply by 1, choosing $\dfrac{5 - \sqrt{3}}{5 - \sqrt{3}}$ for 1.

$$\frac{7}{5 + \sqrt{3}} = \frac{7}{5 + \sqrt{3}} \cdot \frac{5 - \sqrt{3}}{5 - \sqrt{3}} = \frac{7(5 - \sqrt{3})}{5^2 - (\sqrt{3})^2} = \frac{7(5 - \sqrt{3})}{25 - 3}$$

Multiply by 1.

$$= \frac{7(5 - \sqrt{3})}{22} \quad \text{or} \quad \frac{35 - 7\sqrt{3}}{22}.$$

In either form of the answer, there is no radical in the denominator.

Check Point 6 Rationalize the denominator: $\dfrac{8}{4 + \sqrt{5}}$.

6 Evaluate and perform operations with higher roots.

Other Kinds of Roots

We define the **principal nth root** of a real number a, symbolized by $\sqrt[n]{a}$, as follows:

Definition of the Principal nth Root of a Real Number

$$\sqrt[n]{a} = b \text{ means that } b^n = a.$$

If n, the **index**, is even, then a is nonnegative ($a \geq 0$) and b is also nonnegative ($b \geq 0$). If n is odd, a and b can be any real numbers.

For example,

$$\sqrt[3]{64} = 4 \text{ because } 4^3 = 64 \quad \text{and} \quad \sqrt[5]{-32} = -2 \text{ because } (-2)^5 = -32.$$

The same vocabulary that we learned for square roots applies to nth roots. The symbol $\sqrt[n]{a}$ is called a **radical** and a is called the **radicand**.

A number that is the nth power of a rational number is called a **perfect nth power**. For example, 8 is a perfect third power, or perfect cube, because $8 = 2^3$. In general, one of the following rules can be used to find nth roots of perfect nth powers:

Finding nth Roots of Perfect nth Powers

If n is odd, $\sqrt[n]{a^n} = a$.

If n is even, $\sqrt[n]{a^n} = |a|$.

For example,

$$\sqrt[3]{(-2)^3} = -2 \quad \text{and} \quad \sqrt[4]{(-2)^4} = |-2| = 2.$$

Absolute value is not needed with odd roots, but is necessary with even roots.

The Product and Quotient Rules for Other Roots

The product and quotient rules apply to cube roots, fourth roots, and all higher roots.

The Product and Quotient Rules for nth Roots

For all real numbers, where the indicated roots represent real numbers,

$$\sqrt[n]{a} \cdot \sqrt[n]{b} = \sqrt[n]{ab} \quad \text{and} \quad \frac{\sqrt[n]{a}}{\sqrt[n]{b}} = \sqrt[n]{\frac{a}{b}}, b \neq 0.$$

EXAMPLE 7 Simplifying, Multiplying, and Dividing Higher Roots

Simplify: **a.** $\sqrt[3]{24}$ **b.** $\sqrt[4]{8} \cdot \sqrt[4]{4}$ **c.** $\sqrt[4]{\frac{81}{16}}$.

Solution

a. $\sqrt[3]{24} = \sqrt[3]{8 \cdot 3}$ Find the largest perfect cube that is a factor of 24. $\sqrt[3]{8} = 2$, so 8 is a perfect cube and is the largest perfect cube factor of 24.

Study Tip

Some higher even and odd roots occur so frequently that you might want to memorize them.

Cube Roots	
$\sqrt[3]{1} = 1$	$\sqrt[3]{125} = 5$
$\sqrt[3]{8} = 2$	$\sqrt[3]{216} = 6$
$\sqrt[3]{27} = 3$	$\sqrt[3]{1000} = 10$
$\sqrt[3]{64} = 4$	

Fourth Roots	Fifth Roots
$\sqrt[4]{1} = 1$	$\sqrt[5]{1} = 1$
$\sqrt[4]{16} = 2$	$\sqrt[5]{32} = 2$
$\sqrt[4]{81} = 3$	$\sqrt[5]{243} = 3$
$\sqrt[4]{256} = 4$	
$\sqrt[4]{625} = 5$	

$$= \sqrt[3]{8} \cdot \sqrt[3]{3} \qquad \sqrt[n]{ab} = \sqrt[n]{a}\,\sqrt[n]{b}$$
$$= 2\sqrt[3]{3}$$

b. $\sqrt[4]{8} \cdot \sqrt[4]{4} = \sqrt[4]{8 \cdot 4} \qquad \sqrt[n]{a} \cdot \sqrt[n]{b} = \sqrt[n]{ab}$

$$= \sqrt[4]{32}$$

Find the largest perfect fourth power that is a factor of 32.

$$= \sqrt[4]{16 \cdot 2}$$

$\sqrt[4]{16} = 2$, so 16 is a perfect fourth power and is the largest perfect fourth power that is a factor of 32.

$$= \sqrt[4]{16} \cdot \sqrt[4]{2} \qquad \sqrt[n]{ab} = \sqrt[n]{a} \cdot \sqrt[n]{b}$$
$$= 2\sqrt[4]{2}$$

c. $\sqrt[4]{\dfrac{81}{16}} = \dfrac{\sqrt[4]{81}}{\sqrt[4]{16}} \qquad \sqrt[n]{\dfrac{a}{b}} = \dfrac{\sqrt[n]{a}}{\sqrt[n]{b}}$

$$= \dfrac{3}{2}$$

$\sqrt[4]{81} = 3$ because $3^4 = 81$ and $\sqrt[4]{16} = 2$ because $2^4 = 16$.

Check Point 7 Simplify: **a.** $\sqrt[3]{40}$ **b.** $\sqrt[5]{8} \cdot \sqrt[5]{8}$ **c.** $\sqrt[3]{\dfrac{125}{27}}$.

We have seen that adding and subtracting square roots often involves simplifying terms. The same idea applies to adding and subtracting nth roots.

EXAMPLE 8   Combining Cube Roots

Subtract: $5\sqrt[3]{16} - 11\sqrt[3]{2}$.

Solution $5\sqrt[3]{16} - 11\sqrt[3]{2}$

$$= 5\sqrt[3]{8 \cdot 2} - 11\sqrt[3]{2}$$

Because $16 = 8 \cdot 2$ and $\sqrt[3]{8} = 2$, 8 is the largest perfect cube that is a factor of 16.

$$= 5 \cdot 2\sqrt[3]{2} - 11\sqrt[3]{2} \qquad \sqrt[3]{8 \cdot 2} = \sqrt[3]{8}\sqrt[3]{2} = 2\sqrt[3]{2}$$

$$= 10\sqrt[3]{2} - 11\sqrt[3]{2} \qquad \text{Multiply.}$$

$$= (10 - 11)\sqrt[3]{2} \qquad \text{Apply the distributive property.}$$

$$= -1\sqrt[3]{2} \text{ or } -\sqrt[3]{2} \qquad \text{Simplify.}$$

Check Point 8 Subtract: $3\sqrt[3]{81} - 4\sqrt[3]{3}$.

7 Understand and use rational exponents.

Rational Exponents

Animals in the wild have regions to which they confine their movement, called their territorial area. Territorial area, in square miles, is related to an animal's body weight. If an animal weighs W pounds, its territorial area is

$$W^{141/100}$$

square miles.

W to the *what* power?! How can we interpret the information given by this algebraic expression?

In the last part of this section, we turn our attention to rational exponents such as $\frac{141}{100}$ and their relationship to roots of real numbers.

Definition of Rational Exponents

If $\sqrt[n]{a}$ represents a real number and $n \geq 2$ is an integer, then

$$a^{1/n} = \sqrt[n]{a}.$$

Furthermore,

$$a^{-1/n} = \frac{1}{a^{1/n}} = \frac{1}{\sqrt[n]{a}}, a \neq 0.$$

EXAMPLE 9 Using the Definition of $a^{1/n}$

Simplify: **a.** $64^{1/2}$ **b.** $8^{1/3}$ **c.** $64^{-1/3}$.

Solution

   a. $64^{1/2} = \sqrt{64} = 8$

   b. $8^{1/3} = \sqrt[3]{8} = 2$

   c. $64^{-1/3} = \dfrac{1}{64^{1/3}} = \dfrac{1}{\sqrt[3]{64}} = \dfrac{1}{4}$

Check Point 9   Simplify: **a.** $81^{1/2}$ **b.** $27^{1/3}$ **c.** $32^{-1/5}$.

Note that every rational exponent in Example 9 has a numerator of 1 or -1. We now define rational exponents with any integer in the numerator.

Definition of Rational Exponents

If $\sqrt[n]{a}$ represents a real number, $\dfrac{m}{n}$ is a rational number reduced to lowest terms, and $n \geq 2$ is an integer, then

$$a^{m/n} = (\sqrt[n]{a})^m = \sqrt[n]{a^m}.$$

The exponent m/n consists of two parts: the denominator n is the root and the numerator m is the exponent. Furthermore,

$$a^{-m/n} = \frac{1}{a^{m/n}}.$$

EXAMPLE 10 Using the Definition of $a^{m/n}$

Simplify: **a.** $27^{2/3}$ **b.** $9^{3/2}$ **c.** $16^{-3/4}$.

Technology

Here are the calculator keystroke sequences for $27^{2/3}$:

Many Scientific Calculators

27 $\boxed{y^x}$ $\boxed{(}$ 2 $\boxed{\div}$ 3 $\boxed{)}$ $\boxed{=}$

Many Graphing Calculators

27 $\boxed{\wedge}$ $\boxed{(}$ 2 $\boxed{\div}$ 3 $\boxed{)}$ $\boxed{\text{ENTER}}$.

Solution

a. $27^{2/3} = (\sqrt[3]{27})^2 = 3^2 = 9$

The denominator of $\frac{2}{3}$ is the root and the numerator is the exponent.

b. $9^{3/2} = (\sqrt{9})^3 = 3^3 = 27$

c. $16^{-3/4} = \dfrac{1}{16^{3/4}} = \dfrac{1}{(\sqrt[4]{16})^3} = \dfrac{1}{2^3} = \dfrac{1}{8}$

Check Point 10 Simplify: **a.** $4^{3/2}$ **b.** $32^{-2/5}$.

Properties of exponents can be applied to expressions containing rational exponents.

EXAMPLE 11 Simplifying Expressions with Rational Exponents

Simplify using properties of exponents:

a. $(5x^{1/2})(7x^{3/4})$ **b.** $\dfrac{32x^{5/3}}{16x^{3/4}}$.

Solution

a. $(5x^{1/2})(7x^{3/4}) = 5 \cdot 7 x^{1/2} \cdot x^{3/4}$ — Group factors with the same base.

$= 35x^{(1/2)+(3/4)}$ — When multiplying expressions with the same base, add the exponents.

$= 35x^{5/4}$ — $\frac{1}{2} + \frac{3}{4} = \frac{2}{4} + \frac{3}{4} = \frac{5}{4}$

b. $\dfrac{32x^{5/3}}{16x^{3/4}} = \left(\dfrac{32}{16}\right)\left(\dfrac{x^{5/3}}{x^{3/4}}\right)$ — Group factors with the same base.

$= 2x^{(5/3)-(3/4)}$ — When dividing expressions with the same base, subtract the exponents.

$= 2x^{11/12}$ — $\frac{5}{3} - \frac{3}{4} = \frac{20}{12} - \frac{9}{12} = \frac{11}{12}$

Check Point 11 Simplify: **a.** $(2x^{4/3})(5x^{8/3})$ **b.** $\dfrac{20x^4}{5x^{3/2}}$.

Rational exponents are sometimes useful for simplifying radicals by reducing their index.

EXAMPLE 12   Reducing the Index of a Radical

Simplify: $\sqrt[9]{x^3}$.

Solution $\sqrt[9]{x^3} = x^{3/9} = x^{1/3} = \sqrt[3]{x}$

Check Point 12   Simplify: $\sqrt[6]{x^3}$.

EXERCISE SET P.3

Practice Exercises

Evaluate each expression in Exercises 1–6 or indicate that the root is not a real number.

1. $\sqrt{36}$

2. $\sqrt{25}$

3. $\sqrt{-36}$

4. $\sqrt{-25}$

5. $\sqrt{(-13)^2}$

6. $\sqrt{(-17)^2}$

Use the product rule to simplify the expressions in Exercises 7–16. In Exercises 11–16, assume that variables represent nonnegative real numbers.

7. $\sqrt{50}$

8. $\sqrt{27}$

9. $\sqrt{45x^2}$

10. $\sqrt{125x^2}$

11. $\sqrt{2x} \cdot \sqrt{6x}$

12. $\sqrt{10x} \cdot \sqrt{8x}$

13. $\sqrt{x^3}$

14. $\sqrt{y^3}$

15. $\sqrt{2x^2} \cdot \sqrt{6x}$

16. $\sqrt{6x} \cdot \sqrt{3x^2}$

Use the quotient rule to simplify the expressions in Exercises 17–26. Assume that x > 0.

17. $\sqrt{\dfrac{1}{81}}$

18. $\sqrt{\dfrac{1}{49}}$

19. $\sqrt{\dfrac{49}{16}}$

20. $\sqrt{\dfrac{121}{9}}$

21. $\dfrac{\sqrt{48x^3}}{\sqrt{3x}}$

22. $\dfrac{\sqrt{72x^3}}{\sqrt{8x}}$

23. $\dfrac{\sqrt{150x^4}}{\sqrt{3x}}$

24. $\dfrac{\sqrt{24x^4}}{\sqrt{3x}}$

25. $\dfrac{\sqrt{200x^3}}{\sqrt{10x^{-1}}}$

26. $\dfrac{\sqrt{500x^3}}{\sqrt{10x^{-1}}}$

In Exercises 27–38, add or subtract terms whenever possible.

27. $7\sqrt{3} + 6\sqrt{3}$

28. $8\sqrt{5} + 11\sqrt{5}$

29. $6\sqrt{17x} - 8\sqrt{17x}$

30. $4\sqrt{13x} - 6\sqrt{13x}$

31. $\sqrt{8} + 3\sqrt{2}$

32. $\sqrt{20} + 6\sqrt{5}$

33. $\sqrt{50x} - \sqrt{8x}$

34. $\sqrt{63x} - \sqrt{28x}$

35. $3\sqrt{18} + 5\sqrt{50}$

36. $4\sqrt{12} - 2\sqrt{75}$

37. $3\sqrt{8} - \sqrt{32} + 3\sqrt{72} - \sqrt{75}$

38. $3\sqrt{54} - 2\sqrt{24} - \sqrt{96} + 4\sqrt{63}$

In Exercises 39–48, rationalize the denominator.

39. $\dfrac{1}{\sqrt{7}}$

40. $\dfrac{2}{\sqrt{10}}$

41. $\dfrac{\sqrt{2}}{\sqrt{5}}$

42. $\dfrac{\sqrt{7}}{\sqrt{3}}$

43. $\dfrac{13}{3 + \sqrt{11}}$

44. $\dfrac{3}{3 + \sqrt{7}}$

45. $\dfrac{7}{\sqrt{5} - 2}$

46. $\dfrac{5}{\sqrt{3} - 1}$

47. $\dfrac{6}{\sqrt{5} + \sqrt{3}}$

48. $\dfrac{11}{\sqrt{7} - \sqrt{3}}$

Evaluate each expression in Exercises 49–60, or indicate that the root is not a real number.

49. $\sqrt[3]{125}$

50. $\sqrt[3]{8}$

51. $\sqrt[3]{-8}$

52. $\sqrt[3]{-125}$

53. $\sqrt[4]{-16}$

54. $\sqrt[4]{-81}$

55. $\sqrt[4]{(-3)^4}$

56. $\sqrt[4]{(-2)^4}$

57. $\sqrt[5]{(-3)^5}$

58. $\sqrt[5]{(-2)^5}$

59. $\sqrt[5]{-\frac{1}{32}}$

60. $\sqrt[6]{\frac{1}{64}}$

Simplify the radical expressions in Exercises 61–68.

61. $\sqrt[3]{32}$

62. $\sqrt[3]{150}$

63. $\sqrt[3]{x^4}$

64. $\sqrt[3]{x^5}$

65. $\sqrt[3]{9} \cdot \sqrt[3]{6}$

66. $\sqrt[3]{12} \cdot \sqrt[3]{4}$

67. $\dfrac{\sqrt[5]{64x^6}}{\sqrt[5]{2x}}$

68. $\dfrac{\sqrt[4]{162x^5}}{\sqrt[4]{2x}}$ (Assume that x > 0.)

In Exercises 69–76, add or subtract terms whenever possible.

69. $4\sqrt[5]{2} + 3\sqrt[5]{2}$

70. $6\sqrt[5]{3} + 2\sqrt[5]{3}$

71. $5\sqrt[3]{16} + \sqrt[3]{54}$

72. $3\sqrt[3]{24} + \sqrt[3]{81}$

73. $\sqrt[3]{54xy^3} - y\sqrt[3]{128x}$

74. $\sqrt[3]{24xy^3} - y\sqrt[3]{81x}$

75. $\sqrt{2} + \sqrt[3]{8}$

76. $\sqrt{3} + \sqrt[3]{15}$

In Exercises 77–84, evaluate each expression without using a calculator.

77. $36^{1/2}$

78. $121^{1/2}$

79. $8^{1/3}$

80. $27^{1/3}$

81. $125^{2/3}$

82. $8^{2/3}$

83. $32^{-4/5}$ **84.** $16^{-5/2}$

In Exercises 85–94, simplify using properties of exponents.

85. $(7x^{1/3})(2x^{1/4})$ **86.** $(3x^{2/3})(4x^{3/4})$

87. $\dfrac{20x^{1/2}}{5x^{1/4}}$ **88.** $\dfrac{72x^{3/4}}{9x^{1/3}}$

89. $(x^{2/3})^3$ **90.** $(x^{4/5})^5$

91. $(25x^4y^6)^{1/2}$ **92.** $(125x^9y^6)^{1/3}$

93. $\dfrac{(3y^{1/4})^3}{y^{1/12}}$ **94.** $\dfrac{(2y^{1/5})^4}{y^{3/10}}$

In Exercises 95–102, simplify by reducing the index of the radical.

95. $\sqrt[4]{5^2}$ **96.** $\sqrt[4]{7^2}$

97. $\sqrt[3]{x^6}$ **98.** $\sqrt[4]{x^{12}}$

99. $\sqrt[6]{x^4}$ **100.** $\sqrt[9]{x^6}$

101. $\sqrt[9]{x^6y^3}$ **102.** $\sqrt[12]{x^4y^8}$

Application Exercises

103. The algebraic expression $2\sqrt{5L}$ is used to estimate the speed of a car prior to an accident, in miles per hour, based on the length of its skid marks, L, in feet. Find the speed of a car that left skid marks 40 feet long, and write the answer in simplified radical form.

104. The time, in seconds, that it takes an object to fall a distance d, in feet, is given by the algebraic expression $\sqrt{\dfrac{d}{16}}$. Find how long it will take a ball dropped from the top of a building 320 feet tall to hit the ground. Write the answer in simplified radical form.

105. The early Greeks believed that the most pleasing of all rectangles were golden rectangles whose ratio of width to height is

$$\frac{w}{h} = \frac{2}{\sqrt{5} - 1}.$$

Rationalize the denominator for this ratio and then use a calculator to approximate the answer correct to the nearest hundredth.

106. The amount of evaporation, in inches per day, of a large body of water can be described by the algebraic expression

$$\frac{w}{20\sqrt{a}}$$

where

    a = surface area of the water, in square miles

    w = average wind speed of the air over the water, in miles per hour.

Determine the evaporation on a lake whose surface area is 9 square miles on a day when the wind speed over the water is 10 miles per hour.

107. In the Peanuts cartoon shown below, Woodstock appears to be working steps mentally. Fill in the missing steps that show how to go from

$$\frac{7\sqrt{2 \cdot 2 \cdot 3}}{6} \quad \text{to} \quad \frac{7}{3}\sqrt{3}.$$

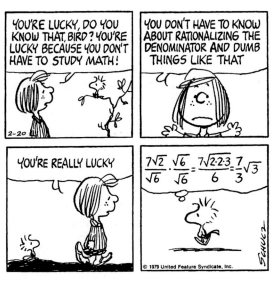

PEANUTS reprinted by permission of United Feature Syndicate, Inc.

108. The algebraic expression $152a^{-1/5}$ describes the percentage of U.S. taxpayers who are a years old who file early. Evaluate the algebraic expression for $a = 32$. Describe what the answer means in practical terms.

109. The algebraic expression $0.07d^{3/2}$ describes the duration of a storm, in hours, whose diameter is d miles. Evaluate the algebraic expression for $d = 9$. Describe what the answer means in practical terms.

Writing in Mathematics

110. Explain how to simplify $\sqrt{10} \cdot \sqrt{5}$.

111. Explain how to add $\sqrt{3} + \sqrt{12}$.

112. Describe what it means to rationalize a denominator. Use both $\dfrac{1}{\sqrt{5}}$ and $\dfrac{1}{5 + \sqrt{5}}$ in your explanation.

113. What difference is there in simplifying $\sqrt[3]{(-5)^3}$ and $\sqrt[4]{(-5)^4}$?

114. What does $a^{m/n}$ mean?

115. Describe the kinds of numbers that have rational fifth roots.

116. Why must a and b represent nonnegative numbers when we write $\sqrt{a} \cdot \sqrt{b} = \sqrt{ab}$? Is it necessary to use this restriction in the case of $\sqrt[3]{a} \cdot \sqrt[3]{b} = \sqrt[3]{ab}$? Explain.

Technology Exercises

117. The algebraic expression

$$\frac{73t^{1/3} - 28t^{2/3}}{t}$$

describes the percentage of people in the United States applying for jobs t years after 1985 who tested positive for illegal drugs. Use a calculator to find the percentage who tested positive from 1986 through 2001. Round answers to the nearest hundredth of a percent. What trend do you observe for the percentage of potential employees testing positive for illegal drugs over time?

118. The territorial area of an animal in the wild is defined to be the area of the region to which the animal confines its movements. The algebraic expression $W^{1.41}$ describes the territorial area, in square miles, of an animal that weighs W pounds. Use a calculator to find the territorial area of animals weighing 25, 50, 150, 200, 250, and 300 pounds. What do the values indicate about the relationship between body weight and territorial area?

Critical Thinking Exercises

119. Which one of the following is true?
  a. Neither $(-8)^{1/2}$ nor $(-8)^{1/3}$ represent real numbers.
  b. $\sqrt{x^2 + y^2} = x + y$
  c. $8^{-1/3} = -2$
  d. $2^{1/2} \cdot 2^{1/2} = 2$

In Exercises 120–121, fill in each box to make the statement true.

120. $(5 + \sqrt{\square})(5 - \sqrt{\square}) = 22$
121. $\sqrt{\square x^{\square}} = 5x^7$

122. Find exact value of $\sqrt{13 + \sqrt{2} + \dfrac{7}{3 + \sqrt{2}}}$ without the use of a calculator.

123. Place the correct symbol, $>$ or $<$, in the box between each of the given numbers. *Do not use a calculator.* Then check your result with a calculator.
  a. $3^{1/2} \square 3^{1/3}$
  b. $\sqrt{7} + \sqrt{18} \square \sqrt{7 + 18}$

SECTION P.4 *Polynomials*

Objectives

1. Understand the vocabulary of polynomials.
2. Add and subtract polynomials.
3. Multiply polynomials.
4. Use FOIL in polynomial multiplication.
5. Use special products in polynomial multiplication.
6. Perform operations with polynomials in several variables.

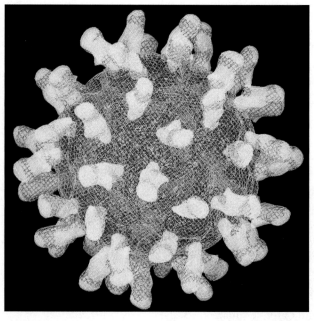

This computer-simulated model of the common cold virus was developed by researchers at Purdue University. Their discovery of how the virus infects human cells could lead to more effective treatment for the illness.

Runny nose? Sneezing? You are probably familiar with the unpleasant onset of a cold. We "catch cold" when the cold virus enters our bodies, where it multiplies. Fortunately, at a certain point the virus begins to die. The algebraic expression $-0.75x^4 + 3x^3 + 5$ describes the billions of viral particles in our bodies after x days of invasion. The expression enables mathematicians to determine the day on which there is a maximum number of viral particles and, consequently, the day we feel sickest.

The algebraic expression $-0.75x^4 + 3x^3 + 5$ is an example of a polynomial. A **polynomial** is a single term or the sum of two or more terms containing variables with whole number exponents. This particular polynomial contains three terms. Equations containing polynomials are used in such diverse areas as science, business, medicine, psychology, and sociology. In this section, we review basic ideas about polynomials and their operations.

1 Understand the vocabulary of polynomials.

The Vocabulary of Polynomials

Consider the polynomial

$$7x^3 - 9x^2 + 13x - 6.$$

We can express this polynomial as

$$7x^3 + (-9x^2) + 13x + (-6).$$

The polynomial contains four terms. It is customary to write the terms in the order of descending powers of the variable. This is the **standard form** of a polynomial.

We begin this section by limiting our discussion to polynomials containing only one variable. Each term of a polynomial in x is of the form ax^n. The **degree** of ax^n is n. For example, the degree of the term $7x^3$ is 3.

Study Tip

We can express 0 in many ways, including $0x$, $0x^2$, and $0x^3$. It is impossible to assign a single exponent on the variable. This is why 0 has no defined degree.

> **The Degree of ax^n**
>
> If $a \neq 0$, the degree of ax^n is n. The degree of a nonzero constant is 0. The constant 0 has no defined degree.

Here is an example of a polynomial and the degree of each of its four terms:

$$6x^4 - 3x^3 + 2x - 5.$$

| degree 4 | degree 3 | degree 1 | degree of non-zero constant: 0 |

Notice that the exponent on x for the term $2x$ is understood to be 1: $2x^1$. For this reason, the degree of $2x$ is 1. You can think of -5 as $-5x^0$; thus, its degree is 0.

A polynomial which when simplified has exactly one term is called a **monomial**. A **binomial** is a simplified polynomial that has two terms, each with a different exponent. A **trinomial** is a simplified polynomial with three terms, each with a different exponent. Simplified polynomials with four or more terms have no special names.

The **degree of a polynomial** is the highest degree of all the terms of the polynomial. For example, $4x^2 + 3x$ is a binomial of degree 2 because the degree of the first term is 2, and the degree of the other term is less than 2. Also, $7x^5 - 2x^2 + 4$ is a trinomial of degree 5 because the degree of the first term is 5, and the degrees of the other terms are less than 5.

Up to now, we have used x to represent the variable in a polynomial. However, any letter can be used. For example,

- $7x^5 - 3x^3 + 8$ is a polynomial (in x) of degree 5.
- $6y^3 + 4y^2 - y + 3$ is a polynomial (in y) of degree 3.
- $z^7 + \sqrt{2}$ is a polynomial (in z) of degree 7.

Not every algebraic expression is a polynomial. Algebraic expressions whose variables do not contain whole number exponents such as

$$3x^{-2} + 7 \quad \text{and} \quad 5x^{3/2} + 9x^{1/2} + 2$$

are not polynomials. Furthermore, a quotient of polynomials such as

$$\frac{x^2 + 2x + 5}{x^3 - 7x^2 + 9x - 3}$$

is not a polynomial because the form of a polynomial involves only addition and subtraction of terms, not division.

We can tie together the threads of our discussion with the formal definition of a polynomial in one variable. In this definition, the coefficients of the terms are represented by a_n (read "a sub n"), a_{n-1} (read "a sub n minus 1"), a_{n-2}, and so on. The small letters to the lower right of each a are called **subscripts** and are *not exponents*. Subscripts are used to distinguish one constant from another when a large and undetermined number of such constants are needed.

Definition of a Polynomial in x

A **polynomial in x** is an algebraic expression of the form

$$a_n x^n + a_{n-1} x^{n-1} + a_{n-2} x^{n-2} + \cdots + a_1 x + a_0,$$

where $a_n, a_{n-1}, a_{n-2}, \ldots, a_1$, and a_0 are real numbers, $a_n \neq 0$, and n is a nonnegative integer. The polynomial is of **degree n**, a_n is the **leading coefficient**, and a_0 is the **constant term**.

2   Add and subtract polynomials.

Adding and Subtracting Polynomials

Polynomials are added and subtracted by combining like terms. For example, we can combine the monomials $-9x^3$ and $13x^3$ using addition as follows:

$$-9x^3 + 13x^3 = (-9 + 13)x^3 = 4x^3.$$

EXAMPLE 1   Adding and Subtracting Polynomials

Perform the indicated operations and simplify:

a. $(-9x^3 + 7x^2 - 5x + 3) + (13x^3 + 2x^2 - 8x - 6)$
b. $(7x^3 - 8x^2 + 9x - 6) - (2x^3 - 6x^2 - 3x + 9)$.

Solution

a. $(-9x^3 + 7x^2 - 5x + 3) + (13x^3 + 2x^2 - 8x - 6)$
$= (-9x^3 + 13x^3) + (7x^2 + 2x^2)$ Group like terms.
$\quad + (-5x - 8x) + (3 - 6)$
$= 4x^3 + 9x^2 + (-13x) + (-3)$ Combine like terms.
$= 4x^3 + 9x^2 - 13x - 3$ Simplify.

Study Tip

You can also arrange like terms in columns and combine vertically:

$$7x^3 - 8x^2 + 9x - 6$$
$$\underline{-2x^3 + 6x^2 + 3x - 9}$$
$$5x^3 - 2x^2 + 12x - 15$$

The like terms can be combined by adding their coefficients and keeping the same variable factor.

b. $(7x^3 - 8x^2 + 9x - 6) - (2x^3 - 6x^2 - 3x + 9)$
$= (7x^3 - 8x^2 + 9x - 6) + (-2x^3 + 6x^2 + 3x - 9)$ Rewrite subtraction as addition of the additive inverse. Be sure to change the sign of each term inside parentheses preceded by the negative sign.

$= (7x^3 - 2x^3) + (-8x^2 + 6x^2)$ Group like terms.
$\quad + (9x + 3x) + (-6 - 9)$
$= 5x^3 + (-2x^2) + 12x + (-15)$ Combine like terms.
$= 5x^3 - 2x^2 + 12x - 15$ Simplify.

Check Point 1   Perform the indicated operations and simplify:

   a. $(-17x^3 + 4x^2 - 11x - 5) + (16x^3 - 3x^2 + 3x - 15)$
   b. $(13x^3 - 9x^2 - 7x + 1) - (-7x^3 + 2x^2 - 5x + 9)$.

3   Multiply polynomials.

Multiplying Polynomials

The product of two monomials is obtained by using properties of exponents. For example,

$$(-8x^6)(5x^3) = -8 \cdot 5x^{6+3} = -40x^9.$$

Multiply coefficients and add exponents.

Furthermore, we can use the distributive property to multiply a monomial and a polynomial that is not a monomial. For example,

$$3x^4(2x^3 - 7x + 3) = 3x^4 \cdot 2x^3 - 3x^4 \cdot 7x + 3x^4 \cdot 3 = 6x^7 - 21x^5 + 9x^4.$$

monomial   trinomial

How do we multiply two polynomials if neither is a monomial? For example, consider

$$(2x + 3)(x^2 + 4x + 5).$$

binomial   trinomial

One way to perform this multiplication is to distribute $2x$ throughout the trinomial

$$2x(x^2 + 4x + 5)$$

and 3 throughout the trinomial

$$3(x^2 + 4x + 5).$$

Then combine the like terms that result.

> **Multiplying Polynomials when Neither is a Monomial**
> Multiply each term of one polynomial by each term of the other polynomial. Then combine like terms.

EXAMPLE 2 Multiplying a Binomial and a Trinomial

Multiply: $(2x + 3)(x^2 + 4x + 5)$.

Solution

$(2x + 3)(x^2 + 4x + 5)$

$= 2x(x^2 + 4x + 5) + 3(x^2 + 4x + 5)$ Multiply the trinomial by each term of the binomial.

$= 2x \cdot x^2 + 2x \cdot 4x + 2x \cdot 5 + 3 \cdot x^2 + 3 \cdot 4x + 3 \cdot 5$ Use the distributive property.

$= 2x^3 + 8x^2 + 10x + 3x^2 + 12x + 15$ Multiply the monomials: multiply coefficients and add exponents.

$= 2x^3 + 11x^2 + 22x + 15$ Combine like terms: $8x^2 + 3x^2 = 11x^2$ and $10x + 12x = 22x$.

Another method for solving Example 2 is to use a vertical format similar to that used for multiplying whole numbers.

$$
\begin{array}{r}
x^2 \;+\; 4x \;+\; 5 \\
2x \;+\; 3 \\
\hline
3x^2 \;+\; 12x \;+\; 15 \\
2x^3 \;+\; 8x^2 \;+\; 10x \\
\hline
2x^3 \;+\; 11x^2 \;+\; 22x \;+\; 15
\end{array}
$$

Write like terms in the same column.

$3(x^2 + 4x + 5)$

$2x(x^2 + 4x + 5)$

Combine like terms.

Check Point 2   Multiply: $(5x - 2)(3x^2 - 5x + 4)$.

4 Use FOIL in polynomial multiplication.

The Product of Two Binomials: FOIL

Frequently we need to find the product of two binomials. We can use a method called FOIL, which is based on the distributive property, to do so. For example, we can find the product of the binomials $3x + 2$ and $4x + 5$ as follows:

$(3x + 2)(4x + 5) = 3x(4x + 5) + 2(4x + 5)$ First, distribute $3x$ over $4x + 5$. Then distribute 2.

$= 3x(4x) + 3x(5) + 2(4x) + 2(5)$

$= 12x^2 + 15x + 8x + 10.$

Two binomials can be quickly multiplied by using the FOIL method, in which F represents the product of the **first** terms in each binomial, O represents the product of the **outside** terms, I represents the product of the two **inside** terms, and L represents the product of the **last**, or second, terms in each binomial.

first   last   F   O   I   L

$(3x + 2)(4x + 5) = 12x^2 + 15x + 8x + 10$

inside

$= 12x^2 + 23x + 10$ Combine like terms.

outside

In general, here is how to use the FOIL method to find the product of $ax + b$ and $cx + d$:

Using the FOIL Method to Multiply Binomials

$$(ax + b)(cx + d) = ax \cdot cx + ax \cdot d + b \cdot cx + b \cdot d$$

first · last · inside · outside

| Product of First terms | Product of Outside terms | Product of Inside terms | Product of Last terms |

EXAMPLE 3 Using the FOIL Method

Multiply: $(3x + 4)(5x - 3)$.

Solution

$$(3x + 4)(5x - 3) = 3x \cdot 5x + 3x(-3) + 4 \cdot 5x + 4(-3)$$

$$= 15x^2 - 9x + 20x - 12$$

$$= 15x^2 + 11x - 12 \qquad \text{Combine like terms.}$$

Check Point 3   Multiply: $(7x - 5)(4x - 3)$.

5 Use special products in polynomial multiplication.

Special Products

There are several products that occur so frequently that it's convenient to memorize the form, or pattern, of these formulas.

Special Products

Let A and B represent real numbers, variables, or algebraic expressions.

Special Product	Example
Sum and Difference of Two Terms	
$(A + B)(A - B) = A^2 - B^2$	$(2x + 3)(2x - 3) = (2x)^2 - 3^2$ $= 4x^2 - 9$
Squaring a Binomial	
$(A + B)^2 = A^2 + 2AB + B^2$	$(y + 5)^2 = y^2 + 2 \cdot y \cdot 5 + 5^2$ $= y^2 + 10y + 25$
$(A - B)^2 = A^2 - 2AB + B^2$	$(3x - 4)^4$ $= (3x)^2 - 2 \cdot 3x \cdot 4 + 4^2$ $= 9x^2 - 24x + 16$
Cubing a Binomial	
$(A + B)^3 = A^3 + 3A^2B + 3AB^2 + B^3$	$(x + 4)^3$ $= x^3 + 3x^2(4) + 3x(4)^2 + 4^3$ $= x^3 + 12x^2 + 48x + 64$
$(A - B)^3 = A^3 - 3A^2B + 3AB^2 - B^3$	$(x - 2)^3$ $= x^3 - 3x^2(2) + 3x(2)^2 - 2^3$ $= x^3 - 6x^2 + 12x - 8$

Study Tip

Although it's convenient to memorize these forms, the FOIL method can be used on all five examples in the box. To cube $x + 4$, you can first square $x + 4$ using FOIL and then multiply this result by $x + 4$. In short, you do not necessarily have to utilize these special formulas. What is the advantage of knowing and using these forms?

6 Perform operations with polynomials in several variables.

Polynomials in Several Variables

The next time you visit the lumber yard and go rummaging through piles of wood, think *polynomials*, although polynomials a bit different from those we have encountered so far. The construction industry uses a polynomial in two variables to determine the number of board feet that can be manufactured from a tree with a diameter of x inches and a length of y feet. This polynomial is

$$\tfrac{1}{4}x^2y - 2xy + 4y.$$

In general, a **polynomial in two variables**, x and y, contains the sum of one or more monomials in the form ax^ny^m. The constant, a, is the **coefficient**. The exponents, n and m, represent whole numbers. The **degree** of the monomial ax^ny^m is $n + m$. We'll use the polynomial from the construction industry to illustrate these ideas.

The coefficients are $\frac{1}{4}$, -2, and 4.

$$\tfrac{1}{4}x^2y \qquad - 2xy \qquad + 4y$$

| Degree of monomial: $2 + 1 = 3$ | Degree of monomial: $1 + 1 = 2$ | Degree of monomial: $0 + 1 = 1$ |

The **degree of a polynomial in two variables** is the highest degree of all its terms. For the preceding polynomial, the degree is 3.

Polynomials containing two or more variables can be added, subtracted, and multiplied just like polynomials that contain only one variable.

EXAMPLE 4   Multiplying Polynomials in Two Variables

Multiply:

a. $(x + 4y)(3x - 5y)$ **b.** $(5x + 3y)^2$.

Solution   We will perform the multiplication in part (a) using the FOIL method. We will multiply in part (b) using the formula for the square of a binomial sum, $(A + B)^2$.

a. $(x + 4y)(3x - 5y)$ *Multiply these binomials using the FOIL method.*

$$
\begin{array}{cccc}
\text{F} & \text{O} & \text{I} & \text{L}
\end{array}
$$

$$= (x)(3x) + (x)(-5y) + (4y)(3x) + (4y)(-5y)$$
$$= 3x^2 - 5xy + 12xy - 20y^2$$
$$= 3x^2 + 7xy - 20y^2 \qquad \text{Combine like terms.}$$

$$(A + B)^2 = A^2 + 2 \cdot A \cdot B + B^2$$

b. $(5x + 3y)^2 = (5x)^2 + 2(5x)(3y) + (3y)^2$
$$= 25x^2 + 30xy + 9y^2$$

Visualizing a Special Products Formula

The formula
$$(A + B)^2 = A^2 + 2AB + B^2$$
can be interpreted geometrically.

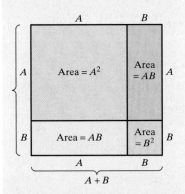

The area of the large rectangle is
$$(A + B)(A + B) = (A + B)^2.$$

The sum of the areas of the four smaller rectangles is
$$A^2 + AB + AB + B^2$$
$$= A^2 + 2AB + B^2.$$

The area of the large rectangle equals the sum of the areas of the four smaller rectangles:
$$(A + B)^2 = A^2 + 2AB + B^2.$$

Check Point 4 Multiply:

    a. $(7x - 6y)(3x - y)$ **b.** $(x^2 + 5y)^2$.

Special products can sometimes be used to find the products of certain trinomials, as illustrated in Example 5.

EXAMPLE 5 Using the Special Products

Multiply:

    a. $(7x + 5 + 4y)(7x + 5 - 4y)$ **b.** $(3x + y + 1)^2$.

Solution

a. By grouping the first two terms within each of the parentheses, we can find the product using the form for the sum and difference of two terms:

$$(A + B) \cdot (A - B) = A^2 - B^2$$

$$\begin{aligned}
[(7x + 5) + 4y] \cdot [(7x + 5) - 4y] &= (7x + 5)^2 - (4y)^2 \\
&= (7x)^2 + 2 \cdot 7x \cdot 5 + 5^2 - (4y)^2 \\
&= 49x^2 + 70x + 25 - 16y^2
\end{aligned}$$

b. We can group the terms so that the formula for the square of a binomial can be applied.

$$(A + B)^2 = A^2 + 2 \cdot A \cdot B + B^2$$

$$\begin{aligned}
[(3x + y) + 1]^2 &= [(3x + y)^2 + 2 \cdot (3x + y) \cdot 1 + 1^2 \\
&= 9x^2 + 6xy + y^2 + 6x + 2y + 1
\end{aligned}$$

Check Point 5 Multiply:

    a. $(3x + 2 + 5y)(3x + 2 - 5y)$ **b.** $(2x + y + 3)^2$.

EXERCISE SET P.4

 Practice Exercises

In Exercises 1–4, is the algebraic expression a polynomia? If it is, write the polynomial in standard form.

1. $2x + 3x^2 - 5$ **2.** $2x + 3x^{-1} - 5$

3. $\dfrac{2x + 3}{x}$ **4.** $x^2 - x^3 + x^4 - 5$

In Exercises 5–8, find the degree of the polynomial.

5. $3x^2 - 5x + 4$ **6.** $-4x^3 + 7x^2 - 11$

7. $x^2 - 4x^3 + 9x - 12x^4 + 63$

8. $x^2 - 8x^3 + 15x^4 + 91$

In Exercises 9–14, perform the indicated operations. Write the resulting polynomial in standard form and indicate its degree.

9. $(-6x^3 + 5x^2 - 8x + 9) + (17x^3 + 2x^2 - 4x - 13)$

10. $(-7x^3 + 6x^2 - 11x + 13) + (19x^3 - 11x^2 + 7x - 17)$

11. $(17x^3 - 5x^2 + 4x - 3) - (5x^3 - 9x^2 - 8x + 11)$

12. $(18x^4 - 2x^3 - 7x + 8) - (9x^4 - 6x^3 - 5x + 7)$

13. $(5x^2 - 7x - 8) + (2x^2 - 3x + 7) - (x^2 - 4x - 3)$

14. $(8x^2 + 7x - 5) - (3x^2 - 4x) - (-6x^3 - 5x^2 + 3)$

In Exercises 15–58, find each product.

15. $(x + 1)(x^2 - x + 1)$ **16.** $(x + 5)(x^2 - 5x + 25)$

17. $(2x - 3)(x^2 - 3x + 5)$ **18.** $(2x - 1)(x^2 - 4x + 3)$

19. $(x + 7)(x + 3)$ **20.** $(x + 8)(x + 5)$

21. $(x - 5)(x + 3)$ **22.** $(x - 1)(x + 2)$
23. $(3x + 5)(2x + 1)$ **24.** $(7x + 4)(3x + 1)$
25. $(2x - 3)(5x + 3)$ **26.** $(2x - 5)(7x + 2)$
27. $(5x^2 - 4)(3x^2 - 7)$ **28.** $(7x^2 - 2)(3x^2 - 5)$
29. $(8x^3 + 3)(x^2 - 5)$ **30.** $(7x^3 + 5)(x^2 - 2)$
31. $(x + 3)(x - 3)$ **32.** $(x + 5)(x - 5)$
33. $(3x + 2)(3x - 2)$ **34.** $(2x + 5)(2x - 5)$
35. $(5 - 7x)(5 + 7x)$ **36.** $(4 - 3x)(4 + 3x)$
37. $(4x^2 + 5x)(4x^2 - 5x)$ **38.** $(3x^2 + 4x)(3x^2 - 4x)$
39. $(1 - y^5)(1 + y^5)$ **40.** $(2 - y^5)(2 + y^5)$
41. $(x + 2)^2$ **42.** $(x + 5)^2$
43. $(2x + 3)^2$ **44.** $(3x + 2)^2$
45. $(x - 3)^2$ **46.** $(x - 4)^2$
47. $(4x^2 - 1)^2$ **48.** $(5x^2 - 3)^2$
49. $(7 - 2x)^2$ **50.** $(9 - 5x)^2$
51. $(x + 1)^3$ **52.** $(x + 2)^3$
53. $(2x + 3)^3$ **54.** $(3x + 4)^3$
55. $(x - 3)^3$ **56.** $(x - 1)^3$
57. $(3x - 4)^3$ **58.** $(2x - 3)^3$

In Exercises 59–82, find each product.

59. $(x + 5y)(7x + 3y)$ **60.** $(x + 9y)(6x + 7y)$
61. $(x - 3y)(2x + 7y)$ **62.** $(3x - y)(2x + 5y)$
63. $(3xy - 1)(5xy + 2)$ **64.** $(7x^2y + 1)(2x^2y - 3)$
65. $(7x + 5y)^2$ **66.** $(9x + 7y)^2$
67. $(x^2y^2 - 3)^2$ **68.** $(x^2y^2 - 5)^2$
69. $(x - y)(x^2 + xy + y^2)$ **70.** $(x + y)(x^2 - xy + y^2)$
71. $(3x + 5y)(3x - 5y)$ **72.** $(7x + 3y)(7x - 3y)$
73. $(x + y + 3)(x + y - 3)$
74. $(x + y + 5)(x + y - 5)$
75. $(3x + 7 - 5y)(3x + 7 + 5y)$
76. $(5x + 7y - 2)(5x + 7y + 2)$
77. $[5y - (2x + 3)][5y + (2x + 3)]$
78. $[8y + (7 - 3x)][8y - (7 - 3x)]$
79. $(x + y + 1)^2$ **80.** $(x + y + 2)^2$
81. $(2x + y + 1)^2$ **82.** $(5x + 1 + 6y)^2$

In Exercises 83–86, a geometric interpretation of a special products formula is illustrated. Select the formula for each illustration from the following options:

$$(A + B)^2 = A^2 + 2AB + B^2,$$
$$(A + B)(A - B) = A^2 - B^2,$$
$$(A + 1)(B + 1) = AB + A + B + 1,$$
$$(A + 1)^2 = A^2 + 2A + 1$$

83.

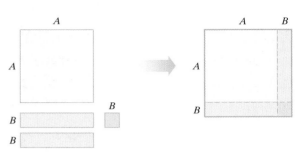

84.

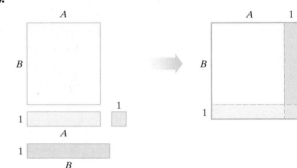

85.

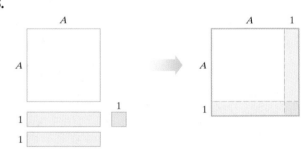

86.

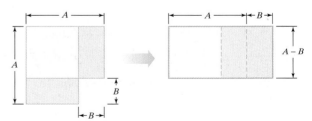

Application Exercises

87. The polynomial $0.018x^2 - 0.757x + 9.047$ describes the amount, in thousands of dollars, that a person earning x thousand dollars a year feels underpaid. Evaluate the polynomial for $x = 40$. Describe what the answer means in practical terms.

88. The polynomial $104.5x^2 - 1501.5x + 6016$ describes the death rate per year, per 100,000 men, for men averaging x hours of sleep each night. Evaluate the polynomial for $x = 10$. Describe what the answer means in practical terms.

89. The polynomial $-1.45x^2 + 38.52x + 470.78$ describes the number of violent crimes in the United States, per 100,000 inhabitants, x years after 1975. Evaluate the polynomial for $x = 25$. Describe what the answer means in practical terms. How well does the polynomial describe the crime rate for the appropriate year shown in the bar graph?

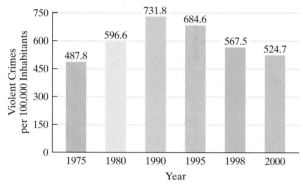

Violent Crime in the United States

Source: F.B.I.

90. The polynomial $-0.02A^2 + 2A + 22$ is used by coaches to get athletes fired up so that they can perform well. The polynomial represents the performance level related to various levels of enthusiasm, from $A = 1$ (almost no enthusiasm) to $A = 100$ (maximum level of enthusiasm). Evaluate the polynomial for $A = 20$, $A = 50$, and $A = 80$. Describe what happens to performance as we get more and more fired up.

91. The number of people who catch a cold t weeks after January 1 is $5t - 3t^2 + t^3$. The number of people who recover t weeks after January 1 is $t - t^2 + \frac{1}{3}t^3$. Write a polynomial in standard form for the number of people who are still ill with a cold t weeks after January 1.

92. The weekly cost, in thousands of dollars, for producing x stereo headphones is $30x + 50$. The weekly revenue, in thousands of dollars, for selling x stereo headphones is $90x^2 - x$. Write a polynomial in standard form for the weekly profit, in thousands of dollars, for producing and selling x stereo headphones.

In Exercises 93–94, write a polynomial in standard form that represents the area of the shaded region of each figure.

93.

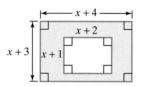

94.

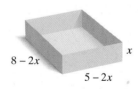

In Exercises 95–96, write a polynomial in standard form that represents the volume of the open box in the figure shown.

95.

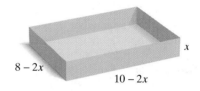

96.

Writing in Mathematics

97. What is a polynomial in x?

98. Explain how to subtract polynomials.

99. Explain how to multiply two binomials using the FOIL method. Give an example with your explanation.

100. Explain how to find the product of the sum and difference of two terms. Give an example with your explanation.

101. Explain how to square a binomial difference. Give an example with your explanation.

102. Explain how to find the degree of a polynomial in two variables.

103. For Exercise 90, explain why performance levels do what they do as we get more and more fired up. If possible, describe an example of a time when you were too enthused and thus did poorly at something you were hoping to do well.

Technology Exercises

104. The common cold is caused by a rhinovirus. The polynomial

$$-0.75x^4 + 3x^3 + 5$$

describes the billions of viral particles in our bodies after x days of invasion. Use a calculator to find the number of viral particles after 0 days (the time of the cold's onset), 1 day, 2 days, 3 days, and 4 days. After how many days is the number of viral particles at a maximum and consequently the day we feel the sickest? By when should we feel completely better?

105. Using data from the National Institute on Drug Abuse, the polynomial

$$0.0032x^3 + 0.0235x^2 - 2.2477x + 61.1998$$

approximately describes the percentage of U.S. high school seniors in the class of x who had ever used marijuana, where x is the number of years after 1980. Use a calculator to find the percentage of high school seniors from the class of 1980 through the class of 2000 who had used marijuana. Round to the nearest tenth of a percent. Describe the trend in the data.

 Critical Thinking Exercises

106. Express the area of the plane figure shown as a polynomial in standard form.

In Exercises 107–108, represent the volume of each figure as a polynomial in standard form.

107.

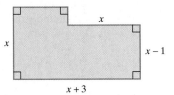

108.

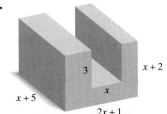

109. Simplify: $(y^n + 2)(y^n - 2) - (y^n - 3)^2$.

SECTION P.5 *Factoring Polynomials*

Objectives

1. Factor out the greatest common factor of a polynomial.
2. Factor by grouping.
3. Factor trinomials.
4. Factor the difference of squares.
5. Factor perfect square trinomials.
6. Factor the sum and difference of cubes.
7. Use a general strategy for factoring polynomials.
8. Factor algebraic expressions containing fractional and negative exponents.

A two-year-old boy is asked, "Do you have a brother?" He answers, "Yes." "What is your brother's name?" "Tom." Asked if Tom has a brother, the two-year-old replies, "No." The child can go in the direction from self to brother, but he cannot reverse this direction and move from brother back to self.

As our intellects develop, we learn to reverse the direction of our thinking. Reversibility of thought is found throughout algebra. For example, we can multiply polynomials and show that

$$(2x + 1)(3x - 2) = 6x^2 - x - 2.$$

We can also reverse this process and express the resulting polynomial as

$$6x^2 - x - 2 = (2x + 1)(3x - 2).$$

Factoring is the process of writing a polynomial as the product of two or more polynomials. The factors of $6x^2 - x - 2$ are $2x + 1$ and $3x - 2$.

In this section, we will be **factoring over the set of integers**, meaning that the coefficients in the factors are integers. Polynomials that cannot be factored using integer coefficients are called **irreducible over the integers**, or **prime**.

The goal in factoring a polynomial is to use one or more factoring techniques until each of the polynomial's factors is prime or irreducible. In this situation, the polynomial is said to be **factored completely**.

We will now discuss basic techniques for factoring polynomials.

1 Factor out the greatest common factor of a polynomial.

Common Factors

In any factoring problem, the first step is to look for the *greatest common factor*. The **greatest common factor**, abbreviated GCF, is an expression of the highest degree that divides each term of the polynomial. The distributive property in the reverse direction

$$ab + ac = a(b + c)$$

can be used to factor out the greatest common factor.

EXAMPLE 1 Factoring out the Greatest Common Factor

Factor: **a.** $18x^3 + 27x^2$ **b.** $x^2(x + 3) + 5(x + 3)$.

Solution

Study Tip

The variable part of the greatest common factor always contains the *smallest* power of a variable or algebraic expression that appears in all terms of the polynomial.

a. We begin by determining the greatest common factor. 9 is the greatest integer that divides 18 and 27. Furthermore, x^2 is the greatest expression that divides x^3 and x^2. Thus, the greatest common factor of the two terms in the polynomial is $9x^2$.

$$18x^3 + 27x^2$$

$$= 9x^2(2x) + 9x^2(3) \qquad \text{Express each term as the product of the greatest common factor and its other factor.}$$

$$= 9x^2(2x + 3) \qquad \text{Factor out the greatest common factor.}$$

b. In this situation, the greatest common factor is the common binomial factor $(x + 3)$. We factor out this common factor as follows:

$$x^2(x + 3) + 5(x + 3) = (x + 3)(x^2 + 5). \qquad \text{Factor out the common binomial factor.}$$

Check Point 1 Factor:

a. $10x^3 - 4x^2$ **b.** $2x(x - 7) + 3(x - 7)$.

2 Factor by grouping.

Factoring by Grouping

Some polynomials have only a greatest common factor of 1. However, by a suitable rearrangement of the terms, it still may be possible to factor. This process, called **factoring by grouping**, is illustrated in Example 2.

EXAMPLE 2   Factoring by Grouping

Factor: $x^3 + 4x^2 + 3x + 12$.

Solution   Group terms that have a common factor:

$$\boxed{x^3 + 4x^2} \quad + \quad \boxed{3x + 12}.$$

Common
factor is x^2.

Common
factor is 3.

We now factor the given polynomial as follows.

$$x^3 + 4x^2 + 3x + 12$$
$$= (x^3 + 4x^2) + (3x + 12) \qquad \text{Group terms with common factors.}$$
$$= x^2(x + 4) + 3(x + 4) \qquad \text{Factor out the greatest common factor from the}$$
$$\qquad\qquad\qquad\qquad\qquad\qquad \text{grouped terms. The remaining two terms have}$$
$$\qquad\qquad\qquad\qquad\qquad\qquad x + 4 \text{ as a common binomial factor.}$$
$$= (x + 4)(x^2 + 3) \qquad \text{Factor } (x + 4) \text{ out of both terms.}$$

Thus, $x^3 + 4x^2 + 3x + 12 = (x + 4)(x^2 + 3)$. Check the factorization by multiplying the right side of the equation using the FOIL method. If the factorization is correct, you will obtain the original polynomial.

Discovery

In Example 2, group the terms as follows:
$$(x^3 + 3x) + (4x^2 + 12).$$
Factor out the greatest common factor from each group and complete the factoring process. Describe what happens. What can you conclude?

**Check
Point
2**   Factor: $x^3 + 5x^2 - 2x - 10$.

3   Factor trinomials.

Factoring Trinomials

To factor a trinomial of the form $ax^2 + bx + c$, a little trial and error may be necessary.

A Strategy for Factoring $ax^2 + bx + c$

(Assume, for the moment, that there is no greatest common factor.)

1. Find two **First** terms whose product is ax^2:

$$(\Box x + \quad)(\Box x + \quad) = ax^2 + bx + c.$$

2. Find two **Last** terms whose product is c:

$$(x + \Box)(x + \Box) = ax^2 + bx + c$$

3. By trial and error, perform steps 1 and 2 until the sum of the **Outside** product and **Inside** product is bx:

$$(\Box x + \Box)(\Box x + \Box) = ax^2 + bx + c.$$

I

O

(sum of O + I)

If no such combinations exist, the polynomial is prime.

EXAMPLE 3 **Factoring Trinomials Whose Leading Coefficients Are 1**

Factor:

a. $x^2 + 6x + 8$ **b.** $x^2 + 3x - 18$.

Solution

a. The factors of the first term are x and x:

$$x^2 + 6x + 8 = (x \quad)(x \quad).$$

To find the second term of each factor, we must find two numbers whose product is 8 and whose sum is 6. From the table in the margin, we see that 4 and 2 are the required integers. Thus,

$$x^2 + 6x + 8 = (x + 4)(x + 2) \quad \text{or} \quad (x + 2)(x + 4).$$

Factors of 8	8, 1	4, 2	−8, −1	−4, −2
Sum of Factors	9	6	−9	−6

This is the desired sum.

b. We begin with

$$x^2 + 3x - 18 = (x \quad)(x \quad).$$

To find the second term of each factor, we must find two numbers whose product is −18 and whose sum is 3. From the table in the margin, we see that 6 and −3 are the required integers. Thus,

$$x^2 + 3x - 18 = (x + 6)(x - 3)$$
$$\text{or } (x - 3)(x + 6).$$

Factors of −18	18, −1	−18, 1	9, −2	−9, 2	6, −3	−6, 3
Sum of Factors	17	−17	7	−7	3	−3

This is the desired sum.

Check Point 3 Factor:

a. $x^2 + 13x + 40$ **b.** $x^2 - 5x - 14$.

EXAMPLE 4 **Factoring a Trinomial Whose Leading Coefficient Is Not 1**

Factor: $8x^2 - 10x - 3$.

Solution

Step 1 Find two *First* terms whose product is $8x^2$.

$$8x^2 - 10x - 3 \stackrel{?}{=} (8x \quad)(x \quad)$$
$$8x^2 - 10x - 3 \stackrel{?}{=} (4x \quad)(2x \quad)$$

Step 2 Find two *Last* terms whose product is −3. The possible factorizations are $1(-3)$ and $-1(3)$.

Step 3 Try various combinations of these factors. The correct factorization of $8x^2 - 10x - 3$ is the one in which the sum of the *Outside* and *Inside* products is equal to $-10x$. Here is a list of the possible factorizations:

Possible Factorizations of $8x^2 - 10x - 3$	Sum of *Outside* and *Inside* Products (Should Equal $-10x$)
$(8x + 1)(x - 3)$	$-24x + x = -23x$
$(8x - 3)(x + 1)$	$8x - 3x = 5x$
$(8x - 1)(x + 3)$	$24x - x = 23x$
$(8x + 3)(x - 1)$	$-8x + 3x = -5x$
$(4x + 1)(2x - 3)$	$-12x + 2x = -10x$
$(4x - 3)(2x + 1)$	$4x - 6x = -2x$
$(4x - 1)(2x + 3)$	$12x - 2x = 10x$
$(4x + 3)(2x - 1)$	$-4x + 6x = 2x$

This is the required middle term.

Thus,

$$8x^2 - 10x - 3 = (4x + 1)(2x - 3) \quad \text{or} \quad (2x - 3)(4x + 1).$$

Show that this factorization is correct by multiplying the factors using the FOIL method. You should obtain the original trinomial.

Check Point 4 Factor: $6x^2 + 19x - 7$.

4 Factor the difference of squares.

Factoring the Difference of Two Squares

A method for factoring the difference of two squares is obtained by reversing the special product for the sum and difference of two terms.

> **The Difference of Two Squares**
> If A and B are real numbers, variables, or algebraic expressions, then
> $$A^2 - B^2 = (A + B)(A - B).$$
> In words: The difference of the squares of two terms factors as the product of a sum and a difference of those terms.

EXAMPLE 5 Factoring the Difference of Two Squares

Factor: **a.** $x^2 - 4$ **b.** $81x^2 - 49$.

Solution We must express each term as the square of some monomial. Then we use the formula for factoring $A^2 - B^2$.

a. $x^2 - 4 = x^2 - 2^2 = (x + 2)(x - 2)$

$$A^2 - B^2 = (A + B)(A - B)$$

b. $81x^2 - 49 = (9x)^2 - 7^2 = (9x + 7)(9x - 7)$

Check Point 5 Factor:
a. $x^2 - 81$ **b.** $36x^2 - 25$.

We have seen that a polynomial is factored completely when it is written as the product of prime polynomials. To be sure that you have factored completely, check to see whether the factors can be factored.

EXAMPLE 6 A Repeated Factorization

Factor completely: $x^4 - 81$.

Solution

$x^4 - 81 = (x^2)^2 - 9^2$ Express as the difference of two squares.

$= (x^2 + 9)(x^2 - 9)$ The factors are the sum and difference of the squared terms.

Study Tip

Factoring $x^4 - 81$ as
$$(x^2 + 9)(x^2 - 9)$$
is not a complete factorization. The second factor, $x^2 - 9$, is itself a difference of two squares and can be factored.

$$= (x^2 + 9)(x^2 - 3^2)$$

The factor $x^2 - 9$ is the difference of two squares and can be factored.

$$= (x^2 + 9)(x + 3)(x - 3)$$

The factors of $x^2 - 9$ are the sum and difference of the squared terms.

Check Point 6   Factor completely: $81x^4 - 16$.

5   Factor perfect square trinomials.

Factoring Perfect Square Trinomials

Our next factoring technique is obtained by reversing the special products for squaring binomials. The trinomials that are factored using this technique are called **perfect square trinomials**.

> **Factoring Perfect Square Trinomials**
>
> Let A and B be real numbers, variables, or algebraic expressions.
>
> **1.** $A^2 + 2AB + B^2 = (A + B)^2$
>
> Same sign
>
> **2.** $A^2 - 2AB + B^2 = (A - B)^2$
>
> Same sign

The two items in the box show that perfect square trinomials come in two forms: one in which the middle term is positive and one in which the middle term is negative. Here's how to recognize a perfect square trinomial:

1. The first and last terms are squares of monomials or integers.

2. The middle term is twice the product of the expressions being squared in the first and last terms.

EXAMPLE 7   Factoring Perfect Square Trinomials

Factor:

   a. $x^2 + 6x + 9$ **b.** $25x^2 - 60x + 36$.

Solution

   a. $x^2 + 6x + 9 = x^2 + 2 \cdot x \cdot 3 + 3^2 = (x + 3)^2$

The middle term has a positive sign.

$$A^2 + 2AB + B^2 = (A + B)^2$$

   b. We suspect that $25x^2 - 60x + 36$ is a perfect square trinomial because $25x^2 = (5x)^2$ and $36 = 6^2$. The middle term can be expressed as twice the product of $5x$ and 6.

$$25x^2 - 60x + 36 = (5x)^2 - 2 \cdot 5x \cdot 6 + 6^2 = (5x - 6)^2$$

$$A^2 - 2AB + B^2 = (A - B)^2$$

Check Point 7 Factor:

a. $x^2 + 14x + 49$ **b.** $16x^2 - 56x + 49$.

6 Factor the sum and difference of cubes.

Factoring the Sum and Difference of Two Cubes

We can use the following formulas to factor the sum or the difference of two cubes:

Factoring the Sum and Difference of Two Cubes

1. Factoring the Sum of Two Cubes
$$A^3 + B^3 = (A + B)(A^2 - AB + B^2)$$

2. Factoring the Difference of Two Cubes
$$A^3 - B^3 = (A - B)(A^2 + AB + B^2)$$

EXAMPLE 8   Factoring Sums and Differences of Two Cubes

Factor:

a. $x^3 + 8$ **b.** $64x^3 - 125$.

Solution

a. $x^3 + 8 = x^3 + 2^3 = (x + 2)(x^2 - x \cdot 2 + 2^2) = (x + 2)(x^2 - 2x + 4)$

$$A^3 + B^3 = (A + B)(A^2 - AB + B^2)$$

b. $64x^3 - 125 = (4x)^3 - 5^3 = (4x - 5)[(4x)^2 + (4x)(5) + 5^2]$

$$A^3 - B^3 = (A - B)(A^2 + AB + B^2)$$

$$= (4x - 5)(16x^2 + 20x + 25)$$

Check Point 8 Factor:

a. $x^3 + 1$ **b.** $125x^3 - 8$.

7 Use a general strategy for factoring polynomials.

A Strategy for Factoring Polynomials

It is important to practice factoring a wide variety of polynomials so that you can quickly select the appropriate technique. The polynomial is factored completely when all its polynomial factors, except possibly for monomial factors, are prime. Because of the commutative property, the order of the factors does not matter.

A Strategy for Factoring a Polynomial

1. If there is a common factor, factor out the GCF.

2. Determine the number of terms in the polynomial and try factoring as follows:

   a. If there are two terms, can the binomial be factored by one of the following special forms?

$$\text{Difference of two squares: } A^2 - B^2 = (A + B)(A - B)$$
$$\text{Sum of two cubes: } A^3 + B^3 = (A + B)(A^2 - AB + B^2)$$
$$\text{Difference of two cubes: } A^3 - B^3 = (A - B)(A^2 + AB + B^2)$$

   b. If there are three terms, is the trinomial a perfect square trinomial? If so, factor by one of the following special forms:

$$A^2 + 2AB + B^2 = (A + B)^2$$
$$A^2 - 2AB + B^2 = (A - B)^2.$$

     If the trinomial is not a perfect square trinomial, try factoring by trial and error.

   c. If there are four or more terms, try factoring by grouping.

3. Check to see if any factors with more than one term in the factored polynomial can be factored further. If so, factor completely.

EXAMPLE 9   Factoring a Polynomial

Factor: $2x^3 + 8x^2 + 8x$.

Solution

Step 1   If there is a common factor, factor out the GCF.   Because $2x$ is common to all terms, we factor it out.

$$2x^3 + 8x^2 + 8x = 2x(x^2 + 4x + 4) \qquad \textit{Factor out the GCF.}$$

Step 2   Determine the number of terms and factor accordingly.   The factor $x^2 + 4x + 4$ has three terms and is a perfect square trinomial. We factor using $A^2 + 2AB + B^2 = (A + B)^2$.

$$2x^3 + 8x^2 + 8x = 2x(x^2 + 4x + 4)$$
$$= 2x(x^2 + 2 \cdot x \cdot 2 + 2^2)$$

$$A^2 + 2AB + B^2$$

$$= 2x(x + 2)^2 \qquad A^2 + 2AB + B^2 = (A + B)^2$$

Step 3   Check to see if factors can be factored further.   In this problem, they cannot. Thus,

$$2x^3 + 8x^2 + 8x = 2x(x + 2)^2.$$

Check Point 9   Factor: $3x^3 - 30x^2 + 75x$.

EXAMPLE 10   Factoring a Polynomial

Factor: $x^2 - 25a^2 + 8x + 16$.

Solution

Step 1   If there is a common factor, factor out the GCF.   Other than 1 or −1, there is no common factor.

Step 2   Determine the number of terms and factor accordingly.   There are four terms. We try factoring by grouping. Grouping into two groups of two terms does not result in a common binomial factor. Let's try grouping as a difference of squares.

$$x^2 - 25a^2 + 8x + 16$$

$$= (x^2 + 8x + 16) - 25a^2 \qquad \text{Rearrange terms and group as a perfect square trinomial minus } 25a^2 \text{ to obtain a difference of squares.}$$

$$= (x + 4)^2 - (5a)^2 \qquad \text{Factor the perfect square trinomial.}$$

$$= (x + 4 + 5a)(x + 4 - 5a) \qquad \text{Factor the difference of squares. The factors are the sum and difference of the expressions being squared.}$$

Step 3   Check to see if factors can be factored further.   In this case, they cannot, so we have factored completely.

Check Point 10   Factor: $x^2 - 36a^2 + 20x + 100$.

---

8 Factor algebraic expressions containing fractional and negative exponents.

Factoring Algebraic Expressions Containing Fractional and Negative Exponents

Algebraic expressions containing radicals and negative exponents occur frequently in calculus. Although these expressions are not polynomials, they can be simplified using factoring techniques.

EXAMPLE 11   Factoring Involving Fractional and Negative Exponents

Factor and simplify: $x(x + 1)^{-3/4} + (x + 1)^{1/4}$.

Solution   The greatest common factor is $x + 1$ with the *smallest exponent* in the two terms. Thus, the greatest common factor is $(x + 1)^{-3/4}$.

$$x(x + 1)^{-3/4} + (x + 1)^{1/4}$$

$$= (x + 1)^{-3/4}x + (x + 1)^{-3/4}(x + 1) \qquad \text{Express each term as the product of the greatest common factor and its other factor.}$$

$$= (x + 1)^{-3/4}[x + (x + 1)] \qquad \text{Factor out the greatest common factor.}$$

$$= \frac{2x + 1}{(x + 1)^{3/4}} \qquad b^{-n} = \frac{1}{b^n}$$

Check
Point
11

Factor and simplify: $x(x - 1)^{-1/2} + (x - 1)^{1/2}$.

EXERCISE SET P.5

Practice Exercises

In Exercises 1–10, factor out the greatest common factor.

1. $18x + 27$ **2.** $16x - 24$

3. $3x^2 + 6x$ **4.** $4x^2 - 8x$

5. $9x^4 - 18x^3 + 27x^2$ **6.** $6x^4 - 18x^3 + 12x^2$

7. $x(x + 5) + 3(x + 5)$ **8.** $x(2x + 1) + 4(2x + 1)$

9. $x^2(x - 3) + 12(x - 3)$ **10.** $x^2(2x + 5) + 17(2x + 5)$

In Exercises 11–16, factor by grouping.

11. $x^3 - 2x^2 + 5x - 10$ **12.** $x^3 - 3x^2 + 4x - 12$

13. $x^3 - x^2 + 2x - 2$ **14.** $x^3 + 6x^2 - 2x - 12$

15. $3x^3 - 2x^2 - 6x + 4$ **16.** $x^3 - x^2 - 5x + 5$

In Exercises 17–30, factor each trinomial, or state that the trinomial is prime.

17. $x^2 + 5x + 6$ **18.** $x^2 + 8x + 15$

19. $x^2 - 2x - 15$ **20.** $x^2 - 4x - 5$

21. $x^2 - 8x + 15$ **22.** $x^2 - 14x + 45$

23. $3x^2 - x - 2$ **24.** $2x^2 + 5x - 3$

25. $3x^2 - 25x - 28$ **26.** $3x^2 - 2x - 5$

27. $6x^2 - 11x + 4$ **28.** $6x^2 - 17x + 12$

29. $4x^2 + 16x + 15$ **30.** $8x^2 + 33x + 4$

In Exercises 31–40, factor the difference of two squares.

31. $x^2 - 100$ **32.** $x^2 - 144$

33. $36x^2 - 49$ **34.** $64x^2 - 81$

35. $9x^2 - 25y^2$ **36.** $36x^2 - 49y^2$

37. $x^4 - 16$ **38.** $x^4 - 1$

39. $16x^4 - 81$ **40.** $81x^4 - 1$

In Exercises 41–48, factor any perfect square trinomials, or state that the polynomial is prime.

41. $x^2 + 2x + 1$ **42.** $x^2 + 4x + 4$

43. $x^2 - 14x + 49$ **44.** $x^2 - 10x + 25$

45. $4x^2 + 4x + 1$ **46.** $25x^2 + 10x + 1$

47. $9x^2 - 6x + 1$ **48.** $64x^2 - 16x + 1$

In Exercises 49–56, factor using the formula for the sum or difference of two cubes.

49. $x^3 + 27$ **50.** $x^3 + 64$

51. $x^3 - 64$ **52.** $x^3 - 27$

53. $8x^3 - 1$ **54.** $27x^3 - 1$

55. $64x^3 + 27$ **56.** $8x^3 + 125$

In Exercises 57–84, factor completely, or state that the polynomial is prime.

57. $3x^3 - 3x$ **58.** $5x^3 - 45x$

59. $4x^2 - 4x - 24$ **60.** $6x^2 - 18x - 60$

61. $2x^4 - 162$ **62.** $7x^4 - 7$

63. $x^3 + 2x^2 - 9x - 18$ **64.** $x^3 + 3x^2 - 25x - 75$

65. $2x^2 - 2x - 112$ **66.** $6x^2 - 6x - 12$

67. $x^3 - 4x$ **68.** $9x^3 - 9x$

69. $x^2 + 64$ **70.** $x^2 + 36$

71. $x^3 + 2x^2 - 4x - 8$ **72.** $x^3 + 2x^2 - x - 2$

73. $y^5 - 81y$ **74.** $y^5 - 16y$

75. $20y^4 - 45y^2$ **76.** $48y^4 - 3y^2$

77. $x^2 - 12x + 36 - 49y^2$ **78.** $x^2 - 10x + 25 - 36y^2$

79. $9b^2x - 16y - 16x + 9b^2y$

80. $16a^2x - 25y - 25x + 16a^2y$

81. $x^2y - 16y + 32 - 2x^2$ **82.** $12x^2y - 27y - 4x^2 + 9$

83. $2x^3 - 8a^2x + 24x^2 + 72x$

84. $2x^3 - 98a^2x + 28x^2 + 98x$

In Exercises 85–94, factor and simplify each algebraic expression.

85. $x^{3/2} - x^{1/2}$ **86.** $x^{3/4} - x^{1/4}$

87. $4x^{-2/3} + 8x^{1/3}$ **88.** $12x^{-3/4} + 6x^{1/4}$

89. $(x + 3)^{1/2} - (x + 3)^{3/2}$

90. $(x^2 + 4)^{3/2} + (x^2 + 4)^{7/2}$

91. $(x + 5)^{-1/2} - (x + 5)^{-3/2}$

92. $(x^2 + 3)^{-2/3} + (x^2 + 3)^{-5/3}$

93. $(4x - 1)^{1/2} - \frac{1}{3}(4x - 1)^{3/2}$

94. $-8(4x + 3)^{-2} + 10(5x + 1)(4x + 3)^{-1}$

Application Exercises

95. Your computer store is having an incredible sale. The price on one model is reduced by 40%. Then the sale price is reduced by another 40%. If x is the computer's original price, the sale price can be represented by

$$(x - 0.4x) - 0.4(x - 0.4x).$$

a. Factor out $(x - 0.4x)$ from each term. Then simplify the resulting expression.

b. Use the simplified expression from part (a) to answer these questions: With a 40% reduction followed by a 40% reduction, is the computer selling at 20% of its original price? If not, at what percentage of the original price is it selling?

96. The polynomial $8x^2 + 20x + 2488$ describes the number, in thousands, of high school graduates in the United States x years after 1993.

    a. According to this polynomial, how many students will graduate from U.S. high schools in 2003?

    b. Factor the polynomial.

    c. Use the factored form of the polynomial in part (b) to find the number of high school graduates in 2003. Do you get the same answer as you did in part (a)? If so, does this prove that your factorization is correct? Explain.

97. A rock is dropped from the top of a 256-foot cliff. The height, in feet, of the rock above the water after t seconds is described by the polynomial $256 - 16t^2$. Factor this expression completely.

98. The amount of sheet metal needed to manufacture a cylindrical tin can, that is, its surface area, S, is $S = 2\pi r^2 + 2\pi rh$. Express the surface area, S, in factored form.

In Exercises 99–100, find the formula for the area of the shaded region and express it in factored form.

99.

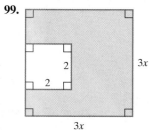

100.

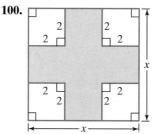

Writing in Mathematics

101. Using an example, explain how to factor out the greatest common factor of a polynomial.

102. Suppose that a polynomial contains four terms. Explain how to use factoring by grouping to factor the polynomial.

103. Explain how to factor $3x^2 + 10x + 8$.

104. Explain how to factor the difference of two squares. Provide an example with your explanation.

105. What is a perfect square trinomial and how is it factored?

106. Explain how to factor $x^3 + 1$.

107. What does it mean to factor completely?

Technology Exercise

108. Computer algebra systems such as *Mathematica, Maple*, and *Derive* will factor polynomials. Graphing calculators, such as the TI-92 with *Derive*, will also allow you to enter a polynomial and use the F2 *Algebra* toolbar menu to display its factored form. Use a computer system or a calculator that can perform symbolic manipulations to verify any five of your factorizations in Exercises 57–76.

Critical Thinking Exercises

109. Which one of the following is true?

    a. Because $x^2 + 1$ is irreducible over the integers, it follows that $x^3 + 1$ is also irreducible.

    b. One correct factored form for $x^2 - 4x + 3$ is $x(x - 4) + 3$.

    c. $x^3 - 64 = (x - 4)^3$

    d. None of the above is true.

In Exercises 110–113, factor completely.

110. $x^{2n} + 6x^n + 8$

111. $-x^2 - 4x + 5$

112. $x^4 - y^4 - 2x^3y + 2xy^3$

113. $(x - 5)^{-1/2}(x + 5)^{-1/2} - (x + 5)^{1/2}(x - 5)^{-3/2}$

In Exercises 114–115, find all integers b so that the trinomial can be factored.

114. $x^2 + bx + 15$ **115.** $x^2 + 4x + b$

Group Exercise

116. Without looking at any factoring problems in the book, create five factoring problems. Make sure that some of your problems require at least two factoring techniques. Next, exchange problems with another person in your group. Work to factor your partner's problems. Evaluate the problems as you work: Are they too easy? Too difficult? Can the polynomials really be factored? Share your response with the person who wrote the problems. Finally, grade each other's work in factoring the polynomials. Each factoring problem is worth 20 points. You may award partial credit. If you take off points, explain why points are deducted and how you decided to take off a particular number of points for the error(s) that you found.

SECTION P.6 *Rational Expressions*

Objectives

1. Specify numbers that must be excluded from the domain of rational expressions.
2. Simplify rational expressions.
3. Multiply rational expressions.
4. Divide rational expressions.
5. Add and subtract rational expressions.
6. Simplify complex rational expressions.
7. Simplify fractional expressions that occur in calculus.
8. Rationalize numerators.

How do we describe the costs of reducing environmental pollution? We often use algebraic expressions involving quotients of polynomials. For example, the algebraic expression

$$\frac{250x}{100 - x}$$

describes the cost, in millions of dollars, to remove x percent of the pollutants that are discharged into a river. Removing a modest percentage of pollutants, say 40%, is far less costly than removing a substantially greater percentage, such as 95%. We see this by evaluating the algebraic expression for $x = 40$ and $x = 95$.

Evaluating $\dfrac{250x}{100 - x}$ for

$x = 40$:	$x = 95$:

Cost is $\dfrac{250(40)}{100 - 40} \approx 167.$ Cost is $\dfrac{250(95)}{100 - 95} = 4750.$

The cost increases from approximately $167 million to a possibly prohibitive $4750 million, or $4.75 billion. Costs spiral upward as the percentage of removed pollutants increases.

Many algebraic expressions that describe costs of environmental projects are examples of rational expressions. First we will define rational expressions. Then we will review how to perform operations with such expressions.

Discovery

What happens if you try substituting 100 for x in

$$\frac{250x}{100 - x}?$$

What does this tell you about the cost of cleaning up all of the river's pollutants?

1 Specify numbers that must be excluded from the domain of rational expressions.

Rational Expressions

A **rational expression** is the quotient of two polynomials. Some examples are

$$\frac{x - 2}{4}, \quad \frac{4}{x - 2}, \quad \frac{x}{x^2 - 1}, \quad \text{and} \quad \frac{x^2 + 1}{x^2 + 2x - 3}.$$

The set of real numbers for which an algebraic expression is defined is the **domain** of the expression. Because rational expressions indicate division and division by zero is undefined, we must exclude numbers from a rational expression's domain that make the denominator zero.

EXAMPLE 1   Excluding Numbers from the Domain

Find all the numbers that must be excluded from the domain of each rational expression:

$$\textbf{a. } \frac{4}{x-2} \qquad \textbf{b. } \frac{x}{x^2-1}.$$

Solution   To determine the numbers that must be excluded from each domain, examine the denominators.

$$\textbf{a. } \frac{4}{x-2} \qquad\qquad \textbf{b. } \frac{x}{x^2-1} = \frac{x}{(x+1)(x-1)}$$

This denominator would equal zero if x = 2.	This factor would equal zero if x = −1.	This factor would equal zero if x = 1.

For the rational expression in part (a), we must exclude 2 from the domain. For the rational expression in part (b), we must exclude both −1 and 1 from the domain. These excluded numbers are often written to the right of a rational expression.

$$\frac{4}{x-2}, x \neq 2 \qquad \frac{x}{x^2-1}, x \neq -1, x \neq 1$$

Check Point 1   Find all the numbers that must be excluded from the domain of each rational expression:

$$\textbf{a. } \frac{7}{x+5} \qquad \textbf{b. } \frac{x}{x^2-36}.$$

2 Simplify rational expressions.

Simplifying Rational Expressions

A rational expression is **simplified** if its numerator and denominator have no common factors other than 1 or −1. The following procedure can be used to simplify rational expressions:

Simplifying Rational Expressions

1. Factor the numerator and denominator completely.
2. Divide both the numerator and denominator by the common factors.

EXAMPLE 2   Simplifying Rational Expressions

Simplify: $\textbf{a. } \dfrac{x^3+x^2}{x+1} \qquad \textbf{b. } \dfrac{x^2+6x+5}{x^2-25}.$

Solution

a. $\dfrac{x^3 + x^2}{x + 1} = \dfrac{x^2(x + 1)}{x + 1}$

Factor the numerator. Because the denominator is $x + 1$, $x \neq -1$.

$= \dfrac{x^2(\overset{1}{\cancel{x + 1}})}{\underset{1}{\cancel{x + 1}}}$

Divide out the common factor, $x + 1$.

$= x^2, x \neq -1$

Denominators of 1 need not be written because $\frac{a}{1} = a$.

b. $\dfrac{x^2 + 6x + 5}{x^2 - 25} = \dfrac{(x + 5)(x + 1)}{(x + 5)(x - 5)}$

Factor the numerator and denominator. Because the denominator is $(x + 5)(x - 5)$, $x \neq -5$ and $x \neq 5$.

$= \dfrac{\overset{1}{\cancel{(x + 5)}}(x + 1)}{\underset{1}{\cancel{(x + 5)}}(x - 5)}$

Divide out the common factor, $x + 5$.

$= \dfrac{x + 1}{x - 5}, \; x \neq -5, x \neq 5$

Check Point 2 Simplify:

a. $\dfrac{x^3 + 3x^2}{x + 3}$ **b.** $\dfrac{x^2 - 1}{x^2 + 2x + 1}$.

Multiplying Rational Expressions

The product of two rational expressions is the product of their numerators divided by the product of their denominators. Here is a step-by-step procedure for multiplying rational expressions:

Multiplying Rational Expressions
1. Factor all numerators and denominators completely.
2. Divide numerators and denominators by common factors.
3. Multiply the remaining factors in the numerator and multiply the remaining factors in the denominator.

EXAMPLE 3   Multiplying Rational Expressions

Multiply and simplify:

$$\dfrac{x - 7}{x - 1} \cdot \dfrac{x^2 - 1}{3x - 21}.$$

Solution

$$\frac{x - 7}{x - 1} \cdot \frac{x^2 - 1}{3x - 21}$$

This is the given multiplication problem.

$$= \frac{x - 7}{x - 1} \cdot \frac{(x + 1)(x - 1)}{3(x - 7)}$$

Factor all numerators and denominators. Because the denominators have factors of $x - 1$ and $x - 7$, $x \neq 1$ and $x \neq 7$.

$$= \frac{\overset{1}{\cancel{x - 7}}}{\underset{1}{\cancel{x - 1}}} \cdot \frac{(x + 1)(\overset{1}{\cancel{x - 1}})}{3(\underset{1}{\cancel{x - 7}})}$$

Divide numerators and denominators by common factors.

$$= \frac{x + 1}{3}, \ x \neq 1, \ x \neq 7$$

Multiply the remaining factors in the numerator and in the denominator.

> These excluded numbers from the domain must also be excluded from the simplified expression's domain.

Check Point 3 Multiply and simplify:

$$\frac{x + 3}{x^2 - 4} \cdot \frac{x^2 - x - 6}{x^2 + 6x + 9}.$$

4 Divide rational expressions.

Dividing Rational Expressions

We find the quotient of two rational expressions by inverting the divisor and multiplying.

EXAMPLE 4 Dividing Rational Expressions

Divide and simplify:

$$\frac{x^2 - 2x - 8}{x^2 - 9} \div \frac{x - 4}{x + 3}.$$

Solution

$$\frac{x^2 - 2x - 8}{x^2 - 9} \div \frac{x - 4}{x + 3}$$

This is the given division problem.

$$= \frac{x^2 - 2x - 8}{x^2 - 9} \cdot \frac{x + 3}{x - 4}$$

Invert the divisor and multiply.

$$= \frac{(x - 4)(x + 2)}{(x + 3)(x - 3)} \cdot \frac{x + 3}{x - 4}$$

Factor throughout. For nonzero denominators, $x \neq -3$, $x \neq 3$, and $x \neq 4$.

$$= \frac{(\overset{1}{\cancel{x - 4}})(x + 2)}{(\cancel{x + 3})(x - 3)} \cdot \frac{(\overset{1}{\cancel{x + 3}})}{(\underset{1}{\cancel{x - 4}})}$$

Divide numerators and denominators by common factors.

$$= \frac{x + 2}{x - 3}, \ x \neq -3, \ x \neq 3, \ x \neq 4$$

Multiply the remaining factors in the numerator and in the denominator.

Check Point 4 Divide and simplify:

$$\frac{x^2 - 2x + 1}{x^3 + x} \div \frac{x^2 + x - 2}{3x^2 + 3}.$$

⑤ Add and subtract rational expressions.

Adding and Subtracting Rational Expressions with the Same Denominator

We add or subtract rational expressions with the same denominator by (1) adding or subtracting the numerators, (2) placing this result over the common denominator, and (3) simplifying, if possible.

EXAMPLE 5 **Subtracting Rational Expressions with the Same Denominator**

Subtract: $\dfrac{5x + 1}{x^2 - 9} - \dfrac{4x - 2}{x^2 - 9}$.

Study Tip

Example 5 shows that when a numerator is being subtracted, we must subtract every term in that expression.

Solution

$$\frac{5x + 1}{x^2 - 9} - \frac{4x - 2}{x^2 - 9} = \frac{5x + 1 - (4x - 2)}{x^2 - 9}$$

Subtract numerators and include parentheses to indicate that both terms are subtracted. Place this difference over the common denominator.

$$= \frac{5x + 1 - 4x + 2}{x^2 - 9}$$

Remove parentheses and then change the sign of each term.

$$= \frac{x + 3}{x^2 - 9}$$

Combine like terms.

$$= \frac{\overset{1}{x + 3}}{\underset{1}{(x + 3)}(x - 3)}$$

Factor and simplify ($x \neq -3$ and $x \neq 3$).

$$= \frac{1}{x - 3}, \quad x \neq -3, x \neq 3$$

Check Point 5 Subtract: $\dfrac{x}{x + 1} - \dfrac{3x + 2}{x + 1}$.

Adding and Subtracting Rational Expressions with Different Denominators

Rational expressions that have no common factors in their denominators can be added or subtracted using one of the following properties:

$$\frac{a}{b} + \frac{c}{d} = \frac{ad + bc}{bd} \qquad \frac{a}{b} - \frac{c}{d} = \frac{ad - bc}{bd}, b \neq 0, d \neq 0.$$

The denominator, bd, is the product of the factors in the two denominators. Because we are looking at rational expressions that have no common factors in their denominators, the product bd gives the least common denominator.

EXAMPLE 6 **Subtracting Rational Expressions Having No Common Factors in Their Denominators**

Subtract: $\dfrac{x + 2}{2x - 3} - \dfrac{4}{x + 3}$.

Solution We need to find the least common denominator. This is the product of the distinct factors in each denominator, namely $(2x - 3)(x + 3)$. We can therefore use the subtraction property given previously as follows:

$$\frac{a}{b} - \frac{c}{d} = \frac{ad - bc}{bd}$$

$$\frac{x + 2}{2x - 3} - \frac{4}{x + 3} = \frac{(x + 2)(x + 3) - (2x - 3)4}{(2x - 3)(x + 3)}$$

Observe that $a = x + 2$, $b = 2x - 3$, $c = 4$, and $d = x + 3$.

$$= \frac{x^2 + 5x + 6 - (8x - 12)}{(2x - 3)(x + 3)}$$

Multiply.

$$= \frac{x^2 + 5x + 6 - 8x + 12}{(2x - 3)(x + 3)}$$

Remove parentheses and then change the sign of each term.

$$= \frac{x^2 - 3x + 18}{(2x - 3)(x + 3)}, x \neq \frac{3}{2}, x \neq -3$$

Combine like terms in the numerator.

Check Point 6 Add: $\dfrac{3}{x + 1} + \dfrac{5}{x - 1}$.

The **least common denominator**, or LCD, of several rational expressions is a polynomial consisting of the product of all prime factors in the denominators, with each factor raised to the greatest power of its occurrence in any denominator. When adding and subtracting rational expressions that have different denominators with one or more common factors in the denominators, it is efficient to find the least common denominator first.

Finding the Least Common Denominator

1. Factor each denominator completely.
2. List the factors of the first denominator.
3. Add to the list in step 2 any factors of the second denominator that do not appear in the list.
4. Form the product of each different factor from the list in step 3. This product is the least common denominator.

EXAMPLE 7 **Finding the Least Common Denominator**

Find the least common denominator of

$$\frac{7}{5x^2 + 15x} \quad \text{and} \quad \frac{9}{x^2 + 6x + 9}.$$

Solution

Step 1 Factor each denominator completely.

$$5x^2 + 15x = 5x(x + 3)$$
$$x^2 + 6x + 9 = (x + 3)^2$$

Step 2 List the factors of the first denominator.

$$5, x, (x + 3)$$

Step 3 Add any unlisted factors from the second denominator. The second denominator is $(x + 3)^2$ or $(x + 3)(x + 3)$. One factor of $x + 3$ is already in our list, but the other factor is not. We add $x + 3$ to the list. We have

$$5, x, (x + 3), (x + 3).$$

Step 4 The least common denominator is the product of all factors in the final list. Thus,

$$5x(x + 3)(x + 3), \quad \text{or} \quad 5x(x + 3)^2$$

is the least common denominator.

Check Point 7 Find the least common denominator of

$$\frac{3}{x^2 - 6x + 9} \quad \text{and} \quad \frac{7}{x^2 - 9}.$$

Finding the least common denominator for two (or more) rational expressions is the first step needed to add or subtract the expressions.

Adding and Subtracting Rational Expressions That Have Different Denominators with Shared Factors

1. Find the least common denominator.
2. Write all rational expressions in terms of the least common denominator. To do so, multiply both the numerator and the denominator of each rational expression by any factor(s) needed to convert the denominator into the least common denominator.
3. Add or subtract the numerators, placing the resulting expression over the least common denominator.
4. If necessary, simplify the resulting rational expression.

EXAMPLE 8   Adding Rational Expressions with Different Denominators

Add: $\dfrac{x + 3}{x^2 + x - 2} + \dfrac{2}{x^2 - 1}.$

Solution

Step 1 Find the least common denominator. Start by factoring the denominators.

$$x^2 + x - 2 = (x + 2)(x - 1)$$
$$x^2 - 1 = (x + 1)(x - 1)$$

The factors of the first denominator are $x + 2$ and $x - 1$. The only factor from the second denominator that is not listed is $x + 1$. Thus, the least common denominator is

$$(x + 2)(x - 1)(x + 1).$$

Step 2   Write all rational expressions in terms of the least common denominator. We do so by multiplying both the numerator and the denominator by any factor(s) needed to convert the denominator into the least common denominator.

$$\frac{x + 3}{x^2 + x - 2} + \frac{2}{x^2 - 1}$$

$$= \frac{x + 3}{(x + 2)(x - 1)} + \frac{2}{(x + 1)(x - 1)}$$

The least common denominator is $(x + 2)(x - 1)(x + 1)$.

$$= \frac{(x + 3)(x + 1)}{(x + 2)(x - 1)(x + 1)} + \frac{2(x + 2)}{(x + 2)(x - 1)(x + 1)}$$

Multiply each numerator and denominator by the extra factor required to form $(x + 2)(x - 1)(x + 1)$, the least common denominator.

Step 3   Add numerators, putting this sum over the least common denominator.

$$= \frac{(x + 3)(x + 1) + 2(x + 2)}{(x + 2)(x - 1)(x + 1)}$$

$$= \frac{x^2 + 4x + 3 + 2x + 4}{(x + 2)(x - 1)(x + 1)}$$

Perform the multiplications in the numerator.

$$= \frac{x^2 + 6x + 7}{(x + 2)(x - 1)(x + 1)}, x \neq -2, x \neq 1, x \neq -1$$

Combine like terms in the numerator.

Step 4   If necessary, simplify.   Because the numerator is prime, no further simplification is possible.

Check Point 8   Subtract: $\dfrac{x}{x^2 - 10x + 25} - \dfrac{x - 4}{2x - 10}$.

6 Simplify complex rational expressions.

Complex Rational Expressions

Complex rational expressions, also called **complex fractions**, have numerators or denominators containing one or more rational expressions. Here are two examples of such expressions:

$$\frac{1 + \dfrac{1}{x}}{1 - \dfrac{1}{x}}$$

Separate rational expressions occur in the numerator and denominator.

$$\frac{\dfrac{1}{x + h} - \dfrac{1}{x}}{h}.$$

Separate rational expressions occur in the numerator.

One method for simplifying a complex rational expression is to combine its numerator into a single expression and combine its denominator into a single expression. Then perform the division by inverting the denominator and multiplying.

EXAMPLE 9 **Simplifying a Complex Rational Expression**

Simplify: $\dfrac{1 + \dfrac{1}{x}}{1 - \dfrac{1}{x}}$.

Solution

$\dfrac{1 + \dfrac{1}{x}}{1 - \dfrac{1}{x}} = \dfrac{\dfrac{x}{x} + \dfrac{1}{x}}{\dfrac{x}{x} - \dfrac{1}{x}}, x \neq 0$

The terms in the numerator and in the denominator are each combined by performing the addition and subtraction. The least common denominator is x.

$= \dfrac{\dfrac{x+1}{x}}{\dfrac{x-1}{x}}$

Perform the addition in the numerator and the subtraction in the denominator.

$= \dfrac{x+1}{x} \div \dfrac{x-1}{x}$

Rewrite the main fraction bar as $\div$.

$= \dfrac{x+1}{x} \cdot \dfrac{x}{x-1}$

Invert the divisor and multiply (x $\neq$ 0 and x $\neq$ 1).

$= \dfrac{x+1}{\overset{}{\underset{1}{\cancel{x}}}} \cdot \dfrac{\overset{1}{\cancel{x}}}{x-1}$

Divide a numerator and denominator by the common factor, x.

$= \dfrac{x+1}{x-1}, x \neq 0, x \neq 1$

Multiply the remaining factors in the numerator and in the denominator.

Check Point 9 Simplify: $\dfrac{\dfrac{1}{x} - \dfrac{3}{2}}{\dfrac{1}{x} + \dfrac{3}{4}}$.

A second method for simplifying a complex rational expression is to find the least common denominator of all the rational expressions in its numerator and denominator. Then multiply each term in its numerator and denominator by this least common denominator. Here we use this method to simplify the complex rational expression in Example 9.

$\dfrac{1 + \dfrac{1}{x}}{1 - \dfrac{1}{x}} = \dfrac{\left(1 + \dfrac{1}{x}\right)}{\left(1 - \dfrac{1}{x}\right)} \cdot \dfrac{x}{x}$

The least common denominator of all the rational expressions is y. Multiply the numerator and denominator by x. Because $\dfrac{x}{x} = 1$, we are not changing the complex fraction (x $\neq$ 0).

$= \dfrac{1 \cdot x + \dfrac{1}{x} \cdot x}{1 \cdot x - \dfrac{1}{x} \cdot x}$

Use the distributive property. Be sure to distribute x to every term.

$= \dfrac{x+1}{x-1}, x \neq 0, x \neq 1$

Multiply. The complex rational expression is now simplified.

EXAMPLE 10 Simplifying a Complex Rational Expression

Simplify: $\dfrac{\dfrac{1}{x+h}-\dfrac{1}{x}}{h}$.

Solution We will use the method of multiplying each of the three terms, $\dfrac{1}{x+h}, \dfrac{1}{x}$, and h by the least common denominator. The least common denominator is $x(x+h)$.

$$\dfrac{\dfrac{1}{x+h}-\dfrac{1}{x}}{h}$$

$$=\dfrac{\left(\dfrac{1}{x+h}-\dfrac{1}{x}\right)x(x+h)}{hx(x+h)}$$

Multiply the numerator and denominator by $x(x+h), h \neq 0, x \neq 0, x \neq -h.$

$$=\dfrac{\dfrac{1}{x+h}\cdot x(x+h)-\dfrac{1}{x}\cdot x(x+h)}{h\cdot x(x+h)}$$

Use the distributive property in the numerator.

$$=\dfrac{x-(x+h)}{hx(x+h)}$$

Simplify: $\dfrac{1}{x+h}\cdot x(x+h) = x$ and $\dfrac{1}{x}\cdot x(x+h) = x+h.$

$$=\dfrac{x-x-h}{hx(x+h)}$$

Subtract in the numerator.

$$=\dfrac{\overset{1}{-h}}{\underset{1}{hx(x+h)}}$$

Simplify: $x - x - h = -h.$

$$=-\dfrac{1}{x(x+h)}, h \neq 0, x \neq 0, x \neq -h$$

Divide the numerator and denominator by h.

Check Point 10 Simplify: $\dfrac{\dfrac{1}{x+7}-\dfrac{1}{x}}{7}$.

7 Simplify fractional expressions that occur in calculus.

Fractional Expressions in Calculus

Fractional expressions containing radicals occur frequently in calculus. Because of the radicals, these expressions are not rational expressions. However, they can often be simplified using the procedure for simplifying complex rational expressions.

EXAMPLE 11 **Simplifying a Fractional Expression Containing Radicals**

Simplify: $\dfrac{\sqrt{9-x^2} + \dfrac{x^2}{\sqrt{9-x^2}}}{9-x^2}$.

Solution

$$\dfrac{\sqrt{9-x^2} + \dfrac{x^2}{\sqrt{9-x^2}}}{9-x^2}$$

The least common denominator of the denominators is $\sqrt{9-x^2}$.

$$= \dfrac{\sqrt{9-x^2} + \dfrac{x^2}{\sqrt{9-x^2}}}{9-x^2} \cdot \dfrac{\sqrt{9-x^2}}{\sqrt{9-x^2}}$$

Multiply the numerator and the denominator by $\sqrt{9-x^2}$.

$$= \dfrac{\sqrt{9-x^2}\sqrt{9-x^2} + \dfrac{x^2}{\sqrt{9-x^2}}\sqrt{9-x^2}}{(9-x^2)\sqrt{9-x^2}}$$

Use the distributive property in the numerator.

$$= \dfrac{(9-x^2) + x^2}{(9-x^2)^{3/2}}$$

In the denominator:
$(9-x^2)^1(9-x^2)^{1/2} = (9-x^2)^{1+(1/2)}$
$= (9-x^2)^{3/2}.$

$$= \dfrac{9}{\sqrt{(9-x^2)^3}}$$

Because the orignal expression was in radical form, write the denominator in radical form.

Check Point 11 Simplify: $\dfrac{\sqrt{x} + \dfrac{1}{\sqrt{x}}}{x}$.

8 Rationalize numerators.

Another fractional expression that you will encounter in calculus is

$$\dfrac{\sqrt{x+h} - \sqrt{x}}{h}.$$

Can you see that this expression is not defined if $h = 0$? However, in calculus, you will ask the following question:

> What happens to the expression as h takes on values that get closer and closer to 0, such as $h = 0.1$, $h = 0.01$, $h = 0.001$, $h = 0.0001$, and so on?

The question is answered by first **rationalizing the numerator**. This process involves rewriting the fractional expression as an equivalent expression in which the numerator no longer contains a radical. **To rationalize a numerator, multiply by 1 to eliminate the radical in the *numerator*.**

EXAMPLE 12   Rationalizing a Numerator

Rationalize the numerator:

$$\frac{\sqrt{x+h} - \sqrt{x}}{h}.$$

Solution If we multiply the numerator and denominator by $\sqrt{x+h} + \sqrt{x}$, the numerator will not contain radicals. Therefore, we multiply by 1, choosing $\dfrac{\sqrt{x+h} + \sqrt{x}}{\sqrt{x+h} + \sqrt{x}}$ for 1.

$$\frac{\sqrt{x+h} - \sqrt{x}}{h} = \frac{\sqrt{x+h} - \sqrt{x}}{h} \cdot \frac{\sqrt{x+h} + \sqrt{x}}{\sqrt{x+h} + \sqrt{x}} \qquad \text{Multiply by 1.}$$

$$= \frac{(\sqrt{x+h})^2 - (\sqrt{x})^2}{h(\sqrt{x+h} + \sqrt{x})} \qquad \begin{array}{l}(\sqrt{a} - \sqrt{b})(\sqrt{a} + \sqrt{b}) = \\ (\sqrt{a})^2 - (\sqrt{b})^2\end{array}$$

$$= \frac{x+h-x}{h(\sqrt{x+h} + \sqrt{x})} \qquad \begin{array}{l}(\sqrt{x+h})^2 = x+h \\ \text{and } (\sqrt{x})^2 = x.\end{array}$$

$$= \frac{h}{h(\sqrt{x+h} + \sqrt{x})} \qquad \text{Simplify: } x+h-x = h.$$

$$= \frac{1}{\sqrt{x+h} + \sqrt{x}}, \quad h \neq 0 \qquad \begin{array}{l}\text{Divide both the numerator} \\ \text{and denominator by h.}\end{array}$$

What happens to $\dfrac{\sqrt{x+h} - \sqrt{x}}{h}$ as h gets closer and closer to 0? In Example 12, we showed that

$$\frac{\sqrt{x+h} - \sqrt{x}}{h} = \frac{1}{\sqrt{x+h} + \sqrt{x}}.$$

As h gets closer to 0, the expression on the right gets closer to $\dfrac{1}{\sqrt{x+0} + \sqrt{x}} = \dfrac{1}{\sqrt{x} + \sqrt{x}}$, or $\dfrac{1}{2\sqrt{x}}$. Thus, the fractional expression $\dfrac{\sqrt{x+h} - \sqrt{x}}{h}$ approaches $\dfrac{1}{2\sqrt{x}}$ as h gets closer to 0.

Calculus Preview

In calculus, you will summarize the discussion on the right using the special notation

$$\lim_{h \to 0} \frac{\sqrt{x+h} - \sqrt{x}}{h} = \frac{1}{2\sqrt{x}}.$$

This is read "the limit of $\dfrac{\sqrt{x+h} - \sqrt{x}}{h}$ as h approaches 0 equals $\dfrac{1}{2\sqrt{x}}$."

Check Point 12   Rationalize the numerator: $\dfrac{\sqrt{x+3} - \sqrt{x}}{3}.$

EXERCISE SET P.6

 Practice Exercises

In Exercises 1–6, find all numbers that must be excluded from the domain of each rational expression.

1. $\dfrac{7}{x - 3}$

2. $\dfrac{13}{x + 9}$

3. $\dfrac{x + 5}{x^2 - 25}$

4. $\dfrac{x + 7}{x^2 - 49}$

5. $\dfrac{x - 1}{x^2 + 11x + 10}$

6. $\dfrac{x - 3}{x^2 + 4x - 45}$

In Exercises 7–12, simplify each rational expression. Find all numbers that must be excluded from the domain of the simplified rational expression.

7. $\dfrac{3x - 9}{x^2 - 6x + 9}$

8. $\dfrac{4x - 8}{x^2 - 4x + 4}$

9. $\dfrac{y^2 + 7y - 18}{y^2 - 3y + 2}$

10. $\dfrac{y^2 - 4y - 5}{y^2 + 5y + 4}$

11. $\dfrac{x^2 + 12x + 36}{x^2 - 36}$

12. $\dfrac{x^2 - 14x + 49}{x^2 - 49}$

In Exercises 13–24, multiply or divide as indicated.

13. $\dfrac{x^2 - 9}{x^2} \cdot \dfrac{x^2 - 3x}{x^2 + x - 12}$

14. $\dfrac{x^2 - 4}{x^2 - 4x + 4} \cdot \dfrac{2x - 4}{x + 2}$

15. $\dfrac{x^2 - 5x + 6}{x^2 - 2x - 3} \cdot \dfrac{x^2 - 1}{x^2 - 4}$

16. $\dfrac{x^2 + 5x + 6}{x^2 + x - 6} \cdot \dfrac{x^2 - 9}{x^2 - x - 6}$

17. $\dfrac{x^3 - 8}{x^2 - 4} \cdot \dfrac{x + 2}{3x}$

18. $\dfrac{x^2 + 6x + 9}{x^3 + 27} \cdot \dfrac{1}{x + 3}$

19. $\dfrac{x^2 - 4}{x} \div \dfrac{x + 2}{x - 2}$

20. $\dfrac{x^2 - 4}{x - 2} \div \dfrac{x + 2}{4x - 8}$

21. $\dfrac{x^2 - 25}{2x - 2} \div \dfrac{x^2 + 10x + 25}{x^2 + 4x - 5}$

22. $\dfrac{x^2 - 4}{x^2 + 3x - 10} \div \dfrac{x^2 + 5x + 6}{x^2 + 8x + 15}$

23. $\dfrac{x^2 + x - 12}{x^2 + x - 30} \cdot \dfrac{x^2 + 5x + 6}{x^2 - 2x - 3} \div \dfrac{x + 3}{x^2 + 7x + 6}$

24. $\dfrac{x^3 - 25x}{4x^2} \cdot \dfrac{2x^2 - 2}{x^2 - 6x + 5} \div \dfrac{x^2 + 5x}{7x + 7}$

In Exercises 25–44, add or subtract as indicated.

25. $\dfrac{4x + 1}{6x + 5} + \dfrac{8x + 9}{6x + 5}$

26. $\dfrac{3x + 2}{3x + 4} + \dfrac{3x + 6}{3x + 4}$

27. $\dfrac{x^2 - 2x}{x^2 + 3x} + \dfrac{x^2 + x}{x^2 + 3x}$

28. $\dfrac{x^2 - 4x}{x^2 - x - 6} + \dfrac{4x - 4}{x^2 - x - 6}$

29. $\dfrac{x^2 + 3x}{x^2 + x - 12} - \dfrac{x^2 - 12}{x^2 + x - 12}$

30. $\dfrac{x^2 - 4x}{x^2 - x - 6} - \dfrac{x - 6}{x^2 - x - 6}$

31. $\dfrac{3}{x + 4} + \dfrac{6}{x + 5}$

32. $\dfrac{8}{x - 2} + \dfrac{2}{x - 3}$

33. $\dfrac{3}{x + 1} - \dfrac{3}{x}$

34. $\dfrac{4}{x} - \dfrac{3}{x + 3}$

35. $\dfrac{2x}{x + 2} + \dfrac{x + 2}{x - 2}$

36. $\dfrac{3x}{x - 3} - \dfrac{x + 4}{x + 2}$

37. $\dfrac{x + 5}{x - 5} + \dfrac{x - 5}{x + 5}$

38. $\dfrac{x + 3}{x - 3} + \dfrac{x - 3}{x + 3}$

39. $\dfrac{4}{x^2 + 6x + 9} + \dfrac{4}{x + 3}$

40. $\dfrac{3}{5x + 2} + \dfrac{5x}{25x^2 - 4}$

41. $\dfrac{3x}{x^2 + 3x - 10} - \dfrac{2x}{x^2 + x - 6}$

42. $\dfrac{x}{x^2 - 2x - 24} - \dfrac{x}{x^2 - 7x + 6}$

43. $\dfrac{4x^2 + x - 6}{x^2 + 3x + 2} - \dfrac{3x}{x + 1} + \dfrac{5}{x + 2}$

44. $\dfrac{6x^2 + 17x - 40}{x^2 + x - 20} + \dfrac{3}{x - 4} - \dfrac{5x}{x + 5}$

In Exercise 45–54, simplify each complex rational expression.

45. $\dfrac{1 + \dfrac{1}{x}}{3 - \dfrac{1}{x}}$

46. $\dfrac{8 + \dfrac{1}{x}}{4 - \dfrac{1}{x}}$

47. $\dfrac{x - \dfrac{x}{x + 3}}{x + 2}$

48. $\dfrac{x - 3}{x - \dfrac{3}{x - 2}}$

49. $\dfrac{\dfrac{3}{x - 2} - \dfrac{4}{x + 2}}{\dfrac{7}{x^2 - 4}}$

50. $\dfrac{\dfrac{x}{x-2}+1}{\dfrac{3}{x^2-4}+1}$

51. $\dfrac{\dfrac{1}{x+1}}{\dfrac{1}{x^2-2x-3}+\dfrac{1}{x-3}}$

52. $\dfrac{\dfrac{6}{x^2+2x-15}-\dfrac{1}{x-3}}{\dfrac{1}{x+5}+1}$

53. $\dfrac{\dfrac{1}{(x+h)^2}-\dfrac{1}{x^2}}{h}$

54. $\dfrac{\dfrac{x+h}{x+h+1}-\dfrac{x}{x+1}}{h}$

Exercises 55–60 contain fractional expressions that occur frequently in calculus. Simplify each expression.

55. $\dfrac{\sqrt{x}-\dfrac{1}{3\sqrt{x}}}{\sqrt{x}}$

56. $\dfrac{\sqrt{x}-\dfrac{1}{4\sqrt{x}}}{\sqrt{x}}$

57. $\dfrac{\dfrac{x^2}{\sqrt{x^2+2}}-\sqrt{x^2+2}}{x^2}$

58. $\dfrac{\sqrt{5-x^2}+\dfrac{x^2}{\sqrt{5-x^2}}}{5-x^2}$

59. $\dfrac{\dfrac{1}{\sqrt{x+h}}-\dfrac{1}{\sqrt{x}}}{h}$

60. $\dfrac{\dfrac{1}{\sqrt{x+3}}-\dfrac{1}{\sqrt{x}}}{3}$

In Exercises 61–64, rationalize the numerator.

61. $\dfrac{\sqrt{x+5}-\sqrt{x}}{5}$

62. $\dfrac{\sqrt{x+7}-\sqrt{x}}{7}$

63. $\dfrac{\sqrt{x}+\sqrt{y}}{x^2-y^2}$

64. $\dfrac{\sqrt{x}-\sqrt{y}}{x^2-y^2}$

Application Exercises

65. The rational expression
$$\frac{130x}{100-x}$$
describes the cost, in millions of dollars, to inoculate x percent of the population against a particular strain of flu.

a. Evaluate the expression for $x=40$, $x=80$, and $x=90$. Describe the meaning of each evaluation in terms of percentage inoculated and cost.

b. For what value of x is the expression undefined?

c. What happens to the cost as x approaches 100%? How can you interpret this observation?

66. Doctors use the rational expression
$$\frac{DA}{A+12}$$
to determine the dosage of a drug prescribed for children. In this expression, A = child's age and D = adult dosage. What is the difference in the child's dosage for a 7-year-old child and a 3-year-old child? Express the answer as a single rational expression in terms of D. Then describe what your answer means in terms of the variables in the rational expression.

67. Anthropologists and forensic scientists classify skulls using
$$\frac{L+60W}{L}-\frac{L-40W}{L}$$
where L is the skull's length and W is its width.

a. Express the classification as a single rational expression.

b. If the value of the rational expression in part (a) is less than 75, a skull is classified as long. A medium skull has a value between 75 and 80, and a round skull has a value over 80. Use your rational expression from part (a) to classify a skull that is 5 inches wide and 6 inches long.

68. The polynomial
$$6t^4-207t^3+2128t^2-6622t+15{,}220$$
describes the annual number of drug convictions in the United States t years after 1984. The polynomial
$$28t^4-711t^3+5963t^2-1695t+27{,}424$$
describes the annual number of drug arrests in the United States t years after 1984. Write a rational expression that describes the conviction rate for drug arrests in the United States t years after 1984.

69. The average speed on a round-trip commute having a one-way distance d is given by the complex rational expression
$$\frac{2d}{\dfrac{d}{r_1}+\dfrac{d}{r_2}}$$
in which r_1 and r_2 are the speeds on the outgoing and return trips, respectively. Simplify the expression. Then find the average speed for a person who drives from home to work at 30 miles per hour and returns on the same route averaging 20 miles per hour. Explain why the answer is not 25 miles per hour.

70. If three resistors with resistances R_1, R_2, and R_3 are connected in parallel, their combined resistance is given by the expression
$$\frac{1}{\dfrac{1}{R_1}+\dfrac{1}{R_2}+\dfrac{1}{R_3}}.$$
Simplify the complex rational expression. Then find the combined resistance when R_1 is 4 ohms, R_2 is 8 ohms, and R_3 is 12 ohms.

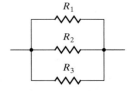

Writing in Mathematics

71. What is a rational expression?

72. Explain how to determine which numbers must be excluded from the domain of a rational expression.

73. Explain how to simplify a rational expression.

74. Explain how to multiply rational expressions.

75. Explain how to divide rational expressions.

76. Explain how to add or subtract rational expressions with the same denominators.

77. Explain how to add rational expressions having no common factors in their denominators. Use $\dfrac{3}{x+5} + \dfrac{7}{x+2}$ in your explanation.

78. Explain how to find the least common denominator for denominators of $x^2 - 100$ and $x^2 - 20x + 100$.

79. Describe two ways to simplify $\dfrac{\dfrac{3}{x} + \dfrac{2}{x^2}}{\dfrac{1}{x^2} + \dfrac{2}{x}}$.

Explain each error in Exercises 80–82. Then rewrite the right side of the equation to correct the error that now exists.

80. $\dfrac{1}{a} + \dfrac{1}{b} = \dfrac{1}{a+b}$

81. $\dfrac{1}{x} + 7 = \dfrac{1}{x+7}$

82. $\dfrac{a}{x} + \dfrac{a}{b} = \dfrac{a}{x+b}$

83. A politician claims that each year the conviction rate for drug arrests in the United States is increasing. Explain how to use the polynomials in Exercise 68 to verify this claim.

Technology Exercise

84. How much are your monthly payments on a loan? If P is the principal, or amount borrowed, i is the monthly interest rate, and n is the number of monthly payments, then the amount, A, of each monthly payment is

$$A = \dfrac{Pi}{1 - \dfrac{1}{(1+i)^n}}.$$

a. Simplify the complex rational expression for the amount of each payment.

b. You purchase a \$20,000 automobile at 1% monthly interest to be paid over 48 months. How much do you pay each month? Use the simplified rational expression from part (a) and a calculator. Round to the nearest dollar.

Critical Thinking Exercises

85. Which one of the following is true?

a. $\dfrac{x^2 - 25}{x - 5} = x - 5$

b. $\dfrac{x}{y} \div \dfrac{y}{x} = 1$, if $x \neq 0$ and $y \neq 0$.

c. The least common denominator needed to find $\dfrac{1}{x} + \dfrac{1}{x+3}$ is $x + 3$.

d. The rational expression

$$\dfrac{x^2 - 16}{x - 4}$$

is not defined for $x = 4$. However, as x gets closer and closer to 4, the value of the expression approaches 8.

In Exercises 86–87, find the missing expression.

86. $\dfrac{3x}{x-5} + \dfrac{\boxed{}}{5-x} = \dfrac{7x+1}{x-5}$

87. $\dfrac{4}{x-2} - \dfrac{\boxed{}}{} = \dfrac{2x+8}{(x-2)(x+1)}$

88. In one short sentence, five words or less, explain what

$$\dfrac{\dfrac{1}{x} + \dfrac{1}{x^2} + \dfrac{1}{x^3}}{\dfrac{1}{x^4} + \dfrac{1}{x^5} + \dfrac{1}{x^6}}$$

does to each number x.

SECTION P.7 *Linear Equations*

Objectives

1. Solve linear equations in one variable.
2. Solve equations with constants in denominators.
3. Solve equations with variables in denominators.
4. Recognize identities, conditional equations, and inconsistent equations.
5. Solve for a variable in a formula.
6. Solve equations involving absolute value.

Unfortunately, many of us have been fined for driving over the speed limit. The amount of the fine depends on how fast we are speeding. Suppose that a highway has a speed limit of 60 miles per hour. The amount that speeders are fined, F, is described by the statement of equality

$$F = 10x - 600$$

where x is the speed, in miles per hour. We can use this statement to determine the fine, F, for a speeder traveling at, say, 70 miles per hour. We substitute 70 for x in the given statement and then find the corresponding value for F.

$$F = 10(70) - 600 = 700 - 600 = 100$$

Thus, a person caught driving 70 miles per hour gets a $100 fine.

A friend, whom we shall call Leadfoot, borrows your car and returns a few hours later with a $400 speeding fine. Leadfoot is furious, protesting that the car was barely driven over the speed limit. Should you believe Leadfoot?

In order to decide if Leadfoot is telling the truth, use $F = 10x - 600$. Leadfoot was fined $400, so substitute 400 for F:

$$400 = 10x - 600.$$

In Example 1, we will find the value for x. This variable represents Leadfoot's speed, which resulted in the $400 fine.

An **equation** consists of two algebraic expressions joined by an equal sign. Thus, $400 = 10x - 600$ is an example of an equation. The equal sign divides the equation into two parts, the left side and the right side:

$$\boxed{400} = \boxed{10x - 600}.$$

Left side   Right side

The two sides of an equation can be reversed. So, we can also express this equation as

$$10x - 600 = 400.$$

Notice that the highest exponent on the variable is 1. Such an equation is called a *linear equation in one variable*. In this section, we will study how to solve linear equations.

1 Solve linear equations in one variable.

Solving Linear Equations in One Variable

We begin with a general definition of a linear equation in one variable.

Definition of a Linear Equation

A **linear equation in one variable** x is an equation that can be written in the form

$$ax + b = 0$$

where a and b are real numbers, and $a \neq 0$.

An example of a linear equation in one variable is $4x + 12 = 0$. **Solving an equation** in x involves determining all values of x that result in a true statement when substituted into the equation. Such values are **solutions**, or **roots**, of the equation. For example, substitute -3 into $4x + 12 = 0$. We obtain $4(-3) + 12 = 0$, or $-12 + 12 = 0$. This simplifies to the true statement $0 = 0$. Thus, -3 is a solution of the equation $4x + 12 = 0$. We also say that -3 **satisfies** the equation $4x + 12 = 0$, because when we substitute -3 for x, a true statement results. The set of all such solutions is called the equation's **solution set**. For example, the solution set of the equation $4x + 12 = 0$ is $\{-3\}$.

Equations that have the same solution set are called **equivalent equations**. For example, the equations $4x + 12 = 0$, $4x = -12$, and $x = -3$ are equivalent equations because the solution set for each is $\{-3\}$. To solve a linear equation in x, we transform the equation into an equivalent equation one or more times. Our final equivalent equation should be in the form $x = d$, where d, is a real number. By inspection, we can see that the solution set of this equation is $\{d\}$.

To generate equivalent equations, we will use the following principles:

Study Tip

We can solve equations such as $3(x - 6) = 5x$ for a variable. However, we cannot solve for a variable in an algebraic expression such as $3(x - 6)$. We *simplify* algebraic expressions.

Correct

Simplify: $3(x - 6)$.
$3(x - 6) = 3x - 18$

Incorrect

Simplify: $3(x - 6)$.
$3(x - 6) = 0$
$3x - 18 = 0$
$3x - 18 + 18 = 0 + 18$
$3x = 18$
$\dfrac{3x}{3} = \dfrac{18}{3}$
$x = 6$

Generating Equivalent Equations

An equation can be transformed into an equivalent equation by one or more of the following operations:

Example

1. Simplify an expression by removing grouping symbols and combining like terms.

$\quad 3(x - 6) = 6x - x$
$\quad 3x - 18 = 5x$

2. Add (or subtract) the same real number or variable expression on *both* sides of the equation.

$\quad 3x - 18 = 5x$

> Subtract 3x from both sides of the equation.

$\quad 3x - 18 - 3x = 5x - 3x$
$\quad\quad\quad -18 = 2x$

3. Multiply (or divide) on *both* sides of the equation by the same *nonzero* quantity.

$\quad -18 = 2x$

> Divide both sides of the equation by 2.

$\quad \dfrac{-18}{2} = \dfrac{2x}{2}$
$\quad -9 = x$

4. Interchange the two sides of the equation.

$\quad -9 = x$
$\quad x = -9$

If you look closely at the equations in the box, you will notice that we have solved the equation $3(x - 6) = 6x - x$. The final equation, $x = -9$, with x isolated by itself on the left side, shows that $\{-9\}$ is the solution set. The idea in solving a linear equation is to get the variable by itself on one side of the equal sign and a number by itself on the other side.

EXAMPLE 1   Solving a Linear Equation (Is Leadfoot Telling the Truth?)

Solve the equation: $10x - 600 = 400$.

Solution   Remember that x represents Leadfoot's speed that resulted in the $400 fine. Our goal is to get x by itself on the left side. We do this by adding 600 to both sides to get $10x$ by itself. Then we isolate x from $10x$ by dividing both sides of the equation by 10.

$$10x - 600 = 400$$ This is the given equation.

$$10x - 600 + 600 = 400 + 600$$ Add 600 to both sides.

$$10x = 1000$$ Combine like terms.

$$\frac{10x}{10} = \frac{1000}{10}$$ Divide both sides by 10.

$$x = 100$$ Simplify.

Can this possibly be correct? Was Leadfoot doing 100 miles per hour in the car he borrowed from you? To find out, check the proposed solution, 100, in the original equation. In other words, evaluate for $x = 100$.

Check 100:

$$10x - 600 = 400$$ This is the original equation.

$$10(100) - 600 \stackrel{?}{=} 400$$ Substitute 100 for x. The question mark indicates that we do not yet know if the two sides are equal.

$$1000 - 600 \stackrel{?}{=} 400$$ Multiply: 10(100) = 1000.

$$400 = 400$$ ⟵ This statement is true.   Subtract: 1000 − 600 = 400.

The true statement $400 = 400$ indicates that 100 is the solution. This verifies that the solution set is $\{100\}$. Leadfoot was doing an outrageous 100 miles per hour, and lied by claiming that your car was barely driven over the speed limit.

Check Point 1   Solve and check: $5x - 8 = 72$.

We now present a step-by-step procedure for solving a linear equation in one variable. Not all of these steps are necessary to solve every equation.

Study Tip

If your proposed solution is incorrect, you will get a false statement when you check your answer. For example, 65 is not a solution of $10x - 600 = 400$. Look what happens when we substitute 65 for x:

$$10x - 600 = 400$$
$$10(65) - 600 \stackrel{?}{=} 400$$
$$650 - 600 \stackrel{?}{=} 400$$
$$50 = 400 \text{ False.}$$

The compact, symbolic notation of algebra enables us to use a clear step-by-step method for solving equations, designed to avoid the confusion shown in the painting.

Solving a Linear Equation

1. Simplify the algebraic expression on each side.
2. Collect the variable terms on one side and the constant terms on the other side.
3. Isolate the variable and solve.
4. Check the proposed solution in the original equation.

EXAMPLE 2   **Solving a Linear Equation**

Solve the equation: $2(x - 3) - 17 = 13 - 3(x + 2)$.

Solution

Step 1   **Simplify the algebraic expression on each side.**

$$2(x - 3) - 17 = 13 - 3(x + 2) \qquad \text{This is given equation.}$$
$$2x - 6 - 17 = 13 - 3x - 6 \qquad \text{Use the distributive property.}$$
$$2x - 23 = -3x + 7 \qquad \text{Combine like terms.}$$

Step 2   **Collect variable terms on one side and constant terms on the other side.** We will collect variable terms on the left by adding $3x$ to both sides. We will collect the numbers on the right by adding 23 to both sides.

$$2x - 23 + 3x = -3x + 7 + 3x \qquad \text{Add 3x to both sides.}$$
$$5x - 23 = 7 \qquad \text{Simplify.}$$
$$5x - 23 + 23 = 7 + 23 \qquad \text{Add 23 to both sides.}$$
$$5x = 30 \qquad \text{Simplify.}$$

> ## Discovery
>
> Solve the equation in Example 2 by collecting terms with the variable on the right and numerical terms on the left. What do you observe?

Step 3   **Isolate the variable and solve.**   We isolate x by dividing both sides by 5.

$$\frac{5x}{5} = \frac{30}{5} \qquad \text{Divide both sides by 5.}$$
$$x = 6 \qquad \text{Simplify.}$$

Step 4   **Check the proposed solution in the original equation.**   Substitute 6 for x in the original equation.

$$2(x - 3) - 17 = 13 - 3(x + 2) \qquad \text{This is the original equation.}$$
$$2(6 - 3) - 17 \overset{?}{=} 13 - 3(6 + 2) \qquad \text{Substitute 6 for x.}$$
$$2(3) - 17 \overset{?}{=} 13 - 3(8) \qquad \text{Simplify inside parentheses.}$$
$$6 - 17 \overset{?}{=} 13 - 24 \qquad \text{Multiply.}$$
$$-11 = -11 \qquad \text{Subtract.}$$

The true statement $-11 = -11$ verifies that the solution set is $\{6\}$.

Check Point 2   Solve and check: $4(2x + 1) - 29 = 3(2x - 5)$.

2 Solve equations with constants in denominators.

Linear Equations with Fractions

Equations are easier to solve when they do not contain fractions. How do we solve equations involving fractions? We begin by multiplying both sides of the equation by the least common denominator of all fractions in the equation. The least common denominator is the smallest number that all the denominators will divide into. Multiplying every term on both sides of the equation by the least common denominator will eliminate the fractions in the equation. Example 3 shows how we "clear an equation of fractions."

EXAMPLE 3   Solving a Linear Equation Involving Fractions

Solve the equation: $\dfrac{3x}{2} = \dfrac{x}{5} - \dfrac{39}{5}$.

Solution   The denominators are 2, 5, and 5. The smallest number that is divisible by 2, 5, and 5 is 10. We begin by multiplying both sides of the equation by 10, the least common denominator.

$$\dfrac{3x}{2} = \dfrac{x}{5} - \dfrac{39}{5}$$ This is the given equation.

$$10 \cdot \dfrac{3x}{2} = 10\left(\dfrac{x}{5} - \dfrac{39}{5}\right)$$ Multiply both sides by 10.

$$10 \cdot \dfrac{3x}{2} = 10 \cdot \dfrac{x}{5} - 10 \cdot \dfrac{39}{5}$$ Use the distributive property and multiply each term by 10.

$$\overset{5}{10} \cdot \dfrac{3x}{\underset{1}{2}} = \overset{2}{10} \cdot \dfrac{x}{\underset{1}{5}} - \overset{2}{10} \cdot \dfrac{39}{\underset{1}{5}}$$ Divide out common factors in each multiplication.

$$15x = 2x - 78$$ Complete the multiplications. The fractions are now cleared.

At this point, we have an equation similar to those we previously solved. Collect the variable terms on one side and the constant terms on the other side.

$$15x - 2x = 2x - 2x - 78$$ Subtract 2x to get the variable terms on the left.

$$13x = -78$$ Simplify.

Isolate x by dividing both sides by 13.

$$\dfrac{13x}{13} = \dfrac{-78}{13}$$ Divide both sides by 13.

$$x = -6$$ Simplify.

Check the proposed solution. Substitute -6 for x in the original equation. You should obtain $-9 = -9$. This true statement verifies that the solution set is $\{-6\}$.

Check Point 3   Solve and check: $\dfrac{x}{4} = \dfrac{2x}{3} + \dfrac{5}{6}$.

3   Solve equations with variables in denominators.

Equations Involving Rational Expressions

In Example 3 we solved a linear equation with constants in denominators. Now, let's consider an equation such as

$$\dfrac{1}{x} = \dfrac{1}{5} + \dfrac{3}{2x}.$$

Can you see how this equation differs from the fractional equation that we solved earlier? The variable, x, appears in two of the denominators. The procedure for solving this equation still involves multiplying each side by the least common denominator. However, we must avoid any values of the variable that make a denominator zero. For example, examine the denominators in the equation

$$\frac{1}{x} = \frac{1}{5} + \frac{3}{2x}.$$

This denominator would equal zero if $x = 0$.

This denominator would equal zero if $x = 0$.

We see that x cannot equal zero. With this in mind, let's solve the equation.

EXAMPLE 4 Solving an Equation Involving Rational Expressions

Solve: $\dfrac{1}{x} = \dfrac{1}{5} + \dfrac{3}{2x}$.

Solution The denominators are x, 5, and $2x$. The least common denominator is $10x$. We begin by multiplying both sides of the equation by $10x$. We will also write the restriction that x cannot equal zero to the right of the equation.

$$\frac{1}{x} = \frac{1}{5} + \frac{3}{2x}, \quad x \neq 0 \qquad \text{This is the given equation.}$$

$$10x \cdot \frac{1}{x} = 10x\left(\frac{1}{5} + \frac{3}{2x}\right) \qquad \text{Multiply both sides by 10x.}$$

$$10x \cdot \frac{1}{x} = 10x \cdot \frac{1}{5} + 10x \cdot \frac{3}{2x} \qquad \text{Use the distributive property and multiply each term by 10x.}$$

$$\overset{1}{10x} \cdot \frac{1}{\underset{1}{x}} = \overset{2}{10x} \cdot \frac{1}{\underset{1}{5}} + \overset{5}{10x} \cdot \frac{3}{\underset{1}{2x}} \qquad \text{Divide out common factors in each multiplication.}$$

$$10 = 2x + 15 \qquad \text{Complete the multiplications.}$$

Observe that the resulting equation,

$$10 = 2x + 15,$$

is now cleared of fractions. With the variable term, $2x$, already on the right, we will collect constant terms on the left by subtracting 15 from both sides.

$$10 - 15 = 2x + 15 - 15 \qquad \text{Subtract 15 from both sides.}$$

$$-5 = 2x \qquad \text{Simplify.}$$

Finally, we isolate the variable, x, in $-5 = 2x$ by dividing both sides by 2.

$$\frac{-5}{2} = \frac{2x}{2} \qquad \text{Divide both sides by 2.}$$

$$-\frac{5}{2} = x \qquad \text{Simplify.}$$

We check our solution by substituting $-\frac{5}{2}$ into the original equation or by using a calculator. With a calculator, evaluate each side of the equation for $x = -\frac{5}{2}$, or for $x = -2.5$. Note that the original restriction that $x \neq 0$ is met. The solution set is $\{-\frac{5}{2}\}$.

Check Point 4 Solve: $\dfrac{5}{2x} = \dfrac{17}{18} - \dfrac{1}{3x}$.

EXAMPLE 5 Solving an Equation Involving Rational Expressions

Solve: $\dfrac{x}{x-3} = \dfrac{3}{x-3} + 9$.

Solution We must avoid any values of the variable x that make a denominator zero.

$$\dfrac{x}{x-3} = \dfrac{3}{x-3} + 9$$

These denominators are zero if $x - 3 = 0$, or equivalently, if $x = 3$.

We see that x cannot equal 3. With denominators of $x - 3$, $x - 3$, and 1, the least common denominator is $x - 3$. We multiply both sides of the equation by $x - 3$. We also write the restriction that x cannot equal 3 to the right of the equation.

$$\dfrac{x}{x-3} = \dfrac{3}{x-3} + 9, \quad x \neq 3$$

This is the given equation.

$$(x-3) \cdot \dfrac{x}{x-3} = (x-3)\left[\dfrac{3}{x-3} + 9 \right]$$

Multiply both sides by $x - 3$.

$$(x-3) \cdot \dfrac{x}{x-3} = (x-3) \cdot \dfrac{3}{x-3} + (x-3) \cdot 9$$

Use the distributive property.

$$(\cancel{x-3}) \cdot \dfrac{x}{\cancel{x-3}} = (\cancel{x-3}) \cdot \dfrac{3}{\cancel{x-3}} + (x-3) \cdot 9$$

Divide out common factors in each multiplication.

$$x = 3 + (x-3) \cdot 9$$

Simplify.

The resulting equation, which can be expressed as

$$x = 3 + 9(x - 3),$$

is cleared of fractions. We now solve for x.

$$x = 3 + 9x - 27$$

Use the distributive property.

$$x = 9x - 24$$

Combine numerical terms.

$$x - 9x = 9x - 24 - 9x \qquad \text{Subtract 9x from both sides.}$$

$$-8x = -24 \qquad \text{Simplify.}$$

$$\frac{-8x}{-8} = \frac{-24}{-8} \qquad \text{Solve for x, dividing both sides by } -8.$$

$$x = 3 \qquad \text{Simplify.}$$

The proposed solution, 3, is *not* a solution because of the restriction that $x \neq 3$. There is *no solution to this equation.* The solution set for this equation contains no elements and is called the empty set, written $\varnothing$.

Study Tip

Reject any proposed solution that causes any denominator in an equation to equal 0.

Check Point 5 Solve: $\dfrac{x}{x - 2} = \dfrac{2}{x - 2} - \dfrac{2}{3}$.

4 Recognize identities, conditional equations, and inconsistent equations.

Types of Equations

We tend to place things in categories, allowing us to order and structure the world. For example, you can categorize yourself by your age group, your ethnicity, your academic major, or your gender. Equations can be placed into categories that depend on their solution sets.

An equation that is true for all real numbers for which both sides are defined is called an **identity**. An example of an identity is

$$x + 3 = x + 2 + 1.$$

Every number plus 3 is equal to that number plus 2 plus 1. Therefore, the solution set to this equation is the set of all real numbers. Another example of an identity is

$$\frac{2x}{x} = 2.$$

Because division by 0 is undefined, this equation is true for all real number values of x except 0. The solution set is the set of nonzero real numbers.

An equation that is not an identity, but that is true for at least one real number, is called a **conditional equation**. The equation $10x - 600 = 400$ is an example of a conditional equation. The equation is not an identity and is true only if x is 100.

An **inconsistent equation** is an equation that is not true for even one real number. An example of an inconsistent equation is

$$x = x + 7.$$

There is no number that is equal to itself plus 7. Some inconsistent equations are less obvious than this. Consider the equation in Example 5,

$$\frac{x}{x - 3} = \frac{3}{x - 3} + 9.$$

This equation is not true for any real number and has no solution. Thus, it is inconsistent.

EXAMPLE 6   Categorizing an Equation

Determine whether the equation

$$2(x + 1) = 2x + 3$$

is an identity, a conditional equation, or an inconsistent equation.

Solution   Let's see what happens if we try solving $2(x + 1) = 2x + 3$. Applying the distributive property on the left side, we obtain

$$2x + 2 = 2x + 3.$$

Does something look strange? Can doubling a number and increasing the product by 2 give the same result as doubling the same number and increasing the product by 3? No. Let's continue solving the equation by subtracting $2x$ from both sides.

$$2x + 2 - 2x = 2x + 3 - 2x$$
$$2 = 3$$

The false statement $2 = 3$ verifies that the given equation is inconsistent.

> **Check Point 6**   Determine whether the equation
>
> $$2(x + 1) = 2x + 2$$
>
> is an identity, a conditional equation, or an inconsistent equation.

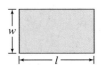

5 Solve for a variable in a formula.

Solving for a Variable in a Formula

A **formula** is an equation that uses letters to express a relationship among two or more variables. For example, the formula

$$P = 2l + 2w$$

expresses the relationship among a rectangle's perimeter, P, its length, l, and its width, w. (See Figure P.7.) A rectangle's perimeter is twice its length plus twice its width.

Figure P.7   The perimeter, P, of a rectangle is given by the formula $P = 2l + 2w$.

When working with formulas, such as the geometric formula for the perimeter of a rectangle, it is often necessary to solve for a specified variable. This is done by isolating the specified variable on one side of the equation. Begin by isolating all terms with the specified variable on one side of the equation and all terms without the specified variable on the other side. The next example shows how to do this.

EXAMPLE 7   Solving for a Variable in a Formula

Solve the formula $2l + 2w = P$ for w.

Solution   First, isolate $2w$ on the left by subtracting $2l$ from both sides. Then solve for w by dividing both sides by 2.

> We need to isolate w.

$$2l + 2w = P \qquad \text{This is the given formula.}$$

$$2l - 2l + 2w = P - 2l \qquad \text{Isolate } 2w \text{ by subtracting } 2l \text{ from both sides.}$$

$$2w = P - 2l \qquad \text{Simplify.}$$

$$\frac{2w}{2} = \frac{P - 2l}{2} \qquad \text{Isolate } w \text{ by dividing both sides by 2.}$$

> You can divide both P and 2l by 2, expressing the answer as $w = \frac{P}{2} - l$.

$$w = \frac{P - 2l}{2} \qquad \text{Simplify.}$$

Check Point 7   Solve $y = mx + b$ for m.

EXAMPLE 8   Solving for a Variable That Occurs Twice in a Formula

Solve the formula $A = P + Prt$ for P.

Solution   Notice that all terms with P already occur on the right side of the equation. Factor P from the two terms on the right to isolate P.

$$A = P + Prt \qquad \text{This is the given formula.}$$

$$A = P(1 + rt) \qquad \text{Factor P on the right side of the equation.}$$

$$\frac{A}{1 + rt} = \frac{P(1 + rt)}{1 + rt} \qquad \text{Divide both sides by } 1 + rt.$$

$$\frac{A}{1 + rt} = P \qquad \text{Simplify: } \frac{P\cancel{(1 + rt)}}{\cancel{(1 + rt)}} = \frac{P}{1} = P.$$

Study Tip

You cannot solve $A = P + Prt$ for P by subtracting Prt from both sides and writing

$$A - Prt = P.$$

When a formula is solved for a specified variable, that variable must be isolated on one side. The variable P occurs on both sides of

$$A - Prt = P.$$

Check Point 8   Solve the formula $P = C + MC$ for C.

6 Solve equations involving absolute value.

Equations Involving Absolute Value

We have seen that the absolute value of x, $|x|$, describes the distance of x from zero on a number line. Now consider **absolute value equations**, such as

$$|x| = 2.$$

Figure P.8   If $|x| = 2$, then $x = 2$ or $x = -2$.

This means that we must determine real numbers whose distance from the origin on the number line is 2. Figure P.8 shows that there are two numbers such that $|x| = 2$, namely, 2 or -2. We write $x = 2$ or $x = -2$. This observation can be generalized as follows:

> **Rewriting an Absolute Value Equation without Absolute Value Bars**
> If c is a positive real number and X represents any algebraic expression, then $|X| = c$ is equivalent to $X = c$ or $X = -c$.

EXAMPLE 9 Solving an Equation Involving Absolute Value

Solve: $|2x - 3| = 11$.

Solution

$\|2x - 3\| = 11$	This is the given equation.
$2x - 3 = 11$　　or　　$2x - 3 = -11$	Rewrite the equation without absolute value bars.
$2x = 14$　　　　　　　$2x = -8$	Add 3 to both sides of each equation.
$x = 7$　　　　　　　$x = -4$	Divide both sides of each equation by 2.

Check 7:	**Check − 4:**	
$\|2x - 3\| = 11$	$\|2x - 3\| = 11$	This is the original equation.
$\|2(7) - 3\| \stackrel{?}{=} 11$	$\|2(-4) - 3\| \stackrel{?}{=} 11$	Substitute the proposed solutions.
$\|14 - 3\| \stackrel{?}{=} 11$	$\|-8 - 3\| \stackrel{?}{=} 11$	Perform operations inside the absolute value bars.
$\|11\| \stackrel{?}{=} 11$	$\|-11\| \stackrel{?}{=} 11$	
$11 = 11 \checkmark$	$11 = 11 \checkmark$	These true statements, indicated by the checks, show that 7 and −4 are solutions.

The solution set is $\{-4, 7\}$.

Check Point 9 Solve: $|2x - 1| = 5$.

The absolute value of a number is never negative. Thus, if X is an algebraic expression and c is a negative number, then $|X| = c$ has no solution. For example, the equation $|3x - 6| = -2$ has no solution because $|3x - 6|$ cannot be negative. The solution set is $\varnothing$, the empty set.

The absolute value of 0 is 0. Thus, if X is an algebraic expression and $|X| = 0$, the solution is found by solving $X = 0$. For example, the solution of $|x - 2| = 0$ is obtained by solving $x - 2 = 0$. The solution is 2 and the solution set is $\{2\}$.

To solve some absolute value equations, it is necessary to first isolate the expression containing the absolute value symbols. For example, consider the equation

$$3|2x - 3| - 8 = 25.$$

We need to isolate $|2x - 3|$.

How can we isolate $|2x - 3|$? Add 8 to both sides of the equation and then divide both sides by 3.

$$3|2x - 3| - 8 = 25 \qquad \text{This is the given equation.}$$
$$3|2x - 3| = 33 \qquad \text{Add 8 to both sides.}$$
$$|2x - 3| = 11 \qquad \text{Divide both sides by 3.}$$

This results in the equation we solved in Example 9.

EXERCISE SET P.7

Practice Exercises

In Exercises 1–16, solve and check each linear equation.

1. $7x - 5 = 72$ **2.** $6x - 3 = 63$

3. $11x - (6x - 5) = 40$ **4.** $5x - (2x - 10) = 35$

5. $2x - 7 = 6 + x$ **6.** $3x + 5 = 2x + 13$

7. $7x + 4 = x + 16$ **8.** $13x + 14 = 12x - 5$

9. $3(x - 2) + 7 = 2(x + 5)$

10. $2(x - 1) + 3 = x - 3(x + 1)$

11. $3(x - 4) - 4(x - 3) = x + 3 - (x - 2)$

12. $2 - (7x + 5) = 13 - 3x$

13. $16 = 3(x - 1) - (x - 7)$

14. $5x - (2x + 2) = x + (3x - 5)$

15. $25 - [2 + 5y - 3(y + 2)] =$
$$- 3(2y - 5) - [5(y - 1) - 3y + 3]$$

16. $45 - [4 - 2y - 4(y + 7)] =$
$$-4(1 + 3y) - [4 - 3(y + 2) - 2(2y - 5)]$$

Exercises 17–30 contain equations with constants in denominators. Solve each equation.

17. $\dfrac{x}{3} = \dfrac{x}{2} - 2$ **18.** $\dfrac{x}{5} = \dfrac{x}{6} + 1$

19. $20 - \dfrac{x}{3} = \dfrac{x}{2}$ **20.** $\dfrac{x}{5} - \dfrac{1}{2} = \dfrac{x}{6}$

21. $\dfrac{3x}{5} = \dfrac{2x}{3} + 1$ **22.** $\dfrac{x}{2} = \dfrac{3x}{4} + 5$

23. $\dfrac{3x}{5} - x = \dfrac{x}{10} - \dfrac{5}{2}$ **24.** $2x - \dfrac{2x}{7} = \dfrac{x}{2} + \dfrac{17}{2}$

25. $\dfrac{x + 3}{6} = \dfrac{3}{8} + \dfrac{x - 5}{4}$ **26.** $\dfrac{x + 1}{4} = \dfrac{1}{6} + \dfrac{2 - x}{3}$

27. $\dfrac{x}{4} = 2 + \dfrac{x - 3}{3}$ **28.** $5 + \dfrac{x - 2}{3} = \dfrac{x + 3}{8}$

29. $\dfrac{x + 1}{3} = 5 - \dfrac{x + 2}{7}$ **30.** $\dfrac{3x}{5} - \dfrac{x - 3}{2} = \dfrac{x + 2}{3}$

*Exercises 31–50 contain equations with variables in denominators. For each equation, **a.** Write the value or values of the variable that make a denominator zero. These are the restrictions on the variable. **b.** Keeping the restrictions in mind, solve the equation.*

31. $\dfrac{4}{x} = \dfrac{5}{2x} + 3$ **32.** $\dfrac{5}{x} = \dfrac{10}{3x} + 4$

33. $\dfrac{2}{x} + 3 = \dfrac{5}{2x} + \dfrac{13}{4}$ **34.** $\dfrac{7}{2x} - \dfrac{5}{3x} = \dfrac{22}{3}$

35. $\dfrac{2}{3x} + \dfrac{1}{4} = \dfrac{11}{6x} - \dfrac{1}{3}$ **36.** $\dfrac{5}{2x} - \dfrac{8}{9} = \dfrac{1}{18} - \dfrac{1}{3x}$

37. $\dfrac{x - 2}{2x} + 1 = \dfrac{x + 1}{x}$ **38.** $\dfrac{4}{x} = \dfrac{9}{5} - \dfrac{7x - 4}{5x}$

39. $\dfrac{1}{x - 1} + 5 = \dfrac{11}{x - 1}$ **40.** $\dfrac{3}{x + 4} - 7 = \dfrac{-4}{x + 4}$

41. $\dfrac{8x}{x + 1} = 4 - \dfrac{8}{x + 1}$ **42.** $\dfrac{2}{x - 2} = \dfrac{x}{x - 2} - 2$

43. $\dfrac{3}{2x - 2} + \dfrac{1}{2} = \dfrac{2}{x - 1}$

44. $\dfrac{3}{x + 3} = \dfrac{5}{2x + 6} + \dfrac{1}{x - 2}$

45. $\dfrac{3}{x + 2} + \dfrac{2}{x - 2} = \dfrac{8}{(x + 2)(x - 2)}$

46. $\dfrac{5}{x + 2} + \dfrac{3}{x - 2} = \dfrac{12}{(x + 2)(x - 2)}$

47. $\dfrac{2}{x + 1} - \dfrac{1}{x - 1} = \dfrac{2x}{x^2 - 1}$

48. $\dfrac{4}{x + 5} + \dfrac{2}{x - 5} = \dfrac{32}{x^2 - 25}$

49. $\dfrac{1}{x - 4} - \dfrac{5}{x + 2} = \dfrac{6}{x^2 - 2x - 8}$

50. $\dfrac{6}{x + 3} - \dfrac{5}{x - 2} = \dfrac{-20}{x^2 + x - 6}$

In Exercises 51–58, determine whether each equation is an identity, a conditional equation, or an inconsistent equation.

51. $4(x - 7) = 4x - 28$

52. $4(x - 7) = 4x + 28$

53. $2x + 3 = 2x - 3$

54. $\dfrac{7x}{x} = 7$

55. $4x + 5x = 8x$

56. $8x + 2x = 9x$

57. $\dfrac{2x}{x - 3} = \dfrac{6}{x - 3} + 4$

58. $\dfrac{3}{x - 3} = \dfrac{x}{x - 3} + 3$

The equations in Exercises 59–70 combine some of the types of equations we have discussed in this section. Solve each equation or state that it is true for all real numbers or no real numbers.

59. $\dfrac{x + 5}{2} - 4 = \dfrac{2x - 1}{3}$

60. $\dfrac{x + 2}{7} = 5 - \dfrac{x + 1}{3}$

61. $\dfrac{2}{x - 2} = 3 + \dfrac{x}{x - 2}$

62. $\dfrac{6}{x + 3} + 2 = \dfrac{-2x}{x + 3}$

63. $8x - (3x + 2) + 10 = 3x$

64. $2(x + 2) + 2x = 4(x + 1)$

65. $\dfrac{2}{x} + \dfrac{1}{2} = \dfrac{3}{4}$

66. $\dfrac{3}{x} - \dfrac{1}{6} = \dfrac{1}{3}$

67. $\dfrac{4}{x - 2} + \dfrac{3}{x + 5} = \dfrac{7}{(x + 5)(x - 2)}$

68. $\dfrac{1}{x - 1} = \dfrac{1}{(2x + 3)(x - 1)} + \dfrac{4}{2x + 3}$

69. $\dfrac{4x}{x + 3} - \dfrac{12}{x - 3} = \dfrac{4x^2 + 36}{x^2 - 9}$

70. $\dfrac{4}{x^2 + 3x - 10} - \dfrac{1}{x^2 + x - 6} = \dfrac{3}{x^2 - x - 12}$

In Exercises 71–90, solve each formula for the specified variable. Do you recognize the formula? If so, what does it describe?

71. $A = lw$ for w

72. $D = RT$ for R

73. $A = \frac{1}{2}bh$ for b

74. $V = \frac{1}{3}Bh$ for B

75. $I = Prt$ for P

76. $C = 2\pi r$ for r

77. $E = mc^2$ for m

78. $V = \pi r^2 h$ for h

79. $T = D + pm$ for p

80. $P = C + MC$ for M

81. $A = \frac{1}{2}h(a + b)$ for a

82. $A = \frac{1}{2}h(a + b)$ for b

83. $S = P + Prt$ for r

84. $S = P + Prt$ for t

85. $B = \dfrac{F}{S - V}$ for S

86. $S = \dfrac{C}{1 - r}$ for r

87. $IR + Ir = E$ for I

88. $A = 2lw + 2lh + 2wh$ for h

89. $\dfrac{1}{p} + \dfrac{1}{q} = \dfrac{1}{f}$ for f

90. $\dfrac{1}{R} = \dfrac{1}{R_1} + \dfrac{1}{R_2}$ for R_1

In Exercises 91–104, solve each absolute value equation or indicate the equation has no solution.

91. $|x| = 8$

92. $|x| = 6$

93. $|x - 2| = 7$

94. $|x + 1| = 5$

95. $|2x - 1| = 5$

96. $|2x - 3| = 11$

97. $2|3x - 2| = 14$

98. $3|2x - 1| = 21$

99. $7|5x| + 2 = 16$

100. $7|3x| + 2 = 16$

101. $|x + 1| + 5 = 3$

102. $|x + 1| + 6 = 2$

103. $|2x - 1| + 3 = 3$

104. $|3x - 2| + 4 = 4$

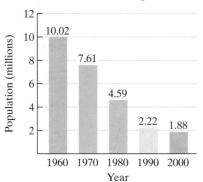

Application Exercises

Growth in human populations and economic activity threatens the continued existence of salmon in the Pacific Northwest. The bar graph shows the Pacific salmon population for various years. The data can be described by the formula

$$P = -0.22t + 9.6$$

in which P is the salmon population, in millions, t years after 1960. Use the formula to solve Exercises 105–106. Round to the nearest year.

Pacific Salmon Population

Source: U.S. Department of the Interior

105. When will the salmon population be reduced to 0.5 million?

106. When will there be no Pacific salmon?

107. The formula

$$\dfrac{W}{2} - 3H = 53$$

describes the recommended weight W, in pounds, for a male, where H represents the man's height, in inches, over 5 feet. What is the recommended weight for a man who is 6 feet, 3 inches tall?

108. The International Panel on Climate Change is a U.N.-sponsored body made up of more than 1500 leading experts from 60 nations. According to their recent findings, increased levels of atmospheric carbon dioxide are affecting our climate. Global warming is under way and the effects could be catastrophic. The formula $C = 1.44t + 280$ describes carbon dioxide concentration, C, in parts per million, t years after 1939. The preindustrial carbon dioxide concentration of 280 parts per million remained fairly constant until World War II, increasing after that due primarily to the burning of fossil fuels related to energy consumption. When will the concentration be double the preindustrial level? Round to the nearest year.

Writing in Mathematics

109. What is a linear equation in one variable? Give an example of this type of equation.

110. What does it mean to solve an equation?

111. What is the solution set of an equation?

112. What are equivalent equations? Give an example.

113. What is the difference between solving an equation such as $2(x - 4) + 5x = 34$ and simplifying an algebraic expression such as $2(x - 4) + 5x$? If there is a difference, which topic should be taught first? Why?

114. Suppose that you solve $\dfrac{x}{5} - \dfrac{x}{2} = 1$ by multiplying both sides by 20, rather than the least common denominator of 5 and 2 (namely, 10). Describe what happens. If you get the correct solution, why do you think we clear the equation of fractions by multiplying by the *least* common denominator?

115. Suppose you are an algebra teacher grading the following solution on an examination:

$$-3(x - 6) = 2 - x$$
$$-3x - 18 = 2 - x$$
$$-2x - 18 = 2$$
$$-2x = -16$$
$$x = 8.$$

You should note that 8 checks, and the solution set is $\{8\}$. The student who worked the problem therefore wants full credit. Can you find any errors in the solution? If full credit is 10 points, how many points should you give the student? Justify your position.

116. Explain how to determine the restrictions on the variable for the equation

$$\frac{3}{x + 5} + \frac{4}{x - 2} = \frac{7}{(x + 5)(x - 2)}.$$

117. What is an identity? Give an example.

118. What is a conditional equation? Give an example.

119. What is an inconsistent equation? Give an example.

120. What is a formula?

121. We discussed formulas in this section after we considered procedures for solving linear equations. Doesn't working with a formula simply mean substituting given numbers into the formula and using the order of operations? Is it necessary to know how to solve equations to work with formulas? Explain.

122. Explain how to solve an equation involving absolute value.

123. Explain why the procedure that you explained in Exercise 122 does not apply to the equation $|x - 2| = -3$. What is the solution set for this equation?

Critical Thinking Exercises

124. Which one of the following is true?

   a. The equation $-7x = x$ has no solution.

   b. The equations $\dfrac{x}{x - 4} = \dfrac{4}{x - 4}$ and $x = 4$ are equivalent.

   c. The equations $3y - 1 = 11$ and $3y - 7 = 5$ are equivalent.

   d. If a and b are any real numbers, then $ax + b = 0$ always has one number in its solution set.

125. Write three equations that are equivalent to $x = 5$.

126. If x represents a number, write an English sentence about the number that results in an inconsistent equation.

127. Find b such that $\dfrac{7x + 4}{b} + 13 = x$ will have a solution set given by $\{-6\}$.

128. Find b such that $\dfrac{4x - b}{x - 5} = 3$ will have a solution set given by $\varnothing$.

129. Solve for C: $\;V = C - \dfrac{C - S}{L}N.$

SECTION P.8 *Quadratic Equations*

Objectives

1. Solve quadratic equations by factoring.
2. Solve quadratic equations by the square root method.
3. Solve quadratic equations by completing the square.
4. Solve quadratic equations using the quadratic formula.
5. Use the discriminant to determine the number and type of solutions.
6. Determine the most efficient method to use when solving a quadratic equation.
7. Solve problems using quadratic equations

Serpico, 1973, starring Al Pacino, is a movie about police corruption.

In 2000, a police scandal shocked Los Angeles. A police officer who had been convicted of stealing cocaine held as evidence described how members of his unit behaved in ways that resembled the gangs they were targeting, assaulting and framing innocent people.

Is police corruption on the rise? The graph in Figure P.9 shows the number of convictions of police officers throughout the United States for seven years.

Convictions of Police Officers

[Bar chart showing Number of Convictions by Year:
1994: 143
1995: 135
1996: 83
1997: 149
1998: 246
1999: 380
2000: 560]

Year

Figure P.9
Source: F.B.I.

The data can be approximated by the formula

$$N = 23.4x^2 - 259.1x + 815.8$$

where N is the number of police officers convicted of felonies x years after 1990. If present trends continue, in which year will 1000 police officers be convicted of felonies? To answer the question, it is necessary to substitute 1000 for N in the formula and solve for x, the number of years after 1990:

$$1000 = 23.4x^2 - 259.1x + 815.8.$$

Do you see how this equation differs from a linear equation? The exponent on x is 2. Solving such an equation involves finding the set of numbers that makes the equation a true statement. In this section, we study a number of methods for solving equations in the form $ax^2 + bx + c = 0$. We also look at applications of these equations.

The General Form of a Quadratic Equation

We begin by defining a quadratic equation.

> **Definition of a Quadratic Equation**
>
> A **quadratic equation** in x is an equation that can be written in the **general form**
> $$ax^2 + bx + c = 0$$
> where a, b, and c are real numbers, with $a \neq 0$. A quadratic equation in x is also called a **second-degree polynomial equation** in x.

An example of a quadratic equation in general form is $x^2 - 7x + 10 = 0$. The coefficient of x^2 is 1 ($a = 1$), the coefficient of x is -7 ($b = -7$), and the constant term is 10 ($c = 10$).

1 Solve quadratic equations by factoring.

Solving Quadratic Equations by Factoring

We can factor the left side of the quadratic equation $x^2 - 7x + 10 = 0$. We obtain $(x - 5)(x - 2) = 0$. If a quadratic equation has zero on one side and a factored expression on the other side, it can be solved using the **zero-product principle.**

> **The Zero-Product Principle**
>
> If the product of two algebraic expressions is zero, then at least one of the factors is equal to zero.
> $$\text{If } AB = 0, \quad \text{then} \quad A = 0 \text{ or } B = 0.$$

For example, consider the equation $(x - 5)(x - 2) = 0$. According to the zero-product principle, this product can be zero only if at least one of the factors is zero. We set each individual factor equal to zero and solve each resulting equation for x.

$$(x - 5)(x - 2) = 0$$
$$x - 5 = 0 \quad \text{or} \quad x - 2 = 0$$
$$x = 5 \qquad\qquad x = 2$$

We can check each of these proposed solutions in the original quadratic equation, $x^2 - 7x + 10 = 0$.

Check 5:	**Check** 2:
$5^2 - 7 \cdot 5 + 10 \stackrel{?}{=} 0$	$2^2 - 7 \cdot 2 + 10 \stackrel{?}{=} 0$
$25 - 35 + 10 \stackrel{?}{=} 0$	$4 - 14 + 10 \stackrel{?}{=} 0$
$0 = 0 \checkmark$	$0 = 0 \checkmark$

The resulting true statements, indicated by the checks, show that the solutions are 5 and 2. The solution set is $\{5, 2\}$. Note that with a quadratic equation, we can have two solutions, compared to the conditional linear equation that had one.

> **Solving a Quadratic Equation by Factoring**
>
> 1. If necessary, rewrite the equation in the form $ax^2 + bx + c = 0$, moving all terms to one side, thereby obtaining zero on the other side.
> 2. Factor.   (continues on the next page)

Solving a Quadratic Equation by Factoring (continued)

3. Apply the zero-product principle, setting each factor equal to zero.
4. Solve the equations in step 3.
5. Check the solutions in the original equation.

EXAMPLE 1   Solving Quadratic Equations by Factoring

Solve by factoring:

a. $4x^2 - 2x = 0$ **b.** $2x^2 + 7x = 4$.

Solution

a. We begin with $4x^2 - 2x = 0$.

Step 1   Move all terms to one side and obtain zero on the other side. All terms are already on the left and zero is on the other side, so we can skip this step.

Step 2   Factor. We factor out $2x$ from the two terms on the left side.

$$4x^2 - 2x = 0 \qquad \text{This is the given equation.}$$
$$2x(2x - 1) = 0 \qquad \text{Factor.}$$

Steps 3 and 4   Set each factor equal to zero and solve the resulting equations.

$$2x = 0 \qquad \text{or} \qquad 2x - 1 = 0$$
$$x = 0 \qquad\qquad\qquad 2x = 1$$
$$x = \tfrac{1}{2}$$

Step 5   Check the solutions in the original equation.

Check 0:

$$4x^2 - 2x = 0$$
$$4 \cdot 0^2 - 2 \cdot 0 \overset{?}{=} 0$$
$$0 - 0 \overset{?}{=} 0$$
$$0 = 0 \checkmark$$

Check $\tfrac{1}{2}$:

$$4x^2 - 2x = 0$$
$$4(\tfrac{1}{2})^2 - 2(\tfrac{1}{2}) \overset{?}{=} 0$$
$$4(\tfrac{1}{4}) - 2(\tfrac{1}{2}) \overset{?}{=} 0$$
$$1 - 1 \overset{?}{=} 0$$
$$0 = 0 \checkmark$$

The solution set is $\{0, \tfrac{1}{2}\}$.

b. Next, we solve $2x^2 + 7x = 4$.

Step 1   Move all terms to one side and obtain zero on the other side.
Subtract 4 from both sides and write the equation in general form.

$$2x^2 + 7x = 4 \qquad \text{This is the given equation.}$$
$$2x^2 + 7x - 4 = 4 - 4 \quad \text{Subtract 4 from both sides.}$$
$$2x^2 + 7x - 4 = 0 \qquad \text{Simplify.}$$

Step 2   Factor.

$$2x^2 + 7x - 4 = 0$$
$$(2x - 1)(x + 4) = 0$$

Steps 3 and 4 Set each factor equal to zero and solve each resulting equation.

$$2x - 1 = 0 \quad \text{or} \quad x + 4 = 0$$
$$2x = 1 \qquad\qquad x = -4$$
$$x = \tfrac{1}{2}$$

Step 5 Check the solutions in the original equation.

Check $\tfrac{1}{2}$:
$$2x^2 + 7x = 4$$
$$2(\tfrac{1}{2})^2 + 7(\tfrac{1}{2}) \overset{?}{=} 4$$
$$\tfrac{1}{2} + \tfrac{7}{2} \overset{?}{=} 4$$
$$4 = 4 \ \checkmark$$

Check -4:
$$2x^2 + 7x = 4$$
$$2(-4)^2 + 7(-4) \overset{?}{=} 4$$
$$32 + (-28) \overset{?}{=} 4$$
$$4 = 4 \ \checkmark$$

The solution set is $\{-4, \tfrac{1}{2}\}$.

Check Point 1 Solve by factoring:

a. $3x^2 - 9x = 0$ **b.** $2x^2 + x = 1$.

2 Solve quadratic equations by the square root method.

Solving Quadratic Equations by the Square Root Method

Quadratic equations of the form $u^2 = d$, where $d > 0$ and u is an algebraic expression, can be solved by the **square root method**. First, isolate the squared expression u^2 on one side of the equation and the number d on the other side. Then take the square root of both sides. Remember, there are two numbers whose square is d. One number is positive and one is negative.

We can use factoring to verify that $u^2 = d$ has two solutions.

$u^2 = d$	This is the given equation.
$u^2 - d = 0$	Move all terms to one side and obtain zero on the other side.
$(u + \sqrt{d})(u - \sqrt{d}) = 0$	Factor.
$u + \sqrt{d} = 0 \quad \text{or} \quad u - \sqrt{d} = 0$	Set each factor equal to zero.
$u = -\sqrt{d} \qquad\qquad u = \sqrt{d}$	Solve the resulting equations.

Because the solutions differ only in sign, we can write them in abbreviated notation as $u = \pm\sqrt{d}$. We read this as "u equals positive or negative the square root of d" or "u equals plus or minus the square root of d."

Now that we have verified these solutions, we can solve $u^2 = d$ directly by taking square roots. This process is called **the square root method**.

The Square Root Method

If u is an algebraic expression and d is a positive real number, then $u^2 = d$ has exactly two solutions:

$$\text{If } u^2 = d, \text{ then } u = \sqrt{d} \text{ or } u = -\sqrt{d}.$$

Equivalently,

$$\text{If } u^2 = d, \text{ then } u = \pm\sqrt{d}.$$

EXAMPLE 2 **Solving Quadratic Equations by the Square Root Method**

Solve by the square root method:

    a. $4x^2 = 20$ **b.** $(x - 2)^2 = 6.$

Solution

a. In order to apply the square root method, we need a squared expression by itself on one side of the equation.

$$4x^2 = 20$$

> We want x^2 by itself.

We can get x^2 by itself if we divide both sides by 4.

$$\frac{4x^2}{4} = \frac{20}{4}$$
$$x^2 = 5$$

Now, we can apply the square root method.

$$x = \pm\sqrt{5}$$

By checking both values in the original equation, we can confirm that the solution set is $\{-\sqrt{5}, \sqrt{5}\}$.

b. $(x - 2)^2 = 6$

> The squared expression is by itself.

With the squared expression by itself, we can apply the square root method.

$$x - 2 = \pm\sqrt{6}$$

We solve for x by adding 2 to both sides.

$$x = 2 \pm \sqrt{6}$$

By checking both values in the original equation, we can confirm that the solution set is $\{2 + \sqrt{6}, 2 - \sqrt{6}\}$ or $\{2 \pm \sqrt{6}\}$.

Check Point 2   Solve by the square root method:

    a. $3x^2 = 21$ **b.** $(x + 5)^2 = 11.$

3 Solve quadratic equations by completing the square.

Completing the Square

How do we solve an equation in the form $ax^2 + bx + c = 0$ if the trinomial $ax^2 + bx + c$ cannot be factored? We cannot use the zero-product principle in such a case. However, we can convert the equation into an equivalent equation that can be solved using the square root method. This is accomplished by **completing the square**.

Completing the Square

If $x^2 + bx$ is a binomial, then by adding $\left(\dfrac{b}{2}\right)^2$, which is the square of half the coefficient of x, a perfect square trinomial will result. That is,

$$x^2 + bx + \left(\frac{b}{2}\right)^2 = \left(x + \frac{b}{2}\right)^2.$$

EXAMPLE 3 Completing the Square

What term should be added to the binomial $x^2 + 8x$ so that it becomes a perfect square trinomial? Then write and factor the trinomial.

Solution The term that should be added is the square of half the coefficient of x. The coefficient of x is 8. Thus, we will add $\left(\frac{8}{2}\right)^2 = 4^2$. A perfect square trinomial is the result.

$$x^2 + 8x + 4^2 = x^2 + 8x + 16 = (x + 4)^2$$

$$\text{(half)}^2$$

Check Point 3 What term should be added to the binomial $x^2 - 14x$ so that it becomes a perfect square trinomial? Then write and factor the trinomial.

We can solve any quadratic equation by completing the square. If the coefficient of the x^2-term is one, we add the square of half the coefficient of x to both sides of the equation. **When you add a constant term to one side of the equation to complete the square, be certain to add the same constant to the other side of the equation.** These ideas are illustrated in Example 4.

EXAMPLE 4 Solving a Quadratic Equation by Completing the Square

Solve by completing the square: $x^2 - 6x + 2 = 0$.

Solution   We begin the procedure of solving $x^2 - 6x + 2 = 0$ by isolating the binomial, $x^2 - 6x$, so that we can complete the square. Thus, we subtract 2 from both sides of the equation.

$$x^2 - 6x + 2 = 0 \qquad \text{This is the given equation.}$$

$$x^2 - 6x + 2 - 2 = 0 - 2 \qquad \text{Subtract 2 from both sides.}$$

$$x^2 - 6x = -2 \qquad \text{Simplify.}$$

> We need to add a constant to this binomial that will make it a perfect square trinomial.

What constant should we add? Add the square of half the coefficient of x.

$$x^2 - 6x = -2$$

> −6 is the coefficient of x.

$$\left(\frac{-6}{2}\right)^2 = (-3)^2 = 9$$

Study Tip

When factoring perfect square trinomials, the constant in the factorization is always half the coefficient of x.

$$x^2 - 6x + 9 = (x - 3)^2$$

> Half the coefficient of x, −6, is −3.

Thus, we need to add 9 to $x^2 - 6x$. In order to obtain an equivalent equation, we must add 9 to both sides.

$$x^2 - 6x = -2 \qquad \text{This is the quadratic equation with the binomial isolated.}$$

$$x^2 - 6x + 9 = -2 + 9 \qquad \text{Add 9 to both sides to complete the square.}$$

$$(x - 3)^2 = 7 \qquad \text{Factor the perfect square trinomial.}$$

$$x - 3 = \pm\sqrt{7} \qquad \text{Apply the square root method.}$$

$$x = 3 \pm \sqrt{7} \qquad \text{Add 3 to both sides.}$$

> In this step we have converted our equation into one that can be solved by the square root method.

The solution set is $\{3 + \sqrt{7}, 3 - \sqrt{7}\}$ or $\{3 \pm \sqrt{7}\}$.

Check Point 4   Solve by completing the square: $x^2 - 2x - 2 = 0$.

If the coefficient of the x^2-term in a quadratic equation is not one, you must divide each side of the equation by this coefficient before completing the square. For example, to solve $3x^2 - 2x - 4 = 0$ by completing the square, first divide every term by 3:

$$\frac{3x^2}{3} - \frac{2x}{3} - \frac{4}{3} = \frac{0}{3}$$

$$x^2 - \frac{2}{3}x - \frac{4}{3} = 0.$$

Now that the coefficient of x^2 is one, we can solve by completing the square using the method of Example 4.

4 Solve quadratic equations using the quadratic formula.

Solving Quadratic Equations Using the Quadratic Formula

We can use the method of completing the square to derive a formula that can be used to solve all quadratic equations. The derivation given here also shows a particular quadratic equation, $3x^2 - 2x - 4 = 0$, to specifically illustrate each of the steps.

Deriving the Quadratic Formula

General Form of a Quadratic Equation	Comment	A Specific Example
$ax^2 + bx + c = 0, \quad a > 0$	This is the given equation.	$3x^2 - 2x - 4 = 0$
$x^2 + \dfrac{b}{a}x + \dfrac{c}{a} = 0$	Divide both sides by the coefficient of x^2.	$x^2 - \dfrac{2}{3}x - \dfrac{4}{3} = 0$
$x^2 + \dfrac{b}{a}x = -\dfrac{c}{a}$	Isolate the binomial by adding $-\dfrac{c}{a}$ on both sides.	$x^2 - \dfrac{2}{3}x = \dfrac{4}{3}$
$x^2 + \dfrac{b}{a}x + \underbrace{\left(\dfrac{b}{2a}\right)^2}_{(\text{half})^2} = -\dfrac{c}{a} + \left(\dfrac{b}{2a}\right)^2$ $x^2 + \dfrac{b}{a}x + \dfrac{b^2}{4a^2} = -\dfrac{c}{a} + \dfrac{b^2}{4a^2}$	Complete the square. Add the square of half the coefficient of x to both sides.	$x^2 - \dfrac{2}{3}x + \underbrace{\left(-\dfrac{1}{3}\right)^2}_{(\text{half})^2} = \dfrac{4}{3} + \left(-\dfrac{1}{3}\right)^2$ $x^2 - \dfrac{2}{3}x + \dfrac{1}{9} = \dfrac{4}{3} + \dfrac{1}{9}$
$\left(x + \dfrac{b}{2a}\right)^2 = -\dfrac{c}{a} \cdot \dfrac{4a}{4a} + \dfrac{b^2}{4a^2}$	Factor on the left side and obtain a common denominator on the right side.	$\left(x - \dfrac{1}{3}\right)^2 = \dfrac{4}{3} \cdot \dfrac{3}{3} + \dfrac{1}{9}$
$\left(x + \dfrac{b}{2a}\right)^2 = \dfrac{-4ac + b^2}{4a^2}$ $\left(x + \dfrac{b}{2a}\right)^2 = \dfrac{b^2 - 4ac}{4a^2}$	Add fractions on the right side.	$\left(x - \dfrac{1}{3}\right)^2 = \dfrac{12 + 1}{9}$ $\left(x - \dfrac{1}{3}\right)^2 = \dfrac{13}{9}$
$x + \dfrac{b}{2a} = \pm\sqrt{\dfrac{b^2 - 4ac}{4a^2}}$	Apply the square root method.	$x - \dfrac{1}{3} = \pm\sqrt{\dfrac{13}{9}}$
$x + \dfrac{b}{2a} = \pm\dfrac{\sqrt{b^2 - 4ac}}{2a}$	Take the square root of the quotient, simplifying the denominator.	$x - \dfrac{1}{3} = \pm\dfrac{\sqrt{13}}{3}$
$x = \dfrac{-b}{2a} \pm \dfrac{\sqrt{b^2 - 4ac}}{2a}$	Solve for x by subtracting $\dfrac{b}{2a}$ from both sides.	$x = \dfrac{1}{3} \pm \dfrac{\sqrt{13}}{3}$
$x = \dfrac{-b \pm \sqrt{b^2 - 4ac}}{2a}$	Combine fractions on the right.	$x = \dfrac{1 \pm \sqrt{13}}{3}$

The formula shown at the bottom of the left column is called the *quadratic formula*. A similar proof shows that the same formula can be used to solve quadratic equations if a, the coefficient of the x^2-term, is negative.

To Die at Twenty

Can the equations

$$7x^5 + 12x^3 - 9x + 4 = 0$$

and

$$8x^6 - 7x^5 + 4x^3 - 19 = 0$$

be solved using a formula similar to the quadratic formula? The first equation has five solutions and the second has six solutions, but they cannot be found using a formula. How do we know? In 1832, a 20-year-old Frenchman, Evariste Galois, wrote down a proof showing that there is no general formula to solve equations when the exponent on the variable is 5 or greater. Galois was jailed as a political activist several times while still a teenager. The day after his brilliant proof he fought a duel over a woman. The duel was a political setup. As he lay dying, Galois told his brother, Alfred, of the manuscript that contained his proof: "Mathematical manuscripts are in my room. On the table. Take care of my work. Make it known. Important. Don't cry, Alfred. I need all my courage—to die at twenty." (Our source is Leopold Infeld's biography of Galois, *Whom the Gods Love.* Some historians, however, dispute the story of Galois's ironic death the very day after his algebraic proof. Mathematical truths seem more reliable than historical ones!)

The Quadratic Formula

The solutions of a quadratic equation in standard form $ax^2 + bx + c = 0$, with $a \neq 0$, are given by the **quadratic formula**

$$x = \frac{-b \pm \sqrt{b^2 - 4ac}}{2a}.$$

x equals negative b, plus or minus the square root of b² – 4ac, all divided by 2a.

To use the quadratic formula, write the quadratic equation in general form if necessary. Then determine the numerical values for a (the coefficient of the squared term), b (the coefficient of the x-term), and c (the constant term). Substitute the values of a, b, and c in the quadratic formula and evaluate the expression. The $\pm$ sign indicates that there are two solutions of the equation.

EXAMPLE 5 Solving a Quadratic Equation Using the Quadratic Formula

Solve using the quadratic formula: $2x^2 - 6x + 1 = 0$.

Solution The given equation is in general form. Begin by identifying the values for a, b, and c.

$$2x^2 - 6x + 1 = 0$$

$a = 2 \qquad b = -6 \qquad c = 1$

$$x = \frac{-b \pm \sqrt{b^2 - 4ac}}{2a}$$
Use the quadratic formula.

$$= \frac{-(-6) \pm \sqrt{(-6)^2 - 4(2)(1)}}{2 \cdot 2}$$
Substitute the values for a, b, and c: $a = 2$, $b = -6$, and $c = 1$.

$$= \frac{6 \pm \sqrt{36 - 8}}{4}$$
$-(-6) = 6$ and $(-6)^2 = (-6)(-6) = 36$.

$$= \frac{6 \pm \sqrt{28}}{2}$$
Complete the subtraction under the radical.

$$= \frac{6 \pm 2\sqrt{7}}{4}$$
$\sqrt{28} = \sqrt{4 \cdot 7} = \sqrt{4}\sqrt{7} = 2\sqrt{7}$

$$= \frac{2(3 \pm \sqrt{7})}{4}$$
Factor out 2 from the numerator.

$$= \frac{3 \pm \sqrt{7}}{2}$$
Divide the numerator and denominator by 2.

The solution set is $\left\{ \dfrac{3 + \sqrt{7}}{2}, \dfrac{3 - \sqrt{7}}{2} \right\}$ or $\left\{ \dfrac{3 \pm \sqrt{7}}{2} \right\}$.

Check Point 5 Solve using the quadratic formula:

$$2x^2 + 2x - 1 = 0.$$

5 Use the discriminant to determine the number and type of solutions.

The Discriminant

The quantity $b^2 - 4ac$, which appears under the radical sign in the quadratic formula, is called the **discriminant**. In Example 5 the discriminant was 28, a positive number that is not a perfect square. The equation had two solutions that were irrational numbers. Table P.3 shows the relationship between the discriminant and the number and type of solutions to a quadratic equation.

Table P.3 The Discriminant and the Kinds of Solutions to $ax^2 + bx + c = 0$

Discriminant $b^2 - 4ac$	Kinds of Solutions to $ax^2 + bx + c = 0$
$b^2 - 4ac > 0$	**Two unequal real solutions**; If a, b, and c are rational numbers and the discriminant is a perfect square, the solutions are rational. If the discriminant is not a perfect square, the solutions are irrational.
$b^2 - 4ac = 0$	**One solution (a repeated solution) that is a real number**; If a, b, and c are rational numbers, the repeated solution is also a rational number.
$b^2 - 4ac < 0$	**No real solutions**

EXAMPLE 6 Using the Discriminant

Compute the discriminant of $4x^2 - 8x + 1 = 0$. What does the discriminant indicate about the number and type of solutions?

Solution Begin by identifying the values for a, b, and c.

$$4x^2 - 8x + 1 = 0$$

$a = 4$ $b = -8$ $c = 1$

Substitute and compute the discriminant:

$$b^2 - 4ac = (-8)^2 - 4 \cdot 4 \cdot 1 = 64 - 16 = 48.$$

The discriminant is 48. Because the discriminant is positive, the equation $4x^2 - 8x + 1 = 0$ has two unequal real solutions.

Check Point 6 Compute the discriminant of $3x^2 - 2x + 5 = 0$. What does the discriminant indicate about the number and type of solutions?

6 Determine the most efficient method to use when solving a quadratic equation.

Determining Which Method to Use

All quadratic equations can be solved by the quadratic formula. However, if an equation is in the form $u^2 = d$, such as $x^2 = 5$ or $(2x + 3)^2 = 8$, it is faster to use the square root method, taking the square root of both sides. If the equation is not in the form $u^2 = d$, write the quadratic equation in general form $(ax^2 + bx + c = 0)$. Try to solve the equation by the factoring method. If $ax^2 + bx + c$ cannot be factored, then solve the quadratic equation by the quadratic formula.

Because we used the method of completing the square to derive the quadratic formula, we no longer need it for solving quadratic equations. However, we will use completing the square later in the book to help graph certain kinds of equations.

Table P.4 summarizes our observations about which technique to use when solving a quadratic equation.

Table P.4   Determining the Most Efficient Technique to Use when Solving a Quadratic Equation

Description and Form of the Quadratic Equation	Most Efficient Solution Method	Example
$ax^2 + bx + c = 0$ and $ax^2 + bx + c$ can be factored easily.	Factor and use the zero-product principle.	$3x^2 + 5x - 2 = 0$ $(3x - 1)(x + 2) = 0$ $3x - 1 = 0$ or $x + 2 = 0$ $x = \dfrac{1}{3}$ $x = -2$
$ax^2 + c = 0$ The quadratic equation has no x-term. $(b = 0)$	Solve for x^2 and apply the square root method.	$4x^2 - 7 = 0$ $4x^2 = 7$ $x^2 = \dfrac{7}{4}$ $x = \pm \dfrac{\sqrt{7}}{2}$
$(ax + c)^2 = d$; $ax + c$ is a first-degree polynomial.	Use the square root method.	$(x + 4)^2 = 5$ $x + 4 = \pm\sqrt{5}$ $x = -4 \pm \sqrt{5}$
$ax^2 + bx + c = 0$ and $ax^2 + bx + c$ cannot be factored or the factoring is too difficult.	Use the quadratic formula: $x = \dfrac{-b \pm \sqrt{b^2 - 4ac}}{2a}$	$x^2 - 2x - 6 = 0$ $a = 1$ $b = -2$ $c = -6$ $x = \dfrac{-(-2) \pm \sqrt{(-2)^2 - 4(1)(-6)}}{2(1)}$ $= \dfrac{2 \pm \sqrt{4 - (-24)}}{2}$ $= \dfrac{2 \pm \sqrt{28}}{2} = \dfrac{2 \pm \sqrt{4}\sqrt{7}}{2}$ $= \dfrac{2 \pm 2\sqrt{7}}{2} = \dfrac{2(1 \pm \sqrt{7})}{2}$ $= 1 \pm \sqrt{7}$

7 Solve problems using quadratic equations.

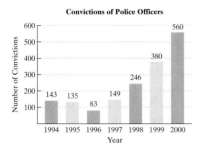

Convictions of Police Officers

Figure P.9, repeated

Source: F.B.I

Applications

We opened this section with a graph (Figure P.9, repeated in the margin) showing the number of convictions of police officers throughout the United States from 1994 through 2000. The data can be approximated by the formula

$$N = 23.4x^2 - 259.1x + 815.8$$

where N is the number of police officers convicted of felonies x years after 1990. Notice that this formula contains an expression in the form $ax^2 + bx + c$ on the right side. If a formula contains such an expression, we can write and solve a quadratic equation to answer questions about the variable x. Our next example shows how this is done.

EXAMPLE 7 Convictions of Police Officers

Use the formula $N = 23.4x^2 - 259.1x + 815.8$ to answer this question: In which year will 1000 police officers be convicted of felonies?

Solution Because we are interested in 1000 convictions, we substitute 1000 for N in the given formula. Then we solve for x, the number of years after 1990.

$$N = 23.4x^2 - 259.1x + 815.8 \qquad \text{This is the given formula.}$$

$$1000 = 23.4x^2 - 259.1x + 815.8 \qquad \text{Substitute 1000 for N.}$$

$$0 = 23.4x^2 - 259.1x - 184.2 \qquad \text{Subtract 1000 from both sides and write the quadratic equation in general form.}$$

$$a = 23.4 \qquad b = -259.1 \qquad c = -184.2$$

Because the trinomial on the right side of the equation is prime, we solve using the quadratic formula.

$$x = \frac{-b \pm \sqrt{b^2 - 4ac}}{2a} \qquad \text{Use the quadratic formula.}$$

$$= \frac{-(-259.1) \pm \sqrt{(-259.1)^2 - 4(23.4)(-184.2)}}{2(23.4)} \qquad \text{Substitute the values for a, b, and c: } a = 23.4, b = -259.1, c = -184.2.$$

$$= \frac{259.1 \pm \sqrt{84,373.93}}{46.8} \qquad \text{Use a calculator to simplify the radicand.}$$

Thus,

$$x = \frac{259.1 + \sqrt{84,373.93}}{46.8} \qquad \text{or} \qquad x = \frac{259.1 - \sqrt{84,373.93}}{46.8}$$

$$x \approx 12 \qquad\qquad\qquad x \approx -1 \quad \text{Use a calculator and round to the nearest integer.}$$

The formula describes the number of convictions x years *after* 1990. Thus, we are interested only in the positive solution, 12. This means that approximately 12 years after 1990, in 2002, 1000 police officers will be convicted of felonies.

Technology

On most calculators, here is how to approximate

$$\frac{259.1 + \sqrt{84,373.93}}{46.8} :$$

Many Scientific Calculators

(259.1 + 84373.93 √

) ÷ 46.8 =

Many Graphing Calculators

(259.1 + √ 84373.93

) ÷ 46.8 ENTER .

Similar keystrokes can be used to approximate the other irrational solution in Example 7.

Check Point 7 Use the formula in Example 7 to answer this question: In which year after 1993 were 250 police officers convicted of felonies? How well does the formula describe the actual number of convictions for that year shown in Figure P.9?

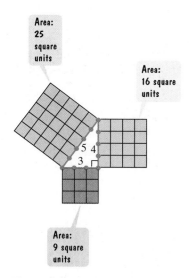

Figure P.10 The area of the large square equals the sum of the areas of the smaller squares.

In our next example, we will be using the *Pythagorean Theorem* to obtain a quadratic equation. The ancient Greek philosopher and mathematician Pythagoras (approximately 582–500 B.C.) founded a school whose motto was "All is number." Pythagoras is best remembered for his work with the **right triangle**, a triangle with one angle measuring 90°. The side opposite the 90° angle is called the **hypotenuse**. The other sides are called **legs**. Pythagoras found that if he constructed squares on each of the legs, as well as a larger square on the hypotenuse, the sum of the areas of the smaller squares is equal to the area of the larger square. This is illustrated in Figure P.10.

This relationship is usually stated in terms of the lengths of the three sides of a right triangle and is called the **Pythagorean Theorem**.

The Pythagorean Theorem

The sum of the squares of the lengths of the legs of a right triangle equals the square of the length of the hypotenuse.

If the legs have lengths a and b, and the hypotenuse has length c, then

$$a^2 + b^2 = c^2.$$

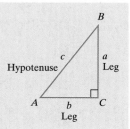

EXAMPLE 8 Using the Pythagorean Theorem

In a 25-inch television set, the length of the screen's diagonal is 25 inches. If the screen's height is 15 inches, what is its width?

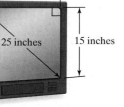

Figure P.11 A right triangle is formed by the television's height, width, and diagonal.

Solution   Figure P.11 shows a right triangle that is formed by the height, width, and diagonal. We can find w, the screen's width, using the Pythagorean Theorem.

$(\text{Leg})^2$ plus $(\text{Leg})^2$ equals $(\text{Hypotenuse})^2$

$$w^2 \quad + \quad 15^2 \quad = \quad 25^2 \qquad \text{This is the equation resulting from the Pythagorean Theorem.}$$

The equation $w^2 + 15^2 = 25^2$ can be solved most efficiently by the square root method.

$$w^2 + 15^2 = 25^2 \qquad \text{This is the equation from the Pythagorean Theorem.}$$

$$w^2 + 225 = 625 \qquad \text{Square 15 and 25.}$$

$$w^2 + 225 - 225 = 625 - 225 \qquad \text{Isolate } w^2 \text{ by subtracting 225 from both sides.}$$

$$w^2 = 400 \qquad \text{Simplify.}$$

$$w = \pm\sqrt{400} \qquad \text{Apply the square root method.}$$

$$w = \pm 20 \qquad \text{Simplify.}$$

Because w represents the width of the television's screen, this dimension must be positive. We reject -20. Thus, the width of the television is 20 inches.

Check Point 8   What is the width in a 15-inch television set whose height is 9 inches?

EXERCISE SET P.8

Practice Exercises

Solve each equation in Exercises 1–14 by factoring.

1. $x^2 - 3x - 10 = 0$

2. $x^2 - 13x + 36 = 0$

3. $x^2 = 8x - 15$

4. $x^2 = -11x - 10$

5. $6x^2 + 11x - 10 = 0$

6. $9x^2 + 9x + 2 = 0$

7. $3x^2 - 2x = 8$

8. $4x^2 - 13x = -3$

9. $3x^2 + 12x = 0$

10. $5x^2 - 20x = 0$

11. $2x(x - 3) = 5x^2 - 7x$

12. $16x(x - 2) = 8x - 25$

13. $7 - 7x = (3x + 2)(x - 1)$

14. $10x - 1 = (2x + 1)^2$

Solve each equation in Exercises 15–26 by the square root method.

15. $3x^2 = 27$

16. $5x^2 = 45$

17. $5x^2 + 1 = 51$

18. $3x^2 - 1 = 47$

19. $(x + 2)^2 = 25$

20. $(x - 3)^2 = 36$

21. $(3x + 2)^2 = 9$

22. $(4x - 1)^2 = 16$

23. $(5x - 1)^2 = 7$

24. $(8x - 3)^2 = 5$

25. $(3x - 4)^2 = 8$

26. $(2x + 8)^2 = 27$

In Exercises 27–38, determine the constant that should be added to the binomial so that it becomes a perfect square trinomial. Then write and factor the trinomial.

27. $x^2 + 12x$

28. $x^2 + 16x$

29. $x^2 - 10x$

30. $x^2 - 14x$

31. $x^2 + 3x$

32. $x^2 + 5x$

33. $x^2 - 7x$

34. $x^2 - 9x$

35. $x^2 - \dfrac{2}{3}x$

36. $x^2 + \dfrac{4}{5}x$

37. $x^2 - \dfrac{1}{3}x$

38. $x^2 - \dfrac{1}{4}x$

Solve each equation in Exercises 39–54 by completing the square.

39. $x^2 + 6x = 7$

40. $x^2 + 6x = -8$

41. $x^2 - 2x = 2$

42. $x^2 + 4x = 12$

43. $x^2 - 6x - 11 = 0$

44. $x^2 - 2x - 5 = 0$

45. $x^2 + 4x + 1 = 0$

46. $x^2 + 6x - 5 = 0$

47. $x^2 + 3x - 1 = 0$

48. $x^2 - 3x - 5 = 0$

49. $2x^2 - 7x + 3 = 0$

50. $2x^2 + 5x - 3 = 0$

51. $4x^2 - 4x - 1 = 0$

52. $2x^2 - 4x - 1 = 0$

53. $3x^2 - 2x - 2 = 0$

54. $3x^2 - 5x - 10 = 0$

Solve each equation in Exercises 55–62 using the quadratic formula.

55. $x^2 + 8x + 15 = 0$

56. $x^2 + 8x + 12 = 0$

57. $x^2 + 5x + 3 = 0$

58. $x^2 + 5x + 2 = 0$

59. $3x^2 - 3x - 4 = 0$

60. $5x^2 + x - 2 = 0$

61. $4x^2 = 2x + 7$

62. $3x^2 = 6x - 1$

Compute the discriminant of each equation in Exercises 63–70. What does the discriminant indicate about the number and type of solutions?

63. $x^2 - 4x - 5 = 0$

64. $4x^2 - 2x + 3 = 0$

65. $2x^2 - 11x + 3 = 0$

66. $2x^2 + 11x - 6 = 0$

67. $x^2 - 2x + 1 = 0$

68. $3x^2 = 2x - 1$

69. $x^2 - 3x - 7 = 0$

70. $3x^2 + 4x - 2 = 0$

Solve each equation in Exercises 71–94 by the method of your choice.

71. $2x^2 - x = 1$

72. $3x^2 - 4x = 4$

73. $5x^2 + 2 = 11x$

74. $5x^2 = 6 - 13x$

75. $3x^2 = 60$

76. $2x^2 = 250$

77. $x^2 - 2x = 1$

78. $2x^2 + 3x = 1$

79. $(2x + 3)(x + 4) = 1$

80. $(2x - 5)(x + 1) = 2$

81. $(3x - 4)^2 = 16$

82. $(2x + 7)^2 = 25$

83. $3x^2 - 12x + 12 = 0$

84. $9 - 6x + x^2 = 0$

85. $4x^2 - 16 = 0$

86. $3x^2 - 27 = 0$

87. $x^2 = 4x - 2$

88. $x^2 = 6x - 7$

89. $2x^2 - 7x = 0$

90. $2x^2 + 5x = 3$

91. $\dfrac{1}{x} + \dfrac{1}{x + 2} = \dfrac{1}{3}$

92. $\dfrac{1}{x} + \dfrac{1}{x + 3} = \dfrac{1}{4}$

93. $\dfrac{2x}{x - 3} + \dfrac{6}{x + 3} = -\dfrac{28}{x^2 - 9}$

94. $\dfrac{3}{x - 3} + \dfrac{5}{x - 4} = \dfrac{x^2 - 20}{x^2 - 7x + 12}$

Application Exercises

A driver's age has something to do with his or her chance of getting into a fatal car crash. The bar graph shows the number of fatal vehicle crashes per 100 million miles driven for drivers of various age groups. For example, 25-year-old drivers are involved in 4.1 fatal crashes per 100 million miles driven. Thus, when a group of 25-year-old Americans have driven a total of 100 million miles, approximately 4 have been in accidents in which someone died.

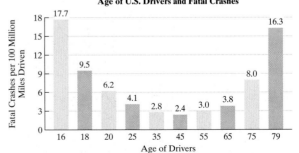

Age of U.S. Drivers and Fatal Crashes

Source: Insurance Institute for Highway Safety

The number of fatal vehicle crashes per 100 million miles, N, for drivers of age x can be approximated by the formula
$$N = 0.013x^2 - 1.19x + 28.24.$$
Use the formula to solve Exercises 95–96.

95. What age groups are expected to be involved in 10 fatal crashes per 100 million miles driven? How well does the formula describe the trend in the actual data shown in the bar graph on the previous page?

96. What age groups are expected to be involved in 3 fatal crashes per 100 million miles driven? How well does the formula describe the trend in the actual data shown in the bar graph on the previous page?

The Food Stamp Program is America's first line of defense against hunger for millions of families. Over half of all participants are children; one out of six is a low-income older adult. The formula
$$y = -\frac{1}{2}x^2 + 4x + 19$$
approximates the number of people, y, in millions, receiving food stamps x years after 1990. Use the formula to solve Exercises 97–98.

97. In which year did 27 million people receive food stamps?

98. In which years did 19 million people receive food stamps?

99. A baseball diamond is actually a square with 90-foot sides. What is the distance from home plate to second base?

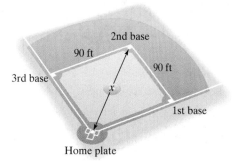

100. A 20-foot ladder is 15 feet from the house. How far up the house does the ladder reach?

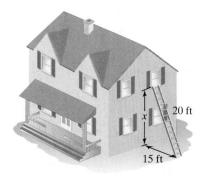

101. An 8-foot tree is supported by two wires that extend from the top of the tree to a point on the ground located 15 feet from the base of the tree. Find the total length of the two support wires.

102. A vertical pole is supported by three wires. Each wire is 13 yards long and is anchored 5 yards from the base of the pole. How far up the pole will the wires be attached?

103. The length of a rectangular garden is 5 feet greater than the width. The area of the garden is 300 square feet. Find the length and the width.

104. A rectangular parking lot has a length that is 3 yards greater than the width. The area of the rectangular lot is 180 square yards. Find the length and the width.

105. A machine produces open boxes using square sheets of metal. The figure illustrates that the machine cuts equal-sized squares measuring 2 inches on a side from the corners and then shapes the metal into an open box by turning up the sides. If each box must have a volume of 200 cubic inches, find the length of the side of the open square-bottom box.

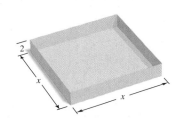

106. A machine produces open boxes using square sheets of metal. The machine cuts equal-sized squares measuring 3 inches on a side from the corners and then shapes the metal into an open box by turning up the sides. If each box must have a volume of 75 cubic inches, find the length of the side of the open square-bottom box.

Writing in Mathematics

107. What is a quadratic equation?

108. Explain how to solve $x^2 + 6x + 8 = 0$ using factoring and the zero-product principle.

109. Explain how to solve $x^2 + 6x + 8 = 0$ by completing the square.

110. Explain how to solve $x^2 + 6x + 8 = 0$ using the quadratic formula.

111. How is the quadratic formula derived?

112. What is the discriminant and what information does it provide about a quadratic equation?

113. If you are given a quadratic equation, how do you determine which method to use to solve it?

114. If $(x + 2)(x - 4) = 0$ indicates that $x + 2 = 0$ or $x - 4 = 0$, explain why $(x + 2)(x - 4) = 6$ does not mean $x + 2 = 6$ or $x - 4 = 6$. Could we solve the equation using $x + 2 = 3$ and $x - 4 = 2$ because $3 \cdot 2 = 6$?

115. Describe the trend shown by the data for the convictions of police officers in the graph in Figure P.9 on page 84. Do you believe that this trend is likely to continue or might something occur that would make it impossible to extend the formula into the future? Explain your answer.

Critical Thinking Exercises

116. Which one of the following is true?

   a. The equation $(2x - 3)^2 = 25$ is equivalent to $2x - 3 = 5$.

   b. Every quadratic equation has two distinct numbers in its solution set.

   c. Some quadratic equations in the form $ax^2 + c = 0$ can be solved by factoring.

   d. The equation $ax^2 + c = 0$ cannot be solved by the quadratic formula.

117. Solve the equation: $x^2 + 2\sqrt{3}x - 9 = 0$.

118. Write a quadratic equation in general form whose solution set is $\{-3, 5\}$.

119. A person throws a rock upward from the edge of an 80-foot cliff. The height, h, in feet, of the rock above the water at the bottom of the cliff after t seconds is described by the formula

$$h = -16t^2 + 64t + 80.$$

How long will it take for the rock to reach the water?

Group Exercise

120. Each group member should find an "intriguing" algebraic formula that contains an expression in the form $ax^2 + bx + c$ on one side. Consult college algebra books or liberal arts mathematics books to do so. Group members should select four of the formulas. For each formula selected, write and solve a problem similar to Exercises 95–98 in this exercise set.

SECTION P.9   *Linear Inequalities*

Objectives

1. Graph an inequality's solution set.
2. Use set-builder and interval notations.
3. Use properties of inequalities to solve inequalities.
4. Solve compound inequalities.
5. Solve inequalities involving absolute value.

You can go online and obtain a list of telecommunication companies that provide residential long-distance phone service. The list contains the monthly fee, the monthly minimum, and the rate per minute for each service provider. You've chosen a plan that has a monthly fee of $15 with a charge of $0.08 per minute for all long-distance calls. Suppose you are limited by how much money you can spend for the month: You can spend at most $35. If we let x represent the number of minutes of long-distance calls in a month, we can write an inequality that describes the given conditions:

The monthly fee of $15	plus	the charge of $0.08 per minute for x minutes	must be less than or equal to	$35.
15	+	0.08x	≤	35.

Using the commutative property of addition, we can express this inequality as

$$0.08x + 15 \le 35.$$

The form of this inequality is $ax + b \le c$, with $a = 0.08$, $b = 15$, and $c = 35$. Any inequality in this form is called **a linear inequality in one variable**. The greatest exponent on the variable in such an inequality is 1. The symbol between $ax + b$ and c can be $\le$ (is less than or equal to), $<$ (is less than), $\ge$ (is greater than or equal to), or $>$ (is greater than).

Study Tip

English phrases such as "at least" and "at most" can be described algebraically using inequalities.

English Sentence	Inequality
x is at least 5.	$x \ge 5$
x is at most 5.	$x \le 5$
x is between 5 and 7.	$5 < x < 7$
x is no more than 5.	$x \le 5$
x is no less than 5.	$x \ge 5$

In this section, we will study how to solve linear inequalities such as $0.08x + 15 \le 35$. **Solving an inequality** is the process of finding the set of numbers that makes the inequality a true statement. These numbers are called the **solutions** of the inequality, and we say that they **satisfy** the inequality. The set of all solutions is called the **solution set** of the inequality. We begin by discussing how to graph and how to represent these solution sets.

1   Graph an inequality's solution set.

Graphs of Inequalities; Interval Notation

There are infinitely many solutions to the inequality $x > -4$, namely all real numbers that are greater than -4. Although we cannot list all the solutions, we can make a drawing on a number line that represents these solutions. Such a drawing is called the **graph of the inequality**.

Graphs of solutions to linear inequalities are shown on a number line by shading all points representing numbers that are solutions. Parentheses indicate endpoints that are not solutions. Square brackets indicate endpoints that are solutions.

EXAMPLE 1   Graphing Inequalities

Graph the solutions of:

     a. $x < 3$ **b.** $x \ge -1$ **c.** $-1 < x \le 3$.

Solution

Study Tip

An inequality symbol points to the smaller number. Thus, another way to express $x < 3$ (x is less than 3) is $3 > x$ (3 is greater than x).

   a. The solutions of $x < 3$ are all real numbers that are less than 3. They are graphed on a number line by shading all points to the left of 3. The parenthesis at 3 indicates that 3 is not a solution, but numbers such as 2.9999 and 2.6 are. The arrow shows that the graph extends indefinitely to the left.

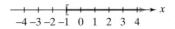

   b. The solutions of $x \ge -1$ are all real numbers that are greater than or equal to -1. We shade all points to the right of -1 and the point for -1 itself. The bracket at -1 shows that -1 is a solution of the given inequality. The arrow shows that the graph extends indefinitely to the right.

   c. The inequality $-1 < x \le 3$ is read "-1 is less than x *and* x is less than or equal to 3," or "x is greater than -1 *and* less than or equal to 3." The solutions of $-1 < x \le 3$ are all real numbers between -1 and 3, not including -1 but including 3. The parenthesis at -1 indicates that -1 is not a solution. By contrast, the bracket at 3 shows that 3 is a solution. Shading indicates the other solutions.

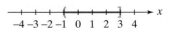

Check Point 1 Graph the solutions of:

  a. $x \leq 2$ **b.** $x > -4$ **c.** $2 \leq x < 6$.

2 Use set-builder and interval notations.

Now that we know how to graph the solution set of an inequality such as $x > -4$, let's see how to represent the solution set. One method is with **set-builder notation**. Using this method, the solution set of $x > -4$ can be expressed as

$$\{x | x > -4\}.$$

The set of all x   such that

We read this as "the set of all real numbers x such that x is greater than -4."

Another method used to represent solution sets of inequalities is **interval notation**. Using this notation, the solution set of $x > -4$ is expressed as $(-4, \infty)$. The parenthesis at -4 indicates that -4 is not included in the interval. The infinity symbol, ∞, does not represent a real number. It indicates that the interval extends indefinitely to the right.

Table P.5 lists nine possible types of intervals used to describe subsets of real numbers.

Table P.5   Intervals on the Real Number Line

Let a and b be real numbers such that $a < b$.

Interval Notation	Set-Builder Notation	Graph
(a,b)	$\{x | a < x < b\}$	
$[a,b]$	$\{x | a \leq x \leq b\}$	
$[a,b)$	$\{x | a \leq x < b\}$	
$(a,b]$	$\{x | a < x \leq b\}$	
(a, ∞)	$\{x | x > a\}$	
$[a, \infty)$	$\{x | x \geq a\}$	
$(-\infty, b)$	$\{x | x < b\}$	
$(-\infty, b]$	$\{x | x \leq b\}$	
$(-\infty, \infty)$	$\mathbb{R}$ (set of all real numbers)	

EXAMPLE 2   Intervals and Inequalities

Express the intervals in terms of inequalities and graph:

  a. $(-1, 4]$ **b.** $[2.5, 4]$ **c.** $(-4, \infty)$.

Solution

  a. $(-1, 4] = \{x | -1 < x \leq 4\}$

  b. $[2.5, 4] = \{x | 2.5 \leq x \leq 4\}$

  c. $(-4, \infty) = \{x | x > -4\}$

Check Point 2 Express the intervals in terms of inequalities and graph:
a. $[-2, 5)$ **b.** $[1, 3.5]$ **c.** $(-\infty, -1)$.

3 Use properties of inequalities to solve inequalities.

Solving Linear Inequalities

Inequalities with the same solution set are called **equivalent inequalities**. We can isolate a variable in a linear inequality the same way we can isolate a variable in a linear equation. The following properties are used to create equivalent inequalities:

Properties of Inequalities

Property	The Property in Words	Example
Addition and Subtraction Properties If $a < b$, then $a + c < b + c$. If $a < b$, then $a - c < b - c$.	If the same quantity is added to or subtracted from both sides of an inequality, the resulting inequality is equivalent to the original one.	$2x + 3 < 7$ Subtract 3: $2x + 3 - 3 < 7 - 3$. Simplify: $2x < 4$.
Positive Multiplication and Division Properties If $a < b$ and c is positive, then $ac < bc$. If $a < b$ and c is positive, then $\dfrac{a}{c} < \dfrac{b}{c}$.	If we multiply or divide both sides of an inequality by the same positive quantity, the resulting inequality is equivalent to the original one.	$2x < 4$ Divide by 2: $\dfrac{2x}{2} < \dfrac{4}{2}$. Simplify: $x < 2$.
Negative Multiplication and Division Properties If $a < b$ and c is negative, then $ac > bc$. If $a < b$ and c is negative, then $\dfrac{a}{c} > \dfrac{b}{c}$.	If we multiply or divide both sides of an inequality by the same negative quantity and reverse the direction of the inequality symbol, the result is an equivalent inequality.	$-4x < 20$ Divide by -4 and reverse the sense of the inequality: $\dfrac{-4x}{-4} > \dfrac{20}{-4}$. Simplify: $x > -5$.

Study Tip

You can solve

$$7x + 15 \geq 13x + 51$$

by isolating x on the right side. Subtract $7x$ from both sides.

$$7x + 15 - 7x \geq 13x + 51 - 7x$$
$$15 \geq 6x + 51$$

Now subtract 51 from both sides.

$$15 - 51 \geq 6x + 51 - 51$$
$$-36 \geq 6x$$

Finally, divide both sides by 6.

$$\frac{-36}{6} \geq \frac{6x}{6}$$
$$-6 \geq x$$

This last inequality means the same thing as

$$x \leq -6.$$

EXAMPLE 3   Solving a Linear Inequality

Solve and graph the solution set: $7x + 15 \geq 13x + 51$.

Solution   We will collect variable terms on the left and constant terms on the right.

$$7x + 15 \geq 13x + 51 \qquad \text{This is the given inequality.}$$

$$7x + 15 - 13x \geq 13x + 51 - 13x \qquad \text{Subtract 13x from both sides.}$$

$$-6x + 15 \geq 51 \qquad \text{Simplify.}$$

$$-6x + 15 - 15 \geq 51 - 15 \qquad \text{Subtract 15 from both sides.}$$

$$-6x \geq 36 \qquad \text{Simplify.}$$

$$\frac{-6x}{-6} \leq \frac{36}{-6} \qquad \text{Divide both sides by } -6 \text{ and reverse the sense of the inequality.}$$

$$x \leq -6 \qquad \text{Simplify.}$$

The solution set consists of all real numbers that are less than or equal to -6, expressed as $\{x \mid x \leq -6\}$. The interval notation for this solution set is $(-\infty, -6]$. The graph of the solution set is shown as follows:

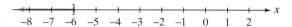

Check Point 3   Solve and graph the solution set: $6 - 3x \leq 5x - 2$.

4 Solve compound inequalities.

Solving Compound Inequalities

We now consider two inequalities such as

$$-3 < 2x + 1 \quad \text{and} \quad 2x + 1 \leq 3$$

expressed as a **compound inequality**

$$-3 < 2x + 1 \leq 3.$$

The word "and" does not appear when the inequality is written in the shorter form, although it is implied. The shorter form enables us to solve both inequalities at once. By performing the same operation on all three parts of the inequality, our goal is to **isolate x in the middle**.

EXAMPLE 4   Solving a Compound Inequality

Solve and graph the solution set:

$$-3 < 2x + 1 \leq 3.$$

Solution   We would like to isolate x in the middle. We can do this by first subtracting 1 from all three parts of the compound inequality. Then we isolate x from $2x$ by dividing all three parts of the inequality by 2.

$$-3 < 2x + 1 \leq 3 \qquad \text{\small This is the given inequality.}$$

$$-3 - 1 < 2x + 1 - 1 \leq 3 - 1 \qquad \text{\small Subtract 1 from all three parts.}$$

$$-4 < 2x \leq 2 \qquad \text{\small Simplify.}$$

$$\frac{-4}{2} < \frac{2x}{2} \leq \frac{2}{2} \qquad \text{\small Divide each part by 2.}$$

$$-2 < x \leq 1 \qquad \text{\small Simplify.}$$

The solution set consists of all real numbers greater than -2 and less than or equal to 1, represented by $\{x \mid -2 < x \leq 1\}$ in set-builder notation and $(-2, 1]$ in interval notation. The graph is shown as follows:

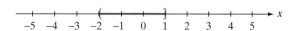

Check Point 4   Solve and graph the solution set: $1 \leq 2x + 3 < 11$.

5 Solve inequalities involving absolute value.

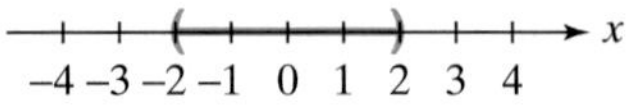

+++(| | | |)+++ → x
−4 −3 −2 −1 0 1 2 3 4

Figure P.12
$|x| < 2$, so $-2 < x < 2$.

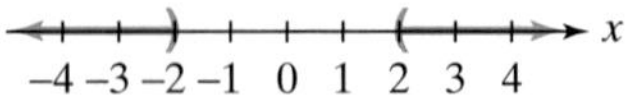

+++)+ | | | (+++ → x
−4 −3 −2 −1 0 1 2 3 4

Figure P.13
$|x| > 2$, so $x < -2$ or $x > 2$.

Solving Inequalities with Absolute Value

We know that $|x|$ describes the distance of x from zero on a real number line. We can use this geometric interpretation to solve an inequality such as

$$|x| < 2.$$

This means that the distance of x from 0 is *less than* 2, as shown in Figure P.12. The interval shows values of x that lie less than 2 units from 0. Thus, x can lie between -2 and 2. That is, x is greater than -2 and less than 2. We write $(-2, 2)$ or $\{x \mid -2 < x < 2\}$.

Some absolute value inequalities use the "greater than" symbol. For example, $|x| > 2$ means that the distance of x from 0 is *greater than* 2, as shown in Figure P.13. Thus, x can be less than -2 *or* greater than 2. We write $x < -2$ or $x > 2$.

These observations suggest the following principles for solving inequalities with absolute value:

Study Tip

In the $|X| < c$ case, we have one compound inequality to solve. In the $|X| > c$ case, we have two separate inequalities to solve.

Solving an Absolute Value Inequality

If X is an algebraic expression and c is a positive number,

1. The solutions of $|X| < c$ are the numbers that satisfy $-c < X < c$.

2. The solutions of $|X| > c$ are the numbers that satisfy $X < -c$ or $X > c$.

These rules are valid if $<$ is replaced by $\leq$ and $>$ is replaced by $\geq$.

EXAMPLE 5 Solving an Absolute Value Inequality with $<$

Solve and graph the solution set: $|x - 4| < 3$.

Solution

$|X| < c$ means $-c < X < c$.

$$|x - 4| < 3 \quad \text{means} \quad -3 < x - 4 < 3.$$

We solve the compound inequality by adding 4 to all three parts.

$$-3 < x - 4 < 3$$
$$-3 + 4 < x - 4 + 4 < 3 + 4$$
$$1 < x < 7$$

The solution set is all real numbers greater than 1 and less than 7, denoted by $\{x \mid 1 < x < 7\}$ or $(1, 7)$. The graph of the solution set is shown as follows:

+ | | | (| | | |) | + → x
−2 −1 0 1 2 3 4 5 6 7 8

Check Point 5 Solve and graph the solution set: $|x - 2| < 5$.

EXAMPLE 6   Solving an Absolute Value Inequality with $\geq$

Solve and graph the solution set: $|2x + 3| \geq 5$.

Solution

$\|X\|$	$\geq$	c means	$X \leq -c$	or	$X \geq c.$

$|2x + 3| \geq 5$ means $2x + 3 \leq -5$ or $2x + 3 \geq 5.$

We solve each of these inequalities separately.

$2x + 3 \geq 5$	or	$2x + 3 \geq 5$	These are the inequalities without absolute value bars.
$2x + 3 - 3 \leq -5 - 3$	or $2x + 3 - 3 \geq 5 - 3$		Subtract 3 from both sides.
$2x \leq -8$	or	$2x \geq 2$	Simplify.
$\dfrac{2x}{2} \leq \dfrac{-8}{2}$	or	$\dfrac{2x}{2} \geq \dfrac{2}{2}$	Divide both sides by 2.
$x \leq -4$	or	$x \geq 1$	Simplify.

Study Tip

The graph of the solution set for $|X| > c$ will be divided into two intervals. The graph of the solution set for $|X| < c$ will be a single interval.

The solution set is $\{x | x \leq -4 \text{ or } x \geq 1\}$, that is, all x in $(-\infty, -4]$ or $[1, \infty)$. The graph of the solution set is shown as follows:

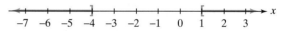

  Solve and graph the solution set: $|2x - 5| \geq 3$.

Applications

Our next example shows how to use an inequality to determine how many minutes of long-distance calls you can make in a month.

EXAMPLE 7   Describing Verbal Conditions with an Inequality

A long-distance telephone plan has a monthly fee of $15 with a charge of $0.08 per minute for all long-distance calls. Suppose that

$x = $ the number of minutes of long-distance calls that you make per month.

What limitations must be placed on x if you can spend at most $35 per month on the plan?

Solution   As we saw in the section opener, the following inequality describes the given conditions:

The monthly fee of $15	plus	the charge of $0.08 per minute for x minutes	must be less than or equal to	$35.
15	+	$0.08x$	$\leq$	35.

We solve $15 + 0.08x \le 35$ to determine the limitations on x, the number of minutes of monthly long-distance calls, with at most \$35 per month to spend.

$$0.08x + 15 \le 35 \qquad \text{This inequality describes the given conditions.}$$

$$0.08x + 15 - 15 \le 35 - 15 \qquad \text{Subtract 15 from both sides.}$$

$$0.08x \le 20 \qquad \text{Simplify.}$$

$$\frac{0.08x}{0.08} \le \frac{20}{0.08} \qquad \text{Divide both sides by 0.08.}$$

$$x \le 250 \qquad \text{Simplify.}$$

You can make no more than 250 minutes of long-distance calls each month.

Check Point 7 A car can be rented from Continental Rental for \$80 per week plus \$0.25 for each mile driven. Suppose that

x = the number of miles you drive in a week in the rented car.

What limitations must be placed on x if you can spend at most \$400 for the week?

EXERCISE SET P.9

Practice Exercises

In Exercises 1–12, graph the solutions of each inequality on a number line.

1. $x > 6$ **2.** $x > -2$

3. $x < -4$ **4.** $x < 0$

5. $x \ge -3$ **6.** $x \ge -5$

7. $x \le 4$ **8.** $x \le 7$

9. $-2 < x \le 5$ **10.** $-3 \le x < 7$

11. $-1 < x < 4$ **12.** $-7 \le x \le 0$

In Exercises 13–26, express each interval in terms of an inequality and graph the interval on a number line.

13. $(1, 6]$ **14.** $(-2, 4]$

15. $[-5, 2)$ **16.** $[-4, 3)$

17. $[-3, 1]$ **18.** $[-2, 5]$

19. $(2, \infty)$ **20.** $(3, \infty)$

21. $[-3, \infty)$ **22.** $[-5, \infty)$

23. $(-\infty, 3)$ **24.** $(-\infty, 2)$

25. $(-\infty, 5.5]$ **26.** $(-\infty, 3.5]$

Solve each linear inequality in Exercises 27–48 and graph the solution set on a number line. Express the solution set using interval notation.

27. $5x + 11 < 26$ **28.** $2x + 5 < 17$

29. $3x - 7 \ge 13$ **30.** $8x - 2 \ge 14$

31. $-9x \ge 36$ **32.** $-5x \le 30$

33. $8x - 11 \le 3x - 13$ **34.** $18x + 45 \le 12x - 8$

35. $4(x + 1) + 2 \ge 3x + 6$

36. $8x + 3 > 3(2x + 1) + x + 5$

37. $2x - 11 < -3(x + 2)$ **38.** $-4(x + 2) > 3x + 20$

39. $1 - (x + 3) \ge 4 - 2x$ **40.** $5(3 - x) \le 3x - 1$

41. $\dfrac{x}{4} - \dfrac{3}{5} \le \dfrac{x}{2} + 1$ **42.** $\dfrac{3x}{10} + 1 \ge \dfrac{1}{5} - \dfrac{x}{10}$

43. $1 - \dfrac{x}{2} > 4$ **44.** $7 - \dfrac{4}{5}x < \dfrac{3}{5}$

45. $\dfrac{x - 4}{6} \ge \dfrac{x - 2}{9} + \dfrac{5}{18}$ **46.** $\dfrac{4x - 3}{6} + 2 \ge \dfrac{2x - 1}{12}$

47. $4(3x - 2) - 3x < 3(1 + 3x) - 7$

48. $3(x - 8) - 2(10 - x) > 5(x - 1)$

Solve each inequality in Exercises 49–56 and graph the solution set on a number line. Express the solution set using interval notation.

49. $6 < x + 3 < 8$ **50.** $7 < x + 5 < 11$

51. $-3 \le x - 2 < 1$ **52.** $-6 < x - 4 \le 1$

53. $-11 < 2x - 1 \le -5$ **54.** $3 \le 4x - 3 < 19$

55. $-3 \le \dfrac{2}{3}x - 5 < -1$ **56.** $-6 \le \dfrac{1}{2}x - 4 < -3$

Solve each inequality in Exercises 57–84 by first rewriting each one as an equivalent inequality without absolute value bars. Graph the solution set on a number line. Express the solution set using interval notation.

57. $|x| < 3$ **58.** $|x| < 5$

59. $|x - 1| \le 2$ **60.** $|x + 3| \le 4$

61. $|2x - 6| < 8$ **62.** $|3x + 5| < 17$

63. $|2(x - 1) + 4| \le 8$ **64.** $|3(x - 1) + 2| \le 20$

65. $\left|\dfrac{2y + 6}{3}\right| < 2$ **66.** $\left|\dfrac{3(x - 1)}{4}\right| < 6$

67. $|x| > 3$

68. $|x| > 5$

69. $|x - 1| \geq 2$

70. $|x + 3| \geq 4$

71. $|3x - 8| > 7$

72. $|5x - 2| > 13$

73. $\left|\dfrac{2x + 2}{4}\right| \geq 2$

74. $\left|\dfrac{3x - 3}{9}\right| \geq 1$

75. $\left|3 - \dfrac{2}{3}x\right| > 5$

76. $\left|3 - \dfrac{3}{4}x\right| > 9$

77. $3|x - 1| + 2 \geq 8$

78. $-2|4 - x| \geq -4$

79. $3 < |2x - 1|$

80. $5 \geq |4 - x|$

81. $12 < \left|-2x + \dfrac{6}{7}\right| + \dfrac{3}{7}$

82. $1 < \left|x - \dfrac{11}{3}\right| + \dfrac{7}{3}$

83. $4 + \left|3 - \dfrac{x}{3}\right| \geq 9$

84. $\left|2 - \dfrac{x}{2}\right| - 1 \leq 1$

Application Exercises

The bar graph shows how we spend our leisure time. Let x represent the percentage of the population regularly participating in an activity. In Exercises 85–92, write the name or names of the activity described by the given inequality or interval.

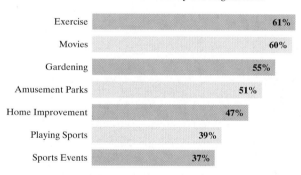

Percentage of U.S. Population Participating in Each Activity on a Regular Basis

Activity	Percentage
Exercise	61%
Movies	60%
Gardening	55%
Amusement Parks	51%
Home Improvement	47%
Playing Sports	39%
Sports Events	37%

Source: U.S. Census Bureau

85. $x < 40\%$

86. $x < 50\%$

87. $[51\%, 61\%]$

88. $[47\%, 60\%]$

89. $(51\%, 61\%)$

90. $(47\%, 60\%)$

91. $(39\%, 55\%]$

92. $(37\%, 47\%]$

The line graph at the top of the next column shows the declining consumption of cigarettes in the United States. The data shown by the graph can be approximated by the formula

$$N = 550 - 9x$$

where N is the number of cigarettes consumed, in billions, x years after 1988. Use this formula to solve Exercises 93–94.

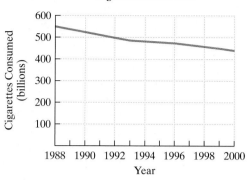

Consumption of Cigarettes in the U.S.

Source: Economic Research Service, USDA

93. How many years after 1988 will cigarette consumption be less than 370 billion cigarettes each year? Which years does this describe?

94. How many years after 1988 will cigarette consumption be less than 325 billion cigarettes each year? Which years does this describe?

95. The formula for converting Fahrenheit temperature, F, to Celsius temperature, C, is

$$C = \frac{5}{9}(F - 32).$$

If Celsius temperature ranges from $15°$ to $35°$, inclusive, what is the range for the Fahrenheit temperature? Use interval notation to express this range.

96. The formula

$$T = 0.01x + 56.7$$

approximates the global mean temperature, T, in degrees Fahrenheit, of Earth x years after 1905. For which range of years was the global mean temperature at least $56.7°$ F and at most $57.2°$ F?

The three television programs viewed by the greatest percentage of U.S. households in the twentieth century are shown in the table. The data are from a random survey of 4000 TV households by Nielsen Media Research. In Exercises 97–98, let x represent the actual viewing percentage in the U.S. population.

TV Programs with the Greatest U.S. Audience Viewing Percentage of the Twentieth Century

Program	Viewing Percentage in Survey
1. "M*A*S*H" Feb. 28, 1983	60.2%
2. "Dallas" Nov. 21, 1980	53.3%
3. "Roots" Part 8 Jan. 30, 1977	51.1%

Source: Nielsen Media Research

(Refer to the discussion at the bottom of the previous page before working Exercises 97–98.)

97. The inequality $|x - 60.2| \leq 1.6$ describes the actual viewing percentage for "M*A*S*H" in the U.S. population. Solve the inequality and interpret the solution. Explain why the survey's *margin of error* is ±1.6%.

98. The inequality $|x - 51.1| \leq 1.6$ describes the actual viewing percentage for "Roots" Part 8 in the U.S. population. Solve the inequality and interpret the solution. Explain why the survey's *margin of error* is ±1.6%.

99. If a coin is tossed 100 times, we would expect approximately 50 of the outcomes to be heads. It can be demonstrated that a coin is unfair if h, the number of outcomes that result in heads, satisfies $\left| \dfrac{h - 50}{5} \right| \geq 1.645$. Describe the number of outcomes that determine an unfair coin that is tossed 100 times.

100. The inequality $|T - 57| \leq 7$ describes the range of monthly average temperature, T, in degrees Fahrenheit, for San Francisco, California. The inequality $|T - 50| \leq 22$ describes the range of monthly average temperature, T, in degrees Fahrenheit, for Albany, New York. Solve each inequality and interpret the solution. Then describe at least three differences between the monthly average temperatures for the two cities.

101. A long-distance telephone plan has a monthly fee of $3 with a charge of $0.12 per minute for all long-distance calls. Let

　　$x =$ the number of minutes of long-distance calls that you make per month.

What limitations must be placed on x if you can spend at most $39 per month on the plan?

102. A car can be rented from Basic Rental for $120 per week plus $0.15 for each mile driven. Let

　　$x =$ the number of miles you drive in a week in the rented car.

What limitations must be placed on x if you can spend at most $420 for the week?

103. On two examinations, you have grades of 86 and 88. There is an optional final examination, which counts as one grade. You decide to take the final in order to get a course grade of A, meaning a final average of at least 90.

　a. What must you get on the final to earn an A in the course?

　b. By taking the final, if you do poorly, you might risk the B that you have in the course based on the first two exam grades. If your final average is less than 80, you will lose your B in the course. Describe the grades on the final that will cause this happen.

104. Parts for an automobile repair cost $175. The mechanic charges $34 per hour. If you receive an estimate for at least $226 and at most $294 for fixing the car, what is the time interval that the mechanic will be working on the job?

Writing in Mathematics

105. When graphing the solutions of an inequality, what does a parenthesis signify? What does a bracket signify?

106. When solving an inequality, when is it necessary to change the sense of the inequality? Give an example.

107. Describe ways in which solving a linear inequality is similar to solving a linear equation.

108. Describe ways in which solving a linear inequality is different than solving a linear equation.

109. What is a compound inequality and how is it solved?

110. Describe how to solve an absolute value inequality involving the symbol $<$. Give an example.

111. Describe how to solve an absolute value inequality involving the symbol $>$. Give an example.

112. Explain why $|x| < -4$ has no solution.

113. Describe the solution set of $|x| > -4$.

114. The formula

$$V = 3.5x + 120$$

describes Super Bowl viewers, V, in millions, x years after 1990. Use the formula to write a word problem that can be solved using a linear inequality. Then solve the problem.

Critical Thinking Exercises

115. Which one of the following is true?

　a. The first step in solving $|2x - 3| > -7$ is to rewrite the inequality as $2x - 3 > -7$ or $2x - 3 < 7$.

　b. The smallest real number in the solution set of $2x > 6$ is 4.

　c. All irrational numbers satisfy $|x - 4| > 0$.

　d. None of these statements is true.

116. What's wrong with this argument? Suppose x and y represent two real numbers, where $x > y$.

$2 > 1$	This is a true statement.
$2(y - x) > 1(y - x)$	Multiply both sides by $y - x$.
$2y - 2x > y - x$	Use the distributive property.
$y - 2x > -x$	Subtract y from both sides.
$y > x$	Add 2x to both sides.

The final inequality, $y > x$, is impossible because we were initially given $x > y$.

117. A city commission has proposed two tax bills. The first bill requires that a homeowner pay $1800 plus 3% of the assessed home value in taxes. The second bill requires taxes of $200 plus 8% of the assessed home value. What price range of home assessment would make the first bill a better deal?

118. The percentage, p, of defective products manufactured by a company is given by $|p - 0.3\%| \leq 0.2\%$. If 100,000 products are manufactured and the company offers a $5 refund for each defective product, describe the company's cost for refunds.

CHAPTER SUMMARY, REVIEW, AND TEST

Summary: Basic Formulas

Definition of Absolute Value

$$|x| = \begin{cases} x & \text{if } x \geq 0 \\ -x & \text{if } x < 0 \end{cases}$$

Distance between Points a and b on a Number Line

$$|a - b| \quad \text{or} \quad |b - a|$$

Properties of Algebra

Commutative	$a + b = b + a, \quad ab = ba$
Associative	$(a + b) + c = a + (b + c)$
	$(ab)c = a(bc)$
Distributive	$a(b + c) = ab + ac$
Identity	$a + 0 = a \quad a \cdot 1 = a$
Inverse	$a + (-a) = 0 \quad a \cdot \dfrac{1}{a} = 1, a \neq 0$

Properties of Exponents

$$b^{-n} = \frac{1}{b^n}, \quad b^0 = 1, \quad b^m \cdot b^n = b^{m+n},$$

$$(b^m)^n = b^{mn}, \quad \frac{b^m}{b^n} = b^{m-n}, \quad (ab)^n = a^n b^n, \quad \left(\frac{a}{b}\right)^n = \frac{a^n}{b^n}$$

Product and Quotient Rules for nth Root

$$\sqrt[n]{a} \cdot \sqrt[n]{b} = \sqrt[n]{ab} \qquad \frac{\sqrt[n]{a}}{\sqrt[n]{b}} = \sqrt[n]{\frac{a}{b}}$$

Rational Exponents

$$a^{1/n} = \sqrt[n]{a}, \quad a^{-1/n} = \frac{1}{a^{1/n}} = \frac{1}{\sqrt[n]{a}},$$

$$a^{m/n} = (\sqrt[n]{a})^m = \sqrt[n]{a^m}, \quad a^{-m/n} = \frac{1}{a^{m/n}}$$

Special Products

$$(A + B)(A - B) = A^2 - B^2$$

$$(A + B)^2 = A^2 + 2AB + B^2$$

$$(A - B)^2 = A^2 - 2AB + B^2$$

$$(A + B)^3 = A^3 + 3A^2B + 3AB^2 + B^3$$

$$(A - B)^3 = A^3 - 3A^2B + 3AB^2 - B^3$$

Factoring Formulas

$$A^2 - B^2 = (A + B)(A - B)$$

$$A^2 + 2AB + B^2 = (A + B)^2$$

$$A^2 - 2AB + B^2 = (A - B)^2$$

$$A^3 + B^3 = (A + B)(A^2 - AB + B^2)$$

$$A^3 - B^3 = (A - B)(A^2 + AB + B^2)$$

The Quadratic Formula

All quadratic equations

$$ax^2 + bx + c = 0, a \neq 0$$

can be solved by the quadratic formula

$$x = \frac{-b \pm \sqrt{b^2 - 4ac}}{2a}.$$

Review Exercises

You can use these review exercises, like the review exercises at the end of each chapter, to test your understanding of the chapter's topics. However, you can also use these exercises as a prerequisite test to check your mastery of the fundamental algebra skills needed in this book.

P.1

1. Consider the set:

$$\{-17, -\tfrac{9}{13}, 0, 0.75, \sqrt{2}, \pi, \sqrt{81}\}.$$

List all numbers from the set that are **a.** natural numbers, **b.** whole numbers, **c.** integers, **d.** rational numbers, **e.** irrational numbers.

In Exercises 2–4, rewrite each expression without absolute value bars.

2. $|-103|$

3. $|\sqrt{2} - 1|$

4. $|3 - \sqrt{17}|$

5. Express the distance between the numbers -17 and 4 using absolute value. Then evaluate the absolute value.

In Exercises 6–7, evaluate each algebraic expression for the given value of the variable.

6. $\dfrac{5}{9}(F - 32); F = 68$

7. $\dfrac{8(x + 5)}{3x + 8}, x = 2$

In Exercises 8–13, state the name of the property illustrated.

8. $3 + 17 = 17 + 3$

9. $(6 \cdot 3) \cdot 9 = 6 \cdot (3 \cdot 9)$

10. $\sqrt{3}(\sqrt{5} + \sqrt{3}) = \sqrt{15} + 3$

11. $(6 \cdot 9) \cdot 2 = 2 \cdot (6 \cdot 9)$

12. $\sqrt{3}(\sqrt{5} + \sqrt{3}) = (\sqrt{5} + \sqrt{3})\sqrt{3}$

13. $(3 \cdot 7) + (4 \cdot 7) = (4 \cdot 7) + (3 \cdot 7)$

In Exercises 14–15, simplify each algebraic expression.

14. $3(7x - 5y) - 2(4y - x + 1)$

15. $\frac{1}{5}(5x) + [(3y) + (-3y)] - (-x)$

P.2

Evaluate each exponential expression in Exercises 16–19.

16. $(-3)^3(-2)^2$

17. $2^{-4} + 4^{-1}$

18. $5^{-3} \cdot 5$

19. $\dfrac{3^3}{3^6}$

Simplify each exponential expression in Exercises 20–23.

20. $(-2x^4y^3)^3$

21. $(-5x^3y^2)(-2x^{-11}y^{-2})$

22. $(2x^3)^{-4}$

23. $\dfrac{7x^5y^6}{28x^{15}y^{-2}}$

In Exercises 24–25, write each number in decimal notation.

24. 3.74×10^4

25. 7.45×10^{-5}

In Exercises 26–27, write each number in scientific notation.

26. 3,590,000

27. 0.00725

In Exercises 28–29, perform the indicated operation and write the answer in decimal notation.

28. $(3 \times 10^3)(1.3 \times 10^2)$

29. $\dfrac{6.9 \times 10^3}{3 \times 10^5}$

30. If you earned \$1 million per year (\$$10^6$), how long would it take to accumulate \$1 billion (\$$10^9$)?

31. If the population of the United States is 2.8×10^8 and each person spends about \$150 per year going to the movies (or renting movies), express the total annual spending on movies in scientific notation.

P.3

Use the product rule to simplify the expressions in Exercises 32–35. In Exercises 34–35, assume that variables represent nonnegative real numbers.

32. $\sqrt{300}$

33. $\sqrt{12x^2}$

34. $\sqrt{10x} \cdot \sqrt{2x}$

35. $\sqrt{r^3}$

Use the quotient rule to simplify the expressions in Exercises 36–37.

36. $\sqrt{\dfrac{121}{4}}$

37. $\dfrac{\sqrt{96x^3}}{\sqrt{2x}}$ (Assume that $x > 0$.)

In Exercises 38–40, add or subtract terms whenever possible.

38. $7\sqrt{5} + 13\sqrt{5}$

39. $2\sqrt{50} + 3\sqrt{8}$

40. $4\sqrt{72} - 2\sqrt{48}$

In Exercises 41–44, rationalize the denominator.

41. $\dfrac{30}{\sqrt{5}}$

42. $\dfrac{\sqrt{2}}{\sqrt{3}}$

43. $\dfrac{5}{6 + \sqrt{3}}$

44. $\dfrac{14}{\sqrt{7} - \sqrt{5}}$

Evaluate each expression in Exercise 45–48 or indicate that the root is not a real number.

45. $\sqrt[3]{125}$

46. $\sqrt[5]{-32}$

47. $\sqrt[4]{-125}$

48. $\sqrt[4]{(-5)^4}$

Simplify the radical expressions in Exercises 49–53.

49. $\sqrt[3]{81}$

50. $\sqrt[3]{y^5}$

51. $\sqrt[4]{8} \cdot \sqrt[4]{10}$

52. $4\sqrt[3]{16} + 5\sqrt[3]{2}$

53. $\dfrac{\sqrt[4]{32x^5}}{\sqrt[4]{16x}}$ (Assume that $x > 0$.)

In Exercises 54–59, evaluate each expression.

54. $16^{1/2}$

55. $25^{-1/2}$

56. $125^{1/3}$

57. $27^{-1/3}$

58. $64^{2/3}$

59. $27^{-4/3}$

In Exercises 60–62, simplify using properties of exponents.

60. $(5x^{2/3})(4x^{1/4})$

61. $\dfrac{15x^{3/4}}{5x^{1/2}}$

62. $(125x^6)^{2/3}$

63. Simplify by reducing the index of the radical: $\sqrt[6]{y^3}$.

P.4

In Exercises 64–65, perform the indicated operations. Write the resulting polynomial in standard form and indicate its degree.

64. $(-6x^3 + 7x^2 - 9x + 3) + (14x^3 + 3x^2 - 11x - 7)$

65. $(13x^4 - 8x^3 + 2x^2) - (5x^4 - 3x^3 + 2x^2 - 6)$

In Exercises 66–72, find each product.

66. $(3x - 2)(4x^2 + 3x - 5)$

67. $(3x - 5)(2x + 1)$

68. $(4x + 5)(4x - 5)$

69. $(2x + 5)^2$

70. $(3x - 4)^2$

71. $(2x + 1)^3$

72. $(5x - 2)^3$

In Exercises 73–79, find each product.

73. $(x + 7y)(3x - 5y)$

74. $(3x - 5y)^2$

75. $(3x^2 + 2y)^2$

76. $(7x + 4y)(7x - 4y)$

77. $(a - b)(a^2 + ab + b^2)$

78. $[5y - (2x + 1)][5y + (2x + 1)]$

79. $(x + 2y + 4)^2$

P.5

In Exercises 80–96, factor completely, or state that the polynomial is prime.

80. $15x^3 + 3x^2$

81. $x^2 - 11x + 28$

82. $15x^2 - x - 2$

83. $64 - x^2$

84. $x^2 + 16$

85. $3x^4 - 9x^3 - 30x^2$

86. $20x^7 - 36x^3$

87. $x^3 - 3x^2 - 9x + 27$

88. $16x^2 - 40x + 25$

89. $x^4 - 16$

90. $y^3 - 8$

91. $x^3 + 64$

92. $3x^4 - 12x^2$

93. $27x^3 - 125$

94. $x^5 - x$

95. $x^3 + 5x^2 - 2x - 10$

96. $x^2 + 18x + 81 - y^2$

In Exercises 97–99, factor and simplify each algebraic expression.

97. $16x^{-3/4} + 32x^{1/4}$

98. $(x^2 - 4)(x^2 + 3)^{1/2} - (x^2 - 4)^2(x^2 + 3)^{3/2}$

99. $12x^{-1/2} + 6x^{-3/2}$

P.6

In Exercises 100–102, simplify each rational expression. Also, list all numbers that must be excluded from the domain.

100. $\dfrac{x^3 + 2x^2}{x + 2}$

101. $\dfrac{x^2 + 3x - 18}{x^2 - 36}$

102. $\dfrac{x^2 + 2x}{x^2 + 4x + 4}$

In Exercises 103–105, multiply or divide as indicated.

103. $\dfrac{x^2 + 6x + 9}{x^2 - 4} \cdot \dfrac{x + 3}{x - 2}$

104. $\dfrac{6x + 2}{x^2 - 1} \div \dfrac{3x^2 + x}{x - 1}$

105. $\dfrac{x^2 - 5x - 24}{x^2 - x - 12} \div \dfrac{x^2 - 10x + 16}{x^2 + x - 6}$

In Exercises 106–109, add or subtract as indicated.

106. $\dfrac{2x - 7}{x^2 - 9} - \dfrac{x - 10}{x^2 - 9}$

107. $\dfrac{3x}{x + 2} + \dfrac{x}{x - 2}$

108. $\dfrac{x}{x^2 - 9} + \dfrac{x - 1}{x^2 - 5x + 6}$

109. $\dfrac{4x - 1}{2x^2 + 5x - 3} - \dfrac{x + 3}{6x^2 + x - 2}$

In Exercises 110–112, simplify each complex rational expression.

110. $\dfrac{3 + \dfrac{12}{x}}{1 - \dfrac{16}{x^2}}$

111. $\dfrac{3 - \dfrac{1}{x + 3}}{3 + \dfrac{1}{x + 3}}$

112. $\dfrac{\sqrt{25 - x^2} + \dfrac{x^2}{\sqrt{25 - x^2}}}{25 - x^2}$

P.7

In Exercises 113–118, solve and check each linear equation.

113. $2x - 5 = 7$

114. $5x + 20 = 3x$

115. $7(x - 4) = x + 2$

116. $1 - 2(6 - x) = 3x + 2$

117. $2(x - 4) + 3(x + 5) = 2x - 2$

118. $2x - 4(5x + 1) = 3x + 17$

Exercises 119–123 contain equations with constants in denominators. Solve each equation and check by the method of your choice.

119. $\dfrac{2x}{3} = \dfrac{x}{6} + 1$ **120.** $\dfrac{x}{2} - \dfrac{1}{10} = \dfrac{x}{5} + \dfrac{1}{2}$

121. $\dfrac{2x}{3} = 6 - \dfrac{x}{4}$ **122.** $\dfrac{x}{4} = 2 + \dfrac{x-3}{3}$

123. $\dfrac{3x+1}{3} - \dfrac{13}{2} = \dfrac{1-x}{4}$

*Exercises 124–127 contain equations with variables in denominators. **a.** List the value or values representing restriction(s) on the variable. **b.** Solve the equation.*

124. $\dfrac{9}{4} - \dfrac{1}{2x} = \dfrac{4}{x}$ **125.** $\dfrac{7}{x-5} + 2 = \dfrac{x+2}{x-5}$

126. $\dfrac{1}{x-1} - \dfrac{1}{x+1} = \dfrac{2}{x^2-1}$

127. $\dfrac{4}{x+2} + \dfrac{2}{x-4} = \dfrac{30}{x^2-2x-8}$

In Exercises 128–130, determine whether each equation is an identity, a conditional equation, or an inconsistent equation.

128. $\dfrac{1}{x+5} = 0$

129. $7x + 13 = 4x - 10 + 3x + 23$

130. $7x + 13 = 3x - 10 + 2x + 23$

131. The percentage, P, of U.S. adults who read the daily newspaper can be described by the formula

$$P = -0.7x + 80$$

where x is the number of years after 1965. In which year will 52% of U.S. adults read the daily newspaper?

In Exercises 132–134, solve each formula for the specified variable.

132. $V = \dfrac{1}{3}Bh$ for h **133.** $F = f(1 - M)$ for M

134. $T = gr + gvt$ for g

Solve the equations containing absolute value in Exercises 135–136.

135. $|2x + 1| = 7$ **136.** $2|x - 3| - 6 = 10$

P.8

Solve each equation in Exercises 137–138 by factoring.

137. $2x^2 + 15x = 8$ **138.** $5x^2 + 20x = 0$

Solve each equation in Exercises 139–140 by the square root method.

139. $2x^2 - 3 = 125$ **140.** $(3x - 4)^2 = 18$

In Exercises 141–142, determine the constant that should be added to the binomial so that it becomes a perfect square trinomial. Then write and factor the trinomial.

141. $x^2 + 20x$ **142.** $x^2 - 3x$

Solve each equation in Exercises 143–144 by completing the square.

143. $x^2 - 12x + 27 = 0$ **144.** $3x^2 - 12x + 11 = 0$

Solve each equation in Exercises 145–146 using the quadratic formula.

145. $x^2 = 2x + 4$ **146.** $2x^2 = 3 - 4x$

Compute the discriminant of each equation in Exercises 147–148. What does the discriminant indicate about the number and type of solutions?

147. $x^2 - 4x + 13 = 0$ **148.** $9x^2 = 2 - 3x$

Solve each equation in Exercises 149–153 by the method of your choice.

149. $2x^2 - 11x + 5 = 0$ **150.** $(3x + 5)(x - 3) = 5$
151. $3x^2 - 7x + 1 = 0$ **152.** $x^2 - 9 = 0$
153. $(x - 3)^2 - 25 = 0$

154. The weight of a human fetus is described by the formula $W = 3t^2$, where W is the weight, in grams, and t is the time, in weeks, $0 \le t \le 39$. After how many weeks does the fetus weigh 1200 grams?

155. The alligator, an endangered species, is the subject of a protection program. The formula

$$P = -10x^2 + 475x + 3500$$

describes the alligator population, P, after x years of the protection program, where $0 \le x \le 12$. After how many years is the population up to 7250?

156. A building casts a shadow that is double the length of its height. If the distance from the end of the shadow to the top of the building is 300 meters, how high is the building? Round to the nearest meter.

P.9

In Exercises 157–159, graph the solutions of each inequality on a number line.

157. $x > 5$ **158.** $x \le 1$
159. $-3 \le x < 0$

In Exercises 160–162, express each interval in terms of an inequality and graph the interval on a number line.

160. $(-2, 3]$ **161.** $[-1.5, 2]$

162. $(-1, \infty)$

Solve each linear inequality in Exercises 163–168 and graph the solution set on a number line. Express each solution set in interval notation.

163. $-6x + 3 \le 15$ **164.** $6x - 9 \ge -4x - 3$

165. $\dfrac{x}{3} - \dfrac{3}{4} - 1 > \dfrac{x}{2}$

166. $6x + 5 > -2(x - 3) - 25$

167. $3(2x - 1) - 2(x - 4) \ge 7 + 2(3 + 4x)$

168. $7 < 2x + 3 \le 9$

Solve each inequality in Exercises 169–171 by first rewriting each one as an equivalent inequality without absolute value bars. Graph the solution set on a number line. Express each solution set in interval notation.

169. $|2x + 3| \le 15$

170. $\left| \dfrac{2x + 6}{3} \right| > 2$

171. $|2x + 5| - 7 \ge -6$

172. Approximately 90% of the population sleeps h hours daily, where h is described by the inequality $|h - 6.5| \le 1$. Write a sentence describing the range for the number of hours that most people sleep. Do *not* use the phrase "absolute value" in your description.

173. The formula for converting Fahrenheit temperature, F, to Celsius temperature, C, is $C = \frac{5}{9}(F - 32)$. If Celsius temperature ranges from $10°$ to $25°$, inclusive, what is the range for the Fahrenheit temperature?

Chapter P Test

1. List all the rational numbers in this set:
$$\{-7, -\tfrac{4}{5}, 0, 0.25, \sqrt{3}, \sqrt{4}, \tfrac{22}{7}, \pi\}.$$

In Exercises 2–3, state the name of the property illustrated.

2. $3(2 + 5) = 3(5 + 2)$

3. $6(7 + 4) = 6 \cdot 7 + 6 \cdot 4$

4. Express in scientific notation: 0.00076.

Simplify each expression in Exercises 5–11.

5. $9(10x - 2y) - 5(x - 4y + 3)$

6. $\dfrac{30x^3 y^4}{6x^9 y^{-4}}$

7. $\sqrt{6r}\sqrt{3r}$ (Assume that $r \ge 0$.)

8. $4\sqrt{50} - 3\sqrt{18}$

9. $\dfrac{3}{5 + \sqrt{2}}$ **10.** $\sqrt[3]{16x^4}$

11. $\dfrac{x^2 + 2x - 3}{x^2 - 3x + 2}$ **12.** Evaluate: $27^{-5/3}$.

In Exercises 13–14, find each product.

13. $(2x - 5)(x^2 - 4x + 3)$ **14.** $(5x + 3y)^2$

In Exercises 15–20, factor completely, or state that the polynomial is prime.

15. $x^2 - 9x + 18$ **16.** $x^3 + 2x^2 + 3x + 6$

17. $25x^2 - 9$ **18.** $36x^2 - 84x + 49$

19. $y^3 - 125$ **20.** $x^2 + 10x + 25 - 9y^2$

21. Factor and simplify:
$$x(x + 3)^{-3/5} + (x + 3)^{2/5}.$$

In Exercises 22–26, perform the operations and simplify, if possible.

22. $\dfrac{2x + 8}{x - 3} \div \dfrac{x^2 + 5x + 4}{x^2 - 9}$

23. $\dfrac{x}{x + 3} + \dfrac{5}{x - 3}$

24. $\dfrac{2x + 3}{x^2 - 7x + 12} - \dfrac{2}{x - 3}$

25. $\dfrac{1 - \dfrac{x}{x + 2}}{1 + \dfrac{1}{x}}$ **26.** $\dfrac{2x\sqrt{x^2 + 5} - \dfrac{2x^3}{\sqrt{x^2 + 5}}}{x^2 + 5}$

Find the solution set for each equation in Exercises 27–33.

27. $7(x - 2) = 4(x + 1) - 21$

28. $\dfrac{2x - 3}{4} = \dfrac{x - 4}{2} - \dfrac{x + 1}{4}$

29. $\dfrac{2}{x - 3} - \dfrac{4}{x + 3} = \dfrac{8}{x^2 - 9}$

30. $2x^2 - 3x - 2 = 0$

31. $(3x - 1)^2 = 75$ **32.** $x(x - 2) = 4$

33. $\left| \dfrac{2}{3}x - 6 \right| = 2$

Solve each inequality in Exercises 34–37. Express the answer in interval notation and graph the solution set on a number line.

34. $3(x + 4) \geq 5x - 12$

35. $\dfrac{x}{6} + \dfrac{1}{8} \leq \dfrac{x}{2} - \dfrac{3}{4}$ **36.** $-3 \leq \dfrac{2x + 5}{3} < 6$

37. $|3x + 2| \geq 3$

In Exercises 38–39, solve each formula for the specified variable.

38. $V = \dfrac{1}{3}lwh$ for h

39. $y - y_1 = m(x - x_1)$ for x

40. The male minority? The graphs show enrollment in U.S. colleges, with projections from 2003 through 2009. The trend indicated by the graphs is among the hottest topics of debate among college-admission officers. Some private liberal arts colleges have quietly begun special efforts to recruit men—including admissions preferences for them.

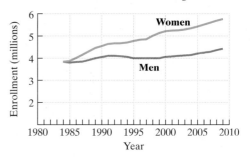

Enrollment in U.S. Colleges

Source: U.S. Department of Education

The data for the men can be approximated by the formula

$$N = 0.01x + 3.9$$

where N represents enrollment, in millions, x years after 1984. According to the formula, when will the projected enrollment for men be 4.1 million? How well does the formula describe enrollment for that year shown by the line graph?

41. A vertical pole is to be supported by a wire that is 26 feet long and anchored 24 feet from the base of the pole. How far up the pole should the wire be attached?

Graphs, Functions, and Models

Chapter 1

The cost of mailing a package depends on its weight. The probability that you and another person in a room share the same birthday depends on the number of people in the room. In both these situations, the relationship between variables can be described by a *function*. Understanding this concept will give you a new perspective on many ordinary situations.

'Tis the season and you've waited until the last minute to mail your holiday gifts. Your only option is overnight express mail. You realize that the cost of mailing a gift depends on its weight, but the mailing costs seem somewhat odd. Your packages that weigh 1.1 pounds, 1.5 pounds, and 2 pounds cost $15.75 each to send overnight. Packages that weigh 2.01 pounds and 3 pounds cost you $18.50 each. Finally, your heaviest gift is barely over 3 pounds and its mailing cost is $21.25. What sort of system is this in which costs increase by $2.75, stepping from $15.75 to $18.50 and from $18.50 to $21.25?

SECTION 1.1 *Graphs and Graphing Utilities*

Objectives

1. Plot points in the rectangular coordinate system.
2. Graph equations in the rectangular coordinate system.
3. Interpret information about a graphing utility's viewing rectangle.
4. Use a graph to determine intercepts.
5. Interpret information given by graphs.

The beginning of the seventeenth century was a time of innovative ideas and enormous intellectual progress in Europe. English theatergoers enjoyed a succession of exciting new plays by Shakespeare. William Harvey proposed the radical notion that the heart was a pump for blood rather than the center of emotion. Galileo, with his new-fangled invention called the telescope, supported the theory of Polish astronomer Copernicus that the sun, not the Earth, was the center of the solar system. Monteverdi was writing the world's first grand operas. French mathematicians Pascal and Fermat invented a new field of mathematics called probability theory.

Into this arena of intellectual electricity stepped French aristocrat René Descartes (1596–1650). Descartes, propelled by the creativity surrounding him, developed a new branch of mathematics that brought together algebra and geometry in a unified way—a way that visualized numbers as points on a graph, equations as geometric figures, and geometric figures as equations. This new branch of mathematics, called *analytic geometry*, established Descartes as one of the founders of modern thought and among the most original mathematicians and philosophers of any age. We begin this section by looking at Descartes's deceptively simple idea, called the **rectangular coordinate system** or (in his honor) the **Cartesian coordinate system.**

1 Plot points in the rectangular coordinate system.

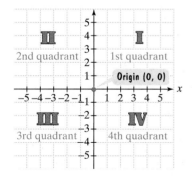

Figure 1.1 The rectangular coordinate system

Points and Ordered Pairs

Descartes used two number lines that intersect at right angles at their zero points, as shown in Figure 1.1. The horizontal number line is the *x*-**axis.** The vertical number line is the *y*-**axis.** The point of intersection of these axes is their zero points, called the **origin.** Positive numbers are shown to the right and above the origin. Negative numbers are shown to the left and below the origin. The axes divide the plane into four quarters, called **quadrants.** The points located on the axes are not in any quadrant.

Each point in the rectangular coordinate system corresponds to an **ordered pair** of real numbers, (x, y). Examples of such pairs are $(4, 2)$ and $(-5, -3)$. The first number in each pair, called the *x*-**coordinate,** denotes the distance and direction from the origin along the *x*-axis. The second number, called the *y*-**coordinate,** denotes vertical distance and direction along a line parallel to the *y*-axis or along the *y*-axis itself.

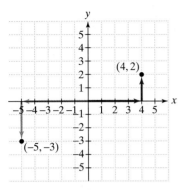

Figure 1.2   Plotting $(4,2)$ and $(-5,-3)$

Figure 1.2 shows how we **plot,** or locate, the points corresponding to the ordered pairs $(4, 2)$ and $(-5, -3)$. We plot $(4, 2)$ by going 4 units from 0 to the right along the x-axis. Then we go 2 units up parallel to the y-axis. We plot $(-5, -3)$ by going 5 units from 0 to the left along the x-axis and 3 units down parallel to the y-axis. The phrase "the point corresponding to the ordered pair $(-5, -3)$" is often abbreviated as "the point $(-5, -3)$."

EXAMPLE 1   Plotting Points in the Rectangular Coordinate System

Plot the points: $A(-3, 5)$, $B(2, -4)$, $C(5, 0)$, $D(-5, -3)$, $E(0, 4)$, and $F(0, 0)$.

Solution   See Figure 1.3. We move from the origin and plot the points in the following way:

$A(-3, 5)$:	3 units left, 5 units up
$B(2, -4)$:	2 units right, 4 units down
$C(5, 0)$:	5 units right, 0 units up or down
$D(-5, -3)$:	5 units left, 3 units down
$E(0, 4)$:	0 units right or left, 4 units up
$F(0, 0)$:	0 units right or left, 0 units up or down

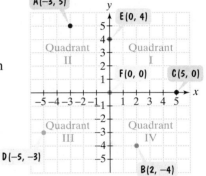

Figure 1.3   Plotting points

Study Tip

Reminder: Answers to all Check Point exercises are given in the answer section. Check your answer before continuing your reading to verify that you understand the concept.

The phrase *ordered pair* is used because **order is important.** For example, the points $(2, 5)$ and $(5, 2)$ are not the same. To plot $(2, 5)$, move 2 units right and 5 units up. To plot $(5, 2)$, move 5 units right and 2 units up. The points $(2, 5)$ and $(5, 2)$ are in different locations. **The order in which coordinates appear makes a difference in a point's location.**

Check Point 1   Plot the points:
$$A(-2, 4), B(4, -2), C(-3, 0), \text{ and } D(0, -3).$$

2 Graph equations in the rectangular coordinate system.

Graphs of Equations

A relationship between two quantities can be expressed as an **equation in two variables,** such as

$$y = x^2 - 4.$$

A **solution** of this equation is an ordered pair of real numbers with the following property: When the x-coordinate is substituted for x and the y-coordinate is substituted for y in the equation, we obtain a true statement. For example, if we let $x = 3$, then $y = 3^2 - 4 = 9 - 4 = 5$. The ordered pair $(3, 5)$ is a solution of the equation $y = x^2 - 4$. We also say that $(3, 5)$ **satisfies** the equation.

We can generate as many ordered-pair solutions as desired of $y = x^2 - 4$ by substituting numbers for x and then finding the values for y. The **graph of the equation** is the set of all points whose coordinates satisfy the equation.

One method for graphing an equation such as $y = x^2 - 4$ is the **point-plotting method.** First, we find several ordered pairs that are solutions of the equation. Next, we plot these ordered pairs as points in the rectangular coordinate system. Finally, we connect the points with a smooth curve or line. This often gives us a picture of all ordered pairs that satisfy the equation.

EXAMPLE 2   Graphing an Equation Using the Point-Plotting Method

Graph $y = x^2 - 4$. Select integers for x, starting with -3 and ending with 3.

Solution   For each value of x we find the corresponding value for y.

x	$y = x^2 - 4$	(x, y)
-3	$y = (-3)^2 - 4 = 9 - 4 = 5$	$(-3, 5)$
-2	$y = (-2)^2 - 4 = 4 - 4 = 0$	$(-2, 0)$
-1	$y = (-1)^2 - 4 = 1 - 4 = -3$	$(-1, -3)$
0	$y = 0^2 - 4 = 0 - 4 = -4$	$(0, -4)$
1	$y = 1^2 - 4 = 1 - 4 = -3$	$(1, -3)$
2	$y = 2^2 - 4 = 4 - 4 = 0$	$(2, 0)$
3	$y = 3^2 - 4 = 9 - 4 = 5$	$(3, 5)$

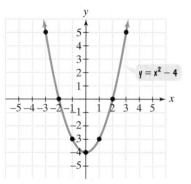

Figure 1.4   The graph of $y = x^2 - 4$

Now we plot the ordered pairs that are solutions of the equation and join the points with a smooth curve, as shown in Figure 1.4. The graph of $y = x^2 - 4$ is a curve where the part of the graph to the right of the y-axis is a reflection of the part to the left of it, and vice versa. The arrows on the left and the right of the curve indicate that it extends indefinitely in both directions.

Check Point 2   Graph $y = 2x - 4$. Select integers for x, starting with -1 and ending with 3.

3 Interpret information about a graphing utility's viewing rectangle.

Graphing Equations Using a Graphing Utility

Graphing calculators or graphing software packages for computers are referred to as **graphing utilities** or graphers. A graphing utility is a powerful tool that quickly generates the graph of an equation in two variables. Figure 1.5 shows two such graphs for the equations in Example 2 and Check Point 2.

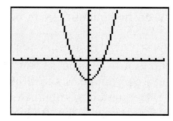

Figure 1.5(a)
The graph of $y = x^2 - 4$

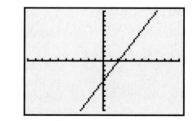

Figure 1.5(b)
The graph of $y = 2x - 4$

Study Tip

Even if you are not using a graphing utility in the course, read this part of the section. Knowing about viewing rectangles will enable you to understand the graphs that we display in the technology boxes throughout the book.

What differences do you notice between these graphs and graphs that we draw by hand? They do seem a bit "jittery." Arrows do not appear on the left and right ends of the graphs. Furthermore, numbers are not given along the axes. For both graphs in Figure 1.5, the x-axis extends from -10 to 10 and the y-axis also extends from -10 to 10. The distance represented by each consecutive tick mark is one unit. We say that the **viewing rectangle** is $[-10, 10, 1]$ by $[-10, 10, 1]$.

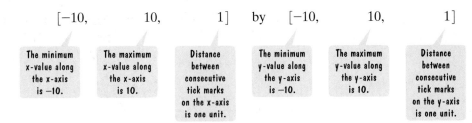

$[-10, \quad 10, \quad 1] \quad$ by $\quad [-10, \quad 10, \quad 1]$

| The minimum x-value along the x-axis is –10. | The maximum x-value along the x-axis is 10. | Distance between consecutive tick marks on the x-axis is one unit. | The minimum y-value along the y-axis is –10. | The maximum y-value along the y-axis is 10. | Distance between consecutive tick marks on the y-axis is one unit. |

To graph an equation in x and y using a graphing utility, enter the equation and specify the size of the viewing rectangle. The size of the viewing rectangle sets minimum and maximum values for both the x- and y-axes. Enter these values, as well as the values between consecutive tick marks on the respective axes. The $[-10, 10, 1]$ by $[-10, 10, 1]$ viewing rectangle used in Figure 1.5 is called the **standard viewing rectangle.**

EXAMPLE 3   Understanding the Viewing Rectangle

What is the meaning of a $[-2, 3, 0.5]$ by $[-10, 20, 5]$ viewing rectangle?

Solution   We begin with $[-2, 3, 0.5]$, which describes the x-axis. The minimum x-value is -2 and the maximum x-value is 3. The distance between consecutive tick marks is 0.5.

Next, consider $[-10, 20, 5]$, which describes the y-axis. The minimum y-value is -10 and the maximum y-value is 20. The distance between consecutive tick marks is 5.

Figure 1.6 illustrates a $[-2, 3, 0.5]$ by $[-10, 20, 5]$ viewing rectangle. To make things clearer, we've placed numbers by each tick mark. These numbers do not appear on the axes when you use a graphing utility to graph an equation.

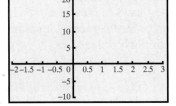

Figure 1.6   A $[-2, 3, 0.5]$ by $[-10, 20, 5]$ viewing rectangle

Check Point 3   What is the meaning of a $[-100, 100, 50]$ by $[-80, 80, 10]$ viewing rectangle? Create a figure like the one in Figure 1.6 that illustrates this viewing rectangle.

On most graphing utilities, the display screen is about two-thirds as high as it is wide. By using a square setting, you can make the distance of one unit along the x-axis the same as the distance of one unit along the y-axis. (This does not occur in the standard viewing rectangle.) Graphing utilities can also *zoom in* and *zoom out.* When you zoom in, you see a smaller portion of the graph, but you see it in greater detail. When you zoom out, you see a larger portion of the graph. Thus, zooming out may help you to develop a better understanding of the overall character of the graph. With practice, you will become more comfortable with graphing equations in two variables using your graphing utility. You will also develop a better sense of the size of the viewing rectangle that will reveal needed information about a particular graph.

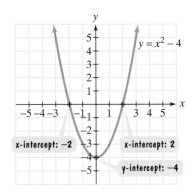

Figure 1.7   Intercepts of $y = x^2 - 4$

Intercepts

An ***x*-intercept** of a graph is the x-coordinate of a point where the graph intersects the x-axis. For example, look at the graph of $y = x^2 - 4$ in Figure 1.7. The graph crosses the x-axis at $(-2, 0)$ and $(2, 0)$. Thus, the x-intercepts are -2 and 2. **The *y*-coordinate corresponding to a graph's *x*-intercept is always zero.**

A ***y*-intercept** of a graph is the y-coordinate of a point where the graph intersects the y-axis. The graph of $y = x^2 - 4$ in Figure 1.7 shows that the graph crosses the y-axis at $(0, -4)$. Thus, the y-intercept is -4. **The *x*-coordinate corresponding to a graph's *y*-intercept is always zero.**

4   Use a graph to determine intercepts.

Figure 1.8 illustrates that a graph may have no intercepts or several intercepts.

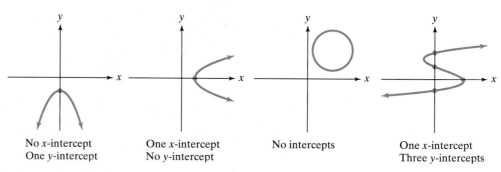

No *x*-intercept
One *y*-intercept

One *x*-intercept
No *y*-intercept

No intercepts

One *x*-intercept
Three *y*-intercepts

Figure 1.8

5 Interpret information given by graphs.

Interpreting Information Given by Graphs

Magazines and newspapers often display information using **line graphs** like the one in Figure 1.9. The graph shows the average age at which women in the United States married for the first time over a 110-year period. The years are listed on the horizontal axis and the ages are listed on the vertical axis.

Like the graph in Figure 1.9, line graphs are often used to illustrate trends over time. Some measure of time, such as months or years, frequently appears on the horizontal axis. Amounts are generally listed on the vertical axis.

Figure 1.9 Average age at which U.S. women married for the first time
Source: U.S. Census Bureau

A line graph displays information in the first quadrant of a rectangular coordinate system. By identifying points on line graphs and their coordinates, you can interpret specific information given by the graph.

For example, Figure 1.10 shows how to find the average age at which women married for the first time in 1930. (Only the part of the graph that reveals what occurred through about 1940 is shown in the margin because we are interested in 1930.)

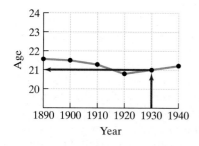

Figure 1.10 In 1930, women were 21 on average when they married for the first time.

Step 1 Locate 1930 on the horizontal axis.

Step 2 Locate the point above 1930.

Step 3 Read across to the corresponding age on the vertical axis.

The age is 21. The coordinates (1930, 21) tell us that in 1930, women in the United States married for the first time at an average age of 21.

EXAMPLE 4   Applying Estimation Techniques to a Line Graph

Use Figure 1.9 to estimate the maximum average age at which U.S. women married for the first time. When did this occur?

Solution   The maximum average age at which U.S. women married for the first time can be found by locating the highest point on the graph. This point lies above 2000 on the horizontal axis. Read across to the corresponding age on the vertical axis. The age falls approximately midway between 23 and 24, at $23\frac{1}{2}$. The coordinates of the point are approximately $(2000, 23\frac{1}{2})$. Thus, according to the graph, the maximum average age at which U.S. women married for the first time is about $23\frac{1}{2}$. This occurred in 2000. Take another look at the complete line graph in Figure 1.9 on page 120 that includes the years 1890 through 2000. Can you see that $23\frac{1}{2}$ is the oldest average age of first marriage over the 110-year period?

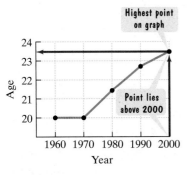
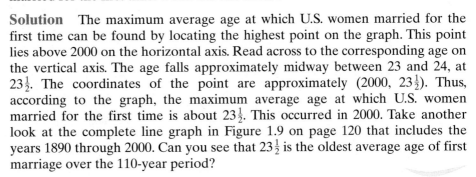

Figure 1.9 Shown again to show only 1960–2000

> **Check Point 4**   Use the complete line graph in Figure 1.9 on page 120 to estimate the maximum average age, for the period from 1900 through 1950, at which U.S. women married for the first time. When did this occur?

EXERCISE SET 1.1

Practice Exercises

In Exercises 1–12, plot the given point in a rectangular coordinate system.

1. $(1, 4)$ **2.** $(2, 5)$

3. $(-2, 3)$ **4.** $(-1, 4)$

5. $(-3, -5)$ **6.** $(-4, -2)$

7. $(4, -1)$ **8.** $(3, -2)$

9. $(-4, 0)$ **10.** $(0, -3)$

11. $\left(\frac{7}{2}, -\frac{3}{2}\right)$ **12.** $\left(-\frac{5}{2}, \frac{3}{2}\right)$

Graph each equation in Exercises 13–28. Let
$x = -3, -2, -1, 0, 1, 2,$ *and* 3.

13. $y = x^2 - 2$ **14.** $y = x^2 + 2$

15. $y = x - 2$ **16.** $y = x + 2$

17. $y = 2x + 1$ **18.** $y = 2x - 4$

19. $y = -\frac{1}{2}x$ **20.** $y = -\frac{1}{2}x + 2$

21. $y = |x|$ **22.** $y = 2|x|$

23. $y = |x| + 1$ **24.** $y = |x| - 1$

25. $y = 4 - x^2$ **26.** $y = 9 - x^2$

27. $y = x^3$ **28.** $y = x^3 - 1$

In Exercises 29–32, match the viewing rectangle with the correct figure shown in the next column. Then label the tick marks in the figure to illustrate this viewing rectangle.

29. $[-5, 5, 1]$ by $[-5, 5, 1]$

30. $[-10, 10, 2]$ by $[-4, 4, 2]$

31. $[-20, 80, 10]$ by $[-30, 70, 10]$

32. $[-40, 40, 20]$ by $[-1000, 1000, 100]$

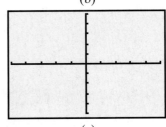

(a)

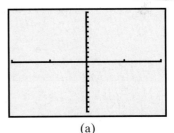

(b)

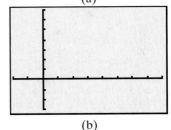

(c)

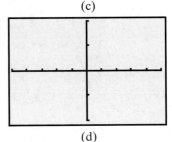

(d)

In Exercises 33–38, use the graph and **a.** *determine the x-intercepts, if any;* **b.** *determine the y-intercepts, if any. For each graph, tick marks along the axes represent one unit each.*

33.   **34.**

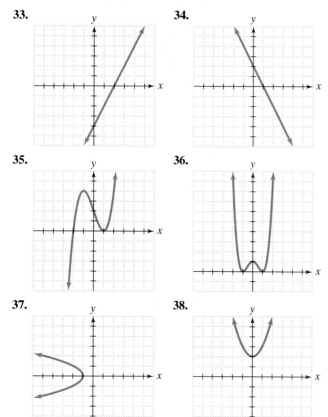

35.   **36.**

37.   **38.**

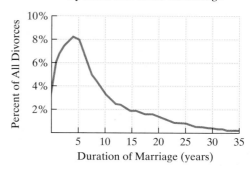

Application Exercises

    A football is thrown by a quarterback to a receiver. The points in the figure show the height of the football, in feet, above the ground in terms of its distance, in yards, from the quarterback. Use this information to solve Exercises 39–44.

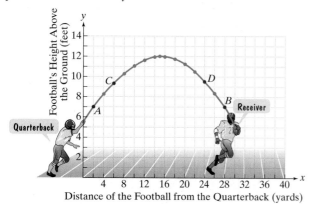

Distance of the Football from the Quarterback (yards)

39. Find the coordinates of point *A*. Then interpret the coordinates in terms of the information given.

40. Find the coordinates of point *B*. Then interpret the coordinates in terms of the information given.

41. Estimate the coordinates of point *C*.

42. Estimate the coordinates of point *D*.

43. What is the football's maximum height? What is its distance from the quarterback when it reaches its maximum height?

44. What is the football's height when it is caught by the receiver? What is the receiver's distance from the quarterback when he catches the football?

The graph shows the percent distribution of divorces in the United States by number of years of marriage. Use the graph to solve Exercises 45–48.

Percent Distribution of Divorces by Number of Years of Marriage

Source: Divorce Center

45. During which years of marriage is the chance of divorce increasing?

46. During which years of marriage is the chance of divorce decreasing?

47. During which year of marriage is the chance of divorce the highest? Estimate, to the nearest percent, the percentage of divorces that occur during this year.

48. During which year of marriage is the chance of divorce the lowest? Estimate, to the nearest percent, the percentage of divorces that occur during this year.

Writing in Mathematics

49. What is the rectangular coordinate system?

50. Explain how to plot a point in the rectangular coordinate system. Give an example with your explanation.

51. Explain why $(5, -2)$ and $(-2, 5)$ do not represent the same point.

52. Explain how to graph an equation in the rectangular coordinate system.

53. What does a $[-20, 2, 1]$ by $[-4, 5, 0.5]$ viewing rectangle mean?

54. Describe the trend shown in the graph for Exercises 45–48. What explanations can you offer for this trend?

Technology Exercises

55. Use a graphing utility to verify each of your hand-drawn graphs in Exercises 13–28. Experiment with the size of the viewing rectangle to make the graph displayed by the graphing utility resemble your hand-drawn graph as much as possible.

56. The stated intent of the 1994 "don't ask, don't tell" policy was to reduce the number of discharges of gay men and lesbians from the military. The equation

$$y = 45.48x^2 - 334.35x + 1237.9$$

describes the number of gay service members, y, discharged from the military for homosexuality x years after 1990. Graph the equation in a $[0, 10, 1]$ by $[0, 2200, 200]$ viewing rectangle. Then describe something about the relationship between x and y that is revealed by looking at the graph that is not obvious from the equation. What does the graph reveal about the success or lack of success of "don't ask, don't tell"?

A graph of an equation is a complete graph if it shows all of the important features of the graph. Use a graphing utility to graph the equations in Exercises 57–59 in each of the given viewing rectangles. Then choose which viewing rectangle gives a complete graph.

57. $y = x^2 + 10$
  a. $[-5, 5, 1]$ by $[-5, 5, 1]$
  b. $[-10, 10, 1]$ by $[-10, 10, 1]$
  c. $[-10, 10, 1]$ by $[-50, 50, 1]$

58. $y = 0.1x^4 - x^3 + 2x^2z$
  a. $[-5, 5, 1]$ by $[-8, 2, 1]$
  b. $[-10, 10, 1]$ by $[-10, 10, 1]$
  c. $[-8, 16, 1]$ by $[-16, 8, 1]$

59. $y = x^3 - 30x + 20$
  a. $[-10, 10, 1]$ by $[-10, 10, 1]$
  b. $[-10, 10, 1]$ by $[-50, 50, 10]$
  c. $[-10, 10, 1]$ by $[-50, 100, 10]$

Critical Thinking Exercises

60. Which one of the following is true?
  a. If the coordinates of a point satisfy the inequality $xy > 0$, then (x, y) must be in quadrant I.
  b. The ordered pair $(2, 5)$ satisfies $3y - 2x = -4$.
  c. If a point is on the x-axis, it is neither up nor down, so $x = 0$.
  d. None of the above is true.

In Exercises 61–64, match the story with the correct figure. The figures are labeled (a), (b), (c), and (d).

61. As the blizzard got worse, the snow fell harder and harder.

62. The snow fell more and more softly.

63. It snowed hard, but then it stopped. After a short time, the snow started falling softly.

64. It snowed softly, and then it stopped. After a short time, the snow started falling hard.

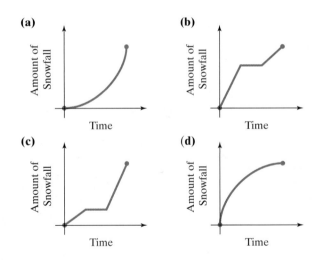

SECTION 1.2 *Lines and Slope*

Objectives

1. Compute a line's slope.
2. Write the point-slope equation of a line.
3. Write and graph the slope-intercept equation of a line.
4. Recognize equations of horizontal and vertical lines.
5. Recognize and use the general form of a line's equation.
6. Find slopes and equations of parallel and perpendicular lines.
7. Model data with linear equations.

Is there a relationship between literacy and child mortality? As the percentage of adult females who are literate increases, does the mortality of children under five decrease? Figure 1.11, based on data from the United Nations, indicates that this is, indeed, the case. Each point in the figure represents one country.

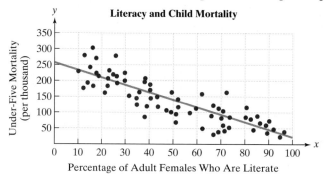

Figure 1.11

Source: United Nations

Data presented in a visual form as a set of points is called a **scatter plot.** Also shown in Figure 1.11 is a line that passes through or near the points. A line that best fits the data points in a scatter plot is called a **regression line.** By writing the equation of this line, we can obtain an algebraic description of the data and make predictions about child mortality based on the percentage of adult females in a country who are literate.

Data often fall on or near a line. In this section we will use equations to describe such data and make predictions. We begin with a discussion of a line's steepness.

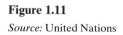 Compute a line's slope.

The Slope of a Line

Mathematicians have developed a useful measure of the steepness of a line, called the *slope* of the line. Slope compares the vertical change (the **rise**) to the horizontal change (the **run**) when moving from one fixed point to another along the line. To calculate the slope of a line, we use a ratio that compares the change in y (the rise) to the corresponding change in x (the run).

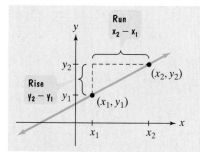

Definition of Slope

The **slope** of the line through the distinct points (x_1, y_1) and (x_2, y_2) is

$$\frac{\text{Change in } y}{\text{Change in } x} = \frac{\text{Rise}}{\text{Run}}$$

$$= \frac{y_2 - y_1}{x_2 - x_1}$$

where $x_2 - x_1 \neq 0$.

It is common notation to let the letter m represent the slope of a line. The letter m is used because it is the first letter of the French verb *monter*, meaning to rise, or to ascend.

Slope and the Streets of San Francisco

San Francisco's Filbert Street has a slope of 0.613, meaning that for every horizontal distance of 100 feet, the street ascends 61.3 feet vertically. With its 31.5° angle of inclination, the street is too steep to pave and is only accessible by wooden stairs.

EXAMPLE 1   Using the Definition of Slope

Find the slope of the line passing through each pair of points:

a. $(-3, -1)$ and $(-2, 4)$ **b.** $(-3, 4)$ and $(2, -2)$.

Solution

a. Let $(x_1, y_1) = (-3, -1)$ and $(x_2, y_2) = (-2, 4)$. We obtain a slope of

$$m = \frac{\text{Change in } y}{\text{Change in } x} = \frac{y_2 - y_1}{x_2 - x_1} = \frac{4 - (-1)}{-2 - (-3)} = \frac{5}{1} = 5.$$

The situation is illustrated in Figure 1.12(a). The slope of the line is 5, indicating that there is a vertical change, a rise, of 5 units for each horizontal change, a run, of 1 unit. The slope is positive, and the line rises from left to right.

Study Tip

When computing slope, it makes no difference which point you call (x_1, y_1) and which point you call (x_2, y_2). If we let $(x_1, y_1) = (-2, 4)$ and $(x_2, y_2) = (-3, -1)$, the slope is still 5:

$$m = \frac{y_2 - y_1}{x_2 - x_1} = \frac{-1 - 4}{-3 - (-2)} = \frac{-5}{-1} = 5.$$

However, you should not subtract in one order in the numerator $(y_2 - y_1)$ and then in a different order in the denominator $(x_1 - x_2)$. The slope is *not*

$$\frac{-1 - 4}{-2 - (-3)} = \frac{-5}{1} = -5. \quad \text{Incorrect}$$

b. We can let $(x_1, y_1) = (-3, 4)$ and $(x_2, y_2) = (2, -2)$. The slope of the line shown in Figure 1.12(b) is computed as follows:

$$m = \frac{-2 - 4}{2 - (-3)} = \frac{-6}{5} = -\frac{6}{5}.$$

The slope of the line is $-\frac{6}{5}$. For every vertical change of -6 units (6 units down), there is a corresponding horizontal change of 5 units. The slope is negative and the line falls from left to right.

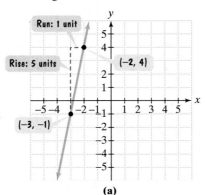

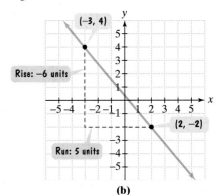

Figure 1.12   Visualizing slope   (a)   (b)

Check Point 1 Find the slope of the line passing through each pair of points:

a. $(-3, 4)$ and $(-4, -2)$ **b.** $(4, -2)$ and $(-1, 5)$.

Example 1 illustrates that a line with a positive slope is rising from left to right and a line with a negative slope is falling from left to right. By contrast, a horizontal line neither rises nor falls and has a slope of zero. A vertical line has no horizontal change, so $x_2 - x_1 = 0$ in the formula for slope. Because we cannot divide by zero, the slope of a vertical line is undefined. This discussion is summarized in Table 1.1.

Table 1.1   Possibilities for a Line's Slope

Positive Slope	Negative Slope	Zero Slope	Undefined Slope
$m > 0$	$m < 0$	$m = 0$	m is undefined.
Line rises from left to right.	Line falls from left to right.	Line is horizontal.	Line is vertical.

2 Write the point-slope equation of a line.

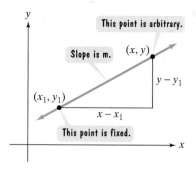

Figure 1.13   A line passing through (x_1, y_1) with slope m

The Point-Slope Form of the Equation of a Line

We can use the slope of a line to obtain various forms of the line's equation. For example, consider a nonvertical line with slope m that contains the point (x_1, y_1). Now, let (x, y) represent any other point on the line, shown in Figure 1.13. Keep in mind that the point (x, y) is arbitrary and is not in one fixed position. By contrast, the point (x_1, y_1) is fixed. Regardless of where the point (x, y) is located, the shape of the triangle in Figure 1.13 remains the same. Thus, the ratio for slope stays a constant m. This means that for all points along the line,

$$m = \frac{y - y_1}{x - x_1}, \quad x \neq x_1.$$

We can clear the fraction by multiplying both sides by $x - x_1$.

$$m(x - x_1) = \frac{y - y_1}{x - x_1} \cdot x - x_1$$

$$m(x - x_1) = y - y_1 \qquad \text{Simplify.}$$

Now, if we reverse the two sides, we obtain the *point-slope form* of the equation of a line.

> **Point-Slope Form of the Equation of a Line**
>
> The **point-slope equation** of a nonvertical line with slope m that passes through the point (x_1, y_1) is
>
> $$y - y_1 = m(x - x_1).$$

For example, an equation of the line passing through $(1, 5)$ with slope 2 $(m = 2)$ is

$$y - 5 = 2(x - 1).$$

After we obtain the point-slope form of a line, it is customary to express the equation with y isolated on one side of the equal sign. Example 2 illustrates how this is done.

EXAMPLE 2   Writing the Point-Slope Equation of a Line

Write the point-slope form of the equation of the line passing through $(-1, 3)$ with slope 4. Then solve the equation for y.

Solution   We use the point-slope equation of a line with $m = 4$, $x_1 = -1$, and $y_1 = 3$.

$$y - y_1 = m(x - x_1) \qquad \text{This is the point-slope form of the equation.}$$
$$y - 3 = 4[x - (-1)] \qquad \text{Substitute the given values.}$$
$$y - 3 = 4(x + 1) \qquad \text{We now have the point-slope form of the equation for the given line.}$$

We can solve this equation for y by applying the distributive property on the right side.

$$y - 3 = 4x + 4$$

Finally, we add 3 to both sides.

$$y = 4x + 7$$

Check Point 2   Write the point-slope form of the equation of the line passing through $(2, -5)$ with slope 6. Then solve the equation for y.

EXAMPLE 3   Writing the Point-Slope Equation of a Line

Write the point-slope form of the equation of the line passing through the points $(4, -3)$ and $(-2, 6)$. (See Figure 1.14.) Then solve the equation for y.

Solution   To use the point-slope form, we need to find the slope. The slope is the change in the y-coordinates divided by the corresponding change in the x-coordinates.

$$m = \frac{6 - (-3)}{-2 - 4} = \frac{9}{-6} = -\frac{3}{2} \qquad \text{This is the definition of slope using } (4, -3) \text{ and } (-2, 6).$$

We can take either point on the line to be (x_1, y_1). Let's use $(x_1, y_1) = (4, -3)$. Now, we are ready to write the point-slope equation.

$$y - y_1 = m(x - x_1) \qquad \text{This is the point-slope form of the equation.}$$
$$y - (-3) = -\tfrac{3}{2}(x - 4) \qquad \text{Substitute: } (x_1, y_1) = (4, -3) \text{ and } m = -\tfrac{3}{2}.$$
$$y + 3 = -\tfrac{3}{2}(x - 4) \qquad \text{Simplify.}$$

We now have the point-slope form of the equation of the line shown in Figure 1.14. Now, we solve this equation for y.

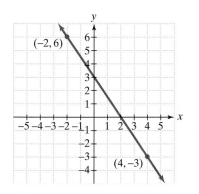

Figure 1.14 Write the point-slope equation of this line.

$$y + 3 = -\tfrac{3}{2}(x - 4) \qquad \text{This is the point-slope form of the equation.}$$

$$y + 3 = -\tfrac{3}{2}x + 6 \qquad \text{Use the distributive property.}$$

$$y = -\tfrac{3}{2}x + 3 \qquad \text{Subtract 3 from both sides.}$$

Discovery

You can use either point for (x_1, y_1) when you write a line's point-slope equation. Rework Example 3 using $(-2, 6)$ for (x_1, y_1). Once you solve for y, you should still obtain

$$y = -\tfrac{3}{2}x + 3.$$

Check Point 3 Write the point-slope form of the equation of the line passing through the points $(-2, -1)$ and $(-1, -6)$. Then solve the equation for y.

3 Write and graph the slope-intercept equation of a line.

The Slope-Intercept Form of the Equation of a Line

Let's write the point-slope form of the equation of a nonvertical line with slope m and y-intercept b. The line is shown in Figure 1.15. Because the y-intercept is b, the line passes through $(0, b)$. We use the point-slope form with $x_1 = 0$ and $y_1 = b$.

$$y - y_1 = m(x - x_1)$$

Let $y_1 = b$.   Let $x_1 = 0$.

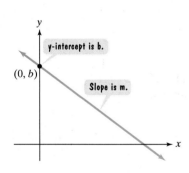

Figure 1.15 A line with slope m and y-intercept b

We obtain

$$y - b = m(x - 0).$$

Simplifying on the right side gives us

$$y - b = mx.$$

Finally, we solve for y by adding b to both sides.

$$y = mx + b$$

Thus, if a line's equation is written with y isolated on one side, the x-coefficient is the line's slope and the constant term is the y-intercept. This form of a line's equation is called the *slope-intercept form* of a line.

Slope-Intercept Form of the Equation of a Line

The **slope-intercept equation** of a nonvertical line with slope m and y-intercept b is

$$y = mx + b.$$

EXAMPLE 4 Graphing by Using the Slope and *y*-Intercept

Graph the line whose equation is $y = \frac{2}{3}x + 2$.

Solution The equation of the line is in the form $y = mx + b$. We can find the slope, m, by identifying the coefficient of x. We can find the y-intercept, b, by identifying the constant term.

$$y = \frac{2}{3}x + 2$$

The slope is $\frac{2}{3}$.

The y-intercept is 2.

We need two points in order to graph the line. We can use the y-intercept, 2, to obtain the first point $(0, 2)$. Plot this point on the y-axis, shown in Figure 1.16.

We know the slope and one point on the line. We can use the slope, $\frac{2}{3}$, to determine a second point on the line. By definition,

$$m = \frac{2}{3} = \frac{\text{Rise}}{\text{Run}}.$$

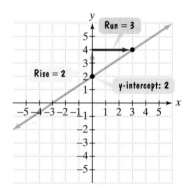

Figure 1.16 The graph of $y = \frac{2}{3}x + 2$

We plot the second point on the line by starting at $(0, 2)$, the first point. Based on the slope, we move 2 units *up* (the rise) and 3 units to the *right* (the run). This puts us at a second point on the line, $(3, 4)$, shown in Figure 1.16.

We use a straightedge to draw a line through the two points. The graph of $y = \frac{2}{3}x + 2$ is shown in Figure 1.16.

Graphing $y = mx + b$ by Using the Slope and y-Intercept

1. Plot the y-intercept on the y-axis. This is the point $(0, b)$.
2. Obtain a second point using the slope, m. Write m as a fraction, and use rise over run, starting at the point containing the y-intercept, to plot this point.
3. Use a straightedge to draw a line through the two points. Draw arrowheads at the ends of the line to show that the line continues indefinitely in both directions.

Check Point 4 Graph the line whose equation is $y = \frac{3}{5}x + 1$.

4 Recognize equations of horizontal and vertical lines.

Equations of Horizontal and Vertical Lines

Some things change very little. For example, Figure 1.17 shows that the percentage of people in the United States satisfied with their lives remains relatively constant for all age groups. Shown in the figure is a horizontal line that passes near most tops of the six bars.

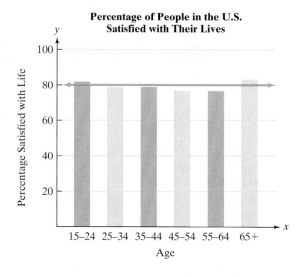

Percentage of People in the U.S. Satisfied with Their Lives

Figure 1.17
Source: *Culture Shift in Advanced Industrial Society*, Princeton University Press

We can use $y = mx + b$, the slope-intercept form of a line's equation, to write the equation of the horizontal line in Figure 1.17. We need the line's slope, m, and its y-intercept, b. Because the line is horizontal, $m = 0$. The line intersects the y-axis at $(0, 80)$, so its y-intercept is 80: $b = 80$.

Thus, an equation in the form $y = mx + b$ that describes the percentage, y, of people at age x satisfied with their lives is

$$y = 0x + 80, \quad \text{or} \quad y = 80.$$

The percentage of people satisfied with their lives remains relatively constant in the United States for all age groups, at approximately 80%.

In general, if a line is horizontal, its slope is zero: $m = 0$. Thus, the equation $y = mx + b$ becomes $y = b$, where b is the y-intercept. All horizontal lines have equations of the form $y = b$.

EXAMPLE 5   Graphing a Horizontal Line

Graph $y = -4$ in the rectangular coordinate system.

Solution   All points on the graph of $y = -4$ have a value of y that is always -4. No matter what the x-coordinate is, the y-coordinate for every point on the line is -4. Let us select three of the possible values for x: $-2, 0,$ and 3. So, three of the points on the graph $y = -4$ are $(-2, -4), (0, -4),$ and $(3, -4)$. Plot each of these points. Drawing a line that passes through the three points gives the horizontal line shown in Figure 1.18.

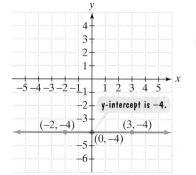

Figure 1.18   The graph of $y = -4$

Check Point 5   Graph $y = 3$ in the rectangular coordinate system.

Next, let's see what we can discover about the graph of an equation of the form $x = a$ by looking at an example.

EXAMPLE 6   Graphing a Vertical Line

Graph $x = 5$ in the rectangular coordinate system.

Solution   All points on the graph of $x = 5$ have a value of x that is always 5. No matter what the y-coordinate is, the corresponding x-coordinate for every point on the line is 5. Let us select three of the possible values of y: -2, 0, and 3. So, three of the points on the graph of $x = 5$ are $(5, -2)$, $(5, 0)$, and $(5, 3)$. Plot each of these points. Drawing a line that passes through the three points gives the vertical line shown in Figure 1.19.

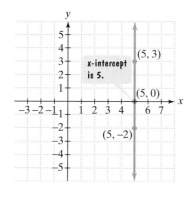

Figure 1.19   The graph $x = 5$

Horizontal and Vertical Lines

The graph of $y = b$ is a horizontal line. The y-intercept is b.

The graph of $x = a$ is a vertical line. The x-intercept is a.

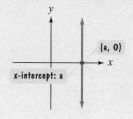

Check Point 6   Graph $x = -1$ in the rectangular coordinate system.

---

5 Recognize and use the general form of a line's equation.

The General Form of the Equation of a Line

The vertical line whose equation is $x = 5$ cannot be written in slope-intercept form, $y = mx + b$, because its slope is undefined. However, every line has an equation that can be expressed in the form $Ax + By + C = 0$. For example, $x = 5$ can be expressed as $1x + 0y - 5 = 0$, or $x - 5 = 0$. The equation $Ax + By + C = 0$ is called the *general form* of the equation of a line.

> **General Form of the Equation of a Line**
> Every line has an equation that can be written in the **general form**
> $$Ax + By + C = 0$$
> where A, B, and C are real numbers, and A and B are not both zero.

If the equation of a line is given in general form, it is possible to find the slope, m, and the y-intercept, b, for the line. We solve the equation for y, transforming it into the slope-intercept form $y = mx + b$. In this form, the coefficient of x is the slope of the line, and the constant term is its y-intercept.

EXAMPLE 7 Finding the Slope and the y-Intercept

Find the slope and the y-intercept of the line whose equation is $2x - 3y + 6 = 0$.

Solution   The equation is given in general form. We begin by rewriting it in the form $y = mx + b$. We need to solve for y.

$$2x - 3y + 6 = 0 \qquad \text{This is the given equation.}$$

$$2x + 6 = 3y \qquad \text{To isolate the } y\text{-term, add } 3y \text{ to both sides.}$$

$$3y = 2x + 6 \qquad \text{Reverse the two sides. (This step is optional.)}$$

$$y = \frac{2}{3}x + 2 \qquad \text{Divide both sides by 3.}$$

The coefficient of x, $\frac{2}{3}$, is the slope and the constant term, 2, is the y-intercept. This is the form of the equation that we graphed in Figure 1.16 on page 129.

> **Check Point 7**   Find the slope and the y-intercept of the line whose equation is $3x + 6y - 12 = 0$. Then use the y-intercept and the slope to graph the equation.

We've covered a lot of territory. Let's take a moment to summarize the various forms for equations of lines.

Equations of Lines

1. Point-slope form: $y - y_1 = m(x - x_1)$
2. Slope-intercept form: $y = mx + b$
3. Horizontal line: $y = b$
4. Vertical line: $x = a$
5. General form: $Ax + By + C = 0$

---

6 Find slopes and equations of parallel and perpendicular lines.

Parallel and Perpendicular Lines

Two nonintersecting lines that lie in the same plane are **parallel.** If two lines do not intersect, the ratio of the vertical change to the horizontal change is the same for each line. Because two parallel lines have the same "steepness," they must have the same slope.

Slope and Parallel Lines

1. If two nonvertical lines are parallel, then they have the same slope.
2. If two distinct nonvertical lines have the same slope, then they are parallel.
3. Two distinct vertical lines, each with undefined slope, are parallel.

EXAMPLE 8 Writing Equations of a Line Parallel to a Given Line

Write an equation of the line passing through $(-3, 2)$ and parallel to the line whose equation is $y = 2x + 1$. Express the equation in point-slope form and slope-intercept form.

Solution The situation is illustrated in Figure 1.20. We are looking for the equation of the line shown on the left. How do we obtain this equation? Notice that the line passes through the point $(-3, 2)$. Using the point-slope form of the line's equation, we have $x_1 = -3$ and $y_1 = 2$.

$$y - y_1 = m(x - x_1)$$

$$y_1 = 2 \qquad x_1 = -3$$

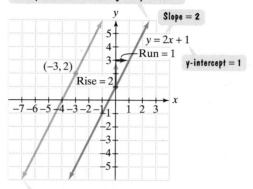

The equation of this line is given: $y = 2x + 1$.

Slope $= 2$

$y = 2x + 1$

Run $= 1$

$(-3, 2)$

y-intercept $= 1$

Rise $= 2$

We must write the equation of this line.

Figure 1.20 Writing equations of a line parallel to a given line

Now, the only thing missing from the equation is m, the slope of the line on the left. Do we know anything about the slope of either line in Figure 1.20? The answer is yes; we know the slope of the line on the right, whose equation is given.

$$y = 2x + 1$$

The slope of the line on the right in Figure 1.20 is 2.

Parallel lines have the same slope. Because the slope of the line with the given equation is 2, $m = 2$ for the line whose equation we must write.

$$y - y_1 = m(x - x_1)$$

$$y_1 = 2 \qquad m = 2 \qquad x_1 = -3$$

The point-slope form of the line's equation is

$$y - 2 = 2[x - (-3)] \text{ or}$$
$$y - 2 = 2(x + 3).$$

Solving for y, we obtain the slope-intercept form of the equation.

$$y - 2 = 2x + 6 \quad \text{Apply the distributive property.}$$
$$y = 2x + 8 \quad \text{Add 2 to both sides. This is the slope-intercept form,}$$
$$y = mx + b, \text{ of the equation.}$$

Check Point 8 Write an equation of the line passing through $(-2, 5)$ and parallel to the line whose equation is $y = 3x + 1$. Express the equation in point-slope form and slope-intercept form.

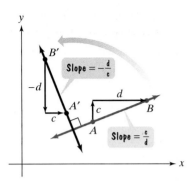

Figure 1.21 Slopes of perpendicular lines

Two lines that intersect at a right angle (90°) are said to be **perpendicular,** shown in Figure 1.21. The relationship between the slopes of perpendicular lines is not as obvious as the relationship between parallel lines. Figure 1.21 shows line AB, with slope $\frac{c}{d}$. Rotate line AB through 90° to the left to obtain line $A'B'$, perpendicular to line AB. The figure indicates that the rise and the run of the new line are reversed from the original line, but the rise is now negative. This means that the slope of the new line is $-\frac{d}{c}$. Notice that the product of the slopes of the two perpendicular lines is -1:

$$\left(\frac{c}{d}\right)\left(-\frac{d}{c}\right) = -1.$$

This relationship holds for all perpendicular lines and is summarized in the following box:

> **Slope and Perpendicular Lines**
> 1. If two nonvertical lines are perpendicular, then the product of their slopes is -1.
> 2. If the product of the slopes of two lines is -1, then the lines are perpendicular.
> 3. A horizontal line having zero slope is perpendicular to a vertical line having undefined slope.

An equivalent way of stating this relationship is to say that one line is perpendicular to another line if its slope is the *negative reciprocal* of the slope of the other. For example, if a line has slope 5, any line having slope $-\frac{1}{5}$ is perpendicular to it. Similarly, if a line has slope $-\frac{3}{4}$, any line having slope $\frac{4}{3}$ is perpendicular to it.

EXAMPLE 9 Finding the Slope of a Line Perpendicular to a Given Line

Find the slope of any line that is perpendicular to the line whose equation is $x + 4y - 8 = 0$.

Solution We begin by writing the equation of the given line, $x + 4y - 8 = 0$, in slope-intercept form. Solve for y.

$x + 4y - 8 = 0$	This is the given equation.
$4y = -x + 8$	To isolate the y-term, subtract x and add 8 on both sides.
$y = -\frac{1}{4}x + 2$	Divide both sides by 4.

The given line has slope $-\frac{1}{4}$. Any line perpendicular to this line has a slope that is the negative reciprocal of $-\frac{1}{4}$. Thus, the slope of any perpendicular line is 4.

Check Point 9 Find the slope of any line that is perpendicular to the line whose equation is $x + 3y - 12 = 0$.

7 Model data with linear equations.

Applications; Mathematical Modeling

Slope is defined as the ratio of a change in y to a corresponding change in x. Our next example shows how slope can be interpreted as a **rate of change** in an applied situation.

Figure 1.22
Source: Forrester Research

EXAMPLE 10 Slope as a Rate of Change

A best guess at the look of our nation in the next decade indicates that the number of men and women living alone will increase each year. Figure 1.22 shows line graphs for the number of U.S. men and women living alone, projected through 2010. Find the slope of the line segment for the women. Describe what the slope represents.

Solution We let x represent a year and y the number of women living alone in that year. The two points shown on the line segment for women have the following coordinates:

$$(1995, 14) \quad \text{and} \quad (2010, 17).$$

In 1995, 14 million U.S. women lived alone.

In 2010, 17 million U.S. women are projected to live alone.

Now we compute the slope:

$$m = \frac{\text{Change in } y}{\text{Change in } x} = \frac{17 - 14}{2010 - 1995}$$

The unit in the numerator is million people.

$$= \frac{3}{15} = \frac{1}{5} = \frac{0.2 \text{ million people}}{\text{year}}.$$

The unit in the denominator is year.

The slope indicates that the number of U.S. women living alone is projected to increase by 0.2 million each year. The rate of change is 0.2 million women per year.

Check Point 10 Use the graph in Figure 1.22 to find the slope of the line segment for the men. Express the slope correct to two decimal places and describe what it represents.

If an equation in slope-intercept form describes a relationship between variables that represent real-world phenomena, then the slope and y-intercept have physical interpretations. For the equation $y = mx + b$, the y-intercept, b, tells us what is happening to y when x is 0. If x represents time, the y-intercept describes the value of y at the beginning, or when time equals 0. The slope, m, represents the rate of change in y per unit change in x.

The process of finding equations and formulas to describe real-world phenomena is called **mathematical modeling.** Such equations and formulas, together with the meanings assigned to the variables, are called **mathematical models.** One method of creating a mathematical model is to use available data and construct an equation that describes the behavior of the data. For example, consider the data for women living alone, shown in Figure 1.22. We let $x =$ the number of years after 1995. At the beginning of our data, or 0 years after 1995, 14 million women lived alone. Thus, $b = 14$. In Example 10, we found that $m = 0.2$ (rate of change is 0.2 million women per year). An equation of the form $y = mx + b$ that models the data is

$$y = 0.2x + 14$$

where y is the number, in millions, of U.S. women living alone x years after 1995.

Linear equations are useful for modeling data in scatter plots that fall on or near a line. For example, Table 1.2 gives the population of the United States, in millions, in the indicated year. The data are displayed in a scatter plot as a set of six points in Figure 1.23.

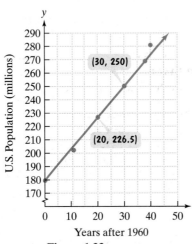

Figure 1.23

Table 1.2

Year	x (Years after 1960)	y (U.S. Population, in millions)
1960	0	179.3
1970	10	203.3
1980	20	226.5
1990	30	250.0
1998	38	268.9
2000	40	281.4

Also shown in Figure 1.23 is a line that passes through or near the six points. By writing the equation of this line, we can obtain a model of the data and make predictions about the population of the United States in the future.

EXAMPLE 11   Modeling U.S. Population

Write the slope-intercept equation of the line shown in Figure 1.23. Use the equation to predict U.S. population in 2010.

Solution   The line in Figure 1.23 passes through $(20, 226.5)$ and $(30, 250)$. We start by finding the slope.

$$m = \frac{\text{Change in } y}{\text{Change in } x} = \frac{250 - 226.5}{30 - 20} = \frac{23.5}{10} = 2.35$$

The slope indicates that the rate of change in the U.S. population is 2.35 million people per year. Now we write the line's slope-intercept equation.

$$y - y_1 = m(x - x_1) \qquad \text{Begin with the point-slope form.}$$

$$y - 250 = 2.35(x - 30) \qquad \text{Either ordered pair can be } (x_1, y_1). \text{Let}$$
$$(x_1, y_1) = (30, 250). \text{From above, } m = 2.35.$$

Technology

You can use a graphing utility to obtain a model for a scatter plot in which the data points fall on or near a straight line. The line that best fits the data is called the **regression line**. After entering the data in Table 1.2, a graphing utility displays a scatter plot of the data and the regression line.

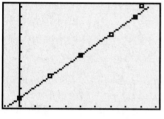

$[-5, 45, 5]$ by $[170, 283, 10]$

Also displayed is the regression line's equation.

```
LinReg
y=ax+b
a=2.45748031496
b=178.377952756
```

$$y - 250 = 2.35x - 70.5 \qquad \text{Apply the distributive property on the right.}$$
$$y = 2.35x + 179.5 \qquad \text{Add 250 to both sides and solve for y.}$$

A linear equation that models U.S. population, y, in millions, x years after 1960 is

$$y = 2.35x + 179.5.$$

Now, let's use this equation to predict U.S. population in 2010. Because 2010 is 50 years after 1960, substitute 50 for x and compute y.

$$y = 2.35(50) + 179.5 = 297$$

Our equation predicts that the population of the United States in the year 2010 will be 297 million. (The projected figure from the U.S. Census Bureau is 297.716 million.)

Check Point 11 Use the data points $(10, 203.3)$ and $(20, 226.5)$ from Table 1.2 to write an equation that models U.S. population x years after 1960. Use the equation to predict U.S. population in 2020.

In creating mathematical models from data, we strive for both accuracy and simplicity. Sometimes a mathematical model gives an estimate that is not a good approximation or is extended too far into the future, resulting in a prediction that does not make sense. In these cases, we say that **model breakdown** has occurred.

Cigarettes and Lung Cancer

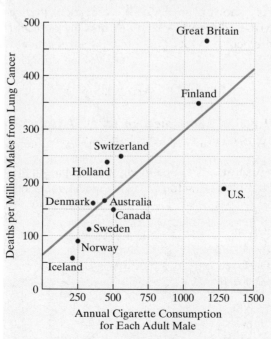

This scatter plot shows a relationship between cigarette consumption among males and deaths due to lung cancer per million males. The data are from 11 countries and date back to a 1964 report by the U.S. Surgeon General. The scatter plot can be modeled by a line whose slope indicates an increasing death rate from lung cancer with increased cigarette consumption. At that time, the tobacco industry argued that in spite of this regression line, tobacco use is not the cause of cancer. Recent data do, indeed, show a causal effect between tobacco use and numerous diseases.

Source: Smoking and Health, Washington, D.C., 1964

EXERCISE SET 1.2

Practice Exercises

In Exercises 1–10, find the slope of the line passing through each pair of points or state that the slope is undefined. Then indicate whether the line through the points rises, falls, is horizontal, or is vertical.

1. $(4, 7)$ and $(8, 10)$ **2.** $(2, 1)$ and $(3, 4)$
3. $(-2, 1)$ and $(2, 2)$ **4.** $(-1, 3)$ and $(2, 4)$
5. $(4, -2)$ and $(3, -2)$ **6.** $(4, -1)$ and $(3, -1)$
7. $(-2, 4)$ and $(-1, -1)$ **8.** $(6, -4)$ and $(4, -2)$
9. $(5, 3)$ and $(5, -2)$ **10.** $(3, -4)$ and $(3, 5)$

In Exercises 11–38, use the given conditions to write an equation for each line in point-slope form and slope-intercept form.

11. Slope $= 2$, passing through $(3, 5)$
12. Slope $= 4$, passing through $(1, 3)$
13. Slope $= 6$, passing through $(-2, 5)$
14. Slope $= 8$, passing through $(4, -1)$
15. Slope $= -3$, passing through $(-2, -3)$
16. Slope $= -5$, passing through $(-4, -2)$
17. Slope $= -4$, passing through $(-4, 0)$
18. Slope $= -2$, passing through $(0, -3)$
19. Slope $= -1$, passing through $\left(-\frac{1}{2}, -2\right)$
20. Slope $= -1$, passing through $\left(-4, -\frac{1}{4}\right)$
21. Slope $= \frac{1}{2}$, passing through the origin
22. Slope $= \frac{1}{3}$, passing through the origin
23. Slope $= -\frac{2}{3}$, passing through $(6, -2)$
24. Slope $= -\frac{3}{5}$, passing through $(10, -4)$
25. Passing through $(1, 2)$ and $(5, 10)$
26. Passing through $(3, 5)$ and $(8, 15)$
27. Passing through $(-3, 0)$ and $(0, 3)$
28. Passing through $(-2, 0)$ and $(0, 2)$
29. Passing through $(-3, -1)$ and $(2, 4)$
30. Passing through $(-2, -4)$ and $(1, -1)$
31. Passing through $(-3, -2)$ and $(3, 6)$
32. Passing through $(-3, 6)$ and $(3, -2)$
33. Passing through $(-3, -1)$ and $(4, -1)$
34. Passing through $(-2, -5)$ and $(6, -5)$
35. Passing through $(2, 4)$ with x-intercept $= -2$
36. Passing through $(1, -3)$ with x-intercept $= -1$
37. x-intercept $= -\frac{1}{2}$ and y-intercept $= 4$
38. x-intercept $= 4$ and y-intercept $= -2$

In Exercises 39–46, give the slope and y-intercept of each line whose equation is given. Then graph the line.

39. $y = 2x + 1$ **40.** $y = 3x + 2$
41. $y = -2x + 1$ **42.** $y = -3x + 2$
43. $y = \frac{3}{4}x - 2$ **44.** $y = \frac{3}{4}x - 3$
45. $y = -\frac{3}{5}x + 7$ **46.** $y = -\frac{2}{5}x + 6$

In Exercises 47–52, graph each equation in the rectangular coordinate system.

47. $y = -2$ **48.** $y = 4$
49. $x = -3$ **50.** $x = 5$
51. $y = 0$ **52.** $x = 0$

In Exercises 53–60,
a. *Rewrite the given equation in slope-intercept form.*
b. *Give the slope and y-intercept.*
c. *Graph the equation.*

53. $3x + y - 5 = 0$ **54.** $4x + y - 6 = 0$
55. $2x + 3y - 18 = 0$ **56.** $4x + 6y + 12 = 0$
57. $8x - 4y - 12 = 0$ **58.** $6x - 5y - 20 = 0$
59. $3x - 9 = 0$ **60.** $4y + 28 = 0$

In Exercises 61–68, use the given conditions to write an equation for each line in point-slope form and slope-intercept form.

61. Passing through $(-8, -10)$ and parallel to the line whose equation is $y = -4x + 3$
62. Passing through $(-2, -7)$ and parallel to the line whose equation is $y = -5x + 4$
63. Passing through $(2, -3)$ and perpendicular to the line whose equation is $y = \frac{1}{5}x + 6$
64. Passing through $(-4, 2)$ and perpendicular to the line whose equation is $y = \frac{1}{3}x + 7$
65. Passing through $(-2, 2)$ and parallel to the line whose equation is $2x - 3y - 7 = 0$
66. Passing through $(-1, 3)$ and parallel to the line whose equation is $3x - 2y - 5 = 0$
67. Passing through $(4, -7)$ and perpendicular to the line whose equation is $x - 2y - 3 = 0$
68. Passing through $(5, -9)$ and perpendicular to the line whose equation is $x + 7y - 12 = 0$

Application Exercises

69. The scatter plot shows that from 1985 through 2001, the number of Americans participating in downhill skiing remained relatively constant. Write an equation that models the number of participants in downhill skiing, y, in millions, for this period.

Number of U.S. Participants in Downhill Skiing

Source: National Ski Areas Association

If talk about a federal budget surplus sounded too good to be true, that's because it probably was. The Congressional Budget Office's estimates for 2010 range from a $1.2 trillion budget surplus to a $286 billion deficit. Use the information provided by the Congressional Budget Office graphs to solve Exercises 70–71.

Federal Budget Projections

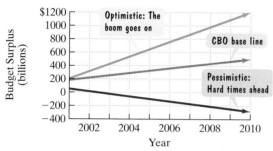

Source: Congressional Budget Office

70. Look at the line that indicates hard times ahead. Find the slope of this line using (2001, 50) and (2010, −286). Use a calculator and round to the nearest whole number. Describe what the slope represents.

71. Look at the line that indicates the boom goes on. Find the slope of this line using (2001, 200) and (2010, 1200). Use a calculator and round to the nearest whole number. Describe what the slope represents.

72. Horrified at the cost the last time you needed a prescription drug? The graph shows that the cost of the average retail prescription has been rising steadily since 1991.

Average Cost of a Retail Prescription

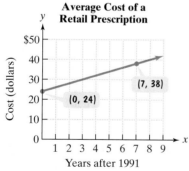

Years after 1991

Source: Newsweek

a. According to the graph, what is the y-intercept? Describe what this represents in this situation.

b. Use the coordinates of the two points shown to compute the slope. What does this mean about the cost of the average retail prescription?

c. Write a linear equation in slope-intercept form that models the cost of the average retail prescription, y, x years after 1991.

d. Use your model from part (c) to predict the cost of the average retail prescription in 2010.

73. For 61 years, Social Security has been a huge success. It is the primary source of income for 66% of Americans over 65 and the only thing that keeps 42% of the elderly from poverty. However, the number of workers per Social Security beneficiary has been declining steadily since 1950.

Number of Workers per Social Security Beneficiary

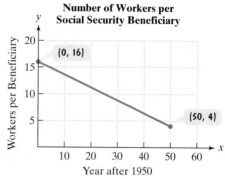

Year after 1950

Source: Social Security Administration

a. According to the graph, what is the y-intercept? Describe what this represents in this situation.

b. Use the coordinates of the two points shown to compute the slope. What does this mean about the number of workers per beneficiary?

c. Write a linear equation in slope-intercept form that models the number of workers per beneficiary, y, x years after 1950.

d. Use your model from part (c) to predict the number of workers per beneficiary in 2010. For every 8 workers, how many beneficiaries will there be?

74. We seem to be fed up with being lectured at about our waistlines. The points in the graph show the average weight of American adults from 1990 through 2000. Also shown is a line that passes through or near the points.

Average Weight of Americans

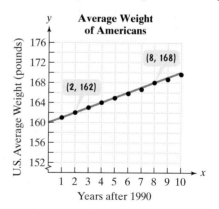

Years after 1990

Source: Diabetes Care

a. Use the two points whose coordinates are shown by the voice balloons to find the point-slope equation of the line that models the average weight of Americans, y, in pounds, x years after 1990.

b. Write the equation in part (a) in slope-intercept form.

c. Use the slope-intercept equation to predict the average weight of Americans in 2008.

75. Films may not be getting any better, but in this era of moviegoing, the number of screens available for new films and the classics has exploded. The points in the graph on the next page show the number of screens in the United

States from 1995 through 2000. Also shown is a line that passes through or near the points.

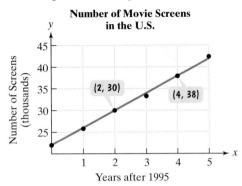

Number of Movie Screens in the U.S.

Source: Motion Picture Association of America

a. Use the two points whose coordinates are shown by the voice balloons to find the point-slope equation of the line that models the number of screens, y, in thousands, x years after 1995.

b. Write the equation in part (a) in slope-intercept form.

c. Use the slope-intercept equation to predict the number of screens, in thousands, in 2008.

76. The scatter plot shows the relationship between the percentage of married women of child-bearing age using contraceptives and the births per woman in selected countries. Also shown is the regression line. Use two points on this line to write both its point-slope and slope-intercept equations. Then find the number of births per woman if 90% of married women of child-bearing age use contraceptives.

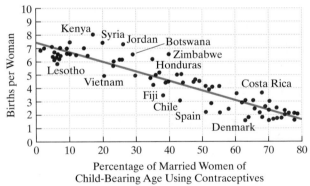

Contraceptive Prevalence and Births per Woman, Selected Countries

Source: Population Reference Bureau

77. Shown, again, at the top of the next column is the scatter plot that indicates a relationship between the percentage of adult females in a country who are literate and the mortality of children under five. Also shown is a line that passes through or near the points. Find a linear equation that models the data by finding the slope-intercept equation of the line. Use the model to make a prediction about child mortality based on the percentage of adult females in a country who are literate.

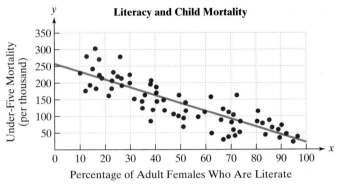

Literacy and Child Mortality

Source: United Nations

In Exercises 78–80, find a linear equation in slope-intercept form that models the given description. Describe what each variable in your model represents. Then use the model to make a prediction.

78. In 1995, the average temperature of Earth was 57.7°F and has increased at a rate of 0.01°F per year since then.

79. In 1995, 60% of U.S. adults read a newspaper and this percentage has decreased at a rate of 0.7% per year since then.

80. A computer that was purchased for $4000 is depreciating at a rate of $950 per year.

81. A business discovers a linear relationship between the number of shirts it can sell and the price per shirt. In particular, 20,000 shirts can be sold at $19 each, and 2000 of the same shirts can be sold at $55 each. Write the slope-intercept equation of the *demand line* that models the number of shirts that can be sold, y, at a price of x dollars. Then determine the number of shirts that can be sold at $50 each.

Writing in Mathematics

82. What is the slope of a line and how is it found?

83. Describe how to write the equation of a line if two points along the line are known.

84. Explain how to derive the slope-intercept form of a line's equation, $y = mx + b$, from the point-slope form
$$y - y_1 = m(x - x_1).$$

85. Explain how to graph the equation $x = 2$. Can this equation be expressed in slope-intercept form? Explain.

86. Explain how to use the general form of a line's equation to find the line's slope and y-intercept.

87. If two lines are parallel, describe the relationship between their slopes.

88. If two lines are perpendicular, describe the relationship between their slopes.

89. If you know a point on a line and you know the equation of a line perpendicular to this line, explain how to write the line's equation.

90. A formula in the form $y = mx + b$ models the cost, y, of a four-year college x years after 2003. Would you expect m to be positive, negative, or zero? Explain your answer.

91. We saw that the percentage of people satisfied with their lives remains relatively constant for all age groups. Exercise 69 showed that the number of skiers in the United States has remained relatively constant over time. Give another example of a real-world phenomenon that has remained relatively constant. Try writing an equation that models this phenomenon.

Technology Exercises

Use a graphing utility to graph each equation in Exercises 92–95. Then use the TRACE *feature to trace along the line and find the coordinates of two points. Use these points to compute the line's slope. Check your result by using the coefficient of x in the line's equation.*

92. $y = 2x + 4$

93. $y = -3x + 6$

94. $y = -\frac{1}{2}x - 5$

95. $y = \frac{3}{4}x - 2$

96. Is there a relationship between alcohol from moderate wine consumption and heart disease death rate? The table gives data from 19 developed countries.

France

Country	A	B	C	D	E	F	G
Liters of alcohol from drinking wine, per person, per year (x)	2.5	3.9	2.9	2.4	2.9	0.8	9.1
Deaths from heart disease, per 100,000 people per year (y)	211	167	131	191	220	297	71

U.S.

Country	H	I	J	K	L	M	N	O	P	Q	R	S
(x)	0.8	0.7	7.9	1.8	1.9	0.8	6.5	1.6	5.8	1.3	1.2	2.7
(y)	211	300	107	167	266	227	86	207	115	285	199	172

Source: New York Times, December 28, 1994

a. Use the statistical menu of your graphing utility to enter the 19 ordered pairs of data items shown in the table.

b. Use the DRAW menu and the scatter plot capability to draw a scatter plot of the data.

c. Select the linear regression option. Use your utility to obtain values for a and b for the equation of the regression line, $y = ax + b$. You may also be given a **correlation coefficient, r**. Values of r close to 1 indicate that the points can be described by a linear relationship and the regression line has a positive slope. Values of r close to -1 indicate that the points can be described by a linear relationship and the regression line has a negative slope. Values of r close to 0 indicate no linear relationship between the variables. In this case, a linear model does not accurately describe the data.

d. Use the appropriate sequence (consult your manual) to graph the regression equation on top of the points in the scatter plot.

Critical Thinking Exercises

97. Which one of the following is true?
a. A linear equation with nonnegative slope has a graph that rises from left to right.
b. The equations $y = 4x$ and $y = -4x$ have graphs that are perpendicular lines.
c. The line whose equation is $5x + 6y - 30 = 0$ passes through the point $(6, 0)$ and has slope $-\frac{5}{6}$.
d. The graph of $y = 7$ in the rectangular coordinate system is the single point $(7, 0)$.

98. Prove that the equation of a line passing through $(a, 0)$ and $(0, b)$ $(a \neq 0, b \neq 0)$ can be written in the form $\frac{x}{a} + \frac{y}{b} = 1$. Why is this called the *intercept form* of a line?

99. Use the figure shown to make the following lists.
a. List the slopes $m_1, m_2, m_3,$ and m_4 in order of decreasing size.
b. List the y-intercepts $b_1, b_2, b_3,$ and b_4 in order of decreasing size.

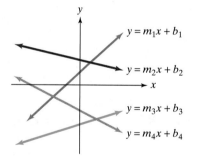

100. Excited about the success of celebrity stamps, post office officials were rumored to have put forth a plan to institute two new types of thermometers. On these new scales, $°E$ represents degrees Elvis and $°M$ represents degrees Madonna. If it is known that $40°E = 25°M$, $280°E = 125°M$, and degrees Elvis is linearly related to degrees Madonna, write an equation expressing E in terms of M.

Group Exercise

101. Group members should consult an almanac, newspaper, magazine, or the Internet to find data that lie approximately on or near a straight line. Working by hand or using a graphing utility, construct a scatter plot for the data. If working by hand, draw a line that approximately fits the data and then write its equation. If using a graphing utility, obtain the equation of the regression line. Then use the equation of the line to make a prediction about what might happen in the future. Are there circumstances that might affect the accuracy of this prediction? List some of these circumstances.

SECTION 1.3 *Distance and Midpoint Formulas; Circles*

Objectives

1. Find the distance between two points.
2. Find the midpoint of a line segment.
3. Write the standard form of a circle's equation.
4. Give the center and radius of a circle whose equation is in standard form.
5. Convert the general form of a circle's equation to standard form.

It's a good idea to know your way around a circle. Clocks, angles, maps, and compasses are based on circles. Circles occur everywhere in nature: in ripples on water, patterns on a butterfly's wings, and cross sections of trees. Some consider the circle to be the most pleasing of all shapes.

The rectangular coordinate system gives us a unique way of knowing a circle. It enables us to translate a circle's geometric definition into an algebraic equation. To do this, we must first develop a formula for the distance between any two points in rectangular coordinates.

---

1 Find the distance between two points.

The Distance Formula

Using the Pythagorean Theorem, we can find the distance between the two points $P_1(x_1, y_1)$ and $P_2(x_2, y_2)$ in the rectangular coordinate system. The two points are illustrated in Figure 1.24.

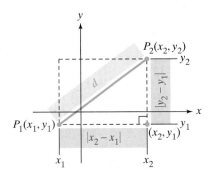

Figure 1.24

The distance that we need to find is represented by d and shown in blue. Notice that the distance between two points on the dashed horizontal line is the absolute value of the difference between the x-coordinates of the two points. This distance, $|x_2 - x_1|$, is shown in pink. Similarly, the distance between two points on the dashed vertical line is the absolute value of the difference between the y-coordinates of the two points. This distance, $|y_2 - y_1|$, is also shown in pink.

Because the dashed lines are horizontal and vertical, a right triangle is formed. Thus, we can use the Pythagorean Theorem to find distance d. By the Pythagorean Theorem,

$$d^2 = |x_2 - x_1|^2 + |y_2 - y_1|^2$$
$$d = \sqrt{|x_2 - x_1|^2 + |y_2 - y_1|^2}$$
$$d = \sqrt{(x_2 - x_1)^2 + (y_2 - y_1)^2}.$$

This result is called the **distance formula.**

The Distance Formula

The distance, d, between the points (x_1, y_1) and (x_2, y_2) in the rectangular coordinate system is

$$d = \sqrt{(x_2 - x_1)^2 + (y_2 - y_1)^2}.$$

When using the distance formula, it does not matter which point you call (x_1, y_1) and which you call (x_2, y_2).

EXAMPLE 1   Using the Distance Formula

Find the distance between $(-1, -3)$ and $(2, 3)$.

Solution   Letting $(x_1, y_1) = (-1, -3)$ and $(x_2, y_2) = (2, 3)$, we obtain

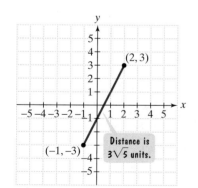

Figure 1.25   Finding the distance between two points

$d = \sqrt{(x_2 - x_1)^2 + (y_2 - y_1)^2}$	Use the distance formula.
$= \sqrt{[2 - (-1)]^2 + [3 - (-3)]^2}$	Substitute the given values.
$= \sqrt{(2 + 1)^2 + (3 + 3)^2}$	Apply the definition of subtraction within the grouping symbols.
$= \sqrt{3^2 + 6^2}$	Perform the resulting additions.
$= \sqrt{9 + 36}$	Square 3 and 6.
$= \sqrt{45}$	Add.
$= 3\sqrt{5} \approx 6.71.$	$\sqrt{45} = \sqrt{9 \cdot 5} = \sqrt{9}\sqrt{5} = 3\sqrt{5}$

The distance between the given points is $3\sqrt{5}$ units, or approximately 6.71 units. The situation is illustrated in Figure 1.25.

Check Point 1   Find the distance between $(2, -2)$ and $(5, 2)$.

2 Find the midpoint of a line segment.

The Midpoint Formula

The distance formula can be used to derive a formula for finding the midpoint of a line segment between two given points. The formula is given as follows:

The Midpoint Formula

Consider a line segment whose endpoints are (x_1, y_1) and (x_2, y_2). The coordinates of the segment's midpoint are

$$\left(\frac{x_1 + x_2}{2}, \frac{y_1 + y_2}{2} \right).$$

To find the midpoint, take the average of the two x-coordinates and the average of the two y-coordinates.

EXAMPLE 2   Using the Midpoint Formula

Find the midpoint of the line segment with endpoints $(1, -6)$ and $(-8, -4)$.

Solution   To find the coordinates of the midpoint, we average the coordinates of the endpoints.

$$\text{Midpoint} = \left(\frac{1 + (-8)}{2}, \frac{-6 + (-4)}{2} \right) = \left(\frac{-7}{2}, \frac{-10}{2} \right) = \left(-\frac{7}{2}, -5 \right)$$

Average the x-coordinates.   **Average the y-coordinates.**

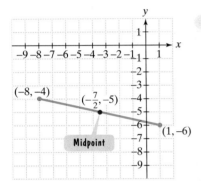

Figure 1.26   Finding a line segment's midpoint

Figure 1.26 illustrates that the point $(-\frac{7}{2}, -5)$ is midway between the points $(1, -6)$ and $(-8, -4)$.

Check Point 2   Find the midpoint of the line segment with endpoints $(1, 2)$ and $(7, -3)$.

Circles

Our goal is to translate a circle's geometric definition into an equation. We begin with this geometric definition.

Definition of a Circle

A **circle** is the set of all points in a plane that are equidistant from a fixed point, called the **center.** The fixed distance from the circle's center to any point on the circle is called the **radius.**

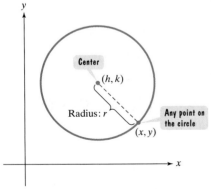

Figure 1.27   A circle centered at (h, k) with radius r

Figure 1.27 is our starting point for obtaining a circle's equation. We've placed the circle into a rectangular coordinate system. The circle's center is (h, k) and its radius is r. We let (x, y) represent the coordinates of any point on the circle.

What does the geometric definition of a circle tell us about point (x, y) in Figure 1.27? The point is on the circle if and only if its distance from the center is r. We can use the distance formula to express this idea algebraically:

The distance between (x, y) and (h, k)   **is always**   **r.**

$$\sqrt{(x - h)^2 + (y - k)^2} \qquad = \qquad r$$

Squaring both sides of $\sqrt{(x-h)^2 + (y-k)^2} = r$ yields the *standard form of the equation of a circle*.

The Standard Form of the Equation of a Circle

The **standard form of the equation of a circle** with center (h, k) and radius r is

$$(x-h)^2 + (y-k)^2 = r^2.$$

---

3 Write the standard form of a circle's equation.

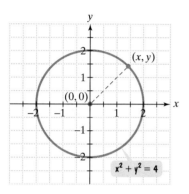

Figure 1.28 The graph of $x^2 + y^2 = 4$

EXAMPLE 3 Finding the Standard Form of a Circle's Equation

Write the standard form of the equation of the circle with center $(0, 0)$ and radius 2. Graph the circle.

Solution   The center is $(0, 0)$. Because the center is represented as (h, k) in the standard form of the equation, $h = 0$ and $k = 0$. The radius is 2, so we will let $r = 2$ in the equation.

$(x-h)^2 + (y-k)^2 = r^2$ *This is the standard form of a circle's equation.*

$(x-0)^2 + (y-0)^2 = 2^2$ *Substitute 0 for h, 0 for k, and 2 for r.*

$\quad\quad\quad x^2 + y^2 = 4$ *Simplify.*

The standard form of the equation of the circle is $x^2 + y^2 = 4$. Figure 1.28 shows the graph.

Check Point 3   Write the standard form of the equation of the circle with center $(0,0)$ and radius 4.

Technology

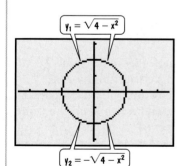

To graph a circle with a graphing utility, first solve the equation for y.

$$x^2 + y^2 = 4$$
$$y^2 = 4 - x^2$$
$$y = \pm\sqrt{4 - x^2}$$

Graph the two equations

$$y_1 = \sqrt{4 - x^2} \quad \text{and} \quad y_2 = -\sqrt{4 - x^2}$$

in the same viewing rectangle. The graph of $y_1 = \sqrt{4 - x^2}$ is the top semicircle because y is always positive. The graph of $y_2 = -\sqrt{4 - x^2}$ is the bottom semicircle because y is always negative. Use a $\boxed{\text{ZOOM SQUARE}}$ setting so that the circle looks like a circle. (Many graphing utilities have problems connecting the two semicircles because the segments directly across from the center become nearly vertical.)

Example 3 and Check Point 3 involved circles centered at the origin. The standard form of the equation of all such circles is $x^2 + y^2 = r^2$, where r is the circle's radius. Now, let's consider a circle whose center is not at the origin.

EXAMPLE 4   Finding the Standard Form of a Circle's Equation

Write the standard form of the equation of the circle with center $(-2, 3)$ and radius 4.

Solution   The center is $(-2, 3)$. Because the center is represented as (h, k) in the standard form of the equation, $h = -2$ and $k = 3$. The radius is 4, so we will let $r = 4$ in the equation.

$$(x - h)^2 + (y - k)^2 = r^2 \quad \text{This is the standard form of a circle's equation.}$$
$$[x - (-2)]^2 + (y - 3)^2 = 4^2 \quad \text{Substitute } -2 \text{ for h, 3 for k, and 4 for r.}$$
$$(x + 2)^2 + (y - 3)^2 = 16 \quad \text{Simplify.}$$

The standard form of the equation of the circle is $(x + 2)^2 + (y - 3)^2 = 16$.

Check Point 4   Write the standard form of the equation of the circle with center $(5, -6)$ and radius 10.

④ Give the center and radius of a circle whose equation is in standard form.

EXAMPLE 5   Using the Standard Form of a Circle's Equation to Graph the Circle

Find the center and radius of the circle whose equation is

$$(x - 2)^2 + (y + 4)^2 = 9$$

and graph the equation.

Solution   In order to graph the circle, we need to know its center, (h, k), and its radius, r. We can find the values for h, k, and r by comparing the given equation to the standard form of the equation of a circle.

$$(x - 2)^2 + (y + 4)^2 = 9$$
$$(x - 2)^2 + [y - (-4)]^2 = 3^2$$

This is $(x - h)^2$, with $h = 2$.   This is $(y - k)^2$, with $k = -4$.   This is r^2, with $r = 3$.

We see that $h = 2, k = -4$, and $r = 3$. Thus, the circle has center $(h, k) = (2, -4)$ and a radius of 3 units. To graph this circle, first plot the center $(2, -4)$. Because the radius is 3, you can locate at least four points on the circle by going out three units to the right, to the left, up, and down from the center.

The points three units to the right and to the left of $(2, -4)$ are $(5, -4)$ and $(-1, -4)$, respectively. The points three units up and down from $(2, -4)$ are $(2, -1)$ and $(2, -7)$, respectively.

Using these points, we obtain the graph in Figure 1.29.

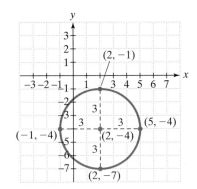

Figure 1.29   The graph of $(x - 2)^2 + (y + 4)^2 = 9$

Check Point 5   Find the center and radius of the circle whose equation is

$$(x + 3)^2 + (y - 1)^2 = 4$$

and graph the equation.

If we square $x - 2$ and $y + 4$ in the standard form of the equation from Example 5, we obtain another form for the circle's equation.

$$(x - 2)^2 + (y + 4)^2 = 9$$ *This is the standard form of the equation from Example 5.*

$$x^2 - 4x + 4 + y^2 + 8y + 16 = 9$$ *Square x − 2 and y + 4.*

$$x^2 + y^2 - 4x + 8y + 20 = 9$$ *Combine numerical terms and rearrange terms.*

$$x^2 + y^2 - 4x + 8y + 11 = 0$$ *Subtract 9 from both sides.*

This result suggests that an equation in the form $x^2 + y^2 + Dx + Ey + F = 0$ can represent a circle. This is called the *general form of the equation of a circle*.

> **The General Form of the Equation of a Circle**
> The **general form of the equation of a circle** is
> $$x^2 + y^2 + Dx + Ey + F = 0.$$

5 Convert the general form of a circle's equation to standard form.

We can convert the general form of the equation of a circle to the standard form $(x - h)^2 + (y - k)^2 = r^2$. We do so by completing the square on x and y. Let's see how this is done.

EXAMPLE 6   **Converting the General Form of a Circle's Equation to Standard Form and Graphing the Circle**

Write in standard form and graph: $x^2 + y^2 + 4x - 6y - 23 = 0$.

Solution   Because we plan to complete the square on both x and y, let's rearrange terms so that x-terms are arranged in descending order, y-terms are arranged in descending order, and the constant term appears on the right.

$$x^2 + y^2 + 4x - 6y - 23 = 0$$ *This is the given equation.*

$$(x^2 + 4x \quad) + (y^2 - 6y \quad) = 23$$ *Rewrite in anticipation of completing the square.*

$$(x^2 + 4x + 4) + (y^2 - 6y + 9) = 23 + 4 + 9$$ *Complete the square on x: $\frac{1}{2} \cdot 4 = 2$ and $2^2 = 4$, so add 4 to both sides. Complete the square on y: $\frac{1}{2}(-6) = -3$ and $(-3)^2 = 9$, so add 9 to both sides.*

> Remember that numbers added on the left side must also be added on the right side.

$$(x + 2)^2 + (y - 3)^2 = 36$$ *Factor on the left and add on the right.*

This last equation is in standard form. We can identify the circle's center and radius by comparing this equation to the standard form of the equation of a circle, $(x - h)^2 + (y - k)^2 = r^2$.

$$(x + 2)^2 + (y - 3)^2 = 36$$
$$[x - (-2)]^2 + (y - 3)^2 = 6^2$$

> This is $(x - h)^2$, with $h = -2$.
> This is $(y - k)^2$, with $k = 3$.
> This is r^2, with $r = 6$.

We use the center, $(h, k) = (-2, 3)$, and the radius, $r = 6$, to graph the circle. The graph is shown in Figure 1.30.

Study Tip

To review completing the square, see Section P.8, pages 89–90.

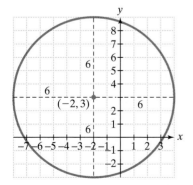

Figure 1.30 The graph of $(x + 2)^2 + (y - 3)^2 = 36$

Technology

To graph $x^2 + y^2 + 4x - 6y - 23 = 0$, rewrite the equation as a quadratic equation in y.

$$y^2 - 6y + (x^2 + 4x - 23) = 0$$

Now solve for y using the quadratic formula, with $a = 1$, $b = -6$, and $c = x^2 + 4x - 23$.

$$y = \frac{-b \pm \sqrt{b^2 - 4ac}}{2a} = \frac{-(-6) \pm \sqrt{(-6)^2 - 4 \cdot 1(x^2 + 4x - 23)}}{2 \cdot 1} = \frac{6 \pm \sqrt{36 - 4(x^2 + 4x - 23)}}{2}$$

Because we will enter these equations, there is no need to simplify. Enter

$$y_1 = \frac{6 + \sqrt{36 - 4(x^2 + 4x - 23)}}{2}$$

and

$$y_2 = \frac{6 - \sqrt{36 - 4(x^2 + 4x - 23)}}{2}.$$

Use a $\boxed{\text{ZOOM SQUARE}}$ setting. The graph is shown on the right.

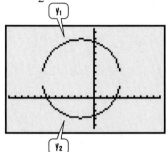

Check Point 6

Write in standard form and graph:

$$x^2 + y^2 + 4x - 4y - 1 = 0.$$

EXERCISE SET 1.3

Practice Exercises

In Exercises 1–18, find the distance between each pair of points. If necessary, round answers to two decimals places.

1. $(2, 3)$ and $(14, 8)$
2. $(5, 1)$ and $(8, 5)$
3. $(4, 1)$ and $(6, 3)$
4. $(2, 3)$ and $(3, 5)$
5. $(0, 0)$ and $(-3, 4)$
6. $(0, 0)$ and $(3, -4)$
7. $(-2, -6)$ and $(3, -4)$
8. $(-4, -1)$ and $(2, -3)$
9. $(0, -3)$ and $(4, 1)$
10. $(0, -2)$ and $(4, 3)$
11. $(3.5, 8.2)$ and $(-0.5, 6.2)$
12. $(2.6, 1.3)$ and $(1.6, -5.7)$
13. $(0, -\sqrt{3})$ and $(\sqrt{5}, 0)$
14. $(0, -\sqrt{2})$ and $(\sqrt{7}, 0)$
15. $(3\sqrt{3}, \sqrt{5})$ and $(-\sqrt{3}, 4\sqrt{5})$
16. $(2\sqrt{3}, \sqrt{6})$ and $(-\sqrt{3}, 5\sqrt{6})$
17. $\left(\frac{7}{3}, \frac{1}{5}\right)$ and $\left(\frac{1}{3}, \frac{6}{5}\right)$
18. $\left(-\frac{1}{4}, -\frac{1}{7}\right)$ and $\left(\frac{3}{4}, \frac{6}{7}\right)$

In Exercises 19–30, find the midpoint of each line segment with the given endpoints.

19. $(6, 8)$ and $(2, 4)$
20. $(10, 4)$ and $(2, 6)$
21. $(-2, -8)$ and $(-6, -2)$
22. $(-4, -7)$ and $(-1, -3)$
23. $(-3, -4)$ and $(6, -8)$
24. $(-2, -1)$ and $(-8, 6)$
25. $\left(-\frac{7}{2}, \frac{3}{2}\right)$ and $\left(-\frac{5}{2}, -\frac{11}{2}\right)$
26. $\left(-\frac{2}{5}, \frac{7}{15}\right)$ and $\left(-\frac{2}{5}, -\frac{4}{15}\right)$
27. $(8, 3\sqrt{5})$ and $(-6, 7\sqrt{5})$
28. $(7\sqrt{3}, -6)$ and $(3\sqrt{3}, -2)$
29. $(\sqrt{18}, -4)$ and $(\sqrt{2}, 4)$
30. $(\sqrt{50}, -6)$ and $(\sqrt{2}, 6)$

In Exercises 31–40, write the standard form of the equation of the circle with the given center and radius.

31. Center $(0, 0)$, $r = 7$
32. Center $(0, 0)$, $r = 8$
33. Center $(3, 2)$, $r = 5$
34. Center $(2, -1)$, $r = 4$
35. Center $(-1, 4)$, $r = 2$

36. Center $(-3, 5)$, $r = 3$

37. Center $(-3, 1)$, $r = \sqrt{3}$

38. Center $(-5, -3)$, $r = \sqrt{5}$

39. Center $(-4, 0)$, $r = 10$

40. Center $(-2, 0)$, $r = 6$

In Exercises 41–48, give the center and radius of the circle described by the equation and graph each equation.

41. $x^2 + y^2 = 16$

42. $x^2 + y^2 = 49$

43. $(x - 3)^2 + (y - 1)^2 = 36$

44. $(x - 2)^2 + (y - 3)^2 = 16$

45. $(x + 3)^2 + (y - 2)^2 = 4$

46. $(x + 1)^2 + (y - 4)^2 = 25$

47. $(x + 2)^2 + (y + 2)^2 = 4$

48. $(x + 4)^2 + (y + 5)^2 = 36$

In Exercises 49–56, complete the square and write the equation in standard form. Then give the center and radius of each circle and graph the equation.

49. $x^2 + y^2 + 6x + 2y + 6 = 0$

50. $x^2 + y^2 + 8x + 4y + 16 = 0$

51. $x^2 + y^2 - 10x - 6y - 30 = 0$

52. $x^2 + y^2 - 4x - 12y - 9 = 0$

53. $x^2 + y^2 + 8x - 2y - 8 = 0$

54. $x^2 + y^2 + 12x - 6y - 4 = 0$

55. $x^2 - 2x + y^2 - 15 = 0$

56. $x^2 + y^2 - 6y - 7 = 0$

Application Exercises

57. A rectangular coordinate system with coordinates in miles is placed on the map in the figure shown. Bangkok has coordinates $(-115, 170)$ and Phnom Penh has coordinates $(65, 70)$. How long will it take a plane averaging 400 miles per hour to fly directly from one city to the other? Round to the nearest tenth of an hour. Approximately how many minutes is the flight?

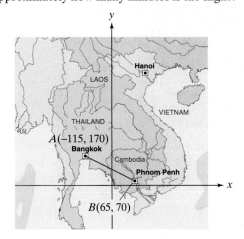

58. We refer to the driveway in the figure shown as being *circular*, meaning that it is bounded by two circles. The figure indicates that the radius of the larger circle is 52 feet and the radius of the smaller circle is 38 feet.

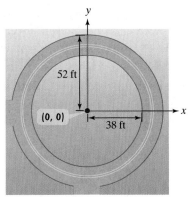

a. Use the coordinate system shown to write the equation of the smaller circle.

b. Use the coordinate system shown to write the equation of the larger circle.

59. The ferris wheel in the figure has a radius of 68 feet. The clearance between the wheel and the ground is 14 feet. The rectangular coordinate system shown has its origin on the ground directly below the center of the wheel. Use the coordinate system to write the equation of the circular wheel.

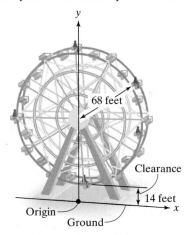

60. The circle formed by the middle lane of a circular running track can be described algebraically by $x^2 + y^2 = 4$, where all measurements are in miles. If you run around the track's middle lane twice, approximately how many miles have you covered?

Writing in Mathematics

61. In your own words, describe how to find the distance between two points in the rectangular coordinate system.

62. In your own words, describe how to find the midpoint of a line segment if its endpoints are known.

63. What is a circle? Without using variables, describe how the definition of a circle can be used to obtain a form of its equation.

64. Give an example of a circle's equation in standard form. Describe how to find the center and radius for this circle.

65. How is the standard form of a circle's equation obtained from its general form?

66. Does $(x - 3)^2 + (y - 5)^2 = 0$ represent the equation of a circle? If not, describe the graph of this equation.

67. Does $(x - 3)^2 + (y - 5)^2 = -25$ represent the equation of a circle? What sort of set is the graph of this equation?

 Technology Exercises

In Exercises 68–70, use a graphing utility to graph each circle whose equation is given.

68. $x^2 + y^2 = 25$

69. $(y + 1)^2 = 36 - (x - 3)^2$

70. $x^2 + 10x + y^2 - 4y - 20 = 0$

Critical Thinking Exercises

71. Which one of the following is true?

a. The equation of the circle whose center is at the origin with radius 16 is $x^2 + y^2 = 16$.

b. The graph of $(x - 3)^2 + (y + 5)^2 = 36$ is a circle with radius 6 centered at $(-3, 5)$.

c. The graph of $(x - 4) + (y + 6) = 25$ is a circle with radius 5 centered at $(4, -6)$.

d. None of the above is true.

72. Show that the points $A(1, 1 + d)$, $B(3, 3 + d)$, and $C(6, 6 + d)$ are collinear (lie along a straight line) by showing that the distance from A to B plus the distance from B to C equals the distance from A to C.

73. Prove the midpoint formula by using the following procedure.

a. Show that the distance between (x_1, y_1) and $\left(\dfrac{x_1 + x_2}{2}, \dfrac{y_1 + y_2}{2}\right)$ is equal to the distance between (x_2, y_2) and $\left(\dfrac{x_1 + x_2}{2}, \dfrac{y_1 + y_2}{2}\right)$.

b. Use the procedure from Exercise 72 and the distances from part (a) to show that the points (x_1, y_1), $\left(\dfrac{x_1 + x_2}{2}, \dfrac{y_1 + y_2}{2}\right)$, and (x_2, y_2) are collinear.

In Exercises 74–75, write the standard form and the general form of the equation of each circle.

74. Center at $(3, -5)$ and passing through the point $(-2, 1)$

75. Passing through $(-7, 2)$ and $(1, 2)$; these points are endpoints of the diameter, the line that passes through the circle's center.

76. Find the area of the donut-shaped region bounded by the graphs of $(x - 2)^2 + (y + 3)^2 = 25$ and $(x - 2)^2 + (y + 3)^2 = 36$.

77. A **tangent line** to a circle is a line that intersects the circle at exactly one point. The tangent line is perpendicular to the radius of the circle at this point of contact. Write the point-slope equation of a line tangent to the circle whose equation is $x^2 + y^2 = 25$ at the point $(3, -4)$.

SECTION 1.4 *Basics of Functions*

Objectives

1. Find the domain and range of a relation.
2. Determine whether a relation is a function.
3. Determine whether an equation represents a function.
4. Evaluate a function.
5. Find and simplify a function's difference quotient.
6. Understand and use piecewise functions.
7. Find the domain of a function.

Jerry Orbach	$34,000
Charles Shaugnessy	$31,800
Andy Richter	$29,400
Norman Schwarzkopf	$28,000
Jon Stewart	$28,000

The answer: See the above list. The question: Who are *Celebrity Jeopardy's* five all-time highest earners? The list indicates a correspondence between the five all-time highest earners and their winnings. We can write this correspondence using a set of ordered pairs:

$$\{(\text{Orbach}, \$34{,}000), (\text{Shaugnessy}, \$31{,}800), (\text{Richter}, \$29{,}400),$$
$$(\text{Schwarzkopf}, \$28{,}000), (\text{Stewart}, \$28{,}000)\}.$$

1 Find the domain and range of a relation.

The mathematical term for a set of ordered pairs is a *relation*.

Definition of a Relation

A **relation** is any set of ordered pairs. The set of all first components of the ordered pairs is called the **domain** of the relation, and the set of all second components is called the **range** of the relation.

EXAMPLE 1   Finding the Domain and Range of a Relation

Find the domain and range of the relation:

$$\{(\text{Orbach, \$34,000}), (\text{Shaugnessy, \$31,800}), (\text{Richter, \$29,400}),$$
$$(\text{Schwarzkopf, \$28,000}), (\text{Stewart, \$28,000})\}.$$

Solution   The domain is the set of all first components. Thus, the domain is

$$\{\text{Orbach, Shaugnessy, Richter, Schwarzkopf, Stewart}\}.$$

The range is the set of all second components. Thus, the range is

$$\{\$34,000, \$31,800, \$29,400, \$28,000\}.$$

Check Point 1   Find the domain and the range of the relation:

$$\{(5, 12.8), (10, 16.2), (15, 18.9), (20, 20.7), (25, 21.8)\}.$$

As you worked Check Point 1 did you wonder if there was a rule that assigned the "inputs" in the domain to the "outputs" in the range? For example, for the ordered pair (15, 18.9), how does the output 18.9 depend on the input 15? Think paid vacation days! The first number in each ordered pair is the number of years a full-time employee has been employed by a medium to large U.S. company. The second number is the average number of paid vacation days each year. Consider, for example, the ordered pair (15, 18.9).

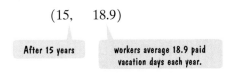

(15,   18.9)

After 15 years   workers average 18.9 paid vacation days each year.

The relation in the vacation-days example of Check Point 1 can be pictured as follows:

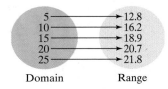

Domain   Range

Average Number of Paid Vacation Days for Full-Time Workers at Medium to Large U.S. Companies

Paid Vacation Days

Years Working for Company

Figure 1.31   The graph of a relation showing a correspondence between years with a company and paid vacation days
Source: Bureau of Labor Statistics

A scatter plot, like the one shown in Figure 1.31, is another way to represent the relation.

2 Determine whether a relation is a function.

Jerry Orbach — $34,000
Charles Shaugnessy — $31,800
Andy Richter — $29,400
Norman Schwarzkopf — $28,000
Jon Stewart — $28,000

Functions

Shown, again, in the margin are *Celebrity Jeopardy's* five all-time highest winners and their winnings. We've used this information to define two relations. Figure 1.32(a) shows a correspondence between winners and their winnings. Figure 1.32(b) shows a correspondence between winnings and winners.

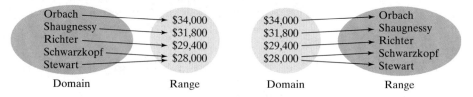

Figure 1.32(a)
Winners correspond to winnings.

Figure 1.32(b)
Winnings correspond to winners.

A relation in which each member of the domain corresponds to exactly one member of the range is a **function.** Can you see that the relation in Figure 1.32(a) is a function? Each winner in the domain corresponds to exactly one winning amount in the range. If we know the winner, we can be sure of the amount won. Notice that more than one element in the domain can correspond to the same element in the range. (Schwarzkopf and Stewart both won $28,000.)

Is the relation in Figure 1.32(b) a function? Does each member of the domain correspond to precisely one member of the range? This relation is not a function because there is a member of the domain that corresponds to two members of the range:

($28,000, Schwarzkopf)   ($28,000, Stewart).

The member of the domain, $28,000, corresponds to both Schwarzkopf and Stewart in the range. If we know the amount won, $28,000, we cannot be sure of the winner. Because **a function is a relation in which no two ordered pairs have the same first component and different second components,** the ordered pairs ($28,000, Schwarzkopf) and ($28,000, Stewart) are not ordered pairs of a function.

Same first component

($28,000, Schwarzkopf)   ($28,000, Stewart)

Different second components

Definition of a Function

A **function** is a correspondence from a first set, called the **domain,** to a second set, called the **range,** such that each element in the domain corresponds to *exactly one* element in the range.

Example 2 illustrates that not every correspondence between sets is a function.

EXAMPLE 2   Determining Whether a Relation Is a Function

Determine whether each relation is a function:

    a. $\{(1, 6), (2, 6), (3, 8), (4, 9)\}$ **b.** $\{(6, 1), (6, 2), (8, 3), (9, 4)\}$.

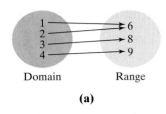

(a)

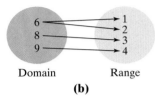

(b)

Figure 1.33

Study Tip

The word "range" can mean many things, from a chain of mountains to a cooking stove. For functions, it means the set of all function values. For graphing utilities, it means the setting used for the viewing rectangle. Try not to confuse these meanings.

Solution We will make a figure for each relation that shows the domain and the range.

a. We begin with the relation $\{(1, 6), (2, 6), (3, 8), (4, 9)\}$. Figure 1.33(a) shows that every element in the domain corresponds to exactly one element in the range. The element 1 in the domain corresponds to the element 6 in the range. Furthermore, 2 corresponds to 6, 3 corresponds to 8, and 4 corresponds to 9. No two ordered pairs in the given relation have the same first component and different second components. Thus, the relation is a function.

b. We now consider the relation $\{(6, 1), (6, 2), (8, 3), (9, 4)\}$. Figure 1.33(b) shows that 6 corresponds to both 1 and 2. If any element in the domain corresponds to more than one element in the range, the relation is not a function. This relation is not a function; two ordered pairs have the same first component and different second components.

Same first components

$(6, 1) \qquad (6, 2)$

Different second components

Look at Figure 1.33(a) again. The fact that 1 and 2 in the domain have the same image, 6, in the range does not violate the definition of a function. **A function can have two different first components with the same second component.** By contrast, a relation is not a function when two different ordered pairs have the same first component and different second components. Thus, the relation pictured in Figure 1.33(b) is not a function.

Check Point 2 Determine whether each relation is a function:

a. $\{(1, 2), (3, 4), (5, 6), (5, 8)\}$
b. $\{(1, 2), (3, 4), (6, 5), (8, 5)\}$.

3 Determine whether an equation represents a function.

Functions as Equations

Functions are usually given in terms of equations rather than as sets of ordered pairs. For example, here is an equation that models paid vacation days each year as a function of years working for a company:

$$y = -0.016x^2 + 0.93x + 8.5.$$

The variable x represents years working for a company. The variable y represents the average number of vacation days each year. The variable y is a function of the variable x. For each value of x, there is one and only one value of y. The variable x is called the **independent variable** because it can be assigned any value from the domain. Thus, x can be assigned any positive integer representing the number of years working for a company. The variable y is called the **dependent variable** because its value depends on x. Paid vacation days depend on years working for a company. The value of the dependent variable, y, is calculated after selecting a value for the independent variable, x.

We have seen that not every set of ordered pairs defines a function. Similarly, not all equations with the variables x and y define a function. If an equation is solved for y and more than one value of y can be obtained for a given x, then the equation does not define y as a function of x.

EXAMPLE 3 Determining Whether an Equation Represents a Function

Determine whether each equation defines y as a function of x:

a. $x^2 + y = 4$ **b.** $x^2 + y^2 = 4$.

Solution Solve each equation for y in terms of x. If two or more values of y can be obtained for a given x, the equation is not a function.

a.

$x^2 + y = 4$	This is the given equation.
$x^2 + y - x^2 = 4 - x^2$	Solve for y by subtracting x^2 from both sides.
$y = 4 - x^2$	Simplify.

From this last equation we can see that for each value of x, there is one and only one value of y. For example, if $x = 1$, then $y = 4 - 1^2 = 3$. The equation defines y as a function of x.

b.

$x^2 + y^2 = 4$	This given equation describes a circle.
$x^2 + y^2 - x^2 = 4 - x^2$	Isolate y^2 by subtracting x^2 from both sides.
$y^2 = 4 - x^2$	Simplify.
$y = \pm\sqrt{4 - x^2}$	Apply the square root method.

The $\pm$ in this last equation shows that for certain values of x (all values between -2 and 2), there are two values of y. For example, if $x = 1$, then $y = \pm\sqrt{4 - 1^2} = \pm\sqrt{3}$. For this reason, the equation does not define y as a function of x.

Check Point 3 Solve each equation for y and then determine whether the equation defines y as a function of x:

a. $2x + y = 6$ **b.** $x^2 + y^2 = 1$.

4 Evaluate a function.

Function Notation

When an equation represents a function, the function is often named by a letter such as f, g, h, F, G, or H. Any letter can be used to name a function. Suppose that f names a function. Think of the domain as the set of the function's inputs and the range as the set of the function's outputs. As shown in Figure 1.34, the input is represented by x and the output by $f(x)$. The special notation $f(x)$, read "f of x" or "f at x," represents the **value of the function at the number x.**

Study Tip

The notation $f(x)$ does *not* mean "f times x." The notation describes the value of the function at x.

Input x

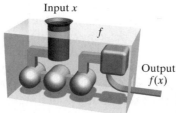

Output $f(x)$

Figure 1.34 A function as a machine with inputs and outputs

Let's make this clearer by considering a specific example. We know that the equation

$$y = -0.016x^2 + 0.93x + 8.5$$

defines y as a function of x. We'll name the function f. Now, we can apply our new function notation.

Input	Output	Equation	We read this equation as "f of x equals $-0.016x^2 + 0.93x + 8.5$."
x	$f(x)$	$f(x) = -0.016x^2 + 0.93x + 8.5$	

Technology

Graphing utilities can be used to evaluate functions. The screens below show the evaluation of

$$f(x) = -0.016x^2 + 0.93x + 8.5$$

at 10 on a T1-83 graphing calculator. The function f is named Y_1.

```
Plot1  Plot2  Plot3
\Y1 = -.016X²+0.93
X+8.5■
·.Y2 =
·.Y3 =
·.Y4 =
·.Y5 =
·.Y6 =
```

```
Y1(10)
                16.2
```

Suppose we are interested in finding $f(10)$, the function's output when the input is 10. To find the value of the function at 10, we substitute 10 for x. We are **evaluating the function** at 10.

$f(x) = -0.016x^2 + 0.93x + 8.5$	This is the given function.
$f(10) = -0.016(10)^2 + 0.93(10) + 8.5$	Replace each occurrence of x with 10.
$= -0.016(100) + 0.93(10) + 8.5$	Evaluate the exponential expression: $10^2 = 100$.
$= -1.6 + 9.3 + 8.5$	Perform the multiplications.
$= 16.2$	Add from left to right.

The statement $f(10) = 16.2$, read "f of 10 equals 16.2," tells us that the value of the function at 10 is 16.2. When the function's input is 10, its output is 16.2 (After 10 years, workers average 16.2 vacation days each year.) To find other function values, such as $f(15)$, $f(20)$, or $f(23)$, substitute the specified input values for x into the function's equation.

If a function is named f and x represents the independent variable, the notation $f(x)$ corresponds to the y-value for a given x. Thus,

$$f(x) = -0.016x^2 + 0.93x + 8.5 \text{ and } y = -0.016x^2 + 0.93x + 8.5$$

define the same function. This function may be written as

$$y = f(x) = -0.016x^2 + 0.93x + 8.5.$$

EXAMPLE 4   Evaluating a Function

If $f(x) = x^2 + 3x + 5$, evaluate:

a. $f(2)$ **b.** $f(x + 3)$ **c.** $f(-x)$.

Solution   We substitute 2, $x + 3$, and $-x$ for x in the definition of f. When replacing x with a variable or an algebraic expression, you might find it helpful to think of the function's equation as

$$f(\boxed{x}) = \boxed{x}^2 + 3\boxed{x} + 5.$$

a. We find $f(2)$ by substituting 2 for x in the equation.

$$f(2) = 2^2 + 3 \cdot 2 + 5 = 4 + 6 + 5 = 15$$

Thus, $f(2) = 15$.

b. We find $f(x + 3)$ by substituting $x + 3$ for x in the equation.

$$f(x + 3) = (x + 3)^2 + 3(x + 3) + 5$$

Equivalently,

$$f(x + 3) = (x + 3)^2 + 3(x + 3) + 5$$
$$= x^2 + 6x + 9 + 3x + 9 + 5 \quad \text{Square } x + 3 \text{ using}$$
$$(A + B)^2 = A^2 + 2AB + B^2.$$
$$\text{Distribute 3 throughout the}$$
$$\text{parentheses.}$$
$$= x^2 + 9x + 23. \quad \text{Combine like terms.}$$

Discovery

Using $f(x) = x^2 + 3x + 5$
and the answers in parts (b)
and (c):
1. Is $f(x + 3)$ equal to
   $f(x) + f(3)$?
2. Is $f(-x)$ equal to $-f(x)$?

c. We find $f(-x)$ by substituting $-x$ for x in the equation $f(x) = x^2 + 3x + 5$.

$$f(-x) = (-x)^2 + 3(-x) + 5$$

Equivalently,

$$f(-x) = (-x)^2 + 3(-x) + 5$$
$$= x^2 - 3x + 5.$$

Check Point 4 If $f(x) = x^2 - 2x + 7$, evaluate:

    a. $f(-5)$ **b.** $f(x + 4)$ **c.** $f(-x)$.

⑤ Find and simplify a function's difference quotient.

Functions and Difference Quotients

We have seen how slope can be interpreted as a rate of change. In the next section, we will be studying the average rate of change of a function. A ratio, called the *difference quotient*, plays an important role in understanding the rate at which functions change.

> **Definition of a Difference Quotient**
> The expression
> $$\frac{f(x + h) - f(x)}{h}$$
> for $h \neq 0$ is called the **difference quotient.**

EXAMPLE 5   Evaluating and Simplifying a Difference Quotient

If $f(x) = x^2 + 3x + 5$, find and simplify:

    a. $f(x + h)$ **b.** $\dfrac{f(x + h) - f(x)}{h}, h \neq 0.$

Solution

a. We find $f(x + h)$ by replacing x with $x + h$ each time that x appears in the equation.

$$f(x) \quad = \quad x^2 \quad + \quad 3x \quad + \quad 5$$

| Replace x with $x + h$. | Replace x with $x + h$. | Replace x with $x + h$. | Copy the 5. There is no x in this term. |

$$f(x + h) \quad = \quad (x + h)^2 \quad + \quad 3(x + h) \quad + \quad 5$$
$$= x^2 + 2xh + h^2 + 3x + 3h \quad + \quad 5$$

b. Using our result from part (a), we obtain the following:

> This is f(x + h) from part (a).

> This is f(x) from the given equation.

$$\frac{f(x + h) - f(x)}{h} = \frac{x^2 + 2xh + h^2 + 3x + 3h + 5 - (x^2 + 3x + 5)}{h}$$

$$= \frac{x^2 + 2xh + h^2 + 3x + 3h + 5 - x^2 - 3x - 5}{h}$$

Remove parentheses and change the sign of each term in the parentheses.

$$= \frac{(x^2 - x^2) + (3x - 3x) + (5 - 5) + 2xh + h^2 + 3h}{h}$$

Group like terms.

$$= \frac{2xh + h^2 + 3h}{h}$$

Simplify.

$$= \frac{h(2x + h + 3)}{h}$$

Factor h from the numerator.

$$= 2x + h + 3, h \neq 0.$$

Cancel identical factors of h in the numerator and denominator.

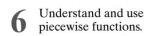 **Check Point 5**
If $f(x) = x^2 - 7x + 3$, find and simplify:

a. $f(x + h)$ **b.** $\dfrac{f(x + h) - f(x)}{h}, h \neq 0.$

Piecewise Functions

⑥ Understand and use piecewise functions.

The early part of the twentieth century was the golden age of immigration in America. More than 13 million people migrated to the United States between 1900 and 1914. By 1910, foreign-born residents accounted for 15% of the total U.S. population. The graph in Figure 1.35 shows the percentage of Americans who were foreign born throughout the twentieth century.

We can model the data from 1910 through 2000 with two equations, one from 1910 through 1970, years in which the percentage was decreasing, and one from 1970 through 2000, years in which the percentage was increasing. These two trends can be approximated by the function

$$P(t) = \begin{cases} -\dfrac{11}{60}t + 15 & \text{if } 0 \leq t < 60 \\ \dfrac{1}{5}t - 8 & \text{if } 60 \leq t \leq 90 \end{cases}$$

in which t represents the number of years after 1910 and $P(t)$ is the percentage of foreign-born Americans. A function that is defined by two (or more) equations over a specified domain is called a **piecewise function**.

Percentage of Americans Who Were Foreign Born in the Twentieth Century

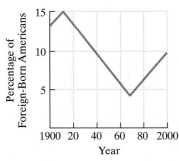

Figure 1.35

Source: U.S. Census Bureau

EXAMPLE 6 Evaluating a Piecewise Function

Use the function $P(t)$, described previously,

$$P(t) = \begin{cases} -\dfrac{11}{60}t + 15 & \text{if } 0 \le t < 60 \\[2mm] \dfrac{1}{5}t - 8 & \text{if } 60 \le t \le 90 \end{cases}$$

to find and interpret:

a. $P(30)$ **b.** $P(80)$.

Solution

a. To find $P(30)$, we let $t = 30$. Because 30 is less than 60, we use the first line of the piecewise function.

$$P(t) = -\tfrac{11}{60}t + 15 \qquad \text{This is the function's equation for } 0 \le t < 60.$$

$$P(30) = -\tfrac{11}{60}\cdot 30 + 15 \qquad \text{Replace t with 30.}$$

$$= 9.5$$

This means that 30 years after 1910, in 1940, 9.5% of Americans were foreign born.

b. To find $P(80)$, we let $t = 80$. Because 80 is between 60 and 90, we use the second line of the piecewise function.

$$P(t) = \frac{1}{5}t - 8 \qquad \text{This is the function's equation for } 60 \le t \le 90.$$

$$P(80) = \tfrac{1}{5}\cdot 80 - 8 \qquad \text{Replace t with 80.}$$

$$= 8$$

This means that 80 years after 1910, in 1990, 8% of Americans were foreign born.

Check Point 6   If $f(x) = \begin{cases} x^2 + 3 & \text{if } x < 0 \\ 5x + 3 & \text{if } x \ge 0 \end{cases}$, find:

a. $f(-5)$ **b.** $f(6)$.

7 Find the domain of a function.

The Domain of a Function

Let's reconsider the function that models the percentage of foreign-born Americans t years after 1910, up through and including 2000. The domain of this function is

$$\{0, \quad 1, \quad 2, \quad 3, \quad \ldots, \quad 90\}.$$

0 years after 1910 is 1910.

3 years after 1910 is 1913.

90 years after 1910 brings the domain up to the year 2000.

Functions that model data often have their domains explicitly given along with the function's equation. However, for most functions, only an equation is given, and the domain is not specified. In cases like these, the domain of f is the largest set of real numbers for which the value of $f(x)$ is a real number. For example, consider the function

$$f(x) = \frac{1}{x - 3}.$$

Because division by 0 is undefined (and not a real number), the denominator $x - 3$ cannot be 0. Thus, x cannot equal 3. The domain of the function consists of all real numbers other than 3, represented by $\{x \mid x \neq 3\}$. We say that f is not defined at 3, or $f(3)$ does not exist.

Just as the domain of a function must exclude real numbers that cause division by zero, it must also exclude real numbers that result in an even root of a negative number. For example, consider the function

$$g(x) = \sqrt{x}.$$

The equation tells us to take the square root of x. Because only nonnegative numbers have real square roots, the expression under the radical sign, x, must be greater than or equal to 0. The domain of g is $\{x \mid x \geq 0\}$, or the interval $[0, \infty)$.

Finding a Function's Domain

If a function f does not model data or verbal conditions, its domain is the largest set of real numbers for which the value of $f(x)$ is a real number. Exclude from a function's domain real numbers that cause division by zero and real numbers that result in an even root of a negative number.

EXAMPLE 7   Finding the Domain of a Function

Find the domain of each function:

a. $f(x) = x^2 - 7x$ **b.** $g(x) = \dfrac{6x}{x^2 - 9}$ **c.** $h(x) = \sqrt{3x + 12}.$

Technology

You can graph a function and often get hints about its domain. For example, $h(x) = \sqrt{3x + 12}$, or $y = \sqrt{3x + 12}$, appears only for $x \geq -4$, verifying $[-4, \infty)$ as the domain.

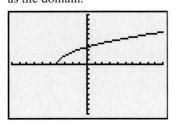

Solution

a. The function $f(x) = x^2 - 7x$ contains neither division nor an even root. The domain of f is the set of all real numbers.

b. The function $g(x) = \dfrac{6x}{x^2 - 9}$ contains division. Because division by 0 is undefined, we must exclude from the domain values of x that cause $x^2 - 9$ to be 0. Thus, x cannot equal -3 or 3. The domain of g is $\{x \mid x \neq -3, x \neq 3\}$.

c. The function $h(x) = \sqrt{3x + 12}$ contains an even root. Because only nonnegative numbers have real square roots, the quantity under the radical sign, $3x + 12$, must be greater than or equal to 0.

$$3x + 12 \geq 0$$
$$3x \geq -12$$
$$x \geq -4$$

The domain of h is $\{x \mid x \geq -4\}$, or the interval $[-4, \infty)$.

Check Point 7   Find the domain of each function:

a. $f(x) = x^2 + 3x - 17$ **b.** $g(x) = \dfrac{5x}{x^2 - 49}$

c. $h(x) = \sqrt{9x - 27}.$

EXERCISE SET 1.4

Practice Exercises

In Exercises 1–8, determine whether each relation is a function. Give the domain and range for each relation.

1. $\{(1, 2), (3, 4), (5, 5)\}$

2. $\{(4, 5), (6, 7), (8, 8)\}$

3. $\{(3, 4), (3, 5), (4, 4), (4, 5)\}$

4. $\{(5, 6), (5, 7), (6, 6), (6, 7)\}$

5. $\{(-3, -3), (-2, -2), (-1, -1), (0, 0)\}$

6. $\{(-7, -7), (-5, -5), (-3, -3), (0, 0)\}$

7. $\{(1, 4), (1, 5), (1, 6)\}$

8. $\{(4, 1), (5, 1), (6, 1)\}$

In Exercises 9–20, determine whether each equation defines y as a function of x.

9. $x + y = 16$

10. $x + y = 25$

11. $x^2 + y = 16$

12. $x^2 + y = 25$

13. $x^2 + y^2 = 16$

14. $x^2 + y^2 = 25$

15. $x = y^2$

16. $4x = y^2$

17. $y = \sqrt{x + 4}$

18. $y = -\sqrt{x + 4}$

19. $x + y^3 = 8$

20. $x + y^3 = 27$

In Exercises 21–32, evaluate each function at the given values of the independent variable and simplify.

21. $f(x) = 4x + 5$

    a. $f(6)$ **b.** $f(x + 1)$ **c.** $f(-x)$

22. $f(x) = 3x + 7$

    a. $f(4)$ **b.** $f(x + 1)$ **c.** $f(-x)$

23. $g(x) = x^2 + 2x + 3$

    a. $g(-1)$ **b.** $g(x + 5)$ **c.** $g(-x)$

24. $g(x) = x^2 - 10x - 3$

    a. $g(-1)$ **b.** $g(x + 2)$ **c.** $g(-x)$

25. $h(x) = x^4 - x^2 + 1$

    a. $h(2)$ **b.** $h(-1)$

    c. $h(-x)$ **d.** $h(3a)$

26. $h(x) = x^3 - x + 1$

    a. $h(3)$ **b.** $h(-2)$

    c. $h(-x)$ **d.** $h(3a)$

27. $f(r) = \sqrt{r + 6} + 3$

    a. $f(-6)$ **b.** $f(10)$ **c.** $f(x - 6)$

28. $f(r) = \sqrt{25 - r} - 6$

    a. $f(16)$ **b.** $f(-24)$ **c.** $f(25 - 2x)$

29. $f(x) = \dfrac{4x^2 - 1}{x^2}$

    a. $f(2)$ **b.** $f(-2)$ **c.** $f(-x)$

30. $f(x) = \dfrac{4x^3 + 1}{x^3}$

    a. $f(2)$ **b.** $f(-2)$ **c.** $f(-x)$

31. $f(x) = \dfrac{x}{|x|}$

    a. $f(6)$ **b.** $f(-6)$ **c.** $f(r^2)$

32. $f(x) = \dfrac{|x + 3|}{x + 3}$

    a. $f(5)$ **b.** $f(-5)$ **c.** $f(-9 - x)$

In Exercises 33–44, find and simplify the difference quotient

$$\frac{f(x + h) - f(x)}{h}, \quad h \neq 0$$

for the given function.

33. $f(x) = 4x$

34. $f(x) = 7x$

35. $f(x) = 3x + 7$

36. $f(x) = 6x + 1$

37. $f(x) = x^2$

38. $f(x) = 2x^2$

39. $f(x) = x^2 - 4x + 3$

40. $f(x) = x^2 - 5x + 8$

41. $f(x) = 6$

42. $f(x) = 7$

43. $f(x) = \dfrac{1}{x}$

44. $f(x) = \dfrac{1}{2x}$

In Exercises 45–50, evaluate each piecewise function at the given values of the independent variable.

45. $f(x) = \begin{cases} 3x + 5 & \text{if } x < 0 \\ 4x + 7 & \text{if } x \geq 0 \end{cases}$

    a. $f(-2)$ **b.** $f(0)$ **c.** $f(3)$

46. $f(x) = \begin{cases} 6x - 1 & \text{if } x < 0 \\ 7x + 3 & \text{if } x \geq 0 \end{cases}$

    a. $f(-3)$ **b.** $f(0)$ **c.** $f(4)$

47. $g(x) = \begin{cases} x + 3 & \text{if } x \geq -3 \\ -(x + 3) & \text{if } x < -3 \end{cases}$

    a. $g(0)$ **b.** $g(-6)$ **c.** $g(-3)$

48. $g(x) = \begin{cases} x + 5 & \text{if } x \geq -5 \\ -(x + 5) & \text{if } x < -5 \end{cases}$

    a. $g(0)$ **b.** $g(-6)$ **c.** $g(-5)$

49. $h(x) = \begin{cases} \dfrac{x^2 - 9}{x - 3} & \text{if } x \neq 3 \\ 6 & \text{if } x = 3 \end{cases}$

    a. $h(5)$ **b.** $h(0)$ **c.** $h(3)$

50. $h(x) = \begin{cases} \dfrac{x^2 - 25}{x - 5} & \text{if } x \neq 5 \\ 10 & \text{if } x = 5 \end{cases}$

    a. $h(7)$ **b.** $h(0)$ **c.** $h(5)$

In Exercises 51–72, find the domain of each function.

51. $f(x) = 4x^2 - 3x + 1$

52. $f(x) = 8x^2 - 5x + 2$

53. $g(x) = \dfrac{3}{x - 4}$

54. $g(x) = \dfrac{2}{x + 5}$

55. $h(x) = \dfrac{7x}{x^2 - 16}$

56. $h(x) = \dfrac{12x}{x^2 - 36}$

57. $f(x) = \dfrac{2}{(x + 3)(x - 7)}$

58. $f(x) = \dfrac{15}{(x + 8)(x - 3)}$

59. $H(r) = \dfrac{4}{r^2 + 11r + 24}$

60. $H(r) = \dfrac{5}{6r^2 + r - 2}$

61. $f(t) = \dfrac{3}{t^2 + 4}$

62. $f(t) = \dfrac{5}{t^2 + 9}$

63. $f(x) = \sqrt{x - 3}$

64. $f(x) = \sqrt{x + 2}$

65. $f(x) = \dfrac{1}{\sqrt{x - 3}}$

66. $f(x) = \dfrac{1}{\sqrt{x + 2}}$

67. $g(x) = \sqrt{5x + 35}$

68. $g(x) = \sqrt{7x - 70}$

69. $f(x) = \sqrt{24 - 2x}$

70. $f(x) = \sqrt{84 - 6x}$

71. $f(x) = \dfrac{\sqrt{x - 2}}{x - 5}$

72. $f(x) = \dfrac{\sqrt{x - 3}}{x - 6}$

Application Exercises

73. The bar graph shows the percentage of people in the United States using the Internet by education level. Write five ordered pairs for

(education level, percentage using Internet)

as a relation. Find the domain and the range of the relation. Is this relation a function? Explain your answer.

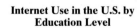

Internet Use in the U.S. by Education Level

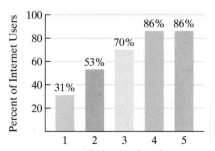

Education Level
(1 = less than high school, 2 = high school graduate, 3 = some college, 4 = college graduate, 5 = advanced degree)

Source: U.C.L.A. Center for Communication Policy

The table shows the ten longest-running television shows of the twentieth century. Use the information in the table to solve Exercises 74–76.

Ten Longest-Running National Network TV Series of the Twentieth Century

Program	Number of Seasons the Show Ran
"Walt Disney"	33
"60 Minutes"	33
"The Ed Sullivan Show"	24
"Gunsmoke"	20
"The Red Skelton Show"	20
"Meet the Press"	18
"What's My Line?"	18
"I've Got a Secret"	17
"Lassie"	17
"The Lawrence Welk Show"	17

Source: Nielsen Media Research

74. Consider the relation for which the domain represents the ten longest-running series and the range represents the number of seasons the series ran. Is this relation a function? Explain your answer.

75. Consider the relation for which the domain represents the number of seasons the ten longest-running series ran and the range represents the ten longest-running series. Is this relation a function? Explain your answer.

76. Use your answers from Exercises 74 and 75 to answer the following question: If the components in a function's ordered pairs are reversed, must the resulting relation also be a function?

77. The function

$$P(x) = 0.72x^2 + 9.4x + 783$$

models the gray wolf population in the United States, $P(x)$, x years after 1960. Find and interpret $P(30)$. How well does the function model the actual value shown in the bar graph?

Gray Wolf Population (to the Nearest Hundred)

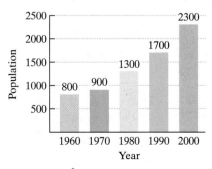

Year

Source: U.S. Department of the Interior

78. As the use of the Internet increases, so has the number of computer infections from viruses. The function

$$N(x) = 0.2x^2 - 1.2x + 2$$

models the number of infections per month for every 1000 computers, $N(x)$, x years after 1990. Find and interpret $N(10)$. How well does the function model the actual value shown in the bar graph?

Computer Infection Rates

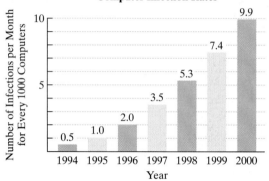

Source: Jupiter Communications

The function

$$P(t) = 6.85\sqrt{t} + 19$$

models the percentage of U.S. households online, $P(t)$, t years after 1997. In Exercises 79–82, use this function to find and interpret the given expression. If necessary, use a calculator and round to the nearest whole percent. How well does the function model the actual data shown in the bar graph?

Percentage of U.S. Households Online

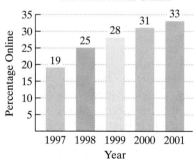

Source: Forrester Research

79. $P(0)$ **80.** $P(4)$

81. $P(3) - P(1)$ **82.** $P(2) - P(1)$

The number of lawyers in the United States can be modeled by the function

$$f(x) = \begin{cases} 6.5x + 200 & \text{if } 0 \le x < 23 \\ 26.2x - 252 & \text{if } x \ge 23 \end{cases}$$

where x represents the number of years after 1951 and $f(x)$ represents the number of lawyers, in thousands. In Exercises 83–86, use this function to find and interpret each of the following.

83. $f(0)$ **84.** $f(10)$

85. $f(50)$ **86.** $f(60)$

During a particular year, the taxes owed, $T(x)$, in dollars, filing separately with an adjusted gross income of x dollars is given by the piecewise function

$$T(x) = \begin{cases} 0.15x & \text{if } 0 \le x < 17{,}900 \\ 0.28(x - 17{,}900) + 2685 & \text{if } 17{,}900 \le x < 43{,}250. \\ 0.31(x - 43{,}250) + 9783 & \text{if } x \ge 43{,}250 \end{cases}$$

In Exercises 87–88, use this function to find and interpret each expression.

87. $T(40{,}000)$

88. $T(70{,}000)$

Writing in Mathematics

89. If a relation is represented by a set of ordered pairs, explain how to determine whether the relation is a function.

90. How do you determine if an equation in x and y defines y as a function of x?

91. A student in introductory algebra hears that functions are studied in subsequent algebra courses. The student asks you what a function is. Provide the student with a clear, relatively concise response.

92. Describe one advantage of using $f(x)$ rather than y in a function's equation.

93. Explain how to find the difference quotient, $\dfrac{f(x + h) - f(x)}{h}$, if a function's equation is given.

94. What is a piecewise function?

95. How is the domain of a function determined?

96. For people filing a single return, federal income tax is a function of adjusted gross income because for each value of adjusted gross income there is a specific tax to be paid. On the other hand, the price of a house is not a function of the lot size on which the house sits because houses on same-sized lots can sell for many different prices.

   a. Describe an everyday situation between variables that is a function.

   b. Describe an everyday situation between variables that is not a function.

Technology Exercises

Use a graphing utility to find the domain of each function in Exercises 97–99. Then verify your observation algebraically.

97. $f(x) = \sqrt{x - 1}$

98. $g(x) = \sqrt{2x + 6}$

99. $h(x) = \sqrt{15 - 3x}$

Critical Thinking Exercises

100. Write a function defined by an equation in x whose domain is $\{x | x \neq -4, x \neq 11\}$.

101. Write a function defined by an equation in x whose domain is $[-6, \infty)$.

102. Give an example of an equation that does not define y as a function of x but that does define x as a function of y.

103. If $f(x) = ax^2 + bx + c$ and $r_1 = \dfrac{-b + \sqrt{b^2 - 4ac}}{2a}$, find $f(r_1)$ without doing any algebra and explain how you arrived at your result.

Group Exercise

104. Almanacs, newspapers, magazines, and the Internet contain bar graphs and line graphs that describe how things are changing over time. For example, the graphs in Exercises 77–82 show how various phenomena are changing over time. Find a bar or line graph showing yearly changes that you find intriguing. Describe to the group what interests you about this data. The group should select their two favorite graphs. For each graph selected:

a. Rewrite the data so that they are presented as a relation in the form of a set of ordered pairs.

b. Determine whether the relation in part (a) is a function. Explain why the relation is a function, or why it is not.

SECTION 1.5 *Graphs of Functions*

Objectives

1. Graph functions by plotting points.
2. Obtain information about a function from its graph.
3. Use the vertical line test to identify functions.
4. Identify intervals on which a function increases, decreases, or is constant.
5. Use graphs to locate relative maxima or minima.
6. Find a function's average rate of change.
7. Identify even or odd functions and recognize their symmetries.
8. Graph step functions.

Have you ever seen a gas-guzzling car from the 1950s, with its huge fins and overstated design? The worst year for automobile fuel efficiency was 1958, when cars averaged a dismal 12.4 miles per gallon. The function

$$f(x) = 0.0075x^2 - 0.2672x + 14.8$$

models the average number of miles per gallon for U.S. automobiles, $f(x)$, x years after 1940. If we could see the graph of the function's equation, we would get a much better idea of the relationship between time and fuel efficiency. In this section, we will learn how to use the graph of a function to obtain useful information about the function.

Graphs of Functions

A graph enables us to visualize a function's behavior. The graph shows the relationship between the function's two variables more clearly than the function's equation does. The **graph of a function** is the graph of its ordered pairs. For example, the graph of $f(x) = \sqrt{x}$ is the set of points (x, y) in the rectangular coordinate system satisfying the equation $y = \sqrt{x}$. Thus, one way to graph a function is by plotting several of its ordered pairs and drawing a line or smooth curve through them. With the function's graph, we can picture its domain on the x-axis and its range on the y-axis. Our first example illustrates how this is done.

1 Graph functions by plotting points.

EXAMPLE 1   Graphing a Function by Plotting Points

Graph $f(x) = x^2 + 1$. To do so, use integer values of x from the set $\{-3, -2, -1, 0, 1, 2, 3\}$ to obtain seven ordered pairs. Plot each ordered pair and draw a smooth curve through the points. Use the graph to specify the function's domain and range.

Solution   The graph of $f(x) = x^2 + 1$ is, by definition, the graph of $y = x^2 + 1$. We begin by setting up a partial table of coordinates.

x	$f(x) = x^2 + 1$	(x, y) or $(x, f(x))$
-3	$f(-3) = (-3)^2 + 1 = 10$	$(-3, 10)$
-2	$f(-2) = (-2)^2 + 1 = 5$	$(-2, 5)$
-1	$f(-1) = (-1)^2 + 1 = 2$	$(-1, 2)$
0	$f(0) = 0^2 + 1 = 1$	$(0, 1)$
1	$f(1) = 1^2 + 1 = 2$	$(1, 2)$
2	$f(2) = 2^2 + 1 = 5$	$(2, 5)$
3	$f(3) = 3^2 + 1 = 10$	$(3, 10)$

Now, we plot the seven points and draw a smooth curve through them, as shown in Figure 1.36. The graph of f has a cuplike shape. The points on the graph of f have x-coordinates that extend indefinitely to the left and to the right. Thus, the domain consists of all real numbers, represented by $(-\infty, \infty)$. By contrast, the points on the graph have y-coordinates that start at 1 and extend indefinitely upward. Thus, the range consists of all real numbers greater than or equal to 1, represented by $[1, \infty)$.

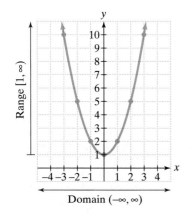

Figure 1.36   The graph of $f(x) = x^2 + 1$

Check Point 1   Graph $f(x) = x^2 - 2$, using integers from -3 to 3 for x in the partial table of coordinates. Use the graph to specify the function's domain and range.

Technology

Does your graphing utility have a TABLE feature? If so, you can use it to create tables of coordinates for a function. You will need to enter the equation of the function and specify the starting value for x, TblStart, and the increment between successive x-values, ΔTbl. For the table of coordinates in Example 1, we start the table at $x = -3$ and increment by 1. Using the up- or down-arrow keys, you can scroll through the table and determine as many ordered pairs of the graph as desired.

2 Obtain information about a function from its graph.

Obtaining Information from Graphs

You can obtain information about a function from its graph. At the right or left of a graph, you will find closed dots, open dots, or arrows.

- A closed dot indicates that the graph does not extend beyond this point and the point belongs to the graph.
- An open dot indicates that the graph does not extend beyond this point and the point does not belong to the graph.
- An arrow indicates that the graph extends indefinitely in the direction in which the arrow points.

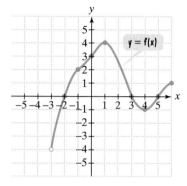

Figure 1.37

EXAMPLE 2   Obtaining Information from a Function's Graph

Use the graph of the function f, shown in Figure 1.37, to answer the following questions:

a. What are the function values $f(-1)$ and $f(1)$?
b. What is the domain of f?
c. What is the range of f?

Solution

a. Because $(-1, 2)$ is a point on the graph of f, the y-coordinate, 2, is the value of the function at the x-coordinate, -1. Thus, $f(-1) = 2$. Similarly, because $(1, 4)$ is also a point on the graph of f, this indicates that $f(1) = 4$.

b. The open dot on the left shows that $x = -3$ is not in the domain of f. By contrast, the closed dot on the right shows that $x = 6$ is in the domain of f. We determine the domain of f by noticing that the points on the graph of f have x-coordinates between -3, excluding -3, and 6, including 6. For each number x between -3 and 6, there is a point $(x, f(x))$ on the graph. Thus, the domain of f is $\{x \mid -3 < x \le 6\}$, or the interval $(-3, 6]$.

c. The points on the graph all have y-coordinates between -4, not including -4, and 4, including 4. The graph does not extend below $y = -4$ or above $y = 4$. Thus, the range of f is $\{y \mid -4 < y \le 4\}$, or the interval $(-4, 4]$.

Check Point 2   Use the graph of function f, shown below, to find $f(4)$, the domain, and the range.

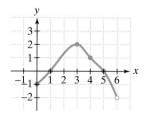

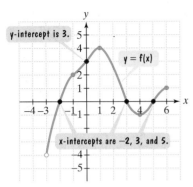

Figure 1.38 Identifying intercepts

Figure 1.38 illustrates how we can identify a graph's intercepts. To find the *x*-intercepts, look for the points at which the graph crosses the *x*-axis. There are three such points: $(-2, 0)$, $(3, 0)$, and $(5, 0)$. Thus, the *x*-intercepts are -2, 3, and 5. We express this in function notation by writing $f(-2) = 0$, $f(3) = 0$, and $f(5) = 0$. We say that -2, 3, and 5 are the *zeros of the function*. The **zeros of a function** *f* are the *x*-values for which $f(x) = 0$.

To find the *y*-intercept, look for the point at which the graph crosses the *y*-axis. This occurs at $(0, 3)$. Thus, the *y*-intercept is 3. We express this in function notation by writing $f(0) = 3$.

By the definition of a function, for each value of *x* we can have at most one value for *y*. What does this mean in terms of intercepts? **A function can have more than one *x*-intercept but at most one *y*-intercept.**

3 Use the vertical line test to identify functions.

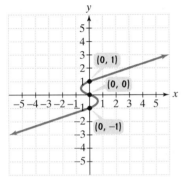

Figure 1.39 *y* is not a function of *x* because 0 is paired with three values of *y*, namely, 1, 0, and −1.

The Vertical Line Test

Not every graph in the rectangular coordinate system is the graph of a function. The definition of a function specifies that no value of *x* can be paired with two or more different values of *y*. Consequently, if a graph contains two or more different points with the same first coordinate, the graph cannot represent a function. This is illustrated in Figure 1.39. Observe that points sharing a common first coordinate are vertically above or below each other.

This observation is the basis of a useful test for determining whether a graph defines *y* as a function of *x*. The test is called the **vertical line test.**

> ### The Vertical Line Test for Functions
>
> If any vertical line intersects a graph in more than one point, the graph does not define *y* as a function of *x*.

EXAMPLE 3   Using the Vertical Line Test

Use the vertical line test to identify graphs in which *y* is a function of *x*.

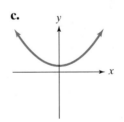

Solution   *y* is a function of *x* for the graphs in **b** and **c.**

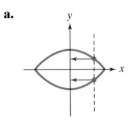

y **is not a function of** *x*.
Two values of *y*
correspond to an *x*-value.

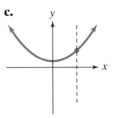

y **is a function of** *x*.

y **is a function of** *x*.

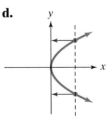

y **is not a function of** *x*.
Two values of *y*
correspond to an *x*-value.

Check Point 3   Use the vertical line test to identify graphs in which y is a function of x.

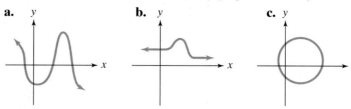

a.   **b.**   **c.**

EXAMPLE 4   Analyzing the Graph of a Function

The function

$$f(x) = -0.016x^2 + 0.93x + 8.5$$

models the average number of paid vacation days each year, $f(x)$, for full-time workers at medium to large U.S. companies after x years. The graph of f is shown in Figure 1.40.

a. Explain why f represents the graph of a function.
b. Use the graph to find a reasonable estimate of $f(5)$.
c. For what value of x is $f(x) = 20$?
d. Describe the general trend shown by the graph.

Average Number of Paid Vacation Days for Full-Time Workers at Medium to Large U.S. Companies

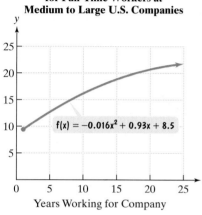

$f(x) = -0.016x^2 + 0.93x + 8.5$

Years Working for Company

Figure 1.40
Source: Bureau of Labor Statistics

Solution

a. No vertical line intersects the graph of f more than once. By the vertical line test, f represents the graph of a function.

b. To find $f(5)$, or f of 5, we locate 5 on the x-axis. The figure shows the point on the graph of f for which 5 is the first coordinate. From this point, we look to the y-axis to find the corresponding y-coordinate. A reasonable estimate of the y-coordinate is 13. Thus, $f(5) \approx 13$. After 5 years, a worker can expect approximately 13 paid vacation days.

c. To find the value of x for which $f(x) = 20$, we locate 20 on the y-axis. The figure shows that there is one point on the graph of f for which 20 is the second coordinate. From this point, we look to the x-axis to find the corresponding x-coordinate. A reasonable estimate of the x-coordinate is 18. Thus, $f(x) = 20$ for $x \approx 18$. A worker with 20 paid vacation days has been with the company approximately 18 years.

Average Number of Paid Vacation Days for Full-Time Workers at Medium to Large U.S. Companies

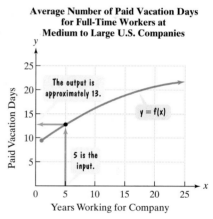

The output is approximately 13.

5 is the input.

$y = f(x)$

Years Working for Company

Average Number of Paid Vacation Days for Full-Time Workers at Medium to Large U.S. Companies

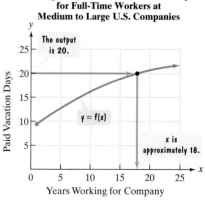

The output is 20.

$y = f(x)$

x is approximately 18.

Years Working for Company

d. The graph of f is rising from left to right. This shows that paid vacation days increase as time with the company increases. However, the rate of increase is slowing down as the graph moves to the right. This means that the increase in paid vacation days takes place more slowly the longer an employee is with the company.

Check Point 4
a. Use the graph of f in Figure 1.40 to find a reasonable estimate of $f(10)$.
b. For what value of x is $f(x) = 15$? Round to the nearest whole number.

4 Identify intervals on which a function increases, decreases, or is constant.

Increasing and Decreasing Functions

A function f is *increasing* when its graph rises, *decreasing* when its graph falls, and remains *constant* when its graph neither rises nor falls. We now provide a more algebraic description for these intuitive concepts.

Study Tip

The open intervals describing where functions increase, decrease, or are constant, use x-coordinates and not the y-coordinates.

Increasing, Decreasing, and Constant Functions

1. A function is **increasing** on an open interval, I, if for any x_1 and x_2 in the interval, where $x_1 < x_2$, then $f(x_1) < f(x_2)$.
2. A function is **decreasing** on an open interval, I, if for any x_1 and x_2 in the interval, where $x_1 < x_2$, then $f(x_1) > f(x_2)$.
3. A function is **constant** on an open interval, I, if for any x_1 and x_2 in the interval, where $x_1 < x_2$, then $f(x_1) = f(x_2)$.

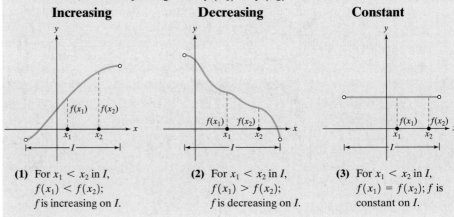

Increasing	Decreasing	Constant

(1) For $x_1 < x_2$ in I, $f(x_1) < f(x_2)$; f is increasing on I.

(2) For $x_1 < x_2$ in I, $f(x_1) > f(x_2)$; f is decreasing on I.

(3) For $x_1 < x_2$ in I, $f(x_1) = f(x_2)$; f is constant on I.

EXAMPLE 5   Intervals on Which a Function Increases, Decreases, or Is Constant

Give the intervals on which each function whose graph is shown is increasing, decreasing, or constant.

a.

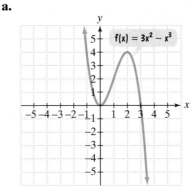

$f(x) = 3x^2 - x^3$

b.

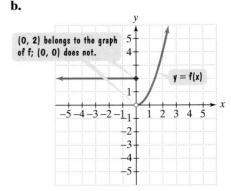

(0, 2) belongs to the graph of f; (0, 0) does not.

$y = f(x)$

Solution

a. The function is decreasing on the interval $(-\infty, 0)$, increasing on the interval $(0, 2)$, and decreasing on the interval $(2, \infty)$.

b. Although the function's equations are not given, the graph indicates that the function is defined in two pieces. The part of the graph to the left of the y-axis shows that the function is constant on the interval $(-\infty, 0)$. The part to the right of the y-axis shows that the function is increasing on the interval $(0, \infty)$.

Check Point 5 Give the intervals on which the function whose graph is shown is increasing, decreasing, or constant.

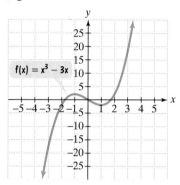

Relative Maxima and Relative Minima

5 Use graphs to locate relative maxima or minima.

The points at which a function changes its increasing or decreasing behavior can be used to find the *relative maximum* or *relative minimum* values of the function. For example, consider the function with which we opened this section:

$$f(x) = 0.0075x^2 - 0.2672x + 14.8.$$

Recall that the function models the average number of miles per gallon of U.S. automobiles, $f(x)$, x years after 1940. The graph of this function is shown as a continuous curve in Figure 1.41. (It can also be shown as a series of points, each point representing a year and miles per gallon for that year.)

The graph of f is decreasing to the left of $x = 18$ and increasing to the right of $x = 18$. Thus, 18 years after 1940, in 1958, fuel efficiency was at a minimum. We say that the relative minimum fuel efficiency is $f(18)$, or approximately 12.4 miles per gallon. Mathematicians use the word "relative" to suggest that relative to an open interval about 18, the value $f(18)$ is smallest.

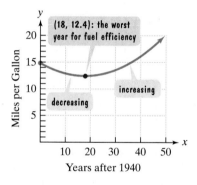

Figure 1.41 Fuel efficiency of U.S. automobiles over time

Definitions of Relative Maximum and Relative Minimum

1. A function value $f(a)$ is a **relative maximum** of f if there exists an open interval about a such that $f(a) > f(x)$ for all x in the open interval.
2. A function value $f(b)$ is a **relative minimum** of f if there exists an open interval about b such that $f(b) < f(x)$ for all x in the open interval.

Study Tip

The word "local" is sometimes used instead of "relative" when describing maxima or minima. If f has a relative, or local, maximum at a, $f(a)$ is greater than the values of f near a. If f has a relative, or local, minimum at b, $f(b)$ is less than the values of f near b.

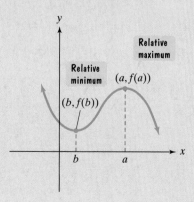

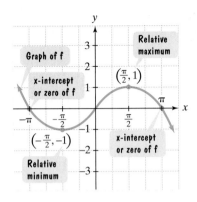

Figure 1.42 Using a graph to locate where *f* has a relative maximum or minimum

If the graph of a function is given, we can often visually locate the number(s) at which the function has a relative maximum or a relative minimum. For example, the graph of *f* in Figure 1.42 shows that

- *f* has a relative maximum at $\dfrac{\pi}{2}$.

   The relative maximum is $f\left(\dfrac{\pi}{2}\right) = 1$.

- *f* has a relative minimum at $-\dfrac{\pi}{2}$.

   The relative minimum is $f\left(-\dfrac{\pi}{2}\right) = -1$.

Notice that *f* does not have a relative maximum or minimum at $-\pi$ and π, the *x*-intercepts, or zeros, of the function.

6 Find a function's average rate of change.

The Average Rate of Change of a Function

We have seen that the slope of a line can be interpreted as its rate of change. If the graph of a function is not a straight line, we speak of an **average rate of change** between any two points on its graph. To find the average rate of change, calculate the slope of the line containing the two points. This line is called a **secant line.**

The Average Rate of Change of a Function

Let $(x_1, f(x_1))$ and $(x_2, f(x_2))$ be distinct points on the graph of a function *f*. (See Figure 1.43.) The **average rate of change of *f*** from x_1 to x_2, denoted by $\dfrac{\Delta y}{\Delta x}$ (read "*delta y divided by delta x*" or "*change in y divided by change in x*"), is

$$\frac{\Delta y}{\Delta x} = \frac{f(x_2) - f(x_1)}{x_2 - x_1}.$$

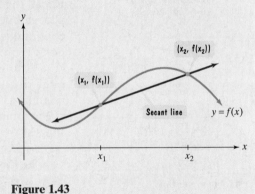

Figure 1.43

EXAMPLE 6 Finding the Average Rate of Change

Find the average rate of change of $f(x) = x^2$ from:

a. $x_1 = 0$ to $x_2 = 1$ **b.** $x_1 = 1$ to $x_2 = 2$ **c.** $x_1 = -2$ to $x_2 = 0$.

Solution

a. The average rate of change of $f(x) = x^2$ from $x_1 = 0$ to $x_2 = 1$ is

$$\frac{\Delta y}{\Delta x} = \frac{f(x_2) - f(x_1)}{x_2 - x_1} = \frac{f(1) - f(0)}{1 - 0} = \frac{1^2 - 0^2}{1} = 1.$$

Figure 1.44(a) shows the secant line of $f(x) = x^2$ from $x_1 = 0$ to $x_2 = 1$. The average rate of change is positive, and the function is increasing on the interval $(0, 1)$.

b. The average rate of change of $f(x) = x^2$ from $x_1 = 1$ to $x_2 = 2$ is

$$\frac{\Delta y}{\Delta x} = \frac{f(x_2) - f(x_1)}{x_2 - x_1} = \frac{f(2) - f(1)}{2 - 1} = \frac{2^2 - 1^2}{1} = 3.$$

Figure 1.44(b) shows the secant line of $f(x) = x^2$ from $x_1 = 1$ to $x_2 = 2$. The average rate of change is positive, and the function is increasing on the interval $(1, 2)$. Can you see that the graph rises more steeply on the interval $(1, 2)$ than on $(0, 1)$? This is because the average rate of change from $x_1 = 1$ to $x_2 = 2$ is greater than the average rate of change from $x_1 = 0$ to $x_2 = 1$.

c. The average rate of change of $f(x) = x^2$ from $x_1 = -2$ to $x_2 = 0$ is

$$\frac{\Delta y}{\Delta x} = \frac{f(x_2) - f(x_1)}{x_2 - x_1} = \frac{f(0) - f(-2)}{0 - (-2)} = \frac{0^2 - (-2)^2}{2} = \frac{-4}{2} = -2.$$

Figure 1.44(c) shows the secant line of $f(x) = x^2$ from $x_1 = -2$ to $x_2 = 0$. The average rate of change is negative, and the function is decreasing on the interval $(-2, 0)$.

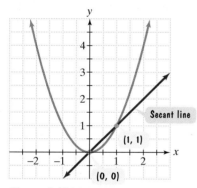

Figure 1.44(a)   The secant line of $f(x) = x^2$ from $x_1 = 0$ to $x_2 = 1$

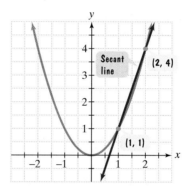

Figure 1.44(b)   The secant line of $f(x) = x^2$ from $x_1 = 1$ to $x_2 = 2$

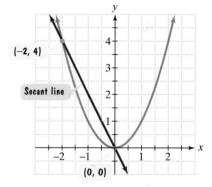

Figure 1.44(c)   The secant line of $f(x) = x^2$ from $x_1 = -2$ to $x_2 = 0$

Check Point 6   Find the average rate of change of $f(x) = x^3$ from:

a. $x_1 = 0$ to $x_2 = 1$
b. $x_1 = 1$ to $x_2 = 2$
c. $x_1 = -2$ to $x_2 = 0$.

Suppose we are interested in the average rate of change of f from $x_1 = x$ to $x_2 = x + h$. In this case, the average rate of change is

$$\frac{\Delta y}{\Delta x} = \frac{f(x_2) - f(x_1)}{x_2 - x_1} = \frac{f(x + h) - f(x)}{x + h - x} = \frac{f(x + h) - f(x)}{h}.$$

Do you recognize the last expression? It is the difference quotient that you used in the previous section to practice evaluating functions. Thus, the difference quotient gives the average rate of change of a function from x to $x + h$. In the difference quotient, h is thought of as a number very close to 0. In this way, the average rate of change can be found for a very short interval.

The **average velocity** of an object is its change in position divided by the change in time between these two positions. If a function expresses an object's position in terms of time, the function's average rate of change describes the object's average velocity.

Average Velocity of an Object

Suppose that a function expresses an object's position, $s(t)$, in terms of time, t. The **average velocity** of the object from t_1 to t_2 is

$$\frac{\Delta s}{\Delta t} = \frac{s(t_2) - s(t_1)}{t_2 - t_1}.$$

EXAMPLE 7   Finding Average Velocity

The distance, $s(t)$, in feet, traveled by a ball rolling down a ramp is given by the function

$$s(t) = 5t^2$$

where t is the time, in seconds, after the ball is released. Find the ball's average velocity from:

    a. $t_1 = 2$ seconds to $t_2 = 3$ seconds
    b. $t_1 = 2$ seconds to $t_2 = 2.5$ seconds
    c. $t_1 = 2$ seconds to $t_2 = 2.01$ seconds.

Solution

    a. The ball's average velocity between 2 and 3 seconds is

$$\frac{\Delta s}{\Delta t} = \frac{s(3) - s(2)}{3 \text{ sec} - 2 \text{ sec}} = \frac{5 \cdot 3^2 - 5 \cdot 2^2}{1 \text{ sec}} = \frac{45 \text{ ft} - 20 \text{ ft}}{1 \text{ sec}} = 25 \text{ ft/sec.}$$

    b. The ball's average velocity between 2 and 2.5 seconds is

$$\frac{\Delta s}{\Delta t} = \frac{s(2.5) - s(2)}{2.5 \text{ sec} - 2 \text{ sec}} = \frac{5(2.5)^2 - 5 \cdot 2^2}{0.5 \text{ sec}} = \frac{31.25 \text{ ft} - 20 \text{ ft}}{0.5 \text{ sec}} = 22.5 \text{ ft/sec.}$$

    c. The ball's average velocity between 2 and 2.01 seconds is

$$\frac{\Delta s}{\Delta t} = \frac{s(2.01) - s(2)}{2.01 \text{ sec} - 2 \text{ sec}} = \frac{5(2.01)^2 - 5 \cdot 2^2}{0.01 \text{ sec}} = \frac{20.2005 \text{ ft} - 20 \text{ ft}}{0.01 \text{ sec}} = 20.05 \text{ ft/sec.}$$

In Example 7, observe that each calculation begins at 2 seconds and involves shorter and shorter time intervals. In calculus, this procedure leads to the concept of *instantaneous*, as opposed to *average, velocity*.

Check Point 7 The distance, $s(t)$, in feet, traveled by a ball rolling down a ramp is given by the function

$$s(t) = 4t^2$$

where t is the time, in seconds, after the ball is released. Find the ball's average velocity from:

a. $t_1 = 1$ second to $t_2 = 2$ seconds

b. $t_1 = 1$ second to $t_2 = 1.5$ seconds

c. $t_1 = 1$ second to $t_2 = 1.01$ seconds.

7 Identify even or odd functions and recognize their symmetries.

Even and Odd Functions and Symmetry

Is beauty in the eye of the beholder? Or are there certain objects (or people) that are so well balanced and proportioned that they are universally pleasing to the eye? What constitutes an attractive human face? In Figure 1.45, we've drawn lines between paired features and marked the midpoints. Notice how the features line up almost perfectly. Each half of the face is a mirror image of the other half through the white vertical line.

Did you know that graphs of some equations exhibit exactly the kind of symmetry shown by the attractive face in Figure 1.45? The word *symmetry* comes from the Greek *symmetria*, meaning "the same measure." We can identify graphs with symmetry by looking at a function's equation and determining if the function is *even* or *odd*.

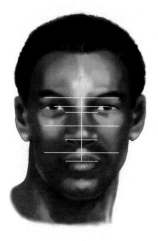

Figure 1.45 To most people, an attractive face is one in which each half is an almost perfect mirror image of the other half.

Definition of Even and Odd Functions

The function f is an **even function** if

$$f(-x) = f(x) \quad \text{for all } x \text{ in the domain of } f.$$

The right side of the equation of an even function does not change if x is replaced with $-x$.

The function f is an **odd function** if

$$f(-x) = -f(x) \quad \text{for all } x \text{ in the domain of } f.$$

Every term in the right side of the equation of an odd function changes sign if x is replaced with $-x$.

EXAMPLE 8 Identifying Even or Odd Functions

Identify each of the following functions as even, odd, or neither:

a. $f(x) = x^3$

b. $g(x) = x^4 - 2x^2$

c. $h(x) = x^2 + 2x + 1.$

Solution In each case, replace x with $-x$ and simplify. If the right side of the equation stays the same, the function is even. If every term on the right changes sign, the function is odd.

a. We use the given function's equation, $f(x) = x^3$, to find $f(-x)$.

Use $f(x) = x^3$.

Replace x with $-x$. Replace x with $-x$.

$$f(-x) = (-x)^3 = (-x)(-x)(-x) = -x^3$$

There is only one term in the equation $f(x) = x^3$, and the term changed signs when we replaced x with $-x$. Because $f(-x) = -f(x)$, f is an odd function.

b. We use the given function's equation, $g(x) = x^4 - 2x^2$, to find $g(-x)$.

Use $g(x) = x^4 - 2x^2$.

Replace x with $-x$.

$$g(-x) = (-x)^4 - 2(-x)^2 = (-x)(-x)(-x)(-x) - 2(-x)(-x)$$
$$= x^4 - 2x^2$$

The right side of the equation of the given function, $g(x) = x^4 - 2x^2$, did not change when we replaced x with $-x$. Because $g(-x) = g(x)$, g is an even function.

c. We use the given function's equation, $h(x) = x^2 + 2x + 1$, to find $h(-x)$.

Use $h(x) = x^2 + 2x + 1$.

Replace x with $-x$.

$$h(-x) = (-x)^2 + 2(-x) + 1 = x^2 - 2x + 1$$

The right side of the equation of the given function, $h(x) = x^2 + 2x + 1$, changed when we replaced x with $-x$. Thus, $h(-x) \neq h(x)$, so h is not an even function. The sign of *each* of the three terms in the equation for $h(x)$ did not change when we replaced x with $-x$. Only the second term changed signs. Thus, $h(-x) \neq -h(x)$, so h is not an odd function. We conclude that h is neither an even nor an odd function.

Check Point 8 Determine whether each of the following functions is even, odd, or neither:

a. $f(x) = x^2 + 6$ **b.** $g(x) = 7x^3 - x$ **c.** $h(x) = x^5 + 1$.

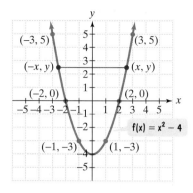

Figure 1.46 *y-axis symmetry with* $f(-x) = f(x)$

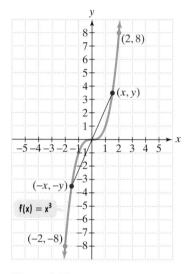

Figure 1.47 Origin symmetry with $f(-x) = -f(x)$

Table 1.3 Cost of First-Class Mail (Effective June 30, 2002)

Weight Not Over	Cost
1 ounce	$0.37
2 ounces	0.60
3 ounces	0.83
4 ounces	1.06
5 ounces	1.29

Source: U.S. Postal Service

Now, let's see what even and odd functions tell us about a function's graph. Begin with the even function $f(x) = x^2 - 4$, shown in Figure 1.46. The function is even because

$$f(-x) = (-x)^2 - 4 = x^2 - 4 = f(x).$$

Examine the pairs of points shown, such as $(3, 5)$ and $(-3, 5)$. Notice that we obtain the same y-coordinate whenever we evaluate the function at a value of x and the value of its opposite, $-x$. Like the attractive face, each half of the graph is a mirror image of the other half through the y-axis. If we were to fold the paper along the y-axis, the two halves of the graph would coincide. This causes the graph to be *symmetric with respect to the y-axis*. A graph is **symmetric with respect to the y-axis** if, for every point (x, y) on the graph, the point $(-x, y)$ is also on the graph. All even functions have graphs with this kind of symmetry.

Even Functions and y-Axis Symmetry

The graph of an even function in which $f(-x) = f(x)$ is symmetric with respect to the y-axis.

Now, consider the graph of the function $f(x) = x^3$. In Example 8, we saw that $f(-x) = -f(x)$, so this is an odd function. Although the graph in Figure 1.47 is not symmetric with respect to the y-axis, it is symmetric in another way. Look at the pairs of points, such as $(2, 8)$ and $(-2, -8)$. For each point (x, y) on the graph, the point $(-x, -y)$ is also on the graph. The points $(2, 8)$ and $(-2, -8)$ are reflections of one another in the origin. This means that

- the points are the same distance from the origin, and
- the points lie on a line through the origin.

A graph is **symmetric with respect to the origin** if, for every point (x, y) on the graph, the point $(-x, -y)$ is also on the graph. Observe that the first- and third-quadrant portions of $f(x) = x^3$ are reflections of one another with respect to the origin. Notice that $f(x)$ and $f(-x)$ have opposite signs, so that $f(-x) = -f(x)$. All odd functions have graphs with origin symmetry.

Odd Functions and Origin Symmetry

The graph of an odd function in which $f(-x) = -f(x)$ is symmetric with respect to the origin.

Step Functions

Have you ever mailed a letter that seemed heavier than usual? Perhaps you worried that the letter would not have enough postage. Costs for mailing a letter weighing up to 5 ounces are given in Table 1.3. If your letter weighs an ounce or less, the cost is $0.37. If your letter weighs 1.05 ounces, 1.50 ounces, 1.90 ounces, or 2.00 ounces, the cost "steps" to $0.60. The cost does not take on any value between $0.37 and $0.60. If your letter weighs 2.05 ounces, 2.50 ounces, 2.90 ounces, or 3 ounces, the cost "steps" to $0.83. Cost increases are $0.23 per step.

8 Graph step functions.

Now, let's see what the graph of the function that models this situation looks like. Let

x = the weight of the letter, in ounces, and

$y = f(x)$ = the cost of mailing a letter weighing x ounces.

The graph is shown in Figure 1.48. Notice how it consists of a series of steps that jump vertically 0.23 unit at each integer. The graph is constant between each pair of consecutive integers.

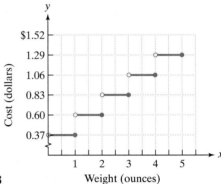

Figure 1.48

Technology

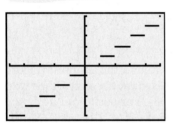

The graph of $f(x) = \text{int}(x)$, shown above, was obtained with a graphing utility. By graphing in "dot" mode, we can see the discontinuities at the integers. By looking at the graph, it is impossible to tell that, for each step, the point on the left is included and the point on the right is not. We must trace along the graph to obtain such information.

Mathematicians have defined functions that describe situations where function values graphically form discontinuous steps. One such function is called the **greatest integer function,** symbolized by $\text{int}(x)$ or $[\![x]\!]$. And what is $\text{int}(x)$?

$\text{int}(x)$ = the greatest integer that is less than or equal to x.

For example,

$$\text{int}(1) = 1, \quad \text{int}(1.3) = 1, \quad \text{int}(1.5) = 1, \quad \text{int}(1.9) = 1.$$

1 is the greatest integer that is less than or equal to 1, 1.3, 1.5, and 1.9.

Here are some additional examples:

$$\text{int}(2) = 2, \quad \text{int}(2.3) = 2, \quad \text{int}(2.5) = 2, \quad \text{int}(2.9) = 2.$$

2 is the greatest integer that is less than or equal to 2, 2.3, 2.5, and 2.9.

Notice how we jumped from 1 to 2 in the function values for $\text{int}(x)$. In particular,

$$\text{If } 1 \le x < 2, \quad \text{then} \quad \text{int}(x) = 1.$$
$$\text{If } 2 \le x < 3, \quad \text{then} \quad \text{int}(x) = 2.$$

The graph of $f(x) = \text{int}(x)$ is shown in Figure 1.49. The graph of the greatest integer function jumps vertically one unit at each integer. However, the graph is constant between each pair of consecutive integers. The rightmost horizontal step shown in the graph illustrates that

$$\text{If } 5 \le x < 6, \quad \text{then} \quad \text{int}(x) = 5.$$

In general,

$$\text{If } n \le x < n + 1, \text{ where } n \text{ is an integer, then } \text{int}(x) = n.$$

By contrast to the graph for the cost of first-class mail, the graph of the greatest integer function includes the point on the left of each horizontal step, but does not include the point on the right. The domain of $f(x) = \text{int}(x)$ is the set of all real numbers, $(-\infty, \infty)$. The range is the set of all integers.

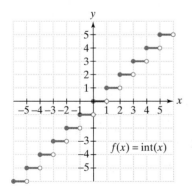

Figure 1.49 The graph of the greatest integer function

EXERCISE SET 1.5

Practice Exercises

Graph each function in Exercises 1–14. Use the integer values of x given to the right of the function to obtain ordered pairs. Use the graph to specify the function's domain and range.

1. $f(x) = x^2 + 2$ $x = -3, -2, -1, 0, 1, 2, 3$
2. $f(x) = x^2 - 1$ $x = -3, -2, -1, 0, 1, 2, 3$
3. $g(x) = \sqrt{x} - 1$ $x = 0, 1, 4, 9$
4. $g(x) = \sqrt{x} + 2$ $x = 0, 1, 4, 9$
5. $h(x) = \sqrt{x - 1}$ $x = 1, 2, 5, 10$
6. $h(x) = \sqrt{x + 2}$ $x = -2, -1, 2, 7$
7. $f(x) = |x| - 1$ $x = -3, -2, -1, 0, 1, 2, 3$
8. $f(x) = |x| + 1$ $x = -3, -2, -1, 0, 1, 2, 3$
9. $g(x) = |x - 1|$ $x = -3, -2, -1, 0, 1, 2, 3$
10. $g(x) = |x + 1|$ $x = -3, -2, -1, 0, 1, 2, 3$
11. $f(x) = 5$ $x = -3, -2, -1, 0, 1, 2, 3$
12. $f(x) = 3$ $x = -3, -2, -1, 0, 1, 2, 3$
13. $f(x) = x^3 - 2$ $x = -2, -1, 0, 1, 2$
14. $f(x) = x^3 + 2$ $x = -2, -1, 0, 1, 2$

*In Exercises 15–30, use the graph to determine **a.** the function's domain; **b.** the function's range; **c.** the x-intercepts, if any; **d.** the y-intercept, if any; and **e.** the function values indicated below some of the graphs.*

15.

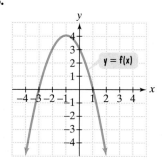

16.

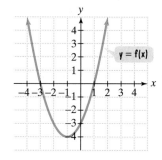

17.

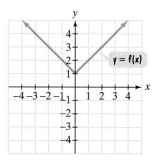

$f(-1) = ?$ $f(3) = ?$

18.

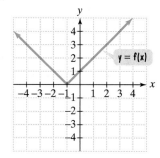

$f(-4) = ?$ $f(3) = ?$

19.

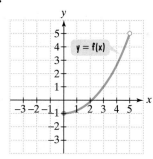

$f(3) = ?$

20.

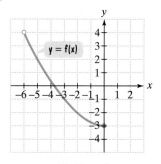

$f(-5) = ?$

21.

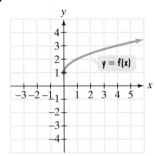

$$f(4) = ?$$

22.

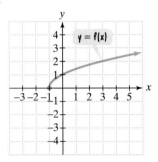

$$f(3) = ?$$

23.

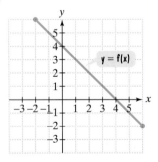

$$f(-1) = ?$$

24.

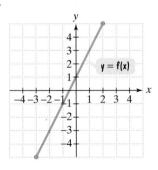

$$f(-2) = ?$$

25.

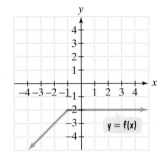

$$f(-4) = ? \quad f(4) = ?$$

26.

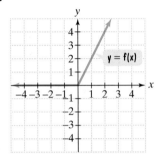

$$f(-2) = ? \quad f(2) = ?$$

27.

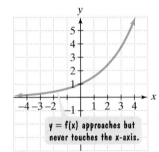

28.

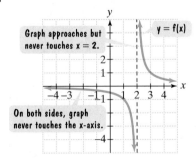

29.

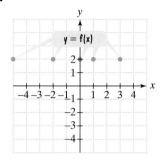

30.

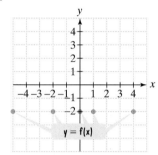

In Exercises 31–38, use the vertical line test to identify graphs in which y is a function of x.

31.

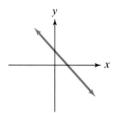

32.

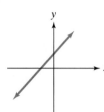

33.

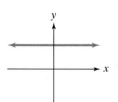

34.

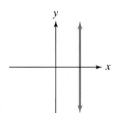

35.

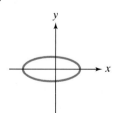

36.

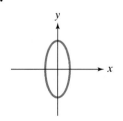

37.

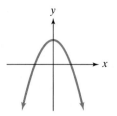

38.

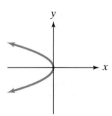

In Exercises 39–50, use the graph to determine:
    a. *intervals on which the function is increasing, if any.*
    b. *intervals on which the function is decreasing, if any.*
    c. *intervals on which the function is constant, if any.*

39. Use the graph in Exercise 15.

40. Use the graph in Exercise 16.

41. Use the graph in Exercise 21.

42. Use the graph in Exercise 22.

43. Use the graph in Exercise 23.

44. Use the graph in Exercise 24.

45. Use the graph in Exercise 25.

46. Use the graph in Exercise 26.

47.

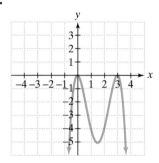

48.

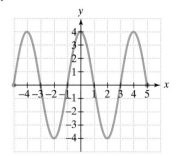

49.

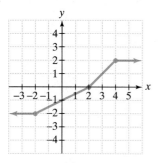

50.

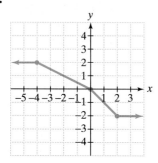

In Exercises 51–54, the graph of a function f is given. Use the graph to find:

    a. *the numbers, if any, at which f has a relative maximum. What are these relative maxima?*

    b. *the numbers, if any, at which f has a relative minimum. What are these relative minima?*

51.

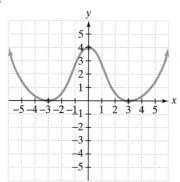

52.

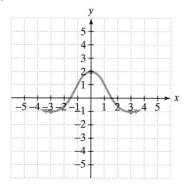

53.

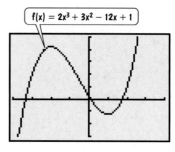

[−4, 4, 1] by [−15, 25, 5]

54.

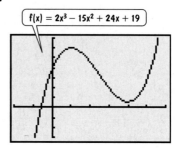

[−2, 6, 1] by [−15, 35, 5]

In Exercises 55–60, find the average rate of change of the function from x_1 to x_2.

55. $f(x) = 3x$ from $x_1 = 0$ to $x_2 = 5$

56. $f(x) = 6x$ from $x_1 = 0$ to $x_2 = 4$

57. $f(x) = x^2 + 2x$ from $x_1 = 3$ to $x_2 = 5$

58. $f(x) = x^2 - 2x$ from $x_1 = 3$ to $x_2 = 6$

59. $f(x) = \sqrt{x}$ from $x_1 = 4$ to $x_2 = 9$

60. $f(x) = \sqrt{x}$ from $x_1 = 9$ to $x_2 = 16$

In Exercises 61–62, suppose that a ball is rolling down a ramp. The distance traveled by the ball is given by the function in each exercise, where t is the time, in seconds, after the ball is released, and s(t) is measured in feet. For each given function, find the ball's average velocity from:
a. $t_1 = 3$ to $t_2 = 4$; **b.** $t_1 = 3$ to $t_2 = 3.5$;
c. $t_1 = 3$ to $t_2 = 3.01$; and **d.** $t_1 = 3$ to $t_2 = 3.001$.

61. $s(t) = 10t^2$

62. $s(t) = 12t^2$

In Exercises 63–74, determine whether each function is even, odd, or neither.

63. $f(x) = x^3 + x$ **64.** $f(x) = x^3 - x$

65. $g(x) = x^2 + x$ **66.** $g(x) = x^2 - x$

67. $h(x) = x^2 - x^4$ **68.** $h(x) = 2x^2 + x^4$

69. $f(x) = x^2 - x^4 + 1$ **70.** $f(x) = 2x^2 + x^4 + 1$

71. $f(x) = \frac{1}{5}x^6 - 3x^2$ **72.** $f(x) = 2x^3 - 6x^5$

73. $f(x) = x\sqrt{1 - x^2}$ **74.** $f(x) = x^2\sqrt{1 - x^2}$

In Exercises 75–78, use possible symmetry to determine whether each graph is the graph of an even function, an odd function, or a function that is neither even nor odd.

75.

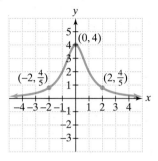

76.

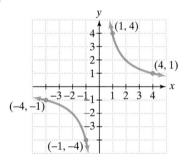

77.

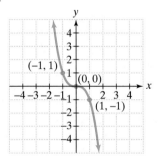

78.

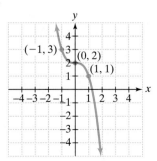

In Exercises 79–84, if $f(x) = \text{int}(x)$, find each function value.

79. $f(1.06)$ **80.** $f(2.99)$

81. $f\left(\frac{1}{3}\right)$ **82.** $f(-1.5)$

83. $f(-2.3)$ **84.** $f(-99.001)$

Application Exercises

The figure shows the percentage of the U.S. population made up of Jewish Americans, $f(x)$, as a function of time, x, where x is the number of years after 1900. Use the graph to solve Exercises 85–92.

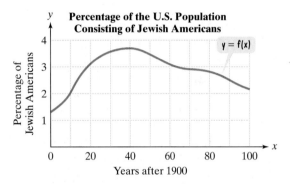

Source: American Jewish Yearbook

85. Use the graph to find a reasonable estimate of $f(60)$. What does this mean in terms of the variables in this situation?

86. Use the graph to find a reasonable estimate of $f(100)$. What does this mean in terms of the variables in this situation?

87. For what value or values of x is $f(x) = 3$? Round to the nearest year. What does this mean in terms of the variables in this situation?

88. For what value or values of x is $f(x) = 2.5$? Round to the nearest year. What does this mean in terms of the variables in this situation?

89. In which year did the percentage of Jewish Americans in the U.S. population reach a maximum? What is a reasonable estimate of the percentage for that year?

90. In which year was the percentage of Jewish Americans in the U.S. population at a minimum? What is a reasonable estimate of the percentage for that year?

91. Explain why f represents the graph of a function.

92. Describe the general trend shown by the graph.

The function

$$f(x) = 0.4x^2 - 36x + 1000$$

models the number of accidents, $f(x)$, per 50 million miles driven as a function of the driver's age, x, in years, where x includes drivers from ages 16 through 74. The graph of f is shown. Use the graph of f, and possibly the equation, to solve Exercises 93–95.

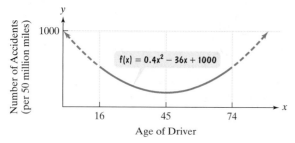

Age of Driver

93. State the intervals on which the function is increasing and decreasing and describe what this means in terms of the variables modeled by the function.

94. For what value of x does the graph reach its lowest point? What is the minimum value of y? Describe the practical significance of this minimum value.

95. Use the graph to identify two different ages for which drivers have the same number of accidents. Use the equation for f to find the number of accidents for drivers at each of these ages.

96. Based on a study by Vance Tucker (*Scientific American*, May 1969), the power expenditure of migratory birds in flight is a function of their flying speed, x, in miles per hour, modeled by $f(x) = 0.67x^2 - 27.74x + 387$. Power expenditure, $f(x)$, is measured in calories, and migratory birds generally fly between 12 and 30 miles per hour. The graph of f is shown in the figure below, with a domain of $[12, 30]$.

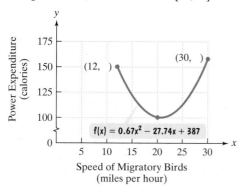

Speed of Migratory Birds
(miles per hour)

a. State the intervals on which the function is increasing and decreasing and describe what this means in terms of the variables modeled by the function.

b. For what approximate value of x does the graph reach its lowest point? What is the minimum value of y? Describe the practical significance of this minimum value.

97. The cost of a telephone call between two cities is $0.10 for the first minute and $0.05 for each additional minute or portion of a minute. Draw a graph of the cost, C, in dollars, of the phone call as a function of time, t, in minutes, on the interval $(0, 5]$.

98. A cargo service charges a flat fee of $4 plus $1 for each pound or fraction of a pound to mail a package. Let $C(x)$ represent the cost to mail a package that weighs x pounds. Graph the cost function on the interval $(0, 5]$.

99. Researchers at Yale University have suggested that levels of passion and commitment in human relations are functions of time. Based on the shapes of the graphs shown, which do you think depicts passion and which represents commitment? Explain how you arrived at your answer.

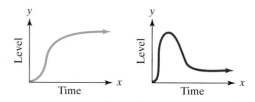

Writing in Mathematics

100. Discuss one disadvantage to using point plotting as a method for graphing functions.

101. Explain how to use a function's graph to find the function's domain and range.

102. Explain how the vertical line test is used to determine whether a graph is a function.

103. What does it mean if a function f is increasing on an interval?

104. Suppose that a function f is increasing on (a, b) and decreasing on (b, c). Describe what occurs at $x = b$. What does the function value $f(b)$ represent?

105. What is a secant line?

106. What is the average rate of change of a function?

107. If you are given a function's equation, how do you determine if the function is even, odd, or neither?

108. If you are given a function's graph, how do you determine if the function is even, odd, or neither?

109. What is a step function? Give an example of an everyday situation that can be modeled using such a function. Do not use the cost-of-mail example.

110. Explain how to find int(-3.000004).

Technology Exercises

111. The function
$$f(x) = -0.00002x^3 + 0.008x^2 - 0.3x + 6.95$$
models the number of annual physician visits, $f(x)$, by a person of age x. Graph the function in a $[0, 100, 5]$ by $[0, 40, 2]$ viewing rectangle. What does the shape of the graph indicate about the relationship between one's age and the number of annual physician visits? Use the $\boxed{\text{TRACE}}$ or minimum function capability to find the coordinates of the minimum point on the graph of the function. What does this mean?

In Exercises 112–117, use a graphing utility to graph each function. Use a $[-5, 5, 1]$ by $[-5, 5, 1]$ viewing rectangle. Then find the intervals on which the function is increasing, decreasing, or constant.

112. $f(x) = x^3 - 6x^2 + 9x + 1$ **113.** $g(x) = |4 - x^2|$

114. $h(x) = |x - 2| + |x + 2|$ **115.** $f(x) = x^{1/3}(x - 4)$

116. $g(x) = x^{2/3}$ **117.** $h(x) = 2 - x^{2/5}$

118. a. Graph the functions $f(x) = x^n$ for $n = 2, 4,$ and 6 in a $[-2, 2, 1]$ by $[-1, 3, 1]$ viewing rectangle.

    b. Graph the functions $f(x) = x^n$ for $n = 1, 3,$ and 5 in a $[-2, 2, 1]$ by $[-2, 2, 1]$ viewing rectangle.

    c. If n is even, where is the graph of $f(x) = x^n$ increasing and where is it decreasing?

    d. If n is odd, what can you conclude about the graph of $f(x) = x^n$ in terms of increasing or decreasing behavior?

    e. Graph all six functions in a $[-1, 3, 1]$ by $[-1, 3, 1]$ viewing rectangle. What do you observe about the graphs in terms of how flat or how steep they are?

Critical Thinking Exercises

119. Which statement at the top of the next column is true based on the graph of f in the figure?

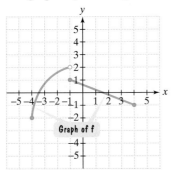

a. The domain of f is $[-4, 1)$ or $(1, 4]$.

b. The range of f is $[-2, 2]$.

c. $f(-1) - f(4) = 2$

d. $f(0) = 2.1$

120. Sketch the graph of f using the following properties. (More than one correct graph is possible.) f is a piecewise function that is decreasing on $(-\infty, 2)$, $f(2) = 0$, f is increasing on $(2, \infty)$, and the range of f is $[0, \infty)$.

121. Define a piecewise function on the intervals $(-\infty, 2]$, $(2, 5)$, and $[5, \infty)$ that does not "jump" at 2 or 5 such that one piece is a constant function, another piece is an increasing function, and the third piece is a decreasing function.

122. Suppose that $h(x) = \dfrac{f(x)}{g(x)}$. The function f can be even, odd, or neither. The same is true for the function g.

    a. Under what conditions is h definitely an even function?

    b. Under what conditions is h definitely an odd function?

123. Take another look at the cost of first-class mail and its graph (Table 1.3 on page 175 and Figure 1.48 on page 176). Change the description of the heading in the left column of Table 1.3 so that the graph includes the point on the left of each horizontal step, but does not include the point on the right.

Group Exercise

124. In Exercise 99, passion and commitment are graphed over time. For this activity, you will be creating a graph of a particular experience that involved your feelings of love, anger, sadness, or any other emotion you choose. The horizontal axis should be labeled time and the vertical axis the emotion you are graphing. You will not be using your algebra skills to create your graph; however, you should try to make the graph as precise as possible. You may use negative numbers on the vertical axis, if appropriate. After each group member has created a graph, pool together all of the graphs and study them to see if there are any similarities in the graphs for a particular emotion or for all emotions.

SECTION 1.6 *Transformations of Functions*

Objectives

1. Recognize graphs of common functions.
2. Use vertical shifts to graph functions.
3. Use horizontal shifts to graph functions.
4. Use reflections to graph functions.
5. Use vertical stretching and shrinking to graph functions.
6. Graph functions involving a sequence of transformations.

Have you seen *Terminator 2, The Mask,* or *The Matrix*? These were among the first films to use spectacular effects in which a character or object having one shape was transformed in a fluid fashion into a quite different shape. The name for such a transformation is **morphing.** The effect allows a real actor to be seamlessly transformed into a computer-generated animation. The animation can be made to perform impossible feats before it is morphed back to the conventionally filmed image.

Like transformed movie images, the graph of one function can be turned into the graph of a different function. To do this, we need to rely on a function's equation. Knowing that a graph is a transformation of a familiar graph makes graphing easier.

1 Recognize graphs of common functions.

Graphs of Common Functions

Table 1.4 below and on page 185 gives names to six frequently encountered functions in algebra. The table shows each function's graph and lists characteristics of the function. Study the shape of each graph and take a few minutes to verify the function's characteristics from its graph. Knowing these graphs is essential for analyzing their transformations into more complicated graphs.

Table 1.4   Algebra's Common Graphs

Constant Function	Indentity Function	Standard Quadratic Function
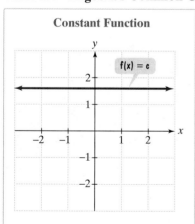	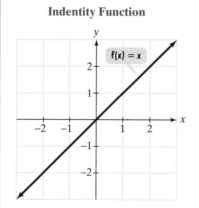	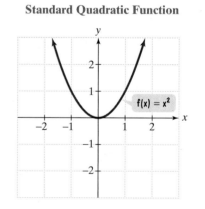
• Domain: $(-\infty, \infty)$ • Range: the single number c • Constant on $(-\infty, \infty)$ • Even function	• Domain: $(-\infty, \infty)$ • Range: $(-\infty, \infty)$ • Increasing on $(-\infty, \infty)$ • Odd function	• Domain: $(-\infty, \infty)$ • Range: $[0, \infty)$ • Decreasing on $(-\infty, 0)$ and increasing on $(0, \infty)$ • Even function

Table 1.4 Algebra's Common Graphs (*continued*)

Standard Cubic Function	Square Root Function	Absolute Value Function
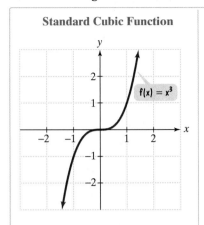	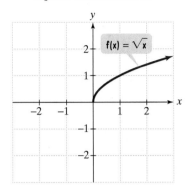	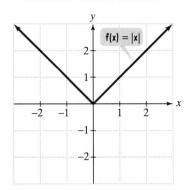
• Domain: $(-\infty, \infty)$ • Range: $(-\infty, \infty)$ • Increasing on $(-\infty, \infty)$ • Odd function	• Domain: $[0, \infty)$ • Range: $[0, \infty)$ • Increasing on $(0, \infty)$ • Neither even nor odd function	• Domain: $(-\infty, \infty)$ • Range: $[0, \infty)$ • Decreasing on $(-\infty, 0)$ and increasing on $(0, \infty)$ • Even function

Discovery

The study of how changing a function's equation can affect its graph can be explored with a graphing utility. Use your graphing utility to verify the hand-drawn graphs as you read this section.

2 Use vertical shifts to graph functions.

Vertical Shifts

Let's begin by looking at three graphs whose shapes are the same. Figure 1.50 shows the graphs. The black graph in the middle is the standard quadratic function, $f(x) = x^2$. Now, look at the blue graph on the top. The equation of this graph, $g(x) = x^2 + 2$, adds 2 to the right side of $f(x) = x^2$. What effect does this have on the graph of f? It shifts the graph vertically up by 2 units.

$$g(x) = x^2 + 2 = f(x) + 2$$

The graph of g shifts the graph of f up **2** units.

Finally, look at the red graph on the bottom of Figure 1.50. The equation of this graph, $h(x) = x^2 - 3$, subtracts 3 from the right side of $f(x) = x^2$. What effect does this have on the graph of f? It shifts the graph vertically down by 3 units.

$$h(x) = x^2 - 3 = f(x) - 3$$

The graph of h shifts the graph of f down **3** units.

Figure 1.50 Vertical shifts
The graph of $f(x) = x^2$ can be gradually morphed into the graph of $g(x) = x^2 + 2$ by using animation to graph $f(x) = x^2 + c$ for $0 \le c \le 2$. By selecting many values for c, we can create an animated sequence in which change appears to occur continuously.

In general, if c is positive, $y = f(x) + c$ shifts the graph of f upward c units and $y = f(x) - c$ shifts the graph of f downward c units. These are called **vertical shifts** of the graph of f.

Vertical Shifts

Let f be a function and c a positive real number.
- The graph of $y = f(x) + c$ is the graph of $y = f(x)$ shifted c units vertically upward.
- The graph of $y = f(x) - c$ is the graph of $y = f(x)$ shifted c units vertically downward.

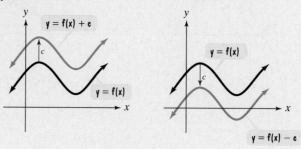

EXAMPLE 1 Vertical Shift Down

Use the graph of $f(x) = |x|$ to obtain the graph of $g(x) = |x| - 4$.

Solution The graph of $g(x) = |x| - 4$ has the same shape as the graph of $f(x) = |x|$. However, it is shifted down vertically 4 units. We have constructed a table showing some of the coordinates for f and g. The graphs of f and g are shown in Figure 1.51.

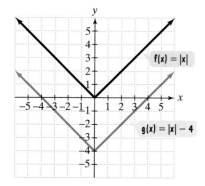

Figure 1.51

x	$y = f(x) = |x|$	$(x, f(x))$	$y = g(x)$ $= |x| - 4 = f(x) - 4$	$(x, g(x))$
-2	$|-2| = 2$	$(-2, 2)$	$|-2| - 4 = -2$	$(-2, -2)$
-1	$|-1| = 1$	$(-1, 1)$	$|-1| - 4 = -3$	$(-1, -3)$
0	$|0| = 0$	$(0, 0)$	$|0| - 4 = -4$	$(0, -4)$
1	$|1| = 1$	$(1, 1)$	$|1| - 4 = -3$	$(1, -3)$
2	$|2| = 2$	$(2, 2)$	$|2| - 4 = -2$	$(2, -2)$

Check Point 1 Use the graph of $f(x) = |x|$ to obtain the graph of $g(x) = |x| + 3$.

3 Use horizontal shifts to graph functions.

Horizontal Shifts

We return to the graph of $f(x) = x^2$, the standard quadratic function. In Figure 1.52 on the next page, the graph of function f is in the middle of the three graphs.

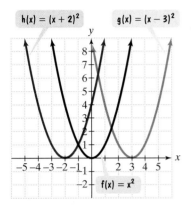

$h(x) = (x + 2)^2$ $g(x) = (x - 3)^2$

$f(x) = x^2$

Figure 1.52   Horizontal shifts

By contrast to the vertical shift situation, this time there are graphs to the left and to the right of the graph of f. Look at the blue graph on the right. The equation of this graph, $g(x) = (x - 3)^2$, subtracts 3 from each value of x in the domain of $f(x) = x^2$. What effect does this have on the graph of f? It shifts the graph horizontally to the right by 3 units.

$$g(x) = (x - 3)^2 = f(x - 3)$$

The graph of g   shifts the graph of f 3 units to the right.

Now, look at the red graph on the left in Figure 1.52. The equation of this graph, $h(x) = (x + 2)^2$, adds 2 to each value of x in the domain of $f(x) = x^2$. What effect does this have on the graph of f? It shifts the graph horizontally to the left by 2 units.

$$h(x) = (x + 2)^2 = f(x + 2)$$

The graph of h   shifts the graph of f 2 units to the left.

In general, if c is positive, $y = f(x + c)$ shifts the graph of f to the left c units and $y = f(x - c)$ shifts the graph of f to the right c units. These are called **horizontal shifts** of the graph of f.

Study Tip

We know that positive numbers are to the right of zero on a number line and negative numbers are to the left of zero. This positive-negative orientation does not apply to horizontal shifts. A *positive* number causes a shift to the *left* and a *negative* number causes a shift to the *right*.

Horizontal Shifts

Let f be a function and c a positive real number.

• The graph of $y = f(x + c)$ is the graph of $y = f(x)$ shifted to the left c units.
• The graph of $y = f(x - c)$ is the graph of $y = f(x)$ shifted to the right c units.

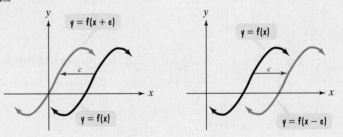

EXAMPLE 2   Horizontal Shift to the Left

Use the graph of $f(x) = \sqrt{x}$ to obtain the graph of $g(x) = \sqrt{x + 5}$.

Solution   Compare the equations for $f(x) = \sqrt{x}$ and $g(x) = \sqrt{x + 5}$. The equation for g adds 5 to each value of x in the domain of f.

$$y = g(x) = \sqrt{x + 5} = f(x + 5)$$

The graph of g   shifts the graph of f 5 units to the left.

The graph of $g(x) = \sqrt{x + 5}$ has the same shape as the graph of $f(x) = \sqrt{x}$. However, it is shifted horizontally to the left 5 units. We have created tables

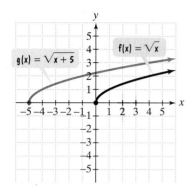

Figure 1.53 Shifting $f(x) = \sqrt{x}$ five units left

showing some of the coordinates for f and g. As shown in Figure 1.53, every point in the graph of g is exactly 5 units to the left of a corresponding point on the graph of f.

x	$y = f(x) = \sqrt{x}$	$(x, f(x))$
0	$\sqrt{0} = 0$	$(0, 0)$
1	$\sqrt{1} = 1$	$(1, 1)$
4	$\sqrt{4} = 2$	$(4, 2)$

x	$y = g(x) = \sqrt{x + 5}$	$(x, g(x))$
-5	$\sqrt{-5 + 5} = \sqrt{0} = 0$	$(-5, 0)$
-4	$\sqrt{-4 + 5} = \sqrt{1} = 1$	$(-4, 1)$
-1	$\sqrt{-1 + 5} = \sqrt{4} = 2$	$(-1, 2)$

Check Point 2 Use the graph of $f(x) = \sqrt{x}$ to obtain the graph of $g(x) = \sqrt{x - 4}$.

Some functions can be graphed by combining horizontal and vertical shifts. These functions will be variations of a function whose equation you know how to graph, such as the standard quadratic function, the standard cubic function, the square root function, or the absolute value function.

In our next example, we will use the graph of the standard quadratic function, $f(x) = x^2$, to obtain the graph of $h(x) = (x + 1)^2 - 3$. We will graph three functions:

$$f(x) = x^2 \qquad g(x) = (x + 1)^2 \qquad h(x) = (x + 1)^2 - 3.$$

> Start by graphing the standard quadratic function.

> Shift the graph of f horizontally one unit to the left.

> Shift the graph of g vertically down 3 units.

EXAMPLE 3 Combining Horizontal and Vertical Shifts

Use the graph of $f(x) = x^2$ to obtain the graph of $h(x) = (x + 1)^2 - 3$.

Solution

Step 1 Graph $f(x) = x^2$. The graph of the standard quadratic function is shown in Figure 1.54(a). We've identified three points on the graph.

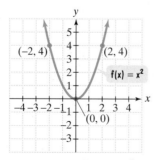

(a) The graph of $f(x) = x^2$

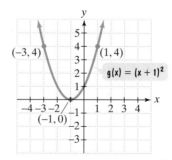

(b) The graph of $g(x) = (x + 1)^2$

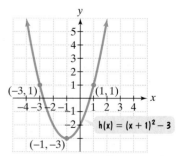

(c) The graph of $h(x) = (x + 1)^2 - 3$

Figure 1.54

Discovery

Work Example 3 by first shifting the graph of $f(x) = x^2$ three units down, graphing $g(x) = x^2 - 3$. Now, shift this graph one unit left to graph $h(x) = (x + 1)^2 - 3$. Did you obtain the graph in Figure 1.54(c)? What can you conclude?

Step 2   Graph $g(x) = (x + 1)^2$. Because we add 1 to each value of x in the domain of the standard quadratic function, $f(x) = x^2$, we shift the graph of f horizontally one unit to the left. This is shown in Figure 1.54(b). Notice that every point in the graph in Figure 1.54(b) has an x-coordinate that is one less than the x-coordinate for the corresponding point in the graph in Figure 1.54(a).

Step 3   Graph $h(x) = (x + 1)^2 - 3$. Because we subtract 3, we shift the graph in Figure 1.54(b) vertically down 3 units. The graph is shown in Figure 1.54(c). Notice that every point in the graph in Figure 1.54(c) has a y-coordinate that is three less than the y-coordinate of the corresponding point in the graph in Figure 1.54(b).

Check Point 3   Use the graph of $f(x) = \sqrt{x}$ to obtain the graph of $h(x) = \sqrt{x - 1} - 2$.

4 Use reflections to graph functions.

Reflections of Graphs

This photograph shows a reflection of an old bridge in a Maryland river. This perfect reflection occurs because the surface of the water is absolutely still. A mild breeze rippling the water's surface would distort the reflection.

Is it possible for graphs to have mirror-like qualities? Yes. Figure 1.55 shows the graphs of $f(x) = x^2$ and $g(x) = -x^2$. The graph of g is a **reflection about the x-axis** of the graph of f. In general, the graph of $y = -f(x)$ reflects the graph of f about the x-axis. Thus, the graph of g is a reflection of the graph of f about the x-axis because

$$g(x) = -x^2 = -f(x).$$

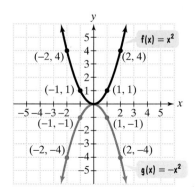

Figure 1.55   Reflections about the x-axis

Reflection about the x-Axis

The graph of $y = -f(x)$ is the graph of $y = f(x)$ reflected about the x-axis.

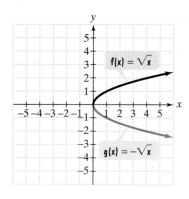

Figure 1.56 Reflecting $f(x) = \sqrt{x}$ about the x-axis

EXAMPLE 4   Reflection about the x-Axis

Use the graph of $f(x) = \sqrt{x}$ to obtain the graph of $g(x) = -\sqrt{x}$.

Solution   Compare the equations for $f(x) = \sqrt{x}$ and $g(x) = -\sqrt{x}$. The graph of g is a reflection about the x-axis of the graph of f because

$$g(x) = -\sqrt{x} = -f(x).$$

We have created a table showing some of the coordinates for f and g. The graphs of f and g are shown in Figure 1.56.

x	$f(x) = \sqrt{x}$	$(x, f(x))$	$g(x) = -\sqrt{x}$	$(x, g(x))$
0	$\sqrt{0} = 0$	$(0, 0)$	$-\sqrt{0} = 0$	$(0, 0)$
1	$\sqrt{1} = 1$	$(1, 1)$	$-\sqrt{1} = -1$	$(1, -1)$
4	$\sqrt{4} = 2$	$(4, 2)$	$-\sqrt{4} = -2$	$(4, -2)$

> **Check Point 4**   Use the graph of $f(x) = |x|$ to obtain the graph of $g(x) = -|x|$.

It is also possible to reflect graphs about the y-axis.

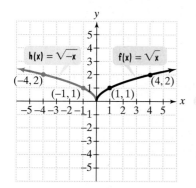

Figure 1.57 Reflecting $f(x) = \sqrt{x}$ about the y-axis

Reflection about the y-Axis
The graph of $y = f(-x)$ is the graph of $y = f(x)$ reflected about the y-axis.

EXAMPLE 5   Reflection about the y-Axis

Use the graph of $f(x) = \sqrt{x}$ to obtain the graph of $h(x) = \sqrt{-x}$.

Solution   Compare the equations for $f(x) = \sqrt{x}$ and $h(x) = \sqrt{-x}$. The graph of h is a reflection about the y-axis of the graph of f because

$$h(x) = \sqrt{-x} = f(-x).$$

We have created tables showing some of the coordinates for f and h. The graphs of f and h are shown in Figure 1.57.

x	$f(x) = \sqrt{x}$	$(x, f(x))$
0	$\sqrt{0} = 0$	$(0, 0)$
1	$\sqrt{1} = 1$	$(1, 1)$
4	$\sqrt{4} = 2$	$(4, 2)$

x	$h(x) = \sqrt{-x}$	$(x, h(x))$
0	$\sqrt{-0} = \sqrt{0} = 0$	$(0, 0)$
-1	$\sqrt{-(-1)} = \sqrt{1} = 1$	$(-1, 1)$
-4	$\sqrt{-(-4)} = \sqrt{4} = 2$	$(-4, 2)$

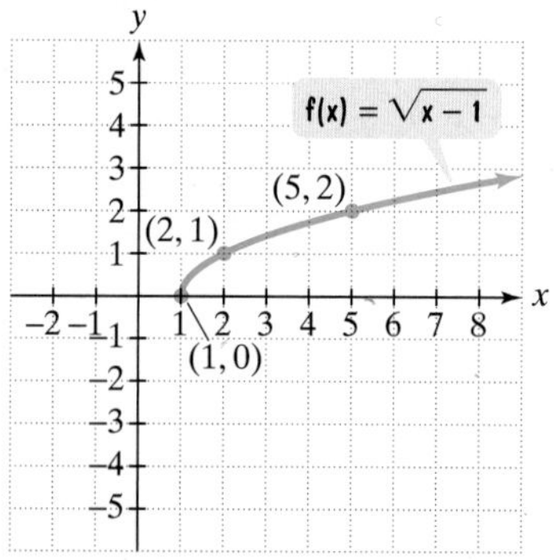

Figure 1.58

> **Check Point 5**   Use the graph of $f(x) = \sqrt{x - 1}$ in Figure 1.58 to obtain the graph of $h(x) = \sqrt{-x - 1}$.

5 Use vertical stretching and shrinking to graph functions.

Vertical Stretching and Shrinking

Morphing does much more than move an image horizontally, vertically, or about an axis. An object having one shape is transformed into a different shape. Horizontal shifts, vertical shifts, and reflections do not change the basic shape of a graph. How can we shrink and stretch graphs, thereby altering their basic shapes?

Look at the three graphs in Figure 1.59. The black graph in the middle is the graph of the standard quadratic function, $f(x) = x^2$. Now, look at the blue graph on the top. The equation of this graph is $g(x) = 2x^2$. Thus, for each x, the y-coordinate of g is 2 times as large as the corresponding y-coordinate on the graph of f. The result is a narrower graph. We say that the graph of g is obtained by vertically *stretching* the graph of f. Now, look at the red graph on the bottom. The equation of this graph is $h(x) = \frac{1}{2}x^2$, or $h(x) = \frac{1}{2}f(x)$. Thus, for each x, the y-coordinate of h is one-half as large as the corresponding y-coordinate on the graph of f. The result is a wider graph. We say that the graph of h is obtained by vertically *shrinking* the graph of f.

These observations can be summarized as follows:

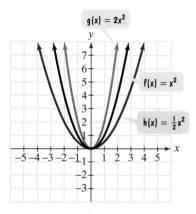

Figure 1.59 Stretching and shrinking $f(x) = x^2$

Stretching and Shrinking Graphs

Let f be a function and c a positive real number.
- If $c > 1$, the graph of $y = cf(x)$ is the graph of $y = f(x)$ vertically stretched by multiplying each of its y-coordinates by c.
- If $0 < c < 1$, the graph of $y = cf(x)$ is the graph of $y = f(x)$ vertically shrunk by multiplying each of its y-coordinates by c.

EXAMPLE 6 Vertically Stretching a Graph

Use the graph of $f(x) = |x|$ to obtain the graph of $g(x) = 2|x|$.

Solution The graph of $g(x) = 2|x|$ is obtained by vertically stretching the graph of $f(x) = |x|$. We have constructed a table showing some of the coordinates for f and g. Observe that the y-coordinate on the graph of g is twice as large as the corresponding y-coordinate on the graph of f. The graphs of f and g are shown in Figure 1.60.

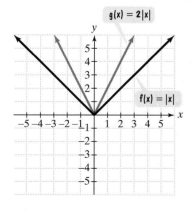

Figure 1.60 Stretching $f(x) = |x|$

x	$f(x) = |x|$	$(x, f(x))$	$g(x) = 2|x| = 2f(x)$	$(x, g(x))$
-2	$|-2| = 2$	$(-2, 2)$	$2|-2| = 4$	$(-2, 4)$
-1	$|-1| = 1$	$(-1, 1)$	$2|-1| = 2$	$(-1, 2)$
0	$|0| = 0$	$(0, 0)$	$2|0| = 0$	$(0, 0)$
1	$|1| = 1$	$(1, 1)$	$2|1| = 2$	$(1, 2)$
2	$|2| = 2$	$(2, 2)$	$2|2| = 4$	$(2, 4)$

Check Point 6 Use the graph of $f(x) = |x|$ to obtain the graph of $g(x) = 3|x|$.

EXAMPLE 7 Vertically Shrinking a Graph

Use the graph of $f(x) = |x|$ to obtain the graph of $h(x) = \frac{1}{2}|x|$.

Solution The graph of $h(x) = \frac{1}{2}|x|$ is obtained by vertically shrinking the graph of $f(x) = |x|$. We have constructed a table showing some of the coordinates for f and h. Observe that the y-coordinate on the graph of h is one-half the corresponding y-coordinate on the graph of f. The graphs of f and h are shown in Figure 1.61.

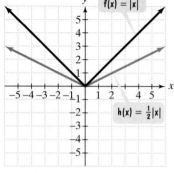

Figure 1.61 Shrinking $f(x) = |x|$

x	$f(x) = |x|$	$(x, f(x))$	$h(x) = \frac{1}{2}|x| = \frac{1}{2}f(x)$	$(x, h(x))$
-2	$|-2| = 2$	$(-2, 2)$	$\frac{1}{2}|-2| = 1$	$(-2, 1)$
-1	$|-1| = 1$	$(-1, 1)$	$\frac{1}{2}|-1| = \frac{1}{2}$	$(-1, \frac{1}{2})$
0	$|0| = 0$	$(0, 0)$	$\frac{1}{2}|0| = 0$	$(0, 0)$
1	$|1| = 1$	$(1, 1)$	$\frac{1}{2}|1| = \frac{1}{2}$	$(1, \frac{1}{2})$
2	$|2| = 2$	$(2, 2)$	$\frac{1}{2}|2| = 1$	$(2, 1)$

Check Point 7 Use the graph of $f(x) = |x|$ to obtain the graph of $h(x) = \frac{1}{4}|x|$.

6 Graph functions involving a sequence of transformations.

Sequences of Transformations

Table 1.5 summarizes the procedures for transforming the graph of $y = f(x)$.

Table 1.5 Summary of Transformations
In each case, c represents a positive real number.

To Graph:	Draw the Graph of f and:	Changes in the Equation of $y = f(x)$
Vertical shifts $y = f(x) + c$ $y = f(x) - c$	Raise the graph of f by c units. Lower the graph of f by c units.	c is added to $f(x)$. c is subtracted from $f(x)$.
Horizontal shifts $y = f(x + c)$ $y = f(x - c)$	Shift the graph of f to the left c units. Shift the graph of f to the right c units.	x is replaced with $x + c$. x is replaced with $x - c$.
Reflection about the x-axis $y = -f(x)$	Reflect the graph of f about the x-axis.	$f(x)$ is multiplied by -1.
Reflection about the y-axis $y = f(-x)$	Reflect the graph of f about the y-axis.	x is replaced with $-x$.
Vertical stretching or shrinking $y = cf(x), c > 1$	Multiply each y-coordinate of $y = f(x)$ by c, vertically stretching the graph of f.	$f(x)$ is multiplied by $c, c > 1$.
$y = cf(x), 0 < c < 1$	Multiply each y-coordinate of $y = f(x)$ by c, vertically shrinking the graph of f.	$f(x)$ is multiplied by $c, 0 < c < 1$.

A function involving more than one transformation can be graphed by performing transformations in the following order:

1. Horizontal shifting
2. Vertical stretching or shrinking
3. Reflecting
4. Vertical shifting.

EXAMPLE 8 Graphing Using a Sequence of Transformations

Use the graph of $f(x) = \sqrt{x}$ to graph $g(x) = \sqrt{1 - x} + 3$.

Solution The following sequence of steps is illustrated in Figure 1.62. We begin with the graph of $f(x) = \sqrt{x}$.

Step 1 Horizontal Shifting Graph $y = \sqrt{x + 1}$. Because x is replaced with $x + 1$, the graph of $f(x) = \sqrt{x}$ is shifted 1 unit to the left.

Step 2 Vertical Stretching or Shrinking Because the equation $y = \sqrt{x + 1}$ is not multiplied by a constant in $g(x) = \sqrt{1 - x} + 3$, no stretching or shrinking is involved.

Step 3 Reflecting We are interested in graphing $y = \sqrt{1 - x} + 3$, or $y = \sqrt{-x + 1} + 3$. We have now graphed $y = \sqrt{x + 1}$. We can graph $y = \sqrt{-x + 1}$ by noting that x is replaced with $-x$. Thus, we graph $y = \sqrt{-x + 1}$ by reflecting the graph of $y = \sqrt{x + 1}$ about the y-axis.

Step 4 Vertical Shifting We can use the graph of $y = \sqrt{1 - x}$ to get the graph of $g(x) = \sqrt{1 - x} + 3$. Because 3 is added, shift the graph of $y = \sqrt{1 - x}$ up 3 units.

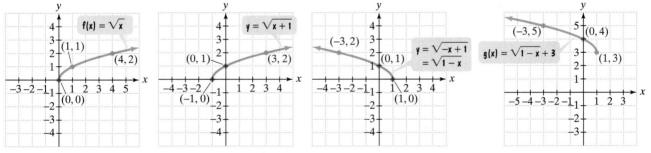

Figure 1.62 Using $f(x) = \sqrt{x}$ to graph $g(x) = \sqrt{1 - x} + 3$

Check Point 8 Use the graph of $f(x) = x^2$ to graph $g(x) = -(x - 2)^2 + 3$.

EXERCISE SET 1.6

Practice Exercises

In Exercises 1–10, begin by graphing the standard quadratic function, $f(x) = x^2$. Then use transformations of this graph to graph the given function.

1. $g(x) = x^2 - 2$
2. $g(x) = x^2 - 1$
3. $g(x) = (x - 2)^2$
4. $g(x) = (x - 1)^2$
5. $h(x) = -(x - 2)^2$
6. $h(x) = -(x - 1)^2$
7. $h(x) = (x - 2)^2 + 1$
8. $h(x) = (x - 1)^2 + 2$
9. $g(x) = 2(x - 2)^2$
10. $g(x) = \frac{1}{2}(x - 1)^2$

In Exercises 11–22, begin by graphing the square root function, $f(x) = \sqrt{x}$. Then use transformations of this graph to graph the given function.

11. $g(x) = \sqrt{x} + 2$

12. $g(x) = \sqrt{x} + 1$

13. $g(x) = \sqrt{x + 2}$

14. $g(x) = \sqrt{x + 1}$

15. $h(x) = -\sqrt{x + 2}$

16. $h(x) = -\sqrt{x + 1}$

17. $h(x) = \sqrt{-x + 2}$

18. $h(x) = \sqrt{-x + 1}$

19. $g(x) = \frac{1}{2}\sqrt{x + 2}$

20. $g(x) = 2\sqrt{x + 1}$

21. $h(x) = \sqrt{x + 2} - 2$

22. $h(x) = \sqrt{x + 1} - 1$

In Exercises 23–34, begin by graphing the absolute value function, $f(x) = |x|$. Then use transformations of this graph to graph the given function.

23. $g(x) = |x| + 4$

24. $g(x) = |x| + 3$

25. $g(x) = |x + 4|$

26. $g(x) = |x + 3|$

27. $h(x) = |x + 4| - 2$

28. $h(x) = |x + 3| - 2$

29. $h(x) = -|x + 4|$

30. $h(x) = -|x + 3|$

31. $g(x) = -|x + 4| + 1$

32. $g(x) = -|x + 4| + 2$

33. $h(x) = 2|x + 4|$

34. $h(x) = 2|x + 3|$

In Exercises 35–44, begin by graphing the standard cubic function, $f(x) = x^3$. Then use transformations of this graph to graph the given function.

35. $g(x) = x^3 - 3$

36. $g(x) = x^3 - 2$

37. $g(x) = (x - 3)^3$

38. $g(x) = (x - 2)^3$

39. $h(x) = -x^3$

40. $h(x) = -(x - 2)^3$

41. $h(x) = \frac{1}{2}x^3$

42. $h(x) = \frac{1}{4}x^3$

43. $r(x) = (x - 3)^3 + 2$

44. $r(x) = (x - 2)^3 + 1$

In Exercises 45–52, use the graph of the function f to sketch the graph of the given function g.

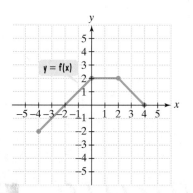

45. $g(x) = f(x) + 1$

46. $g(x) = f(x) + 2$

47. $g(x) = f(x + 1)$

48. $g(x) = f(x + 2)$

49. $g(x) = -f(x)$

50. $g(x) = \frac{1}{2}f(x)$

51. $g(x) = \frac{1}{2}f(x + 1)$

52. $g(x) = -f(x + 2)$

In Exercises 53–56, write a possible equation for the function whose graph is shown. Each graph shows a transformation of a common function.

53.

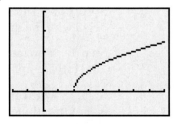

$[-2, 8, 1]$ by $[-1, 4, 1]$

54.

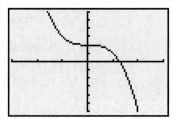

$[-3, 3, 1]$ by $[-6, 6, 1]$

55.

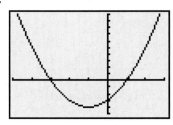

$[-5, 3, 1]$ by $[-5, 10, 1]$

56.

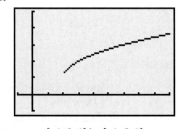

$[-1, 9, 1]$ by $[-1, 5, 1]$

Application Exercises

57. The function $f(x) = 2.9\sqrt{x} + 20.1$ models the median height, $f(x)$, in inches, of boys who are x months of age. The graph of f is shown.

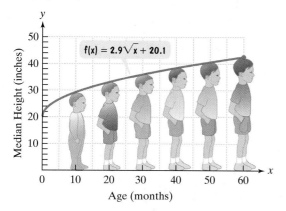

f(x) = 2.9√x + 20.1

a. Describe how the graph can be obtained using transformations of the square root function $f(x) = \sqrt{x}$.
b. According to the model, what is the median height of boys who are 48 months, or four years, old? Use a calculator and round to the nearest tenth of an inch. The actual median height for boys at 48 months is 40.8 inches. How well does the model describe the actual height?
c. Use the model to find the average rate of change, in inches per month, between birth and 10 months. Round to the nearest tenth.
d. Use the model to find the average rate of change, in inches per month, between 50 and 60 months. Round to the nearest tenth. How does this compare with your answer in part (c)? How is this difference shown by the graph?

58. The graph shows the amount of money, in billions of dollars, of new student loans from 1993 through 2000.

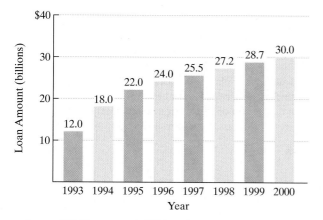

Amount of New Student Loans

Source: U.S. Department of Education

The data shown can be modeled by the function $f(x) = 6.75\sqrt{x} + 12$, where $f(x)$ is the amount, in billions of dollars, of new student loans x years after 1993.

a. Describe how the graph of f can be obtained using transformations of the square root function $f(x) = \sqrt{x}$. Then sketch the graph of f over the interval $0 \le x \le 9$. If applicable, use a graphing utility to verify your hand-drawn graph.
b. According to the model, how much was loaned in 2000? Round to the nearest tenth of a billion. How well does the model describe the actual data?
c. Use the model to find the average rate of change, in billions of dollars per year, between 1993 and 1995. Round to the nearest tenth.
d. Use the model to find the average rate of change, in billions of dollars per year, between 1998 and 2000. Round to the nearest tenth. How does this compare with you answer in part (c)? How is this difference shown by your graph?
e. Rewrite the function so that it represents the amount, $f(x)$, in billions of dollars, of new student loans x years after 1995.

Writing in Mathematics

59. What must be done to a function's equation so that its graph is shifted vertically upward?
60. What must be done to a function's equation so that its graph is shifted horizontally to the right?
61. What must be done to a function's equation so that its graph is reflected about the x-axis?
62. What must be done to a function's equation so that its graph is reflected about the y-axis?
63. What must be done to a function's equation so that its graph is stretched?

Technology Exercises

64. a. Use a graphing utility to graph $f(x) = x^2 + 1$.
b. Graph $f(x) = x^2 + 1$, $g(x) = f(2x)$, $h(x) = f(3x)$, and $k(x) = f(4x)$ in the same viewing rectangle.
c. Describe the relationship among the graphs of f, g, h, and k, with emphasis on different values of x for points on all four graphs that give the same y-coordinate.
d. Generalize by describing the relationship between the graph of f and the graph of g, where $g(x) = f(cx)$ for $c > 1$.
e. Try out your generalization by sketching the graphs of $f(cx)$ for $c = 1, c = 2, c = 3,$ and $c = 4$ for a function of your choice.
65. a. Use a graphing utility to graph $f(x) = x^2 + 1$.
b. Graph $f(x) = x^2 + 1$, $g(x) = f(\frac{1}{2}x)$, and $h(x) = f(\frac{1}{4}x)$ in the same viewing rectangle.
c. Describe the relationship among the graphs of f, g, and h, with emphasis on different values of x for points on all three graphs that give the same y-coordinate.
d. Generalize by describing the relationship between the graph of f and the graph of g, where $g(x) = f(cx)$ for $0 < c < 1$.
e. Try out your generalization by sketching the graphs of $f(cx)$ for $c = 1, c = \frac{1}{2},$ and $c = \frac{1}{4}$ for a function of your choice.

Critical Thinking Exercises

66. Which one of the following is true?

    a. If $f(x) = |x|$ and $g(x) = |x + 3| + 3$, then the graph of g is a translation of three units to the right and three units upward of the graph of f.

    b. If $f(x) = -\sqrt{x}$ and $g(x) = \sqrt{-x}$, then f and g have identical graphs.

    c. If $f(x) = x^2$ and $g(x) = 5(x^2 - 2)$, then the graph of g can be obtained from the graph of f by stretching f five units followed by a downward shift of two units.

    d. If $f(x) = x^3$ and $g(x) = -(x - 3)^3 - 4$, then the graph of g can be obtained from the graph of f by moving f three units to the right, reflecting in the x-axis, and then moving the resulting graph down four units.

In Exercises 67–70, functions f and g are graphed in the same rectangular coordinate system. If g is obtained from f through a sequence of transformations, find an equation for g.

67.

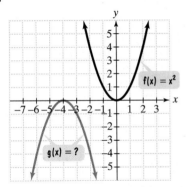

68.

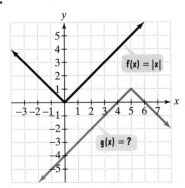

69.

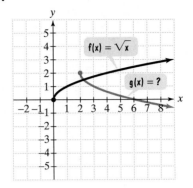

70.

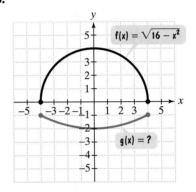

For Exercises 71–74, assume that (a, b) is a point on the graph of f. What is the corresponding point on the graph of each of the following functions?

71. $y = f(-x)$ **72.** $y = 2f(x)$

73. $y = f(x - 3)$ **74.** $y = f(x) - 3$

Group Exercise

75. This activity is a group research project on morphing and should result in a presentation made by group members to the entire class. Be sure to include morphing images that will intrigue class members. You should have no problem finding an array of fascinating images online. Also include a discussion of films using spectacular morphing effects. Rent videos of these films and show appropriate excerpts.

SECTION 1.7   Combinations of Functions; Composite Functions

Objectives

1. Combine functions arithmetically, specifying domains.
2. Form composite functions.
3. Determine domains for composite functions.
4. Write functions as compositions.

They say a fool and his money are soon parted and the rest of us just wait to be taxed. It's hard to believe that the United States was a low-tax country in the early part of the twentieth century. Figure 1.63 shows how the tax burden has grown since then. We can use the information shown to illustrate how two functions can be combined to form a new function. In this section, you will learn how to combine functions to obtain new functions.

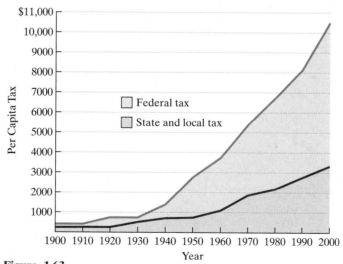

Figure 1.63
Source: Tax Foundation

1 Combine functions arithmetically, specifying domains.

Combinations of Functions

To begin our discussion, take a look at the information shown for the year 2000. The total per capita tax burden is approximately $10,500. The per capita state and local tax is approximately $3400. The per capita federal tax is the difference between these amounts.

$$\text{Per capita federal tax} = \$10,500 - \$3400 = \$7100$$

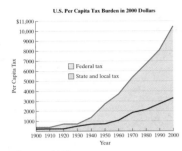

U.S. Per Capita Tax Burden in 2000 Dollars

Figure 1.63, repeated
Source: Tax Foundation

We can think of this subtraction as the subtraction of function values. We do this by introducing the following functions:

Let $T(x)$ = total per capita tax in year x.

Let $S(x)$ = per capita state and local tax in year x.

Using Figure 1.63, we see that

$$T(2000) = \$10{,}500 \quad \text{and} \quad S(2000) = \$3400.$$

We can subtract these function values by introducing a new function, $T - S$, defined by the subtraction of $T(x)$ and $S(x)$. Thus,

$$(T - S)(x) = T(x) - S(x) = \quad \begin{array}{l}\text{total per capita tax in year } x \\ \text{minus state and local per} \\ \text{capita tax in year } x.\end{array}$$

For example,

$$(T - S)(2000) = T(2000) - S(2000) = \$10{,}500 - \$3400 = \$7100.$$

> In 2000, the difference between total tax and state and local tax was $7100. This is the per capita federal tax.

Figure 1.63 illustrates that information involving differences of functions often appears in graphs seen in newspapers and magazines. Like numbers and algebraic expressions, two functions can be added, subtracted multiplied, or divided as long as there are numbers common to the domains of both functions. The common domain for functions T and S in Figure 1.63 is

$$\{1900, 1901, 1902, 1903, \dots, 2000\}.$$

Because functions are usually given as equations, we perform operations by carrying out these operations with the algebraic expressions that appear on the right side of the equations. For example, we can combine the following two functions using addition:

$$f(x) = 2x + 1 \quad \text{and} \quad g(x) = x^2 - 4.$$

To do so, we add the terms to the right of the equal sign for $f(x)$ to the terms to the right of the equal sign for $g(x)$. Here is how it's done:

$$(f + g)(x) = f(x) + g(x)$$
$$= (2x + 1) + (x^2 - 4) \quad \text{Add terms for f(x) and g(x).}$$
$$= 2x - 3 + x^2 \quad \text{Combine like terms.}$$
$$= x^2 + 2x - 3 \quad \text{Arrange terms in descending powers of x.}$$

The name of this new function is $f + g$. Thus, the sum $f + g$ is the function defined by $(f + g)(x) = x^2 + 2x - 3$. The domain of $f + g$ consists of the numbers x that are in the domain of f and in the domain of g. Because neither f nor g contains division or even roots, the domain of each function is the set of all real numbers. Thus, the domain of $f + g$ is also the set of all real numbers.

EXAMPLE 1 Finding the Sum of Two Functions

Let $f(x) = x^2 - 3$ and $g(x) = 4x + 5$. Find

a. $(f + g)(x)$ **b.** $(f + g)(3)$.

Solution

a. $(f + g)(x) = f(x) + g(x) = (x^2 - 3) + (4x + 5) = x^2 + 4x + 2$.
Thus, $(f + g)(x) = x^2 + 4x + 2$.

b. We find $(f + g)(3)$ by substituting 3 for x in the equation for $f + g$.

$(f + g)(x) = x^2 + 4x + 2$ This is the equation for $f + g$.

Substitute 3 for x.

$(f + g)(3) = 3^2 + 4 \cdot 3 + 2 = 9 + 12 + 2 = 23$

Check Point 1 Let $f(x) = 3x^2 + 4x - 1$ and $g(x) = 2x + 7$. Find:

a. $(f + g)(x)$ **b.** $(f + g)(4)$.

Here is a general definition for function addition:

The Sum of Functions

Let f and g be two functions. The **sum $f + g$** is the function defined by

$$(f + g)(x) = f(x) + g(x).$$

The domain of $f + g$ is the set of all real numbers that are common to the domain of f and the domain of g.

EXAMPLE 2 Adding Functions and Determining the Domain

Let $f(x) = \sqrt{x + 3}$ and $g(x) = \sqrt{x - 2}$. Find:

a. $(f + g)(x)$ **b.** the domain of $f + g$.

Solution

a. $(f + g)(x) = f(x) + g(x) = \sqrt{x + 3} + \sqrt{x - 2}$

b. The domain of $f + g$ is the set of all real numbers that are common to the domain of f and the domain of g. Thus, we must find the domains of f and g. We will do so for f first.

Note that $f(x) = \sqrt{x + 3}$ is a function involving the square root of $x + 3$. Because the square root of a negative quantity is not a real number, the value of $x + 3$ must be nonnegative. Thus, the domain of f is all x such that $x + 3 \geq 0$. Equivalently, the domain is $\{x | x \geq -3\}$, or $[-3, \infty)$.

Likewise, $g(x) = \sqrt{x - 2}$ is also a square root function. Because the square root of a negative quantity is not a real number, the value of $x - 2$ must be nonnegative. Thus, the domain of g is all x such that $x - 2 \geq 0$. Equivalently, the domain is $\{x | x \geq 2\}$, or $[2, \infty)$.

Now, we can use a number line to determine the domain of $f + g$. Figure 1.64 shows the domain of f in blue and the domain of g in red. Can you see that all real numbers greater than or equal to 2 are common to both domains? This is shown in purple on the number line. Thus, the domain of $f + g$ is $[2, \infty)$.

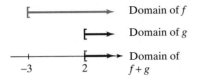

Domain of f

Domain of g

Domain of $f + g$

Figure 1.64 Finding the domain of the sum $f + g$

Technology

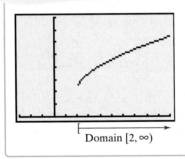

Domain $[2, \infty)$

The graph on the left is the graph of

$$y = \sqrt{x + 3} + \sqrt{x - 2}$$

in a $[-3, 10, 1]$ by $[0, 8, 1]$ viewing rectangle. The graph reveals what we discovered algebraically in Example 2(b). The domain of this function is $[2, \infty)$.

Check Point 2 Let $f(x) = \sqrt{x - 3}$ and $g(x) = \sqrt{x + 1}$. Find:

a. $(f + g)(x)$ **b.** the domain of $f + g$.

We can also combine functions using subtraction, multiplication, and division by performing operations with the algebraic expressions that appear on the right side of the equations.

Definitions: Sum, Difference, Product, and Quotient of Functions

Let f and g be two functions. The **sum** $f + g$, the **difference** $f - g$, the **product** fg, and the **quotient** $\frac{f}{g}$ are functions whose domains are the set of all real numbers common to the domains of f and g, defined as follows:

1. Sum: $(f + g)(x) = f(x) + g(x)$

2. Difference: $(f - g)(x) = f(x) - g(x)$

3. Product: $(fg)(x) = f(x) \cdot g(x)$

4. Quotient: $\left(\dfrac{f}{g}\right)(x) = \dfrac{f(x)}{g(x)}$, provided $g(x) \neq 0$

EXAMPLE 3   Combining Functions

If $f(x) = 2x - 1$ and $g(x) = x^2 + x - 2$, find:

a. $(f - g)(x)$ **b.** $(fg)(x)$ **c.** $\left(\dfrac{f}{g}\right)(x)$.

Determine the domain for each function.

Solution

a. $(f - g)(x) = f(x) - g(x)$

This is the definition of the difference $f - g$.

$= (2x - 1) - (x^2 + x - 2)$ Subtract g(x) from f(x).

$= 2x - 1 - x^2 - x + 2$ Perform the subtraction.

$= -x^2 + x + 1$ Combine like terms and arrange terms in descending powers of x.

b. $(fg)(x) = f(x) \cdot g(x)$

This is the definition of the product fg.

$= (2x - 1)(x^2 + x - 2)$ Multiply f(x) and g(x).

$= 2x(x^2 + x - 2) - 1(x^2 + x - 2)$ Multiply each term in the second factor by 2x and −1, respectively.

$= 2x^3 + 2x^2 - 4x - x^2 - x + 2$ Use the distributive property.

$= 2x^3 + (2x^2 - x^2) + (-4x - x) + 2$ Rearrange terms so that like terms are adjacent.

$= 2x^3 + x^2 - 5x + 2$ Combine like terms.

c. $\left(\dfrac{f}{g}\right)(x) = \dfrac{f(x)}{g(x)}$ This is the definition of the quotient $\dfrac{f}{g}$.

$= \dfrac{2x - 1}{x^2 + x - 2}$ Divide the algebraic expressions for f(x) and g(x).

> **Study Tip**
>
> If the function $\dfrac{f}{g}$ can be simplified, determine the domain *before* simplifying.
>
> **EXAMPLE:**
>
> $f(x) = x^2 - 4$ and
> $g(x) = x - 2$
>
> $\left(\dfrac{f}{g}\right)(x) = \dfrac{x^2 - 4}{x - 2}$
>
> Domain of $\dfrac{f}{g}$ is $\{x | x \neq 2\}$.
>
> $= \dfrac{(x + 2)\overset{1}{\cancel{(x - 2)}}}{\underset{1}{\cancel{(x - 2)}}} = x + 2$

Because the equations for f and g do not involve division or contain even roots, the domain of both f and g is the set of all real numbers. Thus, the domain of $f - g$ and fg is the set of all real numbers. However, for $\frac{f}{g}$, the denominator cannot equal zero. We can factor the denominator as follows:

$$\left(\frac{f}{g}\right)(x) = \frac{2x - 1}{x^2 + x - 2} = \frac{2x - 1}{(x + 2)(x - 1)}.$$

Because x + 2 ≠ 0, x ≠ −2.   Because x − 1 ≠ 0, x ≠ 1.

We see that the domain for $\frac{f}{g}$ is the set of all real numbers except −2 and 1: $\{x | x \neq -2, x \neq 1\}$.

Check Point 3 If $f(x) = x - 5$ and $g(x) = x^2 - 1$, find:

a. $(f - g)(x)$ **b.** $(fg)(x)$ **c.** $\left(\dfrac{f}{g}\right)(x)$.

Determine the domain for each function.

2 Form composite functions.

Composite Functions

There is another way of combining two functions. To help understand this new combination, suppose that your computer store is having a sale. The models that are on sale cost either $300 less than the regular price or 85% of the regular price. If x represents the computer's regular price, both discounts can be described with the following functions:

$$f(x) = x - 300 \qquad g(x) = 0.85x.$$

> The computer is on sale for $300 less than its regular price.

> The computer is on sale for 85% of its regular price.

At the store, you bargain with the salesperson. Eventually, she makes an offer you can't refuse: The sale price is 85% of the regular price followed by a $300 reduction:

$$0.85x - 300.$$

> 85% of the regular price

> followed by a $300 reduction

In terms of functions f and g, this offer can be obtained by taking the output of $g(x) = 0.85x$, namely $0.85x$, and using it as the input of f:

$$f(x) = x - 300$$

> Replace x with 0.85x, the output of g(x) = 0.85x.

$$f(0.85x) = 0.85x - 300.$$

Because $0.85x$ is $g(x)$, we can write this last equation as

$$f(g(x)) = 0.85x - 300.$$

We read this equation as "f of g of x is equal to $0.85x - 300$." We call $f(g(x))$ the *composition of the function f with g*, or a *composite function*. This composite function is written $f \circ g$. Thus,

$$(f \circ g)(x) = f(g(x)) = 0.85x - 300.$$

Like all functions, we can evaluate $f \circ g$ for a specified value of x in the function's domain. For example, here's how to find the value of this function at 1400:

$$(f \circ g)(x) = 0.85x - 300 \qquad \text{This composite function describes the offer you cannot refuse.}$$

> Replace x with 1400.

$$(f \circ g)(1400) = 0.85(1400) - 300 = 1190 - 300 = 890.$$

This means that a computer that regularly sells for $1400 is on sale for $890 subject to both discounts.

Before you run out to buy a new computer, let's generalize our discussion of the computer's double discount and define the composition of any two functions.

The Composition of Functions

The **composition of the function f with g** is denoted by $f \circ g$ and is defined by the equation

$$(f \circ g)(x) = f(g(x)).$$

The domain of the **composite function $f \circ g$** is the set of all x such that

1. x is in the domain of g and

2. $g(x)$ is in the domain of f.

The composition of f with g, $f \circ g$, is pictured as a machine with inputs and outputs in Figure 1.65. The diagram indicates that the output of g, or $g(x)$, becomes the input for "machine" f. If $g(x)$ is not in the domain of f, it cannot be input into machine f, and so $g(x)$ must be discarded.

Inputs, x

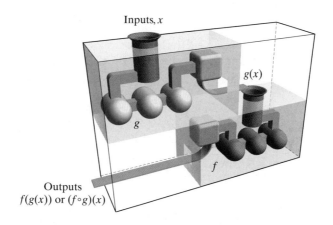

$g(x)$

g

f

Figure 1.65 Inputting one function into a second function

Outputs
$f(g(x))$ or $(f \circ g)(x)$

EXAMPLE 4 Forming Composite Functions

Given $f(x) = 3x - 4$ and $g(x) = x^2 + 6$, find:

    a. $(f \circ g)(x)$ **b.** $(g \circ f)(x)$.

Solution

a. We begin with $(f \circ g)(x)$, the composition of f with g. Because $(f \circ g)(x)$ means $f(g(x))$, we must replace each occurrence of x in the equation for f with $g(x)$.

$$f(x) = 3x - 4 \qquad \text{\small This is the given equation for } f.$$

Replace x with g(x).

$$(f \circ g)(x) = f(g(x)) = 3g(x) - 4$$

$$= 3(x^2 + 6) - 4 \qquad \text{\small Because } g(x) = x^2 + 6, \text{ replace } g(x) \text{ with } x^2 + 6.$$

$$= 3x^2 + 18 - 4 \qquad \text{\small Use the distributive property.}$$

$$= 3x^2 + 14 \qquad \text{\small Simplify.}$$

Thus, $(f \circ g)(x) = 3x^2 + 14$.

b. Next, we find $(g \circ f)(x)$, the composition of g with f. Because $(g \circ f)(x)$ means $g(f(x))$, we must replace each occurrence of x in the equation for g with $f(x)$.

$$g(x) = x^2 + 6 \qquad \text{This is the given equation for } g.$$

Replace x with f(x).

$$
\begin{aligned}
(g \circ f)(x) = g(f(x)) &= (f(x))^2 + 6 \\
&= (3x - 4)^2 + 6 \qquad \text{Because f(x)} = 3x - 4, \text{replace} \\
&\qquad\qquad\qquad\qquad\quad \text{f(x) with } 3x - 4. \\
&= 9x^2 - 24x + 16 + 6 \quad \text{Use} \\
&\qquad\qquad\qquad\qquad\qquad (A - B)^2 = A^2 - 2AB + B^2 \text{ to} \\
&\qquad\qquad\qquad\qquad\qquad \text{square } 3x - 4. \\
&= 9x^2 - 24x + 22 \qquad \text{Simplify.}
\end{aligned}
$$

Thus, $(g \circ f)(x) = 9x^2 - 24x + 22$. Notice that $(f \circ g)(x)$ is not the same function as $(g \circ f)(x)$.

Check Point 4   Given $f(x) = 5x + 6$ and $g(x) = x^2 - 1$, find:

a. $(f \circ g)(x)$ **b.** $(g \circ f)(x)$.

3   Determine domains for composite functions.

We need to be careful in determining the domain for the composite function

$$(f \circ g)(x) = f(g(x)).$$

The following values must be excluded from the input x:

- If x is not in the domain of g, it must not be in the domain of $f \circ g$.
- Any x for which $g(x)$ is not in the domain of f must not be in the domain of $f \circ g$.

EXAMPLE 5   Forming a Composite Function and Finding Its Domain

Given $f(x) = \dfrac{2}{x - 1}$ and $g(x) = \dfrac{3}{x}$, find:

a. $(f \circ g)(x)$ **b.** the domain of $f \circ g$.

Solution

a. Because $(f \circ g)(x)$ means $f(g(x))$, we must replace x in $f(x) = \dfrac{2}{x - 1}$ with $g(x)$.

$$(f \circ g)(x) = f(g(x)) = \frac{2}{g(x) - 1} = \frac{2}{\frac{3}{x} - 1} = \frac{2}{\frac{3}{x} - 1} \cdot \frac{x}{x} = \frac{2x}{3 - x}$$

g(x) = $\frac{3}{x}$ Simplify the complex fraction by multiplying by $\frac{x}{x}$, or 1.

Thus, $(f \circ g)(x) = \dfrac{2x}{3 - x}$.

Study Tip

The procedure for simplifying complex fractions can be found in Section P.6, pages 62-64.

b. We determine the domain of $(f \circ g)(x)$ in two steps.

Rules for Excluding Numbers from the Domain of $(f \circ g)(x) = f(g(x))$	Applying the Rules to $f(x) = \dfrac{2}{x-1}$ and $g(x) = \dfrac{3}{x}$
If x is not in the domain of g, it must not be in the domain of $f \circ g$.	The domain of g is $\{x \mid x \neq 0\}$. Thus, 0 must be excluded from the domain of $f \circ g$.
Any x for which $g(x)$ is not in the domain of f must not be in the domain of $f \circ g$.	The domain of f is $\{x \mid x \neq 1\}$. This means we must exclude from the domain of $f \circ g$ any x for which $g(x) = 1$. $\dfrac{3}{x} = 1$ Set $g(x)$ equal to 1. $3 = x$ Multiply both sides by x. 3 must be excluded from the domain of $f \circ g$.

The domain of $f \circ g$ is $\{x \mid x \neq 0 \text{ and } x \neq 3\}$.

Check Point 5   Given $f(x) = \dfrac{4}{x+2}$ and $g(x) = \dfrac{1}{x}$, find:

a. $(f \circ g)(x)$ **b.** the domain of $f \circ g$.

4 Write functions as compositions.

Decomposing Functions

When you form a composite function, you "compose" two functions to form a new function. It is also possible to reverse this process. That is, you can "decompose" a given function and express it as a composition of two functions. Although there is more than one way to do this, there is often a "natural" selection that comes to mind first. For example, consider the function h defined by

$$h(x) = (3x^2 - 4x + 1)^5.$$

The function h takes $3x^2 - 4x + 1$ and raises it to the power 5. A natural way to write h as a composition of two functions is to raise the function $g(x) = 3x^2 - 4x + 1$ to the power 5. Thus, if we let

$$f(x) = x^5 \text{ and } g(x) = 3x^2 - 4x + 1, \text{ then}$$
$$(f \circ g)(x) = f(g(x)) = f(3x^2 - 4x + 1) = (3x^2 - 4x + 1)^5.$$

Study Tip

Suppose the form of function h is $h(x) = (\text{algebraic expression})^{\text{power}}$. Function h can be expressed as a composition, $f \circ g$, using

$f(x) = x^{\text{power}}$
$g(x) = \text{algebraic expression}$.

EXAMPLE 6   Writing a Function as a Composition

Express as a composition of two functions:
$$h(x) = \sqrt[3]{x^2 + 1}.$$

Solution   The function h takes $x^2 + 1$ and takes its cube root. A natural way to write h as a composition of two functions is to take the cube root of the function $g(x) = x^2 + 1$. Thus, we let

$$f(x) = \sqrt[3]{x} \text{ and } g(x) = x^2 + 1.$$

We can check this composition, namely $f(x) = \sqrt[3]{x}$ and $g(x) = x^2 + 1$, by finding $(f \circ g)(x)$. This should give the original function, namely $h(x) = \sqrt[3]{x^2 + 1}$.

$$(f \circ g)(x) = f(g(x)) = f(x^2 + 1) = \sqrt[3]{x^2 + 1} = h(x)$$

Check Point 6

Express as a composition of two functions:
$$h(x) = \sqrt{x^2 + 5}.$$

EXERCISE SET 1.7

Practice Exercises

1. If $f(x) = 2x^2 - 5$ and $g(x) = 3x + 7$, find:
  a. $(f + g)(x)$ **b.** $(f + g)(4)$.

2. If $f(x) = 3x^2 - 2x + 1$ and $g(x) = 4x - 1$, find:
  a. $(f + g)(x)$ **b.** $(f + g)(5)$.

3. If $f(x) = \sqrt{x - 6}$ and $g(x) = \sqrt{x + 2}$, find:
  a. $(f + g)(x)$ **b.** the domain of $f + g$.

4. If $f(x) = \sqrt{x - 8}$ and $g(x) = \sqrt{x + 5}$, find:
  a. $(f + g)(x)$ **b.** the domain of $f + g$.

In Exercises 5–16, find $f + g, f - g, fg,$ and $\frac{f}{g}$. Determine the domain for each function.

5. $f(x) = 2x + 3, \quad g(x) = x - 1$
6. $f(x) = 3x - 4, \quad g(x) = x + 2$
7. $f(x) = x - 5, \quad g(x) = 3x^2$
8. $f(x) = x - 6, \quad g(x) = 5x^2$
9. $f(x) = 2x^2 - x - 3, \quad g(x) = x + 1$
10. $f(x) = 6x^2 - x - 1, \quad g(x) = x - 1$
11. $f(x) = \sqrt{x}, \quad g(x) = x - 4$
12. $f(x) = \dfrac{1}{x}, \quad g(x) = x - 5$
13. $f(x) = 2 + \dfrac{1}{x}, \quad g(x) = \dfrac{1}{x}$
14. $f(x) = 6 - \dfrac{1}{x}, \quad g(x) = \dfrac{1}{x}$
15. $f(x) = \sqrt{x + 4}, \quad g(x) = \sqrt{x - 1}$
16. $f(x) = \sqrt{x + 6}, \quad g(x) = \sqrt{x - 3}$

In Exercises 17–28, find:
  a. $(f \circ g)(x)$
  b. $(g \circ f)(x)$
  c. $(f \circ g)(2)$.

17. $f(x) = 2x, \quad g(x) = x + 7$
18. $f(x) = 3x, \quad g(x) = x - 5$
19. $f(x) = x + 4, \quad g(x) = 2x + 1$
20. $f(x) = 5x + 2, \quad g(x) = 3x - 4$
21. $f(x) = 4x - 3, \quad g(x) = 5x^2 - 2$
22. $f(x) = 7x + 1, \quad g(x) = 2x^2 - 9$
23. $f(x) = x^2 + 2, \quad g(x) = x^2 - 2$
24. $f(x) = x^2 + 1, \quad g(x) = x^2 - 3$
25. $f(x) = \sqrt{x}, \quad g(x) = x - 1$
26. $f(x) = \sqrt{x}, \quad g(x) = x + 2$
27. $f(x) = 2x - 3, \quad g(x) = \dfrac{x + 3}{2}$
28. $f(x) = 6x - 3, \quad g(x) = \dfrac{x + 3}{6}$

In Exercises 29–38, find:
  a. $(f \circ g)(x)$ **b.** *the domain of $f \circ g$.*

29. $f(x) = \dfrac{2}{x + 3}, \quad g(x) = \dfrac{1}{x}$
30. $f(x) = \dfrac{5}{x + 4}, \quad g(x) = \dfrac{1}{x}$
31. $f(x) = \dfrac{x}{x + 1}, \quad g(x) = \dfrac{4}{x}$
32. $f(x) = \dfrac{x}{x + 5}, \quad g(x) = \dfrac{6}{x}$
33. $f(x) = \sqrt{x}, \quad g(x) = x + 3$
34. $f(x) = \sqrt{x}, \quad g(x) = x - 3$
35. $f(x) = x^2 + 4, \quad g(x) = \sqrt{1 - x}$
36. $f(x) = x^2 + 1, \quad g(x) = \sqrt{2 - x}$
37. $f(x) = 4 - x^2, \quad g(x) = \sqrt{x^2 - 4}$
38. $f(x) = 9 - x^2, \quad g(x) = \sqrt{x^2 - 9}$

In Exercises 39–46, express the given function h as a composition of two functions f and g so that $h(x) = (f \circ g)(x)$.

39. $h(x) = (3x - 1)^4$
40. $h(x) = (2x - 5)^3$
41. $h(x) = \sqrt[3]{x^2 - 9}$
42. $h(x) = \sqrt{5x^2 + 3}$
43. $h(x) = |2x - 5|$
44. $h(x) = |3x - 4|$
45. $h(x) = \dfrac{1}{2x - 3}$
46. $h(x) = \dfrac{1}{4x + 5}$

In Exercises 47–58, use the graphs of f and g to evaluate each function.

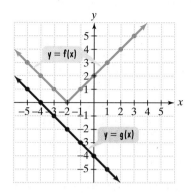

47. $(f + g)(-3)$

48. $(f + g)(-4)$

49. $(f - g)(2)$

50. $(g - f)(2)$

51. $\left(\dfrac{f}{g}\right)(-6)$

52. $\left(\dfrac{f}{g}\right)(-5)$

53. $(fg)(-4)$

54. $(fg)(-2)$

55. $(f \circ g)(2)$

56. $(f \circ g)(1)$

57. $(g \circ f)(0)$

58. $(g \circ f)(-1)$

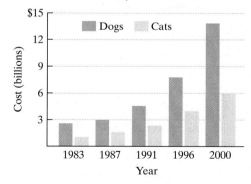

Application Exercises

It seems that Phideau's medical bills are costing us an arm and a paw. The graph shows veterinary costs, in billions of dollars, for dogs and cats in five selected years. Let

$D(x) = $ *veterinary costs, in billions of dollars, for dogs in year x*

$C(x) = $ *veterinary costs, in billions of dollars, for cats in year x*

Use the graph to solve Exercises 59–62.

Veterinary Costs in the U.S.

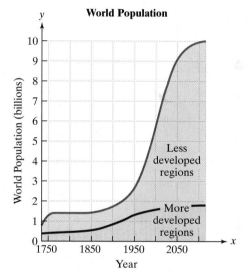

Source: American Veterinary Medical Association

59. Find an estimate of $(D + C)(2000)$. What does this mean in terms of the variables in this situation?

60. Find an estimate of $(D - C)(2000)$. What does this mean in terms of the variables in this situation?

61. Using the information shown in the graph, what is the domain of $D + C$?

62. Using the information shown in the graph, what is the domain of $D - C$?

Consider the following functions:

$f(x) = $ *population of the world's more developed regions in year x*

$g(x) = $ *population of the world's less developed regions in year x*

$h(x) = $ *total world population in year x.*

Use these functions and the graph shown to answer Exercises 63–66.

World Population

Source: Population Reference Bureau

63. What does the function $f + g$ represent?

64. What does the function $h - g$ represent?

65. Use the graph to estimate $(f + g)(2000)$.

66. Use the graph to estimate $(h - g)(2000)$.

67. A company that sells radios has a yearly fixed cost of $600,000. It costs the company $45 to produce each radio. Each radio will sell for $65. The company's costs and revenue are modeled by the following functions:

$C(x) = 600,000 + 45x$ *This function models the company's costs.*

$R(x) = 65x.$ *This function models the company's revenue.*

Find and interpret $(R - C)(20,000)$, $(R - C)(30,000)$, and $(R - C)(40,000)$.

68. A department store has two locations in a city. From 1998 through 2002, the profits for each of the store's two branches are modeled by the functions $f(x) = -0.44x + 13.62$ and $g(x) = 0.51x + 11.14$. In each model, x represents the number of years after 1998 and f and g represent the profit, in millions of dollars.

   a. What is the slope of f? Describe what this means.

   b. What is the slope of g? Describe what this means.

   c. Find $f + g$. What is the slope of this function? What does this mean?

69. The regular price of a computer is x dollars. Let $f(x) = x - 400$ and $g(x) = 0.75x$.

   a. Describe what the functions f and g model in terms of the price of the computer.

   b. Find $(f \circ g)(x)$ and describe what this models in terms of the price of the computer.

   c. Repeat part (b) for $(g \circ f)(x)$.

   d. Which composite function models the greater discount on the computer, $f \circ g$ or $g \circ f$? Explain.

70. The regular price of a pair of jeans is x dollars. Let $f(x) = x - 5$ and $g(x) = 0.6x$.

   a. Describe what functions f and g model in terms of the price of the jeans.

   b. Find $(f \circ g)(x)$ and describe what this models in terms of the price of the jeans.

   c. Repeat part (b) for $(g \circ f)(x)$.

   d. Which composite function models the greater discount on the jeans, $f \circ g$ or $g \circ f$? Explain.

Writing in Mathematics

71. If equations for functions f and g are given, explain how to find $f + g$.

72. If the equations of two functions are given, explain how to obtain the quotient function and its domain.

73. If equations for functions f and g are given, describe two ways to find $(f - g)(3)$.

74. Explain how to use the graphs in Figure 1.63 on page 197 to estimate the per capita federal tax for any one of the years shown on the horizontal axis.

75. Describe a procedure for finding $(f \circ g)(x)$. What is the name of this function?

76. Describe the values of x that must be excluded from the domain of $(f \circ g)(x)$.

Technology Exercises

77. The function $f(t) = -0.14t^2 + 0.51t + 31.6$ models the U.S. population ages 65 and older, $f(t)$, in millions, t years after 1990. The function $g(t) = 0.54t^2 + 12.64t + 107.1$ models the total yearly cost of Medicare, $g(t)$, in billions of dollars, t years after 1990. Graph the function $\frac{g}{f}$ in a $[0, 15, 1]$ by $[0, 60, 10]$ viewing rectangle. What does the shape of the graph indicate about the per capita costs of Medicare for the U.S. population ages 65 and over with increasing time?

78. Graph $y_1 = x^2 - 2x$, $y_2 = x$, and $y_3 = y_1 \div y_2$ in the same $[-10, 10, 1]$ by $[-10, 10, 1]$ viewing rectangle. Then use the $\boxed{\text{TRACE}}$ feature to trace along y_3. What happens at $x = 0$? Explain why this occurs.

79. Graph $y_1 = x^2 - 4$, $y_2 = \sqrt{4 - x^2}$, and $y_3 = y_2^2 - 4$ in the same $[-5, 5, 1]$ by $[-5, 5, 1]$ viewing rectangle. If y_1 represents f and y_2 represents g, use the graph of y_3 to find the domain of $f \circ g$. Then verify your observation algebraically.

Critical Thinking Exercises

80. Which one of the following is true?

   a. If $f(x) = x^2 - 4$ and $g(x) = \sqrt{x^2 - 4}$, then $(f \circ g)(x) = -x^2$ and $(f \circ g)(5) = -25$.

   b. There can never be two functions f and g, where $f \neq g$, for which $(f \circ g)(x) = (g \circ f)(x)$.

   c. If $f(7) = 5$ and $g(4) = 7$ then $(f \circ g)(4) = 35$.

   d. If $f(x) = \sqrt{x}$ and $g(x) = 2x - 1$, then $(f \circ g)(5) = g(2)$.

81. Prove that if f and g are even functions, then fg is also an even function.

82. Define two functions f and g so that $f \circ g = g \circ f$.

83. Use the graphs given in Exercises 63–66 to create a graph that shows the population, in billions, of less developed regions from 1950 through 2050.

Group Exercise

84. Consult an almanac, newspaper, magazine, or the Internet to find data displayed in a graph in the style of Figure 1.63 on page 197. Using the two graphs that group members find most interesting, introduce two functions that are related to the graphs. Then write and solve a problem involving function subtraction for each selected graph. If you are not sure where to begin, reread pages 197–198 or look at Exercises 63–66 in this exercise set.

SECTION 1.8   *Inverse Functions*

Objectives

1. Verify inverse functions.
2. Find the inverse of a function.
3. Use the horizontal line test to determine if a function has an inverse function.
4. Use the graph of a one-to-one function to graph its inverse function.

In most societies, women say they prefer to marry men who are older than themselves, whereas men say they prefer women who are younger. Evolutionary psychologists attribute these preferences to female concern with a partner's material resources and male concern with a partner's fertility (*Source:* David M. Buss, *Psychological Inquiry*, 6, 1–30). When the man is considerably older than the woman, people rarely comment. However, when the woman is older, as in the relationship between actors Susan Sarandon and Tim Robbins, people take notice.

Figure 1.66 shows the preferred age in a mate in five selected countries. We can focus on the data for the women and define a function.

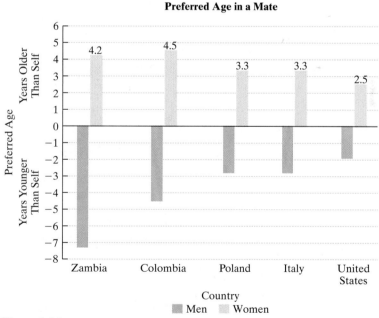

Figure 1.66

Source: Carole Wade and Carol Tavris, *Psychology Sixth Edition*, Prentice Hall, 2000

Let the domain of the function be the set of the five countries shown in the graph. Let the range be the set of the average number of years women in each of the respective countries prefer men who are older than themselves. The function can be written as follows:

f: {(Zambia, 4.2), (Colombia, 4.5), (Poland, 3.3), (Italy, 3.3), (U.S., 2.5)}.

Now let's "undo" f by interchanging the first and second components in each of its ordered pairs. Switching the inputs and outputs of f, we obtain the following relation:

Same first component

Undoing f: {(4.2, Zambia), (4.5, Colombia), (3.3, Poland), (3.3, Italy), (2.5, U.S.)}.

Different second components

Can you see that this relation is not a function? Two of its ordered pairs have the same first component and different second components. This violates the definition of a function.

If a function f is a set of ordered pairs, (x, y), then the changes produced by f can be "undone" by reversing the components of all the ordered pairs. The resulting relation, (y, x), may or may not be a function. In this section, we will develop these ideas by studying functions whose compositions have a special "undoing" relationship.

Inverse Functions

Here are two functions that describe situations related to the price of a computer, x:

$$f(x) = x - 300 \qquad g(x) = x + 300.$$

Function f subtracts \$300 from the computer's price and function g adds \$300 to the computer's price. Let's see what $f(g(x))$ does. Put $g(x)$ into f:

$$f(x) = x - 300 \qquad \text{This is the given equation for f.}$$

Replace x with g(x).

$$f(g(x)) = g(x) - 300$$
$$= x + 300 - 300 \qquad \text{Because g(x) = x + 300, replace g(x) with x + 300.}$$

$$= x. \qquad \text{This is the computer's original price.}$$

Using $f(g(x))$, the computer's price, x, went through two changes: the first, an increase; the second, a decrease:

$$x + 300 - 300.$$

The final price of the computer, x, is identical to its starting price, x.

In general, if the changes made to x by function g are undone by the changes made by function f, then

$$f(g(x)) = x.$$

Assume, also, that this "undoing" takes place in the other direction:

$$g(f(x)) = x.$$

Under these conditions, we say that each function is the *inverse function* of the other. The fact that g is the inverse of f is expressed by renaming g as f^{-1}, read "f-inverse." For example, the inverse functions

$$f(x) = x - 300 \qquad g(x) = x + 300$$

are usually named as follows:

$$f(x) = x - 300 \qquad f^{-1}(x) = x + 300.$$

With these ideas in mind, we present the formal definition of the inverse of a function:

Study Tip

The notation f^{-1} represents the inverse function of f. The -1 is *not* an exponent. The notation f^{-1} does *not* mean $\dfrac{1}{f}$:

$$f^{-1} \neq \frac{1}{f}.$$

Definition of the Inverse of a Function

Let f and g be two functions such that

$$f(g(x)) = x \qquad \text{for every } x \text{ in the domain of } g$$

and

$$g(f(x)) = x \qquad \text{for every } x \text{ in the domain of } f.$$

The function g is the **inverse of the function** f, and is denoted by f^{-1} (read "f-inverse"). Thus, $f(f^{-1}(x)) = x$ and $f^{-1}(f(x)) = x$. The domain of f is equal to the range of f^{-1}, and vice versa.

1   Verify inverse functions.

EXAMPLE 1   Verifying Inverse Functions

Show that each function is an inverse of the other:

$$f(x) = 5x \quad \text{and} \quad g(x) = \frac{x}{5}.$$

Solution   To show that f and g are inverses of each other, we must show that $f(g(x)) = x$ and $g(f(x)) = x$. We begin with $f(g(x))$.

$$f(x) = 5x \qquad \textit{This is the given equation for f.}$$

Replace x with g(x).

$$f(g(x)) = 5g(x) = 5\left(\frac{x}{5}\right) = x$$

Next, we find $g(f(x))$.

$$g(x) = \frac{x}{5} \qquad \textit{This is the given equation for g.}$$

Replace x with f(x).

$$g(f(x)) = \frac{f(x)}{5} = \frac{5x}{5} = x$$

Because g is the inverse of f (and vice versa), we can use inverse notation and write

$$f(x) = 5x \quad \text{and} \quad f^{-1}(x) = \frac{x}{5}.$$

Notice how f^{-1} undoes the change produced by f: f changes x by multiplying by 5 and f^{-1} undoes this by dividing by 5.

Study Tip

The following partial tables of coordinates numerically illustrate that inverse functions reverse each other's coordinates.

x	-5	-2	0	3
$f(x) = 5x$	-25	-10	0	15

x	-25	-10	0	15
$g(x) = \dfrac{x}{5}$	-5	-2	0	3

Using the first two tables, the following table shows how inverse functions undo one another.

x	-5	-2	0	3
$g(f(x))$	$g(f(-5))$ $=g(-25)$ $=-5$	$g(f(-2))$ $=g(-10)$ $=-2$	$g(f(0))$ $=g(0)$ $=0$	$g(f(3))$ $=g(15)$ $=3$

Each output of the composite function is identical to the input.

Check Point 1

Show that each function is an inverse of the other:

$$f(x) = 7x \quad \text{and} \quad g(x) = \frac{x}{7}.$$

EXAMPLE 2 Verifying Inverse Functions

Show that each function is an inverse of the other:

$$f(x) = 3x + 2 \quad \text{and} \quad g(x) = \frac{x - 2}{3}.$$

Solution To show that f and g are inverses of each other, we must show that $f(g(x)) = x$ and $g(f(x)) = x$. We begin with $f(g(x))$.

$$f(x) = 3x + 2 \qquad \text{This is the equation for f.}$$

> Replace x with g(x).

$$f(g(x)) = 3g(x) + 2 = 3\left(\frac{x - 2}{3}\right) + 2 = x - 2 + 2 = x$$

Next, we find $g(f(x))$.

$$g(x) = \frac{x - 2}{3} \qquad \text{This is the equation for g.}$$

> Replace x with f (x).

$$g(f(x)) = \frac{f(x) - 2}{3} = \frac{(3x + 2) - 2}{3} = \frac{3x}{3} = x$$

Because g is the inverse of f (and vice versa), we can use inverse notation and write

$$f(x) = 3x + 2 \quad \text{and} \quad f^{-1}(x) = \frac{x - 2}{3}.$$

Notice how f^{-1} undoes the changes produced by f: f changes x by *multiplying* by 3 and *adding* 2, and f^{-1} undoes this by *subtracting* 2 and *dividing* by 3. This "undoing" process is illustrated in Figure 1.67.

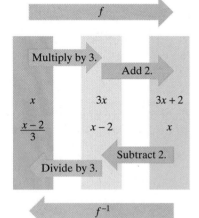

Figure 1.67 f^{-1} undoes the changes produced by f.

Check Point 2 Show that each function is an inverse of the other:

$$f(x) = 4x - 7 \quad \text{and} \quad g(x) = \frac{x + 7}{4}.$$

2 Find the inverse of a function.

Finding the Inverse of a Function

The definition of the inverse of a function tells us that the domain of f is equal to the range of f^{-1}, and vice versa. This means that if the function f is the set of ordered pairs (x, y), then the inverse of f is the set of ordered pairs (y, x). If a function is defined by an equation, we can obtain the equation for f^{-1}, the inverse of f, by interchanging the role of x and y in the equation for the function f.

Study Tip

The procedure for finding a function's inverse uses a *switch-and-solve* strategy. Switch x and y, then solve for y.

> **Finding the Inverse of a Function**
>
> The equation for the inverse of a function f can be found as follows:
>
> 1. Replace $f(x)$ with y in the equation for $f(x)$.
> 2. Interchange x and y.
> 3. Solve for y. If this equation does not define y as a function of x, the function f does not have an inverse function and this procedure ends. If this equation does define y as a function of x, the function f has an inverse function.
> 4. If f has an inverse function, replace y in step 3 by $f^{-1}(x)$. We can verify our result by showing that $f(f^{-1}(x)) = x$ and $f^{-1}(f(x)) = x$.

EXAMPLE 3   Finding the Inverse of a Function

Find the inverse of $f(x) = 7x - 5$.

Solution

Step 1   Replace $f(x)$ with y:

$$y = 7x - 5$$

Step 2   Interchange x and y:

$$x = 7y - 5$$ *This is the inverse function.*

Step 3   Solve for y:

$$x + 5 = 7y$$ *Add 5 to both sides.*

$$\frac{x + 5}{7} = y$$ *Divide both sides by 7.*

Step 4   Replace y with $f^{-1}(x)$:

$$f^{-1}(x) = \frac{x + 5}{7}$$ *The equation is written with f^{-1} on the left.*

Thus, the inverse of $f(x) = 7x - 5$ is $f^{-1}(x) = \dfrac{x + 5}{7}$.

The inverse function, f^{-1}, undoes the changes produced by f. f changes x by multiplying by 7 and subtracting 5. f^{-1} undoes this by adding 5 and dividing by 7.

Discovery

In Example 3, we found that if $f(x) = 7x - 5$, then
$$f^{-1}(x) = \frac{x + 5}{7}.$$
Verify this result by showing that
$$f(f^{-1}(x)) = x$$
and
$$f^{-1}(f(x)) = x.$$

Check Point 3   Find the inverse of $f(x) = 2x + 7$.

EXAMPLE 4   Finding the Equation of the Inverse

Find the inverse of $f(x) = x^3 + 1$.

Solution

Step 1   Replace $f(x)$ with y: $y = x^3 + 1$

Step 2   Interchange x and y: $x = y^3 + 1$

Step 3   Solve for y: $x - 1 = y^3$

$$\sqrt[3]{x - 1} = \sqrt[3]{y^3}$$

$$\sqrt[3]{x - 1} = y$$

Step 4   Replace y with $f^{-1}(x)$: $f^{-1}(x) = \sqrt[3]{x - 1}$.

Thus, the inverse of $f(x) = x^3 + 1$ is $f^{-1}(x) = \sqrt[3]{x - 1}$.

Check Point 4   Find the inverse of $f(x) = 4x^3 - 1$.

3 Use the horizontal line test to determine if a function has an inverse function.

The Horizontal Line Test and One-to-One Functions

Let's see what happens if we try to find the inverse of the standard quadratic function, $f(x) = x^2$.

Step 1 **Replace $f(x)$ with y:** $\quad y = x^2$

Step 2 **Interchange x and y:** $\quad x = y^2$

Step 3 **Solve for y:** We apply the square root method to solve $y^2 = x$ for y.

We obtain

$$y = \pm\sqrt{x}.$$

The $\pm$ in this last equation shows that for certain values of x (all positive real numbers), there are two values of y. Because this equation does not represent y as a function of x, the standard quadratic function does not have an inverse function.

    Can we look at the graph of a function and tell if it represents a function with an inverse? Yes. The graph of the standard quadratic function is shown in Figure 1.68. Four units above the x-axis, a horizontal line is drawn. This line intersects the graph at two of its points, $(-2, 4)$ and $(2, 4)$. Because inverse functions have ordered pairs with the coordinates reversed, let's see what happens if we reverse these coordinates. We obtain $(4, -2)$ and $(4, 2)$. A function provides exactly one output for each input. However, the input 4 is associated with two outputs, -2 and 2. The points $(4, -2)$ and $(4, 2)$ do not define a function.

    If any horizontal line, such as the one in Figure 1.68, intersects a graph at two or more points, these points will not define a function when their coordinates are reversed. This suggests the **horizontal line test** for inverse functions:

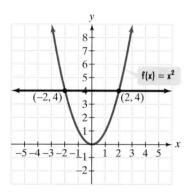

Figure 1.68 The horizontal line intersects the graph twice.

Discovery

How might you restrict the domain of $f(x) = x^2$, graphed in Figure 1.68, so that the remaining portion of the graph passes the horizontal line test?

The Horizontal Line Test for Inverse Functions

A function f has an inverse that is a function, f^{-1}, if there is no horizontal line that intersects the graph of the function f at more than one point.

EXAMPLE 5   Applying the Horizontal Line Test

Which of the following graphs represent functions that have inverse functions?

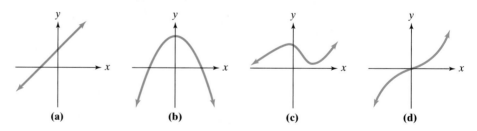

Solution   Can you see that horizontal lines can be drawn in parts (b) and (c) that intersect the graphs more than once? This is illustrated in the figure at the top of the next page. These graphs do not pass the horizontal line test. The graphs in parts (b) and (c) are not the graphs of functions with inverse functions. By contrast, no horizontal line can be drawn in parts (a) and (d) that intersect the graphs more than once. These graphs pass the horizontal line test. Thus, the graphs in parts (a) and (d) represent functions that have inverse functions.

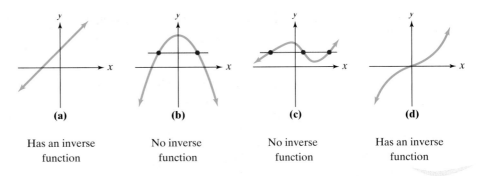

(a)	(b)	(c)	(d)
Has an inverse function	No inverse function	No inverse function	Has an inverse function

Check Point 5 Which of the following graphs represent functions that have inverse functions?

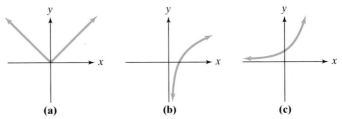

      (a)   (b)   (c)

A function passes the horizontal line test when no two different ordered pairs have the same second component. This means that if $x_1 \neq x_2$, then $f(x_1) \neq f(x_2)$. Such a function is called a **one-to-one function.** Thus, a one-to-one function is a function in which no two different ordered pairs have the same second component. Only one-to-one functions have inverse functions. Any function that passes the horizontal line test is a one-to-one function. Any one-to-one function has a graph that passes the horizontal line test.

4 Use the graph of a one-to-one function to graph its inverse function.

Graphs of f and f^{-1}

There is a relationship between the graph of a one-to-one function, f, and its inverse, f^{-1}. Because inverse functions have ordered pairs with the coordinates reversed, if the point (a, b) is on the graph of f, then the point (b, a) is on the graph of f^{-1}. The points (a, b) and (b, a) are symmetric with respect to the line $y = x$. Thus, **the graph of f^{-1} is a reflection of the graph of f about the line $y = x$.** This is illustrated in Figure 1.69.

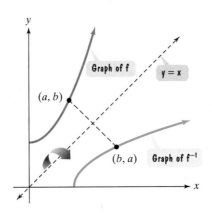

Figure 1.69 The graph of f^{-1} is a reflection of the graph of f about $y = x$.

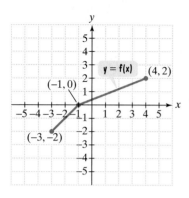

Figure 1.70

EXAMPLE 6   Graphing the Inverse Function

Use the graph of f in Figure 1.70 to draw the graph of its inverse function.

Solution   We begin by noting that no horizontal line intersects the graph of f at more than one point, so f does have an inverse function. Because the points $(-3, -2), (-1, 0)$, and $(4, 2)$ are on the graph of f, the graph of the inverse function, f^{-1}, has points with these ordered pairs reversed. Thus, $(-2, -3), (0, -1)$, and $(2, 4)$ are on the graph of f^{-1}. We can use these points to graph f^{-1}. The graph of f^{-1} is shown in green in Figure 1.71. Note that the graph of f^{-1} is the reflection of the graph of f about the line $y = x$.

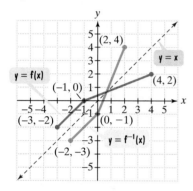

Figure 1.71   The graphs of f and f^{-1}

Check Point 6   Use the graph of f in the following figure to draw the graph of its inverse function.

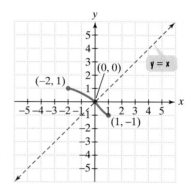

EXERCISE SET 1.8

Practice Exercises

In Exercises 1–10, find $f(g(x))$ and $g(f(x))$ and determine whether each pair of functions f and g are inverses of each other.

1. $f(x) = 4x$ and $g(x) = \dfrac{x}{4}$

2. $f(x) = 6x$ and $g(x) = \dfrac{x}{6}$

3. $f(x) = 3x + 8$ and $g(x) = \dfrac{x - 8}{3}$

4. $f(x) = 4x + 9$ and $g(x) = \dfrac{x - 9}{4}$

5. $f(x) = 5x - 9$ and $g(x) = \dfrac{x + 5}{9}$

6. $f(x) = 3x - 7$ and $g(x) = \dfrac{x + 3}{7}$

7. $f(x) = \dfrac{3}{x - 4}$ and $g(x) = \dfrac{3}{x} + 4$

8. $f(x) = \dfrac{2}{x - 5}$ and $g(x) = \dfrac{2}{x} + 5$

9. $f(x) = -x$ and $g(x) = -x$

10. $f(x) = \sqrt[3]{x - 4}$ and $g(x) = x^3 + 4$

The functions in Exercises 11–28 are all one-to-one. For each function:

a. *Find an equation for $f^{-1}(x)$, the inverse function.*

b. *Verify that your equation is correct by showing that $f(f^{-1}(x)) = x$ and $f^{-1}(f(x)) = x$.*

11. $f(x) = x + 3$ **12.** $f(x) = x + 5$

13. $f(x) = 2x$ **14.** $f(x) = 4x$

15. $f(x) = 2x + 3$ **16.** $f(x) = 3x - 1$

17. $f(x) = x^3 + 2$ **18.** $f(x) = x^3 - 1$

19. $f(x) = (x + 2)^3$ **20.** $f(x) = (x - 1)^3$

21. $f(x) = \dfrac{1}{x}$ **22.** $f(x) = \dfrac{2}{x}$

23. $f(x) = \sqrt{x}$ **24.** $f(x) = \sqrt[3]{x}$

25. $f(x) = x^2 + 1$, for $x \geq 0$

26. $f(x) = x^2 - 1$, for $x \geq 0$

27. $f(x) = \dfrac{2x + 1}{x - 3}$

28. $f(x) = \dfrac{2x - 3}{x + 1}$

Which graphs in Exercises 29–34 represent functions that have inverse functions?

29.

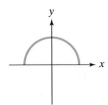

30.

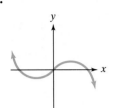

31.

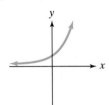

32.

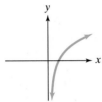

33.

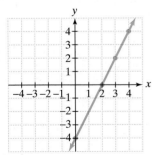

34.

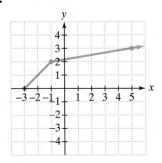

In Exercises 35–38, use the graph of f to draw the graph of its inverse function.

35.

36.

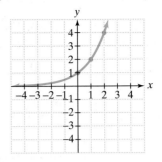

37.

38.

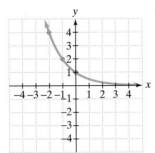

Application Exercises

39. Refer to Figure 1.66 on page 209. Recall that the bar graphs in the figure show the preferred age in a mate in five selected countries.

a. Consider a function, *f*, whose domain is the set of the five countries shown in the graph. Let the range be the set of the average number of years men in each of the respective countries prefer women who are younger than themselves. (You will need to use the graph to estimate these values. Assume that the bars for Poland and Italy have the same length. Round to the nearest tenth of a year.) Write function *f* as a set of ordered pairs.

b. Write the relation that is the inverse of *f* as a set of ordered pairs. Is this relation a function? Explain your answer.

40. The bar graph shows the percentage of land owned by the federal government in western states in which the government owns at least half of the land.

Percentage of Land Owned by the Federal Government

State	Percentage
Nevada	83%
Utah	65%
Idaho	62%
Alaska	62%
Oregon	52%
Wyoming	50%

Source: Bureau of Land Management

(Be sure to turn back a page and refer to the graph at the bottom of the page.)

a. Consider a function, f, whose domain is the set of six states shown. Let the range be the percentage of land owned by the federal government in each of the respective states. Write the function f as a set of ordered pairs.

b. Write the relation that is the inverse of f as a set of ordered pairs. Is this relation a function? Explain your answer.

41. The graph represents the probability of two people in the same room sharing a birthday as a function of the number of people in the room. Call the function f.

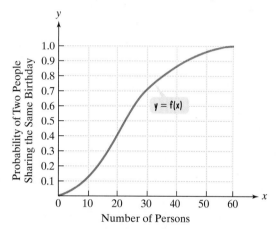

a. Explain why f has an inverse that is a function.

b. Describe in practical terms the meaning of $f^{-1}(0.25)$, $f^{-1}(0.5)$, and $f^{-1}(0.7)$.

42. The graph shows the average age at which women in the United States marry for the first time over a 110-year period.

Average Age at First Marriage for U.S. Women

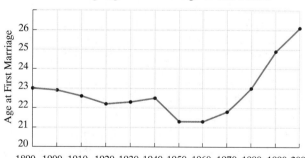

Year

Source: U.S. Census Bureau

a. Does this graph have an inverse that is a function? What does this mean about the average age at which U.S. women marry during the period shown?

b. Identify two or more years in which U.S. women married for the first time at the same average age. What is a reasonable estimate of this average age?

43. The formula

$$y = f(x) = \frac{9}{5}x + 32$$

is used to convert from x degrees Celsius to y degrees Fahrenheit. The formula

$$y = g(x) = \frac{5}{9}(x - 32)$$

is used to convert from x degrees Fahrenheit to y degrees Celsius. Show that f and g are inverse functions.

Writing in Mathematics

44. Explain how to determine if two functions are inverses of each other.

45. Describe how to find the inverse of a one-to-one function.

46. What is the horizontal line test and what does it indicate?

47. Describe how to use the graph of a one-to-one function to draw the graph of its inverse function.

48. How can a graphing utility be used to visually determine if two functions are inverses of each other?

Technology Exercises

In Exercises 49–57, use a graphing utility to graph the function. Use the graph to determine whether the function has an inverse that is a function (that is, whether the function is one-to-one).

49. $f(x) = x^2 - 1$

50. $f(x) = \sqrt[3]{2 - x}$

51. $f(x) = \dfrac{x^3}{2}$

52. $f(x) = \dfrac{x^4}{4}$

53. $f(x) = \text{int}(x - 2)$

54. $f(x) = |x - 2|$

55. $f(x) = (x - 1)^3$

56. $f(x) = -\sqrt{16 - x^2}$

57. $f(x) = x^3 + x + 1$

In Exercises 58–60, use a graphing utility to graph f and g in the same viewing rectangle. In addition, graph the line $y = x$ and visually determine if f and g are inverses.

58. $f(x) = 4x + 4$, $\quad g(x) = 0.25x - 1$

59. $f(x) = \dfrac{1}{x} + 2$, $\quad g(x) = \dfrac{1}{x - 2}$

60. $f(x) = \sqrt[3]{x} - 2$, $\quad g(x) = (x + 2)^3$

Critical Thinking Exercises

61. Which one of the following is true?

a. The inverse of $\{(1, 4), (2, 7)\}$ is $\{(2, 7), (1, 4)\}$.

b. The function $f(x) = 5$ is one-to-one.

c. If $f(x) = 3x$, then $f^{-1}(x) = \dfrac{1}{3x}$.

d. The domain of f is the same as the range of f^{-1}.

62. If $f(x) = 3x$ and $g(x) = x + 5$, find $(f \circ g)^{-1}(x)$ and $(g^{-1} \circ f^{-1})(x)$.

63. Show that
$$f(x) = \frac{3x - 2}{5x - 3}$$
is its own inverse.

64. *Freedom 7* was the spacecraft that carried the first American into space in 1961. Total flight time was 15 minutes, and the spacecraft reached a maximum height of 116 miles. Consider a function, s, that expresses *Freedom 7*'s height, $s(t)$, in miles, after t minutes. Is s a one-to-one function? Explain your answer.

65. If $f(2) = 6$, find x satisfying $8 + f^{-1}(x - 1) = 10$.

Group Exercise

66. In Tom Stoppard's play *Arcadia*, the characters dream and talk about mathematics, including ideas involving graphing, composite functions, symmetry, and lack of symmetry in things that are tangled, mysterious, and unpredictable. Group members should read the play. Present a report on the ideas discussed by the characters that are related to concepts that we studied in this chapter. Bring in a copy of the play and read appropriate excerpts.

SECTION 1.9 *Modeling with Functions*

Objectives

1. Construct functions from verbal descriptions.
2. Construct functions from formulas.

A can of Coca-Cola is sold every six seconds throughout the world.

In 2002, the sale of soft drinks was banned in the public schools of Los Angeles county. Despite the variety of its reputations throughout the world, the soft drink industry has spent far more time reducing the amount of aluminum in its cylindrical cans than addressing the problems of the nutritional disaster floating within its packaging. In the 1960s, one pound of aluminum made fewer than 20 cans; today, almost 30 cans come out of the same amount. The thickness of the can wall is less than five-thousandths of an inch, about the same as a magazine cover.

Many real-world problems involve constructing mathematical models that are functions. The problem of minimizing the amount of aluminum needed to manufacture a 12-ounce soft-drink can first requires that we express the surface area of all such cans as a function of their radius. In constructing such a function, we must be able to translate a verbal description into a mathematical representation—that is, a mathematical model.

Earlier in the chapter, we saw that mathematical models can be obtained from real-world data. In this section, you will learn to obtain mathematical models involving functions from verbal descriptions and formulas.

Study Tip

In calculus, you will solve problems involving maximum or minimum values of functions. Such problems often require creating the function that is to be maximized or minimized. Quite often, the calculus is fairly routine. It is the algebraic setting up of the function that causes difficulty. That is why the material in this section is so important.

1 Construct functions from verbal descriptions.

Functions from Verbal Descriptions

There is no rigid step-by-step procedure that can be used to construct a function from a verbal description. Read the problem carefully. Attempt to write a critical sentence that describes the function's conditions in terms of its independent variable, x. In the following examples, we will use voice balloons that show these critical sentences, or **verbal models.** Then translate the verbal model into the algebraic notation used to represent a function's equation.

EXAMPLE 1   Modeling Costs of Long-Distance Carriers

You are choosing between two long-distance telephone plans. Plan A has a monthly fee of $20 with a charge of $0.05 per minute for all long-distance calls. Plan B has a monthly fee of $5 with a charge of $0.10 per minute for all long-distance calls.

   a. Express the monthly cost for plan A, f, as a function of the number of minutes of long-distance calls in a month, x.
   b. Express the monthly cost for plan B, g, as a function of the number of minutes of long-distance calls in a month, x.
   c. For how many minutes of long-distance calls will the costs for the two plans be the same?

Solution

   a. The monthly cost for plan A is the monthly fee, $20, plus the per-minute charge, $0.05, times the number of minutes of long-distance calls, x.

Monthly cost for plan A	equals	monthly fee	plus	per-minute charge	times	the number of minutes of long-distance calls.
$f(x)$	=	20	+	0.05	·	x

The function $f(x) = 0.05x + 20$ models the monthly cost, f, in dollars, in terms of the number of minutes of long-distance calls, x.

   b. The monthly cost for plan B is the monthly fee, $5, plus the per-minute charge, $0.10, times the number of minutes of long-distance calls, x.

Monthly cost for plan B	equals	monthly fee	plus	per-minute charge	times	the number of minutes of long-distance calls.
$g(x)$	=	5	+	0.10	·	x

The function $g(x) = 0.10x + 5$ models the monthly cost, g, in dollars, in terms of the number of minutes of long-distance calls, x.

   c. We are interested in how many minutes of long-distance calls, x, result in the same monthly costs, f and g, for the two plans. Thus, we must set the equations for f and g equal to each other. We then solve the resulting linear equation for x.

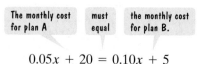

The monthly cost for plan A	must equal	the monthly cost for plan B.

$$0.05x + 20 = 0.10x + 5$$

Technology

The graphs of the functions that model the costs of the two long-distance carriers are shown in a $[0, 600, 100]$ by $[0, 70, 5]$ viewing rectangle. The graphs intersect at $(300, 35)$. This confirms that the costs for the two plans are the same with 300 minutes of long-distance calls. The cost for each plan is $35. To the left of $x = 300$, the graph of plan B's cost lies below that of plan A's cost. This shows that for fewer than 300 minutes of long-distance calls per month, plan B is cheaper. With more than 300 minutes of calls per month, plan A is the better deal.

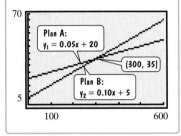

$$0.05x + 20 = 0.10x + 5 \qquad \text{This equation models equal monthly costs.}$$
$$20 = 0.05x + 5 \qquad \text{Subtract 0.05 x from both sides.}$$
$$15 = 0.05x \qquad \text{Subtract 5 from both side}$$
$$300 = x \qquad \text{Divide both sides by 0.05: } \frac{15}{0.05} = 300.$$

The costs for the two plans will be the same with 300 minutes of long-distance calls. Take a moment to verify that $f(300) = g(300) = 35$. Thus, the cost for each plan will be $35.

We have seen that the slope-intercept equation of a nonvertical line with slope m and y-intercept b is $y = mx + b$. This equation can be expressed in function notation by replacing y with $f(x)$:

$$f(x) = mx + b.$$

Functions with equations in this form are called **linear functions.** The functions that we obtained in Example 1 are both linear functions. Based on the meaning of the functions' variables, we can interpret slope and y-intercept as follows:

$$f(x) = 0.05x + 20$$

The slope indicates that the rate of change in the plan's cost is $0.05 per minute.	The y-intercept indicates the starting cost with no long-distance calls is $20.

$$g(x) = 0.10x + 5.$$

The slope indicates that the rate of change in the plan's cost is $0.10 per minute.	The y-intercept indicates the starting cost with no long-distance call is $5.

Check Point 1

You are choosing between two long-distance telephone plans. Plan A has a monthly fee of $15 with a charge of $0.08 per minute for all long-distance calls. Plan B has a monthly fee of $3 with a charge of $0.12 per minute for all long-distance calls.

a. Express the monthly cost for plan A, f, as a function of the number of minutes of long-distance calls in a month, x.

b. Express the monthly cost for plan B, g, as a function of the number of minutes of long-distance calls in a month, x.

c. For how many minutes of long-distance calls will the costs for the two plans be the same?

EXAMPLE 2   Modeling the Number of Customers and Revenue

On a certain route, an airline carries 6000 passengers per month, each paying $200. A market survey indicates that for each $1 increase in the ticket price, the airline will lose 100 passengers.

a. Express the number of passengers per month, N, as a function of the ticket price, x.

b. The airline's monthly revenue for the route is the product of the number of passengers and the ticket price. Express the monthly revenue, R, as a function of the ticket price, x.

Solution

a. The number of passengers, N, depends on the ticket price, x. In particular, the number of passengers is the original number, 6000, minus the number lost to the fare increase. The following table shows how to find the number lost to the fare increase:

English Phrase	Algebraic Translation
Ticket price	x
Amount of fare increase: ticket price minus original ticket price	$x - 200$ *The original ticket price was $200.*
Decrease in passengers due to the fare increase: 100 times the dollar amount of the fare increase	$100(x - 200)$ *100 passengers are lost for each dollar of fare increase.*

The number of passengers per month, N, is the original number of passengers, 6000, minus the decrease due to the fare increase.

$$
\begin{array}{ccccc}
\text{Number of} & & \text{the original} & & \text{the decrease in} \\
\text{passengers per} & \text{equals} & \text{number of} & \text{minus} & \text{passengers due} \\
\text{month} & & \text{passengers} & & \text{to the fare increase.}
\end{array}
$$

$$N(x) \quad = \quad 6000 \quad - \quad 100(x - 200)$$
$$= 6000 - 100x + 20{,}000$$
$$= -100x + 26{,}000$$

The linear function $N(x) = -100x + 26{,}000$ models the number of passengers per month, N, in terms of the price per ticket, x. The linear function's slope, -100, indicates that the rate of change is a loss of 100 passengers per dollar of fare increase.

b. The monthly revenue for the route is the number of passengers, $-100x + 26{,}000$, times the ticket price, x.

$$
\begin{array}{cccc}
\text{Monthly} & & \text{the number of} & \text{the ticket} \\
\text{revenue} & \text{equals} & \text{passengers} & \text{times} & \text{price.}
\end{array}
$$

$$R(x) \quad = (-100x + 26{,}000) \quad \cdot \quad x$$
$$= -100x^2 + 26{,}000x$$

The function $R(x) = -100x^2 + 26{,}000x$ models the airline's monthly revenue for the route, R, in terms of the ticket price, x.

The revenue function in Example 2 is of the form $f(x) = ax^2 + bx + c$. Any function of this form, where $a \neq 0$, is called a **quadratic function.** In the next chapter, you will study quadratic functions and their U-shaped graphs.

Technology

The graph of the function $R(x) = -100x^2 + 26{,}000x$, the model for the airline's revenue, R, in terms of the ticket price, x, is shown in a $[0, 260, 13]$ by $[0, 1{,}700{,}000, 100{,}000]$ viewing rectangle. The graphing utility's maximum function feature indicates that a ticket price of $130 yields the maximum revenue. As the ticket price exceeds $130, the maximum revenue, $1,690,000, starts to decrease.

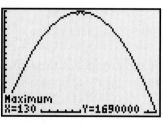

Check Point 2

On a certain route, an airline carries 8000 passengers per month, each paying $100. A market survey indicates that for each $1 increase in ticket price, the airline will lose 100 passengers.

a. Express the number of passengers per month, N, as a function of the ticket price, x.

b. Express the monthly revenue for the route, R, as a function of the ticket price, x.

Functions from Formulas

2 Construct functions from formulas.

A **formula** is an equation that uses letters to express relationships among two or more variables. Finding functions that model geometric situations requires a knowledge of basic geometric formulas. Formulas for area, perimeter, and volume are given in Table 1.6.

Table 1.6 Common Formulas for Area, Perimeter, and Volume

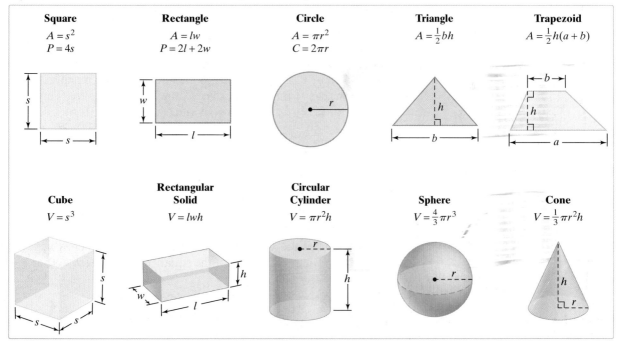

Square	Rectangle	Circle	Triangle	Trapezoid
$A = s^2$	$A = lw$	$A = \pi r^2$	$A = \frac{1}{2}bh$	$A = \frac{1}{2}h(a + b)$
$P = 4s$	$P = 2l + 2w$	$C = 2\pi r$		

Cube	Rectangular Solid	Circular Cylinder	Sphere	Cone
$V = s^3$	$V = lwh$	$V = \pi r^2 h$	$V = \frac{4}{3}\pi r^3$	$V = \frac{1}{3}\pi r^2 h$

Study Tip

When constructing functions in geometric situations, it is helpful to draw a diagram. Label the given and variable quantities on the diagram.

In our next example, we will obtain a function using the formula for the volume of a rectangular solid, $V = lwh$. A rectangular solid's volume is the product of its length, width, and height.

EXAMPLE 3 Obtaining a Function from a Geometric Formula

A machine produces open boxes using square sheets of metal measuring 12 inches on each side. The machine cuts equal-sized squares from each corner. Then it shapes the metal into an open box by turning up the sides.

a. Express the volume of the box, V, in cubic inches, as a function of the length of the side of the square cut from each corner, x, in inches.

b. Find the domain of V.

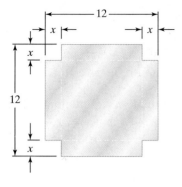

Figure 1.72 Producing open boxes using square sheets of metal

Solution

a. The situation is illustrated in Figure 1.72. The volume of the box in the lower portion of the figure is the product of its length, width, and height. The height of the box is the same as the side of the square cut from each corner, x. Because the 12-inch square has x inches cut from each corner, the length of the resulting box is $12 - x - x$, or $12 - 2x$. Similarly, the width of the resulting box is also $12 - 2x$.

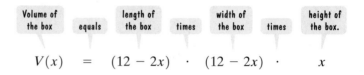

Volume of the box	equals	length of the box	times	width of the box	times	height of the box.

$$V(x) = (12 - 2x) \cdot (12 - 2x) \cdot x$$

The function $V(x) = x(12 - 2x)^2$ models the volume of the box, V, in terms of the length of the side of the square cut from each corner, x.

b. The formula for V involves a polynomial, $x(12 - 2x)^2$, which is defined for any real number, x. However, in the function $V(x) = x(12 - 2x)^2$, x represents the number of inches cut from each corner of the 12-inch square. Thus, $x > 0$. To produce an open box, the machine must cut less than 6 inches from each corner of the 12-inch square. Thus, $x < 6$. The domain of V is $0 < x < 6$, or, in interval notation, $(0, 6)$.

Technology

The graph of the function
$V(x) = x(12 - 2x)^2$,
the model for the volume of the box in Figure 1.72, is shown in a $[0, 6, 1]$ by $[0, 130, 13]$ viewing rectangle. The graphing utility's maximum function feature indicates that the volume of the box is a maximum, 128 cubic inches, when the side of the square cut from each corner of the metal sheet is 2 inches.

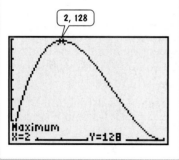

Check Point 3 A machine produces open boxes using rectangular sheets of metal measuring 15 inches by 8 inches. The machine cuts equal-sized squares from each corner. Then it shapes the metal into an open box by turning up the sides.

a. Express the volume of the box, V, in cubic inches, as a function of the length of the side of the square cut from each corner, x, in inches.

b. Find the domain of V.

In many situations, the conditions of the problem result in a function whose equation contains more than one variable. If this occurs, use the given information to write an equation among these variables. Then use this equation to eliminate all but one of the variables in the function's expression.

EXAMPLE 4 Modeling the Area of a Rectangle with a Fixed Perimeter

You have 140 yards of fencing to enclose a rectangular garden. Express the area of the garden, A, as a function of one of its dimensions, x.

Solution Figure 1.73 illustrates three of your options for enclosing the garden. In each case, the perimeter of the rectangle, twice the length plus twice the width, is 140 yards. By contrast, the area, length times width, varies according to the length of a side, x.

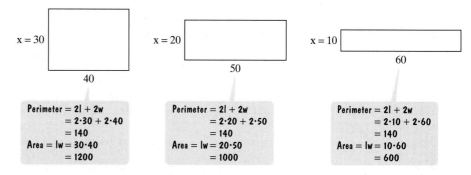

Figure 1.73 Rectangles with fixed perimeters and varying areas

As specified, x represents one of the dimensions of the rectangle. In particular, let

$$x = \text{the length of the garden}$$
$$y = \text{the width of the garden.}$$

The area, A, of the garden is the product of its length and its width:

$$A = xy.$$

There are two variables in this formula—the garden's length, x, and its width, y. We need to transform this into a function in which A is represented by one variable, x, the garden's length. Thus, we must express the width, y, in terms of the length, x. We do this using the information that you have 140 yards of fencing.

$2x + 2y = 140$ — The perimeter, twice the length plus twice the width, is 140 yards.

$2y = 140 - 2x$ — Subtract 2x from both sides.

$y = \dfrac{140 - 2x}{2}$ — Divide both sides by 2.

$y = 70 - x$ — Divide each term in the numerator by 2.

Now we substitute $70 - x$ for y in the formula for area.

$$A = xy = x(70 - x)$$

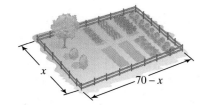

Figure 1.74

The rectangle and its dimensions are illustrated in Figure 1.74. Because A is a function of x, we can write

$$A(x) = x(70 - x) \quad \text{or} \quad A(x) = 70x - x^2.$$

This function models the area, A, of a rectangular garden with a perimeter of 140 yards in terms of the length of a side, x.

Technology

The graph of the function

$$A(x) = x(70 - x)$$

the model for the area of the garden in Example 4, is shown in a $[0, 70, 5]$ by $[0, 1400, 100]$ viewing rectangle. The graph shows that as the length of a side increases, the enclosed area increases, then decreases. The area of the garden is a maximum, 1225 square yards, when the length of one of its sides is 35 yards.

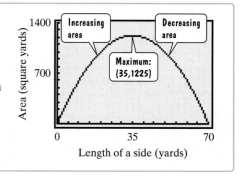

Figure 1.75

Technology

The graph of the function

$$A(r) = 2\pi r^2 + \frac{44}{r}, \text{ or}$$

$$y = 2\pi x^2 + \frac{44}{x},$$

the model for the surface area of the soft-drink can in Figure 1.75, was obtained with a graphing utility.

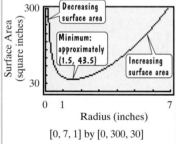

[0, 7, 1] by [0, 300, 30]

As the radius increases, the can's surface area decreases, then increases. Using a graphing utility's minimum function feature, it can be shown that when $x \approx 1.5$, the value of y is smallest ($y \approx 43.5$). Thus, the least amount of material needed to manufacture the can, approximately 43.5 square inches, occurs when its radius is approximately 1.5 inches. In calculus, you will learn techniques that give the exact radius needed to minimize the can's surface area.

Check Point 4 You have 200 feet of fencing to enclose a rectangular garden. Express the area of the garden, A, as a function of one of its dimensions, x.

In order to save on production costs, manufacturers need to use the least amount of material for containers that are required to hold a specified volume of their product. Using the least amount of material involves minimizing the surface area of the container. Formulas for surface area, A, are given in Table 1.7.

Table 1.7   Common Formulas for Surface Area

Cube	Rectangular Solid	Circular Cylinder	Sphere
$A = 6s^2$	$A = 2lw + 2lh + 2wh$	$A = 2\pi r^2 + 2\pi rh$	$A = 4\pi r^2$

EXAMPLE 5   Modeling the Surface Area of a Soft-Drink Can with Fixed Volume

Figure 1.75 shows a cylindrical soft-drink can. The can is to have a volume of 12 fluid ounces, approximately 22 cubic inches. Express the surface area of the can, A, in square inches, as a function of its radius, r, in inches.

Solution   The surface area, A, of the cylindrical can in Figure 1.75 is given by

$$A = 2\pi r^2 + 2\pi rh.$$

There are two variables in this formula—the can's radius, r, and its height, h. We need to transform this into a function in which A is represented by one variable, r, the radius of the can. Thus, we must express the height, h, in terms of the radius, r. We do this using the information that the can's volume, $V = \pi r^2 h$, must be 22 cubic inches.

$$\pi r^2 h = 22 \qquad \text{The volume of the can is 22 cubic inches.}$$

$$h = \frac{22}{\pi r^2} \qquad \text{Divide both sides by } \pi r^2 \text{ and solve for } h.$$

Now we substitute $\frac{22}{\pi r^2}$ for h in the formula for surface area.

$$A = 2\pi r^2 + 2\pi rh = 2\pi r^2 + 2\pi r\left(\frac{22}{\pi r^2}\right) = 2\pi r^2 + \frac{44}{r}$$

Because A is a function of r, the can's radius, we can express the surface area of the can as

$$A(r) = 2\pi r^2 + \frac{44}{r}.$$

Check Point 5 A cylindrical can is to hold 1 liter, or 1000 cubic centimeters, of oil. Express the surface area of the can, A, in square centimeters, as a function of its radius, r, in centimeters.

Our next example involves constructing a function that models simple interest. The annual simple interest that an investment earns is given by the formula

$$I = Pr$$

where I is the simple interest, P is the principal, and r is the simple interest rate, expressed in decimal form. Suppose, for example, that you deposit $2000 ($P = 2000$) in an account that has a simple interest rate of 3% ($r = 0.03$). The annual simple interest is computed as follows:

$$I = Pr = (2000)(0.03) = 60.$$

The annual interest is $60.

EXAMPLE 6   Modeling Simple Interest

You inherit $16,000 with the stipulation that for the first year the money must be placed in two investments expected to pay 6% and 8% annual interest, respectively. Express the expected interest, I, as a function of the amount of money invested at 6%, x.

Solution   As specified, x represents the amount invested at 6%. We will let y represent the amount invested at 8%. The expected interest, I, on the two investments combined is the expected interest on the 6% investment plus the expected interest on the 8% investment.

Expected interest on　　Expected interest on
the 6% investment　　　the 8% investment

$$I = 0.06x + 0.08y$$

Rate times　　Rate times
principal　　　principal

There are two variables in this formula—the amount invested at 6%, x, and the amount invested at 8%, y. We need to transform this into a function in which I is represented by one variable, x, the amount invested at 6%. Thus, we must express the amount invested at 8%, y, in terms of the amount invested at 6%, x. We do this using the information that you have $16,000 to invest.

$$x + y = 16,000 \qquad \text{The sum of the amounts invested at each rate must be \$16,000.}$$

$$y = 16,000 - x \qquad \text{Subtract } x \text{ from both sides and solve for } y.$$

Now we substitute $16,000 - x$ for y in the formula for interest.

$$I = 0.06x + 0.08y = 0.06x + 0.08(16,000 - x).$$

Because I is now a function of x, the amount invested at 6%, the expected interest can be expressed as

$$I(x) = 0.06x + 0.08(16,000 - x).$$

Check Point 6   You place $25,000 in two investments expected to pay 7% and 9% annual interest, respectively. Express the expected interest, I, as a function of the amount of money invested at 7%, x.

Our next example involves constructing a function using the distance formula. Earlier in the chapter, we saw that the distance between two points in the rectangular coordinate system is the square root of the difference between their x-coordinates squared plus the difference between their y-coordinates squared.

EXAMPLE 7 Modeling the Distance from the Origin to a Point on a Graph

Figure 1.76 shows that $P(x, y)$ is a point on the graph of $y = 1 - x^2$. Express the distance, d, from P to the origin as a function of the point's x-coordinate.

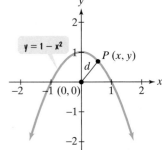

Figure 1.76

Solution   We use the distance formula to find the distance, d, from $P(x, y)$ to the origin, $(0, 0)$.

$$d = \sqrt{(x - 0)^2 + (y - 0)^2} = \sqrt{x^2 + y^2}$$

There are two variables in this formula—the point's x-coordinate and its y-coordinate. We need to transform this into a function in which d is represented by one variable, x, the point's x-coordinate. Thus, we must express y in terms of x. We do this using the information shown in Figure 1.76, namely that $P(x, y)$ is a point on the graph of $y = 1 - x^2$. This means that we can replace y with $1 - x^2$ in our formula for d.

$$d = \sqrt{x^2 + y^2} = \sqrt{x^2 + (1 - x^2)^2} = \sqrt{x^2 + 1 - 2x^2 + x^4} = \sqrt{x^4 - x^2 + 1}$$

Square $1 - x^2$ using
$(A - B)^2 = A^2 - 2AB + B^2.$

The distance, d, from $P(x, y)$ to the origin can be expressed as a function of the point's x-coordinate as

$$d(x) = \sqrt{x^4 - x^2 + 1}.$$

Check Point 7   Let $P(x, y)$ be a point on the graph of $y = x^3$. Express the distance, d, from P to the origin as a function of the point's x-coordinate.

Technology

The graph of the function
$$d(x) = \sqrt{x^4 - x^2 + 1}, \text{ or }$$
$$y = \sqrt{x^4 - x^2 + 1},$$
the model for the distance, d, shown in Figure 1.76, was obtained with a graphing utility.

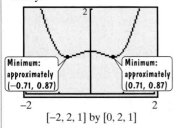

Minimum: approximately $(-0.71, 0.87)$ Minimum: approximately $(0.71, 0.87)$

$[-2, 2, 1]$ by $[0, 2, 1]$

Using a graphing utility's minimum function feature, it can be shown that when $x \approx -0.71$ and when $x \approx 0.71$, the value of d is smallest ($d \approx 0.87$ is a relative minimum). Examine Figure 1.76. The graph's y-axis symmetry indicates that there are two points on $y = 1 - x^2$ whose distance to the origin is smallest.

EXERCISE SET 1.9

Practice and Application Exercises

1. A car rental agency charges $200 per week plus $0.15 per mile to rent a car.

   a. Express the weekly cost to rent the car, f, as a function of the number of miles driven during the week, x.

   b. How many miles did you drive during the week if the weekly cost to rent the car was $320?

2. A car rental agency charges $180 per week plus $0.25 per mile to rent a car.

   a. Express the weekly cost to rent the car, f, as a function of the number of miles driven during the week, x.

   b. How many miles did you drive during the week if the weekly cost to rent the car was $395?

3. One yardstick for measuring how steadily—if slowly—athletic performance has improved is the mile run. In 1954, Roger Bannister of Britain cracked the 4-minute mark, setting the record for running a mile in 3 minutes, 59.4 seconds, or 239.4 seconds. In the half-century since then, the record has decreased by 0.3 second per year.

a. Express the record time for the mile run, M, as a function of the number of years after 1954, x.

b. If this trend continues, in which year will someone run a 3-minute, or 180-second, mile?

4. According to the National Center for Health Statistics, in 1990, 28% of babies in the United States were born to parents who were not married. Throughout the 1990s, this increased by approximately 0.6% per year.

a. Express the percentage of babies born out of wedlock, P, as a function of the number of years after 1990, x.

b. If this trend continues, in which year will 40% of babies be born out of wedlock?

5. The bus fare in a city is $1.25. People who use the bus have the option of purchasing a monthly coupon book for $21.00. With the coupon book, the fare is reduced to $0.50.

a. Express the total monthly cost to use the bus without a coupon book, f, as a function of the number of times in the month the bus is used, x.

b. Express the total monthly cost to use the bus with a coupon book, g, as a function of the number of times in the month the bus is used.

c. Determine the number of times in a month the bus must be used so that the total monthly cost without the coupon book is the same as the total monthly cost with the coupon book. What will be the monthly cost for each option?

6. A coupon book for a bridge costs $21 per month. The toll for the bridge is normally $2.50, but it is reduced to $1 for people who have purchased the coupon book.

a. Express the total monthly cost to use the bridge without a coupon book, f, as a function of the number of times in the month the bridge is crossed, x.

b. Express the total monthly cost to use the bridge with a coupon book, g, as a function of the number of times in the month the bridge is crossed, x.

c. Determine the number of times in a month the bridge must be crossed so that the total monthly cost without the coupon book is the same as the total monthly cost with the coupon book. What will be the monthly cost for each option?

7. You are choosing between two plans at a discount warehouse. Plan A offers an annual membership of $100 and you pay 80% of the manufacturer's recommended list price. Plan B offers an annual membership fee of $40 and you pay 90% of the manufacturer's recommended list price.

a. Express the total yearly amount paid to the warehouse under plan A, f, as a function of the dollars of merchandise purchased during the year, x.

b. Express the total yearly amount paid to the warehouse under plan B, g, as a function of the dollars of merchandise purchased during the year, x.

c. How many dollars of merchandise would you have to purchase in a year to pay the same amount under both plans? What will be the total yearly amount paid to the warehouse for each plan?

8. You are choosing between two plans at a discount warehouse. Plan A offers an annual membership fee of $300 and you pay 70% of the manufacturer's recommended list price. Plan B offers an annual membership fee of $40 and you pay 90% of the manufacturer's recommended list price.

a. Express the total yearly amount paid to the warehouse under plan A, f, as a function of the dollars of merchandise purchased during the year, x.

b. Express the total yearly amount paid to the warehouse under plan B, g, as a function of the dollars of merchandise purchased during the year, x.

c. How many dollars of merchandise would you have to purchase in a year to pay the same amount under both plans? What will be the total yearly amount paid to the warehouse for each plan?

9. A football team plays in a large stadium. With a ticket price of $20, the average attendance at recent games has been 30,000. A market survey indicates that for each $1 increase in the ticket price, attendance decreases by 500.

a. Express the number of spectators at a football game, N, as a function of the ticket price, x.

b. Express the revenue from a football game, R, as a function of the ticket price, x.

10. A baseball team plays in a large stadium. With a ticket price of $15, the average attendance at recent games has been 20,000. A market survey indicates that for each $1 increase in the ticket price, attendance decreases by 400.

a. Express the number of spectators at a baseball game, N, as a function of the ticket price, x.

b. Express the revenue from a baseball game, R, as a function of the ticket price, x.

11. On a certain route, an airline carries 9000 passengers per month, each paying $150. A market survey indicates that for each $1 decrease in the ticket price, the airline will gain 50 passengers.

a. Express the number of passengers per month, N, as a function of the ticket price, x.

b. Express the monthly revenue for the route, R, as a function of the ticket price, x.

12. On a certain route, an airline carries 7000 passengers per month, each paying $90. A market survey indicates that for each $1 decrease in the ticket price, the airline will gain 60 passengers.

    a. Express the number of passengers per month, N, as a function of the ticket price, x.

    b. Express the monthly revenue for the route, R, as a function of the ticket price, x.

13. The annual yield per lemon tree is fairly constant at 320 pounds per tree when the number of trees per acre is 50 or fewer. For each additional tree over 50, the annual yield per tree for all trees on the acre decreases by 4 pounds due to overcrowding.

    a. Express the yield per tree, Y, in pounds, as a function of the number of lemon trees per acre, x.

    b. Express the total yield for an acre, T, in pounds, as a function of the number of lemon trees per acre, x.

14. The annual yield per orange tree is fairly constant at 270 pounds per tree when the number of trees per acre is 30 or fewer. For each additional tree over 30, the annual yield per tree for all trees on the acre decreases by 3 pounds due to overcrowding.

    a. Express the yield per tree, Y, in pounds, as a function of the number of orange trees per acre, x.

    b. Express the total yield for an acre, T, in pounds, as a function of the number of orange trees per acre, x.

15. An open box is made from a square piece of cardboard 24 inches on a side by cutting identical squares from the corners and turning up the sides.

    a. Express the volume of the box, V, as a function of the length of the side of the square cut from each corner, x.

    b. Find and interpret $V(2)$, $V(3)$, $V(4)$, $V(5)$, and $V(6)$. What is happening to the volume of the box as the length of the side of the square cut from each corner increases?

    c. Find the domain of V.

16. An open box is made from a square piece of cardboard 30 inches on a side by cutting identical squares from the corners and turning up the sides.

    a. Express the volume of the box, V, as a function of the length of the side of the square cut from each corner, x.

    b. Find and interpret $V(3)$, $V(4)$, $V(5)$, $V(6)$, and $V(7)$. What is happening to the volume of the box as the length of the side of the square cut from each corner increases?

    c. Find the domain of V.

17. A rain gutter is made from sheets of aluminum that are 20 inches wide. As shown in the figure at the top of the next column, the edges are turned up to form right angles. Express the cross-sectional area of the gutter, A, as a function of its depth, x.

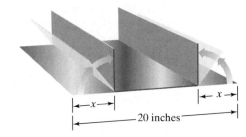

20 inches

18. A piece of wire is 8 inches long. The wire is cut into two pieces and then each piece is bent into a square. Express the sum of the areas of these squares, A, as a function of the length of the cut, x.

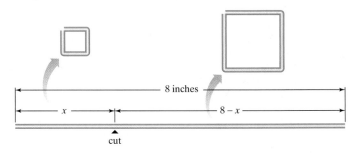

8 inches

x $8 - x$

cut

19. The sum of two numbers is 66. Express the product of the numbers, P, as a function of one of the numbers, x.

20. The sum of two numbers is 50. Express the product of the numbers, P, as a function of one of the numbers, x.

21. You have 800 feet of fencing to enclose a rectangular field. Express the area of the field, A, as a function of one of its dimensions, x.

22. You have 600 feet of fencing to enclose a rectangular field. Express the area of the field, A, as a function of one of its dimensions, x.

23. As in Exercise 21, you have 800 feet of fencing to enclose a rectangular field. However, one side of the field lies along a canal and requires no fencing. Express the area of the field, A, as a function of one of its dimensions, x.

24. As in Exercise 22, you have 600 feet of fencing to enclose a rectangular field. However, one side of the field lies along a canal and requires no fencing. Express the area of the field, A, as a function of one of its dimensions, x.

25. You have 1000 feet of fencing to enclose a rectangular playground and subdivide it into two smaller playgrounds by placing the fencing parallel to one of the sides. Express the area of the playground, A, as a function of one of its dimensions, x.

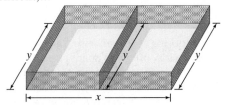

26. You have 1200 feet of fencing to enclose a rectangular region and subdivide it into three smaller rectangular regions by placing two fences parallel to one of the sides. Express the area of the enclosed rectangular region, A, as a function of one of its dimensions, x.

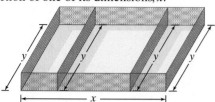

27. A new running track is to be constructed in the shape of a rectangle with semicircles at each end. The track is to be 440 yards long. Express the area of the region enclosed by the track, A, as a function of its radius, r.

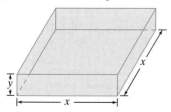

28. Work Exercise 27 if the length of the track is increased to 880 yards.

29. A contractor is to build a warehouse whose rectangular floor will have an area of 4000 square feet. The warehouse will be separated into two rectangular rooms by an interior wall. The cost of the exterior walls is \$175 per linear foot and the cost of the interior wall is \$125 per linear foot. Express the contractor's cost for building the walls, C, as a function of one of the dimensions of the warehouse's rectangular floor, x.

30. The area of a rectangular garden is 125 square feet. The garden is to be enclosed on three sides by a brick wall costing \$20 per foot and on one side by a fence costing \$9 per foot. Express the cost to enclose the garden, C, as a function of one of its dimensions, x.

31. The figure shows an open box with a square base. The box is to have a volume of 10 cubic feet. Express the amount of material needed to construct the box, A, as a function of the length of a side of its square base, x.

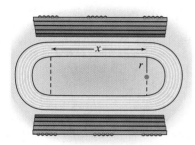

32. The figure shows an open box with a square base and a partition down the middle. The box is to have a volume of 400 cubic inches. Express the amount of material needed to construct the box, A, as a function of the length of a side of its square base, x.

33. The figure shows a package whose front is a square. The length plus girth (the distance around) of the package is 300 inches. (This is the maximum length plus girth permitted by Federal Express for its overnight service.) Express the volume of the package, V, as a function of the length of a side of its square front, x.

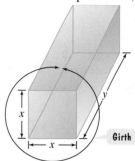

34. Work Exercise 33 if the length plus girth of the package is 108 inches.

35. Your grandmother needs your help. She has \$50,000 to invest. Part of this money is to be invested in noninsured bonds paying 15% annual interest. The rest of this money is to be invested in a government-insured certificate of deposit paying 7% annual interest.

a. Express the interest from both investments, I, as a function of the amount of money invested in noninsured bonds, x.

b. Your grandmother told you that she requires \$6000 per year in extra income from both these investments. How much money should be placed in each investment?

36. You inherit \$18,750 with the stipulation that for the first year the money must be placed in two investments expected to pay 10% and 12% annual interest, respectively.

a. Express the expected interest from both investments, I, as a function of the amount of money invested at 10%, x.

b. If the total interest earned for the year was \$2117, how much money was invested at each rate?

37. You invested \$8000, part of it in a stock that paid 12% annual interest. However, the rest of the money suffered a 5% loss. Express the total annual income from both investments, I, as a function of the amount invested in the 12% stock, x.

38. You invested $12,000, part of it in a stock that paid 14% annual interest. However, the rest of the money suffered a 6% loss. Express the total annual income from both investments, I, as a function of the amount invested in the 14% stock, x.

39. Let $P(x, y)$ be a point on the graph of $y = x^2 - 4$. Express the distance, d, from P to the origin as a function of the point's x-coordinate.

40. Let $P(x, y)$ be a point on the graph of $y = x^2 - 8$. Express the distance, d, from P to the origin as a function of the point's x-coordinate.

41. Let $P(x, y)$ be a point on the graph of $y = \sqrt{x}$. Express the distance, d, from P to $(1, 0)$ as a function of the point's x-coordinate.

42. Let $P(x, y)$ be a point on the graph of $y = \sqrt{x}$. Express the distance, d, from P to $(2, 0)$ as a function of the point's x-coordinate.

43. The figure shows a rectangle with two vertices on a semicircle of radius 2 and two vertices on the x-axis. Let $P(x, y)$ be the vertex that lies in the first quadrant.

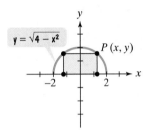

a. Express the area of the rectangle, A, as a function of x.
b. Express the perimeter of the rectangle, P, as a function of x.

44. The figure shows a rectangle with two vertices on a semicircle of radius 3 and two vertices on the x-axis. Let $P(x, y)$ be the vertex that lies in the first quadrant.

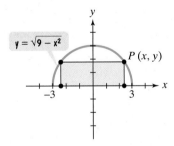

a. Express the area of the rectangle, A, as a function of x.
b. Express the perimeter of the rectangle, P, as a function of x.

45. Two vertical poles of length 6 feet and 8 feet, respectively, stand 10 feet apart. A cable reaches from the top of one pole to some point on the ground between the poles and then to the top of the other pole. Express the amount of cable used, f, as a function of the distance of the cable from the 6-foot pole, x.

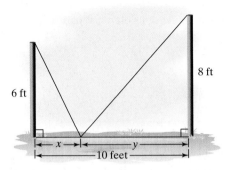

46. Towns A and B are located 6 miles and 3 miles, respectively, from a major expressway. The point on the expressway closest to town A is 12 miles from the point on the expressway closest to town B. Two new roads are to be built from A to the expressway and then to B. Express the combined lengths of the new roads, f, as a function of where the roads are positioned from A, shown as x in the figure.

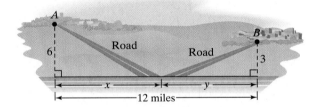

Writing in Mathematics

47. Throughout this section, we started with familiar formulas and created functions by substitution. Describe a specific situation in which we obtained a function using this technique.

48. Describe what should be displayed on the screen of a graphing utility to illustrate the solution that you obtained in Exercise 5(c) or Exercise 6(c).

49. In Exercise 9(b) or Exercise 10(b), describe what important information the team owners could learn from the revenue function.

50. In Exercise 13(b) or 14(b), describe what important information the growers could learn from the total-yield function.

51. In Exercise 31 or 32, describe what important information the box manufacturer could learn from the surface area function.

52. In calculus, you will learn powerful tools that reveal how functions behave. However, before applying these tools, there will be situations in which you are first required to obtain these functions from verbal descriptions. This is why your work in this section is so important. Because there is no rigid step-by-step procedure for modeling from verbal conditions, you might have had some difficulties obtaining functions for the assigned exercises. Discuss what you did if this happened to you. Did your course of action enhance your ability to model with functions?

Technology Exercises

53. Use a graphing utility to graph the function that you obtained in Exercise 1 or Exercise 2. Then use the TRACE or ZOOM feature to verify your answer in part (b) of the exercise.

54. Use a graphing utility to graph the two functions, f and g, that you obtained in any one exercise from Exercises 5–8. Then use the TRACE or INTERSECTION feature to verify your answer in part (c) of the exercise.

55. Use a graphing utility to graph the volume-of-the-box function, V, that you obtained in Exercise 15 or Exercise 16. Then use the TRACE or maximum function feature to find the length of the side of the square that should be cut from each corner of the cardboard to create a box with the greatest possible volume. What is the maximum volume of the open box?

56. Use a graphing utility to graph the area function, A, that you obtained in Exercise 21 or Exercise 22. Then use an appropriate feature on your graphing utility to find the dimensions of the field that result in the greatest possible area. What is the maximum area?

57. Use the maximum or minimum function feature of a graphing utility to provide useful numerical information to any one of the following: the manufacturer of the rain gutters in Exercise 17; the person enclosing the playground in Exercise 25; the contractor in Exercise 29; the manufacturer of the cylindrical cans in Check Point 5 (page 226).

Critical Thinking Exercises

58. A pool measuring 20 meters by 10 meters is surrounded by a path of uniform width, as shown in the figure. Express the area of the path, A, in square meters, as a function of its width, x, in meters.

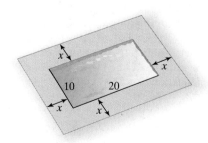

59. You are on an island 2 miles from the nearest point P on a straight shoreline, as shown in the figure. Six miles down the shoreline from point P is a restaurant, shown as point R. To reach the restaurant, you first row from the island to point Q, averaging 2 miles per hour. Then you jog the distance from Q to R, averaging 5 miles per hour. Express the time that, T, it takes to go from the island to the restaurant as a function of the distance, x, from P, where you land the boat.

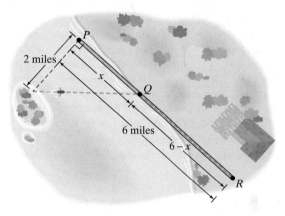

60. The figure shows a Norman window that has the shape of a rectangle with a semicircle attached at the top. The diameter of the semicircle is equal to the width of the rectangle. The window has a perimeter of 12 feet. Express the area of the window, A, as a function of its radius, r.

61. The figure shows water running into a container in the shape of a cone. The radius of the cone is 6 feet and its height is 12 feet. Express the volume of the water in the cone, V, as a function of the height of the water, h.

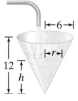

CHAPTER SUMMARY, REVIEW, AND TEST
Summary

DEFINITIONS AND CONCEPTS	EXAMPLES

1.1 Graphs and Graphing Utilities

a. The rectangular coordinate system consists of a horizontal number line, the x-axis, and a vertical number line, the y-axis, intersecting at their zero points, the origin. Each point in the system corresponds to an ordered pair of real numbers (x, y). The first number in the pair is the x-coordinate; the second number is the y-coordinate. See Figure 1.1 on page 116.   Ex. 1, p.117

b. An ordered pair is a solution of an equation in two variables if replacing the variables by the corresponding coordinates results in a true statement. The ordered pair is said to satisfy the equation. The graph of the equation is the set of all points whose coordinates satisfy the equation. One method for graphing an equation is to plot ordered-pair solutions and connect them with a smooth curve or line.   Ex. 2, p.118

c. An x-intercept of a graph is the x-coordinate of a point where the graph intersects the x-axis. The y-coordinate corresponding to a graph's x-intercept is always zero.

d. A y-intercept of a graph is the y-coordinate of a point where the graph intersects the y-axis. The x-coordinate corresponding to a graph's y-intercept is always zero.

1.2 Lines and Slope

a. The slope, m, of the line through (x_1, y_1) and (x_2, y_2) is $m = \dfrac{y_2 - y_1}{x_2 - x_1}$.   Ex. 1, p.125

b. Equations of lines include point-slope form, $y - y_1 = m(x - x_1)$, slope-intercept form, $y = mx + b$, and general form, $Ax + By + C = 0$. The equation of a horizontal line is $y = b$; a vertical line is $x = a$.   Exs. 2 & 3, p.127; Ex. 5, p.130; Ex. 6, p.131

c. Parallel lines have equal slopes. Perpendicular lines have slopes that are negative reciprocals.   Exs. 8 & 9, pp.132–134

1.3 Distance and Midpoint Formulas; Circles

a. The distance, d, between the points (x_1, y_1) and (x_2, y_2) is given by $d = \sqrt{(x_2 - x_1)^2 + (y_2 - y_1)^2}$.   Ex. 1, p.143

b. The midpoint of the line segment whose endpoints are (x_1, y_1) and (x_2, y_2) is the point with coordinates $\left(\dfrac{x_1 + x_2}{2}, \dfrac{y_1 + y_2}{2} \right)$.   Ex. 2, p.144

c. The standard form of the equation of a circle with center (h, k) and radius r is $(x - h)^2 + (y - k)^2 = r^2$.   Ex. 3, p.145; Exs. 4 & 5, p.146

d. The general form of the equation of a circle is $x^2 + y^2 + Dx + Ey + F = 0$.

e. To convert from the general form to the standard form of a circle's equation, complete the square on x and y.   Ex. 6, p.147

1.4 Basics of Functions

a. A relation is any set of ordered pairs. The set of first components is the domain and the set of second components is the range.   Ex. 1, p.151

b. A function is a correspondence from a first set, called the domain, to a second set, called the range, such that each element in the domain corresponds to exactly one element in the range. If any element in a relation's domain corresponds to more than one element in the range, the relation is not a function.   Ex. 2, p.152

c. Functions are usually given in terms of equations involving x and y, in which x is the independent variable and y is the dependent variable. If an equation is solved for y and more than one value of y can be obtained for a given x, then the equation does not define y as a function of x. If an equation defines a function, the value of the function at x, $f(x)$, often replaces y.   Ex. 3, p.154

d. The difference quotient is   Ex. 5, p.156
$$\frac{f(x + h) - f(x)}{h}, h \neq 0.$$

e. If a function f does not model data or verbal conditions, its domain is the largest set of real numbers for which the value of $f(x)$ is a real number. Exclude from the function's domain real numbers that cause division by zero and real numbers that result in an even root of a negative number.   Ex. 7, p.159

DEFINITIONS AND CONCEPTS	EXAMPLES

1.5   Graphs of Functions

a. The graph of a function is the graph of its ordered pairs.

Ex. 1, p.164

b. The vertical line test for functions: If any vertical line intersects a graph in more than one point, the graph does not define y as a function of x.

Ex. 3, p.166

c. A function is increasing on intervals where its graph rises, decreasing on intervals where it falls, and constant on intervals where it neither rises nor falls. Precise definitions are given in the box on page 168.

Ex. 5, p.168

d. If the graph of a function is given, we can often visually locate the number(s) at which the function has a relative maximum or relative minimum. Precise definitions are given in the box on page 169.

Fig 1.42, p.170

e. The average rate of change of f from x_1 to x_2 is
$$\frac{\Delta y}{\Delta x} = \frac{f(x_2) - f(x_1)}{x_2 - x_1}.$$

Ex. 6, p.170

f. If a function expresses an object's position, $s(t)$, in terms of time, t, the average velocity of the object from t_1 to t_2 is
$$\frac{\Delta s}{\Delta t} = \frac{s(t_2) - s(t_1)}{t_2 - t_1}.$$

Ex. 7, p.172

g. The graph of an even function in which $f(-x) = f(x)$ is symmetric with respect to the y-axis. The graph of an odd function in which $f(-x) = -f(x)$ is symmetric with respect to the origin.

Ex. 8, p.173

h. The graph of $f(x) = \text{int}(x)$, where $\text{int}(x)$ is the greatest integer that is less than or equal to x, has function values that form discontinuous steps, shown in Figure 1.49 on page 176. If $n \le x < n + 1$, where n is an integer, then $\text{int}(x) = n$.

1.6   Transformations of Functions

a. Table 1.4 on pages 184–185 shows the graphs of the constant function, $f(x) = c$, the identity function, $f(x) = x$, the standard quadratic function, $f(x) = x^2$, the standard cubic function, $f(x) = x^3$, the square root function, $f(x) = \sqrt{x}$, and the absolute value function, $f(x) = |x|$. The table also lists characteristics of each function.

Ex. 1, p.186;
Ex. 2, p.187;

b. Table 1.5 on page 192 summarizes how to graph a function using vertical shifts, $y = f(x) \pm c$, horizontal shifts, $y = f(x \pm c)$, reflections about the x-axis, $y = -f(x)$, reflections about the y-axis, $y = f(-x)$, vertical stretching, $y = cf(x), c > 1$, and vertical shrinking, $y = cf(x), 0 < c < 1$.

Ex. 3, p.188;
Exs. 4–7,
pp.190–192

c. A function involving more than one transformation can be graphed in the following order: (1) horizontal shifting; (2) vertical stretching or shrinking; (3) reflecting; (4) vertical shifting.

Ex. 8, p.193

1.7   Combinations of Functions; Composite Functions

a. When functions are given as equations, they can be added, subtracted, multiplied, or divided by performing operations with the algebraic expressions that appear on the right side of the equations.

Ex. 1, p.199;
Ex. 2, p.199;

Definitions for the sum $f + g$, the difference $f - g$, the product fg, and the quotient $\dfrac{f}{g}$ functions are given in the box on page 200.

Ex. 3, p.200

b. The composition of functions f and g, $f \circ g$, is defined by $(f \circ g)(x) = f(g(x))$. The domain of the composite function $f \circ g$ is given in the box on page 203. This composite function is obtained by replacing each occurrence of x in the equation for f with $g(x)$.

Ex. 4, p.203;
Ex. 5, p.204

1.8   Inverse Functions

a. If $f(g(x)) = x$ and $g(f(x)) = x$, function g is the inverse of function f, denoted f^{-1} and read "f inverse." Thus, to show that f and g are inverses of each other, one must show $f(g(x)) = x$ and $g(f(x)) = x$.

Ex. 1, p.211;
Ex. 2, p.212

b. The procedure for finding a function's inverse uses a switch-and-solve strategy. Switch x and y, then solve for y. The procedure is given in the box on page 212.

Exs. 3 & 4, p.213

c. The horizontal line test for inverse functions: A function f has an inverse that is a function, f^{-1}, if there is no horizontal line that intersects the graph of the function f at more than one point.

Ex. 5, p.214

DEFINITIONS AND CONCEPTS	**EXAMPLES**

d. A one-to-one function is one in which no two different ordered pairs have the same second component. Only one-to-one functions have inverse functions.

e. If the point (a, b) is on the graph of f, then the point (b, a) is on the graph of f^{-1}. The graph of f^{-1} is a reflection of the graph of f about the line $y = x$.

Ex. 6, p.216

1.9 Modeling with Functions

a. Verbal models are often helpful in obtaining functions from verbal descriptions.

Ex. 1, p.220; Ex. 2, p.221

b. Functions can be constructed from formulas, such as formulas for area, perimeter, and volume (Table 1.6 on page 223) and formulas for surface area (Table 1.7 on page 226).

Ex. 3, p. 223

c. If a problem's conditions are modeled by a function whose equation contains more than one variable, use the given information to write an equation among these variables. Then use this equation to eliminate all but one of the variables in the function's expression.

Ex. 4, p.224; Ex. 5, p.226; Ex. 6, p.227; Ex. 7, p.228

Review Exercises

1.1

Graph each equation in Exercises 1–4.
Let $x = -3, -2, -1, 0, 1, 2,$ and 3.

1. $y = 2x - 2$ **2.** $y = x^2 - 3$

3. $y = x$ **4.** $y = |x| - 2$

5. What does a $[-20, 40, 10]$ by $[-5, 5, 1]$ viewing rectangle mean? Draw axes with tick marks and label the tick marks to illustrate this viewing rectangle.

In Exercises 6–8, use the graph and determine the x-intercepts, if any, and the y-intercepts, if any. For each graph, tick marks along the axes represent one unit each.

6.

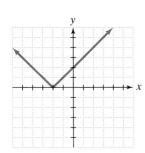

7.

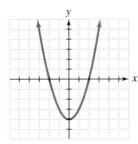

8.

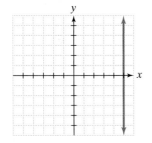

The caseload of Alzheimer's disease in the United States is expected to explode as baby boomers head into their later years. The graph shows the percentage of Americans with the disease, by age. Use the graph to solve Exercises 9–11.

Alzheimer's Prevalence in the U.S., by Age

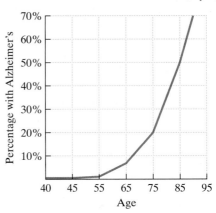

Source: Centers for Disease Control

9. What percentage of Americans who are 75 have Alzheimer's disease?

10. What age represents 50% prevalence of Alzheimer's disease?

11. Describe the trend shown by the graph.

1.2

In Exercises 12–15, find the slope of the line passing through each pair of points or state that the slope is undefined. Then indicate whether the line through the points rises, falls, is horizontal, or is vertical.

12. $(3, 2)$ and $(5, 1)$ **13.** $(-1, -2)$ and $(-3, -4)$

14. $(-3, \frac{1}{4})$ and $(6, \frac{1}{4})$ **15.** $(-2, 5)$ and $(-2, 10)$

In Exercises 16–17, use the given conditions to write an equation for each line in point-slope form and slope-intercept form.

16. Passing through $(-3, 2)$ with slope -6

17. Passing through $(1, 6)$ and $(-1, 2)$

In Exercises 18–21, give the slope and y-intercept of each line whose equation is given. Then graph the line.

18. $y = \frac{2}{5}x - 1$

19. $y = -4x + 5$

20. $2x + 3y + 6 = 0$

21. $2y - 8 = 0$

22. Corporations in the United States are doing quite well, thank you. The scatter plot below shows corporate profits, in billions of dollars, from 1990 through 2000. Also shown is a line that passes through or near the points.

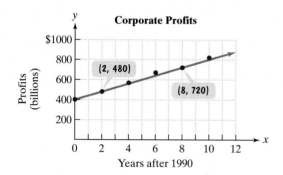

Corporate Profits

(2, 480)

(8, 720)

Source: U.S. Department of Labor

a. Use the two points whose coordinates are shown by the voice balloons to find the point-slope equation of the line that models corporate profits, y, in billions of dollars, x years after 1990.

b. Write the equation in part (a) in slope-intercept form.

c. Use the linear model to predict corporate profits in 2010.

23. The scatter plot shows the number of minutes each that 16 people exercise per week and the number of headaches per month each person experiences.

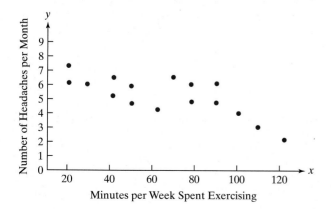

a. Draw a line that fits the data so that the spread of the data points around the line is as small as possible.

b. Use the coordinates of two points along your line to write its point-slope and slope-intercept equations.

c. Use the equation in part (b) to predict the number of headaches per month for a person exercising 130 minutes per week.

In Exercises 24–25, use the given conditions to write an equation for each line in point-slope form and slope-intercept form.

24. Passing through $(4, -7)$ and parallel to the line whose equation is $3x + y - 9 = 0$

25. Passing through $(-3, 6)$ and perpendicular to the line whose equation is $y = \frac{1}{3}x + 4$

1.3

In Exercises 26–27, find the distance between each pair of points. If necessary, round answers to two decimal places.

26. $(-2, -3)$ and $(3, 9)$ **27.** $(-4, 3)$ and $(-2, 5)$

In Exercises 28–29, find the midpoint of each line segment with the given endpoints.

28. $(2, 6)$ and $(-12, 4)$ **29.** $(4, -6)$ and $(-15, 2)$

In Exercises 30–31, write the standard form of the equation of the circle with the given center and radius.

30. Center $(0, 0), r = 3$ **31.** Center $(-2, 4), r = 6$

In Exercises 32–34, give the center and radius of each circle and graph its equation.

32. $x^2 + y^2 = 1$

33. $(x + 2)^2 + (y - 3)^2 = 9$

34. $x^2 + y^2 - 4x + 2y - 4 = 0$

1.4

In Exercises 35–37, determine whether each relation is a function. Give the domain and range for each relation.

35. $\{(2, 7), (3, 7), (5, 7)\}$ **36.** $\{(1, 10)(2, 500), (13, \pi)\}$

37. $\{(12, 13), (14, 15), (12, 19)\}$

In Exercises 38–40, determine whether each equation defines y as a function of x.

38. $2x + y = 8$ **39.** $3x^2 + y = 14$

40. $2x + y^2 = 6$

In Exercises 41–44, evaluate each function at the given values of the independent variable and simplify.

41. $f(x) = 5 - 7x$

   a. $f(4)$ **b.** $f(x + 3)$ **c.** $f(-x)$

42. $g(x) = 3x^2 - 5x + 2$

   a. $g(0)$ **b.** $g(-2)$

   c. $g(x-1)$ **d.** $g(-x)$

43. $g(x) = \begin{cases} \sqrt{x-4} & \text{if } x \geq 4 \\ 4-x & \text{if } x < 4 \end{cases}$

   a. $g(13)$ **b.** $g(0)$ **c.** $g(-3)$

44. $f(x) = \begin{cases} \dfrac{x^2-1}{x-1} & \text{if } x \neq 1 \\ 12 & \text{if } x = 1 \end{cases}$

   a. $f(-2)$ **b.** $f(1)$ **c.** $f(2)$

In Exercises 45–46, find and simplify the difference quotient

$$\frac{f(x+h) - f(x)}{h}, \quad h \neq 0$$

for the given function.

45. $f(x) = 8x - 11$ **46.** $f(x) = x^2 - 13x + 5$

In Exercises 47–51, find the domain of each function.

47. $f(x) = x^2 + 6x - 3$ **48.** $g(x) = \dfrac{4}{x-7}$

49. $h(x) = \sqrt{8 - 2x}$

50. $f(x) = \dfrac{x}{x^2 - 1}$

51. $g(x) = \dfrac{\sqrt{x-2}}{x-5}$

1.5

Graph the functions in Exercises 52–53. Use the integer values of x given to the right of the function to obtain the ordered pairs. Use the graph to specify the function's domain and range.

52. $f(x) = x^2 - 4x + 4$ $x = -1, 0, 1, 2, 3, 4$

53. $f(x) = |2 - x|$ $x = -1, 0, 1, 2, 3, 4$

In Exercises 54–56, use the graph to determine **a.** *the function's domain;* **b.** *the function's range;* **c.** *the x-intercepts, if any;* **d.** *the y-intercept, if any;* **e.** *intervals on which the function is increasing, decreasing, or constant; and* **f.** *the function values indicated below the graphs.*

54.

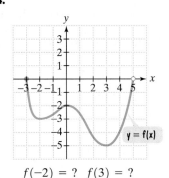

$f(-2) = ?$ $f(3) = ?$

55.

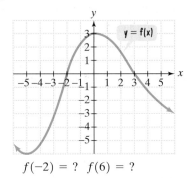

$f(-2) = ?$ $f(6) = ?$

56.

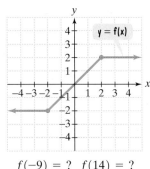

$f(-9) = ?$ $f(14) = ?$

In Exercises 57–58, find:

   a. *the numbers, if any, at which f has a relative maximum. What are these relative maxima?*

   b. *the numbers, if any, at which f has a relative minimum. What are these relative minima?*

57. Use the graph in Exercise 54.

58. Use the graph in Exercise 55.

In Exercises 59–62, use the vertical line test to identify graphs in which y is a function of x.

59.

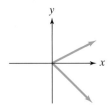

60.

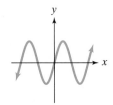

61.

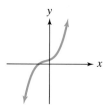

62.

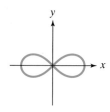

63. Find the average rate of change of $f(x) = x^2 - 4x$ from $x_1 = 5$ to $x_2 = 9$.

64. A person standing on the roof of a building throws a ball directly upward. The ball misses the rooftop on its way down and eventually strikes the ground. The function

$$s(t) = -16t^2 + 64t + 80$$

describes the ball's height above the ground, $s(t)$, in feet, t seconds after it was thrown.
  a. Find the ball's average velocity between the time it was thrown and 2 seconds later.
  b. Find the ball's average velocity between 2 and 4 seconds after it was thrown.
  c. What do the signs in your answers to parts (a) and (b) mean in terms of the direction of the ball's motion?

65. The graph shows annual spending per uniformed member of the U.S. military in inflation-adjusted dollars. Find the average rate of change of spending per year from 1955 through 2000. Round to the nearest dollar per year.

Spending per Uniformed Member of the U.S. Military

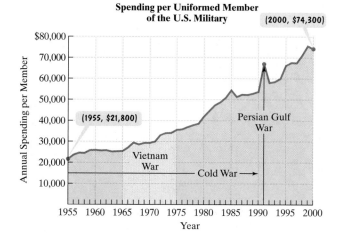

Source: Centers for Strategic and Budgetary Assessments

In Exercises 66–68, determine whether each function is even, odd, or neither. State each function's symmetry. If you are using a graphing utility, graph the function and verify its possible symmetry.

66. $f(x) = x^3 - 5x$ **67.** $f(x) = x^4 - 2x^2 + 1$

68. $f(x) = 2x\sqrt{1 - x^2}$

69. The graph shows the height, in meters, of a vulture in terms of its time, in seconds, in flight.

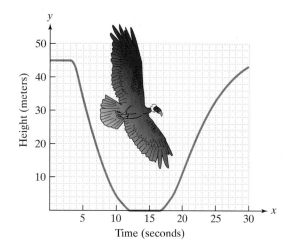

  a. Is the vulture's height a function of time? Use the graph to explain why or why not.
  b. On which interval is the function decreasing? Describe what this means in practical terms.
  c. On which intervals is the function constant? What does this mean for each of these intervals?
  d. On which interval is the function increasing? What does this mean?

70. A cargo service charges a flat fee of $5 plus $1.50 for each pound or fraction of a pound. Graph shipping cost, $C(x)$, in dollars, as a function of weight, x, in pounds, for $0 < x \le 5$.

1.6

In Exercises 71–73, begin by graphing the standard quadratic function, $f(x) = x^2$. Then use transformations of this graph to graph the given function.

71. $g(x) = x^2 + 2$ **72.** $h(x) = (x + 2)^2$

73. $r(x) = -(x + 1)^2$

In Exercises 74–76, begin by graphing the square root function, $f(x) = \sqrt{x}$. Then use transformations of this graph to graph the given function.

74. $g(x) = \sqrt{x + 3}$ **75.** $h(x) = \sqrt{3 - x}$

76. $r(x) = 2\sqrt{x + 2}$

In Exercises 77–79, begin by graphing the absolute value function, $f(x) = |x|$. Then use transformations of this graph to graph the given function.

77. $g(x) = |x + 2| - 3$

78. $h(x) = -|x - 1| + 1$

79. $r(x) = \frac{1}{2}|x + 2|$

In Exercises 80–82, begin by graphing the standard cubic function, $f(x) = x^3$. Then use transformations of this graph to graph the given function.

80. $g(x) = \frac{1}{2}(x - 1)^3$

81. $h(x) = -(x + 1)^3$

82. $r(x) = \frac{1}{4}x^3 - 1$

In Exercises 83–85, use the graph of the function f to sketch the graph of the given function g.

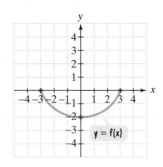

83. $g(x) = f(x + 2) + 3$

84. $g(x) = \frac{1}{2}f(x - 1)$

85. $g(x) = -2 + 2f(x + 2)$

1.7

In Exercises 86–88, find $f + g$, $f - g$, fg, and $\frac{f}{g}$. Determine the domain for each function.

86. $f(x) = 3x - 1$, $g(x) = x - 5$

87. $f(x) = x^2 + x + 1$, $g(x) = x^2 - 1$

88. $f(x) = \sqrt{x + 7}$, $g(x) = \sqrt{x - 2}$

In Exercises 89–90, find:

   a. $(f \circ g)(x)$

   b. $(g \circ f)(x)$

   c. $(f \circ g)(3)$.

89. $f(x) = x^2 + 3$, $g(x) = 4x - 1$

90. $f(x) = \sqrt{x}$, $g(x) = x + 1$

In Exercises 91–92, find:

   a. $(f \circ g)(x)$

   b. the domain of $(f \circ g)$.

91. $f(x) = \dfrac{x + 1}{x - 2}$, $g(x) = \dfrac{1}{x}$

92. $f(x) = \sqrt{x - 1}$, $g(x) = x + 3$

In Exercises 93–94, express the given function h as a composition of two functions f and g so that $h(x) = (f \circ g)(x)$.

93. $h(x) = (x^2 + 2x - 1)^4$

94. $h(x) = \sqrt[3]{7x + 4}$

1.8

In Exercises 95–96, find $f(g(x))$ and $g(f(x))$ and determine whether each pair of functions f and g are inverses of each other.

95. $f(x) = \dfrac{3}{5}x + \dfrac{1}{2}$ and $g(x) = \dfrac{5}{3}x - 2$

96. $f(x) = 2 - 5x$ and $g(x) = \dfrac{2 - x}{5}$

The functions in Exercises 97–99 are all one-to-one. For each function:

   a. *Find an equation for $f^{-1}(x)$, the inverse function.*

   b. *Verify that your equation is correct by showing that $f(f^{-1}(x)) = x$ and $f^{-1}(f(x)) = x$.*

97. $f(x) = 4x - 3$

98. $f(x) = \sqrt{x + 2}$

99. $f(x) = 8x^3 + 1$

Which graphs in Exercises 100–103 represent functions that have inverse functions?

100.

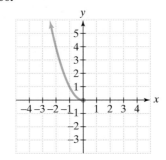

101.

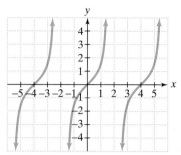

102.

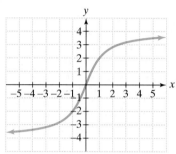

103.

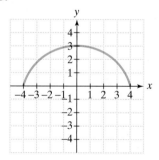

104. Use the graph of f in the figure shown to draw the graph of its inverse function.

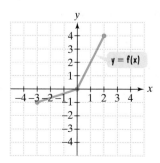

1.9

105. In 2000, the average weekly salary for workers in the United States was $567. This amount is increasing by approximately $15 per year.
   a. Express the average weekly salary for U.S. workers, W, as a function of the number of years after 2000, x.
   b. If this trend continues, in which year will the average weekly salary be $702?

106. You are choosing between two long-distance telephone plans. Plan A has a monthly fee of $15 with a charge of $0.05 per minute. Plan B has a monthly fee of $5 with a charge of $0.07 per minute.
   a. Express the monthly cost for plan A, f, as a function of the number of minutes of long-distance calls in a month, x.
   b. Express the monthly cost for plan B, g, as a function of the number of minutes of long-distance calls in a month, x.
   c. For how many minutes of long-distance calls will the costs for the two plans be the same?

107. A 400-r.oom hotel can rent every one of its rooms at $120 per room. For each $1 increase in rent, two fewer rooms are rented.
   a. Express the number of rooms rented, N, as a function of the rent, x.
   b. Express the hotel's revenue, R, as a function of the rent, x.

108. An open box is made by cutting identical squares from the corners of a 16-inch by 24-inch piece of cardboard and then turning up the sides.
   a. Express the volume of the box, V, as a function of the length of the side of the square cut from each corner, x.
   b. Find the domain of V.

109. You have 400 feet of fencing to enclose a rectangular lot and divide it in two by another fence that is parallel to one side of the lot. Express the area of the rectangular lot, A, as a function of the length of the fence that divides the rectangular lot, x.

110. The figure shows a box with a square base and a square top. The box is to have a volume of 8 cubic feet. Express the surface area of the box, A, as a function of the length of a side of its square base, x.

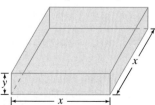

111. You inherit $10,000 with the stipulation that for the first year the money must be placed in two investments expected to earn 8% and 12% annual interest, respectively. Express the expected interest from both investments, I, as a function of the amount of money invested at 8%, x.

Chapter 1 Test

1. Graph $y = x^2 - 4$ by letting x equal integers from -3 through 3.

2. The graph of $y = -\frac{3}{2}x + 3$ is shown in a $[-6, 6, 1]$ by $[-6, 6, 1]$ viewing rectangle: Determine the x-intercepts, if any, and the y-intercepts, if any.

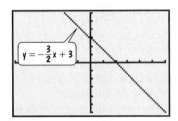

$$y = -\frac{3}{2}x + 3$$

3. The graph shows the unemployment rate in the United States from 1990 through 2000. For the period shown, during which year did the unemployment rate reach a maximum? Estimate the percentage of the work force unemployed, to the nearest tenth of a percent, at that time.

U.S. Unemployment Rate

Source: Bureau of Labor Statistics

In Exercises 4–5, use the given conditions to write an equation for each line in point-slope form and slope-intercept form.

4. Passing through $(2, 1)$ and $(-1, -8)$

5. Passing through $(-4, 6)$ and perpendicular to the line whose equation is $y = -\frac{1}{4}x + 5$

6. Strong demand plus higher fuel and labor costs are driving up the price of flying. The graph at the top of the next column shows the national averages for one-way fares. Also shown is a line that models that data.

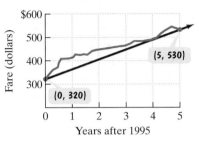

National Averages for One-Way Airline Fares: Business Travel

(5, 530)

(0, 320)

Years after 1995

Source: American Express

a. Use the two points whose coordinates are shown by the voice balloons to write the slope-intercept equation of the line that models the average one-way fare, y, in dollars, x years after 1995.

b. According to the model, what will the national average for one-way fares be in 2008?

7. Give the center and radius of the circle whose equation is $x^2 + y^2 + 4x - 6y - 3 = 0$ and graph the equation.

8. List by letter all relations that are not functions.

   a. $\{(7, 5), (8, 5), (9, 5)\}$

   b. $\{(5, 7), (5, 8), (5, 9)\}$

   c.

   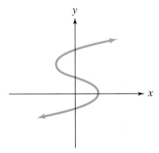

   d. $x^2 + y^2 = 100$

   e.

   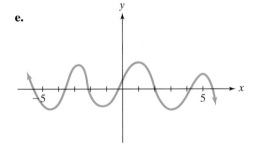

9. If $f(x) = x^2 - 2x + 5$, find $f(x - 1)$ and simplify.

10. If $g(x) = \begin{cases} \sqrt{x - 3} & \text{if } x \geq 3 \\ 3 - x & \text{if } x < 3 \end{cases}$, find $g(-1)$ and $g(7)$.

11. If $f(x) = \sqrt{12 - 3x}$, find the domain of f.

12. If $f(x) = x^2 + 11x - 7$, find and simplify the difference quotient $\dfrac{f(x + h) - f(x)}{h}$.

13. Use the graph of function f to answer the following questions.

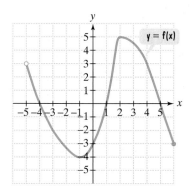

a. What is $f(4) - f(-3)$?

b. What is the domain of f?

c. What is the range of f?

d. On which interval or intervals is f increasing?

e. On which interval or intervals is f decreasing?

f. For what number does f have a relative maximum? What is the relative maximum?

g. For what number does f have a relative minimum? What is the relative minimum?

h. What are the x-intercepts?

i. What is the y-intercept?

14. Find the average rate of change of $f(x) = 3x^2 - 5$ from $x_1 = 6$ to $x_2 = 10$.

15. Determine whether $f(x) = x^4 - x^2$ is even, odd, or neither. Use your answer to explain why the graph in the figure shown at the top of the next cannot be the graph of f.

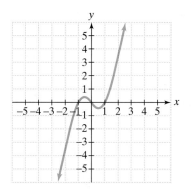

16. The figure below shows how the graph of $h(x) = -2(x - 3)^2$ is obtained from the graph of $f(x) = x^2$. Describe this process, using the graph of g in your description.

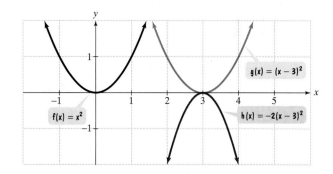

17. Begin by graphing the absolute value function, $f(x) = |x|$. Then use transformations of this graph to graph $g(x) = \frac{1}{2}|x + 1| + 3$.

If $f(x) = x^2 + 3x - 4$ and $g(x) = 5x - 2$, find each function or function value in Exercises 18–22.

18. $(f - g)(x)$

19. $\left(\dfrac{f}{g}\right)(x)$ and its domain

20. $(f \circ g)(x)$

21. $(g \circ f)(x)$

22. $f(g(2))$

23. If $f(x) = \dfrac{7}{x-4}$ and $g(x) = \dfrac{2}{x}$, find $(f \circ g)(x)$ and the domain of $f \circ g$.

24. Express $h(x) = (2x + 13)^7$ as a composition of two functions f and g so that $h(x) = (f \circ g)(x)$.

25. If $f(x) = \sqrt{x - 2}$, find the equation for $f^{-1}(x)$. Then verify that your equation is correct by showing that $f(f^{-1}(x)) = x$ and $f^{-1}(f(x)) = x$.

26. A function f models the amount given to charity as a function of income. The graph of f is shown in the figure.

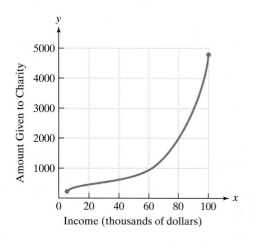

Income (thousands of dollars)

a. Explain why f has an inverse that is a function.
b. Find $f(80)$.
c. Describe in practical terms the meaning of $f^{-1}(2000)$.

27. Use a graphing utility to graph

$$f(x) = \frac{x^3}{3} + x^2 - 15x + 3 \text{ in a } [-10, 10, 1] \text{ by}$$

$[-30, 70, 10]$ viewing rectangle. Use the graph to answer the following questions.

a. Is f one-to-one? Explain.
b. Is f even, odd, or neither? Explain.
c. What is the range of f?
d. On which interval or intervals is f increasing?
e. On which interval or intervals is f decreasing?
f. For what number does f have a relative maximum? What is the relative maximum?
g. For what number does f have a relative minimum? What is the relative minimum?

28. In 1980, the winning time for women in the Olympic 500-meter speed skating event was 41.78 seconds. The average rate of decrease in the winning time has been about 0.19 second per year.

a. Express the winning time, T, in this event as a function of the number of years after 1980, x.
b. According to the function, when will the winning time be 37.22 seconds?

29. The annual yield per walnut tree is fairly constant at 50 pounds per tree when the number of trees per acre is 30 or fewer. For each additional tree over 30, the annual yield per tree for all trees on the acre decreases by 1.5 pounds due to overcrowding.

a. Express the yield per tree, Y, in pounds, as a function of the number of walnut trees per acre, x.
b. Express the total yield for an acre, T, in pounds, as a function of the number of walnut trees per acre, x.

30. You have 600 yards of fencing to enclose a rectangular field. Express the area of the field, A, as a function of one of its dimensions, x.

31. A closed rectangular box with a square base has a volume of 8000 cubic centimeters. Express the surface area of the box, A, as a function of the length of a side of its square base, x.

Polynomial and Rational Functions

Chapter 2

There is a function that models the age in human years, $H(x)$, of a dog that is x years old:

$$H(x) = -0.001618x^4 + 0.077326x^3$$
$$-1.2367x^2 + 11.460x + 2.914.$$

The function contains variables to powers that are whole numbers and is an example of a **polynomial function.** In this chapter, we study polynomial functions and functions that consist of quotients of polynomials, called **rational functions.**

One of the joys of your life is your dog, your very special buddy. Lately, however, you've noticed that your companion is slowing down a bit. He's now 8 years old and you wonder how this translates into human years. You remember something about every year of a dog's life being equal to seven years for a human. Is there a more accurate description?

SECTION 2.1 *Complex Numbers*

Objectives

1. Add and subtract complex numbers.
2. Multiply complex numbers.
3. Divide complex numbers.
4. Perform operations with square roots of negative numbers.
5. Solve quadratic equations with complex imaginary solutions.

THE KID WHO LEARNED ABOUT MATH ON THE STREET

If you divide 6,973 by 0, you die.

Once, this guy tried to find the square root of -9, and his eyeballs turned black.

This girl my brother knows found out exactly what π equals, but she went nuts.

© 2000 Roz Chast from Cartoonbank.com. All rights reserved.

Who is this kid warning us about our eyeballs turning black if we attempt to find the square root of −9? Don't believe what you hear on the street. Although square roots of negative numbers are not real numbers, they do play a significant role in algebra. In this section, we move beyond the real numbers and discuss square roots with negative radicands.

The Imaginary Unit *i*

In this chapter, we'll be studying equations whose solutions involve the square roots of negative numbers. Because the square of a real number is never negative, there is no real number x such that $x^2 = -1$. To provide a setting in which such equations have solutions, mathematicians invented an expanded system of numbers, the complex numbers. The *imaginary number i*, defined to be a solution of the equation $x^2 = -1$, is the basis of this new set.

The Imaginary Unit *i*

The **imaginary unit** *i* is defined as

$$i = \sqrt{-1}, \quad \text{where} \quad i^2 = -1.$$

Using the imaginary unit *i*, we can express the square root of any negative number as a real multiple of *i*. For example,

$$\sqrt{-25} = i\sqrt{25} = 5i.$$

We can check this result by squaring *5i* and obtaining −25.

$$(5i)^2 = 5^2 i^2 = 25(-1) = -25$$

A new system of numbers, called *complex numbers*, is based on adding multiples of i, such as $5i$, to the real numbers.

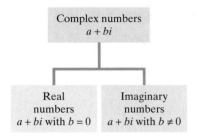

Figure 2.1 The complex number system

Complex Numbers

The set of all numbers in the form

$$a + bi$$

with real numbers a and b, and i, the imaginary unit, is called the set of **complex numbers.** The real number a is called the **real part,** and the real number b is called the **imaginary part,** of the complex number $a + bi$. If $b \neq 0$, then the complex number is called an **imaginary number** (Figure 2.1). An imaginary number in the form bi is called a **pure imaginary number.**

Here are some examples of complex numbers. Each number can be written in the form $a + bi$.

$$-4 + 6i \qquad\qquad 2i = 0 + 2i \qquad\qquad 3 = 3 + 0i$$

| a, the real part, is −4. | b, the imaginary part, is 6. | a, the real part, is 0. | b, the imaginary part, is 2. | a, the real part, is 3. | b, the imaginary part, is 0. |

Can you see that b, the imaginary part, is not zero in the first two complex numbers? Because $b \neq 0$, these complex numbers are imaginary numbers. Furthermore, the imaginary number $2i$ is a pure imaginary number. By contrast, the imaginary part of the complex number on the right is zero. This complex number is not an imaginary number. The number 3, or $3 + 0i$, is a real number.

A complex number is said to be **simplified** if it is expressed in the **standard form** $a + bi$. If b is a radical, we usually write i before b. For example, we write $7 + i\sqrt{5}$ rather than $7 + \sqrt{5}i$, which could easily be confused with $7 + \sqrt{5i}$.

Expressed in standard form, two complex numbers are equal if and only if their real parts are equal and their imaginary parts are equal.

Equality of Complex Numbers

$a + bi = c + di$ if and only if $a = c$ and $b = d$.

1 Add and subtract complex numbers.

Operations with Complex Numbers

The form of a complex number $a + bi$ is like the binomial $a + bx$. Consequently, we can add, subtract, and multiply complex numbers using the same methods we used for binomials, remembering that $i^2 = -1$.

Adding and Subtracting Complex Numbers

1. $(a + bi) + (c + di) = (a + c) + (b + d)i$

In words, this says that you add complex numbers by adding their real parts, adding their imaginary parts, and expressing the sum as a complex number.

2. $(a + bi) - (c + di) = (a - c) + (b - d)i$

In words, this says that you subtract complex numbers by subtracting their real parts, subtracting their imaginary parts, and expressing the difference as a complex number.

EXAMPLE 1 Adding and Subtracting Complex Numbers

Perform the indicated operations, writing the result in standard form:

    a. $(5 - 11i) + (7 + 4i)$ **b.** $(-5 + 7i) - (-11 - 6i)$.

Study Tip

The following examples, using the same integers as in Example 1, show how operations with complex numbers are just like operations with polynomials.

a. $(5 - 11x) + (7 + 4x)$
    $= 12 - 7x$

b. $(-5 + 7x) - (-11 - 6x)$
    $= -5 + 7x + 11 + 6x$
    $= 6 + 13x$

Solution

a. $(5 - 11i) + (7 + 4i)$

    $= 5 - 11i + 7 + 4i$ Remove the parentheses.

    $= 5 + 7 - 11i + 4i$ Group real and imaginary terms.

    $= (5 + 7) + (-11 + 4)i$ Add real parts and add imaginary parts.

    $= 12 - 7i$ Simplify.

b. $(-5 + 7i) - (-11 - 6i)$

    $= -5 + 7i + 11 + 6i$ Remove the parentheses.

    $= -5 + 11 + 7i + 6i$ Group real and imaginary terms.

    $= (-5 + 11) + (7 + 6)i$ Add real parts and add imaginary parts.

    $= 6 + 13i$ Simplify.

Check Point 1   Add or subtract as indicated:

    a. $(5 - 2i) + (3 + 3i)$ **b.** $(2 + 6i) - (12 - 4i)$.

2   Multiply complex numbers.

    Multiplication of complex numbers is performed the same way as multiplication of polynomials, using the distributive property and the FOIL method. After completing the multiplication, we replace i^2 with -1. This idea is illustrated in the next example.

EXAMPLE 2 Multiplying Complex Numbers

Find the products:

    a. $4i(3 - 5i)$ **b.** $(7 - 3i)(-2 - 5i)$.

Solution

a. $4i(3 - 5i) = 4i(3) - 4i(5i)$ Distribute 4i throughout the parentheses.

                $= 12i - 20i^2$ Multiply.

                $= 12i - 20(-1)$ Replace i² with −1.

                $= 20 + 12i$ Simplify to 12i + 20 and write in standard form.

b. $(7 - 3i)(-2 - 5i)$

         F   O   I   L

    $= -14 - 35i + 6i + 15i^2$ Use the FOIL method.

    $= -14 - 35i + 6i + 15(-1)$ i² = −1

    $= -14 - 15 - 35i + 6i$ Group real and imaginary terms.

    $= -29 - 29i$ Combine real and imaginary terms.

Check Point 2   Find the products:

    a. $7i(2 - 9i)$ **b.** $(5 + 4i)(6 - 7i)$.

3 Divide complex numbers.

Complex Conjugates and Division

It is possible to multiply complex numbers and obtain a real number. This occurs when we multiply $a + bi$ and $a - bi$.

$$
\begin{aligned}
(a + bi)(a - bi) &= a^2 - abi + abi - b^2i^2 && \text{Use the FOIL method.} \\
&= a^2 - b^2(-1) && i^2 = -1 \\
&= a^2 + b^2 && \text{Notice that this product eliminates } i.
\end{aligned}
$$

(with F O I L labels above the first line)

For the complex number $a + bi$, we define its *complex conjugate* to be $a - bi$. The multiplication of complex conjugates results in a real number.

> ### Conjugate of a Complex Number
> The **complex conjugate** of the number $a + bi$ is $a - bi$, and the complex conjugate of $a - bi$ is $a + bi$. The multiplication of complex conjugates gives a real number.
> $$(a + bi)(a - bi) = a^2 + b^2$$
> $$(a - bi)(a + bi) = a^2 + b^2$$

Complex conjugates are used when dividing complex numbers. By multiplying the numerator and the denominator of the division by the complex conjugate of the denominator, you will obtain a real number in the denominator.

EXAMPLE 3 Using Complex Conjugates to Divide Complex Numbers

Divide and express the result in standard form: $\dfrac{7 + 4i}{2 - 5i}$.

Solution The complex conjugate of the denominator, $2 - 5i$, is $2 + 5i$. Multiplication of both the numerator and the denominator by $2 + 5i$ will eliminate i from the denominator.

$$
\frac{7 + 4i}{2 - 5i} = \frac{(7 + 4i)}{(2 - 5i)} \cdot \frac{(2 + 5i)}{(2 + 5i)}
$$

Multiply the numerator and the denominator by the complex conjugate of the denominator.

$$
= \frac{14 + 35i + 8i + 20i^2}{2^2 + 5^2}
$$

(with F O I L labels above)

Use the FOIL method in the numerator and $(a - bi)(a + bi) = a^2 + b^2$ in the denominator.

$$
= \frac{14 + 43i + 20(-1)}{29}
$$

Combine imaginary terms and replace i^2 with -1.

$$
= \frac{-6 + 43i}{29}
$$

Combine real terms in the numerator.

$$
= -\frac{6}{29} + \frac{43}{29}i
$$

Express the answer in standard form.

Observe that the quotient is expressed in the standard form $a + bi$, with $a = -\frac{6}{29}$ and $b = \frac{43}{29}$.

Complex Numbers on a Postage Stamp

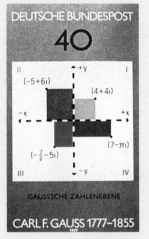

DEUTSCHE BUNDESPOST

40

GAUSSSCHE ZAHLENEBENE

CARL F. GAUSS 1777–1855

This stamp honors the work done by the German mathematician Carl Friedrich Gauss (1777–1855) with complex numbers. Gauss represented complex numbers as points in the plane.

Check Point 3 Divide and express the result in standard form: $\dfrac{5 + 4i}{4 - 2i}$.

4 Perform operations with square roots of negative numbers.

Roots of Negative Numbers

The square of $4i$ and the square of $-4i$ both result in -16.

$$(4i)^2 = 16i^2 = 16(-1) = -16 \qquad (-4i)^2 = 16i^2 = 16(-1) = -16$$

Consequently, in the complex number system -16 has two square roots, namely, $4i$ and $-4i$. We call $4i$ the **principal square root** of -16.

> **Principal Square Root of a Negative Number**
> For any positive number real number b, the **principal square root** of the negative number $-b$ is defined by
> $$\sqrt{-b} = i\sqrt{b}.$$

EXAMPLE 4   Operations Involving Square Roots of Negative Numbers

Perform the indicated operations and write the result in standard form:

a. $\sqrt{-18} - \sqrt{-8}$ **b.** $(-1 + \sqrt{-5})^2$ **c.** $\dfrac{-25 + \sqrt{-50}}{15}$.

Solution   Begin by expressing all square roots of negative numbers in terms of i.

a. $\sqrt{-18} - \sqrt{-8} = i\sqrt{18} - i\sqrt{8} = i\sqrt{9 \cdot 2} - i\sqrt{4 \cdot 2}$
$$= 3i\sqrt{2} - 2i\sqrt{2} = i\sqrt{2}$$

$$(A + B)^2 = A^2 + 2AB + B^2$$

b. $(-1 + \sqrt{-5})^2 = (-1 + i\sqrt{5})^2 = (-1)^2 + 2(-1)(i\sqrt{5}) + (i\sqrt{5})^2$
$$= 1 - 2i\sqrt{5} + 5i^2$$
$$= 1 - 2i\sqrt{5} + 5(-1)$$
$$= -4 - 2i\sqrt{5}$$

c. $\dfrac{-25 + \sqrt{-50}}{15}$

$= \dfrac{-25 + i\sqrt{50}}{15}$ $\sqrt{-b} = i\sqrt{b}$

$= \dfrac{-25 + 5i\sqrt{2}}{15}$ $\sqrt{50} = \sqrt{25 \cdot 2} = 5\sqrt{2}$

$= \dfrac{-25}{15} + \dfrac{5i\sqrt{2}}{15}$ Write the complex number in standard form.

$= -\dfrac{5}{3} + i\dfrac{\sqrt{2}}{3}$ Simplify.

Study Tip

Do not apply the properties
$$\sqrt{b}\sqrt{c} = \sqrt{bc}$$
and
$$\frac{\sqrt{b}}{\sqrt{c}} = \sqrt{\frac{b}{c}}$$
to the pure imaginary numbers because these properties can only be used when b and c are positive.

Correct:
$$\sqrt{-25}\sqrt{-4} = i\sqrt{25}\, i\sqrt{4}$$
$$= (5i)(2i)$$
$$= 10i^2$$
$$= 10(-1)$$
$$= -10$$

Incorrect:
$$\sqrt{-25}\sqrt{-4} = \sqrt{(-25)(-4)}$$
$$= \sqrt{100}$$
$$= 10$$

One way to avoid confusion is to represent square roots of negative numbers in terms of i before performing any operations.

Check Point 4 Perform the indicated operations and write the result in standard form:

a. $\sqrt{-27} + \sqrt{-48}$ **b.** $(-2 + \sqrt{-3})^2$ **c.** $\dfrac{-14 + \sqrt{-12}}{2}$.

⑤ Solve quadratic equations with complex imaginary solutions.

Quadratic Equations with Complex Imaginary Solutions

We have seen that a quadratic equation can be expressed in the general form

$$ax^2 + bx + c = 0, \quad a \neq 0.$$

All quadratic equations can be solved by the quadratic formula:

$$x = \frac{-b \pm \sqrt{b^2 - 4ac}}{2a}.$$

Study Tip

If you need to review quadratic equations and how to solve them, read Section P.8 beginning on page 84.

Recall that the quantity $b^2 - 4ac$, which appears under the radical sign in the quadratic formula, is called the discriminant. If the discriminant is negative, a quadratic equation has no real solutions. However, quadratic equations with negative discriminants do have two solutions. These solutions are imaginary numbers that are complex conjugates.

EXAMPLE 5 A Quadratic Equation with Imaginary Solutions

Solve using the quadratic formula: $3x^2 - 2x + 4 = 0$.

Solution The given equation is in general form. Begin by identifying the values for a, b, and c.

$$3x^2 - 2x + 4 = 0$$

$$a = 3 \qquad b = -2 \qquad c = 4$$

$$x = \frac{-b \pm \sqrt{b^2 - 4ac}}{2a} \qquad \text{Use the quadratic formula.}$$

$$= \frac{-(-2) \pm \sqrt{(-2)^2 - 4(3)(4)}}{2(3)} \qquad \begin{array}{l}\text{Substitute the values for } a, b, \text{ and } c: \\ a = 3, b = -2, \text{ and } c = 4.\end{array}$$

$$= \frac{2 \pm \sqrt{4 - 48}}{6} \qquad -(-2) = 2 \text{ and } (-2)^2 = (-2)(-2) = 4.$$

$$= \frac{2 \pm \sqrt{-44}}{6} \qquad \begin{array}{l}\text{Subtract under the radical. Because the} \\ \text{discriminant, } -44, \text{ is negative, the} \\ \text{solutions will not be real numbers.}\end{array}$$

$$= \frac{2 \pm 2i\sqrt{11}}{6} \qquad \begin{array}{l}\sqrt{-44} = \sqrt{4(11)(-1)} \\ \qquad = 2i\sqrt{11}\end{array}$$

$$= \frac{2(1 \pm i\sqrt{11})}{6} \qquad \text{Factor 2 from the numerator.}$$

$$= \frac{1 \pm i\sqrt{11}}{3} \qquad \text{Divide numerator and denominator by 2.}$$

$$= \frac{1}{3} \pm i\frac{\sqrt{11}}{3} \qquad \begin{array}{l}\text{Write the complex numbers in standard} \\ \text{form.}\end{array}$$

The solutions are complex conjugates, and the solution set is $\left\{ \frac{1}{3} + i\frac{\sqrt{11}}{3}, \frac{1}{3} - i\frac{\sqrt{11}}{3} \right\}$ or $\left\{ \frac{1}{3} \pm i\frac{\sqrt{11}}{3} \right\}$.

Check Point 5 Solve using the quadratic formula:

$$x^2 - 2x + 2 = 0.$$

EXERCISE SET 2.1

Practice Exercises

In Exercises 1–8, add or subtract as indicated and write the result in standard form.

1. $(7 + 2i) + (1 - 4i)$ **2.** $(-2 + 6i) + (4 - i)$
3. $(3 + 2i) - (5 - 7i)$ **4.** $(-7 + 5i) - (-9 - 11i)$
5. $6 - (-5 + 4i) - (-13 - 11i)$
6. $7 - (-9 + 2i) - (-17 - 6i)$
7. $8i - (14 - 9i)$ **8.** $15i - (12 - 11i)$

In Exercises 9–20, find each product and write the result in standard form.

9. $-3i(7i - 5)$ **10.** $-8i(2i - 7)$
11. $(-5 + 4i)(3 + 7i)$ **12.** $(-4 - 8i)(3 + 9i)$
13. $(7 - 5i)(-2 - 3i)$ **14.** $(8 - 4i)(-3 + 9i)$
15. $(3 + 5i)(3 - 5i)$ **16.** $(2 + 7i)(2 - 7i)$
17. $(-5 + 3i)(-5 - 3i)$ **18.** $(-7 - 4i)(-7 + 4i)$
19. $(2 + 3i)^2$ **20.** $(5 - 2i)^2$

In Exercises 21–28, divide and express the result in standard form.

21. $\dfrac{2}{3 - i}$ **22.** $\dfrac{3}{4 + i}$

23. $\dfrac{2i}{1 + i}$ **24.** $\dfrac{5i}{2 - i}$

25. $\dfrac{8i}{4 - 3i}$ **26.** $\dfrac{-6i}{3 + 2i}$

27. $\dfrac{2 + 3i}{2 + i}$ **28.** $\dfrac{3 - 4i}{4 + 3i}$

In Exercises 29–44, perform the indicated operations and write the result in standard form.

29. $\sqrt{-64} - \sqrt{-25}$ **30.** $\sqrt{-81} - \sqrt{-144}$
31. $5\sqrt{-16} + 3\sqrt{-81}$ **32.** $5\sqrt{-8} + 3\sqrt{-18}$
33. $(-2 + \sqrt{-4})^2$ **34.** $(-5 - \sqrt{-9})^2$
35. $(-3 - \sqrt{-7})^2$ **36.** $(-2 + \sqrt{-11})^2$
37. $\dfrac{-8 + \sqrt{-32}}{24}$ **38.** $\dfrac{-12 + \sqrt{-28}}{32}$
39. $\dfrac{-6 - \sqrt{-12}}{48}$ **40.** $\dfrac{-15 - \sqrt{-18}}{33}$
41. $\sqrt{-8}(\sqrt{-3} - \sqrt{5})$ **42.** $\sqrt{-12}(\sqrt{-4} - \sqrt{2})$
43. $(3\sqrt{-5})(-4\sqrt{-12})$ **44.** $(3\sqrt{-7})(2\sqrt{-8})$

In Exercises 45–50, solve each quadratic equation using the quadratic formula. Express solutions in standard form.

45. $x^2 - 6x + 10 = 0$ **46.** $x^2 - 2x + 17 = 0$
47. $4x^2 + 8x + 13 = 0$ **48.** $2x^2 + 2x + 3 = 0$
49. $3x^2 = 8x - 7$ **50.** $3x^2 = 4x - 6$

Writing in Mathematics

51. What is i?
52. Explain how to add complex numbers. Provide an example with your explanation.
53. Explain how to multiply complex numbers and give an example.
54. What is the complex conjugate of $2 + 3i$? What happens when you multiply this complex number by its complex conjugate?
55. Explain how to divide complex numbers. Provide an example with your explanation.
56. A stand-up comedian uses algebra in some jokes, including one about a telephone recording that announces "You have just reached an imaginary number. Please multiply by i and dial again." Explain the joke.

Explain the error in Exercises 57–58.
57. $\sqrt{-9} + \sqrt{-16} = \sqrt{-25} = i\sqrt{25} = 5i$
58. $(\sqrt{-9})^2 = \sqrt{-9} \cdot \sqrt{-9} = \sqrt{81} = 9$

Critical Thinking Exercises

59. Which one of the following is true?
  a. Some irrational numbers are not complex numbers.
  b. $(3 + 7i)(3 - 7i)$ is an imaginary number.
  c. $\dfrac{7 + 3i}{5 + 3i} = \dfrac{7}{5}$
  d. In the complex number system, $x^2 + y^2$ (the sum of two squares) can be factored as $(x + yi)(x - yi)$.

In Exercises 60–62, perform the indicated operations and write the result in standard form.
60. $(8 + 9i)(2 - i) - (1 - i)(1 + i)$
61. $\dfrac{4}{(2 + i)(3 - i)}$ **62.** $\dfrac{1 + i}{1 + 2i} + \dfrac{1 - i}{1 - 2i}$
63. Evaluate $x^2 - 2x + 2$ for $x = 1 + i$.

SECTION 2.2 Quadratic Functions

Objectives

1. Recognize characteristics of parabolas.
2. Graph parabolas.
3. Solve problems involving minimizing or maximizing quadratic functions.

The Food Stamp Program is the first line of defense against hunger for millions of American families. The program provides benefits for eligible participants to purchase approved food items at approved food stores. Over half of all participants are children; one out of six is a low-income older adult. The function

$$f(x) = -0.5x^2 + 4x + 19$$

models the number of people, $f(x)$, in millions, receiving food stamps x years after 1990. For example, to find the number of food stamp recipients in 2000, substitute 10 for x because 2000 is 10 years after 1990:

$$f(10) = -0.5(10)^2 + 4(10) + 19 = 9.$$

Thus, in 2000, there were 9 million food stamp recipients.

The function $f(x) = -0.5x^2 + 4x + 19$ is an example of a *quadratic function*. A **quadratic function** is any function of the form

$$f(x) = ax^2 + bx + c$$

where a, b, and c are real numbers and $a \neq 0$. A quadratic function is a polynomial function whose highest power is 2. In this section, we will study quadratic functions and their graphs.

1 Recognize characteristics of parabolas.

Graphs of Quadratic Functions

The graph of any quadratic function is called a **parabola.** Parabolas are shaped like cups, as shown in Figure 2.2. If the coefficient of x^2 (the value of a in $ax^2 + bx + c$) is positive, the parabola opens upward. If the coefficient of x^2 is negative, the graph opens downward. The **vertex** (or turning point) of the parabola is the minimum point on the graph when it opens upward, and the maximum point on the graph when it opens downward.

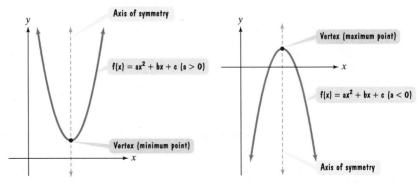

Figure 2.2 Characteristics of parabolas

$a > 0$: Parabola opens upward. $a < 0$: Parabola opens downward.

Look at the unusual image of the word "mirror" shown below. The artist, Scott Kim, has created the image so that the two halves of the whole are mirror images of each other. A parabola shares this kind of symmetry, in which a line through the vertex divides the figure in half. Parabolas are symmetric with respect to this line, called the **axis of symmetry.** The movements of gymnasts, divers, and swimmers can approximate this symmetry. If a parabola is folded along its axis of symmetry, the two halves match exactly.

2 Graph parabolas.

Graphing Quadratic Functions in Standard Form

In Section 1.6, we applied a series of transformations to the graph of $f(x) = x^2$. The graph of this function is a parabola. The vertex for this parabola is $(0, 0)$. In Figure 2.3(a), the graph of $f(x) = ax^2$ for $a > 0$ is shown in black; it opens *upward*. In Figure 2.3(b), the graph of $f(x) = ax^2$ for $a < 0$ is shown in black; it opens *downward*.

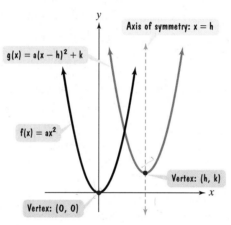

(a) $a > 0$: Parabola opens upward.

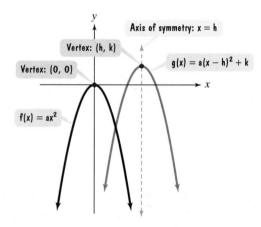

(b) $a < 0$: Parabola opens downward.

Figure 2.3   Transformations of $f(x) = ax^2$

Figure 2.3 also shows the graphs of $g(x) = a(x - h)^2 + k$ in blue. Compare these graphs to those of $f(x) = ax^2$. Observe that h determines the horizontal shift and k determines the vertical shift of the graph of $f(x) = ax^2$:

$$g(x) = a(x - h)^2 + k.$$

If $h > 0$, it shifts the graph of $f(x) = ax^2$ h units to the right.

If $k > 0$, it shifts the graph of $y = a(x-h)^2$ k units up.

Consequently, the vertex $(0, 0)$ on the black graph of $f(x) = ax^2$ moves to the point (h, k) on the blue graph of $g(x) = a(x - h)^2 + k$. The axis of symmetry is the vertical line whose equation is $x = h$.

The form of the expression for g is convenient because it immediately identifies the vertex of the parabola as (h, k). This is the **standard form** of a quadratic function.

The Standard Form of a Quadratic Function

The quadratic function

$$f(x) = a(x - h)^2 + k, \quad a \neq 0$$

is in **standard form.** The graph of f is a parabola whose vertex is the point (h, k). The parabola is symmetric with respect to the line $x = h$. If $a > 0$, the parabola opens upward; if $a < 0$, the parabola opens downward.

The sign of a in $f(x) = a(x - h)^2 + k$ determines whether the parabola opens upward or downward. Furthermore, if $|a|$ is small, the parabola opens more widely than if $|a|$ is large. Here is a general procedure for graphing parabolas whose equations are in standard form:

Graphing Quadratic Functions with Equations in Standard Form

To graph $f(x) = a(x - h)^2 + k$,

1. Determine whether the parabola opens upward or downward. If $a > 0$, it opens upward. If $a < 0$, it opens downward.
2. Determine the vertex of the parabola. The vertex is (h, k).
3. Find any x-intercepts by replacing $f(x)$ with 0. Solve the resulting quadratic equation for x.
4. Find the y-intercept by replacing x with 0.
5. Plot the intercepts and vertex. Connect these points with a smooth curve that is shaped like a cup. Draw a dashed vertical line for the axis of symmetry.

EXAMPLE 1 Graphing a Quadratic Function in Standard Form

Graph the quadratic function $f(x) = -2(x - 3)^2 + 8$.

Solution   We can graph this function by following the steps in the preceding box. We begin by identifying values for a, h, and k.

Standard form $f(x) = a(x - h)^2 + k$

$a = -2$ $h = 3$ $k = 8$

Given equation $f(x) = -2(x - 3)^2 + 8$

Step 1   Determine how the parabola opens.   Note that a, the coefficient of x^2, is -2. Thus, $a < 0$; this negative value tells us that the parabola opens downward.

Step 2   Find the vertex.   The vertex of the parabola is (h, k). Because $h = 3$ and $k = 8$, the parabola's vertex is $(3, 8)$.

Step 3 Find the x-intercepts. Replace $f(x)$ with 0 in $f(x) = -2(x - 3)^2 + 8$.

$$0 = -2(x - 3)^2 + 8$$ Find x-intercepts, setting f(x) equal to 0.

$$2(x - 3)^2 = 8$$ Solve for x. Add $2(x - 3)^2$ to both sides of the equation.

$$(x - 3)^2 = 4$$ Divide both sides by 2.

$$(x - 3) = \pm\sqrt{4}$$ Apply the square root method. If $(x - c)^2 = d$, then $x - c = \pm\sqrt{d}$.

$$x - 3 = -2 \quad \text{or} \quad x - 3 = 2$$ Express as two separate equations.

$$x = 1 \quad \text{or} \quad x = 5$$ Add 3 to both sides in each equation.

The x-intercepts are 1 and 5. The parabola passes through $(1, 0)$ and $(5, 0)$.

Step 4 Find the y-intercept. Replace x with 0 in $f(x) = -2(x - 3)^2 + 8$.

$$f(0) = -2(0 - 3)^2 + 8 = -2(-3)^2 + 8 = -2(9) + 8 = -10$$

The y-intercept is -10. The parabola passes through $(0, -10)$.

Step 5 Graph the parabola. With a vertex at $(3, 8)$, x-intercepts at 1 and 5, and a y-intercept at -10, the graph of f is shown in Figure 2.4. The axis of symmetry is the vertical line whose equation is $x = 3$.

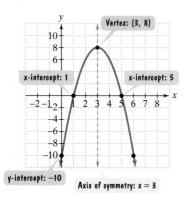

Figure 2.4 The graph of $f(x) = -2(x - 3)^2 + 8$

Check Point 1 Graph the quadratic function $f(x) = -(x - 1)^2 + 4$.

EXAMPLE 2 Graphing a Quadratic Function in Standard Form

Graph the quadratic function $f(x) = (x + 3)^2 + 1$.

Solution We begin by finding values for a, h, and k.

Standard form $\qquad f(x) = a(x - h)^2 + k$

Given equation $\qquad f(x) = (x + 3)^2 + 1$

$\qquad\qquad\qquad$ or $\quad f(x) = 1(x - (-3))^2 + 1$

$a = 1 \qquad h = -3 \qquad k = 1$

Step 1 Determine how the parabola opens. Note that a, the coefficient of x^2, is 1. Thus, $a > 0$; this positive value tells us that the parabola opens upward.

Step 2 Find the vertex. The vertex of the parabola is (h, k). Because $h = -3$ and $k = 1$, the parabola's vertex is $(-3, 1)$.

Step 3 **Find the *x*-intercepts.** Replace $f(x)$ with 0 in $f(x) = (x + 3)^2 + 1$. Because the vertex is $(-3, 1)$, which lies above the *x*-axis, and the parabola opens upward, it appears that this parabola has no *x*-intercepts. We can verify this observation algebraically.

$$0 = (x + 3)^2 + 1 \quad \text{\small Find possible x-intercepts, setting f(x) equal to 0.}$$

$$-1 = (x + 3)^2 \quad \text{\small Solve for x. Subtract 1 from both sides.}$$

$$x + 3 = \pm\sqrt{-1} \quad \text{\small Apply the square root method.}$$

$$x + 3 = \pm i \quad \text{\small Recall that } \sqrt{-1} = i, \text{ an imaginary number.}$$

$$x = -3 \pm i \quad \text{\small Subtract 3 from both sides.}$$

Because this equation has no real solutions, the parabola has no *x*-intercepts.

Step 4 **Find the *y*-intercept.** Replace *x* with 0 in $f(x) = (x + 3)^2 + 1$.

$$f(0) = (0 + 3)^2 + 1 = 3^2 + 1 = 9 + 1 = 10$$

The *y*-intercept is 10. The parabola passes through $(0, 10)$.

Step 5 **Graph the parabola.** With a vertex at $(-3, 1)$, no *x*-intercepts, and a *y*-intercept at 10, the graph of *f* is shown in Figure 2.5. The axis of symmetry is the vertical line whose equation is $x = -3$.

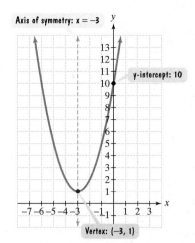

Figure 2.5 The graph of $g(x) = (x + 3)^2 + 1$

Check Point 2   Graph the quadratic function $f(x) = (x - 2)^2 + 1$.

Graphing Quadratic Functions in the Form $f(x) = ax^2 + bx + c$

Quadratic functions are frequently expressed in the form $f(x) = ax^2 + bx + c$. How can we identify the vertex of a parabola whose equation is in this form? By completing the square, we can find a way to describe the vertex in terms of *a* and *b*.

$$f(x) = ax^2 + bx + c$$

$$= a\left(x^2 + \frac{b}{a}x\right) + c \quad \text{\small Factor out a from } ax^2 + bx.$$

$$= a\left(x^2 + \frac{b}{a}x + \frac{b^2}{4a^2}\right) + c - a\left(\frac{b^2}{4a^2}\right)$$

Complete the square by adding the square of half the coefficient of x.

By completing the square, we added $a \cdot \frac{b^2}{4a^2}$. To avoid changing the function's equation, we must subtract this term.

$$= a\left(x + \frac{b}{2a}\right)^2 + c - \frac{b^2}{4a} \quad \text{\small Write the trinomial as the square of a binomial and simplify the constant term.}$$

Compare the form of the equation at the bottom of the previous page with a quadratic function's standard form.

Standard form $\qquad\qquad f(x) = a(x - h)^2 + k$

$$h = -\frac{b}{2a} \qquad k = c - \frac{b^2}{4a}$$

Equation under discussion $\qquad f(x) = a\left(x - \left(-\frac{b}{2a}\right)\right)^2 + c - \frac{b^2}{4a}$

The important part of this observation is that h, the x-coordinate of the vertex, is $-\dfrac{b}{2a}$. The y-coordinate can be found by evaluating the function at $-\dfrac{b}{2a}$.

The Vertex of a Parabola Whose Equation is $f(x) = ax^2 + bx + c$

Consider the parabola defined by the quadratic function $f(x) = ax^2 + bx + c$. The parabola's vertex is $\left(-\dfrac{b}{2a}, f\left(-\dfrac{b}{2a}\right)\right)$.

We can apply our five-step procedure and graph parabolas in the form $f(x) = ax^2 + bx + c$. The only step that is different is how we determine the vertex.

EXAMPLE 3 Graphing a Quadratic Function in the Form $f(x) = ax^2 + bx + c$

Graph the quadratic function $f(x) = -x^2 + 4x - 1$.

Solution

Step 1   Determine how the parabola opens.   Note that a, the coefficient of x^2, is -1. Thus, $a < 0$; this negative value tells us that the parabola opens downward.

Step 2   Find the vertex.   We know that the x-coordinate of the vertex is $x = -\dfrac{b}{2a}$. We identify a, b, and c in $f(x) = ax^2 + bx + c$.

$$f(x) = -x^2 + 4x - 1$$

$$a = -1 \qquad b = 4 \qquad c = -1$$

Substitute the values of a and b into the equation for the x-coordinate:

$$x = -\frac{b}{2a} = -\frac{4}{2(-1)} = \frac{-4}{-2} = 2.$$

The x-coordinate of the vertex is 2. We substitute 2 for x in $f(x) = -x^2 + 4x - 1$, the equation of the function, to find the y-coordinate:

$$f(2) = -2^2 + 4 \cdot 2 - 1 = -4 + 8 - 1 = 3.$$

The vertex is $(2, 3)$.

Step 3 Find the x-intercepts. Replace $f(x)$ with 0 in $f(x) = -x^2 + 4x - 1$. We obtain $0 = -x^2 + 4x - 1$ or $-x^2 + 4x - 1 = 0$. This equation cannot be solved by factoring. We will use the quadratic formula to solve it.

$$a = -1, \quad b = 4, \quad c = -1$$

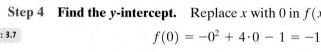

$$x = \frac{-b \pm \sqrt{b^2 - 4ac}}{2a} = \frac{-4 \pm \sqrt{4^2 - 4(-1)(-1)}}{2(-1)} = \frac{-4 \pm \sqrt{16 - 4}}{-2}$$

$$x = \frac{-4 - \sqrt{12}}{-2} \approx 3.7 \quad \text{or} \quad x = \frac{-4 + \sqrt{12}}{-2} \approx 0.3$$

The x-intercepts are approximately 0.3 and 3.7. The parabola passes through the corresponding points, which we approximate as $(0.3, 0)$ and $(3.7, 0)$.

Step 4 Find the y-intercept. Replace x with 0 in $f(x) = -x^2 + 4x - 1$.

$$f(0) = -0^2 + 4 \cdot 0 - 1 = -1$$

The y-intercept is -1. The parabola passes through $(0, -1)$.

Step 5 Graph the parabola. With a vertex at $(2, 3)$, x-intercepts at approximately 0.3 and 3.7, and a y-intercept at -1, the graph of f is shown in Figure 2.6. The axis of symmetry is the vertical line whose equation is $x = 2$.

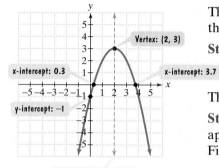

Axis of symmetry: x = 2

Figure 2.6 The graph of $f(x) = -x^2 + 4x - 1$

Check Point 3 Graph the quadratic function $f(x) = x^2 - 2x - 3$.

3 Solve problems involving minimizing or maximizing quadratic functions.

Applications of Quadratic Functions

When did the maximum number of Americans participate in the food stamp program? What is the age of a driver having the least number of car accidents? How do people launching fireworks know when they should explode to be viewed at the greatest possible height? The answers to these questions involve finding the maximum or minimum value of quadratic functions.

Consider the quadratic function $f(x) = ax^2 + bx + c$. If $a > 0$, the parabola opens upward and the vertex is its lowest point. If $a < 0$, the parabola opens downward and the vertex is its highest point. The x-coordinate of the vertex is $-\dfrac{b}{2a}$. Thus, we can find the minimum or maximum value of f by evaluating the quadratic function at $x = -\dfrac{b}{2a}$.

Minimum and Maximum: Quadratic Functions
Consider $f(x) = ax^2 + bx + c$.

1. If $a > 0$, then f has a minimum that occurs at $x = -\dfrac{b}{2a}$.
   This minimum value is $f\left(-\dfrac{b}{2a}\right)$.

2. If $a < 0$, then f has a maximum that occurs at $x = -\dfrac{b}{2a}$.
   This maximum value is $f\left(-\dfrac{b}{2a}\right)$.

EXAMPLE 4   An Application: The Food Stamp Program

The function

$$f(x) = -0.5x^2 + 4x + 19$$

models the number of people, $f(x)$, in millions, receiving food stamps x years after 1990. (*Source: New York Times*) In which year was this number at a maximum? How many food stamp recipients were there for that year?

Solution   The quadratic function is in the form $f(x) = ax^2 + bx + c$ with $a = -0.5$ and $b = 4$. Because $a < 0$, the function has a maximum value that occurs at $x = -\dfrac{b}{2a}$.

$$x = -\frac{b}{2a} = -\frac{4}{2(-0.5)} = \frac{-4}{-1} = 4$$

This means that the number of people receiving food stamps was at a maximum 4 years after 1990, in 1994. The number of recipients, in millions, for that year was

$$f(4) = -0.5(4)^2 + 4(4) + 19 = -8 + 16 + 19 = 27.$$

In 1994, the number of people receiving food stamps reached a maximum of 27 million.

Technology

The graph of the function modeling the millions of food stamp recipients

$$f(x) = -0.5x^2 + 4x + 19$$

is shown in a $[0, 10, 1]$ by $[0, 35, 5]$ viewing rectangle. The maximum function feature verifies that 4 years after 1990, the number of recipients reached a maximum of 27 million. Notice that x gives the location of the maximum and y gives the maximum value.

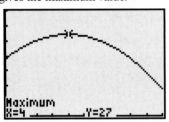

Check Point 4   The function $f(x) = 0.4x^2 - 36x + 1000$ models the number of accidents, $f(x)$, per 50 million miles driven, in terms of a driver's age, x, in years, where $16 \le x \le 74$. What is the age of a driver having the least number of car accidents? What is the minimum number of car accidents per 50 million miles driven?

Visualizing Irritability by Age

The quadratic function

$$P(x) = -0.05x^2 + 4.2x - 26$$

models the percentage of coffee drinkers, $P(x)$, who are x years old who become irritable if they do not have coffee at their regular time. Figure 2.7 shows the graph of the function. The vertex reveals that 62.2% of 42-year-old coffee drinkers become irritable. This is the maximum percentage for any age, x, in the function's domain.

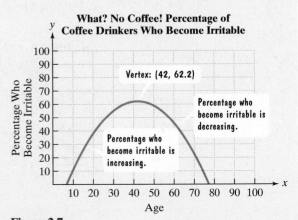

Figure 2.7

Source: LMK Associates

In Section 1.9, we constructed functions from verbal conditions. If the conditions of a problem can be modeled with a quadratic function, we can use the x-coordinate of the vertex to determine the function's maximum or minimum value. Here is a step-by-step strategy for solving these kinds of problems:

Study Tip

The hardest part in solving applied maximum or minimum problems often involves constructing the function to be maximized or minimized. Review Section 1.9, beginning on page 219, if you need more practice modeling functions from verbal conditions and formulas.

Strategy for Solving Problems Involving Maximizing or Minimizing Quadratic Functions

1. Read the problem carefully and decide which quantity is to be maximized or minimized.
2. Use the conditions of the problem to express the quantity as a function in one variable.
3. Rewrite the function in the form $f(x) = ax^2 + bx + c$.
4. Calculate $-\dfrac{b}{2a}$. If $a > 0$, f has a minimum at $x = -\dfrac{b}{2a}$. This minimum value is $f\left(-\dfrac{b}{2a}\right)$. If $a < 0$, f has a maximum at $x = -\dfrac{b}{2a}$. This maximum value is $f\left(-\dfrac{b}{2a}\right)$.
5. Answer the question posed in the problem.

EXAMPLE 5   Solving a Number Problem

Among all pairs of numbers whose sum is 40, find a pair whose product is as large as possible. What is the maximum product?

Solution

Step 1   Decide what must be maximized or minimized.   We must maximize the product of two numbers. Calling the numbers x and y, and calling the product P, we must maximize

$$P = xy.$$

Step 2   Express this quantity as a function in one variable.   In the formula $P = xy$, P is expressed in terms of two variables, x and y. However, because the sum of the numbers is 40, we can write

$$x + y = 40.$$

We can solve this equation for y in terms of x (or vice versa), substitute the result into $P = xy$, and obtain P as a function of one variable

$$y = 40 - x. \qquad \text{Subtract 40 from both sides of } x + y = 40.$$

Now we substitute $40 - x$ for y in $P = xy$.

$$P = xy = x(40 - x)$$

Because P is now a function of x, we can write

$$P(x) = x(40 - x).$$

This function models the product, P, of two numbers whose sum is 40 in terms of one of the numbers, x.

Step 3   Write the function in the form $f(x) = ax^2 + bx + c$.   We apply the distributive property to obtain

$$P(x) = x(40 - x) = 40x - x^2 = -x^2 + 40x.$$

$a = -1 \qquad b = 40$

Technology

The graph of the product function

$$P(x) = x(40 - x)$$

was obtained with a graphing utility using a $[0, 40, 4]$ by $[0, 400, 20]$ viewing rectangle. The graphing utility's maximum function feature verifies that a maximum product of 400 occurs when one of the numbers is 20.

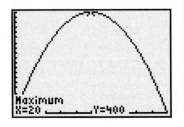

Step 4 Calculate $-\dfrac{b}{2a}$. If $a < 0$, the function has a maximum at this value.
The voice balloons on the previous page show that $a = -1$ and $b = 40$ in $P(x) = -x^2 + 40x$.

$$x = -\frac{b}{2a} = -\frac{40}{2(-1)} = 20$$

This means that the product, P, of two numbers whose sum is 40 is a maximum when one of the numbers, x, is 20.

Step 5 Answer the question posed by the problem. The problem asks for the two numbers and the maximum product. We found that one of the numbers, x, is 20. Now we must find the second number, y: $y = 40 - x = 40 - 20 = 20$. The number pair whose sum is 40 and whose product is as large as possible is $20, 20$. The maximum product is $20 \cdot 20$, or 400.

Check Point 5 Among all pairs of numbers whose sum is 8, find a pair whose product is as large as possible. What is the maximum product?

EXAMPLE 6 Maximizing Area

You have 100 yards of fencing to enclose a rectangular region. Find the dimensions of the rectangle that maximize the enclosed area. What is the maximum area?

Solution

Step 1 Decide what must be maximized or minimized. We must maximize area. What we do not know are the rectangle's dimensions, x and y.

Step 2 Express this quantity as a function in one variable. Because we must maximize area, we have $A = xy$. We need to transform this into a function in which A is represented by one variable. Because you have 100 yards of fencing, the perimeter of the rectangle is 100 yards. This means that

$$2x + 2y = 100.$$

We can solve this equation for y in terms of x, substitute the result into $A = xy$, and obtain A as a function in one variable. Solving for y, we get

$$2y = 100 - 2x \qquad \text{Subtract 2x from both sides of}$$
$$\text{2x + 2y = 100.}$$

$$y = \frac{100 - 2x}{2} \qquad \text{Divide both sides by 2.}$$

$$y = 50 - x. \qquad \text{Divide each term in the numerator by 2.}$$

Now we substitute $50 - x$ for y in $A = xy$.

$$A = xy = x(50 - x).$$

The rectangle and its dimensions are illustrated in Figure 2.8. Because A is now a function of x, we can write

$$A(x) = x(50 - x).$$

This function models the area, A, of any rectangle whose perimeter is 100 yards in terms of one of its dimensions, x.

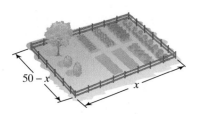

Figure 2.8 What value of x will maximize the rectangle's area?

Step 3   Write the function in the form $f(x) = ax^2 + bx + c$.   We apply the distributive property to obtain

$$A(x) = x(50 - x) = 50x - x^2 = -x^2 + 50x.$$

$a = -1$ $b = 50$

Step 4   Calculate $-\dfrac{b}{2a}$. If $a < 0$, the function has a maximum at this value.
The voice balloons show that $a = -1$ and $b = 50$.

$$x = -\frac{b}{2a} = -\frac{50}{2(-1)} = 25$$

This means that the area, A, of a rectangle with perimeter 100 yards is a maximum when one of the rectangle's dimensions, x, is 25 yards.

Step 5   Answer the question posed by the problem.   We found that $x = 25$. Figure 2.8 shows that the rectangle's other dimension is $50 - x = 50 - 25 = 25$. The dimensions of the rectangle that maximize the enclosed area are 25 yards by 25 yards. The rectangle that gives the maximum area is actually a square with an area of 25 yards · 25 yards, or 625 square yards.

Technology

The graph of the area function

$$A(x) = x(50 - x)$$

was obtained with a graphing utility using a [0, 50, 2] by [0, 700, 25] viewing rectangle. The maximum function feature verifies that a maximum area of 625 square yards occurs when one of the dimensions is 25 yards.

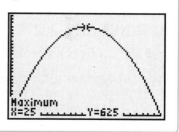

Check Point 6   You have 120 feet of fencing to enclose a rectangular region. Find the dimensions of the rectangle that maximize the enclosed area. What is the maximum area?

The ability to express a quantity to be maximized or minimized as a function in one variable plays a critical role in solving max-min problems. In calculus, you will learn a technique for maximizing or minimizing all functions, not only quadratic functions.

Technology

We've come a long way from the small nation of "embattled farmers" who launched the American Revolution. In the early days of our Republic, 95% of the population was involved in farming. The graph in Figure 2.9 shows the number of farms in the United States from 1850 through 2010 (projected). Because the graph is shaped like a cup, with an increasing number of farms from 1850 to 1910 and a decreasing number of farms from 1910 to 2010, a quadratic function is an appropriate model for the data. You can use the statistical menu of a graphing utility to enter the data in Figure 2.9. We entered the data using (number of decades after 1850, millions of U.S. farms). The data are shown to the right of Figure 2.9.

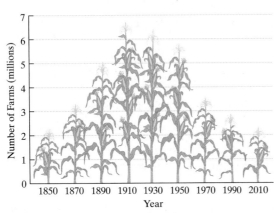

Number of U. S. Farms, 1850–2010

Data:
$(0, 2.3), (2, 3.3), (4, 5.1),$
$(6, 6.7), (8, 6.4), (10, 5.8)$
$(12, 3.6), (14, 2.9), (16, 2.3)$

```
QuadReg
 y=ax²+bx+c
 a=⁻.0643668831
 b=.9873701299
 c=2.203636364
```

Figure 2.9

Source: U.S. Bureau of the Census

Upon entering the QUADratic REGression program, we obtain the results shown in the screen. Thus, the quadratic function of best fit is

$$f(x) = -0.064x^2 + 0.99x + 2.2$$

where x represents the number of decades after 1850 and $f(x)$ represents the number of U.S. farms, in millions.

EXERCISE SET 2.2

Practice Exercises

In Exercises 1–4, the graph of a quadratic function is given. Write the function's equation, selecting from the following options.

$f(x) = (x + 1)^2 - 1 \qquad g(x) = (x + 1)^2 + 1$
$h(x) = (x - 1)^2 + 1 \qquad j(x) = (x - 1)^2 - 1$

1.

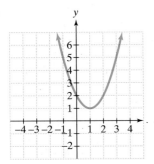

2.

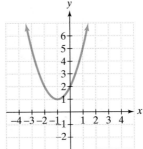

3.

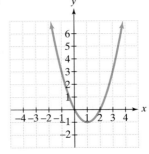

4.

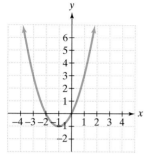

8.

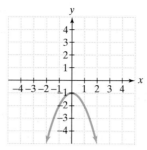

In Exercises 5–8, the graph of a quadratic function is given. Write the function's equation, selecting from the following options.

$$f(x) = x^2 + 2x + 1 \qquad g(x) = x^2 - 2x + 1$$
$$h(x) = x^2 - 1 \qquad\qquad j(x) = -x^2 - 1$$

In Exercises 9–16, find the coordinates of the vertex for the parabola defined by the given quadratic function.

9. $f(x) = 2(x - 3)^2 + 1$ **10.** $f(x) = -3(x - 2)^2 + 12$

11. $f(x) = -2(x + 1)^2 + 5$ **12.** $f(x) = -2(x + 4)^2 - 8$

13. $f(x) = 2x^2 - 8x + 3$ **14.** $f(x) = 3x^2 - 12x + 1$

15. $f(x) = -x^2 - 2x + 8$ **16.** $f(x) = -2x^2 + 8x - 1$

5.

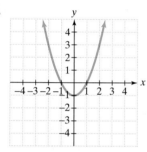

In Exercises 17–34, use the vertex and intercepts to sketch the graph of each quadratic function. Give the equation of the parabola's axis of symmetry. Use the graph to determine the function's domain and range.

17. $f(x) = (x - 4)^2 - 1$ **18.** $f(x) = (x - 1)^2 - 2$

19. $f(x) = (x - 1)^2 + 2$ **20.** $f(x) = (x - 3)^2 + 2$

21. $y - 1 = (x - 3)^2$ **22.** $y - 3 = (x - 1)^2$

23. $f(x) = 2(x + 2)^2 - 1$ **24.** $f(x) = \frac{5}{4} - (x - \frac{1}{2})^2$

25. $f(x) = 4 - (x - 1)^2$ **26.** $f(x) = 1 - (x - 3)^2$

6.

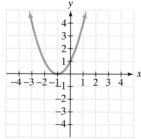

27. $f(x) = x^2 - 2x - 3$ **28.** $f(x) = x^2 - 2x - 15$

29. $f(x) = x^2 + 3x - 10$ **30.** $f(x) = 2x^2 - 7x - 4$

31. $f(x) = 2x - x^2 + 3$ **32.** $f(x) = 5 - 4x - x^2$

33. $f(x) = 2x - x^2 - 2$ **34.** $f(x) = 6 - 4x + x^2$

In Exercises 35–40, determine, without graphing, whether the given quadratic function has a minimum value or a maximum value. Then find the coordinates of the minimum or the maximum point.

7.

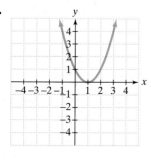

35. $f(x) = 3x^2 - 12x - 1$ **36.** $f(x) = 2x^2 - 8x - 3$

37. $f(x) = -4x^2 + 8x - 3$ **38.** $f(x) = -2x^2 - 12x + 3$

39. $f(x) = 5x^2 - 5x$ **40.** $f(x) = 6x^2 - 6x$

Application Exercises

Cigarette Consumption per U.S. Adult

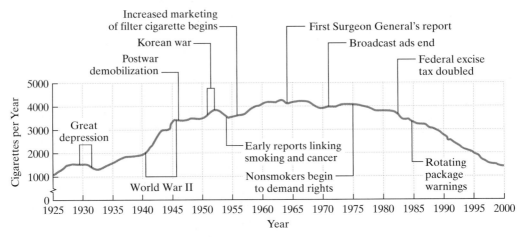

Source: U.S. Department of Health and Human Services

41. Please see the graph above. The function

$$f(x) = -3.1x^2 + 51.4x + 4024.5$$

models the average annual per capita consumption of cigarettes, $f(x)$, by Americans 18 and older x years after 1960. According to this model, in which year did cigarette consumption per capita reach a maximum? What was the consumption for that year? Does this accurately model what actually occurred as shown by the graph above?

42. The function

$$f(x) = 104.5x^2 - 1501.5x + 6016$$

models the death rate per year per 100,000 males, $f(x)$, for U.S. men who average x hours of sleep each night. How many hours of sleep, to the nearest tenth of an hour, corresponds to the minimum death rate? What is this minimum death rate, to the nearest whole number?

43. Fireworks are launched into the air. The quadratic function

$$s(t) = -16t^2 + 200t + 4$$

models the fireworks' height, $s(t)$, in feet, t seconds after they are launched. When should the fireworks explode so that they go off at the greatest height? What is that height?

44. A football is thrown by a quarterback to a receiver 40 yards away. The quadratic function

$$s(t) = -0.025t^2 + t + 5$$

models the football's height above the ground, $s(t)$, in feet, when it is t yards from the quarterback. How many yards from the quarterback does the football reach its greatest height? What is that height?

In the United States, HCV, or hepatitis C virus, is four times as widespread as HIV. Few of the nation's three to four million carriers have any idea they are infected. The graph shows the projected mortality, in thousands, from the virus. Use the information in the graph to solve Exercises 45–46.

Projected Hepatitis C Deaths in the United States

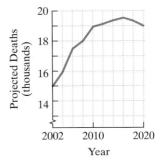

Source: American Journal of Public Health

45. Why is a quadratic function an appropriate model for the data shown in the graph?

46. Suppose that a quadratic function is used to model the data shown with ordered pairs representing (number of years after 2002, thousands of hepatitis C deaths). Determine, without obtaining an actual quadratic function that models the data, the approximate coordinates of the vertex for the function's graph. Describe what this means in practical terms. Use the word "maximum" in your description.

47. Among all pairs of numbers whose sum is 16, find a pair whose product is as large as possible. What is the maximum product?

48. Among all pairs of numbers whose sum is 9, find a pair whose product is as large as possible. What is the maximum product?

49. Among all pairs of numbers whose difference is 10, find a pair whose product is as small as possible. What is the minimum product?

50. Among all pairs of numbers whose difference is 8, find a pair whose product is as small as possible. What is the minimum product?

51. You have 20 yards of fencing to enclose a rectangular region. Find the dimensions of the rectangle that maximize the enclosed area. What is the maximum area?

52. You have 80 feet of fencing to enclose a rectangular region. Find the dimensions of the rectangle that maximize the enclosed area. What is the maximum area?

53. You have 120 feet of fencing to enclose a rectangular plot that borders on a river. If you do not fence the side along the river, find the length and width of the plot that will maximize the area. What is the largest area that can be enclosed?

54. You have 600 feet of fencing to enclose a rectangular plot that borders on a river. If you do not fence the side along the river, find the length and width of the plot that will maximize the area. What is the largest area that can be enclosed?

55. A rectangular playground is to be fenced off and divided in two by another fence parallel to one side of the playground. Six hundred feet of fencing is used. Find the dimensions of the playground that maximize the total enclosed area. What is the maximum area?

56. A rectangular playground is to be fenced off and divided into two by another fence parallel to one side of the playground. Four hundred feet of fencing is used. Find the dimensions of the playground that maximize the total enclosed area. What is the maximum area?

57. A rain gutter is made from sheets of aluminum that are 12 inches wide by turning up the edges to form right angles. Determine the depth of the gutter that will maximize its cross-sectional area and allow the greatest amount of water to flow.

58. A rain gutter is made from sheets of aluminum that are 20 inches wide by turning up the edges to form right angles. Determine the depth of the gutter that will maximize its cross-sectional area and allow the greatest amount of water to flow.

If you have difficulty obtaining the functions to be maximized in Exercises 59–62, read Example 2 on pages 221–222.

59. On a certain route, an airline carries 8000 passengers per month, each paying $50. A market survey indicates that for each $1 increase in the ticket price, the airline will lose 100 passengers. Find the ticket price that will maximize the airline's monthly revenue for the route. What is the maximum monthly revenue?

60. A car rental agency can rent every one of its 200 cars at $30 per day. For each $1 increase in rate, five fewer cars are rented. Find the rental amount that will maximize the agency's daily revenue. What is the maximum daily revenue?

61. The annual yield per walnut tree is fairly constant at 60 pounds per tree when the number of trees per acre is 20 or fewer. For each additional tree over 20, the annual yield per tree for all trees on the acre decreases by 2 pounds due to overcrowding. How many walnut trees should be planed per acre to maximize the annual yield for the acre? What is the maximum number of pounds of walnuts per acre?

62. The annual yield per cherry tree is fairly constant at 50 pounds per tree when the number of trees per acre is 30 or fewer. For each additional tree over 30, the annual yield per tree for all trees on the acre decreases by 1 pound due to overcrowding. How many cherry trees should be planted per acre to maximize the annual yield for the acre? What is the maximum number of pounds of cherries per acre?

Writing in Mathematics

63. What is a quadratic function?

64. What is a parabola? Describe its shape.

65. Explain how to decide whether a parabola opens upward or downward.

66. Describe how to find a parabola's vertex if its equation is expressed in standard form. Give an example.

67. Describe how to find a parabola's vertex if its equation is in the form $f(x) = ax^2 + bx + c$. Use $f(x) = x^2 - 6x + 8$ as an example.

68. A parabola that opens upward has its vertex at $(1, 2)$. Describe as much as you can about the parabola based on this information. Include in your discussion the number of x-intercepts (if any) for the parabola.

69. The quadratic function

$$f(x) = -0.018x^2 + 1.93x - 25.34$$

describes the miles per gallon, $f(x)$, of a Ford Taurus driven at x miles per hour. Suppose that you own a Ford Taurus. Describe how you can use this function to save money.

70. Write two different word problems that can be solved by maximizing or minimizing quadratic functions. The problems may be similar to Exercises 47–62 in this exercise set. Then solve each of your problems.

Technology Exercises

71. Use a graphing utility to verify any five of your hand-drawn graphs in Exercises 17–34.

72. a. Use a graphing utility to graph $y = 2x^2 - 82x + 720$ in a standard viewing rectangle. What do you observe?

b. Find the coordinates of the vertex for the given quadratic function.

c. The answer to part (b) is $(20.5, -120.5)$. Because the leading coefficient of the given function, 2, is positive, the vertex is a minimum point on the graph. Use this fact to help find a viewing rectangle that will give a relatively complete picture of the parabola. With an axis of symmetry at $x = 20.5$, the setting for x should extend past this, so try Xmin = 0 and Xmax = 30. The setting for y should include (and probably go below) the y-coordinate of the graph's minimum point, so try Ymin = -130. Experiment with Ymax until your utility shows the parabola's major features.

d. In general, explain how knowing the coordinates of a parabola's vertex can help determine a reasonable viewing rectangle on a graphing utility for obtaining a complete picture of the parabola.

In Exercises 73–76, find the vertex for each parabola. Then determine a reasonable viewing rectangle on your graphing utility and use it to graph the quadratic function.

73. $y = -0.25x^2 + 40x$ **74.** $y = -4x^2 + 20x + 160$

75. $y = 5x^2 + 40x + 600$ **76.** $y = 0.01x^2 + 0.6x + 100$

77. The function $y = 0.011x^2 - 0.097x + 4.1$ models the number of people in the United States, y, in millions, holding more than one job x years after 1970. Use a graphing utility to graph the function in a $[0, 20, 1]$ by $[3, 6, 1]$ viewing rectangle. TRACE along the curve or use your utility's minimum value feature to approximate the coordinates of the parabola's vertex. Describe what this represents in practical terms.

78. The following data show fuel efficiency, in miles per gallon, for all U.S. automobiles in the indicated year.

x (Years after 1940)	y (Average Number of Miles per Gallon for U.S. Automobiles)
1940: 0	14.8
1950: 10	13.9
1960: 20	13.4
1970: 30	13.5
1980: 40	15.9
1990: 50	20.2
1998: 58	21.8

Source: U.S. Department of Transportation

a. Use a graphing utility to draw a scatter plot of the data. Explain why a quadratic function is appropriate for modeling these data.

b. Use the quadratic regression feature to find the quadratic function that best fits the data.

c. Use the model in part (b) to determine the worst year for automobile fuel efficiency. What was the average number of miles per gallon for that year?

d. Use a graphing utility to draw a scatter plot of the data and graph the quadratic function of best fit on the scatter plot.

79. Use a graphing utility to verify your algebraic work in any three exercises from Exercises 47–62. Begin by graphing the quadratic function that models the quantity to be maximized or minimized. Then use the maximum or minimum function feature, or possibly the TRACE feature, to verify your answer.

Critical Thinking Exercises

80. Which one of the following is true?

a. No quadratic functions have a range of $(-\infty, \infty)$.

b. The vertex of the parabola described by $f(x) = 2(x - 5)^2 - 1$ is $(5, 1)$.

c. The graph of $f(x) = -2(x + 4)^2 - 8$ has one y-intercept and two x-intercepts.

d. The maximum value of y for the quadratic function $f(x) = -x^2 + x + 1$ is 1.

81. What explanations can you offer for your answer to Exercise 41? Use a graphing utility to graph f. Do you agree with the long-term predictions made by the graph? Explain.

In Exercises 82–83, find the axis of symmetry for each parabola whose equation is given. Use the axis of symmetry to find a second point on the parabola whose y-coordinate is the same as the given point.

82. $f(x) = 3(x + 2)^2 - 5$; $(-1, -2)$

83. $f(x) = (x - 3)^2 + 2$; $(6, 11)$

84. Find the point on the line whose equation is $2x + y = 2$ that is closest to the origin. *Hint*: Minimize the distance function by minimizing the expression under the square root.

85. A 300-room hotel can rent every one of its rooms at $80 per room. For each $1 increase in rent, three fewer rooms are rented. Each rented room costs the hotel $10 to service per day. How much should the hotel charge for each room to maximize its daily profit? What is the maximum daily profit?

86. A track and field area is to be constructed in the shape of a rectangle with semicircles at each end. The inside perimeter of the track is to be 440 yards. Find the dimensions of the rectangle that maximize the area of the rectangular portion of the field.

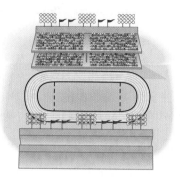

Group Exercise

87. Each group member should consult an almanac, newspaper, magazine, or the Internet to find data that can be modeled by a quadratic function. Group members should select the two sets of data that are most interesting and relevant. For each data set selected:

a. Use the quadratic regression feature of a graphing utility to find the quadratic function that best fits the data.

b. Use the equation of the quadratic function to make a prediction from the data. What circumstances might affect the accuracy of your prediction?

c. Use the equation of the quadratic function to write and solve a problem involving maximizing or minimizing the function.

SECTION 2.3 *Polynomial Functions and Their Graphs*

Objectives

1. Recognize characteristics of graphs of polynomial functions.

2. Determine end behavior.

3. Use factoring to find zeros of polynomial functions.

4. Identify the multiplicity of a zero.

5. Use the Intermediate Value Theorem.

6. Understand the relationship between degree and turning points.

7. Graph polynomial functions.

Magnified 6000 times, this color-scanned image shows a T-lymphocyte blood cell (green) infected with the HIV virus (red). Depletion of the number of T-cells causes destruction of the immune system.

In 1980, U.S. doctors diagnosed 41 cases of a rare form of cancer, Kaposi's sarcoma, that involved skin lesions, pneumonia, and severe immunological deficiencies. All cases involved gay men ranging in age from 26 to 51. By the end of 2000, approximately 775,000 Americans, straight and gay, male and female, old and young, were infected with the HIV virus.

Modeling AIDS-related data and making predictions about the epidemic's havoc is serious business. Changing circumstances and unforeseen events have resulted in models that are not particularly useful over long periods of time. For example, the function

$$f(x) = -143x^3 + 1810x^2 - 187x + 2331$$

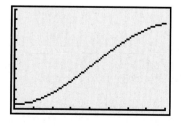

Figure 2.10 The graph of a function modeling the number of new AIDS cases in the U.S. from 1983 through 1991

models the number of AIDS cases diagnosed in the United States x years after 1983. The model was obtained using cases diagnosed from 1983 through 1991. Figure 2.10 shows the graph of f from 1983 through 1991 in a $[0, 8, 1]$ by $[0, 50,000, 5000]$ viewing rectangle. The function used to describe the number of new AIDS cases in the United States over a limited period of time is an example of a *polynomial function*.

Definition of a Polynomial Function

Let n be a nonnegative integer and let $a_n, a_{n-1}, \ldots, a_2, a_1, a_0$, be real numbers, with $a_n \neq 0$. The function defined by

$$f(x) = a_n x^n + a_{n-1} x^{n-1} + \cdots + a_2 x^2 + a_1 x + a_0$$

is called a **polynomial function of x of degree n.** The number a_n, the coefficient of the variable to the highest power, is called the **leading coefficient.**

A constant function $f(x) = c$, where $c \neq 0$, is a polynomial function of degree 0. A linear function $f(x) = mx + b$, where $m \neq 0$, is a polynomial function of degree 1. A quadratic function $f(x) = ax^2 + bx + c$, where $a \neq 0$, is a polynomial function of degree 2. In this section, we focus on polynomial functions of degree 3 or higher.

1 Recognize characteristics of graphs of polynomial functions.

Smooth, Continuous Graphs

Polynomial functions of degree 2 or less have graphs that are either parabolas or lines. We can graph such functions by plotting points. We can also graph polynomial functions of degree 3 or higher by plotting points. However, the process is rather tedious: Many points must be plotted. It may be easier to use a graphing utility for such functions. Regardless of the graphing method you use, you will find an ability to recognize the basic features of polynomial functions helpful. For example, they may help you choose an appropriate viewing rectangle for a graphing utility.

Two important features of the graphs of polynomial functions are that they are *smooth* and *continuous*. By **smooth,** we mean that the graph contains only rounded curves with no sharp corners. By **continuous,** we mean that the graph has no breaks and can be drawn without lifting your pencil from the rectangular coordinate system. These ideas are illustrated in Figure 2.11.

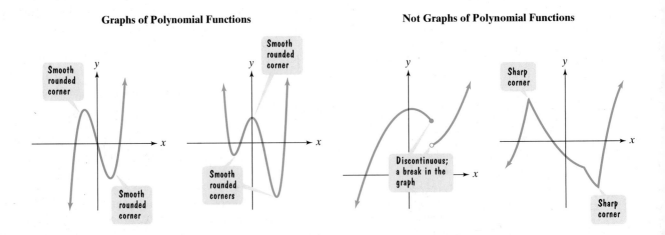

Figure 2.11   Recognizing graphs of polynomial functions

2 Determine end behavior.

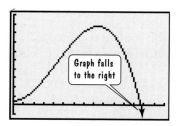

Figure 2.12 By extending the viewing rectangle, y is eventually negative and the function no longer models the number of AIDS cases.

End Behavior of Polynomial Functions

Figure 2.12 shows the graph of the function

$$f(x) = -143x^3 + 1810x^2 - 187x + 2331,$$

which models U.S. AIDS cases from 1983 through 1991. Look what happens to the graph when we extend the year up through 1998 with a $[0, 15, 1]$ by $[-5000, 50,000, 5000]$ viewing rectangle. By year 13 (1996), the values of y are negative and the function no longer models AIDS cases. We've added an arrow to the graph at the far right to emphasize that it continues to decrease without bound. It is this far-right *end behavior* of the graph that makes it inappropriate for modeling AIDS cases into the future.

The behavior of a graph of a function to the far left or the far right is called its **end behavior.** Although the graph of a polynomial function may have intervals where it increases or decreases, the graph will eventually rise or fall without bound as it moves far to the left or far to the right.

How can you determine whether the graph of a polynomial function goes up or down at each end? The end behavior of a polynomial function

$$f(x) = a_nx^n + a_{n-1}x^{n-1} + \cdots + a_1x + a_0$$

depends upon the leading term a_nx^n. In particular, the sign of the leading coefficient, a_n, and the degree, n, of the polynomial function reveal its end behavior. In terms of end behavior, only the term of highest degree counts, summarized by the **Leading Coefficient Test.**

The Leading Coefficient Test

As x increases or decreases without bound the graph of the polynomial function

$$f(x) = a_nx^n + a_{n-1}x^{n-1} + a_{n-2}x^{n-2} + \cdots + a_1x + a_0 \quad (a_n \neq 0)$$

eventually rises or falls. In particular,

1. For n odd:

If the leading coefficient is positive, the graph falls to the left and rises to the right.

If the leading coefficient is negative, the graph rises to the left and falls to the right.

2. For n even:

If the leading coefficient is positive, the graph rises to the left and to the right.

If the leading coefficient is negative, the graph falls to the left and to the right.

$a_n > 0$

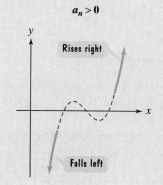

$a_n < 0$

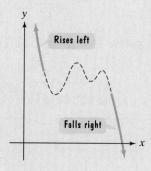

$a_n > 0$

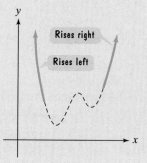

$a_n < 0$

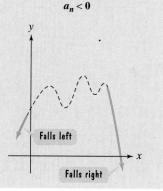

Study Tip

Odd-degree polynomial functions have graphs with opposite behavior at each end. Even-degree polynomial functions have graphs with the same behavior at each end.

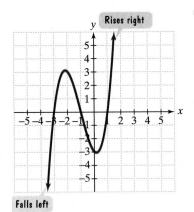

Figure 2.13 The graph of $f(x) = x^3 + 3x^2 - x - 3$

EXAMPLE 1   Using the Leading Coefficient Test

Use the Leading Coefficient Test to determine the end behavior of the graph of

$$f(x) = x^3 + 3x^2 - x - 3.$$

Solution   Because the degree is odd ($n = 3$) and the leading coefficient, 1, is positive, the graph falls to the left and rises to the right, as shown in Figure 2.13.

Check Point 1   Use the Leading Coefficient Test to determine the end behavior of the graph of $f(x) = x^4 - 4x^2$.

EXAMPLE 2   Using the Leading Coefficient Test

Use end behavior to explain why

$$f(x) = -143x^3 + 1810x^2 - 187x + 2331$$

is only an appropriate model for AIDS cases for a limited time period.

Solution   Because the degree is odd ($n = 3$) and the leading coefficient, -143, is negative, the graph rises to the left and falls to the right. The fact that it falls to the right indicates at some point the number of AIDS cases will be negative, an impossibility. If a function has a graph that decreases without bound over time, it will not be capable of modeling nonnegative phenomena over long time periods.

Check Point 2   The polynomial function

$$f(x) = -0.27x^3 + 9.2x^2 - 102.9x + 400$$

models the ratio of students to computers, $f(x)$, in U.S. public schools x years after 1980. Use end behavior to determine whether this function could be an appropriate model for computers in the classroom well into the twenty-first century. Explain your answer.

If you use a graphing utility to graph a polynomial function, it is important to select a viewing rectangle that accurately reveals the graph's end behavior. If the viewing rectangle is too small, it may not accurately show the end behavior.

EXAMPLE 3   Using the Leading Coefficient Test

The graph of $f(x) = -x^4 + 8x^3 + 4x^2 + 2$ was obtained with a graphing utility using a $[-8, 8, 1]$ by $[-10, 10, 1]$ viewing rectangle. The graph is shown in Figure 2.14(a). Does the graph show the end behavior of the function?

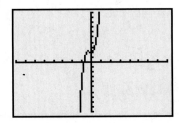 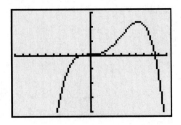

Figure 2.14   **(a)** $[-8, 8, 1]$ by $[-10, 10, 1]$ **(b)** $[-10, 10, 1]$ by $[-1000, 750, 250]$

Solution Note that the degree is even ($n = 4$) and the leading coefficient, -1, is negative. Even-degree polynomial functions have graphs with the same behavior at each end. The Leading Coefficient Test indicates that the graph should fall to the left and fall to the right. The graph in Figure 2.14(a) is falling to the left, but it is not falling to the right. Therefore, the graph is not complete enough to show end behavior. A more complete graph of the function is shown in a larger viewing rectangle in Figure 2.14(b).

Figure 2.15

Check Point 3 The graph of $f(x) = x^3 + 13x^2 + 10x - 4$ is shown in a standard viewing rectangle in Figure 2.15. Use the Leading Coefficient Test to determine whether the graph shows the end behavior of the function. Explain your answer.

3 Use factoring to find zeros of polynomial functions.

Zeros of Polynomial Functions

If f is a polynomial function, then the values of x for which $f(x)$ is equal to 0 are called the **zeros** of f. These values of x are the **roots,** or **solutions,** of the polynomial equation $f(x) = 0$. Each real root of the polynomial equation appears as an x-intercept of the graph of the polynomial function.

EXAMPLE 4 Finding Zeros of a Polynomial Function

Find all zeros of $f(x) = x^3 + 3x^2 - x - 3$.

Solution By definition, the zeros are the values of x for which $f(x)$ is equal to 0. Thus, we set $f(x)$ equal to 0:

$$f(x) = x^3 + 3x^2 - x - 3 = 0.$$

We solve the polynomial equation $x^3 + 3x^2 - x - 3 = 0$ for x as follows:

$x^3 + 3x^2 - x - 3 = 0$	This is the equation needed to find the function's zeros.
$x^2(x + 3) - 1(x + 3) = 0$	Factor x^2 from the first two terms and -1 from the last two terms.
$(x + 3)(x^2 - 1) = 0$	A common factor of $x + 3$ is factored from the expression.
$x + 3 = 0$ or $x^2 - 1 = 0$	Set each factor equal to 0.
$x = -3$ $\qquad$ $x^2 = 1$	Solve for x.
$x = \pm 1$	Remember that if $x^2 = d$, then $x = \pm\sqrt{d}$.

The zeros of f are $-3, -1$, and 1. The graph of f in Figure 2.16 shows that each zero is an x-intercept.

Technology

The graph of

$$f(x) = x^3 + 3x^2 - x - 3$$

is shown in a $[-6, 6, 1]$ by $[-6, 6, 1]$ viewing rectangle. The x-intercepts indicate that f has three real zeros,

$$-3, -1, \text{ and } 1.$$

The table provides numerical support for the three real zeros shown by the graph.

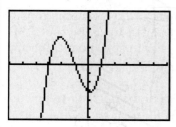

X	Y1	
-4	-15	
-3	0	
-2	3	-3, -1 and
-1	0	1 are the
0	-3	real zeros.
1	0	
2	15	

Y1⊟X3+3X2-X-3

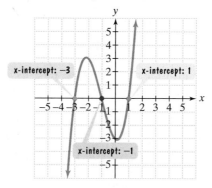

Figure 2.16 The real zeros of $f(x) = x^3 + 3x^2 - x - 3$, namely $-3, -1$, and 1, are the x-intercepts for the graph of f.

Check Point 4   Find all zeros of $f(x) = x^3 + 2x^2 - 4x - 8$.

EXAMPLE 5   Finding Zeros of a Polynomial Function

Find all zeros of $f(x) = -x^4 + 4x^3 - 4x^2$.

Solution   We find the zeros of f by setting $f(x)$ equal to 0 and solving the resulting equation.

$$-x^4 + 4x^3 - 4x^2 = 0 \qquad \textit{We now have a polynomial equation.}$$

$$x^4 - 4x^3 + 4x^2 = 0 \qquad \textit{Multiply both sides by } -1. \textit{ This step is optional.}$$

$$x^2(x^2 - 4x + 4) = 0 \qquad \textit{Factor out } x^2.$$

$$x^2(x - 2)^2 = 0 \qquad \textit{Factor completely.}$$

$$x^2 = 0 \quad \text{or} \quad (x - 2)^2 = 0 \qquad \textit{Set each factor equal to 0.}$$

$$x = 0 \qquad\qquad x = 2 \qquad \textit{Solve for x.}$$

The zeros of $f(x) = -x^4 + 4x^3 - 4x^2$ are 0 and 2. The graph of f, shown in Figure 2.17, has x-intercepts at 0 and 2.

Check Point 5   Find all zeros of $f(x) = x^4 - 4x^2$.

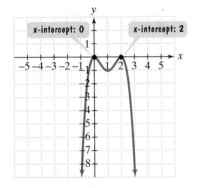

x-intercept: 0   x-intercept: 2

Figure 2.17   The zeros of $f(x) = -x^4 + 4x^3 - 4x^2$, namely 0 and 2, are the x-intercepts for the graph of f.

We can use the results of factoring to express a polynomial as a product of factors. For instance, in Example 5, we can use our factoring to express the function's equation as follows:

$$f(x) = -x^4 + 4x^3 - 4x^2 = -(x^4 - 4x^3 + 4x^2) = -x^2(x - 2)^2.$$

The factor x occurs twice: $x^2 = x \cdot x$.

The factor $(x - 2)$ occurs twice: $(x - 2)^2 = (x - 2)(x - 2)$.

4 Identify the multiplicity of a zero.

Notice that each factor occurs twice. In factoring the equation for the polynomial function f, if the same factor $x - r$ occurs k times, but not $k + 1$ times, we call r a **repeated zero with multiplicity k.** For the polynomial function

$$f(x) = -x^2(x - 2)^2,$$

0 and 2 are both repeated zeros with multiplicity 2. For the polynomial function

$$f(x) = 4(x - 5)(x + 2)^3(x - \tfrac{1}{4})^4,$$

5 is a zero with multiplicity 1, -2 is a repeated zero with multiplicity 3, and $\tfrac{1}{4}$ is a repeated zero with multiplicity 4.

The multiplicity of a zero tells us if the graph of a polynomial function touches the x-axis at the zero and turns around, or crosses the x-axis at the zero. For example, look again at the graph of $f(x) = -x^4 + 4x^3 - 4x^2$ in Figure 2.17. Each zero, 0 and 2, is a repeated zero with multiplicity 2. The graph of f touches, but does not cross, the x-axis at each of these zeros of even multiplicity. By contrast, a graph crosses the x-axis at zeros of odd multiplicity.

Multiplicity and x-Intercepts

If r is a zero of even multiplicity, then the graph **touches** the x-axis and turns around at r. If r is a zero of odd multiplicity, then the graph **crosses** the x-axis at r. Regardless of whether a zero is even or odd, graphs tend to flatten out at zeros with multiplicity greater than one.

5 Use the Intermediate Value Theorem.

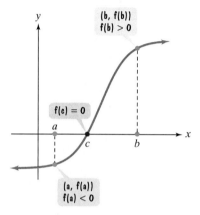

Figure 2.18 The graph must cross the x-axis at some value between a and b.

The Intermediate Value Theorem

We can find decimal approximations for real zeros of polynomial functions using a graphing utility. The *Intermediate Value Theorem* tells us of the existence of real zeros. The idea behind the theorem is illustrated in Figure 2.18. The figure shows that if $(a, f(a))$ lies below the x-axis and $(b, f(b))$ lies above the x-axis, the smooth, continuous graph of a polynomial function f must cross the x-axis at some value c between a and b. This value is a real zero for the function.

These observations are summarized in the **Intermediate Value Theorem.**

The Intermediate Value Theorem for Polynomials

Let f be a polynomial function with real coefficients. If $f(a)$ and $f(b)$ have opposite signs, then there is at least one value of c between a and b for which $f(c) = 0$. Equivalently, the equation $f(x) = 0$ has at least one real root between a and b.

EXAMPLE 6   Using the Intermediate Value Theorem

Show that the polynomial function $f(x) = x^3 - 2x - 5$ has a real zero between 2 and 3.

Solution   Let us evaluate f at 2 and 3. If $f(2)$ and $f(3)$ have opposite signs, then there is a real zero between 2 and 3. Using $f(x) = x^3 - 2x - 5$, we obtain

$$f(2) = 2^3 - 2 \cdot 2 - 5 = 8 - 4 - 5 = -1$$

and

> f(2) is negative.

$$f(3) = 3^3 - 2 \cdot 3 - 5 = 27 - 6 - 5 = 16.$$

> f(3) is positive.

This sign change shows that the polynomial function has a real zero between 2 and 3.

Technology

Most graphing utilities have a zero or root feature that gives zeros to a great degree of accuracy. Using the graph shown on the right, the real zero of

$$f(x) = x^3 - 2x - 5,$$

correct to the nearest thousandth, is 2.095.

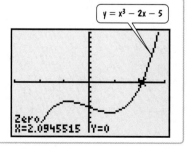

Check Point 6   Show that the polynomial function $f(x) = 3x^3 - 10x + 9$ has a real zero between -3 and -2.

6   Understand the relationship between degree and turning points.

Turning Points of Polynomial Functions

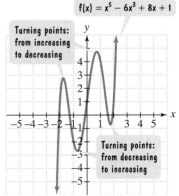

Figure 2.19   Graph with four turning points

The graph of $f(x) = x^5 - 6x^3 + 8x + 1$ is shown in Figure 2.19. The graph has four smooth **turning points.** At each turning point, the graph changes direction from increasing to decreasing or vice versa. The function value at a turning point is either a relative maximum of f or a relative minimum of f. The given equation has 5 as its greatest exponent and is therefore a polynomial function of degree 5. Notice that the graph has four turning points. In general, **if f is a polynomial of degree n, then the graph of f has at most $n - 1$ turning points**.

7   Graph polynomial functions.

A Strategy for Graphing Polynomial Functions

Here's a general strategy for graphing a polynomial function. A graphing utility is a valuable complement to this strategy. Some of the steps listed in the following box will help you to select a viewing rectangle that shows the important parts of the graph.

Graphing a Polynomial Function

$$f(x) = a_n x^n + a_{n-1} x^{n-1} + a_{n-2} x^{n-2} + \cdots + a_1 x + a_0, \quad a_n \neq 0$$

1. Use the Leading Coefficient Test to determine the graph's end behavior.
2. Find x-intercepts by setting $f(x) = 0$ and solving the resulting polynomial equation. If there is an x-intercept at r as a result of $(x - r)^k$ in the complete factorization of $f(x)$, then
   a. If k is even, the graph touches the x-axis at r and turns around.
   b. If k is odd, the graph crosses the x-axis at r.
   c. If $k > 1$, the graph flattens out at $(r, 0)$.
3. Find the y-intercept by computing $f(0)$.
4. Use symmetry, if applicable, to help draw the graph.
   a. y-axis symmetry: $f(-x) = f(x)$
   b. Origin symmetry: $f(-x) = -f(x)$
5. Use the fact that the maximum number of turning points of the graph is $n - 1$ to check whether it is drawn correctly.

EXAMPLE 7   Graphing a Polynomial Function

Graph: $f(x) = x^4 - 2x^2 + 1$.

Solution

Step 1   Determine end behavior.   Because the degree of $f(x) = 1x^4 - 2x^2 + 1$ is even $(n = 4)$ and the leading coefficient, 1, is positive, the graph rises to the left and rises to the right.

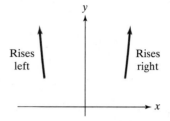

Step 2   Find x-intercepts (zeros of the function) by setting $f(x) = 0$.

$$x^4 - 2x^2 + 1 = 0$$
$$(x^2 - 1)(x^2 - 1) = 0 \quad \text{Factor.}$$
$$(x + 1)(x - 1)(x + 1)(x - 1) = 0 \quad \text{Factor completely.}$$
$$(x + 1)^2(x - 1)^2 = 0 \quad \text{Express the factoring in a more compact notation.}$$
$$(x + 1)^2 = 0 \quad \text{or} \quad (x - 1)^2 = 0 \quad \text{Set each factor equal to 0.}$$
$$x = -1 \qquad\qquad x = 1 \quad \text{Solve for x.}$$

We see that -1 and 1 are both repeated zeros with multiplicity 2. Because of the even multiplicity, the graph touches the x-axis at -1 and 1 and turns around. Furthermore, the graph tends to flatten out at these zeros with multiplicity greater than one.

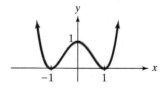

Step 3   Find the y-intercept by computing $f(0)$.   We use $f(x) = x^4 - 2x^2 + 1$ and compute $f(0)$.

$$f(0) = 0^4 - 2 \cdot 0^2 + 1 = 1$$

There is a y-intercept at 1, so the graph passes through $(0, 1)$:

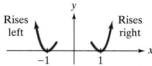

Step 4   Use possible symmetry to help draw the graph.   Our partial graph suggests y-axis symmetry. Let's verify this by finding $f(-x)$.

$$f(x) = \quad x^4 \quad - \quad 2x^2 + 1$$

Replace x with −x.

$$f(-x) = (-x)^4 - 2(-x)^2 + 1 = x^4 - 2x^2 + 1$$

Because $f(-x) = f(x)$, the graph of f is symmetric with respect to the y-axis. Figure 2.20 shows the graph of $f(x) = x^4 - 2x^2 + 1$.

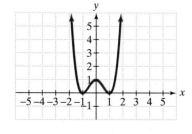

Figure 2.20   The graph of $f(x) = x^4 - 2x^2 + 1$

Step 5 Use the fact that the maximum number of turning points of the graph is $n - 1$ to check whether it is drawn correctly. Because we are graphing $f(x) = x^4 - 2x^2 + 1$, with $n = 4$, the maximum number of turning points is $4 - 1$, or 3. Because the graph in Figure 2.20 has three turning points, we have not violated the maximum number possible. (Take a moment to turn back the page and verify this observation.)

Check Point 7 Use the five-step strategy to graph $f(x) = x^3 - 3x^2$.

EXERCISE SET 2.3

Practice Exercises

In Exercises 1–10, determine which functions are polynomial functions. For those that are, identify the degree.

1. $f(x) = 5x^2 + 6x^3$

2. $f(x) = 7x^2 + 9x^4$

3. $g(x) = 7x^5 - \pi x^3 + \frac{1}{5}x$

4. $g(x) = 6x^7 + \pi x^5 + \frac{2}{3}x$

5. $h(x) = 7x^3 + 2x^2 + \frac{1}{x}$

6. $h(x) = 8x^3 - x^2 + \frac{2}{x}$

7. $f(x) = x^{1/2} - 3x^2 + 5$

8. $f(x) = x^{1/3} - 4x^2 + 7$

9. $f(x) = \frac{x^2 + 7}{x^3}$

10. $f(x) = \frac{x^2 + 7}{3}$

In Exercises 11–14, identify which graphs are not those of polynomial functions.

11.

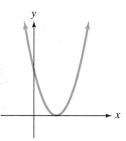

12.

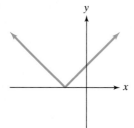

13.

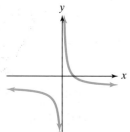

14.

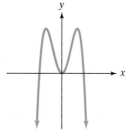

In Exercises 15–20, use the Leading Coefficient Test to determine the end behavior of the graph of the given polynomial function. Then use this end behavior to match the polynomial function with its graph. [The graphs are labeled (a) through (f).]

15. $f(x) = -x^4 + x^2$ c

16. $f(x) = x^3 - 4x^2$

17. $f(x) = (x - 3)^2$ b

18. $f(x) = -x^3 - x^2 + 5x - 3$

19. $f(x) = x - 3$ A

20. $f(x) = (x + 1)^2(x - 1)^2$

(a)

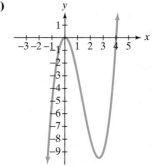

(b)

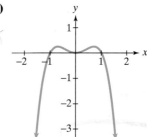

(c)

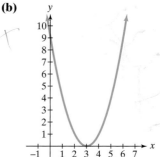

(d)

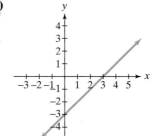

(e)

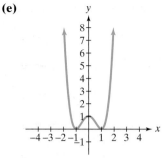

(f)

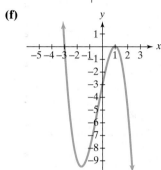

In Exercises 21–26, use the Leading Coefficient Test to determine the end behavior of the graph of the polynomial function.

21. $f(x) = 5x^3 + 7x^2 - x + 9$

22. $f(x) = 11x^3 - 6x^2 + x + 3$

23. $f(x) = 5x^4 + 7x^2 - x + 9$

24. $f(x) = 11x^4 - 6x^2 + x + 3$

25. $f(x) = -5x^4 + 7x^2 - x + 9$

26. $f(x) = -11x^4 - 6x^2 + x + 3$

In Exercises 27–34, find the zeros for each polynomial function and give the multiplicity for each zero. State whether the graph crosses the x-axis, or touches the x-axis and turns around, at each zero.

27. $f(x) = 2(x - 5)(x + 4)^2$

28. $f(x) = 3(x + 5)(x + 2)^2$

29. $f(x) = 4(x - 3)(x + 6)^3$

30. $f(x) = -3(x + \frac{1}{2})(x - 4)^3$

31. $f(x) = x^3 - 2x^2 + x$

32. $f(x) = x^3 + 4x^2 + 4x$

33. $f(x) = x^3 + 7x^2 - 4x - 28$

34. $f(x) = x^3 + 5x^2 - 9x - 45$

In Exercises 35–42, use the Intermediate Value Theorem to show that each polynomial has a real zero between the given integers.

35. $f(x) = x^3 - x - 1$; between 1 and 2

36. $f(x) = x^3 - 4x^2 + 2$; between 0 and 1

37. $f(x) = 2x^4 - 4x^2 + 1$; between −1 and 0

38. $f(x) = x^4 + 6x^3 - 18x^2$; between 2 and 3

39. $f(x) = x^3 + x^2 - 2x + 1$; between −3 and −2

40. $f(x) = x^5 - x^3 - 1$; between 1 and 2

41. $f(x) = 3x^3 - 10x + 9$; between −3 and −2

42. $f(x) = 3x^3 - 8x^2 + x + 2$; between 2 and 3

In Exercises 43–58,

> **a.** *Use the Leading Coefficient Test to determine the graph's end behavior.*
>
> **b.** *Find x-intercepts by setting $f(x) = 0$ and solving the resulting polynomial equation. State whether the graph crosses the x-axis, or touches the x-axis and turns around, at each intercept.*
>
> **c.** *Find the y-intercept by setting x equal to 0 and computing $f(0)$.*
>
> **d.** *Determine whether the graph has y-axis symmetry, origin symmetry, or neither.*
>
> **e.** *If necessary, find a few additional points and graph the function. Use the fact that the maximum number of turning points of the graph is $n - 1$ to check whether it is drawn correctly.*

43. $f(x) = x^3 + 2x^2 - x - 2$

44. $f(x) = x^3 + x^2 - 4x - 4$

45. $f(x) = x^4 - 9x^2$ **46.** $f(x) = x^4 - x^2$

47. $f(x) = -x^4 + 16x^2$ **48.** $f(x) = -x^4 + 4x^2$

49. $f(x) = x^4 - 2x^3 + x^2$ **50.** $f(x) = x^4 - 6x^3 + 9x^2$

51. $f(x) = -2x^4 + 4x^3$ **52.** $f(x) = -2x^4 + 2x^3$

53. $f(x) = 6x^3 - 9x - x^5$ **54.** $f(x) = 6x - x^3 - x^5$

55. $f(x) = 3x^2 - x^3$ **56.** $f(x) = \frac{1}{2} - \frac{1}{2}x^4$

57. $f(x) = -3(x - 1)^2(x^2 - 4)$

58. $f(x) = -2(x - 4)^2(x^2 - 25)$

Application Exercises

59. A herd of 100 elk is introduced to a small island. The number of elk, $N(t)$, after t years is described by the polynomial function $N(t) = -t^4 + 21t^2 + 100$.

    a. Use the Leading Coefficient Test to determine the graph's end behavior to the right. What does this mean about what will eventually happen to the elk population?

    b. Graph the function.

    c. Graph only the portion of the function that serves as a realistic model for the elk population over time. When does the population become extinct?

60. The common cold is caused by a rhinovirus. After x days of invasion by the viral particles, the number of particles in our bodies, $f(x)$, in billions, can be modeled by the polynomial function

$$f(x) = -0.75x^4 + 3x^3 + 5.$$

Use the Leading Coefficient Test to determine the graph's end behavior to the right. What does this mean about the number of viral particles in our bodies over time?

61. The polynomial function

$$f(x) = -0.87x^3 + 0.35x^2 + 81.62x + 7684.94$$

models the number of thefts, $f(x)$, in thousands, in the United States x years after 1987. Will this function be useful in modeling the number of thefts over an extended period of time? Explain your answer.

62. The graphs show the percentage of husbands and wives with one or more children who said their marriage was going well "all the time" at various stages in their relationships.

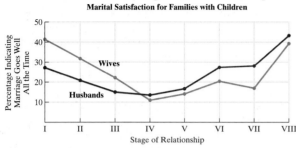

Marital Satisfaction for Families with Children

Source: Rollins, B,. & Feldman, H. (1970), Marital satisfaction over the family life cycle. *Journal of Marriage and the Family*, 32, 20–28.

Stage   I: Beginning families
Stage   II: Child-bearing families
Stage   III: Families with preschool children
Stage   IV: Families with school-age children
Stage   V: Families with teenagers
Stage   VI: Families with adult children leaving home
Stage VII: Families in the middle years
Stage VIII: Aging families

  a. Between which stages was marital satisfaction for wives decreasing?
  b. Between which stages was marital satisfaction for wives increasing?
  c. How many turning points (from decreasing to increasing or from increasing to decreasing) are shown in the graph for wives?
  d. Suppose that a polynomial function is used to model the data shown in the graph for wives using

    (stage in the relationship, percentage indicating that the marriage was going well all the time).

    Use the number of turning points to determine the degree of the polynomial function of best fit.
  e. For the model in part (d), should the leading coefficient of the polynomial function be positive or negative? Explain your answer.

The United States has more people in prison, as well as more people in prison per capita, than any other western industrialized nation. The bar graph shows the number of inmates in U.S. state and federal prisons in seven selected years from 1985 through 2000.

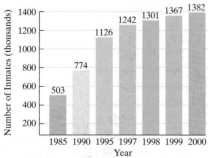

Number of Inmates in U.S. State and Federal Prisons

Source: U.S. Justice Department

The data in the graph can be modeled by

a linear function, $f(x) = 61.3x + 495$
a quadratic function,
$$g(x) = -0.131x^2 + 63.27x + 491.6$$
a third-degree polynomial function,
$$h(x) = -0.219x^3 + 4.885x^2 + 35.14x + 503.14.$$

For each of these functions, x represents the number of years after 1985 and the function value represents the number of inmates, in thousands. Use this information to solve Exercises 63–64.

63. The graph indicates that in 2000, there were 1382 thousand inmates. Substitute 1382 for $f(x)$ and $g(x)$ in the linear and quadratic models. Then solve each resulting equation to find how many years after 1985, to the nearest tenth of a year, inmate population was 1382 thousand. How well do the linear and quadratic functions serve as a model for 2000?

64. The graph indicates that in 2000, there were 1382 thousand inmates. Substitute 1382 for $h(x)$ in the third-degree model. Subtract 1382 from both sides of the equation to obtain a polynomial equal to 0. Use the Intermediate Value Theorem to show that this equation has a real root between 14 and 15. Then use a graphing utility's zero feature to find an approximation, to the nearest tenth, for this root. How well does the third-degree polynomial function serve as a model for 2000?

Writing in Mathematics

65. What is a polynomial function?

66. What do we mean when we describe the graph of a polynomial function as smooth and continuous?

67. What is meant by the end behavior of a polynomial function?

68. Explain how to use the Leading Coefficient Test to determine the end behavior of a polynomial function.

69. Why is a third-degree polynomial function with a negative leading coefficient not appropriate for modeling non-negative real-world phenomena over a long period of time?

70. What are the zeros of a polynomial function and how are they found?

71. Explain the relationship between the multiplicity of a zero and whether or not the graph crosses or touches the *x*-axis at that zero.

72. How do you show that a polynomial function has a real zero between two given numbers?

73. Explain the relationship between the degree of a polynomial function and the number of turning points on its graph.

74. Can the graph of a polynomial function have no *x*-intercepts? Explain.

75. Can the graph of a polynomial function have no *y*-intercept? Explain.

76. Describe a strategy for graphing a polynomial function. In your description, mention intercepts, the polynomial's degree, and turning points.

77. In a favorable habitat and without natural predators, a population of reindeer is introduced to an island preserve. The reindeer population, $f(t)$, t years after their introduction is modeled by the polynomial function $f(t) = -0.125t^5 + 3.125t^4 + 4000$. Discuss the growth and decline of the reindeer population. Describe the factors that might contribute to this population model.

78. The graphs shown in Exercise 62 indicate that marital satisfaction tends to be greatest at the beginning and at the end of the stages in the relationship, with a decline occurring in the middle. What explanations can you offer for this trend?

 Technology Exercises

79. Use a graphing utility to verify any five of the graphs that you drew by hand in Exercises 43–58.

Write a polynomial function that imitates the end behavior of each graph in Exercises 80–83. The dashed portions of the graphs indicate that you should focus only on imitating the left and right behavior of the graph and can be flexible about what occurs between the left and right ends. Then use your graphing utility to graph the polynomial function and verify that you imitated the end behavior shown in the given graph.

80. 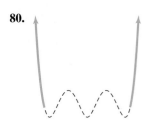   **81.**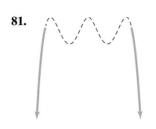

82. 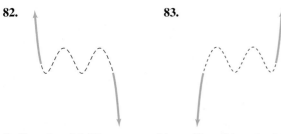   **83.**

In Exercises 84–87, use a graphing utility with a viewing rectangle large enough to show end behavior to graph each polynomial function.

84. $f(x) = x^3 + 13x^2 + 10x - 4$
85. $f(x) = -2x^3 + 6x^2 + 3x - 1$
86. $f(x) = -x^4 + 8x^3 + 4x^2 + 2$
87. $f(x) = -x^5 + 5x^4 - 6x^3 + 2x + 20$

In Exercises 88–89, use a graphing utility to graph f and g in the same viewing rectangle. Then use the $\boxed{\text{ZOOM OUT}}$ feature to show that f and g have identical end behavior.

88. $f(x) = x^3 - 6x + 1$, $g(x) = x^3$
89. $f(x) = -x^4 + 2x^3 - 6x$, $g(x) = -x^4$

 Critical Thinking Exercises

90. Which one of the following is true?

 a. If $f(x) = -x^3 + 4x$, then the graph of f falls to the left and to the right.

 b. A mathematical model that is a polynomial of degree n whose leading term is $a_n x^n$, n odd and $a_n < 0$, is ideally suited to describe nonnegative phenomena over unlimited periods of time.

 c. There is more than one third-degree polynomial function with the same three x-intercepts.

 d. The graph of a function with origin symmetry can rise to the left and to the right.

Use the descriptions in Exercises 91–92 to write an equation of a polynomial function with the given characteristics. Use a graphing utility to graph your function to see if you are correct. If not, modify the function's equation and repeat this process.

91. Crosses the x-axis at -4, 0, and 3; lies above the x-axis between -4 and 0; lies below the x-axis between 0 and 3

92. Touches the x-axis at 0 and crosses the x-axis at 2; lies below the x-axis between 0 and 2

93. Give an example of a function that is not subject to the Intermediate Value Theorem.

 Group Exercise

94. This exercise is based on the group's work in Exercise 87 of Exercise Set 2.2. For the two data sets that the group selected:

 a. Use the polynomial regression feature of a graphing utility to find the third-degree polynomial function that best fits the data.

 b. Use this function to repeat the predictions that you made with the quadratic function. How do these predictions compare with those that you obtained previously?

 c. For each data set, describe whether the quadratic function or the third-degree function is a better fit. Use a graphing utility, a scatter plot of the data, and the function of best fit drawn on the scatter plot to help determine which function is the better fit.

SECTION 2.4   *Dividing Polynomials; Remainder and Factor Theorems*

Objectives

1. Use long division to divide polynomials.
2. Use synthetic division to divide polynomials.
3. Evaluate a polynomial using the Remainder Theorem.
4. Use the Factor Theorem to solve a polynomial equation.

For those of you who are dog lovers, you might still be thinking of the polynomial function that models the age in human years, $H(x)$, of a dog that is x years old, namely

$$H(x) = -0.001618x^4 + 0.077326x^3 - 1.2367x^2 + 11.460x + 2.914.$$

Suppose that you are in your twenties, say 25. What is Fido's equivalent age? To answer this question, we must substitute 25 for $H(x)$ and solve the resulting polynomial equation for x:

$$25 = -0.001618x^4 + 0.077326x^3 - 1.2367x^2 + 11.460x + 2.914.$$

How can we solve such an equation? You might begin by subtracting 25 from both sides to obtain zero on one side. But then what? The factoring that we used in the previous section will not work in this situation.

In the next section, we will present techniques for solving certain kinds of polynomial equations. These techniques will further enhance your ability to manipulate algebraically the polynomial functions that model your world. Because these techniques are based on understanding polynomial division, in this section we look at two methods for dividing polynomials.

1 Use long division to divide polynomials.

Long Division of Polynomials and the Division Algorithm

We begin by looking at division by a polynomial containing more than one term, such as

$$x + 3 \overline{)x^2 + 10x + 21}$$

> Divisor has two terms.   Dividend has three terms.

When a divisor has more than one term, the four steps used to divide whole numbers—**divide, multiply, subtract, bring down the next term**—form the repetitive procedure for polynomial long division.

EXAMPLE 1   Long Division of Polynomials

Divide $x^2 + 10x + 21$ by $x + 3$.

Solution   The following steps illustrate how polynomial division is very similar to numerical division.

$$x + 3 \overline{)x^2 + 10x + 21}$$ Arrange the terms of the dividend ($x^2 + 10x + 21$) and the divisor ($x + 3$) in descending powers of x.

$$\begin{array}{r} x \\ x + 3 \overline{)x^2 + 10x + 21} \end{array}$$ Divide x^2 (the first term in the dividend) by x (the first term in the divisor): $\dfrac{x^2}{x} = x$. Align like terms.

$$\overset{\text{times}}{\underset{\text{equals}}{x + 3 \overline{\smash{)}\, x^2 + 10x + 21}}}$$
$$\,x^2 + 3x$$
$$\qquad\qquad x$$

Multiply each term in the divisor x + 3 by x, aligning terms of the product under like terms in the dividend.

$$\overset{x}{x + 3 \overline{\smash{)}\, x^2 + 10x + 21}}$$
$$\ominus\,x^2 \ominus 3x$$
$$\overline{7x}$$

Change signs of the polynomial being subtracted.

Subtract x² + 3x from x² + 10x by changing the sign of each term in the lower expression and adding.

$$\overset{x}{x + 3 \overline{\smash{)}\, x^2 + 10x + 21}}$$
$$x^2 + 3x \downarrow$$
$$\overline{7x + 21}$$

Bring down 21 from the original dividend and add algebraically to form a new dividend.

Is It Hot in Here, or Is It Just Me?

$$\overset{x + 7}{x + 3 \overline{\smash{)}\, x^2 + 10x + 21}}$$
$$x^2 + 3x \downarrow$$
$$\overline{7x + 21}$$

Find the second term of the quotient. Divide the first term of 7x + 21 by x, the first term of

the divisor: $\dfrac{7x}{x} = 7$.

$$\overset{\text{times}}{\overset{x + 7}{x + 3 \overline{\smash{)}\, x^2 + 10x + 21}}}$$
$$x^2 + 3x$$
$$\underset{\text{equals}}{\ominus\,7x \ominus 21}$$
$$\overline{7x + 21}$$
$$0 \quad \boxed{\text{Remainder}}$$

Multiply the divisor x + 3 by 7, aligning under like terms in the new dividend. Then subtract to obtain the remainder of 0.

In the 1980s, a rising trend in global surface temperature was observed, and the term "global warming" was coined. Scientists are more convinced than ever that burning coal, oil, and gas results in a buildup of gases and particles that trap heat and raise the planet's temperature. The average increase in global surface temperature, $T(x)$, in degrees Centigrade, x years after 1980 can be modeled by the polynomial function.

$$T(x) = \frac{21}{5,000,000}x^3$$
$$- \frac{127}{1,000,000}x^2 + \frac{1293}{50,000}x.$$

Use your graphing utility to graph the function in a [0, 60, 3] by [0, 2, 0.1] viewing rectangle. (Place parentheses around each fractional coefficient when you enter the equation). What do you observe about global warming through the year 2040?

The quotient is $x + 7$. Because the remainder is 0, we can conclude that $x + 3$ is a factor of $x^2 + 10x + 21$ and

$$\frac{x^2 + 10x + 21}{x + 3} = x + 7.$$

Check Point 1 Divide $x^2 + 14x + 45$ by $x + 9$.

Before considering additional examples, let's summarize the general procedure for dividing one polynomial by another.

Long Division of Polynomials

1. **Arrange the terms** of both the dividend and the divisor in descending powers of the variable.
2. **Divide** the first term in the dividend by the first term in the divisor. The result is the first term of the quotient.
3. **Multiply** every term in the divisor by the first term in the quotient. Write the resulting product beneath the dividend with like terms lined up.
4. **Subtract** the product from the dividend.
5. **Bring down** the next term in the original dividend and write it next to the remainder to form a new dividend.
6. Use this new expression as the dividend and repeat this process until the remainder can no longer be divided. This will occur when the degree of the remainder (the highest exponent on a variable in the remainder) is less than the degree of the divisor.

In our next long division, we will obtain a nonzero remainder.

EXAMPLE 2   Long Division of Polynomials

Divide $4 - 5x - x^2 + 6x^3$ by $3x - 2$.

Solution   We begin by writing the divisor and dividend in descending powers of x.

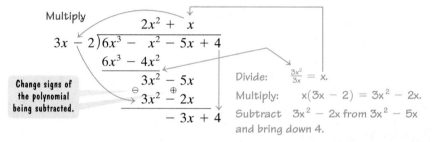

Divide: $\frac{6x^3}{3x} = 2x^2$.

Multiply: $2x^2(3x - 2) = 6x^3 - 4x^2$.

Subtract $6x^3 - 4x^2$ from $6x^3 - x^2$ and bring down $-5x$.

Now we divide $3x^2$ by $3x$ to obtain x, multiply x and the divisor, and subtract.

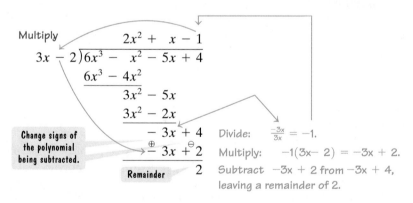

Divide: $\frac{3x^2}{3x} = x$.

Multiply: $x(3x - 2) = 3x^2 - 2x$.

Subtract $3x^2 - 2x$ from $3x^2 - 5x$ and bring down 4.

Now we divide $-3x$ by $3x$ to obtain -1, multiply -1 and the divisor, and subtract.

$$
\begin{array}{r}
2x^2 + x - 1 \\
3x - 2 \overline{)6x^3 - x^2 - 5x + 4} \\
\underline{6x^3 - 4x^2} \\
3x^2 - 5x \\
\underline{3x^2 - 2x} \\
-3x + 4 \\
\underline{-3x + 2} \\
2
\end{array}
$$

Change signs of the polynomial being subtracted.

Remainder

Divide: $\frac{-3x}{3x} = -1$.

Multiply: $-1(3x - 2) = -3x + 2$.

Subtract $-3x + 2$ from $-3x + 4$, leaving a remainder of 2.

In Example 2, the quotient is $2x^2 + x - 1$ and the remainder is 2. When there is a nonzero remainder, as in this example, list the quotient, plus the remainder above the divisor. Thus,

Dividend   Quotient   Remainder

$$
\frac{6x^3 - x^2 - 5x + 4}{3x - 2} = 2x^2 + x - 1 + \frac{2}{3x - 2}.
$$

Divisor   Divisor

Multiplying both sides of this equation by $3x - 2$ results in the following equation:

$$
6x^3 - x^2 - 5x + 4 = (3x - 2)(2x^2 + x - 1) + 2.
$$

Dividend   Divisor   Quotient   Remainder

Polynomial long division is checked by multiplying the divisor with the quotient and then adding the remainder. This should give the dividend. The process illustrates the **Division Algorithm.**

> **The Division Algorithm**
>
> If $f(x)$ and $d(x)$ are polynomials, with $d(x) \neq 0$, and the degree of $d(x)$ is less than or equal to the degree of $f(x)$, then there exist unique polynomials $q(x)$ and $r(x)$ such that
>
> $$f(x) \quad = \quad d(x) \quad \cdot \quad q(x) \quad + \quad r(x).$$
>
>   Dividend   Divisor   Quotient   Remainder
>
> The remainder, $r(x)$, equals 0 or it is of degree less than the degree of $d(x)$. If $r(x) = 0$, we say that $d(x)$ **divides evenly** into $f(x)$ and that $d(x)$ and $q(x)$ are **factors** of $f(x)$.

Check Point 2

Divide $7 - 11x - 3x^2 + 2x^3$ by $x - 3$. Use the remainder to express your result in fractional form.

If a power of x is missing in either a dividend or a divisor, add that power of x with a coefficient of 0 and then divide. In this way, like terms will be aligned as you carry out the long division.

EXAMPLE 3   Long Division of Polynomials

Divide $6x^4 + 5x^3 + 3x - 5$ by $3x^2 - 2x$.

Solution   We write the dividend, $6x^4 + 5x^3 + 3x - 5$,   as $6x^4 + 5x^3 + 0x^2 + 3x - 5$ to keep all like terms aligned.

The division process is finished because the degree of $7x - 5$, which is 1, is less than the degree of the divisor $3x^2 - 2x$, which is 2. The answer is

$$\frac{6x^4 + 5x^3 + 3x - 5}{3x^2 - 2x} = 2x^2 + 3x + 2 + \frac{7x - 5}{3x^2 - 2x}.$$

Check Point 3

Divide $2x^4 + 3x^3 - 7x - 10$ by $x^2 - 2x$.

2 Use synthetic division to divide polynomials.

Dividing Polynomials Using Synthetic Division

We can use **synthetic division** to divide polynomials if the divisor is of the form $x - c$. This method provides a quotient more quickly than long division. Let's compare the two methods showing $x^3 + 4x^2 - 5x + 5$ divided by $x - 3$.

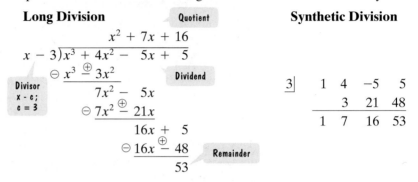

Long Division

Quotient

$$x^2 + 7x + 16$$
$$x - 3 \overline{)x^3 + 4x^2 - 5x + 5}$$

Divisor
$x - c$;
$c = 3$

$$\ominus \underline{x^3 \overset{\oplus}{-} 3x^2}$$ Dividend
$$7x^2 - 5x$$
$$\ominus \underline{7x^2 \overset{\oplus}{-} 21x}$$
$$16x + 5$$
$$\ominus \underline{16x \overset{\oplus}{-} 48}$$ Remainder
$$53$$

Synthetic Division

$$
\begin{array}{r|rrrr}
3 & 1 & 4 & -5 & 5 \\
  &   & 3 & 21 & 48 \\
\hline
  & 1 & 7 & 16 & 53
\end{array}
$$

Notice the relationship between the polynomials in the long division process and the numbers that appear in synthetic division.

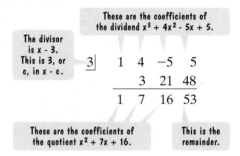

The divisor
is $x - 3$.
This is 3, or
c, in $x - c$.

These are the coefficients of
the dividend $x^3 + 4x^2 - 5x + 5$.

$$
\begin{array}{r|rrrr}
3 & 1 & 4 & -5 & 5 \\
  &   & 3 & 21 & 48 \\
\hline
  & 1 & 7 & 16 & 53
\end{array}
$$

These are the coefficients of
the quotient $x^2 + 7x + 16$.

This is the
remainder.

Now let's look at the steps involved in synthetic division.

Synthetic Division

To divide a polynomial by $x - c$,

1. Arrange polynomials in descending powers, with a 0 coefficient for any missing term.

2. Write c for the divisor, $x - c$. To the right, write the coefficients of the dividend.

3. Write the leading coefficient of the dividend on the bottom row.

4. Multiply c (in this case, 3) times the value just written on the bottom row. Write the product in the next column in the second row.

5. Add the values in this new column, writing the sum in the bottom row.

Example

$$x - 3 \overline{)x^3 + 4x^2 - 5x + 5}$$

$$
\begin{array}{r|rrrr}
3 & 1 & 4 & -5 & 5
\end{array}
$$

$$
\begin{array}{r|rrrr}
3 & 1 & 4 & -5 & 5 \\
\end{array}
$$
Bring down 1.
$$1$$

$$
\begin{array}{r|rrrr}
3 & 1 & 4 & -5 & 5 \\
  &   & 3 & & \\
\hline
  & 1 &   &   &
\end{array}
$$
Multiply by 3: $3 \cdot 1 = 3$.

$$
\begin{array}{r|rrrr}
3 & 1 & 4 & -5 & 5 \\
  &   & 3 & & \\
\hline
  & 1 & 7 &   &
\end{array}
$$
Add.

6. Repeat this series of multiplications and additions until all columns are filled in.

$$3\ |\ 1\quad 4\quad -5\quad 5$$
$$\underline{\qquad 3\quad 21}\ \text{Add.}$$
$$1\quad 7\quad 16$$

Multiply by 3: 3 · 7 = 21.

$$3\ |\ 1\quad 4\quad -5\quad 5$$
$$\underline{\qquad 3\quad 21\quad 48}\ \text{Add.}$$
$$1\quad 7\quad 16\quad 53$$

Multiply by 3: 3 · 16 = 48.

7. Use the numbers in the last row to write the quotient, plus the remainder above the divisor. **The degree of the first term of the quotient is one less than the degree of the first term of the dividend.** The final value in this row is the remainder.

Written from the last row of the synthetic division

$$1x^2 + 7x + 16 + \frac{53}{x-3}$$

$$x - 3 \overline{)x^3 + 4x^2 - 5x + 5}$$

EXAMPLE 4 Using Synthetic Division

Use synthetic division to divide $5x^3 + 6x + 8$ by $x + 2$.

Solution The divisor must be in the form $x - c$. Thus, we write $x + 2$ as $x - (-2)$. This means that $c = -2$. Writing a 0 coefficient for the missing x^2-term in the dividend, we can express the division as follows:

$$x - (-2)\overline{)5x^3 + 0x^2 + 6x + 8}.$$

Now we are ready to set up the problem so that we can use synthetic division.

Use the coefficients of the dividend $5x^3 + 0x^2 + 6x + 8$ in descending powers of x.

This is c in x−(−2).

$$-2\ |\ 5\quad 0\quad 6\quad 8$$

We begin the synthetic division process by bringing down 5. This is followed by a series of multiplications and additions.

1. Bring down 5.

$$-2\ |\ 5\quad 0\quad 6\quad 8$$
$$5$$

2. Multiply: −2(5) = −10.

$$-2\ |\ 5\quad 0\quad 6\quad 8$$
$$\underline{\quad -10}$$
$$5$$

Multiply by −2.

3. Add: 0 + (−10) = −10.

$$-2\ |\ 5\quad 0\quad 6\quad 8$$
$$\underline{\quad -10}\ \text{Add.}$$
$$5\quad -10$$

4. Multiply: −2(−10) = 20.

$$-2\ |\ 5\quad 0\quad 6\quad 8$$
$$\underline{\quad -10\quad 20}$$
$$5\quad -10$$

Multiply by −2.

5. Add: 6 + 20 = 26.

$$-2\ |\ 5\quad 0\quad 6\quad 8$$
$$\underline{\quad -10\quad 20}\ \text{Add.}$$
$$5\quad -10\quad 26$$

6. Multiply: −2(26) = −52.

$$-2\ |\ 5\quad 0\quad 6\quad 8$$
$$\underline{\quad -10\quad 20\quad -52}$$
$$5\quad -10\quad 26$$

Multiply by −2.

7. Add: 8 + (−52) = −44.

$$-2\ |\ 5\quad 0\quad 6\quad 8$$
$$\underline{\quad -10\quad 20\quad -52}\ \text{Add.}$$
$$5\quad -10\quad 26\quad -44$$

$$\begin{array}{r|rrrr} -2 & 5 & 0 & 6 & 8 \\ & & -10 & 20 & -52 \\ \hline & 5 & -10 & 26 & -44 \end{array}$$

The quotient is $5x^2 - 10x + 26$.   The remainder is -44.

The numbers in the last row represent the coefficients of the quotient and the remainder. The degree of the first term of the quotient is one less than that of the dividend. Because the degree of the dividend, $5x^3 + 6x + 8$, is 3, the degree of the quotient is 2. This means that the 5 in the last row represents $5x^2$.
Thus,

$$x + 2 \overline{)5x^3 + 6x + 8} \quad \begin{array}{c} 5x^2 - 10x + 26 - \dfrac{44}{x + 2} \end{array}$$

Check Point 4   Use synthetic division to divide $x^3 - 7x - 6$ by $x + 2$.

3 Evaluate a polynomial using the Remainder Theorem.

The Remainder Theorem

Let's consider the Division Algorithm when the dividend, $f(x)$, is divided by $x - c$. In this case, the remainder must be a constant because its degree is less than one, the degree of $x - c$.

$$f(x) = (x - c)q(x) + r$$

Dividend   Divisor   Quotient   The remainder, r, is a constant when dividing by x - c.

Now let's evaluate f at c.

$f(c) = (c - c)q(c) + r$ Find $f(c)$ by letting $x = c$ in $f(x) = (x - c)q(x) + r$. This will give an expression for r.

$f(c) = 0 \cdot q(c) + r$ $c - c = 0$

$f(c) = r$ $0 \cdot q(c) = 0$ and $0 + r = r$.

What does this last equation mean? If a polynomial is divided by $x - c$, the remainder is the value of the polynomial at c. This result is called the **Remainder Theorem.**

The Remainder Theorem

If the polynomial $f(x)$ is divided by $x - c$, then the remainder is $f(c)$.

Example 5 shows how we can use the Remainder Theorem to evaluate a polynomial function at 2. Rather than substituting 2 for x, we divide the function by $x - 2$. The remainder is $f(2)$.

EXAMPLE 5 **Using the Remainder Theorem**
to Evaluate a Polynomial Function

Given $f(x) = x^3 - 4x^2 + 5x + 3$, use the Remainder Theorem to find $f(2)$.

Solution By the Remainder Theorem, if $f(x)$ is divided by $x - 2$, then the remainder is $f(2)$. We'll use synthetic division to divide.

$$\begin{array}{r|rrrr} 2 & 1 & -4 & 5 & 3 \\ & & 2 & -4 & 2 \\ \hline & 1 & -2 & 1 & 5 \end{array} \quad \text{Remainder}$$

The remainder, 5, is the value of $f(2)$. Thus, $f(2) = 5$. We can verify that this is correct by evaluating $f(2)$ directly. Using $f(x) = x^3 - 4x^2 + 5x + 3$, we obtain

$$f(2) = 2^3 - 4 \cdot 2^2 + 5 \cdot 2 + 3 = 8 - 16 + 10 + 3 = 5.$$

> **Check**
> **Point**
> **5** Given $f(x) = 3x^3 + 4x^2 - 5x + 3$, use the Remainder Theorem to find $f(-4)$.

4 Use the Factor Theorem to solve a polynomial equation.

The Factor Theorem

Let's look again at the Division Algorithm when the divisor is of the form $x - c$.

$$f(x) = (x - c)q(x) + r$$

Dividend   Divisor   Quotient   Constant
remainder

By the Remainder Theorem, the remainder r is $f(c)$, so we can substitute $f(c)$ for r:

$$f(x) = (x - c)q(x) + f(c).$$

Notice that if $f(c) = 0$, then

$$f(x) = (x - c)q(x)$$

so that $x - c$ is a factor of $f(x)$. This means that for the polynomial function $f(x)$, if $f(c) = 0$, then $x - c$ is a factor of $f(x)$.

Let's reverse directions and see what happens if $x - c$ is a factor of $f(x)$. This means that

$$f(x) = (x - c)q(x).$$

If we replace x with c, we obtain

$$f(c) = (c - c)q(c) = 0.$$

Thus, if $x - c$ is a factor of $f(x)$, then $f(c) = 0$.
We have proved a result known as the **Factor Theorem.**

The Factor Theorem

Let $f(x)$ be a polynomial.
  a. If $f(c) = 0$, then $x - c$ is a factor of $f(x)$.
  b. If $x - c$ is a factor of $f(x)$, then $f(c) = 0$.

The example that follows shows how the Factor Theorem can be used to solve a polynomial equation.

EXAMPLE 6   Using the Factor Theorem

Solve the equation $2x^3 - 3x^2 - 11x + 6 = 0$ given that 3 is a zero of $f(x) = 2x^3 - 3x^2 - 11x + 6$.

Solution   We are given that $f(3) = 0$. The Factor Theorem tells us that $x - 3$ is a factor of $f(x)$. We'll use synthetic division to divide $f(x)$ by $x - 3$.

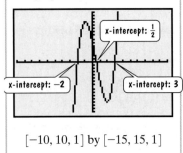

$$\underline{3}\begin{array}{rrrr} 2 & -3 & -11 & 6 \\ & 6 & 9 & -6 \\ \hline 2 & 3 & -2 & 0 \end{array}$$

$$\begin{array}{r} 2x^2 + 3x - 2 \\ x - 3\overline{)2x^3 - 3x^2 - 11x + 6} \end{array}$$

> The remainder, 0, verifies that x − 3 is a factor of $2x^3 - 3x^2 - 11x + 6.$

Equivalently,
$$2x^3 - 3x^2 - 11x + 6 = (x - 3)(2x^2 + 3x - 2).$$

Now we can solve the polynomial equation.

$$
\begin{array}{ll}
2x^3 - 3x^2 - 11x + 6 = 0 & \text{This is the given equation.} \\
(x - 3)(2x^2 + 3x - 2) = 0 & \text{Factor using the result from the synthetic division.} \\
(x - 3)(2x - 1)(x + 2) = 0 & \text{Factor the trinomial.} \\
x - 3 = 0 \quad \text{or} \quad 2x - 1 = 0 \quad \text{or} \quad x + 2 = 0 & \text{Set each factor equal to 0.} \\
\quad x = 3 \qquad\qquad\quad x = \tfrac{1}{2} \qquad\qquad x = -2 & \text{Solve for x.}
\end{array}
$$

The solution set is $\{-2, \tfrac{1}{2}, 3\}$.

Based on the Factor Theorem, the following statements are useful in solving polynomial equations:

1. If $f(x)$ is divided by $x - c$ and the remainder is zero, then c is a zero of f and c is a root of the polynomial equation $f(x) = 0$.

2. If $f(x)$ is divided by $x - c$ and the remainder is zero, then $x - c$ is a factor of $f(x)$.

> **Check Point 6**   Solve the equation $15x^3 + 14x^2 - 3x - 2 = 0$ given that -1 is a zero of $f(x) = 15x^3 + 14x^2 - 3x - 2$.

Technology

Because the solution set of
$$2x^3 - 3x^2 - 11x + 6 = 0$$
is $\{-2, \tfrac{1}{2}, 3\}$, this implies that the polynomial function
$$f(x) = 2x^3 - 3x^2 - 11x + 6$$
has x-intercepts (or zeros) at $-2, \tfrac{1}{2}$, and 3. This is verified by the graph of f.

$[-10, 10, 1]$ by $[-15, 15, 1]$

EXERCISE SET 2.4

Practice Exercises

In Exercises 1–16, divide using long division. State the quotient, $q(x)$, and the remainder, $r(x)$.

1. $(x^2 + 8x + 15) \div (x + 5)$
2. $(x^2 + 3x - 10) \div (x - 2)$
3. $(x^3 + 5x^2 + 7x + 2) \div (x + 2)$
4. $(x^3 - 2x^2 - 5x + 6) \div (x - 3)$
5. $(6x^3 + 7x^2 + 12x - 5) \div (3x - 1)$
6. $(6x^3 + 17x^2 + 27x + 20) \div (3x + 4)$
7. $(12x^2 + x - 4) \div (3x - 2)$

8. $(4x^2 - 8x + 6) \div (2x - 1)$
9. $\dfrac{2x^3 + 7x^2 + 9x - 20}{x + 3}$
10. $\dfrac{3x^2 - 2x + 5}{x - 3}$
11. $\dfrac{4x^4 - 4x^2 + 6x}{x - 4}$
12. $\dfrac{x^4 - 81}{x - 3}$
13. $\dfrac{6x^3 + 13x^2 - 11x - 15}{3x^2 - x - 3}$
14. $\dfrac{x^4 + 2x^3 - 4x^2 - 5x - 6}{x^2 + x - 2}$
15. $\dfrac{18x^4 + 9x^3 + 3x^2}{3x^2 + 1}$
16. $\dfrac{2x^5 - 8x^4 + 2x^3 + x^2}{2x^3 + 1}$

In Exercises 17–32, divide using synthetic division.

17. $(2x^2 + x - 10) \div (x - 2)$

18. $(x^2 + x - 2) \div (x - 1)$

19. $(3x^2 + 7x - 20) \div (x + 5)$

20. $(5x^2 - 12x - 8) \div (x + 3)$

21. $(4x^3 - 3x^2 + 3x - 1) \div (x - 1)$

22. $(5x^3 - 6x^2 + 3x + 11) \div (x - 2)$

23. $(6x^5 - 2x^3 + 4x^2 - 3x + 1) \div (x - 2)$

24. $(x^5 + 4x^4 - 3x^2 + 2x + 3) \div (x - 3)$

25. $(x^2 - 5x - 5x^3 + x^4) \div (5 + x)$

26. $(x^2 - 6x - 6x^3 + x^4) \div (6 + x)$

27. $\dfrac{x^5 + x^3 - 2}{x - 1}$

28. $\dfrac{x^7 + x^5 - 10x^3 + 12}{x + 2}$

29. $\dfrac{x^4 - 256}{x - 4}$

30. $\dfrac{x^7 - 128}{x - 2}$

31. $\dfrac{2x^5 - 3x^4 + x^3 - x^2 + 2x - 1}{x + 2}$

32. $\dfrac{x^5 - 2x^4 - x^3 + 3x^2 - x + 1}{x - 2}$

33. Given $f(x) = 2x^3 - 11x^2 + 7x - 5$, use the Remainder Theorem to find $f(4)$.

34. Given $f(x) = x^3 - 7x^2 + 5x - 6$, use the Remainder Theorem to find $f(3)$.

35. Given $f(x) = 7x^4 - 3x^3 + 6x + 9$, use the Remainder Theorem to find $f(-5)$.

36. Given $f(x) = 3x^4 + 6x^3 - 2x + 4$, use the Remainder Theorem to find $f(-4)$.

37. Use synthetic division to divide $f(x) = x^3 - 4x^2 + x + 6$ by $x + 1$. Use the result to find all zeros of f.

38. Use synthetic division to divide $f(x) = x^3 - 2x^2 - x + 2$ by $x + 1$. Use the result to find all zeros of f.

39. Solve the equation $2x^3 - 5x^2 + x + 2 = 0$ given that 2 is a zero of $f(x) = 2x^3 - 5x^2 + x + 2$.

40. Solve the equation $2x^3 - 3x^2 - 11x + 6 = 0$ given that -2 is a zero of $f(x) = 2x^3 - 3x^2 - 11x + 6$.

41. Solve the equation $12x^3 + 16x^2 - 5x - 3 = 0$ given that $-\frac{3}{2}$ is a root.

42. Solve the equation $3x^3 + 7x^2 - 22x - 8 = 0$ given that $-\frac{1}{3}$ is a root.

Application Exercises

43. A rectangle with length $2x + 5$ inches has an area of $2x^4 + 15x^3 + 7x^2 - 135x - 225$ square inches. Write a polynomial that represents its width.

44. If you travel a distance of $x^3 + 3x^2 + 5x + 3$ miles at a rate of $x + 1$ miles per hour, write a polynomial that represents the number of hours you traveled.

During the 1980s, the controversial economist Arthur Laffer promoted the idea that tax increases lead to a reduction in government revenue. Called supply-side economics, the theory uses function such as

$$f(x) = \frac{80x - 8000}{x - 110}, \quad 30 \le x \le 100.$$

This function models the government tax revenue, $f(x)$, in tens of billions of dollars, in terms of the tax rate, x. The graph of the function is shown. It illustrates tax revenue decreasing quite dramatically as the tax rate increases. At a tax rate of (gasp) 100%, the government takes all our money and no one has an incentive to work. With no income earned, zero dollars in tax revenue is generated.

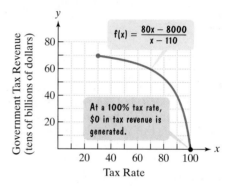

Use function f and its graph to solve Exercises 45–46.

45. a. Find and interpret $f(30)$. Identify the solution as a point on the graph of the function.

b. Rewrite the function by using long division to perform $(80x - 8000) \div (x - 110)$. Then use this new form of the function to find $f(30)$. Do you obtain the same answer as you did in part (a)?

c. Is f a polynomial function? Explain your answer.

46. a. Find and interpret $f(40)$. Identify the solution as a point on the graph of the function.

b. Rewrite the function by using long division to perform $(80x - 8000) \div (x - 110)$. Then use this new form of the function to find $f(40)$. Do you obtain the same answer as you did in part (a)?

c. Is f a polynomial function? Explain your answer.

Writing in Mathematics

47. Explain how to perform long division of polynomials. Use $2x^3 - 3x^2 - 11x + 7$ divided by $x - 3$ in your explanation.

48. In your own words, state the Division Algorithm.

49. How can the Division Algorithm be used to check the quotient and remainder in a long division problem?

50. Explain how to perform synthetic division. Use the division problem in Exercise 47 to support your explanation.

51. State the Remainder Theorem.

52. Explain how the Remainder Theorem can be used to find $f(-6)$ if $f(x) = x^4 + 7x^3 + 8x^2 + 11x + 5$. What advantage is there to using the Remainder Theorem in this situation rather than evaluating $f(-6)$ directly?

53. How can the Factor Theorem be used to determine if $x - 1$ is a factor of $x^3 - 2x^2 - 11x + 12$?

54. If you know that -2 is a zero of
$$f(x) = x^3 + 7x^2 + 4x - 12,$$
explain how to solve the equation
$$x^3 + 7x^2 + 4x - 12 = 0.$$

55. The idea of supply-side economics (see Exercises 45–46) is that an increase in the tax rate may actually reduce government revenue. What explanation can you offer for this theory?

Technology Exercises

In Exercises 56–59, use a graphing utility to determine if the division has been performed correctly. Graph the function on each side of the equation in the same viewing rectangle. If the graphs do not coincide, correct the expression on the right side by using polynomial long division. Then verify your correction using the graphing utility.

56. $(6x^2 + 16x + 8) \div (3x + 2) = 2x + 4, x \neq -\frac{2}{3}$

57. $\dfrac{x^4 + 6x^3 + 6x^2 - 10x - 3}{x^2 + 2x - 3} = x^2 + 4x + 1,$
$x \neq -3, x \neq 1$

58. $\dfrac{2x^3 - 3x^2 - 3x + 4}{x - 1} = 2x^2 - x + 4, x \neq 1$

59. $\dfrac{3x^4 + 4x^3 - 32x^2 - 5x - 20}{x + 4} = 3x^3 + 8x^2 - 5,$
$x \neq -4$

Critical Thinking Exercises

60. Which one of the following is true?
   a. If a trinomial in x of degree 6 is divided by a trinomial in x of degree 3, the degree of the quotient is 2.
   b. Synthetic division could not be used to find the quotient of $10x^3 - 6x^2 + 4x - 1$ and $x - \frac{1}{2}$.
   c. Any problem that can be done by synthetic division can also be done by the method for long division of polynomials.
   d. If a polynomial long-division problem results in a remainder that is a whole number, then the divisor is a factor of the dividend.

61. Find k so that $4x + 3$ is a factor of
$$20x^3 + 23x^2 - 10x + k.$$

62. When $2x^2 - 7x + 9$ is divided by a polynomial, the quotient is $2x - 3$ and the remainder is 3. Find the polynomial.

63. Find the quotient of $x^{3n} + 1$ and $x^n + 1$.

64. Synthetic division is a process for dividing a polynomial by $x - c$. The coefficient of x is 1. How might synthetic division be used if you are dividing by $2x - 4$?

SECTION 2.5 *Zeros of Polynomial Functions*

Objectives

1. Use the Rational Zero Theorem to find possible rational zeros.
2. Find zeros of a polynomial function.
3. Solve polynomial equations.
4. Use the Linear Factorization Theorem to factor a polynomial.
5. Factor polynomials with given zeros.
6. Use Descartes's Rule of Signs.

You stole my formula!

Tartaglia's Secret Formula for One Solution of $x^3 + mx = n$

$$x = \sqrt[3]{\sqrt{\left(\frac{n}{2}\right)^2 + \left(\frac{m}{3}\right)^3} + \frac{n}{2}} - \sqrt[3]{\sqrt{\left(\frac{n}{2}\right)^2 + \left(\frac{m}{3}\right)^3} - \frac{n}{2}}$$

Popularizers of mathematics are sharing bizarre stories that are giving math a secure place in popular culture. One episode, able to compete with the wildest fare served up by television talk shows and the tabloids, involves three Italian mathematicians and, of all things, zeros of polynomial functions.

Tartaglia (1499–1557), poor and starving, has found a formula that gives a root for a third-degree polynomial equation. Cardano (1501–1576) begs Tartaglia to reveal the secret formula, wheedling it from him with the promise he will find the impoverished Tartaglia a patron. Then Cardano publishes his famous work *Ars Magna*, in which he presents Tartaglia's formula as his own. Cardano uses his most talented student, Ferrari (1522–1565), who derived a formula for a root of a fourth-degree polynomial equation, to falsely accuse Tartaglia of plagiarism. The dispute becomes violent and Tartaglia is fortunate to escape alive.

The noise from this "You Stole My Formula" episode is quieted by the work of French mathematician Evariste Galois (1811–1832). Galois proved that there is no general formula for finding roots of polynomial equations of degree 5 or higher. There are, however, methods for finding roots. In this section, we study methods for finding zeros of polynomial functions. We begin with a theorem that plays an important role in this process.

1 Use the Rational Zero Theorem to find possible rational zeros.

The Rational Zero Theorem

The Rational Zero Theorem gives a list of all possible rational zeros of a polynomial function. Equivalently, the theorem gives all possible rational roots of a polynomial equation. Not every number in the list will be a zero of the function, but every rational zero of the polynomial function will appear somewhere in the list.

The Rational Zero Theorem

If $f(x) = a_n x^n + a_{n-1} x^{n-1} + \cdots + a_1 x + a_0$ has *integer* coefficients and $\dfrac{p}{q}$ (where $\dfrac{p}{q}$ is reduced) is a rational zero, then p is a factor of the constant term, a_0, and q is a factor of the leading coefficient, a_n.

You can explore the "why" behind the Rational Zero Theorem in Exercise 88 of Exercise Set 2.5. For now, let's see if we can figure out what the theorem tells us about possible rational zeros. In order to use the theorem, list all the integers that are factors of the constant term, a_0. Then list all the integers that are factors of the leading coefficient, a_n. Finally list all possible rational zeros:

$$\text{Possible rational zeros} = \frac{\text{Factors of the constant term}}{\text{Factors of the leading coefficient}}.$$

EXAMPLE 1   Using the Rational Zero Theorem

List all possible rational zeros of $f(x) = -x^4 + 3x^2 + 4$.

Solution   The constant term is 4. We list all of its factors: $\pm 1, \pm 2, \pm 4$. The leading coefficient is -1. Its factors are ± 1.

Factors of the constant term:	$\pm 1, \pm 2, \pm 4$
Factors of the leading coefficient:	± 1

Because

$$\text{Possible rational zeros} = \frac{\text{Factors of the constant term}}{\text{Factors of the leading coefficient}},$$

we must take each number in the first row, $\pm 1, \pm 2, \pm 4$, and divide by each number in the second row, ± 1.

$$\text{Possible rational zeros} = \frac{\text{Factors of 4}}{\text{Factors of } -1} = \frac{\pm 1, \pm 2, \pm 4}{\pm 1} = \pm 1, \ \pm 2, \ \pm 4$$

| Divide ± 1 by ± 1. | Divide ± 2 by ± 1. | Divide ± 4 by ± 1. |

Study Tip

Always keep in mind the relationship among zeros, roots, and x-intercepts. The zeros of function f are the roots, or solutions, of the equation $f(x) = 0$. Furthermore, the real zeros, or real roots, are the x-intercepts of the graph of f.

There are six possible rational zeros of $f(x) = -x^4 + 3x^2 + 4$, namely $\pm 1, \pm 2$, and ± 4. The graph of f is shown in Figure 2.21. The x-intercepts are -2 and 2. Thus, -2 and 2 are the actual rational zeros.

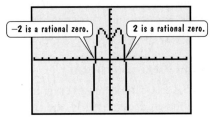

Figure 2.21 The graph of $f(x) = -x^4 + 3x^2 + 4$ shows that -2 and 2 are rational zeros.

Check Point 1 List all possible rational zeros of

$$f(x) = x^3 + 2x^2 - 5x - 6.$$

EXAMPLE 2 Using the Rational Zero Theorem

List all possible rational zeros of $f(x) = 15x^3 + 14x^2 - 3x - 2$.

Solution The constant term is -2 and the leading coefficient is 15.

$$\text{Possible rational zeros} = \frac{\text{Factors of the constant term, } -2}{\text{Factors of the leading coefficient, } 15}$$

$$= \frac{\pm 1, \pm 2}{\pm 1, \pm 3, \pm 5, \pm 15}$$

$$= \pm 1, \quad \pm 2, \quad \pm\tfrac{1}{3}, \quad \pm\tfrac{2}{3}, \quad \pm\tfrac{1}{5}, \quad \pm\tfrac{2}{5}, \quad \pm\tfrac{1}{15}, \quad \pm\tfrac{2}{15}$$

Divide ± 1 and ± 2 by ± 1.   Divide ± 1 and ± 2 by ± 3.   Divide ± 1 and ± 2 by ± 5.   Divide ± 1 and ± 2 by ± 15.

There are 16 possible rational zeros. The actual solution set of

$$15x^3 + 14x^2 - 3x - 2 = 0$$

is $\{-1, -\tfrac{1}{3}, \tfrac{2}{5}\}$, which contains 3 of the 16 possible zeros.

Check Point 2 List all possible rational zeros of

$$f(x) = 4x^5 + 12x^4 - x - 3.$$

2 Find zeros of a polynomial function.

How do we determine which (if any) of the possible rational zeros are rational zeros of the polynomial function? To find the first rational zero, we can use a trial-and-error process involving synthetic division: If $f(x)$ is divided by $x - c$ and the remainder is zero, then c is a zero of f. After we identify the first rational zero, we use the result of the synthetic division to factor the original polynomial. Then we set each factor equal to zero to identify any additional rational zeros.

EXAMPLE 3   Finding Zeros of a Polynomial Function

Find all rational zeros of $f(x) = x^3 + 2x^2 - 5x - 6$.

Solution   We begin by listing all possible rational zeros.

Possible rational zeros

$$= \frac{\text{Factors of the constant term, } -6}{\text{Factors of the leading coefficient, } 1} = \frac{\pm 1, \pm 2, \pm 3, \pm 6}{\pm 1} = \pm 1, \pm 2, \pm 3, \pm 6$$

> Divide the eight numbers in the numerator by ±1.

Now we will use synthetic division to see if we can find a rational zero among the possible rational zeros $\pm 1, \pm 2, \pm 3, \pm 6$. Keep in mind that if $f(x)$ is divided by $x - c$ and the remainder is zero, then c is a zero of f. Let's start by testing 1. If 1 is not a rational zero, then we will test other possible rational zeros.

Test 1

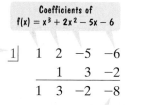

Possible rational zero

Coefficients of
$f(x) = x^3 + 2x^2 - 5x - 6$

$$\begin{array}{r|rrrr} 1 & 1 & 2 & -5 & -6 \\ & & 1 & 3 & -2 \\ \hline & 1 & 3 & -2 & -8 \end{array}$$

The nonzero remainder shows that 1 is not a zero.

Test 2

Possible rational zero

Coefficients of
$f(x) = x^3 + 2x^2 - 5x - 6$

$$\begin{array}{r|rrrr} 2 & 1 & 2 & -5 & -6 \\ & & 2 & 8 & 6 \\ \hline & 1 & 4 & 3 & 0 \end{array}$$

The zero remainder shows that 2 is a zero.

The zero remainder tells us that 2 is a zero of the polynomial function $f(x) = x^3 + 2x^2 - 5x - 6$. Equivalently, 2 is a solution, or root, of the polynomial equation $x^3 + 2x^2 - 5x - 6 = 0$. Thus, $x - 2$ is a factor of the polynomial.

$$x^3 + 2x^2 - 5x - 6 = 0$$
Finding the zeros of $f(x) = x^3 + 2x^2 - 5x - 6$ is the same as finding the roots of this equation.

$$(x - 2)(x^2 + 4x + 3) = 0$$
Factor using the result from the synthetic division.

$$(x - 2)(x + 3)(x + 1) = 0$$
Factor completely.

$$x - 2 = 0 \quad \text{or} \quad x + 3 = 0 \quad \text{or} \quad x + 1 = 0$$
Set each factor equal to zero.

$$x = 2 \qquad\qquad x = -3 \qquad\qquad x = -1$$
Solve for x.

The solution set is $\{-3, -1, 2\}$. The rational zeros of f are $-3, -1$, and 2.

Check Point 3   Find all rational zeros of

$$f(x) = x^3 + 8x^2 + 11x - 20.$$

Our work in Example 3 involved solving a third-degree equation. We found one factor by synthetic division and then factored the remaining quadratic factor. If the degree of a polynomial function or equation is 4 or higher, it is often necessary to find more than one linear factor by synthetic division.

One way to speed up the process of finding the first zero is to graph the function. Any *x*-intercept is a zero.

3 Solve polynomial equations.

EXAMPLE 4 Solving a Polynomial Equation

Solve: $x^4 - 6x^2 - 8x + 24 = 0$.

Solution   Recall that we refer to the *zeros* of a polynomial function and the *roots* of a polynomial equation. Because we are given an equation, we will use the word "roots," rather than "zeros," in the solution process. We begin by listing all possible rational roots.

Possible rational roots

$$= \frac{\text{Factors of the constant term, 24}}{\text{Factors of the leading coefficient, 1}}$$

$$= \frac{\pm 1, \pm 2, \pm 3, \pm 4, \pm 6, \pm 8, \pm 12, \pm 24}{\pm 1} = \pm 1, \pm 2, \pm 3, \pm 4, \pm 6, \pm 8, \pm 12, \pm 24$$

The graph of $f(x) = x^4 - 6x^2 - 8x + 24$ is shown in Figure 2.22. Because the *x*-intercept is 2, we will test 2 by synthetic division and show that it is a root of the given equation.

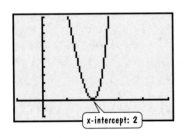

$$
\begin{array}{r|rrrrr}
2 & 1 & 0 & -6 & -8 & 24 \\
  &   & 2 & 4 & -4 & -24 \\
\hline
  & 1 & 2 & -2 & -12 & 0
\end{array}
$$

Careful!
$x^4 - 6x^2 - 8x + 24 = x^4 + 0x^3 - 6x^2 - 8x + 24$

The zero remainder indicates that 2 is a root of $x^4 - 6x^2 - 8x + 24 = 0$.

Figure 2.22   The graph of $f(x) = x^4 - 6x^2 - 8x + 24$ in a $[-1, 5, 1]$ by $[-2, 10, 1]$ viewing rectangle

Now we can rewrite the given equation in factored form.

$$x^4 - 6x^2 - 8x + 24 = 0 \qquad \text{This is the given equation.}$$

$$(x - 2)(x^3 + 2x^2 - 2x - 12) = 0 \qquad \text{This is the result obtained from the synthetic division.}$$

$$x - 2 = 0 \quad \text{or} \quad x^3 + 2x^2 - 2x - 12 = 0 \qquad \text{Set each factor equal to 0.}$$

We can use the same approach to look for rational roots of the polynomial equation $x^3 + 2x^2 - 2x - 12 = 0$, listing all possible rational roots. However, take a second look at the graph in Figure 2.22. Because the graph turns around at 2, this means that 2 is a root of even multiplicity. Thus, 2 must also be a root of $x^3 + 2x^2 - 2x - 12 = 0$, confirmed by the following synthetic division.

$$
\begin{array}{r|rrrr}
2 & 1 & 2 & -2 & -12 \\
  &   & 2 & 8 & 12 \\
\hline
  & 1 & 4 & 6 & 0
\end{array}
$$

These are the coefficients of $x^3 + 2x^2 - 2x - 12 = 0$.

The zero remainder indicates that 2 is a root of $x^3 + 2x^2 - 2x - 12 = 0$.

Now we can solve the original equation as follows:

$$x^4 - 6x^2 - 8x + 24 = 0 \qquad \text{This is the given equation.}$$

$$(x - 2)(x^3 + 2x^2 - 2x - 12) = 0 \qquad \text{This factorization is obtained from the first synthetic division.}$$

$$(x - 2)(x - 2)(x^2 + 4x + 6) = 0 \qquad \text{This factorization is obtained from the second synthetic division.}$$

$$x - 2 = 0 \quad \text{or} \quad x - 2 = 0 \quad \text{or} \quad x^2 + 4x + 6 = 0 \qquad \text{Set each factor equal to 0.}$$

$$x = 2 \qquad\qquad\quad x = 2 \qquad\qquad x^2 + 4x + 6 = 0 \qquad \text{Solve.}$$

We can use the quadratic formula to solve $x^2 + 4x + 6 = 0$.

$$x = \frac{-b \pm \sqrt{b^2 - 4ac}}{2a}$$
    We use the quadratic formula because $x^2 + 4x + 6$ cannot be factored.

$$= \frac{-4 \pm \sqrt{4^2 - 4(1)(6)}}{2(1)}$$
    Let $a = 1$, $b = 4$, and $c = 6$.

$$= \frac{-4 \pm \sqrt{-8}}{2}$$
    Multiply and subtract under the radical.

$$= \frac{-4 \pm 2i\sqrt{2}}{2}$$
    $\sqrt{-8} = \sqrt{4(2)(-1)} = 2i\sqrt{2}$

$$= -2 \pm i\sqrt{2}$$
    Simplify.

The solution set of the original equation, $x^4 - 6x^2 - 8x + 24 = 0$, is $\{2, -2 - i\sqrt{2}, -2 + i\sqrt{2}\}$. The graph in Figure 2.22 illustrates that a graphing utility does not reveal the two imaginary roots.

In Example 4, 2 is a repeated root of the equation with multiplicity 2. The example illustrates two general properties:

Properties of Polynomial Equations

1. If a polynomial equation is of degree n, then counting multiple roots separately, the equation has n roots.
2. If $a + bi$ is a root of a polynomial equation ($b \neq 0$), then the complex imaginary number $a - bi$ is also a root. Complex imaginary roots, if they exist, occur in conjugate pairs.

Check Point 4   Solve: $x^4 - 6x^3 + 22x^2 - 30x + 13 = 0$.

The Fundamental Theorem of Algebra

The fact that a polynomial equation of degree n has n roots is a consequence of a theorem proved in 1799 by a 22-year-old student named Carl Friedrich Gauss in his doctoral dissertation. His result is called the **Fundamental Theorem of Algebra.**

The Fundamental Theorem of Algebra
If $f(x)$ is a polynomial of degree n, where $n \geq 1$, then the equation $f(x) = 0$ has at least one complex root.

Suppose, for example, that $f(x) = 0$ represents a polynomial equation of degree n. By the Fundamental Theorem of Algebra, we know that this equation has at least one complex root; we'll call it c_1. By the Factor Theorem, we know that $x - c_1$ is a factor of $f(x)$. Therefore, we obtain

$$(x - c_1)q_1(x) = 0$$ The degree of the polynomial $q_1(x)$ is $n - 1$.
$$x - c_1 = 0 \quad \text{or} \quad q_1(x) = 0.$$ Set each factor equal to 0.

If the degree of $q_1(x)$ is at least 1, by the Fundamental Theorem of Algebra the equation $q_1(x) = 0$ has at least one complex root. We'll call it c_2. The Factor Theorem gives us

$$q_1(x) = 0 \qquad \text{The degree of } q_1(x) \text{ is } n - 1.$$
$$(x - c_2)q_2(x) = 0 \qquad \text{The degree of } q_2(x) \text{ is } n - 2.$$
$$x - c_2 = 0 \quad \text{or} \quad q_2(x) = 0. \qquad \text{Set each factor equal to } 0.$$

Let's see what we have up to this point, and then continue the process.

$$f(x) = 0 \qquad \text{This is the original polynomial equation of degree } n.$$
$$(x - c_1)q_1(x) = 0 \qquad \text{This is the result from our first application of the Fundamental Theorem.}$$
$$(x - c_1)(x - c_2)q_2(x) = 0 \qquad \text{This is the result from our second application of the Fundamental Theorem.}$$

By continuing this process, we will obtain the product of n linear factors. Setting each of these linear factors equal to zero results in n complex roots. Thus, if $f(x)$ is a polynomial of degree n, where $n \geq 1$, then $f(x) = 0$ has exactly n roots, where roots are counted according to their multiplicity.

4 Use the Linear Factorization Theorem to factor a polynomial.

The Linear Factorization Theorem

In Example 4, we found that $x^4 - 6x^2 - 8x + 24 = 0$ has $\{2, -2 \pm i\sqrt{2}\}$ as a solution set, where 2 is a repeated root with multiplicity 2. The polynomial can be factored over the complex nonreal numbers as follows:

$$f(x) = x^4 - 6x^2 - 8x + 24$$

These are the four zeros.

$$= [x - (-2 + i\sqrt{2})][x - (-2 - i\sqrt{2})](x - 2)(x - 2).$$

These are the four linear factors.

This fourth-degree polynomial has four linear factors. Just as an nth-degree polynomial equation has n roots, an nth-degree polynomial has n linear factors. This is formally stated as the **Linear Factorization Theorem.**

The Linear Factorization Theorem

If $f(x) = a_n x^n + a_{n-1} x^{n-1} + \cdots + a_1 x + a_0$, where $n \geq 1$ and $a_n \neq 0$, then

$$f(x) = a_n(x - c_1)(x - c_2) \cdots (x - c_n),$$

where $c_1, c_2, \ldots, c_n$ are complex numbers (possibly real and not necessarily distinct). In words: An nth-degree polynomial can be expressed as the product of a nonzero constant and n linear factors.

The Linear Factorization Theorem involves factors somewhat different than those you are used to seeing. For example, the polynomial $x^2 - 3$ is irreducible over the rational numbers. However, it can be factored over the real numbers as follows:

$$x^2 - 3 = (x + \sqrt{3})(x - \sqrt{3}). \qquad \text{Use } a^2 - b^2 = (a + b)(a - b) \text{ with } a = x \text{ and } b = \sqrt{3}.$$

The polynomial $x^2 + 1$ is irreducible over the real numbers, but reducible over the complex imaginary numbers:

$$x^2 + 1 = (x + i)(x - i).$$

Study Tip

The sum of squares, irreducible over the real numbers, can be factored over the imaginary numbers as

$$a^2 + b^2 = (a + bi)(a - bi).$$

EXAMPLE 5   Factoring a Polynomial

Factor $x^4 - 3x^2 - 28$:

   a. As the product of factors that are irreducible over the rational numbers.
   b. As the product of factors that are irreducible over the real numbers.
   c. In completely factored form involving complex imaginary number.

Solution

   a. $x^4 - 3x^2 - 28 = (x^2 - 7)(x^2 + 4)$ *Both quadratic factors are irreducible over the rational numbers.*

   b. $= (x + \sqrt{7})(x - \sqrt{7})(x^2 + 4)$ *The third factor is still irreducible over the real numbers.*

   c. $= (x + \sqrt{7})(x - \sqrt{7})(x + 2i)(x - 2i)$ *This is the completely factored form using complex imaginary numbers.*

> **Check Point 5**   Factor $x^4 - 4x^2 - 5$ as the product of factors that are irreducible over **a.** the rational numbers; **b.** the real numbers; **c.** the complex imaginary numbers.

5   Find polynomials with given zeros.

Reversing Things: Finding Polynomials when the Zeros Are Given

Many of our problems involving polynomial functions and polynomial equations dealt with the process of finding zeros and roots. The Linear Factorization Theorem enables us to reverse this process, finding a polynomial function when the zeros are given.

EXAMPLE 6   Finding a Polynomial Function with Given Zeros

Find a fourth-degree polynomial function $f(x)$ with real coefficients that has $-2, 2,$ and i as zeros and such that $f(3) = -150$.

Solution   Because i is a zero and the polynomial has real coefficients, the conjugate, $-i$, must also be a zero. We can now use the Linear Factorization Theorem.

$f(x) = a_n(x - c_1)(x - c_2)(x - c_3)(x - c_4)$ *This is the linear factorization for a fourth-degree polynomial.*

$ = a_n(x + 2)(x - 2)(x - i)(x + i)$ *Use the given zeros: $c_1 = -2, c_2 = 2, c_3 = i$, and, including the conjugate $c_4 = -i$.*

$ = a_n(x^2 - 4)(x^2 + 1)$ *Multiply.*

$f(x) = a_n(x^4 - 3x^2 - 4)$ *Complete the multiplication.*

$f(3) = a_n(3^4 - 3 \cdot 3^2 - 4) = -150$ *To find a_n, use the fact that $f(3) = -150$.*

Technology

The graph of

$$f(x) = -3x^4 + 9x^2 + 12,$$

shown in a $[-3, 3, 1]$ by $[-200, 20, 20]$ viewing rectangle, verifies that -2 and 2 are real zeros. By tracing along the curve, we can check that $f(3) = -150$.

-2 is a zero.   **2 is a zero.**

$$a_n(81 - 27 - 4) = -150 \qquad \text{Solve for } a_n.$$

$$50a_n = -150 \qquad \text{Simplify: } 81 - 27 - 4 = 50.$$

$$a_n = -3 \qquad \text{Divide both sides by 50.}$$

Substituting -3 for a_n in the formula for $f(x)$, namely, $f(x) = a_n(x^4 - 3x^2 - 4)$, we obtain

$$f(x) = -3(x^4 - 3x^2 - 4).$$

Equivalently,

$$f(x) = -3x^4 + 9x^2 + 12.$$

Check Point 6 Find a third-degree polynomial function $f(x)$ with real coefficients that has -3 and i as zeros and such that $f(1) = 8$.

6 Use Descartes's Rule of Signs.

Descartes's Rule of Signs

Because an nth-degree polynomial equation might have roots that are imaginary numbers, we should note that such an equation can have *at most n* real roots. **Descartes's Rule of Signs** provides even more specific information about the number of real zeros that a polynomial can have. The rule is based on considering *variations in sign* between consecutive coefficients. For example, the function

$$f(x) = 3x^7 - 2x^5 - x^4 + 7x^2 + x - 3$$

has three sign changes.

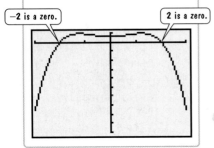

An equation can have as many true [positive] roots as it contains changes of sign, from plus to minus or from minus to plus. ... René Descartes (1596–1650) in *La Géométrie* (1637)

Descartes's Rule of Signs

Let $f(x) = a_nx^n + a_{n-1}x^{n-1} + \cdots + a_2x^2 + a_1x + a_0$ be polynomial with real coefficients.

1. The number of *positive real zeros* of f is either equal to the number of sign changes of $f(x)$ or is less than that number by an even integer. If there is only one variation in sign, there is exactly one positive real zero.

2. The number of *negative real zeros* of f is either equal to the number of sign changes of $f(-x)$ or is less than that number by an even integer. If $f(-x)$ has only one variation in sign, then f has exactly one negative real zero.

EXAMPLE 7   Using Descartes's Rule of Signs

Determine the possible number of positive and negative real zeros of $f(x) = x^3 + 2x^2 + 5x + 4$.

Solution

1. To find possibilities for positive real zeros, count the number of sign changes in the equation for $f(x)$. Because all the terms are positive, there are no variations in sign. Thus, there are no positive real zeros.

2. To find possibilities for negative real zeros, count the number of sign changes in the equation for $f(-x)$. We obtain this equation by replacing x with $-x$ in the given function.

$$f(x) = x^3 + 2x^2 + 5x + 4 \qquad \text{This is the given polynomial function.}$$

Replace x with −x.

$$f(-x) = (-x)^3 + 2(-x)^2 + 5(-x) + 4$$
$$= -x^3 + 2x^2 - 5x + 4$$

Now count the sign changes.

$$f(-x) = -x^3 + 2x^2 - 5x + 4$$

$$\underbrace{}_{1} \underbrace{}_{2} \underbrace{}_{3}$$

There are three variations in sign. The number of negative real zeros of f is either equal to the number of sign changes, 3, or is less than this number by an even integer. This means that there are either 3 negative real zeros or $3 - 2 = 1$ negative real zero.

What do the results of Example 7 mean in terms of solving

$$x^3 + 2x^2 + 5x + 4 = 0?$$

Without using Descartes's Rule of Signs, we list possible rational roots as follows:

Possible rational roots

$$= \frac{\text{Factors of the constant term, 4}}{\text{Factors of the leading coefficient, 1}} = \frac{\pm 1, \pm 2, \pm 4}{\pm 1} = \pm 1, \pm 2, \pm 4.$$

However, Descartes's Rule of Signs informed us that $f(x) = x^3 + 2x^2 + 5x + 4$ has no positive real zeros. Thus, the polynomial equation $x^3 + 2x^2 + 5x + 4 = 0$ has no positive real roots. This means that we can eliminate the positive numbers from our list of possible rational roots. Possible rational roots include only $-1, -2$, and -4. We can use synthetic division to test the three possible rational roots. Our test on two of the three rational roots is shown below.

-1	1	2	5	4
		-1	-1	-4
	1	1	4	0

-2	1	2	5	4
		-2	0	-10
	1	0	5	-6

The zero remainder shows that −1 is a root.

The nonzero remainder shows that −2 is not a root.

By solving the equation $x^3 + 2x^2 + 5x + 4 = 0$, you will find that this equation of degree 3 has three roots. One root is -1, and the other two roots are imaginary numbers in a conjugate pair. Verify this by completing the solution process.

Check Point 7 Determine the possible number of positive and negative real zeros of $f(x) = x^4 - 14x^3 + 71x^2 - 154x + 120$.

EXERCISE SET 2.5

Practice Exercises

In Exercises 1–8, use the Rational Zero Theorem to list all possible rational zeros for each given function.

1. $f(x) = x^3 + x^2 - 4x - 4$

2. $f(x) = x^3 + 3x^2 - 6x - 8$

3. $f(x) = 3x^4 - 11x^3 - x^2 + 19x + 6$

4. $f(x) = 2x^4 + 3x^3 - 11x^2 - 9x + 15$

5. $f(x) = 4x^4 - x^3 + 5x^2 - 2x - 6$

6. $f(x) = 3x^4 - 11x^3 - 3x^2 - 6x + 8$

7. $f(x) = x^5 - x^4 - 7x^3 + 7x^2 - 12x - 12$

8. $f(x) = 4x^5 - 8x^4 - x + 2$

In Exercises 9–14,

   a. List all possible rational zeros.

   b. Use synthetic division to test the possible rational zeros and find an actual zero.

   c. Use the zero from part (b) to find all the zeros of the polynomial function.

9. $f(x) = x^3 + x^2 - 4x - 4$

10. $f(x) = x^3 - 2x^2 - 11x + 12$

11. $f(x) = 2x^3 - 3x^2 - 11x + 6$

12. $f(x) = 2x^3 - 5x^2 + x + 2$

13. $f(x) = 3x^3 + 7x^2 - 22x - 8$

14. $f(x) = 3x^3 + 8x^2 - 15x + 4$

In Exercises 15–22,

   a. List all possible rational roots.

   b. Use synthetic division to test the possible rational roots and find an actual root.

   c. Use the root from part (b) and solve the equation.

15. $x^3 - 2x^2 - 11x + 12 = 0$

16. $x^3 - 2x^2 - 7x - 4 = 0$

17. $x^3 - 10x - 12 = 0$

18. $x^3 - 5x^2 + 17x - 13 = 0$

19. $6x^3 + 25x^2 - 24x + 5 = 0$

20. $2x^3 - 5x^2 - 6x + 4 = 0$

21. $x^4 - 2x^3 - 5x^2 + 8x + 4 = 0$

22. $x^4 - 2x^2 - 16x - 15 = 0$

In Exercises 23–28, factor each polynomial:

   a. *as the product of factors that are irreducible over the rational numbers.*

   b. *as the product of factors that are irreducible over the real numbers.*

   c. *in completely factored form involving complex nonreal, or imaginary, numbers.*

23. $x^4 - x^2 - 20$ **24.** $x^4 + 6x^2 - 27$

25. $x^4 + x^2 - 6$ **26.** $x^4 - 9x^2 - 22$

27. $x^4 - 2x^3 + x^2 - 8x - 12$

   (*Hint:* One factor is $x^2 + 4$.)

28. $x^4 - 4x^3 + 14x^2 - 36x + 45$

   (*Hint:* One factor is $x^2 + 9$.)

In Exercises 29–36, find an nth-degree polynomial function with real coefficients satisfying the given conditions. If you are using a graphing utility, use it to graph the function and verify the real zeros and the given function value.

29. $n = 3$; 1 and $5i$ are zeros; $f(-1) = -104$

30. $n = 3$; 4 and $2i$ are zeros; $f(-1) = -50$

31. $n = 3$; -5 and $4 + 3i$ are zeros; $f(2) = 91$

32. $n = 3$; 6 and $-5 + 2i$ are zeros; $f(2) = -636$

33. $n = 4$; i and $3i$ are zeros; $f(-1) = 20$

34. $n = 4$; $-2, -\frac{1}{2}$, and i are zeros; $f(1) = 18$

35. $n = 4$; $-2, 5$, and $3 + 2i$ are zeros; $f(1) = -96$

36. $n = 4$; $-4, \frac{1}{3}$, and $2 + 3i$ are zeros; $f(1) = 100$

In Exercises 37–42, use Descartes's Rule of Signs to determine the possible number of positive and negative real zeros for each given function.

37. $f(x) = x^3 + 2x^2 + 5x + 4$

38. $f(x) = x^3 + 7x^2 + x + 7$

39. $f(x) = 5x^3 - 3x^2 + 3x - 1$

40. $f(x) = -2x^3 + x^2 - x + 7$

41. $f(x) = 2x^4 - 5x^3 - x^2 - 6x + 4$

42. $f(x) = 4x^4 - x^3 + 5x^2 - 2x - 6$

In Exercises 43–56, find all zeros of the polynomial function or solve the given polynomial equation. Use the Rational Zero Theorem, Descartes's Rule of Signs, and possibly the graph of the polynomial function shown by a graphing utility as an aid in obtaining the first zero or the first root.

43. $f(x) = x^3 - 4x^2 - 7x + 10$

44. $f(x) = x^3 + 12x^2 + 21x + 10$

45. $2x^3 - x^2 - 9x - 4 = 0$

46. $3x^3 - 8x^2 - 8x + 8 = 0$

47. $f(x) = x^4 - 2x^3 + x^2 + 12x + 8$

48. $f(x) = x^4 - 4x^3 - x^2 + 14x + 10$

49. $x^4 - 3x^3 - 20x^2 - 24x - 8 = 0$

50. $x^4 - x^3 + 2x^2 - 4x - 8 = 0$

51. $f(x) = 3x^4 - 11x^3 - x^2 + 19x + 6$

52. $f(x) = 2x^4 + 3x^3 - 11x^2 - 9x + 15$

53. $4x^4 - x^3 + 5x^2 - 2x - 6 = 0$

54. $3x^4 - 11x^3 - 3x^2 - 6x + 8 = 0$

55. $2x^5 + 7x^4 - 18x^2 - 8x + 8 = 0$

56. $4x^5 + 12x^4 - 41x^3 - 99x^2 + 10x + 24 = 0$

Application Exercises

The graphs are based on a study of the percentage of professional works completed in each age decade of life by 738 people who lived to be at least 79. Use the graphs to solve Exercises 57–58.

Age Trends in Professional Productivity

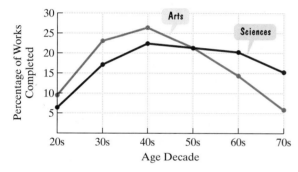

Source: Dennis, W. (1966), Creative productivity between the ages of 20 and 80 years. *Journal of Gerontology*, 21, 1–8.

57. Suppose that a polynomial function f is used to model the data shown in the graph for the arts using

  (age decade, percentage of works completed).

  a. Use the graph to solve the polynomial equation $f(x) = 27$. Describe what this means in terms of an age decade and productivity.

  b. Describe the degree and the leading coefficient of a function f that can be used to model the data in the graph.

58. Suppose that a polynomial function g is used to model the data shown in the graph for the sciences using

  (age decade, percentage of works completed).

  a. Use the graph to solve the polynomial equation $g(x) = 20$. Find only the meaningful value of x and then describe what this means in terms of an age decade and productivity.

  b. Describe the degree and the leading coefficient of a function g that can be used to model the data in the graph.

59. The number of eggs, N, in a female moth is a function of her abdominal width, W, in millimeters, modeled by $N = 14W^3 - 17W^2 - 16W + 34$, for $1.5 \le W \le 3.5$. What is the abdominal width when there are 211 eggs?

60. The concentration of a drug, $f(x)$, in parts per million, in a patient's blood x hours after the drug is administered is given by the function

$$f(x) = -x^4 + 12x^3 - 58x^2 + 132x.$$

How many hours after the drug is administered will it be eliminated from the bloodstream?

61. The width of a rectangular box is twice the height and the length is 7 inches more than the height. If the volume is 72 cubic inches, find the dimensions of the box.

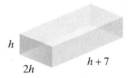

62. A box with an open top is formed by cutting squares out of the corners of a rectangular piece of cardboard 10 inches by 8 inches and then folding up the sides. If x represents the length of the side of the square cut from each corner of the rectangle, what size square must be cut if the volume of the box is to be 48 cubic inches?

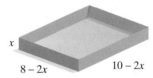

We have seen that the polynomial function

$$H(x) = -0.001618x^4 + 0.077326x^3 - 1.2367x^2 + 11.460x + 2.914$$

models the age in human years, $H(x)$, of a dog that is x years old, where $x \ge 1$. Although the coefficients make it difficult to solve equations algebraically using this function, a graph of the function makes approximate solutions possible.

Use the graph shown to solve Exercises 63–64. Round all answers to the nearest year.

Dog's Age in Human Years

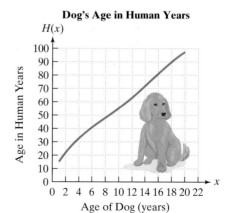

Source: U.C. Davis

(Refer to the graph on the previous page in solving Exercises 63–64.)

63. If you are 25, what is the equivalent age for dogs?

64. If you are 90, what is the equivalent age for dogs?

65. Use the function H given on the previous page and set up an equation to answer the question in either Exercise 63 or 64. Bring all terms to one side and obtain zero on the other side. What are some of the difficulties involved in solving this equation? Explain how the Intermediate Value Theorem can be used to verify the approximate solution that you obtained from the graph.

Writing in Mathematics

66. Describe how to find the possible rational zeros of a polynomial function.

67. How does the linear factorization of $f(x)$, that is,

$$f(x) = a_n(x - c_1)(x - c_2) \cdots (x - c_n),$$

show that a polynomial equation of degree n has n roots?

68. Describe how to use Descartes's Rule of Signs to determine the possible number of positive real zeros of a polynomial function.

69. Describe how to use Descartes's Rule of Signs to determine the possible number of negative roots of a polynomial equation.

70. Why must every polynomial equation of degree 3 have at least one real root?

71. Explain why the equation $x^4 + 6x^2 + 2 = 0$ has no rational roots.

72. Suppose $\frac{3}{4}$ is a root of a polynomial equation. What does this tell us about the leading coefficient and the constant term in the equation?

73. Use the graphs for Exercises 57–58 to describe one similarity and one difference between age trends in professional productivity in the arts and the sciences.

Technology Exercises

The equations in Exercises 74–77 have real roots that are rational. Use the Rational Zero Theorem to list all possible rational roots. Then graph the polynomial function in the given viewing rectangle to determine which possible rational roots are actual roots of the equation.

74. $2x^3 - 15x^2 + 22x + 15 = 0; [-1, 6, 1]$ by $[-50, 50, 10]$

75. $6x^3 - 19x^2 + 16x - 4 = 0; [0, 2, 1]$ by $[-3, 2, 1]$

76. $2x^4 + 7x^3 - 4x^2 - 27x - 18 = 0; [-4, 3, 1]$ by $[-45, 45, 15]$

77. $4x^4 + 4x^3 + 7x^2 - x - 2 = 0; [-2, 2, 1]$ by $[-5, 5, 1]$

78. Use Descartes's Rule of Signs to determine the possible number of positive and negative real zeros of $f(x) = 3x^4 + 5x^2 + 2$. What does this mean in terms of the graph of f? Verify your result by using a graphing utility to graph f.

79. Use Descartes's Rule of Signs to determine the possible number of positive and negative real zeros of $f(x) = x^5 - x^4 + x^3 - x^2 + x - 8$. Verify your result by using a graphing utility to graph f.

80. Determine a number of polynomial functions of odd degree and graph each function. Is it possible for the graph to have no real zeros? Explain. Try doing the same thing for polynomial functions of even degree. Now is it possible to have no real zeros?

Use a graphing utility to obtain a complete graph for each polynomial function in Exercises 81–84. Then determine the number of real zeros and the number of nonreal complex zeros for each function.

81. $f(x) = x^3 - 6x - 9$

82. $f(x) = 3x^5 - 2x^4 + 6x^3 - 4x^2 - 24x + 16$

83. $f(x) = 3x^4 + 4x^3 - 7x^2 - 2x - 3$

84. $f(x) = x^6 - 64$

Critical Thinking Exercises

85. Which one of the following is true?
   a. The equation $x^3 + 5x^2 + 6x + 1 = 0$ has one positive real root.
   b. Descartes's Rule of Signs gives the exact number of positive and negative real roots for a polynomial equation.
   c. Every polynomial equation of degree 3 has at least one rational root.
   d. None of the above is true.

86. Give an example of a polynomial equation that has no real roots. Describe how you obtained the equation.

87. If the volume of the solid shown in the figure is 208 cubic inches, find the value of x.

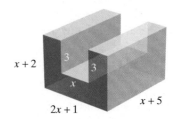

88. In this exercise, we lead you through the steps involved in the proof of the Rational Zero Theorem. Consider the polynomial equation

$$a_n x^n + a_{n-1} x^{n-1} + a_{n-2} x^{n-2} + \cdots + a_1 x + a_0 = 0$$

where $\dfrac{p}{q}$ is a rational root reduced to lowest terms.

a. Substitute $\dfrac{p}{q}$ for x in the equation and show that the equation can be written as

$$a_n p^n + a_{n-1} p^{n-1} q + a_{n-2} p^{n-2} q^2 + \cdots + a_1 p q^{n-1} = -a_0 q^n.$$

b. Why is p a factor of the left side of the equation?

c. Because p divides the left side, it must also divide the right side. However, because $\dfrac{p}{q}$ is reduced to lowest terms, p cannot divide q. Thus, p and q have no common factors other than -1 and 1. Because p does divide the right side and it is not a factor of q^n, what can you conclude?

d. Rewrite the equation from part (a) with all terms containing q on the left and the term that does not have a factor of q on the right. Use an argument that parallels parts (b) and (c) to conclude that q is a factor of a_n.

In Exercises 89–92, what is the smallest degree that each polynomial could have?

89.

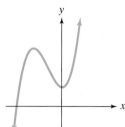

90.

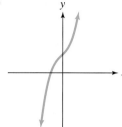

91.

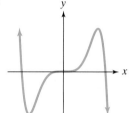

92.

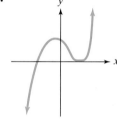

93. Explain why nonreal complex zeros are gained or lost in pairs in terms of graphs of polynomial functions.

94. Explain why a polynomial function of degree 20 cannot cross the x-axis exactly once.

SECTION 2.6   *Rational Functions and Their Graphs*

Objectives

1. Find the domain of rational functions.
2. Use arrow notation.
3. Identify vertical asymptotes.
4. Identify horizontal asymptotes.
5. Graph rational functions.
6. Identify slant asymptotes.
7. Solve applied problems involving rational functions.

Technology is now promising to bring light, fast, and beautiful wheelchairs to millions of disabled people. The cost of manufacturing these radically different wheelchairs can be modeled by rational functions. In this section, we will see how graphs of these functions illustrate that low prices are possible with high production levels, urgently needed in this situation. There are more than half a billion people with disabilities in developing countries; an estimated 20 million need wheelchairs right now.

1 Find the domain of rational functions.

Rational Functions

Rational functions are quotients of polynomial functions. This means that rational functions can be expressed as

$$f(x) = \frac{p(x)}{q(x)}$$

where p and q are polynomial functions and $q(x) \neq 0$. The **domain** of a rational function is the set of all real numbers except the x-values that make the denominator zero. For example, the domain of the rational function

$$f(x) = \frac{x^2 + 7x + 9}{x(x - 2)(x + 5)}$$

This is $p(x)$.

This is $q(x)$.

is the set of all real numbers except $0, 2$, and -5.

EXAMPLE 1 Finding the Domain of a Rational Function

Find the domain of each rational function:

a. $f(x) = \dfrac{x^2 - 9}{x - 3}$ **b.** $g(x) = \dfrac{x}{x^2 - 9}$ **c.** $h(x) = \dfrac{x + 3}{x^2 + 9}$.

Solution   Rational functions contain division. Because division by 0 is undefined, we must exclude from the domain of each function values of x that cause the polynomial function in the denominator to be 0.

a. The denominator of $f(x) = \dfrac{x^2 - 9}{x - 3}$ is 0 if $x = 3$. Thus, x cannot equal 3. The domain of f consists of all real numbers except 3. We can express the domain in set-builder or interval notation:

$$\text{Domain of } f = \{x | x \neq 3\}$$
$$\text{Domain of } f = (-\infty, 3) \text{ or } (3, \infty).$$

b. The denominator of $g(x) = \dfrac{x}{x^2 - 9}$ is 0 if $x = -3$ or $x = 3$. Thus, the domain of g consists of all real numbers except -3 and 3. We can express the domain in set-builder or interval notation:

$$\text{Domain of } g = \{x | x \neq -3, x \neq 3\}$$
$$\text{Domain of } g = (-\infty, -3) \text{ or } (-3, 3) \text{ or } (3, \infty).$$

c. No real numbers cause the denominator of $h(x) = \dfrac{x + 3}{x^2 + 9}$ to equal 0. The domain of h consists of all real numbers.

$$\text{Domain of } h = (-\infty, \infty)$$

Study Tip

Because the domain of a rational function is the set of all real numbers except those for which the denominator is 0, you can identify such numbers by setting the denominator equal to 0 and solving for x. Exclude the resulting real values of x from the domain.

Check Point 1   Find the domain of each rational function:

a. $f(x) = \dfrac{x^2 - 25}{x - 5}$ **b.** $g(x) = \dfrac{x}{x^2 - 25}$ **c.** $h(x) = \dfrac{x + 5}{x^2 + 25}$.

(Ask your professor if a particular notation is preferred.)

2 Use arrow notation.

The most basic rational function is the **reciprocal function,** defined by $f(x) = \dfrac{1}{x}$. The denominator of the reciprocal function is zero when $x = 0$, so the domain of f is the set of all real numbers except 0.

Let's look at the behavior of f near the excluded value 0. We start by evaluating $f(x)$ to the left of 0.

x approaches 0 from the left.					
x	-1	-0.5	-0.1	-0.01	-0.001
$f(x) = \dfrac{1}{x}$	-1	-2	-10	-100	-1000

Mathematically, we say that "x approaches 0 from the left." From the table and the accompanying graph, it appears that as x approaches 0 from the left, the function values, $f(x)$, decrease without bound. We say that "$f(x)$ approaches negative infinity." We use a special arrow notation to describe this situation symbolically:

$$\text{As } x \to 0^{-}, f(x) \to -\infty.$$

> As x approaches 0 from the left, f(x) approaches negative infinity (that is, the graph falls).

Observe that the minus $(-)$ superscript on the 0 $(x \to 0^{-})$ is read "from the left."

Next, we evaluate $f(x)$ to the right of 0.

x approaches 0 from the right.					
x	0.001	0.01	0.1	0.5	1
$f(x) = \dfrac{1}{x}$	1000	100	10	2	1

Mathematically, we say that "x approaches 0 from the right." From the table and the accompanying graph, it appears that as x approaches 0 from the right, the function values, $f(x)$, increase without bound. We say that "$f(x)$ approaches infinity." We again use a special arrow notation to describe this situation symbolically:

$$\text{As } x \to 0^{+}, f(x) \to \infty.$$

> As x approaches 0 from the right, f(x) approaches infinity (that is, the graph rises).

Observe that the plus $(+)$ superscript on the 0 $(x \to 0^{+})$ is read "from the right."

Now let's see what happens to the function values, $f(x)$, as x gets farther away from the origin. The following tables suggest that as x increases or decreases without bound, the function values are getting closer to 0.

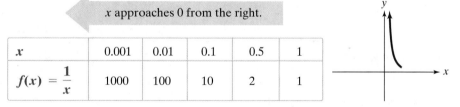

x increases without bound:				
x	1	10	100	1000
$f(x) = \dfrac{1}{x}$	1	0.1	0.01	0.001

x decreases without bound:				
x	-1	-10	-100	-1000
$f(x) = \dfrac{1}{x}$	-1	-0.1	-0.01	-0.001

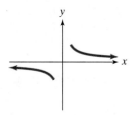

Figure 2.23 $f(x)$ approaches 0 as x increases or decreases without bound.

Figure 2.23 illustrates the end behavior of $f(x) = \frac{1}{x}$ as x increases or decreases without bound. The function values, $f(x)$, are getting progressively closer to 0. This means that as x increases or decreases without bound, the graph of f is approaching the horizontal line $y = 0$ (that is, the x-axis). We use the arrow notation to describe this situation:

$$\text{As } x \to \infty, f(x) \to 0 \qquad \text{and} \qquad \text{as } x \to -\infty, f(x) \to 0.$$

As x approaches infinity (that is, increases without bound), f(x) approaches 0.

As x approaches negative infinity (that is, decreases without bound), f(x) approaches 0.

Thus, as x approaches infinity $(x \to \infty)$ or as x approaches negative infinity $(x \to -\infty)$, the function values are approaching zero: $f(x) \to 0$.

The graph of the reciprocal function $f(x) = \frac{1}{x}$ is shown in Figure 2.24. Unlike the graph of a polynomial function, the graph of the reciprocal function has a break in it and is composed of two distinct branches.

The arrow notation used throughout our discussion of the reciprocal function is summarized in the following box:

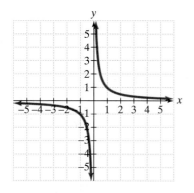

Figure 2.24 The graph of the reciprocal function $f(x) = \frac{1}{x}$

Arrow Notation

Symbol	Meaning
$x \to a^{+}$	x approaches a from the right.
$x \to a^{-}$	x approaches a from the left.
$x \to \infty$	x approaches infinity; that is, x increases without bound.
$x \to -\infty$	x approaches negative infinity; that is, x decreases without bound.

In calculus, you will use **limits** to convey ideas involving a function's end behavior or its possible asymptotic behavior. For example, examine the graph of $f(x) = \frac{1}{x}$ in Figure 2.24 and its end behavior to the right. As $x \to \infty$, the values of $f(x)$ approach 0: $f(x) \to 0$. In calculus, this is symbolized by

$$\lim_{x \to \infty} f(x) = 0. \qquad \text{This is read "the limit of f(x) as x approaches infinity equals zero."}$$

3 Identify vertical asymptotes.

Vertical Asymptotes of Rational Functions

Look again at the graph of $f(x) = \frac{1}{x}$ in Figure 2.24. The curve approaches, but does not touch, the y-axis. The y-axis, or $x = 0$, is said to be a *vertical asymptote* of the graph. A rational function may have no vertical asymptotes, one vertical asymptote, or several vertical asymptotes. The graph of a rational function never intersects a vertical asymptote. We will use dashed lines to show asymptotes.

Definition of a Vertical Asymptote

The line $x = a$ is a **vertical asymptote** of the graph of a function f if $f(x)$ increases or decreases without bound as x approaches a.

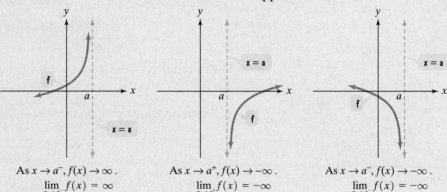

As $x \to a^+$, $f(x) \to \infty$.
$\lim\limits_{x \to a^+} f(x) = \infty$

As $x \to a^-$, $f(x) \to \infty$.
$\lim\limits_{x \to a^-} f(x) = \infty$

As $x \to a^+$, $f(x) \to -\infty$.
$\lim\limits_{x \to a^+} f(x) = -\infty$

As $x \to a^-$, $f(x) \to -\infty$.
$\lim\limits_{x \to a^-} f(x) = -\infty$

Thus, as x approaches a from either the left or the right, $f(x) \to \infty$ or $f(x) \to -\infty$.

If the graph of a rational function has vertical asymptotes, they can be located using the following theorem:

Locating Vertical Asymptotes

If $f(x) = \dfrac{p(x)}{q(x)}$ is a rational function in which $p(x)$ and $q(x)$ have no common factors and a is a zero of $q(x)$, the denominator, then $x = a$ is a vertical asymptote of the graph of f.

EXAMPLE 2 Finding the Vertical Asymptotes of a Rational Function

Find the vertical asymptotes, if any, of the graph of each rational function:

a. $f(x) = \dfrac{x}{x^2 - 9}$ **b.** $g(x) = \dfrac{x + 3}{x^2 - 9}$ **c.** $h(x) = \dfrac{x + 3}{x^2 + 9}$.

Solution   Factoring is usually helpful in identifying zeros of denominators.

a. $f(x) = \dfrac{x}{x^2 - 9} = \dfrac{x}{(x + 3)(x - 3)}$

This factor is 0 if $x = -3$.

This factor is 0 if $x = 3$.

There are no common factors in the numerator and the denominator. The zeros of the denominator are -3 and 3. Thus, the lines $x = -3$ and $x = 3$ are the vertical asymptotes for the graph of f.

b. We will use factoring to see if there are common factors.

$$g(x) = \frac{x + 3}{x^2 - 9} = \frac{(x + 3)}{(x + 3)(x - 3)} = \frac{1}{x - 3}$$

There is a common factor, x + 3, so simplify.

This denominator is 0 if x = 3.

The only zero of the denominator of $g(x)$ in simplified form is 3. Thus, the line $x = 3$ is the only vertical asymptote of the graph of g.

c. We cannot factor the denominator of $h(x)$ over the real numbers.

$$h(x) = \frac{x + 3}{x^2 + 9}$$

No real numbers make this denominator 0.

The denominator has no real zeros. Thus, the graph of h has no vertical asymptotes.

Check Point 2 Find the vertical asymptotes, if any, of the graph of each rational function:

a. $f(x) = \dfrac{x}{x^2 - 1}$ **b.** $g(x) = \dfrac{x - 1}{x^2 - 1}$ **c.** $h(x) = \dfrac{x - 1}{x^2 + 1}$.

A value where the denominator of a function is zero does not necessarily result in a vertical asymptote. There is a hole corresponding to $x = a$, and not a vertical asymptote, in the graph of a function under the following conditions: The value a causes the denominator to be zero, but there is a reduced form of the function's equation in which a does not cause the denominator to be zero.

Consider, for example, the function

$$f(x) = \frac{x^2 - 9}{x - 3}.$$

Because the denominator is zero when $x = 3$, the function's domain is all real numbers except 3. However, there is a reduced form of the equation in which 3 does not cause the denominator to be zero:

$$f(x) = \frac{x^2 - 9}{x - 3} = \frac{(x + 3)(x - 3)}{x - 3} = x + 3, \; x \neq 3$$

Denominator is zero at x = 3.

In this reduced form, 3 does not result in a zero denominator.

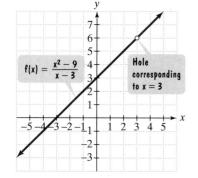

Figure 2.25 A graph with a hole corresponding to the denominator's zero

Figure 2.25 shows that the graph has a hole corresponding to $x = 3$. Graphing utilities do not generally show this feature of the graph.

4 Identify horizontal asymptotes.

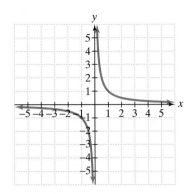

Figure 2.24 The graph of the reciprocal function $f(x) = \dfrac{1}{x}$, repeated

Horizontal Asymptotes of Rational Functions

Figure 2.24 shows the graph of the reciprocal function $f(x) = \dfrac{1}{x}$. As $x \to \infty$ and as $x \to -\infty$, the function values are approaching 0: $f(x) \to 0$. The line $y = 0$ (that is, the x-axis) is a *horizontal asymptote* of the graph. Many, but not all, rational functions have horizontal asymptotes.

Definition of a Horizontal Asymptote
The line $y = b$ is a **horizontal asymptote** of the graph of a function f if $f(x)$ approaches b as x increases or decreases without bound.

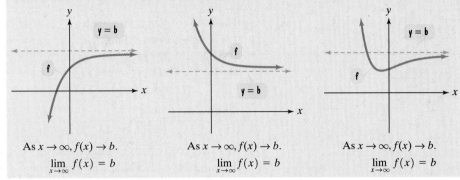

As $x \to \infty$, $f(x) \to b$.
$$\lim_{x \to \infty} f(x) = b$$

As $x \to \infty$, $f(x) \to b$.
$$\lim_{x \to \infty} f(x) = b$$

As $x \to \infty$, $f(x) \to b$.
$$\lim_{x \to \infty} f(x) = b$$

Recall that a rational function may have several vertical asymptotes. By contrast, it can have at most one horizontal asymptote. Although a graph can never intersect a vertical asymptote, it may cross its horizontal asymptote.

If the graph of a rational function has a horizontal asymptote, it can be located using the following theorem:

Locating Horizontal Asymptotes
Let f be the rational function given by
$$f(x) = \frac{a_n x^n + a_{n-1} x^{n-1} + \cdots + a_1 x + a_0}{b_m x^m + b_{m-1} x^{m-1} + \cdots + b_1 x + b_0}, \ a_n \neq 0, b_m \neq 0.$$

The degree of the numerator is n. The degree of the denominator is m.
1. If $n < m$, the x-axis, or $y = 0$, is the horizontal asymptote of the graph of f.
2. If $n = m$, the line $y = \dfrac{a_n}{b_m}$ is the horizontal asymptote of the graph of f.
3. If $n > m$, the graph of f has no horizontal asymptote.

EXAMPLE 3 Finding the Horizontal Asymptote of a Rational Function

Find the horizontal asymptote, if any, of the graph of each rational function:

a. $f(x) = \dfrac{4x}{2x^2 + 1}$ **b.** $g(x) = \dfrac{4x^2}{2x^2 + 1}$ **c.** $h(x) = \dfrac{4x^3}{2x^2 + 1}.$

Solution

a. $f(x) = \dfrac{4x}{2x^2 + 1}$

The degree of the numerator, 1, is less than the degree of the denominator, 2. Thus, the graph of f has the x-axis as a horizontal asymptote [see Figure 2.26(a)]. The equation of the horizontal asymptote is $y = 0$.

b. $g(x) = \dfrac{4x^2}{2x^2 + 1}$

The degree of the numerator, 2, is equal to the degree of the denominator, 2. The leading coefficients of the numerator and denominator, 4 and 2, are used to obtain the equation of the horizontal asymptote. The equation of the horizontal asymptote is $y = \frac{4}{2}$ or $y = 2$ [see Figure 2.26(b)].

c. $h(x) = \dfrac{4x^3}{2x^2 + 1}$

The degree of the numerator, 3, is greater than the degree of the denominator, 2. Thus, the graph of h has no horizontal asymptote [see Figure 2.26(c)].

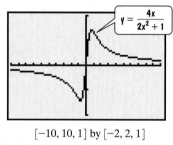

$[-10, 10, 1]$ by $[-2, 2, 1]$

(a) The horizontal asymptote of the graph is $y = 0$.

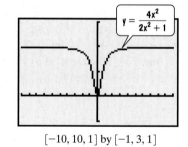

$[-10, 10, 1]$ by $[-1, 3, 1]$

(b) The horizontal asymptote of the graph is $y = 2$.

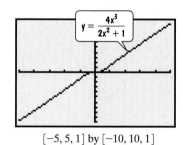

$[-5, 5, 1]$ by $[-10, 10, 1]$

(c) The graph has no horizontal asymptote.

Figure 2.26

Check Point 3

Find the horizontal asymptote, if any, of the graph of each rational function:

a. $f(x) = \dfrac{9x^2}{3x^2 + 1}$ **b.** $g(x) = \dfrac{9x}{3x^2 + 1}$ **c.** $h(x) = \dfrac{9x^3}{3x^2 + 1}$.

5 Graph rational functions.

Graphing Rational Functions

Here are some suggestions for graphing rational functions:

Strategy for Graphing a Rational Function

Suppose that

$$f(x) = \frac{p(x)}{q(x)}$$

where p and q are polynomial functions with no common factors.

1. Determine whether the graph of f has symmetry.
$$f(-x) = f(x): y\text{-axis symmetry.}$$
$$f(-x) = -f(x): \text{origin symmetry}$$
2. Find the y-intercept (if there is one) by evaluating $f(0)$.
3. Find the x-intercepts (if there are any) by solving the equation $p(x) = 0$.
4. Find any vertical asymptote(s) by solving the equation $q(x) = 0$.
5. Find the horizontal asymptote (if there is one) using the rule for determining the horizontal asymptote of a rational function.
6. Plot at least one point between and beyond each x-intercept and vertical asymptote.
7. Use the information obtained previously to graph the function between and beyond the vertical asymptotes.

EXAMPLE 4   Graphing a Rational Function

Graph: $f(x) = \dfrac{2x}{x - 1}$.

Solution

Step 1   Determine symmetry.

$$f(-x) = \frac{2(-x)}{-x - 1} = \frac{-2x}{-x - 1} = \frac{2x}{x + 1}$$

Because $f(-x)$ does not equal $f(x)$ or $-f(x)$, the graph has neither y-axis nor origin symmetry.

Step 2   Find the y-intercept.   Evaluate $f(0)$.

$$f(0) = \frac{2 \cdot 0}{0 - 1} = \frac{0}{-1} = 0$$

The y-intercept is 0, and so the graph passes through the origin.

Step 3   Find x-intercept(s).   This is done by solving $p(x) = 0$.

$$2x = 0 \qquad \text{Set the numerator equal to 0.}$$
$$x = 0$$

There is only one x-intercept. This verifies that the graph passes through the origin.

The function to be graphed,

$$f(x) = \frac{2x}{x - 1}, \text{ repeated}$$

Step 4 Find the vertical asymptote(s). Solve $q(x) = 0$, thereby finding zeros of the denominator.

$$x - 1 = 0 \qquad \textit{Set the denominator equal to 0.}$$
$$x = 1$$

The equation of the vertical asymptote is $x = 1$.

Step 5 Find the horizontal asymptote. Because the numerator and denominator have the same degree, the leading coefficients of the numerator and denominator, 2 and 1, are used to obtain the equation of the horizontal asymptote.

$$y = \frac{2}{1} = 2$$

The equation of the horizontal asymptote is $y = 2$.

Step 6 Plot points between and beyond each x-intercept and vertical asymptote. With an x-intercept at 0 and a vertical asymptote at $x = 1$, we evaluate the function at $-2, -1, \frac{1}{2}, 2,$ and 4.

x	-2	-1	$\dfrac{1}{2}$	2	4
$f(x) = \dfrac{2x}{x - 1}$	$\dfrac{4}{3}$	1	-2	4	$\dfrac{8}{3}$

Figure 2.27 shows these points, the y-intercept, the x-intercept, and the asymptotes.

Step 7 Graph the function. The graph of $f(x) = \dfrac{2x}{x - 1}$ is shown in Figure 2.28.

Technology

The graph of $y = \dfrac{2x}{x - 1}$, obtained using the $\boxed{\text{DOT}}$ mode in a $[-6, 6, 1]$ by $[-6, 6, 1]$ viewing rectangle verifies that our hand-drawn graph in Figure 2.28 is correct.

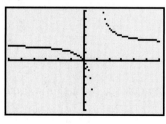

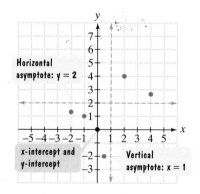

Figure 2.27 Preparing to graph the rational function $f(x) = \dfrac{2x}{x - 1}$

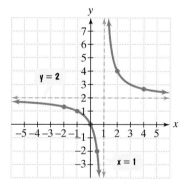

Figure 2.28 The graph of $f(x) = \dfrac{2x}{x - 1}$

Check Point 4 Graph: $f(x) = \dfrac{3x}{x - 2}$.

EXAMPLE 5   Graphing a Rational Function

Graph: $f(x) = \dfrac{3x^2}{x^2 - 4}$.

Solution

Step 1   Determine symmetry: $f(-x) = \dfrac{3(-x)^2}{(-x)^2 - 4} = \dfrac{3x^2}{x^2 - 4} = f(x)$: The graph of f is symmetric with respect to the y-axis.

Step 2   Find the y-intercept: $f(0) = \dfrac{3 \cdot 0^2}{0^2 - 4} = \dfrac{0}{-4} = 0$: The y-intercept is 0.

Step 3   Find the x-intercept: $3x^2 = 0$, so $x = 0$: The x-intercept is 0.

Step 4   Find the vertical asymptotes: Set $q(x) = 0$.

$$x^2 - 4 = 0 \qquad \text{Set the denominator equal to 0.}$$
$$x^2 = 4 \qquad \text{Add 4 to both sides.}$$
$$x = \pm 2 \qquad \text{Use the square root method.}$$

The vertical asymptotes are $x = -2$ and $x = 2$.

Step 5   Find the horizontal asymptote: The horizontal asymptote is $y = \dfrac{3}{1} = 3$.

Step 6   Plot points between and beyond the x-intercept and the vertical asymptotes. With an x-intercept at 0 and vertical asymptotes at $x = -2$ and $x = 2$, we evaluate the function at $-3, -1, 1, 3,$ and 4.

x	-3	-1	1	3	4
$f(x) = \dfrac{3x^2}{x^2 - 4}$	$\dfrac{27}{5}$	-1	-1	$\dfrac{27}{5}$	4

Figure 2.29 shows these points, the y-intercept, the x-intercept, and the asymptotes.

Step 7   Graph the function. The graph of $f(x) = \dfrac{3x^2}{x^2 - 4}$ is shown in Figure 2.30. The y-axis symmetry is now obvious.

Study Tip

Because the graph has y-axis symmetry, it is not necessary to evaluate the even function at -3 and again at 3.

$$f(-3) = f(3) = \tfrac{27}{5}$$

This also applies to evaluation at -1 and 1.

Technology

The graph of $y = \dfrac{3x^2}{x^2 - 4}$, generated by a graphing utility, verifies that our hand-drawn graph in Figure 2.30 is correct.

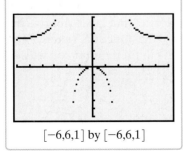

$[-6,6,1]$ by $[-6,6,1]$

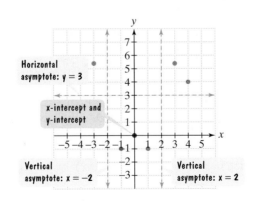

Figure 2.29   Preparing to graph $f(x) = \dfrac{3x^2}{x^2 - 4}$

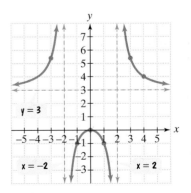

Figure 2.30   The graph of $f(x) = \dfrac{3x^2}{x^2 - 4}$

Check Point 5 Graph: $f(x) = \dfrac{2x^2}{x^2 - 9}$.

Example 6 illustrates that not every rational function has vertical and horizontal asymptotes.

EXAMPLE 6 Graphing a Rational Function

Graph: $f(x) = \dfrac{x^4}{x^2 + 1}$.

Solution

Step 1 Determine symmetry: $f(-x) = \dfrac{(-x)^4}{(-x)^2 + 1} = \dfrac{x^4}{x^2 + 1} = f(x)$: The graph of f is symmetric with respect to the y-axis.

Step 2 Find the y-intercept: $f(0) = \dfrac{0^4}{0^2 + 1} = \dfrac{0}{1} = 0$: The y-intercept is 0.

Step 3 Find the x-intercept: $x^4 = 0$, so $x = 0$: The x-intercept is 0.

Step 4 Find the vertical asymptote: Set $q(x) = 0$.

$$x^2 + 1 = 0 \qquad \text{Set the denominator equal to 0.}$$
$$x^2 = -1$$

Although this equation has imaginary roots ($x = \pm i$), there are no real roots. Thus, there is no vertical asymptote.

Step 5 Find the horizontal asymptote: Because the degree of the numerator, 4, is greater than the degree of the denominator, 2, there is no horizontal asymptote.

Step 6 Plot points between and beyond the x-intercept and the vertical asymptotes. With an x-intercept at 0 and no vertical asymptotes, let's look at function values at $-2, -1, 1,$ and 2. You can evaluate the function at 1 and 2. Use y-axis symmetry to obtain function values at -1 and -2:

$$f(-1) = f(1) \text{ and } f(-2) = f(2).$$

x	-2	-1	1	2
$f(x) = \dfrac{x^4}{x^2 + 1}$	$\dfrac{16}{5}$	$\dfrac{1}{2}$	$\dfrac{1}{2}$	$\dfrac{16}{5}$

Step 7 Graph the function. Figure 2.31 shows the graph of f using the points obtained from the table and y-axis symmetry. Notice that as x approaches infinity or negative infinity ($x \to \infty$ or $x \to -\infty$), the function values, $f(x)$, are getting larger without bound [$f(x) \to \infty$]. Using limit notation you will see in calculus, we can write $\displaystyle\lim_{x \to \infty} f(x) = \infty$ and $\displaystyle\lim_{x \to -\infty} f(x) = \infty$.

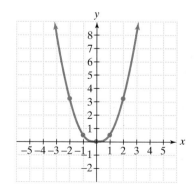

Figure 2.31 The graph of $f(x) = \dfrac{x^4}{x^2 + 1}$

Check Point 6 Graph: $f(x) = \dfrac{x^4}{x^2 + 2}$.

6 Identify slant asymptotes.

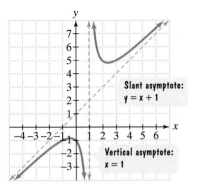

Figure 2.32 The graph of $f(x) = \dfrac{x^2 + 1}{x - 1}$ with a slant asymptote

Slant Asymptotes

Examine the graph of

$$f(x) = \frac{x^2 + 1}{x - 1}$$

shown in Figure 2.32. Note that the degree of the numerator, 2, is greater than the degree of the denominator, 1. Thus, the graph of this function has no horizontal asymptote. However, the graph has a **slant asymptote,** $y = x + 1$.

The graph of a rational function has a slant asymptote if the degree of the numerator is one more than the degree of the denominator. The equation of the slant asymptote can be found by division. For example, to find the slant asymptote for the graph of $f(x) = \dfrac{x^2 + 1}{x - 1}$, divide $x - 1$ into $x^2 + 1$:

$$
\begin{array}{r}
\underline{1|} \quad 1 \quad 0 \quad 1 \\
\quad\ \underline{1 \quad 1} \\
\quad 1 \quad 1 \quad 2
\end{array}
\qquad
\begin{array}{r}
1x + 1 + \dfrac{2}{x-1} \\[2pt]
x - 1 \overline{)\,x^2 + 0x + 1}
\end{array}
$$

> Remainder

Observe that

$$f(x) = \frac{x^2 + 1}{x - 1} = \underbrace{x + 1}_{\substack{\text{Slant asymptote:}\\ y = x + 1}} + \frac{2}{x - 1}$$

If $|x| \to \infty$, the value of $\dfrac{2}{x - 1}$ is approximately 0. Thus, when $|x|$ is large, the function is very close to $y = x + 1 + 0$. This means that as $x \to \infty$ or as $x \to -\infty$, the graph of f gets closer and closer to the line whose equation is $y = x + 1$. The line $y = x + 1$ is a slant asymptote of the graph.

In general, if $f(x) = \dfrac{p(x)}{q(x)}$, p and q have no common factors, and the degree of p is one greater than the degree of q, find the slant asymptote by dividing $q(x)$ into $p(x)$. The division will take the form

$$\frac{p(x)}{q(x)} = mx + b + \frac{\text{remainder}}{q(x)}.$$

> Slant asymptote:
> $y = mx + b$

The equation of the slant asymptote is obtained by dropping the term with the remainder. Thus, the equation of the slant asymptote is $y = mx + b$.

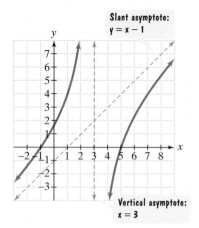

Figure 2.33 The graph of
$f(x) = \dfrac{x^2 - 4x - 5}{x - 3}$

EXAMPLE 7 Finding the Slant Asymptote of a Rational Function

Find the slant asymptote of $f(x) = \dfrac{x^2 - 4x - 5}{x - 3}$.

Solution Because the degree of the numerator, 2, is exactly one more than the degree of the denominator, 1, and $x - 3$ is not a factor of $x^2 - 4x - 5$, the graph of f has a slant asymptote. To find the equation of the slant asymptote, divide $x - 3$ into $x^2 - 4x - 5$:

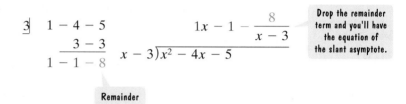

The equation of the slant asymptote is $y = x - 1$. Using our strategy for graphing rational functions, the graph of $f(x) = \dfrac{x^2 - 4x - 5}{x - 3}$ is shown in Figure 2.33.

> **Check Point 7** Find the slant asymptote of $f(x) = \dfrac{2x^2 - 5x + 7}{x - 2}$.

7 Solve applied problems involving rational functions.

Applications

There are numerous examples of asymptotic behavior in functions that describe real-world phenomena. Let's consider an example from the business world. The **cost function, C**, for a business is the sum of its fixed and variable costs:

$$C(x) = (\text{fixed cost}) + cx.$$
Cost per unit times the number of units produced, x

The **average cost** per unit for a company to produce x units is the sum of its fixed and variable costs divided by the number of units produced. The **average cost function** is a rational function that is denoted by $\overline{C}$. Thus,

$$\overline{C}(x) = \dfrac{(\text{fixed cost}) + cx}{x}.$$
Cost of producing x units: fixed plus variable costs

Number of units produced

EXAMPLE 8 Average Cost of Producing a Wheelchair

A company is planning to manufacture wheelchairs that are light, fast, and beautiful. Fixed monthly cost will be $500,000, and it will cost $400 to produce each radically innovative chair.

    a. Write the cost function, C, of producing x wheelchairs.
    b. Write the average cost function, $\overline{C}$, of producing x wheelchairs.
    c. Find and interpret $\overline{C}(1000)$, $\overline{C}(10,000)$, and $\overline{C}(100,000)$.
    d. What is the horizontal asymptote for the average cost function, $\overline{C}$? Describe what this represents for the company.

Solution

a. The cost function of producing x wheelchairs, C, is the sum of the fixed cost and the variable cost.

> Fixed cost is $500,000.
>
> Variable cost: $400 for each wheelchair produced

$$C(x) = 500{,}000 + 400x$$

b. The average cost function of producing x wheelchairs, $\overline{C}$, is the sum of fixed and variable costs divided by the number of wheelchairs produced.

$$\overline{C}(x) = \frac{500{,}000 + 400x}{x} \quad \text{or} \quad \overline{C}(x) = \frac{400x + 500{,}000}{x}$$

c. We evaluate $\overline{C}$ at 1000, 10,000, and 100,000, interpreting the results.

$$\overline{C}(1000) = \frac{400(1000) + 500{,}000}{1000} = 900$$

The average cost per wheelchair of producing 1000 wheelchairs per month is $900.

$$\overline{C}(10{,}000) = \frac{400(10{,}000) + 500{,}000}{10{,}000} = 450$$

The average cost per wheelchair of producing 10,000 wheelchairs per month is $450.

$$\overline{C}(100{,}000) = \frac{400(100{,}000) + 500{,}000}{100{,}000} = 405$$

The average cost per wheelchair of producing 100,000 wheelchairs per month is $405. Notice that with higher production levels, the cost of producing each wheelchair decreases.

d. We developed the average cost function

$$\overline{C}(x) = \frac{400x + 500{,}000}{x}$$

in which the degree of the numerator, 1, is equal to the degree of the denominator, 1. The leading coefficients of the numerator and denominator, 400 and 1, are used to obtain the equation of the horizontal asymptote. The equation of the horizontal asymptote is

$$y = \frac{400}{1} \quad \text{or} \quad y = 400.$$

The horizontal asymptote is shown in Figure 2.34. This means that the more wheelchairs produced per month, the closer the average cost per wheelchair for the company comes to $400. The least possible cost per wheelchair is approaching $400. Competitively low prices take place with high production levels, posing a major problem for small businesses.

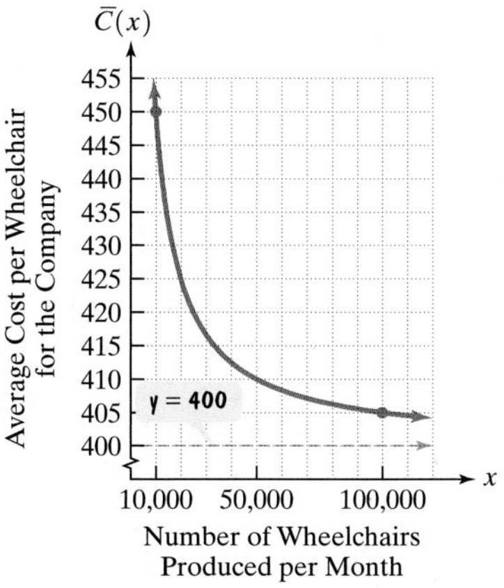

Figure 2.34 As production level increases, the average cost per wheelchair approaches $400:

$$\lim_{x \to \infty} \overline{C}(x) = 400.$$

Check Point 8   The time: the not-too-distant future. A new company is hoping to replace traditional computers and two-dimensional monitors with its virtual reality system. The fixed monthly cost will be $600,000, and it will cost $500 to produce each system.

a. Write the cost function, C, of producing x virtual reality systems.

b. Write the average cost function, $\overline{C}$, of producing x virtual reality systems.

c. Find and interpret $\overline{C}(1000)$, $\overline{C}(10{,}000)$, and $\overline{C}(100{,}000)$.

d. What is the horizontal asymptote for the average cost function, $\overline{C}$? Describe what this represents for the company.

If an object moves at an average velocity v, the distance, s, covered in time t is given by the formula

$$s = vt.$$

Thus, *distance = velocity · time*. Objects that move in accordance with this formula are said to be in **uniform motion**. In Example 9, we use a rational function to model time, t, in uniform motion. Solving the uniform motion formula for t, we obtain

$$t = \frac{s}{v}.$$

Thus, time is the quotient of distance and average velocity.

EXAMPLE 9 Time Involved in Uniform Motion

Two commuters drove to work a distance of 40 miles and then returned again on the same route. The average velocity on the return trip was 30 miles per hour faster than the average velocity on the outgoing trip. Express the total time required to complete the round trip, T, as a function of the average velocity on the outgoing trip, x.

Solution As specified, the average velocity on the outgoing trip is represented by x. Because the average velocity on the return trip was 30 miles per hour faster than the average velocity on the outgoing trip, let

$$x + 30 = \text{the average velocity on the return trip.}$$

The sentence that we use as a verbal model to write our rational function is

Total time on the round trip	equals	time on the outgoing trip	plus	time on the return trip.
$T(x)$	$=$	$\dfrac{40}{x}$	$+$	$\dfrac{40}{x + 30}.$

This is outgoing distance, 40 miles, divided by outgoing velocity, x.

This is return distance, 40 miles, divided by return velocity, x + 30.

The function that expresses the total time required to complete the round trip is

$$T(x) = \frac{40}{x} + \frac{40}{x + 30}.$$

Once you have modeled a problem's conditions with a function, you can use a graphing utility to explore the function's behavior. For example let's graph the function in Example 9. Because it seems unlikely that an average outgoing velocity exceeds 60 miles per hour with an average return velocity that is 30 miles per hour faster, we graph the function for $0 \le x \le 60$. Figure 2.35 shows the graph of $T(x) = \dfrac{40}{x} + \dfrac{40}{x + 30}$ in a $[0, 60, 3]$ by $[0, 10, 1]$ viewing rectangle. Notice that the function is decreasing. This shows decreasing times with increasing average velocities. Can you see that the vertical asymptote is $x = 0$, or the y-axis? This indicates that close to an outgoing average velocity of zero miles per hour, the round trip will take nearly forever: $\lim\limits_{x \to 0^+} T(x) = \infty$.

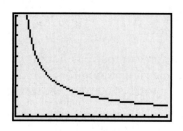

Figure 2.35 The graph of $T(x) = \dfrac{40}{x} + \dfrac{40}{x + 30}$. As average velocity increases, time for the trip decreases: $\lim\limits_{x \to \infty} T(x) = 0$.

Check Point 9
Two commuters drove to work a distance of 20 miles and then returned again on the same route. The average velocity on the return trip was 10 miles per hour slower than the average velocity on the outgoing trip. Express the total time required to complete the round trip, T, as a function of the average velocity on the outgoing trip, x.

EXERCISE SET 2.6

Practice Exercises

In Exercises 1–8, find the domain of each rational function.

1. $f(x) = \dfrac{5x}{x - 4}$

2. $f(x) = \dfrac{7x}{x - 8}$

3. $g(x) = \dfrac{3x^2}{(x - 5)(x + 4)}$

4. $g(x) = \dfrac{2x^2}{(x - 2)(x + 6)}$

5. $h(x) = \dfrac{x + 7}{x^2 - 49}$

6. $h(x) = \dfrac{x + 8}{x^2 - 64}$

7. $f(x) = \dfrac{x + 7}{x^2 + 49}$

8. $f(x) = \dfrac{x + 8}{x^2 + 64}$

Use the graph of the rational function in the figure shown to complete each statement in Exercises 9–14.

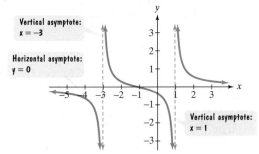

9. As $x \to -3^-$, $f(x) \to$ _____.

10. As $x \to -3^+$, $f(x) \to$ _____.

11. As $x \to 1^-$, $f(x) \to$ _____.

12. As $x \to 1^+$, $f(x) \to$ _____.

13. As $x \to -\infty$, $f(x) \to$ _____.

14. As $x \to \infty$, $f(x) \to$ _____.

Use the graph of the rational function in the figure shown to complete each statement in Exercises 15–20.

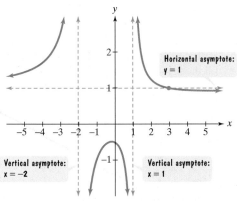

15. As $x \to 1^+$, $f(x) \to$ _____.

16. As $x \to 1^-$, $f(x) \to$ _____.

17. As $x \to -2^+$, $f(x) \to$ _____.

18. As $x \to -2^-$, $f(x) \to$ _____.

19. As $x \to \infty$, $f(x) \to$ _____.

20. As $x \to -\infty$, $f(x) \to$ _____.

In Exercises 21–28, find the vertical asymptotes, if any, of the graph of each rational function.

21. $f(x) = \dfrac{x}{x + 4}$

22. $f(x) = \dfrac{x}{x - 3}$

23. $g(x) = \dfrac{x + 3}{x(x + 4)}$

24. $g(x) = \dfrac{x + 3}{x(x - 3)}$

25. $h(x) = \dfrac{x}{x(x + 4)}$

26. $h(x) = \dfrac{x}{x(x - 3)}$

27. $r(x) = \dfrac{x}{x^2 + 4}$

28. $r(x) = \dfrac{x}{x^2 + 3}$

In Exercises 29–36, find the horizontal asymptote, if any, of the graph of each rational function.

29. $f(x) = \dfrac{12x}{3x^2 + 1}$

30. $f(x) = \dfrac{15x}{3x^2 + 1}$

31. $g(x) = \dfrac{12x^2}{3x^2 + 1}$

32. $g(x) = \dfrac{15x^2}{3x^2 + 1}$

33. $h(x) = \dfrac{12x^3}{3x^2 + 1}$

34. $h(x) = \dfrac{15x^3}{3x^2 + 1}$

35. $f(x) = \dfrac{-2x + 1}{3x + 5}$

36. $f(x) = \dfrac{-3x + 7}{5x - 2}$

In Exercises 37–58, follow the seven steps on page 313 to graph each rational function.

37. $f(x) = \dfrac{4x}{x - 2}$

38. $f(x) = \dfrac{3x}{x - 1}$

39. $f(x) = \dfrac{2x}{x^2 - 4}$

40. $f(x) = \dfrac{4x}{x^2 - 1}$

41. $f(x) = \dfrac{2x^2}{x^2 - 1}$

42. $f(x) = \dfrac{4x^2}{x^2 - 9}$

43. $f(x) = \dfrac{-x}{x + 1}$

44. $f(x) = \dfrac{-3x}{x + 2}$

45. $f(x) = -\dfrac{1}{x^2 - 4}$

46. $f(x) = -\dfrac{2}{x^2 - 1}$

47. $f(x) = \dfrac{2}{x^2 + x - 2}$

48. $f(x) = \dfrac{-2}{x^2 - x - 2}$

49. $f(x) = \dfrac{2x^2}{x^2 + 4}$

50. $f(x) = \dfrac{4x^2}{x^2 + 1}$

51. $f(x) = \dfrac{x + 2}{x^2 + x - 6}$

52. $f(x) = \dfrac{x - 4}{x^2 - x - 6}$

53. $f(x) = \dfrac{x^4}{x^2 + 2}$

54. $f(x) = \dfrac{2x^4}{x^2 + 1}$

55. $f(x) = \dfrac{x^2 + x - 12}{x^2 - 4}$

56. $f(x) = \dfrac{x^2}{x^2 + x - 6}$

57. $f(x) = \dfrac{3x^2 + x - 4}{2x^2 - 5x}$

58. $f(x) = \dfrac{x^2 - 4x + 3}{(x + 1)^2}$

In Exercises 59–66, **a.** *Find the slant asymptote of the graph of each rational function and* **b.** *Follow the seven-step strategy and use the slant asymptote to graph each rational function.*

59. $f(x) = \dfrac{x^2 - 1}{x}$

60. $f(x) = \dfrac{x^2 - 4}{x}$

61. $f(x) = \dfrac{x^2 + 1}{x}$

62. $f(x) = \dfrac{x^2 + 4}{x}$

63. $f(x) = \dfrac{x^2 + x - 6}{x - 3}$

64. $f(x) = \dfrac{x^2 - x + 1}{x - 1}$

65. $f(x) = \dfrac{x^3 + 1}{x^2 + 2x}$

66. $f(x) = \dfrac{x^3 - 1}{x^2 - 9}$

Application Exercises

67. A company is planning to manufacture mountain bikes. Fixed monthly cost will be $100,000 and it will cost $100 to produce each bicycle.

   a. Write the cost function, C, of producing x mountain bikes.

   b. Write the average cost function, $\overline{C}$, of producing x mountain bikes.

   c. Find and interpret $\overline{C}(500), \overline{C}(1000), \overline{C}(2000),$ and $\overline{C}(4000)$.

   d. What is the horizontal asymptote for the function, $\overline{C}$? Describe what this means in practical terms.

68. A company that manufactures running shoes has a fixed monthly cost of $300,000. It costs $30 to produce each pair of shoes.

   a. Write the cost function, C, of producing x pairs of shoes.

   b. Write the average cost function, $\overline{C}$, of producing x pairs of shoes.

   c. Find and interpret $\overline{C}(1000),$ $\overline{C}(10,000),$ and $\overline{C}(100,000)$.

   d. What is the horizontal asymptote for the average cost function, $\overline{C}$? Describe what this represents for the company.

69. Textbook sales at college stores have increased during the past two decades. The function

$$B(x) = 190.9x + 2413.99$$

models textbook sales, $B(x)$, in millions of dollars, x years after 1985. College enrollment has also increased. The function

$$E(x) = 0.234x + 12.54$$

models total college enrollment, $E(x)$, in millions, x years after 1985.

   a. Write a rational function that models the average amount of money spent on textbooks per college student, $M(x)$, in dollars per student, x years after 1985. The graph of M is shown in the figure.

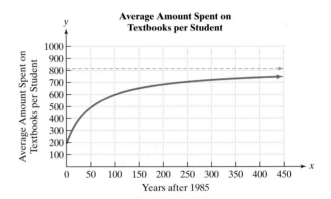

Average Amount Spent on Textbooks per Student

Years after 1985

   b. Use the function from part (a) to find the average amount of money that will be spent on textbooks per college student in 2004. How is this shown on the graph of M?

   c. What is the horizontal asymptote for the function that models the average amount of money spent on textbooks per college student? Describe what this represents in practical terms.

70. The rational function

$$C(x) = \dfrac{130x}{100 - x}, \quad 0 \le x < 100,$$

describes the cost, $C(x)$, in millions of dollars, to inoculate $x\%$ of the population against a particular strain of flu.

   a. Find and interpret $C(20), C(40), C(60), C(80),$ and $C(90)$.

   b. What is the equation of the vertical asymptote? What does this mean in terms of the variables in the function?

   c. Graph the function.

*Among all deaths from a particular disease, the percentage that are smoking related (21–39 cigarettes per day) is a function of the disease's **incidence ratio**. The incidence ratio describes the number of times more likely smokers are than nonsmokers to die from the disease. The following table shows the incidence ratios for heart disease and lung cancer for two age groups.*

Incidence Ratios

	Heart Disease	Lung Cancer
Ages 55–64	1.9	10
Ages 65–74	1.7	9

Source: Alexander M. Walker, *Observations and Inference,* page 20.

For example, the incidence ratio of 9 in the table means that smokers between the ages of 65 and 74 are 9 times more likely than nonsmokers in the same group to die from lung cancer. The rational function

$$P(x) = \frac{100(x - 1)}{x}$$

models the percentage of smoking-related deaths among all deaths from a disease, P(x), in terms of the disease's incidence ratio, x. The graph of the rational function is shown. Use this function to solve Exercises 71–74.

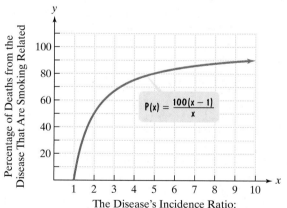

The Disease's Incidence Ratio:
The number of times more likely smokers are than nonsmokers to die from the disease

71. Find $P(10)$. Describe what this means in terms of the incidence ratio, 10, given in the table. Identify your solution as a point on the graph.

72. Find $P(9)$. Round to the nearest percent. Describe what this means in terms of the incidence ratio, 9, given in the table. Identify your solution as a point on the graph.

73. What is the horizontal asymptote of the graph? Describe what this means about the percentage of deaths caused by smoking with increasing incidence ratios.

74. According to the model and its graph, is there a disease for which all deaths are caused by smoking? Explain your answer.

75. Rational functions are often used to model how much we remember over time. In an experiment on memory, students in a language class are asked to memorize 40 vocabulary words in Latin, a language with which the students are not familiar. After studying the words for one day, the class is tested each day thereafter to see how many words they remember. The class average is taken and the results are graphed below.

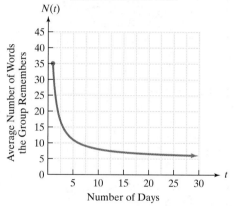

a. Use the graph to find a reasonable estimate of the number of Latin words remembered after 1 day, 5 days, and 15 days.

b. The function that models the number of Latin words remembered by the students after t days is given by

$$N(t) = \frac{5t + 30}{t}, \quad \text{where } t \geq 1.$$

Find $N(1)$, $N(5)$, and $N(15)$, comparing these values with your estimates from part (a).

c. What does the graph indicate about the number of Latin words remembered by the group over time?

d. Use the function in part (b) to find the horizontal asymptote for the graph. Describe what this horizontal asymptote means in terms of the variables modeled in this situation.

Exercises 76–79 involve writing functions that model a problem's conditions.

76. You drive from your home to a vacation resort 600 miles away. You return on the same highway. The average velocity on the return trip is 10 miles per hour slower than the average velocity on the outgoing trip. Express the total time required to complete the round trip, T, as a function of the average velocity on the outgoing trip, x.

77. A tourist drives 90 miles along a scenic highway and then takes a 5-mile walk along a hiking trail. The average velocity driving is nine times that while hiking. Express the total time for driving and hiking, T, as a function of the average velocity on the hike, x.

78. A contractor is constructing the house shown in the figure. The cross section up to the roof is in the shape of a rectangle. The area of the rectangular floor of the house is 2500 square feet. Express the perimeter of the rectangular floor, P, as a function of the width of the rectangle, x.

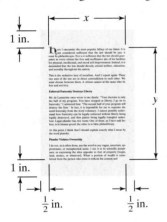

Length

Width: x

79. The figure shows a page with 1-inch margins at the top and the bottom and half-inch side margins. A publishing company is willing to vary the page dimensions subject to the condition that the printed area of the page is 50 square inches. Express the total area of the page, A, as a function of the width of the rectangle containing the print, x.

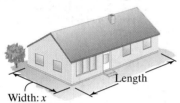

x

1 in.

y

1 in.

$\frac{1}{2}$ in. $\frac{1}{2}$ in.

Writing in Mathematics

80. What is a rational function?

81. Use everyday language to describe the graph of a rational function f such that as $x \to -\infty$, $f(x) \to 3$.

82. Use everyday language to describe the behavior of a graph near its vertical asymptote if $f(x) \to \infty$ as $x \to -2^-$ and $f(x) \to -\infty$ as $x \to -2^+$.

83. If you are given the equation of a rational function, explain how to find the vertical asymptotes, if any, of the function's graph.

84. If you are given the equation of a rational function, explain how to find the horizontal asymptote, if any, of the function's graph.

85. Describe how to graph a rational function.

86. If you are given the equation of a rational function, how can you tell if the graph has a slant asymptote? If it does, how do you find its equation?

87. Is every rational function a polynomial function? Why or why not? Does a true statement result if the two adjectives *rational* and *polynomial* are reversed? Explain.

88. Although your friend has a family history of heart disease, he smokes, on average, 25 cigarettes per day. He sees the table showing incidence ratios for heart disease (see Exercises 71–74) and feels comfortable that they are less than 2, compared to 9 and 10 for lung cancer. He claims that all family deaths have been from heart disease, and decides not to give up smoking. Use the given function and its graph to describe some additional information not given in the table that might influence his decision.

Technology Exercises

89. Use a graphing utility to verify any five of your hand-drawn graphs in Exercises 37–66.

90. Use a graphing utility to verify your hand-drawn graph in Exercise 70.

91. Use a graphing utility to graph $y = \dfrac{1}{x}$, $y = \dfrac{1}{x^3}$, and $\dfrac{1}{x^5}$ in the same viewing rectangle. For odd values of n, how does changing n affect the graph of $y = \dfrac{1}{x^n}$?

92. Use a graphing utility to graph $y = \dfrac{1}{x^2}$, $y = \dfrac{1}{x^4}$, and $y = \dfrac{1}{x^6}$ in the same viewing rectangle. For even values of n, how does changing n affect the graph of $y = \dfrac{1}{x^n}$?

93. Use a graphing utility to graph $f(x) = \dfrac{x^2 - 4x + 3}{x - 2}$ and $g(x) = \dfrac{x^2 - 5x + 6}{x - 2}$. What differences do you observe between the graph of f and g? How do you account for these differences?

94. The rational function

$$f(x) = \frac{27,725(x - 14)}{x^2 + 9} - 5x$$

models the number of arrests, $f(x)$, per 100,000 drivers, for driving under the influence of alcohol, as a function of a driver's age, x.

a. Graph the function in a [0, 70, 5] by [0, 400, 20] viewing rectangle.

b. Describe the trend shown by the graph.

c. Use the ZOOM and TRACE features or the maximum function feature of your graphing utility to find the age that corresponds to the greatest number of arrests. How many arrests, per 100,000 drivers, are there for this age group?

Critical Thinking Exercises

95. Which one of the following is true?

a. The graph of a rational function cannot have both a vertical and a horizontal asymptote.

b. It is not possible to have a rational function whose graph has no y-intercept.

c. The graph of a rational function can have three horizontal asymptotes.

d. The graph of a rational function can never cross a vertical asymptote.

96. Which one of the following is true?

a. The function $f(x) = \dfrac{1}{\sqrt{x - 3}}$ is a rational function.

b. The x-axis is a horizontal asymptote for the graph of $f(x) = \dfrac{4x - 1}{x + 3}$.

c. The number of televisions that a company can produce per week after t weeks of production is given by

$$N(t) = \frac{3000t^2 + 30,000t}{t^2 + 10t + 25}.$$

Using this model, the company will eventually be able to produce 30,000 televisions in a single week.

d. None of the given statements is true.

In Exercises 97–100, write the equation of a rational function
$f(x) = \dfrac{p(x)}{q(x)}$ *having the indicated properties, in which the*
degrees of p and q are as small as possible. More than one
correct function may be possible. Graph your function using
a graphing utility to verify that it has the required properties.

97. f has a vertical asymptote given by $x = 3$, a horizontal asymptote $y = 0$, y-intercept at -1, and no x-intercept.

98. f has vertical asymptotes given by $x = -2$ and $x = 2$, a horizontal asymptote $y = 2$, y-intercept at $\frac{9}{2}$, x-intercepts at -3 and 3, and y-axis symmetry.

99. f has a vertical asymptote given by $x = 1$, a slant asymptote whose equation is $y = x$, y-intercept at 2, and x-intercepts at -1 and 2.

100. f has no vertical, horizontal, or slant asymptotes, and no x-intercepts.

Group Exercise

101. Group members form the sales team for a company that makes computer video games. It has been determined that the rational function

$$f(x) = \frac{200x}{x^2 + 100}$$

models the monthly sales, $f(x)$, in thousands of games, of a new video game as a function of the number of months, x, after the game is introduced. The figure shows the graph of the function. What are the team's recommendations to the company in terms of how long the video game should be on the market before another new video game is introduced? What other factors might members want to take into account in terms of the recommendations? What will eventually happen to sales, and how is this indicated by the graph? What does this have to do with a horizontal asymptote? What could the company do to change the behavior of this function and continue generating sales? Would this be cost effective?

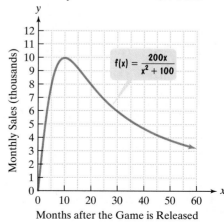

Monthly Sales of a New Video Game

$f(x) = \dfrac{200x}{x^2 + 100}$

Months after the Game is Released

SECTION 2.7 *Polynomial and Rational Inequalities*

Objectives

1. Solve polynomial inequalities.
2. Solve rational inequalities.
3. Solve problems modeled by polynomial or rational inequalities.

Not afraid of heights and cutting-edge excitement? How about sky diving? Behind your exhilarating experience is the world of algebra. After you jump from the airplane, your height above the ground at every instant of your fall can be described by a formula involving a variable that is squared. At some point, you'll need to open your parachute. How can you determine when you must do so? Let x represent the number of seconds you are falling. You can compute when to open the parachute by solving an inequality that takes on the form $ax^2 + bx + c < 0$. Such an inequality is called a *polynomial inequality*.

> ## Study Tip
>
> If you need to review basic principles of inequality and procedures for solving linear inequalities, refer to Section P.9, beginning on page 99.

> ### Definition of a Polynomial Inequality
> A **polynomial inequality** is any inequality that can be put in one of the forms
> $$f(x) < 0, \quad f(x) > 0, \quad f(x) \leq 0, \quad \text{or} \quad f(x) \geq 0$$
> where f is a polynomial function.

In this section, we establish the basic techniques for solving polynomial inequalities. We will use these techniques to solve inequalities involving rational functions. Finally, we will consider a function that models the position of any free-falling object. As a sky diver, you could be that free-falling object!

1 Solve polynomial inequalities.

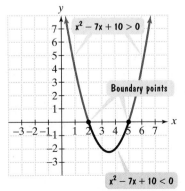

Figure 2.36

Solving Polynomial Inequalities

Graphs can help us visualize the solutions of polynomial inequalities. For example, the graph of $f(x) = x^2 - 7x + 10$ is shown in Figure 2.36. The zeros, 2 and 5, are **boundary points** between where the graph lies above the x-axis, shown in blue, and where the graph lies below the x-axis, shown in red.

Locating the zeros of a polynomial function, f, is an important step in finding the solution set for polynomial inequalities in the form $f(x) < 0$ or $f(x) > 0$. Use the zeros of f as boundary points that divide the real number line into intervals. On each interval, the graph of f is either above the x-axis $[f(x) > 0]$ or below the x-axis $[f(x) < 0]$. For this reason, zeros play a fundamental role in solving polynomial inequalities.

Procedure for Solving Polynomial Inequalities

1. Express the inequality in the form
$$f(x) < 0 \quad \text{or} \quad f(x) > 0$$
where f is a polynomial function.
2. Find the zeros of f. The real zeros are the **boundary points.**
3. Locate these boundary points on a number line, thereby dividing the number line into intervals.
4. Choose one representative number within each interval and evaluate f at that number.
   a. If the value of f is positive, then $f(x) > 0$ for all numbers, x, in the interval.
   b. If the value of f is negative, then $f(x) < 0$ for all numbers, x, in the interval.
5. Write the solution set, selecting the interval(s) that satisfy the given inequality.

This procedure is valid if $<$ is replaced by $\leq$ or $>$ is replaced by $\geq$. However, if the inequality involves $\leq$ or $\geq$, include the boundary points [the solutions of $f(x) = 0$] in the solution set.

EXAMPLE 1 Solving a Polynomial Inequality

Solve and graph the solution set on a real number line: $2x^2 + x > 15$.

Solution

Step 1 Express the inequality in the form $f(x) < 0$ or $f(x) > 0$. We begin by rewriting the inequality so that 0 is on the right side.

$2x^2 + x > 15$	This is the given inequality.
$2x^2 + x - 15 > 15 - 15$	Subtract 15 from both sides.
$2x^2 + x - 15 > 0$	Simplify.

This inequality is equivalent to the one we wish to solve. It is in the form $f(x) > 0$, where $f(x) = 2x^2 + x - 15$.

Step 2 Find the zeros of f. We find the zeros of $f(x) = 2x^2 + x - 15$ by solving the equation $2x^2 + x - 15 = 0$.

$2x^2 + x - 15 = 0$	This polynomial equation is a quadratic equation.
$(2x - 5)(x + 3) = 0$	Factor.
$2x - 5 = 0 \quad \text{or} \quad x + 3 = 0$	Set each factor equal to 0.
$x = \frac{5}{2} \qquad\qquad x = -3$	Solve for x.

The zeros of f are -3 and $\frac{5}{2}$. We will use these zeros as boundary points on a number line.

Step 3 Locate the boundary points on a number line and separate the line into intervals. The number line with the boundary points is shown as follows:

The boundary points divide the number line into three intervals:
$$(-\infty, -3) \quad (-3, \tfrac{5}{2}) \quad (\tfrac{5}{2}, \infty).$$

Step 4   Choose one representative number within each interval and evaluate f at that number.

Interval	Representative Number	Substitute into $f(x) = 2x^2 + x - 15$	Conclusion
$(-\infty, -3)$	-4	$f(-4) = 2(-4)^2 + (-4) - 15$ $= 13$, positive	$f(x) > 0$ for all x in $(-\infty, -3)$.
$\left(-3, \frac{5}{2}\right)$	0	$f(0) = 2 \cdot 0^2 + 0 - 15$ $= -15$, negative	$f(x) < 0$ for all x in $\left(-3, \frac{5}{2}\right)$.
$\left(\frac{5}{2}, \infty\right)$	3	$f(3) = 2 \cdot 3^2 + 3 - 15$ $= 6$, positive	$f(x) > 0$ for all x in $\left(\frac{5}{2}, \infty\right)$.

Step 5   Write the solution set, selecting the interval(s) that satisfy the given inequality.   Based on our work in step 4, we see that $f(x) > 0$ for all x in $(-\infty, -3)$ or $(\frac{5}{2}, \infty)$. Thus, the solution set of the given inequality, $2x^2 + x > 15$, or, equivalently, $2x^2 + x - 15 > 0$, is

$$(-\infty, -3) \quad \text{or} \quad (\tfrac{5}{2}, \infty).$$

The graph of the solution set on a number line is shown as follows:

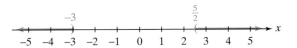

Technology

The solution set for
$$2x^2 + x > 15$$
or, equivalently,
$$2x^2 + x - 15 > 0$$
can be verified with a graphing utility. The graph of $f(x) = 2x^2 + x - 15$ was obtained using a $[-10, 10, 1]$ by $[-16, 6, 1]$ viewing rectangle. The graph lies above the x-axis, representing $>$, for all x in $(-\infty, -3)$ or $(\frac{5}{2}, \infty)$.

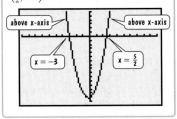

Check Point 1   Solve and graph the solution set: $x^2 - x > 20$.

EXAMPLE 2   Solving a Polynomial Inequality

Solve and graph the solution set on a real number line: $x^3 + x^2 \leq 4x + 4$.

Solution

Step 1   Express the inequality in the form $f(x) \leq 0$ or $f(x) \geq 0$. We begin by rewriting the inequality so that 0 is on the right side.

$$x^3 + x^2 \leq 4x + 4 \qquad \text{This is the given inequality.}$$
$$x^3 + x^2 - 4x - 4 \leq 4x + 4 - 4x - 4 \qquad \text{Subtract } 4x + 4 \text{ from both sides.}$$
$$x^3 + x^2 - 4x - 4 \leq 0 \qquad \text{Simplify.}$$

This inequality is equivalent to the one we wish to solve. It is in the form $f(x) \leq 0$, where $f(x) = x^3 + x^2 - 4x - 4$.

Step 2   Find the zeros of f. We find the zeros of $f(x) = x^3 + x^2 - 4x - 4$ by solving the equation $x^3 + x^2 - 4x - 4 = 0$.

$$x^3 - x^2 - 4x - 4 = 0 \qquad \text{This polynomial equation of degree 3 has three zeros.}$$
$$x^2(x + 1) - 4(x + 1) = 0 \qquad \text{Factor } x^2 \text{ from the first two terms and } -4 \text{ from the last two terms.}$$

$$(x + 1)(x^2 - 4) = 0 \qquad \text{A common factor of } x + 1 \text{ is factored from the expression.}$$

$$x + 1 = 0 \quad \text{or} \quad x^2 - 4 = 0 \qquad \text{Set each factor equal to 0.}$$

$$x = -1 \qquad\qquad x^2 = 4 \qquad \text{Solve for x.}$$

$$x = \pm 2 \qquad \text{Remember that if } x^2 = d, \text{ then } x = \pm\sqrt{d}.$$

The zeros of f are $-2, -1$, and 2. We will use these zeros as boundary points on a number line.

Step 3 Locate the boundary points on a number line and separate the line into intervals. The number line with the boundary points is shown as follows:

The boundary points divide the number line into four intervals:

$$(-\infty, -2) \quad (-2, -1) \quad (-1, 2) \quad (2, \infty).$$

Step 4 Choose one representative number within each interval and evaluate f at that number.

Interval	Representative Number	Substitute into $f(x) = x^3 + x^2 - 4x - 4$	Conclusion
$(-\infty, -2)$	-3	$f(-3) = (-3)^3 + (-3)^2 - 4(-3) - 4$ $= -10,$ negative	$f(x) < 0$ for all x in $(-\infty, -2)$.
$(-2, -1)$	-1.5	$f(-1.5) = (-1.5)^3 + (-1.5)^2 - 4(-1.5) - 4$ $= 0.875,$ positive	$f(x) > 0$ for all x in $(-2, -1)$.
$(-1, 2)$	0	$f(0) = 0^3 + 0^2 - 4 \cdot 0 - 4$ $= -4,$ negative	$f(x) < 0$ for all x in $(-1, 2)$.
$(2, \infty)$	3	$f(3) = 3^3 + 3^2 - 4 \cdot 3 - 4$ $= 20,$ positive	$f(x) > 0$ for all x in $(2, \infty)$.

Technology

The solution set for
$$x^3 + x^2 \le 4x + 4$$
or, equivalently,
$$x^3 + x^2 - 4x - 4 \le 0$$
can be verified with a graphing utility. The graph of $f(x) = x^3 + x^2 - 4x - 4$ lies on or below the x-axis, representing $\le$, for all x in $(-\infty, -2]$ or $[-1, 2]$.

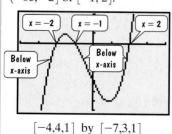

$[-4, 4, 1]$ by $[-7, 3, 1]$

Step 5 Write the solution set, selecting the interval(s) that satisfy the given inequality. Based on our work in step 4, we see that $f(x) < 0$ for all x in $(-\infty, -2)$ or $(-1, 2)$. However, because the inequality involves $\le$ (less than or *equal to*), we must also include the solutions of $x^3 + x^2 - 4x - 4 = 0$, namely $-2, -1$, and 2, in the solution set. Thus, the solution set of the given inequality, $x^3 + x^2 \le 4x + 4$, or, equivalently, $x^3 + x^2 - 4x - 4 \le 0$, is

$$(-\infty, -2] \quad \text{or} \quad [-1, 2].$$

The graph of the solution set on a number line is shown as follows:

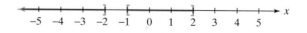

Check Point 2 Solve and graph the solution set on a real number line: $x^3 + 3x^2 \le x + 3$.

2 Solve rational inequalities.

Solving Rational Inequalities

A **rational inequality** is any inequality that can be put in one of the forms

$$f(x) < 0, \quad f(x) > 0, \quad f(x) \leq 0, \quad \text{or } f(x) \geq 0$$

where f is a rational function. An example of a rational inequality is

$$\frac{3x + 3}{2x + 4} > 0.$$

This inequality is in the form $f(x) > 0$, where f is the rational function given by

$$f(x) = \frac{3x + 3}{2x + 4}.$$

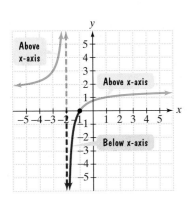

Figure 2.37 The graph of $f(x) = \dfrac{3x + 3}{2x + 4}$

The graph of f is shown in Figure 2.37.

The zero of the numerator of f, $3x + 3 = 0$, or $x = -1$, shows that the function's x-intercept is at -1. The zero of the denominator of f, $2x + 4 = 0$, or $x = -2$, shows that the function's vertical asymptote is $x = -2$. The zeros of the numerator and the denominator, -2 and -1, separate the x-axis into three intervals: $(-\infty, -2)$, $(-2, -1)$, and $(-1, \infty)$. On each interval, the graph of f is either above the x-axis $[f(x) > 0]$ or below the x-axis $[f(x) < 0]$.

Examine the graph in Figure 2.37 carefully. Can you see that it is above the x-axis for all x in $(-\infty, -2)$ or $(-1, \infty)$? Thus, the solution set of $\dfrac{3x + 3}{2x + 4} > 0$ is $(-\infty, -2)$ or $(-1, \infty)$. By contrast, the graph of f lies below the x-axis for all x in $(-2, -1)$. Thus, the solution set of $\dfrac{3x + 3}{2x + 4} < 0$ is $(-2, -1)$.

The first step in solving a rational inequality is to bring all terms to one side, obtaining zero on the other side. Then express the rational function on the nonzero side as a single quotient. The second step is to find the zeros of the rational function's numerator and denominator. These zeros serve as boundary points that separate the real number line into intervals. At this point, the procedure is the same as the one we used for solving polynomial inequalities.

EXAMPLE 3   Solving a Rational Inequality

Solve and graph the solution set: $\dfrac{x + 1}{x + 3} \geq 2$.

Solution

Step 1   Express the inequality so that one side is zero and the other side is a single quotient.   We subtract 2 from both sides to obtain zero on the right.

$$\frac{x + 1}{x + 3} \geq 2 \qquad \text{\small This is the given inequality.}$$

$$\frac{x + 1}{x + 3} - 2 \geq 0 \qquad \text{\small Subtract 2 from both sides, obtaining 0 on the right.}$$

$$\frac{x + 1}{x + 3} - \frac{2(x + 3)}{x + 3} \geq 0 \qquad \text{\small The least common denominator is } x + 3\text{. Express 2 in terms of this denominator.}$$

Study Tip

Do not begin solving

$$\frac{x + 1}{x + 3} \geq 2$$

by multiplying both sides by $x + 3$. We do not know if $x + 3$ is positive or negative. Thus, we do not know whether or not to reverse the sense of the inequality.

$$\frac{x + 1 - 2(x + 3)}{x + 3} \geq 0 \qquad \text{Subtract rational expressions.}$$

$$\frac{x + 1 - 2x - 6}{x + 3} \geq 0 \qquad \text{Apply the distributive property.}$$

$$\frac{-x - 5}{x + 3} \geq 0 \qquad \text{Simplify.}$$

This inequality is equivalent to the one we wish to solve. It is in the form $f(x) \geq 0$, where $f(x) = \dfrac{-x - 5}{x + 3}$.

Step 2   Find the zeros of the numerator and the denominator of f.

$$-x - 5 = 0 \qquad x + 3 = 0 \qquad \text{Set the numerator and denominator equal to 0. These are the values that make the previous quotient zero or undefined.}$$

$$x = -5 \qquad x = -3 \qquad \text{Solve for } x.$$

The zero of the numerator of f is -5 and the zero of the denominator of f is -3. We will use these zeros as boundary points on a number line.

Step 3   Locate the boundary points on a number line and separate the line into intervals.   The number line with the boundary points is shown as follows:

The boundary points divide the number line into three intervals:
$$(-\infty, -5) \quad (-5, -3) \quad (-3, \infty).$$

Step 4   Choose one representative number within each interval and evaluate f at that number.

Interval	Representative Number	Substitute into $f(x) = \dfrac{-x - 5}{x + 3}$	Conclusion
$(-\infty, -5)$	-6	$f(-6) = \dfrac{-(-6) - 5}{-6 + 3}$ $= -\frac{1}{3}$, negative	$f(x) < 0$ for all x in $(-\infty, -5)$.
$(-5, -3)$	-4	$f(-4) = \dfrac{-(-4) - 5}{-4 + 3}$ $= 1$, positive	$f(x) > 0$ for all x in $(-5, -3)$.
$(-3, \infty)$	0	$f(0) = \dfrac{-0 - 5}{0 + 3}$ $= -\frac{5}{3}$, negative	$f(x) < 0$ for all x in $(-3, \infty)$.

Step 5   Write the solution set, selecting the interval(s) that satisfy the given inequality.   Based on our work in step 4, we see that $f(x) > 0$ for all x in $(-5, -3)$. However, because the inequality involves $\geq$ (greater than or *equal to*), we must also include the solution of $f(x) = 0$, namely the zero of the numerator of f. Thus, we must include -5 in the solution set. The solution set of the given inequality is

$$[-5, -3).$$

The graph of the solution set on a number line is shown as follows:

Study Tip

Never include the zeros of a rational function's denominator in the solution set of a rational inequality. Division by zero is undefined.

Technology

The solution set for
$$\frac{x + 1}{x + 3} \geq 2$$
or, equivalently,
$$\frac{-x - 5}{x + 3} \geq 0$$
can be verified with a graphing utility. The graph of
$$f(x) = \frac{-x - 5}{x + 3}$$ lies on or above the x-axis, representing $\geq$, for all x in $[-5, -3)$.

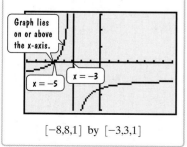

$[-8, 8, 1]$ by $[-3, 3, 1]$

Discovery

Because $(x + 3)^2$ is positive, it is possible to solve

$$\frac{x + 1}{x + 3} \geq 2$$

by first multiplying both sides by $(x + 3)^2$ (where $x \neq -3$). This will not reverse the sense of the inequality and will clear the fraction. Try using this solution method and compare it to the one shown on pages 330–331.

Check Point 3 Solve and graph the solution set: $\dfrac{2x}{x + 1} \geq 1$.

3 Solve problems modeled by polynomial or rational inequalities.

Applications

We are surrounded by evidence that the world is profoundly mathematical. For example, did you know that every time you throw an object vertically upward, its changing height above the ground can be modeled by a quadratic function? The same function can be used to describe objects that are falling, such as the sky divers shown in the opening to this section.

> **The Position Function for a Free-Falling Object Near Earth's Surface**
>
> An object that is falling or vertically projected into the air has its height above the ground, $s(t)$, in feet, after t seconds given by the **position function**
>
> $$s(t) = -16t^2 + v_0 t + s_0$$
>
> where v_0 is the original velocity (initial velocity) of the object, in feet per second, and s_0 is the original height (initial height) of the object, in feet.

In Example 4, we solve a polynomial inequality in a problem about the position of a free-falling object.

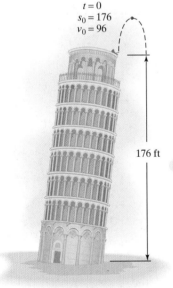

$t = 0$
$s_0 = 176$
$v_0 = 96$

176 ft

Figure 2.38 Throwing a ball from 176 feet with a velocity of 96 feet per second

EXAMPLE 4 Using the Position Model

A ball is thrown vertically upward from the top of the Leaning Tower of Pisa (176 feet high) with an initial velocity of 96 feet per second (Figure 2.38). During which time period will the ball's height exceed that of the tower?

Solution

$$s(t) = -16t^2 + v_0 t + s_0 \qquad \text{This is the position function for a free-falling object.}$$

$$s(t) = -16t^2 + 96t + 176 \qquad \text{Because } v_0 \text{ (initial velocity)} = 96 \text{ and } s_0 \text{ (initial position)} = 176, \text{ substitute these values into the formula.}$$

| When will s(t), the ball's height | exceed that | of the tower? |

$$-16t^2 + 96t + 176 \quad > \quad 176$$

$-16t^2 + 96t + 176 > 176$ This is the inequality that models the problem's question. We must find t.

$-16t^2 + 96t > 0$ Subtract 176 from both sides. This inequality is in the form f(t) > 0, where f(t) = −16t² + 96t.

$-16t^2 + 96t = 0$ Find the zeros of f(t) = −16t² + 96t.

$-16t(t - 6) = 0$ Factor.

$-16t = 0 \quad \text{or} \quad t - 6 = 0$ Set each factor equal to 0.

$t = 0 \qquad\qquad t = 6$ Solve for t. The boundary points are 0 and 6.

```
 ┼──┼──┼──●──┼──┼──┼──┼──●──┼──┼──→ t
-2 -1  0  1  2  3  4  5  6  7  8
```

Locate these values on a number line, with t ≥ 0.

Technology

The graphs of
$$y_1 = -16x^2 + 96x + 176$$
and
$$y_2 = 176$$
are shown in a

$$[0, 8, 1] \text{ by } [0, 320, 32]$$

| seconds in motion | height, in feet |

viewing rectangle. The graphs show that the ball's height exceeds that of the tower between 0 and 6 seconds.

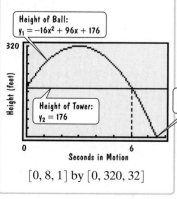

Height of Ball:
$y_1 = -16x^2 + 96x + 176$

320

Height (feet)

Height of Tower:
$y_2 = 176$

Ball hits ground after 7.47 seconds.

0   6
Seconds in Motion

$$[0, 8, 1] \text{ by } [0, 320, 32]$$

The intervals are $(-\infty, 0)$, $(0, 6)$ and $(6, \infty)$. For our purposes, the mathematical model is useful only from $t = 0$ until the ball hits the ground. (By setting $-16t^2 + 96t + 176$ equal to zero, we find $t \approx 7.47$; the ball hits the ground after approximately 7.47 seconds.) Thus, we use $(0, 6)$ and $(6, 7.47)$ for our intervals.

Interval	Representative Number	Substitute into $f(t) = -16t^2 + 96t$	Conclusion
$(0, 6)$	1	$f(1) = -16 \cdot 1^2 + 96 \cdot 1$ $= 80$, positive	$f(t) > 0$ for all t in $(0, 6)$.
$(6, 7.47)$	7	$f(7) = -16 \cdot 7^2 + 96 \cdot 7$ $= -112$, negative	$f(t) < 0$ for all t in $(6, 7.47)$.

We see that $f(t) > 0$ for all t in $(0, 6)$. This means that the ball's height exceeds that of the tower between 0 and 6 seconds.

Check Point 4   A toy rocket is propelled straight up from ground level with an initial velocity of 80 feet per second. During which time period will the rocket be more than 64 feet above the ground?

EXERCISE SET 2.7

Practice Exercises

Solve each inequality in Exercises 1–30, and graph the solution set on a real number line. Express each solution set in interval notation.

1. $(x - 4)(x + 2) > 0$

2. $(x + 3)(x - 5) > 0$

3. $(x - 7)(x + 3) \leq 0$

4. $(x + 1)(x - 7) \leq 0$

5. $x^2 - 5x + 4 > 0$

6. $x^2 - 4x + 3 < 0$

7. $x^2 + 5x + 4 > 0$

8. $x^2 + x - 6 > 0$

9. $3x^2 + 10x - 8 \leq 0$

10. $9x^2 + 3x - 2 \geq 0$

11. $2x^2 + x < 15$

12. $6x^2 + x > 1$

13. $4x^2 + 7x < -3$

14. $3x^2 + 16x < -5$

15. $2x^2 + 3x > 0$

16. $3x^2 - 5x \leq 0$

17. $x^2 - 6x + 9 < 0$

18. $4x^2 - 4x + 1 \geq 0$

19. $(x - 1)(x - 2)(x - 3) \geq 0$

20. $(x + 1)(x + 2)(x + 3) \geq 0$

21. $x^3 + 2x^2 - x - 2 \geq 0$

22. $x^3 + 2x^2 - 4x - 8 \geq 0$

23. $x^3 - 3x^2 - 9x + 27 < 0$

24. $x^3 + 7x^2 - x - 7 < 0$

25. $x^3 + x^2 + 4x + 4 > 0$

26. $x^3 - x^2 + 9x - 9 > 0$

27. $x^3 \geq 9x^2$

28. $x^3 \leq 4x^2$

29. $|x^2 + 2x - 36| > 12$

30. $|x^2 + 6x + 1| > 8$

Solve each rational inequality in Exercises 31–44, and graph the solution set on a real number line. Express each solution set in interval notation.

31. $\dfrac{x - 4}{x + 3} > 0$

32. $\dfrac{x + 5}{x - 2} > 0$

33. $\dfrac{x + 3}{x + 4} < 0$

34. $\dfrac{x + 5}{x + 2} < 0$

35. $\dfrac{x + 1}{x + 3} < 2$

36. $\dfrac{x}{x - 1} > 2$

37. $\dfrac{x - 2}{x + 2} \leq 2$

38. $\dfrac{x}{x + 2} \geq 2$

39. $\dfrac{3}{x + 3} > \dfrac{3}{x - 2}$

40. $\dfrac{1}{x + 1} > \dfrac{2}{x - 1}$

41. $\dfrac{x^2 - x - 2}{x^2 - 4x + 3} > 0$

42. $\dfrac{x^2 - 3x + 2}{x^2 - 2x - 3} > 0$

43. $\dfrac{(x + 4)(2 - x)}{(x - 1)^2} \geq 0$

44. $\dfrac{(x + 5)(3 - x)}{(x - 1)^2} \geq 0$

In Exercises 45–46, use the graph of the polynomial function to solve each inequality.

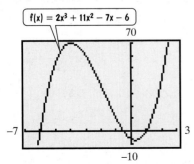

45. $2x^3 + 11x^2 \geq 7x + 6$

46. $2x^3 + 11x^2 < 7x + 6$

In Exercises 47–48, use the graph of the rational function to solve each inequality.

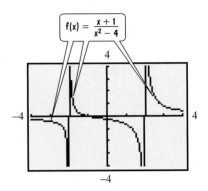

47. $\dfrac{1}{4(x + 2)} \leq -\dfrac{3}{4(x - 2)}$

48. $\dfrac{1}{4(x + 2)} > -\dfrac{3}{4(x - 2)}$

Application Exercises

Use the position function

$$s(t) = -16t^2 + v_0 t + s_0$$

$$(v_0 = \text{initial velocity}, s_0 = \text{initial position}, t = \text{time})$$

to answer Exercises 49–52. If necessary, round answers to the nearest hundredth of a second.

49. A projectile is fired straight upward from ground level with an initial velocity of 80 feet per second. During which interval of time will the projectile's height exceed 96 feet?

50. A projectile is fired straight upward from ground level with an initial velocity of 128 feet per second. During which interval of time will the projectile's height exceed 128 feet?

51. A ball is thrown vertically upward with a velocity of 64 feet per second from the top edge of a building 80 feet high. For how long is the ball higher than 96 feet?

52. A diver leaps into the air at 20 feet per second from a diving board that is 10 feet above the water. For how many seconds is the diver at least 12 feet above the water?

53. The function

$$H(x) = \frac{15}{8}x^2 - 30x + 200$$

models heart rate, $H(x)$, in beats per minute, x minutes after a strenuous workout.

a. What is the heart rate immediately following the workout?

b. According to the model, during which intervals of time after the strenuous workout does the heart rate exceed 110 beats per minute? For which of these intervals has model breakdown occurred? Which interval provides a more realistic answer? How did you determine this?

The bar graph shows the cost of Medicare, in billions of dollars, projected through 2005. The data can be modeled by

a linear function, $f(x) = 27x + 163$;

a quadratic function, $g(x) = 1.2x^2 + 15.2x + 181.4$.

In each function x represent the number of years after 1995 and $f(x)$ or $g(x)$ represents Medicare spending, in billions of dollars. Use these formulas to solve Exercises 54–56.

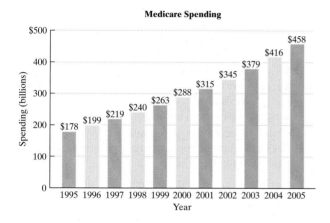

Medicare Spending

Source: Congressional Budget Office

54. The graph indicates that Medicare spending will reach $458 billion in 2005. Find the amount predicted by each of the functions for that year. How well do the functions model the value in the graph? Which function serves as a better model for that year?

55. For which years does the quadratic model indicate that Medicare spending will exceed $536.6 billion?

56. For which years does the quadratic model indicate that Medicare spending will exceed $629.4 billion?

In Exercises 57–60, use a polynomial or rational inequality to solve each problem.

In the previous section, we presented an example involving a company that manufactures light, fast, and beautiful wheelchairs. The company's fixed monthly cost is $500,000, and it costs $400 to produce each wheelchair. Use this information to solve Exercises 57–58.

57. Describe the company's production level so that the average cost of producing each wheelchair does not exceed $425.

58. Describe the company's production level so that the average cost of producing each wheelchair does not exceed $410.

59. An open box is made from a rectangular piece of cardboard measuring 11 inches by 8 inches by cutting identical squares from the corners and turning up the sides. Describe the possible lengths of the sides of the removed squares if the area of the bottom of the open box is not to exceed 50 square inches.

60. The perimeter of a rectangle is 50 feet. Describe the possible lengths of a side if the area of the rectangle is not to exceed 114 square feet.

Writing in Mathematics

61. What is a polynomial inequality?

62. What is a rational inequality?

63. If f is a polynomial or rational function, explain how the graph of f can be used to visualize the solution set of the inequality $f(x) < 0$.

Technology Exercises

64. Use a graphing utility to verify your solution sets to any three of the inequalities that you solved algebraically in Exercises 1–30.

65. Use a graphing utility to verify your solutions sets to any three of the rational inequalities that you solved algebraically in Exercises 31–44.

Solve each inequality in Exercises 66–69 using a graphing utility.

66. $x^2 + 3x - 10 > 0$

67. $2x^2 + 5x - 3 \leq 0$

68. $x^3 + x^2 - 4x - 4 > 0$

69. $\dfrac{x - 4}{x - 1} \leq 0$

Critical Thinking Exercises

70. Which one of the following is true?

a. The solution set of $x^2 > 25$ is $(5, \infty)$.

b. The inequality $\dfrac{x - 2}{x + 3} < 2$ can be solved by multiplying both sides by $x + 3$, resulting in the equivalent inequality $x - 2 < 2(x + 3)$.

c. $(x + 3)(x - 1) \geq 0$ and $\dfrac{x + 3}{x - 1} \geq 0$ have the same solution set.

d. None of these statements is true.

71. Write a polynomial inequality whose solution set is $[-3, 5]$.

72. Write a rational inequality whose solution set is $(-\infty, -4)$ or $[3, \infty)$.

In Exercises 73–76, use inspection to describe each inequality's solution set. Do not solve any of the inequalities.

73. $(x - 2)^2 > 0$

74. $(x - 2)^2 \leq 0$

75. $(x - 2)^2 < -1$

76. $\dfrac{1}{(x - 2)^2} > 0$

77. The graphing utility screen shows the graph of $y = 4x^2 - 8x + 7$.

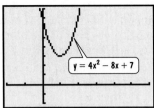

$y = 4x^2 - 8x + 7$

a. Use the graph to describe the solution set of $4x^2 - 8x + 7 > 0$.

b. Use the graph to describe the solution set of $4x^2 - 8x + 7 < 0$.

c. Use an algebraic approach to verify each of your descriptions in parts (a) and (b).

78. The graphing utility screen shows the graph of $y = \sqrt{27 - 3x^2}$. Write and solve a quadratic inequality that explains why the graph only appears for $-3 \le x \le 3$.

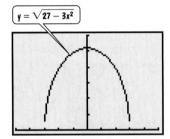

$y = \sqrt{27 - 3x^2}$

Group Exercise

79. This exercise is intended as a group learning experience and is appropriate for groups of three to five people. Before working on the various parts of the problem, reread the description of the position function on page 332.

a. Drop a ball from a height of 3 feet, 6 feet, and 12 feet. Record the number of seconds it takes for the ball to hit the ground.

b. For each of the three initial positions, use the position function to determine the time required for the ball to hit the ground.

c. What factors might result in differences between the times that you recorded and the times indicated by the function?

d. What appears to be happening to the time required for a free-falling object to hit the ground as its initial height is doubled? Verify this observation algebraically and with a graphing utility.

e. Repeat part (a) using a sheet of paper rather than a ball. What differences do you observe? What factor seems to be ignored in the position function?

f. What is meant by the acceleration of gravity and how does this number appear in the position function for a free-falling object?

SECTION 2.8 *Modeling Using Variation*

Objectives

1. Solve direct variation problems.

2. Solve inverse variation problems.

3. Solve combined variation problems.

4. Solve problems involving joint variation.

Have you ever wondered how telecommunication companies estimate the number of phone calls expected per day between two cities? The formula

$$N = \frac{400P_1P_2}{d^2}$$

shows that the daily number of phone calls, N, increases as the populations of the cities, P_1 and P_2, in thousands, increase and decreases as the distance, d, between the cities increases.

Certain formulas occur so frequently in applied situations that they are given special names. Variation formulas show how one quantity changes in relation to other quantities. Quantities can vary *directly, inversely,* or *jointly.* In this section, we look at situations that can be modeled by each of these kinds of variation. And think of this: The next time you get one of those "all-circuits-are-busy" messages, you will be able to use a variation formula to estimate how many other callers you're competing with for those precious 8-cent minutes.

1 Solve direct variation problems.

Direct Variation

Because light travels faster than sound, during a thunderstorm we see lightning before we hear thunder. The formula

$$d = 1080t$$

describes the distance, d, in feet, of the storm's center if it takes t seconds to hear thunder after seeing lightning. Thus,

If $t = 1$, $d = 1080 \cdot 1 = 1080.$ If it takes 1 second to hear thunder, the storm's center is 1080 feet away.

If $t = 2$, $d = 1080 \cdot 2 = 2160.$ If it takes 2 seconds to hear thunder, the storm's center is 2160 feet away.

If $t = 3$, $d = 1080 \cdot 3 = 3240.$ If it takes 3 seconds to hear thunder, the storm's center is 3240 feet away.

As the formula $d = 1080t$ illustrates, the distance to the storm's center is a constant multiple of how long it takes to hear the thunder. When the time is doubled, the storm's distance is doubled; when the time is tripled, the storm's distance is tripled; and so on. Because of this, the distance is said to *vary directly* as the time. The *equation of variation* is

$$d = 1080t.$$

Generalizing, we obtain the following statement:

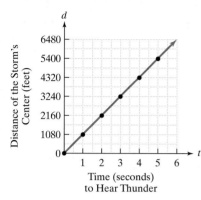

The graph of $d = 1080t$. Distance to a storm's center varies directly as the time it takes to hear thunder.

> ### Direct Variation
> If a situation is described by an equation in the form
> $$y = kx$$
> where k is a nonzero constant, we say that **y varies directly as x** or **y is directly proportional to x.** The number k is called the **constant of variation** or the **constant of proportionality.**

Problems involving direct variation can be solved using the following procedure. This procedure applies to direct variation problems, as well as to the other kinds of variation problems that we will discuss.

> ### Solving Variation Problems
> **1.** Write an equation that describes the given English statement.
> **2.** Substitute the given pair of values into the equation in step 1 and find the value of k.
> **3.** Substitute the value of k into the equation in step 1.
> **4.** Use the equation from step 3 to answer the problem's question.

EXAMPLE 1   Solving a Direct Variation Problem

The amount of garbage, G, varies directly as the population, P. Allegheny County, Pennsylvania, has a population of 1.3 million and creates 26 million pounds of garbage each week. Find the weekly garbage produced by New York City with a population of 7.3 million.

Solution

Step 1   **Write an equation.**   We know that y varies directly as x is expressed as

$$y = kx.$$

By changing letters, we can write an equation that describes the following English statement: Garbage production, G, varies directly as the population, P.

$$G = kP$$

Equivalently, garbage production is directly proportional to the population.

Step 2 Use the given values to find k. Allegheny County has a population of 1.3 million and creates 26 million pounds of garbage weekly. Substitute 26 for G and 1.3 for P in the direct variation equation. Then solve for k.

$$G = kP \qquad \text{This is the direct variation equation.}$$
$$26 = k \cdot 1.3 \qquad G = 26 \text{ and } P = 1.3.$$
$$\frac{26}{1.3} = \frac{k \cdot 1.3}{1.3} \qquad \text{Divide both sides by 1.3.}$$
$$20 = k \qquad \text{Simplify.}$$

Step 3 Substitute the value of k into the equation.

$$G = kP \qquad \text{Use the direct variation equation from step 1.}$$
$$G = 20P \qquad \text{Replace } k, \text{ the constant of variation, with 20.}$$

Step 4 Answer the problem's question. New York City has a population of 7.3 million. To find its weekly garbage production, substitute 7.3 for P in $G = 20P$ and solve for G.

$$G = 20P \qquad \text{Use the equation from step 3.}$$
$$G = 20(7.3) \qquad \text{Substitute 7.3 for } P.$$
$$G = 146$$

The weekly garbage produced by New York City weighs approximately 146 million pounds.

 Check Point 1 The pressure, P, of water on an object below the surface varies directly as its distance, D, below the surface. If a submarine experiences a pressure of 25 pounds per square inch 60 feet below the surface, how much pressure will it experience 330 feet below the surface?

The direct variation equation $y = kx$, or $f(x) = kx$, is a linear function. If $k > 0$, then the slope of the line is positive. Consequently, as x increases, y also increases.

A direct variation situation can involve variables to higher powers. For example, y can vary directly as x^2 ($y = kx^2$) or as x^3 ($y = kx^3$).

Direct Variation with Powers

y **varies directly as the nth power of x** if there exists some nonzero constant k such that

$$y = kx^n.$$

We also say that y **is directly proportional to the nth power of x.**

Direct variation with powers is modeled by polynomial functions. In our next example, the graph of the variation equation is the familiar parabola.

EXAMPLE 2 Solving a Direct Variation Problem

The distance, s, that a body falls from rest varies directly as the square of the time, t, of the fall. If skydivers fall 64 feet in 2 seconds, how far will they fall in 4.5 seconds?

Solution

Step 1 Write an equation. We know that y varies directly as the square of x is expressed as

$$y = kx^2.$$

By changing letters, we can write an equation that describes the following English statement: Distance, s, varies directly as the square of time, t, of the fall.

$$s = kt^2$$

> Equivalently, distance is directly proportional to the square of time.

Step 2 Use the given values to find k. Skydivers fall 64 feet in 2 seconds. Substitute 64 for s and 2 for t in the direct variation equation. Then solve for k.

$s = kt^2$	This is the direct variation equation.
$64 = k \cdot 2^2$	$s = 64$ and $t = 2$.
$64 = 4k$	Simplify.
$\dfrac{64}{4} = \dfrac{4k}{4}$	Divide both sides by 4.
$16 = k$	Simplify.

Step 2 Substitute the value of k into the equation.

$s = kt^2$	Use the direct variation equation from step 1.
$s = 16t^2$	Replace k, the constant of variation, with 16.

Step 4 Answer the problem's question. How far will the skydivers fall in 4.5 seconds? Substitute 4.5 for t in $s = 16t^2$ and solve for s.

$$s = 16(4.5)^2 = 16(20.25) = 324$$

Thus, in 4.5 seconds, skydivers will fall 324 feet.

We can express the variation equation from Example 2 in function notation, writing

$$s(t) = 16t^2.$$

The distance that a body falls from rest is a function of the time, t, of the fall. The parabola that is the graph of this quadratic function is shown in Figure 2.39 for $t \geq 0$. The graph increases rapidly from left to right, showing the effects of the acceleration of gravity.

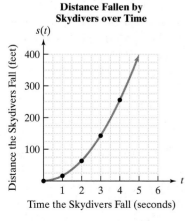

Distance Fallen by Skydivers over Time

Distance the Skydivers Fall (feet)

Time the Skydivers Fall (seconds)

Figure 2.39 The graph of $s(t) = 16t^2$

Check Point 2 The distance required to stop a car varies directly as the square of its speed. If 200 feet are required to stop a car traveling 60 miles per hour, how many feet are required to stop a car traveling 100 miles per hour?

2 Solve inverse variation problems.

Inverse Variation

The distance from Atlanta, Georgia, to Orlando, Florida, is 450 miles. The time that it takes to drive from Atlanta to Orlando depends on the rate at which one drives and is given by

$$\text{Time} = \frac{450}{\text{Rate}}.$$

For example, if you average 45 miles per hour, the time for the drive is

$$\text{Time} = \frac{450}{45} = 10,$$

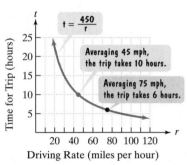

Figure 2.40

or 10 hours. If you ignore speed limits and average 75 miles per hour, the time for the drive is

$$\text{Time} = \frac{450}{75} = 6,$$

or 6 hours. As your rate (or speed) increases, the time for the trip decreases and vice versa. This is illustrated in Figure 2.40.

We can express the time for the Atlanta–Orlando trip using t for time and r for rate:

$$t = \frac{450}{r}.$$

This equation is an example of an *inverse variation* equation. Time, *t, varies inversely* as rate, *r.* When two quantities vary inversely, and the constant of variation is positive, such as 450, one quantity increases as the other decreases, and vice versa.

Generalizing, we obtain the following statement:

> ### Inverse Variation
> If a situation is described by an equation in the form
>
> $$y = \frac{k}{x}$$
>
> where k is a nonzero constant, we say that **y varies inversely as x** or **y is inversely proportional to x.** The number k is called the **constant of variation.**

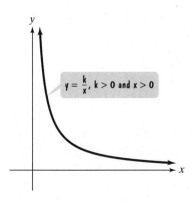

Figure 2.41 The graph of the inverse variation equation

Notice that the inverse variation equation

$$y = \frac{k}{x}, \quad \text{or} \quad f(x) = \frac{k}{x},$$

is a rational function. For $k > 0$ and $x > 0$, the graph of the function takes on the shape shown in Figure 2.41.

We use the same procedure to solve inverse variation problems as we did to solve direct variation problems. Example 3 illustrates this procedure.

EXAMPLE 3 Solving an Inverse Variation Problem

P $2P$

$2V$ V

Doubling the pressure halves the volume.

When you use a spray can and press the valve at the top, you decrease the pressure of the gas in the can. This decrease of pressure causes the volume of the gas in the can to increase. Because the gas needs more room than is provided in the can, it expands in spray form through the small hole near the valve. In general, if the temperature is constant, the pressure, P, of a gas in a container varies inversely as the volume, V, of the container. The pressure of a gas sample in a container whose volume is 8 cubic inches is 12 pounds per square inch. If the sample expands to a volume of 22 cubic inches, what is the new pressure of the gas?

Solution

Step 1   Write an equation.   We know that y varies inversely as x is expressed as

$$y = \frac{k}{x}.$$

By changing letters, we can write an equation that describes the following English statement: The pressure, P, of a gas in a container varies inversely as the volume, V.

$$P = \frac{k}{V}$$

> Equivalently, pressure is inversely proportional to volume.

Step 2   Use the given values to find k.   The pressure of a gas sample in a container whose volume is 8 cubic inches is 12 pounds per square inch. Substitute 12 for P and 8 for V in the inverse variation equation. Then solve for k.

$$P = \frac{k}{V} \qquad \text{This is the inverse variation equation.}$$

$$12 = \frac{k}{8} \qquad P = 12 \text{ and } V = 8.$$

$$12 \cdot 8 = \frac{k}{8} \cdot 8 \qquad \text{Multiply both sides by 8.}$$

$$96 = k \qquad \text{Simplify.}$$

Step 1   Substitute the value of k into the equation.

$$P = \frac{k}{V} \qquad \text{Use the inverse variation equation from step 1.}$$

$$P = \frac{96}{V} \qquad \text{Replace } k, \text{ the constant of variation, with 96.}$$

Step 4   Answer the problem's question.   We need to find the pressure when the volume expands to 22 cubic inches. Substitute 22 for V and solve for P.

$$P = \frac{96}{V} = \frac{96}{22} = 4\frac{4}{11}$$

When the volume is 22 cubic inches, the pressure of the gas is $4\frac{4}{11}$ pounds per square inch.

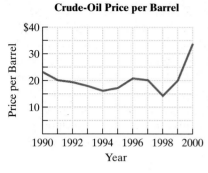

Crude-Oil Price per Barrel

Source: U.S. Department of Commerce

3 Solve combined variation problems.

Check Point 3

The price, P, of oil varies inversely as the supply, S. An OPEC nation sells oil for $19.50 per barrel when its daily production level is 4 million barrels. At what price will it sell oil if the daily production level is decreased to 3 million barrels?

Combined Variation

In a **combined variation** situation, direct and inverse variation occur at the same time. For example, as the advertising budget, A, of a company increases, its monthly sales, S, also increase. Monthly sales vary directly as the advertising budget:

$$S = kA.$$

By contrast, as the price of the company's product, P, increases, its monthly sales, S, decrease. Monthly sales vary inversely as the price of the product:

$$S = \frac{k}{P}.$$

We can combine these two variation equations into one equation:

$$S = \frac{kA}{P}.$$

The following example illustrates the application of combined variation.

EXAMPLE 4 Solving a Combined Variation Problem

The owners of Rollerblades Now determine that the monthly sales, S, of its skates vary directly as its advertising budget, A, and inversely as the price of the skates, P. When $60,000 is spent on advertising and the price of the skates is $40, the monthly sales are 12,000 pairs of rollerblades.

  a. Write an equation of variation that describes this situation.
  b. Determine monthly sales if the amount of the advertising budget is increased to $70,000.

Solution

  a. Write an equation.

$$S = \frac{kA}{P}$$ Translate "sales vary directly as the advertising budget and inversely as the skates' price."

Use the given values to find k.

$$12,000 = \frac{k(60,000)}{40}$$ When $60,000 is spent on advertising ($A = 60,000$) and the price is $40 ($P = 40$), monthly sales are 12,000 units ($S = 12,000$).

$$12,000 = k \cdot 1500$$ Divide 60,000 by 40.

$$\frac{12,000}{1500} = \frac{k \cdot 1500}{1500}$$ Divide both sides of the equation by 1500.

$$8 = k$$ Simplify

Using $k = 8$ and $S = \dfrac{kA}{P}$, the equation of variation that describes monthly sales is

$$S = \dfrac{8A}{P}.$$

b. The advertising budget is increased to $70,000, so $A = 70,000$. The skates' price is still $40, so $P = 40$.

$$S = \dfrac{8A}{P} \qquad \text{\textit{This is the combined variation equation from part (a).}}$$

$$S = \dfrac{8(70,000)}{40} \qquad \text{\textit{Substitute 70,000 for A and 40 for P.}}$$

$$S = 14,000 \qquad \text{\textit{Simplify.}}$$

With a $70,000 advertising budget and $40 price, the company can expect to sell 14,000 pairs of rollerblades in a month (up from 12,000).

> **Check Point 4**
> The number of minutes needed to solve an exercise set of variation problems varies directly as the number of problems and inversely as the number of people working to solve the problems. It takes 4 people 32 minutes to solve 16 problems. How many minutes will it take 8 people to solve 24 problems?

4 Solve problems involving joint variation.

Joint Variation

Joint variation is a variation in which a variable varies directly as the product of two or more other variables. Thus, the equation $y = kxz$ is read "y varies jointly as x and z."

Joint variation plays a critical role in Isaac Newton's formula for gravitation:

$$F = G\dfrac{m_1 m_2}{d^2}.$$

The formula states that the force of gravitation, F, between two bodies varies jointly as the product of their masses, m_1 and m_2, and inversely as the square of the distance between them, d. (G is the gravitational constant.) The formula indicates that gravitational force exists between any two objects in the universe, increasing as the distance between the bodies decreases. One practical result is that the pull of the moon on the oceans is greater on the side of the Earth closer to the moon. This gravitational imbalance is what produces tides.

EXAMPLE 5 Modeling Centrifugal Force

The centrifugal force, C, of a body moving in a circle varies jointly with the radius of the circular path, r, and the body's mass, m, and inversely with the square of the time, t, it takes to move about one full circle. A 6-gram body moving in a circle with radius 100 centimeters at a rate of 1 revolution in 2 seconds has a centrifugal force of 6000 dynes. Find the centrifugal force of an 18-gram body moving in a circle with radius 100 centimeters at a rate of 1 revolution in 3 seconds.

Solution

$$C = \frac{krm}{t^2}$$

Translate "Centrifugal force, C, varies jointly with radius, r, and mass, m, and inversely with the square of time, t."

$$6000 = \frac{k(100)(6)}{2^2}$$

If $r = 100$, $m = 6$, and $t = 2$, then $C = 6000$.

$$40 = k$$

Solve for k.

$$C = \frac{40rm}{t^2}$$

Substitute 40 for k in the model for centrifugal force.

$$= \frac{40(100)(18)}{3^2}$$

Find C when $r = 100$, $m = 18$, and $t = 3$.

$$= 8000$$

The centrifugal force is 8000 dynes.

 Check Point 5
The volume of a cone, V, varies jointly as its height, h, and the square of its radius, r. A cone with a radius measuring 6 feet and a height measuring 10 feet has a volume of 120π cubic feet. Find the volume of a cone having a radius of 12 feet and a height of 2 feet.

EXERCISE SET 2.8

 Practice Exercises

Use the four-step procedure for solving variation problems given on page 337 to solve Exercises 1–8.

1. y varies directly as x. $y = 35$ when $x = 5$. Find y when $x = 12$.

2. y varies directly as x. $y = 55$ when $x = 5$. Find y when $x = 13$.

3. y varies inversely as x. $y = 10$ when $x = 5$. Find y when $x = 2$.

4. y varies inversely as x. $y = 5$ when $x = 3$. Find y when $x = 9$.

5. y is directly proportional to x and inversely proportional to the square of z. $y = 20$ when $x = 50$ and $z = 5$. Find y when $x = 3$ and $z = 6$.

6. a is directly proportional to b and inversely proportional to the square of c. $a = 7$ when $b = 9$ and $c = 6$. Find a when $b = 4$ and $c = 8$.

7. y varies jointly as x and z. $y = 25$ when $x = 2$ and $z = 5$. Find y when $x = 8$ and $z = 12$.

8. C varies jointly as A and T. $C = 175$ when $A = 2100$ and $T = 4$. Find C when $A = 2400$ and $T = 6$.

 Application Exercises

9. A person's fingernail growth, G, in inches, varies directly as the number of weeks it has been growing, W.
   a. Write an equation that expresses this relationship.
   b. Fingernails grow at a rate of about 0.02 inch per week. Substitute 0.02 for k, the constant of variation, in the equation in part (a) and write the equation for fingernail growth.
   c. Substitute 52 for W to determine your fingernail length at the end of one year if for some bizarre reason you decided not to cut them and they did not break.

10. A person's wages, W, vary directly as the number of hours worked, h.
    a. Write an equation that expresses this relationship.
    b. For a 40-hour work week, Gloria earned $1400. Substitute 1400 for W and 40 for h in the equation from part (a) and find k, the constant of variation.
    c. Substitute the value of k into your equation in part (a) and write the equation that describes Gloria's wages in terms of the number of hours she works.
    d. Use the equation from part (c) to find Gloria's wages for 25 hours of work.

Use the four-step procedure for solving variation problems given on page 337 to solve Exercises 11–30.

11. The cost, C, of an airplane ticket varies directly as the number of miles, M, in the trip. A 3000-mile trip costs $400. What is the cost of a 450-mile trip?

12. An object's weight on the moon, M, varies directly as its weight on Earth, E. A person who weighs 55 kilograms on Earth weights 8.8 kilograms on the moon. What is the moon weight of a person who weighs 90 kilograms on Earth?

13. The Mach number is a measurement of speed named after the man who suggested it, Ernst Mach (1838–1916). The speed of an aircraft is directly proportional to its Mach number. Shown here are two aircraft. Use the figures for the Concorde to determine the Blackbird's speed.

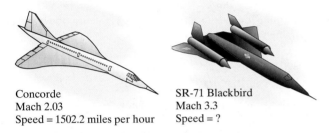

Concorde
Mach 2.03
Speed = 1502.2 miles per hour

SR-71 Blackbird
Mach 3.3
Speed = ?

14. Do you still own records, or are you strictly a CD person? Record owners claim that the quality of sound on good vinyl surpasses that of a CD, although this is up for debate. This, however, is not debatable: The number of revolutions a record makes as it is being played is directly proportional to the time that it is on the turntable. A record that lasted 3 minutes made 135 revolutions. If a record takes 2.4 minutes to play, how many revolutions does it make?

15. If all men had identical body types, their weight would vary directly as the cube of their height. Shown is Robert Wadlow, who reached a record height of 8 feet 11 inches (107 inches) before his death at age 22. If a man who is 5 feet 10 inches tall (70 inches) with the same body type as Mr. Wadlow weighs 170 pounds, what was Robert Wadlow's weight shortly before his death?

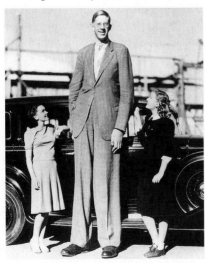

16. The distance that an object falls varies directly as the square of the time it has been falling. An object falls 144 feet in 3 seconds. Find how far it will fall in 7 seconds.

17. The time that it takes you to get to campus varies inversely as your driving rate. Averaging 20 miles per hour in terrible traffic, it takes you 1.5 hours to get to campus. How long would the trip take averaging 60 miles per hour?

18. The weight that can be supported by a 2-inch by 4-inch piece of pine (called a 2-by-4) varies inversely as its length. A 10-foot 2-by-4 can support 500 pounds. What weight can be supported by a 5-foot 2-by-4?

19. The volume of a gas in a container at a constant temperature is inversely proportional to the pressure. If the volume is 32 cubic centimeters at a pressure of 8 pounds, find the pressure when the volume is 40 cubic centimeters.

20. The current in a circuit is inversely proportional to the resistance. The current is 20 amperes when the resistance is 5 ohms. Find the current for a resistance of 16 ohms.

21. A person's body-mass index is used to assess levels of fatness, with an index from 20 to 26 considered in the desirable range. The index varies directly as one's weight, in pounds, and inversely as one's height, in inches. A person who weighs 150 pounds and is 70 inches tall has an index of 21. What is the body-mass index of a person who weighs 240 pounds and is 74 inches tall? Because the index is rounded to the nearest whole number, do so and then determine if this person's level of fatness is in the desirable range.

22. The volume of a gas varies directly as its temperature and inversely as its pressure. At a temperature of 100 Kelvin and a pressure of 15 kilograms per square meter, the gas occupies a volume of 20 cubic meters. Find the volume at a temperature of 150 Kelvin and a pressure of 30 kilograms per square meter.

23. The intensity of illumination on a surface varies inversely as the square of the distance of the light source from the surface. The illumination from a source is 25 foot-candles at a distance of 4 feet. What is the illumination when the distance is 6 feet?

24. The gravitational force with which Earth attracts an object varies inversely with the square of the distance from the center of Earth. A gravitational force of 0.4 pound acts on an object 8000 miles from Earth's center. Find the force of attraction on an object 6000 miles from the center of Earth.

25. Kinetic energy varies jointly as the mass and the square of the velocity. A mass of 8 grams and velocity of 3 centimeters per second has a kinetic energy of 36 ergs. Find the kinetic energy for a mass of 4 grams and velocity of 6 centimeters per second.

26. The electrical resistance of a wire varies directly as its length and inversely as the square of its diameter. A wire of 720 feet with $\frac{1}{4}$-inch diameter has a resistance of $1\frac{1}{2}$ ohms. Find the resistance for 960 feet of the same kind of wire if its diameter is doubled.

27. The average number of phone calls between two cities in a day varies jointly as the product of their populations and inversely as the square of the distance between them. The population of Minneapolis is 2538 thousand and the population of Cincinnati is 1818 thousand. Separated by 108 miles, the average number of telephone calls per day between the two cities is 158,233. Find the average number of telephone calls per day between Orlando, Florida (population 1225 thousand) and Seattle, Washington (population 2970 thousand), two cities that are 3403 miles apart.

28. The force of attraction between two bodies varies jointly as the product of their masses and inversely as the square of the distance between them. Two 1-kilogram masses separated by 1 meter exert a force of attraction of 6.67×10^{-11} newton. What is the gravitational force exerted by Earth on a 1000-kilogram satellite orbiting at an altitude of 300 kilometers? (The mass of Earth is 5.98×10^{24} kilograms and its radius is 6400 kilometers. Consequently, the distance between Earth and the satellite is $6400 + 300$, or 6700 kilometers.)

29. The force of wind blowing on a window positioned at a right angle to the direction of the wind varies jointly as the area of the window and the square of the wind's speed. It is known that a wind of 30 miles per hour blowing on a window measuring 4 feet by 5 feet exerts a force of 150 pounds. During a storm with winds of 60 miles per hour, should hurricane shutters be placed on a window that measures 3 feet by 4 feet and is capable of withstanding 300 pounds of force?

30. The table of values shows the values for the current, I, in an electric circuit and the resistance, R, of the circuit.

I (amperes)	0.5	1.0	1.5	2.0	2.5	3.0	4.0	5.0
R (ohms)	12	6.0	4.0	3.0	2.4	2.0	1.5	1.2

   a. Graph the ordered pairs in the table of values, with values of I along the x-axis and values of R along the y-axis. Connect the eight points with a smooth curve.

   b. Does current vary directly or inversely as resistance? Use your graph and explain how you arrived at your answer.

   c. Write an equation of variation for I and R, using one of the ordered pairs in the table to find the constant of variation. Then use your variation equation to verify the other seven ordered pairs in the table.

Writing in Mathematics

31. What does it mean if two quantities vary directly?

32. In your own words, explain how to solve a variation problem.

33. What does it mean if two quantities vary inversely?

34. Explain what is meant by combined variation. Give an example with your explanation.

35. Explain what is meant by joint variation. Give an example with your explanation.

In Exercises 36–37, describe in words the variation shown by the given equation.

36. $z = \dfrac{k\sqrt{x}}{y^2}$ **37.** $z = kx^2\sqrt{y}$

38. We have seen that the daily number of phone calls between two cities varies jointly as their populations and inversely as the square of the distance between them. This model, used by telecommunication companies to estimate the line capacities needed between various cities, is called the *gravity model*. Compare the model to Newton's formula for gravitation on page 343 and describe why the name *gravity model* is appropriate.

Technology Exercise

39. Use a graphing utility to graph any three of the variation equations in Exercises 11–20. Then $\boxed{\text{TRACE}}$ along each curve and identify the point that corresponds to the problem's solution.

Critical Thinking Exercises

40. In a hurricane, the wind pressure varies directly as the square of the wind velocity. If wind pressure is a measure of a hurricane's destructive capacity, what happens to this destructive power when the wind speed doubles?

41. The illumination from a light source varies inversely as the square of the distance from the light source. If you raise a lamp from 15 inches to 30 inches over your desk, what happens to the illumination?

42. The heat generated by a stove element varies directly as the square of the voltage and inversely as the resistance. If the voltage remains constant, what needs to be done to triple the amount of heat generated?

43. Galileo's telescope brought about revolutionary changes in astronomy. A comparable leap in our ability to observe the universe took place as a result of the Hubble Space Telescope. The space telescope can see stars and galaxies whose brightness is $\frac{1}{50}$ of the faintest objects now observable using ground-based telescopes. Use the fact that the brightness of a point source, such as a star, varies inversely as the square of its distance from an observer to show that the space telescope can see about seven times farther than a ground-based telescope.

Group Exercise

44. Begin by deciding on a product that interests the group because you are now in charge of advertising this product. Members were told that the demand for the product varies directly as the amount spent on advertising and inversely as the price of the product. However, as more money is spent on advertising, the price of your product rises.

Under what conditions would members recommend an increased expense in advertising? Once you've determined what your product is, write formulas for the given conditions and experiment with hypothetical numbers. What other factors might you take into consideration in terms of your recommendation? How do these factors affect the demand for your product?

CHAPTER SUMMARY, REVIEW, AND TEST

Summary

DEFINITIONS AND CONCEPTS	EXAMPLES

2.1 Complex Numbers

a. The imaginary unit i is defined as
$$i = \sqrt{-1}, \text{ where } i^2 = -1.$$
The set of numbers in the form $a + bi$ is called the set of complex numbers; a is the real part and b is the imaginary part. If $b = 0$, the complex number is a real number. If $b \neq 0$, the complex number is an imaginary number. Complex numbers in the form bi are called pure imaginary numbers.

b. Rules for adding and subtracting complex numbers are given in the box on page 247. Ex. 1, p. 248

c. To multiply complex numbers, multiply as if they are polynomials. After completing the multiplication, replace i^2 with -1. Ex. 2, p. 248

d. The complex conjugate of $a + bi$ is $a - bi$ and vice versa. The multiplication of complex conjugates gives a real number:
$$(a + bi)(a - bi) = a^2 + b^2.$$

e. To divide complex numbers, multiply the numerator and the denominator by the complex conjugate of the denominator. Ex. 3, p. 249

f. When performing operations with square roots of negative numbers, begin by expressing all square roots in terms of i. The principal square root of $-b$ is defined by
$$\sqrt{-b} = i\sqrt{b}.$$ Ex. 4, p. 250

g. Quadratic equations ($ax^2 + bx + c = 0, a \neq 0$) with negative discriminants ($b^2 - 4ac < 0$) have imaginary solutions that are complex conjugates. Ex. 5, p. 251

2.2 Quadratic Functions

a. A quadratic function is of the form $f(x) = ax^2 + bx + c, a \neq 0$.

b. The standard form of a quadratic function is $f(x) = a(x - h)^2 + k, a \neq 0$.

c. The graph of a quadratic function is a parabola. The vertex is (h, k) or $\left(-\dfrac{b}{2a}, f\left(-\dfrac{b}{2a}\right)\right)$. Ex. 1, p. 255; Ex. 2, p. 256; Ex. 3, p. 258

A procedure for graphing a quadratic function is given in the box on page 255.

d. See the box on page 259 for minimum or maximum values of quadratic functions. Ex. 4, p. 260

e. A strategy for solving problems involving maximizing or minimizing quadratic functions is given in box on page 261. Ex. 5, p. 261; Ex. 6, p. 262

DEFINITIONS AND CONCEPTS	**EXAMPLES**

2.3 Polynomial Functions and Their Graphs

a. Polynomial Function of x of Degree n: $f(x) = a_nx^n + a_{n-1}x^{n-1} + \cdots + a_2x^2 + a_1x + a_0, a_n \neq 0$

b. The graphs of polynomial functions are smooth and continuous.

c. The end behavior of the graph of a polynomial function depends on the leading term, given by the Leading Coefficient Test in the box on page 271.
Exs. 1&2, p. 272; Ex. 3, p. 272

d. The values of x for which $f(x)$ is equal to 0 are the zeros of the polynomial function f. These values are the roots, or solutions, of the polynomial equation $f(x) = 0$.
Ex. 4, p. 273; Ex. 5, p. 274

e. If $x - r$ occurs k times in a polynomial function's factorization, r is a repeated zero with multiplicity k. If k is even, the graph touches the x-axis at r; if odd, it crosses the x-axis at r.

f. The Intermediate Value Theorem: If $f(a)$ and $f(b)$ have opposite signs, there is at least one value of c between a and b for which $f(c) = 0$.
Ex. 6, p. 275

g. If f is a polynomial of degree n, the graph of f has at most $n - 1$ turning points.

h. A strategy for graphing a polynomial function is given in the box on page 276.
Ex. 7, p. 276

2.4 Dividing Polynomials; Remainder and Factor Theorems

a. Long division of polynomials is performed by dividing, multiplying, subtracting, bringing down the next term, and repeating this process until the degree of the remainder is less than the degree of the divisor. The details are given in the box on page 283.
Ex. 1, p. 282; Ex. 2, p. 284; Ex. 3, p. 285

b. The Division Algorithm: $f(x) = d(x)q(x) + r(x)$. The dividend is the product of the divisor and the quotient plus the remainder.

c. Synthetic division is used to divide a polynomial by $x - c$. The details are given in the box on pages 286–287.
Ex. 4, p. 287

d. The Remainder Theorem: If polynomial $f(x)$ is divided by $x - c$, then the remainder is $f(c)$.
Ex. 5, p. 289

e. The Factor Theorem: If $x - c$ is a factor of a polynomial function $f(x)$, c is a zero of f and a root of $f(x) = 0$. If c is a zero of f or a root of $f(x) = 0$, then $x - c$ is a factor of $f(x)$.
Ex. 6, p. 290

2.5 Zeros of Polynomial Functions

a. The Rational Zero Theorem states that possible rational zeros of a polynomial
function $= \dfrac{\text{Factors of the constant term}}{\text{Factors of the leading coefficient}}$. The theorem is stated in the box on page 293.
Ex. 1, p. 293; Ex. 2, p. 294; Ex. 3, p. 295; Ex. 4, p. 296

b. Number of roots: If $f(x)$ is a polynomial of degree $n \geq 1$, then, counting multiple roots separately, the equation $f(x) = 0$ has n roots.

c. If $a + bi$ is a root of $f(x) = 0$, then $a - bi$ is also a root.

d. The Linear Factorization Theorem: An nth-degree polynomial can be expressed as the product of n linear factors. Thus, $f(x) = a_n(x - c_1)(x - c_2) \cdots (x - c_n)$.
Ex. 5, p. 299; Ex. 6, p. 299

e. Descartes's Rule of Signs: The number of positive real zeros of f equals the number of sign changes of $f(x)$ or is less than that number by an even integer. The number of negative real zeros of f applies a similar statement to $f(-x)$.
Ex. 7, p. 300

2.6 Rational Functions and Their Graphs

a. Rational function: $f(x) = \dfrac{p(x)}{q(x)}$; p and q are polynomial functions and $q(x) \neq 0$. The
Ex. 1, p. 306

domain of f is the set of all real numbers excluding values of x that make $q(x)$ zero.

b. Arrow notation is summarized in the box on page 308.

c. The line $x = a$ is a vertical asymptote of the graph of f if $f(x)$ increases or decreases without bound as x approaches a. Vertical asymptotes are identified using the location theorem in the box on page 309.
Ex. 2, p. 309

DEFINITIONS AND CONCEPTS	EXAMPLES

d. The line $y = b$ is a horizontal asymptote of the graph of f if $f(x)$ approaches b as x increases or decreases without bound. Horizontal asymptotes are identified using the location theorem in the box on page 311.

Ex. 3, p. 311

e. A strategy for graphing rational functions is given in the box on page 313.

Ex. 4, p. 313;
Ex. 5, p. 315;
Ex. 6, p. 316

f. The graph of a rational function has a slant asymptote when the degree of the numerator is one more than the degree of the denominator. The equation of the slant asymptote is found using division and dropping the remainder term.

Ex. 7, p. 318

2.7 Polynomial and Rational Inequalities

a. A polynomial inequality can be expressed as $f(x) < 0$, $f(x) > 0$, $f(x) \le 0$, or $f(x) \ge 0$, where f is a polynomial function.

b. A procedure for solving polynomial inequalities is given in the box on page 327.

Ex. 1, p. 327;
Ex. 2, p. 328

c. A rational inequality can be expressed as $f(x) < 0$, $f(x) > 0$, $f(x) \le 0$, or $f(x) \ge 0$, where f is a rational function.

d. To solve a rational inequality, bring all terms to one side, obtaining zero on the other side. Express the rational function on the nonzero side as a single quotient. Find boundary points by determining the zeros of the rational function's numerator and denominator. Then follow a procedure similar to that for solving polynomial inequalities.

Ex. 3, p. 330

2.8 Modeling Using Variation

a.

English Statement	Equation	
y varies directly as x. y is directly proportional to x.	$y = kx$	Ex. 1, p. 337
y varies directly as x^n. y is directly proportional to x^n.	$y = kx^n$	Ex. 2, p. 339
y varies inversely as x. y is inversely proportional to x.	$y = \dfrac{k}{x}$	Ex. 3, p. 341; Ex. 4, p. 342
y varies inversely as x^n. y is inversely proportional to x^n.	$y = \dfrac{k}{x^n}$	
y varies jointly as x and z.	$y = kxz$	Ex. 5, p. 343

b. A procedure for solving variation problems is given in the box on page 337.

Review Exercises

2.1

In Exercises 1–10, perform the indicated operations and write the result in standard form

1. $(8 - 3i) - (17 - 7i)$ **2.** $4i(3i - 2)$
3. $(7 - 5i)(2 + 3i)$ **4.** $(3 - 4i)^2$
5. $(7 + 8i)(7 - 8i)$ **6.** $\dfrac{6}{5 + i}$
7. $\dfrac{3 + 4i}{4 - 2i}$ **8.** $\sqrt{-32} - \sqrt{-18}$
9. $(-2 + \sqrt{-100})^2$ **10.** $\dfrac{4 + \sqrt{-8}}{2}$

In Exercises 11–12, solve each quadratic equation using the quadratic formula. Express solutions in standard form.

11. $x^2 - 2x + 4 = 0$ **12.** $2x^2 - 6x + 5 = 0$

2.2

In Exercises 13–16, use the vertex and intercepts to sketch the graph of each quadratic function. Give the equation for the parabola's axis of symmetry.

13. $f(x) = -2(x - 1)^2 + 3$ **14.** $f(x) = (x + 4)^2 - 2$
15. $f(x) = -x^2 + 2x + 3$ **16.** $f(x) = 2x^2 - 4x - 6$

17. A person standing close to the edge on the top of an 80-foot building throws a ball vertically upward with an initial velocity of 64 feet per second. The function $s(t) = -16t^2 + 64t + 80$ describes the ball's height above the ground, $s(t)$, in feet, t seconds after it is thrown. After how many seconds does the ball reach its maximum height? What is the maximum height?

18. Suppose that a quadratic function is used to model the data shown in the graph using

(number of years after 1960, divorce rate per 1000 population).

U.S. Divorce Rate

Source: U.S. Department of Health and Human Services

Determine, without obtaining an actual quadratic function that models the data, the approximate coordinates of the vertex for the function's graph. Describe what this means in practical terms. Use the word "maximum" in your description.

19. A field bordering a straight stream is to be enclosed. The side bordering the stream is not to be fenced. If 1000 yards of fencing material is to be used, what are the dimensions of the largest rectangular field that can be fenced? What is the maximum area?

20. You have 1000 feet of fencing to construct six corrals, as shown in the figure. Find the dimensions that maximize the enclosed area.

21. The annual yield per fruit tree is fairly constant at 150 pounds per tree when the number of trees per acre is 35 or fewer. For each additional tree over 35, the annual yield per tree for all trees on the acre decreases by 4 pounds due to overcrowding. How many fruit trees should be planted per acre to maximize the annual yield for the acre? What is the maximum number of pounds of fruit per acre?

2.3

In Exercises 22–25, use the Leading Coefficient Test to determine the end behavior of the graph of the given polynomial function. Then use this end behavior to match the polynomial function with its graph. [The graphs are labeled (a) through (d).]

22. $f(x) = -x^3 + 12x^2 - x$

23. $f(x) = x^6 - 6x^4 + 9x^2$

24. $f(x) = x^5 - 5x^3 + 4x$

25. $f(x) = -x^4 + 1$

a.

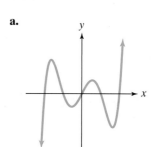

b.

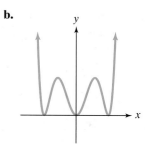

c.

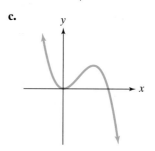

d.

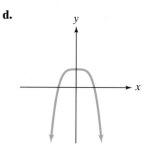

26. The function $f(x) = -0.0013x^3 + 0.78x^2 - 1.43x + 18.1$ models the percentage of U.S. families below the poverty level, $f(x)$, x years after 1960. Use end behavior to explain why the model is valid only for a limited period of time.

27. Despite a combination of drugs used to inhibit the growth of the HIV virus, a patient dies as a result of the virus overwhelming his body. Could the function $N(t) = -\frac{3}{4}t^4 + 3t^3 + 5$ model the number of viral particles, $N(t)$, in billions, in this patient's body over time? Use the graph's end behavior to the right to answer the question. Explain your answer.

In Exercises 28–29, find the zeros for each polynomial function and give the multiplicity of each zero. State whether the graph crosses or touches the x-axis at each zero.

28. $f(x) = -2(x - 1)(x + 2)^2(x + 5)^3$

29. $f(x) = x^3 - 5x^2 - 25x + 125$

In Exercises 30–31, use the Intermediate Value Theorem to show that the polynomial has a zero between the given integers.

30. $f(x) = x^3 - 2x - 1$; between 1 and 2

31. $f(x) = 3x^3 + 2x^2 - 8x + 7$; between -3 and -2

In Exercises 32–37,
    a. *Use the Leading Coefficient Test to determine the graph's end behavior.*
    b. *Determine whether the graph has y-axis symmetry, origin symmetry, or neither.*
    c. *Graph the function.*

32. $f(x) = x^3 - x^2 - 9x + 9$ **33.** $f(x) = 4x - x^3$
34. $f(x) = 2x^3 + 3x^2 - 8x - 12$
35. $f(x) = -x^4 + 25x^2$
36. $f(x) = -x^4 + 6x^3 - 9x^2$ **37.** $f(x) = 3x^4 - 15x^3$

2.4

In Exercises 38–40, divide using long division.
38. $(4x^3 - 3x^2 - 2x + 1) \div (x + 1)$
39. $(10x^3 - 26x^2 + 17x - 13) \div (5x - 3)$
40. $(4x^4 + 6x^3 + 3x - 1) \div (2x^2 + 1)$

In Exercises 41–42, divide using synthetic division.
41. $(3x^4 + 11x^3 - 20x^2 + 7x + 35) \div (x + 5)$
42. $(3x^4 - 2x^2 - 10x) \div (x - 2)$
43. Given $f(x) = 2x^3 - 7x^2 + 9x - 3$, use the Remainder Theorem to find $f(-13)$.
44. Use synthetic division to divide $f(x) = 2x^3 + x^2 - 13x + 6$ by $x - 2$. Use the result to find all zeros of f.
45. Solve the equation $x^3 - 17x + 4 = 0$ given that 4 is a root.

2.5

In Exercises 46–47, use the Rational Zero Theorem to list all possible rational zeros for each given function.
46. $f(x) = x^4 - 6x^3 + 14x^2 - 14x + 5$
47. $f(x) = 3x^5 - 2x^4 - 15x^3 + 10x^2 + 12x - 8$

In Exercises 48–49, use Descartes's Rule of Signs to determine the possible number of positive and negative real zeros for each given function.
48. $f(x) = 3x^4 - 2x^3 - 8x + 5$
49. $f(x) = 2x^5 - 3x^3 - 5x^2 + 3x - 1$
50. Use Descartes's Rule of Signs to explain why $2x^4 + 6x^2 + 8 = 0$ has no real roots.

For Exercises 51–56,
    a. *List all possible rational roots or rational zeros.*
    b. *Use Descartes's Rule of Signs to determine the possible number of positive and negative real roots or real zeros.*
    c. *Use synthetic division to test the possible rational roots or zeros and find an actual root or zero.*
    d. *Use the root or zero from part (c) to find all the zeros or roots.*

51. $f(x) = x^3 + 3x^2 - 4$
52. $f(x) = 6x^3 + x^2 - 4x + 1$
53. $8x^3 - 36x^2 + 46x - 15 = 0$
54. $x^4 - x^3 - 7x^2 + x + 6 = 0$

55. $4x^4 + 7x^2 - 2 = 0$
56. $f(x) = 2x^4 + x^3 - 9x^2 - 4x + 4$

In Exercises 57–58, find an nth-degree polynomial function with real coefficients satisfying the given conditions. If you are using a graphing utility, graph the function and verify the real zeros and the given function value.
57. $n = 3$; 2 and $2 - 3i$ are zeros; $f(1) = -10$
58. $n = 4$; i is a zero; -3 is a zero of multiplicity 2; $f(-1) = 16$

In Exercises 59–60, find all the zeros of each polynomial function and write the polynomial as a product of linear factors.
59. $f(x) = 2x^4 + 3x^3 + 3x - 2$
60. $g(x) = x^4 - 6x^3 + x^2 + 24x + 16$

In Exercises 61–64, graphs of fifth-degree polynomial functions are shown. In each case, specify the number of real zeros and the number of imaginary zeros. Indicate whether there are any real zeros with multiplicity other than 1.

61.
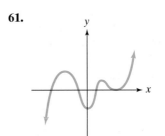

62.
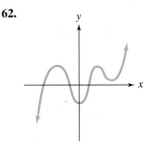

63.
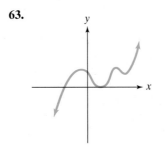

64.
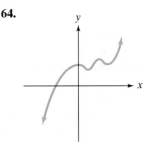

2.6

In Exercises 65–72, find the vertical asymptotes, if any, the horizontal asymptote, if there is one, and the slant asymptote, if there is one, of the graph of each rational function. Then graph the rational function.

65. $f(x) = \dfrac{2x}{x^2 - 9}$ **66.** $g(x) = \dfrac{2x - 4}{x + 3}$

67. $h(x) = \dfrac{x^2 - 3x - 4}{x^2 - x - 6}$ **68.** $r(x) = \dfrac{x^2 + 4x + 3}{(x + 2)^2}$

69. $y = \dfrac{x^2}{x + 1}$ **70.** $y = \dfrac{x^2 + 2x - 3}{x - 3}$

71. $f(x) = \dfrac{-2x^3}{x^2 + 1}$ **72.** $g(x) = \dfrac{4x^2 - 16x + 16}{2x - 3}$

73. A company is planning to manufacture affordable graphing calculators. Fixed monthly cost will be $50,000, and it will cost $25 to produce each calculator.
  a. Write the cost function, C, of producing x graphing calculators.
  b. Write the average cost function, $\overline{C}$, of producing x graphing calculators.
  c. Find and interpret $\overline{C}(50)$, $\overline{C}(100)$, $\overline{C}(1000)$, and $\overline{C}(100,000)$.
  d. What is the horizontal asymptote for this function, and what does it represent?

74. In Palo Alto, California, a government agency ordered computer-related companies to contribute to a monetary pool to clean up underground water supplies. (The companies had stored toxic chemicals in leaking underground containers.) The rational function

$$C(x) = \frac{200x}{100 - x}$$

models the cost, $C(x)$, in tens of thousands of dollars, for removing x percent of the contaminants.
  a. Find and interpret $C(90) - C(50)$.
  b. What is the equation for the vertical asymptote? What does this mean in terms of the variables given by the function?

Exercises 75–76 involve rational functions that model the given situations. In each case, find the horizontal asymptote as $x \to \infty$ and then describe what this means in practical terms.

75. $f(x) = \dfrac{150x + 120}{0.05x + 1}$; the number of bass, $f(x)$, after x months in a lake that was stocked with 120 bass

76. $P(x) = \dfrac{72,900}{100x^2 + 729}$; the percentage, $P(x)$, of people in the United States with x years of education who are unemployed.

77. The function $p(x) = 1.96x + 3.14$ models the number of nonviolent prisoners, $p(x)$, in thousands, in New York State prisons x years after 1980. The function $q(x) = 3.04x + 21.79$ models the total number of prisoners, $q(x)$, in thousands, in New York State prisons x years after 1980.
  a. Write a function that models the fraction of nonviolent prisoners in New York prisons x years after 1980.
  b. What is the equation of the horizontal asymptote associated with the function in part (a)? Describe what this means about the percentage, to the nearest tenth of a percent, of nonviolent prisoners in New York prisons over time.
  c. Use your equation in part (b) to explain why, in 1998, New York implemented a strategy where more nonviolent offenders are granted parole and more violent offenders are denied parole.

78. A jogger ran 4 miles and then walked 2 miles. The average velocity running was 3 miles per hour faster than the average velocity walking. Express the total time for running and walking, T, as a function of the average velocity walking, x.

79. The area of a rectangular floor is 1000 square feet. Express the perimeter of the floor, P, as a function of the width of the rectangle, x.

2.7

Solve each polynomial inequality in Exercises 80–82, and graph the solution set on a real number line. Express each solution set in interval notation.

80. $2x^2 + 7x \le 4$ **81.** $2x^2 > 6x - 3$

82. $x^3 + 2x^2 - 3x > 0$

Solve each rational inequality in Exercises 83–84, and graph the solution set on a real number line. Express each solution set in interval notation.

83. $\dfrac{x - 6}{x + 2} > 0$ **84.** $\dfrac{x + 3}{x - 4} \le 5$

85. Use the position function

$$s(t) = -16t^2 + v_0 t + s_0$$

initial velocity   initial height

to solve this problem. A projectile is fired vertically upward from ground level with an initial velocity of 48 feet per second. During which time period will the projectile's height exceed 32 feet?

2.8

Solve the variation problems in Exercises 86–91.

86. An electric bill varies directly as the amount of electricity used. The bill for 1400 kilowatts of electricity is $98. What is the bill for 2200 kilowatts of electricity?

87. The distance that a body falls from rest is directly proportional to the square of the time of the fall. If skydivers fall 144 feet in 3 seconds, how far will they fall in 10 seconds?

88. The time it takes to drive a certain distance is inversely proportional to the rate of travel. If it takes 4 hours at 50 miles per hour to drive the distance, how long will it take at 40 miles per hour?

89. The loudness of a stereo speaker, measured in decibels, varies inversely as the square of your distance from the speaker. When you are 8 feet from the speaker, the loudness is 28 decibels. What is the loudness when you are 4 feet from the speaker?

90. The time required to assemble computers varies directly as the number of computers assembled and inversely as the number of workers. If 30 computers can be assembled by 6 workers in 10 hours, how long would it take 5 workers to assemble 40 computers?

91. The volume of a pyramid varies jointly as its height and the area of its base. A pyramid with a height of 15 feet and a base with an area of 35 square feet has a volume of 175 cubic feet. Find the volume of a pyramid with a height of 20 feet and a base with an area of 120 square feet.

Chapter 2 Test

In Exercises 1–3, perform the indicated operations and write the result in standard form.

1. $(6 - 7i)(2 + 5i)$ **2.** $\dfrac{5}{2 - i}$

3. $2\sqrt{-49} + 3\sqrt{-64}$

4. Solve and express solutions in standard form:
$x^2 = 4x - 8$.

In Exercises 5–6, use the vertex and intercepts to sketch the graph of each quadratic function. Give the equation for the parabola's axis of symmetry.

5. $f(x) = (x + 1)^2 + 4$ **6.** $f(x) = x^2 - 2x - 3$

7. Determine, without graphing, whether the quadratic function $f(x) = -2x^2 + 12x - 16$ has a minimum value or a maximum value. Then find the coordinates of the minimum or the maximum point.

8. The function $f(x) = -x^2 + 46x - 360$ models the daily profit, $f(x)$, in hundreds of dollars, for a company that manufactures x computers daily. How many computers should be manufactured each day to maximize profit? What is the maximum daily profit?

9. Among all pairs of numbers whose difference is 14, find a pair whose product is as small as possible. What is the minimum product?

10. Consider the function $f(x) = x^3 - 5x^2 - 4x + 20$.

    a. Use factoring to find all zeros of f.

    b. Use the Leading Coefficient Test and the zeros of f to graph the function.

11. Use end behavior to explain why the graph cannot be the graph of $f(x) = x^5 - x$. Then use intercepts to explain why the graph cannot represent $f(x) = x^5 - x$.

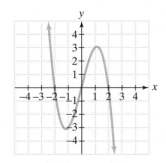

12. The graph of $f(x) = 6x^3 - 19x^2 + 16x - 4$ is shown in the figure at the top of the next column.

    a. Based on the graph of f, find the root of the equation $6x^3 - 19x^2 + 16x - 4 = 0$ that is an integer.

    b. Use synthetic division to find the other two roots of $6x^3 - 19x^2 + 16x - 4 = 0$.

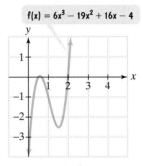

$f(x) = 6x^3 - 19x^2 + 16x - 4$

13. Use the Rational Zero Theorem to list all possible rational zeros of $f(x) = 2x^3 + 11x^2 - 7x - 6$.

14. Use Descartes's Rule of Signs to determine the possible number of positive and negative real zeros of $f(x) = 3x^5 - 2x^4 - 2x^2 + x - 1$.

15. Solve: $x^3 + 6x^2 - x - 30 = 0$.

16. Consider the function whose equation is given by $f(x) = 2x^4 - x^3 - 13x^2 + 5x + 15$.

    a. List all possible rational zeros.

    b. Use the graph of f in the figure shown and synthetic division to find all zeros of the function.

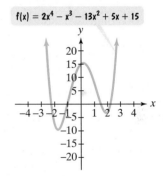

$f(x) = 2x^4 - x^3 - 13x^2 + 5x + 15$

17. Use the graph of $f(x) = x^3 + 3x^2 - 4$ in the figure shown to factor $x^3 + 3x^2 - 4$.

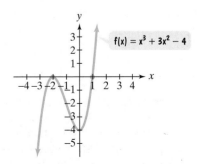

$f(x) = x^3 + 3x^2 - 4$

In Exercises 18–21, find the domain of each rational function and graph the function.

18. $f(x) = \dfrac{x}{x^2 - 16}$ **19.** $f(x) = \dfrac{x^2 - 9}{x - 2}$

20. $f(x) = \dfrac{x + 1}{x^2 + 2x - 3}$ **21.** $f(x) = \dfrac{4x^2}{x^2 + 3}$

22. Rational functions can be used to model learning. Many of these functions model the proportion of correct responses as a function of the number of trials of a particular task. One such model, called a learning curve, is

$$f(x) = \dfrac{0.9x - 0.4}{0.9x + 0.1}$$

where $f(x)$ is the proportion of correct responses after x trials. If $f(x) = 0$, there are no correct responses. If $f(x) = 1$, all responses are correct. The graph of the rational function is shown.

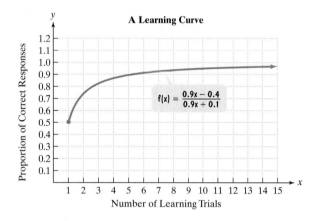

A Learning Curve

$f(x) = \dfrac{0.9x - 0.4}{0.9x + 0.1}$

Number of Learning Trials

Proportion of Correct Responses

a. According to the graph, what proportion of responses are correct after 5 learning trials?

b. According to the graph, how many learning trials are necessary for 0.95 of the responses to be correct?

c. Use the function's equation to write the equation of the horizontal asymptote. What does this mean in terms of the variables modeled by the learning curve?

Simplify each inequality in Exercises 23–24 and graph the solution set on a real number line. Express each solution set in interval notation.

23. $x^2 < x + 12$

24. $\dfrac{2x + 1}{x - 3} > 3$

25. The intensity of light received at a source varies inversely as the square of the distance from the source. A particular light has an intensity of 20 foot-candles at 15 feet. What is the light's intensity at 10 feet?

Cumulative Review Exercises (Chapters P–2)

Simplify each expression in Exercises 1–3.

1. $\dfrac{1}{2 - \sqrt{3}}$

2. $3\left(x^2 - 3x + 1\right) - 2\left(3x^2 + x - 4\right)$

3. $3\sqrt{8} + 5\sqrt{50} - 4\sqrt{32}$

4. Factor completely: $x^7 - x^5$.

Solve each equation in Exercises 5–8.

5. $|2x - 1| = 3$ **6.** $3x^2 - 5x + 1 = 0$

7. $9 + \dfrac{3}{x} = \dfrac{2}{x^2}$ **8.** $x^3 + 2x^2 - 5x - 6 = 0$

Solve each inequality in Exercises 9–10. Express the answer in interval notation.

9. $|2x - 5| > 3$ **10.** $3x^2 > 2x + 5$

11. Give the center and radius. Then graph the equation: $x^2 + y^2 - 2x + 4y - 4 = 0$.

12. Solve for t: $V = C(1 - t)$.

13. If $f(x) = \sqrt{45 - 9x}$, find the domain of f.

If $f(x) = x^2 + 2x - 5$ and $g(x) = 4x - 1$, find each function or function value in Exercises 14–16.

14. $(f - g)(x)$ **15.** $(f \circ g)(x)$

16. $g\left(f(-3)\right)$

17. Consider the function $f(x) = x^3 - 4x^2 - x + 4$.
   a. Use factoring to find all zeros of f.
   b. Use the Leading Coefficient Test and the zeros of f to graph the function.

Graph each function in Exercises 18–20.

18. $f(x) = x^2 + 2x - 8$ **19.** $f(x) = x^2(x - 3)$

20. $f(x) = \dfrac{x - 1}{x - 2}$

Exponential and Logarithmic Functions

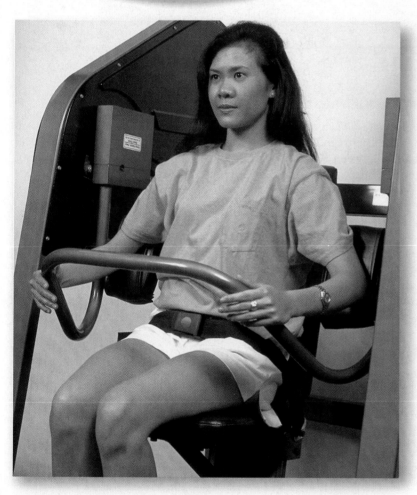

What went wrong on the space shuttle *Challenger*? Will population growth lead to a future without comfort or individual choice? Can I put aside a small amount of money and have millions for early retirement? Why did I feel I was walking too slowly on my visit to New York City? Why are people in California at far greater risk from drunk drivers than from earthquakes? What is the difference between earthquakes measuring 6 and 7 on the Richter scale? And what can I hope to accomplish in weightlifting?

The functions that you will be learning about in this chapter will provide you with the mathematics for answering these questions. You will see how these remarkable functions enable us to predict the future and rediscover the past.

You've recently taken up weightlifting, recording the maximum number of pounds you can lift at the end of each week. At first your weight limit increases rapidly, but now you notice that this growth is beginning to level off. You wonder about a function that would serve as a mathematical model to predict the number of pounds you can lift as you continue the sport.

SECTION 3.1 *Exponential Functions*

Objectives

1. Evaluate exponential functions.
2. Graph exponential functions.
3. Evaluate functions with base *e*.
4. Use compound interest formulas.

The space shuttle *Challenger* exploded approximately 73 seconds into flight on January 28, 1986. The tragedy involved damage to O-rings, which were used to seal the connections between different sections of the shuttle engines. The number of O-rings damaged increases dramatically as temperature falls.

The function

$$f(x) = 13.49(0.967)^x - 1$$

models the number of O-rings expected to fail when the temperature is $x°$F. Can you see how this function is different from polynomial functions? The variable x is in the exponent. Functions whose equations contain a variable in the exponent are called **exponential functions.** Many real-life situations, including population growth, growth of epidemics, radioactive decay, and other changes that involve rapid increase or decrease, can be described using exponential functions.

Definition of the Exponential Function

The **exponential function** f **with base** b is defined by

$$f(x) = b^x \quad \text{or} \quad y = b^x$$

where b is a positive constant other than 1 ($b > 0$ and $b \neq 1$) and x is any real number.

Here are some examples of exponential functions:

$$f(x) = 2^x \qquad g(x) = 10^x \qquad h(x) = 3^{x+1}.$$

> Base is 2. Base is 10. Base is 3.

Each of these functions has a constant base and a variable exponent. By contrast, the following functions are not exponential:

$$F(x) = x^2 \qquad G(x) = 1^x \qquad H(x) = x^x.$$

> Variable is the base and not the exponent.
>
> The base of an exponential function must be a positive constant other than 1.
>
> Variable is both the base and the exponent.

Why is $G(x) = 1^x$ not classified as an exponential function? The number 1 raised to any power is 1. Thus, the function G can be written as $G(x) = 1$, which is a constant function.

1 Evaluate exponential functions.

You will need a calculator to evaluate exponential expressions. Most scientific calculators have a $\boxed{y^x}$ key. Graphing calculators have a $\boxed{\wedge}$ key. To evaluate expressions of the form b^x, enter the base b, press $\boxed{y^x}$ or $\boxed{\wedge}$, enter the exponent x, and finally press $\boxed{=}$ or $\boxed{\text{ENTER}}$.

EXAMPLE 1 Evaluating an Exponential Function

The exponential function $f(x) = 13.49(0.967)^x - 1$ describes the number of O-rings expected to fail, $f(x)$, when the temperature is $x°F$. On the morning the *Challenger* was launched, the temperature was $31°F$, colder than any previous experience. Find the number of O-rings expected to fail at this temperature.

Solution Because the temperature was $31°F$, substitute 31 for x and evaluate the function.

$$f(x) = 13.49(0.967)^x - 1 \quad \text{This is the given function.}$$
$$f(31) = 13.49(0.967)^{31} - 1 \quad \text{Substitute 31 for } x.$$

Use a scientific or graphing calculator to evaluate $f(31)$. Press the following keys on your calculator to do this:

Scientific calculator: $13.49 \boxed{\times} .967 \boxed{y^x} 31 \boxed{-} 1 \boxed{=}$

Graphing calculator: $13.49 \boxed{\times} .967 \boxed{\wedge} 31 \boxed{-} 1 \boxed{\text{ENTER}}$.

The display should be approximately 3.7668627.

$$f(31) = 13.49(0.967)^{31} - 1 \approx 3.8 \approx 4$$

Thus, four O-rings are expected to fail at a temperature of $31°F$.

Check Point 1 Use the function in Example 1 to find the number of O-rings expected to fail at a temperature of $60°F$. Round to the nearest whole number.

2 Graph exponential functions.

Graphing Exponential Functions

We are familiar with expressions involving b^x, where x is a rational number. For example,

$$b^{1.7} = b^{17/10} = \sqrt[10]{b^{17}} \quad \text{and} \quad b^{1.73} = b^{173/100} = \sqrt[100]{b^{173}}.$$

However, note that the definition of $f(x) = b^x$ includes all real numbers for the domain x. You may wonder what b^x means when x is an irrational number, such as $b^{\sqrt{3}}$ or b^{π}. Using closer and closer approximations for $\sqrt{3}$ ($\sqrt{3} \approx 1.73205$), we can think of $b^{\sqrt{3}}$ as the value that has the successively closer approximations

$$b^{1.7}, b^{1.73}, b^{1.732}, b^{1.73205}, \dots.$$

In this way, we can graph the exponential function with no holes, or points of discontinuity, at the irrational domain values.

EXAMPLE 2   Graphing an Exponential Function

Graph: $f(x) = 2^x$.

Solution   We begin by setting up a table of coordinates.

x	$f(x) = 2^x$
-3	$f(-3) = 2^{-3} = \frac{1}{8}$
-2	$f(-2) = 2^{-2} = \frac{1}{4}$
-1	$f(-1) = 2^{-1} = \frac{1}{2}$
0	$f(0) = 2^0 = 1$
1	$f(1) = 2^1 = 2$
2	$f(2) = 2^2 = 4$
3	$f(3) = 2^3 = 8$

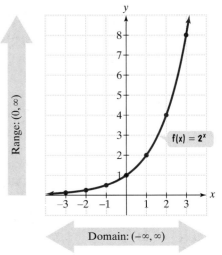

Figure 3.1 The graph of $f(x) = 2^x$

We plot these points, connecting them with a continuous curve. Figure 3.1 shows the graph of $f(x) = 2^x$. Observe that the graph approaches, but never touches, the negative portion of the x-axis. Thus, the x-axis is a horizontal asymptote. The range is the set of all positive real numbers. Although we used integers for x in our table of coordinates, you can use a calculator to find additional points. For example, $f(0.3) = 2^{0.3} \approx 1.231$, $f(0.95) = 2^{0.95} \approx 1.932$. The points $(0.3, 1.231)$ and $(0.95, 1.932)$ approximately fit the graph.

Check Point 2   Graph: $f(x) = 3^x$.

Four exponential functions have been graphed in Figure 3.2. Compare the black and green graphs, where $b > 1$, to those in blue and red, where $b < 1$. When $b > 1$, the value of y increases as the value of x increases. When $b < 1$, the value of y decreases as the value of x increases. Notice that all four graphs pass through $(0, 1)$.

Study Tip

The graph of $y = \left(\frac{1}{2}\right)^x$, meaning $y = 2^{-x}$, is the graph of $y = 2^x$ reflected about the y-axis.

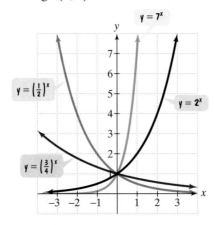

Figure 3.2 Graphs of four exponential functions

The graphs on the previous page illustrate the following general characteristics of exponential functions:

Characteristics of Exponential Functions of the Form $f(x) = b^x$

1. The domain of $f(x) = b^x$ consists of all real numbers. The range of $f(x) = b^x$ consists of all positive real numbers.
2. The graphs of all exponential functions of the form $f(x) = b^x$ pass through the point $(0, 1)$ because $f(0) = b^0 = 1 (b \neq 0)$. The y-intercept is 1.
3. If $b > 1$, $f(x) = b^x$ has a graph that goes up to the right and is an increasing function. The greater the value of b, the steeper the increase.
4. If $0 < b < 1$, $f(x) = b^x$ has a graph that goes down to the right and is a decreasing function. The smaller the value of b, the steeper the decrease.
5. $f(x) = b^x$ is one-to-one and has an inverse that is a function.
6. The graph of $f(x) = b^x$ approaches, but does not cross, the x-axis. The x-axis is a horizontal asymptote.

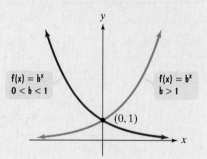

Transformations of Exponential Functions   The graphs of exponential functions can be translated vertically or horizontally, reflected, stretched, or shrunk. We use the ideas of Section 1.6 to do so, as summarized in Table 3.1.

Table 3.1   Transformations Involving Exponential Functions
In each case, c represents a positive real number.

Transformation	Equation	Description
Vertical translation	$g(x) = b^x + c$	• Shifts the graph of $f(x) = b^x$ upward c units.
	$g(x) = b^x - c$	• Shifts the graph of $f(x) = b^x$ downward c units.
Horizontal translation	$g(x) = b^{x+c}$	• Shifts the graph of $f(x) = b^x$ to the left c units.
	$g(x) = b^{x-c}$	• Shifts the graph of $f(x) = b^x$ to the right c units.
Reflecting	$g(x) = -b^x$	• Reflects the graph of $f(x) = b^x$ about the x-axis.
	$g(x) = b^{-x}$	• Reflects the graph of $f(x) = b^x$ about the y-axis.
Vertical stretching or shrinking	$g(x) = cb^x$	• Stretches the graph of $f(x) = b^x$ if $c > 1$. • Shrinks the graph of $f(x) = b^x$ if $0 < c < 1$.

Using the information in Table 3.1 and a table of coordinates, you will obtain relatively accurate graphs that can be verified using a graphing utility.

EXAMPLE 3 Transformations Involving Exponential Functions

Use the graph of $f(x) = 3^x$ to obtain the graph of $g(x) = 3^{x+1}$.

Solution Examine Table 3.1. Note that the function $g(x) = 3^{x+1}$ has the general form $g(x) = b^{x+c}$, where $c = 1$. Thus, we graph $g(x) = 3^{x+1}$ by shifting the graph of $f(x) = 3^x$ *one* unit to the *left*. We construct a table showing some of the coordinates for f and g, selecting integers from –2 to 2 for x. The graphs of f and g are shown in Figure 3.3.

x	$f(x) = 3^x$	$g(x) = 3^{x+1}$
−2	$f(-2) = 3^{-2} = \frac{1}{9}$	$g(-2) = 3^{-2+1} = 3^{-1} = \frac{1}{3}$
−1	$f(-1) = 3^{-1} = \frac{1}{3}$	$g(-1) = 3^{-1+1} = 3^0 = 1$
0	$f(0) = 3^0 = 1$	$g(0) = 3^{0+1} = 3^1 = 3$
1	$f(1) = 3^1 = 3$	$g(1) = 3^{1+1} = 3^2 = 9$
2	$f(2) = 3^2 = 9$	$g(2) = 3^{2+1} = 3^3 = 27$

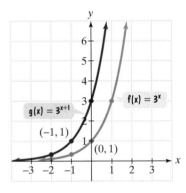

Figure 3.3 The graph of $g(x) = 3^{x+1}$ is the graph of $f(x) = 3^x$ shifted one unit to the left.

Check Point 3 Use the graph of $f(x) = 3^x$ to obtain the graph of $g(x) = 3^{x-1}$.

If an exponential function is translated upward or downward, the horizontal asymptote is shifted by the amount of the vertical shift.

EXAMPLE 4 Transformations Involving Exponential Functions

Use the graph of $f(x) = 2^x$ to obtain the graph of $g(x) = 2^x - 3$.

Solution The function $g(x) = 2^x - 3$ has the general form $g(x) = b^x - c$, where $c = 3$. Thus, we graph $g(x) = 2^x - 3$ by shifting the graph of $f(x) = 2^x$ *down three* units. We construct a table showing some of the coordinates for f and g, selecting integers from –2 to 2 for x.

The graphs of f and g are shown in Figure 3.4. Notice that the horizontal asymptote for f, the x-axis, is shifted down three units for the horizontal asymptote for g. As a result, $y = -3$ is the horizontal asymptote for g.

x	$f(x) = 2^x$	$g(x) = 2^x - 3$
-2	$f(-2) = 2^{-2} = \frac{1}{4}$	$g(-2) = 2^{-2} - 3 = \frac{1}{4} - 3 = -2\frac{3}{4}$
-1	$f(-1) = 2^{-1} = \frac{1}{2}$	$g(-1) = 2^{-1} - 3 = \frac{1}{2} - 3 = -2\frac{1}{2}$
0	$f(0) = 2^0 = 1$	$g(0) = 2^0 - 3 = 1 - 3 = -2$
1	$f(1) = 2^1 = 2$	$g(1) = 2^1 - 3 = 2 - 3 = -1$
2	$f(2) = 2^2 = 4$	$g(2) = 2^2 - 3 = 4 - 3 = 1$

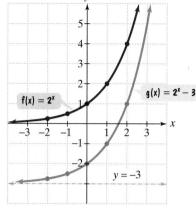

Figure 3.4 The graph of $g(x) = 2^x - 3$ is the graph of $f(x) = 2^x$ shifted down three units.

Check Point 4   Use the graph of $f(x) = 2^x$ to obtain the graph of $g(x) = 2^x + 1$.

3   Evaluate functions with base e.

The Natural Base e

An irrational number, symbolized by the letter e, appears as the base in many applied exponential functions. This irrational number is approximately equal to 2.72. More accurately,

$$e \approx 2.71828\ldots.$$

The number e is called the **natural base.** The function $f(x) = e^x$ is called the **natural exponential function.**

Use a scientific or graphing calculator with an $\boxed{e^x}$ key to evaluate e to various powers. For example, to find e^2, press the following keys on most calculators:

Scientific calculator: 2 $\boxed{e^x}$

Graphing calculator: $\boxed{e^x}$ 2 $\boxed{\text{ENTER}}$.

The display should be approximately 7.389.

$$e^2 \approx 7.389$$

The number e lies between 2 and 3. Because $2^2 = 4$ and $3^2 = 9$, it makes sense that e^2, approximately 7.389, lies between 4 and 9.

Because $2 < e < 3$, the graph of $y = e^x$ lies between the graphs of $y = 2^x$ and $y = 3^x$, shown in Figure 3.5.

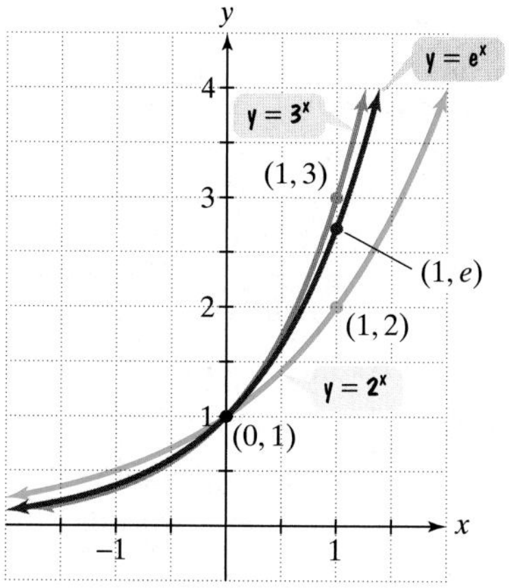

Figure 3.5 Graphs of three exponential functions

EXAMPLE 5   World Population

In a report entitled *Resources and Man*, the U.S. National Academy of Sciences concluded that a world population of 10 billion "is close to (if not above) the maximum that an intensely managed world might hope to support with some degree of comfort and individual choice." At the time the report was issued in 1969, world population was approximately 3.6 billion, with a growth rate of 2% per year. The function

$$f(x) = 3.6e^{0.02x}$$

describes world population, $f(x)$, in billions, x years after 1969. Use the function to find world population in the year 2020. Is there cause for alarm?

World Population, in Billions

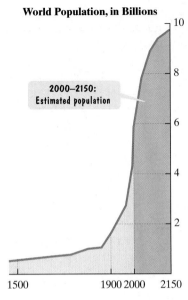

Source: U.N. Population Division

Solution Because 2020 is 51 years after 1969, we substitute 51 for x in $f(x) = 3.6e^{0.02x}$:

$$f(51) = 3.6e^{0.02(51)}.$$

Perform this computation on your calculator.

Scientific calculator: 3.6 × (.02 × 51) e^x =

Graphing calculator: 3.6 × e^x (.02 × 51) ENTER

The display should be approximately 9.9835012. Thus,
$$f(51) = 3.6e^{0.02(51)} \approx 9.98.$$

This indicates that world population in the year 2020 will be approximately 9.98 billion. Because this number is quite close to 10 billion, the given function suggests that there may be cause for alarm.

World population in 2000 was approximately 6 billion, but the growth rate was no longer 2%. It had slowed down to 1.3%. Using this current growth rate, exponential functions now predict a world population of 7.8 billion in the year 2020. Experts think the population may stabilize at 10 billion after 2200 if the deceleration in growth rate continues.

Check Point 5 The function $f(x) = 6e^{0.013x}$ describes world population, $f(x)$, in billions, x years after 2000 subject to a growth rate of 1.3% annually. Use the function to find world population in 2050.

Compound Interest

4 Use compound interest formulas.

We all want a wonderful life with fulfilling work, good health, and loving relationships. And let's be honest: Financial security wouldn't hurt! Achieving this goal depends on understanding how money in savings accounts grows in remarkable ways as a result of *compound interest*. **Compound interest** is interest computed on your original investment as well as on any accumulated interest.

Suppose a sum of money, called the **principal,** P, is invested at an annual percentage rate r, in decimal form, compounded once per year. Because the interest is added to the principal at year's end, the accumulated value, A, is
$$A = P + Pr = P(1 + r).$$

The accumulated amount of money follows this pattern of multiplying the previous principal by $(1 + r)$ for each successive year, as indicated in Table 3.2.

Table 3.2

Time in Years	Accumulated Value after Each Compounding
0	$A = P$
1	$A = P(1 + r)$
2	$A = P(1 + r)(1 + r) = P(1 + r)^2$
3	$A = P(1 + r)^2(1 + r) = P(1 + r)^3$
4	$A = P(1 + r)^3(1 + r) = P(1 + r)^4$
⋮	⋮
t	$A = P(1 + r)^t$

This formula gives the balance, A, that a principal, P, is worth after t years at interest rate r, compounded once a year.

n	$\left(1 + \dfrac{1}{n}\right)^n$
1	2
2	2.25
5	2.48832
10	2.59374246
100	2.704813829
1000	2.716923932
10,000	2.718145927
100,000	2.718268237
1,000,000	2.718280469
1,000,000,000	2.718281827

As n takes on increasingly large values, the expression $\left(1 + \dfrac{1}{n}\right)^n$ approaches e:

$$\lim_{n \to \infty}\left(1 + \frac{1}{n}\right)^n = e.$$

Most savings institutions have plans in which interest is paid more than once a year. If compound interest is paid twice a year, the compounding period is six months. We say that the interest is **compounded semiannually.** When compound interest is paid four times a year, the compounding period is three months and the interest is said to be **compounded quarterly.** Some plans allow for monthly compounding or daily compounding.

In general, when compound interest is paid n times a year, we say that there are n **compounding periods per year.** The formula $A = P(1 + r)^t$ can be adjusted to take into account the number of compounding periods in a year. If there are n compounding periods per year, the formula becomes

$$A = P\left(1 + \frac{r}{n}\right)^{nt}.$$

Some banks use **continuous compounding,** where the number of compounding periods increases infinitely (compounding interest every trillionth of a second, every quadrillionth of a second, etc.). As n, the number of compounding periods in a year, increases without bound, the expression $\left(1 + \dfrac{1}{n}\right)^n$ approaches e. As a result, the formula for continuous compounding is $A = Pe^{rt}$. Although continuous compounding sounds terrific, it yields only a fraction of a percent more interest over a year than daily compounding.

Formulas for Compound Interest

After t years, the balance, A, in an account with principal P and annual interest rate r (in decimal form) is given by the following formulas:

1. For n compoundings per year: $A = P\left(1 + \dfrac{r}{n}\right)^{nt}$

2. For continuous compounding: $A = Pe^{rt}$.

EXAMPLE 6   Choosing between Investments

You want to invest \$8000 for 6 years, and you have a choice between two accounts. The first pays 7% per year, compounded monthly. The second pays 6.85% per year, compounded continuously. Which is the better investment?

Solution   The better investment is the one with the greater balance in the account after 6 years. Let's begin with the account with monthly compounding. We use the compound interest model with $P = 8000$, $r = 7\% = 0.07$, $n = 12$ (monthly compounding means 12 compoundings per year), and $t = 6$.

$$A = P\left(1 + \frac{r}{n}\right)^{nt} = 8000\left(1 + \frac{0.07}{12}\right)^{12 \cdot 6} \approx 12{,}160.84$$

The balance in this account after 6 years is \$12,160.84. For the second investment option, we use the model for continuous compounding with $P = 8000$, $r = 6.85\% = 0.0685$, and $t = 6$.

$$A = Pe^{rt} = 8000e^{0.0685(6)} \approx 12{,}066.60$$

Technology

The graphs illustrate that as x increases, $\left(1 + \dfrac{1}{x}\right)^x$ approaches e.

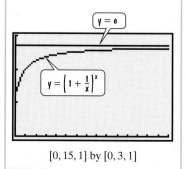

$y = e$

$y = \left(1 + \dfrac{1}{x}\right)^x$

$[0, 15, 1]$ by $[0, 3, 1]$

The balance in this account after 6 years is $12,066.60, slightly less than the previous amount. Thus, the better investment is the 7% monthly compounding option.

Check Point 6 A sum of $10,000 is invested at an annual rate of 8%. Find the balance in the account after 5 years subject to **a.** quarterly compounding and **b.** continuous compounding.

EXERCISE SET 3.1

 Practice Exercises

In Exercises 1–10, approximate each number using a calculator. Round your answer to three decimal places.

1. $2^{3.4}$ **2.** $3^{2.4}$ **3.** $3^{\sqrt{5}}$ **4.** $5^{\sqrt{3}}$ **5.** $4^{-1.5}$

6. $6^{-1.2}$ **7.** $e^{2.3}$ **8.** $e^{3.4}$ **9.** $e^{-0.95}$ **10.** $e^{-0.75}$

In Exercises 11–18, graph each function by making a table of coordinates. If applicable, use a graphing utility to confirm your hand-drawn graph.

11. $f(x) = 4^x$ **12.** $f(x) = 5^x$

13. $g(x) = \left(\frac{3}{2}\right)^x$ **14.** $g(x) = \left(\frac{4}{3}\right)^x$

15. $h(x) = \left(\frac{1}{2}\right)^x$ **16.** $h(x) = \left(\frac{1}{3}\right)^x$

17. $f(x) = (0.6)^x$ **18.** $f(x) = (0.8)^x$

In Exercises 19–24, the graph of an exponential function is given. Select the function for each graph from the following options:

$$f(x) = 3^x, g(x) = 3^{x-1}, h(x) = 3^x - 1,$$
$$F(x) = -3^x, G(x) = 3^{-x}, H(x) = -3^{-x}.$$

19.

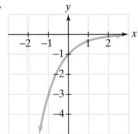

20.

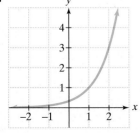

21.

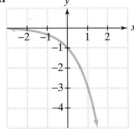

22.

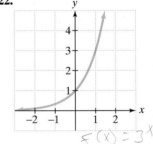

23.

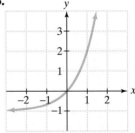

24.

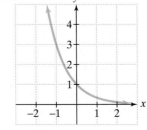

In Exercises 25–34, begin by graphing $f(x) = 2^x$. Then use transformations of this graph and a table of coordinates to graph the given function. If applicable, use a graphing utility to confirm your hand-drawn graphs.

25. $g(x) = 2^{x+1}$ **26.** $g(x) = 2^{x+2}$

27. $g(x) = 2^x - 1$ **28.** $g(x) = 2^x + 2$

29. $h(x) = 2^{x+1} - 1$ **30.** $h(x) = 2^{x+2} - 1$

31. $g(x) = -2^x$ **32.** $g(x) = 2^{-x}$

33. $g(x) = 2 \cdot 2^x$ **34.** $g(x) = \frac{1}{2} \cdot 2^x$

In Exercises 35–40, graph functions f and g in the same rectangular coordinate system. If applicable, use a graphing utility to confirm your hand-drawn graphs.

35. $f(x) = 3^x$ and $g(x) = 3^{-x}$

36. $f(x) = 3^x$ and $g(x) = -3^x$

37. $f(x) = 3^x$ and $g(x) = \frac{1}{3} \cdot 3^x$

38. $f(x) = 3^x$ and $g(x) = 3 \cdot 3^x$

39. $f(x) = (\frac{1}{2})^x$ and $g(x) = (\frac{1}{2})^{x-1} + 1$

40. $f(x) = (\frac{1}{2})^x$ and $g(x) = (\frac{1}{2})^{x-1} + 2$

Use the compound interest formulas $A = P\left(1 + \frac{r}{n}\right)^{nt}$ and $A = Pe^{rt}$ to solve Exercises 41–44. Round answers to the nearest cent.

41. Find the accumulated value of an investment of $10,000 for 5 years at an interest rate of 5.5% if the money is **a.** compounded semiannually; **b.** compounded quarterly; **c.** compounded monthly; **d.** compounded continuously.

42. Find the accumulated value of an investment of $5000 for 10 years at an interest rate of 6.5% if the money is **a.** compounded semiannually; **b.** compounded quarterly; **c.** compounded monthly; **d.** compounded continuously.

43. Suppose that you have $12,000 to invest. Which investment yields the greatest return over 3 years: 7% compounded monthly or 6.85% compounded continuously?

44. Suppose that you have $6000 to invest. Which investment yields the greatest return over 4 years: 8.25% compounded quarterly or 8.3% compounded semiannually?

Application Exercises

Use a calculator with a $\boxed{y^x}$ key or a $\boxed{\wedge}$ key to solve Exercises 45–52.

45. The exponential function $f(x) = 67.38(1.026)^x$ describes the population of Mexico, $f(x)$, in millions, x years after 1980.

   a. Substitute 0 for x and, without using a calculator, find Mexico's population in 1980.

   b. Substitute 27 for x and use your calculator to find Mexico's population in the year 2007 as predicted by this function.

c. Find Mexico's population in the year 2034 as predicted by this function.

d. Find Mexico's population in the year 2061 as predicted by this function.

e. What appears to be happening to Mexico's population every 27 years?

46. The 1986 explosion at the Chernobyl nuclear power plant in the former Soviet Union sent about 1000 kilograms of radioactive cesium-137 into the atmosphere. The function $f(x) = 1000(0.5)^{x/30}$ describes the amount, $f(x)$, in kilograms, of cesium-137 remaining in Chernobyl x years after 1986. If even 100 kilograms of cesium-137 remain in Chernobyl's atmosphere, the area is considered unsafe for human habitation. Find $f(80)$ and determine if Chernobyl will be safe for human habitation by 2066.

It is 8:00 P.M. and West Side Story is scheduled to begin. When the curtain does not go up, a rumor begins to spread through the 400-member audience: The lead roles of Tony and Maria might be understudied by Anthony Hopkins and Jodie Foster. The function

$$f(x) = \frac{400}{1 + 399(0.67)^x}$$

models the number of people in the audience, $f(x)$, who have heard the rumor x minutes after 8:00. Use this function to solve Exercises 47–48.

47. Evaluate $f(10)$ and describe what this means in practical terms.

48. Evaluate $f(20)$ and describe what this means in practical terms.

The formula $S = C(1 + r)^t$ models inflation, where C = the value today, r = the annual inflation rate, and S = the inflated value t years from now. Use this formula to solve Exercises 49–50.

49. If the inflation rate is 6%, how much will a house now worth $65,000 be worth in 10 years?

50. If the inflation rate is 3%, how much will a house now worth $110,000 be worth in 5 years?

51. A decimal approximation for $\sqrt{3}$ is 1.7320508. Use a calculator to find $2^{1.7}, 2^{1.73}, 2^{1.732}, 2^{1.73205}$, and $2^{1.7320508}$. Now find $2^{\sqrt{3}}$. What do you observe?

52. A decimal approximation for π is 3.141593. Use a calculator to find $2^3, 2^{3.1}, 2^{3.14}, 2^{3.141}, 2^{3.1415}, 2^{3.14159}$, and $2^{3.141593}$. Now find 2^π. What do you observe?

The graph on the next page shows the number of Americans enrolled in HMOs, in millions, from 1992 through 2000. The data can be modeled by the exponential function

$$f(x) = 36.1e^{0.113x}$$

which describes enrollment in HMOs, $f(x)$, in millions, x years after 1992. Use this function to solve Exercises 53–54.

Enrollment in HMOs

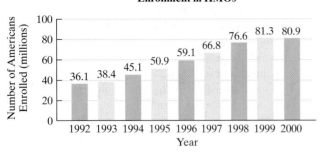

Source: Department of Health and Human Services

53. According to the model, how many Americans will be enrolled in HMOs in the year 2006? Round to the nearest tenth of a million.

54. According to the model, how many Americans will be enrolled in HMOs in the year 2008? Round to the nearest tenth of a million.

55. In college, we study large volumes of information—information that, unfortunately, we do not often retain for very long. The function

$$f(x) = 80e^{-0.5x} + 20$$

describes the percentage of information, $f(x)$, that a particular person remembers x weeks after learning the information.

a. Substitute 0 for x and, without using a calculator, find the percentage of information remembered at the moment it is first learned.

b. Substitute 1 for x and find the percentage of information that is remembered after 1 week.

c. Find the percentage of information that is remembered after 4 weeks.

d. Find the percentage of information that is remembered after one year (52 weeks).

56. In 1626, Peter Minuit convinced the Wappinger Indians to sell him Manhattan Island for $24. If the Native Americans had put the $24 into a bank account paying 5% interest, how much would the investment be worth in the year 2000 if interest were compounded

a. monthly? **b.** continuously?

57. The function

$$N(t) = \frac{30,000}{1 + 20e^{-1.5t}}$$

describes the number of people, $N(t)$, who become ill with influenza t weeks after its initial outbreak in a town with 30,000 inhabitants. The horizontal asymptote in the graph at the top of the next column indicates that there is a limit to the epidemic's growth.

a. How many people became ill with the flu when the epidemic began? (When the epidemic began, $t = 0$.)

b. How many people were ill by the end of the third week?

c. Why can't the spread of an epidemic simply grow indefinitely? What does the horizontal asymptote shown in the graph indicate about the limiting size of the population that becomes ill?

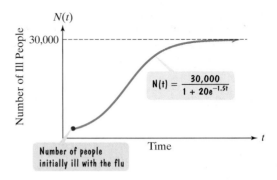

Writing in Mathematics

58. What is an exponential function?

59. What is the natural exponential function?

60. Use a calculator to evaluate $\left(1 + \dfrac{1}{x}\right)^x$ for $x = 10, 100, 1000, 10,000, 100,000,$ and $1,000,000$. Describe what happens to the expression as x increases.

61. Write an example similar to Example 6 on page 363 in which continuous compounding at a slightly lower yearly interest rate is a better investment than compounding n times per year.

62. Describe how you could use the graph of $f(x) = 2^x$ to obtain a decimal approximation for $\sqrt{2}$.

63. The exponential function $y = 2^x$ is one-to-one and has an inverse function. Try finding the inverse function by exchanging x and y and solving for y. Describe the difficulty that you encounter in this process. What is needed to overcome this problem?

64. In 2000, world population was approximately 6 billion with an annual growth rate of 1.3%. Discuss two factors that would cause this growth rate to slow down over the next ten years.

Technology Exercises

65. Graph $y = 13.49(0.967)^x - 1$, the function for the number of O-rings expected to fail at $x°F$, in a $[0, 90, 10]$ by $[0, 20, 5]$ viewing rectangle. If NASA engineers had used this function and its graph, is it likely they would have allowed the *Challenger* to be launched when the temperature was $31°F$? Explain.

66. You have $10,000 to invest. One bank pays 5% interest compounded quarterly and the other pays 4.5% interest compounded monthly.

a. Use the formula for compound interest to write a function for the balance in each account at any time t.

b. Use a graphing utility to graph both functions in an appropriate viewing rectangle. Based on the graphs, which bank offers the better return on your money?

67. a. Graph $y = e^x$ and $y = 1 + x + \dfrac{x^2}{2}$ in the same viewing rectangle.

   b. Graph $y = e^x$ and $y = 1 + x + \dfrac{x^2}{2} + \dfrac{x^3}{6}$ in the same viewing rectangle.

   c. Graph $y = e^x$ and $y = 1 + x + \dfrac{x^2}{2} + \dfrac{x^3}{6} + \dfrac{x^4}{24}$ in the same viewing rectangle.

   d. Describe what you observe in parts (a)–(c). Try generalizing this observation.

Critical Thinking Exercises

68. Which one of the following is true?

   a. As the number of compounding periods increases on a fixed investment, the amount of money in the account over a fixed interval of time will increase without bound.

   b. The functions $f(x) = 3^{-x}$ and $g(x) = -3^x$ have the same graph.

   c. $e = 2.718$

   d. The functions $f(x) = \left(\frac{1}{3}\right)^x$ and $g(x) = 3^{-x}$ have the same graph.

69. The graphs labeled (a)–(d) in the figure represent $y = 3^x$, $y = 5^x$, $y = \left(\frac{1}{3}\right)^x$, and $y = \left(\frac{1}{5}\right)^x$, but not necessarily in that order. Which is which? Describe the process that enables you to make this decision.

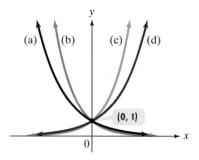

70. Graph $f(x) = 2^x$ and its inverse function in the same rectangular coordinate system.

71. The *hyperbolic cosine* and *hyperbolic sine* functions are defined by

$$\cosh x = \frac{e^x + e^{-x}}{2} \quad \text{and} \quad \sinh x = \frac{e^x - e^{-x}}{2}.$$

Prove that $(\cosh x)^2 - (\sinh x)^2 = 1$.

SECTION 3.2   *Logarithmic Functions*

Objectives

1. Change from logarithmic to exponential form.
2. Change from exponential to logarithmic form.
3. Evaluate logarithms.
4. Use basic logarithmic properties.
5. Graph logarithmic functions.
6. Find the domain of a logarithmic function.
7. Use common logarithms.
8. Use natural logarithms.

The earthquake that ripped through northern California on October 17, 1989, measured 7.1 on the Richter scale, killed more than 60 people, and injured more than 2400. Shown here is San Francisco's Marina district, where shock waves tossed houses off their foundations and into the street.

A higher measure on the Richter scale is more devastating than it seems because for each increase in one unit on the scale, there is a tenfold increase in the intensity of an earthquake. In this section, our focus is on the inverse of the exponential function, called the logarithmic function. The logarithmic function will help you to understand diverse phenomena, including earthquake intensity, human memory, and the pace of life in large cities.

Study Tip

In case you need to review inverse functions, they are discussed in Section 1.8 on pages 209–216. The horizontal line test appears on page 214.

The Definition of Logarithmic Functions

No horizontal line can be drawn that intersects the graph of an exponential function at more than one point. This means that the exponential function is one-to-one and has an inverse. The inverse function of the exponential function with base b is called the *logarithmic function with base b*.

> **Definition of the Logarithmic Function**
> For $x > 0$ and $b > 0, b \neq 1$,
> $$y = \log_b x \text{ is equivalent to } b^y = x.$$
> The function $f(x) = \log_b x$ is the **logarithmic function with base b**.

Study Tip

The inverse function of $y = b^x$ is $x = b^y$. Logarithms give us a way to express this inverse function for y in terms of x.

The equations

$$y = \log_b x \quad \text{and} \quad b^y = x$$

are different ways of expressing the same thing. The first equation is in **logarithmic form** and the second equivalent equation is in **exponential form.**

Notice that a **logarithm, y, is an exponent.** You should learn the location of the base and exponent in each form.

> **Location of Base and Exponent in Exponential and Logarithmic Forms**
>
> Exponent
>
> Logarithmic Form: $y = \log_b x$ Exponential Form: $b^y = x$
>
> Base   Base

1 Change from logarithmic to exponential form.

EXAMPLE 1   Changing from Logarithmic to Exponential Form

Write each equation in its equivalent exponential form:

   a. $2 = \log_5 x$ **b.** $3 = \log_b 64$ **c.** $\log_3 7 = y.$

Solution   We use the fact that $y = \log_b x$ means $b^y = x$.

   a. $2 = \log_5 x$ means $5^2 = x.$ **b.** $3 = \log_b 64$ means $b^3 = 64.$

   Logarithms are exponents.   Logarithms are exponents.

   c. $\log_3 7 = y$ or $y = \log_3 7$ means $3^y = 7.$

Check Point 1   Write each equation in its equivalent exponential form:

   a. $3 = \log_7 x$ **b.** $2 = \log_b 25$ **c.** $\log_4 26 = y.$

2 Change from exponential to logarithmic form.

EXAMPLE 2   Changing from Exponential to Logarithmic Form

Write each equation in its equivalent logarithmic form:

   a. $12^2 = x$ **b.** $b^3 = 8$ **c.** $e^y = 9.$

Solution   We use the fact that $b^y = x$ means $y = \log_b x$.

a. $12^2 = x$ means $2 = \log_{12} x$.   **b.** $b^3 = 8$ means $3 = \log_b 8$.

Exponents are logarithms.   Exponents are logarithms.

c. $e^y = 9$ means $y = \log_e 9$.

Check
Point
2

Write each equation in its equivalent logarithmic form:

a. $2^5 = x$ **b.** $b^3 = 27$ **c.** $e^y = 33$.

3 Evaluate logarithms.

Remembering that logarithms are exponents makes it possible to evaluate some logarithms by inspection. The logarithm of x with base b, $\log_b x$, is the exponent to which b must be raised to get x. For example, suppose we want to evaluate $\log_2 32$. We ask, 2 to what power gives 32? Because $2^5 = 32$, $\log_2 32 = 5$.

EXAMPLE 3   Evaluating Logarithms

Evaluate:

    a. $\log_2 16$ **b.** $\log_3 9$ **c.** $\log_{25} 5$.

Solution

Logarithmic Expression	Question Needed for Evaluation	Logarithmic Expression Evaluated
a. $\log_2 16$	2 to what power gives 16?	$\log_2 16 = 4$ because $2^4 = 16$.
b. $\log_3 9$	3 to what power gives 9?	$\log_3 9 = 2$ because $3^2 = 9$.
c. $\log_{25} 5$	25 to what power gives 5?	$\log_{25} 5 = \frac{1}{2}$ because $25^{1/2} = \sqrt{25} = 5$.

Check
Point
3

Evaluate:

    a. $\log_{10} 100$ **b.** $\log_3 3$ **c.** $\log_{36} 6$.

4 Use basic logarithmic properties.

Basic Logarithmic Properties

Because logarithms are exponents, they have properties that can be verified using properties of exponents.

Basic Logarithmic Properties Involving One

1. $\log_b b = 1$ because 1 is the exponent to which b must be raised to obtain b. $\left(b^1 = b\right)$

2. $\log_b 1 = 0$ because 0 is the exponent to which b must be raised to obtain 1. $\left(b^0 = 1\right)$

EXAMPLE 4   Using Properties of Logarithms

Evaluate:

a. $\log_7 7$ **b.** $\log_5 1.$

Solution

a. Because $\log_b b = 1$, we conclude $\log_7 7 = 1$.

b. Because $\log_b 1 = 0$, we conclude $\log_5 1 = 0$.

Check Point 4   Evaluate:

a. $\log_9 9$ **b.** $\log_8 1.$

The inverse of the exponential function is the logarithmic function. Thus, if $f(x) = b^x$, then $f^{-1}(x) = \log_b x$. In Chapter 1, we saw how inverse functions "undo" one another. In particular,

$$f(f^{-1}(x)) = x \text{ and } f^{-1}(f(x)) = x.$$

Applying these relationships to exponential and logarithmic functions, we obtain the following **inverse properties of logarithms:**

> **Inverse Properties of Logarithms**
> For $b > 0$ and $b \neq 1$,
>
> $$\log_b b^x = x \qquad \text{The logarithm with base } b \text{ of } b \text{ raised to a power equals that power.}$$
>
> $$b^{\log_b x} = x \qquad b \text{ raised to the logarithm with base } b \text{ of a number equals that number.}$$

EXAMPLE 5   Using Inverse Properties of Logarithms

Evaluate:

a. $\log_4 4^5$ **b.** $6^{\log_6 9}.$

Solution

a. Because $\log_b b^x = x$, we conclude $\log_4 4^5 = 5$.

b. Because $b^{\log_b x} = x$, we conclude $6^{\log_6 9} = 9$.

Check Point 5   Evaluate:

a. $\log_7 7^8$ **b.** $3^{\log_3 17}.$

5 Graph logarithmic functions.

Graphs of Logarithmic Functions

How do we graph logarithmic functions? We use the fact that the logarithmic function is the inverse of the exponential function. This means that the logarithmic function reverses the coordinates of the exponential function. It also means that the graph of the logarithmic function is a reflection of the graph of the exponential function about the line $y = x$.

EXAMPLE 6 Graphs of Exponential and Logarithmic Functions

Graph $f(x) = 2^x$ and $g(x) = \log_2 x$ in the same rectangular coordinate system.

Solution   We first set up a table of coordinates for $f(x) = 2^x$. Reversing, these coordinates gives the coordinates for the inverse function $g(x) = \log_2 x$.

x	-2	-1	0	1	2	3
$f(x) = 2^x$	$\frac{1}{4}$	$\frac{1}{2}$	1	2	4	8

x	$\frac{1}{4}$	$\frac{1}{2}$	1	2	4	8
$g(x) = \log_2 x$	-2	-1	0	1	2	3

Reverse coordinates.

We now plot the ordered pairs in both tables, connecting them with smooth curves. Figure 3.6 shows the graphs of $f(x) = 2^x$ and its inverse function $g(x) = \log_2 x$. The graph of the inverse can also be drawn by reflecting the graph of $f(x) = 2^x$ about the line $y = x$.

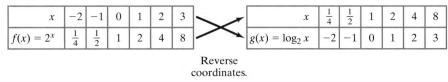

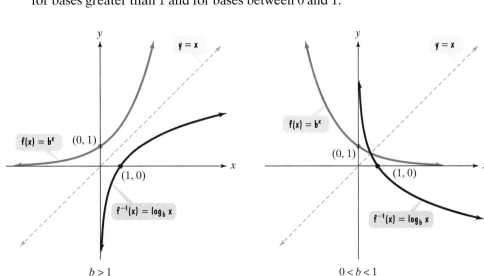

Figure 3.6 The graphs of $f(x) = 2^x$ and its inverse function

Check Point 6   Graph $f(x) = 3^x$ and $g(x) = \log_3 x$ in the same rectangular coordinate system.

Figure 3.7 illustrates the relationship between the graph of the exponential function, shown in blue, and its inverse, the logarithmic function, shown in red, for bases greater than 1 and for bases between 0 and 1.

Figure 3.7 Graphs of exponential and logarithmic functions

Discovery

Verify each of the four characteristics in the box for the red graphs in Figure 3.7.

Characteristics of the Graphs of Logarithmic Functions of the Form $f(x) = \log_b x$

- The x-intercept is 1. There is no y-intercept.
- The y-axis is a vertical asymptote.
- If $b > 1$, the function is increasing. If $0 < b < 1$, the function is decreasing.
- The graph is smooth and continuous. It has no sharp corners or gaps.

The graphs of logarithmic functions can be translated vertically or horizontally, reflected, stretched, or shrunk. We use the ideas of Section 1.6 to do so, as summarized in Table 3.3.

Table 3.3   Transformations Involving Logarithmic Functions
In each case, c represents a positive real number.

Transformation	Equation	Description
Vertical translation	$g(x) = \log_b x + c$	• Shifts the graph of $f(x) = \log_b x$ upward c units.
	$g(x) = \log_b x - c$	• Shifts the graph of $f(x) = \log_b x$ downward c units.
Horizontal translation	$g(x) = \log_b (x + c)$	• Shifts the graph of $f(x) = \log_b x$ to the left c units. Vertical asymptote: $x = -c$.
	$g(x) = \log_b (x - c)$	• Shifts the graph of $f(x) = \log_b x$ to the right c units. Vertical asymptote: $x = c$.
Reflecting	$g(x) = -\log_b x$	• Reflects the graph of $f(x) = \log_b x$ about the x-axis.
	$g(x) = \log_b (-x)$	• Reflects the graph of $f(x) = \log_b x$ about the y-axis.
Vertical stretching or shrinking	$g(x) = c \log_b x$	• Stretches the graph of $f(x) = \log_b x$ if $c > 1$. • Shrinks the graph of $f(x) = \log_b x$ if $0 < c < 1$.

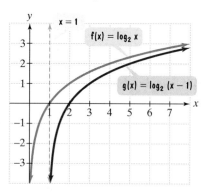

Figure 3.8 Shifting $f(x) = \log_2 x$ one unit to the right

For example, Figure 3.8 illustrates that the graph of $g(x) = \log_2 (x - 1)$ is the graph of $f(x) = \log_2 x$ shifted one unit to the right. If a logarithmic function is translated to the left or to the right, both the x-intercept and the vertical asymptote are shifted by the amount of the horizontal shift. In Figure 3.8, the x-intercept of f is 1. Because g is shifted one unit to the right, its x-intercept is 2. Also observe that the vertical asymptote for f, the y-axis, is shifted one unit to the right for the vertical asymptote for g. Thus, $x = 1$ is the vertical asymptote for g.

Here are some other examples of transformations of graphs of logarithmic functions:

- The graph of $g(x) = 3 + \log_4 x$ is the graph of $f(x) = \log_4 x$ shifted up three units, shown in Figure 3.9.
- The graph of $h(x) = -\log_2 x$ is the graph of $f(x) = \log_2 x$ reflected about the x-axis, shown in Figure 3.10.
- The graph of $r(x) = \log_2 (-x)$ is the graph of $f(x) = \log_2 x$ reflected about the y-axis, shown in Figure 3.11.

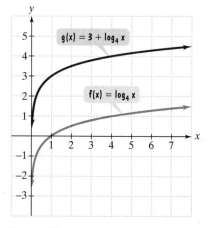

Figure 3.9 Shifting vertically up three units

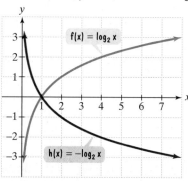

Figure 3.10 Reflection about the x-axis

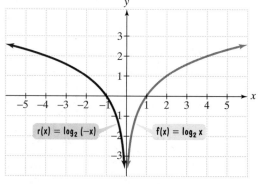

Figure 3.11 Reflection about the y-axis

6 Find the domain of a logarithmic function.

The Domain of a Logarithmic Function

In Section 3.1, we learned that the domain of an exponential function of the form $f(x) = b^x$ includes all real numbers and its range is the set of positive real numbers. Because the logarithmic function reverses the domain and the range of the exponential function, the **domain of a logarithmic function of the form** $f(x) = \log_b x$ **is the set of all positive real numbers.** Thus, $\log_2 8$ is defined because the value of x in the logarithmic expression, 8, is greater than zero and therefore is included in the domain of the logarithmic function $f(x) = \log_2 x$. However, $\log_2 0$ and $\log_2 (-8)$ are not defined because 0 and -8 are not positive real numbers and therefore are excluded from the domain of the logarithmic function $f(x) = \log_2 x$. In general, the domain of $f(x) = \log_b (x + c)$ consists of all x for which $x + c > 0$.

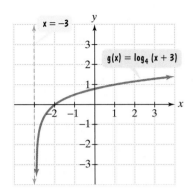

Figure 3.12 The domain of $g(x) = \log_4 (x + 3)$ is $(-3, \infty)$.

EXAMPLE 7 Finding the Domain of a Logarithmic Function

Find the domain of $g(x) = \log_4 (x + 3)$.

Solution   The domain of g consists of all x for which $x + 3 > 0$. Solving this inequality for x, we obtain $x > -3$. Thus, the domain of g is $(-3, \infty)$. This is illustrated in Figure 3.12. The vertical asymptote is $x = -3$, and all points on the graph of g have x-coordinates that are greater than -3.

Check Point 7   Find the domain of $h(x) = \log_4 (x - 5)$.

7 Use common logarithms.

Common Logarithms

The logarithmic function with base 10 is called the **common logarithmic function.** The function $f(x) = \log_{10} x$ is usually expressed as $f(x) = \log x$. A calculator with a $\boxed{\text{LOG}}$ key can be used to evaluate common logarithms. Here are some examples:

Logarithm	Most Scientific Calculator Keystrokes	Most Graphing Calculator Keystrokes	Display (or Approximate Display)
$\log 1000$	1000 $\boxed{\text{LOG}}$	$\boxed{\text{LOG}}$ 1000 $\boxed{\text{ENTER}}$	3
$\log \dfrac{5}{2}$	$\boxed{(}$ 5 $\boxed{\div}$ 2 $\boxed{)}$ $\boxed{\text{LOG}}$	$\boxed{\text{LOG}}$ $\boxed{(}$ 5 $\boxed{\div}$ 2 $\boxed{)}$ $\boxed{\text{ENTER}}$	0.39794
$\dfrac{\log 5}{\log 2}$	5 $\boxed{\text{LOG}}$ $\boxed{\div}$ 2 $\boxed{\text{LOG}}$ $\boxed{=}$	$\boxed{\text{LOG}}$ 5 $\boxed{\div}$ $\boxed{\text{LOG}}$ 2 $\boxed{\text{ENTER}}$	2.32193
$\log (-3)$	3 $\boxed{+/-}$ $\boxed{\text{LOG}}$	$\boxed{\text{LOG}}$ $\boxed{(-)}$ 3 $\boxed{\text{ENTER}}$	$\boxed{\text{ERROR}}$

The error message given by many calculators for $\log (-3)$ is a reminder that the domain of every logarithmic function, including the common logarithmic function, is the set of positive real numbers.

   Many real-life phenomena start with rapid growth, and then the growth begins to level off. This type of behavior can be modeled by logarithmic functions.

EXAMPLE 8 Modeling Height of Children

The percentage of adult height attained by a boy who is x years old can be modeled by

$$f(x) = 29 + 48.8 \log (x + 1)$$

where x represents the boy's age and $f(x)$ represents the percentage of his adult height. Approximately what percent of his adult height is a boy at age eight?

Solution We substitute the boy's age, 8, for x and evaluate the function.

$$f(x) = 29 + 48.8 \log (x + 1) \qquad \text{This is the given function.}$$
$$f(8) = 29 + 48.8 \log (8 + 1) \qquad \text{Substitute 8 for x.}$$
$$= 29 + 48.8 \log 9 \qquad \text{Graphing calculator keystrokes:}$$
$$\approx 76 \qquad \qquad\qquad \text{29 } \boxed{+}\text{ 48.8 } \boxed{\text{LOG}}\text{ 9 } \boxed{\text{ENTER}}.$$

Thus, an 8-year-old boy is approximately 76% of his adult height.

> **Check Point 8** Use the function in Example 8 to answer this question: Approximately what percent of his adult height is a boy at age 10?

The basic properties of logarithms that were listed earlier in this section can be applied to common logarithms.

Properties of Common Logarithms	
General Properties	**Common Logarithm Properties**
1. $\log_b 1 = 0$	**1.** $\log 1 = 0$
2. $\log_b b = 1$	**2.** $\log 10 = 1$
3. $\log_b b^x = x$	**3.** $\log 10^x = x$
4. $b^{\log_b x} = x$	**4.** $10^{\log x} = x$

The property $\log 10^x = x$ can be used to evaluate common logarithms involving powers of 10. For example,

$$\log 100 = \log 10^2 = 2, \quad \log 1000 = \log 10^3 = 3, \quad \text{and} \quad \log 10^{7.1} = 7.1.$$

EXAMPLE 9 Earthquake Intensity

The magnitude, R, on the Richter scale of an earthquake of intensity I is given by

$$R = \log \frac{I}{I_0}$$

where I_0 is the intensity of a barely felt zero-level earthquake. The earthquake that destroyed San Francisco in 1906 was $10^{8.3}$ times as intense as a zero-level earthquake. What was its magnitude on the Richter scale?

Solution Because the earthquake was $10^{8.3}$ times as intense as a zero-level earthquake, the intensity, I, is $10^{8.3}I_0$.

$$R = \log \frac{I}{I_0} \qquad \text{This is the formula for magnitude on the Richter scale.}$$

$$R = \log \frac{10^{8.3}I_0}{I_0} \qquad \text{Substitute } 10^{8.3}I_0 \text{ for } I.$$

$$= \log 10^{8.3} \qquad \text{Simplify.}$$

$$= 8.3 \qquad \text{Use the property } \log 10^x = x.$$

San Francisco's 1906 earthquake registered 8.3 on the Richter scale.

Check Point 9 Use the formula in Example 9 to solve this problem. If an earthquake is 10,000 times as intense as a zero-level quake ($I = 10,000I_0$), what is its magnitude on the Richter scale?

---

8 Use natural logarithms.

Natural Logarithms

The logarithmic function with base e is called the **natural logarithmic function.** The function $f(x) = \log_e x$ is usually expressed as $f(x) = \ln x$, read "el en of x." A calculator with an $\boxed{\text{LN}}$ key can be used to evaluate natural logarithms.

Like the domain of all logarithmic functions, the domain of the natural logarithmic function is the set of all positive real numbers. Thus, the domain of $f(x) = \ln(x + c)$ consists of all x for which $x + c > 0$.

EXAMPLE 10   Finding Domains of Natural Logarithmic Functions

Find the domain of each function:

a. $f(x) = \ln(3 - x)$ **b.** $g(x) = \ln(x - 3)^2$.

Solution

a. The domain of f consists of all x for which $3 - x > 0$. Solving this inequality for x, we obtain $x < 3$. Thus, the domain of f is $\{x | x < 3\}$, or $(-\infty, 3)$. This is verified by the graph in Figure 3.13.

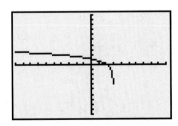

Figure 3.13 The domain of $f(x) = \ln(3 - x)$ is $(-\infty, 3)$.

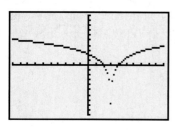

Figure 3.14 3 is excluded from the domain of $g(x) = \ln(x - 3)^2$.

b. The domain of g consists of all x for which $(x - 3)^2 > 0$. It follows that the domain of g is the set of all real numbers except 3. Thus, the domain of g is $\{x \mid x \neq 3\}$, or, in interval notation, $(-\infty, 3)$ or $(3, \infty)$. This is shown by the graph in Figure 3.14. To make it more obvious that 3 is excluded from the domain, we changed the $\boxed{\text{MODE}}$ to Dot.

Check Point 10

Find the domain of each function:

a. $f(x) = \ln(4 - x)$ **b.** $g(x) = \ln x^2$.

The basic properties of logarithms that were listed earlier in this section can be applied to natural logarithms.

Properties of Natural Logarithms

General Properties	Natural Logarithm Properties
1. $\log_b 1 = 0$	**1.** $\ln 1 = 0$
2. $\log_b b = 1$	**2.** $\ln e = 1$
3. $\log_b b^x = x$	**3.** $\ln e^x = x$
4. $b^{\log_b x} = x$	**4.** $e^{\ln x} = x$

The property $\ln e^x = x$ can be used to evaluate natural logarithms involving powers of e. For example,

$$\ln e^2 = 2, \quad \ln e^3 = 3, \quad \ln e^{7.1} = 7.1, \quad \text{and} \quad \ln \frac{1}{e} = \ln e^{-1} = -1.$$

EXAMPLE 11   Using Inverse Properties

Use inverse properties to simplify:

a. $\ln e^{7x}$ **b.** $e^{\ln 4x^2}$.

Solution

a. Because $\ln e^x = x$, we conclude that $\ln e^{7x} = 7x$.

b. Because $e^{\ln x} = x$, we conclude $e^{\ln 4x^2} = 4x^2$.

Check Point 11

Use inverse properties to simplify:

a. $\ln e^{25x}$ **b.** $e^{\ln \sqrt{x}}$.

EXAMPLE 12   Walking Speed and City Population

As the population of a city increases, the pace of life also increases. The formula

$$W = 0.35 \ln P + 2.74$$

models average walking speed, W, in feet per second, for a resident of a city whose population is P thousand. Find the average walking speed for people living in New York City with a population of 7323 thousand.

Solution We use the formula and substitute 7323 for P, the population in thousands.

$$W = 0.35 \ln P + 2.74 \qquad \text{This is the given formula.}$$
$$W = 0.35 \ln 7323 + 2.74 \qquad \text{Substitute 7323 for P.}$$
$$\approx 5.9 \qquad \text{Graphing calculator keystrokes:}$$
$$0.35 \boxed{\text{LN}} \ 7323 \ \boxed{+} \ 2.74 \ \boxed{\text{ENTER}}.$$

The average walking speed in New York City is approximately 5.9 feet per second.

Check Point 12 Use the formula $W = 0.35 \ln P + 2.74$ to find the average walking speed in Jackson, Mississippi, with a population of 197 thousand.

EXERCISE SET 3.2

Practice Exercises

In Exercises 1–8, write each equation in its equivalent exponential form.

1. $4 = \log_2 16$

2. $6 = \log_2 64$

3. $2 = \log_3 x$

4. $2 = \log_9 x$

5. $5 = \log_b 32$

6. $3 = \log_b 27$

7. $\log_6 216 = y$

8. $\log_5 125 = y$

In Exercises 9–20, write each equation in its equivalent logarithmic form.

9. $2^3 = 8$

10. $5^4 = 625$

11. $2^{-4} = \frac{1}{16}$

12. $5^{-3} = \frac{1}{125}$

13. $\sqrt[3]{8} = 2$

14. $\sqrt[3]{64} = 4$

15. $13^2 = x$

16. $15^2 = x$

17. $b^3 = 1000$

18. $b^3 = 343$

19. $7^y = 200$

20. $8^y = 300$

In Exercises 21–38, evaluate each expression without using a calculator.

21. $\log_4 16$

22. $\log_7 49$

23. $\log_2 64$

24. $\log_3 27$

25. $\log_7 \sqrt{7}$

26. $\log_6 \sqrt{6}$

27. $\log_2 \frac{1}{8}$

28. $\log_3 \frac{1}{9}$

29. $\log_{64} 8$

30. $\log_{81} 9$

31. $\log_5 5$

32. $\log_{11} 11$

33. $\log_4 1$

34. $\log_6 1$

35. $\log_5 5^7$

36. $\log_4 4^6$

37. $8^{\log_8 19}$

38. $7^{\log_7 23}$

39. Graph $f(x) = 4^x$ and $g(x) = \log_4 x$ in the same rectangular coordinate system.

40. Graph $f(x) = 5^x$ and $g(x) = \log_5 x$ in the same rectangular coordinate system.

41. Graph $f(x) = \left(\frac{1}{2}\right)^x$ and $g(x) = \log_{1/2} x$ in the same rectangular coordinate system.

42. Graph $f(x) = \left(\frac{1}{4}\right)^x$ and $g(x) = \log_{1/4} x$ in the same rectangular coordinate system.

In Exercises 43–48, the graph of a logarithmic function is given. Select the function for each graph from the following options:

$$f(x) = \log_3 x, \ g(x) = \log_3 (x - 1), \ h(x) = \log_3 x - 1,$$

$$F(x) = -\log_3 x, \ G(x) = \log_3 (-x), \ H(x) = 1 - \log_3 x.$$

43.

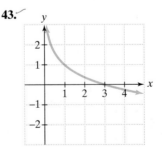

44.

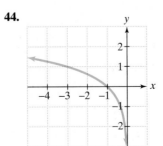

45.

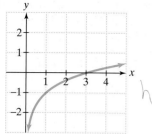

46.

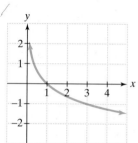

47.

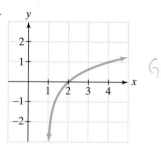

48.

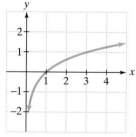

In Exercises 49–54, begin by graphing $f(x) = \log_2 x$. Then use transformations of this graph to graph the given function. What is the graph's x-intercept? What is the vertical asymptote?

49. $g(x) = \log_2 (x + 1)$ **50.** $g(x) = \log_2 (x + 2)$

51. $h(x) = 1 + \log_2 x$ **52.** $h(x) = 2 + \log_2 x$

53. $g(x) = \frac{1}{2} \log_2 x$ **54.** $g(x) = -2 \log_2 x$

In Exercises 55–60, find the domain of each logarithmic function.

55. $f(x) = \log_5 (x + 4)$ **56.** $f(x) = \log_5 (x + 6)$

57. $f(x) = \log (2 - x)$ **58.** $f(x) = \log (7 - x)$

59. $f(x) = \ln (x - 2)^2$ **60.** $f(x) = \ln (x - 7)^2$

In Exercises 61–74, evaluate each expression without using a calculator.

61. $\log 100$ **62.** $\log 1000$ **63.** $\log 10^7$

64. $\log 10^8$ **65.** $10^{\log 33}$ **66.** $10^{\log 53}$

67. $\ln 1$ **68.** $\ln e$ **69.** $\ln e^6$

70. $\ln e^7$ **71.** $\ln \dfrac{1}{e^6}$ **72.** $\ln \dfrac{1}{e^7}$

73. $e^{\ln 125}$ **74.** $e^{\ln 300}$

In Exercises 75–80, use inverse properties of logarithms to simplify each expression.

75. $\ln e^{9x}$ **76.** $\ln e^{13x}$ **77.** $e^{\ln 5x^2}$

78. $e^{\ln 7x^2}$ **79.** $10^{\log \sqrt{x}}$ **80.** $10^{\log \sqrt[3]{x}}$

Application Exercises

The percentage of adult height attained by a girl who is x years old can be modeled by

$$f(x) = 62 + 35 \log (x - 4)$$

where x represents the girl's age (from 5 to 15) and $f(x)$ represents the percentage of her adult height. Use the function to solve Exercises 81–82.

81. Approximately what percent of her adult height is a girl at age 13?

82. Approximately what percent of her adult height is a girl at age ten?

83. The annual amount that we spend to attend sporting events can be modeled by

$$f(x) = 2.05 + 1.3 \ln x$$

where x represents the number of years after 1984 and $f(x)$ represents the total annual expenditures for admission to spectator sports, in billions of dollars. In 2000, approximately how much was spent on admission to spectator sports?

84. The percentage of U.S. households with cable television can be modeled by

$$f(x) = 18.32 + 15.94 \ln x$$

where x represents the number of years after 1979 and $f(x)$ represents the percentage of U.S. households with cable television. What percentage of U.S. households had cable television in 1990?

The loudness level of a sound, D, in decibels, is given by the formula

$$D = 10 \log (10^{12} I)$$

where I is the intensity of the sound, in watts per meter^2. Decibel levels range from 0, a barely audible sound, to 160, a sound resulting in a ruptured eardrum. Use the formula to solve Exercises 85–86.

85. The sound of a blue whale can be heard 500 miles away, reaching an intensity of 6.3×10^6 watts per meter^2. Determine the decibel level of this sound. At close range, can the sound of a blue whale rupture the human eardrum?

86. What is the decibel level of a normal conversation, 3.2×10^{-6} watt per meter^2?

87. Students in a psychology class took a final examination. As part of an experiment to see how much of the course content they remembered over time, they took equivalent forms of the exam in monthly intervals thereafter. The average score for the group, $f(t)$, after t months was modeled by the function

$$f(t) = 88 - 15 \ln (t + 1), \qquad 0 \le t \le 12.$$

  a. What was the average score on the original exam?

  b. What was the average score after 2 months? 4 months? 6 months? 8 months? 10 months? one year?

  c. Sketch the graph of f (either by hand or with a graphing utility). Describe what the graph indicates in terms of the material retained by the students.

Writing in Mathematics

88. Describe the relationship between an equation in logarithmic form and an equivalent equation in exponential form.

89. What question can be asked to help evaluate $\log_3 81$?

90. Explain why the logarithm of 1 with base b is 0.

91. Describe the following property using words: $\log_b b^x = x$.

92. Explain how to use the graph of $f(x) = 2^x$ to obtain the graph of $g(x) = \log_2 x$.

93. Explain how to find the domain of a logarithmic function.

94. New York City is one of the world's great walking cities. Use the formula in Example 12 on page 376 to describe what frequently happens to tourists exploring the city by foot.

95. Logarithmic models are well suited to phenomena in which growth is initially rapid but then begins to level off. Describe something that is changing over time that can be modeled using a logarithmic function.

96. Suppose that a girl is 4' 6" at age 10. Explain how to use the function in Exercises 81–82 to determine how tall she can expect to be as an adult.

Technology Exercises

In Exercises 97–100, graph f and g in the same viewing rectangle. Then describe the relationship of the graph of g to the graph of f.

97. $f(x) = \ln x, g(x) = \ln (x + 3)$

98. $f(x) = \ln x, g(x) = \ln x + 3$

99. $f(x) = \log x, g(x) = -\log x$

100. $f(x) = \log x, g(x) = \log (x - 2) + 1$

101. Students in a mathematics class took a final examination. They took equivalent forms of the exam in monthly intervals thereafter. The average score, $f(t)$, for the group after t months was modeled by the human memory function $f(t) = 75 - 10 \log (t + 1)$, where $0 \le t \le 12$.

Use a graphing utility to graph the function. Then determine how many months will elapse before the average score falls below 65.

102. Graph f and g in the same viewing rectangle.

  a. $f(x) = \ln (3x), g(x) = \ln 3 + \ln x$

  b. $f(x) = \log (5x^2), g(x) = \log 5 + \log x^2$

  c. $f(x) = \ln (2x^3), g(x) = \ln 2 + \ln x^3$

  d. Describe what you observe in parts (a)–(c). Generalize this observation by writing an equivalent expression for $\log_b (MN)$, where $M > 0$ and $N > 0$.

  e. Complete this statement: The logarithm of a product is equal to _____.

103. Graph each of the following functions in the same viewing rectangle and then place the functions in order from the one that increases most slowly to the one that increases most rapidly.

$$y = x, y = \sqrt{x}, y = e^x, y = \ln x, y = x^x, y = x^2$$

Critical Thinking Exercises

104. Which one of the following is true?

  a. $\dfrac{\log_2 8}{\log_2 4} = \dfrac{8}{4}$

  b. $\log (-100) = -2$

  c. The domain of $f(x) = \log_2 x$ is $(-\infty, \infty)$.

  d. $\log_b x$ is the exponent to which b must be raised to obtain x.

105. Without using a calculator, find the exact value of

$$\frac{\log_3 81 - \log_\pi 1}{\log_{2\sqrt{2}} 8 - \log 0.001}.$$

106. Solve for x: $\log_4[\log_3(\log_2 x)] = 0$.

107. Without using a calculator, determine which is the greater number: $\log_4 60$ or $\log_3 40$.

Group Exercise

108. This group exercise involves exploring the way we grow. Group members should create a graph for the function that models the percentage of adult height attained by a boy who is x years old, $f(x) = 29 + 48.8 \log (x + 1)$. Let $x = 1, 2, 3, \ldots, 12$, find function values, and connect the resulting points with a smooth curve. Then create a function that models the percentage of adult height attained by a girl who is x years old, $g(x) = 62 + 35 \log (x - 4)$. Let $x = 5, 6, 7, \ldots, 15$, find function values, and connect the resulting points with a smooth curve. Group members should then discuss similarities and differences in the growth patterns for boys and girls based on the graphs.

SECTION 3.3 *Properties of Logarithms*

Objectives

1. Use the product rule.
2. Use the quotient rule.
3. Use the power rule.
4. Expand logarithmic expressions.
5. Condense logarithmic expressions.
6. Use the change-of-base property.

We all learn new things in different ways. In this section, we consider important properties of logarithms. What would be the most effective way for you to learn about these properties? Would it be helpful to use your graphing utility and discover one of these properties for yourself? To do so, work Exercise 102 in Exercise Set 3.2 before continuing. Would the properties become more meaningful if you could see exactly where they come from? If so, you will find details of the proofs of many of these properties in the appendix. The remainder of our work in this chapter will be based on the properties of logarithms that you learn in this section.

1 Use the product rule.

The Product Rule

Properties of exponents correspond to properties of logarithms. For example, when we multiply with the same base, we add exponents:

$$b^m \cdot b^n = b^{m+n}.$$

This property of exponents, coupled with an awareness that a logarithm is an exponent, suggests the following property, called the **product rule:**

Discovery

We know that log 100,000 = 5. Show that you get the same result by writing 100,000 as 1000 · 100 and then using the product rule. Then verify the product rule by using other numbers whose logarithms are easy to find.

> **The Product Rule**
>
> Let b, M, and N be positive real numbers with $b \neq 1$.
>
> $$\log_b (MN) = \log_b M + \log_b N$$
>
> The logarithm of a product is the sum of the logarithms.

When we use the product rule to write a single logarithm as the sum of two logarithms, we say that we are **expanding a logarithmic expression.** For example, we can use the product rule to expand $\ln(4x)$:

$$\ln (7x) = \ln 7 + \ln x.$$

The logarithm of a product   is   the sum of the logarithms.

EXAMPLE 1 Using the Product Rule

Use the product rule to expand each logarithmic expression:

    a. $\log_4 (7 \cdot 5)$ **b.** $\log (10x)$.

Solution

a. $\log_4 (7 \cdot 5) = \log_4 7 + \log_4 5$ *The logarithm of a product is the sum of the logarithms.*

b. $\log (10x) = \log 10 + \log x$ *The logarithm of a product is the sum of the logarithms. These are common logarithms with base 10 understood.*

$= 1 + \log x$ *Because $\log_b b = 1$, then $\log 10 = 1$.*

Check Point 1   Use the product rule to expand each logarithmic expression:

a. $\log_6 (7 \cdot 11)$ **b.** $\log (100x)$.

2 Use the quotient rule.

The Quotient Rule

When we divide with the same base, we subtract exponents:

$$\frac{b^m}{b^n} = b^{m-n}.$$

This property suggests the following property of logarithms, called the **quotient rule:**

Discovery

We know that $\log_2 16 = 4$. Show that you get the same result by writing 16 as $\frac{32}{2}$ and then using the quotient rule. Then verify the quotient rule using other numbers whose logarithms are easy to find.

The Quotient Rule

Let b, M, and N be positive real numbers with $b \neq 1$.

$$\log_b \left(\frac{M}{N}\right) = \log_b M - \log_b N$$

The logarithm of a quotient is the difference of the logarithms.

When we use the quotient rule to write a single logarithm as the difference of two logarithms, we say that we are **expanding a logarithmic expression.** For example, we can use the quotient rule to expand $\log \frac{x}{2}$:

$$\log \frac{x}{2} = \log x - \log 2.$$

The logarithm of a quotient   is   the difference of the logarithms.

EXAMPLE 2   Using the Quotient Rule

Use the quotient rule to expand each logarithmic expression:

a. $\log_7 \left(\frac{19}{x}\right)$ **b.** $\ln \left(\frac{e^3}{7}\right)$.

Solution

a. $\log_7 \left(\frac{19}{x}\right) = \log_7 19 - \log_7 x$ *The logarithm of a quotient is the difference of the logarithms.*

b. $\ln \left(\frac{e^3}{7}\right) = \ln e^3 - \ln 7$ *The logarithm of a quotient is the difference of the logarithms. These are natural logarithms with base e understood.*

$= 3 - \ln 7$ *Because $\ln e^x = x$, then $\ln e^3 = 3$.*

Check Point 2 Use the quotient rule to expand each logarithmic expression:

a. $\log_8\left(\dfrac{23}{x}\right)$ **b.** $\ln\left(\dfrac{e^5}{11}\right)$.

3 Use the power rule.

The Power Rule

When an exponential expression is raised to a power, we multiply exponents:

$$\left(b^m\right)^n = b^{mn}.$$

This property suggests the following property of logarithms, called the **power rule:**

> ### The Power Rule
>
> Let b and M be positive real numbers with $b \neq 1$, and let p be any real number.
>
> $$\log_b M^p = p \log_b M$$
>
> The logarithm of a number with an exponent is the product of the exponent and the logarithm of that number.

When we use the power rule to "pull the exponent to the front," we say that we are **expanding a logarithmic expression.** For example, we can use the power rule to expand $\ln x^2$:

$$\ln x^2 = 2 \ln x.$$

The logarithm of a number with an exponent   is   the product of the exponent and the logarithm of that number.

Figure 3.15 shows the graphs of $y = \ln x^2$ and $y = 2 \ln x$. Are $\ln x^2$ and $2 \ln x$ the same? The graphs illustrate that $y = \ln x^2$ and $y = 2 \ln x$ have different domains. The graphs are only the same if $x > 0$. Thus, we should write

$$\ln x^2 = 2 \ln x \text{ for } x > 0.$$

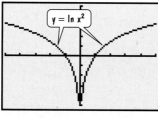

Domain: $(-\infty, 0)$ or $(0, \infty)$

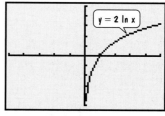

Domain: $(0, \infty)$

Figure 3.15 $\ln x^2$ and $2 \ln x$ have different domains.

When expanding a logarithmic expression, you might want to determine whether the rewriting has changed the domain of the expression. For the rest of this section, assume that all variable and variable expressions represent positive numbers.

EXAMPLE 3   Using the Power Rule

Use the power rule to expand each logarithmic expression:

a. $\log_5 7^4$ **b.** $\ln \sqrt{x}$.

Solution

a. $\log_5 7^4 = 4 \log_5 7$ The logarithm of a number with an exponent is the exponent times the logarithm of the number.

b. $\ln \sqrt{x} = \ln x^{1/2}$ — Rewrite the radical using a rational exponent.

$= \frac{1}{2} \ln x$ — Use the power rule to bring the exponent to the front.

> **Check Point 3** — Use the power rule to expand each logarithmic expression:
>
> **a.** $\log_6 3^9$ **b.** $\ln \sqrt[3]{x}$.

④ Expand logarithmic expressions.

Expanding Logarithmic Expressions

It is sometimes necessary to use more than one property of logarithms when you expand a logarithmic expression. Properties for expanding logarithmic expressions are as follows:

Properties for Expanding Logarithmic Expressions

For $M > 0$ and $N > 0$:

1. $\log_b (MN) = \log_b M + \log_b N$ Product rule

2. $\log_b \left(\dfrac{M}{N} \right) = \log_b M - \log_b N$ Quotient rule

3. $\log_b M^p = p \log_b M$ Power rule

EXAMPLE 4   Expanding Logarithmic Expressions

Use logarithmic properties to expand each expression as much as possible:

a. $\log_b (x^2 \sqrt{y})$ **b.** $\log_6 \left(\dfrac{\sqrt[3]{x}}{36 y^4} \right)$.

Solution   We will have to use two or more of the properties for expanding logarithms in each part of this example.

a. $\log_b (x^2 \sqrt{y}) = \log_b (x^2 y^{1/2})$ — Use exponential notation.

$= \log_b x^2 + \log_b y^{1/2}$ — Use the product rule.

$= 2 \log_b x + \dfrac{1}{2} \log_b y$ — Use the power rule.

b. $\log_6 \left(\dfrac{\sqrt[3]{x}}{36 y^4} \right) = \log_6 \left(\dfrac{x^{1/3}}{36 y^4} \right)$ — Use exponential notation.

$= \log_6 x^{1/3} - \log_6 (36 y^4)$ — Use the quotient rule.

$= \log_6 x^{1/3} - (\log_6 36 + \log_6 y^4)$ — Use the product rule on $\log_6 (36 y^4)$.

$= \dfrac{1}{3} \log_6 x - (\log_6 36 + 4 \log_6 y)$ — Use the power rule.

$= \dfrac{1}{3} \log_6 x - \log_6 36 - 4 \log_6 y$ — Apply the distributive property.

$= \dfrac{1}{3} \log_6 x - 2 - 4 \log_6 y$ — $\log_6 36 = 2$ because 2 is the power to which we must raise 6 to get 36. $(6^2 = 36)$

> **Check Point 4** — Use logarithmic properties to expand each expression as much as possible:
>
> **a.** $\log_b (x^4 \sqrt[3]{y})$ **b.** $\log_5 \dfrac{\sqrt{x}}{25 y^3}$.

Study Tip

The graphs show

$$y_1 = \ln (x + 3)$$

and

$$y_2 = \ln x + \ln 3.$$

The graphs are not the same. The graph of y_1 is the graph of the natural logarithmic function shifted 3 units to the left. By contrast, the graph of y_2 is the graph of the natural logarithmic function shifted upward by $\ln 3$, or about 1.1 units. Thus we see that

$$\ln (x + 3) \neq \ln x + \ln 3.$$

In general,

$$\log_b (M + N) \neq \log_b M + \log_b N.$$

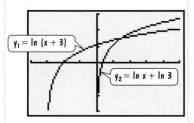

Try to avoid the following errors:

INCORRECT

$\log_b (M + N) = \log_b M + \log_b N$

$\log_b (M - N) = \log_b M - \log_b N$

$\log_b (M \cdot N) = \log_b M \cdot \log_b N$

$\log_b \left(\dfrac{M}{N} \right) = \dfrac{\log_b M}{\log_b N}$

$\dfrac{\log_b M}{\log_b N} = \log_b M - \log_b N$

5 Condense logarithmic expressions.

Condensing Logarithmic Expressions

To **condense a logarithmic expression,** we write the sum or difference of two or more logarithmic expressions as a single logarithmic expression. We use the properties of logarithms to do so.

Study Tip

These properties are the same as those in the box on page 383. The only difference is that we've reversed the sides in each property from the previous box.

Properties for Condensing Logarithmic Expressions

For $M > 0$ and $N > 0$:

1. $\log_b M + \log_b N = \log_b(MN)$ Product rule

2. $\log_b M - \log_b N = \log_b\left(\dfrac{M}{N}\right)$ Quotient rule

3. $p\log_b M = \log_b M^p$ Power rule

EXAMPLE 5   Condensing Logarithmic Expressions

Write as a single logarithm:

   a. $\log_4 2 + \log_4 32$ **b.** $\log(4x - 3) - \log x.$

Solution

  a. $\log_4 2 + \log_4 32 = \log_4(2 \cdot 32)$ Use the product rule.

                         $= \log_4 64$ We now have a single logarithm. However, we can simplify.

                         $= 3$ $\log_4 64 = 3$ because $4^3 = 64$.

  b. $\log(4x - 3) - \log x = \log\left(\dfrac{4x - 3}{x}\right)$ Use the quotient rule.

Check Point 5   Write as a single logarithm:

   a. $\log 25 + \log 4$ **b.** $\log(7x + 6) - \log x.$

Coefficients of logarithms must be 1 before you can condense them using the product and quotient rules. For example, to condense

$$2\ln x + \ln(x + 1),$$

the coefficient of the first term must be 1. We use the power rule to rewrite the coefficient as an exponent:

> 1. Use the power rule to make the number in front an exponent.

$$2\ln x + \ln(x + 1) = \ln x^2 + \ln(x + 1) = \ln[x^2(x + 1)].$$

> 2. Use the product rule. The sum of logarithms with coefficients 1 is the logarithm of the product.

EXAMPLE 6   Condensing Logarithmic Expressions

Write as a single logarithm:

   a. $\frac{1}{2}\log x + 4\log(x - 1)$ **b.** $3\ln(x + 7) - \ln x$

   c. $4\log_b x - 2\log_b 6 + \frac{1}{2}\log_b y.$

Solution

a. $\frac{1}{2}\log x + 4\log(x-1)$

$\quad = \log x^{1/2} + \log(x-1)^4$ *Use the power rule so that all coefficients are 1.*

$\quad = \log[x^{1/2}(x-1)^4]$ *Use the product rule. The condensation can be expressed as $\log[\sqrt{x}\,(x-1)^4]$.*

b. $3\ln(x+7) - \ln x$

$\quad = \ln(x+7)^3 - \ln x$ *Use the power rule so that all coefficients are 1.*

$\quad = \ln\left[\dfrac{(x+7)^3}{x}\right]$ *Use the quotient rule.*

c. $4\log_b x - 2\log_b 6 + \frac{1}{2}\log_b y$

$\quad = \log_b x^4 - \log_b 6^2 + \log_b y^{1/2}$ *Use the power rule so that all coefficients are 1.*

$\quad = (\log_b x^4 - \log_b 36) + \log_b y^{1/2}$ *This optional step emphasizes the order of operations.*

$\quad = \log_b\left(\dfrac{x^4}{36}\right) + \log_b y^{1/2}$ *Use the quotient rule.*

$\quad = \log_b\left(\dfrac{x^4}{36} \cdot y^{1/2}\right)$ or $\log\left(\dfrac{x^4\sqrt{y}}{36}\right)$ *Use the product rule.*

Check Point 6 Write as a single logarithm:

    a. $2\ln x + \frac{1}{3}\ln(x+5)$ **b.** $2\log(x-3) - \log x$

    c. $\frac{1}{4}\log_b x - 2\log_b 5 + 10\log_b y$.

6 Use the change-of-base property.

The Change-of-Base Property

We have seen that calculators give the values of both common logarithms (base 10) and natural logarithms (base e). To find a logarithm with any other base, we can use the following change-of-base property:

The Change-of-Base Property

For any logarithmic bases a and b, and any positive number M,

$$\log_b M = \frac{\log_a M}{\log_a b}.$$

The logarithm of M with base b is equal to the logarithm of M with any new base divided by the logarithm of b with that new base.

    In the change-of-base property, base b is the base of the original logarithm. Base a is a new base that we introduce. Thus, the change-of-base property allows us to change from base b to *any* new base a, as long as the newly introduced base is a positive number not equal to 1.

    The change-of-base property is used to write a logarithm in terms of quantities that can be evaluated with a calculator. Because calculators contain keys for common (base 10) and natural (base e) logarithms, we will frequently introduce base 10 or base e.

Change-of-Base Property	**Introducing Common Logarithms**	**Introducing Natural Logarithms**
$\log_b M = \dfrac{\log_a M}{\log_a b}$	$\log_b M = \dfrac{\log_{10} M}{\log_{10} b}$	$\log_b M = \dfrac{\log_e M}{\log_e b}$
a is the new introduced base.	*10 is the new introduced base.*	*e is the new introduced base.*

Using the notations for common logarithms and natural logarithms, we have the following results:

> ### The Change-of-Base Property: Introducing Common and Natural Logarithms
>
> **Introducing Common Logarithms** **Introducing Natural Logarithms**
>
> $$\log_b M = \frac{\log M}{\log b} \qquad\qquad \log_b M = \frac{\ln M}{\ln b}$$

EXAMPLE 7 Changing Base to Common Logarithms

Use common logarithms to evaluate $\log_5 140$.

Solution Because $\log_b M = \dfrac{\log M}{\log b}$,

$$\log_5 140 = \frac{\log 140}{\log 5}$$

$$\approx 3.07.$$

Use a calculator: 140 LOG ÷ 5 LOG = or LOG 140 ÷ LOG 5 ENTER.

This means that $\log_5 140 \approx 3.07$.

Check Point 7 Use common logarithms to evaluate $\log_7 2506$.

Discovery

Find a reasonable estimate of $\log_5 140$ to the nearest whole number. 5 to what power is 140? Compare your estimate to the value obtained in Example 7.

EXAMPLE 8 Changing Base to Natural Logarithms

Use natural logarithms to evaluate $\log_5 140$.

Solution Because $\log_b M = \dfrac{\ln M}{\ln b}$,

$$\log_5 140 = \frac{\ln 140}{\ln 5}$$

$$\approx 3.07.$$

Use a calculator: 140 LN ÷ 5 LN = or LN 140 ÷ LN 5 ENTER.

We have again shown that $\log_5 140 \approx 3.07$.

Check Point 8 Use natural logarithms to evaluate $\log_7 2506$.

We can use the change-of-base property to graph logarithmic functions with bases other than 10 or e on a graphing utility. For example, Figure 3.16 shows the graphs of

$$y = \log_2 x \quad \text{and} \quad y = \log_{20} x$$

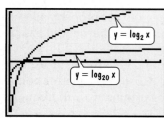

Figure 3.16 Using the change-of-base property to graph logarithmic functions

in a $[0, 10, 1]$ by $[-3, 3, 1]$ viewing rectangle. Because $\log_2 x = \dfrac{\ln x}{\ln 2}$ and $\log_{20} x = \dfrac{\ln x}{\ln 20}$, the functions can be entered as

$$y_1 = \boxed{\text{LN}}\ x\ \boxed{\div}\ \boxed{\text{LN}}\ 2$$

and $y_2 = \boxed{\text{LN}}\ x\ \boxed{\div}\ \boxed{\text{LN}}\ 20.$

The Curious Number e

You will learn more about each curiosity mentioned below when you take calculus.

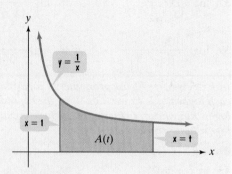

- The number e was named by the Swiss mathematician Leonhard Euler (1707–1783), who proved that it is the limit as $n \to \infty$ of $\left(1 + \dfrac{1}{n}\right)^n$.

- e features in Euler's remarkable relationship $e^{i\pi} = -1$, in which $i = \sqrt{-1}$.

- The first few decimal places of e are fairly easy to remember: $e = 2.7\ 1828\ 1828\ 45\ 90\ 45 \ldots$.

- The best approximation of e using numbers less than 1000 is also easy to remember: $e \approx \dfrac{878}{323} \approx 2.71826 \ldots$.

Figure 3.17

- Isaac Newton (1642–1727), one of the cofounders of calculus, showed that $e^x = 1 + x + \dfrac{x^2}{2!} + \dfrac{x^3}{3!} + \dfrac{x^4}{4!} + \ldots$, from which we obtain $e = 1 + 1 + \dfrac{1}{2!} + \dfrac{1}{3!} + \dfrac{1}{4!} \ldots$, an infinite sum suitable for calculation because its terms decrease so rapidly. (*Note:* $n!$ (n factorial) is the product of all the consecutive integers from n down to 1: $n! = n(n-1)(n-2)(n-3) \cdot \ldots \cdot 3 \cdot 2 \cdot 1$.)

- The area of the region bounded by $y = \dfrac{1}{x}$, the x-axis, $x = 1$ and $x = t$ (shaded in Figure 3.17) is a function of t, designated by $A(t)$. Grégoire de Saint-Vincent, a Belgian Jesuit (1584–1667), spent his entire professional life attempting to find a formula for $A(t)$. With his student, he showed that $A(t) = \ln t$, becoming one of the first mathematicians to make use of the logarithmic function for something other than a computational device.

EXERCISE SET 3.3

Practice Exercises

In Exercises 1–40, use properties of logarithms to expand each logarithmic expression as much as possible. Where possible, evaluate logarithmic expressions without using a calculator.

1. $\log_5 (7 \cdot 3)$

2. $\log_8 (13 \cdot 7)$

3. $\log_7 (7x)$

4. $\log_9 (9x)$

5. $\log (1000x)$

6. $\log (10{,}000x)$

7. $\log_7 \left(\dfrac{7}{x}\right)$

8. $\log_9 \left(\dfrac{9}{x}\right)$

9. $\log \left(\dfrac{x}{100}\right)$

10. $\log \left(\dfrac{x}{1000}\right)$

11. $\log_4 \left(\dfrac{64}{y}\right)$

12. $\log_5 \left(\dfrac{125}{y}\right)$

13. $\ln \left(\dfrac{e^2}{5}\right)$

14. $\ln \left(\dfrac{e^4}{8}\right)$

15. $\log_b x^3$

16. $\log_b x^7$

17. $\log N^{-6}$

18. $\log M^{-8}$

19. $\ln \sqrt[5]{x}$

20. $\ln \sqrt[7]{x}$

21. $\log_b (x^2 y)$

22. $\log_b (xy^3)$

23. $\log_4 \left(\dfrac{\sqrt{x}}{64}\right)$

24. $\log_5 \left(\dfrac{\sqrt{x}}{25}\right)$

25. $\log_6 \left(\dfrac{36}{\sqrt{x+1}}\right)$

26. $\log_8 \left(\dfrac{64}{\sqrt{x+1}}\right)$

27. $\log_b \left(\dfrac{x^2 y}{z^2}\right)$

28. $\log_b \left(\dfrac{x^3 y}{z^2}\right)$

29. $\log \sqrt{100x}$

30. $\ln \sqrt{ex}$

31. $\log \sqrt[3]{\dfrac{x}{y}}$

32. $\log \sqrt[5]{\dfrac{x}{y}}$

33. $\log_b \left(\dfrac{\sqrt{x}\, y^3}{z^3} \right)$

34. $\log_b \left(\dfrac{\sqrt[3]{x}\, y^4}{z^5} \right)$

35. $\log_5 \sqrt[3]{\dfrac{x^2 y}{25}}$

36. $\log_2 \sqrt[5]{\dfrac{x y^4}{16}}$

37. $\ln \left[\dfrac{x^3 \sqrt{x^2 + 1}}{(x + 1)^4} \right]$

38. $\ln \left[\dfrac{x^4 \sqrt{x^2 + 3}}{(x + 3)^5} \right]$

39. $\log \left[\dfrac{10 x^2 \sqrt[3]{1 - x}}{7(x + 1)^2} \right]$

40. $\log \left[\dfrac{100 x^3 \sqrt[3]{5 - x}}{3(x + 7)^2} \right]$

In Exercises 41–70, use properties of logarithms to condense each logarithmic expression. Write the expression as a single logarithm whose coefficient is 1. Where possible, evaluate logarithmic expressions.

41. $\log 5 + \log 2$

42. $\log 250 + \log 4$

43. $\ln x + \ln 7$

44. $\ln x + \ln 3$

45. $\log_2 96 - \log_2 3$

46. $\log_3 405 - \log_3 5$

47. $\log (2x + 5) - \log x$

48. $\log (3x + 7) - \log x$

49. $\log x + 3 \log y$

50. $\log x + 7 \log y$

51. $\frac{1}{2} \ln x + \ln y$

52. $\frac{1}{3} \ln x + \ln y$

53. $2 \log_b x + 3 \log_b y$

54. $5 \log_b x + 6 \log_b y$

55. $5 \ln x - 2 \ln y$

56. $7 \ln x - 3 \ln y$

57. $3 \ln x - \frac{1}{3} \ln y$

58. $2 \ln x - \frac{1}{2} \ln y$

59. $4 \ln (x + 6) - 3 \ln x$

60. $8 \ln (x + 9) - 4 \ln x$

61. $3 \ln x + 5 \ln y - 6 \ln z$

62. $4 \ln x + 7 \ln y - 3 \ln z$

63. $\frac{1}{2} (\log x + \log y)$

64. $\frac{1}{3} (\log_4 x - \log_4 y)$

65. $\frac{1}{2} (\log_5 x + \log_5 y) - 2 \log_5 (x + 1)$

66. $\frac{1}{3} (\log_4 x - \log_4 y) + 2 \log_4 (x + 1)$

67. $\frac{1}{3} \left[2 \ln (x + 5) - \ln x - \ln (x^2 - 4) \right]$

68. $\frac{1}{3} \left[5 \ln (x + 6) - \ln x - \ln (x^2 - 25) \right]$

69. $\log x + \log 7 + \log (x^2 - 1) - \log (x + 1)$

70. $\log x + \log 15 + \log (x^2 - 4) - \log (x + 2)$

In Exercises 71–78, use common logarithms or natural logarithms and a calculator to evaluate to four decimal places.

71. $\log_5 13$

72. $\log_6 17$

73. $\log_{14} 87.5$

74. $\log_{16} 57.2$

75. $\log_{0.1} 17$

76. $\log_{0.3} 19$

77. $\log_\pi 63$

78. $\log_\pi 400$

In Exercises 79–82, use a graphing utility and the change-of-base property to graph each function.

79. $y = \log_3 x$

80. $y = \log_{15} x$

81. $y = \log_2 (x + 2)$

82. $y = \log_3 (x - 2)$

Application Exercises

83. The loudness level of a sound can be expressed by comparing the sound's intensity to the intensity of a sound barely audible to the human ear. The formula

$$D = 10(\log I - \log I_0)$$

describes the loudness level of a sound, D, in decibels, where I is the intensity of the sound, in watts per meter^2, and I_0 is the intensity of a sound barely audible to the human ear.

a. Express the formula so that the expression in parentheses is written as a single logarithm.

b. Use the form of the formula from part (a) to answer this question: If a sound has an intensity 100 times the intensity of a softer sound, how much larger on the decibel scale is the loudness level of the more intense sound?

84. The formula

$$t = \frac{1}{c} \left[\ln A - \ln (A - N) \right]$$

describes the time, t, in weeks, that it takes to achieve mastery of a portion of a task, where A is the maximum learning possible, N is the portion of the learning that is to be achieved, and c is a constant used to measure an individual's learning style.

a. Express the formula so that the expression in brackets is written as a single logarithm.

b. The formula is also used to determine how long it will take chimpanzees and apes to master a task. For example, a typical chimpanzee learning sign language can master a maximum of 65 signs. Use the form of the formula from part (a) to answer this question: How many weeks will it take a chimpanzee to master 30 signs if c for that chimp is 0.03?

Writing in Mathematics

85. Describe the product rule for logarithms and give an example.

86. Describe the quotient rule for logarithms and give an example.

87. Describe the power rule for logarithms and give an example.

88. Without showing the details, explain how to condense $\ln x - 2 \ln (x + 1)$.

89. Describe the change-of-base property and give an example.

90. Explain how to use your calculator to find $\log_{14} 283$.

91. You overhear a student talking about a property of logarithms in which division becomes subtraction. Explain what the student means by this.

92. Find $\ln 2$ using a calculator. Then calculate each of the following: $1 - \frac{1}{2}$; $1 - \frac{1}{2} + \frac{1}{3}$; $1 - \frac{1}{2} + \frac{1}{3} - \frac{1}{4}$; $1 - \frac{1}{2} + \frac{1}{3} - \frac{1}{4} + \frac{1}{5}$; .... Describe what you observe.

Technology Exercises

93. a. Use a graphing utility (and the change-of-base property) to graph $y = \log_3 x$.

b. Graph $y = 2 + \log_3 x$, $y = \log_3 (x + 2)$, and $y = -\log_3 x$ in the same viewing rectangle as $y = \log_3 x$. Then describe the change or changes that need to be made to the graph of $y = \log_3 x$ to obtain each of these three graphs.

94. Graph $y = \log x$, $y = \log (10x)$, and $y = \log (0.1x)$ in the same viewing rectangle. Describe the relationship among the three graphs. What logarithmic property accounts for this relationship?

95. Use a graphing utility and the change-of-base property to graph $y = \log_3 x$, $y = \log_{25} x$, and $y = \log_{100} x$ in the same viewing rectangle.

a. Which graph is on the top in the interval $(0, 1)$? Which is on the bottom?

b. Which graph is on the top in the interval $(1, \infty)$? Which is on the bottom?

c. Generalize by writing a statement about which graph is on top, which is on the bottom, and in which intervals, using $y = \log_b x$ where $b > 1$.

Disprove each statement in Exercises 96–100 by
a. *letting y equal a positive constant of your choice.*
b. *using a graphing utility to graph the function on each side of the equal sign. The two functions should have different graphs, showing that the equation is not true in general.*

96. $\log(x + y) = \log x + \log y$ **97.** $\log \dfrac{x}{y} = \dfrac{\log x}{\log y}$

98. $\ln(x - y) = \ln x - \ln y$ **99.** $\ln(xy) = (\ln x)(\ln y)$

100. $\dfrac{\ln x}{\ln y} = \ln x - \ln y$

Critical Thinking Exercises

101. Which one of the following is true?

a. $\dfrac{\log_7 49}{\log_7 7} = \log_7 49 - \log_7 7$

b. $\log_b(x^3 + y^3) = 3 \log_b x + 3 \log_b y$

c. $\log_b(xy)^5 = (\log_b x + \log_b y)^5$

d. $\ln \sqrt{2} = \dfrac{\ln 2}{2}$

102. Use the change-of-base property to prove that
$$\log e = \frac{1}{\ln 10}.$$

103. If $\log 3 = A$ and $\log 7 = B$, find $\log_7 9$ in terms of A and B.

104. Write as a single term that does not contain a logarithm:
$$e^{\ln 8x^5 - \ln 2x^2}.$$

105. If $f(x) = \log_b x$, show that
$$\frac{f(x + h) - f(x)}{h} = \log_b \left(1 + \frac{h}{x}\right)^{1/h}, h \neq 0.$$

SECTION 3.4 Exponential and Logarithmic Equations

Objectives
1. Solve exponential equations.
2. Solve logarithmic equations.
3. Solve applied problems involving exponential and logarithmic equations.

Is an early retirement awaiting you?

You inherited $30,000. You'd like to put aside $25,000 and eventually have over half a million dollars for early retirement. Is this possible? In this section, you will see how techniques for solving equations with variable exponents provide an answer to the question.

1 Solve exponential equations.

Exponential Equations

An **exponential equation** is an equation containing a variable in an exponent. Examples of exponential equations include

$$3^x = 81, \quad 4^x = 15, \quad \text{and} \quad 40e^{0.6x} = 240.$$

Each side of the first equation can be expressed with the same base. Can you see that we can rewrite

$$3^x = 81 \quad \text{as} \quad 3^x = 3^4?$$

All exponential functions are one-to-one—that is, if b is a positive number other than 1 and $b^M = b^N$, then $M = N$. Because we have expressed $3^x = 81$ as $3^x = 3^4$, we conclude that $x = 4$. The equation's solution set is {4}.

Most exponential equations cannot be rewritten so that each side has the same base. Logarithms are extremely useful in solving such equations. The solution begins with isolating the exponential expression and taking the natural logarithm on both sides. Why can we do this? All logarithmic relations are functions. Thus, if M and N are positive real numbers and $M = N$, then $\log_b M = \log_b N$.

Using Natural Logarithms to Solve Exponential Equations

1. Isolate the exponential expression.
2. Take the natural logarithm on both sides of the equation.
3. Simplify using one of the following properties:

$$\ln b^x = x \ln b \quad \text{or} \quad \ln e^x = x.$$

4. Solve for the variable.

EXAMPLE 1 Solving an Exponential Equation

Solve: $4^x = 15$.

Solution   Because the exponential expression, 4^x, is already isolated on the left, we begin by taking the natural logarithm on both sides of the equation.

$4^x = 15$	This is the given equation.
$\ln 4^x = \ln 15$	Take the natural logarithm on both sides.
$x \ln 4 = \ln 15$	Use the power rule and bring the variable exponent to the front: $\ln b^x = x \ln b$.
$x = \dfrac{\ln 15}{\ln 4}$	Solve for x by dividing both sides by $\ln 4$.

We now have an exact value for x. We use the exact value for x in the equation's solution set. Thus, the equation's solution is $\dfrac{\ln 15}{\ln 4}$ and the solution set is $\left\{ \dfrac{\ln 15}{\ln 4} \right\}$. We can obtain a decimal approximation by using a calculator: $x \approx 1.95$. Because $4^2 = 16$, it seems reasonable that the solution to $4^x = 15$ is approximately 1.95.

Discovery

The base that is used when taking the logarithm on both sides of an equation can be any base at all. Solve $4^x = 15$ by taking the common logarithm on both sides. Solve again, this time taking the logarithm with base 4 on both sides. Use the change-of-base property to show that the solutions are the same as the one obtained in Example 1.

> **Check Point 1** Solve: $5^x = 134$. Find the solution set and then use a calculator to obtain a decimal approximation to two decimal places for the solution.

EXAMPLE 2 Solving an Exponential Equation

Solve: $40e^{0.6x} = 240$.

Solution We begin by dividing both sides by 40 to isolate the exponential expression, $e^{0.6x}$. Then we take the natural logarithm on both sides of the equation.

$$40e^{0.6x} = 240 \qquad \text{This is the given equation.}$$

$$e^{0.6x} = 6 \qquad \text{Isolate the exponential factor by dividing both sides by 40.}$$

$$\ln e^{0.6x} = \ln 6 \qquad \text{Take the natural logarithm on both sides.}$$

$$0.6x = \ln 6 \qquad \text{Use the inverse property } \ln e^x = x \text{ on the left.}$$

$$x = \frac{\ln 6}{0.6} \approx 2.99 \qquad \text{Divide both sides by 0.6.}$$

Thus, the solution of the equation is $\dfrac{\ln 6}{0.6} \approx 2.99$. Try checking this approximate solution in the original equation to verify that $\left\{ \dfrac{\ln 6}{0.6} \right\}$ is the solution set.

> **Check Point 2** Solve: $7e^{2x} = 63$. Find the solution set and then use a calculator to obtain a decimal approximation to two decimal places for the solution.

EXAMPLE 3 Solving an Exponential Equation

Solve: $5^{4x-7} - 3 = 10$

Solution We begin by adding 3 to both sides to isolate the exponential expression, 5^{4x-7}. Then we take the natural logarithm on both sides of the equation.

$$5^{4x-7} - 3 = 10 \qquad \text{This is the given equation.}$$

$$5^{4x-7} = 13 \qquad \text{Add 3 to both sides.}$$

$$\ln 5^{4x-7} = \ln 13 \qquad \text{Take the natural logarithm on both sides.}$$

$$(4x - 7)\ln 5 = \ln 13 \qquad \text{Use the power rule to bring the exponent to the front: } \ln M^p = p \ln M.$$

$$4x\ln 5 - 7\ln 5 = \ln 13 \qquad \text{Use the distributive property and distribute } \ln 5 \text{ to both terms in parentheses.}$$

$$4x\ln 5 = \ln 13 + 7\ln 5 \qquad \text{Isolate the variable term by adding } 7\ln 5 \text{ to both sides.}$$

$$x = \frac{\ln 13 + 7\ln 5}{4\ln 5} \qquad \text{Isolate x by dividing both sides by } 4\ln 5.$$

The solution set is $\left\{ \dfrac{\ln 13 + 7\ln 5}{4\ln 5} \right\}$. The solution is approximately 2.15.

> **Check Point 3**
> Solve: $6^{3x-4} - 7 = 2081$. Find the solution set and then use a calculator to obtain a decimal approximation to two decimal places for the solution.

EXAMPLE 4 Solving an Exponential Equation

Solve: $e^{2x} - 4e^x + 3 = 0$.

Solution The given equation is quadratic in form. If $t = e^x$, the equation can be expressed as $t^2 - 4t + 3 = 0$. Because this equation can be solved by factoring, we factor to isolate the exponential term.

Technology

Shown below is the graph of $y = e^{2x} - 4e^x + 3$. There are two x-intercepts, one at 0 and one at approximately 1.10. These intercepts verify our algebraic solution.

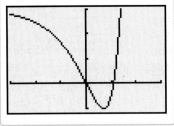

$$e^{2x} - 4e^x + 3 = 0 \qquad \text{This is the given equation.}$$

$$(e^x - 3)(e^x - 1) = 0 \qquad \text{Factor on the left. Notice that if } t = e^x,$$
$$t^2 - 4t + 3 = (t - 3)(t - 1).$$

$$e^x - 3 = 0 \quad \text{or} \quad e^x - 1 = 0 \qquad \text{Set each factor equal to 0.}$$

$$e^x = 3 \qquad\qquad e^x = 1 \qquad \text{Solve for } e^x.$$

$$\ln e^x = \ln 3 \qquad\qquad x = 0 \qquad \text{Take the natural logarithm on both sides of the first equation. The equation on the right can be solved by inspection.}$$

$$x = \ln 3 \qquad\qquad\qquad \text{ln } e^x = x$$

The solution set is $\{0, \ln 3\}$. The solutions are 0 and approximately 1.10.

> **Check Point 4**
> Solve: $e^{2x} - 8e^x + 7 = 0$. Find the solution set and then use a calculator to obtain a decimal approximation to two decimal places, if necessary, for the solutions.

② Solve logarithmic equations.

Logarithmic Equations

A **logarithmic equation** is an equation containing a variable in a logarithmic expression. Examples of logarithmic equations include

$$\log_4 (x + 3) = 2 \quad \text{and} \quad \ln (2x) = 3.$$

If a logarithmic equation is in the form $\log_b x = c$, we can solve the equation by rewriting it in its equivalent exponential form $b^c = x$. Example 5 illustrates how this is done.

EXAMPLE 5 Solving a Logarithmic Equation

Solve: $\log_4 (x + 3) = 2$.

Solution We first rewrite the equation as an equivalent equation in exponential form using the fact that $\log_b x = c$ means $b^c = x$.

$$\log_4(x + 3) = 2 \qquad \text{means} \qquad 4^2 = x + 3$$

> Logarithms are exponents.

Technology

The graphs of
$y_1 = \log_4(x + 3)$ and $y_2 = 2$
have an intersection point
whose x-coordinate is 13. This
verifies that {13} is the solution
set for $\log_4(x + 3) = 2$.

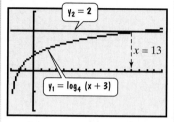

$[-3, 17, 1]$ by $[-2, 3, 1]$

Note:
Because

$$\log_b x = \frac{\ln x}{\ln b}$$

(change-of-base property),
we entered y_1 using

$$y_1 = \frac{\ln(x + 3)}{\ln 4}.$$

Now we solve the equivalent equation for x.

$$4^2 = x + 3 \qquad \text{This is the equation equivalent to } \log_4(x + 3) = 2.$$
$$16 = x + 3 \qquad \text{Square 4.}$$
$$13 = x \qquad \text{Subtract 3 from both sides.}$$

Check 13:

$$\log_4(x + 3) = 2 \qquad \text{This is the given logarithmic equation.}$$
$$\log_4(13 + 3) \overset{?}{=} 2 \qquad \text{Substitute 13 for } x.$$
$$\log_4 16 \overset{?}{=} 2$$
$$2 = 2 \;\checkmark \qquad \log_4 16 = 2 \text{ because } 4^2 = 16.$$

This true statement indicates that the solution set is {13}.

Check Point 5   Solve: $\log_2(x - 4) = 3$.

Logarithmic expressions are defined only for logarithms of positive real numbers. Always check proposed solutions of a logarithmic equation in the original equation. Exclude from the solution set any proposed solution that produces the logarithm of a negative number or the logarithm of 0.

To rewrite the logarithmic equation $\log_b x = c$ in the equivalent exponential form $b^c = x$, we need a single logarithm whose coefficient is one. It is sometimes necessary to use properties of logarithms to condense logarithms into a single logarithm. In the next example, we use the product rule for logarithms to obtain a single logarithmic expression on the left side.

EXAMPLE 6   Using the Product Rule to Solve a Logarithmic Equation

Solve: $\log_2 x + \log_2(x - 7) = 3$.

Solution

$$\log_2 x + \log_2(x - 7) = 3 \qquad \text{This is the given equation.}$$
$$\log_2[x(x - 7)] = 3 \qquad \text{Use the product rule to obtain a single}$$
$$\text{logarithm: } \log_b M + \log_b N = \log_b(MN).$$
$$2^3 = x(x - 7) \qquad \log_b x = c \text{ means } b^c = x.$$
$$8 = x^2 - 7x \qquad \text{Apply the distributive property}$$
$$\text{on the right and evaluate } 2^3 \text{ on the left.}$$
$$0 = x^2 - 7x - 8 \qquad \text{Set the equation equal to 0.}$$
$$0 = (x - 8)(x + 1) \qquad \text{Factor.}$$
$$x - 8 = 0 \quad \text{or} \quad x + 1 = 0 \qquad \text{Set each factor equal to 0.}$$
$$x = 8 \qquad\qquad x = -1 \qquad \text{Solve for } x.$$

Check 8:

$$\log_2 x + \log_2(x - 7) = 3$$
$$\log_2 8 + \log_2(8 - 7) \overset{?}{=} 3$$
$$\log_2 8 + \log_2 1 \overset{?}{=} 3$$
$$3 + 0 \overset{?}{=} 3$$
$$3 = 3 \;\checkmark$$

Check -1:

$$\log_2 x + \log_2(x - 7) = 3$$
$$\log_2(-1) + \log_2(-1 - 7) \overset{?}{=} 3$$

The number -1 does not check.
Negative numbers do not have logarithms.

The solution set is {8}.

Study Tip

You can also solve $\ln x = 5$, meaning $\log_e x = 5$, by rewriting the equation in exponential form:

$\log_e x = 5$ means

A logarithm is an exponent.

$e^5 = x$.

Check Point 6 Solve: $\log x + \log (x - 3) = 1$.

Equations involving natural logarithms can be solved using the inverse property $e^{\ln x} = x$. For example, to solve

$$\ln x = 5$$

we write both sides of the equation as exponents on base e:

$$e^{\ln x} = e^5.$$

This is called **exponentiating both sides** of the equation. Using the inverse property $e^{\ln x} = x$, we simplify the left side of the equation and obtain the solution:

$$x = e^5.$$

EXAMPLE 7   Solving an Equation with a Natural Logarithm

Solve: $3 \ln (2x) = 12$.

Solution

$3 \ln(2x) = 12$	This is the given equation.
$\ln (2x) = 4$	Divide both sides by 3.
$e^{\ln (2x)} = e^4$	Exponentiate both sides.
$2x = e^4$	Use the inverse property to simplify the left side: $e^{\ln x} = x$.
$x = \dfrac{e^4}{2} \approx 27.30$	Divide both sides by 2.

Check $\dfrac{e^4}{2}$:

$3 \ln (2x) = 12$	This is the given logarithmic equation.
$3 \ln \left[2 \left(\dfrac{e^4}{2} \right) \right] \stackrel{?}{=} 12$	Substitute $\dfrac{e^4}{2}$ for x.
$3 \ln e^4 \stackrel{?}{=} 12$	Simplify: $\dfrac{2}{1} \cdot \dfrac{e^4}{2} = e^4$.
$3 \cdot 4 \stackrel{?}{=} 12$	Because $\ln e^x = x$, we conclude $\ln e^4 = 4$.
$12 = 12$ ✓	

This true statement indicates that the solution set is $\left\{ \dfrac{e^4}{2} \right\}$.

Check Point 7 Solve: $4 \ln 3x = 8$.

3 Solve applied problems involving exponential and logarithmic equations.

Applications

Our first applied example provides a mathematical perspective on the old slogan "Alcohol and driving don't mix." In California, where 38% of fatal traffic crashes involve drinking drivers, it is illegal to drive with a blood alcohol concentration of 0.08 or higher. At these levels, drivers may be arrested and charged with driving under the influence.

EXAMPLE 8   Alcohol and Risk of a Car Accident

Medical research indicates that the risk of having a car accident increases exponentially as the concentration of alcohol in the blood increases. The risk is modeled by

$$R = 6e^{12.77x}$$

where x is the blood alcohol concentration and R, given as a percent, is the risk of having a car accident. What blood alcohol concentration corresponds to a 17% risk of a car accident?

Solution   For a risk of 17%, we let $R = 17$ in the equation and solve for x, the blood alcohol concentration.

$$R = 6e^{12.77x} \qquad \text{This is the given equation.}$$

$$6e^{12.77x} = 17 \qquad \text{Substitute 17 for R and (optional) reverse the two sides of the equation.}$$

$$e^{12.77x} = \frac{17}{6} \qquad \text{Isolate the exponential factor by dividing both sides by 6.}$$

$$\ln e^{12.77x} = \ln\left(\frac{17}{6}\right) \qquad \text{Take the natural logarithm on both sides.}$$

$$12.77x = \ln\left(\frac{17}{6}\right) \qquad \text{Use the inverse property } \ln e^x = x \text{ on the left.}$$

$$x = \frac{\ln\left(\dfrac{17}{6}\right)}{12.77} \approx 0.08 \qquad \text{Divide both sides by 12.77.}$$

For a blood alcohol concentration of 0.08, the risk of a car accident is 17%. In many states, it is illegal to drive at this blood alcohol concentration.

Visualizing the Relationship between Blood Alcohol Concentration and the Risk of a Car Accident

A blood alcohol concentration of 0.22 corresponds to near certainty, or a 100% probability, of a car accident.

$R = 6e^{12.77x}$

Risk of a Car Accident

Blood Alcohol Concentration

Check Point 8   Use the formula in Example 8 to answer this question: What blood alcohol concentration corresponds to a 7% risk of a car accident? (In many states, drivers under the age of 21 can lose their license for driving at this level.)

Suppose that you inherit $30,000. Is it possible to invest $25,000 and have over half a million dollars for early retirement? Our next example illustrates the power of compound interest.

EXAMPLE 9   Revisiting the Formula for Compound Interest

The formula

$$A = P\left(1 + \frac{r}{n}\right)^{nt}$$

describes the accumulated value, A, of a sum of money, P, the principal, after t years at annual percentage rate r (in decimal form) compounded n times a year. How long will it take $25,000 to grow to $500,000 at 9% annual interest compounded monthly?

Solution

$$A = P\left(1 + \frac{r}{n}\right)^{nt}$$
This is the given formula.

$$500,000 = 25,000\left(1 + \frac{0.09}{12}\right)^{12t}$$
A (the desired accumulated value) = $500,000,
P (the principal) = $25,000,
r (the interest rate) = 9% = 0.09, and n = 12
(monthly compounding).

Our goal is to solve the equation for t. Let's reverse the two sides of the equation and then simplify within parentheses.

$$25,000\left(1 + \frac{0.09}{12}\right)^{12t} = 500,000$$
Reverse the two sides of the previous equation.

$$25,000(1 + 0.0075)^{12t} = 500,000$$
Divide within parentheses: $\frac{0.09}{12} = 0.0075$.

$$25,000(1.0075)^{12t} = 500,000$$
Add within parentheses.

$$(1.0075)^{12t} = 20$$
Divide both sides by 25,000.

$$\ln(1.0075)^{12t} = \ln 20$$
Take the natural logarithm on both sides.

$$12t \ln(1.0075) = \ln 20$$
Use the power rule to bring the exponent to the front: $\ln M^p = p\ln M$.

$$t = \frac{\ln 20}{12 \ln 1.0075}$$
Solve for t, dividing both sides by 12 ln 1.0075.

$$\approx 33.4$$
Use a calculator.

After approximately 33.4 years, the $25,000 will grow to an accumulated value of $500,000. If you set aside the money at age 20, you can begin enjoying a life of leisure at about age 53.

Check Point 9 How long, to the nearest tenth of a year, will it take $1000 to grow to $3600 at 8% annual interest compounded quarterly?

Yogi Berra, catcher and renowned hitter for the New York Yankees (1946–1963), said it best: "Prediction is very hard, especially when it's about the future." At the start of the twenty-first century, we are plagued by questions about the environment. Will we run out of gas? How hot will it get? Will there be neighborhoods where the air is pristine? Can we make garbage disappear? Will there be any wilderness left? Which wild animals will become extinct? These concerns have led to the growth of the environmental industry in the United States.

EXAMPLE 10 The Growth of the Environmental Industry

The formula

$$N = 461.87 + 299.4\ln x$$

models the thousands of workers, N, in the environmental industry in the United States x years after 1979. By which year will there be 1,500,000, or 1500 thousand, U.S. workers in the environmental industry?

Playing Doubles: Interest Rates and Doubling Time

One way to calculate what your savings will be worth at some point in the future is to consider doubling time. Shown below is how long it takes for your money to double at different annual interest rates subject to continuous compounding.

Annual Interest Rate	Years to Double
5%	13.9 years
7%	9.9 years
9%	7.7 years
11%	6.3 years

Of course, the first problem is collecting some money to invest. The second problem is finding a reasonably safe investment with a return of 9% or more.

Solution   We substitute 1500 for N and solve for x, the number of years after 1979.

$$N = 461.87 + 299.4 \ln x$$ This is the given formula.

$$461.87 + 299.4 \ln x = 1500$$ Substitute 1500 for N and reverse the two sides of the equation.

Our goal is to isolate $\ln x$. We can then find x by exponentiating both sides of the equation, using the inverse property $e^{\ln x} = x$.

$$299.4 \ln x = 1038.13$$ Subtract 461.87 from both sides.

$$\ln x = \frac{1038.13}{299.4}$$ Divide both sides by 299.4.

$$e^{\ln x} = e^{1038.13/299.4}$$ Exponentiate both sides.

$$x = e^{1038.13/299.4}$$ $e^{\ln x} = x$

$$\approx 32$$ Use a calculator.

Approximately 32 years after 1979, in the year 2011, there will be 1.5 million U.S. workers in the environmental industry.

Check Point 10   Use the formula in Example 10 to find by which year there will be two million, or 2000 thousand, U.S. workers in the environmental industry.

EXERCISE SET 3.4

Practice Exercises

Solve each exponential equation in Exercises 1–26. Express the solution set in terms of natural logarithms. Then use a calculator to obtain a decimal approximation, correct to two decimal places, for the solution.

1. $10^x = 3.91$

2. $10^x = 8.07$

3. $e^x = 5.7$

4. $e^x = 0.83$

5. $5^x = 17$

6. $19^x = 143$

7. $5e^x = 23$

8. $9e^x = 107$

9. $3e^{5x} = 1977$

10. $4e^{7x} = 10,273$

11. $e^{1-5x} = 793$

12. $e^{1-8x} = 7957$

13. $e^{5x-3} - 2 = 10,476$

14. $e^{4x-5} - 7 = 11,243$

15. $7^{x+2} = 410$

16. $5^{x-3} = 137$

17. $7^{0.3x} = 813$

18. $3^{x/7} = 0.2$

19. $5^{2x+3} = 3^{x-1}$

20. $7^{2x+1} = 3^{x+2}$

21. $e^{2x} - 3e^x + 2 = 0$

22. $e^{2x} - 2e^x - 3 = 0$

23. $e^{4x} + 5e^{2x} - 24 = 0$

24. $e^{4x} - 3e^{2x} - 18 = 0$

25. $3^{2x} + 3^x - 2 = 0$

26. $2^{2x} + 2^x - 12 = 0$

Solve each logarithmic equation in Exercises 27–44. Be sure to reject any value of x that produces the logarithm of a negative number or the logarithm of 0.

27. $\log_3 x = 4$

28. $\log_5 x = 3$

29. $\log_4(x + 5) = 3$

30. $\log_5(x - 7) = 2$

31. $\log_3(x - 4) = -3$

32. $\log_7(x + 2) = -2$

33. $\log_4(3x + 2) = 3$

34. $\log_2(4x + 1) = 5$

35. $\log_5 x + \log_5(4x - 1) = 1$

36. $\log_6(x + 5) + \log_6 x = 2$

37. $\log_3(x - 5) + \log_3(x + 3) = 2$

38. $\log_2(x - 1) + \log_2(x + 1) = 3$

39. $\log_2(x + 2) - \log_2(x - 5) = 3$

40. $\log_4(x + 2) - \log_4(x - 1) = 1$

41. $2 \log_3(x + 4) = \log_3 9 + 2$

42. $3 \log_2(x - 1) = 5 - \log_2 4$

43. $\log_2(x - 6) + \log_2(x - 4) - \log_2 x = 2$

44. $\log_2(x - 3) + \log_2 x - \log_2(x + 2) = 2$

Exercises 45–52 involve equations with natural logarithms. Solve each equation by isolating the natural logarithm and exponentiating both sides. Express the answer in terms of e. Then use a calculator to obtain a decimal approximation, correct to two decimal places, for the solution.

45. $\ln x = 2$

46. $\ln x = 3$

47. $5 \ln(2x) = 20$

48. $6 \ln(2x) = 30$

49. $6 + 2 \ln x = 5$

50. $7 + 3 \ln x = 6$

51. $\ln \sqrt{x + 3} = 1$

52. $\ln \sqrt{x + 4} = 1$

Application Exercises

Use the formula $R = 6e^{12.77x}$, where x is the blood alcohol concentration and R, given as a percent, is the risk of having a car accident, to solve Exercises 53–54.

53. What blood alcohol concentration corresponds to a 25% risk of a car accident?

54. What blood alcohol concentration corresponds to a 50% risk of a car accident?

55. The formula $A = 18.9e^{0.0055t}$ models the population of New York State, A, in millions, t years after 2000.
  a. What was the population of New York in 2000?
  b. When will the population of New York reach 19.6 million?

56. The formula $A = 15.9e^{0.0235t}$ models the population of Florida, A, in millions, t years after 2000.
  a. What was the population of Florida in 2000?
  b. When will the population of Florida reach 17.5 million?

In Exercises 57–60, complete the table for a savings account subjected to n compoundings yearly $\left[A = P\left(1 + \dfrac{r}{n}\right)^{nt} \right]$. Round answers to one decimal place.

Amount Invested	Number of Compounding Periods	Annual Interest Rate	Accumulated Amount	Time t in Years
57. $12,500	4	5.75%	$20,000	
58. $7250	12	6.5%	$15,000	
59. $1000	360		$1400	2
60. $5000	360		$9000	4

In Exercises 61–64, complete the table for a savings account subjected to continuous compounding ($A = Pe^{rt}$). Round answers to one decimal place.

Amount Invested	Annual Interest Rate	Accumulated Amount	Time t in Years
61. $8000	8%	Double the amount invested	
62. $8000		$12,000	2
63. $2350		Triple the amount invested	7
64. $17,425	4.25%	$25,000	

65. The function $f(x) = 15{,}557 + 5259 \ln x$ models the average cost of a new car, $f(x)$, in dollars, x years after 1989. When was the average cost of a new car $25,000?

66. The function $f(x) = 68.41 + 1.75 \ln x$ models the life expectancy, $f(x)$, in years, for African-American females born x years after 1969. In which birth year was life expectancy 73.7 years? Round to the nearest year.

The function $P(x) = 95 - 30\log_2 x$ models the percentage, $P(x)$, of students who could recall the important features of a classroom lecture as a function of time, where x represents the number of days that have elapsed since the lecture was given. The figure shows the graph of the function. Use this information to solve Exercises 67–68. Round answers to one decimal place.

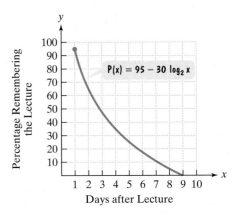

67. After how many days do only half the students recall the important features of the classroom lecture? (Let $P(x) = 50$ and solve for x.) Locate the point on the graph that conveys this information.

68. After how many days have all students forgotten the important features of the classroom lecture? (Let $P(x) = 0$ and solve for x.) Locate the point on the graph on the previous page that conveys this information.

The pH of a solution ranges from 0 to 14. An acid solution has a pH less than 7. Pure water is neutral and has a pH of 7. Normal, unpolluted rain has a pH of about 5.6. The pH of a solution is given by

$$pH = -\log x$$

where x represents the concentration of the hydrogen ions in the solution, in moles per liter. Use the formula to solve Exercises 69–70.

69. An environmental concern involves the destructive effects of acid rain. The most acidic rainfall ever had a pH of 2.4. What was the hydrogen ion concentration? Express the answer as a power of 10, and then round to the nearest thousandth.

70. The figure shows very acidic rain in the northeast United States. What is the hydrogen ion concentration of rainfall with a pH of 4.2? Express the answer as a power of 10, and then round to the nearest hundred-thousandth.

Acid Rain over Canada and the United States

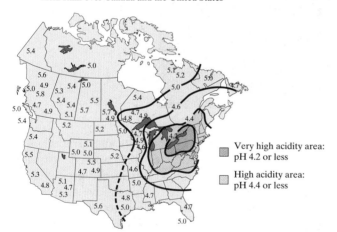

Very high acidity area: pH 4.2 or less

High acidity area: pH 4.4 or less

Source: National Atmospheric Program

 Writing in Mathematics

71. Explain how to solve an exponential equation. Use $3^x = 140$ in your explanation.

72. Explain how to solve a logarithmic equation. Use $\log_3(x - 1) = 4$ in your explanation.

73. In many states, a 17% risk of a car accident with a blood alcohol concentration of 0.08 is the lowest level for charging a motorist with driving under the influence. Do you agree with the 17% risk as a cutoff percentage, or do you feel that the percentage should be lower or higher?

Explain your answer. What blood alcohol concentration corresponds to what you believe is an appropriate percentage?

74. Have you purchased a new or used car recently? If so, describe if the function in Exercise 65 accurately models what you paid for your car. If there is a big difference between the figure given by the formula and the amount that you paid, how can you explain this difference?

Technology Exercises

In Exercises 75–82, use your graphing utility to graph each side of the equation in the same viewing rectangle. Then use the x-coordinate of the intersection point to find the equation's solution set. Verify this value by direct substitution into the equation.

75. $2^{x+1} = 8$
76. $3^{x+1} = 9$
77. $\log_3(4x - 7) = 2$
78. $\log_3(3x - 2) = 2$
79. $\log(x + 3) + \log x = 1$
80. $\log(x - 15) + \log x = 2$
81. $3^x = 2x + 3$
82. $5^x = 3x + 4$

Hurricanes are one of nature's most destructive forces. These low-pressure areas often have diameters of over 500 miles. The function $f(x) = 0.48 \ln(x + 1) + 27$ models the barometric air pressure, $f(x)$, in inches of mercury, at a distance of x miles from the eye of a hurricane. Use this function to solve Exercises 83–84.

83. Graph the function in a $[0, 500, 50]$ by $[27, 30, 1]$ viewing rectangle. What does the shape of the graph indicate about barometric air pressure as the distance from the eye increases?

84. Use an equation to answer this question: How far from the eye of a hurricane is the barometric air pressure 29 inches of mercury? Use the TRACE and ZOOM features or the intersect command of your graphing utility to verify your answer.

85. The function $P(t) = 145e^{-0.092t}$ models a runner's pulse, $P(t)$, in beats per minute, t minutes after a race, where $0 \le t \le 15$. Graph the function using a graphing utility. TRACE along the graph and determine after how many minutes the runner's pulse will be 70 beats per minute. Round to the nearest tenth of a minute. Verify your observation algebraically.

86. The function $W(t) = 2600(1 - 0.51e^{-0.075t})^3$ models the weight, $W(t)$, in kilograms, of a female African elephant at age t years. (1 kilogram $\approx$ 2.2 pounds) Use a graphing utility to graph the function. Then TRACE along the curve to estimate the age of an adult female elephant weighing 1800 kilograms.

Critical Thinking Exercises

87. Which one of the following is true?
   a. If $\log(x + 3) = 2$, then $e^2 = x + 3$.
   b. If $\log(7x + 3) - \log(2x + 5) = 4$, then in exponential form $10^4 = (7x + 3) - (2x + 5)$.

c. If $x = \dfrac{1}{k} \ln y$, then $y = e^{kx}$.

d. Examples of exponential equations include $10^x = 5.71$, $e^x = 0.72$, and $x^{10} = 5.71$.

88. If \$4000 is deposited into an account paying 3% interest compounded annually and at the same time \$2000 is deposited into an account paying 5% interest compounded annually, after how long will the two accounts have the same balance?

Solve each equation in Exercises 89–91. Check each proposed solution by direct substitution or with a graphing utility.

89. $(\ln x)^2 = \ln x^2$

90. $(\log x)(2 \log x + 1) = 6$

91. $\ln(\ln x) = 0$

Group Exercise

92. Research applications of logarithmic functions as mathematical models and plan a seminar based on your group's research. Each group member should research one of the following areas or any other area of interest: pH (acidity of solutions), intensity of sound (decibels), brightness of stars, consumption of natural resources, human memory, progress over time in a sport, profit over time. For the area that you select, explain how logarithmic functions are used and provide examples.

SECTION 3.5 *Modeling with Exponential and Logarithmic Functions*

Objectives

1. Model exponential growth and decay.
2. Use logistic growth models.
3. Use Newton's Law of Cooling.
4. Model data with exponential and logarithmic functions.
5. Express an exponential model in base e.

The most casual cruise on the Internet shows how people disagree when it comes to making predictions about the effects of the world's growing population. Some argue that there is a recent slowdown in the growth rate, economies remain robust, and famines in Biafra and Ethiopia are aberrations rather than signs of the future. Others say that the 6 billion people on Earth is twice as many as can be supported in middle-class comfort, and the world is running out of arable land and fresh water. Debates about entities that are growing exponentially can be approached mathematically: We can create functions that model data and use these functions to make predictions. In this section we will show you how this is done.

1 Model exponential growth and decay.

Exponential Growth and Decay

One of algebra's many applications is to predict the behavior of variables. This can be done with *exponential growth* and *decay models*. With exponential growth or decay, quantities grow or decay at a rate directly proportional to their size. Populations that are growing exponentially grow extremely rapidly as they get larger because there are more adults to have offspring. For example, the **growth rate** for world population is 1.3%, or 0.013. This means that each year world population is 1.3% more than what it was in the previous year. In 2001, world population was approximately 6.2 billion. Thus, we compute the world population in 2002 as follows:

6.2 billion + 1.3% of 6.2 billion = 6.2 + (0.013)(6.2) = 6.2806.

This computation suggests that 6.2806 billion people will populate the world in 2002. The 0.0806 billion represents an increase of 80.6 million people from 2001 to 2002, the equivalent of the population of Germany. Using 1.3% as the annual growth rate, world population for 2003 is found in a similar manner:

6.2806 + 1.3% of 6.2806 = 6.2806 + (0.013)(6.2806) ≈ 6.3622.

This computation suggests that approximately 6.3622 billion people will populate the world in 2003.

The explosive growth of world population may remind you of the growth of money in an account subject to compound interest. Just as the growth rate for world population is multiplied by the population plus any increase in the population, a compound interest rate is multiplied by your original investment plus any accumulated interest. The balance in an account subject to continuous compounding and world population are special cases of an *exponential growth model*.

Study Tip

You have seen the formula for exponential growth before, but with different letters. It is the formula for compound interest with continous compounding.

$$A = Pe^{rt}$$

Amount at time t Principal is the original amount.   Interest rate is the growth rate.

$$A = A_o e^{kt}$$

Exponential Growth and Decay Models

The mathematical model for **exponential growth** or **decay** is given by
$$f(t) = A_0 e^{kt} \quad \text{or} \quad A = A_0 e^{kt}.$$

- **If $k > 0$, the function models the amount, or size, of a *growing* entity.** A_0 is the original amount, or size, of the growing entity at time $t = 0$, A is the amount at time t, and k is a constant representing the growth rate.
- **If $k < 0$, the function models the amount, or size, of a *decaying* entity.** A_0 is the original amount, or size, of the decaying entity at time $t = 0$, A is the amount at time t, and k is a constant representing the decay rate.

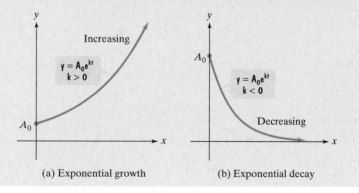

(a) Exponential growth   (b) Exponential decay

Sometimes we need to use given data to determine k, the rate of growth or decay. After we compute the value of k, we can use the formula $A = A_0 e^{kt}$ to make predictions. This idea is illustrated in our first two examples.

EXAMPLE 1 Modeling the Growth of the Minimum Wage

The graph in Figure 3.18 on the next page shows the growth of the minimum wage from 1970 through 2000. In 1970, the minimum wage was $1.60 per hour. By 2000, it had grown to $5.15 per hour.

Federal Minimum Wages, 1970–2000

Figure 3.18 *Source:* U.S. Employment Standards Administration

a. Find the exponential growth function that models the data for 1970 through 2000.

b. By which year will the minimum wage reach $7.50 per hour?

Solution

a. We use the exponential growth model

$$A = A_0 e^{kt}$$

in which t is the number of years after 1970. This means that 1970 corresponds to $t = 0$. At that time the minimum wage was $1.60, so we substitute 1.6 for A_0 in the growth model:

$$A = 1.6e^{kt}.$$

We are given that $5.15 is the minimum wage in 2000. Because 2000 is 30 years after 1970, when $t = 30$ the value of A is 5.15. Substituting these numbers into the growth model will enable us to find k, the growth rate. We know that $k > 0$ because the problem involves growth.

$A = 1.6e^{kt}$	Use the growth model, $A = A_0 e^{kt}$, with $A_0 = 1.6$.
$5.15 = 1.6e^{k \cdot 30}$	When $t = 30$, $A = 5.15$. Substitute these numbers into the model.
$e^{30k} = \dfrac{5.15}{1.6}$	Isolate the exponential factor by dividing both sides by 1.6. We also reversed the sides.
$\ln e^{30k} = \ln \dfrac{5.15}{1.6}$	Take the natural logarithm on both sides.
$30k = \ln \dfrac{5.15}{1.6}$	Simplify the left side using $\ln e^x = x$.
$k = \dfrac{\ln \dfrac{5.15}{1.6}}{30} \approx 0.039$	Divide both sides by 30 and solve for k.

We substitute 0.039 for k in the growth model to obtain the exponential growth function for the minimum wage. It is

$$A = 1.6e^{0.039t}$$

where t is measured in years after 1970.

b. To find the year in which the minimum wage will reach $7.50 per hour, we substitute 7.5 for A in the model from part (a) and solve for t.

$$A = 1.6e^{0.039t}$$ This is the model from part (a).

$$7.5 = 1.6e^{0.039t}$$ Substitute 7.5 for A.

$$e^{0.039t} = \frac{7.5}{1.6}$$ Divide both sides by 1.6. We also reversed the sides.

$$\ln e^{0.039t} = \ln \frac{7.5}{1.6}$$ Take the natural logarithm on both sides.

$$0.039t = \ln \frac{7.5}{1.6}$$ Simplify on the left using $\ln e^x = x$.

$$t = \frac{\ln \dfrac{7.5}{1.6}}{0.039} \approx 40$$ Solve for t by dividing both sides by 0.039.

Because 40 is the number of years after 1970, the model indicates that the minimum wage will reach $7.50 by 1970 + 40, or in the year 2010.

Check Point 1 In 1990, the population of Africa was 643 million and by 2000 it had grown to 813 million.

a. Use the exponential growth model $A = A_0 e^{kt}$, in which t is the number of years after 1990, to find the exponential growth function that models the data.

b. By which year will Africa's population reach 2000 million, or two billion?

Lying with Statistics

Benjamin Disraeli, Queen Victoria's prime minister, stated that there are "lies, damned lies, and statistics." The problem is not that data lie, but rather that liars use data. For example, the data in Example 1 create the impression that wages are on the rise and workers are better off each year. The graph in Figure 3.19 is more effective in creating an accurate picture. Why? It is adjusted for inflation and measured in constant 1996 dollars. Something else to think about: In predicting a minimum wage of $7.50 by 2010, are we using the best possible model for the data? We return to this issue in the exercise set.

Figure 3.19
Source: U.S. Employment Standards Administration

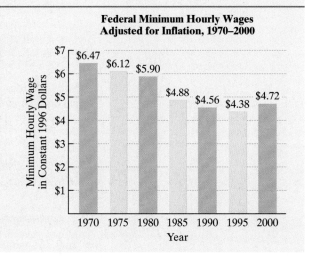

Federal Minimum Hourly Wages Adjusted for Inflation, 1970–2000

Carbon Dating and Artistic Development

The artistic community was electrified by the discovery in 1995 of spectacular cave paintings in a limestone cavern in France. Carbon dating of the charcoal from the site showed that the images, created by artists of remarkable talent, were 30,000 years old, making them the oldest cave paintings ever found. The artists seemed to have used the cavern's natural contours to heighten a sense of perspective. The quality of the painting suggests that the art of early humans did not mature steadily from primitive to sophisticated in any simple linear fashion.

Our next example involves exponential decay and its use in determining the age of fossils and artifacts. The method is based on considering the percentage of carbon-14 remaining in the fossil or artifact. Carbon-14 decays exponentially with a *half-life* of approximately 5715 years. The **half-life** of a substance is the time required for half of a given sample to disintegrate. Thus, after 5715 years a given amount of carbon-14 will have decayed to half the original amount. Carbon dating is useful for artifacts or fossils up to 80,000 years old. Older objects do not have enough carbon-14 left to date age accurately.

EXAMPLE 2   Carbon-14 Dating: The Dead Sea Scrolls

a. Use the fact that after 5715 years a given amount of carbon-14 will have decayed to half the original amount to find the exponential decay model for carbon-14.

b. In 1947, earthenware jars containing what are known as the Dead Sea Scrolls were found by an Arab Bedouin herdsman. Analysis indicated that the scroll wrappings contained 76% of their original carbon-14. Estimate the age of the Dead Sea Scrolls.

Solution   We begin with the exponential decay model $A = A_0 e^{kt}$. We know that $k < 0$ because the problem involves the decay of carbon-14. After 5715 years ($t = 5715$), the amount of carbon-14 present, A, is half the original amount A_0. Thus, we can substitute $\dfrac{A_0}{2}$ for A in the exponential decay model. This will enable us to find k, the decay rate.

a.

$$A = A_0 e^{kt}$$

Begin with the exponential decay model .

$$\frac{A_0}{2} = A_0 e^{k \cdot 5715}$$

After 5715 years ($t = 5715$), $A = \dfrac{A_0}{2}$ (because the amount present, A, is half the original amount, A_0).

$$\frac{1}{2} = e^{5715k}$$

Divide both sides of the equation by A_0.

$$\ln \frac{1}{2} = \ln e^{5715k}$$

Take the natural logarithm on both sides.

$$\ln \frac{1}{2} = 5715k$$

Simplify the right side using $\ln e^x = x$.

$$k = \frac{\ln \frac{1}{2}}{5715} \approx -0.000121$$

Divide both sides by 5715 and solve for k.

Substituting for k in the decay model, $A = A_0 e^{kt}$, the model for carbon-14 is

$$A = A_0 e^{-0.000121t}.$$

b.

$$A = A_0 e^{-0.000121t}$$

This is the decay model for carbon-14.

$$0.76A_0 = A_0 e^{-0.000121t}$$

A, the amount present, is 76% of the original amount, so A = 0.76A_0.

$$0.76 = e^{-0.000121t}$$

Divide both sides of the equation by A_0.

$$\ln 0.76 = \ln e^{-0.000121t}$$

Take the natural logarithm on both sides.

$$\ln 0.76 = \ln e^{-0.000121t}$$

We've repeated this equation from the bottom of the previous page.

$$\ln 0.76 = -0.000121t$$

Simplify the right side using $\ln e^x = x$.

$$t = \frac{\ln 0.76}{-0.000121} \approx 2268$$

Divide both sides by -0.000121 and solve for t.

The Dead Sea Scrolls are approximately 2268 years old plus the number of years between 1947 and the current year.

Check Point 2 Strontium-90 is a waste product from nuclear reactors. As a consequence of fallout from atmospheric nuclear tests, we all have a measurable amount of strontium-90 in our bones.

a. Use the fact that after 28 years a given amount of strontium-90 will have decayed to half the original amount to find the exponential decay model for strontium-90.

b. Suppose that a nuclear accident occurs and releases 60 grams of strontium-90 into the atmosphere. How long will it take for strontium-90 to decay to a level of 10 grams?

2 Use logistic growth models.

Logistic Growth Models

From population growth to the spread of an epidemic, nothing on Earth can grow exponentially indefinitely. Growth is always limited. This is shown in Figure 3.20 by the horizontal asymptote. The **logistic growth model** is an exponential function used to model situations in which growth is limited.

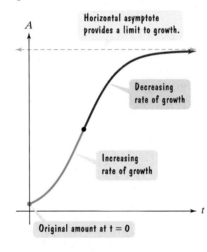

Figure 3.20 The logistic growth curve has a horizontal asymptote that limits the growth of A over time.

Logistic Growth Model

The mathematical model for limited logistic growth is given by

$$f(t) = \frac{c}{1 + ae^{-bt}} \quad \text{or} \quad A = \frac{c}{1 + ae^{-bt}}$$

where a, b, and c are constants, with $c > 0$ and $b > 0$.

As time increases $(t \to \infty)$, the expression ae^{-bt} in the model approaches 0, and A gets closer and closer to c. This means that $y = c$ is a horizontal asymptote for the graph of the function. Thus, the value of A can never exceed c and c represents the limiting size that A can attain.

EXAMPLE 3   Modeling the Spread of the Flu

The function

$$f(t) = \frac{30,000}{1 + 20e^{-1.5t}}$$

describes the number of people, $f(t)$, who have become ill with influenza t weeks after its initial outbreak in a town with 30,000 inhabitants.

a. How many people became ill with the flu when the epidemic began?

b. How many people were ill by the end of the fourth week?

c. What is the limiting size of $f(t)$, the population that becomes ill?

Solution

a. The time at the beginning of the flu epidemic is $t = 0$. Thus, we can find the number of people who were ill at the beginning of the epidemic by substituting 0 for t.

$$f(t) = \frac{30,000}{1 + 20e^{-1.5t}} \qquad \text{This is the given logistic growth function.}$$

$$f(0) = \frac{30,000}{1 + 20e^{-1.5(0)}} \qquad \text{When the epidemic began, } t = 0.$$

$$= \frac{30,000}{1 + 20} \qquad e^{-1.5(0)} = e^0 = 1$$

$$\approx 1429$$

Approximately 1429 people were ill when the epidemic began.

b. We find the number of people who were ill at the end of the fourth week by substituting 4 for t in the logistic growth function.

$$f(t) = \frac{30,000}{1 + 20e^{-1.5t}} \qquad \text{Use the given logistic growth function.}$$

$$f(4) = \frac{30,000}{1 + 20e^{-1.5(4)}} \qquad \text{To find the number of people ill by the end of week four, let } t = 4.$$

$$= 28,583 \qquad \text{Use a calculator.}$$

Approximately 28,583 people were ill by the end of the fourth week. Compared with the number of people who were ill initially, 1429, this illustrates the virulence of the epidemic.

c. Recall that in the logistic growth model, $f(t) = \dfrac{c}{1 + ae^{-bt}}$, the constant c represents the limiting size that $f(t)$ can attain. Thus, the number in the numerator, 30,000, is the limiting size of the population that becomes ill.

Technology

The graph of the logistic growth function for the flu epidemic

$$y = \frac{30,000}{1 + 20e^{-1.5x}}$$

can be obtained using a graphing utility. We started x at 0 and ended at 10. This takes us to week 10. (In Example 3, we found that by week 4 approximately 28,583 people were ill.) We also know that 30,000 is the limiting size, so we took values of y up to 30,000. Using a $[0, 10, 1]$ by $[0, 30,000, 3000]$ viewing rectangle, the graph of the logistic growth function is shown below.

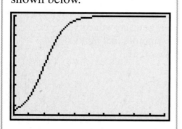

Check Point 3   In a learning theory project, psychologists discovered that

$$f(t) = \frac{0.8}{1 + e^{-0.2t}}$$

is a model for describing the proportion of correct responses, $f(t)$, after t learning trials.

a. Find the proportion of correct responses prior to learning trials taking place.

b. Find the proportion of correct responses after 10 learning trials.

c. What is the limiting size of $f(t)$, the proportion of correct responses, as continued learning trials take place?

3 Use Newton's Law of Cooling

Modeling Cooling

Over a period of time, a cup of hot coffee cools to the temperature of the surrounding air. **Newton's Law of Cooling**, named after Sir Isaac Newton, states that the temperature of a heated object decreases exponentially over time toward the temperature of the surrounding medium.

Study Tip

Newton's Law of Cooling applies to any situation in which an object's temperature is different from that of the surrounding medium. Thus, it can be used to model a heated object cooling to room temperature as well as a frozen object thawing to room temperature.

Newton's Law of Cooling

The temperature, T, of a heated object at time t is given by
$$T = C + (T_0 - C)e^{kt}$$
where C is the constant temperature of the surrounding medium, T_0 is the initial temperature of the heated object, and k is a negative constant that is associated with the cooling object.

EXAMPLE 4 Using Newton's Law of Cooling

A cake removed from the oven has a temperature of 210°F. It is left to cool in a room that has a temperature of 70°F. After 30 minutes, the temperature of the cake is 140°F.

a. Use Newton's Law of Cooling to find a model for the temperature of the cake, T, after t minutes.

b. What is the temperature of the cake after 40 minutes?

c. When will the temperature of the cake be 90°F?

Solution

a. We use Newton's Law of Cooling
$$T = C + (T_0 - C)e^{kt}.$$

When the cake is removed from the oven, its temperature is 210°F. This is its initial temperature: $T_0 = 210$. The constant temperature of the room is 70°F: $C = 70$. Substitute these values into Newton's Law of Cooling. Thus, the temperature of the cake, T, in degrees Fahrenheit, at time t, in minutes, is
$$T = 70 + (210 - 70)e^{kt} = 70 + 140e^{kt}$$

After 30 minutes, the temperature of the cake is 140°F. This means that when $t = 30$, $T = 140$. Substituting these numbers into Newton's Law of

Cooling will enable us to find k, a negative constant.

$$T = 70 + 140e^{kt}$$ Use Newton's Law of Cooling from pag. 000.

$$140 = 70 + 140e^{k \cdot 30}$$ When $t = 30$, $T = 140$. Substitute these numbers into the cooling model.

$$70 = 140e^{30k}$$ Subtract 70 from both sides.

$$e^{30k} = \tfrac{1}{2}$$ Isolate the exponential factor by dividing both sides by 140. We also reversed the sides.

$$\ln e^{30k} = \ln \tfrac{1}{2}$$ Take the natural logarithm on both sides.

$$30k = \ln \tfrac{1}{2}$$ Simplify the left side using $\ln e^x = x$.

$$k = \frac{\ln \tfrac{1}{2}}{30} \approx -0.0231$$ Divide both sides by 30 and solve for k.

We substitute -0.0231 for k in Newton's Law of Cooling. The temperature of the cake, T, in degrees Fahrenheit, after t minutes is modeled by

$$T = 70 + 140e^{-0.0231t}.$$

b. To find the temperature of the cake after 40 minutes, we substitute 40 for t in the cooling model from part (a) and evaluate to find T.

$$T = 70 + 140e^{-0.0231(40)} \approx 126$$

After 40 minutes, the temperature of the cake will be approximately 126°F.

c. To find when the temperature of the cake will be 90°F, we substitute 90 for T in the cooling model from part (a) and solve for t.

$$T = 70 + 140e^{-0.0231t}$$ This is the cooling model from part (a).

$$90 = 70 + 140e^{-0.0231t}$$ Substitute 90 for T.

$$20 = 140e^{-0.0231t}$$ Subtract 70 from both sides.

$$e^{-0.0231t} = \tfrac{1}{7}$$ Divide both sides by 140.

$$\ln e^{-0.0231t} = \ln \tfrac{1}{7}$$ Take the natural logarithm on both sides.

$$-0.0231t = \ln \tfrac{1}{7}$$ Simplify the left side using $\ln e^x = x$.

$$t = \frac{\ln \tfrac{1}{7}}{-0.0231} \approx 84$$ Solve for t by dividing both sides by -0.0231.

The temperature of the cake will be 90°F after approximately 84 minutes.

Technology

The graphs illustrate how the temperature of the cake decreases exponentially over time toward the 70°F room temperature.

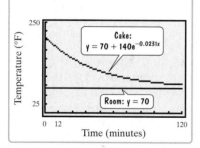

Check Point 4

An object is heated to 100°C. It is left to cool in a room that has a temperature of 30°C. After 5 minutes, the temperature of the object is 80°C.

a. Use Newton's Law of Cooling to find a model for the temperature of the object, T, after t minutes.

b. What is the temperature of the object after 20 minutes?

c. When will the temperature of the object be 35°C?

4 Model data with exponential and logarithmic functions.

The Art of Modeling

Throughout this chapter, we have been working with models that were given. However, we can create functions that model data by observing patterns in scatter plots. Figure 3.21 shows scatter plots for data that are exponential or logarithmic.

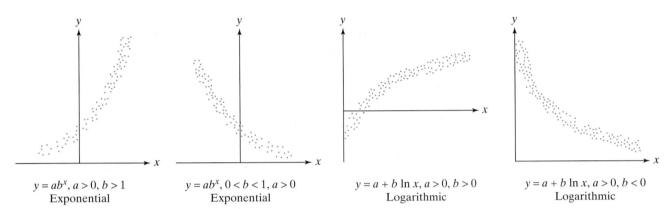

$y = ab^x, a > 0, b > 1$
Exponential

$y = ab^x, 0 < b < 1, a > 0$
Exponential

$y = a + b \ln x, a > 0, b > 0$
Logarithmic

$y = a + b \ln x, a > 0, b < 0$
Logarithmic

Figure 3.21 Scatter plots for exponential or logarithmic models

Graphing utilities can be used to find the equation of a function that is derived from data. For example, earlier in the chapter we encountered a function that modeled the size of a city and the average walking speed, in feet per second, of pedestrians. The function was derived from the data in Table 3.4. The scatter plot is shown in Figure 3.22.

Table 3.4

x, Population (thousands)	y, Walking Speed (feet per second)
5.5	3.3
14	3.7
71	4.3
138	4.4
342	4.8

Source: Mark and Helen Bornstein, "The Pace of Life"

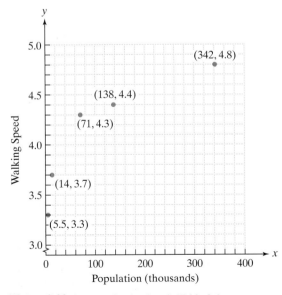

Figure 3.22 Scatter plot for data in Table 3.4

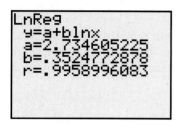

Figure 3.23 A logarithmic model for the data in Table 3.4

Because the data in this scatter plot increase rapidly at first and then begin to level off a bit, the shape suggests that a logarithmic model might be a good choice. A graphing utility fits the data in Table 3.4 to a logarithmic model of the form $y = a + b \ln x$ by using the Natural Logarithmic REGression (LnReg) option (see Figure 3.23). From the figure, we see that the logarithmic model of the data, with numbers rounded to three decimal places, is

$$y = 2.735 + 0.352 \ln x.$$

The number r that appears in Figure 3.23 is called the **correlation coefficient** and is a measure of how well the model fits the data. The value of r is such that $-1 \le r \le 1$. A positive r means that as the x-values increase, so do the y-values. A negative r means that as the x-values increase, the y-values decrease. **The closer that r is to -1 or 1, the better the model fits the data.** Because r is approximately 0.996, the model

$$y = 2.735 + 0.352 \ln x$$

fits the data very well.

Now let's look at data whose scatter plot suggests an exponential model. The data in Table 3.5 indicate world population for six years. The scatter plot is shown in Figure 3.24.

Table 3.5

x, Year	y, World Population (billions)
1950	2.6
1960	3.1
1970	3.7
1980	4.5
1989	5.3
2001	6.2

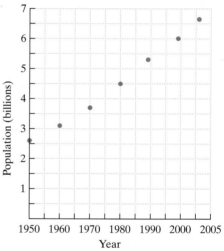

Figure 3.24 A scatter plot for data in Table 4.5

Because the data in this scatter plot have a rapidly increasing pattern, the shape suggests that an exponential model might be a good choice. (You might also want to try a linear model.) If you select the exponential option, you will use a graphing utility's Exponential REGression option. With this feature, a graphing utility fits the data to an exponential model of the form $y = ab^x$.

When computing an exponential model of the form $y = ab^x$, many graphing utilities rewrite the equation using logarithms. Because the domain of the logarithmic function is the set of positive numbers, **zero must not be a value for x** when using such utilities. What does this mean in terms of our data for world population that starts in the year 1950? We must start values of x after 0. Thus, we'll assign x to represent the number of years after 1949.

This gives us the data shown in Table 3.6. Using the Exponential REGression option, we obtain the equation in Figure 3.25.

Table 3.6

x, Numbers of Years after 1949		y, World Population (billions)
1	(1950)	2.6
11	(1960)	3.1
21	(1970)	3.7
31	(1980)	4.5
40	(1989)	5.3
52	(2001)	6.2

```
ExpReg
y=a*b^x
a=2.5731257405
b=1.01759434422
r=.998331654439
```

Figure 3.25 An exponential model for the data in Table 4.6

From Figure 3.25, we see that the exponential model of the data for world population, y, in billions, x years after 1949, with numbers rounded to three decimal places, is

$$y = 2.573(1.018)^x.$$

The correlation coefficient, r, is close to 1, indicating that the model fits the data very well.

Because $b = e^{\ln b}$, we can rewrite any model in the form $y = ab^x$ in terms of base e.

5 Express an exponential model in base e.

> **Expressing an Exponential Model in Base e**
>
> $y = ab^x$ is equivalent to $y = ae^{(\ln b) \cdot x}$.

EXAMPLE 5 Rewriting an Exponential Model in Base e

Rewrite $y = 2.573(1.018)^x$ in terms of base e.

Solution

$y = ab^x$ is equivalent to $y = ae^{(\ln b) \cdot x}$.

$y = 2.573(1.018)^x$ is equivalent to $y = 2.573e^{(\ln 1.018) \cdot x}$.

Using $\ln 1.018 \approx 0.018$, the exponential growth model for world population, y, in billions, x years after 1949 is

$$y = 2.573e^{0.018x}.$$

In Example 5, we can replace y with A and x with t so that the model has the same letters as those in the exponential growth model $A = A_0 e^{kt}$.

$A = A_0 e^{kt}$ This is the exponential growth model.

$A = 2.573e^{0.018t}$ This is the model for world population.

The value of k, 0.018, indicates a growth rate of 1.8%. Although this is an excellent model for the data, we must be careful about making projections about world population using this growth function. Why? World population growth rate is now 1.3%, not 1.8%, so our model will overestimate future populations.

Check Point 5 Rewrite $y = 4(7.8)^x$ in terms of base e. Express the answer in terms of a natural logarithm, and then round to three decimal places.

When using a graphing utility to model data, begin with a scatter plot, drawn either by hand or with the graphing utility, to obtain a general picture for the shape of the data. It might be difficult to determine which model best fits the data—linear, logarithmic, exponential, quadratic, or something else. If necessary, use your graphing utility to fit several models to the data. The best model is the one that yields the value r, the correlation coefficient, closest to 1 or -1. Finding a proper fit for data can be almost as much art as it is mathematics. In this era of technology, the process of creating models that best fit data is one that involves more decision making than computation.

EXERCISE SET 3.5

Practice and Application Exercises

The exponential growth model $A = 203e^{0.011t}$ describes the population of the United States, A, in millions, t years after 1970. Use this model to solve Exercises 1–4.

1. What was the population of the United States in 1970?
2. By what percentage is the population of the United States increasing each year?
3. When will the U.S. population be 300 million?
4. When will the U.S. population be 350 million?

India is currently one of the world's fastest-growing countries. By 2040, the population of India will be larger than the population of China; by 2050, nearly one-third of the world's population will live in these two countries alone. The exponential growth model $A = 574e^{0.026t}$ describes the population of India, A, in millions, t years after 1974. Use this model to solve Exercises 5–8.

5. By what percentage is the population of India increasing each year?
6. What was the population of India in 1974?
7. When will India's population be 1624 million?
8. When will India's population be 2732 million?
9. Low interest rates, easy credit, and strong demand from new immigrants have driven up the average sales price of new one-family houses in the United States. In 1995, the average sales price was $158,700 and by 2000 it had increased to $207,200.
   a. Use the exponential growth model $A = A_0e^{kt}$, in which t is the number of years after 1995, to find the exponential growth function that models the data.
   b. According to your model, by which year will the average sales price of a new one-family house reach $300,000?

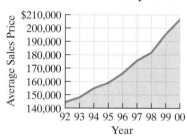

Average Sales Prices of New One-Family Houses

Source: U.S. Census Bureau

About the size of New Jersey, Israel has seen its population soar to more than 6 million since it was established. With the help of U.S. aid, the country now has a diversified economy rivaling those of other developed Western nations. By contrast, the Palestinians, living under Israeli occupation and a corrupt regime, endure bleak conditions. The graphs show that by 2050, Palestinians in the West Bank, Gaza Strip, and East Jerusalem will outnumber Israelis. Exercises 10–12 involve the projected growth of these two populations.

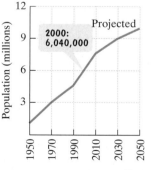

Population of Israel

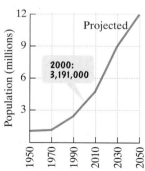

Palestinian Population in West Bank, Gaza, and East Jerusalem

Source: Newsweek

10. In 2000, the population of the Palestinians in the West Bank, Gaza Strip, and East Jerusalem was approximately 3.2 million and by 2050 it is projected to grow to 12 million. Use the exponential growth model $A = A_0e^{kt}$, in which t is the number of years after 2000, to find the exponential growth function that models the data.

11. In 2000, the population of Israel was approximately 6.04 million and by 2050 it is projected to grow to 10 million. Use the exponential growth model $A = A_0e^{kt}$, in which t is the number of years after 2000, to find an exponential growth function that models the data.

12. Use the growth models in Exercises 10 and 11 to determine the year in which the two populations will be the same.

An artifact originally had 16 grams of carbon-14 present. The decay model $A = 16e^{-0.000121t}$ describes the amount of carbon-14 present, A, in grams, after t years. Use this model to solve Exercises 13–14.

13. How many grams of carbon-14 will be present after 5715 years?

14. How many grams of carbon-14 will be present after 11,430 years?

15. The half-life of the radioactive element krypton-91 is 10 seconds. If 16 grams of krypton-91 are initially present, how many grams are present after 10 seconds? 20 seconds? 30 seconds? 40 seconds? 50 seconds?

16. The half-life of the radioactive element plutonium-239 is 25,000 years. If 16 grams of plutonium-239 are initially present how many grams are present after 25,000 years? 50,000 years? 75,000 years? 100,000 years? 125,000 years?

Use the exponential decay model for carbon-14, $A = A_0e^{-0.000121t}$, to solve Exercises 17–18.

17. Prehistoric cave paintings were discovered in a cave in France. The paint contained 15% of the original carbon-14. Estimate the age of the paintings.

18. Skeletons were found at a construction site in San Francisco in 1989. The skeletons contained 88% of the expected amount of carbon-14 found in a living person. In 1989, how old were the skeletons?

19. The August 1978 issue of *National Geographic* described the 1964 find of dinosaur bones of a newly discovered dinosaur weighing 170 pounds, measuring 9 feet, with a 6-inch claw on one toe of each hind foot. The age of the dinosaur was estimated using potassium-40 dating of rocks surrounding the bones.

   a. Potassium-40 decays exponentially with a half-life of approximately 1.31 billion years. Use the fact that after 1.31 billion years a given amount of potassium-40 will have decayed to half the original amount to show that the decay model for potassium-40 is given by $A = A_0e^{-0.52912t}$, where t is in billions of years.

   b. Analysis of the rocks surrounding the dinosaur bones indicated that 94.5% of the original amount of

potassium-40 was still present. Let $A = 0.945A_0$ in the model in part (a) and estimate the age of the bones of the dinosaur.

20. A bird species in danger of extinction has a population that is decreasing exponentially $(A = A_0e^{kt})$. Five years ago the population was at 1400 and today only 1000 of the birds are alive. Once the population drops below 100, the situation will be irreversible. When will this happen?

21. Use the exponential growth model, $A = A_0e^{kt}$, to show that the time it takes a population to double (to grow from A_0 to $2A_0$) is given by $t = \dfrac{\ln 2}{k}$.

22. Use the exponential growth model, $A = A_0e^{kt}$, to show that the time it takes a population to triple (to grow from A_0 to $3A_0$) is given by $t = \dfrac{\ln 3}{k}$.

Use the formula $t = \dfrac{\ln 2}{k}$ that gives the time for a population with a growth rate k to double to solve Exercises 23–24. Express each answer to the nearest whole year.

23. China is growing at a rate of 1.1% per year. How long will it take China to double its population?

24. Japan is growing at a rate of 0.3% per year. How long will it take Japan to double its population?

25. The logistic growth function

$$f(t) = \frac{100,000}{1 + 5000e^{-t}}$$

describes the number of people, $f(t)$, who have become ill with influenza t weeks after its initial outbreak in a particular community.

   a. How many people became ill with the flu when the epidemic began?

   b. How many people were ill by the end of the fourth week?

   c. What is the limiting size of the population that becomes ill?

26. The logistic growth function

$$f(t) = \frac{500}{1 + 83.3e^{-0.162t}}$$

describes the population, $f(t)$, of an endangered species of birds t years after they are introduced to a nonthreatening habitat.

   a. How many birds were initially introduced to the habitat?

   b. How many birds are expected in the habitat after 10 years?

   c. What is the limiting size of the bird population that the habitat will sustain?

The logistic growth function

$$P(x) = \frac{90}{1 + 271e^{-0.122x}}$$

models the percentage, $P(x)$, of Americans who are x years old with some coronary heart disease. Use the function to solve Exercises 27–30.

27. What percentage of 20-year-olds have some coronary heart disease?

28. What percentage of 80-year-olds have some coronary heart disease?

29. At what age is the percentage of some coronary heart disease 50%?

30. At what age is the percentage of some coronary heart disease 70%?

Use Newton's Law of Cooling, $T = C + (T_0 - C)e^{kt}$, to solve Exercises 31–34.

31. A bottle of juice initially has a temperature of 70°F. It is left to cool in a refrigerator that has a temperature of 45°F. After 10 minutes, the temperature of the juice is 55°F.
  a. Use Newton's Law of Cooling to find a model for the temperature of the juice, T, after t minutes.
  b. What is the temperature of the juice after 15 minutes?
  c. When will the temperature of the juice be 50°F?

32. A pizza removed from the oven has a temperature of 450°F. It is left sitting in a room that has a temperature of 70°F. After 5 minutes, the temperature of the pizza is 300°F.
  a. Use Newton's Law of Cooling to find a model for the temperature of the pizza, T, after t minutes.
  b. What is the temperature of the pizza after 20 minutes?
  c. When will the temperature of the pizza be 140°F?

33. A frozen steak initially has a temperature of 28°F. It is left to thaw in a room that has a temperature of 75°F. After 10 minutes, the temperature of the steak has risen to 38°F. After how many minutes will the temperature of the steak be 50°F?

34. A frozen steak initially has a temperature of 24°F. It is left to thaw in a room that has a temperature of 65°F. After 10 minutes, the temperature of the steak has risen to 30°F. After how many minutes will the temperature of the steak be 45°F?

In Exercises 35–38, rewrite the equation in terms of base e. Express the answer in terms of a natural logarithm, and then round to three decimal places.

35. $y = 100(4.6)^x$

36. $y = 1000(7.3)^x$

37. $y = 2.5(0.7)^x$

38. $y = 4.5(0.6)^x$

Writing in Mathematics

39. Nigeria has a growth rate of 0.031 or 3.1%. Describe what this means.

40 How can you tell if an exponential model describes exponential growth or exponential decay?

41. Suppose that a population that is growing exponentially increases from 800,000 people in 2003 to 1,000,000 people in 2006. Without showing the details, describe how to obtain the exponential growth function that models the data.

42. What is the half-life of a substance?

43. Describe a difference between exponential growth and logistic growth.

44. Describe the shape of a scatter plot that suggests modeling the data with an exponential function.

45. Based on the graphs in Exercises 10–12, for which population is an exponential growth function a better model? Explain your answer.

46. You take up weightlifting and record the maximum number of pounds you can lift at the end of each week. You start off with rapid growth in terms of the weight you can lift from week to week, but then the growth begins to level off. Describe how to obtain a function that models the number of pounds you can lift at the end of each week. How can you use this function to predict what might happen if you continue the sport?

47. Would you prefer that your salary be modeled exponentially or logarithmically? Explain your answer.

48. One problem with all exponential growth models is that nothing can grow exponentially forever. Describe factors that might limit the size of a population.

Technology Exercises

In Example 1 on pages 401–402, we used two data points and an exponential function to model federal minimum wages that were not adjusted for inflation from 1970 through 2000. The data are shown again in the table. Use all seven data points to solve Exercises 49–53.

x, Number of Years after 1969	y, Federal Minimum Wage
1	1.60
6	2.10
11	3.10
16	3.35
21	3.80
26	4.25
31	5.15

49. Use your graphing utility's Exponential REGression option to obtain a model of the form $y = ab^x$ that fits the data. How well does the correlation coefficient, r, indicate that the model fits the data?

50. Use your graphing utility's Logarithmic REGression option to obtain a model of the form $y = a + b \ln x$ that fits the data. How well does the correlation coefficient, r, indicate that the model fits the data?

51. Use your graphing utility's Linear REGression option to obtain a model of the form $y = ax + b$ that fits the data. How well does the correlation coefficient, r, indicate that the model fits the data?

52. Use your graphing utility's Power REGression option to obtain a model of the form $y = ax^b$ that fits the data. How well does the correlation coefficient, r, indicate that the model fits the data?

53. Use the value of r in Exercises 49–52 to select the model of best fit. Use this model to predict by which year the minimum wage will reach $7.50. How does this answer compare to the year we found in Example 1, namely 2010? If you obtained a different year, how do you account for this difference?

54. In Exercises 27–30, you worked with the logistic growth function

$$P(x) = \frac{90}{1 + 271e^{-0.122x}}$$

which models the percentage, $P(x)$, of Americans who are x years old with some coronary heart disease. Use your graphing utility to graph the function in a $[0, 100, 10]$ by $[0, 100, 10]$ viewing rectangle. Describe as specifically as possible what the logistic curve indicates about aging and the percentage of Americans with coronary heart disease.

In Exercises 55–56, use a graphing utility to find the model that best fits the given data. Then use the model to make a reasonable prediction for a value that exceeds those shown on the graph's horizontal axis.

55.

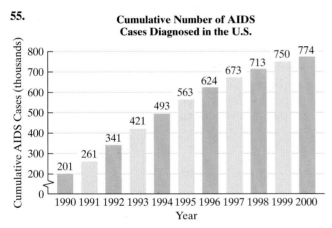

Cumulative Number of AIDS Cases Diagnosed in the U.S.

Source: Centers for Disease Control

56.

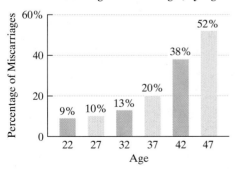

Percentage of Miscarriages, by Age

Source: Time

Critical Thinking Exercises

57. The World Health Organization makes predictions about the number of AIDS cases based on a compromise between a linear model and an exponential growth model. Explain why the World Health Organization does this.

58. Use Newton's Law of Cooling, $T = C + (T_0 - C)e^{kt}$, to solve this exercise. At 9:00 A.M., a coroner arrived at the home a person who had died during the night. The temperature of the room was 70°F, and at the time of death the person had a body temperature of 98.6°F. The coroner took the body's temperature at 9:30 A.M., at which time it was 85.6°F, and again at 10:00 A.M., when it was 82.7°F. At what time did the person die?

Group Exercises

59. This activity is intended for three or four people who would like to take up weightlifting. Each person in the group should record the maximum number of pounds that he or she can lift at the end of each week for the first 10 consecutive weeks. Use the Logarithmic REGression option of a graphing utility to obtain a model showing the amount of weight that group members can lift from week 1 through week 10. Graph each of the models in the same viewing rectangle to observe similarities and differences among weight-growth patterns of each member. Use the functions to predict the amount of weight that group members will be able to lift in the future. If the group continues to work out together, check the accuracy of these predictions.

60. Each group member should consult an almanac, newspaper, magazine, or the Internet to find data that can be modeled by exponential or logarithmic functions. Group members should select the two sets of data that are most interesting and relevant. For each data set selected, find a model that best fits the data. Each group member should make one prediction based on the model and then discuss a consequence of this prediction. What factors might change the accuracy of each prediction?

CHAPTER SUMMARY, REVIEW, AND TEST

Summary

DEFINITIONS AND CONCEPTS	EXAMPLES

3.1 Exponential Functions

a. The exponential function with base b is defined by $f(x) = b^x$, where $b > 0$ and $b \neq 1$. — Ex. 1, p. 357

b. Characteristics of exponential functions and graphs for $0 < b < 1$ and $b > 1$ are shown in the box on page 359. — Ex. 2, p. 358

c. Transformations involving exponential functions are summarized in Table 3.1 on page 359. — Exs. 3 & 4, p. 360

d. The natural exponential function is $f(x) = e^x$. The irrational number e is called the natural base, where $e \approx 2.7183$. — Ex. 5, p. 361

e. Formulas for compound interest: After t years, the balance, A, in an account with principal P and annual interest rate r (in decimal form) is given by one of the following formulas: — Ex. 6, p. 363

1. For n compoundings per year: $A = P\left(1 + \dfrac{r}{n}\right)^{nt}$

2. For continuous compounding: $A = Pe^{rt}$

3.2 Logarithmic Functions

a. Definition of the logarithmic function: For $x > 0$ and $b > 0$, $b \neq 1$, $y = \log_b x$ is equivalent to $b^y = x$. The function $f(x) = \log_b x$ is the logarithmic function with base b. This function is the inverse function of the exponential function with base b. — Ex. 1, p. 368; Ex. 2, p. 368; Ex. 3, p. 369

b. Graphs of logarithmic functions for $b > 1$ and $0 < b < 1$ are shown in Figure 3.7 on page 371. Characteristics of the graphs are summarized in the box that follows the figure. — Ex. 6, p. 371

c. Transformations involving logarithmic functions are summarized in Table 3.3 on page 372.

d. The domain of a logarithmic function of the form $f(x) = \log_b x$ is the set of all positive real numbers. The domain of $f(x) = \log_b(x + c)$ consists of all x for which $x + c > 0$. — Ex. 7, p. 373; Ex. 10, p. 375

e. Common and natural logarithms: $f(x) = \log x$ means $f(x) = \log_{10} x$ and is the common logarithmic function. $f(x) = \ln x$ means $f(x) = \log_e x$ and is the natural logarithmic function. — Ex. 8, p. 374; Ex. 9, p. 374

f. Basic Logarithmic Properties

Base b $(b > 0, b \neq 1)$	Base 10 (Common Logarithms)	Base e (Natural Logarithms)	
$\log_b 1 = 0$	$\log 1 = 0$	$\ln 1 = 0$	Ex. 4, p. 370;
$\log_b b = 1$	$\log 10 = 1$	$\ln e = 1$	Ex. 5, p. 370;
$\log_b b^x = x$	$\log 10^x = 1$	$\ln e^x = x$	Ex. 11, p. 376
$b^{\log_b x} = x$	$10^{\log x} = x$	$e^{\ln x} = x$	

3.3 Properties of Logarithms

a. *The Product Rule:* $\log_b(MN) = \log_b M + \log_b N$ — Ex. 1, p. 380

b. *The Quotient Rule:* $\log_b\left(\dfrac{M}{N}\right) = \log_b M - \log_b N$ — Ex. 2, p. 381

c. *The Power Rule:* $\log_b M^p = p \log_b M$ — Ex. 3, p. 382

d. *The Change-of-Base Property:*

The General Property	Introducing Common Logarithms	Introducing Natural Logarithms	
$\log_b M = \dfrac{\log_a M}{\log_a b}$	$\log_b M = \dfrac{\log M}{\log b}$	$\log_b M = \dfrac{\ln M}{\ln b}$	Ex. 7, p. 386; Ex. 8, p. 386

DEFINITIONS AND CONCEPTS   **EXAMPLES**

3.4 Exponential and Logarithmic Equations

a. An exponential equation is an equation containing a variable in an exponent. The solution procedure involves isolating the exponential expression and taking the natural logarithm on both sides. The box on page 390 provides the details.   Ex. 1, p. 390; Ex. 2, p. 391; Ex. 3, p. 391; Ex. 4, p. 392

b. A logarithmic equation is an equation containing a variable in a logarithmic expression. Logarithmic equations in the form $\log_b x = c$ can be solved by rewriting as $b^c = x$.   Ex. 5, p. 392

c. When checking logarithmic equations, reject proposed solutions that produce the logarithm of a negative number or the logarithm of 0 in the original equation.   Ex. 6, p. 393

d. Equations involving natural logarithms are solved by isolating the natural logarithm with coefficient 1 on one side and exponentiating both sides. Simplify using $e^{\ln x} = x$.   Ex. 7, p. 394

3.5 Modeling with Exponential and Logarithmic Functions

a. Exponential growth and decay models are given by $A = A_0 e^{kt}$ in which t represents time, A_0 is the amount present at $t = 0$, and A is the amount present at time t. If $k > 0$, the model describes growth and k is the growth rate. If $k < 0$, the model describes decay and k is the decay rate.   Ex. 1, p. 401; Ex. 2, p. 404

b. The logistic growth model, given by $A = \dfrac{c}{1 + ae^{-bt}}$, describes situations in which growth is limited. $y = c$ is a horizontal asymptote for the graph, and growth, A, can never exceed c.   Ex. 3, p. 406

c. Newton's Law of Cooling: The temperature, T, of a heated object at time t is given by   Ex. 4, p. 407

$$T = C + (T_0 - C)e^{kt}$$

where C is the constant temperature of the surrounding medium, T_0 is the initial temperature of the heated object, and k is a negative constant.

d. Scatter plots for exponential and logarithmic models are shown in Figure 3.21 on page 409. When using a graphing utility to model data, the closer that the correlation coefficient, r, is to -1 or 1, the better the model fits the data.

e. Expressing an Exponential Model in Base e: $y = ab^x$ is equivalent to $y = ae^{(\ln b) \cdot x}$.   Ex. 5, p. 411

Review Exercises

3.1

In Exercises 1–4, the graph of an exponential function is given. Select the function for each graph from the following options:

$$f(x) = 4^x, \; g(x) = 4^{-x},$$
$$h(x) = -4^{-x}, \; r(x) = -4^{-x} + 3.$$

1.

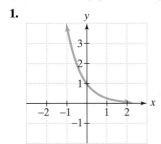

2.

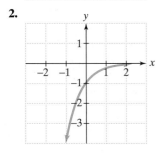

3.

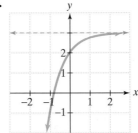

4.

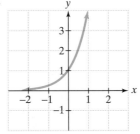

In Exercises 5–8, sketch by hand the graphs of the two functions in the same rectangular coordinate system. Use a table of coordinates to sketch the first function and transformations of this function with a table of coordinates to graph the second function.

5. $f(x) = 2^x$ and $g(x) = 2^{x-1}$

6. $f(x) = 3^x$ and $g(x) = 3^x - 1$

7. $f(x) = 3^x$ and $g(x) = -3^x$

8. $f(x) = \left(\frac{1}{2}\right)^x$ and $g(x) = \left(\frac{1}{2}\right)^{-x}$

Use the compound interest formulas to solve Exercises 9–10.

9. Suppose that you have $5000 to invest. Which investment yields the greater return over 5 years: 5.5% compounded semiannually or 5.25% compounded monthly?

10. Suppose that you have $14,000 to invest. Which investment yields the greater return over 10 years: 7% compounded monthly or 6.85% compounded continuously?

11. A cup of coffee is taken out of a microwave oven and placed in a room. The temperature, T, in degrees Fahrenheit, of the coffee after t minutes is modeled by the function $T = 70 + 130e^{-0.04855t}$. The graph of the function is shown in the figure.

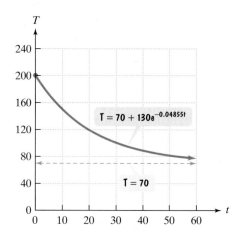

Use the graph to answer each of the following questions.

a. What was the temperature of the coffee when it was first taken out of the microwave?

b. What is a reasonable estimate of the temperature of the coffee after 20 minutes? Use your calculator to verify this estimate.

c. What is the limit of the temperature to which the coffee will cool? What does this tell you about the temperature of the room?

3.2

In Exercises 12–14, write each equation in its equivalent exponential form.

12. $\frac{1}{2} = \log_{49} 7$ **13.** $3 = \log_4 x$ **14.** $\log_3 81 = y$

In Exercises 15–17, write each equation in its equivalent logarithmic form.

15. $6^3 = 216$ **16.** $b^4 = 625$ **17.** $13^y = 874$

In Exercises 18–25, evaluate each expression without using a calculator. If evaluation is not possible, state the reason.

18. $\log_4 64$ **19.** $\log_5 \frac{1}{25}$ **20.** $\log_3(-9)$

21. $\log_{16} 4$ **22.** $\log_{17} 17$ **23.** $\log_3 3^8$

24. $\ln e^5$ **25.** $\log_3(\log_8 8)$

26. Graph $f(x) = 2^x$ and $g(x) = \log_2 x$ in the same rectangular coordinate system.

27. Graph $f(x) = \left(\frac{1}{3}\right)^x$ and $g(x) = \log_{1/3} x$ in the same rectangular coordinate system.

In Exercises 28–31, the graph of a logarithmic function is given. Select the function for each graph from the following options:

$$f(x) = \log x, g(x) = \log(-x),$$
$$h(x) = \log(2 - x), r(x) = 1 + \log(2 - x).$$

28.

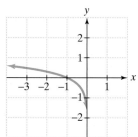

29.

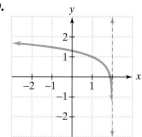

30.

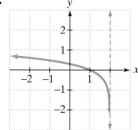

31.

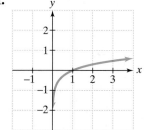

In Exercises 32–34, begin by graphing $f(x) = \log_2 x$. Then use transformations of this graph to graph the given function. What is the graph's x-intercept? What is the vertical asymptote?

32. $g(x) = \log_2(x - 2)$ **33.** $h(x) = -1 + \log_2 x$

34. $r(x) = \log_2(-x)$

In Exercises 35–37, find the domain of each logarithmic function.

35. $f(x) = \log_8(x + 5)$ **36.** $f(x) = \log(3 - x)$

37. $f(x) = \ln(x - 1)^2$

In Exercises 38–40, use inverse properties of logarithms to simplify each expression.

38. $\ln e^{6x}$ **39.** $e^{\ln \sqrt{x}}$ **40.** $10^{\log 4x^2}$

41. On the Richter scale, the magnitude, R, of an earthquake of intensity I is given by $R = \log \dfrac{I}{I_0}$, where I_0 is the intensity of a barely felt zero-level earthquake. If the intensity of an earthquake is $1000I_0$, what is its magnitude on the Richter scale?

42. Students in a psychology class took a final examination. As part of an experiment to see how much of the course content they remembered over time, they took equivalent forms of the exam in monthly intervals thereafter. The average score, $f(t)$, for the group after t months is modeled by the function $f(t) = 76 - 18 \log(t + 1)$, where $0 \le t \le 12$.

  a. What was the average score when the exam was first given?

  b. What was the average score after 2 months? 4 months? 6 months? 8 months? one year?

  c. Use the results from parts (a) and (b) to graph f. Describe what the shape of the graph indicates in terms of the material retained by the students.

43. The formula

$$t = \frac{1}{c} \ln\left(\frac{A}{A - N}\right)$$

describes the time, t, in weeks, that it takes to achieve mastery of a portion of a task. In the formula, A represents maximum learning possible, N is the portion of the learning that is to be achieved, and c is a constant used to measure an individual's learning style. A 50-year-old man decides to start running as a way to maintain good health. He feels that the maximum rate he could ever hope to achieve is 12 miles per hour. How many weeks will it take before the man can run 5 miles per hour if $c = 0.06$ for this person?

3.3

In Exercises 44–47, use properties of logarithms to expand each logarithmic expression as much as possible. Where possible, evaluate logarithmic expressions without using a calculator.

44. $\log_6(36x^3)$ **45.** $\log_4\left(\dfrac{\sqrt{x}}{64}\right)$

46. $\log_2\left(\dfrac{xy^2}{64}\right)$ **47.** $\ln\sqrt[3]{\dfrac{x}{e}}$

In Exercises 48–51, use properties of logarithms to condense each logarithmic expression. Write the expression as a single logarithm whose coefficient is 1.

48. $\log_b 7 + \log_b 3$ **49.** $\log 3 - 3 \log x$

50. $3 \ln x + 4 \ln y$ **51.** $\frac{1}{2} \ln x - \ln y$

In Exercises 52–53, use common logarithms or natural logarithms and a calculator to evaluate to four decimal places.

52. $\log_6 72{,}348$ **53.** $\log_4 0.863$

3.4

Solve each exponential equation in Exercises 54–58. Express the answer in terms of natural logarithms. Then use a calculator to obtain a decimal approximation, correct to two decimal places, for the solution.

54. $8^x = 12{,}143$ **55.** $9e^{5x} = 1269$

56. $e^{12-5x} - 7 = 123$ **57.** $5^{4x+2} = 37{,}500$

58. $e^{2x} - e^x - 6 = 0$

Solve each logarithmic equation in Exercises 59–63.

59. $\log_4(3x - 5) = 3$

60. $\log_2(x + 3) + \log_2(x - 3) = 4$

61. $\log_3(x - 1) - \log_3(x + 2) = 2$

62. $\ln x = -1$ **63.** $3 + 4 \ln(2x) = 15$

64. The formula $A = 10.1e^{0.005t}$ models the population of Los Angeles, California, A, in millions, t years after 1992. If the growth rate continues into the future, when will the population reach 13 million?

65. The amount of carbon dioxide in the atmosphere, measured in parts per million, has been increasing as a result of the burning of oil and coal. The buildup of gases and particles traps heat and raises the planet's temperature, a phenomenon called the *greenhouse effect*. Carbon dioxide accounts for about half of the warming. The function $f(t) = 364(1.005)^t$ projects carbon dioxide concentration, $f(t)$, in parts per million, t years after 2000. Using the projections given by the function, when will the carbon dioxide concentration be double the preindustrial level of 280 parts per million?

66. The formula $\overline{C}(x) = 15{,}557 + 5259 \ln x$ models the average cost of a new car, $\overline{C}(x)$, x years after 1989. When will the average cost of a new car reach $30,000?

67. Use the formula for compound interest with n compoundings each year to solve this problem. How long, to the nearest tenth of a year, will it take $12,500 to grow to $20,000 at 6.5% annual interest compounded quarterly?

Use the formula for continuous compounding to solve Exercises 68–69.

68. How long, to the nearest tenth of a year, will it take $50,000 to triple in value at 7.5% annual interest compounded continuously?

69. What interest rate is required for an investment subject to continuous compounding to triple in 5 years?

3.5

70. According to the U.S. Bureau of the Census, in 1990 there were 22.4 million residents of Hispanic origin living in the United States. By 2000, the number had increased to 35.3 million. The exponential growth function $A = 22.4e^{kt}$ describes the U.S. Hispanic population, A, in millions, t years after 1990.

 a. Find k, correct to three decimal places.

 b. Use the resulting model to project the Hispanic resident population in 2010.

 c. In which year will the Hispanic resident population reach 60 million?

71. Use the exponential decay model for carbon-14, $A = A_0e^{-0.000121t}$, to solve this exercise. Prehistoric cave paintings were discovered in the Lascaux cave in France. The paint contained 15% of the original carbon-14. Estimate the age of the paintings at the time of the discovery.

72. The function

$$f(t) = \frac{500{,}000}{1 + 2499e^{-0.92t}}$$

models the number of people, $f(t)$, in a city who have become ill with influenza t weeks after its initial outbreak.

 a. How many people became ill with the flu when the epidemic began?

 b. How many people were ill by the end of the sixth week?

 c. What is the limiting size of $f(t)$, the population that becomes ill?

73. Use Newton's Law of Cooling, $T = C + (T_0 - C)e^{kt}$, to solve this exercise. You are served a cup of coffee that has a temperature of 185°F. The room temperature is 65°F. After 2 minutes, the temperature of the coffee is 155°F.

 a. Write a model for the temperature of the coffee, T, after t minutes.

 b. When will the temperature of the coffee be 105°F?

In Exercises 74–75, rewrite the equation in terms of base e. Express the answer in terms of a natural logarithm, and then round to three decimal places.

74. $y = 73(2.6)^x$ **75.** $y = 6.5(0.43)^x$

76. The figure shows world population projections through the year 2150. The data are from the United Nations Family Planning Program and are based on optimistic or pessimistic expectations for successful control of human population growth. Suppose that you are interested in modeling these data using exponential, logarithmic, linear, and quadratic functions. Which function would you use to model each of the projections? Explain your choices. For the choice corresponding to a quadratic model, would your formula involve one with a positive or negative leading coefficient? Explain.

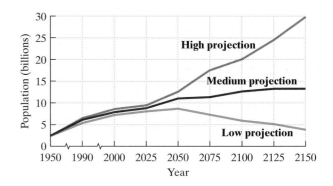

77. The figure shows the number of people in the United States age 65 and over, with projected figures for the year 2010 and beyond.

U. S. Population Age 65 and Over

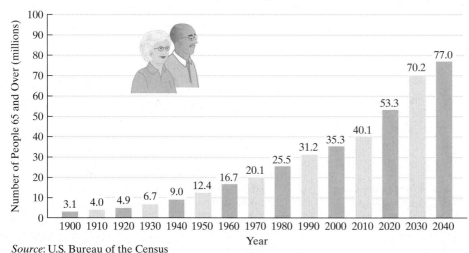

Source: U.S. Bureau of the Census

Let x represent the number of years after 1899 and let y represent the U.S. population, in millions, age 65 and over. Use your graphing utility to find the model that best fits the data in the bar graph. Then use the model to find the projected U.S. population age 65 and over in 2050.

Chapter 3 Test

1. Graph $f(x) = 2^x$ and $g(x) = 2^{x+1}$ in the same rectangular coordinate system.
2. Graph $f(x) = \log_2 x$ and $g(x) = \log_2(x-1)$ in the same rectangular coordinate system.
3. Write in exponential form: $\log_5 125 = 3$.
4. Write in logarithmic form: $\sqrt{36} = 6$.
5. Find the domain of $f(x) = \ln(3-x)$.

In Exercises 6–7, use properties of logarithms to expand each logarithmic expression as much as possible. Where possible, evaluate logarithmic expressions without using a calculator.

6. $\log_4(64x^5)$
7. $\log_3\left(\dfrac{\sqrt[3]{x}}{81}\right)$

In Exercises 8–9, write each expression as a single logarithm.

8. $6\log x + 2\log y$
9. $\ln 7 - 3\ln x$
10. Use a calculator to evaluate $\log_{15} 71$ to four decimal places.

In Exercises 11–16, solve each equation.

11. $5^x = 1.4$
12. $400e^{0.005x} = 1600$
13. $e^{2x} - 6e^x + 5 = 0$
14. $\log_6(4x-1) = 3$
15. $\log x + \log(x+15) = 2$
16. $2\ln(3x) = 8$

17. Suppose you have \$3000 to invest. Which investment yields the greater return over 10 years: 6.5% compounded semiannually or 6% compounded continuously? How much more (to the nearest dollar) is yielded by the better investment?

18. On the decibel scale, the loudness of a sound, D, in decibels, is given by $D = 10\log\dfrac{I}{I_0}$, where I is the intensity of the sound, in watts per meter^2, and I_0 is the intensity of a sound barely audible to the human ear. If the intensity of a sound is $10^{12}I_0$, what is its loudness in decibels? (Such a sound is potentially damaging to the ear.)

19. The function
$$P(t) = 89.18e^{-0.004t}$$
models the percentage, $P(t)$, of married men in the United States who were employed t years after 1959.
 a. What percentage of married men were employed in 1959?
 b. Is the percentage of married men who are employed increasing or decreasing? Explain.
 c. In what year were 77% of U.S. married men employed?

20. The 1990 population of Europe was 509 million; in 2000, it was 729 million. Write the exponential growth function that describes the population of Europe, in millions, t years after 1990.

21. Use the exponential decay model for carbon-14, $A = A_0 e^{-0.000121t}$, to solve this exercise. Bones of a prehistoric man were discovered and contained 5% of the original amount of carbon-14. How long ago did the man die?

22. The logistic growth function
$$f(t) = \frac{140}{1 + 9e^{-0.165t}}$$
describes the population, $f(t)$, of an endangered species of elk t years after they were introduced to a nonthreatening habitat.
 a. How many elk were initially introduced to the habitat?
 b. How many elk are expected in the habitat after 10 years?
 c. What is the limiting size of the elk population that the habitat will sustain?

Cumulative Review Exercises (Chapters P–3)

Solve each equation in Exercises 1–5.

1. $|3x - 4| = 2$
2. $x^2 + 2x + 5 = 0$
3. $x^4 + x^3 - 3x^2 - x + 2 = 0$
4. $e^{5x} - 32 = 96$
5. $\log_2(x+5) + \log_2(x-1) = 4$

Solve each inequality in Exercises 6–7. Express the answer in interval notation.

6. $14 - 5x \geq -6$
7. $|2x - 4| \leq 2$
8. Write the point-slope form and the slope-intercept form of the line passing through $(1,3)$ and $(3,-3)$.
9. If $f(x) = x^2$ and $g(x) = x + 2$, find $(f \circ g)(x)$ and $(g \circ f)(x)$.
10. If $f(x) = 2x - 7$, find $f^{-1}(x)$.
11. Divide $x^3 + 5x^2 + 3x - 10$ by $x + 2$.
12. Use the Rational Zero Theorem to list all possible rational zeros for $f(x) = 4x^3 - 7x - 3$.

13. The value of y varies directly as the square of x. If $x = 3$ when $y = 12$, find y when $x = 15$.
14. Show that $f(x) = x^5 - x^3 - 1$ has a zero between 1 and 2.

In Exercises 15–18, graph each equation.

15. $(x-3)^2 + (y+2)^2 = 4$
16. $f(x) = (x-2)^2 - 1$
17. $f(x) = \dfrac{x^2 - 1}{x^2 - 4}$
18. $f(x) = (x-2)^2(x+1)$

19. You have 2000 yards of fence to enclose a rectangular field. What are the dimensions of the rectangle that encloses the most area? What is the maximum area that can be enclosed?
20. The function $F(t) = 1 - k\ln(t+1)$ models the fraction of people, $F(t)$, who remember all the words in a list of nonsense words t hours after memorizing the list. After 3 hours, only half the people could remember all the words. Determine the value of k and then predict the fraction of people in the group who will remember all the words after 6 hours. Round to three decimal places and then express the fraction with a denominator of 1000.

Trigonometric Functions

Chapter 4

Have you had days where your physical, intellectual, and emotional potentials were all at their peak? Then there are those other days when we feel we should not even bother getting out of bed. Do our potentials run in oscillating cycles like the tides? Can they be described mathematically? In this chapter you will encounter functions that enable us to model phenomena that occur in cycles.

What a day! It started when you added two miles to your morning run. You've experienced a feeling of peak physical well-being ever since. College was wonderful: You actually enjoyed two difficult lectures and breezed through a math test that had you worried. Now you're having dinner with an old group of friends. You experience the warmth from bonds of friendship filling the room.

SECTION 4.1   *Angles and Their Measure*

Objectives

1. Recognize and use the vocabulary of angles.
2. Use degree measure.
3. Draw angles in standard position.
4. Find coterminal angles.
5. Find complements and supplements.
6. Use radian measure.
7. Convert between degrees and radians.
8. Find the length of a circular arc.
9. Use linear and angular speed to describe motion on a circular path.

The San Francisco Museum of Modern Art was constructed in 1995 to illustrate how art and architecture can enrich one another. The exterior involves geometric shapes, symmetry, and unusual facades. Although there are no windows, natural light streams in through a truncated cylindrical skylight that crowns the building. The architect worked with a scale model of the museum at the site and observed how light hit it during different times of the day. These observations were used to cut the cylindrical skylight at an angle that maximizes sunlight entering the interior.

Angles play a critical role in creating modern architecture. They are also fundamental in trigonometry. In this section, we begin our study of trigonometry by looking at angles and methods for measuring them.

1 Recognize and use the vocabulary of angles.

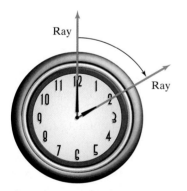

Figure 4.1 Clock with hands forming an angle

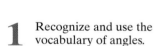

Figure 4.2 An angle; two rays with a common endpoint

Angles

The hour hand of a clock suggests a **ray,** a part of a line that has only one endpoint and extends forever in the opposite direction. An **angle** is formed by two rays that have a common endpoint. One ray is called the **initial side** and the other the **terminal side.**

A rotating ray is often a useful way to think about angles. The ray in Figure 4.1 rotates from 12 to 2. The ray pointing to 12 is the **initial side** and the ray pointing to 2 is the **terminal side.** The common endpoint of an angle's initial side and terminal side is the **vertex** of the angle.

Figure 4.2 shows an angle. The arrow near the vertex shows the direction and the amount of rotation from the initial side to the terminal side. Several methods can be used to name an angle. Lowercase Greek letters, such as α (alpha), β (beta), γ (gamma), and θ (theta), are often used.

An angle is in **standard position** if

- its vertex is at the origin of a rectangular coordinate system

and

- its initial side lies along the positive *x*-axis.

The angles in Figure 4.3 are both in standard position.

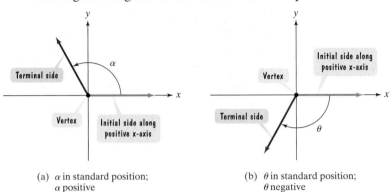

(a) α in standard position;
α positive

(b) θ in standard position;
θ negative

Figure 4.3 Two angles in standard position

When we see an initial side and a terminal side in place, there are two kinds of rotation that could have generated it. The arrow in Figure 4.3(a) indicates that the rotation from the initial side to the terminal side is in the counterclockwise direction. **Positive angles** are generated by counterclockwise rotation. Thus, angle α is positive. By contrast, the arrow in Figure 4.3(b) shows that the rotation from the initial side to the terminal side is in the clockwise direction. **Negative angles** are generated by clockwise rotation. Thus, angle θ is negative.

When an angle is in standard position, its terminal side can lie in a quadrant. We say that the angle **lies in that quadrant.** For example, in Figure 4.3(a), the terminal side of angle α lies in quadrant II. Thus, angle α lies in quadrant II. By contrast, in Figure 4.3(b), the terminal side of angle θ lies in quadrant III. Thus, angle θ lies in quadrant III.

Must all angles in standard position lie in a quadrant? The answer is no. The terminal side can lie on the x-axis or the y-axis. For example, angle β in Figure 4.4 has a terminal side that lies on the negative y-axis. An angle is called a **quadrantal angle** if its terminal side lies on the x-axis or the y-axis. Angle β in Figure 4.4 is an example of a quadrantal angle.

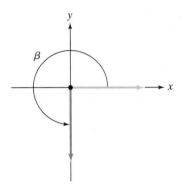

Figure 4.4 β is a quadrantal angle.

2 Use degree measure.

A complete 360° rotation

Figure 4.5 Classifying angles by their degree measurement

Measuring Angles Using Degrees

Angles are measured by determining the amount of rotation from the initial side to the terminal side. One way to measure angles is in **degrees,** symbolized by a small, raised circle °. Think of the hour hand of a clock. From 12 noon to 12 midnight, the hour hand moves around in a complete circle. By definition, the ray has rotated through 360 degrees, or 360°. Using 360° as the amount of rotation of a ray back onto itself, a degree, 1°, is $\frac{1}{360}$ of a complete rotation.

Figure 4.5 shows angles classified by their degree measurement. An **acute angle** measures less than 90° [see Figure 4.5(a)]. A **right angle,** one quarter of a complete rotation, measures 90° [Figure 4.5(b)]. Examine the right angle—do you see a small square at the vertex? This symbol is used to indicate a right angle. An **obtuse angle** measures more than 90°, but less than 180° [Figure 4.5(c)]. Finally, a **straight angle,** one-half a complete rotation, measures 180° [Figure 4.5(d)].

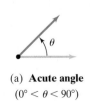 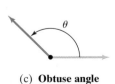 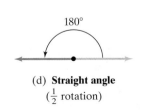

(a) **Acute angle**
($0° < \theta < 90°$)

(b) **Right angle**
($\frac{1}{4}$ rotation)

(c) **Obtuse angle**
($90° < \theta < 180°$)

(d) **Straight angle**
($\frac{1}{2}$ rotation)

3 Draw angles in standard position.

We will be using notation such as $\theta = 60°$ to refer to an angle θ whose measure is $60°$. We also refer to *an angle of 60°* or a *60° angle*, rather than using the more precise (but cumbersome) phrase *an angle whose measure is 60°*.

Technology

Fractional parts of degrees are measured in minutes and seconds. One minute, written $1'$, is $\frac{1}{60}$ degree: $1' = \frac{1}{60}°$.

One second, written $1''$, is $\frac{1}{3600}$ degree: $1'' = \frac{1}{3600}°$.

For example,

$31° \, 47' \, 12''$

$= \left(31 + \dfrac{47}{60} + \dfrac{12}{3600}\right)^{\circ}$

$\approx 31.787°$.

Many calculators have keys for changing an angle from degree, minute, second notation (D°M′S″) to a decimal form and vice versa.

EXAMPLE 1 Drawing Angles in Standard Position

Draw each angle in standard position:

    a. a $45°$ angle   **b.** a $225°$ angle   **c.** a $-135°$ angle   **d.** a $405°$ angle.

Solution Because we are drawing angles in standard position, each vertex is at the origin and each initial side lies along the positive x-axis.

    a. A $45°$ angle is half of a right angle. The angle lies in quadrant I and is shown in Figure 4.6(a).

    b. A $225°$ angle is a positive angle. It has a counterclockwise rotation of $180°$ followed by a counterclockwise rotation of $45°$. The angle lies in quadrant III and is shown in Figure 4.6(b).

    c. A $-135°$ angle is negative angle. It has a clockwise rotation of $90°$ followed by a clockwise rotation of $45°$. The angle lies in quadrant III and is shown in Figure 4.6(c).

    d. A $405°$ angle is a positive angle. It has a counterclockwise rotation of $360°$, one complete rotation, followed by a counterclockwise rotation of $45°$. The angle lies in quadrant I and is shown in Figure 4.6(d).

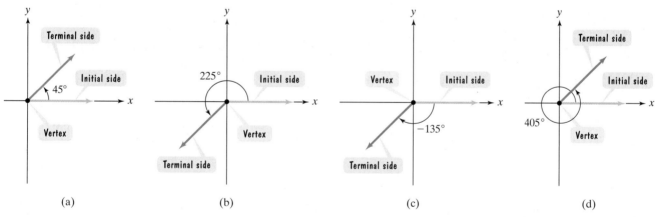

    (a)   (b)   (c)   (d)

Figure 4.6 Four angles in standard position

Check Point 1 Draw each angle in standard position:

    a. a $30°$ angle   **b.** a $210°$ angle

    c. a $-120°$ angle   **d.** a $390°$ angle.

4 Find coterminal angles.

Look at Figure 4.6 again. The $45°$ and $405°$ angles in parts (a) and (d) have the same initial and terminal sides. Similarly, the $225°$ and $-135°$ angles in parts (b) and (c) have the same initial and terminal sides. Two angles with the same initial and terminal sides are called **coterminal angles.**

Every angle has infinitely many coterminal angles. Why? Think of an angle in standard position. One or more complete rotations of 360°, clockwise or counterclockwise, result in angles with the same initial and terminal sides as the original angle.

> **Coterminal Angles**
>
> An angle of $x°$ is coterminal with angles of
>
> $$x° + k \cdot 360°$$
>
> where k is an integer.

Two coterminal angles for an angle of $x°$ can be found by adding 360° to $x°$ and subtracting 360° from $x°$.

Counterclockwise Clocks

The counterclockwise rotation associated with positive angles was used in England to manufacture counterclockwise clocks. They ran backward but told the time perfectly correctly.

EXAMPLE 2 Finding Coterminal Angles

Assume the following angles are in standard position. Find a positive angle less than 360° that is coterminal with:

a. a 420° angle   **b.** a −120° angle.

Solution   We obtain the coterminal angle by adding or subtracting 360°. The requirement to obtain a positive angle less than 360° determines whether we should add or subtract.

a. For a 420° angle, subtract 360° to find a positive coterminal angle.

$$420° − 360° = 60°$$

A 60° angle is coterminal with a 420° angle. Figure 4.7(a) illustrates that these angles have the same initial and terminal sides.

b. For a −120° angle, add 360° to find a positive coterminal angle.

$$−120° + 360° = 240°$$

A 240° angle is coterminal with a −120° angle. Figure 4.7(b) illustrates that these angles have the same initial and terminal sides.

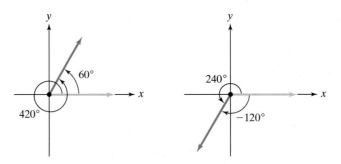

(a) Angles of 420° and 60° are coterminal.

(b) Angles of −120° and 240° are coterminal.

Figure 4.7 Pairs of coterminal angles

Check Point 2 Find a positive angle less than 360° that is coterminal with:
a. a 400° angle   **b.** a −135° angle.

5 Find complements and supplements.

Two positive angles are **complements** if their sum is 90°. For example, angles of 70° and 20° are complements because 70° + 20° = 90°.

Two positive angles are **supplements** if their sum is 180°. For example, angles of 130° and 50° are supplements because 130° + 50° = 180°.

Finding Complements and Supplements

- For an $x°$ angle, the complement is a $90° - x°$ angle. Thus, the complement's measure is found by subtracting the angle's measure from 90°.
- For an $x°$ angle, the supplement is a $180° - x°$ angle. Thus, the supplement's measure is found by subtracting the angle's measure from 180°.

Because we use only positive angles for complements and supplements, some angles do not have complements and supplements.

EXAMPLE 3   Complements and Supplements

If possible, find the complement and the supplement of the given angle:

a. $\theta = 62°$ **b.** $\alpha = 123°$.

Solution   We find the complement by subtracting the angle's measure from 90°. We find the supplement by subtracting the angle's measure from 180°.

a. We begin with $\theta = 62°$.

$$\text{complement} = 90° - 62° = 28°$$
$$\text{supplement} = 180° - 62° = 118°$$

For a 62° angle, the complement is a 28° angle and the supplement is a 118° angle.

b. Now we turn to $\alpha = 123°$. For the angle's complement, we consider subtracting 123° from 90°. The difference is negative. Because we use only positive angles for complements, a 123° angle has no complement. It does, however, have a supplement.

$$\text{supplement} = 180° - 123° = 57°$$

The supplement of a 123° angle is a 57° angle.

Check Point 3 If possible, find the complement and the supplement of the given angle:
a. $\theta = 78°$ **b.** $\alpha = 150°$.

6 Use radian measure.

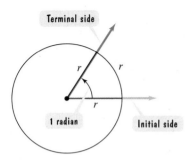

Figure 4.8 For a 1-radian angle, the intercepted arc and the radius are equal.

Measuring Angles Using Radians

Another way to measure angles is in *radians*. Let's first define an angle measuring **1 radian.** We use a circle of radius *r*. In Figure 4.8, we've constructed an angle whose vertex is at the center of the circle. Such an angle is called a **central angle.** Notice that this central angle intercepts an arc along the circle measuring *r* units. The radius of the circle is also *r* units. The measure of such an angle is 1 radian.

Definition of a Radian

One radian is the measure of the central angle of a circle that intercepts an arc equal in length to the radius of the circle.

The **radian measure** of any central angle is the length of the intercepted arc divided by the circle's radius. In Figure 4.9(a), the length of the arc intercepted by angle β is double the radius, *r*. We find the measure of angle β in radians by dividing the length of the intercepted arc by the radius.

$$\beta = \frac{\text{length of the intercepted arc}}{\text{radius}} = \frac{2r}{r} = 2$$

Thus, angle β measures 2 radians.

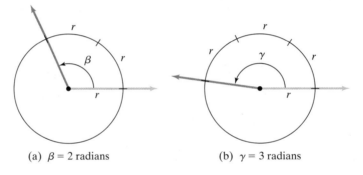

(a) $\beta = 2$ radians   (b) $\gamma = 3$ radians

Figure 4.9 Two central angles measured in radians

In Figure 4.9(b), the length of the intercepted arc is triple the radius, *r*. Let us find the measure of angle γ:

$$\gamma = \frac{\text{length of the intercepted arc}}{\text{radius}} = \frac{3r}{r} = 3.$$

Thus, angle γ measures 3 radians.

Radian Measure

Consider an arc of length *s* on a circle of radius *r*. The measure of the central angle, θ, that intercepts the arc is

$$\theta = \frac{s}{r} \text{ radians.}$$

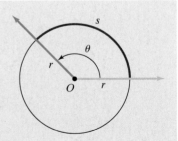

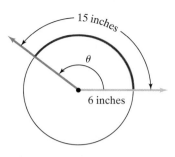

Figure 4.10

EXAMPLE 4   Computing Radian Measure

A central angle, θ, in a circle of radius 6 inches intercepts an arc of length 15 inches. What is the radian measure of θ?

Solution   Angle θ is shown in Figure 4.10. The radian measure of a central angle is the length of the intercepted arc, s, divided by the circle's radius, r. The length of the intercepted arc is 15 inches: $s = 15$ inches. The circle's radius is 6 inches: $r = 6$ inches. Now we use the formula for radian measure to find the radian measure of θ.

$$\theta = \frac{s}{r} = \frac{15 \text{ inches}}{6 \text{ inches}} = 2.5$$

Thus, the radian measure of θ is 2.5.

Study Tip

Before applying the formula for radian measure, be sure that the same unit of length is used for the intercepted arc, s, and the radius, r.

In Example 4, notice that the units (inches) cancel when we use the formula for radian measure. We are left with a number with no units. Thus, if an angle θ has a measure of 2.5 radians, we can write $\theta = 2.5$ radians or $\theta = 2.5$. We will often include the word *radians* simply for emphasis. There should be no confusion as to whether radian or degree measure is being used. Why is this so? If θ has a degree measure of, say, 2.5°, we must include the degree symbol and write $\theta = 2.5°$, and *not* $\theta = 2.5$.

> **Check Point 4**   A central angle, θ, in a circle of radius 12 feet intercepts an arc of length 42 feet. What is the radian measure of θ?

7   Convert between degrees and radians.

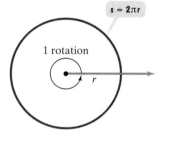

Figure 4.11 A complete rotation

Relationship between Degrees and Radians

How can we obtain a relationship between degrees and radians? We compare the number of degrees and the number of radians in one complete rotation, shown in Figure 4.11. We know that 360° is the amount of rotation of a ray back onto itself. The length of the intercepted arc is equal to the circumference of the circle. Thus, the radian measure of this central angle is the circumference of the circle divided by the circle's radius, r. The circumference of a circle of radius r is $2\pi r$. We use the formula for radian measure to find the radian measure of the 360° angle.

$$\theta = \frac{s}{r} = \frac{\text{the circle's circumference}}{r} = \frac{2\pi r}{r} = 2\pi$$

Because one complete rotation measures 360° and 2π radians,

$$360° = 2\pi \text{ radians.}$$

Dividing both sides by 2, we have

$$180° = \pi \text{ radians.}$$

Dividing this last equation by 180° or π gives the conversion rules that appear on the next page.

Study Tip

The unit you are converting *to* appears in the *numerator* of the conversion factor.

Conversion between Degrees and Radians

Using the basic relationship π radians $= 180°$,

1. To convert degrees to radians, multiply degrees by $\dfrac{\pi \text{ radians}}{180°}$.

2. To convert radians to degrees, multiply radians by $\dfrac{180°}{\pi \text{ radians}}$.

Angles that are fractions of a complete rotation are usually expressed in radian measure as fractional multiples of π, rather than as decimal approximations. For example, we write $\theta = \dfrac{\pi}{2}$ rather than using the decimal approximation $\theta \approx 1.57$.

EXAMPLE 5   Converting from Degrees to Radians

Convert each angle in degrees to radians:

a. $30°$ **b.** $90°$ **c.** $-135°$.

Solution   To convert degrees to radians, multiply by $\dfrac{\pi \text{ radians}}{180°}$. Observe how the degree units cancel.

a. $30° = 30° \cdot \dfrac{\pi \text{ radians}}{180°} = \dfrac{30\pi}{180} \text{ radians} = \dfrac{\pi}{6} \text{ radians}$

b. $90° = 90° \cdot \dfrac{\pi \text{ radians}}{180°} = \dfrac{90\pi}{180} \text{ radians} = \dfrac{\pi}{2} \text{ radians}$

c. $-135° = -135° \cdot \dfrac{\pi \text{ radians}}{180°} = -\dfrac{135\pi}{180} \text{ radians} = -\dfrac{3\pi}{4} \text{ radians}$

Divide the numerator and denominator by 45.

Check Point 5   Convert each angle in degrees to radians:

a. $60°$ **b.** $270°$ **c.** $-300°$.

EXAMPLE 6   Converting from Radians to Degrees

Convert each angle in radians to degrees:

a. $\dfrac{\pi}{3}$ radians   **b.** $-\dfrac{5\pi}{3}$ radians   **c.** 1 radian.

Solution   To convert radians to degrees, multiply by $\dfrac{180°}{\pi \text{ radians}}$. Observe how the radian units cancel.

Study Tip

In Example 6(c), we see that 1 radian is approximately 57°. Keep in mind that a radian is much larger than a degree.

a. $\dfrac{\pi}{3}$ radians $= \dfrac{\pi \text{ radians}}{3} \cdot \dfrac{180°}{\pi \text{ radians}} = \dfrac{180°}{3} = 60°$

b. $-\dfrac{5\pi}{3}$ radians $= -\dfrac{5\pi \text{ radians}}{3} \cdot \dfrac{180°}{\pi \text{ radians}} = -\dfrac{5 \cdot 180°}{3} = -300°$

c. 1 radian $= 1 \text{ radian} \cdot \dfrac{180°}{\pi \text{ radians}} = \dfrac{180°}{\pi} \approx 57.3°$

Check Point 6 Convert each angle in radians to degrees:

a. $\dfrac{\pi}{4}$ radians   **b.** $-\dfrac{4\pi}{3}$ radians   **c.** 6 radians.

Figure 4.12 illustrates the degree and radian measures of angles that you will commonly see in trigonometry. Each angle is in standard position, so that the initial side lies along the positive *x*-axis. We will be using both degree and radian measure for these angles.

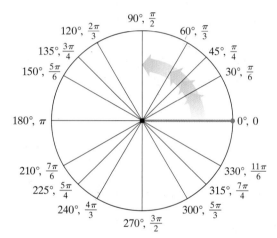

Figure 4.12 Degree and radian measures of selected angles

8 Find the length of a circular arc.

The Length of a Circular Arc

We can use the radian measure formula, $\theta = \dfrac{s}{r}$, to find the length of the arc of a circle. How do we do this? Remember that *s* represents the length of the arc intercepted by the central angle θ. Thus, by solving the formula for *s*, we have an equation for arc length.

The Length of a Circular Arc

Let *r* be the radius of a circle and θ the nonnegative radian measure of a central angle of the circle. The length of the arc intercepted by the central angle is

$$s = r\theta.$$

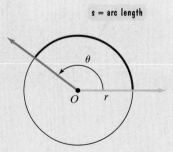

s = arc length

EXAMPLE 7   Finding the Length of a Circular Arc

A circle has a radius of 10 inches. Find the length of the arc intercepted by a central angle of 120°.

Study Tip

> The unit used to describe the length of a circular arc is the same unit that is given in the circle's radius.

Solution   The formula $s = r\theta$ can be used only when θ is expressed in radians. Thus, we begin by converting 120° to radians. Multiply by $\dfrac{\pi \text{ radians}}{180°}$.

$$120° = 120° \cdot \frac{\pi \text{ radians}}{180°} = \frac{120\pi}{180} \text{ radians} = \frac{2\pi}{3} \text{ radians}$$

Now we can use the formula $s = r\theta$ to find the length of the arc. The circle's radius is 10 inches: $r = 10$ inches. The measure of the central angle, in radians, is $\dfrac{2\pi}{3}$: $\theta = \dfrac{2\pi}{3}$. The length of the arc intercepted by this central angle is

$$s = r\theta = (10 \text{ inches})\left(\frac{2\pi}{3}\right) = \frac{20\pi}{3} \text{ inches} \approx 20.94 \text{ inches}.$$

Check Point 7   A circle has a radius of 6 inches. Find the length of the arc intercepted by a central angle of 45°. Express arc length in terms of π. Then round your answer to two decimal places.

9   Use linear and angular speed to describe motion on a circular path.

Linear and Angular Speed

A carousel contains four circular rows of animals. As the carousel revolves, the animals in the outer row travel a greater distance per unit of time than those in the inner rows. These animals have a greater *linear speed* than those in the inner rows. By contrast, all animals, regardless of the row, complete the same number of revolutions per unit of time. All animals in the four circular rows travel at the same *angular speed*.

Using v for linear speed and ω (omega) for angular speed, we define these two kinds of speeds along a circular path as follows:

Definitions of Linear and Angular Speed

If a point is in motion on a circle of radius r through an angle of θ radians in time t, then its **linear speed** is

$$v = \frac{s}{t}$$

where s is the arc length given by $s = r\theta$, and its **angular speed** is

$$\omega = \frac{\theta}{t}.$$

The hard drive in a computer rotates at 3600 revolutions per minute. This angular speed, expressed in revolutions per minute, can also be expressed in revolutions per second, radians per minute, and radians per second. Using 2π radians $= 1$ revolution, we express the angular speed of a hard drive in radians per minute as follows:

3600 revolutions per minute

$$= \frac{3600 \text{ revolutions}}{1 \text{ minute}} \cdot \frac{2\pi \text{ radians}}{1 \text{ revolution}} = \frac{7200\pi \text{ radians}}{1 \text{ minute}}$$

$= 7200\pi$ radians per minute.

We can establish a relationship between the two kinds of speed by dividing both sides of the arc length formula, $s = r\theta$, by t:

$$\frac{s}{t} = \frac{r\theta}{t} = r\frac{\theta}{t}.$$

This expression defines linear speed.

This expression defines angular speed.

Thus, linear speed is the product of the radius and the angular speed.

Linear Speed in Terms of Angular Speed

The linear speed, v, of a point a distance r from the center of rotation is given by

$$v = r\omega$$

where ω is the angular speed in radians per unit of time.

EXAMPLE 8 Finding Linear Speed

A wind machine used to generate electricity has blades that are 10 feet in length (see Figure 4.13). The propeller is rotating at four revolutions per second. Find the linear speed, in feet per second, of the tips of the blades.

10 feet

Figure 4.13

Figure 4.13, repeated

Solution   We are given ω, the angular speed.

$$\omega = 4 \text{ revolutions per second}$$

We use the formula $v = r\omega$ to find v, the linear speed. Before applying the formula, we must express ω in radians per second.

$$\omega = \frac{4 \text{ revolutions}}{1 \text{ second}} \cdot \frac{2\pi \text{ radians}}{1 \text{ revolution}} = \frac{8\pi \text{ radians}}{1 \text{ second}} \quad \text{or} \quad \frac{8\pi}{1 \text{ second}}$$

The angular speed of the propeller is 8π radians per second. The linear speed is

$$v = r\omega = 10 \text{ feet} \cdot \frac{8\pi}{1 \text{ second}} = \frac{80\pi \text{ feet}}{\text{second}}.$$

The linear speed of the tips of the blades is 80π feet per second, which is approximately 251 feet per second.

> **Check Point 8**   A 45-rpm record has an angular speed of 45 revolutions per minute. Find the linear speed, in inches per minute, at the point where the needle is 1.5 inches from the record's center.

EXERCISE SET 4.1

Practice Exercises

In Exercises 1–6, each angle is in standard position. Determine the quadrant in which the angle lies.

1. 145°
2. 285°
3. −100°
4. −110°
5. 362°
6. 364°

In Exercises 7–10, classify the angle as acute, right, straight, or obtuse.

7.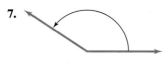

8.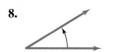

9.

10.

In Exercises 11–18, draw each angle in standard position.

11. 135°
12. 120°
13. −150°
14. −240°
15. 420°
16. 450°
17. −90°
18. −270°

In Exercises 19–24, find a positive angle less than 360° that is coterminal with the given angle.

19. 395°
20. 415°
21. −150°
22. −160°
23. −45°
24. −40°

In Exercises 25–30, if possible, find the complement and the supplement of the given angle.

25. 52°
26. 85°
27. 37.4°
28. 47.6°
29. 111°
30. 95°

In Exercises 31–36, find the radian measure of the central angle of a circle of radius r that intercepts an arc of length s.

Radius, r	Arc length, s
31. 10 inches	40 inches
32. 5 feet	30 feet
33. 6 yards	8 yards
34. 8 yards	18 yards
35. 1 meter	400 centimeters
36. 1 meter	600 centimeters

In Exercises 37–44, convert each angle in degrees to radians. Express your answer as a multiple of π.

37. $45°$ **38.** $18°$

39. $135°$ **40.** $150°$

41. $300°$ **42.** $330°$

43. $-225°$ **44.** $-270°$

In Exercises 45–52, convert each angle in radians to degrees.

45. $\dfrac{\pi}{2}$ **46.** $\dfrac{\pi}{9}$

47. $\dfrac{2\pi}{3}$ **48.** $\dfrac{3\pi}{4}$

49. $\dfrac{7\pi}{6}$ **50.** $\dfrac{11\pi}{6}$

51. -3π **52.** -4π

In Exercises 53–58, convert each angle in degrees to radians. Round to two decimal places.

53. $18°$ **54.** $76°$

55. $-40°$ **56.** $-50°$

57. $200°$ **58.** $250°$

In Exercises 59–64, convert each angle in radians to degrees. Round to two decimal places.

59. 2 radians   **60.** 3 radians

61. $\dfrac{\pi}{13}$ radians   **62.** $\dfrac{\pi}{17}$ radians

63. -4.8 radians   **64.** -5.2 radians

In Exercises 65–68, find the length of the arc on a circle of radius r intercepted by a central angle θ. Express arc length in terms of π. Then round your answer to two decimal places.

Radius, r	Central angle, θ
65. 12 inches	$\theta = 45°$
66. 16 inches	$\theta = 60°$
67. 8 feet	$\theta = 225°$
68. 9 yards	$\theta = 315°$

In Exercises 69–70, express each angular speed in radians per second.

69. 6 revolutions per second

70. 20 revolutions per second

 Application Exercises

71. The minute hand of a clock moves from 12 to 2 o'clock, or $\frac{1}{6}$ of a complete revolution. Through how many degrees does it move? Through how many radians does it move?

72. The minute hand of a clock moves from 12 to 4 o'clock, or $\frac{1}{3}$ of a complete revolution. Through how many degrees does it move? Through how many radians does it move?

73. The minute hand of a clock is 8 inches long and moves from 12 to 2 o'clock. How far does the tip of the minute hand move? Express your answer in terms of π and then round to two decimal places.

74. The minute hand of a clock is 6 inches long and moves from 12 to 4 o'clock. How far does the tip of the minute hand move? Express your answer in terms of π and then round to two decimal places.

75. The figure shows a highway sign that warns of a railway crossing. The lines that form the cross pass through the circle's center and intersect at right angles. If the radius of the circle is 24 inches, find the length of each of the four arcs formed by the cross. Express your answer in terms of π and then round to two decimal places.

76. The radius of a wheel is 80 centimeters. If the wheel rotates through an angle of $60°$, how many centimeters does it move? Express your answer in terms of π and then round to two decimal places.

How do we measure the distance between two points, A and B, on Earth? We measure along a circle with a center, C, at the center of Earth. The radius of the circle is equal to the distance from C to the surface. Use the fact that Earth is a sphere of radius equal to approximately 4000 miles to solve Exercises 77–80.

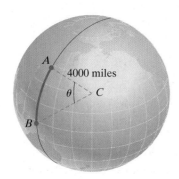

77. If two points, *A* and *B,* are 8000 miles apart, express angle *θ* in radians and in degrees.

78. If two points, *A* and *B,* are 10,000 miles apart, express angle *θ* in radians and in degrees.

79. If *θ* = 30°, find the distance between *A* and *B* to the nearest mile.

80. If *θ* = 10°, find the distance between *A* and *B* to the nearest mile.

81. The angular speed of a point on Earth is $\frac{\pi}{12}$ radians per hour. The Equator lies on a circle of radius approximately 4000 miles. Find the linear velocity, in miles per hour, of a point on the Equator.

82. A ferris wheel has a radius of 25 feet. The wheel is rotating at three revolutions per minute. Find the linear speed, in feet per minute, of this ferris wheel.

83. A water wheel has a radius of 12 feet. The wheel is rotating at 20 revolutions per minute. Find the linear speed, in feet per minute, of the water.

84. On a carousel, the outer row of animals is 20 feet from the center. The inner row of animals is 10 feet from the center. The carousel is rotating at 2.5 revolutions per minute. What is the difference, in feet per minute, in the linear speeds of the animals in the outer and inner rows? Round to the nearest foot per minute.

Writing in Mathematics

85. What is an angle?

86. What determines the size of an angle?

87. Describe an angle in standard position.

88. Explain the difference between positive and negative angles. What are coterminal angles?

89. Explain what is meant by one radian.

90. Explain how to find the radian measure of a central angle.

91. Describe how to convert an angle in degrees to radians.

92. Explain how to convert an angle in radians to degrees.

93. Explain how to find the length of a circular arc.

94. If a carousel is rotating at 2.5 revolutions per minute, explain how to find the linear speed of a child seated on one of the animals.

95. The angular velocity of a point on Earth is $\frac{\pi}{12}$ radians per hour. Describe what happens every 24 hours.

96. Have you ever noticed that we use the vocabulary of angles in everyday speech? Here is an example:

> My opinion about art museums took a 180° turn after visiting the San Francisco Museum of Modern Art.

Explain what this means. Then give another example of the vocabulary of angles in everyday use.

Technology Exercises

In Exercises 97–100, use the keys on your calculator or graphing utility for converting an angle in degrees, minutes, and seconds (D°M'S″) into decimal form, and vice versa.

In Exercises 97–98, convert each angle to a decimal in degrees. Round your answer to two decimal places.

97. 30°15′10″   **98.** 65°45′20″

In Exercises 99–100, convert each angle to D°M'S″ form. Round your answer to the nearest second.

99. 30.42°   **100.** 50.42°

Critical Thinking Exercises

101. If $\theta = \frac{3}{2}$, is this angle larger or smaller than a right angle?

102. A railroad curve is laid out on a circle. What radius should be used if the track is to change direction by 20° in a distance of 100 miles? Round your answer to the nearest mile.

103. Assuming Earth to be a sphere of radius 4000 miles, how many miles north of the Equator is Miami, Florida, if it is 26° north from the Equator? Round your answer to the nearest mile.

SECTION 4.2   *Trigonometric Functions: The Unit Circle*

Objectives

1. Use a unit circle to define trigonometric functions of real numbers.
2. Use a unit circle to find values of the trogonometric functions.
3. Recognize the domain and range of sine and cosine functions.
4. Find exact values of the trigonometric functions at $\frac{\pi}{4}$.
5. Use even and odd trigonometric functions.
6. Recognize and use fundamental identities.
7. Use periodic properties.
8. Evaluate trigonometric functions with a calculator.

There is something comforting in the repetition of some of nature's patterns. The ocean level at a beach varies between high and low tide approximately every 12 hours. The number of hours of daylight oscillates from a maximum on the summer solstice, June 21, to a minimum on the winter solstice, December 21. Then it increases to the same maximum the following June 21. Some believe that cycles, called biorhythms, represent physical, emotional, and intellectual aspects of our lives. In this chapter, we study six functions, the six *trigonometric functions*, that are used to model phenomena that occur again and again.

Calculus and the Unit Circle

The word *trigonometry* means measurement of triangles. Trigonometric functions, with domains consisting of sets of angles, were first defined using right triangles. By contrast, problems in calculus are solved using functions whose domains are sets of real numbers. Therefore, we introduce the trigonometric functions using unit circles and radians, rather than right triangles and degrees.

A **unit circle** is a circle of radius 1, with its center at the origin of a rectangular coordinate system. The equation of this unit circle is $x^2 + y^2 = 1$. Figure 4.14 shows a unit circle in which the central angle measures t radians. We can use the formula for the length of a circular arc, $s = r\theta$, to find the length of the intercepted arc.

$$s = r\theta = 1t = t$$

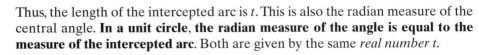

The radius of a unit circle is 1.   The radian measure of the central angle is t.

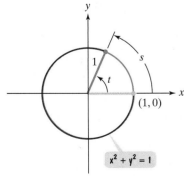

Figure 4.14 A unit circle with a central angle measuring t radians

Thus, the length of the intercepted arc is t. This is also the radian measure of the central angle. **In a unit circle, the radian measure of the angle is equal to the measure of the intercepted arc.** Both are given by the same *real number t*.

In Figure 4.15, the radian measure of the angle and the length of the intercepted arc are both shown by t. Let $P = (x, y)$ denote the point on the unit circle that has arc length t from $(1, 0)$. Figure 4.15(a) shows that if t is positive, point P is reached by moving counterclockwise along the unit circle from $(1, 0)$. Figure 4.15(b) shows that if t is negative, point P is reached by moving clockwise along the unit circle from $(1, 0)$. For each real number t, there corresponds a point $P = (x, y)$ on the unit circle.

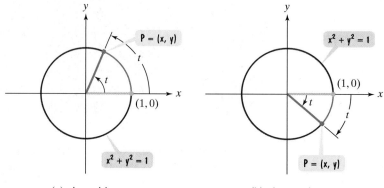

Figure 4.15   (a) t is positive.   (b) t is negative.

1 Use a unit circle to define trigonometric functions of real numbers.

The Six Trigonometric Functions

We begin the study of trigonometry by defining the six trigonometric functions. The inputs of these functions are real numbers, represented by t in Figure 4.15. The outputs involve point $P = (x, y)$ on the unit circle that corresponds to t and the coordinates of this point.

The trigonometric functions have names that are words, rather than single letters such as f, g, and h. For example, the **sine of t** is the y-coordinate of point P on the unit circle:

$$\sin t = y.$$

Input is the real number t.

Output is the y-coordinate of a point on the unit circle.

The value of y depends on the real number t and thus is a function of t. The expression $\sin t$ really means $\sin(t)$, where sine is the name of the function and t, a real number, is an input.

For example, a point $P = (x, y)$ on the unit circle corresponding to a real number t is shown in Figure 4.16, for $\pi < t < \dfrac{3\pi}{2}$. We see that the coordinates of $P = (x, y)$ are $x = -\dfrac{3}{5}$ and $y = -\dfrac{4}{5}$. Because the sine function is the y-coordinate of P, the value of this trigonometric function at the real number t is

$$\sin t = -\frac{4}{5}.$$

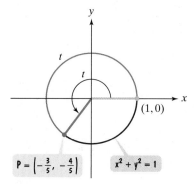

Figure 4.16

Here are the names of the six trigonometric functions, along with their abbreviations.

Name	Abbreviation	Name	Abbreviation
sine	sin	cosecant	csc
cosine	cos	secant	sec
tangent	tan	cotangent	cot

Definitions of the Trigonometric Functions in Terms of a Unit Circle

If t is a real number and $P = (x, y)$ is a point on the unit circle that corresponds to t, then

$$\sin t = y \qquad\qquad \cos t = x \qquad\qquad \tan t = \frac{y}{x}, x \neq 0$$

$$\csc t = \frac{1}{y}, y \neq 0 \qquad \sec t = \frac{1}{x}, x \neq 0 \qquad \cot t = \frac{x}{y}, y \neq 0.$$

Because this definition expresses function values in terms of coordinates of a point on a unit circle, the trigonometric functions are sometimes called the **circular functions**. Observe that the functions in the second row in the box are the reciprocals of the corresponding functions in the first row.

2 Use a unit circle to find values of the trigonometric functions.

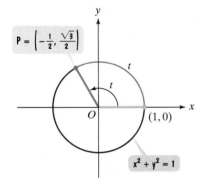

Figure 4.17

EXAMPLE 1 Finding Values of the Trigonometric Functions

In Figure 4.17, t is a real number and $P = \left(-\frac{1}{2}, \frac{\sqrt{3}}{2}\right)$ is a point on the unit circle that corresponds to t. Use the figure to find the values of the trigonometric functions at t.

Solution The point P on the unit circle that corresponds to t has coordinates $\left(-\frac{1}{2}, \frac{\sqrt{3}}{2}\right)$. We use $x = -\frac{1}{2}$ and $y = \frac{\sqrt{3}}{2}$ to find the values of the trigonometric functions.

$$\sin t = y = \frac{\sqrt{3}}{2} \qquad \cos t = x = -\frac{1}{2} \qquad \tan t = \frac{y}{x} = \frac{\dfrac{\sqrt{3}}{2}}{-\dfrac{1}{2}} = -\sqrt{3}$$

$$\csc t = \frac{1}{y} = \frac{1}{\dfrac{\sqrt{3}}{2}} = \frac{2}{\sqrt{3}} \qquad \sec t = \frac{1}{x} = \frac{1}{-\dfrac{1}{2}} = -2 \qquad \cot t = \frac{x}{y} = \frac{-\dfrac{1}{2}}{\dfrac{\sqrt{3}}{2}} = -\frac{1}{\sqrt{3}}$$

You can rationalize the denominator:
$$\frac{2}{\sqrt{3}} \cdot \frac{\sqrt{3}}{\sqrt{3}} = \frac{2\sqrt{3}}{3}.$$

You can rationalize the denominator:
$$-\frac{1}{\sqrt{3}} \cdot \frac{\sqrt{3}}{\sqrt{3}} = -\frac{\sqrt{3}}{3}.$$

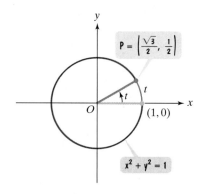

Check Point 1 Use the figure on the right to find the values of the trigonometric functions at t.

$P = \left(\frac{\sqrt{3}}{2}, \frac{1}{2}\right)$

$(1, 0)$

$x^2 + y^2 = 1$

EXAMPLE 2 Finding Values of the Trigonometric Functions

Use Figure 4.18 to find the values of the trigonometric functions at $t = \dfrac{\pi}{2}$.

Solution The point P on the unit circle that corresponds to $t = \dfrac{\pi}{2}$ has coordinates $(0, 1)$. We use $x = 0$ and $y = 1$ to find the values of the trigonometric functions.

$$\sin \frac{\pi}{2} = y = 1 \qquad\qquad \cos \frac{\pi}{2} = x = 0$$

$$\csc \frac{\pi}{2} = \frac{1}{y} = \frac{1}{1} = 1 \qquad \cot \frac{\pi}{2} = \frac{x}{y} = \frac{0}{1} = 0$$

By definition, $\tan t = \dfrac{y}{x}$ and $\sec t = \dfrac{1}{x}$. Because $x = 0$, $\tan \dfrac{\pi}{2}$ and $\sec \dfrac{\pi}{2}$ are undefined.

$P = (0, 1)$

$\dfrac{\pi}{2}$

$\dfrac{\pi}{2}$

$(1, 0)$

$x^2 + y^2 = 1$

Figure 4.18

Check Point 2 Use the figure on the right to find the values of the trigonometric functions at $t = \pi$.

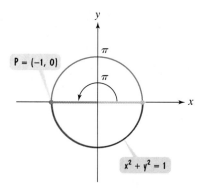

$P = (-1, 0)$

π

π

$x^2 + y^2 = 1$

3 Recognize the domain and range of sine and cosine functions

Domain and Range of Sine and Cosine Functions

The domain and range of each trigonometric function can be found from the unit circle definition. At this point, let's look only at the sine and cosine functions,

$$\sin t = y \quad \text{and} \quad \cos t = x.$$

Because t can be the radian measure of any angle, or equivalently, the measure of any intercepted arc, t can be any real number. Thus, the domain of the sine function and the cosine function is the set of all real numbers. Because the radius of the unit circle is 1, we have

$$-1 \le x \le 1 \quad \text{and} \quad -1 \le y \le 1.$$

Therefore, with $x = \cos t$ and $y = \sin t$, we obtain

$$-1 \le \cos t \le 1 \quad \text{and} \quad -1 \le \sin t \le 1.$$

The range of the cosine and sine functions is $[-1, 1]$.

The Domain and Range of the Sine and Cosine Functions

The domain of the sine function and the cosine function is the set of all real numbers. The range of these functions is the set of all real numbers from -1 to 1, inclusive.

---

4 Find exact values of the trigonometric functions at $\frac{\pi}{4}$.

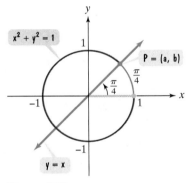

Figure 4.19

Exact Values of Trigonometric Functions at $t = \dfrac{\pi}{4}$

Trigonometric functions at $t = \dfrac{\pi}{4}$ occur frequently. How do we use the unit circle to find values of the trigonometric functions at $t = \dfrac{\pi}{4}$? Look at Figure 4.19. We must find the coordinates of point $P = (a, b)$ on the unit circle that correspond to $t = \dfrac{\pi}{4}$. Can you see that P lies on the line $y = x$? Thus, point P has equal x- and y-coordinates: $a = b$. We find these coordinates as follows:

$x^2 + y^2 = 1$	This is the equation of the unit circle.
$a^2 + b^2 = 1$	Point $P = (a, b)$ lies on the unit circle. Thus, its coordinates satisfy the circle's equation.
$a^2 + a^2 = 1$	Because $a = b$, substitute a^2 for b^2 in the equation.
$2a^2 = 1$	Add like terms.
$a^2 = \frac{1}{2}$	Divide both sides of the equation by 2.
$a = \sqrt{\frac{1}{2}}$	Because $a > 0$, take the positive square root of both sides.

We see that $a = \dfrac{1}{\sqrt{2}}$. Because $a = b$, we also have $b = \dfrac{1}{\sqrt{2}}$. Thus, if $t = \dfrac{\pi}{4}$, point

$$P = \left(\frac{1}{\sqrt{2}}, \frac{1}{\sqrt{2}} \right)$$ is the point on the unit circle that corresponds to t. We use these

coordinates to find the values of the trigonometric functions at $t = \dfrac{\pi}{4}$.

EXAMPLE 3 **Finding Values of the Trigonometric Functions at** $t = \dfrac{\pi}{4}$

Find $\sin \dfrac{\pi}{4}$, $\cos \dfrac{\pi}{4}$, and $\tan \dfrac{\pi}{4}$.

Solution The point P on the unit circle that corresponds to $t = \dfrac{\pi}{4}$ has coordinates $\left(\dfrac{1}{\sqrt{2}}, \dfrac{1}{\sqrt{2}} \right)$. We use $x = \dfrac{1}{\sqrt{2}}$ and $y = \dfrac{1}{\sqrt{2}}$ to find the values the three trigonometric functions at $\dfrac{\pi}{4}$.

$$\sin \frac{\pi}{4} = y = \frac{1}{\sqrt{2}} \qquad \cos \frac{\pi}{4} = x = \frac{1}{\sqrt{2}} \qquad \tan \frac{\pi}{4} = \frac{y}{x} = \frac{\dfrac{1}{\sqrt{2}}}{\dfrac{1}{\sqrt{2}}} = 1$$

Check Point 3 Find $\csc \dfrac{\pi}{4}$, $\sec \dfrac{\pi}{4}$, and $\cot \dfrac{\pi}{4}$.

When you worked Check Point 3, did you use the x- and y-coordinates of $P = \left(\dfrac{1}{\sqrt{2}}, \dfrac{1}{\sqrt{2}} \right)$, or did you use reciprocals to find the values?

$$\csc \frac{\pi}{4} = \sqrt{2} \qquad \sec \frac{\pi}{4} = \sqrt{2} \qquad \cot \frac{\pi}{4} = 1$$

| Take the reciprocal of $\sin \frac{\pi}{4} = \frac{1}{\sqrt{2}}$. | Take the reciprocal of $\cos \frac{\pi}{4} = \frac{1}{\sqrt{2}}$. | Take the reciprocal of $\tan \frac{\pi}{4} = 1$. |

We found that $\sin \dfrac{\pi}{4} = \dfrac{1}{\sqrt{2}}$ and $\cos \dfrac{\pi}{4} = \dfrac{1}{\sqrt{2}}$. This value is often expressed by rationalizing the denominator:

$$\frac{1}{\sqrt{2}} = \frac{1}{\sqrt{2}} \cdot \frac{\sqrt{2}}{\sqrt{2}} = \frac{\sqrt{2}}{2}.$$

We are multiplying by 1 and not changing the value of $\dfrac{1}{\sqrt{2}}$.

Thus, $\sin \dfrac{\pi}{4} = \dfrac{\sqrt{2}}{2}$ and $\cos \dfrac{\pi}{4} = \dfrac{\sqrt{2}}{2}$.

Because you will often see the trigonometric functions at $\dfrac{\pi}{4}$, it is a good idea to memorize the values shown in the following box. In the next section, you will learn to use a right triangle to obtain these values.

Trigonometric Functions at $\dfrac{\pi}{4}$

$$\sin \frac{\pi}{4} = \frac{\sqrt{2}}{2} \qquad \cos \frac{\pi}{4} = \frac{\sqrt{2}}{2} \qquad \tan \frac{\pi}{4} = 1$$

$$\csc \frac{\pi}{4} = \sqrt{2} \qquad \sec \frac{\pi}{4} = \sqrt{2} \qquad \cot \frac{\pi}{4} = 1$$

5 Use even and odd trigonometric functions.

Even and Odd Trigonometric Functions

In Chapter 1, we saw that a function is even if $f(-t) = f(t)$ and odd if $f(-t) = -f(t)$. We can use Figure 4.20 to show that the cosine function is an even function and the sine function is an odd function. By definition, the coordinates of the points P and Q in Figure 4.20 are as follows:

$$P: (\cos t, \sin t)$$
$$Q: (\cos(-t), \sin(-t))$$

In Figure 4.20, the x-coordinates of P and Q are the same. Thus,

$$\cos(-t) = \cos t.$$

This shows that cosine function is an even function. By contrast, the y-coordinates of P and Q are negatives of each other. Thus,

$$\sin(-t) = -\sin t.$$

This shows that the sine function is an odd function.

This argument is valid regardless of the length of t. Thus, the arc may terminate in any of the four quadrants or on any axis. Using the unit circle definition of the trigonometric functions, we obtain the following results:

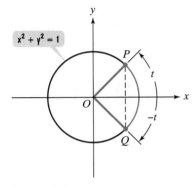

Figure 4.20

Even and Odd Trigonometric Functions

The cosine and secant functions are *even*.

$$\cos(-t) = \cos t \qquad \qquad \sec(-t) = \sec t$$

The sine, cosecant, tangent, and cotangent functions are *odd*.

$$\sin(-t) = -\sin t \qquad \qquad \csc(-t) = -\csc t$$
$$\tan(-t) = -\tan t \qquad \qquad \cot(-t) = -\cot t$$

EXAMPLE 4 **Using Even and Odd Functions to Find Values of the Trigonometric Functions**

Table 4.1 shows the values of the trigonometric functions at $t = \dfrac{\pi}{4}$. Use these values to find:

a. $\cos\left(-\dfrac{\pi}{4}\right)$ **b.** $\tan\left(-\dfrac{\pi}{4}\right)$.

Table 4.1

t	$\sin t$	$\cos t$	$\tan t$
$\dfrac{\pi}{4}$	$\dfrac{\sqrt{2}}{2}$	$\dfrac{\sqrt{2}}{2}$	1

t	$\csc t$	$\sec t$	$\cot t$
$\dfrac{\pi}{4}$	$\sqrt{2}$	$\sqrt{2}$	1

Solution

a. $\cos\left(-\dfrac{\pi}{4}\right) = \cos \dfrac{\pi}{4} = \dfrac{\sqrt{2}}{2}$ **b.** $\tan\left(-\dfrac{\pi}{4}\right) = -\tan \dfrac{\pi}{4} = -1$

Check Point 4 Use the values in Table 4.1 to find:

a. $\sec\left(-\dfrac{\pi}{4}\right)$ **b.** $\sin\left(-\dfrac{\pi}{4}\right)$.

6 Recognize and use fundamental identities.

Fundamental Identities

Many relationships exist among the six trigonometric functions. These relationships are described using **trigonometric identities**. Trigonometric identities are equations that are true for all real numbers for which the trigonometric expressions in the equations are defined. For example, the definitions of the cosine and secant functions are given by

$$\cos t = x \quad \text{and} \quad \sec t = \frac{1}{x}, x \neq 0.$$

Substituting $\cos t$ for x in the equation on the right, we see that

$$\sec t = \frac{1}{\cos t}, \quad \cos t \neq 0.$$

This identity is one of six **reciprocal identities**.

Reciprocal Identities

$$\sin t = \frac{1}{\csc t} \qquad \cos t = \frac{1}{\sec t} \qquad \tan t = \frac{1}{\cot t}$$

$$\csc t = \frac{1}{\sin t} \qquad \sec t = \frac{1}{\cos t} \qquad \cot t = \frac{1}{\tan t}$$

Two other relationships that follow from the definitions of the trigonometric functions are called the **quotient identities**.

Quotient Identities

$$\tan t = \frac{\sin t}{\cos t} \qquad \cot t = \frac{\cos t}{\sin t}$$

If $\sin t$ and $\cos t$ are known, a quotient identity and three reciprocal identities make it possible to find the value of each of the four remaining trigonometric functions.

EXAMPLE 5   Using Quotient and Reciprocal Identities

Given $\sin t = \dfrac{1}{2}$ and $\cos t = \dfrac{\sqrt{3}}{2}$, find the value of each of the four remaining trigonometric functions.

Solution   We can find $\tan t$ by using the quotient identity that describes $\tan t$ as the quotient of $\sin t$ and $\cos t$.

$$\tan t = \frac{\sin t}{\cos t} = \frac{\dfrac{1}{2}}{\dfrac{\sqrt{3}}{2}} = \frac{1}{2} \cdot \frac{2}{\sqrt{3}} = \frac{1}{\sqrt{3}} = \frac{1}{\sqrt{3}} \cdot \frac{\sqrt{3}}{\sqrt{3}} = \frac{\sqrt{3}}{3}$$

Rationalize the denominator.

We use the reciprocal identities to find the value of each of the remaining three functions.

$$\csc t = \frac{1}{\sin t} = \frac{1}{\dfrac{1}{2}} = 2$$

$$\sec t = \frac{1}{\cos t} = \frac{1}{\dfrac{\sqrt{3}}{2}} = \frac{2}{\sqrt{3}} = \frac{2}{\sqrt{3}} \cdot \frac{\sqrt{3}}{\sqrt{3}} = \frac{2\sqrt{3}}{3}$$

Rationalize the denominator.

$$\cot t = \frac{1}{\tan t} = \frac{1}{\dfrac{1}{\sqrt{3}}} = \sqrt{3} \qquad \text{We found } \tan t = \frac{1}{\sqrt{3}}. \text{ We could use } \tan t = \frac{\sqrt{3}}{3},$$
$$\text{but then we would have to rationalize the denominator.}$$

Check Point 5   Given $\sin t = \dfrac{2}{3}$ and $\cos t = \dfrac{\sqrt{5}}{3}$, find the value of each of the four remaining trigonometric functions.

Other relationships among trigonometric functions follow from the equation of the unit circle

$$x^2 + y^2 = 1.$$

Because $\cos t = x$ and $\sin t = y$, we see that

$$(\cos t)^2 + (\sin t)^2 = 1.$$

We will eliminate the parentheses in this identity by writing $\cos^2 t$ instead of $(\cos t)^2$ and $\sin^2 t$ instead of $(\sin t)^2$. With this notation, we can write the identity as

$$\cos^2 t + \sin^2 t = 1$$

or

$$\sin^2 t + \cos^2 t = 1. \qquad \text{The identity usually appears in this form.}$$

Two additional identities can be obtained from $x^2 + y^2 = 1$ by dividing both sides by x^2 and y^2, respectively. The three identities are called the **Pythagorean identities**.

Pythagorean Identities

$$\sin^2 t + \cos^2 t = 1 \qquad 1 + \tan^2 t = \sec^2 t \qquad 1 + \cot^2 t = \csc^2 t$$

EXAMPLE 6   Using a Pythagorean Identity

Given that $\sin t = \dfrac{3}{5}$ and $0 \le t < \dfrac{\pi}{2}$, find the value of $\cos t$ using a trigonometric identity.

Solution   We can find the value of $\cos t$ by using the Pythagorean identity

$$\sin^2 t + \cos^2 t = 1.$$

$$\left(\frac{3}{5}\right)^2 + \cos^2 t = 1 \qquad \text{We are given that } \sin t = \frac{3}{5}.$$

$$\frac{9}{25} + \cos^2 t = 1 \qquad \text{Square } \frac{3}{5}: \left(\frac{3}{5}\right)^2 = \frac{3^2}{5^2} = \frac{9}{25}.$$

$$\cos^2 t = 1 - \frac{9}{25} \qquad \text{Subtract } \frac{9}{25} \text{ from both sides.}$$

$$\cos^2 t = \frac{16}{25} \qquad \text{Simplify: } 1 - \frac{9}{25} = \frac{25}{25} - \frac{9}{25} = \frac{16}{25}.$$

$$\cos t = \sqrt{\frac{16}{25}} = \frac{4}{5} \qquad \text{Because } 0 \le t < \frac{\pi}{2}, \cos t \text{ is positive.}$$

Thus, $\cos t = \dfrac{4}{5}$.

Check Point 6   Given that $\sin t = \dfrac{1}{2}$ and $0 \le t < \dfrac{\pi}{2}$, find the value of $\cos t$ using a trigonometric identity.

7 Use periodic properties.

Periodic Functions

Certain patterns in nature repeat again and again. For example, the ocean level at a beach varies from low tide to high tide and then back to low tide approximately every 12 hours. If low tide occurs at noon, then high tide will be around 6 P.M. and low tide will occur again around midnight, and so on infinitely. If $f(t)$ represents the ocean level at the beach at any time t, then the level is the same 12 hours later. Thus,

$$f(t + 12) = f(t).$$

The word *periodic* means that this tidal behavior repeats infinitely. The *period*, 12 hours, is the time it takes to complete one full cycle.

Definition of a Periodic Function

A function f is **periodic** if there exists a positive number p such that

$$f(t + p) = f(t)$$

for all t in the domain of f. The smallest number p for which f is periodic is called the **period** of f.

The trigonometric functions are used to model periodic phenomena. Why? If we begin at any point P on the unit circle and travel a distance of 2π units along the perimeter, we will return to the same point P. Because the trigonometric functions are defined in terms of the coordinates of that point P, we obtain the following results:

Periodic Properties of the Sine and Cosine Functions

$$\sin(t + 2\pi) = \sin t \quad \text{and} \quad \cos(t + 2\pi) = \cos t$$

The sine and cosine functions are periodic functions and have period 2π.

EXAMPLE 7   Using Periodic Properties to Find Values of the Trigonometric Functions

Table 4.1, repeated

t	$\sin t$	$\cos t$	$\tan t$
$\dfrac{\pi}{4}$	$\dfrac{\sqrt{2}}{2}$	$\dfrac{\sqrt{2}}{2}$	1
t	$\csc t$	$\sec t$	$\cot t$
$\dfrac{\pi}{4}$	$\sqrt{2}$	$\sqrt{2}$	1

The values of the trigonometric functions at $t = \dfrac{\pi}{4}$ are repeated in Table 4.1. Use a value in the table to find $\sin \dfrac{9\pi}{4}$.

Solution

reverse

$$\sin \frac{9\pi}{4} = \sin\left(\frac{\pi}{4} + 2\pi\right) = \sin \frac{\pi}{4} = \frac{\sqrt{2}}{2}$$

$$\sin(t + 2\pi) = \sin t$$

Check Point 7 Use a value in Table 4.1 to find $\tan \dfrac{9\pi}{4}$.

Like the sine and cosine functions, the secant and cosecant functions have period 2π. However, the tangent and cotangent functions have a smaller period. If we begin at any point $P = (x, y)$ on the unit circle and travel a distance of π units along the perimeter, we arrive at the point $(-x, -y)$. The tangent function, defined in terms of the coordinates of a point, is the same at (x, y) and $(-x, -y)$.

Tangent function at (x, y) $\dfrac{y}{x} = \dfrac{-y}{-x}$ Tangent function π radians later

We see that $\tan(t + \pi) = \tan t$. The same observations apply to the cotangent function.

Periodic Properties of the Tangent and Cotangent Functions

$$\tan(t + \pi) = \tan t \quad \text{and} \quad \cot(t + \pi) = \cot t$$

The tangent and cotangent functions are periodic functions and have period π.

Why do the trigonometric functions model phenomena that repeat *infinitely*? By starting at point P on the unit circle and traveling a distance of 2π units, 4π units, 6π units, and so on, we return to the starting point P. Because the trigonometric functions are defined in terms of the coordinates of that point P, if we add (or subtract) multiples of 2π, the trigonometric values do not change. Furthermore, the trigonometric values for the tangent and cotangent functions do not change if we add (or subtract) multiples of π.

Repetitive Behavior of the Sine, Cosine, and Tangent Functions

For any integer n and real number t,

$$\sin(t + 2\pi n) = \sin t, \quad \cos(t + 2\pi n) = \cos t, \quad \text{and} \quad \tan(t + \pi n) = \tan t.$$

---

8 Evaluate trigonometric functions with a calculator.

Using a Calculator to Evaluate Trigonometric Functions

We used a unit circle to find values of the trigonometric functions at $\frac{\pi}{4}$. These are exact values. We can find approximate values of the trigonometric functions using a calculator.

The first step in using a calculator to evaluate trigonometric functions is to set the calculator to the correct *mode*, degrees or radians. The domains of the trigonometric functions in the unit circle are sets of real numbers. Therefore, we use the radian mode.

Most calculators have keys marked $\boxed{\text{SIN}}$, $\boxed{\text{COS}}$, and $\boxed{\text{TAN}}$. For example, to find the value of sin 1.2, set the calculator to the radian mode and enter 1.2 $\boxed{\text{SIN}}$ on most scientific calculators and $\boxed{\text{SIN}}$ 1.2 $\boxed{\text{ENTER}}$ on most graphing calculators. Consult the manual for your calculator.

To evaluate the cosecant, secant, and cotangent functions, use the key for the respective reciprocal function, $\boxed{\text{SIN}}$, $\boxed{\text{COS}}$, or $\boxed{\text{TAN}}$, and then use the reciprocal key. The reciprocal key is $\boxed{^1/_x}$ on many scientific calculators and $\boxed{x^{-1}}$ on many graphing calculators. For example, we can evaluate $\sec\frac{\pi}{12}$ using the following reciprocal relationship:

$$\sec\frac{\pi}{12} = \frac{1}{\cos\dfrac{\pi}{12}}.$$

Using the radian mode, enter one of the following keystroke sequences:

Many Scientific Calculators

$\boxed{\pi} \boxed{\div} 12 \boxed{=} \boxed{\text{COS}} \boxed{^1/_x}$

Many Graphing Calculators

$\boxed{(} \boxed{\text{COS}} \boxed{(} \boxed{\pi} \boxed{\div} 12 \boxed{)} \boxed{)} \boxed{x^{-1}} \boxed{\text{ENTER}}$.

Rounding the display to four decimal places, we obtain $\sec \dfrac{\pi}{12} = 1.0353$.

EXAMPLE 8 **Evaluating Trigonometric Functions with a Calculator**

Use a calculator to find the value to four decimal places of:

    a. $\cos \dfrac{\pi}{4}$ **b.** $\cot 1.2$.

Solution

Scientific Calculator Solution

Function	Mode	Keystrokes	Display, rounded to four decimal places
a. $\cos \dfrac{\pi}{4}$	Radian	$\boxed{\pi} \boxed{\div} \boxed{4} \boxed{=} \boxed{\text{COS}}$	0.7071
b. $\cot 1.2$	Radian	$1.2 \boxed{\text{TAN}} \boxed{^1/_x}$	0.3888

Graphing Calculator Solution

Function	Mode	Keystrokes	Display, rounded to four decimal places
a. $\cos \dfrac{\pi}{4}$	Radian	$\boxed{\text{COS}} \boxed{(} \boxed{\pi} \boxed{\div} \boxed{4} \boxed{)} \boxed{\text{ENTER}}$	0.7071
b. $\cot 1.2$	Radian	$\boxed{(} \boxed{\text{TAN}} 1.2 \boxed{)} \boxed{x^{-1}} \boxed{\text{ENTER}}$	0.3888

Check Point 8   Use a calculator to find the value to four decimal places of:

    a. $\sin \dfrac{\pi}{4}$ **b.** $\csc 1.5$.

EXERCISE SET 4.2

Practice Exercises

In Exercises 1–4, a point $P(x, y)$ is shown on the unit circle corresponding to a real number t. Find the values of the trigonometric functions at t.

1.

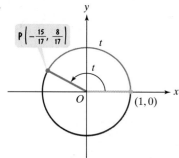

2.

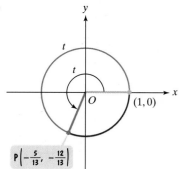

3.

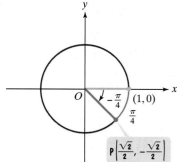

4.

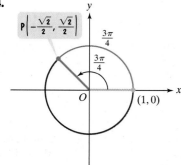

In Exercises 5–18, the unit circle has been divided into twelve equal arcs, corresponding to t-values of

$$0, \frac{\pi}{6}, \frac{\pi}{3}, \frac{\pi}{2}, \frac{2\pi}{3}, \frac{5\pi}{6}, \pi, \frac{7\pi}{6}, \frac{4\pi}{3}, \frac{3\pi}{2}, \frac{5\pi}{3}, \frac{11\pi}{6}, \text{ and } 2\pi.$$

Use the (x, y) coordinates in the figure to find the value of each trigonometric function at the indicated real number, t, or state that the expression is undefined.

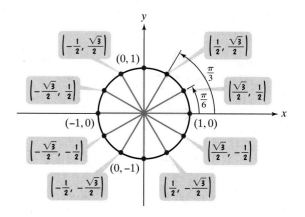

5. $\sin \dfrac{\pi}{6}$ **6.** $\sin \dfrac{\pi}{3}$

7. $\cos \dfrac{5\pi}{6}$ **8.** $\cos \dfrac{2\pi}{3}$

9. $\tan \pi$ **10.** $\tan 0$

11. $\csc \dfrac{7\pi}{6}$ **12.** $\csc \dfrac{4\pi}{3}$

13. $\sec \dfrac{11\pi}{6}$ **14.** $\sec \dfrac{5\pi}{3}$

15. $\sin \dfrac{3\pi}{2}$ **16.** $\cos \dfrac{3\pi}{2}$

17. $\sec \dfrac{3\pi}{2}$ **18.** $\tan \dfrac{3\pi}{2}$

In Exercises 19–24:

a. Use the unit circle shown for Exercises 5–18 to find the value of the trigonometric function.

b. Use even and odd trigonometric functions and your answer from part (a) to find the value of the same trigonometric function at the indicated real number.

19. a. $\cos \dfrac{\pi}{6}$ **b.** $\cos\left(-\dfrac{\pi}{6}\right)$

20. a. $\cos \dfrac{\pi}{3}$ **b.** $\cos\left(-\dfrac{\pi}{3}\right)$

21. a. $\sin \dfrac{5\pi}{6}$ **b.** $\sin\left(-\dfrac{5\pi}{6}\right)$

22. a. $\sin \dfrac{2\pi}{3}$ **b.** $\sin\left(-\dfrac{2\pi}{3}\right)$

23. a. $\tan \dfrac{5\pi}{3}$ **b.** $\tan\left(-\dfrac{5\pi}{3}\right)$

24. a. $\tan \dfrac{11\pi}{6}$ **b.** $\tan\left(-\dfrac{11\pi}{6}\right)$

In Exercises 25–28, $\sin t$ and $\cos t$ are given. Use identities to find $\tan t$, $\csc t$, $\sec t$, and $\cot t$. Where necessary, rationalize denominators.

25. $\sin t = \dfrac{8}{17},\ \cos t = \dfrac{15}{17}$ **26.** $\sin t = \dfrac{3}{5},\ \cos t = \dfrac{4}{5}$

27. $\sin t = \dfrac{1}{3},\ \cos t = \dfrac{2\sqrt{2}}{3}$ **28.** $\sin t = \dfrac{2}{3},\ \cos t = \dfrac{\sqrt{5}}{3}$

In Exercises 29–32, $0 \le t < \dfrac{\pi}{2}$ and $\sin t$ is given. Use the Pythagorean identity $\sin^2 t + \cos^2 t = 1$ to find $\cos t$.

29. $\sin t = \dfrac{6}{7}$ **30.** $\sin t = \dfrac{7}{8}$

31. $\sin t = \dfrac{\sqrt{39}}{8}$ **32.** $\sin t = \dfrac{\sqrt{21}}{5}$

In Exercises 33–38, use an identity to find the value of each expression. Do not use a calculator.

33. $\sin 1.7 \csc 1.7$ **34.** $\cos 2.3 \sec 2.3$

35. $\sin^2 \dfrac{\pi}{6} + \cos^2 \dfrac{\pi}{6}$ **36.** $\sin^2 \dfrac{\pi}{3} + \cos^2 \dfrac{\pi}{3}$

37. $\sec^2 \dfrac{\pi}{3} - \tan^2 \dfrac{\pi}{3}$ **38.** $\csc^2 \dfrac{\pi}{6} - \cot^2 \dfrac{\pi}{6}$

In Exercises 39–46, the unit circle has been divided into eight equal arcs, corresponding to t-values of

$$0,\ \frac{\pi}{4},\ \frac{\pi}{2},\ \frac{3\pi}{4},\ \pi,\ \frac{5\pi}{4},\ \frac{3\pi}{2},\ \frac{7\pi}{4},\ \text{and } 2\pi.$$

a. *Use the (x, y) coordinates in the figure to find the value of the trigonometric function.*

b. *Use periodic properties and your answer from part (a) to find the value of the same trigonometric function at the indicated real number.*

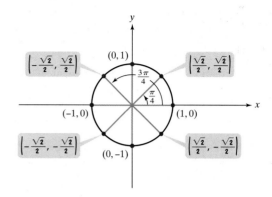

39. a. $\sin \dfrac{3\pi}{4}$ **b.** $\sin \dfrac{11\pi}{4}$

40. a. $\cos \dfrac{3\pi}{4}$ **b.** $\cos \dfrac{11\pi}{4}$

41. a. $\cos \dfrac{\pi}{2}$ **b.** $\cos \dfrac{9\pi}{2}$

42. a. $\sin \dfrac{\pi}{2}$ **b.** $\sin \dfrac{9\pi}{2}$

43. a. $\tan \pi$ **b.** $\tan 17\pi$

44. a. $\cot \dfrac{\pi}{2}$ **b.** $\cot \dfrac{15\pi}{2}$

45. a. $\sin \dfrac{7\pi}{4}$ **b.** $\sin \dfrac{47\pi}{4}$

46. a. $\cos \dfrac{7\pi}{4}$ **b.** $\cos \dfrac{47\pi}{4}$

In Exercises 47–56, use a calculator to find the value of the trigonometric function to four decimal places.

47. $\sin 0.8$ **48.** $\cos 0.6$

49. $\tan 3.4$ **50.** $\tan 3.7$

51. $\csc 1$ **52.** $\sec 1$

53. $\cos \dfrac{\pi}{10}$ **54.** $\sin \dfrac{3\pi}{10}$

55. $\cot \dfrac{\pi}{12}$ **56.** $\cot \dfrac{\pi}{18}$

Application Exercises

57. The number of hours of daylight, H, on day t of any given year (on January 1, $t = 1$) in Fairbanks, Alaska, can be modeled by the function

$$H(t) = 12 + 8.3 \sin\left[\frac{2\pi}{365}(t - 80)\right].$$

a. March 21, the 80th day of the year, is the spring equinox. Find the number of hours of daylight in Fairbanks on this day.

b. June 21, the 172nd day of the year, is the summer solstice, the day with the maximum number of hours of daylight. To the nearest tenth of an hour, find the number of hours of daylight in Fairbanks on this day.

c. December 21, the 355th day of the year, is the winter solstice, the day with the minimum number of hours of daylight. Find, to the nearest tenth of an hour, the number of hours of daylight in Fairbanks on this day.

58. The number of hours of daylight, H, on day t of any given year (on January 1, $t = 1$) in San Diego, California, can be modeled by the function

$$H(t) = 12 + 2.4 \sin \left[\frac{2\pi}{365} (t - 80) \right].$$

a. March 21, the 80th day of the year, is the spring equinox. Find the number of hours of daylight in San Diego on this day.

b. June 21, the 172nd day of the year, is the summer solstice, the day with the maximum number of hours of daylight. Find, to the nearest tenth of an hour, the number of hours of daylight in San Diego on this day.

c. December 21, the 355th day of the year, is the winter solstice, the day with the minimum number of hours of daylight. To the nearest tenth of an hour, find the number of hours of daylight in San Diego on this day.

59. People who believe in biorhythms claim that there are three cycles that rule our behavior—the physical, emotional, and mental. Each is a sine function of a certain period. The function for our emotional fluctuations is

$$E = \sin \frac{\pi}{14} t$$

where t is measured in days starting at birth. Emotional fluctuations, E, are measured from -1 to 1, inclusive, with 1 representing peak emotional well-being, -1 representing the low for emotional well-being, and 0 representing feeling neither emotionally high or low.

a. Find E corresponding to $t = 7, 14, 21, 28,$ and 35. Describe what you observe.

b. What is the period of the emotional cycle?

60. The height of the water, H in feet, at a boat dock t hours after 6 A.M. is given by

$$H(t) = 10 + 4 \sin \frac{\pi}{6} t.$$

a. Find the height of the water at the dock at 6 A.M., 9 A.M., noon, 6 P.M., midnight, and 3 A.M.

b. When is low tide and when is high tide?

c. What is the period of this function and what does this mean about the tides?

Writing in Mathematics

61. Why are the trigonometric functions sometimes called circular functions?

62. Define the sine of t.

63. Given a point on the unit circle that corresponds to t, explain how to find $\tan t$.

64. What is the range of the sine function? Use the unit circle to explain where this range comes from.

65. Explain how to use the unit circle to find values of the trigonometric functions at $\frac{\pi}{4}$.

66. What do we mean by even trigonometric functions? Which of the six functions fall into this category?

67. Use words (not an equation) to describe one of the reciprocal identities.

68. Use words (not an equation) to describe one of the quotient identities.

69. Use words (not an equation) to describe one of the Pythagorean identities.

70. What is a periodic function? Why are the sine and cosine functions periodic?

71. Explain how you can use the function for emotional fluctuations in Exercise 59 to determine good days for having dinner with your moody boss.

72. Describe a phenomenon that repeats again and again. What is its period?

Critical Thinking Exercises

73. If $f(x) = \sin x$ and $f(a) = \frac{1}{4}$, find the value of
$$f(a) + f(a + 2\pi) + f(a + 4\pi) + f(a + 6\pi).$$

74. If $f(x) = \sin x$ and $f(a) = \frac{1}{4}$, find the value of $f(a) + 2f(-a)$.

75. The seats of a ferris wheel are 40 feet from the wheel's center. When you get on the ride, your seat is 5 feet above the ground. How far above the ground are you after rotating through an angle of $\frac{17\pi}{4}$ radians?

SECTION 4.3 *Right Triangle Trigonometry*

Objectives

1. Use right triangles to evaluate trigonometric functions.

2. Find function values for $30° \left(\frac{\pi}{6} \right)$, $45° \left(\frac{\pi}{4} \right)$, and $60° \left(\frac{\pi}{3} \right)$.

3. Use equal cofunctions of complements.

4. Use right triangle trigonometry to solve applied problems.

In the last century, Ang Rita Sherpa climbed Mount Everest eight times, all without the use of bottled oxygen.

Mountain climbers have forever been fascinated by reaching the top of Mount Everest, sometimes with tragic results. The mountain, on Asia's Tibet-Nepal border, is Earth's highest, peaking at an incredible 29,035 feet. The heights of mountains can be found using trigonometric functions. Remember that the word *trigonometry* means *measurement of triangles*. Trigonometry is used in navigation, building, and engineering. For centuries, Muslims have used trigonometry and the stars to navigate across the Arabian desert to Mecca, the birthplace of the prophet Muhammad, the founder of Islam. The ancient Greeks used trigonometry to record the locations of thousands of stars and worked out the motion of the Moon relative to Earth. These applications involve looking at the trigonometric functions from the perspective of a right triangle.

1 Use right triangles to evaluate trigonometric functions.

Right Triangle Definitions of Trigonometric Functions

We have seen that in a unit circle, the radian measure of a central angle is equal to the measure of the intercepted arc. Thus, the value of a trigonometric function at the real number t is its value at an angle of t radians.

Figure 4.21(a) shows a central angle that measures $\frac{\pi}{3}$ radians and an intercepted arc of length $\frac{\pi}{3}$. Interpret $\frac{\pi}{3}$ as the measure of the central angle. In Figure 4.21(b) we construct a right triangle by dropping a perpendicular line

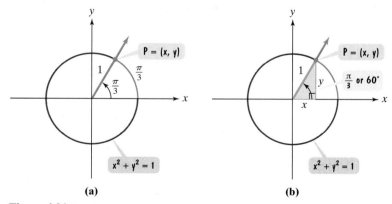

(a)   **(b)**

Figure 4.21 Interpreting trigonometric functions using a unit circle and a right triangle

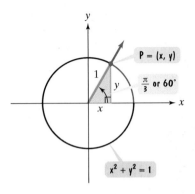

Figure 4.21(b), repeated

segment from point P to the x-axis. Now we can think of $\dfrac{\pi}{3}$, or $60°$, as the measure of an acute angle in this triangle. Because $\sin t$ is the second coordinate of point P and $\cos t$ is the first coordinate of point P, we see that

$$\sin \frac{\pi}{3} = \sin 60° = y = \frac{y}{1}$$

This is the length of the side opposite the 60° angle in the right triangle.

This is the length of the hypotenuse in the right triangle.

$$\cos \frac{\pi}{3} = \cos 60° = x = \frac{x}{1}$$

This is the length of the side adjacent to the 60° angle.

This is the length of the hypotenuse.

In solving certain kinds of problems, it is helpful to interpret trigonometric functions in right triangles where angles are limited to acute angles. Figure 4.22 shows a right triangle with one of its acute angles labeled θ. The side opposite the right angle, the hypotenuse, has length c. The other sides of the triangle are described by their position relative to the acute angle θ. One side is opposite θ. The length of this side is a. One side is adjacent to θ. The length of this side is b.

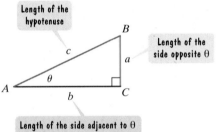

Length of the hypotenuse

Length of the side opposite θ

Length of the side adjacent to θ

Figure 4.22

Right Triangle Definitions of Trigonometric Functions

See Figure 4.22. The six **trigonometric functions of the acute angle θ** are defined as follows:

$$\sin \theta = \frac{\text{length of side opposite angle } \theta}{\text{length of hypotenuse}} = \frac{a}{c} \qquad \csc \theta = \frac{\text{length of hypotenuse}}{\text{length of side opposite angle } \theta} = \frac{c}{a}$$

$$\cos \theta = \frac{\text{length of side adjacent to angle } \theta}{\text{length of hypotenuse}} = \frac{b}{c} \qquad \sec \theta = \frac{\text{length of hypotenuse}}{\text{length of side adjacent angle } \theta} = \frac{c}{b}$$

$$\tan \theta = \frac{\text{length of side opposite angle } \theta}{\text{length of side adjacent to angle } \theta} = \frac{a}{b} \qquad \cot \theta = \frac{\text{length of side adjacent to angle } \theta}{\text{length of side opposite angle } \theta} = \frac{b}{a}$$

Each of the trigonometric functions of the acute angle θ is positive. Observe that the functions in the second column in the box are the reciprocals of the corresponding functions in the first column.

Figure 4.23 on the next page shows four right triangles of varying sizes. In each of the triangles, θ is the same acute angle, measuring approximately $56.3°$. All four of these similar triangles have the same shape and the lengths of corresponding sides are in the same ratio. In each triangle, the tangent function has the same value: $\tan \theta = \frac{3}{2}$.

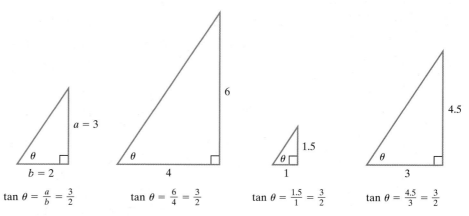

$$\tan \theta = \frac{a}{b} = \frac{3}{2} \qquad \tan \theta = \frac{6}{4} = \frac{3}{2} \qquad \tan \theta = \frac{1.5}{1} = \frac{3}{2} \qquad \tan \theta = \frac{4.5}{3} = \frac{3}{2}$$

Figure 4.23 A particular acute angle always gives the same ratio of opposite to adjacent sides.

In general, **the trigonometric function values of θ depend only on the size of angle θ, and not on the size of the triangle.**

EXAMPLE 1 Evaluating Trigonometric Functions

Find the value of each of the six trigonometric functions of θ in Figure 4.24.

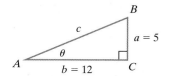

Figure 4.24

Solution We need to find the values of the six trigonometric functions of θ. However, we must know the lengths of all three sides of the triangle (a, b, and c) to evaluate all six functions. The values of a and b are given. We can use the Pythagorean Theorem, $c^2 = a^2 + b^2$, to find c.

$$\boxed{a = 5} \quad \boxed{b = 12}$$

$$c^2 = a^2 + b^2 = 5^2 + 12^2 = 25 + 144 = 169$$

$$c = \sqrt{169} = 13$$

Now that we know the lengths of the three sides of the triangle, we apply the definitions of the six trigonometric functions of θ. Referring to these lengths as opposite, adjacent, and hypotenuse, we have

$$\sin \theta = \frac{\text{opposite}}{\text{hypotenuse}} = \frac{5}{13} \qquad \csc \theta = \frac{\text{hypotenuse}}{\text{opposite}} = \frac{13}{5}$$

$$\cos \theta = \frac{\text{adjacent}}{\text{hypotenuse}} = \frac{12}{13} \qquad \sec \theta = \frac{\text{hypotenuse}}{\text{adjacent}} = \frac{13}{12}$$

$$\tan \theta = \frac{\text{opposite}}{\text{adjacent}} = \frac{5}{12} \qquad \cot \theta = \frac{\text{adjacent}}{\text{opposite}} = \frac{12}{5}.$$

Study Tip

The functions in the second column are reciprocals of those in the first column. You can obtain their values by exchanging the numerator and denominator of the corresponding ratios in the first column.

Check Point 1 Find the value of each of the six trigonometric functions of θ in the figure.

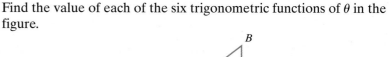
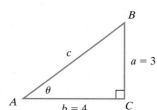

2 Find function values for $30° \left(\dfrac{\pi}{6}\right)$, $45° \left(\dfrac{\pi}{4}\right)$, and $60° \left(\dfrac{\pi}{3}\right)$.

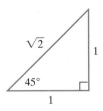

Figure 4.25 An isosceles right triangle

Function Values for Some Special Angles

In Section 4.2, we used the unit circle to find values of the trigonometric functions at $\dfrac{\pi}{4}$. How can we find the values of the trigonometric functions at $\dfrac{\pi}{4}$, or $45°$, using a right triangle? We construct a right triangle with a $45°$ angle, shown in Figure 4.25. The triangle actually has two $45°$ angles. Thus, the triangle is isosceles—that is, it has two sides of the same length. Assume that each leg of the triangle has a length equal to 1. We can find the length of the hypotenuse using the Pythagorean Theorem.

$$(\text{length of hypotenuse})^2 = 1^2 + 1^2 = 2$$
$$\text{length of hypotenuse} = \sqrt{2}$$

With Figure 4.25, we can determine the trigonometric function values for $45°$.

EXAMPLE 2 **Evaluating Trigonometric Functions of 45°**

Use Figure 4.25 to find $\sin 45°$, $\cos 45°$, and $\tan 45°$.

Solution We apply the definitions of these three trigonometric functions.

$$\sin 45° = \frac{\text{length of side opposite } 45°}{\text{length of hypotenuse}} = \frac{1}{\sqrt{2}} \quad \text{or} \quad \frac{\sqrt{2}}{2}$$

$$\cos 45° = \frac{\text{length of side adjacent to } 45°}{\text{length of hypotenuse}} = \frac{1}{\sqrt{2}} \quad \text{or} \quad \frac{\sqrt{2}}{2}$$

$$\tan 45° = \frac{\text{length of side opposite } 45°}{\text{length of side adjacent to } 45°} = \frac{1}{1} = 1$$

Check Point 2 Use Figure 4.25 to find $\csc 45°$, $\sec 45°$, and $\cot 45°$.

Two other angles that occur frequently in trigonometry are $30°$, or $\dfrac{\pi}{6}$ radian, and $60°$, or $\dfrac{\pi}{3}$ radian, angles. We can find the values of the trigonometric functions of $30°$ and $60°$ by using a right triangle. To form this right triangle, draw an equilateral triangle—that is a triangle with all sides the same length. Assume that each side has a length equal to 2. Now take half of the equilateral triangle. We obtain the solid blue triangle in Figure 4.26. This right triangle has a hypotenuse of length 2 and a leg of length 1. The other leg has length a, which can be found using the Pythagorean Theorem.

$$a^2 + 1^2 = 2^2$$
$$a^2 + 1 = 4$$
$$a^2 = 3$$
$$a = \sqrt{3}$$

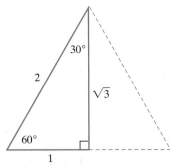

Figure 4.26 A $30°$-$60°$-$90°$ triangle

With the blue right triangle in Figure 4.26, we can determine the trigonometric functions for $30°$ and $60°$.

EXAMPLE 3 **Evaluating Trigonometric Functions of 30° and 60°**

Use Figure 4.26 to find $\sin 60°$, $\cos 60°$, $\sin 30°$, and $\cos 30°$.

Solution We begin with 60°. Use the angle on the lower left in Figure 4.26.

$$\sin 60° = \frac{\text{length of side opposite } 60°}{\text{length of hypotenuse}} = \frac{\sqrt{3}}{2}$$

$$\cos 60° = \frac{\text{length of side adjacent to } 60°}{\text{length of hypotenuse}} = \frac{1}{2}$$

To find $\sin 30°$ and $\cos 30°$, use the angle on the upper right in Figure 4.26.

$$\sin 30° = \frac{\text{length of side opposite } 30°}{\text{length of hypotenuse}} = \frac{1}{2}$$

$$\cos 30° = \frac{\text{length of side adjacent to } 30°}{\text{length of hypotenuse}} = \frac{\sqrt{3}}{2}$$

Check Point 3 Use Figure 4.26 to find $\tan 60°$ and $\tan 30°$. If necessary, express the value without a square root in the denominator by rationalizing the denominator.

Because we will often use the function values of 30°, 45°, and 60°, you should learn to construct the right triangles shown in Figures 4.25 and 4.26. With sufficient practice, you will memorize the values shown in the following box:

Sines, Cosines, and Tangents of Special Angles

$$\sin 30° = \sin \frac{\pi}{6} = \frac{1}{2} \qquad \cos 30° = \cos \frac{\pi}{6} = \frac{\sqrt{3}}{2} \qquad \tan 30° = \tan \frac{\pi}{6} = \frac{\sqrt{3}}{3}$$

$$\sin 45° = \sin \frac{\pi}{4} = \frac{\sqrt{2}}{2} \qquad \cos 45° = \cos \frac{\pi}{4} = \frac{\sqrt{2}}{2} \qquad \tan 45° = \tan \frac{\pi}{4} = 1$$

$$\sin 60° = \sin \frac{\pi}{3} = \frac{\sqrt{3}}{2} \qquad \cos 60° = \cos \frac{\pi}{3} = \frac{1}{2} \qquad \tan 60° = \tan \frac{\pi}{3} = \sqrt{3}$$

---

③ Use equal cofunctions of complements.

Trigonometric Functions and Complements

In Section 4.2, we used the unit circle to establish fundamental trigonometric identities. Another relationship among trigonometric functions is based on angles that are complements. We use a right triangle to establish this relationship. Refer to Figure 4.27. Because the sum of the angles of any triangle is 180°, in a right triangle the sum of the acute angles is 90°. Thus, the acute angles are complements. If the degree measure of one acute angle is θ, then the degree measure of the other acute angle is $(90° - \theta)$. This angle is shown on the upper right in Figure 4.27.

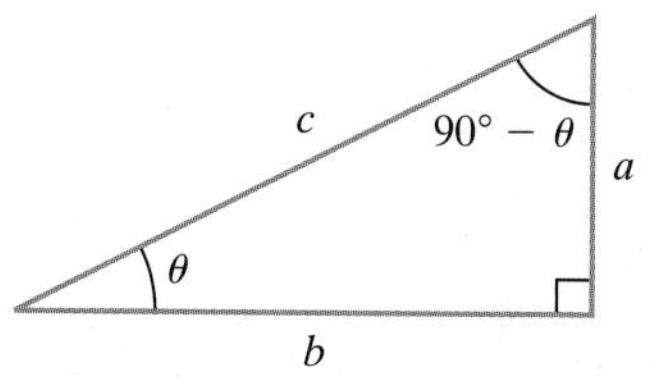

Let's use Figure 4.27 to compare $\sin \theta$ and $\cos (90° - \theta)$.

$$\sin \theta = \frac{\text{length of side opposite } \theta}{\text{length of hypotenuse}} = \frac{a}{c}$$

$$\cos (90° - \theta) = \frac{\text{length of side adjacent to } (90° - \theta)}{\text{length of hypotenuse}} = \frac{a}{c}$$

Figure 4.27

Thus, $\sin \theta = \cos (90° - \theta)$. If two angles are complements, the sine of one equals the cosine of the other. Because of this relationship, the sine and cosine are called *cofunctions* of each other. The name *cosine* is a shortened form of the phrase *complement's sine*.

Any pair of trigonometric functions f and g for which

$$f(\theta) = g(90° - \theta) \quad \text{and} \quad g(\theta) = f(90° - \theta)$$

are called **cofunctions**. Using Figure 4.27, we can show that the tangent and cotangent are cofunctions of each other. So are the secant and cosecant.

Cofunction Identities

The value of a trigonometric function of θ is equal to the cofunction of the complement of θ.

$$\sin \theta = \cos (90° - \theta) \qquad \cos \theta = \sin (90° - \theta)$$
$$\tan \theta = \cot (90° - \theta) \qquad \cot \theta = \tan (90° - \theta)$$
$$\sec \theta = \csc (90° - \theta) \qquad \csc \theta = \sec (90° - \theta)$$

If θ is in radians, replace $90°$ with $\dfrac{\pi}{2}$.

EXAMPLE 4

Find a cofunction with the same value as the given expression.

a. $\sin 72°$ **b.** $\csc \dfrac{\pi}{3}$

Solution Because the value of a trigonometric function of θ is equal to the cofunction of the complement of θ, we need to find the complement of each angle. We do this by subtracting the angle's measure from $90°$ or its radian equivalent, $\dfrac{\pi}{2}$.

a. $\sin 72° = \cos (90° - 72°) = \cos 18°$

We have a function and its cofunction.

b. $\csc \dfrac{\pi}{3} = \sec \left(\dfrac{\pi}{2} - \dfrac{\pi}{3} \right) = \sec \left(\dfrac{3\pi}{6} - \dfrac{2\pi}{6} \right) = \sec \dfrac{\pi}{6}$

We have a cofunction and its function.

Perform the subtraction using the least common denominator, 6.

Check Point 4 Find a cofunction with the same value as the given expression.

a. $\sin 46°$ **b.** $\cot \dfrac{\pi}{12}$

---

4 Use right triangle trigonometry to solve applied problems.

Applications

Many applications of right triangle trigonometry involve the angle made with an imaginary horizontal line. As shown in Figure 4.28, an angle formed by a horizontal line and the line of sight to an object that is above the horizontal line

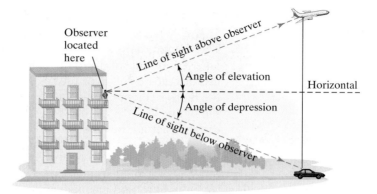

Figure 4.28

is called the **angle of elevation**. The angle formed by a horizontal line and the line of sight to an object that is below the horizontal line is called the **angle of depression**. Transits and sextants are instruments used to measure such angles.

EXAMPLE 5   Problem Solving Using an Angle of Elevation

Sighting the top of a building, a surveyor measured the angle of elevation to be 22°. The transit is 5 feet above the ground and 300 feet from the building. Find the building's height.

Solution   The situation is illustrated in Figure 4.29. Let a be the height of the portion of the building that lies above the transit. The height of the building is the transit's height, 5 feet, plus a. Thus, we need to identify a trigonometric function that will make it possible to find a. In terms of the 22° angle, we are looking for the side opposite the angle. The transit is 300 feet from the building, so the side adjacent to the 22° angle is 300 feet. Because we have a known angle, an unknown opposite side, and a known adjacent side, we select the tangent function.

$$\tan 22° = \frac{a}{300}$$

Length of side opposite the 22° angle
Length of side adjacent to the 22° angle

$$a = 300 \tan 22°$$ Multiply both sides of the equation by 300.

$$a \approx 300(0.4040) \approx 121$$ Find $\tan 22°$ with a calculator in the degree mode.

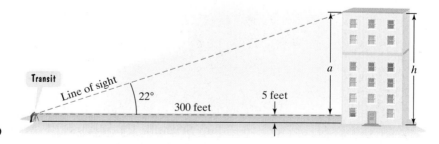

Figure 4.29

The height of the part of the building above the transit is 121 feet. Thus, the height of the building is determined by adding the transit's height, 5 feet, to 121 feet.

$$h \approx 5 + 121 = 126$$

The building's height is approximately 126 feet.

Check Point 5 The irregular blue shape in Figure 4.30 represents a lake. The distance across the lake, a, is unknown. To find this distance, a surveyor took the measurements shown in the figure. What is the distance across the lake?

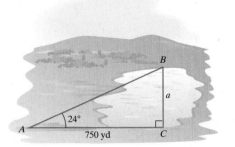

Figure 4.30

If two sides of a right triangle are known, an appropriate trigonometric function can be used to find an acute angle θ in the triangle. You will also need to use the *inverse key* on a calculator. This key uses a function value to display the acute angle θ. For example, suppose that $\sin\theta = 0.866$. We can find θ in the degree mode by using the secondary *inverse sine* key, usually labeled $\boxed{\text{SIN}^{-1}}$.

Study Tip

$\boxed{\text{SIN}^{-1}}$ is not a button you will actually press. It is the secondary function for the button labeled $\boxed{\text{SIN}}$.

Many Scientific Calculators:	**Many Graphing Calculators:**
.866 $\boxed{\text{2nd}}$ $\boxed{\text{SIN}^{-1}}$	$\boxed{\text{2nd}}$ $\boxed{\text{SIN}^{-1}}$ .866 $\boxed{\text{ENTER}}$

The display shows approximately 59.99, which we can round to 60. Thus, if $\sin\theta = 0.866$, then $\theta \approx 60°$.

EXAMPLE 6 Determining the Angle of Elevation

A building that is 21 meters tall casts a shadow 25 meters long. Find the angle of elevation of the sun to the nearest degree.

Solution The situation is illustrated in Figure 4.31. We are asked to find θ. We begin with the tangent function.

$$\tan\theta = \frac{\text{side opposite } \theta}{\text{side adjacent to } \theta} = \frac{21}{25}$$

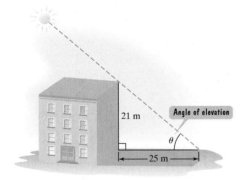

Figure 4.31

We use a calculator in the degree mode to find θ.

Many Scientific Calculators:	**Many Graphing Calculators:**
$\boxed{(}$21 $\boxed{\div}$ 25$\boxed{)}$ $\boxed{\text{2nd}}$ $\boxed{\text{TAN}^{-1}}$	$\boxed{\text{2nd}}$ $\boxed{\text{TAN}^{-1}}$ $\boxed{(}$21 $\boxed{\div}$ 25$\boxed{)}$ $\boxed{\text{ENTER}}$

The display should show approximately 40. Thus, the angle of elevation of the sun is approximately 40°.

Check Point 6 A flagpole that is 14 meters tall casts a shadow 10 meters long. Find the angle of elevation of the sun to the nearest degree.

The Mountain Man

In the 1930s, a *National Geographic* team headed by Brad Washburn used trigonometry to create a map of the 5000-square-mile region of the Yukon, near the Canadian border. The team started with aerial photography. By drawing a network of angles on the photographs, the approximate locations of the major mountains and their rough heights were determined. The expedition then spent three months on foot to find the exact heights. Team members established two base points a known distance apart, one directly under the mountain's peak. By measuring the angle of elevation from one of the base points to the peak, the tangent function was used to determine the peak's height. The Yukon expedition was a major advance in the way maps are made.

EXERCISE SET 4.3

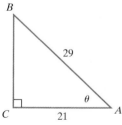 **Practice Exercises**

In Exercises 1–8, use the Pythagorean Theorem to find the length of the missing side of each right triangle. Then find the value of each of the six trigonometric functions of θ.

1.

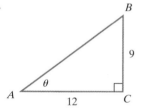

2.

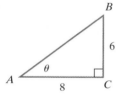

3. B

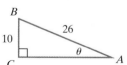

4. B

5. B

6. B

7.

8.

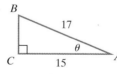

In Exercises 9–20, use the given triangles to evaluate each expression. If necessary, express the value without a square root in the denominator by rationalizing the denominator.

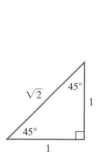 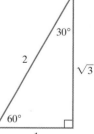

9. $\cos 30°$

10. $\tan 30°$

11. $\sec 45°$

12. $\csc 45°$

13. $\tan \dfrac{\pi}{3}$

14. $\cot \dfrac{\pi}{3}$

15. $\sin \dfrac{\pi}{4} - \cos \dfrac{\pi}{4}$

16. $\tan \dfrac{\pi}{4} + \csc \dfrac{\pi}{6}$

17. $\sin \dfrac{\pi}{3} \cos \dfrac{\pi}{4} - \tan \dfrac{\pi}{4}$ **18.** $\cos \dfrac{\pi}{3} \sec \dfrac{\pi}{3} - \cot \dfrac{\pi}{3}$

19. $2 \tan \dfrac{\pi}{3} + \cos \dfrac{\pi}{4} \tan \dfrac{\pi}{6}$ **20.** $6 \tan \dfrac{\pi}{4} + \sin \dfrac{\pi}{3} \sec \dfrac{\pi}{6}$

In Exercises 21–28, find a cofunction with the same value as the given expression.

21. $\sin 7°$

22. $\sin 19°$

23. $\csc 25°$

24. $\csc 35°$

25. $\tan \dfrac{\pi}{9}$

26. $\tan \dfrac{\pi}{7}$

27. $\cos \dfrac{2\pi}{5}$

28. $\cos \dfrac{3\pi}{8}$

In Exercises 29–34, find the measure of the side of the right triangle whose length is designated by a lowercase letter. Round answers to the nearest whole number.

29.

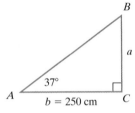

30.

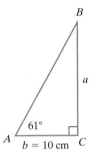

31.

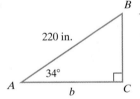

32.

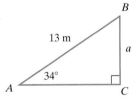

33.

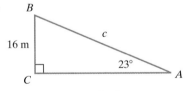

34.

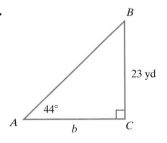

In Exercises 35–38, use a calculator to find the value of the acute angle θ to the nearest degree.

35. $\sin \theta = 0.2974$ **36.** $\cos \theta = 0.8771$

37. $\tan \theta = 4.6252$ **38.** $\tan \theta = 26.0307$

In Exercises 39–42, use a calculator to find the value of the acute angle θ in radians, rounded to three decimal places.

39. $\cos \theta = 0.4112$ **40.** $\sin \theta = 0.9499$

41. $\tan \theta = 0.4169$ **42.** $\tan \theta = 0.5117$

 Application Exercises

43. To find the distance across a lake, a surveyor took the measurements in the figure shown. Use these measurements to determine how far it is across the lake. Round to the nearest yard.

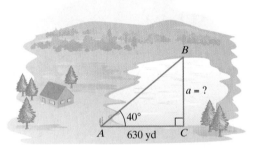

44. At a certain time of day, the angle of elevation of the sun is 40°. To the nearest foot, find the height of a tree whose shadow is 35 feet long.

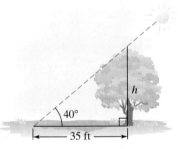

45. A tower that is 125 feet tall casts a shadow 172 feet long. Find the angle of elevation of the sun to the nearest degree.

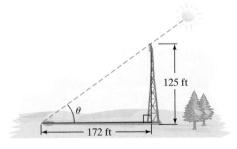

46. The Washington Monument is 555 feet high. If you stand one quarter of a mile, or 1320 feet, from the base of the monument and look to the top, find the angle of elevation to the nearest degree.

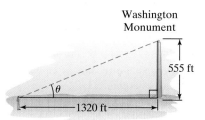

Washington
Monument

555 ft

θ

1320 ft

47. A plane rises from a take-off and flies at an angle of 10° with the horizontal runway. When it has gained 500 feet, find the distance, to the nearest foot, the plane has flown.

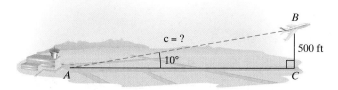

B

c = ?

500 ft

10°

A

C

48. A road is inclined at an angle of 5°. After driving 5000 feet along this road, find the driver's increase in altitude. Round to the nearest foot.

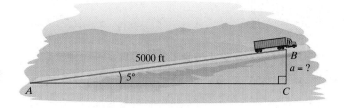

5000 ft

B
a = ?

5°

A

C

49. A telephone pole is 60 feet tall. A guy wire 75 feet long is attached from the ground to the top of the pole. Find the angle between the wire and the pole to the nearest degree.

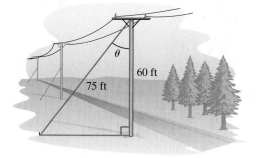

θ

60 ft

75 ft

50. A telephone pole is 55 feet tall. A guy wire 80 feet long is attached from the ground to the top of the pole. Find the angle between the wire and the pole to the nearest degree.

Writing in Mathematics

51. If you are given the lengths of the sides of a right triangle, describe how to find the sine of either acute angle.

52. Describe one similarity and one difference between the definitions of $\sin\theta$ and $\cos\theta$, where θ is an acute angle of a right triangle.

53. Describe the triangle used to find the trigonometric functions of 45°.

54. Describe the triangle used to find the trigonometric functions of 30° and 60°.

55. Describe a relationship among trigonometric functions that is based on angles that are complements.

56. Describe what is meant by an angle of elevation and an angle of depression.

57. Stonehenge, the famous "stone circle" in England, was built between 2750 B.C. and 1300 B.C. using solid stone blocks weighing over 99,000 pounds each. It required 550 people to pull a single stone up a ramp inclined at a 9° angle. Describe how right triangle trigonometry can be used to determine the distance the 550 workers had to drag a stone in order to raise it to a height of 30 feet.

Technology Exercises

58. Use a calculator in the radian mode to fill in the values in the following table. Then draw a conclusion about $\dfrac{\sin\theta}{\theta}$ as θ approaches 0.

θ	0.4	0.3	0.2	0.1	0.01	0.001	0.0001	0.00001
$\sin\theta$								
$\dfrac{\sin\theta}{\theta}$								

59. Use a calculator in the radian mode to fill in the values in the following table. Then draw a conclusion about $\dfrac{\cos\theta - 1}{\theta}$ as θ approaches 0.

θ	0.4	0.3	0.2	0.1	0.01	0.001	0.0001	0.00001
$\cos\theta$								
$\dfrac{\cos\theta - 1}{\theta}$								

Critical Thinking Exercises

60. Which one of the following is true?

a. $\dfrac{\tan 45°}{\tan 15°} = \tan 3°$ **b.** $\tan^2 15° - \sec^2 15° = -1$

c. $\sin 45° + \cos 45° = 1$ **d.** $\tan^2 5° = \tan 25°$

61. Explain why the sine or cosine of an acute angle cannot be greater than or equal to 1.

62. Describe what happens to the tangent of an acute angle as the angle gets close to 90°. What happens at 90°?

63. From the top of a 250-foot lighthouse, a plane is sighted overhead and a ship is observed directly below the plane. The angle of elevation of the plane is 22° and the angle of depression of the ship is 35°. Find **a.** the distance of the ship from the lighthouse; **b.** the plane's height above the water. Round to the nearest foot.

SECTION 4.4 *Trigonometric Functions of Any Angle*

Objectives

1. Use the definitions of trigonometric functions of any angle.

2. Use the signs of the trigonometric functions.

3. Find reference angles.

4. Use reference angles to evaluate trigonometric functions.

Cycles govern many aspects of life—heartbeats, sleep patterns, seasons, and tides all follow regular, predictable cycles. Because of their periodic nature, trigonometric functions are used to model phenomena that occur in cycles. It is helpful to apply these models regardless of whether we think of the domains of trigonometric functions as sets of real numbers or sets of angles. In order to understand and use models for cyclic phenomena from an angle perspective, we need to move beyond right triangles.

1 Use the definitions of trigonometric functions of any angle.

Trigonometric Functions of Any Angle

In the last section, we evaluated trigonometric functions of acute angles, such as that shown in Figure 4.32(a). Note that this angle is in standard position. The point $P = (x, y)$ is a point r units from the origin on the terminal side of θ. A right triangle is formed by drawing a perpendicular from $P = (x, y)$ to the x-axis. Note that y is the length of the side opposite θ and x is the length of the side adjacent to θ.

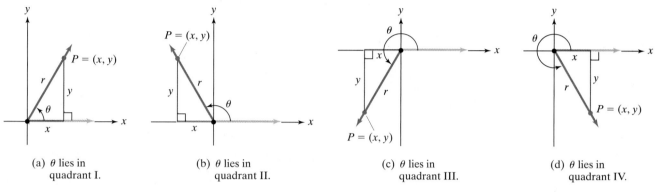

(a) θ lies in quadrant I.

(b) θ lies in quadrant II.

(c) θ lies in quadrant III.

(d) θ lies in quadrant IV.

Figure 4.32

Figures 4.32(b), (c), and (d) show angles in standard position, but they are not acute. We can extend our definitions of the six trigonometric functions to include such angles, as well as quadrantal angles. (Recall that a quadrantal angle has its terminal side on the x-axis or y-axis; such angles are *not* shown in Figure 4.32.) The point $P = (x, y)$ may be any point on the terminal side of the angle θ other than the origin, $(0, 0)$.

Study Tip

If θ is acute, we have the right triangle shown in Figure 4.32(a). In this situation, the definitions in the box are the right triangle definitions of the trigonometric functions. This should make it easier for you to remember the six definitions.

Definitions of Trigonometric Functions of Any Angle

Let θ be any angle in standard position, and let $P = (x, y)$ be a point on the terminal side of θ. If $r = \sqrt{x^2 + y^2}$ is the distance from $(0, 0)$ to (x, y), as shown in Figure 4.32, the **six trigonometric functions of θ** are defined by the following ratios:

$$\sin\theta = \frac{y}{r} \qquad \cos\theta = \frac{x}{r} \qquad \tan\theta = \frac{y}{x}, x \neq 0$$

$$\csc\theta = \frac{r}{y}, y \neq 0 \qquad \sec\theta = \frac{r}{x}, x \neq 0 \qquad \cot\theta = \frac{x}{y}, y \neq 0.$$

Because the point $P = (x, y)$ is any point on the terminal side of θ other than the origin, $(0, 0)$, $r = \sqrt{x^2 + y^2}$ cannot be zero. Examine the six trigonometric functions defined previously. Note that the denominator of the sine and cosine functions is r. Because $r \neq 0$, the sine and cosine functions are defined for any real value of the angle θ. This is not true for the other four trigonometric functions. Note that the denominator of the tangent and secant functions is x. These functions are not defined if $x = 0$. If the point $P = (x, y)$ is on the y-axis, then $x = 0$. Thus, the tangent and secant functions are undefined for all quadrantal angles with terminal sides on the positive or negative y-axis. Likewise, if $P = (x, y)$ is on the x-axis, then $y = 0$, and the cotangent and cosecant functions are undefined. The cotangent and cosecant functions are undefined for all quadrantal angles with terminal sides on the positive or negative x-axis.

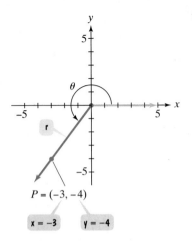

Figure 4.33

EXAMPLE 1 Evaluating Trigonometric Functions

Let $P = (-3, -4)$ be a point on the terminal side of θ. Find each of the six trigonometric functions of θ.

Solution   The situation is shown in Figure 4.33. We need values for x, y, and r to evaluate all six trigonometric functions. We are given the values of x and y. Because $P = (-3, -4)$ is a point on the terminal side of θ, $x = -3$ and $y = -4$. Furthermore,

$$r = \sqrt{x^2 + y^2} = \sqrt{(-3)^2 + (-4)^2} = \sqrt{9 + 16} = \sqrt{25} = 5.$$

Now that we know x, y, and r, we can find the six trigonometric functions of θ.

$$\sin\theta = \frac{y}{r} = \frac{-4}{5} = -\frac{4}{5}, \quad \cos\theta = \frac{x}{r} = \frac{-3}{5} = -\frac{3}{5}, \quad \tan\theta = \frac{y}{x} = \frac{-4}{-3} = \frac{4}{3}$$

$$\csc\theta = \frac{r}{y} = \frac{5}{-4} = -\frac{5}{4}, \quad \sec\theta = \frac{r}{x} = \frac{5}{-3} = -\frac{5}{3}, \quad \cot\theta = \frac{x}{y} = \frac{-3}{-4} = \frac{3}{4}$$

These ratios are the reciprocals of those shown directly above.

Check Point 1   Let $P = (4, -3)$ be a point on the terminal side of θ. Find each of the six trigonometric functions of θ.

How do we find the values of the trigonometric functions for a quadrantal angle? First, draw the angle in standard position. Second, choose a point P on the angle's terminal side. The trigonometric function values of θ depend only on the size of θ and not on the distance of point P from the origin. Thus, we choose a point that is 1 unit from the origin. Finally, apply the definition of the appropriate trigonometric function.

EXAMPLE 2 Trigonometric Functions of Quadrantal Angles

Evaluate, if possible, the sine function and the tangent function at the following four quadrantal angles:

a. $\theta = 0° = 0$ **b.** $\theta = 90° = \dfrac{\pi}{2}$ **c.** $\theta = 180° = \pi$ **d.** $\theta = 270° = \dfrac{3\pi}{2}$.

Solution

a. If $\theta = 0° = 0$ radians, then the terminal side of the angle is on the positive x-axis. Let us select the point $P = (1, 0)$ with $x = 1$ and $y = 0$. This point is 1 unit from the origin, so $r = 1$. Now that we know x, y, and r, we can apply the definitions of the sine and tangent functions.

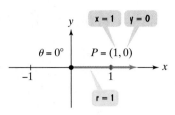

$$\sin 0° = \sin 0 = \frac{y}{r} = \frac{0}{1} = 0$$

$$\tan 0° = \tan 0 = \frac{y}{x} = \frac{0}{1} = 0$$

b. If $\theta = 90° = \dfrac{\pi}{2}$ radians, then the terminal side of the angle is on the positive y-axis. Let us select the point $P = (0, 1)$ with $x = 0$ and $y = 1$. This point is 1 unit from the origin, so $r = 1$. Now that we know x, y, and r, we can apply the definitions of the sine and tangent functions.

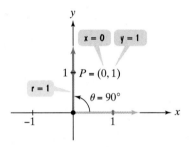

$$\sin 90° = \sin \frac{\pi}{2} = \frac{y}{r} = \frac{1}{1} = 1$$

$$\tan 90° = \tan \frac{\pi}{2} = \frac{y}{x} = \frac{1}{0}$$

Because division by 0 is undefined, $\tan 90°$ is undefined.

c. If $\theta = 180° = \pi$ radians, then the terminal side of the angle is on the negative x-axis. Let us select the point $P = (-1, 0)$ with $x = -1$ and $y = 0$. This point is 1 unit from the origin, so $r = 1$. Now that we know x, y, and r, we can apply the definitions of the sine and tangent functions.

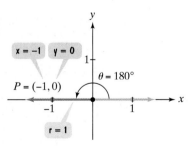

$$\sin 180° = \sin \pi = \frac{y}{r} = \frac{0}{1} = 0$$

$$\tan 180° = \tan \pi = \frac{y}{x} = \frac{0}{-1} = 0$$

Discovery

Try finding tan 90° and tan 270° with your calculator. Describe what occurs.

d. If $\theta = 270° = \dfrac{3\pi}{2}$ radians, then the terminal side of the angle is on the negative y-axis. Let us select the point $P = (0, -1)$ with $x = 0$ and $y = -1$. This point is 1 unit from the origin, so $r = 1$. Now that we know x, y, and r, we can apply the definitions of the sine and tangent functions.

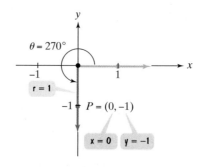

$$\sin 270° = \sin \dfrac{3\pi}{2} = \dfrac{y}{r} = \dfrac{-1}{1} = -1$$

$$\tan 270° = \tan \dfrac{3\pi}{2} = \dfrac{y}{x} = \dfrac{-1}{0}$$

Because division by 0 is undefined, tan 270° is undefined.

> **Check Point 2**
> Evaluate, if possible, the cosine function and the cosecant function at the following four quadrantal angles:
>
> **a.** $\theta = 0° = 0$ **b.** $\theta = 90° = \dfrac{\pi}{2}$
>
> **c.** $\theta = 180° = \pi$ **d.** $\theta = 270° = \dfrac{3\pi}{2}$.

2 Use the signs of the trigonometric functions.

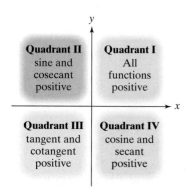

Figure 4.34 The signs of the trigonometric functions

The Signs of the Trigonometric Functions

In Example 2, we evaluated trigonometric functions of quadrantal angles. However, we will now return to the trigonometric functions of nonquadrantal angles. **If θ is not a quadrantal angle, the sign of a trigonometric function depends on the quadrant in which θ lies.** In all four quadrants, r is positive. However, x and y can be positive or negative. For example, if θ lies in quadrant II, x is negative and y is positive. Thus, the only positive ratios in this quadrant are $\dfrac{y}{r}$ and its reciprocal, $\dfrac{r}{y}$. These ratios are the function values for the sine and cosecant, respectively. In short, if θ lies in quadrant II, $\sin \theta$ and $\csc \theta$ are positive. The other four trigonometric functions are negative.

Figure 4.34 summarizes the signs of the trigonometric functions. If θ lies in quadrant I, all six functions are positive. If θ lies in quadrant II, only $\sin \theta$ and $\csc \theta$ are positive. If θ lies in quadrant III, only $\tan \theta$ and $\cot \theta$ are positive. Finally, if θ lies in quadrant IV, only $\cos \theta$ and $\sec \theta$ are positive. Observe that the positive functions in each quadrant occur in reciprocal pairs.

EXAMPLE 3 **Finding the Quadrant in Which an Angle Lies**

If $\tan \theta < 0$ and $\cos \theta > 0$, name the quadrant in which angle θ lies.

Solution Because $\tan \theta < 0$, θ cannot lie in quadrant I; all the functions are positive in quadrant I. Furthermore, θ cannot lie in quadrant III; $\tan \theta$ is positive in quadrant III. Thus, with $\tan \theta < 0$, θ lies in quadrant II or quadrant IV. We are also given that $\cos \theta > 0$. Because quadrant IV is the only quadrant in which the cosine is positive and the tangent is negative, we conclude that θ lies in quadrant IV.

> **Check Point 3** If $\sin \theta < 0$ and $\cos \theta < 0$, name the quadrant in which angle θ lies.

EXAMPLE 4 **Evaluating Trigonometric Functions**

Given $\tan \theta = -\frac{2}{3}$ and $\cos \theta > 0$, find $\cos \theta$ and $\csc \theta$.

Solution Because the tangent is negative and the cosine is positive, θ lies in quadrant IV. This will help us to determine whether the negative sign in $\tan \theta = -\frac{2}{3}$ should be associated with the numerator or the denominator. Keep in mind that in quadrant IV, x is positive and y is negative. Thus,

In quadrant IV, y is negative.

$$\tan \theta = -\frac{2}{3} = \frac{y}{x} = \frac{-2}{3}.$$

(See Figure 4.35.) Thus, $x = 3$ and $y = -2$. Furthermore,

$$r = \sqrt{x^2 + y^2} = \sqrt{3^2 + (-2)^2} = \sqrt{9 + 4} = \sqrt{13}.$$

Now that we know x, y, and r, we can find $\cos \theta$ and $\csc \theta$.

$$\cos \theta = \frac{x}{r} = \frac{3}{\sqrt{13}} = \frac{3}{\sqrt{13}} \cdot \frac{\sqrt{13}}{\sqrt{13}} = \frac{3\sqrt{13}}{13} \qquad \csc \theta = \frac{r}{y} = \frac{\sqrt{13}}{-2} = -\frac{\sqrt{13}}{2}$$

> **Check Point 4** Given $\tan \theta = -\frac{1}{3}$ and $\cos \theta < 0$, find $\sin \theta$ and $\sec \theta$.

The figure in the left margin:

Figure 4.35 $\tan \theta = -\frac{2}{3}$ and $\cos \theta > 0$

$r = \sqrt{13}$

$P = (3, -2)$

$x = 3$ $y = -2$

3 Find reference angles.

Reference Angles

We will often evaluate trigonometric functions of positive angles greater than 90° and all negative angles by making use of a positive acute angle. This positive acute angle is called a *reference angle*.

> **Definition of a Reference Angle**
>
> Let θ be a nonacute angle in standard position that lies in a quadrant. Its **reference angle** is the positive acute angle θ' formed by the terminal side of θ and the x-axis.

Figure 4.36 shows the reference angle for θ lying in quadrants II, III, and IV. Notice that the formula used to find θ', the reference angle, varies according to the quadrant in which θ lies. You may find it easier to find the reference angle for a given angle by making a figure that shows the angle in standard position. The acute angle formed by the terminal side of this angle and the x-axis is the reference angle.

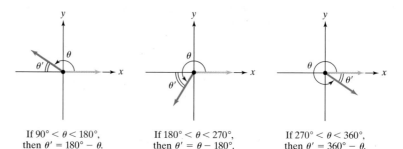

Figure 4.36 Reference angles, θ', for positive angles, θ, in quadrants II, III, and IV

If $90° < \theta < 180°$, then $\theta' = 180° - \theta$.

If $180° < \theta < 270°$, then $\theta' = \theta - 180°$.

If $270° < \theta < 360°$, then $\theta' = 360° - \theta$.

EXAMPLE 5   Finding Reference Angles

Find the reference angle, θ', for each of the following angles:

a. $\theta = 345°$ **b.** $\theta = \dfrac{5\pi}{6}$ **c.** $\theta = -135°$ **d.** $\theta = 2.5$.

Solution

a. A $345°$ angle in standard position is shown in Figure 4.37. Because $345°$ lies in quadrant IV, the reference angle is

$$\theta' = 360° - 345° = 15°.$$

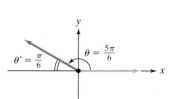

Figure 4.37

b. Because $\dfrac{5\pi}{6}$ lies between $\dfrac{\pi}{2} = \dfrac{3\pi}{6}$ and

$\pi = \dfrac{6\pi}{6}$, $\theta = \dfrac{5\pi}{6}$ lies in quadrant II.

The angle is shown in Figure 4.38. The reference angle is

$$\theta' = \pi - \dfrac{5\pi}{6} = \dfrac{6\pi}{6} - \dfrac{5\pi}{6} = \dfrac{\pi}{6}.$$

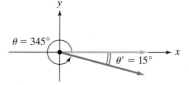

Figure 4.38

c. A $-135°$ angle in standard position is shown in Figure 4.39. The figure indicates that the positive acute angle formed by the terminal side of θ and the x-axis is $45°$. The reference angle is

$$\theta' = 45°.$$

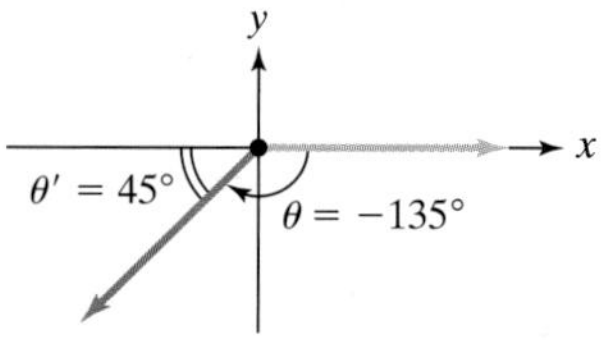

Figure 4.39

d. The angle $\theta = 2.5$ lies between $\dfrac{\pi}{2} \approx 1.57$

and $\pi \approx 3.14$. This means that $\theta = 2.5$ is in quadrant II, shown in Figure 4.40. The reference angle is

$$\theta' = \pi - 2.5 \approx 0.64.$$

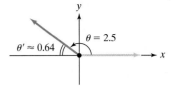

Figure 4.40

Check Point 5 Find the reference angle, θ', for each of the following angles:

a. $\theta = 210°$ **b.** $\theta = \dfrac{7\pi}{4}$ **c.** $\theta = -240°$ **d.** $\theta = 3.6$.

The way that reference angles are defined makes them useful in evaluating trigonometric functions.

> **4** Use reference angles to evaluate trigonometric functions.

Using Reference Angles to Evaluate Trigonometric Functions

The values of the trigonometric functions of a given angle, θ, are the same as the values of the trigonometric functions of the reference angle, θ', except possibly for the sign. A function value of the acute reference angle, θ', is always positive. However, the same function value for θ may be positive or negative.

For example, we can use a reference angle, θ', to obtain an exact value for $\tan 120°$. The reference angle for $\theta = 120°$ is $\theta' = 180° - 120° = 60°$. We know the exact value of the tangent function of the reference angle: $\tan 60° = \sqrt{3}$. We also know that the value of a trigonometric function of a given angle, θ, is the same as that of its reference angle, θ', except possibly for the sign. Thus, we can conclude that $\tan 120°$ equals $-\sqrt{3}$ or $\sqrt{3}$.

What sign should we attach to $\sqrt{3}$? A $120°$ angle lies in quadrant II, where only the sine and cosecant are positive. Thus, the tangent function is negative for a $120°$ angle. Therefore,

> Prefix by a negative sign to show tangent is negative in quadrant II.

$$\tan 120° = -\tan 60° = -\sqrt{3}.$$

> The reference angle for $120°$ is $60°$.

In the previous section, we used two right triangles to find exact trigonometric values of $30°$, $45°$, and $60°$. Using a procedure similar to finding $\tan 120°$, we can now find the function values of all angles for which $30°$, $45°$, or $60°$ are reference angles.

A Procedure for Using Reference Angles to Evaluate Trigonometric Functions

The value of a trigonometric function of any angle θ is found as follows:

1. Find the associated reference angle, θ', and the function value for θ'.

2. Use the quadrant in which θ lies to prefix the appropriate sign to the function value in step 1.

Discovery

Draw the two right triangles involving 30°, 45°, and 60°. Indicate the length of each side. Use these lengths to verify the function values for the reference angles in the solution to Example 6.

EXAMPLE 6 Using Reference Angles to Evaluate Trigonometric Functions

Use reference angles to find the exact value of each of the following trigonometric functions:

a. $\sin 135°$ **b.** $\cos \dfrac{4\pi}{3}$ **c.** $\cot\left(-\dfrac{\pi}{3}\right)$.

Solution

a. We use our two-step procedure to find $\sin 135°$.

Step 1 Find the reference angle, θ', and $\sin\theta'$. Figure 4.41 shows 135° lies in quadrant II. The reference angle is

$$\theta' = 180° - 135° = 45°.$$

The function value for the reference angle is $\sin 45° = \dfrac{\sqrt{2}}{2}$.

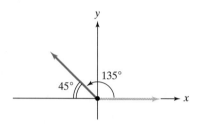

Figure 4.41 Reference angle for 135°

Step 2 Use the quadrant in which θ lies to prefix the appropriate sign to the function value in step 1. The angle $\theta = 135°$ lies in quadrant II. Because the sine is positive in quadrant II, we put a + sign before the function value of the reference angle. Thus,

> The sine is positive in quadrant II.

$$\sin 135° = +\sin 45° = \dfrac{\sqrt{2}}{2}.$$

> The reference angle for 135° is 45°.

b. We use our two-step procedure to find $\cos\dfrac{4\pi}{3}$.

Step 1 Find the reference angle, θ', and $\cos\theta'$. Figure 4.42 shows that $\theta = \dfrac{4\pi}{3}$ lies in quadrant III. The reference angle is

$$\theta' = \frac{4\pi}{3} - \pi = \frac{4\pi}{3} - \frac{3\pi}{3} = \frac{\pi}{3}.$$

The function value for the reference angle is

$$\cos\frac{\pi}{3} = \frac{1}{2}.$$

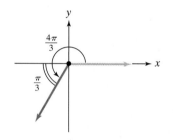

Figure 4.42 Reference angle for $\dfrac{4\pi}{3}$

Step 2   Use the quadrant in which θ lies to prefix the appropriate sign to the function value in step 1.   The angle $\theta = \dfrac{4\pi}{3}$ lies in quadrant III. Because only the tangent and cotangent are positive in quadrant III, the cosine is negative in this quadrant. We put a $-$ sign before the function value of the reference angle. Thus,

The cosine is negative in quadrant III.

$$\cos \frac{4\pi}{3} = -\cos \frac{\pi}{3} = -\frac{1}{2}.$$

The reference angle for $\frac{4\pi}{3}$ is $\frac{\pi}{3}$.

c. We use our two-step procedure to find $\cot\left(-\dfrac{\pi}{3}\right)$.

Step 1   Find the reference angle, θ', and $\cot\theta'$.   Figure 4.43 shows that $\theta = -\dfrac{\pi}{3}$ lies in quadrant IV. The reference angle is $\theta' = \dfrac{\pi}{3}$. The function value for the reference angle is $\cot\dfrac{\pi}{3} = \dfrac{\sqrt{3}}{3}$.

Step 2   Use the quadrant in which θ lies to prefix the appropriate sign to the function value in step 1.   The angle $\theta = -\dfrac{\pi}{3}$ lies in quadrant IV. Because only the cosine and secant are positive in quadrant IV, the cotangent is negative in this quadrant. We put a $-$ sign before the function value of the reference angle. Thus,

The cotangent is negative in quadrant IV.

$$\cot\left(-\frac{\pi}{3}\right) = -\cot\frac{\pi}{3} = -\frac{\sqrt{3}}{3}.$$

The reference angle for $-\frac{\pi}{3}$ is $\frac{\pi}{3}$.

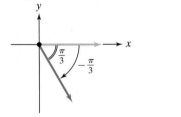

Figure 4.43 Reference angle for $-\dfrac{\pi}{3}$

Check Point 6   Use reference angles to find the exact value of the following trigonometric functions:

a. $\sin 300°$ **b.** $\tan \dfrac{5\pi}{4}$ **c.** $\sec\left(-\dfrac{\pi}{6}\right)$.

EXERCISE SET 4.4

Practice Exercises

In Exercises 1–8, a point on the terminal side of angle θ is given. Find the exact value of each of the six trigonometric functions of θ.

1. $(-4, 3)$ **2.** $(-12, 5)$
3. $(2, 3)$ **4.** $(3, 7)$
5. $(3, -3)$ **6.** $(5, -5)$
7. $(-2, -5)$ **8.** $(-1, -3)$

In Exercises 9–16, evaluate the trigonometric function at the quadrantal angle, or state that the expression is undefined.

9. $\cos \pi$ **10.** $\tan \pi$
11. $\sec \pi$ **12.** $\csc \pi$
13. $\tan \dfrac{3\pi}{2}$ **14.** $\cos \dfrac{3\pi}{2}$
15. $\cot \dfrac{\pi}{2}$ **16.** $\tan \dfrac{\pi}{2}$

In Exercises 17–22, let θ be an angle in standard position. Name the quadrant in which θ lies.

17. $\sin \theta > 0, \quad \cos \theta > 0$
18. $\sin \theta < 0, \quad \cos \theta > 0$
19. $\sin \theta < 0, \quad \cos \theta < 0$
20. $\tan \theta < 0, \quad \sin \theta < 0$
21. $\tan \theta < 0, \quad \cos \theta < 0$
22. $\cot \theta > 0, \quad \sec \theta < 0$

In Exercises 23–34, find the exact value of each of the remaining trigonometric functions of θ.

23. $\cos \theta = -\frac{3}{5}, \quad \theta$ in quadrant III
24. $\sin \theta = -\frac{12}{13}, \quad \theta$ in quadrant III
25. $\sin \theta = \frac{5}{13}, \quad \theta$ in quadrant II
26. $\cos \theta = \frac{4}{5}, \quad \theta$ in quadrant IV
27. $\cos \theta = \frac{8}{17}, \quad 270° < \theta < 360°$
28. $\cos \theta = \frac{1}{3}, \quad 270° < \theta < 360°$
29. $\tan \theta = -\frac{2}{3}, \quad \sin \theta > 0$
30. $\tan \theta = -\frac{1}{3}, \quad \sin \theta > 0$
31. $\tan \theta = \frac{4}{3}, \quad \cos \theta < 0$
32. $\tan \theta = \frac{5}{12}, \quad \cos \theta < 0$
33. $\sec \theta = -3, \quad \tan \theta > 0$
34. $\csc \theta = -4, \quad \tan \theta > 0$

In Exercises 35–50, find the reference angle for each angle.

35. $160°$ **36.** $170°$
37. $205°$ **38.** $210°$
39. $355°$ **40.** $351°$

41. $\dfrac{7\pi}{4}$ **42.** $\dfrac{5\pi}{4}$
43. $\dfrac{5\pi}{6}$ **44.** $\dfrac{5\pi}{7}$
45. $-150°$ **46.** $-250°$
47. $-335°$ **48.** $-359°$
49. 4.7 **50.** 5.5

In Exercises 51–66, use reference angles to find the exact value of each expression. Do not use a calculator.

51. $\cos 225°$ **52.** $\sin 300°$
53. $\tan 210°$ **54.** $\sec 240°$
55. $\tan 420°$ **56.** $\tan 405°$
57. $\sin \dfrac{2\pi}{3}$ **58.** $\cos \dfrac{3\pi}{4}$
59. $\csc \dfrac{7\pi}{6}$ **60.** $\cot \dfrac{7\pi}{4}$
61. $\tan \dfrac{9\pi}{4}$ **62.** $\tan \dfrac{9\pi}{2}$
63. $\sin(-240°)$ **64.** $\sin(-225°)$
65. $\tan\left(-\dfrac{\pi}{4}\right)$ **66.** $\tan\left(-\dfrac{\pi}{6}\right)$

 ### Writing in Mathematics

67. If you are given a point on the terminal side of angle θ, explain how to find $\sin \theta$.
68. Explain why $\tan 90°$ is undefined.
69. If $\cos \theta > 0$ and $\tan \theta < 0$, explain how to find the quadrant in which θ lies.
70. What is a reference angle? Give an example with your description.
71. Explain how reference angles are used to evaluate trigonometric functions. Give an example with your description.

SECTION 4.5   *Graphs of Sine and Cosine Functions*

Objectives

1. Understand the graph of $y = \sin x$.
2. Graph variations of $y = \sin x$.
3. Understand the graph of $y = \cos x$.
4. Graph variations of $y = \cos x$.
5. Use vertical shifts of sine and cosine curves.
6. Model periodic behavior.

Take a deep breath and relax. Many relaxation exercises involve slowing down our breathing. Some people suggest that the way we breathe affects every part of our lives. Did you know that graphs of trigonometric functions can be used to analyze the breathing cycle, which is our closest link to both life and death?

In this section, we use graphs of sine and cosine functions to visualize their properties. We use the traditional symbol x, rather than θ or t, to represent the independent variable. We use the symbol y for the dependent variable, or the function's value at x. Thus, we will be graphing $y = \sin x$ and $y = \cos x$ in rectangular coordinates. In all graphs of trigonometric functions, the independent variable, x, is measured in radians.

1 Understand the graph of $y = \sin x$.

The Graph of $y = \sin x$

The trigonometric functions can be graphed in a rectangular coordinate system by plotting points whose coordinates satisfy the function. Thus, we graph $y = \sin x$ by listing some points on the graph. Because the period of the sine function is 2π, we will graph the function on the interval $[0, 2\pi]$. The rest of the graph is made up of repetitions of this portion.

Table 4.2 lists some values of (x, y) on the graph of $y = \sin x, 0 \le x \le 2\pi$.

Table 4.2   Values of (x, y) on $y = \sin x$

x	0	$\dfrac{\pi}{6}$	$\dfrac{\pi}{3}$	$\dfrac{\pi}{2}$	$\dfrac{2\pi}{3}$	$\dfrac{5\pi}{6}$	π	$\dfrac{7\pi}{6}$	$\dfrac{4\pi}{3}$	$\dfrac{3\pi}{2}$	$\dfrac{5\pi}{3}$	$\dfrac{11\pi}{6}$	2π
$y = \sin x$	0	$\dfrac{1}{2}$	$\dfrac{\sqrt{3}}{2}$	1	$\dfrac{\sqrt{3}}{2}$	$\dfrac{1}{2}$	0	$-\dfrac{1}{2}$	$-\dfrac{\sqrt{3}}{2}$	-1	$-\dfrac{\sqrt{3}}{2}$	$-\dfrac{1}{2}$	0

As x increases from 0 to $\frac{\pi}{2}$, y increases from 0 to 1.

As x increases from $\frac{\pi}{2}$ to π, y decreases from 1 to 0.

As x increases from π to $\frac{3\pi}{2}$, y decreases from 0 to -1.

As x increases from $\frac{3\pi}{2}$ to 2π, y increases from -1 to 0.

In plotting the points obtained in Table 4.2, we will use the approximation $\dfrac{\sqrt{3}}{2} \approx 0.87$. Rather than approximating π, we will mark off units on the x-axis in terms of π. If we connect these points with a smooth curve, we obtain the graph shown in Figure 4.44 on the next page. The figure shows one period of the graph of $y = \sin x$.

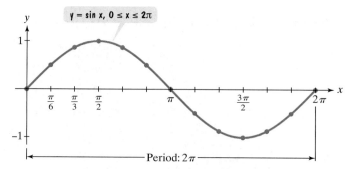

Figure 4.44 One period of the graph of $y = \sin x$

We can obtain a more complete graph of $y = \sin x$ by continuing the portion shown in Figure 4.44 to the left and to the right. The graph of the sine function, called a **sine curve,** is shown in Figure 4.45. Any part of the graph that corresponds to one period (2π) is one cycle of the graph of $y = \sin x$.

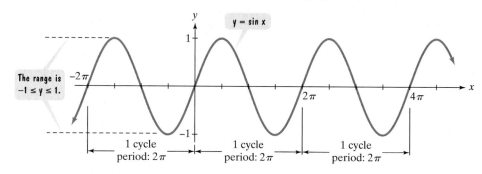

Figure 4.45 The graph of $y = \sin x$

The graph of $y = \sin x$ allows us to visualize some of the properties of the sine function.

- The domain is the set of all real numbers. The graph extends indefinitely to the left and to the right with no gaps or holes.
- The range consists of all numbers between -1 and 1, inclusive. The graph never rises above 1 or falls below -1.
- The period is 2π. The graph's pattern repeats in every interval of length 2π.
- The function is an odd function: $\sin(-x) = -\sin x$. This can be seen by observing that the graph is symmetric with respect to the origin.

2 Graph variations of $y = \sin x$.

Graphing Variations of $y = \sin x$

To graph variations of $y = \sin x$ by hand, it is helpful to find x-intercepts, maximum points, and minimum points. One complete cycle of the sine curve includes three x-intercepts, one maximum point, and one minimum point. The graph of $y = \sin x$ has x-intercepts at the beginning, middle, and end of its full period, shown in Figure 4.46. The curve reaches its maximum point $\frac{1}{4}$ of the way through the period. It reaches its minimum point $\frac{3}{4}$ of the way through the period. Thus, key points in graphing sine functions are obtained by dividing the period into four equal parts.

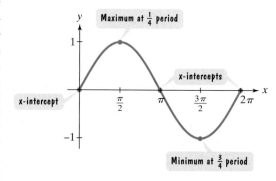

Figure 4.46 Key points in graphing the sine function

The x-coordinates of the five key points are as follows:

$$x_1 = \text{value of } x \text{ where the cycle begins}$$

$$x_2 = x_1 + \frac{\text{period}}{4}$$

$$x_3 = x_2 + \frac{\text{period}}{4}$$

$$x_4 = x_3 + \frac{\text{period}}{4}$$

$$x_5 = x_4 + \frac{\text{period}}{4}.$$

Add "quarter-periods" to find successive value of x.

The y-coordinates of the five key points are obtained by evaluating the given function at each of these values of x.

The graph of $y = \sin x$ forms the basis for graphing functions of the form

$$y = A \sin x.$$

For example, consider $y = 2 \sin x$, in which $A = 2$. We can obtain the graph of $y = 2 \sin x$ from that of $y = \sin x$ if we multiply each y-coordinate on the graph of $y = \sin x$ by 2. Figure 4.47 shows the graphs. The basic sine curve is *stretched* and ranges between -2 and 2, rather than between -1 and 1. However, both $y = \sin x$ and $y = 2 \sin x$ have a period of 2π.

In general, the graph of $y = A \sin x$ ranges between $-|A|$ and $|A|$. Thus, the range of the function is $-|A| \le y \le |A|$. If $|A| > 1$, the basic sine curve is *stretched*, as in Figure 4.47. If $|A| < 1$, the basic sine curve is *shrunk*. We call $|A|$ the **amplitude** of $y = A \sin x$. The maximum value of y on the graph of $y = A \sin x$ is $|A|$, the amplitude.

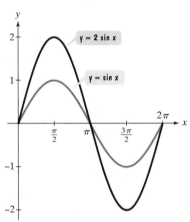

Figure 4.47 Comparing the graphs of $y = \sin x$ and $y = 2 \sin x$

Graphing Variations of $y = \sin x$

1. Identify the amplitude and the period.
2. Find the values of x for the five key points—the three x-intercepts, the maximum point, and the minimum point. Start with the value of x where the cycle begins and add quarter-periods—that is, $\dfrac{\text{period}}{4}$—to find successive values of x.
3. Find the values of y for the five key points by evaluating the function at each value of x from step 2.
4. Connect the five key points with a smooth curve and graph one complete cycle of the given function.
5. Extend the graph in step 4 to the left or right as desired.

EXAMPLE 1   Graphing a Variation of $y = \sin x$

Determine the amplitude of $y = \frac{1}{2} \sin x$. Then graph $y = \sin x$ and $y = \frac{1}{2} \sin x$ for $0 \le x \le 2\pi$.

Solution

Step 1   Identify the amplitude and the period.   The equation $y = \frac{1}{2} \sin x$ is of the form $y = A \sin x$ with $A = \frac{1}{2}$. Thus, the amplitude is $|A| = \frac{1}{2}$. This means that the maximum value of y is $\frac{1}{2}$ and the minimum value of y is $-\frac{1}{2}$. The period for both $y = \frac{1}{2} \sin x$ and $y = \sin x$ is 2π.

Step 2   Find the values of x for the five key points.   We need to find the three x-intercepts, the maximum point, and the minimum point on the interval $[0, 2\pi]$. To do so, we begin by dividing the period, 2π, by 4.

$$\frac{\text{period}}{4} = \frac{2\pi}{4} = \frac{\pi}{2}$$

We start with the value of x where the cycle begins: $x = 0$. Now we add quarter-periods, $\frac{\pi}{2}$, to generate x-values for each of the key points. The five x-values are

$$x = 0, \quad x = 0 + \frac{\pi}{2} = \frac{\pi}{2}, \quad x = \frac{\pi}{2} + \frac{\pi}{2} = \pi,$$

$$x = \pi + \frac{\pi}{2} = \frac{3\pi}{2}, \quad x = \frac{3\pi}{2} + \frac{\pi}{2} = 2\pi.$$

Step 3   Find the values of y for the five key points.   We evaluate the function at each value of x from step 2.

Value of x	Value of y: $y = \frac{1}{2}\sin x$	Coordinates of key point	
0	$y = \frac{1}{2}\sin 0 = \frac{1}{2} \cdot 0 = 0$	$(0, 0)$	
$\dfrac{\pi}{2}$	$y = \frac{1}{2}\sin\frac{\pi}{2} = \frac{1}{2} \cdot 1 = \frac{1}{2}$	$\left(\dfrac{\pi}{2}, \dfrac{1}{2}\right)$	maximum point
π	$y = \frac{1}{2}\sin\pi = \frac{1}{2} \cdot 0 = 0$	$(\pi, 0)$	
$\dfrac{3\pi}{2}$	$y = \frac{1}{2}\sin\frac{3\pi}{2} = \frac{1}{2}(-1) = -\frac{1}{2}$	$\left(\dfrac{3\pi}{2}, -\dfrac{1}{2}\right)$	minimum point
2π	$y = \frac{1}{2}\sin 2\pi = \frac{1}{2} \cdot 0 = 0$	$(2\pi, 0)$	

There are x-intercepts at 0, π, and 2π. The maximum and minimum points are indicated by the voice balloons.

Step 4   Connect the five key points with a smooth curve and graph one complete cycle of the given function.   The five key points for $y = \frac{1}{2}\sin x$ are shown in Figure 4.48. By connecting the points with a smooth curve, the figure shows one complete cycle of $y = \frac{1}{2}\sin x$. Also shown is the graph of $y = \sin x$. The graph of $y = \frac{1}{2}\sin x$ shrinks the graph of $y = \sin x$.

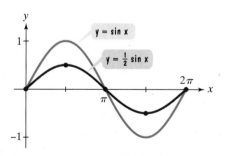

Figure 4.48 The graphs of $y = \sin x$ and $y = \frac{1}{2}\sin x$, $0 \le x \le 2\pi$

Check Point 1 Determine the amplitude of $y = 3 \sin x$. Then graph $y = \sin x$ and $y = 3 \sin x$ for $0 \le x \le 2\pi$.

EXAMPLE 2 Graphing a Variation of $y = \sin x$

Determine the amplitude of $y = -2 \sin x$. Then graph $y = \sin x$ and $y = -2 \sin x$ for $-\pi \le x \le 3\pi$.

Solution

Step 1   Identify the amplitude and the period.   The equation $y = -2 \sin x$ is of the form $y = A \sin x$ with $A = -2$. Thus, the amplitude is $|A| = |-2| = 2$. This means that the maximum value of y is 2 and the minimum value of y is -2. Both $y = \sin x$ and $y = -2 \sin x$ have a period of 2π.

Step 2   Find the x-values for the five key points.   Begin by dividing the period, 2π, by 4.

$$\frac{\text{period}}{4} = \frac{2\pi}{4} = \frac{\pi}{2}$$

Start with the value of x where the cycle begins: $x = 0$. Adding quarter-periods, $\frac{\pi}{2}$, the five x-values for the key points are

$$x = 0, \quad x = 0 + \frac{\pi}{2} = \frac{\pi}{2}, \quad x = \frac{\pi}{2} + \frac{\pi}{2} = \pi,$$

$$x = \pi + \frac{\pi}{2} = \frac{3\pi}{2}, \quad x = \frac{3\pi}{2} + \frac{\pi}{2} = 2\pi.$$

Step 3   Find the values of y for the five key points.   We evaluate the function at each value of x from step 2.

Value of x	Value of y: $y = -2 \sin x$	Coordinates of key point	
0	$y = -2 \sin 0 = -2 \cdot 0 = 0$	$(0, 0)$	
$\dfrac{\pi}{2}$	$y = -2 \sin \dfrac{\pi}{2} = -2 \cdot 1 = -2$	$\left(\dfrac{\pi}{2}, -2\right)$	minimum point
π	$y = -2 \sin \pi = -2 \cdot 0 = 0$	$(\pi, 0)$	
$\dfrac{3\pi}{2}$	$y = -2 \sin \dfrac{3\pi}{2} = -2(-1) = 2$	$\left(\dfrac{3\pi}{2}, 2\right)$	maximum point
2π	$y = -2 \sin 2\pi = -2 \cdot 0 = 0$	$(2\pi, 0)$	

There are x-intercepts at 0, π, and 2π. The minimum and maximum points are indicated by the voice balloons.

Step 4   Connect the five key points with a smooth curve and graph one complete cycle of the given function.   The five key points for $y = -2 \sin x$ are shown in Figure 4.49. By connecting the points with a smooth curve, the dark red portion shows one complete cycle of $y = -2 \sin x$. Also shown in dark blue is one complete cycle of the graph of $y = \sin x$.

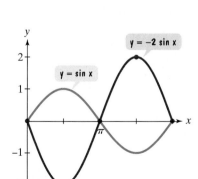

Figure 4.49 The graphs of $y = \sin x$ and $y = -2 \sin x$, $0 \le x \le 2\pi$

Step 5   Extend the graph in step 4 to the left or right as desired.   The dark red and dark blue portions of the graphs in Figure 4.49 are from 0 to 2π. In order to graph for $-\pi \le x \le 3\pi$, continue the pattern of each graph to the left and to the right. These extensions are shown by the lighter colors in Figure 4.49.

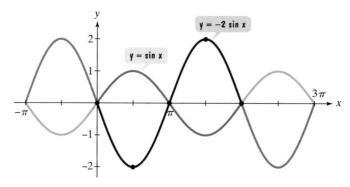

Figure 4.49, extended
The graphs of $y = \sin x$ and
$y = -2 \sin x$, $-\pi \le x \le 3\pi$

Check Point 2   Determine the amplitude of $y = -\frac{1}{2}\sin x$. Then graph $y = \sin x$ and $y = -\frac{1}{2}\sin x$ for $-\pi \le x \le 3\pi$.

Now let us examine the graphs of functions of the form $y = A \sin Bx$, where B is the coefficient of x. How do such graphs compare to those of functions of the form $y = A \sin x$? We know that $y = A \sin x$ completes one cycle from $x = 0$ to $x = 2\pi$. Thus, $y = A \sin Bx$ completes one cycle from $Bx = 0$ to $Bx = 2\pi$. Solve each of these equations for x.

$$Bx = 0 \qquad\qquad Bx = 2\pi$$
$$x = 0 \qquad\qquad x = \frac{2\pi}{B} \qquad \text{Divide both sides of each equation by } B.$$

This means that $y = A \sin Bx$ completes one cycle from 0 to $\frac{2\pi}{B}$. The period is $\frac{2\pi}{B}$.

Amplitudes and Periods
The graph of $y = A \sin Bx$ has

$$\text{amplitude} = |A|$$

$$\text{period} = \frac{2\pi}{B}.$$

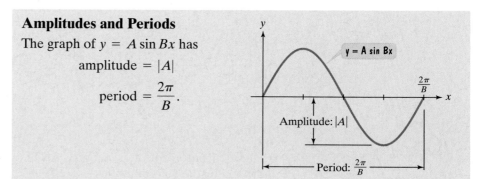

EXAMPLE 3   Graphing a Function of the Form $y = A \sin Bx$

Determine the amplitude and period of $y = 3 \sin 2x$. Then graph the function for $0 \le x \le 2\pi$.

Solution

Step 1 Identify the amplitude and the period. The equation $y = 3 \sin 2x$ is of the form $y = A \sin Bx$ with $A = 3$ and $B = 2$.

$$\text{amplitude:} \quad |A| = |3| = 3$$

$$\text{period:} \quad \frac{2\pi}{B} = \frac{2\pi}{2} = \pi$$

The amplitude, 3, tells us that the maximum value of y is 3 and the minimum value of y is -3.

Step 2 Find the x-values for the five key points. Begin by dividing the period, π, by 4.

$$\frac{\text{period}}{4} = \frac{\pi}{4}$$

Start with the value of x where the cycle begins: $x = 0$. Adding quarter-periods, $\frac{\pi}{4}$, the five x-values for the key points are

$$x = 0, \quad x = 0 + \frac{\pi}{4} = \frac{\pi}{4}, \quad x = \frac{\pi}{4} + \frac{\pi}{4} = \frac{\pi}{2},$$

$$x = \frac{\pi}{2} + \frac{\pi}{4} = \frac{3\pi}{4}, \quad x = \frac{3\pi}{4} + \frac{\pi}{4} = \pi.$$

Step 3 Find the values of y for the five key points. We evaluate the function at each value of x from step 2.

Value of x	Value of y: $y = 3 \sin 2x$	Coordinates of key point	
0	$y = 3 \sin 2 \cdot 0$ $= 3 \sin 0 = 3 \cdot 0 = 0$	$(0, 0)$	
$\dfrac{\pi}{4}$	$y = 3 \sin 2 \cdot \dfrac{\pi}{4}$ $= 3 \sin \dfrac{\pi}{2} = 3 \cdot 1 = 3$	$\left(\dfrac{\pi}{4}, 3\right)$	maximum point
$\dfrac{\pi}{2}$	$y = 3 \sin 2 \cdot \dfrac{\pi}{2}$ $= 3 \sin \pi = 3 \cdot 0 = 0$	$\left(\dfrac{\pi}{2}, 0\right)$	
$\dfrac{3\pi}{4}$	$y = 3 \sin 2 \cdot \dfrac{3\pi}{4}$ $= 3 \sin \dfrac{3\pi}{2} = 3(-1) = -3$	$\left(\dfrac{3\pi}{4}, -3\right)$	minimum point
π	$y = 3 \sin 2 \cdot \pi$ $= 3 \sin 2\pi = 3 \cdot 0 = 0$	$(\pi, 0)$	

In the interval $[0, \pi]$, there are x-intercepts at 0, $\frac{\pi}{2}$, and π. The maximum and minimum points are indicated by the voice balloons.

Step 4 Connect the five key points with a smooth curve and graph one complete cycle of the given function. The five key points for $y = 3 \sin 2x$ are shown in Figure 4.50. By connecting the points with a smooth curve, the blue portion shows one complete cycle of $y = 3 \sin 2x$ from 0 to π.

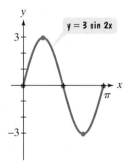

Figure 4.50 The graph of $y = 3 \sin 2x$, $0 \le x \le \pi$

Technology

The graph of $y = 3 \sin 2x$ in a $\left[0, 2\pi, \dfrac{\pi}{2}\right]$ by $[-4, 4, 1]$ viewing rectangle verifies our hand-drawn graph in Figure 4.50.

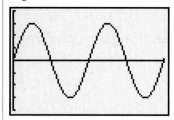

Step 5 Extend the graph in step 4 to the left or right as desired. The blue portion of the graph in Figure 4.50 is from 0 to π. In order to graph for $0 \le x \le 2\pi$, we continue this portion and extend the graph another full period to the right. This extension is shown in black in Figure 4.50.

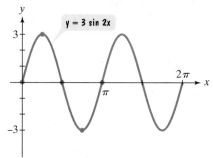

Figure 4.50, extended

Check Point 3   Determine the amplitude and period of $y = 2 \sin \frac{1}{2} x$. Then graph the function for $0 \le x \le 8\pi$.

Now let us examine the graphs of functions of the form $y = A \sin(Bx - C)$. How do such graphs compare to those of functions of the form $y = A \sin Bx$?

In both cases, the amplitude is $|A|$ and the period is $\dfrac{2\pi}{B}$. One complete cycle occurs if $Bx - C$ increases from 0 to 2π. This means that we can find an interval containing one cycle by solving the equations

$$Bx - C = 0 \quad \text{and} \quad Bx - C = 2\pi.$$

$$Bx = C \qquad\qquad Bx = C + 2\pi \qquad \text{Add } C \text{ to both sides in each equation.}$$

$$x = \frac{C}{B} \qquad\qquad x = \frac{C}{B} + \frac{2\pi}{B}. \qquad \text{Divide both sides by } B \text{ in each equation.}$$

This is the x-coordinate on the left where the cycle begins.

This is the x-coordinate on the right where the cycle ends. $\dfrac{2\pi}{B}$ is the period.

The voice balloon on the left indicates that $y = A \sin(Bx - C)$ shifts the graph of $y = A \sin Bx$ horizontally by $\dfrac{C}{B}$. Thus, the number $\dfrac{C}{B}$ is the **phase shift** associated with the graph.

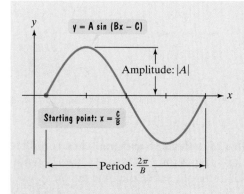

The Graph of $y = A \sin(Bx - C)$

The graph of $y = A \sin(Bx - C)$ is obtained by horizontally shifting the graph of $y = A \sin Bx$ so that the starting point of the cycle is shifted from $x = 0$ to $x = \dfrac{C}{B}$. If $\dfrac{C}{B} > 0$, the shift is to the right. If $\dfrac{C}{B} < 0$, the shift is to the left. The number $\dfrac{C}{B}$ is called the **phase shift.**

$$\text{amplitude} = |A|$$

$$\text{period} = \frac{2\pi}{B}$$

EXAMPLE 4 Graphing a Function of the Form $y = A \sin(Bx - C)$

Determine the amplitude, period, and phase shift of $y = 4\sin\left(2x - \dfrac{2\pi}{3}\right)$. Then graph one period of the function.

Solution

Step 1 Identify the amplitude, the period, and the phase shift. We must first identify values for A, B, and C.

> This equation is of the form
> $y = A \sin(Bx - C)$.

$$y = 4\sin\left(2x - \frac{2\pi}{3}\right)$$

Using the voice balloon, we see that $A = 4$, $B = 2$, and $C = \dfrac{2\pi}{3}$.

amplitude: $|A| = |4| = 4$ *The maximum y is 4 and the minimum is −4.*

period: $\dfrac{2\pi}{B} = \dfrac{2\pi}{2} = \pi$ *Each cycle is completed in π radians.*

phase shift: $\dfrac{C}{B} = \dfrac{\frac{2\pi}{3}}{2} = \dfrac{2\pi}{3} \cdot \dfrac{1}{2} = \dfrac{\pi}{3}$ *A cycle starts at $x = \frac{\pi}{3}$.*

Step 2 Find the x-values for the five key points. Begin by dividing the period, π, by 4.

$$\frac{\text{period}}{4} = \frac{\pi}{4}$$

Start with the value of x where the cycle begins: $x = \dfrac{\pi}{3}$. Adding quarter-periods, $\dfrac{\pi}{4}$, the five x-values for the key points are

$$x = \frac{\pi}{3}, \quad x = \frac{\pi}{3} + \frac{\pi}{4} = \frac{4\pi}{12} + \frac{3\pi}{12} = \frac{7\pi}{12},$$

$$x = \frac{7\pi}{12} + \frac{\pi}{4} = \frac{7\pi}{12} + \frac{3\pi}{12} = \frac{10\pi}{12} = \frac{5\pi}{6},$$

$$x = \frac{5\pi}{6} + \frac{\pi}{4} = \frac{10\pi}{12} + \frac{3\pi}{12} = \frac{13\pi}{12},$$

$$x = \frac{13\pi}{12} + \frac{\pi}{4} = \frac{13\pi}{12} + \frac{3\pi}{12} = \frac{16\pi}{12} = \frac{4\pi}{3}.$$

Study Tip

You can speed up the additions on the right by first writing the starting point, $\frac{\pi}{3}$, and the quarter-period, $\frac{\pi}{4}$, with a common denominator, 12.

starting point
$$= \frac{\pi}{3} = \frac{4\pi}{12}$$

quarter-period
$$= \frac{\pi}{4} = \frac{3\pi}{12}$$

Step 3 Find the values of y for the five key points. We evaluate the function at each value of x from step 2.

Value of x	Value of y: $y = 4 \sin\left(2x - \dfrac{2\pi}{3}\right)$	Coordinates of key point
$\dfrac{\pi}{3}$	$y = 4 \sin\left(2 \cdot \dfrac{\pi}{3} - \dfrac{2\pi}{3}\right)$ $= 4 \sin 0 = 4 \cdot 0 = 0$	$\left(\dfrac{\pi}{3}, 0\right)$
$\dfrac{7\pi}{12}$	$y = 4 \sin\left(2 \cdot \dfrac{7\pi}{12} - \dfrac{2\pi}{3}\right)$ $= 4 \sin\left(\dfrac{7\pi}{6} - \dfrac{2\pi}{3}\right)$ $= 4 \sin \dfrac{3\pi}{6} = 4 \sin \dfrac{\pi}{2} = 4 \cdot 1 = 4$	$\left(\dfrac{7\pi}{12}, 4\right)$
$\dfrac{5\pi}{6}$	$y = 4 \sin\left(2 \cdot \dfrac{5\pi}{6} - \dfrac{2\pi}{3}\right)$ $= 4 \sin\left(\dfrac{5\pi}{3} - \dfrac{2\pi}{3}\right)$ $= 4 \sin \dfrac{3\pi}{3} = 4 \sin \pi = 4 \cdot 0 = 0$	$\left(\dfrac{5\pi}{6}, 0\right)$
$\dfrac{13\pi}{12}$	$y = 4 \sin\left(2 \cdot \dfrac{13\pi}{12} - \dfrac{2\pi}{3}\right)$ $= 4 \sin\left(\dfrac{13\pi}{6} - \dfrac{4\pi}{6}\right)$ $= 4 \sin \dfrac{9\pi}{6} = 4 \sin \dfrac{3\pi}{2} = 4(-1) = -4$	$\left(\dfrac{13\pi}{12}, -4\right)$
$\dfrac{4\pi}{3}$	$y = 4 \sin\left(2 \cdot \dfrac{4\pi}{3} - \dfrac{2\pi}{3}\right)$ $= 4 \sin \dfrac{6\pi}{3} = 4 \sin 2\pi = 4 \cdot 0 = 0$	$\left(\dfrac{4\pi}{3}, 0\right)$

maximum point (pointing to $\left(\dfrac{7\pi}{12}, 4\right)$)

minimum point (pointing to $\left(\dfrac{13\pi}{12}, -4\right)$)

In the interval $\left[\dfrac{\pi}{3}, \dfrac{4\pi}{3}\right]$, there are x-intercepts at $\dfrac{\pi}{3}, \dfrac{5\pi}{6}$, and $\dfrac{4\pi}{3}$. The maximum and minimum points are indicated by the voice balloons.

Step 4 Connect the five key points with a smooth curve and graph one complete cycle of the given function. The key points and the graph of $y = 4 \sin\left(2x - \dfrac{2\pi}{3}\right)$ are shown in Figure 4.51.

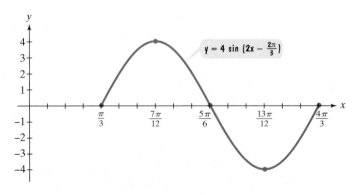

Figure 4.51

Check Point 4 Determine the amplitude, period, and phase shift of $y = 3 \sin\left(2x - \dfrac{\pi}{3}\right)$. Then graph one period of the function.

3 Understand the graph of $y = \cos x$.

The Graph of $y = \cos x$

We graph $y = \cos x$ by listing some points on the graph. Because the period of the cosine function is 2π, we will concentrate on the graph of the basic cosine curve on the interval $[0, 2\pi]$. The rest of the graph is made up of repetitions of this portion. Table 4.3 lists some values of (x, y) on the graph of $y = \cos x$.

Table 4.3 **Values of (x, y) on $y = \cos x$**

x	0	$\dfrac{\pi}{6}$	$\dfrac{\pi}{3}$	$\dfrac{\pi}{2}$	$\dfrac{2\pi}{3}$	$\dfrac{5\pi}{6}$	π	$\dfrac{7\pi}{6}$	$\dfrac{4\pi}{3}$	$\dfrac{3\pi}{2}$	$\dfrac{5\pi}{3}$	$\dfrac{11\pi}{6}$	2π
$y = \cos x$	1	$\dfrac{\sqrt{3}}{2}$	$\dfrac{1}{2}$	0	$-\dfrac{1}{2}$	$-\dfrac{\sqrt{3}}{2}$	-1	$-\dfrac{\sqrt{3}}{2}$	$-\dfrac{1}{2}$	0	$\dfrac{1}{2}$	$\dfrac{\sqrt{3}}{2}$	1

As x increases from 0 to $\dfrac{\pi}{2}$, y decreases from 1 to 0.

As x increases from $\dfrac{\pi}{2}$ to π, y decreases from 0 to -1.

As x increases from π to $\dfrac{3\pi}{2}$, y increases from -1 to 0.

As x increases from $\dfrac{3\pi}{2}$ to 2π, y increases from 0 to 1.

Plotting the points in Table 4.3 and connecting them with a smooth curve, we obtain the graph shown in Figure 4.52. The portion of the graph in dark blue shows one complete period. We can obtain a more complete graph of $y = \cos x$ by extending this dark blue portion to the left and to the right.

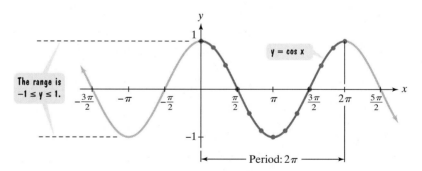

The range is $-1 \le y \le 1$.

y = cos x

Period: 2π

Figure 4.52 The graph of $y = \cos x$

The graph of $y = \cos x$ allows us to visualize some of the properties of the cosine function.

- The domain is the set of all real numbers. The graph extends indefinitely to the left and to the right with no gaps or holes.
- The range consists of all numbers between -1 and 1, inclusive. The graph never rises above 1 or falls below -1.
- The period is 2π. The graph's pattern repeats in every interval of length 2π.
- The function is an even function: $\cos(-x) = \cos x$. This can be seen by observing that the graph is symmetric with respect to the y-axis.

Turn back the page and take a second look at Figure 4.52. Can you see that the graph of $y = \cos x$ is the graph of $y = \sin x$ with a phase shift of $-\dfrac{\pi}{2}$ radians? If you trace along the curve from $x = -\dfrac{\pi}{2}$ to $x = \dfrac{3\pi}{2}$, you are tracing one complete cycle of the sine curve. This can be expressed as an identity:

$$\cos x = \sin\left(x + \frac{\pi}{2}\right).$$

Because of this similarity, the graphs of sine functions and cosine functions are called **sinusoidal graphs.**

4 Graph variations of $y = \cos x$.

Graphing Variations of $y = \cos x$

We use the same steps to graph variations of $y = \cos x$ as we did for graphing variations of $y = \sin x$. We will continue finding key points by dividing the period into four equal parts. Amplitudes, periods, and phase shifts play an important role when graphing by hand.

The Graph of $y = A \cos Bx$

The graph of $y = A \cos Bx$ has

$$\text{amplitude} = |A|$$

$$\text{period} = \frac{2\pi}{B}.$$

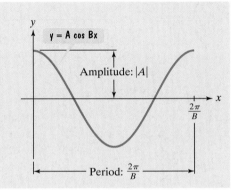

EXAMPLE 5   Graphing a Function of the Form $y = A \cos Bx$

Determine the amplitude and period of $y = -3 \cos \dfrac{\pi}{2} x$. Then graph the function for $-4 \le x \le 4$.

Solution

Step 1   Identify the amplitude and the period.   The equation $y = -3 \cos \dfrac{\pi}{2} x$ is of the form $y = A \cos Bx$ with $A = -3$ and $B = \dfrac{\pi}{2}$.

amplitude: $|A| = |-3| = 3$ The maximum y is 3 and the minimum is -3.

period: $\dfrac{2\pi}{B} = \dfrac{2\pi}{\dfrac{\pi}{2}} = 2\pi \cdot \dfrac{2}{\pi} = 4$ Each cycle is completed in 4 radians.

Step 2   Find the x-values for the five key points.   Begin by dividing the period, 4, by 4.

$$\frac{\text{period}}{4} = \frac{4}{4} = 1$$

Start with the value of x where the cycle begins: $x = 0$. Adding quarter-periods, 1, the five x-values for the key points are

$$x = 0, \quad x = 0 + 1 = 1, \quad x = 1 + 1 = 2, \quad x = 2 + 1 = 3, \quad x = 3 + 1 = 4$$

Step 3 **Find the values of y for the five key points.** We evaluate the function at each value of x from step 2.

Value of x	Value of y: $y = -3 \cos \dfrac{\pi}{2} x$	Coordinates of key point	
0	$y = -3 \cos \dfrac{\pi}{2} \cdot 0$ $= -3 \cos 0 = -3 \cdot 1 = -3$	$(0, -3)$	minimum point
1	$y = -3 \cos \dfrac{\pi}{2} \cdot 1$ $= -3 \cos \dfrac{\pi}{2} = -3 \cdot 0 = 0$	$(1, 0)$	
2	$y = -3 \cos \dfrac{\pi}{2} \cdot 2$ $= -3 \cos \pi = -3(-1) = 3$	$(2, 3)$	maximum point
3	$y = -3 \cos \dfrac{\pi}{2} \cdot 3$ $= -3 \cos \dfrac{3\pi}{2} = -3(0) = 0$	$(3, 0)$	
4	$y = -3 \cos \dfrac{\pi}{2} \cdot 4$ $= -3 \cos 2\pi = -3(1) = -3$	$(4, -3)$	minimum point

In the interval $[0, 4]$, there are x-intercepts at 1 and 3. The minimum and maximum points are indicated by the voice balloons.

Step 4 **Connect the five key points with a smooth curve and graph one complete cycle of the given function.** The five key points for $y = -3 \cos \dfrac{\pi}{2} x$ are shown in Figure 4.53. By connecting the points with a smooth curve, the blue portion shows one complete cycle of $y = -3 \cos \dfrac{\pi}{2} x$ from 0 to 4.

Technology

The graph of $y = -3 \cos \dfrac{\pi}{2} x$ in a $[-4, 4, 1]$ by $[-4, 4, 1]$ viewing rectangle verifies our hand-drawn graph in Figure 4.53.

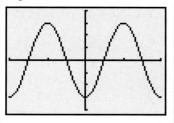

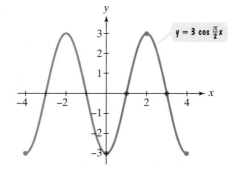

$y = 3 \cos \frac{\pi}{2} x$

Figure 4.53

Step 5 **Extend the graph in step 4 to the left or right as desired.** The blue portion of the graph in Figure 4.53 is for x from 0 to 4. In order to graph for $-4 \le x \le 4$, we continue this portion and extend the graph another full period to the left. This extension is shown in black in Figure 4.53.

Check Point 5 Determine the amplitude and period of $y = -4\cos \pi x$. Then graph the function for $-2 \le x \le 2$.

Finally, let us examine the graphs of functions of the form $y = A\cos(Bx - C)$.

Graphs of these functions shift the graph of $y = A\cos Bx$ horizontally by $\dfrac{C}{B}$.

The Graph of $y = A\cos(Bx - C)$

The graph of $y = A\cos(Bx - C)$ is obtained by horizontally shifting the graph of $y = A\cos Bx$ so that the starting point of the cycle is shifted from $x = 0$ to $x = \dfrac{C}{B}$. If $\dfrac{C}{B} > 0$, the shift is to the right. If $\dfrac{C}{B} < 0$, the shift is to the left. The number $\dfrac{C}{B}$ is called the **phase shift.**

$$\text{amplitude} = |A|$$

$$\text{period} = \frac{2\pi}{B}$$

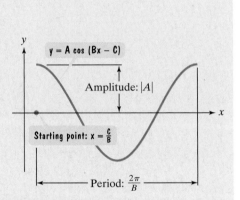

EXAMPLE 6 **Graphing a Function of the Form $y = A\cos(Bx - C)$**

Determine the amplitude, period, and phase shift of $y = \frac{1}{2}\cos(4x + \pi)$. Then graph one period of the function.

Solution

Step 1 Identify the amplitude, the period, and the phase shift. We must first identify values for A, B, and C. To do this, we need to express the equation in the form $y = A\cos(Bx - C)$. Thus, we write $y = \frac{1}{2}\cos(4x + \pi)$ as $y = \frac{1}{2}\cos[4x - (-\pi)]$. Now we can identify values for A, B, and C.

> This equation is of the form $y = A\cos(Bx - C)$.

$$y = \frac{1}{2}\cos[4x - (-\pi)]$$

Using the voice balloon, we see that $A = \frac{1}{2}$, $B = 4$, and $C = -\pi$.

amplitude: $|A| = \left|\dfrac{1}{2}\right| = \dfrac{1}{2}$ $\quad$ The maximum y is $\frac{1}{2}$ and the minimum is $-\frac{1}{2}$.

period: $\dfrac{2\pi}{B} = \dfrac{2\pi}{4} = \dfrac{\pi}{2}$ $\quad$ Each cycle is completed in $\frac{\pi}{2}$ radians.

phase shift: $\dfrac{C}{B} = -\dfrac{\pi}{4}$ $\quad$ A cycle starts at $x = -\frac{\pi}{4}$.

Step 2 Find the x-values for the five key points. Begin by dividing the period, $\frac{\pi}{2}$, by 4.

$$\frac{\text{period}}{4} = \frac{\frac{\pi}{2}}{4} = \frac{\pi}{8}$$

Start with the value of x where the cycle begins: $x = -\frac{\pi}{4}$. Adding quarter-periods, $\frac{\pi}{8}$, the five x-values for the key points are

$$x = -\frac{\pi}{4}, \quad x = -\frac{\pi}{4} + \frac{\pi}{8} = -\frac{2\pi}{8} + \frac{\pi}{8} = -\frac{\pi}{8}, \quad x = -\frac{\pi}{8} + \frac{\pi}{8} = 0,$$

$$x = 0 + \frac{\pi}{8} = \frac{\pi}{8}, \quad x = \frac{\pi}{8} + \frac{\pi}{8} = \frac{2\pi}{8} = \frac{\pi}{4}.$$

Step 3 Find the values of y for the five key points. Take a few minutes and use your calculator to evaluate the function at each value of x from step 2. Show that the key points are

$$\left(-\frac{\pi}{4}, \frac{1}{2}\right), \quad \left(-\frac{\pi}{8}, 0\right), \quad \left(0, -\frac{1}{2}\right), \quad \left(\frac{\pi}{8}, 0\right), \quad \text{and} \quad \left(\frac{\pi}{4}, \frac{1}{2}\right).$$

maximum point	x-intercept at $-\frac{\pi}{8}$	minimum point	x-intercept at $\frac{\pi}{8}$	maximum point

Technology

The graph of

$$y = \frac{1}{2}\cos(4x + \pi)$$

in a $\left[-\frac{\pi}{4}, \frac{\pi}{4}, \frac{\pi}{8}\right]$ by $[-1, 1, 1]$ viewing rectangle verifies our hand-drawn graph in Figure 4.54.

Step 4 Connect the five key points with a smooth curve and graph one complete cycle of the given function. The key points and the graph of $y = \frac{1}{2}\cos(4x + \pi)$ are shown in Figure 4.54.

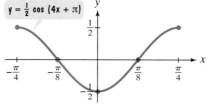

Figure 4.54

> **Check Point 6** Determine the amplitude, period, and phase shift of $y = \frac{3}{2}\cos(2x + \pi)$. Then graph one period of the function.

5 Use vertical shifts of sine and cosine curves.

Vertical Shifts of Sinusoidal Graphs

We now look at sinusoidal graphs of

$$y = A\sin(Bx - C) + D \quad \text{and} \quad y = A\cos(Bx - C) + D.$$

The constant D causes vertical shifts in the graphs of $y = A\sin(Bx - C)$ and $y = A\cos(Bx - C)$. If D is positive, the shift is D units upward. If D is negative, the shift is D units downward. These vertical shifts result in sinusoidal graphs oscillating about the horizontal line $y = D$ rather than about the x-axis. Thus, the maximum y is $D + |A|$ and the minimum y is $D - |A|$.

EXAMPLE 7 A Vertical Shift

Graph one period of the function $y = \frac{1}{2}\cos x - 1$.

Solution   The graph of $y = \frac{1}{2}\cos x - 1$ is the graph of $y = \frac{1}{2}\cos x$ shifted one unit downward. The period of $y = \frac{1}{2}\cos x$ is 2π, which is also the period for the vertically shifted graph. The key points on the interval $[0, 2\pi]$ for $y = \frac{1}{2}\cos x - 1$ are found by first determining their x-coordinates. The quarter-period is $\dfrac{2\pi}{4}$, or $\dfrac{\pi}{2}$.

The cycle begins at $x = 0$. As always, we add quarter-periods to generate x-values for each of the key points. The five x-values are

$$x = 0, \quad x = 0 + \frac{\pi}{2} = \frac{\pi}{2}, \quad x = \frac{\pi}{2} + \frac{\pi}{2} = \pi,$$

$$x = \pi + \frac{\pi}{2} = \frac{3\pi}{2}, \quad x = \frac{3\pi}{2} + \frac{\pi}{2} = 2\pi.$$

The values of y for the five key points and their coordinates are determined as follows.

Value of x	Value of y: $y = \dfrac{1}{2}\cos x - 1$	Coordinates of key point
0	$y = \dfrac{1}{2}\cos 0 - 1$ $= \dfrac{1}{2}\cdot 1 - 1 = -\dfrac{1}{2}$	$\left(0, -\dfrac{1}{2}\right)$
$\dfrac{\pi}{2}$	$y = \dfrac{1}{2}\cos\dfrac{\pi}{2} - 1$ $= \dfrac{1}{2}\cdot 0 - 1 = -1$	$\left(\dfrac{\pi}{2}, -1\right)$
π	$y = \dfrac{1}{2}\cos\pi - 1$ $= \dfrac{1}{2}(-1) - 1 = -\dfrac{3}{2}$	$\left(\pi, -\dfrac{3}{2}\right)$
$\dfrac{3\pi}{2}$	$y = \dfrac{1}{2}\cos\dfrac{3\pi}{2} - 1$ $= \dfrac{1}{2}\cdot 0 - 1 = -1$	$\left(\dfrac{3\pi}{2}, -1\right)$
2π	$y = \dfrac{1}{2}\cos 2\pi - 1$ $= \dfrac{1}{2}\cdot 1 - 1 = -\dfrac{1}{2}$	$\left(2\pi, -\dfrac{1}{2}\right)$

The five key points for $y = \frac{1}{2}\cos x - 1$ are shown in Figure 4.55. By connecting the points with a smooth curve, we obtain one period of the graph.

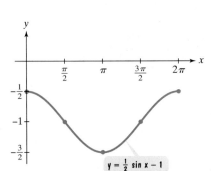

$y = \frac{1}{2}\sin x - 1$

Figure 4.55

Check Point 7   Graph one period of the function $y = 2\cos x + 1$.

6 Model periodic behavior.

Modeling Periodic Behavior

Our breathing consists of alternating periods of inhaling and exhaling. Each complete pumping cycle of the human heart can be described using a sine function. Our brain waves during deep sleep are sinusoidal. Viewed in this way, trigonometry becomes an intimate experience.

Some graphing utilities have a SINe REGression feature. This feature gives the sine function in the form $y = A \sin(Bx + C) + D$ of best fit for wavelike data. At least four data points must be used. However, it is not always necessary to use technology. In our next example, we use our understanding of sinusoidal graphs to model the process of breathing.

EXAMPLE 8   A Trigonometric Breath of Life

The graph in Figure 4.56 shows one complete normal breathing cycle. The cycle consists of inhaling and exhaling. It takes place every 5 seconds. Velocity of air flow is positive when we inhale and negative when we exhale. It is measured in liters per second. If y represents velocity of air flow after x seconds, find a function of the form $y = A \sin Bx$ that models air flow in a normal breathing cycle.

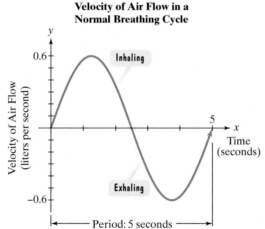

Velocity of Air Flow in a Normal Breathing Cycle

Figure 4.56

Solution   We need to determine values for A and B in the equation $y = A \sin Bx$. A, the amplitude, is the maximum value of y. Figure 4.56 shows that this maximum value is 0.6. Thus, $A = 0.6$.

The value of B in $y = A \sin Bx$ can be found using the formula for the period: period $= \dfrac{2\pi}{B}$. The period of our breathing cycle is 5 seconds. Thus,

$$5 = \frac{2\pi}{B} \qquad \textit{Our goal is to solve this equation for B.}$$

$$5B = 2\pi \qquad \textit{Multiply both sides of the equation by B.}$$

$$B = \frac{2\pi}{5}. \qquad \textit{Divide both sides of the equation by 5.}$$

We see that $A = 0.6$ and $B = \dfrac{2\pi}{5}$. Substitute these values into $y = A \sin Bx$. The breathing cycle is modeled by

$$y = 0.6 \sin \frac{2\pi}{5} x.$$

Check Point 8 Find an equation of the form $y = A \sin Bx$ that produces the graph shown in the figure on the right.

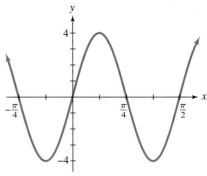

EXAMPLE 9 Modeling a Tidal Cycle

Figure 4.57 shows that the depth of water at a boat dock varies with the tides. The depth is 5 feet at low tide and 13 feet at high tide. On a certain day, low tide occurs at 4 A.M. and high tide at 10 A.M. If y represents the depth of the water, in feet, x hours after midnight, use a sine function of the form $y = A \sin(Bx - C) + D$ to model the water's depth.

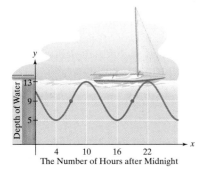

Figure 4.57

Solution We need to determine values for A, B, C, and D in the equation $y = A \sin(Bx - C) + D$. We can find these values using Figure 4.57. We begin with D.

To find D, we use the vertical shift. Because the water's depth ranges from a minimum of 5 feet to a maximum of 13 feet, the curve oscillates about the middle value, 9 feet. Thus, $D = 9$, which is the vertical shift.

At maximum depth, the water is 4 feet above 9 feet. Thus, A, the amplitude, is 4: $A = 4$.

To find B, we use the period. The blue portion of the graph shows that one complete tidal cycle occurs in $19 - 7$, or 12 hours. The period is 12. Thus,

$$12 = \frac{2\pi}{B} \qquad \text{Our goal is to solve this equation for B.}$$

$$12B = 2\pi \qquad \text{Multiply both sides by B.}$$

$$B = \frac{2\pi}{12} = \frac{\pi}{6}. \qquad \text{Divide both sides by 12.}$$

To find C, we use the phase shift. The blue portion of the graph shows that the starting point of the cycle is shifted from 0 to 7. The phase shift, $\frac{C}{B}$, is 7.

$$7 = \frac{C}{B} \qquad \text{The phase shift of } y = A \sin(Bx - C) \text{ is } \frac{C}{B}.$$

$$7 = \frac{C}{\frac{\pi}{6}} \qquad \text{From above, we have } B = \frac{\pi}{6}.$$

$$\frac{7\pi}{6} = C \qquad \text{Multiply both sides of the equation by } \frac{\pi}{6}.$$

Technology

We can use a graphing utility to verify that the model in Example 9

$$y = 4 \sin \left(\frac{\pi}{6} x - \frac{7\pi}{6} \right) + 9$$

is correct. The graph of the function is shown in a $[0, 28, 4]$ by $[0, 15, 5]$ viewing rectangle.

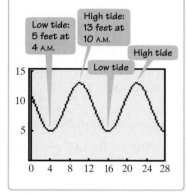

Low tide: 5 feet at 4 A.M.

High tide: 13 feet at 10 A.M.

High tide

Low tide

We see that $A = 4$, $B = \dfrac{\pi}{6}$, $C = \dfrac{7\pi}{6}$, and $D = 9$. Substitute these values into $y = A \sin(Bx - C) + D$. The water's depth, in feet, x hours after midnight is modeled by

$$y = 4 \sin \left(\frac{\pi}{6} x - \frac{7\pi}{6} \right) + 9.$$

Check Point 9 The figure shows the number of hours of daylight for a region that is 30° north of the equator. Hours of daylight are at a minimum of 10 hours in December. Hours of daylight are at a maximum of 14 hours in June. Let x represent the month of the year, with 1 for January, 2 for February, 3 for March, and 12 for December. If y represents the number of hours of daylight in month x, use a sine function of the form $y = A \sin(Bx - C) + D$ to model the hours of daylight.

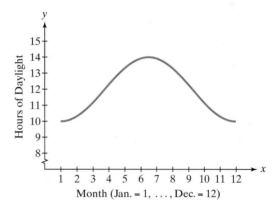

EXERCISE SET 4.5

Practice Exercises

In Exercises 1–6, determine the amplitude of each function. Then graph the function and $y = \sin x$ in the same rectangular coordinate system for $0 \leq x \leq 2\pi$.

1. $y = 4 \sin x$

2. $y = 5 \sin x$

3. $y = \frac{1}{3} \sin x$

4. $y = \frac{1}{4} \sin x$

5. $y = -3 \sin x$

6. $y = -4 \sin x$

In Exercises 7–16, determine the amplitude and period of each function. Then graph one period of the function.

7. $y = \sin 2x$

8. $y = \sin 4x$

9. $y = 3 \sin \frac{1}{2} x$

10. $y = 2 \sin \frac{1}{4} x$

11. $y = 4 \sin \pi x$

12. $y = 3 \sin 2\pi x$

13. $y = -3 \sin 2\pi x$

14. $y = -2 \sin \pi x$

15. $y = -\sin \frac{2}{3} x$

16. $y = -\sin \frac{4}{3} x$

In Exercises 17–30, determine the amplitude, period, and phase shift of each function. Then graph one period of the function.

17. $y = \sin(x - \pi)$

18. $y = \sin \left(x - \dfrac{\pi}{2} \right)$

19. $y = \sin(2x - \pi)$

20. $y = \sin \left(2x - \dfrac{\pi}{2} \right)$

21. $y = 3 \sin(2x - \pi)$ ✗ **22.** $y = 3 \sin\left(2x - \dfrac{\pi}{2}\right)$

23. $y = \frac{1}{2} \sin\left(x + \dfrac{\pi}{2}\right)$ **24.** $y = \frac{1}{2} \sin(x + \pi)$

25. $y = -2 \sin\left(2x + \dfrac{\pi}{2}\right)$ **26.** $y = -3 \sin\left(2x + \dfrac{\pi}{2}\right)$

✗ **27.** $y = 3 \sin(\pi x + 2)$ **28.** $y = 3 \sin(2\pi x + 4)$

29. $y = -2 \sin(2\pi x + 4\pi)$ **30.** $y = -3 \sin(2\pi x + 4\pi)$

In Exercises 31–34, determine the amplitude of each function. Then graph the function and $y = \cos x$ in the same rectangular coordinate system for $0 \le x \le 2\pi$.

31. $y = 2 \cos x$ **32.** $y = 3 \cos x$

33. $y = -2 \cos x$ **34.** $y = -3 \cos x$

In Exercises 35–42, determine the amplitude and period of each function. Then graph one period of the function.

35. $y = \cos 2x$ **36.** $y = \cos 4x$

> **37.** $y = 4 \cos 2\pi x$ **38.** $y = 5 \cos 2\pi x$

39. $y = -4 \cos \frac{1}{2} x$ **40.** $y = -3 \cos \frac{1}{3} x$

41. $y = -\frac{1}{2} \cos \dfrac{\pi}{3} x$ **42.** $y = -\frac{1}{2} \cos \dfrac{\pi}{4} x$

In Exercises 43–50, determine the amplitude, period, and phase shift of each function. Then graph one period of the function.

43. $y = 3 \cos(2x - \pi)$ **44.** $y = 4 \cos(2x - \pi)$

45. $y = \frac{1}{2} \cos\left(3x + \dfrac{\pi}{2}\right)$ ✗ **46.** $y = \frac{1}{2} \cos(2x + \pi)$

47. $y = -3 \cos\left(2x - \dfrac{\pi}{2}\right)$ **48.** $y = -4 \cos\left(2x - \dfrac{\pi}{2}\right)$

49. $y = 2 \cos(2\pi x + 8\pi)$ **50.** $y = 3 \cos(2\pi x + 4\pi)$

In Exercises 51–58, use a vertical shift to graph one period of the function.

51. $y = \sin x + 2$ **52.** $y = \sin x - 2$

53. $y = \cos x - 3$ **54.** $y = \cos x + 3$

55. $y = 2 \sin \frac{1}{2} x + 1$ **56.** $y = 2 \cos \frac{1}{2} x + 1$

57. $y = -3 \cos 2\pi x + 2$ **58.** $y = -3 \sin 2\pi x + 2$

Application Exercises

In the theory of biorhythms, sine functions are used to measure a person's potential. You can obtain your biorhythm chart online by simply entering your date of birth, the date you want your biorhythm chart to begin, and the number of months you wish to be included in the plot. At the top of the next column is your author's chart, beginning January 25, 2003, when he was 21,093 days old. We all have cycles with the same amplitudes and periods as those shown here. Each of our three basic cycles begins at birth. Use the

biorhythm chart shown to solve Exercises 59–66. The longer tick marks correspond to the dates shown.

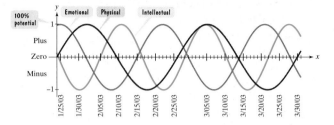

59. What is the period of the physical cycle?

60. What is the period of the emotional cycle?

61. What is the period of the intellectual cycle?

62. For the period shown, what is the worst day in February for your author to run in a marathon?

63. For the period shown, what is the best day in March for your author to meet an online friend for the first time?

64. For the period shown, what is the best day in February for your author to begin writing this trigonometry chapter?

65. If you extend these sinusoidal graphs to the end of the year, is there a day when your author should not even bother getting out of bed?

66. If you extend these sinusoidal graphs to the end of the year, are there any days where your author is at near-peak physical, emotional, and intellectual potential?

67. Rounded to the nearest hour, Los Angeles averages 14 hours of daylight in June, 10 hours in December, and 12 hours in March and September. Let x represent the number of months after June and y represent the number of hours of daylight in month x. Make a graph that displays the information from June of one year to June of the following year.

68. A clock with an hour hand that is 15 inches long is hanging on a wall. At noon, the distance between the tip of the hour hand and the ceiling is 23 inches. At 3 P.M., the distance is 38 inches; at 6 P.M., 53 inches; at 9 P.M., 38 inches; and at midnight the distance is again 23 inches. If y represents the distance between the tip of the hour hand and the ceiling x hours after noon, make a graph that displays the information for $0 \le x \le 24$.

69. The number of hours of daylight in Boston is given by

$$y = 3 \sin \dfrac{2\pi}{365} (x - 79) + 12$$

where x is the number of days after January 1.
a. What is the amplitude of this function?
b. What is the period of this function?

c. How many hours of daylight are there on the longest day of the year?

d. How many hours of daylight are there on the shortest day of the year?

e. Graph the function for one period, starting on January 1.

70. The average monthly temperature, y, in degrees Fahrenheit, for Juneau, Alaska, can be modeled by

$$y = 16 \sin\left(\frac{\pi}{6}x - \frac{2\pi}{3}\right) + 40,$$ where x is the month of

the year (January $= 1$, February $= 2, \ldots$ December $= 12$). Graph the function for $1 \le x \le 12$. What is the highest average monthly temperature? In which month does this occur?

71. The figure shows the depth of water at the end of a boat dock. The depth is 6 feet at low tide and 12 feet at high tide. On a certain day, low tide occurs at 6 A.M. and high tide at noon. If y represents the depth of the water x hours after midnight, use a cosine function of the form $y = A \cos Bx + D$ to model the water's depth.

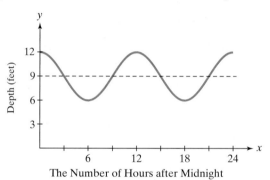

The Number of Hours after Midnight

72. The figure shows the depth of water at the end of a boat dock. The depth is 5 feet at high tide and 3 feet at low tide. On a certain day, high tide occurs at noon and low tide at 6 P.M. If y represents the depth of the water x hours after noon, use a cosine function of the form $y = A \cos Bx + D$ to model the water's depth.

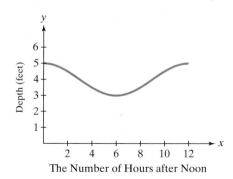

The Number of Hours after Noon

Writing in Mathematics

73. Without drawing a graph, describe the behavior of the basic sine curve.

74. What is the amplitude of the sine function? What does this tell you about the graph?

75. If you are given the equation of a sine function, how do you determine the period?

76. What does a phase shift indicate about the graph of a sine function? How do you determine the phase shift from the function's equation?

77. Describe a general procedure for obtaining the graph of $y = A \sin(Bx - C)$.

78. Without drawing a graph, describe the behavior of the basic cosine curve.

79. Describe a relationship between the graphs of $y = \sin x$ and $y = \cos x$.

80. Describe the relationship between the graphs of $y = A \cos(Bx - C)$ and $y = A \cos(Bx - C) + D$.

81. Biorhythm cycles provide interesting applications of sinusoidal graphs. But do you believe in the validity of biorhythms? Write a few sentences explaining why or why not.

Technology Exercises

82. Use a graphing utility to verify any five of the sine curves that you drew by hand in Exercises 7–30. The amplitude, period, and phase shift should help you to determine appropriate range settings.

83. Use a graphing utility to verify any five of the cosine curves that you drew by hand in Exercises 31–50.

84. Use a graphing utility to verify any two of the sinusoidal curves with vertical shifts that you drew in Exercises 51–58.

In Exercises 85–88, use a graphing utility to graph two periods of the function.

85. $y = 3 \sin(2x + \pi)$ **86.** $y = -2 \cos\left(2\pi x - \frac{\pi}{2}\right)$

87. $y = 0.2 \sin\left(\frac{\pi}{10}x + \pi\right)$ **88.** $y = 3 \sin(2x - \pi) + 5$

89. Use a graphing utility to graph $y = \sin x$ and $y = x - \frac{x^3}{6} + \frac{x^5}{120}$ in a $\left[-\pi, \pi, \frac{\pi}{2}\right]$ by $[-2, 2, 1]$ viewing rectangle. How do the graphs compare?

90. Use a graphing utility to graph $y = \cos x$ and $y = 1 - \dfrac{x^2}{2} + \dfrac{x^4}{24}$ in a $\left[-\pi, \pi, \dfrac{\pi}{2}\right]$ by $[-2, 2, 1]$ viewing rectangle. How do the graphs compare?

91. Use a graphing utility to graph

$$y = \sin x + \frac{\sin 2x}{2} + \frac{\sin 3x}{3} + \frac{\sin 4x}{4}$$

in a $\left[-2\pi, 2\pi, \dfrac{\pi}{2}\right]$ by $[-2, 2, 1]$ viewing rectangle. How do these waves compare to the smooth rolling waves of the basic sine curve?

92. Use a graphing utility to graph

$$y = \sin x - \frac{\sin 3x}{9} + \frac{\sin 5x}{25}$$

in a $\left[-2\pi, 2\pi, \dfrac{\pi}{2}\right]$ by $[-2, 2, 1]$ viewing rectangle. How do these waves compare to the smooth rolling waves of the basic sine curve?

93. The data show the average monthly temperatures for Washington, D.C.

  a. Use your graphing utility to draw a scatter plot of the data from $x = 1$ through $x = 12$.

  b. Use the SINe REGression feature to find the sinusoidal function of the form $y = A \sin(Bx + C) + D$ that best fits the data.

  c. Use your graphing utility to draw the sinusoidal function of best fit on the scatter plot.

x Month		Average Monthly Temperature, °F
1	(January)	34.6
2	(February)	37.5
3	(March)	47.2
4	(April)	56.5
5	(May)	66.4
6	(June)	75.6
7	(July)	80.0
8	(August)	78.5
9	(September)	71.3
10	(October)	59.7
11	(November)	49.8
12	(December)	39.4

Source: U.S. National Oceanic and Atmospheric Administration

94. Repeat Exercise 93 for data of your choice. The data can involve the average monthly temperatures for the region where you live or any data whose scatter plot takes the form of a sinusoidal function.

Critical Thinking Exercises

Graph each function in Exercises 95–96 by hand.

95. $y = \sin x + \cos x$ for $0 \le x \le 2\pi$

96. $y = x + \cos x$ for $0 \le x \le \dfrac{5\pi}{2}$

97. Use the cosine function to find an equation of the graph in the figure shown.

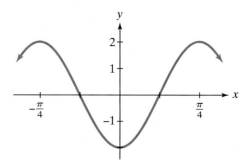

Group Exercise

98. This exercise is intended to provide some fun with biorhythms, regardless of whether you believe they have any validity. We will use each member's chart to determine biorhythmic compatibility. Before meeting, each group member should go online and obtain his or her biorhythm chart. The date of the group meeting is the date on which your chart should begin. Include 12 months in the plot. At the meeting, compare differences and similarities among the intellectual sinusoidal curves. Using these comparisons, each person should find the one other person with whom he or she would be most intellectually compatible.

SECTION 4.6   *Graphs of Other Trigonometric Functions*

Objectives

1. Understand the graph of $y = \tan x$.
2. Graph variations of $y = \tan x$.
3. Understand the graph of $y = \cot x$.
4. Graph variations of $y = \cot x$.
5. Understand the graphs of $y = \csc x$ and $y = \sec x$.
6. Graph variations of $y = \csc x$ and $y = \sec x$.

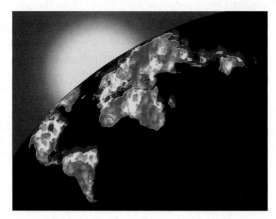

Recent advances in our understanding of climate have changed global warming from a subject for a disaster movie (the Statue of Liberty up to its chin in water) to a serious but manageable scientific and policy issue. Global warming is related to the burning of fossil fuels, which adds carbon dioxide to the atmosphere. In the new millennium, we will see whether our use of fossil fuels will add enough carbon dioxide to the atmosphere to change it (and our climate) in significant ways. In this section's exercise set, you will see how trigonometric graphs reveal interesting patterns in carbon dioxide concentration from 1990 through 2005. In the section itself, trigonometric graphs will reveal patterns involving the tangent, cotangent, secant, and cosecant functions.

1 Understand the graph of $y = \tan x$.

The Graph of $y = \tan x$

The properties of the tangent function discussed in Section 4.2 will help us determine its graph. Because the tangent function has properties that are different from sinusoidal functions, its graph differs significantly from those of sine and cosine. Properties of the tangent function include the following:

- The period is π. It is only necessary to graph $y = \tan x$ over an interval of length π. The remainder of the graph consists of repetitions of that graph at intervals of π.
- The tangent function is an odd function: $\tan(-x) = -\tan x$. The graph is symmetric with respect to the origin.
- The tangent function is undefined at $\frac{\pi}{2}$. The graph of $y = \tan x$ has a vertical asymptote at $x = \frac{\pi}{2}$.

We obtain the graph of $y = \tan x$ using some points on the graph and origin symmetry. Table 4.4 lists some values of (x, y) on the graph of $y = \tan x$ on the interval $\left[0, \frac{\pi}{2} \right)$.

Table 4.4   Values of (x, y) on $y = \tan x$

x	0	$\frac{\pi}{6}$	$\frac{\pi}{4}$	$\frac{\pi}{3}$	$\frac{5\pi}{12}$ $(75°)$	$\frac{17\pi}{36}$ $(85°)$	$\frac{89\pi}{180}$ $(89°)$	1.57	$\frac{\pi}{2}$
$y = \tan x$	0	$\frac{\sqrt{3}}{3} \approx 0.6$	1	$\sqrt{3} \approx 1.7$	3.7	11.4	57.3	1255.8	undefined

As x increases from 0 to $\frac{\pi}{2}$, y increases slowly at first, then more and more rapidly.

The graph in Figure 4.58(a) is based on our observation that as x increases from 0 to $\frac{\pi}{2}$, y increases slowly at first, then more and more rapidly. Notice that y increases without bound as x approaches $\frac{\pi}{2}$. As the figure shows, the graph of $y = \tan x$ has a vertical asymptote at $x = \frac{\pi}{2}$.

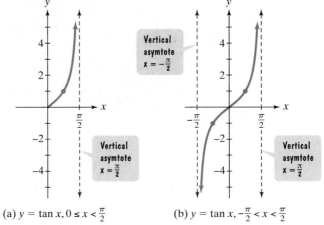

Figure 4.58 Graphing the tangent function

(a) $y = \tan x, 0 \leq x < \frac{\pi}{2}$ (b) $y = \tan x, -\frac{\pi}{2} < x < \frac{\pi}{2}$

The graph of $y = \tan x$ can be completed on the interval $\left(-\frac{\pi}{2}, \frac{\pi}{2}\right)$ by using origin symmetry. Figure 4.58(b) shows the result of reflecting the graph in Figure 4.58(a) about the origin. The graph of $y = \tan x$ has another vertical asymptote at $x = -\frac{\pi}{2}$. Notice that y decreases without bound as x approaches $-\frac{\pi}{2}$.

Because the period of the tangent function is π radians, the graph in Figure 4.58(b) shows one complete period of $y = \tan x$. We obtain the complete graph of $y = \tan x$ by repeating the graph in Figure 4.58(b) to the left and right over intervals of π. The resulting graph and its main characteristics are shown in the following box:

The Tangent Curve: The Graph of $y = \tan x$ and Its Characteristics

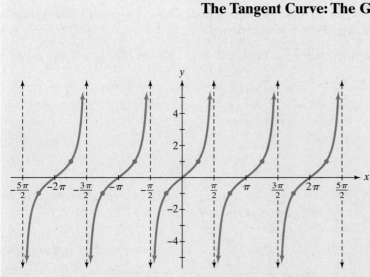

Characteristics

- **Period**: π
- **Domain**: All real numbers except odd multiples of $\frac{\pi}{2}$
- **Range**: All real numbers
- **Vertical asymptotes** at odd multiples of $\frac{\pi}{2}$
- **An x-intercept** occurs midway between each pair of consecutive asymptotes.
- **Odd function** with origin symmetry
- Points on the graph $\frac{1}{4}$ and $\frac{3}{4}$ of the way between consecutive asymptotes have y-coordinates of -1 and 1.

2 Graph variations of $y = \tan x$.

Graphing Variations of $y = \tan x$

We use the characteristics of the tangent curve to graph tangent functions of the form $y = A \tan(Bx - C)$.

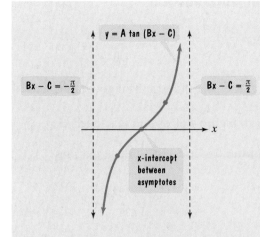

Graphing $y = A \tan(Bx - C)$

1. Find two consecutive asymptotes by setting the variable expression in the tangent equal to $-\dfrac{\pi}{2}$ and $\dfrac{\pi}{2}$ and solving

$$Bx - C = -\frac{\pi}{2} \quad \text{and} \quad Bx - C = \frac{\pi}{2}.$$

2. Identify an x-intercept, midway between the consecutive asymptotes.

3. Find the points on the graph $\dfrac{1}{4}$ and $\dfrac{3}{4}$ of the way between the consecutive asymptotes. These points have y-coordinates of $-A$ and A.

4. Use steps 1–3 to graph one full period of the function. Add additional cycles to the left or right as needed.

EXAMPLE 1 Graphing a Tangent Function

Graph $y = 2 \tan \dfrac{x}{2}$ for $-\pi < x < 3\pi$.

Solution Refer to Figure 4.59 as you read each step.

Step 1 Find two consecutive asymptotes. We must solve the equations

$$\frac{x}{2} = -\frac{\pi}{2} \quad \text{and} \quad \frac{x}{2} = \frac{\pi}{2}. \quad \textit{Set the variable expression in the tangent equal to } -\frac{\pi}{2} \textit{ and } \frac{\pi}{2}.$$

$$x = -\pi \qquad\qquad x = \pi \quad \textit{Multiply both sides of each equation by 2.}$$

Thus, two consecutive asymptotes occur at $x = -\pi$ and $x = \pi$.

Step 2 Identify an x-intercept, midway between the consecutive asymptotes. Midway between $x = -\pi$ and $x = \pi$ is $x = 0$. An x-intercept is 0 and the graph passes through $(0, 0)$.

Step 3 Find points on the graph $\dfrac{1}{4}$ and $\dfrac{3}{4}$ of the way between the consecutive asymptotes. These points have y-coordinates of $-A$ and A. Because A, the coefficient of the tangent in $y = 2 \tan \dfrac{x}{2}$ is 2, these points have y-coordinates of -2 and 2. The graph passes through $\left(-\dfrac{\pi}{2}, -2\right)$ and $\left(\dfrac{\pi}{2}, 2\right)$.

Step 4 Use steps 1–3 to graph one full period of the function. We use the two consecutive asymptotes, $x = -\pi$ and $x = \pi$, an x-intercept of 0, and points midway between the x-intercept and asymptotes with y-coordinates of -2 and 2. We graph one period of $y = 2 \tan \dfrac{\pi}{2}$ from $-\pi$ to π. In order to graph for $-\pi < x < 3\pi$, we continue the pattern and extend the graph another full period to the right. The graph is shown in Figure 4.59.

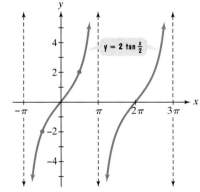

Figure 4.59 The graph is shown for two full periods.

Check Point 1 Graph $y = 3 \tan 2x$ for $-\dfrac{\pi}{4} < x < \dfrac{3\pi}{4}$.

EXAMPLE 2 Graphing a Tangent Function

Graph two full periods of $y = \tan\left(x + \dfrac{\pi}{4}\right)$.

Solution The graph of $y = \tan\left(x + \dfrac{\pi}{4}\right)$ is the graph of $y = \tan x$ shifted horizontally to the left $\dfrac{\pi}{4}$ units. Refer to Figure 4.60 as you read each step.

Step 1 Find two consecutive asymptotes. We must solve the equations

$$x + \frac{\pi}{4} = -\frac{\pi}{2} \quad \text{and} \quad x + \frac{\pi}{4} = \frac{\pi}{2}.$$

Set the variable expression in the tangent equal to $-\dfrac{\pi}{2}$ and $\dfrac{\pi}{2}$.

$$x = -\frac{\pi}{4} - \frac{\pi}{2} \qquad\qquad x = -\frac{\pi}{4} + \frac{\pi}{2}$$

Subtract $\dfrac{\pi}{4}$ from both sides in each equation.

$$x = -\frac{3\pi}{4} \qquad\qquad\qquad x = \frac{\pi}{4}$$

Simplify.

Thus, two consecutive asymptotes occur at $x = -\dfrac{3\pi}{4}$ and $x = \dfrac{\pi}{4}$.

Step 2 Identify an x-intercept, midway between the consecutive asymptotes.

$$x\text{-intercept} = \frac{-\dfrac{3\pi}{4} + \dfrac{\pi}{4}}{2} = \frac{-\dfrac{2\pi}{4}}{2} = -\frac{2\pi}{8} = -\frac{\pi}{4}$$

An x-intercept is $-\dfrac{\pi}{4}$ and the graph passes through $\left(-\dfrac{\pi}{4}, 0\right)$.

Step 3 Find points on the graph $\dfrac{1}{4}$ and $\dfrac{3}{4}$ of the way between the consecutive asymptotes. These points have y-coordinates of $-A$ and A. Because A, the coefficient of the tangent in $y = \tan\left(x + \dfrac{\pi}{4}\right)$ is 1, these points have y-coordinates of -1 and 1. They are shown as blue dots in Figure 4.60.

Step 4 Use steps 1–3 to graph one full period of the function. We use the two consecutive asymptotes, $x = -\dfrac{3\pi}{4}$ and $x = \dfrac{\pi}{4}$, to graph one full period of $y = \tan\left(x + \dfrac{\pi}{4}\right)$ from $-\dfrac{3\pi}{4}$ to $\dfrac{\pi}{4}$. We graph two full periods by continuing the pattern and extending the graph another full period to the right. The graph is shown in Figure 4.60.

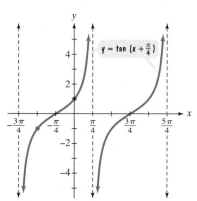

Figure 4.60 The graph is shown for two full periods.

Check Point 2 Graph two full periods of $y = \tan\left(x - \dfrac{\pi}{2}\right)$.

3 Understand the graph of $y = \cot x$.

The Graph of $y = \cot x$

Like the tangent function, the cotangent function, $y = \cot x$, has a period of π. The graph and its main characteristics are shown in the following box:

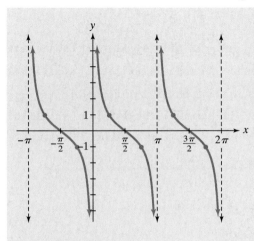

The Cotangent Curve: The Graph of $y = \cot x$ and Its Characteristics

Characteristics

- **Period:** π
- **Domain:** All real numbers except integral multiples of π
- **Range:** All real numbers
- **Vertical asymptotes** at integral multiples of π
- An **x-intercept** occurs midway between each pair of consecutive asymptotes.
- **Odd function** with origin symmetry
- Points on the graph $\dfrac{1}{4}$ and $\dfrac{3}{4}$ of the way between consecutive asymptotes have y-coordinates of 1 and -1.

4 Graph variations of $y = \cot x$.

Graphing Variations of $y = \cot x$

We use the characteristics of the cotangent curve to graph cotangent functions of the form $y = A \cot(Bx - C)$.

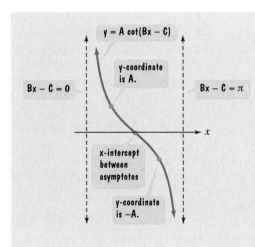

Graphing $y = A \cot(Bx - C)$

1. Find two consecutive asymptotes by setting the variable expression in the cotangent equal to 0 and π and solving

$$Bx - C = 0 \text{ and } Bx - C = \pi.$$

2. Identify an x-intercept, midway between the consecutive asymptotes.

3. Find the points on the graph $\dfrac{1}{4}$ and $\dfrac{3}{4}$ of the way between the consecutive asymptotes. These points have y-coordinates of A and $-A$.

4. Use steps 1–3 to graph one full period of the function. Add additional cycles to the left or right as needed.

EXAMPLE 3   Graphing a Cotangent Function

Graph $y = 3 \cot 2x$.

Solution   Refer to Figure 4.61, shown on the next page, as you read each step.

Step 1   Find two consecutive asymptotes. We must solve the equations

$$2x = 0 \quad \text{and} \quad 2x = \pi. \quad \text{Set the variable expression in the cotangent equal to 0 and } \pi.$$

$$x = 0 \qquad x = \frac{\pi}{2} \quad \text{Divide both sides of each equation by 2.}$$

Two consecutive asymptotes occur at $x = 0$ and $x = \dfrac{\pi}{2}$.

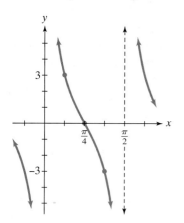

Figure 4.61 The graph of $y = 3 \cot 2x$

Step 2   Identify an *x*-intercept, midway between the consecutive asymptotes. Midway between $x = 0$ and $x = \dfrac{\pi}{2}$ is $x = \dfrac{\pi}{4}$. An *x*-intercept is $\dfrac{\pi}{4}$ and the graph passes through $\left(\dfrac{\pi}{4}, 0 \right)$.

Step 3   Find points on the graph $\dfrac{1}{4}$ and $\dfrac{3}{4}$ of the way between consecutive asymptotes. These points have *y*-coordinates of A and $-A$. Because A, the coefficient of the cotangent in $y = 3 \cot 2x$ is 3, these points have *y*-coordinates of 3 and -3.

Step 4   Use steps 1–3 to graph one full period of the function. We use the two consecutive asymptotes, $x = 0$ and $x = \dfrac{\pi}{2}$, to graph one full period of $y = 3 \cot 2x$. This curve is repeated to the left and right, as shown in Figure 4.61.

> **Check Point 3**   Graph $y = \dfrac{1}{2} \cot \dfrac{\pi}{2} x$.

5 Understand the graphs of $y = \csc x$ and $y = \sec x$.

The Graphs of $y = \csc x$ and $y = \sec x$

We obtain the graphs of the cosecant and secant curves by using the reciprocal identities

$$\csc x = \frac{1}{\sin x} \quad \text{and} \quad \sec x = \frac{1}{\cos x}.$$

The identity on the left tells us that the value of the cosecant function $y = \csc x$ at a given value of x equals the reciprocal of the corresponding value of the sine function, provided that the value of the sine function is not 0. If the value of $\sin x$ is 0, then at each of these values of x, the cosecant function is not defined. A vertical asymptote is associated with each of these values on the graph of $y = \csc x$.

We obtain the graph of $y = \csc x$ by taking reciprocals of the *y*-values in the graph of $y = \sin x$. Vertical asymptotes of $y = \csc x$ occur at the *x*-intercepts of $y = \sin x$. Likewise, we obtain the graph of $y = \sec x$ by taking the reciprocal of $y = \cos x$. Vertical asymptotes of $y = \sec x$ occur at the *x*-intercepts of $y = \cos x$. The graphs of $y = \csc x$ and $y = \sec x$ and their key characteristics are shown in the following boxes. We have used dashed red lines to first graph $y = \sin x$ and $y = \cos x$, drawing vertical asymptotes through the *x*-intercepts.

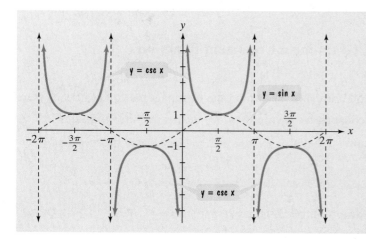

The Cosecant Curve: The Graph of $y = \csc x$ and Its Characteristics

Characteristics

- **Period:** 2π
- **Domain:** All real numbers except integral multiples of π
- **Range:** All real numbers y such that $y \le -1$ or $y \ge 1$
- **Vertical asymptotes** at integral multiples of π
- **Odd function,** $\csc(-x) = -\csc x$, with origin symmetry

The Secant Curve: The Graph of $y = \sec x$ and Its Characteristics

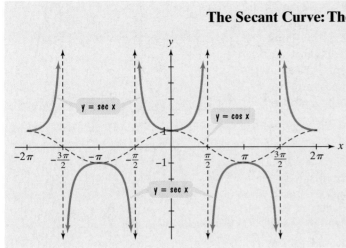

Characteristics

- **Period:** 2π
- **Domain:** All real numbers except odd multiples of $\dfrac{\pi}{2}$
- **Range:** All real numbers y such that $y \leq -1$ or $y \geq 1$
- **Vertical asymptotes** at odd multiples of $\dfrac{\pi}{2}$
- **Even function,** $\sec(-x) = \sec x$, with y-axis symmetry

6 Graph variations of $y = \csc x$ and $y = \sec x$.

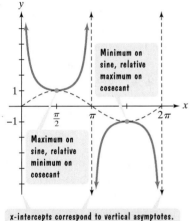

x-intercepts correspond to vertical asymptotes.

Figure 4.62

Graphing Variations of $y = \csc x$ and $y = \sec x$

We use graphs of reciprocal functions to obtain graphs of cosecant and secant functions. To graph a cosecant or secant curve, begin by graphing the reciprocal function. For example, to graph $y = 2 \csc 2x$, we use the graph of $y = 2 \sin 2x$. Likewise, to graph $y = -3 \sec \dfrac{x}{2}$, we use the graph of $y = -3 \cos \dfrac{x}{2}$.

Figure 4.62 illustrates how we use a sine curve to obtain a cosecant curve. Notice that

- x-intercepts on the red sine curve correspond to vertical asymptotes of the blue cosecant curve.
- A maximum point on the red sine curve corresponds to a minimum point on a continuous portion of the blue cosecant curve.
- A minimum point on the red sine curve corresponds to a maximum point on a continuous portion of the blue cosecant curve.

EXAMPLE 4 Using a Sine Curve to Obtain a Cosecant Curve

Use the graph of $y = 2 \sin 2x$ in Figure 4.63 to obtain the graph of $y = 2 \csc 2x$.

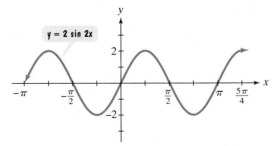

Figure 4.63

Solution On the next page, we begin our work by showing the given graph, the graph of $y = 2 \sin 2x$, using dashed red lines.

The x-intercepts of $y = 2 \sin 2x$ correspond to the vertical asymptotes of $y = 2 \csc 2x$. Thus, we draw vertical asymptotes through the x-intercepts, shown in Figure 4.64. Using the asymptotes as guides, we sketch the graph of $y = 2 \csc 2x$ in Figure 4.64.

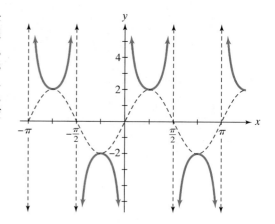

Figure 4.64 Using a sine curve to graph $y = 2 \csc 2x$

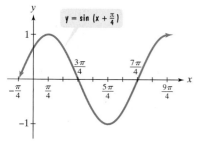

Check Point 4 Use the graph of $y = \sin\left(x + \dfrac{\pi}{4}\right)$, shown on the left, to obtain the graph of $y = \csc\left(x + \dfrac{\pi}{4}\right)$.

We use a cosine curve to obtain a secant curve in exactly the same way we used a sine curve to obtain a cosecant curve. Thus,

- x-intercepts on the cosine curve correspond to vertical asymptotes on the secant curve.
- A maximum point on the cosine curve corresponds to a minimum point on a continuous portion of the secant curve.
- A minimum point on the cosine curve corresponds to a maximum point on a continuous portion of the secant curve.

EXAMPLE 5 Graphing a Secant Function

Graph $y = -3 \sec \dfrac{x}{2}$ for $-\pi < x < 5\pi$.

Solution We begin by graphing the reciprocal cosine function, $y = -3 \cos \dfrac{x}{2}$. This equation is of the form $y = A \cos Bx$ with $A = -3$ and $B = \frac{1}{2}$.

amplitude: $|A| = |-3| = 3$ — The maximum y is 3 and the minimum is -3.

period: $\dfrac{2\pi}{B} = \dfrac{2\pi}{\frac{1}{2}} = 4\pi$ — Each cycle is completed in 4π radians.

We use quarter-periods, $\dfrac{4\pi}{4}$, or π, to find the x-values for the five key points. Starting with $x = 0$, the x-values are $0, \pi, 2\pi, 3\pi,$ and 4π. Evaluating the function at each of these values of x, the key points are

$$(0, -3), (\pi, 0), (2\pi, 3), (3\pi, 0), \text{ and } (4\pi, -3).$$

We use these key points to graph $y = -3 \cos \dfrac{x}{2}$ from 0 to 4π. In order to graph for $-\pi \le x \le 5\pi$, extend the graph π units to the left and π units to the right. The graph is shown using a dashed red line in Figure 4.65. Now use this dashed red graph to obtain the graph of the reciprocal function. Draw vertical asymptotes through the x-intercepts. Using these asymptotes as guides, the graph of $y = -3 \sec \dfrac{x}{2}$ is shown in blue in Figure 4.65.

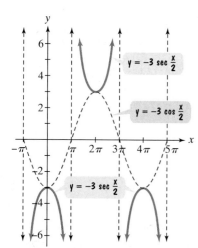

Figure 4.65 Using a cosine curve to graph $y = -3 \sec \dfrac{x}{2}$

Check Point 5 Graph $y = 2 \sec 2x$ for $-\dfrac{3\pi}{4} < x < \dfrac{3\pi}{4}$.

The Six Curves of Trigonometry

Table 4.5 summarizes the graphs of the six trigonometric functions. Below each of the graphs is a description of the domain, range, and period of the function.

Table 4.5 Graphs of the Six Trigonometric Functions

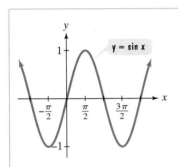

Domain: all real numbers

Range: $[-1, 1]$
Period: 2π

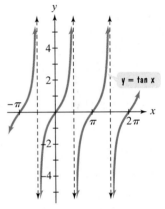

Domain: all real numbers

Range: $[-1, 1]$
Period: 2π

Domain: all real numbers
except odd multiples of $\dfrac{\pi}{2}$

Range: all real numbers
Period: π

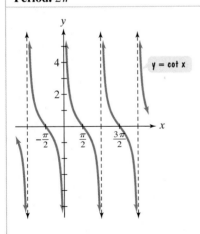

Domain: all real numbers
except integral multiples of π

Range: all real numbers
Period: π

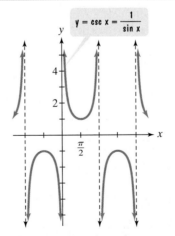

Domain: all real numbers
except integral multiples of π

Range: $(-\infty, -1]$ or $[1, \infty)$
Period: 2π

Domain: all real numbers
except odd multiples of $\dfrac{\pi}{2}$

Range: $(-\infty, -1]$ or $[1, \infty)$
Period: 2π

EXERCISE SET 4.6

Practice Exercises

In Exercises 1–4, the graph of a tangent function is given.
Select the equation for each graph from the following options:

$$y = \tan\left(x + \frac{\pi}{2}\right), \quad y = \tan(x + \pi), \quad y = -\tan x, \quad y = -\tan\left(x - \frac{\pi}{2}\right).$$

1.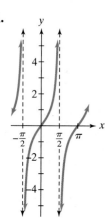

2.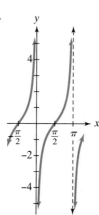

3.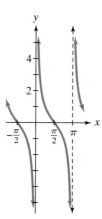

4.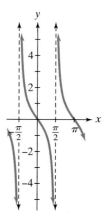

In Exercises 5–12, graph two periods of the given tangent function.

5. $y = 3\tan\dfrac{x}{4}$

6. $y = 2\tan\dfrac{x}{4}$

7. $y = \dfrac{1}{2}\tan 2x$

8. $y = 3\tan 2x$

9. $y = -2\tan\dfrac{1}{2}x$

10. $y = -3\tan\dfrac{1}{2}x$

11. $y = \tan(x - \pi)$

12. $y = \tan\left(x + \dfrac{\pi}{2}\right)$

In Exercises 13–16, the graph of a cotangent function is given. Select the equation for each graph from the following options:

$$y = \cot\left(x + \frac{\pi}{2}\right), \quad y = \cot(x + \pi), \quad y = -\cot x, \quad y = -\cot\left(x - \frac{\pi}{2}\right).$$

13.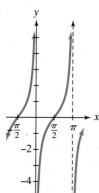

14.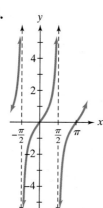

15.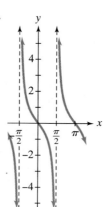

16.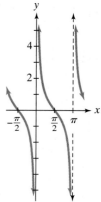

In Exercises 17–24, graph two periods of the given cotangent function.

17. $y = 2\cot x$

18. $y = \dfrac{1}{2}\cot x$

19. $y = \dfrac{1}{2}\cot 2x$

20. $y = 2\cot 2x$

21. $y = -3 \cot \dfrac{\pi}{2} x$

22. $y = -2 \cot \dfrac{\pi}{4} x$

23. $y = 3 \cot\left(x + \dfrac{\pi}{2}\right)$

24. $y = 3 \cot\left(x + \dfrac{\pi}{4}\right)$

In Exercises 25–28, use each graph to obtain the graph of the reciprocal function. Give the equation of the function for the graph that you obtain.

25.

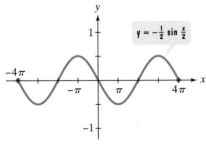

$y = -\dfrac{1}{2}\sin\dfrac{x}{2}$

26.

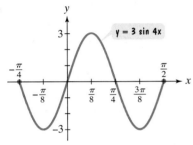

$y = 3\sin 4x$

27.

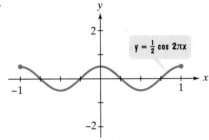

$y = \dfrac{1}{2}\cos 2\pi x$

28.

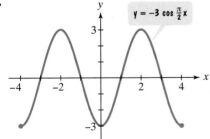

$y = -3\cos\dfrac{\pi}{2}x$

In Exercises 29–44, graph two periods of the given cosecant or secant function.

29. $y = 3 \csc x$

30. $y = 2 \csc x$

31. $y = \dfrac{1}{2}\csc\dfrac{x}{2}$

32. $y = \dfrac{3}{2}\csc\dfrac{x}{4}$

33. $y = 2\sec x$

34. $y = 3\sec x$

35. $y = \sec\dfrac{x}{3}$

36. $y = \sec\dfrac{x}{2}$

37. $y = -2\csc \pi x$

38. $y = -\dfrac{1}{2}\csc \pi x$

39. $y = -\dfrac{1}{2}\sec \pi x$

40. $y = -\dfrac{3}{2}\sec \pi x$

41. $y = \csc(x - \pi)$

42. $y = \csc\left(x - \dfrac{\pi}{2}\right)$

43. $y = 2\sec(x + \pi)$

44. $y = 2\sec\left(x + \dfrac{\pi}{2}\right)$

Application Exercises

45. An ambulance with a rotating beacon of light is parked 12 feet from a building. The function

$$d = 12 \tan 2\pi t$$

describes the distance, d, in feet, of the rotating beacon from point C after t seconds.

a. Graph the function on the interval $[0, 2]$.

b. For what values of t in $[0, 2]$ is the function undefined? What does this mean in terms of the rotating beacon in the figure shown?

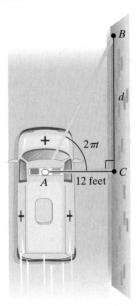

46. The angle of elevation from the top of a house to a jet flying 2 miles above the house is x radians. If d represents the horizontal distance, in miles, of the jet from the house, express d in terms of a trigonometric function of x. Then graph the function for $0 < x < \pi$.

47. Your best friend is marching with a band and has asked you to film her. The figure below shows that you have set yourself up 10 feet from the street where your friend will be passing from left to right. If d represents your distance, in feet, from your friend and x is the radian measure of the angle shown, express d in terms of a trigonometric function of x. Then graph the function for $-\frac{\pi}{2} < x < \frac{\pi}{2}$. Negative angles indicate that your marching buddy is on your left.

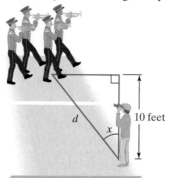

d x 10 feet

In Exercises 48–50, sketch a reasonable graph that models the given situation.

48. The number of hours of daylight per day in your hometown over a two-year period

49. The motion of a diving board vibrating 10 inches in each direction per second just after someone has dived off

50. The distance of a rotating beacon of light from a point on a wall (See the figure for Exercise 45.)

Writing in Mathematics

51. Without drawing a graph, describe the behavior of the basic tangent curve.

52. If you are given the equation of a tangent function, how do you find consecutive asymptotes?

53. If you are given the equation of a tangent function, how do you identify an x-intercept?

54. Without drawing a graph, describe the behavior of the basic cotangent curve.

55. If you are given the equation of a cotangent function, how do you find consecutive asymptotes?

56. Explain how to determine the range of $y = \csc x$ from the graph. What is the range?

57. Explain how to use a sine curve to obtain a cosecant curve. Why can the same procedure be used to obtain a secant curve from a cosine curve?

58. Scientists record brain activity by attaching electrodes to the scalp and then connecting these electrodes to a machine. The record of brain activity recorded with this machine is shown in the three graphs at the top of the next column. Which trigonometric functions would be most appropriate for describing the oscillations in brain

activity? Describe similarities and differences among these functions when modeling brain activity when awake, during dreaming sleep, and during nondreaming sleep.

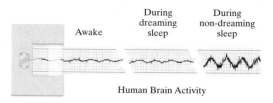

Awake   During dreaming sleep   During non-dreaming sleep

Human Brain Activity

Technology Exercises

In working Exercises 59–62, describe what happens at the asymptotes on the graphing utility. Compare the graphs in the connected and dot modes.

59. Use a graphing utility to verify any two of the tangent curves that you drew by hand in Exercises 5–12.

60. Use a graphing utility to verify any two of the cotangent curves that you drew by hand in Exercises 17–24.

61. Use a graphing utility to verify any two of the cosecant curves that you drew by hand in Exercises 29–44.

62. Use a graphing utility to verify any two of the secant curves that you drew by hand in Exercises 29–44.

In Exercises 63–68, use a graphing utility to graph each function. Use a range setting so that the graph is shown for at least two periods.

63. $y = \tan \frac{x}{4}$

64. $y = \tan 4x$

65. $y = \cot 2x$

66. $y = \cot \frac{x}{2}$

67. $y = \frac{1}{2} \tan \pi x$

68. $y = \frac{1}{2} \tan(\pi x + 1)$

In Exercises 69–72, use a graphing utility to graph each pair of functions in the same viewing rectangle. Use a range setting so that the graphs are shown for at least two periods.

69. $y = 0.8 \sin \frac{x}{2}$ and $y = 0.8 \csc \frac{x}{2}$

70. $y = -2.5 \sin \frac{\pi}{3} x$ and $y = -2.5 \csc \frac{\pi}{3} x$

71. $y = 4 \cos \left(2x - \frac{\pi}{6} \right)$ and $y = 4 \sec \left(2x - \frac{\pi}{6} \right)$

72. $y = -3.5 \cos \left(\pi x - \frac{\pi}{6} \right)$ and $y = -3.5 \sec \left(\pi x - \frac{\pi}{6} \right)$

73. Carbon dioxide particles in our atmosphere trap heat and raise the planet's temperature. The resultant gradually increasing temperature is called the greenhouse effect. Carbon dioxide accounts for about half of global warming. The function

$$y = 2.5 \sin 2\pi x + 0.0216x^2 + 0.654x + 316$$

models carbon dioxide concentration, y, in parts per million, where $x = 0$ represents January 1960; $x = \frac{1}{12}$, February 1960; $x = \frac{2}{12}$, March 1960; $\cdots$, $x = 1$, January 1961; $x = \frac{13}{12}$, February 1961; and so on. Use a graphing utility to graph the function in a $[30, 45, 5]$ by $[310, 420, 5]$

viewing rectangle. Describe what the graph reveals about carbon dioxide concentration from 1990 through 2005.

74. Graph $y = \sin\dfrac{1}{x}$ in a $[-0.2, 0.2, 0.01]$ by $[-1.2, 1.2, 0.01]$

viewing rectangle. What is happening as x approaches 0 from the left or the right? Explain this behavior.

 Critical Thinking Exercises

In Exercises 75–76, write an equation for each blue graph.

75.

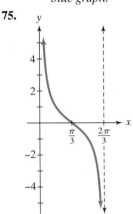

76.

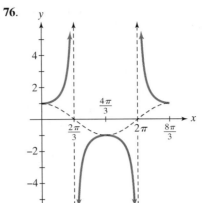

77. For $x > 0$, what effect does 2^{-x} in $y = 2^{-x} \sin x$ have on the graph of $y = \sin x$? What kind of behavior can be modeled by a function such as $y = 2^{-x} \sin x$?

SECTION 4.7 Inverse Trigonometric Functions

Objectives

1. Understand and use the inverse sine function.
2. Understand and use the inverse cosine function.
3. Understand and use the inverse tangent function.
4. Use a calculator to evaluate inverse trigonometric functions.
5. Find exact values of composite functions with inverse trigonometric functions.

You watched *The Matrix* on video and were impressed by the elaborate computer-generated effects. The movie is being shown again at a local theater, where you can experience its stunning visual force on a large screen. Where in the theater should you sit to maximize the film's visual impact? In this section's exercise set, you will see how an inverse trigonometric function can enhance your movie-going experiences.

Study Tip

Here are some helpful things to remember from our discussion of inverse functions in Section 1.8.

- If no horizontal line intersects the graph of a function more than once, the function is one-to-one and has an inverse function.
- If the point (a, b) is on the graph of f, then the point (b, a) is on the graph of the inverse function, denoted f^{-1}. The graph of f^{-1} is a reflection of the graph of f about the line $y = x$.

1 Understand and use the inverse sine function.

The Inverse Sine Function

Figure 4.66 shows the graph of $y = \sin x$. Can you see that every horizontal line that can be drawn between -1 and 1 intersects the graph infinitely many times? Thus, the sine function is not one-to-one and has no inverse function.

In Figure 4.67, we have taken a portion of the sine curve, restricting the domain of the sine function to $-\dfrac{\pi}{2} \le x \le \dfrac{\pi}{2}$. With this restricted domain, every horizontal line that can be drawn between -1 and 1 intersects the graph exactly once. Thus, the restricted function passes the horizontal line test and is one-to-one.

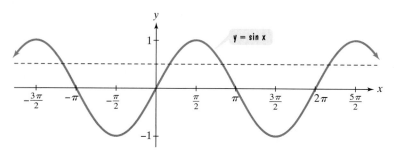

Figure 4.66 The horizontal line test shows that the sine function is not one-to-one and has no inverse function.

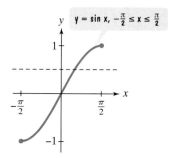

Figure 4.67 The restricted sine function passes the horizontal line test. It is one-to-one and has an inverse function.

On the restricted domain $-\dfrac{\pi}{2} \le x \le \dfrac{\pi}{2}$, $y = \sin x$ has an inverse function.

The inverse of the restricted sine function is called the **inverse sine function.** Two notations are commonly used to denote the inverse sine function:

$$y = \sin^{-1} x \quad \text{or} \quad y = \arcsin x.$$

In this book, we will use $y = \sin^{-1} x$. This notation has the same symbol as the inverse function notation $f^{-1}(x)$.

The Inverse Sine Function

The **inverse sine function,** denoted by $\sin^{-1}$, is the inverse of the restricted sine function $y = \sin x, -\dfrac{\pi}{2} \le x \le \dfrac{\pi}{2}$. Thus,

$$y = \sin^{-1} x \quad \text{means} \quad \sin y = x$$

where $-\dfrac{\pi}{2} \le y \le \dfrac{\pi}{2}$ and $-1 \le x \le 1$. We read $y = \sin^{-1} x$ as "y equals the inverse sine at x."

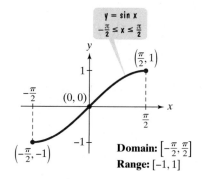

Figure 4.68 The restricted sine function

One way to graph $y = \sin^{-1} x$ is to take points on the graph of the restricted sine function and reverse the order of the coordinates. For example, Figure 4.68 shows that $\left(-\dfrac{\pi}{2}, -1\right)$, $(0, 0)$, and $\left(\dfrac{\pi}{2}, 1\right)$ are on the graph of the restricted sine function. Reversing the order of the coordinates gives $\left(-1, -\dfrac{\pi}{2}\right)$, $(0, 0)$, and $\left(1, \dfrac{\pi}{2}\right)$. We now use these three points to sketch the inverse sine function. The graph of $y = \sin^{-1} x$ is shown in Figure 4.69.

Another way to obtain the graph of $y = \sin^{-1} x$ is to reflect the graph of the restricted sine function about the line $y = x$, shown in Figure 4.70. The red graph is the restricted sine function and the blue graph is the graph of $y = \sin^{-1} x$.

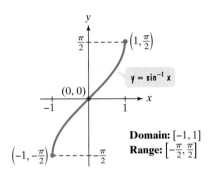

Domain: $[-1, 1]$
Range: $\left[-\dfrac{\pi}{2}, \dfrac{\pi}{2}\right]$

Figure 4.69 The graph of the inverse sine function

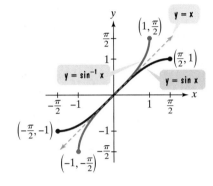

Figure 4.70 Using a reflection to obtain the graph of the inverse sine function

Exact values of $\sin^{-1} x$ can be found by thinking of **$\sin^{-1} x$ as the angle in the interval $\left[-\dfrac{\pi}{2}, \dfrac{\pi}{2}\right]$ whose sine is x.** For example, we can use the two points on the blue graph of the inverse sine function in Figure 4.70 and write

$$\sin^{-1}(-1) = -\frac{\pi}{2} \quad \text{and} \quad \sin^{-1} 1 = \frac{\pi}{2}.$$

The angle whose sine is -1 is $-\dfrac{\pi}{2}$.

The angle whose sine is 1 is $\dfrac{\pi}{2}$.

Because we are thinking of $\sin^{-1} x$ in terms of an angle, we will represent such an angle by θ.

Finding Exact Values of $\sin^{-1} x$.

1. Let $\theta = \sin^{-1} x$.
2. Rewrite $\theta = \sin^{-1} x$ as $\sin \theta = x$.
3. Use the exact values in Table 4.6, shown on the next page, to find the value of θ in $\left[-\dfrac{\pi}{2}, \dfrac{\pi}{2}\right]$ that satisfies $\sin \theta = x$.

Table 4.6   Exact Values for
$$\sin\theta, -\frac{\pi}{2} \le \theta \le \frac{\pi}{2}$$

θ	$\sin\theta$
$-\dfrac{\pi}{2}$	-1
$-\dfrac{\pi}{3}$	$-\dfrac{\sqrt{3}}{2}$
$-\dfrac{\pi}{4}$	$-\dfrac{\sqrt{2}}{2}$
$-\dfrac{\pi}{6}$	$-\dfrac{1}{2}$
0	0
$\dfrac{\pi}{6}$	$\dfrac{1}{2}$
$\dfrac{\pi}{4}$	$\dfrac{\sqrt{2}}{2}$
$\dfrac{\pi}{3}$	$\dfrac{\sqrt{3}}{2}$
$\dfrac{\pi}{2}$	1

EXAMPLE 1   Finding the Exact Value of an Inverse Sine Function

Find the exact value of $\sin^{-1}\dfrac{\sqrt{2}}{2}$.

Solution

Step 1   Let $\theta = \sin^{-1}x$. Thus,

$$\theta = \sin^{-1}\frac{\sqrt{2}}{2}.$$

We must find the angle θ, $-\dfrac{\pi}{2} \le \theta \le \dfrac{\pi}{2}$, whose sine equals $\dfrac{\sqrt{2}}{2}$.

Step 2   Rewrite $\theta = \sin^{-1}x$ as $\sin\theta = x$. Using the definition of the inverse sine function, we rewrite $\theta = \sin^{-1}\dfrac{\sqrt{2}}{2}$ as

$$\sin\theta = \frac{\sqrt{2}}{2}.$$

Step 3   Use the exact values in Table 4.6 to find the value of θ in $\left[-\dfrac{\pi}{2}, \dfrac{\pi}{2}\right]$ that satisfies $\sin\theta = x$. Table 4.6 shows that the only angle in the interval $\left[-\dfrac{\pi}{2}, \dfrac{\pi}{2}\right]$ that satisfies $\sin\theta = \dfrac{\sqrt{2}}{2}$ is $\dfrac{\pi}{4}$. Thus, $\theta = \dfrac{\pi}{4}$. Because θ, in step 1, represents $\sin^{-1}\dfrac{\sqrt{2}}{2}$, we conclude that

$$\sin^{-1}\frac{\sqrt{2}}{2} = \frac{\pi}{4}. \quad \text{The angle in } \left[-\frac{\pi}{2}, \frac{\pi}{2}\right] \text{ whose sine is } \frac{\sqrt{2}}{2} \text{ is } \frac{\pi}{4}.$$

Check Point 1   Find the exact value of $\sin^{-1}\dfrac{\sqrt{3}}{2}$.

EXAMPLE 2   Finding the Exact Value of an Inverse Sine Function

Find the exact value of $\sin^{-1}\left(-\dfrac{1}{2}\right)$.

Solution

Step 1   Let $\theta = \sin^{-1}x$. Thus,

$$\theta = \sin^{-1}\left(-\frac{1}{2}\right).$$

We must find the angle θ, $-\dfrac{\pi}{2} \le \theta \le \dfrac{\pi}{2}$, whose sine equals $-\dfrac{1}{2}$.

Step 2   Rewrite $\theta = \sin^{-1}x$ as $\sin\theta = x$. We rewrite $\theta = \sin^{-1}\left(-\dfrac{1}{2}\right)$ and obtain

$$\sin\theta = -\frac{1}{2}.$$

Step 3   Use the exact values in Table 4.6 to find the value of θ in $\left[-\dfrac{\pi}{2}, \dfrac{\pi}{2}\right]$ that satisfies $\sin\theta = x$. Table 4.6, on page 000, shows that the only angle in the interval $\left[-\dfrac{\pi}{2}, \dfrac{\pi}{2}\right]$ that satisfies $\sin\theta = -\dfrac{1}{2}$ is $-\dfrac{\pi}{6}$. Thus,

$$\sin^{-1}\left(-\frac{1}{2}\right) = -\frac{\pi}{6}$$

Check Point 2   Find the exact value of $\sin^{-1}\left(-\dfrac{\sqrt{2}}{2}\right)$.

Some inverse sine expressions cannot be evaluated. Because the domain of the inverse sine function is $[-1, 1]$, it is only possible to evaluate $\sin^{-1}x$ for values of x in this domain. Thus, $\sin^{-1}3$ cannot be evaluated. There is no angle whose sine is 3.

2   Understand and use the inverse cosine function.

The Inverse Cosine Function

Figure 4.71 shows how we restrict the domain of the cosine function so that it becomes one-to-one and has an inverse function. Restrict the domain to the interval $[0, \pi]$, shown by the dark blue graph. Over this interval, the restricted cosine function passes the horizontal line test and has an inverse function.

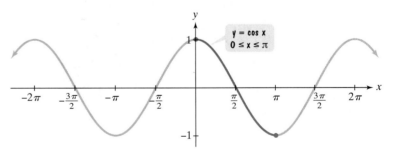

Figure 4.71 $y = \cos x$ is one-to-one on the interval $[0, \pi]$.

The Inverse Cosine Function

The **inverse cosine function,** denoted by $\cos^{-1}$, is the inverse of the restricted cosine function $y = \cos x, 0 \le x \le \pi$. Thus,

$$y = \cos^{-1}x \quad \text{means} \quad \cos y = x$$

where $0 \le y \le \pi$ and $-1 \le x \le 1$.

One way to graph $y = \cos^{-1} x$ is to take points on the graph of the restricted cosine function and reverse the order of the coordinates. For example, Figure 4.72 shows that $(0, 1)$, $\left(\dfrac{\pi}{2}, 0\right)$, and $(\pi, -1)$ are on the graph of the restricted cosine function. Reversing the order of the coordinates gives $(1, 0)$, $\left(0, \dfrac{\pi}{2}\right)$, and $(-1, \pi)$.

We now use these three points to sketch the inverse cosine function. The graph of $y = \cos^{-1} x$ is shown in Figure 4.73. You can also obtain this graph by reflecting the graph of the restricted cosine function about the line $y = x$.

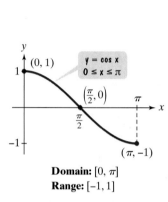

Domain: $[0, \pi]$
Range: $[-1, 1]$

Figure 4.72 The restricted cosine function

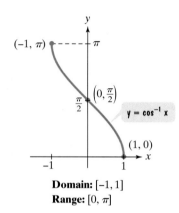

Domain: $[-1, 1]$
Range: $[0, \pi]$

Figure 4.73 The graph of the inverse cosine function

Table 4.7 Exact Values for $\cos\theta$, $0 \le \theta \le \pi$

θ	$\cos\theta$
0	1
$\dfrac{\pi}{6}$	$\dfrac{\sqrt{3}}{2}$
$\dfrac{\pi}{4}$	$\dfrac{\sqrt{2}}{2}$
$\dfrac{\pi}{3}$	$\dfrac{1}{2}$
$\dfrac{\pi}{2}$	0
$\dfrac{2\pi}{3}$	$-\dfrac{1}{2}$
$\dfrac{3\pi}{4}$	$-\dfrac{\sqrt{2}}{2}$
$\dfrac{5\pi}{6}$	$-\dfrac{\sqrt{3}}{2}$
π	-1

Exact values of $\cos^{-1} x$ can be found by thinking of $\cos^{-1} x$ as **the angle in the interval $[0, \pi]$ whose cosine is x.** This time we will use Table 4.7, which shows exact values for $\cos\theta$ for θ in the interval $[0, \pi]$.

EXAMPLE 3   Finding the Exact Value of an Inverse Cosine Function

Find the exact value of $\cos^{-1}\left(-\dfrac{\sqrt{3}}{2}\right)$.

Solution

Step 1   Let $\theta = \cos^{-1} x$. Thus,

$$\theta = \cos^{-1}\left(-\dfrac{\sqrt{3}}{2}\right).$$

We must find the angle θ, $0 \le \theta \le \pi$, whose cosine equals $-\dfrac{\sqrt{3}}{2}$.

Step 2   Rewrite $\theta = \cos^{-1} x$ as $\cos\theta = x$. We obtain

$$\cos\theta = -\dfrac{\sqrt{3}}{2}.$$

Step 3   **Use the exact values in Table 4.7 to find the value of θ in $[0, \pi]$ that satisfies $\cos \theta = x$.** The table shows that the only angle in the interval $[0, \pi]$ that satisfies $\cos \theta = -\dfrac{\sqrt{3}}{2}$ is $\dfrac{5\pi}{6}$. Thus, $\theta = \dfrac{5\pi}{6}$ and

$$\cos^{-1}\left(-\frac{\sqrt{3}}{2}\right) = \frac{5\pi}{6}. \qquad \text{The angle in } [0, \pi] \text{ whose cosine is } -\frac{\sqrt{3}}{2} \text{ is } \frac{5\pi}{6}.$$

Check Point 3   Find the exact value of $\cos^{-1}\left(-\dfrac{1}{2}\right)$.

3 Understand and use the inverse tangent function.

The Inverse Tangent Function

Figure 4.74 shows how we restrict the domain of the tangent function so that it becomes one-to-one and has an inverse function. Restrict the domain to the interval $\left(-\dfrac{\pi}{2}, \dfrac{\pi}{2}\right)$, shown by the solid blue graph. Over this interval, the restricted tangent function passes the horizontal line test and has an inverse function.

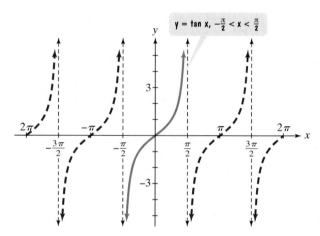

Figure 4.74 $y = \tan x$ is one-to-one on the interval $\left(-\dfrac{\pi}{2}, \dfrac{\pi}{2}\right)$.

The Inverse Tangent Function

The **inverse tangent function,** denoted by $\tan^{-1}$, is the inverse of the restricted tangent function $y = \tan x, -\dfrac{\pi}{2} < x < \dfrac{\pi}{2}$. Thus,

$$y = \tan^{-1} x \quad \text{means} \quad \tan y = x$$

where $-\dfrac{\pi}{2} < y < \dfrac{\pi}{2}$ and $-\infty < x < \infty$.

We graph $y = \tan^{-1} x$ by taking points on the graph of the restricted function and reversing the order of the coordinates. Figure 4.75 shows that $\left(-\frac{\pi}{4}, -1\right)$, $(0, 0)$, and $\left(\frac{\pi}{4}, 1\right)$ are on the graph of the restricted tangent function. Reversing the order gives $\left(-1, -\frac{\pi}{4}\right)$, $(0, 0)$, and $\left(1, \frac{\pi}{4}\right)$. We now use these three points to graph the inverse tangent function. The graph of $y = \tan^{-1} x$ is shown in Figure 4.76. Notice that the vertical asymptotes become horizontal asymptotes for the graph of the inverse function.

Domain: $\left(-\frac{\pi}{2}, \frac{\pi}{2}\right)$
Range: $(-\infty, \infty)$

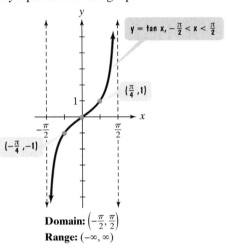

Figure 4.75 The restricted tangent function

Domain: $(-\infty, \infty)$
Range: $\left(-\frac{\pi}{2}, \frac{\pi}{2}\right)$

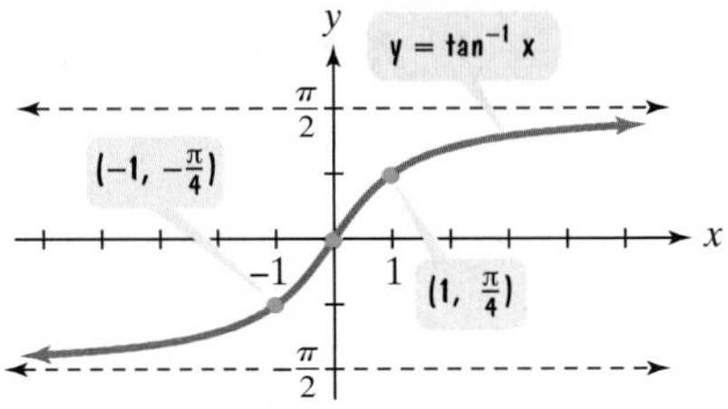

Figure 4.76 The graph of the inverse tangent function

Exact values of $\tan^{-1} x$ can be found by thinking of **$\tan^{-1} x$ as the angle in the interval $\left(-\frac{\pi}{2}, \frac{\pi}{2}\right)$ whose tangent is x.** We use Table 4.8, which shows exact values for $\tan\theta$ for θ in the interval $\left(-\frac{\pi}{2}, \frac{\pi}{2}\right)$.

Table 4.8 Exact Values for $\tan\theta, -\frac{\pi}{2} < \theta < \frac{\pi}{2}$

θ	$\tan\theta$
$-\frac{\pi}{2}$	Undefined
$-\frac{\pi}{3}$	$-\sqrt{3}$
$-\frac{\pi}{4}$	-1
$-\frac{\pi}{6}$	$-\frac{\sqrt{3}}{3}$
0	0
$\frac{\pi}{6}$	$\frac{\sqrt{3}}{3}$
$\frac{\pi}{4}$	1
$\frac{\pi}{3}$	$\sqrt{3}$
$\frac{\pi}{2}$	Undefined

EXAMPLE 4 Finding the Exact Value of an Inverse Tangent Function

Find the exact value of $\tan^{-1}\sqrt{3}$.

Solution

Step 1 Let $\theta = \tan^{-1} x$. Thus,

$$\theta = \tan^{-1}\sqrt{3}.$$

We must find the angle θ, $-\frac{\pi}{2} < \theta < \frac{\pi}{2}$, whose tangent equals $\sqrt{3}$.

Step 2 Rewrite $\theta = \tan^{-1} x$ as $\tan\theta = x$. We obtain $\tan\theta = \sqrt{3}$.

Step 3 Use the exact values in Table 4.8 to find the value of θ in $\left(-\frac{\pi}{2}, \frac{\pi}{2}\right)$ that satisfies $\tan\theta = x$. The table shows that the only angle in the interval $\left(-\frac{\pi}{2}, \frac{\pi}{2}\right)$ that satisfies $\tan\theta = \sqrt{3}$ is $\frac{\pi}{3}$. Thus, $\theta = \frac{\pi}{3}$ and

$$\tan^{-1}\sqrt{3} = \frac{\pi}{3}.$$ The angle in $\left(-\frac{\pi}{2}, \frac{\pi}{2}\right)$ whose tangent is $\sqrt{3}$ is $\frac{\pi}{3}$.

Check Point 4 Find the exact value of $\tan^{-1}(-1)$.

Table 4.9 summarizes the graphs of the three basic inverse trigonometric functions. Below each of the graphs is a description of the function's domain and range.

Table 4.9   Graphs of the Three Basic Inverse Trigonometric Functions

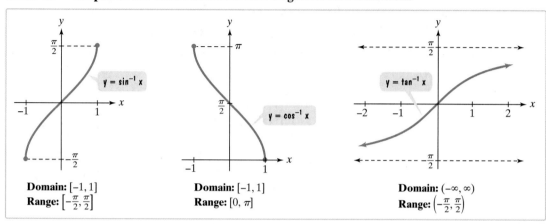

Domain: $[-1, 1]$	**Domain:** $[-1, 1]$	**Domain:** $(-\infty, \infty)$
Range: $\left[-\frac{\pi}{2}, \frac{\pi}{2}\right]$	**Range:** $[0, \pi]$	**Range:** $\left(-\frac{\pi}{2}, \frac{\pi}{2}\right)$

4 Use a calculator to evaluate inverse trigonometric functions.

Using a Calculator to Evaluate Inverse Trigonometric Functions

Calculators give approximate values of inverse trigonometric functions. Use the secondary keys marked $\boxed{\text{SIN}^{-1}}$, $\boxed{\text{COS}^{-1}}$, and $\boxed{\text{TAN}^{-1}}$. These keys are not buttons that you actually press. They are the secondary functions for the buttons labeled $\boxed{\text{SIN}}$, $\boxed{\text{COS}}$, and $\boxed{\text{TAN}}$, respectively. Consult your manual for the location of this feature.

EXAMPLE 5   Calculators and Inverse Trigonometric Functions

Use a calculator to find the value to four decimal places of:

    a. $\sin^{-1}\dfrac{1}{4}$ **b.** $\tan^{-1}(-9.65)$.

Solution

Scientific Calculator Solution

Function	Mode	Keystrokes	Display, rounded to four places
a. $\sin^{-1}\dfrac{1}{4}$	Radian	$1 \boxed{\div} 4 \boxed{=} \boxed{\text{SIN}^{-1}}$	0.2527
b. $\tan^{-1}(-9.65)$	Radian	$9.65 \boxed{{}^+/_-} \boxed{\text{TAN}^{-1}}$	-1.4675

Graphing Calculator Solution

Function	Mode	Keystrokes	Display, rounded to four places
a. $\sin^{-1}\dfrac{1}{4}$	Radian	$\boxed{\text{SIN}^{-1}} \boxed{(} 1 \boxed{\div} 4 \boxed{)} \boxed{\text{ENTER}}$	0.2527
b. $\tan^{-1}(-9.65)$	Radian	$\boxed{\text{TAN}^{-1}} \boxed{(-)} 9.65 \boxed{\text{ENTER}}$	-1.4675

Check Point 5 Use a calculator to find the value to four decimal places of:

a. $\cos^{-1}\dfrac{1}{3}$ **b.** $\tan^{-1}(-35.85)$.

What happens if you attempt to evaluate an inverse trigomometric function at a value that is not in its domain? In real number mode, most calculators will display an error message. For example, an error message can result if you attempt to approximate $\cos^{-1}3$. There is no angle whose cosine is 3. The domain of the inverse cosine function is $[-1, 1]$, and 3 does not belong to this domain.

5 Find exact values of composite functions with inverse trigonometric functions.

Composition of Functions Involving Inverse Trigonometric Functions

In our discussion of functions and their inverses in Section 1.8, we saw that

$$f(f^{-1}(x)) = x \quad \text{and} \quad f^{-1}(f(x)) = x.$$

x must be in the domain of f^{-1}.

x must be in the domain of f.

We apply these properties to the sine, cosine, tangent, and their inverse functions to obtain the following properties:

Inverse Properties

The Sine Function and Its Inverse

$\sin(\sin^{-1}x) = x$ for every x in the interval $[-1, 1]$

$\sin^{-1}(\sin x) = x$ for every x in the interval $\left[-\dfrac{\pi}{2}, \dfrac{\pi}{2}\right]$

The Cosine Function and Its Inverse

$\cos(\cos^{-1}x) = x$ for every x in the interval $[-1, 1]$

$\cos^{-1}(\cos x) = x$ for every x in the interval $[0, \pi]$

The Tangent Function and Its Inverse

$\tan(\tan^{-1}x) = x$ for every real number x

$\tan^{-1}(\tan x) = x$ for every x in the interval $\left(-\dfrac{\pi}{2}, \dfrac{\pi}{2}\right)$

The restrictions on x in the inverse properties are a bit tricky. For example,

$$\sin^{-1}\left(\sin\frac{\pi}{4}\right) = \frac{\pi}{4}$$

$\sin^{-1}(\sin x) = x$ for x in $\left[-\dfrac{\pi}{2}, \dfrac{\pi}{2}\right]$.
Observe that $\dfrac{\pi}{4}$ is in this interval.

Can we use $\sin^{-1}(\sin x) = x$ to find the exact value of $\sin^{-1}\left(\sin\dfrac{5\pi}{4}\right)$? Is $\dfrac{5\pi}{4}$ in the interval $\left[-\dfrac{\pi}{2}, \dfrac{\pi}{2}\right]$? No. Thus, to evaluate $\sin^{-1}\left(\sin\dfrac{5\pi}{4}\right)$, we must first find $\sin\dfrac{5\pi}{4}$.

$\dfrac{5\pi}{4}$ is in quadrant III, where the sine is negative.

$$\sin\frac{5\pi}{4} = -\sin\frac{\pi}{4} = -\frac{\sqrt{2}}{2}$$

The reference angle for $\dfrac{5\pi}{4}$ is $\dfrac{\pi}{4}$.

We evaluate $\sin^{-1}\left(\sin\dfrac{5\pi}{4}\right)$ as follows:

$$\sin^{-1}\left(\sin\frac{5\pi}{4}\right) = \sin^{-1}\left(-\frac{\sqrt{2}}{2}\right) = -\frac{\pi}{4} \qquad \text{If necessary, see Table 4.6 on p. 512.}$$

To determine how to evaluate the composition of functions involving inverse trigonometric functions, first examine the value of x. You can use the inverse properties in the box on the previous page only if x is in the specified interval.

EXAMPLE 6   Evaluating Compositions of Functions and Their Inverses

Find the exact value, if possible, of:

a. $\cos(\cos^{-1}0.6)$ **b.** $\sin^{-1}\left(\sin\dfrac{3\pi}{2}\right)$ **c.** $\cos(\cos^{-1}2\pi)$.

Solution

a. The inverse property $\cos(\cos^{-1}x) = x$ applies for every x in $[-1, 1]$. To evaluate $\cos(\cos^{-1}0.6)$, observe that $x = 0.6$. This value of x lies in $[-1, 1]$, which is the domain of the inverse cosine function. This means that we can use the inverse property $\cos(\cos^{-1}x) = x$. Thus,

$$\cos(\cos^{-1}0.6) = 0.6.$$

b. The inverse property $\sin^{-1}(\sin x) = x$ applies for every x in $\left[-\dfrac{\pi}{2}, \dfrac{\pi}{2}\right]$. To evaluate $\sin^{-1}\left(\sin\dfrac{3\pi}{2}\right)$, observe that $x = \dfrac{3\pi}{2}$. This value of x does not lie in $\left[-\dfrac{\pi}{2}, \dfrac{\pi}{2}\right]$. To evaluate this expression, we first find $\sin\dfrac{3\pi}{2}$.

$$\sin^{-1}\left(\sin\frac{3\pi}{2}\right) = \sin^{-1}(-1) = -\frac{\pi}{2} \qquad \text{The angle in } \left[-\frac{\pi}{2}, \frac{\pi}{2}\right] \text{ whose sine is } -1 \text{ is } -\frac{\pi}{2}.$$

c. The inverse property $\cos(\cos^{-1}x) = x$ applies for every x in $[-1, 1]$. To attempt to evaluate $\cos(\cos^{-1}2\pi)$, observe that $x = 2\pi$. This value of x does not lie in $[-1, 1]$, which is the domain of the inverse cosine function. Thus, the expression $\cos(\cos^{-1}2\pi)$ is not defined because $\cos^{-1}2\pi$ is not defined.

Check Point 6 Find the exact value, if possible, of:
a. $\cos(\cos^{-1}0.7)$ **b.** $\sin^{-1}(\sin\pi)$ **c.** $\cos(\cos^{-1}\pi)$.

We can use points on terminal sides of angles in standard position to find exact values of expressions involving the composition of a function and a different inverse function. Here are two examples:

$$\cos\left(\tan^{-1}\frac{5}{12}\right) \qquad \cot\left[\sin^{-1}\left(-\frac{1}{3}\right)\right].$$

Inner part involves the angle in $\left(-\frac{\pi}{2},\frac{\pi}{2}\right)$ whose tangent is $\frac{5}{12}$.

Inner part involves the angle in $\left[-\frac{\pi}{2},\frac{\pi}{2}\right]$ whose sine is $-\frac{1}{3}$.

The inner part of each expression involves an angle. To evaluate such expressions, we represent such angles by θ. Then we use a sketch that illustrates our representation. Examples 7 and 8 show how to carry out such evaluations.

EXAMPLE 7 Evaluating a Composite Trigonometric Expression

Find the exact value of $\cos\left(\tan^{-1}\frac{5}{12}\right)$.

Solution We let θ represent the angle in $\left(-\frac{\pi}{2},\frac{\pi}{2}\right)$ whose tangent is $\frac{5}{12}$. Thus,

$$\theta = \tan^{-1}\frac{5}{12}.$$

Using the definition of the inverse tangent function, we can rewrite this as

$$\tan\theta = \frac{5}{12}.$$

Because $\tan\theta$ is positive, θ must be an angle in $\left(0,\frac{\pi}{2}\right)$. Thus, θ is a first-quadrant angle. Figure 4.77 shows a right triangle in quadrant I with

$$\tan\theta = \frac{5}{12}. \quad \text{Side opposite } \theta, \text{ or } y$$

$$\text{Side adjacent to } \theta, \text{ or } x$$

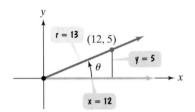

Figure 4.77 Representing $\tan\theta = \frac{5}{12}$

The hypotenuse of the triangle, r, or the distance from the origin to $(12, 5)$, is found using $r = \sqrt{x^2 + y^2}$.

$$r = \sqrt{x^2 + y^2} = \sqrt{12^2 + 5^2} = \sqrt{144 + 25} = \sqrt{169} = 13$$

We use the values for x and r to find the exact value of $\cos\left(\tan^{-1}\frac{5}{12}\right)$.

$$\cos\left(\tan^{-1}\frac{5}{12}\right) = \cos\theta = \frac{\text{side adjacent to } \theta, \text{ or } x}{\text{hypotenuse, or } r} = \frac{12}{13}$$

Check Point 7 Find the exact value of $\sin\left(\tan^{-1}\frac{3}{4}\right)$.

EXAMPLE 8 Evaluating a Composite Trigonometric Expression

Find the exact value of $\cot\left[\sin^{-1}\left(-\frac{1}{3}\right)\right]$.

Solution We let θ represent the angle in $\left[-\frac{\pi}{2},\frac{\pi}{2}\right]$ whose sine is $-\frac{1}{3}$. Thus,

$$\theta = \sin^{-1}\left(-\frac{1}{3}\right) \quad \text{and} \quad \sin\theta = -\frac{1}{3}.$$

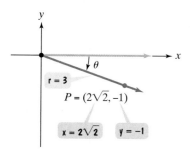

Figure 4.78 Representing $\sin\theta = -\frac{1}{3}$

Because $\sin\theta$ is negative in $\sin\theta = -\frac{1}{3}$, θ must be an angle in $\left[-\frac{\pi}{2}, 0\right)$. Thus, θ is a negative angle that lies in quadrant IV. Figure 4.78 shows angle θ in quadrant IV with

In quadrant IV, y is negative.

$$\sin\theta = -\frac{1}{3} = \frac{y}{r} = \frac{-1}{3}.$$

Thus, $y = -1$ and $r = 3$. The value of x can be found using $r = \sqrt{x^2 + y^2}$, or $x^2 + y^2 = r^2$.

$$x^2 + (-1)^2 = 3^2$$
$$x^2 + 1 = 9$$
$$x^2 = 8$$
$$x = \sqrt{8} = \sqrt{4\cdot 2} = 2\sqrt{2} \qquad \text{Remember that x is positive in quadrant IV.}$$

We use values for x and y to find the exact value of $\cot\left[\sin^{-1}\left(-\frac{1}{3}\right)\right]$.

$$\cot\left[\sin^{-1}\left(-\frac{1}{3}\right)\right] = \cot\theta = \frac{x}{y} = \frac{2\sqrt{2}}{-1} = -2\sqrt{2}$$

Check Point 8 Find the exact value of $\cos\left[\sin^{-1}\left(-\frac{1}{2}\right)\right]$.

Some composite functions with inverse trigonometric functions can be simplified to algebraic expressions. To simplify such an expression, we represent the inverse trigonometric function in the expression by θ. Then we use a right triangle.

EXAMPLE 9 Simplifying an Expression Involving $\sin^{-1}x$

If $0 < x \le 1$, write $\cos(\sin^{-1}x)$ as an algebraic expression in x.

Solution We let θ represent the angle in $\left[-\frac{\pi}{2}, \frac{\pi}{2}\right]$ whose sine is x. Thus,

$$\theta = \sin^{-1}x, \quad \text{and} \quad \sin\theta = x.$$

Because $0 < x \le 1$, $\sin\theta$ is positive. Thus, θ is a first-quadrant angle and can be represented as an acute angle of a right triangle. Figure 4.79 shows a right triangle with

$$\sin\theta = x = \frac{x}{1}. \qquad \text{Side opposite } \theta$$

Hypotenuse

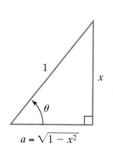

Figure 4.79 Representing $\sin\theta = x$

The third side, a in Figure 4.79, can be found using the Pythagorean Theorem.

$$a^2 + x^2 = 1^2 \qquad \text{Apply the Pythagorean Theorem to the right triangle in Figure 4.79.}$$

$$a^2 = 1 - x^2 \qquad \text{Subtract } x^2 \text{ from both sides.}$$

$$a = \sqrt{1 - x^2} \qquad \text{Solve for a.}$$

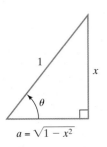

Figure 4.79, repeated

We use the right triangle in Figure 4.79 to write $\cos(\sin^{-1} x)$ as an algebraic expression.

$$\cos(\sin^{-1} x) = \cos\theta = \frac{\text{side adjacent to } \theta}{\text{hypotenuse}} = \frac{\sqrt{1 - x^2}}{1} = \sqrt{1 - x^2}$$

Check Point 9 If $x > 0$, write $\sec(\tan^{-1} x)$ as an algebraic expression in x.

The inverse secant function, $y = \sec^{-1} x$, is used in calculus. However, inverse cotangent and inverse cosecant functions are rarely used. Two of these remaining inverse trigonometric functions are briefly developed in the exercise set that follows.

EXERCISE SET 4.7

Practice Exercises

In Exercises 1–18, find the exact value of each expression.

1. $\sin^{-1}\dfrac{1}{2}$ **2.** $\sin^{-1} 0$ **3.** $\sin^{-1}\dfrac{\sqrt{2}}{2}$

4. $\sin^{-1}\dfrac{\sqrt{3}}{2}$ **5.** $\sin^{-1}\left(-\dfrac{1}{2}\right)$ **6.** $\sin^{-1}\left(-\dfrac{\sqrt{3}}{2}\right)$

7. $\cos^{-1}\dfrac{\sqrt{3}}{2}$ **8.** $\cos^{-1}\dfrac{\sqrt{2}}{2}$ **9.** $\cos^{-1}\left(-\dfrac{\sqrt{2}}{2}\right)$

10. $\cos^{-1}\left(-\dfrac{\sqrt{3}}{2}\right)$ **11.** $\cos^{-1} 0$ **12.** $\cos^{-1} 1$

13. $\tan^{-1}\dfrac{\sqrt{3}}{3}$ **14.** $\tan^{-1} 1$ **15.** $\tan^{-1} 0$

16. $\tan^{-1}(-1)$ **17.** $\tan^{-1}(-\sqrt{3})$ **18.** $\tan^{-1}\left(-\dfrac{\sqrt{3}}{3}\right)$

In Exercises 19–30, use a calculator to find the value of each expression rounded to two decimal places.

19. $\sin^{-1} 0.3$ **20.** $\sin^{-1} 0.47$

21. $\sin^{-1}(-0.32)$ **22.** $\sin^{-1}(-0.625)$

23. $\cos^{-1}\dfrac{3}{8}$ **24.** $\cos^{-1}\dfrac{4}{9}$

25. $\cos^{-1}\dfrac{\sqrt{5}}{7}$ **26.** $\cos^{-1}\dfrac{\sqrt{7}}{10}$

27. $\tan^{-1}(-20)$ **28.** $\tan^{-1}(-30)$

29. $\tan^{-1}(-\sqrt{473})$ **30.** $\tan^{-1}(-\sqrt{5061})$

In Exercises 31–46, find the exact value of each expression, if possible. Do not use a calculator.

31. $\sin(\sin^{-1} 0.9)$ **32.** $\cos(\cos^{-1} 0.57)$

33. $\sin^{-1}\left(\sin\dfrac{\pi}{3}\right)$ **34.** $\cos^{-1}\left(\cos\dfrac{2\pi}{3}\right)$

35. $\sin^{-1}\left(\sin\dfrac{5\pi}{6}\right)$ **36.** $\cos^{-1}\left(\cos\dfrac{4\pi}{3}\right)$

37. $\tan(\tan^{-1} 125)$ **38.** $\tan(\tan^{-1} 380)$

39. $\tan^{-1}\left[\tan\left(-\dfrac{\pi}{6}\right)\right]$ **40.** $\tan^{-1}\left[\tan\left(-\dfrac{\pi}{3}\right)\right]$

41. $\tan^{-1}\left(\tan\dfrac{2\pi}{3}\right)$ **42.** $\tan^{-1}\left(\tan\dfrac{3\pi}{4}\right)$

43. $\sin^{-1}(\sin\pi)$ **44.** $\cos^{-1}(\cos 2\pi)$

45. $\sin(\sin^{-1}\pi)$ **46.** $\cos(\cos^{-1} 3\pi)$

In Exercises 47–60, use a sketch to find the exact value of each expression.

47. $\cos\left(\sin^{-1}\frac{4}{5}\right)$ **48.** $\sin\left(\tan^{-1}\frac{7}{24}\right)$

49. $\tan\left(\cos^{-1}\frac{5}{13}\right)$ **50.** $\cot\left(\sin^{-1}\frac{5}{13}\right)$

51. $\tan\left[\sin^{-1}\left(-\frac{3}{5}\right)\right]$ **52.** $\cos\left[\sin^{-1}\left(-\frac{4}{5}\right)\right]$

53. $\sin\left(\cos^{-1}\dfrac{\sqrt{2}}{2}\right)$ **54.** $\cos\left(\sin^{-1}\frac{1}{2}\right)$

55. $\sec\left[\sin^{-1}\left(-\frac{1}{4}\right)\right]$ **56.** $\sec\left[\sin^{-1}\left(-\frac{1}{2}\right)\right]$

57. $\tan\left[\cos^{-1}\left(-\frac{1}{3}\right)\right]$ **58.** $\tan\left[\cos^{-1}\left(-\frac{1}{4}\right)\right]$

59. $\csc\left[\cos^{-1}\left(-\dfrac{\sqrt{3}}{2}\right)\right]$ **60.** $\sec\left[\sin^{-1}\left(-\dfrac{\sqrt{2}}{2}\right)\right]$

In Exercises 61–70, use a right triangle to write each expression as an algebraic expression. Assume that x is positive and in the domain of the given inverse trigonometric function.

61. $\tan(\cos^{-1} x)$ **62.** $\sin(\tan^{-1} x)$

63. $\cos(\sin^{-1} 2x)$ **64.** $\sin(\cos^{-1} 2x)$

65. $\cos\left(\sin^{-1}\dfrac{1}{x}\right)$ **66.** $\sec\left(\cos^{-1}\dfrac{1}{x}\right)$

67. $\cot\left(\tan^{-1}\dfrac{x}{\sqrt{3}}\right)$ **68.** $\cot\left(\tan^{-1}\dfrac{x}{\sqrt{2}}\right)$

69. $\sec\left(\sin^{-1}\dfrac{x}{\sqrt{x^2+4}}\right)$ **70.** $\cot\left(\sin^{-1}\dfrac{\sqrt{x^2-9}}{x}\right)$

71. a. Graph the restricted secant function, $y = \sec x$, by restricting x to the intervals $\left[0, \dfrac{\pi}{2}\right)$ and $\left(\dfrac{\pi}{2}, \pi\right]$.

     b. Use the horizontal line test to explain why the restricted secant function has an inverse function.

     c. Use the graph of the restricted secant function to graph $y = \sec^{-1} x$.

72. a. Graph the restricted cotangent function, $y = \cot x$, by restricting x to the interval $(0, \pi)$.

     b. Use the horizontal line test to explain why the restricted cotangent function has an inverse function.

     c. Use the graph of the restricted cotangent function to graph $y = \cot^{-1} x$.

 Application Exercises

73. Your neighborhood movie theater has a 25-foot-high screen located 8 feet above your eye level. If you sit too close to the screen, your viewing angle is too small, resulting in a distorted picture. By contrast, if you sit too far back, the image is quite small, diminishing the movie's visual impact. If you sit x feet back from the screen, your viewing angle, θ, is given by

$$\theta = \tan^{-1}\dfrac{33}{x} - \tan^{-1}\dfrac{8}{x}.$$

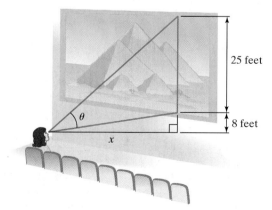

25 feet

8 feet

Find the viewing angle, in radians, at distances of 5 feet, 10 feet, 15 feet, 20 feet, and 25 feet.

74. The function $\theta = \tan^{-1}\dfrac{33}{x} - \tan^{-1}\dfrac{8}{x}$, described in Exercise 73, is graphed below in a $[0, 50, 10]$ by $[0, 1, 0.1]$ viewing rectangle. Use the graph to describe what happens to your viewing angle as you move farther back from the screen. How far back from the screen, to the nearest foot, should you sit to maximize your viewing angle? Verify this observation by finding the viewing angle one foot closer to the screen and one foot farther from the screen for this ideal viewing distance.

The formula

$$\theta = 2\tan^{-1}\dfrac{21.634}{x}$$

gives the viewing angle, θ, in radians, for a camera whose lens is x millimeters wide. Use this formula to solve Exercises 75–76.

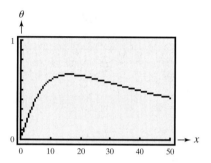

75. Find the viewing angle, in radians and in degrees (to the nearest tenth of a degree), of a 28-millimeter lens.

76. Find the viewing angle, in radians and degrees (to the nearest tenth of a degree), of a 300-millimeter telephoto lens.

For years, mathematicians were challenged by the following problem: What is the area of a region under a curve between two values of x? The problem was solved in the seventeenth century with the development of integral calculus. Using calculus, the area of the region under $y = \dfrac{1}{x^2+1}$, above the x-axis, and between $x = a$ and $x = b$ is $\tan^{-1} b - \tan^{-1} a$. Use this result, shown in the figure on the next page, to find *the area of the region under* $y = \dfrac{1}{x^2 + 1}$, *above the x-axis, and between the values of a and b given in Exercises 77–78.*

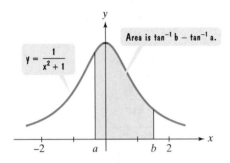

Area is $\tan^{-1} b - \tan^{-1} a$.

$y = \dfrac{1}{x^2 + 1}$

77. $a = 0$ and $b = 2$

78. $a = -2$ and $b = 1$

Writing in Mathematics

79. Explain why, without restrictions, no trigonometric function has an inverse function.

80. Describe the restriction on the sine function so that it has an inverse function.

81. How can the graph of $y = \sin^{-1} x$ be obtained from the graph of the restricted sine function?

82. Without drawing a graph, describe the behavior of the graph of $y = \sin^{-1} x$. Mention the function's domain and range in your description.

83. Describe the restriction on the cosine function so that it has an inverse function.

84. Without drawing a graph, describe the behavior of the graph of $y = \cos^{-1} x$. Mention the function's domain and range in your description.

85. Describe the restriction on the tangent function so that it has an inverse function.

86. Without drawing a graph, describe the behavior of the graph of $y = \tan^{-1} x$. Mention the function's domain and range in your description.

87. If $\sin^{-1}\left(\sin \dfrac{\pi}{3}\right) = \dfrac{\pi}{3}$, is $\sin^{-1}\left(\sin \dfrac{5\pi}{6}\right) = \dfrac{5\pi}{6}$? Explain your answer.

88. Explain how a right triangle can be used to find the exact value of $\sec\left(\sin^{-1}\frac{4}{5}\right)$.

89. Find the height of the screen and the number of feet that it is located above eye level in your favorite movie theater. Modify the formula given in Exercise 73 so that it applies to your theater. Then describe where in the theater you should sit so that a movie creates the greatest visual impact.

Technology Exercises

In Exercises 90–93, graph each pair of functions in the same viewing rectangle. Use your knowledge of the domain and range for the inverse trigonometric functions to select an appropriate viewing rectangle. What does the graph of the second equation in each exercise do to the graph of the first equation?

90. $y = \sin^{-1} x$ and $y = \sin^{-1} x + 2$

91. $y = \cos^{-1} x$ and $y = \cos^{-1}(x - 1)$

92. $y = \tan^{-1} x$ and $y = -2 \tan^{-1} x$

93. $y = \sin^{-1} x$ and $y = \sin^{-1}(x + 2) + 1$

94 Graph $y = \tan^{-1} x$ and its two horizontal asymptotes in a $[-3, 3, 1]$ by $\left[-\pi, \pi, \dfrac{\pi}{2}\right]$ viewing rectangle. Then change the range setting to $[-50, 50, 5]$ by $\left[-\pi, \pi, \dfrac{\pi}{2}\right]$. What do you observe?

95. Graph $y = \sin^{-1} x + \cos^{-1} x$ in a $[-2, 2, 1]$ by $[0, 3, 1]$ viewing rectangle. What appears to be true about the sum of the inverse sine and inverse cosine for values between -1 and 1 inclusive?

Critical Thinking Exercises

96. Solve $y = 2 \sin^{-1}(x - 5)$ for x in terms of y.

97. Solve for x: $2 \sin^{-1} x = \dfrac{\pi}{4}$.

98. Prove that if $x > 0$, $\tan^{-1} x + \tan^{-1} \dfrac{1}{x} = \dfrac{\pi}{2}$.

99. Derive the formula for θ, your viewing angle at the movie theater, in Exercise 73. *Hint:* Use the figure shown and represent the acute angle on the left in the smaller right triangle by α. Find expressions for $\tan \alpha$ and $\tan(\alpha + \theta)$.

SECTION 4.8 *Applications of Trigonometric Functions*

Objectives

1. Solve a right triangle.
2. Solve problems involving bearings.
3. Model simple harmonic motion.

In the late 1960s, popular musicians were searching for new sounds. Film composers were looking for ways to create unique sounds as well. From these efforts, synthesizers that electronically reproduce musical sounds were born. From providing the backbone of today's most popular music to providing the strange sounds for the most experimental music, synthesizers are at the forefront of music technology.

If we did not understand the periodic nature of sinusoidal functions, the synthesizers used in almost all forms of music would not exist. In this section, we look at applications of trigonometric functions in right triangles and in modeling periodic phenomena such as sound.

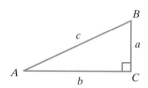

1 Solve a right triangle.

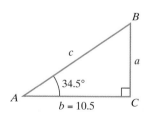

Figure 4.80 Labeling right triangles

Solving Right Triangles

Solving a right triangle means finding the missing lengths of its sides and the measurements of its angles. We will label right triangles so that side a is opposite angle A, side b is opposite angle B, and side c, the hypotenuse, is opposite right angle C. Figure 4.80 illustrates this labeling.

When solving a right triangle, we will use the sine, cosine, and tangent functions, rather than their reciprocals. Example 1 shows how to solve a right triangle when we know the length of a side and the measure of an acute angle.

EXAMPLE 1 Solving a Right Triangle

Solve the right triangle shown in Figure 4.81.

Solution We begin by finding the measure of angle B. We do not need a trigonometric function to do so. Because $C = 90°$ and the sum of a triangle's angles is $180°$, we see that $A + B = 90°$. Thus,

$$B = 90° - A = 90° - 34.5° = 55.5°.$$

Now we need to find a. Because we have a known angle, an unknown opposite side, and a known adjacent side, we use the tangent function.

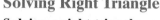

Figure 4.81 Find B, a, and c.

$$\tan 34.5° = \frac{a}{10.5} \quad \begin{array}{l} \text{Side opposite the 34.5° angle} \\ \\ \text{Side adjacent to the 34.5° angle} \end{array}$$

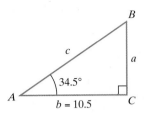

Figure 4.81, repeated

Now we multiply both sides of this equation by 10.5 and solve for a.

$$a = 10.5 \tan 34.5° \approx 7.22$$

Finally, we need to find c. Because we have a known angle, a known adjacent side, and an unknown hypotenuse, we use the cosine function.

$$\cos 34.5° = \frac{10.5}{c}$$

Side adjacent to the 34.5° angle

hypotenuse

Discovery

There is often more than one correct way to solve a right triangle. In Example 1, find a using angle $B = 55.5°$. Find c using the Pythagorean Theorem.

Now we multiply both sides of this equation by c and then solve for c.

$$c \cos 34.5° = 10.5 \qquad \text{Multiply both sides by } c.$$

$$c = \frac{10.5}{\cos 34.5°} \approx 12.74 \qquad \text{Divide both sides by } \cos 34.5° \text{ and solve for } c.$$

In summary, $B = 55.5°$, $a \approx 7.22$, and $c \approx 12.74$.

> **Check Point 1** In Figure 4.80, let $A = 62.7°$ and $a = 8.4$. Solve the right triangle, rounding lengths to two decimal places.

Trigonometry was first developed to measure heights and distances that were inconvenient or impossible to measure. In solving application problems, begin by making a sketch involving a right triangle that illustrates the problem's conditions. Then put your knowledge of solving right triangles to work and find the required distance or height.

EXAMPLE 2   Finding the Side of a Triangle

From a point on level ground 125 feet from the base of a tower, the angle of elevation is 57.2°. Approximate the height of the tower to the nearest foot.

Solution   A sketch is shown in Figure 4.82, where a represents the height of the tower. In the right triangle, we have a known angle, an unknown opposite side, and a known adjacent side. Therefore, we use the tangent function.

$$\tan 57.2° = \frac{a}{125}$$

Side opposite the 57.2° angle

Side adjacent to the 57.2° angle

Now we multiply both sides of this equation by 125 and solve for a.

$$a = 125 \tan 57.2° \approx 194$$

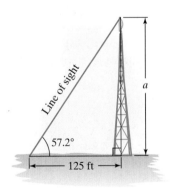

Figure 4.82 Determining height without using direct measurement

The tower is approximately 194 feet high.

> **Check Point 2** From a point on level ground 80 feet from the base of the Eiffel Tower, the angle of elevation is 85.4°. Approximate the height of the Eiffel Tower to the nearest foot.

Example 3 illustrates how to find the measure of an acute angle of a right triangle if the lengths of two sides are known.

EXAMPLE 3   Finding the Angle of a Triangle

A kite flies at a height of 30 feet when 65 feet of string is out. If the string is in a straight line, find the angle that it makes with the ground. Round to the nearest tenth of a degree.

Solution   A sketch is shown in Figure 4.83, where A represents the angle the string makes with the ground. In the right triangle, we have an unknown angle, a known opposite side, and a known hypotenuse. Therefore, we use the sine function.

$$\sin A = \frac{30}{65} \quad \text{Side opposite A} \\ \text{hypotenuse}$$

$$A = \sin^{-1}\frac{30}{65} \approx 27.5°$$

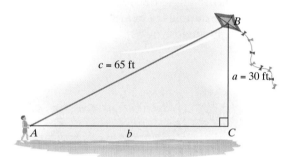

Figure 4.83 Flying a kite

The string makes an angle of approximately 27.5° with the ground.

Check Point 3   A guy wire is 13.8 yards long and is attached from the ground to a pole 6.7 yards above the ground. Find the angle, to the nearest tenth of a degree, that the wire makes with the ground.

EXAMPLE 4   Using Two Right Triangles to Solve a Problem

You are taking your first hot-air balloon ride. Your friend is standing on level ground, 100 feet away from your point of launch, making a video of the terrified look on your rapidly ascending face. How rapidly? At one instant, the angle of elevation from the video camera to your face is 31.7°. One minute later, the angle of elevation is 76.2°. How far did you travel during that minute?

Solution   A sketch that illustrates the problem is shown in Figure 4.84. We need to determine $b - a$, the distance traveled during the one-minute period. We find a using the small right triangle. Because we have a known angle, an unknown opposite side, and a known adjacent side, we use the tangent function.

$$\tan 31.7° = \frac{a}{100} \quad \text{Side opposite the 31.7° angle} \\ \text{Side adjacent to the 31.7° angle}$$

$$a = 100 \tan 31.7° \approx 61.8$$

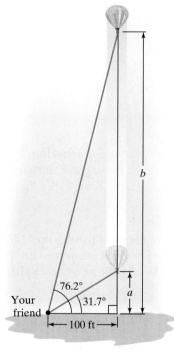

Figure 4.84 Ascending in a hot-air balloon

We find b using the tangent function in the large right triangle. Turn back the page and verify this observation in Figure 4.84.

$$\tan 76.2° = \frac{b}{100} \quad \text{Side opposite the 76.2° angle} \\ \text{Side adjacent to the 76.2° angle}$$

$$b = 100 \tan 76.2° \approx 407.1$$

The balloon traveled $407.1 - 61.8$, or approximately 345.3 feet, during the minute.

Check Point 4 You are standing on level ground 800 feet from Mt. Rushmore, looking at the sculpture of Abraham Lincoln's face. The angle of elevation to the bottom of the sculpture is 32° and the angle of elevation to the top is 35°. Find the height of the sculpture of Lincoln's face to the nearest tenth of a foot.

2 Solve problems involving bearings.

Trigonometry and Bearings

In navigation and surveying problems, the term *bearing* is used to specify the location of one point relative to another. The **bearing** from point O to point P is the acute angle between ray OP and a north-south line. Figure 4.85 illustrates some examples of bearings. The north-south line and the east-west line intersect at right angles.

Study Tip

The bearing from O to P can also be described using the phrase "the bearing of P from O."

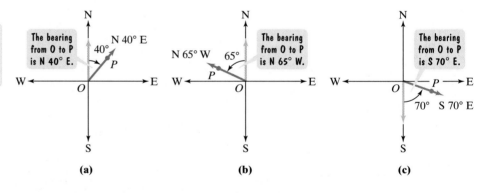

Figure 4.85 An illustration of three bearings

(a)   (b)   (c)

Each bearing has three parts: a letter (N or S), the measure of an acute angle, and a letter (E or W). Here's how we write a bearing:

- If the acute angle is measured from the *north side* of the north-south line, then we write N first. [See Figure 4.85(a).] If the acute angle is measured from the *south side* of the north-south line, then we write S first. [See Figure 4.85(c).]
- Second, we write the measure of the acute angle.
- If the acute angle is measured on the *east side* of the north-south line, then we write E last. [See Figure 4.85(a)]. If the acute angle is measured on the *west side* of the north-south line, then we write W last. [See Figure 4.85(b).]

EXAMPLE 5 Understanding Bearings

Use Figure 4.86 on the next page to find:

a. the bearing from O to B.

b. the bearing from O to A.

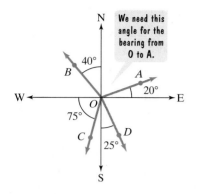

Figure 4.86 Finding bearings

Solution

a. To find the bearing from O to B, we need the acute angle between the ray OB and the north-south line through O. The measurement of this angle is given to be 40°. Figure 4.86 shows that the angle is measured from the north side of the north-south line and lies west of the north-south line. Thus, the bearing from O to B is N 40° W.

b. To find the bearing from O to A, we need the acute angle between the ray OA and the north-south line through O. This angle is specified by the voice balloon in Figure 4.86. Because of the given 20° angle, this angle measures 90° − 20°, or 70°. This angle is measured from the north side of the north-south line. This angle is also east of the north-south line. Thus, the bearing from O to A is N 70° E.

> **Check Point 5** Use Figure 4.86 to find:
> **a.** the bearing from O to D.
> **b.** the bearing from O to C.

EXAMPLE 6 Finding the Bearing of a Boat

A boat leaves the entrance to a harbor and travels 25 miles on a bearing of N 42° E. Figure 4.87 shows that the captain then turns the boat 90°clockwise and travels 18 miles on a bearing of S 48° E. At that time:

a. How far is the boat from the harbor entrance?

b. What is the bearing of the boat from the harbor entrance?

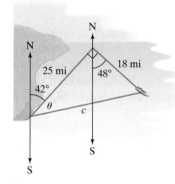

Figure 4.87 Finding a boat's bearing from the harbor entrance

Study Tip

When making a diagram showing bearings, draw a north-south line through each point at which a change in course occurs. The north side of the line lies above each point. The south side of the line lies below each point.

Solution

a. The boat's distance from the harbor entrance is represented by c in Figure 4.87. Because we know the length of two sides of the right triangle, we find c using the Pythagorean Theorem. We have

$$c^2 = a^2 + b^2 = 25^2 + 18^2 = 949$$
$$c = \sqrt{949} \approx 30.8.$$

The boat is approximately 30.8 miles from the harbor entrance.

b. The bearing of the boat from the harbor entrance means the bearing from the entrance to the boat. Look at the north-south line passing through the harbor entrance on the left in Figure 4.87. The acute angle from this line to the ray on which the boat lies is $42° + \theta$. Because we are measuring the angle from the north side of the line and the boat is east of the harbor, its bearing from the harbor entrance is N$(42° + \theta)$E. To find θ, we use the right triangle shown in Figure 4.87 and the tangent function.

$$\tan \theta = \frac{\text{side opposite } \theta}{\text{side adjacent to } \theta} = \frac{18}{25}$$

$$\theta = \tan^{-1} \frac{18}{25}$$

We can use a calculator in degree mode to find the value of θ: $\theta \approx 35.8°$. Thus, $42° + \theta = 42° + 35.8° = 77.8°$. The bearing of the boat from the harbor entrance is N 77.8° E.

Check Point 6 You leave the entrance to a system of hiking trails and hike 2.3 miles on a bearing of S 31° W. Then the trail turns 90° clockwise and you hike 3.5 miles on a bearing of N 59° W. At that time:

a. How far are you from the entrance to the trail system?

b. What is your bearing from the entrance to the trail system?

3 Model simple harmonic motion.

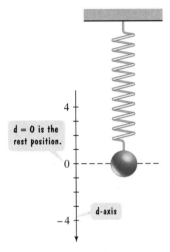

d = 0 is the rest position.

d-axis

Figure 4.88 Using a d-axis to describe a ball's distance from its rest position

Simple Harmonic Motion

Because of their periodic nature, trigonometric functions are used to model phenomena that occur again and again. This includes vibratory or oscillatory motion, such as the motion of a vibrating guitar string, the swinging of a pendulum, or the bobbing of an object attached to a spring. Trigonometric functions are also used to describe radio waves from your favorite FM station, television waves from your not-to-be-missed weekly sitcom, and sound waves from your most-prized CDs.

To see how trigonometric functions are used to model vibratory motion, consider this: A ball is attached to a spring hung from the ceiling. You pull the ball down 4 inches and then release it. If we neglect the effects of friction and air resistance, the ball will continue bobbing up and down on the end of the spring. These up-and-down oscillations are called **simple harmonic motion.**

To better understand this motion, we use a d-axis, where d represents distance. This axis is shown in Figure 4.88. On this axis, the position of the ball before you pull it down is $d = 0$. This rest position is called the **equilibrium position.** Now you pull the ball down 4 inches to $d = -4$ and release it. Figure 4.89 shows a sequence of "photographs" taken at one-second time intervals illustrating the distance of the ball from its rest position, d.

The curve in Figure 4.89 shows how the ball's distance from its rest position changes over time. The curve is sinusoidal and the motion can be described using a cosine or a sine function.

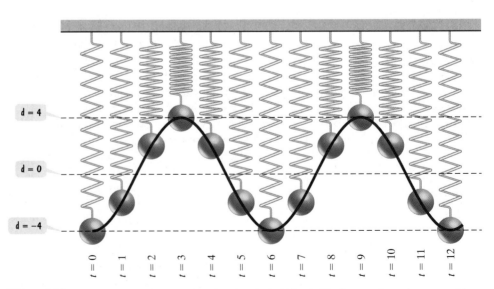

d = 4

d = 0

d = -4

$t = 0$ $t = 1$ $t = 2$ $t = 3$ $t = 4$ $t = 5$ $t = 6$ $t = 7$ $t = 8$ $t = 9$ $t = 10$ $t = 11$ $t = 12$

Figure 4.89 A sequence of "photographs" showing the bobbing ball's distance from the rest position, taken at one-second intervals

Simple Harmonic Motion

An object that moves on a coordinate axis is in **simple harmonic motion** if its distance from the origin, d, at time t is given by either

$$d = a \cos \omega t \ \text{ or } \ d = a \sin \omega t.$$

The motion has **amplitude** $|a|$, the maximum displacement of the object from its rest position. The **period** of the motion is $\dfrac{2\pi}{\omega}$, where $\omega > 0$. The period gives the time it takes for the motion to go through one complete cycle.

In describing simple harmonic motion, the equation with the cosine function is used if the object is at its greatest distance from rest position, the origin, at $t = 0$. By contrast, the equation with the sine function is used if the object is at its rest position, the origin, at $t = 0$.

Diminishing Motion with Increasing Time

Due to friction and other resistive forces, the motion of an oscillating object decreases over time. The function

$$d = 3e^{-0.1t} \cos 2t$$

models this type of motion. The graph of the function is shown in a $t = [0, 10, 1]$ by $d = [-3, 3, 1]$ viewing rectangle. Notice how the amplitude is decreasing with time as the moving object loses energy.

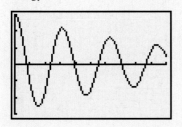

EXAMPLE 7   Finding an Equation for an Object in Simple Harmonic Motion

A ball on a spring is pulled 4 inches below its rest position and then released. The period of the motion is 6 seconds. Write the equation for the ball's simple harmonic motion.

Solution   We need to write an equation that describes d, the distance of the ball from its rest position, after t seconds. (The motion is illustrated by the "photo" sequence in Figure 4.89 on page 530.) When the object is released ($t = 0$), the ball's distance from its rest position is 4 inches down. Because it is *down* 4 inches, d is negative: When $t = 0$, $d = -4$. Notice the greatest distance from rest position occurs at $t = 0$. Thus, we will use the equation with the cosine function,

$$d = a \cos \omega t,$$

to model the ball's simple harmonic motion.

Now we determine values for a and ω. Recall that $|a|$ is the maximum displacement. Because the ball initially moves down, $a = -4$.

The value of ω in $d = a \cos \omega t$ can be found using the formula for the period.

$$\text{period} = \frac{2\pi}{\omega} = 6 \qquad \textit{We are given that the period of the motion is 6 seconds.}$$

$$2\pi = 6\omega \qquad \textit{Multiply both sides by } \omega.$$

$$\omega = \frac{2\pi}{6} = \frac{\pi}{3} \qquad \textit{Divide both sides by 6 and solve for } \omega.$$

We see that $a = -4$ and $\omega = \dfrac{\pi}{3}$. Substitute these values into $d = a \cos \omega t$. The equation for the ball's simple harmonic motion is

$$d = -4 \cos \frac{\pi}{3} t.$$

Modeling Music

Sounds are caused by vibrating objects that result in variations in pressure in the surrounding air. Areas of high and low pressure moving through the air are modeled by the harmonic motion formulas. When these vibrations reach our eardrums, the eardrums' vibrations send signals to our brains which create the sensation of hearing.

  French mathematician John Fourier (1768–1830) proved that all musical sounds—instrumental and vocal—could be modeled by sums involving sine functions. Modeling musical sounds with sinusoidal functions is used by synthesizers to electronically produce sounds unobtainable from ordinary musical instruments.

 Check Point 7 A ball on a spring is pulled 6 inches below its rest position and then released. The period for the motion is 4 seconds. Write the equation for the ball's simple harmonic motion.

  The period of the harmonic motion in Example 7 was 6 seconds. It takes 6 seconds for the moving object to complete one cycle. Thus, $\frac{1}{6}$ of a cycle is completed every second. We call $\frac{1}{6}$ the *frequency* of the moving object. **Frequency** describes the number of complete cycles per unit time and is the reciprocal of the period.

Frequency of an Object in Simple Harmonic Motion

An object in simple harmonic motion given by

$$d = a \cos \omega t \ \text{ or } \ d = a \sin \omega t$$

has **frequency** f given by

$$f = \frac{\omega}{2\pi}, \ \omega > 0.$$

Equivalently,

$$f = \frac{1}{\text{period}}.$$

EXAMPLE 8 Analyzing Simple Harmonic Motion

Figure 4.90 shows a mass on a smooth table attached to a spring. The mass moves in simple harmonic motion described by

$$d = 10 \cos \frac{\pi}{6} t$$

with t measured in seconds and d in centimeters. Find **a.** the maximum displacement, **b.** the frequency, and **c.** the time required for one cycle.

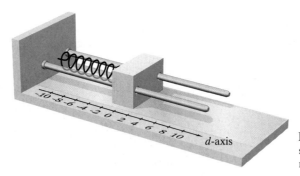

Figure 4.90 A mass attached to a spring, moving in simple harmonic motion

Solution We begin by identifying values for a and ω.

$$d = 10 \cos \frac{\pi}{6} t$$

The form of this equation is
$d = a \cos \omega t$
with $a = 10$ and $\omega = \frac{\pi}{6}$.

a. The maximum displacement from the rest position is the amplitude. Because $a = 10$, the maximum displacement is 10 centimeters.

b. The frequency, f, is

$$f = \frac{\omega}{2\pi} = \frac{\dfrac{\pi}{6}}{2\pi} = \frac{\pi}{6} \cdot \frac{1}{2\pi} = \frac{1}{12}.$$

The frequency is $\frac{1}{12}$ cycle (or oscillation) per second.

c. The time required for one cycle is the period.

$$\text{period} = \frac{2\pi}{\omega} = \frac{2\pi}{\dfrac{\pi}{6}} = 2\pi \cdot \frac{6}{\pi} = 12$$

The time required for one cycle is 12 seconds.

Check Point 8

An object moves in simple harmonic motion described by $d = 12 \cos \dfrac{\pi}{4} t$, where t is measured in seconds and d in centimeters. Find **a.** the maximum displacement, **b.** the frequency, and **c.** the time required for one cycle.

Resisting Damage of Simple Harmonic Motion

Simple harmonic motion from an earthquake caused this highway in Oakland, California, to collapse. By studying the harmonic motion of the soil under the highway, engineers learn to build structures that can resist damage.

EXERCISE SET 4.8

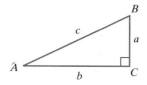

Practice Exercises

In Exercises 1–12, solve the right triangle shown in the figure. Round lengths to two decimal places and express angles to the nearest tenth of a degree.

(figure of right triangle with vertices A, B, C; sides a, b, c and right angle at C)

1. $A = 23.5°, b = 10$ **2.** $A = 41.5°, b = 20$

✗ **3.** $A = 52.6°, c = 54$ **4.** $A = 54.8°, c = 80$

5. $B = 16.8°, b = 30.5$ **6.** $B = 23.8°, b = 40.5$

7. $a = 30.4, c = 50.2$ **8.** $a = 11.2, c = 65.8$

✗ **9.** $a = 10.8, b = 24.7$ **10.** $a = 15.3, b = 17.6$

11. $b = 2, c = 7$ **12.** $b = 4, c = 9$

Use the figure shown to solve Exercises 13–16.

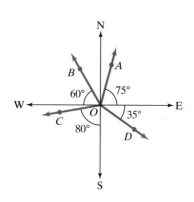

13. Find the bearing from O to A.

14. Find the bearing from O to B.

15. Find the bearing from O to C.

16. Find the bearing from O to D.

In Exercises 17–20, an object is attached to a coiled spring. The object is pulled down (negative direction from the rest position) and then released. Write an equation for the distance of the object from its rest position after t seconds.

	Distance from rest position at $t = 0$	Amplitude	Period
17.	6 centimeters	6 centimeters	4 seconds
18.	8 inches	8 inches	2 seconds
19.	0	3 inches	1.5 seconds
20.	0	5 centimeters	2.5 seconds

In Exercises 21–28, an object moves in simple harmonic motion described by the given equation, where t is measured in seconds and d in inches. In each exercise, find:

a. *the maximum displacement*

b. *the frequency*

c. *the time required for one cycle.*

21. $d = 5 \cos \dfrac{\pi}{2} t$ **22.** $d = 10 \cos 2\pi t$

23. $d = -6 \cos 2\pi t$ **24.** $d = -8 \cos \dfrac{\pi}{2} t$

25. $d = \frac{1}{2} \sin 2t$ **26.** $d = \frac{1}{3} \sin 2t$

27. $d = -5 \sin \dfrac{2\pi}{3} t$ **28.** $d = -4 \sin \dfrac{3\pi}{2} t$

Application Exercises

✗ **29.** The tallest television transmitting tower in the world is in North Dakota. From a point on level ground 5280 feet (one mile) from the base of the tower, the angle of elevation is 21.3°. Approximate the height of the tower to the nearest foot.

30. From a point on level ground 30 yards from the base of a building, the angle of elevation is 38.7°. Approximate the height of the building to the nearest foot.

31. The Statue of Liberty is approximately 305 feet tall. If the angle of elevation of a ship to the top of the statue is 23.7°, how far, to the nearest foot, is the ship from the statue's base?

32. A 200-foot cliff drops vertically into the ocean. If the angle of elevation of a ship to the top of the cliff is 22.3°, how far off shore, to the nearest foot, is the ship?

33. A helicopter hovers 1000 feet above a small island. The figure shows that the angle of depression from the helicopter to point P is 36°. How far off the coast, to the nearest foot, is the island?

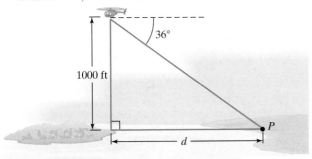

34. A police helicopter is flying at 800 feet. A stolen car is sighted at an angle of depression of 72°. Find the distance of the stolen car, to the nearest foot, from a point directly below the helicopter.

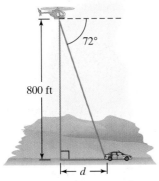

35. A wheelchair ramp is to be built beside the steps to the campus library. Find the angle of elevation of the 23-foot ramp, to the nearest tenth of a degree, if its final height is 6 feet.

36. A building that is 250 feet high casts a shadow 40 feet long. Find the angle of elevation, to the nearest tenth of a degree, of the sun at this time.

37. A hot-air balloon is rising vertically. The angle of elevation from a point on level ground 125 feet from the balloon to a point directly under the passenger compartment changes from 19.2° to 31.7°. How far, to the nearest tenth of a foot, does the balloon rise during this period?

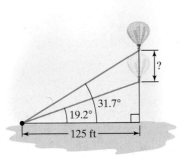

38. A flagpole is situated on top of a building. The angle of elevation from a point on level ground 330 feet from the building to the top of the flagpole is 63°. The angle of elevation from the same point to the bottom of the flagpole is 53°. Find the height of the flagpole to the nearest tenth of a foot.

39. A boat leaves the entrance to a harbor and travels 150 miles on a bearing of N 53° E. How many miles north and how many miles east from the harbor has the boat traveled?

40. A boat leaves the entrance to a harbor and travels 40 miles on a bearing of S 64° E. How many miles south and how many miles east from the harbor has the boat traveled?

41. A forest ranger sights a fire directly to the south. A second ranger, 7 miles east of the first ranger, also sights the fire. The bearing from the second ranger to the fire is S 28° W. How far, to the nearest tenth of a mile, is the first ranger from the fire?

42. A ship sights a lighthouse directly to the south. A second ship, 9 miles east of the first ship, also sights the lighthouse. The bearing from the second ship to the lighthouse is S 34° W. How far, to the nearest tenth of a mile, is the first ship from the lighthouse?

43. You leave your house and run 2 miles due west followed by 1.5 miles due north. At that time, what is your bearing from your house?

44. A ship is 9 miles east and 6 miles south of a harbor. What bearing should be taken to sail directly to the harbor?

45. A jet leaves a runway whose bearing is N 35° E from the control tower. After flying 5 miles, the jet turns 90° and flies on a bearing of S 55° E for 7 miles. At that time, what is the bearing of the jet from the control tower?

46. A ship leaves port with a bearing of S 40° W. After traveling 7 miles, the ship turns 90° and travels on a bearing of N 50° W for 11 miles. At that time, what is the bearing of the ship from port?

47. An object in simple harmonic motion has a frequency of $\frac{1}{2}$ oscillation per minute and an amplitude of 6 feet. Write an equation in the form $d = a \sin \omega t$ for the object's simple harmonic motion.

48. An object in simple harmonic motion has a frequency of $\frac{1}{4}$ oscillation per minute and an amplitude of 8 feet. Write an equation in the form $d = a \sin \omega t$ for the object's simple harmonic motion.

49. A piano tuner uses a tuning fork. If middle C has a frequency of 264 vibrations per second, write an equation in the form $d = \sin \omega t$ for the simple harmonic motion.

50. A radio station, 98.1 on the FM dial, has radio waves with a frequency of 98.1 million cycles per second. Write an equation in the form $d = \sin \omega t$ for the simple harmonic motion of the radio waves.

Writing in Mathematics

51. What does it mean to solve a right triangle?

52. Explain how to find one of the acute angles of a right triangle if two sides are known.

53. Describe a situation in which a right triangle and a trigonometric function are used to measure a height or distance that would otherwise be inconvenient or impossible to measure.

54. What is meant by the bearing from point O to point P? Give an example with your description.

55. What is simple harmonic motion? Give an example with your description.

56. Explain the period and the frequency of simple harmonic motion. How are they related?

57. Explain how the photograph of the damaged highway on page 533 illustrates simple harmonic motion.

Technology Exercises

The functions in Exercises 58–59 model motion in which the amplitude decreases with time due to friction or other resistive forces. Graph each function in the given viewing rectangle. How many complete oscillations occur on the time interval $0 \le x \le 10$?

58. $y = 4e^{-0.1} \cos 2x$ $[0, 10, 1]$ by $[-4, 4, 1]$

59. $y = -6e^{-0.09x} \cos 2\pi x$ $[0, 10, 1]$ by $[-6, 6, 1]$

Critical Thinking Exercises

60. The figure at the top of the next column shows a satellite circling 112 miles above Earth. When the satellite is directly above point B, angle A measures $76.6°$. Find Earth's radius to the nearest mile.

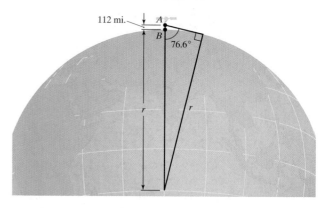

61. The figure shows that the angle of elevation to the top of the building changes from $20°$ to $40°$ as an observer advances 75 feet toward the building. Find the height of the building to the nearest foot.

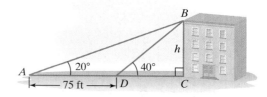

Group Exercise

62. Music and mathematics have been linked over the centuries. Group members should research and present a seminar to the class on music and mathematics. Be sure to include the role of trigonometric functions in the music-mathematics link.

Chapter Summary, Review, and Test

Summary

DEFINITIONS AND CONCEPTS   EXAMPLES

4.1 Angles and Their Measure

a. An angle consists of two rays with a common endpoint, the vertex.

b. An angle is in standard position if its vertex is at the origin and its initial side lies along   Ex. 1, p. 425
the positive x-axis. Figure 4.3 on page 424 shows positive and negative angles in standard position.

c. A quadrantal angle is one with its terminal side on the x-axis or the y-axis.

d. Angles can be measured in degrees. $1°$ is $\frac{1}{360}$ of a complete rotation.

e. Acute angles measure less than $90°$, right angles $90°$, obtuse angles more than $90°$ but less
than $180°$, and straight angles $180°$.

f. Two angles with the same initial and terminal sides are called coterminal angles.   Ex. 2, p. 426

g. Two angles are complements if their sum is $90°$ and supplements if their sum is $180°$. Only   Ex. 3, p. 427
positive angles are used.

DEFINITIONS AND CONCEPTS   **EXAMPLES**

h. Angles can be measured in radians. One radian is the measure of the central angle when the   Ex. 4, p. 429
intercepted arc and radius have the same length. In general, the radian measure of a central angle

is the length of the intercepted arc divided by the circle's radius: $\theta = \dfrac{s}{r}$.

i. To convert from degrees to radians, multiply degrees by $\dfrac{\pi \text{ radians}}{180°}$. To convert from radians to   Ex. 5, p. 430;
Ex. 6, p. 430

degrees, multiply radians by $\dfrac{180°}{\pi \text{ radians}}$.

j. The arc length formula, $s = r\theta$, is described in the box on page 431.   Ex. 7, p. 432

k. The definitions of linear speed, $v = \dfrac{s}{t}$, and angular speed, $\omega = \dfrac{\theta}{t}$, are given in the box on page 432.

l. Linear speed is expressed in terms of angular speed by $v = r\omega$, where v is the linear speed of a   Ex. 8, p. 433
point a distance r from the center of rotation and ω is the angular speed in radians per unit of time.

4.2 Trigonometric Functions: The Unit Circle

a. Definitions of the trigonometric functions in terms of a unit circle are given in the box on page 439.   Ex. 1, p. 439;
Ex. 2, p. 440

b. The cosine and secant functions are even: $\cos(-t) = \cos t$, $\sec(-t) = \sec t$.   Ex. 4, p. 443
The other trigonometric functions are odd: $\sin(-t) = -\sin t$, $\tan(-t) = -\tan t$,
$\cot(-t) = -\cot t$, $\csc(-t) = -\csc t$.

c. Fundamental Identities

1. Reciprocal Identities

$$\sin t = \frac{1}{\csc t} \quad \cos t = \frac{1}{\sec t} \quad \tan t = \frac{1}{\cot t}$$

$$\csc t = \frac{1}{\sin t} \quad \sec t = \frac{1}{\cos t} \quad \cot t = \frac{1}{\tan t}$$

2. Quotient Identities   Ex. 5, p. 444

$$\tan t = \frac{\sin t}{\cos t} \quad \cot t = \frac{\cos t}{\sin t}$$

3. Pythagorean Identities   Ex. 6, p. 446

$$\sin^2 t + \cos^2 t = 1 \quad 1 + \tan^2 t = \sec^2 t \quad 1 + \cot^2 t = \csc^2 t$$

d. If $f(t + p) = f(t)$, function f is periodic. The smallest p for which f is periodic is the period of f.   Ex. 7, p. 447
The tangent and cotangent functions have period π. The other four trigonometric functions have
period 2π.

4.3 Right Triangle Trigonometry

a. The right triangle definitions of the six trigonometric functions are given in the box on page 454.   Ex. 1, p. 455

b. Function values for 30°, 45°, and 60° can be obtained using these special triangles.   Ex. 2, p. 456;
Ex. 3, p. 457

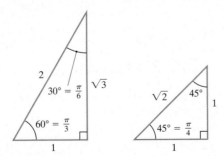

c. The value of a trigonometric function of θ is equal to the cofunction of the complement of θ.   Ex. 4, p. 458
Cofunction identities are listed in the box on page 458.

DEFINITIONS AND CONCEPTS **EXAMPLES**

4.4 Trigonometric Functions of Any Angle

a. Definitions of the trigonometric functions of any angle are given in the box on page 465.

Ex. 1, p. 466;
Ex. 2, p. 466

b. Signs of the trigonometric functions: All functions are positive in quadrant I. If θ lies in quadrant II, $\sin\theta$ and $\csc\theta$ are positive. If θ lies in quadrant III, $\tan\theta$ and $\cot\theta$ are positive. If θ lies in quadrant IV, $\cos\theta$ and $\sec\theta$ are positive.

Ex. 3, p. 469;
Ex. 4, p. 469

c. If θ is a nonacute angle in standard position that lies in a quadrant, its reference angle is the positive acute angle θ' formed by the terminal side of θ and the x-axis. The reference angle for a given angle can be found by making a sketch that shows the angle in standard position. Figure 4.36 on page 470 shows reference angles for θ in quadrants II, III, and IV.

Ex. 5, p. 470

d. The values of the trigonometric functions of a given angle are the same as the values of the functions of the reference angle, except possibly for the sign. A procedure for using reference angles to evaluate trigonometric functions is given in the box on page 471.

Ex. 6, p. 472

4.5 and 4.6 Graphs of the Trigonometric Functions

a. Graphs of the six trigonometric functions, with a description of the domain, range, and period of each function, are given in Table 4.5 on page 505.

b. The graph of $y = A\sin(Bx - C)$ can be obtained using amplitude $= |A|$, period $= \dfrac{2\pi}{B}$ and phase shift $= \dfrac{C}{B}$. See the illustration in the box on page 482.

Ex. 1, p. 477;
Ex. 2, p. 479;
Ex. 3, p. 480;
Ex. 4, p. 483
Ex. 5, p. 486;
Ex. 6, p. 488

c. The graph of $y = A\cos(Bx - C)$ can be obtained using amplitude $= |A|$, period $= \dfrac{2\pi}{B}$, and phase shift $= \dfrac{C}{B}$. See the illustration in the box on page 488.

d. The constant D in $y = A\sin(Bx - C) + D$ and $y = A\cos(Bx - C) + D$ causes vertical shifts in the graphs in the preceding items (b) and (c). If $D > 0$, the shift is D units upward and if $D < 0$, the shift is D units downward. Oscillation is about $y = D$.

Ex. 7, p. 490

e. The graph of $y = A\tan(Bx - C)$ is obtained using the procedure in the box on page 499. Consecutive asymptotes $\left(\text{solve } Bx - C = -\dfrac{\pi}{2} \text{ and } Bx - C = \dfrac{\pi}{2}\right)$ and an x-intercept midway between them play a key role in the graphing process.

Ex. 1, p. 499;
Ex. 2, p. 500

f. The graph of $y = A\cot(Bx - C)$ is obtained using the procedure in the box on page 501. Consecutive asymptotes (solve $Bx - C = 0$ and $Bx - C = \pi$) and an x-intercept midway between them play a key role in the graphing process.

Ex. 3, p. 501

g. To graph a cosecant curve, begin by graphing the reciprocal sine curve. Draw vertical asymptotes through x-intercepts, using asymptotes as guides to sketch the graph. To graph a secant curve, first graph the reciprocal cosine curve and use the same procedure.

Ex. 4, p. 503;
Ex. 5, p. 504

4.7 Inverse Trigonometric Functions

a. On the restricted domain $-\dfrac{\pi}{2} \le x \le \dfrac{\pi}{2}$, $y = \sin x$ has an inverse function, defined in the box on page 510. Think of $\sin^{-1} x$ as the angle in $\left[-\dfrac{\pi}{2}, \dfrac{\pi}{2}\right]$ whose sine is x.

Ex. 1, p. 512;
Ex. 2, p. 512

b. On the restricted domain $0 \le x \le \pi$, $y = \cos x$ has an inverse function, defined in the box on page 513. Think of $\cos^{-1} x$ as the angle in $[0, \pi]$ whose cosine is x.

Ex. 3, p. 514

DEFINITIONS AND CONCEPTS   **EXAMPLES**

c. On the restricted domain $-\dfrac{\pi}{2} < x < \dfrac{\pi}{2}$, $y = \tan x$ has an inverse function, defined in the box on   Ex. 4, p. 516

page 515. Think of $\tan^{-1} x$ as the angle in $\left(-\dfrac{\pi}{2}, \dfrac{\pi}{2}\right)$ whose tangent is x.

d. Graphs of the three basic inverse trigonometric functions, with a description of the domain and range of each function, are given in Table 4.9 on page 517.

e. Inverse properties are given in the box on page 518. Points on terminal sides of angles in standard   Ex. 6, p. 519;
position and right triangles are used to find exact values of the composition of a function and a   Ex. 7, p. 520;
different inverse function.   Ex. 8, p. 520;
                                                                                                            Ex. 9, p. 521

4.8 Applications of Trigonometric Functions

a. Solving a right triangle means finding the missing lengths of its sides and the measurements of its   Ex. 1, p. 525;
angles. The Pythagorean Theorem, two acute angles whose sum is 90°, and appropriate trigonometric   Ex. 2, p. 526;
functions are used in this process.   Ex. 3, p. 527;
                                                                                                            Ex. 4, p. 527

b. The bearing from point O to point P is the acute angle between ray OP and a north-south line.   Ex. 5, p. 528;
                                                                                                            Ex. 6, p. 529

c. Simple harmonic motion, described in the box on page 531, is modeled by $d = a \cos \omega t$ or   Ex. 7, p. 531;
                                                                                                            Ex. 8, p. 532
$d = a \sin \omega t$, with amplitude $= |a|$, period $= \dfrac{2\pi}{\omega}$ and frequency $= \dfrac{\omega}{2\pi} = \dfrac{1}{\text{period}}$.

Review Exercises

4.1

In Exercises 1–4, draw each angle in standard position.

1. 190°   **2.** −135°

3. $\dfrac{5\pi}{6}$ **4.** $-\dfrac{2\pi}{3}$

In Exercises 5–6, find a positive angle less than 360° that is coterminal with the given angle.

5. 400°   **6.** −85°

In Exercises 7–8, if possible, find the complement and the supplement of the given angle.

7. 73°   **8.** $\dfrac{2\pi}{3}$

9. Find the radian measure of the central angle of a circle of radius 6 centimeters that intercepts an arc of length 27 centimeters.

In Exercises 10–12, convert each angle in degrees to radians. Express your answer as a multiple of π.

10. 15°   **11.** 120°   **12.** 315°

In Exercises 13–15, convert each angle in radians to degrees.

13. $\dfrac{5\pi}{3}$ **14.** $\dfrac{7\pi}{5}$ **15.** $-\dfrac{5\pi}{6}$

16. Find the length of the arc on a circle of radius 10 feet intercepted by a 135° central angle. Express arc length in terms of π. Then round your answer to two decimal places.

17. The angular speed of a propeller on a wind generator is 10.3 revolutions per minute. Express this angular speed in radians per minute.

18. The propeller of an airplane has a radius of 3 feet. The propeller is rotating at 2250 revolutions per minute. Find the linear speed, in feet per minute, of the tip of the propeller.

4.2

In Exercises 19–20, a point P(x, y) is shown on the unit circle corresponding to a real number t. Find the values of the trigonometric functions at t.

19.

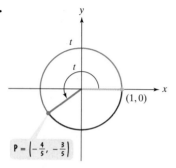

$$P = \left(-\frac{4}{5}, -\frac{3}{5}\right)$$

20.

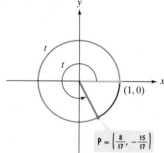

$$P = \left(\frac{8}{17}, -\frac{15}{17}\right)$$

In Exercises 21–24, use the figure shown to find the value of each trigonometric function at the indicated real number or state that the expression is undefined.

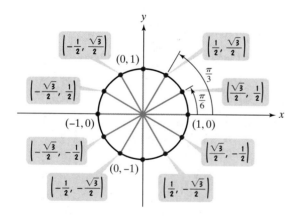

21. $\sec \frac{5\pi}{6}$ **22.** $\tan \frac{4\pi}{3}$ **23.** $\sec \frac{\pi}{2}$ **24.** $\cot \pi$

25. If $\sin t = \frac{2}{\sqrt{7}}$, $0 \leq t < \frac{\pi}{2}$, use identities to find the remaining trigonometric functions.

In Exercises 26–28, evaluate each expression without using a calculator.

26. $\tan 4.7 \cot 4.7$

27. $\sin^2 \frac{\pi}{17} + \cos^2 \frac{\pi}{17}$

28. $\cot^2 1.4 - \csc^2 1.4$

4.3

29. Use the triangle to find each of the six trigonometric functions of θ.

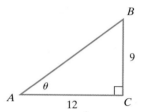

In Exercises 30–31, find the exact value of each expression.

30. $4 \cot \frac{\pi}{4} + \cos \frac{\pi}{3} \csc \frac{\pi}{6}$ **31.** $\cos \frac{\pi}{6} \sin \frac{\pi}{4} - \tan \frac{\pi}{4}$

In Exercises 32–33, find a cofunction with the same value as the given expression.

32. $\sin 70°$

33. $\cos \frac{\pi}{2}$

In Exercises 34–36, find the measure of the side of the right triangle whose length is designated by a lowercase letter. Round answers to the nearest whole number.

34.

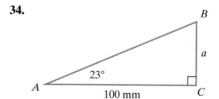

35.

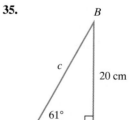

36.

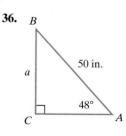

37. A hiker climbs for a half mile up a slope whose inclination is 17°. How many feet of altitude, to the nearest foot, does the hiker gain?

38. To find the distance across a lake, a surveyor took the measurements in the figure shown. What is the distance across the lake? Round to the nearest meter.

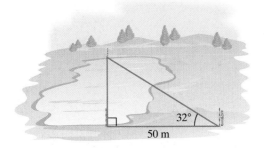

32°

50 m

39. When a six-foot pole casts a four-foot shadow, what is the angle of elevation of the sun? Round to the nearest whole degree.

4.4

In Exercises 40–41, a point on the terminal side of angle θ is given. Find the exact value of each of the six trigonometric functions of θ, or state that the function is undefined.

40. $(-1, -5)$

41. $(0, -1)$

In Exercises 42–43, let θ be an angle in standard position. Name the quadrant in which θ lies.

42. $\tan \theta > 0$ and $\sec \theta > 0$

43. $\tan \theta > 0$ and $\cos \theta < 0$

In Exercises 44–45, find the exact value of each of the remaining trigonometric functions of θ.

44. $\cos \theta = \frac{2}{5}, \sin \theta < 0$

45. $\tan \theta = -\frac{1}{3}, \sin \theta > 0$

In Exercises 46–48, find the reference angle for each angle.

46. $265°$

47. $\frac{5\pi}{8}$

48. $-410°$

In Exercises 49–57, find the exact value of each expression. Do not use a calculator.

49. $\sin 240°$

50. $\tan 120°$

51. $\sec \frac{7\pi}{4}$

52. $\cos \frac{11\pi}{6}$

53. $\cot (-210°)$

54. $\csc \left(-\frac{2\pi}{3} \right)$

55. $\sin \left(-\frac{\pi}{3} \right)$

56. $\sin 495°$

57. $\tan \frac{13\pi}{4}$

4.5

In Exercises 58–63, determine the amplitude and period of each function. Then graph one period of the function.

58. $y = 3 \sin 4x$

59. $y = -2 \cos 2x$

60. $y = 2 \cos \frac{1}{2} x$

61. $y = \frac{1}{2} \sin \frac{\pi}{3} x$

62. $y = -\sin \pi x$

63. $y = 3 \cos \frac{x}{3}$

In Exercises 64–68, determine the amplitude, period, and phase shift of each function. Then graph one period of the function.

64. $y = 2 \sin (x - \pi)$

65. $y = -3 \cos (x + \pi)$

66. $y = \frac{3}{2} \cos \left(2x + \frac{\pi}{4} \right)$

67. $y = \frac{5}{2} \sin \left(2x + \frac{\pi}{2} \right)$

68. $y = -3 \sin \left(\frac{\pi}{3} x - 3\pi \right)$

In Exercises 69–70, use a vertical shift to graph one period of the function.

69. $y = \sin 2x + 1$

70. $y = 2 \cos \frac{1}{3} x - 2$

71. The equation

$$y = 98.6 + 0.3 \sin \left(\frac{\pi}{12} x - \frac{11\pi}{12} \right)$$

models variation in body temperature, y, in °F, x hours after midnight.
   a. What is body temperature at midnight?
   b. What is the period of the body temperature cycle?
   c. When is body temperature highest? What is the body temperature at this time?
   d. When is body temperature lowest? What is the body temperature at this time?
   e. Graph one period of the body temperature function.

4.6

In Exercises 72–78, graph two full periods of the given tangent or cotangent function.

72. $y = 4 \tan 2x$

73. $y = -2 \tan \frac{\pi}{4} x$

74. $y = \tan (x + \pi)$

75. $y = -\tan \left(x - \frac{\pi}{4} \right)$

76. $y = 2 \cot 3x$

77. $y = -\frac{1}{2} \cot \frac{\pi}{2} x$

78. $y = 2 \cot \left(x + \frac{\pi}{2} \right)$

In Exercises 79–82, graph two full periods of the given cosecant or secant function.

79. $y = 3 \sec 2\pi x$

80. $y = -2 \csc \pi x$

81. $y = 3 \sec (x + \pi)$

82. $y = \frac{5}{2} \csc (x - \pi)$

4.7

In Exercises 83–100, find the exact value of each expression. Do not use a calculator.

83. $\sin^{-1} 1$

84. $\cos^{-1} 1$

85. $\tan^{-1} 1$

86. $\sin^{-1} \left(-\frac{\sqrt{3}}{2} \right)$

87. $\cos^{-1} \left(-\frac{1}{2} \right)$

88. $\tan^{-1} \left(-\frac{\sqrt{3}}{3} \right)$

89. $\cos \left(\sin^{-1} \frac{\sqrt{2}}{2} \right)$

90. $\sin (\cos^{-1} 0)$

91. $\tan \left[\sin^{-1} \left(-\frac{1}{2} \right) \right]$

92. $\tan \left[\cos^{-1} \left(-\frac{\sqrt{3}}{2} \right) \right]$

93. $\csc\left(\tan^{-1}\dfrac{\sqrt{3}}{3}\right)$

94. $\cos\left(\tan^{-1}\frac{3}{4}\right)$

95. $\sin\left(\cos^{-1}\frac{3}{5}\right)$

96. $\tan\left[\sin^{-1}\left(-\frac{3}{5}\right)\right]$

97. $\tan\left[\cos^{-1}\left(-\frac{4}{5}\right)\right]$

98. $\sin^{-1}\left(\sin\dfrac{\pi}{3}\right)$

99. $\sin^{-1}\left(\sin\dfrac{2\pi}{3}\right)$

100. $\sin^{-1}\left(\cos\dfrac{2\pi}{3}\right)$

In Exercises 101–102, use a right triangle to write each expression as an algebraic expression. Assume that x is positive and in the domain of the given inverse trigonometric function.

101. $\cos\left(\tan^{-1}\dfrac{x}{2}\right)$

102. $\sec\left(\sin^{-1}\dfrac{1}{x}\right)$

4.8

In Exercises 103–106, solve the right triangle shown in the figure. Round lengths to two decimal places and express angles to the nearest tenth of a degree.

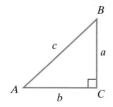

103. $A = 22.3°, c = 10$

104. $B = 37.4°, b = 6$

105. $a = 2, c = 7$

106. $a = 1.4, b = 3.6$

107. From a point on level ground 80 feet from the base of a building, the angle of elevation is 25.6°. Approximate the height of the building to the nearest foot.

108. Two buildings with flat roofs are 60 yards apart. The height of the shorter building is 40 yards. From its roof, the angle of elevation to the edge of the roof of the taller building is 40°. Find the height of the taller building to the nearest yard.

109. You want to measure the height of an antenna on the top of a 125-foot building. From a point in front of the building, you measure the angle of elevation to the top of the building to be 68° and the angle of elevation to the top of the antenna to be 71°. How tall is the antenna, to the nearest tenth of a foot?

In Exercises 110–111, use the figure shown to find the bearing from O to A.

110.   **111.**

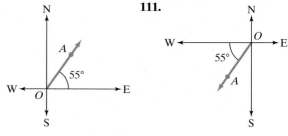

112. A ship is due west of a lighthouse. A second ship is 12 miles south of the first ship. The bearing from the second ship to the lighthouse is N 64° E. How far, to the nearest tenth of a mile, is the first ship from the lighthouse?

113. From city A to city B, a plane flies 850 miles at a bearing of N 58° E. From city B to city C, the plane flies 960 miles at a bearing of S 72° E.
   a. Find, to the nearest tenth of a mile, the distance from city A to city C.
   b. What is the bearing from city A to city C?

In Exercises 114–115, an object moves in simple harmonic motion described by the given equation, where t is measured in seconds and d in centimeters. In each exercise, find

   a. *the maximum displacement.*
   b. *the frequency.*
   c. *the time required for one cycle.*

114. $d = 20\cos\dfrac{\pi}{4}t$

115. $d = \frac{1}{2}\sin 4t$

In Exercises 116–117, an object is attached to a coiled spring. The object is pulled down (negative direction from the rest position) and then released. Write an equation for the distance of the object from its rest position after t seconds.

	Distance from rest position at $t = 0$	Amplitude	Period
116.	30 inches	30 inches	2 seconds
117.	0	$\frac{1}{4}$ inches	5 seconds

Chapter 4 Test

1. Convert 135° to exact radian measure.

2. Find the supplement of the angle whose radian measure is $\dfrac{9\pi}{13}$. Express the answer in terms of π.

3. Find the length of the arc on a circle of radius 20 feet intercepted by a 75° central angle. Express arc length in terms of π. Then round your answer to two decimal places.

4. If $(-2, 5)$ is a point on the terminal side of angle θ, find the exact value of each of the six trigonometric functions of θ.

5. Determine the quadrant in which θ lies if $\cos \theta < 0$ and $\cot \theta > 0$.

6. If $\cos \theta = \frac{1}{3}$ and $\tan \theta < 0$, find the exact value of each of the remaining trigonometric functions of θ.

In Exercises 7–9, find the exact value of each expression. Do not use a calculator.

7. $\tan \dfrac{\pi}{6} \cos \dfrac{\pi}{3} - \cos \dfrac{\pi}{2}$ **8.** $\tan 300°$

9. $\sin \dfrac{7\pi}{4}$

In Exercises 10–13, graph one period of each function.

10. $y = 3 \sin 2x$ **11.** $y = -2 \cos \left(x - \dfrac{\pi}{2} \right)$

12. $y = 2 \tan \dfrac{x}{2}$ **13.** $y = -\dfrac{1}{2} \csc \pi x$

14. Find the exact value of $\tan \left[\cos^{-1} \left(-\dfrac{1}{2} \right) \right]$.

15. Solve the right triangle in the figure shown. Round lengths to one decimal place.

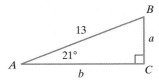

16. The angle of elevation of a building from a point on the ground 30 yards from its base is 37°. Find the height of the building to the nearest yard.

17. A 73-foot rope from the top of a circus tent pole is anchored to the flat ground 43 feet from the bottom of the pole. Find the angle, to the nearest tenth of a degree, that the rope makes with the pole.

18. Use the figure to find the bearing from O to P.

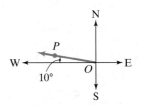

19. An object moves in simple harmonic motion described by $d = -6 \cos \pi t$, where t is measured in seconds and d in inches. Find **a.** the maximum displacement, **b.** the frequency, and **c.** the time required for one oscillation.

20. Why are trigonometric functions ideally suited to model phenomena that repeat in cycles?

Cumulative Review Exercises (Chapters P–4)

Solve each equation or inequality in Exercises 1–6.

1. $x^2 = 18 + 3x$ **2.** $x^3 + 5x^2 - 4x - 20 = 0$

3. $\log_2 x + \log_2 (x - 2) = 3$ **4.** $e^{0.04x} = 4500$

5. $x^3 - 4x^2 + x + 6 = 0$ **6.** $|2x - 5| \le 11$

7. If $f(x) = \sqrt{x - 6}$, find $f^{-1}(x)$.

8. Divide $20x^3 - 6x^2 - 9x + 10$ by $5x + 2$.

9. Write as a single logarithm and evaluate: $\log 25 + \log 40$.

10. Convert $\dfrac{14\pi}{9}$ radians to degrees.

11. Find the maximum number of positive and negative real roots of the equation $3x^4 - 2x^3 + 5x^2 + x - 9 = 0$.

In Exercises 12–16, graph each equation.

12. $f(x) = \dfrac{x}{x^2 - 1}$ **13.** $(x - 2)^2 + y^2 = 1$

14. $y = (x - 1)(x + 2)^2$

15. $y = \sin \left(2x + \dfrac{\pi}{2} \right)$, from 0 to 2π

16. $y = 2 \tan 3x$; graph two complete cycles.

17. The height of a triangle is half the base. Express the area of the triangle, A, as function of the base, x.

18. Use the exponential growth model $A = A_0 e^{kt}$ to solve this exercise. Data from the Federal Communication Commission show that the use of toll-free 800 numbers has grown exponentially. In 1991 there were 10.2 billion such calls and by 1998, there were 86.7 billion.

  a. Find the exponential function that models the data.

  b. By which year will the number of toll-free 800 numbers reach 200 billion?

19. The rate of heat lost through insulation varies inversely as the thickness of the insulation. The rate of heat lost through a 3.5-inch thickness of insulation is 2200 Btu per hour. What is the rate of heat lost through a 5-inch thickness of the same insulation?

20. A tower is 200 feet tall. To the nearest degree, find the angle of elevation from a point 50 feet from the base of the tower to the top of the tower.

Analytic Trigonometry

Chapter 5

This chapter emphasizes the algebraic aspects of trigonometry. We derive important categories of identities involving trigonometric functions. These identities are used to simplify and analyze expressions that model phenomena as diverse as the distance achieved when throwing an object and musical sounds on a touch-tone phone. For example, we can find out critical information about an athlete's performance by using an identity to analyze an expression involving throwing distance. You will learn how to use trigonometric identities to better understand your periodic world.

You enjoy watching your friend participate in the shot put at college track and field events. After a few full turns in a circle, she throws ("puts") an 8-pound, 13-ounce shot from the shoulder. The range of her throwing distance continues to improve. Knowing that you are studying trigonometry, she asks if there is some way that a trigonometric expression might help achieve her best possible distance in the event.

SECTION 5.1   *Verifying Trigonometric Identities*

Objective

1. Use the fundamental trigonometric identities to verify identities.

Do you enjoy solving puzzles? The process is a natural way to develop problem-solving skills that are important in every area of our lives. Engaging in problem solving for sheer pleasure releases chemicals in the brain that enhance our feeling of well-being. Perhaps this is why puzzles have fascinated people for over 12,000 years.

Thousands of relationships exist among the six trigonometric functions. Verifying these relationships is like solving a puzzle. Why? There are no rigid rules for the process. Thus, proving a trigonometric relationship requires you to be creative in your approach to problem solving. By learning to establish these relationships, you will become a better, more confident problem solver. Furthermore, you may enjoy the feeling of satisfaction that accompanies solving each "puzzle."

The Fundamental Identities

In Chapter 4, we used right triangles to establish relationships among the trigonometric functions. Although we limited domains to acute angles, the fundamental identities listed in the following box are true for all values of x for which the expressions are defined.

Study Tip

Memorize the identities in the box. You may need to use variations of these fundamental identities. For example, instead of

$$\sin^2 x + \cos^2 x = 1,$$

you might want to use

$$\sin^2 x = 1 - \cos^2 x$$

or

$$\cos^2 x = 1 - \sin^2 x.$$

Therefore, it is important to know each relationship well so that mental algebraic manipulation is possible.

Fundamental Trigonometric Identities

Reciprocal Identities

$$\sin x = \frac{1}{\csc x} \quad \cos x = \frac{1}{\sec x} \quad \tan x = \frac{1}{\cot x}$$

$$\csc x = \frac{1}{\sin x} \quad \sec x = \frac{1}{\cos x} \quad \cot x = \frac{1}{\tan x}$$

Quotient Identities

$$\tan x = \frac{\sin x}{\cos x} \quad \cot x = \frac{\cos x}{\sin x}$$

Pythagorean Identities

$$\sin^2 x + \cos^2 x = 1 \quad 1 + \tan^2 x = \sec^2 x \quad 1 + \cot^2 x = \csc^2 x$$

Even-Odd Identities

$$\sin(-x) = -\sin x \quad \cos(-x) = \cos x \quad \tan(-x) = -\tan x$$

$$\csc(-x) = -\csc x \quad \sec(-x) = \sec x \quad \cot(-x) = -\cot x$$

1 Use the fundamental trigonometric identities to verify identities.

Using Fundamental Identities to Verify Other Identities

The fundamental trigonometric identities are used to establish other relationships among trigonometric functions. To **verify an identity**, we show that one side of the identity can be simplified so that it is identical to the other side. Each side of the equation is manipulated independently of the other side of the equation. Start with the side containing the more complicated expression. If you substitute one or more fundamental identities on the more complicated side, you will often be able to rewrite it in a form identical to that of the other side.

No one method can be used to verify every identity. Some identities can be verified by rewriting the more complicated side so that it contains only sines and cosines.

EXAMPLE 1 Changing to Sines and Cosines to Verify an Identity

Verify the identity: $\sec x \cot x = \csc x$.

<section name="Technology">
Technology

You can use a graphing utility to provide evidence of an identity. For example, consider the equation in Example 1,

$$\sec x \cot x = \csc x.$$

Graph

$$y = \sec x \cot x$$

and

$$y = \csc x$$

in the same viewing rectangle. The screen shown indicates that the graphs appear to be the same. Example 1 shows the equivalence algebraically.

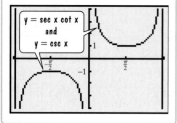
</section>

Solution   The left side of the equation contains the more complicated expression. Thus, we work with the left side. Let us express this side of the identity in terms of sines and cosines. Perhaps this strategy will enable us to transform the left side into $\csc x$, the expression on the right.

$$\sec x \cot x = \frac{1}{\cos x} \cdot \frac{\cos x}{\sin x}$$

Apply a reciprocal identity: $\sec x = \dfrac{1}{\cos x}$ and a quotient identity: $\cot x = \dfrac{\cos x}{\sin x}$.

$$= \frac{1}{\cancel{\cos x}} \cdot \frac{\overset{1}{\cancel{\cos x}}}{\sin x}$$

Divide both the numerator and the denominator by $\cos x$, the common factor.

$$= \frac{1}{\sin x}$$

Multiply the remaining factors in the numerator and denominator.

$$= \csc x$$

Apply a reciprocal identity: $\csc x = \dfrac{1}{\sin x}$.

By working with the left side and simplifying it so that it is identical to the right side, we have verified the given identity.

Check Point 1   Verify the identity: $\csc x \tan x = \sec x$.

In verifying an identity, stay focused on your goal. When manipulating one side of the equation, continue to look at the other side to keep the desired form of the result in mind.

Study Tip

Verifying that an equation is an identity is different from solving an equation. You do not verify an identity by adding, subtracting, multiplying, or dividing each side by the same expression. If you do this, you have already assumed that the given statement is true. You do not know that it is true until after you have verified it.

EXAMPLE 2 **Changing to Sines and Cosines to Verify an Identity**

Verify the identity: $\sin x \tan x + \cos x = \sec x$.

Solution The left side is more complicated, so we start with it. Let us express this side of the identity so that it contains only sines and cosines. Thus, we apply a quotient identity and replace $\tan x$ by $\dfrac{\sin x}{\cos x}$. Perhaps this strategy will enable us to transform the left side into $\sec x$, the expression on the right.

$$\sin x \tan x + \cos x = \sin x \left(\frac{\sin x}{\cos x} \right) + \cos x \qquad \text{Apply a quotient identity:} \quad \tan x = \frac{\sin x}{\cos x}.$$

$$= \frac{\sin^2 x}{\cos x} + \cos x. \qquad \text{Multiply.}$$

$$= \frac{\sin^2 x}{\cos x} + \cos x \cdot \frac{\cos x}{\cos x} \qquad \text{The least common denominator is } \cos x. \text{ Write the second expression with a denominator of } \cos x.$$

$$= \frac{\sin^2 x}{\cos x} + \frac{\cos^2 x}{\cos x} \qquad \text{Multiply.}$$

$$= \frac{\sin^2 x + \cos^2 x}{\cos x} \qquad \text{Add numerators, putting this sum over the least common denominator.}$$

$$= \frac{1}{\cos x} \qquad \text{Apply a Pythagorean identity:} \quad \sin^2 x + \cos^2 x = 1.$$

$$= \sec x \qquad \text{Apply a reciprocal identity:} \quad \sec x = \frac{1}{\cos x}.$$

By working with the left side and arriving at the right side, the identity is verified.

Check Point 2 Verify the identity: $\cos x \cot x + \sin x = \csc x$.

Some identities are verified using factoring to simplify a trigonometric expression.

EXAMPLE 3 **Using Factoring to Verify an Identity**

Verify the identity: $\cos x - \cos x \sin^2 x = \cos^3 x$.

Solution We start with the more complicated side, the left side. Factor out the greatest common factor, $\cos x$, from each of the two terms.

$$\cos x - \cos x \sin^2 x = \cos x (1 - \sin^2 x) \qquad \text{Factor } \cos x \text{ from the two terms.}$$

$$= \cos x \cdot \cos^2 x \qquad \text{Use a variation of } \sin^2 x + \cos^2 x = 1.$$
$$\text{Solving for } \cos^2 x, \text{ we obtain}$$
$$\cos^2 x = 1 - \sin^2 x.$$

$$= \cos^3 x \qquad \text{Multiply.}$$

We worked with the left side and arrived at the right side. Thus, the identity is verified.

Check Point 3 Verify the identity: $\sin x - \sin x \cos^2 x = \sin^3 x$.

How do we verify identities in which sums or differences of fractions with trigonometric functions appear on one side? Use the least common denominator and combine the fractions. This technique is especially useful when the other side of the identity contains only one term.

EXAMPLE 4 Combining Fractional Expressions to Verify an Identity

Verify the identity: $\dfrac{\cos x}{1 + \sin x} + \dfrac{1 + \sin x}{\cos x} = 2 \sec x$.

Solution We start with the more complicated side, the left side. The least common denominator of the fractions is $(1 + \sin x)(\cos x)$. We express each fraction in terms of this least common denominator by multiplying the numerator and denominator by the extra factor needed to form $(1 + \sin x)(\cos x)$.

<table>
<tr>
<td>

$\dfrac{\cos x}{1 + \sin x} + \dfrac{1 + \sin x}{\cos x}$

</td>
<td>

The least common denominator is $(1 + \sin x)(\cos x)$.

</td>
</tr>
<tr>
<td>

$= \dfrac{\cos x (\cos x)}{(1 + \sin x)(\cos x)} + \dfrac{(1 + \sin x)(1 + \sin x)}{(1 + \sin x)(\cos x)}$

</td>
<td>

Rewrite each fraction with the least common denominator.

</td>
</tr>
<tr>
<td>

$= \dfrac{\cos^2 x}{(1 + \sin x)(\cos x)} + \dfrac{1 + 2 \sin x + \sin^2 x}{(1 + \sin x)(\cos x)}$

</td>
<td>

Use the FOIL method to multiply $(1 + \sin x)(1 + \sin x)$.

</td>
</tr>
<tr>
<td>

$= \dfrac{\cos^2 x + 1 + 2 \sin x + \sin^2 x}{(1 + \sin x)(\cos x)}$

</td>
<td>

Add numerators. Put this sum over the least common denominator.

</td>
</tr>
<tr>
<td>

$= \dfrac{(\sin^2 x + \cos^2 x) + 1 + 2 \sin x}{(1 + \sin x)(\cos x)}$

</td>
<td>

Regroup terms to apply a Pythagorean identity.

</td>
</tr>
<tr>
<td>

$= \dfrac{1 + 1 + 2 \sin x}{(1 + \sin x)(\cos x)}$

</td>
<td>

Apply a Pythagorean identity: $\sin^2 x + \cos^2 x = 1$.

</td>
</tr>
<tr>
<td>

$= \dfrac{2 + 2 \sin x}{(1 + \sin x)(\cos x)}$

</td>
<td>

Add constant terms in the numerator: $1 + 1 = 2$.

</td>
</tr>
<tr>
<td>

$= \dfrac{2 (1 + \sin x)}{(1 + \sin x) (\cos x)}$

</td>
<td>

Factor and simplify.

</td>
</tr>
<tr>
<td>

$= \dfrac{2}{\cos x}$

</td>
<td></td>
</tr>
<tr>
<td>

$= 2 \sec x$

</td>
<td>

Apply a reciprocal identity: $\sec x = \dfrac{1}{\cos x}$.

</td>
</tr>
</table>

Study Tip

Some students have difficulty verifying identities due to problems working with fractions. If this applies to you, review the material on rational expressions in Chapter P, Section P.6.

We worked with the left side and arrived at the right side. Thus, the identity is verified.

Check Point 4   Verify the identity: $\dfrac{\sin x}{1 + \cos x} + \dfrac{1 + \cos x}{\sin x} = 2 \csc x.$

Some identities are verified using a technique that may remind you of rationalizing a denominator.

EXAMPLE 5   **Multiplying the Numerator and Denominator by the Same Factor to Verify an Identity**

Verify the identity: $\dfrac{\sin x}{1 + \cos x} = \dfrac{1 - \cos x}{\sin x}.$

Solution   The suggestions given in the previous examples do not apply here. Everything is already expressed in terms of sines and cosines. Furthermore, there are no fractions to combine and neither side looks more complicated than the other. Let's solve the puzzle by working with the left side and making it look like the expression on the right. The expression on the right contains $1 - \cos x$ in the numerator. This suggests multiplying the numerator and denominator of the left side by $1 - \cos x$. By doing this, we obtain a factor of $1 - \cos x$ in the numerator, as in the numerator on the right.

Discovery

Verify the identity in Example 5 by making the right side look like the left side. Start with the expression on the right. Multiply the numerator and denominator by $1 + \cos x$.

$$\dfrac{\sin x}{1 + \cos x} = \dfrac{\sin x}{1 + \cos x} \cdot \dfrac{1 - \cos x}{1 - \cos x} \qquad \text{Multiply numerator and denominator by } 1 - \cos x.$$

$$= \dfrac{\sin x (1 - \cos x)}{1 - \cos^2 x} \qquad \text{Multiply. Use } (A + B)(A - B) = A^2 - B^2, \text{ with } A = 1 \text{ and } B = \cos x, \text{ to multiply denominators.}$$

$$= \dfrac{\sin x (1 - \cos x)}{\sin^2 x} \qquad \text{Use a variation of } \sin^2 x + \cos^2 x = 1. \text{ Solving for } \sin^2 x, \text{ we obtain } \sin^2 x = 1 - \cos^2 x.$$

$$= \dfrac{1 - \cos x}{\sin x} \qquad \text{Simplify: } \dfrac{\sin x}{\sin^2 x} = \dfrac{\cancel{\sin x}}{\cancel{\sin x} \cdot \sin x} = \dfrac{1}{\sin x}.$$

We worked with the left side and arrived at the right side. Thus, the identity is verified.

Check Point 5   Verify the identity: $\dfrac{\cos x}{1 + \sin x} = \dfrac{1 - \sin x}{\cos x}.$

EXAMPLE 6 Changing to Sines and Cosines to Verify an Identity

Verify the identity: $\dfrac{\tan x - \sin(-x)}{1 + \cos x} = \tan x.$

Solution We begin with the left side. Our goal is to obtain $\tan x$, the expression on the right.

$$\frac{\tan x - \sin(-x)}{1 + \cos x} = \frac{\tan x - (-\sin x)}{1 + \cos x} \qquad \text{The sine function is odd: } \sin(-x) = -\sin x.$$

$$= \frac{\tan x + \sin x}{1 + \cos x} \qquad \text{Simplify.}$$

$$= \frac{\dfrac{\sin x}{\cos x} + \sin x}{1 + \cos x} \qquad \text{Apply a quotient identity: } \tan x = \frac{\sin x}{\cos x}.$$

$$= \frac{\dfrac{\sin x}{\cos x} + \dfrac{\sin x \cos x}{\cos x}}{1 + \cos x} \qquad \text{Express the terms in the numerator with the least common denominator, } \cos x.$$

$$= \frac{\dfrac{\sin x + \sin x \cos x}{\cos x}}{1 + \cos x} \qquad \text{Add in the numerator.}$$

$$= \frac{\sin x + \sin x \cos x}{\cos x} \div \frac{1 + \cos x}{1} \qquad \text{Rewrite the main fraction bar as } \div.$$

$$= \frac{\sin x + \sin x \cos x}{\cos x} \cdot \frac{1}{1 + \cos x} \qquad \text{Invert the divisor and multiply.}$$

$$= \frac{\sin x \cancel{(1 + \cos x)}^{1}}{\cos x} \cdot \frac{1}{\cancel{1 + \cos x}_{1}} \qquad \text{Factor and simplify.}$$

$$= \frac{\sin x}{\cos x} \qquad \text{Multiply the remaining factors in the numerator and the denominator.}$$

$$= \tan x \qquad \text{Apply a quotient identity.}$$

The left side simplifies to $\tan x$, the right side. Thus, the identity is verified.

Discovery

Try simplifying

$$\frac{\dfrac{\sin x}{\cos x} + \sin x}{1 + \cos x}$$

by multiplying the two terms in the numerator and the two terms in the denominator by $\cos x$. This method for simplifying the complex fraction involves multiplying the numerator and the denominator by the least common denominator of all fractions in the expression. Do you prefer this simplification procedure over the method used on the right?

Check Point 6 Verify the identity: $\dfrac{\sec x + \csc(-x)}{\sec x \csc x} = \sin x - \cos x.$

Is every identity verified by working with only one side? No. You can sometimes work with each side separately and show that both sides are equal to the same trigonometric expression. This is illustrated in Example 7.

EXAMPLE 7 **Working with Both Sides Separately to Verify an Identity**

Verify the identity: $\dfrac{1}{1 + \cos\theta} + \dfrac{1}{1 - \cos\theta} = 2 + 2\cot^2\theta.$

Solution We begin by working with the left side.

$$\dfrac{1}{1 + \cos\theta} + \dfrac{1}{1 - \cos\theta}$$

The least common denominator is $(1 + \cos\theta)(1 - \cos\theta).$

$$= \dfrac{1(1 - \cos\theta)}{(1 + \cos\theta)(1 - \cos\theta)} + \dfrac{1(1 + \cos\theta)}{(1 + \cos\theta)(1 - \cos\theta)}$$

Rewrite each fraction with the least common denominator.

$$= \dfrac{1 - \cos\theta + 1 + \cos\theta}{(1 + \cos\theta)(1 - \cos\theta)}$$

Add numerators. Put this sum over the least common denominator.

$$= \dfrac{2}{(1 + \cos\theta)(1 - \cos\theta)}$$

Simplify the numerator: $-\cos\theta + \cos\theta = 0$ and $1 + 1 = 2.$

$$= \dfrac{2}{1 - \cos^2\theta}$$

Multiply the factors in the denominator.

Now we work with the right side. Our goal is to transform this side into the simplified form attained for the left side, $\dfrac{2}{1 - \cos^2\theta}.$

$$2 + 2\cot^2\theta = 2 + 2\left(\dfrac{\cos^2\theta}{\sin^2\theta}\right)$$

Use a quotient identity: $\cot\theta = \dfrac{\cos\theta}{\sin\theta}.$

$$= \dfrac{2\sin^2\theta}{\sin^2\theta} + \dfrac{2\cos^2\theta}{\sin^2\theta}$$

Rewrite each fraction with the least common denominator, $\sin^2\theta.$

$$= \dfrac{2\sin^2\theta + 2\cos^2\theta}{\sin^2\theta}$$

Add numerators. Put this sum over the least common denominator.

$$= \dfrac{2(\sin^2\theta + \cos^2\theta)}{\sin^2\theta}$$

Factor out the greatest common factor, 2.

$$= \dfrac{2}{\sin^2\theta}$$

Apply a Pythagorean identity: $\sin^2\theta + \cos^2\theta = 1.$

$$= \dfrac{2}{1 - \cos^2\theta}$$

Use a variation of $\sin^2\theta + \cos^2\theta = 1$ and solve for $\sin^2\theta$: $\sin^2\theta = 1 - \cos^2\theta.$

The identity is verified because both sides are equal to $\dfrac{2}{1 - \cos^2\theta}.$

> **Check Point 7** Verify the identity: $\dfrac{1}{1 + \sin\theta} + \dfrac{1}{1 - \sin\theta} = 2 + 2\tan^2\theta$.

Guidelines for Verifying Trigonometric Identities

There is often more than one correct way to solve a puzzle, although one method may be shorter and more efficient than another. The same is true for verifying an identity. For example, how would you verify

$$\frac{\csc^2 x - 1}{\csc^2 x} = \cos^2 x?$$

One approach is to use a Pythagorean identity, $1 + \cot^2 x = \csc^2 x$, on the left side. Then change the resulting expression to sines and cosines.

$$\frac{\csc^2 x - 1}{\csc^2 x} = \frac{(1 + \cot^2 x) - 1}{\csc^2 x} = \frac{\cot^2 x}{\csc^2 x} = \frac{\dfrac{\cos^2 x}{\sin^2 x}}{\dfrac{1}{\sin^2 x}} = \frac{\cos^2 x}{\sin^2 x} \cdot \frac{\sin^2 x}{1} = \cos^2 x$$

Apply a Pythagorean identity: $1 + \cot^2 x = \csc^2 x$.

Use $\cot x = \dfrac{\cos x}{\sin x}$ and $\csc x = \dfrac{1}{\sin x}$ to change to sines and cosines.

Invert the divisor and multiply.

A more efficient strategy for verifying this identity may not be apparent at first glance. Work with the left side and divide each term in the numerator by the denominator, $\csc^2 x$.

$$\frac{\csc^2 x - 1}{\csc^2 x} = \frac{\csc^2 x}{\csc^2 x} - \frac{1}{\csc^2 x} = 1 - \sin^2 x = \cos^2 x$$

Apply a reciprocal identity: $\sin x = \dfrac{1}{\csc x}$.

Use $\sin^2 x + \cos^2 x = 1$ and solve for $\cos^2 x$.

With this strategy, we again obtain $\cos^2 x$, the expression on the right side, and it takes fewer steps than the first approach.

An even longer strategy, but one that works, is to replace each of the two occurrences of $\csc^2 x$ on the left side by $\dfrac{1}{\sin^2 x}$. This may be the approach that you first consider, particularly if you become accustomed to rewriting the more complicated side in terms of sines and cosines. The selection of an appropriate fundamental identity to solve the puzzle most efficiently is learned through lots of practice.

The more identities you prove, the more confident and efficient you will become. Although practice is the only way to learn how to verify identities, there are some guidelines developed throughout the section that should help you get started.

Guidelines for Verifying Trigonometric Identities

1. Work with each side of the equation independently of the other side. Start with the more complicated side and transform it in a step-by-step fashion until it looks exactly like the other side.

2. Analyze the identity and look for opportunities to apply the fundamental identities. Rewriting the more complicated side of the equation in terms of sines and cosines is often helpful.

3. If sums or differences of fractions appear on one side, use the least common denominator and combine the fractions.

4. Don't be afraid to stop and start over again if you are not getting anywhere. Creative puzzle solvers know that strategies leading to dead ends often provide good problem-solving ideas.

EXERCISE SET 5.1

Practice Exercises

In Exercises 1–60, verify each identity.

1. $\sin x \sec x = \tan x$

2. $\cos x \csc x = \cot x$

3. $\tan(-x) \cos x = -\sin x$

4. $\cot(-x) \sin x = -\cos x$

5. $\tan x \csc x \cos x = 1$

6. $\cot x \sec x \sin x = 1$

7. $\sec x - \sec x \sin^2 x = \cos x$

8. $\csc x - \csc x \cos^2 x = \sin x$

9. $\cos^2 x - \sin^2 x = 1 - 2\sin^2 x$

10. $\cos^2 x - \sin^2 x = 2\cos^2 x - 1$

11. $\csc \theta - \sin \theta = \cot \theta \cos \theta$

12. $\tan \theta + \cot \theta = \sec \theta \csc \theta$

13. $\dfrac{\tan \theta \cot \theta}{\csc \theta} = \sin \theta$

14. $\dfrac{\cos \theta \sec \theta}{\cot \theta} = \tan \theta$

15. $\sin^2 \theta (1 + \cot^2 \theta) = 1$

16. $\cos^2 \theta (1 + \tan^2 \theta) = 1$

17. $\sin t \tan t = \dfrac{1 - \cos^2 t}{\cos t}$

18. $\cos t \cot t = \dfrac{1 - \sin^2 t}{\sin t}$

19. $\dfrac{\csc^2 t}{\cot t} = \csc t \sec t$

20. $\dfrac{\sec^2 t}{\tan t} = \sec t \csc t$

21. $\dfrac{\tan^2 t}{\sec t} = \sec t - \cos t$

22. $\dfrac{\cot^2 t}{\csc t} = \csc t - \sin t$

23. $\dfrac{\sin t}{\csc t} + \dfrac{\cos t}{\sec t} = 1$

24. $\dfrac{\sin t}{\tan t} + \dfrac{\cos t}{\cot t} = \sin t + \cos t$

25. $\tan t + \dfrac{\cos t}{1 + \sin t} = \sec t$

26. $\cot t + \dfrac{\sin t}{1 + \cos t} = \csc t$

27. $1 - \dfrac{\sin^2 x}{1 + \cos x} = \cos x$

28. $1 - \dfrac{\cos^2 x}{1 + \sin x} = \sin x$

29. $\dfrac{\cos x}{1 - \sin x} + \dfrac{1 - \sin x}{\cos x} = 2\sec x$

30. $\dfrac{\sin x}{\cos x + 1} + \dfrac{\cos x - 1}{\sin x} = 0$

31. $\sec^2 x \csc^2 x = \sec^2 x + \csc^2 x$

32. $\csc^2 x \sec x = \sec x + \csc x \cot x$

33. $\dfrac{\sec x - \csc x}{\sec x + \csc x} = \dfrac{\tan x - 1}{\tan x + 1}$

34. $\dfrac{\csc x - \sec x}{\csc x + \sec x} = \dfrac{\cot x - 1}{\cot x + 1}$

35. $\dfrac{\sin^2 x - \cos^2 x}{\sin x + \cos x} = \sin x - \cos x$

36. $\dfrac{\tan^2 x - \cot^2 x}{\tan x + \cot x} = \tan x - \cot x$

37. $\tan^2 2x + \sin^2 2x + \cos^2 2x = \sec^2 2x$

38. $\cot^2 2x + \cos^2 2x + \sin^2 2x = \csc^2 2x$

39. $\dfrac{\tan 2\theta + \cot 2\theta}{\csc 2\theta} = \sec 2\theta$

40. $\dfrac{\tan 2\theta + \cot 2\theta}{\sec 2\theta} = \csc 2\theta$

41. $\dfrac{\tan x + \tan y}{1 - \tan x \tan y} = \dfrac{\sin x \cos y + \cos x \sin y}{\cos x \cos y - \sin x \sin y}$

42. $\dfrac{\cot x + \cot y}{1 - \cot x \cot y} = \dfrac{\cos x \sin y + \sin x \cos y}{\sin x \sin y - \cos x \cos y}$

43. $(\sec x - \tan x)^2 = \dfrac{1 - \sin x}{1 + \sin x}$

44. $(\csc x - \cot x)^2 = \dfrac{1 - \cos x}{1 + \cos x}$

45. $\dfrac{\sec t + 1}{\tan t} = \dfrac{\tan t}{\sec t - 1}$

46. $\dfrac{\csc t - 1}{\cot t} = \dfrac{\cot t}{\csc t + 1}$

47. $\dfrac{1 + \cos t}{1 - \cos t} = (\csc t + \cot t)^2$

48. $\dfrac{1 - \sin t}{1 + \sin t} = (\sec t - \tan t)^2$

49. $\cos^4 t - \sin^4 t = 1 - 2\sin^2 t$

50. $\sin^4 t - \cos^4 t = 1 - 2\cos^2 t$

51. $\dfrac{\sin \theta - \cos \theta}{\sin \theta} + \dfrac{\cos \theta - \sin \theta}{\cos \theta} = 2 - \sec \theta \csc \theta$

52. $\dfrac{\sin \theta}{1 - \cot \theta} - \dfrac{\cos \theta}{\tan \theta - 1} = \sin \theta + \cos \theta$

53. $(\tan^2 \theta + 1)(\cos^2 \theta + 1) = \tan^2 \theta + 2$

54. $(\cot^2 \theta + 1)(\sin^2 \theta + 1) = \cot^2 \theta + 2$

55. $(\cos \theta - \sin \theta)^2 + (\cos \theta + \sin \theta)^2 = 2$

56. $(3\cos \theta - 4\sin \theta)^2 + (4\cos \theta + 3\sin \theta)^2 = 25$

57. $\dfrac{\cos^2 x - \sin^2 x}{1 - \tan^2 x} = \cos^2 x$

58. $\dfrac{\sin x + \cos x}{\sin x} - \dfrac{\cos x - \sin x}{\cos x} = \sec x \csc x$

59. $(\sec x - \tan x)^2 = \dfrac{1 - \sin x}{1 + \sin x}$

60. $(\cot x - \csc x)^2 = \dfrac{1 - \cos x}{1 + \cos x}$

Writing in Mathematics

61. Explain how to verify an identity.

62. Describe two strategies that can be used to verify identities.

63. Describe how you feel when you successfully verify a difficult identity. What other activities do you engage in that evoke the same feelings?

64. A 10-point question on a quiz asks students to verify the identity

$$\dfrac{\sin^2 x - \cos^2 x}{\sin x + \cos x} = \sin x - \cos x.$$

One student begins with the left side and obtains the right side as shown in the next column:

$$\dfrac{\sin^2 x - \cos^2 x}{\sin x + \cos x} = \dfrac{\sin^2 x}{\sin x} - \dfrac{\cos^2 x}{\cos x} = \sin x - \cos x.$$

How many points (out of 10) would you give this student? Explain your answer.

Technology Exercises

In Exercises 65–73, graph each side of the equation in the same viewing rectangle. If the graphs appear to coincide, verify that the equation is an identity. If the graphs do not appear to coincide, this indicates the equation is not an identity. In these exercises, find a value of x for which both sides are defined but not equal.

65. $\tan x = \sec x(\sin x - \cos x) + 1$

66. $\sin x = -\cos x \tan(-x)$

67. $\sin\left(x + \dfrac{\pi}{4}\right) = \sin x + \sin \dfrac{\pi}{4}$

68. $\cos\left(x + \dfrac{\pi}{4}\right) = \cos x + \cos \dfrac{\pi}{4}$

69. $\cos(x + \pi) = \cos x$

70. $\sin(x + \pi) = \sin x$

71. $\dfrac{\sin x}{1 - \cos^2 x} = \csc x$

72. $\sin x - \sin x \cos^2 x = \sin^3 x$

73. $\sqrt{\sin^2 x + \cos^2 x} = \sin x + \cos x$

Critical Thinking Exercises

In Exercises 74–76, verify each identity.

74. $\dfrac{\sin^3 x - \cos^3 x}{\sin x - \cos x} = 1 + \sin x \cos x$

75. $\dfrac{\sin x - \cos x + 1}{\sin x + \cos x - 1} = \dfrac{\sin x + 1}{\cos x}$

76. $\ln|\sec x| = -\ln|\cos x|$

77. Use one of the fundamental identities in the box on page 545 to create an original identity.

Group Exercise

78. Group members are to write a helpful list of items for a pamphlet called "The Underground Guide to Verifying Identities." The pamphlet will be used primarily by students who sit, stare, and freak out every time they are asked to verify an identity. List easy ways to remember the fundamental identities. What helpful guidelines can you offer from the perspective of a student that you probably won't find in math books? If you have your own strategies that work particularly well, include them in the pamphlet.

SECTION 5.2   *Sum and Difference Formulas*

Objectives

1. Use the formula for the cosine of the difference of two angles.
2. Use sum and difference formulas for cosines and sines.
3. Use sum and difference formulas for tangents.

Listen to the same note played on a piano and a violin. The notes have a different quality or "tone." Tone depends on the way an instrument vibrates. However, the less than 1% of the population with amusia, or true tone deafness, cannot tell the two sounds apart. Even simple, familiar tunes such as *Happy Birthday* and *Jingle Bells* are mystifying to amusics.

When a note is played, it vibrates at a specific fundamental frequency and has a particular amplitude. Amusics cannot tell the difference between two sounds from tuning forks modeled by $p = 3 \sin 2t$ and $p = 2 \sin(2t + \pi)$, respectively. However, they can recognize the difference between the two equations. Notice that the second equation contains the sine of the sum of two angles. In this section, we will be developing identities involving the sums or differences of two angles. These formulas are called the **sum and difference formulas.** We begin with $\cos(\alpha - \beta)$, the cosine of the difference of two angles.

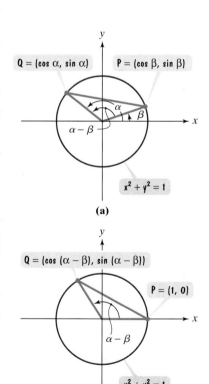

Figure 5.1 Using the unit circle and QP to develop a formula for $\cos(\alpha - \beta)$

The Cosine of the Difference of Two Angles

> **The Cosine of the Difference of Two Angles**
>
> $$\cos(\alpha - \beta) = \cos \alpha \cos \beta + \sin \alpha \sin \beta$$
>
> The cosine of the difference of two angles equals the cosine of the first angle times the cosine of the second angle plus the sine of the first angle times the sine of the second angle.

We use Figure 5.1 to prove the identity in the box. The graph in Figure 5.1(a) shows a unit circle, $x^2 + y^2 = 1$. The figure uses the definitions of the cosine and sine functions as the x- and y-coordinates of points along the unit circle. For example, point P corresponds to angle β. By definition, the x-coordinate of P is $\cos \beta$ and the y-coordinate is $\sin \beta$. Similarly, point Q corresponds to angle α. By definition, the x-coordinate of Q is $\cos \alpha$ and the y-coordinate is $\sin \alpha$.

Note that if we draw a line segment between points P and Q, a triangle is formed. Angle $\alpha - \beta$ is one of the angles of this triangle. What happens if we rotate this triangle so that point P falls on the x-axis at $(1, 0)$? The result is shown in Figure 5.1(b). This rotation changes the coordinates of points P and Q. However, it has no effect on the length of line segment PQ.

We can use the distance formula, $d = \sqrt{(x_2 - x_1)^2 + (y_2 - y_1)^2}$, to find an expression for PQ in Figure 5.1(a) and in Figure 5.1(b). By equating the two expressions for PQ, we will obtain the identity for the cosine of the difference of two angles, $\alpha - \beta$. We first apply the distance formula in Figure 5.1(a).

$PQ = \sqrt{(\cos\alpha - \cos\beta)^2 + (\sin\alpha - \sin\beta)^2}$

Apply the distance formula, $d = \sqrt{(x_2 - x_1)^2 + (y_2 - y_1)^2}$, to find the distance between $(\cos\beta, \sin\beta)$ and $(\cos\alpha, \sin\alpha)$.

$= \sqrt{\cos^2\alpha - 2\cos\alpha\cos\beta + \cos^2\beta + \sin^2\alpha - 2\sin\alpha\sin\beta + \sin^2\beta}$

Square each expression using $(A - B)^2 = A^2 - 2AB + B^2$.

$= \sqrt{(\sin^2\alpha + \cos^2\alpha) + (\sin^2\beta + \cos^2\beta) - 2\cos\alpha\cos\beta - 2\sin\alpha\sin\beta}$

Regroup terms to apply a Pythagorean identity.

$= \sqrt{1 + 1 - 2\cos\alpha\cos\beta - 2\sin\alpha\sin\beta}$

Because $\sin^2 x + \cos^2 x = 1$, each expression in parentheses equals 1.

$= \sqrt{2 - 2\cos\alpha\cos\beta - 2\sin\alpha\sin\beta}$

Simplify.

Next, we apply the distance formula in Figure 5.1(b) to obtain a second expression for PQ. We let $(x_1, y_1) = (1, 0)$ and $(x_2, y_2) = (\cos(\alpha - \beta), \sin(\alpha - \beta))$.

$PQ = \sqrt{[\cos(\alpha - \beta) - 1]^2 + [\sin(\alpha - \beta) - 0]^2}$

Apply the distance formula to find the distance between $(1, 0)$ and $(\cos(\alpha - \beta), \sin(\alpha - \beta))$.

$= \sqrt{\cos^2(\alpha - \beta) - 2\cos(\alpha - \beta) + 1 + \sin^2(\alpha - \beta)}$

Square each expression.

Using a Pythagorean identity,
$\sin^2(\alpha - \beta) + \cos^2(\alpha - \beta) = 1$

$= \sqrt{1 - 2\cos(\alpha - \beta) + 1}$

Use a Pythagorean identity.

$= \sqrt{2 - 2\cos(\alpha - \beta)}$

Simplify.

Now we equate the two expressions for PQ.

$\sqrt{2 - 2\cos(\alpha - \beta)} = \sqrt{2 - 2\cos\alpha\cos\beta - 2\sin\alpha\sin\beta}$

The rotation does not change the length of PQ.

$2 - 2\cos(\alpha - \beta) = 2 - 2\cos\alpha\cos\beta - 2\sin\alpha\sin\beta$

Square both sides to eliminate radicals.

$-2\cos(\alpha - \beta) = -2\cos\alpha\cos\beta - 2\sin\alpha\sin\beta$

Subtract 2 from both sides of the equation.

$\cos(\alpha - \beta) = \cos\alpha\cos\beta + \sin\alpha\sin\beta$

Divide both sides of the equation by -2.

This proves the identity for the cosine of the difference of two angles.

Now that we see where the identity for the cosine of the difference of two angles comes from, let's look at some applications of this result.

1 Use the formula for the cosine of the difference of two angles.

EXAMPLE 1 Using the Difference Formula for Cosines to Find Exact Values

Find the exact value of cos 15°.

Solution We know exact values for trigonometric functions of 60° and 45°. Thus, we write 15° as 60° − 45° and use the difference formula for cosines.

$$\cos 15° = \cos(60° − 45°)$$

$$= \cos 60° \cos 45° + \sin 60° \sin 45° \quad \text{cos}(\alpha - \beta) = \cos \alpha \cos \beta + \sin \alpha \sin \beta$$

$$= \frac{1}{2} \cdot \frac{\sqrt{2}}{2} + \frac{\sqrt{3}}{2} \cdot \frac{\sqrt{2}}{2} \quad \text{Substitute exact values from memory or use special right triangles.}$$

$$= \frac{\sqrt{2}}{4} + \frac{\sqrt{6}}{4} \quad \text{Multiply.}$$

$$= \frac{\sqrt{2} + \sqrt{6}}{4} \quad \text{Add.}$$

Check Point 1 We know that $\cos 30° = \frac{\sqrt{3}}{2}$. Obtain this exact value using $\cos 30° = \cos(90° − 60°)$ and the difference formula for cosines.

EXAMPLE 2 Using the Difference Formula for Cosines to Find Exact Values

Find the exact value of cos 80° cos 20° + sin 80° sin 20°.

Solution The given expression is the right side of the formula for $\cos(\alpha − \beta)$ with $\alpha = 80°$ and $\beta = 20°$.

$$\cos(\alpha - \beta) = \cos \alpha \cos \beta + \sin \alpha \sin \beta$$

$$\cos 80° \cos 20° + \sin 80° \sin 20° = \cos(80° − 20°) = \cos 60° = \frac{1}{2}$$

Check Point 2 Find the exact value of
$$\cos 70° \cos 40° + \sin 70° \sin 40°.$$

Sound Quality and Amusia

People with true tone deafness cannot hear the difference among tones produced by a tuning fork, a flute, an oboe, and a violin. They cannot dance or tell the difference between harmony and dissonance. People with amusia appear to have been born without the wiring necessary to process music. Intriguingly, they show no overt signs of brain damage and their brain scans appear normal. Thus, they can visually recognize the difference among sound waves that produce varying sound qualities.

Varying Sound Qualities

• Tuning fork: Sound waves are rounded and regular, giving a pure and gentle tone.

• Flute: Sound waves are smooth and give a fluid tone.

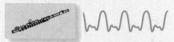

• Oboe: Rapid wave changes give a richer tone.

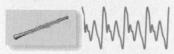

• Violin: Jagged waves give a brighter harsher tone.

EXAMPLE 3 Verifying an Identity

Verify the identity: $\dfrac{\cos(\alpha - \beta)}{\sin\alpha\cos\beta} = \cot\alpha + \tan\beta$.

Solution We work with the left side.

$$\dfrac{\cos(\alpha - \beta)}{\sin\alpha\cos\beta} = \dfrac{\cos\alpha\cos\beta + \sin\alpha\sin\beta}{\sin\alpha\cos\beta}$$ Use the formula for $\cos(\alpha - \beta)$.

$$= \dfrac{\cos\alpha}{\sin\alpha}\dfrac{\cos\beta}{\cos\beta} + \dfrac{\sin\alpha}{\sin\alpha}\dfrac{\sin\beta}{\cos\beta}$$ Divide each term in the numerator by $\sin\alpha\cos\beta$.

$$= \cot\alpha \cdot 1 + 1 \cdot \tan\beta$$ Use quotient identities.

$$= \cot\alpha + \tan\beta$$ Simplify.

We worked with the left side and arrived at the right side. Thus, the identity is verified.

Check Point 3 Verify the identity: $\dfrac{\cos(\alpha - \beta)}{\cos\alpha\cos\beta} = 1 + \tan\alpha\tan\beta$.

The difference formula for cosines is used to establish other identities. For example, in our work with right triangles we noted that cofunctions of complements are equal. Thus, because $\dfrac{\pi}{2} - \theta$ and θ are complements,

$$\cos\left(\dfrac{\pi}{2} - \theta\right) = \sin\theta.$$

We can use the formula for $\cos(\alpha - \beta)$ to prove this cofunction identity.

Apply $\cos(\alpha - \beta)$ with $\alpha = 90°$ and $\theta = \beta$.
$$\cos(\alpha - \beta) = \cos\alpha\cos\beta + \sin\alpha\sin\beta$$

$$\cos\left(\dfrac{\pi}{2} - \theta\right) = \cos\dfrac{\pi}{2}\cos\theta + \sin\dfrac{\pi}{2}\sin\theta$$

$$= 0 \cdot \cos\theta + 1 \cdot \sin\theta$$

$$= \sin\theta$$

Technology

The graphs of

$$y = \cos\left(\dfrac{\pi}{2} - x\right)$$

and

$$y = \sin x$$

are shown in the same viewing rectangle. The graphs are the same. The displayed math on the right shows the equivalence algebraically.

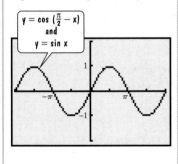

$y = \cos\left(\dfrac{\pi}{2} - x\right)$
and
$y = \sin x$

2 Use sum and difference formulas for cosines and sines.

Sum and Difference Formulas for Cosines and Sines

Our formula for $\cos(\alpha - \beta)$ can be used to verify an identity for a sum involving cosines, as well as identities for a sum and a difference for sines.

Sum and Difference Formulas for Cosines and Sines

1. $\cos(\alpha + \beta) = \cos\alpha\cos\beta - \sin\alpha\sin\beta$

2. $\cos(\alpha - \beta) = \cos\alpha\cos\beta + \sin\alpha\sin\beta$

3. $\sin(\alpha + \beta) = \sin\alpha\cos\beta + \cos\alpha\sin\beta$

4. $\sin(\alpha - \beta) = \sin\alpha\cos\beta - \cos\alpha\sin\beta$

Up to now, we have concentrated on the second formula in the box. The first identity gives a formula for the cosine of the sum of two angles. It is proved as follows:

$$\cos(\alpha + \beta) = \cos[\alpha - (-\beta)] \quad \text{Express addition as subtraction of an inverse.}$$

$$= \cos\alpha\cos(-\beta) + \sin\alpha\sin(-\beta) \quad \text{Use the difference formula for cosines.}$$

$$= \cos\alpha\cos\beta + \sin\alpha(-\sin\beta) \quad \text{Cosine is even: } \cos(-\beta) = \cos\beta. \text{ Sine is odd: } \sin(-\beta) = -\sin\beta.$$

$$= \cos\alpha\cos\beta - \sin\alpha\sin\beta. \quad \text{Simplify.}$$

Thus, the cosine of the sum of two angles equals the cosine of the first angle times the cosine of the second angle minus the sine of the first angle times the sine of the second angle.

The third identity in the box gives a formula for $\sin(\alpha + \beta)$, the sine of the sum of two angles. It is proved as follows:

$$\sin(\alpha + \beta) = \cos\left[\frac{\pi}{2} - (\alpha + \beta)\right] \quad \text{Use a cofunction identity: } \sin\theta = \cos\left(\frac{\pi}{2} - \theta\right).$$

$$= \cos\left[\left(\frac{\pi}{2} - \alpha\right) - \beta\right] \quad \text{Regroup.}$$

$$= \cos\left(\frac{\pi}{2} - \alpha\right)\cos\beta + \sin\left(\frac{\pi}{2} - \alpha\right)\sin\beta \quad \text{Use the difference formula for cosines.}$$

$$= \sin\alpha\cos\beta + \cos\alpha\sin\beta. \quad \text{Use cofunction identities.}$$

Thus, the sine of the sum of two angles equals the sine of the first angle times the cosine of the second angle plus the cosine of the first angle times the sine of the second angle.

The final identity in the box gives a formula for $\sin(\alpha - \beta)$, the sine of the difference of two angles. It is proved by writing $\sin(\alpha - \beta)$ as $\sin[\alpha + (-\beta)]$ and then using the formula for the sine of a sum.

EXAMPLE 4 Using the Sine of a Sum to Find an Exact Value

Find the exact value of $\sin\dfrac{7\pi}{12}$ using the fact that $\dfrac{7\pi}{12} = \dfrac{\pi}{3} + \dfrac{\pi}{4}$.

Solution   We apply the formula for the sine of a sum.

$$\sin\frac{7\pi}{12} = \sin\left(\frac{\pi}{3} + \frac{\pi}{4}\right)$$

$$= \sin\frac{\pi}{3}\cos\frac{\pi}{4} + \cos\frac{\pi}{3}\sin\frac{\pi}{4} \quad \sin(\alpha + \beta) = \sin\alpha\cos\beta + \cos\alpha\sin\beta$$

$$= \frac{\sqrt{3}}{2}\cdot\frac{\sqrt{2}}{2} + \frac{1}{2}\cdot\frac{\sqrt{2}}{2} \quad \text{Substitute exact values.}$$

$$= \frac{\sqrt{6} + \sqrt{2}}{4} \quad \text{Simplify.}$$

Check Point 4 Find the exact value of $\sin \dfrac{5\pi}{12}$ using the fact that

$$\frac{5\pi}{12} = \frac{\pi}{6} + \frac{\pi}{4}.$$

EXAMPLE 5 Finding Exact Values

Suppose that $\sin \alpha = \frac{12}{13}$ for a quadrant II angle α and $\sin \beta = \frac{3}{5}$ for a quadrant I angle β. Find the exact value of:

a. $\cos \alpha$ **b.** $\cos \beta$ **c.** $\cos (\alpha + \beta)$ **d.** $\sin (\alpha + \beta)$.

Solution

a. We find $\cos \alpha$ using a sketch that illustrates

$$\sin \alpha = \frac{12}{13} = \frac{y}{r}.$$

Figure 5.2 shows a quadrant II angle α with $\sin \alpha = \frac{12}{13}$. We find x using $x^2 + y^2 = r^2$. Because α lies in quadrant II, x is negative.

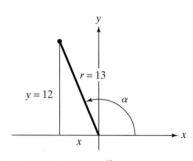

Figure 5.2 $\sin \alpha = \frac{12}{13}$: α lies in quadrant II

$$x^2 + 12^2 = 13^2 \qquad \text{\small{$x^2 + y^2 = r^2$}}$$
$$x^2 + 144 = 169 \qquad \text{\small{Square 12 and 13, respectively.}}$$
$$x^2 = 25 \qquad \text{\small{Subtract 144 from both sides.}}$$
$$x = -\sqrt{25} = -5 \qquad \text{\small{In quadrant II, x is negative.}}$$

Thus,

$$\cos \alpha = \frac{x}{r} = \frac{-5}{13} = -\frac{5}{13}.$$

b. We find $\cos \beta$ using a sketch that illustrates

$$\sin \beta = \frac{3}{5} = \frac{y}{r}.$$

Figure 5.3 shows a quadrant I angle β with $\sin \beta = \frac{3}{5}$. We find x using $x^2 + y^2 = r^2$.

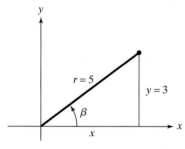

Figure 5.3 $\sin \beta = \frac{3}{5}$: β lies in quadrant I

$$x^2 + 3^2 = 5^2$$
$$x^2 = 25 - 9 = 16$$
$$x = \sqrt{16} = 4 \qquad \text{\small{In quadrant I, x is positive.}}$$

Thus,

$$\cos \beta = \frac{x}{r} = \frac{4}{5}.$$

We use the given values and the exact values that we determined to find exact values for $\cos (\alpha + \beta)$ and $\sin (\alpha + \beta)$.

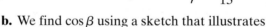

These values are given.   These are the values we found.

$$\sin \alpha = \frac{12}{13}, \sin \beta = \frac{3}{5} \qquad \cos \alpha = -\frac{5}{13}, \cos \beta = \frac{4}{5}$$

c. We use the formula for the cosine of a sum.

$$\cos(\alpha + \beta) = \cos\alpha\cos\beta - \sin\alpha\sin\beta$$

$$= -\frac{5}{13}\left(\frac{4}{5}\right) - \frac{12}{13}\left(\frac{3}{5}\right) = -\frac{56}{65}$$

d. We use the formula for the sine of a sum.

$$\sin(\alpha + \beta) = \sin\alpha\cos\beta + \cos\alpha\sin\beta$$

$$= \frac{12}{13}\cdot\frac{4}{5} + \left(-\frac{5}{13}\right)\cdot\frac{3}{5} = \frac{33}{65}$$

> **Check Point 5**
>
> Suppose that $\sin\alpha = \frac{4}{5}$ for a quadrant II angle α and $\sin\beta = \frac{1}{2}$ for a quadrant I angle β. Find the exact value of:
> **a.** $\cos\alpha$ **b.** $\cos\beta$
> **c.** $\cos(\alpha + \beta)$ **d.** $\sin(\alpha + \beta)$.

EXAMPLE 6 Verifying Observations on a Graphing Utility

Figure 5.4 shows the graph of $y = \sin\left(x - \dfrac{3\pi}{2}\right)$ in a $\left[0, 2\pi, \dfrac{\pi}{2}\right]$ by $[-2, 2, 1]$ viewing rectangle.

a. Describe the graph using another equation.

b. Verify that the two equations are equivalent.

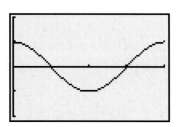

Figure 5.4 The graph of
$y = \sin\left(x - \dfrac{3\pi}{2}\right)$ in a $\left[0, 2\pi, \dfrac{\pi}{2}\right]$ by
$[-2, 2, 1]$ viewing rectangle

Solution

a. The graph appears to be the cosine curve $y = \cos x$. It cycles through maximum, intercept, minimum, intercept, and back to maximum. Thus, $y = \cos x$ also describes the graph.

b. We must show that

$$\sin\left(x - \frac{3\pi}{2}\right) = \cos x.$$

We apply the formula for the sine of a difference on the left side.

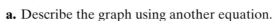

$$\sin\left(x - \frac{3\pi}{2}\right) = \sin x\cos\frac{3\pi}{2} - \cos x\sin\frac{3\pi}{2} \qquad \begin{array}{l} \sin(\alpha - \beta) = \\ \sin\alpha\cos\beta - \cos\alpha\sin\beta \end{array}$$

$$= \sin x\cdot 0 - \cos x(-1) \qquad \cos\frac{3\pi}{2} = 0 \text{ and } \sin\frac{3\pi}{2} = -1$$

$$= \cos x \qquad\qquad \text{Simplify.}$$

This verifies our observation that $y = \sin\left(x - \dfrac{3\pi}{2}\right)$ and $y = \cos x$ describe the same graph.

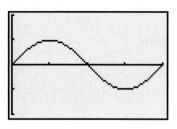

Figure 5.5

Check Point 6

Figure 5.5 shows the graph of $y = \cos\left(x + \dfrac{3\pi}{2}\right)$ in a $\left[0, 2\pi, \dfrac{\pi}{2}\right]$ by $[-2, 2, 1]$ viewing rectangle.

a. Describe the graph using another equation.

b. Verify that the two equations are equivalent.

3 Use sum and difference formulas for tangents.

Sum and Difference Formulas for Tangents

By writing $\tan(\alpha + \beta)$ as the quotient of $\sin(\alpha + \beta)$ and $\cos(\alpha + \beta)$ we can develop a formula for the tangent of a sum. Writing subtraction as addition of an inverse leads to a formula for the tangent of a difference.

Discovery

Derive the sum and difference formulas for tangents by working Exercises 55 and 56 in Exercise Set 5.2.

Sum and Difference Formulas for Tangents

$$\tan(\alpha + \beta) = \frac{\tan\alpha + \tan\beta}{1 - \tan\alpha \tan\beta}$$

The tangent of the sum of two angles equals the tangent of the first angle plus the tangent of the second angle divided by 1 minus their product.

$$\tan(\alpha - \beta) = \frac{\tan\alpha - \tan\beta}{1 + \tan\alpha \tan\beta}$$

The tangent of the difference of two angles equals the tangent of the first angle minus the tangent of the second angle divided by 1 plus their product.

EXAMPLE 7   Verifying an Identity

Verify the identity: $\tan\left(x - \dfrac{\pi}{4}\right) = \dfrac{\tan x - 1}{\tan x + 1}$.

Solution   We work with the left side.

$$\tan\left(x - \frac{\pi}{4}\right) = \frac{\tan x - \tan\dfrac{\pi}{4}}{1 + \tan x \tan\dfrac{\pi}{4}} \qquad \tan(\alpha - \beta) = \frac{\tan\alpha - \tan\beta}{1 + \tan\alpha \tan\beta}$$

$$= \frac{\tan x - 1}{1 + \tan x \cdot 1} \qquad \tan\frac{\pi}{4} = 1$$

$$= \frac{\tan x - 1}{1 + \tan x}$$

Check Point 7   Verify the identity: $\tan(x + \pi) = \tan x$.

EXERCISE SET 5.2

Practice Exercises

Use the formula for the cosine of the difference of two angles to solve Exercises 1–12.

In Exercises 1–4, find the exact value of each expression.

1. $\cos(45° - 30°)$

2. $\cos(120° - 45°)$

3. $\cos\left(\dfrac{3\pi}{4} - \dfrac{\pi}{6}\right)$

4. $\cos\left(\dfrac{2\pi}{3} - \dfrac{\pi}{6}\right)$

In Exercises 5–8, each expression is the right side of the formula for $\cos(\alpha - \beta)$ with particular values for α and β.

a. *Identify α and β in each expression.*

b. *Write the expression as the cosine of an angle.*

c. *Find the exact value of the expression.*

5. $\cos 50° \cos 20° + \sin 50° \sin 20°$

6. $\cos 50° \cos 5° + \sin 50° \sin 5°$

7. $\cos\dfrac{5\pi}{12}\cos\dfrac{\pi}{12} + \sin\dfrac{5\pi}{12}\sin\dfrac{\pi}{12}$

8. $\cos\dfrac{5\pi}{18}\cos\dfrac{\pi}{9} + \sin\dfrac{5\pi}{18}\sin\dfrac{\pi}{9}$

In Exercises 9–12, verify each identity.

9. $\dfrac{\cos(\alpha - \beta)}{\cos\alpha \sin\beta} = \tan\alpha + \cot\beta$

10. $\dfrac{\cos(\alpha - \beta)}{\sin\alpha \sin\beta} = \cot\alpha \cot\beta + 1$

11. $\cos\left(x - \dfrac{\pi}{4}\right) = \dfrac{\sqrt{2}}{2}(\cos x + \sin x)$

12. $\cos\left(x - \dfrac{5\pi}{4}\right) = -\dfrac{\sqrt{2}}{2}(\cos x + \sin x)$

Use one or more of the six sum and difference identities to solve Exercises 13–54.

In Exercises 13–24, find the exact value of each expression.

13. $\sin(45° - 30°)$

14. $\sin(60° - 45°)$

15. $\sin 105°$

16. $\sin 75°$

17. $\tan(30° + 45°)$

18. $\tan(60° + 45°)$

19. $\tan(240° - 45°)$

20. $\tan(300° - 45°)$

21. $\cos\left(\dfrac{3\pi}{4} + \dfrac{\pi}{6}\right)$

22. $\cos\left(\dfrac{4\pi}{3} + \dfrac{\pi}{4}\right)$

23. $\cos\dfrac{5\pi}{12}$

24. $\cos\dfrac{7\pi}{12}$

In Exercises 25–32, write each expression as the sine, cosine, or tangent of an angle. Then find the exact value of the expression.

25. $\sin 25° \cos 5° + \cos 25° \sin 5°$

26. $\sin 40° \cos 20° + \cos 40° \sin 20°$

27. $\dfrac{\tan 10° + \tan 35°}{1 - \tan 10° \tan 35°}$

28. $\dfrac{\tan 50° - \tan 20°}{1 + \tan 50° \tan 20°}$

29. $\sin\dfrac{5\pi}{12}\cos\dfrac{\pi}{4} - \cos\dfrac{5\pi}{12}\sin\dfrac{\pi}{4}$

30. $\sin\dfrac{7\pi}{12}\cos\dfrac{\pi}{12} - \cos\dfrac{7\pi}{12}\sin\dfrac{\pi}{12}$

31. $\dfrac{\tan\dfrac{\pi}{5} - \tan\dfrac{\pi}{30}}{1 + \tan\dfrac{\pi}{5}\tan\dfrac{\pi}{30}}$

32. $\dfrac{\tan\dfrac{\pi}{5} + \tan\dfrac{4\pi}{5}}{1 - \tan\dfrac{\pi}{5}\tan\dfrac{4\pi}{5}}$

In Exercises 33–54, verify each identity.

33. $\sin\left(x + \dfrac{\pi}{2}\right) = \cos x$

34. $\sin\left(x + \dfrac{3\pi}{2}\right) = -\cos x$

35. $\cos\left(x - \dfrac{\pi}{2}\right) = \sin x$

36. $\cos(\pi - x) = -\cos x$

37. $\tan(2\pi - x) = -\tan x$

38. $\tan(\pi - x) = -\tan x$

39. $\sin(\alpha + \beta) + \sin(\alpha - \beta) = 2\sin\alpha\cos\beta$

40. $\cos(\alpha + \beta) + \cos(\alpha - \beta) = 2\cos\alpha\cos\beta$

41. $\dfrac{\sin(\alpha - \beta)}{\cos\alpha\cos\beta} = \tan\alpha - \tan\beta$

42. $\dfrac{\sin(\alpha + \beta)}{\cos\alpha\cos\beta} = \tan\alpha + \tan\beta$

43. $\tan\left(\theta + \dfrac{\pi}{4}\right) = \dfrac{\cos\theta + \sin\theta}{\cos\theta - \sin\theta}$

44. $\tan\left(\dfrac{\pi}{4} - \theta\right) = \dfrac{\cos\theta - \sin\theta}{\cos\theta - \sin\theta}$

45. $\cos(\alpha + \beta)\cos(\alpha - \beta) = \cos^2\beta - \sin^2\alpha$

46. $\sin(\alpha + \beta)\sin(\alpha - \beta) = \cos^2\beta - \cos^2\alpha$

47. $\dfrac{\sin(\alpha + \beta)}{\sin(\alpha - \beta)} = \dfrac{\tan\alpha + \tan\beta}{\tan\alpha - \tan\beta}$

48. $\dfrac{\cos(\alpha + \beta)}{\cos(\alpha - \beta)} = \dfrac{1 - \tan\alpha\tan\beta}{1 + \tan\alpha\tan\beta}$

49. $\dfrac{\cos(x + h) - \cos x}{h} = \cos x\,\dfrac{\cos h - 1}{h} - \sin x\,\dfrac{\sin h}{h}$

50. $\dfrac{\sin(x+h) - \sin x}{h} = \cos x \dfrac{\sin h}{h} + \sin x \dfrac{\cos h - 1}{h}$

51. $\sin 2\alpha = 2 \sin \alpha \cos \alpha$
Hint: Write $\sin 2\alpha$ as $\sin(\alpha + \alpha)$.

52. $\cos 2\alpha = \cos^2\alpha - \sin^2\alpha$
Hint: Write $\cos 2\alpha$ as $\cos(\alpha + \alpha)$.

53. $\tan 2\alpha = \dfrac{2\tan\alpha}{1 - \tan^2\alpha}$
Hint: Write $\tan 2\alpha$ as $\tan(\alpha + \alpha)$.

54. $\tan\left(\dfrac{\pi}{4} + \alpha\right) - \tan\left(\dfrac{\pi}{4} - \alpha\right) = 2\tan 2\alpha$
Hint: Use the result in Exercise 53.

55. Derive the identity for $\tan(\alpha + \beta)$ using

$$\tan(\alpha + \beta) = \dfrac{\sin(\alpha + \beta)}{\cos(\alpha + \beta)}.$$

After applying the formulas for sums of sines and cosines, divide the numerator and denominator by $\cos\alpha \cos\beta$.

56. Derive the identity for $\tan(\alpha - \beta)$ using

$$\tan(\alpha - \beta) = \tan\left[\alpha + (-\beta)\right].$$

After applying the formula for the tangent of the sum of two angles, use the fact that the tangent is an odd function.

In Exercises 57–62, find the exact value of the following under the given conditions:

a. $\cos(\alpha + \beta)$ **b.** $\sin(\alpha + \beta)$ **c.** $\tan(\alpha + \beta)$

57. $\sin\alpha = \frac{3}{5}$, α lies in quadrant I, and $\sin\beta = \frac{5}{13}$, β lies in quadrant II.

58. $\sin\alpha = \frac{4}{5}$, α lies in quadrant I, and $\sin\beta = \frac{7}{25}$, β lies in quadrant II.

59. $\tan\alpha = -\frac{3}{4}$, α lies in quadrant II, and $\cos\beta = \frac{1}{3}$, β lies in quadrant I.

60. $\tan\alpha = -\frac{4}{3}$, α lies in quadrant II, and $\cos\beta = \frac{2}{3}$, β lies in quadrant I.

61. $\cos\alpha = \frac{8}{17}$, α lies in quadrant IV, and $\sin\beta = -\frac{1}{2}$, β lies in quadrant III.

62. $\cos\alpha = \frac{1}{2}$, α lies in quadrant IV, and $\sin\beta = -\frac{1}{3}$, β lies in quadrant III.

In Exercises 63–66, the graph with the given equation is shown in a $\left[0, 2\pi, \dfrac{\pi}{2}\right]$ by $[-2, 2, 1]$ viewing rectangle.

a. *Describe the graph using another equation.*
b. *Verify that the two equations are equivalent.*

63.

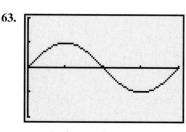

$y = \sin(\pi - x)$

64.

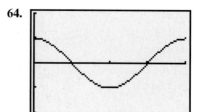

$y = \cos(x - 2\pi)$

65.

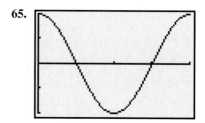

$y = \sin\left(x + \dfrac{\pi}{2}\right) + \sin\left(\dfrac{\pi}{2} - x\right)$

66.

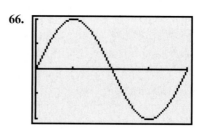

$y = \cos\left(x - \dfrac{\pi}{2}\right) - \cos\left(x + \dfrac{\pi}{2}\right)$

Application Exercises

67. A ball attached to a spring is raised 2 feet and released with an initial vertical velocity of 3 feet per second. The distance of the ball from its rest position after t seconds is given by $d = 2\cos t + 3\sin t$. Show that

$$2\cos t + 3\sin t = \sqrt{13}\cos(t - \theta)$$

where θ lies in quadrant I and $\tan\theta = \frac{3}{2}$. Use the identity to find the amplitude and the period of the ball's motion.

68. A tuning fork is held a certain distance from your ears and struck. Your eardrums' vibrations after t seconds are given by $p = 3\sin 2t$. When a second tuning fork is struck, the formula $p = 2\sin(2t + \pi)$ describes the effects of the sound on the eardrums' vibrations. The total vibrations are given by $p = 3\sin 2t + 2\sin(2t + \pi)$.
a. Express p using a single occurrence of the sine.
b. If the amplitude of p is zero, no sound is heard. Based on your equation in part (a), does this occur with the two tuning forks in this exercise? Explain your answer.

Writing in Mathematics

In Exercises 69–74, use words to describe the formula for:

69. the cosine of the difference of two angles.
70. the cosine of the sum of two angles.
71. the sine of the sum of two angles.
72. the sine of the difference of two angles.
73. the tangent of the difference of two angles.
74. the tangent of the sum of two angles.
75. The distance formula and the definitions for cosine and sine are used to prove the formula for the cosine of the difference of two angles. This formula logically leads the way to the other sum and difference identities. Using this development of ideas and formulas, describe a characteristic of mathematical logic.

Technology Exercises

In Exercises 76–81, graph each side of the equation in the same viewing rectangle. If the graphs appear to coincide, verify that the equation is an identity. If the graphs do not appear to coincide, this indicates that the equation is not an identity. In these exercises, find a value of x for which both sides are defined but not equal.

76. $\cos\left(\dfrac{3\pi}{2} - x\right) = -\sin x$

77. $\tan(\pi - x) = -\tan x$

78. $\sin\left(x + \dfrac{\pi}{2}\right) = \sin x + \sin\dfrac{\pi}{2}$

79. $\cos\left(x + \dfrac{\pi}{2}\right) = \cos x + \cos\dfrac{\pi}{2}$

80. $\cos 1.2x \cos 0.8x - \sin 1.2x \sin 0.8x = \cos 2x$

81. $\sin 1.2x \cos 0.8x + \cos 1.2x \sin 0.8x = \sin 2x$

Critical Thinking Exercises

82. Graph $y = \sin 5x \cos 3x - \cos 5x \sin 3x$ from 0 to 2π.
83. Verify the identity:
$$\frac{\sin(x - y)}{\cos x \cos y} + \frac{\sin(y - z)}{\cos y \cos z} + \frac{\sin(z - x)}{\cos z \cos x} = 0.$$

In Exercises 84–87, find the exact value of each expression. Do not use a calculator.

84. $\sin\left(\cos^{-1}\dfrac{1}{2} + \sin^{-1}\dfrac{3}{5}\right)$

85. $\sin\left[\sin^{-1}\dfrac{3}{5} - \cos^{-1}\left(-\dfrac{4}{5}\right)\right]$

86. $\cos\left(\tan^{-1}\dfrac{4}{3} + \cos^{-1}\dfrac{5}{13}\right)$

87. $\cos\left[\cos^{-1}\left(-\dfrac{\sqrt{3}}{2}\right) - \sin^{-1}\left(-\dfrac{1}{2}\right)\right]$

In Exercises 88–90, write each trigonometric expression as an algebraic expression (that is, without any trigonometric functions). Assume that x and y are positive and in the domain of the given inverse trigonometric function.

88. $\cos(\sin^{-1} x - \cos^{-1} y)$

89. $\sin(\tan^{-1} x - \sin^{-1} y)$

90. $\tan(\sin^{-1} x + \cos^{-1} y)$

Group Exercise

91. Remembering the six sum and difference identities can be difficult. Did you have problems with some exercises because the identity you were using in your head turned out to be an incorrect formula? Are there easy ways to remember the six new identities presented in this section? Group members should address this question, considering one identity at a time. For each formula, list ways to make it easier to remember.

SECTION 5.3   Double-Angle and Half-Angle Formulas

Objectives

1. Use the double-angle formulas.
2. Use the power-reducing formulas.
3. Use the half-angle formulas.

We have a long history of throwing things. Prior to 400 B.C., the Greeks competed in games that included discus throwing. In the seventeenth century,

English soldiers organized cannonball-throwing competitions. In 1827, a Yale University student, disappointed over failing an exam, took out his frustrations at the passing of a collection plate in chapel. Seizing the money tray, he flung it in the direction of a large open space on campus. Yale students see this act of frustration as the origin of the Frisbee.

In this section, we develop other important classes of identities, called the double-angle and half-angle formulas. We will see how one of these formulas can be used by athletes to increase throwing distance.

1 Use the double-angle formulas.

Double-Angle Formulas

A number of basic identities follow from the sum formulas for sine, cosine, and tangent. The first category of identities involves **double-angle formulas.**

Study Tip

The 2 that appears in each of the double-angle expressions cannot be pulled to the front and written as a coefficient.

INCORRECT

$$\sin 2\theta = 2 \sin \theta$$
$$\cos 2\theta = 2 \cos \theta$$
$$\tan 2\theta = 2 \tan \theta$$

The figure shows, in a

$$\left[0, 2\pi, \frac{\pi}{2} \right] \text{ by } [-3, 3, 1]$$

viewing rectangle, that the graphs of

$$y = \sin 2x$$

and

$$y = 2 \sin x$$

do not coincide.

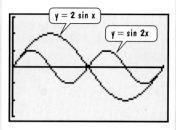

> **Double-Angle Formulas**
>
> $$\sin 2\theta = 2 \sin \theta \cos \theta$$
> $$\cos 2\theta = \cos^2 \theta - \sin^2 \theta$$
> $$\tan 2\theta = \frac{2 \tan \theta}{1 - \tan^2 \theta}$$

To prove each of these formulas, we replace α and β by θ in the sum formulas for $\sin (\alpha + \beta)$, $\cos (\alpha + \beta)$, and $\tan (\alpha + \beta)$.

• $\sin 2\theta = \sin (\theta + \theta) = \sin \theta \cos \theta + \cos \theta \sin \theta = 2 \sin \theta \cos \theta$

> We use
> $\sin(\alpha + \beta) = \sin \alpha \cos \beta + \cos \alpha \sin \beta.$

• $\cos 2\theta = \cos (\theta + \theta) = \cos \theta \cos \theta - \sin \theta \sin \theta = \cos^2 \theta - \sin^2 \theta$

> We use
> $\cos(\alpha + \beta) = \cos \alpha \cos \beta - \sin \alpha \sin \beta.$

• $\tan 2\theta = \tan (\theta + \theta) = \dfrac{\tan \theta + \tan \theta}{1 - \tan \theta \tan \theta} = \dfrac{2 \tan \theta}{1 - \tan^2 \theta}$

> We use
> $\tan(\alpha + \beta) = \dfrac{\tan \alpha + \tan \beta}{1 - \tan \alpha \tan \beta}.$

EXAMPLE 1 Using Double-Angle Formulas to Find Exact Values

If $\sin \theta = \frac{5}{13}$ and θ lies in quadrant II, find the exact value of:

a. $\sin 2\theta$ **b.** $\cos 2\theta$ **c.** $\tan 2\theta$.

Solution We begin with a sketch that illustrates

$$\sin \theta = \frac{5}{13} = \frac{y}{r}.$$

Figure 5.6 shows a quadrant II angle θ for which $\sin \theta = \frac{5}{13}$. We find x using $x^2 + y^2 = r^2$. Because θ lies in quadrant II, x is negative.

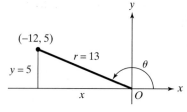

Figure 5.6 $\sin \theta = \frac{5}{13}$ and θ lies in quadrant II

$$x^2 + 5^2 = 13^2 \qquad x^2 + y^2 = r^2$$

$$x^2 = 13^2 - 5^2 = 144 \qquad \text{Solve for } x^2.$$

$$x = -\sqrt{144} = -12 \qquad \text{In quadrant II, } x \text{ is negative.}$$

Now we can use values for x, y, and r to find the required values. We will use $\cos\theta = \dfrac{x}{r} = -\dfrac{12}{13}$ and $\tan\theta = \dfrac{y}{x} = -\dfrac{5}{12}$. We were given $\sin\theta = \dfrac{5}{13}$.

a. $\sin 2\theta = 2\sin\theta\cos\theta = 2\left(\dfrac{5}{13}\right)\left(-\dfrac{12}{13}\right) = -\dfrac{120}{169}$

b. $\cos 2\theta = \cos^2\theta - \sin^2\theta = \left(-\dfrac{12}{13}\right)^2 - \left(\dfrac{5}{13}\right)^2 = \dfrac{144}{169} - \dfrac{25}{169} = \dfrac{119}{169}$

c. $\tan 2\theta = \dfrac{2\tan\theta}{1 - \tan^2\theta} = \dfrac{2\left(-\dfrac{5}{12}\right)}{1 - \left(-\dfrac{5}{12}\right)^2} = \dfrac{-\dfrac{5}{6}}{1 - \dfrac{25}{144}} = \dfrac{-\dfrac{5}{6}}{\dfrac{119}{144}} = \left(-\dfrac{5}{6}\right)\left(\dfrac{144}{119}\right) = -\dfrac{120}{119}$

Check Point 1 If $\sin\theta = \frac{4}{5}$ and θ lies in quadrant II, find the exact value of:

    a. $\sin 2\theta$ **b.** $\cos 2\theta$ **c.** $\tan 2\theta$.

EXAMPLE 2   Using the Double-Angle Formula for Tangents to Find an Exact Value

Find the exact value of $\dfrac{2\tan 15°}{1 - \tan^2 15°}$.

Solution   The given expression is the right side of the formula for $\tan 2\theta$ with $\theta = 15$.

$$\tan 2\theta = \dfrac{2\tan\theta}{1 - \tan^2\theta}$$

$$\dfrac{2\tan 15°}{1 - \tan^2 15°} = \tan(2 \cdot 15°) = \tan 30° = \dfrac{\sqrt{3}}{3}$$

Check Point 2   Find the exact value of $\cos^2 15° - \sin^2 15°$.

There are three forms of the double-angle formula for $\cos 2\theta$. The form we have seen involves both the cosine and the sine:

$$\cos 2\theta = \cos^2\theta - \sin^2\theta.$$

Using the Pythagorean identity $\sin^2\theta + \cos^2\theta = 1$, we can write this last formula in terms of the cosine only. We substitute $1 - \cos^2\theta$ for $\sin^2\theta$.

$$\cos 2\theta = \cos^2\theta - \sin^2\theta = \cos^2\theta - (1 - \cos^2\theta)$$

$$= \cos^2\theta - 1 + \cos^2\theta = 2\cos^2\theta - 1$$

We can also use a Pythagorean identity to write $\cos 2\theta$ in terms of sine only. We substitute $1 - \sin^2 \theta$ for $\cos^2 \theta$.

$$\cos 2\theta = \cos^2 \theta - \sin^2 \theta = 1 - \sin^2 \theta - \sin^2 \theta = 1 - 2 \sin^2 \theta$$

Three Forms of the Double-Angle Formula for $\cos 2\theta$

$$\cos 2\theta = \cos^2 \theta - \sin^2 \theta$$
$$\cos 2\theta = 2 \cos^2 \theta - 1$$
$$\cos 2\theta = 1 - 2 \sin^2 \theta$$

EXAMPLE 3   Verifying an Identity

Verify the identity: $\cos 3\theta = 4 \cos^2 \theta - 3 \cos \theta$.

Solution   We begin by working with the left side. In order to obtain an expression for $\cos 3\theta$, we use the sum formula and write 3θ as $2\theta + \theta$.

$\cos 3\theta = \cos(2\theta + \theta)$ *Write 3θ as $2\theta + \theta$.*

$\quad = \cos 2\theta \cos \theta - \sin 2\theta \sin \theta$ *$\cos(\alpha + \beta)$*
$\quad\quad\quad\quad\quad\quad\quad\quad\quad\quad\quad\quad\quad\quad\quad = \cos \alpha \cos \beta - \sin \alpha \sin \beta$

$\quad\quad\quad\quad\quad\quad\quad\quad\underset{2\cos^2\theta - 1}{\underbrace{\quad\quad}}\quad\underset{2\sin\theta\cos\theta}{\underbrace{\quad\quad}}$

$\quad = (2 \cos^2 \theta - 1) \cos \theta - 2 \sin \theta \cos \theta \sin \theta$ *Substitute double-angle formulas. Because the right side of the given equation involves cosines only, use this form for $\cos 2\theta$.*

$\quad = 2 \cos^3 \theta - \cos \theta - 2 \sin^2 \theta \cos \theta$ *Multiply.*

$\quad\quad\quad\quad\quad\quad\quad\quad\quad\quad\underset{1 - \cos^2\theta}{\underbrace{\quad\quad}}$

$\quad = 2 \cos^3 \theta - \cos \theta - 2(1 - \cos^2 \theta) \cos \theta$ *To get cosines only, use $\sin^2 \theta + \cos^2 \theta = 1$ and substitute $1 - \cos^2 \theta$ for $\sin^2 \theta$.*

$\quad = 2 \cos^3 \theta - \cos \theta - 2 \cos \theta + 2 \cos^3 \theta$ *Multiply.*
$\quad = 4 \cos^3 \theta - 3 \cos \theta$ *Simplify: $2 \cos^3 \theta + 2 \cos^3 \theta = 4 \cos^3 \theta$ and $-\cos \theta - 2 \cos \theta = -3 \cos \theta$.*

By working with the left side and expressing it in a form identical to the right side, we have verified the identity.

Check Point 3   Verify the identity: $\sin 3\theta = 3 \sin \theta - 4 \sin^3 \theta$.

2 Use the power-reducing formulas.

Power-Reducing Formulas

The double-angle formulas are used to derive the **power-reducing formulas:**

Power-Reducing Formulas

$$\sin^2 \theta = \frac{1 - \cos 2\theta}{2} \quad \cos^2 \theta = \frac{1 + \cos 2\theta}{2} \quad \tan^2 \theta = \frac{1 - \cos 2\theta}{1 + \cos 2\theta}$$

We can prove the first two formulas in the box by working with two forms of the double-angle formula for cos 2θ.

> This is the form with sine only.

> This is the form with cosine only.

$$\cos 2\theta = 1 - 2 \sin^2 \theta \qquad \cos 2\theta = 2 \cos^2 \theta - 1$$

Solve the formula on the left for $\sin^2 \theta$. Solve the formula on the right for $\cos^2 \theta$.

$$2 \sin^2 \theta = 1 - \cos 2\theta \qquad 2 \cos^2 \theta = 1 + \cos 2\theta$$

$$\sin^2 \theta = \frac{1 - \cos 2\theta}{2} \qquad \cos^2 \theta = \frac{1 + \cos 2\theta}{2}$$

Divide both sides of each equation by 2.

These are the first two formulas in the box. The third formula in the box is proved by writing the tangent as the quotient of the sine and the cosine.

$$\tan^2 \theta = \frac{\sin^2 \theta}{\cos^2 \theta} = \frac{\dfrac{1 - \cos 2\theta}{2}}{\dfrac{1 + \cos 2\theta}{2}} = \frac{1 - \cos 2\theta}{\overset{1}{\cancel{2}}} \cdot \frac{\overset{1}{\cancel{2}}}{1 + \cos 2\theta} = \frac{1 - \cos 2\theta}{1 + \cos 2\theta}$$

Power-reducing formulas are quite useful in calculus. By reducing the power of trigonometric functions, calculus can better explore the relationship between a function and how it is changing at every single instant in time.

EXAMPLE 4 Reducing the Power of a Trigonometric Function

Write an equivalent expression for $\cos^4 x$ that does not contain powers of trigonometric functions greater than 1.

Solution We will apply the formula for $\cos^2 \theta$ twice.

$$\cos^4 x = \left(\cos^2 x \right)^2$$

$$= \left(\frac{1 + \cos 2x}{2} \right)^2$$

Use $\cos^2 \theta = \dfrac{1 + \cos 2\theta}{2}$ with $\theta = x$.

$$= \frac{1 + 2 \cos 2x + \cos^2 2x}{4}$$

Square the numerator:
$(A + B)^2 = A^2 + 2AB + B^2$.
Square the denominator.

$$= \frac{1}{4} + \frac{1}{2} \cos 2x + \frac{1}{4} \cos^2 2x$$

Divide each term in the numerator by 4.

> We can reduce the power of $\cos^2 2x$ using
> $$\cos^2 \theta = \frac{1 + \cos 2\theta}{2}$$
> with $\theta = 2x$.

$$= \frac{1}{4} + \frac{1}{2} \cos 2x + \frac{1}{4} \left[\frac{1 + \cos 2(2x)}{2} \right]$$

Use the power-reducing formula for $\cos^2 \theta$ with $\theta = 2x$.

$$\cos^4 x = \frac{1}{4} + \frac{1}{2}\cos 2x + \frac{1}{4}\left[\frac{1 + \cos 2(2x)}{2}\right]$$

This is our expression for $\cos^4 x$ from the previous page.

$$= \frac{1}{4} + \frac{1}{2}\cos 2x + \frac{1}{8}(1 + \cos 4x)$$

Multiply.

$$= \frac{1}{4} + \frac{1}{2}\cos 2x + \frac{1}{8} + \frac{1}{8}\cos 4x$$

Distribute $\frac{1}{8}$ throughout parentheses.

$$= \frac{3}{8} + \frac{1}{2}\cos 2x + \frac{1}{8}\cos 4x$$

Simplify: $\frac{1}{4} + \frac{1}{8} = \frac{2}{8} + \frac{1}{8} = \frac{3}{8}$.

Thus, $\cos^4 x = \frac{3}{8} + \frac{1}{2}\cos 2x + \frac{1}{8}\cos 4x$. The expression for $\cos^4 x$ does not contain powers of trigonometric functions greater than 1.

> **Check Point 4** Write an equivalent expression for $\sin^4 x$ that does not contain powers of trigonometric functions greater than 1.

3 Use the half-angle formulas.

Half-Angle Formulas

Useful equivalent forms of the power-reducing formulas can be obtained by replacing θ by $\frac{\alpha}{2}$. Then solve for the trigonometric function on the left sides of the equations. The resulting identities are called the **half-angle formulas:**

Half-Angle Formulas

$$\sin\frac{\alpha}{2} = \pm\sqrt{\frac{1 - \cos\alpha}{2}}$$

$$\cos\frac{\alpha}{2} = \pm\sqrt{\frac{1 + \cos\alpha}{2}}$$

$$\tan\frac{\alpha}{2} = \pm\sqrt{\frac{1 - \cos\alpha}{1 + \cos\alpha}}$$

The $+$ or $-$ in each formula is determined by the quadrant in which $\frac{\alpha}{2}$ lies.

If we know the exact value for the sine, cosine, or tangent of an angle, we can use the half-angle formulas to find exact values for half that angle. For example, we know that $\cos 225° = -\frac{\sqrt{2}}{2}$. In the next example, we find the exact value of the cosine of half of $225°$, or $\cos 112.5°$.

EXAMPLE 5 Using a Half-Angle Formula to Find an Exact Value

Find the exact value of $\cos 112.5°$.

Solution Because $112.5° = \frac{225°}{2}$, we use the half-angle formula for $\cos\frac{\alpha}{2}$ with $\alpha = 225°$. What sign should we use when we apply the formula? Because $112.5°$ lies in quadrant II, where only the sine and cosecant are positive, $\cos 112.5° < 0$. Thus, we use the $-$ sign in the half-angle formula.

Study Tip

The $\frac{1}{2}$ that appears in each of the half-angle formulas cannot be pulled to the front and written as a coefficient.

INCORRECT

$$\sin\frac{\theta}{2} = \frac{1}{2}\sin\theta$$

$$\cos\frac{\theta}{2} = \frac{1}{2}\cos\theta$$

$$\tan\frac{\theta}{2} = \frac{1}{2}\tan\theta$$

The figure shows, in a $\left[0, 2\pi, \frac{\pi}{2}\right]$ by $[-2, 2, 1]$ viewing rectangle, that the graphs of

$$y = \sin\frac{x}{2}$$

and

$$y = \frac{1}{2}\sin x$$

do not coincide.

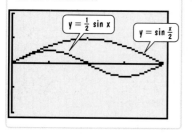

$$\cos 112.5° = \cos \frac{225°}{2}$$

$$= -\sqrt{\frac{1 + \cos 225°}{2}}$$ Use $\cos \frac{\alpha}{2} = -\sqrt{\frac{1 + \cos \alpha}{2}}$ with $\alpha = 225°$.

$$= -\sqrt{\frac{1 + \left(-\dfrac{\sqrt{2}}{2}\right)}{2}}$$ $\cos 225° = -\dfrac{\sqrt{2}}{2}$

$$= -\sqrt{\frac{2 - \sqrt{2}}{4}}$$ Multiply the radicand by $\frac{2}{2}$:

$$\frac{1 + \left(-\dfrac{\sqrt{2}}{2}\right)}{2} \cdot \frac{2}{2} = \frac{2 - \sqrt{2}}{4}.$$

$$= -\frac{\sqrt{2 - \sqrt{2}}}{2}$$ Simplify: $\sqrt{4} = 2$.

Discovery

Use your calculator to find approximations for
$$-\frac{\sqrt{2 - \sqrt{2}}}{2}$$
and cos 112.5°. What do you observe?

Check Point 5   Use $\cos 210° = -\dfrac{\sqrt{3}}{2}$ to find the exact value of $\cos 105°$.

There are alternate formulas for $\tan \dfrac{\alpha}{2}$ that do not require us to determine what sign to use when applying the formula. These formulas are logically connected to the identities in Example 6 and Check Point 6.

EXAMPLE 6   Verifying an Identity

Verify the identity: $\tan \theta = \dfrac{1 - \cos 2\theta}{\sin 2\theta}$.

Solution   We work with the right side.

$$\frac{1 - \cos 2\theta}{\sin 2\theta} = \frac{1 - (1 - 2\sin^2 \theta)}{2 \sin \theta \cos \theta}$$ The form $\cos 2\theta = 1 - 2\sin^2 \theta$ is used because it produces only one term in the numerator. Use the double-angle formula for sine in the denominator.

$$= \frac{2 \sin^2 \theta}{2 \sin \theta \cos \theta}$$ Simplify the numerator.

$$= \frac{\sin \theta}{\cos \theta}$$ Divide the numerator and denominator by 2 sin θ.

$$= \tan \theta$$ Use a quotient identity: $\tan \theta = \dfrac{\sin \theta}{\cos \theta}$.

The right side simplifies to $\tan \theta$, the expression on the left side. Thus, the identity is verified.

Check Point 6 Verify the identity: $\tan\theta = \dfrac{\sin 2\theta}{1 + \cos 2\theta}$.

$$\tan\theta = \frac{1 - \cos 2\theta}{\sin 2\theta}$$

Example 6 identity, repeated

Half-angle formulas for $\tan\dfrac{\alpha}{2}$ can be obtained using the identities in Example 6 and Check Point 6. Do you see how to do this? Replace each occurrence of θ with $\dfrac{\alpha}{2}$. This results in the following identities:

Half-Angle Formulas for $\tan\dfrac{\alpha}{2}$

$$\tan\frac{\alpha}{2} = \frac{1 - \cos\alpha}{\sin\alpha}$$

$$\tan\frac{\alpha}{2} = \frac{\sin\alpha}{1 + \cos\alpha}$$

EXAMPLE 7   Verifying an Identity

Verify the identity: $\tan\dfrac{\alpha}{2} = \csc\alpha - \cot\alpha$.

Solution   We begin with the right side.

$$\csc\alpha - \cot\alpha = \frac{1}{\sin\alpha} - \frac{\cos\alpha}{\sin\alpha} = \frac{1 - \cos\alpha}{\sin\alpha} = \tan\frac{\alpha}{2}$$

Express functions in terms of sines and cosines.

This is the first of the two half-angle formulas in the preceding box.

We worked with the right side and arrived at the left side. Thus, the identity is verified.

Check Point 7 Verify the identity: $\tan\dfrac{\alpha}{2} = \dfrac{\sec\alpha}{\sec\alpha\csc\alpha + \csc\alpha}$.

We conclude with a summary of the principal trigonometric identities developed in this section and the previous section. The fundamental identities can be found in the box on page 545.

Principal Trigonometric Identities

Sum and Difference Formulas

$$\sin(\alpha + \beta) = \sin \alpha \cos \beta + \cos \alpha \sin \beta \qquad \sin(\alpha - \beta) = \sin \alpha \cos \beta - \cos \alpha \sin \beta$$

$$\cos(\alpha + \beta) = \cos \alpha \cos \beta - \sin \alpha \sin \beta \qquad \cos(\alpha - \beta) = \cos \alpha \cos \beta + \sin \alpha \sin \beta$$

$$\tan(\alpha + \beta) = \frac{\tan \alpha + \tan \beta}{1 - \tan \alpha \tan \beta} \qquad \tan(\alpha - \beta) = \frac{\tan \alpha - \tan \beta}{1 + \tan \alpha \tan \beta}$$

Double-Angle Formulas

$$\sin 2\theta = 2 \sin \theta \cos \theta$$

$$\cos 2\theta = \cos^2 \theta - \sin^2 \theta = 2 \cos^2 \theta - 1 = 1 - 2 \sin^2 \theta$$

$$\tan 2\theta = \frac{2 \tan \theta}{1 - \tan^2 \theta}$$

Power-Reducing Formulas

$$\sin^2 \theta = \frac{1 - \cos 2\theta}{2} \qquad \cos^2 \theta = \frac{1 + \cos 2\theta}{2} \qquad \tan^2 \theta = \frac{1 - \cos 2\theta}{1 + \cos 2\theta}$$

Half-Angle Formulas

$$\sin \frac{\alpha}{2} = \pm \sqrt{\frac{1 - \cos \alpha}{2}} \qquad \cos \frac{\alpha}{2} = \pm \sqrt{\frac{1 + \cos \alpha}{2}}$$

$$\tan \frac{\alpha}{2} = \pm \sqrt{\frac{1 - \cos \alpha}{1 + \cos \alpha}} = \frac{1 - \cos \alpha}{\sin \alpha} = \frac{\sin \alpha}{1 + \cos \alpha}$$

Study Tip

To help remember the correct sign in the numerator in the first two power-reducing formulas and the first two half-angle formulas, remember *sinus-minus*–the sine is minus.

EXERCISE SET 5.3

Practice Exercises

In Exercises 1–6, use the figures to find the exact value of each trigonometric function.

1. $\sin 2\theta$ **2.** $\cos 2\theta$ **3.** $\tan 2\theta$

4. $\sin 2\alpha$ **5.** $\cos 2\alpha$ **6.** $\tan 2\alpha$

In Exercises 7–14, use the given information to find the exact value of:

  a. $\sin 2\theta$ **b.** $\cos 2\theta$ **c.** $\tan 2\theta$

7. $\sin \theta = \frac{15}{17}$, θ lies in quadrant II.

8. $\sin \theta = \frac{12}{13}$, θ lies in quadrant II.

9. $\cos \theta = \frac{24}{25}$, θ lies in quadrant IV.

10. $\cos \theta = \frac{40}{41}$, θ lies in quadrant IV.

11. $\cot \theta = 2$, θ lies in quadrant III.

12. $\cot \theta = 3$, θ lies in quadrant III.

13. $\sin \theta = -\frac{9}{41}$, θ lies in quadrant III.

14. $\sin \theta = -\frac{2}{3}$, θ lies in quadrant III.

In Exercises 15–22, write each expression as the sine, cosine, or tangent of a double angle. Then find the exact value of the expression.

15. $2 \sin 15° \cos 15°$ **16.** $2 \sin 22.5° \cos 22.5°$

17. $\cos^2 75° - \sin^2 75°$ **18.** $\cos^2 105° - \sin^2 105°$

19. $2 \cos^2 \frac{\pi}{8} - 1$ **20.** $1 - 2 \sin^2 \frac{\pi}{12}$

21. $\dfrac{2 \tan \dfrac{\pi}{12}}{1 - \tan^2 \dfrac{\pi}{12}}$ **22.** $\dfrac{2 \tan \dfrac{\pi}{8}}{1 - \tan^2 \dfrac{\pi}{8}}$

In Exercises 23–34, verify each identity.

23. $\sin 2\theta = \dfrac{2 \tan \theta}{1 + \tan^2 \theta}$ **24.** $\sin 2\theta = \dfrac{2 \cot \theta}{1 + \cot^2 \theta}$

25. $(\sin \theta + \cos \theta)^2 = 1 + \sin 2\theta$

26. $(\sin \theta - \cos \theta)^2 = 1 - \sin 2\theta$

27. $\sin^2 x + \cos 2x = \cos^2 x$

28. $1 - \tan^2 x = \dfrac{\cos 2x}{\cos^2 x}$

29. $\cot x = \dfrac{\sin 2x}{1 - \cos 2x}$

30. $\cot x = \dfrac{1 + \cos 2x}{\sin 2x}$

31. $\sin 2t - \tan t = \tan t \cos 2t$

32. $\sin 2t - \cot t = -\cot t \cos 2t$

33. $\sin 4t = 4 \sin t \cos^3 t - 4 \sin^3 t \cos t$

34. $\cos 4t = 8 \cos^4 t - 8 \cos^2 t + 1$

In Exercises 35–38, use the power-reducing formulas to rewrite each expression as an equivalent expression that does not contain powers of trigonometric functions greater than 1.

35. $6 \sin^4 x$ **36.** $10 \cos^4 x$

37. $\sin^2 x \cos^2 x$ **38.** $8 \sin^2 x \cos^2 x$

In Exercises 39–46, use a half-angle formula to find the exact value of each expression.

39. $\sin 15°$ **40.** $\cos 22.5°$

41. $\cos 157.5°$ **42.** $\sin 105°$

43. $\tan 75°$ **44.** $\tan 112.5°$

45. $\tan \dfrac{7\pi}{8}$ **46.** $\tan \dfrac{3\pi}{8}$

In Exercises 47–54, use the figures to find the exact value of each trigonometric function.

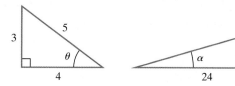

47. $\sin \dfrac{\theta}{2}$ **48.** $\cos \dfrac{\theta}{2}$ **49.** $\tan \dfrac{\theta}{2}$

50. $\sin \dfrac{\alpha}{2}$ **51.** $\cos \dfrac{\alpha}{2}$ **52.** $\tan \dfrac{\alpha}{2}$

53. $2 \sin \dfrac{\theta}{2} \cos \dfrac{\theta}{2}$ **54.** $2 \sin \dfrac{\alpha}{2} \cos \dfrac{\alpha}{2}$

In Exercises 55–58, use the given information to find the exact value of:

a. $\sin \dfrac{\alpha}{2}$ **b.** $\cos \dfrac{\alpha}{2}$ **c.** $\tan \dfrac{\alpha}{2}$.

55. $\tan \alpha = \frac{4}{3}$, α lies in quadrant III.

56. $\tan \alpha = \frac{8}{15}$, α lies in quadrant III.

57. $\sec \alpha = -\frac{13}{5}$, α lies in quadrant II.

58. $\sec \alpha = -3$, α lies in quadrant II.

In Exercises 59–68, verify each identity.

59. $\sin^2 \dfrac{\theta}{2} = \dfrac{\sec \theta - 1}{2 \sec \theta}$

60. $\sin^2 \dfrac{\theta}{2} = \dfrac{\csc \theta - \cot \theta}{2 \csc \theta}$

61. $\cos^2 \dfrac{\theta}{2} = \dfrac{\sin \theta + \tan \theta}{2 \tan \theta}$

62. $\cos^2 \dfrac{\theta}{2} = \dfrac{\sec \theta + 1}{2 \sec \theta}$

63. $\tan \dfrac{\alpha}{2} = \dfrac{\tan \alpha}{\sec \alpha + 1}$

64. $2 \tan \dfrac{\alpha}{2} = \dfrac{\sin^2 \alpha + 1 - \cos^2 \alpha}{\sin \alpha(1 + \cos \alpha)}$

65. $\cot \dfrac{x}{2} = \dfrac{\sin x}{1 - \cos x}$ **66.** $\cot \dfrac{x}{2} = \dfrac{1 + \cos x}{\sin x}$

67. $\tan \dfrac{x}{2} + \cot \dfrac{x}{2} = 2 \csc x$

68. $\tan \dfrac{x}{2} - \cot \dfrac{x}{2} = -2 \cot x$

Application Exercises

69. Throwing events in track and field include the shot put, the discus throw, the hammer throw, and the javelin throw. The distance that the athlete can achieve depends on the initial speed of the object thrown and the angle above the horizontal at which the object leaves the hand. This angle is represented by θ in the figure shown. The distance, d, in feet, that the athlete throws is modeled by the formula

$$d = \dfrac{v_0^2}{16} \sin \theta \cos \theta$$

in which v_0 is the initial speed of the object thrown, in feet per second, and θ is the angle, in degrees, at which the object leaves the hand.

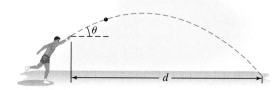

a. Use an identity to express the formula so that it contains the sine function only.

b. Use your formula from part (a) to find the angle, θ, that produces the maximum distance, d, for a given initial speed, v_0.

Use this information to solve Exercises 70–71.

The speed of a supersonic aircraft is usually represented by a Mach number, named after Austrian physicist Ernst Mach (1838–1916). A Mach number is the speed of the aircraft, in miles per hour, divided by the speed of sound, approximately 740 miles per hour. Thus, a plane flying at twice the speed of sound has a speed, M, of Mach 2. If an aircraft has a speed greater than Mach 1, a sonic boom is heard, created by sound waves that form a cone with a vertex angle θ, shown in the figure.

The relationship between the cone's vertex angle, θ, and the Mach speed, M, of an aircraft that is flying faster than the speed of sound is given by

$$\sin \frac{\theta}{2} = \frac{1}{M}.$$

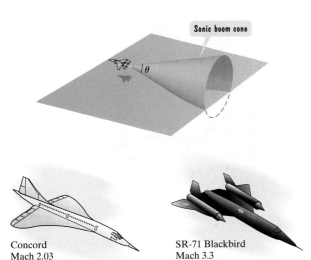

Sonic boom cone

Concord
Mach 2.03

SR-71 Blackbird
Mach 3.3

70. If $\theta = \frac{\pi}{6}$, determine the Mach speed, M, of the aircraft. Express the speed as an exact value and as a decimal to the nearest tenth.

71. If $\theta = \frac{\pi}{4}$, determine the Mach speed, M, of the aircraft. Express the speed as an exact value and as a decimal to the nearest tenth.

Writing in Mathematics

In Exercises 72–79, use words to describe the formula for:

72. the sine of double an angle.

73. the cosine of double an angle. (Describe one of the three formulas.)

74. the tangent of double an angle.

75. the power-reducing formula for the sine squared of an angle.

76. the power-reducing formula for the cosine squared of an angle.

77. the sine of half an angle.

78. the cosine of half an angle.

79. the tangent of half an angle. (Describe one of the two formulas that does not involve a square root.)

80. Explain how the double-angle formulas are derived.

81. How can there be three forms of the double-angle formula for cos 2θ?

82. Without showing algebraic details, describe in words how to reduce the power of $\cos^4 x$.

83. Describe one or more of the techniques you use to help remember the identities in the box on page 573.

84. Your friend is about to compete as a shot-putter in a college field event. Using Exercise 69(b), write a short description to your friend on how to achieve the best distance possible in the throwing event.

Technology Exercises

In Exercises 85–88, graph each side of the equation in the same viewing rectangle. If the graphs appear to coincide, verify that the equation is an identity. If the graphs do not appear to coincide, find a value of x for which both sides are defined but not equal.

85. $3 - 6 \sin^2 x = 3 \cos 2x$ **86.** $4 \cos^2 \frac{x}{2} = 2 + 2 \cos x$

87. $\sin \frac{x}{2} = \frac{1}{2} \sin x$ **88.** $\cos \frac{x}{2} = \frac{1}{2} \cos x$

*In Exercises 89–91, graph each equation in a $\left[-2\pi, 2\pi, \frac{\pi}{2} \right]$ by $[-3, 3, 1]$ viewing rectangle. Then **a.** Describe the graph using another equation, and **b.** Verify that the two equations are equivalent.*

89. $y = \dfrac{1 - 2 \cos 2x}{2 \sin x - 1}$

90. $y = \dfrac{2 \tan \dfrac{x}{2}}{1 + \tan^2 \dfrac{x}{2}}$

91. $y = \csc x - \cot x$

Critical Thinking Exercises

92. Verify the identity:

$$\sin^3 x + \cos^3 x = (\sin x + \cos x)\left(1 - \frac{\sin 2x}{2} \right).$$

93. Use the power-reducing formulas to rewrite $\sin^6 x$ as an equivalent expression that does not contain powers of trigonometric functions greater than 1.

94. Use a right triangle to write $\sin(2 \sin^{-1} x)$ as an algebraic expression. Assume that x is positive and in the domain of the given inverse trigonometric function.

SECTION 5.4 *Product-to-Sum and Sum-to-Product Formulas*

Objectives

1. Use the product-to-sum formulas.
2. Use the sum-to-product formulas.

James K. Polk
Born November 2, 1795

Warren G. Harding
Born November 2, 1865

Of the 43 U.S. presidents, two share a birthday (same month and day). The probability of two or more people in a group sharing a birthday rises sharply as the group's size increases. Above 50 people, the probability approaches certainty. (You can verify the mathematics of this surprising result by studying Sections 10.6 and 10.7, and working Exercise 66 in Exercise Set 10.7.) So, come November 2, we salute Presidents Polk and Harding with

$$112, 163\text{-}, 112, 196\text{-}, 110, 8521\text{-}, 008, 121\text{-}.$$

Were you aware that each button on your touch-tone phone produces a unique sound? If we treat the commas as pauses and the hyphens as held notes, this sequence of numbers is *Happy Birthday* on a touch-tone phone.

Although *Happy Birthday* isn't Mozart or Sondheim, it is sinusoidal. Each of its touch-tone musical sounds can be described by the sum of two sine functions or the product of sines and cosines. In this section, we develop identities that enable us to use both descriptions. They are called the product-to-sum and sum-to-product formulas.

1 Use the product-to-sum formulas.

The Product-to-Sum Formulas

How do we write the products of sines and/or cosines as sums or differences? We use the following identities, which are called **product-to-sum formulas:**

Study Tip

You may not need to memorize the formulas in this section. When you need them, you can either refer to one of the two boxes in the section or perhaps even derive them using the methods shown.

Product-to-Sum Formulas

$$\sin\alpha\sin\beta = \tfrac{1}{2}\big[\cos(\alpha - \beta) - \cos(\alpha + \beta)\big]$$

$$\cos\alpha\cos\beta = \tfrac{1}{2}\big[\cos(\alpha - \beta) + \cos(\alpha + \beta)\big]$$

$$\sin\alpha\cos\beta = \tfrac{1}{2}\big[\sin(\alpha + \beta) + \sin(\alpha - \beta)\big]$$

$$\cos\alpha\sin\beta = \tfrac{1}{2}\big[\sin(\alpha + \beta) - \sin(\alpha - \beta)\big]$$

Although these formulas are difficult to remember, they are fairly easy to derive. For example, let's derive the first identity in the box,

$$\sin \alpha \sin \beta = \tfrac{1}{2}\left[\cos (\alpha - \beta) - \cos (\alpha + \beta)\right].$$

We begin with the difference and sum formulas for the cosine, and subtract the second identity from the first:

$$
\begin{aligned}
\cos (\alpha - \beta) &= \cos \alpha \cos \beta + \sin \alpha \sin \beta \\
-[\cos (\alpha + \beta) &= \cos \alpha \cos \beta - \sin \alpha \sin \beta] \quad \text{Subtract the identities.} \\
\cos (\alpha - \beta) - \cos (\alpha + \beta) &= 0 + 2 \sin \alpha \sin \beta \,.
\end{aligned}
$$

Subtract terms on the left side.

Subtract terms on the right side: $\cos \alpha \cos \beta - \cos \alpha \cos \beta = 0.$

Subtract terms on the right side: $\sin \alpha \sin \beta - (-\sin \alpha \sin \beta) = 2 \sin \alpha \sin \beta.$

Now we use this result to derive the product-to-sum formula for $\sin \alpha \sin \beta$.

$$2 \sin \alpha \sin \beta = \cos (\alpha - \beta) - \cos (\alpha + \beta) \quad \text{Reverse the sides in the preceding equation.}$$

$$\sin \alpha \sin \beta = \tfrac{1}{2}\left[\cos (\alpha - \beta) - \cos (\alpha + \beta)\right] \quad \text{Multiply each side by } \tfrac{1}{2}.$$

This last equation is the desired formula. Likewise, we can derive the product-to-sum formula for cosine, $\cos \alpha \cos \beta = \tfrac{1}{2}\left[\cos (\alpha - \beta) + \cos (\alpha + \beta)\right]$. As we did for the previous derivation, begin with the difference and sum formulas for cosine. However, we *add* the formulas rather than subtracting them. Reversing both sides of this result and multiplying each side by $\tfrac{1}{2}$ produces the formula for $\cos \alpha \cos \beta$. The last two product-to-sum formulas, $\sin \alpha \cos \beta$ and $\cos \alpha \sin \beta$, are derived using the sum and difference formulas for sine in a similar manner.

Technology

The graphs of

$$y = \sin 8x \sin 3x$$

and

$$y = \tfrac{1}{2}(\cos 5x - \cos 11x)$$

are shown in a $\left[-2\pi, 2\pi, \dfrac{\pi}{2}\right]$ by $[-1, 1, 1]$ viewing rectangle. The graphs coincide. This supports our algebraic work in Example 1(a).

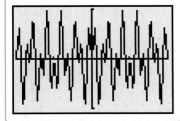

EXAMPLE 1 Using the Product-to-Sum Formulas

Express each of the following products as a sum or difference:

a. $\sin 8x \sin 3x$ **b.** $\sin 4x \cos x.$

Solution The product-to-sum formula that we are using is shown in each of the voice balloons.

a. $\sin \alpha \sin \beta = \tfrac{1}{2}[\cos(\alpha - \beta) - \cos(\alpha + \beta)]$

$$\sin 8x \sin 3x = \tfrac{1}{2}\left[\cos (8x - 3x) - \cos (8x + 3x)\right] = \tfrac{1}{2}(\cos 5x - \cos 11x)$$

b. $\sin \alpha \cos \beta = \tfrac{1}{2}[\sin(\alpha + \beta) + \sin(\alpha - \beta)]$

$$\sin 4x \cos x = \tfrac{1}{2}\left[\sin (4x + x) + \sin (4x - x)\right] = \tfrac{1}{2}(\sin 5x + \sin 3x)$$

Check Point 1 Express each of the following products as a sum or difference:
a. $\sin 5x \sin 2x$ **b.** $\cos 7x \cos x.$

2 Use the sum-to-product formulas.

The Sum-to-Product Formulas

How do we write the sum or difference of sines and/or cosines as products? We use the following identities, which are called the **sum-to-product formulas:**

Sum-to-Product Formulas

$$\sin \alpha + \sin \beta = 2 \sin \frac{\alpha + \beta}{2} \cos \frac{\alpha - \beta}{2}$$

$$\sin \alpha - \sin \beta = 2 \sin \frac{\alpha - \beta}{2} \cos \frac{\alpha + \beta}{2}$$

$$\cos \alpha + \cos \beta = 2 \cos \frac{\alpha + \beta}{2} \cos \frac{\alpha - \beta}{2}$$

$$\cos \alpha - \cos \beta = -2 \sin \frac{\alpha + \beta}{2} \sin \frac{\alpha - \beta}{2}$$

We verify these formulas using the product-to-sum formulas. Let's verify the first sum-to-product formula

$$\sin \alpha + \sin \beta = 2 \sin \frac{\alpha + \beta}{2} \cos \frac{\alpha - \beta}{2}.$$

We start with the right side of the formula, the side with the product. We can apply the product-to-sum formula for $\sin \alpha \cos \beta$ to this expression. By doing so, we obtain the left side of the formula, $\sin \alpha + \sin \beta$. Here's how:

$$\sin \alpha \cos \beta = \tfrac{1}{2} [\sin(\alpha + \beta) + \sin(\alpha - \beta)]$$

$$2 \left| \sin \frac{\alpha + \beta}{2} \cos \frac{\alpha - \beta}{2} \right| = 2 \cdot \left[\frac{1}{2} \left[\sin \left(\frac{\alpha + \beta}{2} + \frac{\alpha - \beta}{2} \right) + \sin \left(\frac{\alpha + \beta}{2} - \frac{\alpha - \beta}{2} \right) \right] \right]$$

$$= \sin \left(\frac{\alpha + \beta + \alpha - \beta}{2} \right) + \sin \left(\frac{\alpha + \beta - \alpha + \beta}{2} \right)$$

$$= \sin \frac{2\alpha}{2} + \sin \frac{2\beta}{2} = \sin \alpha + \sin \beta.$$

The three other sum-to-product formulas in the box are verified in a similar manner. Start with the right side and obtain the left side using an appropriate product-to-sum formula.

EXAMPLE 2   Using the Sum-to-Product Formulas

Express each sum or difference as a product:

  a. $\sin 9x + \sin 5x$ **b.** $\cos 4x - \cos 3x$.

Solution   The sum-to-product formula that we are using is shown in each of the voice balloons.

a.
$$\sin\alpha + \sin\beta = 2\sin\frac{\alpha+\beta}{2}\cos\frac{\alpha-\beta}{2}$$

$$\sin 9x + \sin 5x = 2\sin\frac{9x+5x}{2}\cos\frac{9x-5x}{2}$$

$$= 2\sin\frac{14x}{2}\cos\frac{4x}{2}$$

$$= 2\sin 7x\cos 2x$$

b.
$$\cos\alpha - \cos\beta = -2\sin\frac{\alpha+\beta}{2}\sin\frac{\alpha-\beta}{2}$$

$$\cos 4x - \cos 3x = -2\sin\frac{4x+3x}{2}\sin\frac{4x-3x}{2}$$

$$= -2\sin\frac{7x}{2}\sin\frac{x}{2}$$

> **Check Point 2** Express each sum or difference as a product:
> **a.** $\sin 7x + \sin 3x$ **b.** $\cos 3x + \cos 2x$.

Some identities contain a fraction on one side with sums and differences of sines and/or cosines. Applying the sum-to-product formulas in the numerator and the denominator is often helpful in verifying these identities.

EXAMPLE 3 Using Sum-to-Product Formulas to Verify an Identity

Verify the identity: $\dfrac{\cos 3x - \cos 5x}{\sin 3x + \sin 5x} = \tan x.$

Solution Because the left side is more complicated, we will work with it. We use sum-to-product formulas for the numerator and the denominator of the fraction on this side.

$$\frac{\cos 3x - \cos 5x}{\sin 3x + \sin 5x}$$

$$\cos\alpha - \cos\beta = -2\sin\frac{\alpha+\beta}{2}\sin\frac{\alpha-\beta}{2}$$

$$= \frac{-2\sin\dfrac{3x+5x}{2}\sin\dfrac{3x-5x}{2}}{\sin 3x + \sin 5x}$$

$$\sin\alpha + \sin\beta = 2\sin\frac{\alpha+\beta}{2}\cos\frac{\alpha-\beta}{2}$$

$$= \frac{-2\sin\dfrac{3x+5x}{2}\sin\dfrac{3x-5x}{2}}{2\sin\dfrac{3x+5x}{2}\cos\dfrac{3x-5x}{2}}$$

$$\frac{\cos 3x - \cos 5x}{\sin 3x + \sin 5x} = \frac{-2 \sin \dfrac{3x + 5x}{2} \sin \dfrac{3x - 5x}{2}}{2 \sin \dfrac{3x + 5x}{2} \cos \dfrac{3x - 5x}{2}}$$

This is the expression from the previous page.

$$= \frac{-2 \sin \dfrac{8x}{2} \sin \left(\dfrac{-2x}{2} \right)}{2 \sin \dfrac{8x}{2} \cos \left(\dfrac{-2x}{2} \right)}$$

Perform the indicated additions.

$$= \frac{-2 \, \sin 4x \, \sin (-x)}{2 \, \sin 4x \, \cos (-x)}$$

Simplify.

$$= \frac{-(-\sin x)}{\cos x}$$

The sine function is odd: $\sin (-x) = -\sin x$. The cosine function is even: $\cos (-x) = \cos x$.

$$= \frac{\sin x}{\cos x}$$

Simplify.

$$= \tan x$$

Apply a quotient identity: $\tan x = \dfrac{\sin x}{\cos x}$.

We worked with the left side and arrived at the right side. Thus, the identity is verified.

Check Point 3 Verify the identity: $\dfrac{\cos 3x - \cos x}{\sin 3x + \sin x} = -\tan x.$

Sinusoidal Sounds

Music is all around us. A mere snippet of a song from the past can trigger vivid memories, inducing emotions ranging from unabashed joy to deep sorrow. Trigonometric functions can explain how sound travels from its source and describe its pitch, loudness, and quality. Still unexplained is the remarkable influence music has on the brain, including the deepest question of all: Why do we appreciate music?

When a note is played, it disturbs nearby air molecules, creating regions of higher-than-normal pressure and regions of lower-than-normal pressure. If we graph pressure, y, versus time, t, we get a sine wave that represents the note. The frequency of the sine wave is the number of high-low disturbances, or vibrations, per second. The greater the frequency, the higher the pitch; the lesser the frequency, the lower the pitch.

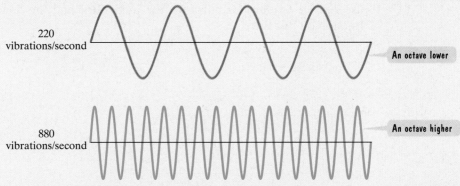

220 vibrations/second — An octave lower

880 vibrations/second — An octave higher

The amplitude of a note's sine wave is related to its loudness. The amplitude for the two sine waves shown above is the same. Thus, the notes have the same loudness, although they differ in pitch. The greater the amplitude, the louder the sound; the lesser the amplitude, the softer the sound. The amplitude and frequency are characteristic of every note—and thus of its graph—until the note dissipates.

EXERCISE SET 5.4

Practice Exercises

In Exercises 1–8, express each product as a sum or difference.

1. $\sin 6x \sin 2x$

2. $\sin 8x \sin 4x$

3. $\cos 7x \cos 3x$

4. $\cos 9x \cos 2x$

5. $\sin x \cos 2x$

6. $\sin 2x \cos 3x$

7. $\cos \dfrac{3x}{2} \sin \dfrac{x}{2}$

8. $\cos \dfrac{5x}{2} \sin \dfrac{x}{2}$

In Exercises 9–22, express each sum or difference as a product. If possible, find this product's exact value.

9. $\sin 6x + \sin 2x$

10. $\sin 8x + \sin 2x$

11. $\sin 7x - \sin 3x$

12. $\sin 11x - \sin 5x$

13. $\cos 4x + \cos 2x$

14. $\cos 9x - \cos 7x$

15. $\sin x + \sin 2x$

16. $\sin x - \sin 2x$

17. $\cos \dfrac{3x}{2} + \cos \dfrac{x}{2}$

18. $\sin \dfrac{3x}{2} + \sin \dfrac{x}{2}$

19. $\sin 75° + \sin 15°$

20. $\cos 75° - \cos 15°$

21. $\sin \dfrac{\pi}{12} - \sin \dfrac{5\pi}{12}$

22. $\cos \dfrac{\pi}{12} - \cos \dfrac{5\pi}{12}$

In Exercises 23–30, verify each identity.

23. $\dfrac{\sin 3x - \sin x}{\cos 3x - \cos x} = -\cot 2x$

24. $\dfrac{\sin x + \sin 3x}{\cos x + \cos 3x} = \tan 2x$

25. $\dfrac{\sin 2x + \sin 4x}{\cos 2x + \cos 4x} = \tan 3x$

26. $\dfrac{\cos 4x - \cos 2x}{\sin 2x - \sin 4x} = \tan 3x$

27. $\dfrac{\sin x - \sin y}{\sin x + \sin y} = \tan \dfrac{x - y}{2} \cot \dfrac{x + y}{2}$

28. $\dfrac{\sin x + \sin y}{\sin x - \sin y} = \tan \dfrac{x + y}{2} \cot \dfrac{x - y}{2}$

29. $\dfrac{\sin x + \sin y}{\cos x + \cos y} = \tan \dfrac{x + y}{2}$

30. $\dfrac{\sin x - \sin y}{\cos x - \cos y} = -\cot \dfrac{x + y}{2}$

In Exercises 31–34, the graph with the given equation is shown in a $\left[0, 2\pi, \dfrac{\pi}{2} \right]$ by $[-2, 2, 1]$ viewing rectangle.

a. *Describe the graph using another equation.*

b. *Verify that the two equations are equivalent.*

31.

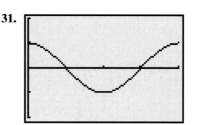

$$y = \frac{\sin x + \sin 3x}{2 \sin 2x}$$

32.

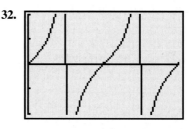

$$y = \frac{\cos x - \cos 3x}{\sin x + \sin 3x}$$

33.

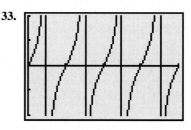

$$y = \frac{\cos x - \cos 5x}{\sin x + \sin 5x}$$

34.

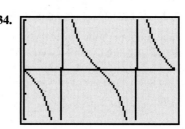

$$y = \frac{\cos 5x - \cos 3x}{\sin 5x + \sin 3x}$$

Application Exercises

Use this information to solve Exercises 35–36. The sound produced by touching each button on a touch-tone phone is described by

$$y = \sin 2\pi l t + \sin 2\pi h t,$$

where l and h are the low and high frequencies in the figure shown. For example, what sound is produced by touching 5? The low frequency is l = 770 cycles per second and the high frequency is h = 1336 cycles per second. The sound produced by touching 5 is described by

$$y = \sin 2\pi(770)t + \sin 2\pi(1336)t.$$

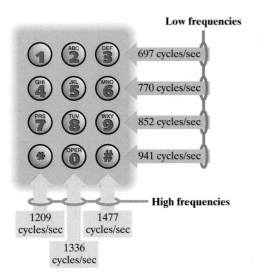

35. The touch-tone phone sequence for that most naive of melodies is given as follows:

Mary Had A Little Lamb

3212333,222,399,3212333322321.

a. Many numbers do not appear in this sequence, including 7. If you accidently touch 7 for one of the notes, describe this sound as the sum of sines.

b. Describe this accidental sound as a product of sines and cosines.

36. The touch-tone phone sequence for *Jingle Bells* is given as follows:

Jingle Bells

333,333,39123,666-663333322329,333,333,39123,666-6633,399621.

a. The first six notes of the song are produced by repeatedly touching 3. Describe this repeated sound as the sum of sines.

b. Describe the repeated sound as a product of sines and cosines.

Writing in Mathematics

In Exercises 37–40, use words to describe the given formula.

37. $\sin \alpha \sin \beta = \frac{1}{2}\left[\cos(\alpha - \beta) - \cos(\alpha + \beta)\right]$

38. $\cos \alpha \cos \beta = \frac{1}{2}\left[\cos(\alpha - \beta) + \cos(\alpha + \beta)\right]$

39. $\sin \alpha + \sin \beta = 2 \sin \dfrac{\alpha + \beta}{2} \cos \dfrac{\alpha - \beta}{2}$

40. $\cos \alpha + \cos \beta = 2 \cos \dfrac{\alpha + \beta}{2} \cos \dfrac{\alpha - \beta}{2}$

41. Describe identities that can be verified using the sum-to-product formulas.

42. Why do the sounds produced by touching each button on a touch-tone phone have the same loudness? Answer the question using the equation described for Exercises 35 and 36, $y = \sin 2\pi l t + \sin 2\pi h t$, and determine the maximum value of y for each sound.

Technology Exercises

In Exercises 43–46, graph each side of the equation in the same viewing rectangle. If the graphs appear to coincide, verify that the equation is an identity. If the graphs do not appear to coincide, find a value of x for which both sides are defined but not equal.

43. $\sin x + \sin 2x = \sin 3x$

44. $\cos x + \cos 2x = \cos 3x$

45. $\sin x + \sin 3x = 2 \sin 2x \cos x$

46. $\cos x + \cos 3x = 2 \cos 2x \cos x$

47. In Exercise 35(a), you wrote an equation for the sound produced by touching 7 on a touch-tone phone. Graph the equation in a $[0, 0.01, 0.001]$ by $[-2, 2, 1]$ viewing rectangle.

48. In Exercise 36(a), you wrote an equation for the sound produced by touching 3 on a touch-tone phone. Graph the equation in a $[0, 0.01, 0.001]$ by $[-2, 2, 1]$ viewing rectangle.

49. In this section, we saw how sums could be expressed as products. Sums of trigonometric functions can also be used to describe functions that are not trigonometric. French mathematician Jean Fourier (1768–1830) showed that *any function* can be described by a series of trigonometric functions. For example, the basic linear function $f(x) = x$ can also be represented by

$$f(x) = 2\left(\frac{\sin x}{1} - \frac{\sin 2x}{2} + \frac{\sin 3x}{3} - \frac{\sin 4x}{4} + \cdots\right).$$

a. Graph

$$y = 2\left(\frac{\sin x}{1}\right),$$

$$y = 2\left(\frac{\sin x}{1} - \frac{\sin 2x}{2}\right),$$

$$y = 2\left(\frac{\sin x}{1} - \frac{\sin 2x}{2} + \frac{\sin 3x}{3}\right)$$

and

$$y = 2\left(\frac{\sin x}{1} - \frac{\sin 2x}{2} + \frac{\sin 3x}{3} - \frac{\sin 4x}{4}\right)$$

in a $\left[-\pi, \pi, \frac{\pi}{2}\right]$ by $[-3, 3, 1]$ viewing rectangle. What patterns do you observe?

b. Graph

$$y = 2\left(\frac{\sin x}{1} - \frac{\sin 2x}{2} + \frac{\sin 3x}{3} - \frac{\sin 4x}{4} + \frac{\sin 5x}{5} - \frac{\sin 6x}{6}\right.$$
$$\left. + \frac{\sin 7x}{7} - \frac{\sin 8x}{8} + \frac{\sin 9x}{9} - \frac{\sin 10x}{10}\right)$$

in a $\left[-\pi, \pi, \frac{\pi}{2}\right]$ by $[-3, 3, 1]$ viewing rectangle. Is a portion of the graph beginning to look like the graph of $f(x) = x$? Obtain a better approximation for the line by graphing functions that contain more and more terms involving sines of multiple angles.

c. Use

$$x = 2\left(\frac{\sin x}{1} - \frac{\sin 2x}{2} + \frac{\sin 3x}{3} - \frac{\sin 4x}{4} + \cdots\right)$$

and substitute $\frac{\pi}{2}$ for x to obtain a formula for $\frac{\pi}{2}$. Show at least four nonzero terms. Then multiply both sides of your formula by 2 to write a nonending series of subtractions and additions that approaches π. Use this series to obtain an approximation for π that is more accurate than the one given by your graphing utility.

Critical Thinking Exercises

Use the identities for sin (α + β) and sin (α − β) to solve Exercises 50–51.

50. Add the left and right sides of the identities and derive the product-to-sum formula for $\sin \alpha \cos \beta$.

51. Subtract the left and right sides of the identities and derive the product-to-sum formula for $\cos \alpha \sin \beta$.

In Exercises 52–53, verify the given sum-to-product formula. Start with the right side and obtain the expression on the left side by using an appropriate product-to-sum formula.

52. $\sin \alpha - \sin \beta = 2 \sin \dfrac{\alpha - \beta}{2} \cos \dfrac{\alpha + \beta}{2}$

53. $\cos \alpha + \cos \beta = 2 \cos \dfrac{\alpha + \beta}{2} \cos \dfrac{\alpha - \beta}{2}$

In Exercises 54–55, verify each identity.

54. $\dfrac{\sin 2x + (\sin 3x + \sin x)}{\cos 2x + (\cos 3x + \cos x)} = \tan 2x$

55. $4 \cos x \cos 2x \sin 3x = \sin 2x + \sin 4x + \sin 6x$

Group Exercise

56. This activity should result in an unusual group display entitled "*Frere Jacques*, a New Perspective." Here is the touch-tone phone sequence:

Frere Jacques

4564,4564,69#,69#,#*#964,#*#964,414,414.

Group members should write every sound in the sequence as both the sum of sines and the product of sines and cosines. Use the sum of sines form and a graphing utility with a $[0, 0.01, 0.001]$ by $[-2, 2, 1]$ viewing rectangle to obtain a graph for every sound. Download these graphs. Use the graphs and equations to create your display in such a way that adults find the trigonometry of this naive melody interesting.

SECTION 5.5 *Trigonometric Equations*

Objectives

1. Find all solutions of a trigonometric equation.
2. Solve equations with multiple angles.
3. Solve trigonometric equations quadratic in form.
4. Use factoring to separate different functions in trigonometric equations.
5. Use identities to solve trigonometric equations.

Exponential functions display the manic energies of uncontrolled growth. By contrast, trigonometric functions repeat their behavior. Do they embody in their regularity some basic rhythm of the universe? The cycles of periodic phenomena provide events that we can comfortably count on. When will the moon look just as it does at this moment? When can I count on 13.5 hours of daylight? When will my breathing be exactly as it is right now? Models with trigonometric functions embrace the periodic rhythms of our world. Equations containing trigonometric functions are used to answer questions about these models.

Trigonometric Equations and Their Solutions

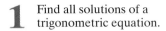

Find all solutions of a trigonometric equation.

A **trigonometric equation** is an equation that contains a trigonometric expression with a variable, such as $\sin x$. We have seen that some trigonometric equations are identities, such as $\sin^2 x + \cos^2 x = 1$. These equations are true for every value of the variable for which the expressions are defined. In this section, we consider trigonometric equations that are true for only some values of the variable. The values that satisfy the equation are its **solutions.** (There are trigonometric equations that have no solution.)

An example of a trigonometric equation is

$$\sin x = \tfrac{1}{2}.$$

A solution of this equation is $\frac{\pi}{6}$ because $\sin \frac{\pi}{6} = \frac{1}{2}$. By contrast, π is not a solution because $\sin \pi = 0 \neq \frac{1}{2}$.

Is $\frac{\pi}{6}$ the only solution of $\sin x = \frac{1}{2}$? The answer is no. Because of the periodic nature of the sine function, there are infinitely many values of x for which $\sin x = \frac{1}{2}$. Figure 5.7 shows five of the solutions, including $\frac{\pi}{6}$, for $-\frac{3\pi}{2} \leq x \leq \frac{7\pi}{2}$. Notice that the x-coordinates of the points where the graph of $y = \sin x$ intersects the line $y = \frac{1}{2}$ are the solutions of the equation $\sin x = \frac{1}{2}$.

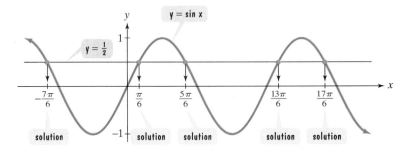

Figure 5.7 The equation $\sin x = \frac{1}{2}$ has five solutions when x is restricted to the interval $\left[-\frac{3\pi}{2}, \frac{7\pi}{2} \right]$

How do we represent all solutions of $\sin x = \frac{1}{2}$? Because the period of the sine function is 2π, first find all solutions in $[0, 2\pi)$. The solutions are

$$x = \frac{\pi}{6} \quad \text{and} \quad x = \pi - \frac{\pi}{6} = \frac{5\pi}{6}.$$

> The sine is positive in quadrants I and II.

Any multiple of 2π can be added to these values and the sine is still $\frac{1}{2}$. Thus, all solutions of $\sin x = \frac{1}{2}$ are given by

$$x = \frac{\pi}{6} + 2n\pi \quad \text{or} \quad x = \frac{5\pi}{6} + 2n\pi$$

where n is any integer. By choosing any two integers, such as $n = 0$ and $n = 1$, we can find some solutions of $\sin x = \frac{1}{2}$. Thus, four of the solutions are:

> Let $n = 0$.

$$x = \frac{\pi}{6} + 2 \cdot 0\pi \qquad x = \frac{5\pi}{6} + 2 \cdot 0\pi$$

$$= \frac{\pi}{6} \qquad\qquad = \frac{5\pi}{6}$$

> Let $n = 1$.

$$x = \frac{\pi}{6} + 2 \cdot 1\pi \qquad x = \frac{5\pi}{6} + 2 \cdot 1\pi$$

$$= \frac{\pi}{6} + 2\pi \qquad\qquad = \frac{5\pi}{6} + 2\pi$$

$$= \frac{\pi}{6} + \frac{12\pi}{6} = \frac{13\pi}{6} \qquad = \frac{5\pi}{6} + \frac{12\pi}{6} = \frac{17\pi}{6}.$$

These four solutions are shown among the five solutions in Figure 5.7.

Equations Involving a Single Trigonometric Function

To solve an equation containing a single trigonometric function:

- Isolate the function on one side of the equation.
- Solve for the variable.

EXAMPLE 1 Finding All Solutions of a Trigonometric Equation

Solve the equation: $3 \sin x - 2 = 5 \sin x - 1$.

Solution The equation contains a single trigonometric function, $\sin x$.

Step 1 Isolate the function on one side of the equation. We can solve for $\sin x$ by collecting terms with $\sin x$ on the left side and constant terms on the right side.

$3 \sin x - 2 = 5 \sin x - 1$	This is the given equation.
$3 \sin x - 5 \sin x - 2 = 5 \sin x - 5 \sin x - 1$	Subtract 5 sin x from both sides.
$-2 \sin x - 2 = -1$	Simplify.
$-2 \sin x = 1$	Add 2 to both sides.
$\sin x = -\frac{1}{2}$	Divide both sides by −2 and solve for sin x.

Urban Canyons

A city's tall buildings and narrow streets reduce the amount of sunlight. If h is the average height of the buildings and w is the width of the street, the angle of elevation from the street to the top of the buildings is given by the trigonometric equation

$$\tan \theta = \frac{h}{w}.$$

A value of $\theta = 63°$ can result in an 85% loss of illumination.

Step 2 Solve for the variable. We must solve for x in $\sin x = -\dfrac{1}{2}$. Because $\sin \dfrac{\pi}{6} = \dfrac{1}{2}$, the solutions of $\sin x = -\dfrac{1}{2}$ in $[0, 2\pi)$ are

$$x = \pi + \frac{\pi}{6} = \frac{6\pi}{6} + \frac{\pi}{6} = \frac{7\pi}{6} \qquad x = 2\pi - \frac{\pi}{6} = \frac{12\pi}{6} - \frac{\pi}{6} = \frac{11\pi}{6}.$$

> The sine is negative
> in quadrant III.

> The sine is negative
> in quadrant IV.

Because the period of the sine function is 2π, the solutions of the equation are given by

$$x = \frac{7\pi}{6} + 2n\pi \quad \text{and} \quad x = \frac{11\pi}{6} + 2n\pi$$

where n is any integer.

Check Point 1 Solve the equation: $5 \sin x = 3 \sin x + \sqrt{3}$.

Now we will concentrate on finding solutions of trigonometric equations for $0 \le x < 2\pi$. You can use a graphing utility to check the solutions of these equations. Graph the left side and graph the right side. The solutions are the x-coordinates of the points where the graphs intersect.

2 Solve equations with multiple angles.

Equations Involving Multiple Angles

Here are examples of two equations that include multiple angles:

$$\tan 3x = 1 \qquad \sin \frac{x}{2} = \frac{\sqrt{3}}{2}.$$

> This angle is a
> multiple of 3.

> This angle is a
> multiple of $\frac{1}{2}$.

We will solve each equation for $0 \le x < 2\pi$. The period of the function plays an important role in ensuring that we do not leave out any solutions.

EXAMPLE 2 Solving an Equation with a Multiple Angle

Solve the equation: $\tan 3x = 1, \quad 0 \le x < 2\pi$.

Solution The period of the tangent function is π. In the interval $[0, \pi)$, the only value for which the tangent function is 1 is $\dfrac{\pi}{4}$. This means that $3x = \dfrac{\pi}{4}$. Because the period is π, all the solutions to $\tan 3x = 1$ are given by

$$3x = \frac{\pi}{4} + n\pi \qquad \text{\textit{n is any integer.}}$$

$$x = \frac{\pi}{12} + \frac{n\pi}{3} \qquad \text{\textit{Divide both sides by 3 and solve for x.}}$$

Technology

Shown below are the graphs of

$$y = \tan 3x$$

and

$$y = 1$$

in a $\left[0, 2\pi, \dfrac{\pi}{2}\right]$ by $[-3, 3, 1]$ viewing rectangle. The solutions of

$$\tan 3x = 1$$

in $[0, 2\pi)$ are shown by the x-coordinates of the six intersection points.

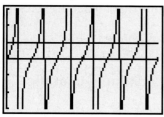

In the interval $[0, 2\pi)$, we obtain the solutions of $\tan 3x = 1$ as follows:

Let $n = 0$.

$$x = \dfrac{\pi}{12} + \dfrac{0\pi}{3}$$

$$= \dfrac{\pi}{12}$$

Let $n = 1$.

$$x = \dfrac{\pi}{12} + \dfrac{1\pi}{3}$$

$$= \dfrac{\pi}{12} + \dfrac{4\pi}{12} = \dfrac{5\pi}{12}$$

Let $n = 2$.

$$x = \dfrac{\pi}{12} + \dfrac{2\pi}{3}$$

$$= \dfrac{\pi}{12} + \dfrac{8\pi}{12} = \dfrac{9\pi}{12} = \dfrac{3\pi}{4}$$

Let $n = 3$.

$$x = \dfrac{\pi}{12} + \dfrac{3\pi}{3}$$

$$= \dfrac{\pi}{12} + \dfrac{12\pi}{12} = \dfrac{13\pi}{12}$$

Let $n = 4$.

$$x = \dfrac{\pi}{12} + \dfrac{4\pi}{3}$$

$$= \dfrac{\pi}{12} + \dfrac{16\pi}{12} = \dfrac{17\pi}{12}$$

Let $n = 5$.

$$x = \dfrac{\pi}{12} + \dfrac{5\pi}{3}$$

$$= \dfrac{\pi}{12} + \dfrac{20\pi}{12} = \dfrac{21\pi}{12} = \dfrac{7\pi}{4}.$$

If you let $n = 6$, you will obtain $x = \dfrac{25\pi}{12}$. This value exceeds 2π. In the interval $[0, 2\pi)$, the solutions of $\tan 3x = 1$ are $\dfrac{\pi}{12}, \dfrac{5\pi}{12}, \dfrac{3\pi}{4}, \dfrac{13\pi}{12}, \dfrac{17\pi}{12}$, and $\dfrac{7\pi}{4}$. These solutions are illustrated by the six intersection points in the technology box.

Check Point 2   Solve the equation: $\tan 2x = \sqrt{3}, 0 \leq x < 2\pi$.

EXAMPLE 3   Solving an Equation with a Multiple Angle

Solve the equation: $\sin \dfrac{x}{2} = \dfrac{\sqrt{3}}{2},\ 0 \leq x < 2\pi$.

Solution   The period of the sine function is 2π. In the interval $[0, 2\pi)$, there are two values at which the sine function is $\dfrac{\sqrt{3}}{2}$. One of these values is $\dfrac{\pi}{3}$. The sine is positive in quadrant II; thus, the other value is $\pi - \dfrac{\pi}{3}$, or $\dfrac{2\pi}{3}$. This means that $\dfrac{x}{2} = \dfrac{\pi}{3}$ or $\dfrac{x}{2} = \dfrac{2\pi}{3}$. Because the period is 2π, all the solutions of $\sin \dfrac{x}{2} = \dfrac{\sqrt{3}}{2}$ are given by

$$\dfrac{x}{2} = \dfrac{\pi}{3} + 2n\pi \quad \text{or} \quad \dfrac{x}{2} = \dfrac{2\pi}{3} + 2n\pi. \quad \text{\small n is any integer.}$$

$$x = \dfrac{2\pi}{3} + 4n\pi \qquad x = \dfrac{4\pi}{3} + 4n\pi. \quad \text{\small Multiply both sides by 2 and solve for x.}$$

If $n = 0$, we obtain $x = \dfrac{2\pi}{3}$ from the equation on the left and $x = \dfrac{4\pi}{3}$ from the equation on the right. If we let $n = 1$, we are adding $4 \cdot 1 \cdot \pi$, or 4π, to each of these expressions. These values of x exceed 2π. Thus, in the interval $[0, 2\pi)$, the only solutions of $\sin \dfrac{x}{2} = \dfrac{\sqrt{3}}{2}$ are $\dfrac{2\pi}{3}$ and $\dfrac{4\pi}{3}$.

> **Check Point 3**
>
> Solve the equation: $\sin \dfrac{x}{3} = \dfrac{1}{2}, 0 \le x < 2\pi.$

③ Solve trigonometric equations quadratic in form.

Trigonometric Equations Quadratic in Form

Some trigonometric equations are in the form of a quadratic equation $at^2 + bt + c = 0$, where t is a trigonometric function. Here are two examples of trigonometric equations that are quadratic in form:

$$2\cos^2 x + \cos x - 1 = 0 \qquad 2\sin^2 x - 3\sin x + 1 = 0.$$

> The form of this equation is $2t^2 + t - 1 = 0$ with $t = \cos x.$

> The form of this equation is $2t^2 - 3t + 1 = 0$ with $t = \sin x.$

To solve this kind of equation, try using factoring. If the trigonometric expression does not factor, use the quadratic formula.

EXAMPLE 4   Solving a Trigonometric Equation Quadratic in Form

Solve the equation: $2\cos^2 x + \cos x - 1 = 0, \quad 0 \le x < 2\pi.$

Solution The given equation is in quadratic form $2t^2 + t - 1 = 0$ with $t = \cos x.$ Let us attempt to solve the equation by factoring.

$$2\cos^2 x + \cos x - 1 = 0 \qquad \text{This is the given equation.}$$

$$(2\cos x - 1)(\cos x + 1) = 0 \qquad \text{Factor. Notice that } 2t^2 + t - 1 \text{ factors as } (2t - 1)(t + 1)$$

$$2\cos x - 1 = 0 \quad \text{or} \quad \cos x + 1 = 0 \qquad \text{Set each factor equal to 0.}$$

$$2\cos x = 1 \qquad\qquad \cos x = -1 \qquad \text{Solve for cos x.}$$

$$\cos x = \tfrac{1}{2}$$

$$x = \frac{\pi}{3} \qquad x = 2\pi - \frac{\pi}{3} = \frac{5\pi}{3} \qquad x = \pi \qquad \begin{array}{l}\text{Solve each equation for x,}\\ 0 \le x < 2\pi.\end{array}$$

> The cosine is positive in quadrants I and IV.

The solutions in the interval $[0, 2\pi)$ are $\dfrac{\pi}{3}$, π, and $\dfrac{5\pi}{3}$.

Technology

The graph of

$$y = 2\cos^2 x + \cos x - 1$$

is shown in a

$$\left[0, 2\pi, \frac{\pi}{2}\right] \text{ by } [-3, 3, 1]$$

viewing rectangle. The x-intercepts,

$$\frac{\pi}{3}, \pi, \text{ and } \frac{5\pi}{3},$$

verify the three solutions of

$$2\cos^2 x + \cos x - 1 = 0$$

in $[0, 2\pi).$

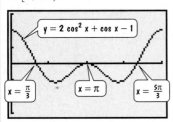

> **Check Point 4**
>
> Solve the equation: $2\sin^2 x - 3\sin x + 1 = 0, 0 \le x < 2\pi.$

4 Use factoring to separate different functions in trigonometric equations.

Using Factoring to Separate Two Different Trigonometric Functions in an Equation

We have seen that factoring is used to solve some trigonometric equations that are quadratic in form. Factoring can also be used to solve some trigonometric equations that contain two different functions such as

$$\tan x \sin^2 x = 3 \tan x.$$

In such a case, move all terms to one side and obtain zero on the other side. Then try to use factoring to separate the different functions. Example 5 shows how this is done.

EXAMPLE 5   Using Factoring to Separate Different Functions

Solve the equation: $\tan x \sin^2 x = 3 \tan x$, $0 \le x < 2\pi$.

Study Tip

In solving

$$\tan x \sin^2 x = 3 \tan x,$$

do not begin by dividing both sides by $\tan x$. Division by zero is undefined. If you divide by $\tan x$, you lose the two solutions for which $\tan x = 0$, namely 0 and π.

Solution
Move all terms to one side and obtain zero on the other side.

$$\tan x \sin^2 x = 3 \tan x \quad \text{This is the given equation.}$$
$$\tan x \sin^2 x - 3 \tan x = 0 \qquad \text{Subtract 3 tan x from both sides.}$$

Use factoring to separate the two functions.

$$\tan x(\sin^2 x - 3) = 0 \qquad \begin{array}{l}\text{Factor out tan x from the two terms}\\ \text{on the left side.}\end{array}$$

$$\tan x = 0 \quad \text{or} \quad \sin^2 x - 3 = 0 \qquad \text{Set each factor equal to 0.}$$
$$x = 0 \quad x = \pi \qquad\qquad \sin^2 x = 3 \qquad \text{Solve for x.}$$
$$\sin x = \pm\sqrt{3}$$

> This equation has no solution because sin x cannot be greater than 1 or less than –1.

The solutions in the interval $[0, 2\pi)$ are 0 and π.

Check Point 5   Solve the equation: $\sin x \tan x = \sin x$, $0 \le x < 2\pi$.

5 Use identities to solve trigonometric equations.

Using Identities to Solve Trigonometric Equations

Some trigonometric equations contain more than one function on the same side and these functions cannot be separated by factoring. For example, consider the equation

$$2 \sin^2 x - \cos x - 1 = 0.$$

How can we obtain an equivalent equation that has only one trigonometric function? We use the identity $\sin^2 x + \cos^2 x = 1$ and substitute $1 - \cos^2$ for $\sin^2 x$.

$$2 \sin^2 x - \cos x - 1 = 0 \qquad \text{This is the given equation.}$$
$$2(1 - \cos^2 x) - \cos x - 1 = 0 \qquad \sin^2 x = 1 - \cos^2 x$$
$$2 - 2 \cos^2 x - \cos x - 1 = 0 \qquad \text{Use the distributive property.}$$
$$-2 \cos^2 x - \cos x + 1 = 0 \qquad \text{Combine like terms.}$$

Multiplying both sides of $-2\cos^2 x - \cos x + 1 = 0$ by -1, we obtain

$$2\cos^2 x + \cos x - 1 = 0.$$

This equivalent equation contains only the cosine function. This is the equation that we solved in Example 4.

EXAMPLE 6 Using an Identity to Solve a Trigonometric Equation

Solve the equation: $\cos 2x + 3\sin x - 2 = 0, \ 0 \le x < 2\pi$.

Solution The given equation contains a cosine function and a sine function. The cosine is a function of $2x$ and the sine is a function of x. We want one trigonometric function of the same angle. This can be accomplished by using the double-angle identity $\cos 2x = 1 - 2\sin^2 x$ to obtain an equivalent equation involving $\sin x$ only.

$$\cos 2x + 3\sin x - 2 = 0 \qquad \text{This is the given equation.}$$
$$1 - 2\sin^2 x + 3\sin x - 2 = 0 \qquad \cos 2x = 1 - 2\sin^2 x$$
$$-2\sin^2 x + 3\sin x - 1 = 0 \qquad \text{Combine like terms.}$$
$$2\sin^2 x - 3\sin x + 1 = 0 \qquad \text{Multiply both sides by } -1.$$

The equation is now in quadratic form $2t^2 - 3t + 1 = 0$ with $t = \sin x$. We solve using factoring.

$$(2\sin x - 1)(\sin x - 1) = 0 \qquad \begin{array}{l}\text{Factor. Notice that} \\ 2t^2 - 3t + 1 \text{ factors as} \\ (2t - 1)(t - 1).\end{array}$$

$$2\sin x - 1 = 0 \qquad \text{or} \qquad \sin x - 1 = 0 \qquad \text{Set each factor equal to 0.}$$
$$\sin x = \tfrac{1}{2} \qquad\qquad\qquad \sin x = 1 \qquad \text{Solve for } \sin x.$$
$$x = \frac{\pi}{6} \quad x = \pi - \frac{\pi}{6} = \frac{5\pi}{6} \qquad x = \frac{\pi}{2} \qquad \begin{array}{l}\text{Solve each equation for } x, \\ 0 \le x < 2\pi.\end{array}$$

> The sine is positive in quadrants I and II.

The solutions in the interval $[0, 2\pi)$ are $\dfrac{\pi}{6}, \dfrac{\pi}{2}$, and $\dfrac{5\pi}{6}$.

Check Point 6 Solve the equation: $\cos 2x + \sin x = 0, \ 0 \le x < 2\pi$.

Sometimes it is necessary to do something to both sides of a trigonometric equation to substitute an identity. For example, consider the equation

$$\sin x \cos x = \tfrac{1}{2}.$$

This equation contains both a sine and a cosine function. How can we obtain a single function? Multiply both sides by 2. In this way, we can use the double-angle identity $\sin 2x = 2\sin x \cos x$ and obtain $\sin 2x$, a single function, on the left side.

Technology

Shown below are the graphs of

$$y = \sin x \cos x$$

and

$$y = \tfrac{1}{2}$$

in a $\left[0, 2\pi, \dfrac{\pi}{2}\right]$ by $[-1, 1, 1]$ viewing rectangle.

The solutions of

$$\sin x \cos x = \tfrac{1}{2}$$

are shown by the x-coordinates of the two intersection points.

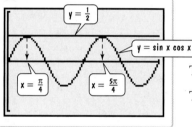

EXAMPLE 7 Using an Identity to Solve a Trigonometric Equation

Solve the equation: $\sin x \cos x = \tfrac{1}{2}, \ 0 \le x < 2\pi.$

Solution

$\sin x \cos x = \tfrac{1}{2}$	This is the given equation.
$2 \sin x \cos x = 1$	Multiply both sides by 2 in anticipation of using $\sin 2x = 2 \sin x \cos x$.
$\sin 2x = 1$	Use a double-angle identity.

Notice that we have an equation with $2x$, a multiple angle. The period of the sine function is 2π. In the interval $[0, 2\pi)$, the only value for which the sine function is 1 is $\dfrac{\pi}{2}$. This means that $2x = \dfrac{\pi}{2}$. Because the period is 2π, all the solutions of $\sin 2x = 1$ are given by

$$2x = \dfrac{\pi}{2} + 2n\pi \qquad \text{n is any integer.}$$

$$x = \dfrac{\pi}{4} + n\pi \qquad \text{Divide both sides by 2 and solve for x.}$$

The solutions in the interval $[0, 2\pi)$ are obtained by letting $n = 0$ and $n = 1$. The solutions are $\dfrac{\pi}{4}$ and $\dfrac{5\pi}{4}$.

Check Point 7 Solve the equation: $\sin x \cos x = -\tfrac{1}{2}, \ 0 \le x < 2\pi.$

Let's look at another equation that contains two different functions, $\sin x - \cos x = 1$. Can you think of an identity that can be used to produce only one function? Perhaps $\sin^2 x + \cos^2 x = 1$ might be helpful. The next example shows how we use this identity by squaring both sides of the given equation. However, if we raise both sides of an equation to an even power, we have the possibility of introducing extra solutions. Such solutions, which are *not solutions of the given equation*, are called **extraneous solutions**. Thus, we must check each proposed solution in the given equation. Alternatively, we can use a graphing utility to verify actual solutions.

Study Tip

When solving equations by raising both sides to an even power, don't forget to check for extraneous solutions. Here is a simple example:

$x = 4$	
$x^2 = 16$	Square both sides.
$x = \pm\sqrt{16}$	Use the square root method.
$x = \pm 4.$	

However, -4 does not check in $x = 4$. Thus, -4 is an extraneous solution.

EXAMPLE 8 Using an Identity to Solve a Trigonometric Equation

Solve the equation: $\sin x - \cos x = 1, \ 0 \le x < 2\pi.$

Solution We square both sides of the equation in anticipation of using $\sin^2 x + \cos^2 x = 1$.

$\sin x - \cos x = 1$	This is the given equation.
$(\sin x - \cos x)^2 = 1^2$	Square both sides.
$\sin^2 x - 2 \sin x \cos x + \cos^2 x = 1$	Square the left side using $(A - B)^2 = A^2 - 2AB + B^2$.

Technology

A graphing utility can be used instead of the algebraic check on the right. Shown are the graphs of

$$y = \sin x - \cos x$$

and

$$y = 1$$

in a $\left[0, 2\pi, \dfrac{\pi}{2}\right]$ by $[-2, 2, 1]$ viewing rectangle. The actual solutions of

$$\sin x - \cos x = 1$$

are shown by the x-coordinates of the two intersection points, $\dfrac{\pi}{2}$ and π.

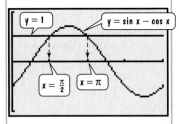

$\sin^2 x - 2 \sin x \cos x + \cos^2 x = 1$	We have repeated the equation from the previous page.
$\sin^2 x + \cos^2 x - 2 \sin x \cos x = 1$	Rearrange terms.
$1 - 2 \sin x \cos x = 1$	Apply a Pythagorean identity: $\sin^2 x + \cos^2 x = 1$.
$-2 \sin x \cos x = 0$	Subtract 1 from both sides of the equation.
$\sin x \cos x = 0$	Divide both sides of the equation by -2.
$\sin x = 0 \quad$ or $\quad \cos x = 0$	Set each factor equal to 0.
$x = 0 \quad x = \pi \quad x = \dfrac{\pi}{2} \quad x = \dfrac{3\pi}{2}$	Solve for x in $[0, 2\pi)$.

We check these proposed solutions to see if any are extraneous.

Check 0:

$\sin x - \cos x = 1$

$\sin 0 - \cos 0 \overset{?}{=} 1$

$0 - 1 \overset{?}{=} 1$

$-1 = 1$, False

Check $\dfrac{\pi}{2}$:

$\sin x - \cos x = 1$

$\sin \dfrac{\pi}{2} - \cos \dfrac{\pi}{2} \overset{?}{=} 1$

$1 - 0 \overset{?}{=} 1$

$1 = 1$, True

Check π:

$\sin x - \cos x = 1$

$\sin \pi - \cos \pi \overset{?}{=} 1$

$0 - (-1) \overset{?}{=} 1$

$1 = 1$, True

Check $\dfrac{3\pi}{2}$:

$\sin x - \cos x = 1$

$\sin \dfrac{3\pi}{2} - \cos \dfrac{3\pi}{2} \overset{?}{=} 1$

$-1 - 0 \overset{?}{=} 1$

$-1 = 1$, F

0 and $\dfrac{3\pi}{2}$ are extraneous.

The false statement $-1 = 1$ indicates 0 and $\dfrac{3\pi}{2}$ are not solutions. They are extraneous solutions brought about by squaring both sides of the given equation. The actual solutions in the interval $[0, 2\pi)$ are $\dfrac{\pi}{2}$ and π.

Check Point 8   Solve the equation: $\cos x - \sin x = -1$, $0 \leq x < 2\pi$.

EXERCISE SET 5.5

Practice Exercises

In Exercises 1–10, use substitution to determine whether the given x-value is a solution of the equation.

1. $\cos x = \dfrac{\sqrt{2}}{2}, \quad x = \dfrac{\pi}{4}$

2. $\tan x = \sqrt{3}, \quad x = \dfrac{\pi}{3}$

3. $\sin x = \dfrac{\sqrt{3}}{2}, \quad x = \dfrac{\pi}{6}$

4. $\sin x = \dfrac{\sqrt{2}}{2}, \quad x = \dfrac{\pi}{3}$

5. $\cos x = -\dfrac{1}{2}, \quad x = \dfrac{2\pi}{3}$

6. $\cos x = -\dfrac{1}{2}, \quad x = \dfrac{4\pi}{3}$

7. $\tan 2x = -\dfrac{\sqrt{3}}{3}, \quad x = \dfrac{5\pi}{12}$

8. $\cos \dfrac{2x}{3} = -\dfrac{1}{2}, \quad x = \pi$ **9.** $\cos x = \sin 2x, x = \dfrac{\pi}{3}$

10. $\cos x + 2 = \sqrt{3} \sin x, \quad x = \dfrac{\pi}{6}$

In Exercises 11–24, find all solutions of each equation.

11. $\sin x = \dfrac{\sqrt{3}}{2}$

12. $\cos x = \dfrac{\sqrt{3}}{2}$

13. $\tan x = 1$

14. $\tan x = \sqrt{3}$

15. $\cos x = -\dfrac{1}{2}$

16. $\sin x = -\dfrac{\sqrt{2}}{2}$

17. $\tan x = 0$

18. $\sin x = 0$

19. $2 \cos x + \sqrt{3} = 0$

20. $2 \sin x + \sqrt{3} = 0$

21. $4 \sin \theta - 1 = 2 \sin \theta$

22. $5 \sin \theta + 1 = 3 \sin \theta$

23. $3 \sin \theta + 5 = -2 \sin \theta$

24. $7 \cos \theta + 9 = -2 \cos \theta$

Exercises 25–38 involve equations with multiple angles. Solve each equation on the interval $[0, 2\pi)$.

25. $\sin 2x = \dfrac{\sqrt{3}}{2}$

26. $\cos 2x = \dfrac{\sqrt{2}}{2}$

27. $\cos 4x = -\dfrac{\sqrt{3}}{2}$

28. $\sin 4x = -\dfrac{\sqrt{2}}{2}$

29. $\tan 3x = \dfrac{\sqrt{3}}{3}$

30. $\tan 3x = \sqrt{3}$

31. $\tan \dfrac{x}{2} = \sqrt{3}$

32. $\tan \dfrac{x}{2} = \dfrac{\sqrt{3}}{3}$

33. $\sin \dfrac{2\theta}{3} = -1$

34. $\cos \dfrac{2\theta}{3} = -1$

35. $\sec \dfrac{3\theta}{2} = -2$

36. $\cot \dfrac{3\theta}{2} = -\sqrt{3}$

37. $\sin \left(2x + \dfrac{\pi}{6} \right) = \dfrac{1}{2}$

38. $\sin \left(2x - \dfrac{\pi}{4} \right) = \dfrac{\sqrt{2}}{2}$

Exercises 39–46 involve trigonometric equations quadratic in form. Solve each equation on the interval $[0, 2\pi)$.

39. $2 \sin^2 x - \sin x - 1 = 0$

40. $2 \sin^2 x + \sin x - 1 = 0$

41. $2 \cos^2 x + 3 \cos x + 1 = 0$

42. $\cos^2 x + 2 \cos x - 3 = 0$

43. $2 \sin^2 x = \sin x + 3$ **44.** $2 \sin^2 x = 4 \sin x + 6$

45. $\sin^2 \theta - 1 = 0$ **46.** $\cos^2 \theta - 1 = 0$

In Exercises 47–56, solve each equation on the interval $[0, 2\pi)$.

47. $(\tan x - 1)(\cos x + 1) = 0$

48. $(\tan x + 1)(\sin x - 1) = 0$

49. $(2 \cos x + \sqrt{3})(2 \sin x + 1) = 0$

50. $(2 \cos x - \sqrt{3})(2 \sin x - 1) = 0$

51. $\cot x(\tan x - 1) = 0$ **52.** $\cot x(\tan x + 1) = 0$

53. $\sin x + 2 \sin x \cos x = 0$ **54.** $\cos x - 2 \sin x \cos x = 0$

55. $\tan^2 x \cos x = \tan^2 x$ **56.** $\cot^2 x \sin x = \cot^2 x$

In Exercises 57–78, use an identity to solve each equation on the interval $[0, 2\pi)$.

57. $2 \cos^2 x + \sin x - 1 = 0$

58. $2 \cos^2 x - \sin x - 1 = 0$

59. $\sin^2 x - 2 \cos x - 2 = 0$

60. $4 \sin^2 x + 4 \cos x - 5 = 0$

61. $4 \cos^2 x = 5 - 4 \sin x$ **62.** $3 \cos^2 x = \sin^2 x$

63. $\sin 2x = \cos x$ **64.** $\sin 2x = \sin x$

65. $\cos 2x = \cos x$ **66.** $\cos 2x = \sin x$

67. $\cos 2x + 5 \cos x + 3 = 0$

68. $\cos 2x + \cos x + 1 = 0$

69. $\sin x \cos x = \dfrac{\sqrt{2}}{4}$ **70.** $\sin x \cos x = \dfrac{\sqrt{3}}{4}$

71. $\sin x + \cos x = 1$ **72.** $\sin x + \cos x = -1$

73. $\sin \left(x + \dfrac{\pi}{4} \right) + \sin \left(x - \dfrac{\pi}{4} \right) = 1$

74. $\sin \left(x + \dfrac{\pi}{3} \right) + \sin \left(x - \dfrac{\pi}{3} \right) = 1$

75. $\sin 2x \cos x + \cos 2x \sin x = \dfrac{\sqrt{2}}{2}$

76. $\sin 3x \cos 2x + \cos 3x \sin 2x = 1$

77. $\tan x + \sec x = 1$ **78.** $\tan x - \sec x = 1$

Application Exercises

Use this information to solve Exercises 79–80. Our cycle of normal breathing takes place every 5 seconds. Velocity of air flow, y, measured in liters per second, after x seconds is modeled by

$$y = 0.6 \sin \dfrac{2\pi}{5} x.$$

Velocity of air flow is positive when we inhale and negative when we exhale.

79. Within each breathing cycle, when are we inhaling at 0.3 liter per second? Round to the nearest tenth of a second.

80. Within each breathing cycle, when are we exhaling at 0.3 liter per second? Round to the nearest tenth of a second.

Use this information to solve Exercises 81–82. The number of hours of daylight in Boston is given by

$$y = 3 \sin \left[\dfrac{2\pi}{365} (x - 79) \right] + 12$$

where x is the number of days after January 1.

81. Within a year, when does Boston have 10.5 hours of daylight? Give your answer in days after January 1 and round to the nearest day.

82. Within a year, when does Boston have 13.5 hours of daylight? Give your answer in days after January 1 and round to the nearest day.

Use this information to solve Exercises 83–84. A ball on a spring is pulled 4 inches below its rest position and then released. After t seconds, the ball's distance, d, in inches from its rest position is given by

$$d = -4 \cos \dfrac{\pi}{3} t.$$

83. Find all values of t for which the ball is 2 inches above its rest position.

84. Find all values of t for which the ball is 2 inches below its rest position.

Use this information to solve Exercises 85–86. When throwing an object, the distance achieved depends on its initial velocity, v_0, and the angle above the horizontal at which the object is thrown, θ. The distance, d, in feet, that describes the range covered is given by

$$d = \frac{v_0^2}{16} \sin \theta \cos \theta$$

where v_0 is measured in feet per second.

85. You and your friend are throwing a baseball back and forth. If you throw the ball with an initial velocity of $v_0 = 90$ feet per second, at what angle of elevation, θ, to the nearest degree, should you direct your throw so that it can be easily caught by your friend located 170 feet away?

86. In Exercise 85, you increase the distance between you and your friend to 200 feet. With this increase, at what angle of elevation, θ, to the nearest degree, should you direct your throw?

Writing in Mathematics

87. What are the solutions of a trigonometric equation?

88. Describe the difference between verifying a trigonometric identity and solving a trigonometric equation.

89. Without actually solving the equation, describe how to solve

$$3 \tan x - 2 = 5 \tan x - 1.$$

90. In the interval $[0, 2\pi)$, the solutions of $\sin x = \cos 2x$ are $\frac{\pi}{6}, \frac{5\pi}{6}$, and $\frac{3\pi}{2}$. Explain how to use graphs generated by a graphing utility to check these solutions.

91. Suppose you are solving equations in the interval $[0, 2\pi)$. Without actually solving equations, what is the difference between the number of solutions of $\sin x = \frac{1}{2}$ and $\sin 2x = \frac{1}{2}$? How do you account for this difference?

In Exercises 92–93, describe a general strategy for solving each equation. Do not solve the equation.

92. $2 \sin^2 x + 5 \sin x + 3 = 0$

93. $\sin 2x = \sin x$

94. Describe a natural periodic phenomenon. Give an example of a question that can be answered by a trigonometric equation in the study of this phenomenon.

95. Some people experience depression with loss of sunlight. Use the essay on page 585 to determine whether such a person should live on a city street that is 80 feet wide with buildings whose heights average 400 feet. Explain your answer and include θ, to the nearest degree, in your argument.

 Technology Exercises

96. Use a graphing utility to verify the solutions of any five equations that you solved in Exercises 57–78.

In Exercises 97–101, use a graphing utility to approximate the solutions of each equation in the interval $[0, 2\pi)$. Round to the nearest hundredth of a radian.

97. $15 \cos^2 x + 7 \cos x - 2 = 0$

98. $\cos x = x$

99. $2 \sin^2 x = 1 - 2 \sin x$

100. $\sin 2x = 2 - x^2$

101. $\sin x + \sin 2x + \sin 3x = 0$

 Critical Thinking Exercises

102. Which one of the following is true?
   a. The equation $(\sin x - 3)(\cos x + 2) = 0$ has no solution.
   b. The equation $\tan x = \frac{\pi}{2}$ has no solution.
   c. A trigonometric equation with an infinite number of solutions is an identity.
   d. The equations $\sin 2x = 1$ and $\sin 2x = \frac{1}{2}$ have the same number of solutions on the interval $[0, 2\pi)$.

In Exercises 103–105, solve each equation on the interval $[0, 2\pi)$. Do not use a calculator.

103. $2 \cos x - 1 + 3 \sec x = 0$

104. $\sin 3x + \sin x + \cos x = 0$

105. $\sin x + 2 \sin \frac{x}{2} = \cos \frac{x}{2} + 1$

CHAPTER SUMMARY, REVIEW, AND TEST

Summary

DEFINITIONS AND CONCEPTS	EXAMPLES
5.1 Verifying Trigonometric Identities	
a. Identities are trigonometric equations that are true for all values of the variable for which the expressions are defined.	Ex. 1, p. 546;
	Exs. 2 & 3, p. 547;
b. Fundamental trigonometric identities are given in the box on page 545.	Ex. 4, p. 548;
c. Guidelines for verifying trigonometric identities are given in the box on page 553.	Ex. 5, p. 549;
	Ex. 6, p. 550;
	Ex. 7, p. 551

DEFINITIONS AND CONCEPTS	EXAMPLES

5.2 and 5.3 Sum, Difference, Double-Angle, and Half-Angle Formulas

a. Sum and difference formulas, double-angle formulas, power-reducing formulas, and half-angle formulas are given in the box on page 573.

Exs. 1–7, pp. 557–562;
Exs. 1–7, pp. 566–572

5.4 Product-to-Sum and Sum-to-Product Formulas

a. The product-to-sum formulas are given in the box on page 576.

Ex. 1, p. 577

b. The sum-to-product formulas are given in the box on page 578. These formulas are useful to verify identities with fractions that contain sums and differences of sines and/or cosines.

Ex. 2, p. 578;
Ex. 3, p. 579

5.5 Trigonometric Equations

a. The values that satisfy a trigonometric equation are its solutions.

Exs. 1–8, pp. 585–591

b. Algebraic techniques such as isolating an expression on one side of the equation and factoring are useful in solving trigonometric equations. Identities are also used to solve some trigonometric equations.

Review Exercises

5.1

In Exercises 1–12, verify each identity.

1. $\sec x - \cos x = \tan x \sin x$

2. $\cos x + \sin x \tan x = \sec x$

3. $\sin^2 \theta (1 + \cot^2 \theta) = 1$

4. $(\sec \theta - 1)(\sec \theta + 1) = \tan^2 \theta$

5. $\dfrac{1}{\sin t - 1} + \dfrac{1}{\sin t + 1} = -2 \tan t \sec t$

6. $\dfrac{1 + \sin t}{\cos^2 t} = \tan^2 t + 1 + \tan t \sec t$

7. $\dfrac{\cos x}{1 - \sin x} = \dfrac{1 + \sin x}{\cos x}$

8. $1 - \dfrac{\cos^2 x}{1 + \sin x} = \sin x$

9. $(\tan \theta + \cot \theta)^2 = \sec^2 \theta + \csc^2 \theta$

10. $\dfrac{1}{\sin \theta + \cos \theta} + \dfrac{1}{\sin \theta - \cos \theta} = \dfrac{2 \sin \theta}{\sin^4 \theta - \cos^4 \theta}$

11. $\dfrac{\cos t}{\cot t - 5 \cos t} = \dfrac{1}{\csc t - 5}$

12. $\dfrac{1 - \cos t}{1 + \cos t} = (\csc t - \cot t)^2$

5.2 and 5.3

In Exercises 13–18, use a sum or difference formula to find the exact value of each expression.

13. $\cos(45° + 30°)$

14. $\sin 195°$

15. $\tan\left(\dfrac{4\pi}{3} - \dfrac{\pi}{4}\right)$

16. $\tan \dfrac{5\pi}{12}$

17. $\cos 65° \cos 5° + \sin 65° \sin 5°$

18. $\sin 80° \cos 50° - \cos 80° \sin 50°$

In Exercises 19–30, verify each identity.

19. $\sin\left(x + \dfrac{\pi}{6}\right) - \cos\left(x + \dfrac{\pi}{3}\right) = \sqrt{3} \sin x$

20. $\tan\left(x + \dfrac{3\pi}{4}\right) = \dfrac{\tan x - 1}{1 + \tan x}$

21. $\sec(\alpha + \beta) = \dfrac{\sec \alpha \sec \beta}{1 - \tan \alpha \tan \beta}$

22. $\dfrac{\cos(\alpha - \beta)}{\cos \alpha \cos \beta} = 1 + \tan \alpha \tan \beta$

23. $\cos^4 t - \sin^4 t = \cos 2t$

24. $\sin t - \cos 2t = (2 \sin t - 1)(\sin t + 1)$

25. $\dfrac{\sin 2\theta - \sin \theta}{\cos 2\theta + \cos \theta} = \dfrac{1 - \cos \theta}{\sin \theta}$

26. $\dfrac{\sin 2\theta}{1 - \sin^2 \theta} = 2 \tan \theta$

27. $\tan 2t = 2 \sin t \cos t \sec 2t$

28. $\cos 4t = 1 - 8 \sin^2 t \cos^2 t$

29. $\tan \dfrac{x}{2} (1 + \cos x) = \sin x$

30. $\tan \dfrac{x}{2} = \dfrac{\sec x - 1}{\tan x}$

In Exercises 31–33, the graph with the given equation is shown on the next page in a $\left[0, 2\pi, \dfrac{\pi}{2}\right]$ by $[-2, 2, 1]$ viewing rectangle.

a. *Describe the graph using another equation.*

b. *Verify that the two equations are equivalent.*

31.

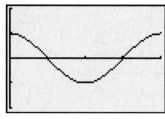

$$y = \sin\left(x - \frac{3\pi}{2}\right)$$

32.

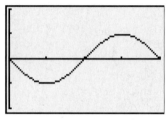

$$y = \cos\left(x + \frac{\pi}{2}\right)$$

33.

$$y = \frac{\tan x - 1}{1 - \cot x}$$

In Exercises 34–37, find the exact value of the following under the given conditions:

 a. $\sin(\alpha + \beta)$ **b.** $\cos(\alpha - \beta)$ **c.** $\tan(\alpha + \beta)$

 d. $\sin 2\alpha$ **e.** $\cos\dfrac{\beta}{2}$

34. $\sin \alpha = \frac{3}{5}$, α lies in quadrant I, and $\sin \beta = \frac{12}{13}$, β lies in quadrant II.

35. $\tan \alpha = \frac{4}{3}$, α lies in quadrant III, and $\tan \beta = \frac{5}{12}$, β lies in quadrant I.

36. $\tan \alpha = -3$, α lies in quadrant II, and $\cot \beta = -3$, β lies in quadrant IV.

37. $\sin \alpha = -\frac{1}{3}$, α lies in quadrant III, and $\cos \beta = -\frac{1}{3}$, β lies in quadrant III.

In Exercises 38–41, use double- and half-angle formulas to find the exact value of each expression.

38. $\cos^2 15° - \sin^2 15°$ **39.** $\dfrac{2\tan\dfrac{5\pi}{12}}{1 - \tan^2\dfrac{5\pi}{12}}$

40. $\sin 22.5°$ **41.** $\tan\dfrac{\pi}{12}$

5.4

In Exercises 42–43, express each product as a sum or difference.

42. $\sin 6x \sin 4x$ **43.** $\sin 7x \cos 3x$

In Exercises 44–45, express each sum or difference as a product. If possible, find this product's exact value.

44. $\sin 2x - \sin 4x$ **45.** $\cos 75° + \cos 15°$

In Exercises 46–47, verify each identity.

46. $\dfrac{\cos 3x + \cos 5x}{\cos 3x - \cos 5x} = \cot x \cot 4x$

47. $\dfrac{\sin 2x + \sin 6x}{\sin 2x - \sin 6x} = -\tan 4x \cot 2x$

48. The graph with the given equation is shown in a $\left[0, 2\pi, \dfrac{\pi}{2}\right]$ by $[-2, 2, 1]$ viewing rectangle.

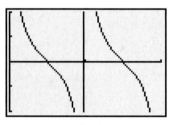

$$y = \frac{\cos 3x + \cos x}{\sin 3x - \sin x}$$

 a. Describe the graph using another equation.

 b. Verify that the two equations are equivalent.

5.5

In Exercises 49–52, find all solutions of each equation.

49. $\cos x = -\dfrac{1}{2}$ **50.** $\sin x = \dfrac{\sqrt{2}}{2}$

51. $2\sin x + 1 = 0$ **52.** $\sqrt{3}\tan x - 1 = 0$

In Exercises 53–62, solve each equation on the interval $[0, 2\pi)$.

53. $\cos 2x = -1$ **54.** $\sin 3x = 1$

55. $\tan\dfrac{x}{2} = -1$ **56.** $\tan x = 2\cos x \tan x$

57. $\cos^2 x - 2\cos x = 3$ **58.** $2\cos^2 x - \sin x = 1$

59. $4\sin^2 x = 1$ **60.** $\cos 2x - \sin x = 1$

61. $\sin 2x = \sqrt{3}\sin x$ **62.** $\sin x = \tan x$

63. A ball on a spring is pulled 6 inches below its rest position and then released. After t seconds, the ball's distance, d, in inches from its rest position is given by

$$d = -6\cos\frac{\pi}{2}t.$$

Find all values of t for which the ball is 3 inches below its rest position.

64. You are playing catch with a friend located 100 feet away. If you throw the ball with an initial velocity of $v_0 = 90$ feet per second, at what angle of elevation, θ, to the nearest degree should you direct your throw so that it can be caught easily?

Use the formula

$$d = \frac{v_0^2}{16} \sin \theta \cos \theta.$$

Chapter 5 Test

Use the following conditions to solve Exercises 1–4:

$$\sin \alpha = \tfrac{4}{5}, \alpha \text{ lies in quadrant II}$$

$$\cos \beta = \tfrac{5}{13}, \beta \text{ lies in quadrant I.}$$

Find the exact value of each of the following.

1. $\cos(\alpha + \beta)$ **2.** $\tan(\alpha - \beta)$

3. $\sin 2\alpha$ **4.** $\cos \dfrac{\beta}{2}$

5. Use $105° = 135° - 30°$ to find the exact value of $\sin 105°$.

In Exercises 6–11, verify each identity.

6. $\cos x \csc x = \cot x$ **7.** $\dfrac{\sec x}{\cot x + \tan x} = \sin x$

8. $1 - \dfrac{\cos^2 x}{1 + \sin x} = \sin x$ **9.** $\cos\left(\theta + \dfrac{\pi}{2}\right) = -\sin \theta$

10. $\dfrac{\sin(\alpha - \beta)}{\sin \alpha \cos \beta} = 1 - \cot \alpha \tan \beta$

11. $\sin t \cos t (\tan t + \cot t) = 1$

In Exercises 12–15, solve each equation on the interval $[0, 2\pi)$.

12. $\sin 3x = -\tfrac{1}{2}$ **13.** $\sin 2x + \cos x = 0$

14. $2 \cos^2 x - 3 \cos x + 1 = 0$

15. $2 \sin^2 x + \cos x = 1$

Cumulative Review Exercises (Chapters P–5)

Solve each equation or inequality in Exercises 1–4.

1. $x^3 + x^2 - x + 15 = 0$ **2.** $11^{x-1} = 125$

3. $x^2 + 2x - 8 > 0$

4. $\cos 2x + 3 = 5 \cos x, \quad 0 \le x < 2\pi$

In Exercises 5–10, graph each equation.

5. $y = \sqrt{x + 2} - 1$; Use transformations of the graph of $y = \sqrt{x}$.

6. $(x - 1)^2 + (y + 2)^2 = 9$

7. $y + 2 = \tfrac{1}{3}(x - 1)$

8. $y = 3 \cos 2x, \quad -2\pi \le x \le 2\pi$

9. $y = 2 \sin \dfrac{x}{2} + 1, \quad -2\pi \le x \le 2\pi$

10. $f(x) = (x - 1)^2(x - 3)$

11. If $f(x) = x^2 + 3x - 1$, find $\dfrac{f(a + h) - f(a)}{h}$.

12. Find the exact value of $\sin 225°$.

13. Verify the identity: $\sec^4 x - \sec^2 x = \tan^4 x + \tan^2 x$.

14. Convert $320°$ to radians.

15. How long would it take for any amount of money, compounded continuously at 5.75% per year, to triple? Round to the nearest tenth of a year.

16. If $f(x) = \dfrac{2x + 1}{x - 3}$, find $f^{-1}(x)$.

17. If C is a right angle in triangle ABC with $A = 23°$ and $a = 12$, solve the triangle.

18. A formula for calculating an infant's dosage for medication is

$$\text{Infant's dose} = \frac{\text{age of infant in months}}{150} \times \text{adult dose.}$$

If a 12-month-old infant is to receive 8.5 mg of medication, find the equivalent adult dose to the nearest milligram.

19. From a point on the ground 12 feet from the base of a flagpole, the angle of elevation to the top of the pole is $53°$. Approximate the height of the flagpole to the nearest tenth of a foot.

20. In *A Tour of the Calculus*, David Berlinski describes trigonometric identities in the following way: "An invisible inner connection exists among the trigonometric functions, one revealed in various identities, strange places where the trigonometric functions appear fluidly to exchange identities or to resolve themselves into unlikely numbers." What does Berlinski mean by this? Use two fundamental trigonometric identities to illustrate your answer.

Additional Topics in Trigonometry

Chapter 6

These days, computers and trigonometric functions are everywhere. Trigonometry plays a critical role in analyzing the forces that surround your every move. Using trigonometry to understand how forces are measured is one of the topics in this chapter that focuses on additional applications of trigonometry.

You enjoy running, although lately you experience discomfort at various points of impact. Your doctor suggests a computer analysis. By attaching sensors to your running shoes as you jog along a treadmill, the computer provides a printout of the magnitude and direction of the forces as your feet hit the ground. Based on this analysis, customized orthotics can be made to fit inside your shoes to minimize the impact.

SECTION 6.1   *The Law of Sines*

Objectives

1. Use the Law of Sines to solve oblique triangles.
2. Use the Law of Sines to solve, if possible, the triangle or triangles in the ambiguous case.
3. Find the area of an oblique triangle using the sine function.
4. Solve applied problems using the Law of Sines.

Point Reyes National Seashore, 40 miles north of San Francisco, consists of 75,000 acres with miles of pristine surf-pummeled beaches, forested ridges, and bays flanked by white cliffs. A few people, inspired by nature in the raw, live on private property adjoining the National Seashore. In 1995, a fire in the park burned 12,350 acres and destroyed 45 homes.

   Fire is a necessary part of the life cycle in many wilderness areas. It is also an ongoing threat to those who choose to live surrounded by nature's unspoiled beauty. In this section, we see how trigonometry can be used to locate small wilderness fires before they become raging infernos. To do this, we begin by considering triangles other than right triangles.

The Law of Sines and Its Derivation

An **oblique triangle** is a triangle that does not contain a right angle. Figure 6.1 shows that an oblique triangle has either three acute angles or two acute angles and one obtuse angle. Notice that the angles are labeled A, B, and C. The sides opposite each angle are labeled as a, b, and c, respectively.

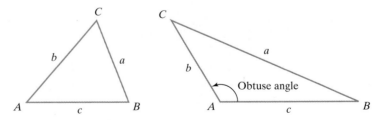

Figure 6.1 Oblique triangles

   Many relationships exist among the sides and angles in an oblique triangle. One such relationship is called the **Law of Sines.**

Study Tip

The Law of Sines can be expressed with the sines in the numerator:

$$\frac{\sin A}{a} = \frac{\sin B}{b} = \frac{\sin C}{c}.$$

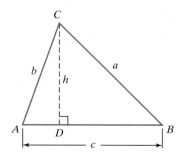

Figure 6.2 Drawing an altitude to prove the Law of Sines

The Law of Sines

If A, B, and C are the measures of the angles of a triangle, and a, b, and c are the lengths of the sides opposite these angles, then

$$\frac{a}{\sin A} = \frac{b}{\sin B} = \frac{c}{\sin C}.$$

The ratio of the length of the side of any triangle to the sine of the angle opposite that side is the same for all three sides of the triangle.

To prove the Law of Sines, we draw an altitude of length h from one of the vertices of the triangle. In Figure 6.2, the altitude is drawn from vertex C. Two smaller triangles are formed, triangles ACD and BCD. Note that both are right triangles. Thus, we can use the definition of the sine of an angle of a right triangle.

$$\sin B = \frac{h}{a} \qquad \sin A = \frac{h}{b} \qquad \sin \theta = \frac{\text{opposite}}{\text{hypotenuse}}$$

$$h = a \sin B \qquad h = b \sin A \qquad \text{Solve each equation for } h.$$

Because we have found two expressions for h, we can set these expressions equal to each other.

$$a \sin B = b \sin A \qquad \text{Equate the expressions for } h.$$

$$\frac{a \sin B}{\sin A \sin B} = \frac{b \sin A}{\sin A \sin B} \qquad \text{Divide both sides by } \sin A \sin B.$$

$$\frac{a}{\sin A} = \frac{b}{\sin B} \qquad \text{Simplify.}$$

This proves part of the Law of Sines. If we use the same process and draw an altitude of length h from vertex A, we obtain the following result:

$$\frac{b}{\sin B} = \frac{c}{\sin C}.$$

When this equation is combined with the previous equation, we obtain the Law of Sines. Because the sine of an angle is equal to the sine of $180°$ minus that angle, the Law of Sines is derived in a similar manner if the oblique triangle contains an obtuse angle.

1 Use the Law of Sines to solve oblique triangles.

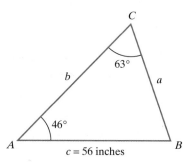

Figure 6.3 Solving an oblique SAA triangle

Solving Oblique Triangles

Solving an oblique triangle means finding the lengths of its sides and the measurement of its angles. The Law of Sines can be used to solve a triangle in which one side and two angles are known. The three known measurements can be abbreviated using SAA (a side and two angles are known) or ASA (two angles and the side between them are known).

EXAMPLE 1   Solving an SAA Triangle Using the Law of Sines

Solve the triangle shown in Figure 6.3 with $A = 46°$, $C = 63°$, and $c = 56$ inches.

Solution   We begin by finding B, the third angle of the triangle. We do not need the Law of Sines to do this. Instead, we use the fact that the sum of the measures of the interior angles of a triangle is $180°$.

$$A + B + C = 180°$$
$$46° + B + 63° = 180° \quad \text{Substitute the given values:} \\ A = 46° \text{ and } C = 63°.$$
$$109° + B = 180° \quad \text{Add.}$$
$$B = 71° \quad \text{Subtract 109° from both sides.}$$

When we use the Law of Sines, we must be given one of the three ratios. In this example, we are given c and C: $c = 56$ and $C = 63°$. Thus, we use the ratio $\dfrac{c}{\sin C}$, or $\dfrac{56}{\sin 63°}$, to find the other two sides. Use the Law of Sines to find a.

$$\dfrac{a}{\sin A} = \dfrac{c}{\sin C} \quad \text{The ratio of any side to the sine of its opposite angle equals the ratio of any other side to the sine of its opposite angle.}$$

$$\dfrac{a}{\sin 46°} = \dfrac{56}{\sin 63°} \quad A = 46°, c = 56, \text{ and } C = 63°.$$

$$a = \dfrac{56 \sin 46°}{\sin 63°} \quad \text{Multiply both sides by } \sin 46° \text{ and solve for } a.$$

$$a \approx 45 \text{ inches} \quad \text{Use a calculator.}$$

Use the Law of Sines again, this time to find b.

$$\dfrac{b}{\sin B} = \dfrac{c}{\sin C} \quad \text{We use the given ratio, } \dfrac{c}{\sin C}, \text{ to find } b.$$

$$\dfrac{b}{\sin 71°} = \dfrac{56}{\sin 63°} \quad \text{We found that } B = 71°. \text{ We are given } c = 56 \text{ and } C = 63°.$$

$$b = \dfrac{56 \sin 71°}{\sin 63°} \quad \text{Multiply both sides by } \sin 71° \text{ and solve for } b.$$

$$b \approx 59 \text{ inches} \quad \text{Use a calculator.}$$

The solution is $B = 71°$, $a \approx 45$ inches, and $b \approx 59$ inches.

> **Check Point 1**   Solve the triangle shown in Figure 6.4 with $A = 64°$, $C = 82°$, and $c = 14$ centimeters.

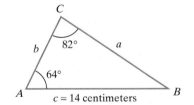

Figure 6.4

EXAMPLE 2   Solving an ASA Triangle Using the Law of Sines

Solve triangle ABC if $A = 50°$, $C = 33.5°$, and $b = 76$.

Solution   We begin by drawing a picture of triangle ABC and labeling it with the given information. Figure 6.5 shows the triangle that we must solve. We begin by finding B.

$$A + B + C = 180° \quad \text{The sum of the measures of a triangle's interior angles is 180°.}$$
$$50° + B + 33.5° = 180° \quad A = 50° \text{ and } C = 33.5°.$$
$$83.5° + B = 180° \quad \text{Add.}$$
$$B = 96.5° \quad \text{Subtract 83.5° from both sides.}$$

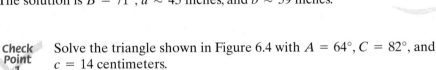

Figure 6.5 Solving an ASA triangle

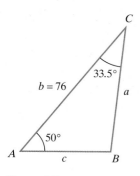

Figure 6.5, repeated

Keep in mind that we must be given one of the three ratios to apply the Law of Sines. In this example, we are given that $b = 76$ and we found that $B = 96.5°$. Thus, we use the ratio $\dfrac{b}{\sin B}$, or $\dfrac{76}{\sin 96.5°}$, to find the other two sides. Use the Law of Sines to find a and c.

Find a:

This is the known ratio.

$$\frac{a}{\sin A} = \frac{b}{\sin B}$$

$$\frac{a}{\sin 50°} = \frac{76}{\sin 96.5°}$$

$$a = \frac{76 \sin 50°}{\sin 96.5°} \approx 59$$

Find c:

$$\frac{c}{\sin C} = \frac{b}{\sin B}$$

$$\frac{c}{\sin 33.5°} = \frac{76}{\sin 96.5°}$$

$$c = \frac{76 \sin 33.5°}{\sin 96.5°} \approx 42$$

The solution is $B = 96.5°$, $a \approx 59$, and $c \approx 42$.

2 Use the Law of Sines to solve, if possible, the triangle or triangles in the ambiguous case.

Check Point 2 Solve triangle ABC if $A = 40°$, $C = 22.5°$, and $b = 12$.

The Ambiguous Case (SSA)

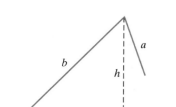

Figure 6.6 Given SSA, no triangle may result.

If we are given two sides and an angle opposite one of them (SSA), does this determine a unique triangle? Can we solve this case using the Law of Sines? Such a case is called the **ambiguous case** because the given information may result in one triangle, two triangles, or no triangle at all. For example, in Figure 6.6, we are given a, b, and A. Because a is shorter than h, it is not long enough to form a triangle. The number of possible triangles, if any, that can be formed in the SSA case depends on h, the length of the altitude, where $h = b \sin A$.

The Ambiguous Case (SSA)

Consider a triangle in which a, b, and A are given. This information may result in

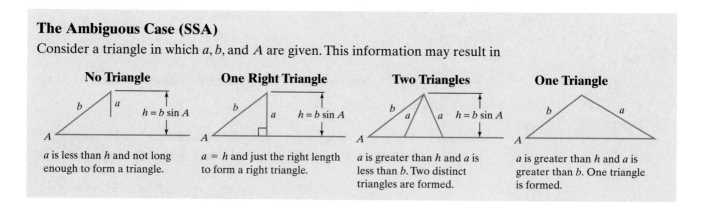

No Triangle	**One Right Triangle**	**Two Triangles**	**One Triangle**
a is less than h and not long enough to form a triangle.	$a = h$ and just the right length to form a right triangle.	a is greater than h and a is less than b. Two distinct triangles are formed.	a is greater than h and a is greater than b. One triangle is formed.

In an SSA situation, it is not necessary to draw an accurate sketch like those shown in the box. The Law of Sines determines the number of triangles, if any, and gives the solution for each triangle.

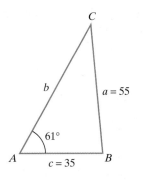

Figure 6.7 Solving an SSA triangle; the ambiguous case

EXAMPLE 3   Solving an SSA Triangle Using the Law of Sines (One Solution)

Solve triangle ABC if $A = 61°$, $a = 55$, and $c = 35$.

Solution   We begin with the sketch in Figure 6.7. The known ratio is $\dfrac{a}{\sin A}$, or $\dfrac{55}{\sin 61°}$. Because side c is given, we use the Law of Sines to find angle C.

$$\frac{a}{\sin A} = \frac{c}{\sin C} \qquad \text{Apply the Law of Sines.}$$

$$\frac{55}{\sin 61°} = \frac{35}{\sin C} \qquad A = 55, c = 35, \text{ and } A = 61°.$$

$$55 \sin C = 35 \sin 61° \qquad \text{Cross multiply: If } \frac{a}{b} = \frac{c}{d}, \text{ then } ad = bc.$$

$$\sin C = \frac{35 \sin 61°}{55} \qquad \text{Divide both sides by 55 and solve for } \sin C.$$

$$\sin C \approx 0.5566 \qquad \text{Use a calculator.}$$

There are two angles C between $0°$ and $180°$ for which $\sin C \approx 0.5566$.

$$C_1 \approx 34° \qquad\qquad C_2 \approx 180° - 34° \approx 146°$$

Obtain the acute angle with your calculator: $\sin^{-1} 0.5566$

The sine is positive in quadrant II.

Look at Figure 6.7. Given that $A = 61°$, can you see that $C_2 \approx 146°$ is impossible? By adding $146°$ to the given angle, $61°$, we exceed a $180°$ sum:

$$61° + 146° = 207°.$$

Thus, the only possibility is that $C_1 \approx 34°$. We find B using this approximation for C_1 and the measure that was given for A: $A = 61°$.

$$B = 180° - C_1 - A \approx 180° - 34° - 61° = 85°$$

Side b that lies opposite this $85°$ angle can now be found using the Law of Sines.

$$\frac{b}{\sin B} = \frac{a}{\sin A} \qquad \text{Apply the Law of Sines.}$$

$$\frac{b}{\sin 85°} = \frac{55}{\sin 61°} \qquad a = 55, B \approx 85°, \text{ and } A = 61°.$$

$$b = \frac{55 \sin 85°}{\sin 61°} \approx 63 \qquad \text{Multiply both sides by } \sin 85° \text{ and solve for } b.$$

There is one triangle and the solution is C_1 (or C) $\approx 34°$, $B \approx 85°$, and $b \approx 63$.

Check Point 3   Solve triangle ABC if $A = 123°$, $a = 47$, and $c = 23$.

EXAMPLE 4 Solving an SSA Triangle Using the Law of Sines (No Solution)

Solve triangle ABC if $A = 75°$, $a = 51$, and $b = 71$.

Solution   The known ratio is $\dfrac{a}{\sin A}$, or $\dfrac{51}{\sin 75°}$. Because side b is given, we use the Law of Sines to find angle B.

$$\frac{a}{\sin A} = \frac{b}{\sin B}$$ Use the Law of Sines.

$$\frac{51}{\sin 75°} = \frac{71}{\sin B}$$ Substitute the given values.

$$51 \sin B = 71 \sin 75°$$ Cross multiply: If $\dfrac{a}{b} = \dfrac{c}{d}$, then $ad = bc$.

$$\sin B = \frac{71 \sin 75°}{51} \approx 1.34$$ Divide by 51 and solve for sin B.

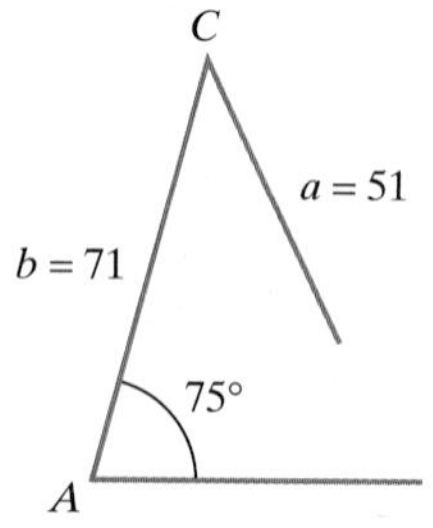

Figure 6.8 a is not long enough to form a triangle.

Because the sine can never exceed 1, there is no angle B for which $\sin B \approx 1.34$. There is no triangle with the given measurements, as illustrated in Figure 6.8.

> **Check Point 4**   Solve triangle ABC if $A = 50°$, $a = 10$, and $b = 20$.

EXAMPLE 5 Solving an SSA Triangle Using the Law of Sines (Two Solutions)

Solve triangle ABC if $A = 40°$, $a = 54$, and $b = 62$.

Solution   The known ratio is $\dfrac{a}{\sin A}$, or $\dfrac{54}{\sin 40°}$. We use the Law of Sines to find angle B.

$$\frac{a}{\sin A} = \frac{b}{\sin B}$$ Use the Law of Sines.

$$\frac{54}{\sin 40°} = \frac{62}{\sin B}$$ Substitute the given values.

$$54 \sin B = 62 \sin 40°$$ Cross multiply: If $\dfrac{a}{b} = \dfrac{c}{d}$, then $ad = bc$.

$$\sin B = \frac{62 \sin 40°}{54} \approx 0.7380$$ Divide by 54 and solve for sin B.

There are two angles B between $0°$ and $180°$ for which $\sin B \approx 0.7380$.

$$B_1 \approx 48° \qquad\qquad B_2 \approx 180° - 48° = 132°$$

Find $\sin^{-1} 0.7380$ with your calculator.   The sine is positive in quadrant II.

If you add either angle to the given angle, $40°$, the sum does not exceed $180°$. Thus, there are two triangles with the given conditions, shown in Figure 6.9(a). The triangles, $AB_1 C_1$ and $AB_2 C_2$, are shown separately in Figures 6.9(b) and (c).

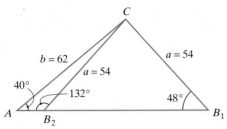

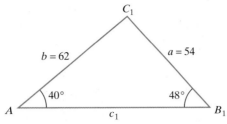

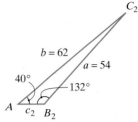

(a) Two triangles are possible with $A = 40°$, $a = 54$, and $b = 62$.

(b) In one possible triangle, $B_1 = 48°$.

(c) In the second possible triangle, $B_2 = 132°$.

Figure 6.9

Study Tip

The two triangles shown in Figure 6.9 are helpful in organizing the solutions. However, if you keep track of the two triangles, one with the given information and $B_1 = 48°$, and the other with the given information and $B_2 = 132°$, you do not have to draw the figure to solve the triangles.

We find angles C_1 and C_2 using a $180°$ angle sum in each of the two triangles.

$$C_1 = 180° - A - B_1 \qquad\qquad C_2 = 180° - A - B_2$$
$$\approx 180° - 40° - 48° \qquad\qquad \approx 180° - 40° - 132°$$
$$= 92° \qquad\qquad\qquad\qquad = 8°$$

We use the Law of Sines to find c_1 and c_2.

$$\frac{c_1}{\sin C_1} = \frac{a}{\sin A} \qquad\qquad \frac{c_2}{\sin C_2} = \frac{a}{\sin A}$$

$$\frac{c_1}{\sin 92°} = \frac{54}{\sin 40°} \qquad\qquad \frac{c_2}{\sin 8°} = \frac{54}{\sin 40°}$$

$$c_1 = \frac{54 \sin 92°}{\sin 40°} \approx 84 \qquad\qquad c_2 = \frac{54 \sin 8°}{\sin 40°} \approx 12$$

There are two triangles. In one triangle, the solution is $B_1 \approx 48°$, $C_1 \approx 92°$, and $c_1 \approx 84$. In the other triangle, $B_2 \approx 132°$, $C_2 \approx 8°$, and $c_2 \approx 12$.

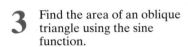

Check Point 5 Solve triangle ABC if $A = 35°$, $a = 12$, and $b = 16$.

3 Find the area of an oblique triangle using the sine function.

The Area of an Oblique Triangle

A formula for the area of an oblique triangle can be obtained using the procedure for proving the Law of Sines. We draw an altitude of length h from one of the vertices of the triangle, as shown in Figure 6.10. We apply the definition of the sine of angle A, $\dfrac{\text{opposite}}{\text{hypotenuse}}$, in right triangle ACD:

$$\sin A = \frac{h}{b} \quad \text{or} \quad h = b \sin A.$$

The area of a triangle is $\frac{1}{2}$ the product of any side and the altitude drawn to that side. Using the altitude h in Figure 6.10, we have

$$\text{Area} = \tfrac{1}{2}ch = \tfrac{1}{2}cb \sin A.$$

Use the result from above: $h = b \sin A$.

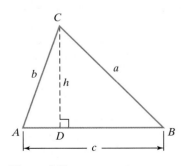

Figure 6.10

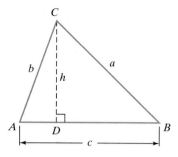

Figure 6.10, repeated

This result, Area $= \frac{1}{2} cb \sin A$, or $\frac{1}{2} bc \sin A$, indicates that the area of the triangle is one-half the product of b and c times the sine of their included angle. If we draw altitudes from the other two vertices, we can use any two sides to compute the area.

> ### Area of An Oblique Triangle
>
> The area of a triangle equals one-half the product of the lengths of two sides times the sine of their included angle. In Figure 6.10, this wording can be expressed by the formulas
>
> $$\text{Area} = \tfrac{1}{2} bc \sin A = \tfrac{1}{2} ab \sin C = \tfrac{1}{2} ac \sin B.$$

EXAMPLE 6   Finding the Area of an Oblique Triangle

Find the area of a triangle having two sides of lengths 24 meters and 10 meters and an included angle of 62°.

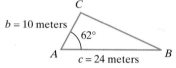

Figure 6.11 Finding the area of an SAS triangle

Solution   The triangle is shown in Figure 6.11. Its area is half the product of the lengths of the two sides times the sine of the included angle.

$$\text{Area} = \tfrac{1}{2}(24)(10)(\sin 62°) \approx 106$$

The area of the triangle is approximately 106 square meters.

> **Check Point 6**   Find the area of a triangle having two sides of lengths 8 meters and 12 meters and an included angle of 135°.

4 Solve applied problems using the Law of Sines.

Applications of the Law of Sines

We have seen how the trigonometry of right triangles can be used to solve many different kinds of applied problems. The Law of Sines enables us to work with triangles that are not right triangles. As a result, this law can be used to solve problems involving surveying, engineering, astronomy, navigation, and the environment. Example 7 illustrates the use of the Law of Sines in detecting potentially devastating fires.

EXAMPLE 7   An Application of the Law of Sines

Two fire-lookout stations are 20 miles apart, with station B directly east of station A. Both stations spot a fire on a mountain to the north. The bearing from station A to the fire is N50°E (50° east of north). The bearing from station B to the fire is N36°W (36° west of north). How far is the fire from station A?

Solution   Figure 6.12 at the top of the next page shows the information given in the problem. The distance from station A to the fire is represented by b. Notice that the angles describing the bearing from each station to the fire, 50° and 36°, are not interior angles of triangle ABC. Using a north-south line, the interior angles are found as follows:

$$A = 90° - 50° = 40° \qquad B = 90° - 36° = 54°.$$

To find b using the Law of Sines, we need a known side and an angle opposite that side. Because $c = 20$ miles, we find angle C using a 180° angle sum in the triangle. Thus,

$$C = 180° - A - B = 180° - 40° - 54° = 86°.$$

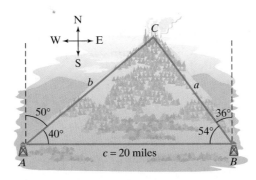

Figure 6.12

The ratio $\dfrac{c}{\sin C}$, or $\dfrac{20}{\sin 86°}$, is now known. We use this ratio and the Law of Sines to find b.

$$\frac{b}{\sin B} = \frac{c}{\sin C} \qquad \text{Use the Law of Sines.}$$

$$\frac{b}{\sin 54°} = \frac{20}{\sin 86°} \qquad c = 20, B = 54°, \text{ and } C = 86°.$$

$$b = \frac{20 \sin 54°}{\sin 86°} \approx 16 \qquad \text{Multiply both sides by } \sin 54° \text{ and solve for } b.$$

The fire is approximately 16 miles from station A.

Check Point 7 Two fire-lookout stations are 13 miles apart, with station B directly east of station A. Both stations spot a fire. The bearing of the fire from station A is N35°E, and the bearing of the fire from station B is N49°W. How far, to the nearest mile, is the fire from station B?

EXERCISE SET 6.1

Practice Exercises

In Exercises 1–8, solve each triangle. Round lengths of sides to the nearest tenth and angle measures to the nearest degree.

1.

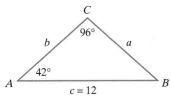

2.

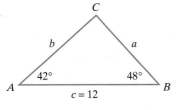

3.

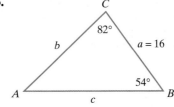

4.

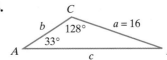

5.

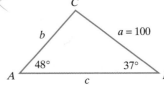

6.

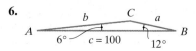

7.

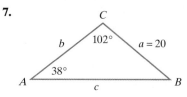

8.

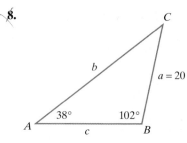

In Exercises 9–16, solve each triangle. Round lengths to the nearest tenth and angle measures to the nearest degree.

9. $A = 44°, B = 25°, a = 12$

10. $A = 56°, C = 24°, a = 22$

11. $B = 85°, C = 15°, b = 40$

12. $A = 85°, B = 35°, c = 30$

13. $A = 115°, C = 35°, c = 200$

14. $B = 5°, C = 125°, b = 200$

15. $A = 65°, B = 65°, c = 6$

16. $B = 80°, C = 10°, a = 8$

In Exercises 17–32, two sides and an angle (SSA) of a triangle are given. Determine whether the given measurements produce one triangle, two triangles, or no triangle at all. Solve each triangle that results. Round to the nearest tenth and the nearest degree for sides and angles, respectively.

17. $a = 20, b = 15, A = 40°$

18. $a = 30, b = 20, A = 50°$

19. $a = 10, c = 8.9, A = 63°$

20. $a = 57.5, c = 49.8, A = 136°$

21. $a = 42.1, c = 37, A = 112°$

22. $a = 6.1, b = 4, A = 162°$

23. $a = 10, b = 40, A = 30°$

24. $a = 10, b = 30, A = 150°$

25. $a = 16, b = 18, A = 60°$

26. $a = 30, b = 40, A = 20°$

27. $a = 12, b = 16.1, A = 37°$

28. $a = 7, b = 28, A = 12°$

29. $a = 22, c = 24.1, A = 58°$

30. $a = 95, c = 125, A = 49°$

31. $a = 9.3, b = 41, A = 18°$

32. $a = 1.4, b = 2.9, A = 142°$

In Exercises 33–38, find the area of the triangle having the given measurements. Round to the nearest square unit.

33. $A = 48°, b = 20$ feet, $c = 40$ feet

34. $A = 22°, b = 20$ feet, $c = 50$ feet

35. $B = 36°, a = 3$ yards, $c = 6$ yards

36. $B = 125°, a = 8$ yards, $c = 5$ yards

37. $C = 124°, a = 4$ meters, $b = 6$ meters

38. $C = 102°, a = 16$ meters, $b = 20$ meters

Application Exercises

39. Two fire-lookout stations are 10 miles apart, with station B directly east of station A. Both stations spot a fire. The bearing of the fire from station A is N25°E and the bearing of the fire from station B is N56°W. How far, to the nearest mile, is the fire from each lookout station?

40. The Federal Communications Commission is attempting to locate an illegal radio station. It sets up two monitoring stations, A and B, with station B 40 miles east of station A. Station A measures the illegal signal from the radio station as coming from a direction of 48° east of north. Station B measures the signal as coming from a point 34° west of north. How far is the illegal radio station from monitoring stations A and B?

41. The figure shows a 1200-yard-long sand beach and an oil platform in the ocean. The angle made with the platform from one end of the beach is 85° and from the other end is 76°. Find the distance of the oil platform, to the nearest yard, from each end of the beach.

42. A surveyor needs to determine the distance between two points that lie on opposite banks of a river. The figure shows that 300 yards are measured along one bank. The angles from each end of this line segment to a point on the opposite bank are 62° and 53°. Find the distance between A and B to the nearest foot.

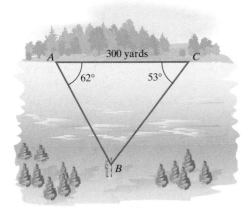

43. The Leaning Tower of Pisa in Italy leans at an angle of about 84.7°. The figure shows that 171 feet from the base of the tower, the angle of elevation to the top is 50°. Find the distance, to the nearest foot, from the base to the top of the tower.

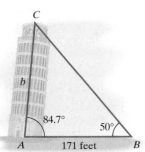

44. A pine tree growing on a hillside makes a 75° angle with the hill. From a point 80 feet up the hill, the angle of elevation to the top of the tree is 62° and the angle of depression to the bottom is 23°. Find, to the nearest foot, the height of the tree.

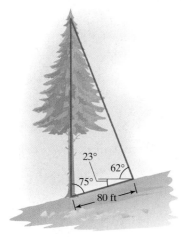

45. The figure shows a shot-put ring. The shot is tossed from A and lands at B. Using modern electronic equipment, the distance of the toss can be measured without the use of measuring tapes. When the shot lands at B, an electronic transmitter placed at B sends a signal to a device in the official's booth above the track. The device determines the angles at B and C. At a track meet, the distance from the official's booth to the shot-put ring is 562 feet. If $B = 85.3°$ and $C = 5.7°$, determine the length of the toss to the nearest foot.

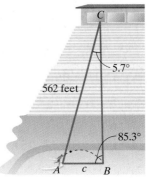

46. A pier forms an 85° angle with a straight shore. At a distance of 100 feet from the pier, the line of sight to the tip forms a 37° angle. Find the length of the pier to the nearest foot.

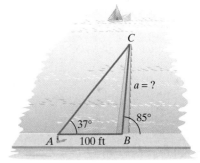

47. When the angle of elevation of the sun is 62°, a telephone pole that is tilted at an angle of 8° directly away from the sun casts a shadow 20 feet long. Determine the length of the pole to the nearest foot.

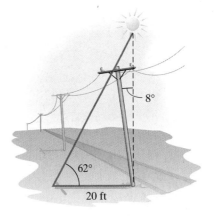

48. A leaning wall is inclined 6° from the vertical. At a distance of 40 feet from the wall, the angle of elevation to the top is 22°. Find the height of the wall to the nearest foot.

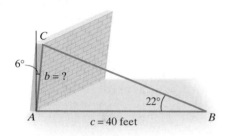

49. Redwood trees in California's Redwood National Park are hundreds of feet tall. The height of one of these trees is represented by h in the figure shown.

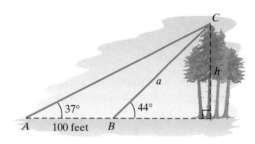

a. Use the measurements shown to find a, to the nearest foot, in oblique triangle ABC.

b. Use the right triangle shown to find the height, to the nearest foot, of a typical redwood tree in the park.

50. The figure shows a cable car that carries passengers from A to C. Point A is 1.6 miles from the base of the mountain. The angles of elevation from A and B to the mountain's peak are 22° and 66°, respectively.

a. Determine, to the nearest foot, the distance covered by the cable car.

b. Find a, to the nearest foot, in oblique triangle ABC.

c. Use the right triangle to find the height of the mountain to the nearest foot.

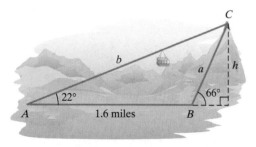

51. Lighthouse B is 7 miles west of lighthouse A. A boat leaves A and sails 5 miles. At this time, it is sighted from B. If the bearing of the boat from B is N62°E, how far from B is the boat? Round to the nearest tenth of a mile.

52. After a wind storm, you notice that your 16-foot flagpole may be leaning, but you are not sure. From a point on the ground 15 feet from the base of the flagpole, you find that the angle of elevation to the top is 48°. Is the flagpole leaning? If so, find the acute angle, to the nearest degree, that the flagpole makes with the ground.

Writing in Mathematics

53. What is an oblique triangle?

54. Without using symbols, state the Law of Sines in your own words.

55. Briefly describe how the Law of Sines is proved.

56. What does it mean to solve an oblique triangle?

57. What do the abbreviations SAA and ASA mean?

58. Why is SSA called the ambiguous case?

59. How is the sine function used to find the area of an oblique triangle?

60. Write an original problem that can be solved using the Law of Sines. Then solve the problem.

61. Use Exercise 45 to describe how the Law of Sines is used for throwing events at track and field meets. Why aren't tape measures used to determine tossing distance?

62. You are cruising in your boat parallel to the coast, looking at a lighthouse. Explain how you can use your boat's speed and a device for measuring angles to determine the distance at any instant from your boat to the lighthouse.

Critical Thinking Exercises

63. If you are given two sides of a triangle and their included angle, you can find the triangle's area. Can the Law of Sines be used to solve the triangle with this given information? Explain your answer.

64. Two buildings of equal height are 800 feet apart. An observer on the street between the buildings measures the angles of elevation to the tops of the buildings as 27° and 41°, respectively. How high, to the nearest foot, are the buildings?

65. The figure shows the design for the top of the wing of a jet fighter. The fuselage is 5 feet wide. Find the wing span CC' to the nearest foot.

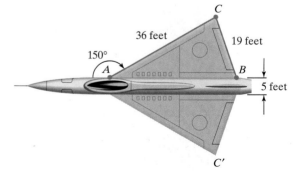

SECTION 6.2   *The Law of Cosines*

Objectives

1. Use the Law of Cosines to solve oblique triangles.
2. Solve applied problems using the Law of Cosines.
3. Use Heron's formula to find the area of a triangle.

Baseball was developed in the United States in the mid-1800s and became our national sport. Little League baseball is the largest youth sports program in the world. Three million boys and girls ages 5 to 18 play on 200,000 Little League teams in 90 countries. Although there are differences between Major League and Little League baseball diamonds, trigonometry can be used to find angles and distances in these fields of dreams. To see how this is done, we turn to the Law of Cosines.

The Law of Cosines and Its Derivation

We now look at another relationship that exists among the sides and angles in an oblique triangle. **The Law of Cosines** is used to solve triangles in which two sides and the included angle (SAS) are known, or those in which three sides (SSS) are known.

Discovery

What happens to the Law of Cosines

$$c^2 = a^2 + b^2 - 2ab \cos C$$

if $C = 90°$? What familiar theorem do you obtain?

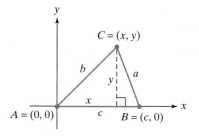

Figure 6.13

The Law of Cosines

If A, B, and C are the measures of the angles of a triangle, and a, b, and c are the lengths of the sides opposite these angles, then

$$a^2 = b^2 + c^2 - 2bc \cos A$$
$$b^2 = a^2 + c^2 - 2ac \cos B$$
$$c^2 = a^2 + b^2 - 2ab \cos C.$$

The square of a side of a triangle equals the sum of the squares of the other two sides minus twice their product times the cosine of their included angle.

To prove the Law of Cosines, we place triangle ABC in a rectangular coordinate system. Figure 6.13 shows a triangle with three acute angles. The vertex A is at the origin and side c lies along the positive x-axis. The coordinates of C are (x, y). Using the right triangle that contains angle A, we apply the definitions of the cosine and the sine.

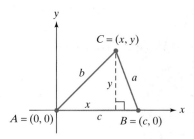

Figure 6.13, repeated

$$\cos A = \frac{x}{b} \qquad \sin A = \frac{y}{b}$$

$$x = b \cos A \qquad y = b \sin A \qquad \text{Multiply both sides of each equation by } b \text{ and solve for x and y, respectively.}$$

Thus, the coordinates of C are $(x, y) = (b \cos A, b \sin A)$. Although triangle ABC in Figure 6.13 shows angle A as an acute angle, if A is obtuse, the coordinates of C are still $(b \cos A, b \sin A)$. This means that our proof applies to both kinds of oblique triangles.

We now apply the distance formula to the side of the triangle with length a. Notice that a is the distance from (x, y) to $(c, 0)$.

$$a = \sqrt{(x - c)^2 + (y - 0)^2} \qquad \text{Use the distance formula.}$$

$$a^2 = (x - c)^2 + y^2 \qquad \text{Square both sides of the equation.}$$

$$a^2 = (b \cos A - c)^2 + (b \sin A)^2 \qquad x = b \cos A \text{ and } y = b \sin A.$$

$$a^2 = b^2 \cos^2 A - 2bc \cos A + c^2 + b^2 \sin^2 A \qquad \text{Square the two expressions.}$$

$$a^2 = b^2 \sin^2 A + b^2 \cos^2 A + c^2 - 2bc \cos A \qquad \text{Rearrange terms.}$$

$$a^2 = b^2(\sin^2 A + \cos^2 A) + c^2 - 2bc \cos A \qquad \text{Factor } b^2 \text{ from the first two terms.}$$

$$a^2 = b^2 + c^2 - 2bc \cos A \qquad \sin^2 A + \cos^2 A = 1$$

The resulting equation is one of the three formulas for the Law of Cosines. The other two formulas are derived in a similar manner.

1 Use the Law of Cosines to solve oblique triangles.

Solving Oblique Triangles

If you are given two sides and an included angle (SAS) of an oblique triangle, none of the three ratios in the Law of Sines is known. This means that we do not begin solving the triangle using the Law of Sines. Instead, we apply the Law of Cosines and the following procedure:

Solving an SAS Triangle

1. Use the Law of Cosines to find the side opposite the given angle.

2. Use the Law of Sines to find the angle opposite the shorter of the two given sides. This angle is always acute.

3. Find the third angle by subtracting the measure of the given angle and the angle found in step 2 from 180°.

EXAMPLE 1   Solving an SAS Triangle

Solve the triangle in Figure 6.14 with $A = 60°$, $b = 20$, and $c = 30$.

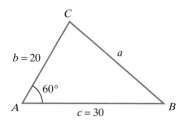

Figure 6.14 Solving an SAS triangle

Solution   We are given two sides and an included angle. Therefore, we apply the three-step procedure for solving an SAS triangle.

Step 1   Use the Law of Cosines to find the side opposite the given angle.   Thus, we will find a.

$$a^2 = b^2 + c^2 - 2bc \cos A$$ Apply the Law of Cosines to find a.

$$a^2 = 20^2 + 30^2 - 2(20)(30) \cos 60°$$ $b = 20, c = 30,$ and $A = 60°$

$$= 400 + 900 - 1200(0.5)$$ Perform the indicated operations.

$$= 700$$

$$a = \sqrt{700} \approx 26$$ Take the square root of both sides and solve for a.

Step 2   Use the Law of Sines to find the angle opposite the shorter of the two given sides. This angle is always acute.   The shorter of the two given sides is $b = 20$. Thus, we will find acute angle B.

$$\frac{b}{\sin B} = \frac{a}{\sin A}$$ Apply the Law of Sines.

$$\frac{20}{\sin B} = \frac{\sqrt{700}}{\sin 60°}$$ We are given $b = 20$ and $A = 60°$. Use the exact value of a, $\sqrt{700}$, from step 1.

$$\sqrt{700} \sin B = 20 \sin 60°$$ Cross multiply: If $\dfrac{a}{b} = \dfrac{c}{d}$, then $ad = bc$.

$$\sin B = \frac{20 \sin 60°}{\sqrt{700}} \approx 0.6547$$ Divide by $\sqrt{700}$ and solve for $\sin B$.

$$B \approx 41°$$ Find $\sin^{-1} 0.6547$ using a calculator.

Step 3   Find the third angle. Subtract the measure of the given angle and the angle found in step 2 from 180°.

$$C = 180° - A - B \approx 180° - 60° - 41° = 79°$$

The solution is $a \approx 26$, $B \approx 41°$, and $C \approx 79°$.

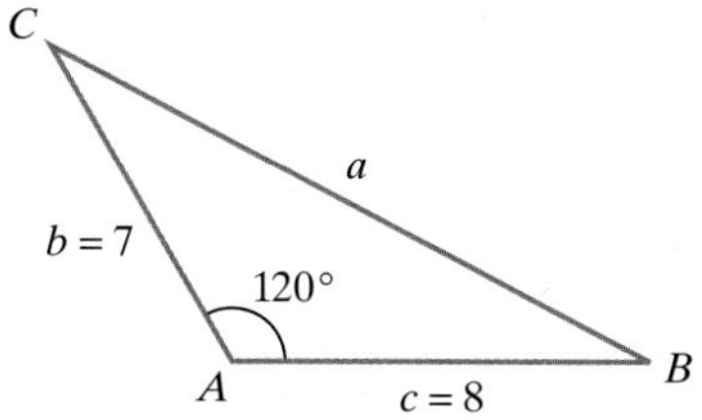

Check Point 1   Solve the triangle shown in Figure 6.15 with $A = 120°$, $b = 7$, and $c = 8$.

Figure 6.15

If you are given three sides of a triangle (SSS), solving the triangle involves finding the three angles. We use the following procedure:

Solving an SSS Triangle

1. Use the Law of Cosines to find the angle opposite the longest side.

2. Use the Law of Sines to find either of the two remaining acute angles.

3. Find the third angle by subtracting the measures of the angles found in steps 1 and 2 from 180°.

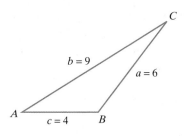

Figure 6.16 Solving an SSS triangle

EXAMPLE 2 Solving an SSS Triangle

Solve triangle ABC if $a = 6$, $b = 9$, and $c = 4$. Round to the nearest tenth of a degree.

Solution We are given three sides. Therefore, we apply the three-step procedure for solving an SSS triangle. The triangle is shown in Figure 6.16.

Step 1 Use the Law of Cosines to find the angle opposite the longest side. The longest side is $b = 9$. Thus, we will find angle B.

$$b^2 = a^2 + c^2 - 2ac \cos B \qquad \text{Apply the Law of Cosines to find } B.$$

$$2ac \cos B = a^2 + c^2 - b^2 \qquad \text{Solve for } \cos B.$$

$$\cos B = \frac{a^2 + c^2 - b^2}{2ac}$$

$$\cos B = \frac{6^2 + 4^2 - 9^2}{2 \cdot 6 \cdot 4} = -\frac{29}{48} \qquad a = 6, b = 9, \text{ and } c = 4.$$

Using a calculator, $\cos^{-1}\left(\frac{29}{48}\right) \approx 52.8°$. Because $\cos B$ is negative, B is an obtuse angle. Thus,

$$B \approx 180° - 52.8° = 127.2°. \qquad \text{Because the domain of } y = \cos^{-1}x \text{ is } [0,\pi], \text{ you can use a calculator to find } \cos^{-1}(-\frac{29}{48}) \approx 127.2°.$$

Study Tip

You can use the Law of Cosines in step 2 to find either of the remaining angles. However, it is simpler to use the Law of Sines. Because the largest angle has been found, the remaining angles must be acute. Thus, there is no need to be concerned about two possible triangles or an ambiguous case.

Step 2 Use the Law of Sines to find either of the two remaining acute angles. We will find angle A.

$$\frac{a}{\sin A} = \frac{b}{\sin B} \qquad \text{Apply the Law of Sines.}$$

$$\frac{6}{\sin A} = \frac{9}{\sin 127.2°} \qquad \text{We are given } a = 6 \text{ and } b = 9. \text{ We found that } B \approx 127.2°.$$

$$9 \sin A = 6 \sin 127.2° \qquad \text{Cross multiply.}$$

$$\sin A = \frac{6 \sin 127.2°}{9} \approx 0.5310 \qquad \text{Divide by 9 and solve for } \sin A.$$

$$A \approx 32.1° \qquad \text{Find } \sin^{-1} 0.5310 \text{ using a calculator.}$$

Step 3 Find the third angle. Subtract the measures of the angles found in steps 1 and 2 from 180°.

$$C = 180° - B - A \approx 180° - 127.2° - 32.1° = 20.7°$$

The solution is $B \approx 127.2°$, $A \approx 32.1°$, and $C \approx 20.7°$.

Check Point 2 Solve triangle ABC if $a = 8$, $b = 10$, and $c = 5$. Round to the nearest tenth of a degree.

2 Solve applied problems using the Law of Cosines.

Applications of the Law of Cosines

Applied problems involving SAS and SSS triangles can be solved using the Law of Cosines.

EXAMPLE 3 An Application of the Law of Cosines

Two airplanes leave an airport at the same time on different runways. One flies on a bearing of N66°W at 325 miles per hour. The other airplane flies on a bearing of S26°W at 300 miles per hour. How far apart will the airplanes be after two hours?

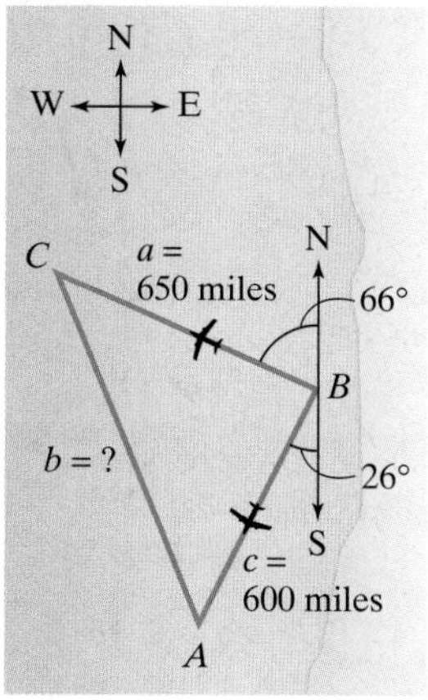

Solution After two hours, the plane flying at 325 miles per hour travels $325 \cdot 2$ miles, or 650 miles. Similarly, the plane flying at 300 miles per hour travels 600 miles. The situation is illustrated in Figure 6.17.

Let b = the distance between the planes after two hours. We can use a north-south line to find angle B in triangle ABC. Thus,

$$B = 180° - 66° - 26° = 88°.$$

Figure 6.17

We now have $a = 650$, $c = 600$, and $B = 88°$. We use the Law of Cosines to find b in this SAS situation.

$$b^2 = a^2 + c^2 - 2ac \cos B \qquad \text{Apply the Law of Cosines.}$$

$$b^2 = 650^2 + 600^2 - 2(650)(600) \cos 88° \qquad \text{Substitute: } a = 650, c = 600, \text{ and } b = 88°.$$

$$\approx 755{,}278 \qquad \text{Use a calculator.}$$

$$b \approx \sqrt{755{,}278} \approx 869 \qquad \text{Take the square root and solve for } b.$$

After two hours, the planes are approximately 869 miles apart.

> **Check Point 3** Two airplanes leave an airport at the same time on different runways. One flies directly north at 400 miles per hour. The other airplane flies on a bearing of N75°E at 350 miles per hour. How far apart will the airplanes be after two hours?

3 Use Heron's formula to find the area of a triangle.

Heron's Formula

Approximately 2000 years ago, the Greek mathematician Heron of Alexandria derived a formula for the area of a triangle in terms of the lengths of its sides. A more modern derivation uses the Law of Cosines and can be found in the appendix.

> **Heron's Formula for the Area of a Triangle**
> The area of a triangle with sides a, b, and c is
> $$\text{Area} = \sqrt{s(s-a)(s-b)(s-c)}$$
> where s is one-half the perimeter: $s = \frac{1}{2}(a + b + c)$.

EXAMPLE 4   Using Heron's Formula

Find the area of the triangle with $a = 12$ yards, $b = 16$ yards, and $c = 24$ yards.

Solution   Begin by calculating one-half the perimeter:

$$s = \tfrac{1}{2}(a + b + c) = \tfrac{1}{2}(12 + 16 + 24) = 26.$$

Use Heron's formula to find the area:

$$\begin{aligned}
\text{Area} &= \sqrt{s(s - a)(s - b)(s - c)} \\
&= \sqrt{26(26 - 12)(26 - 16)(26 - 24)} \\
&= \sqrt{7280} \approx 85.
\end{aligned}$$

The area of the triangle is approximately 85 square yards.

Check Point 4   Find the area of the triangle with $a = 6$ meters, $b = 16$ meters, and $c = 18$ meters. Round to the nearest square meter.

EXERCISE SET 6.2

Practice Exercises

In Exercises 1–8, solve each triangle. Round lengths of sides to the nearest tenth and angle measures to the nearest degree.

1.

C, $b = 4$, a, $46°$, A, $c = 8$, B

2.

C, b, $a = 6$, $32°$, A, $c = 8$, B

3.

C, $96°$, $b = 4$, $a = 6$, A, c, B

4.

$b = 6$, C, a, $22°$, A, $c = 15$, B

5.

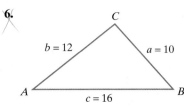

C, $b = 8$, $a = 6$, A, $c = 8$, B

6.

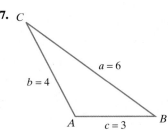

C, $b = 12$, $a = 10$, A, $c = 16$, B

7.

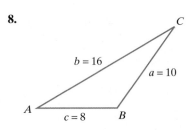

C, $a = 6$, $b = 4$, A, $c = 3$, B

8.

C, $b = 16$, $a = 10$, A, $c = 8$, B

In Exercises 9–24, solve each triangle. Round lengths to the nearest tenth and angle measures to the nearest degree.

9. $a = 5, b = 7, C = 42°$ **10.** $a = 10, b = 3, C = 15°$

11. $b = 5, c = 3, A = 102°$ **12.** $b = 4, c = 1, A = 100°$

13. $a = 6, c = 5, B = 50°$ **14.** $a = 4, c = 7, B = 55°$

15. $a = 5, c = 2, B = 90°$ **16.** $a = 7, c = 3, B = 90°$

17. $a = 5, b = 7, c = 10$ **18.** $a = 4, b = 6, c = 9$

19. $a = 3, b = 9, c = 8$ **20.** $a = 4, b = 7, c = 6$

21. $a = 3, b = 3, c = 3$ **22.** $a = 5, b = 5, c = 5$

23. $a = 73, b = 22, c = 50$ **24.** $a = 66, b = 25, c = 45$

In Exercises 25–30, use Heron's formula to find the area of each triangle. Round to the nearest square unit.

25. $a = 4$ feet, $b = 4$ feet, $c = 2$ feet

26. $a = 5$ feet, $b = 5$ feet, $c = 4$ feet

27. $a = 14$ meters, $b = 12$ meters, $c = 4$ meters

28. $a = 16$ meters, $b = 10$ meters, $c = 8$ meters

29. $a = 11$ yards, $b = 9$ yards, $c = 7$ yards

30. $a = 13$ yards, $b = 9$ yards, $c = 5$ yards

 Application Exercises

31. Two ships leave a harbor at the same time. One ship travels on a bearing of S12°W at 14 miles per hour. The other ship travels on a bearing of N75°E at 10 miles per hour. How far apart will the ships be after three hours? Round to the nearest tenth of a mile.

32. A plane leaves airport A and travels 580 miles to airport B on a bearing of N34°E. The plane later leaves airport B and travels to airport C 400 miles away on a bearing of S74°E. Find the distance from airport A to airport C to the nearest tenth of a mile.

33. Find the distance across the lake from A to C, to the nearest yard, using the measurements shown in the figure.

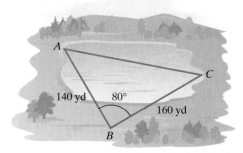

34. To find the distance across a protected cove at a lake, a surveyor makes the measurements shown in the figure at the top of the next column.

Use these measurements to find the distance from A to B to the nearest yard.

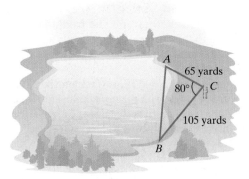

The diagram shows three islands in Florida Bay. You rent a boat and plan to visit each of these remote islands. Use the diagram to solve Exercises 35–36.

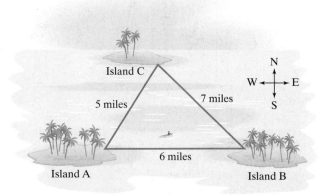

35. If you are on island A, on what bearing should you navigate to go to island C?

36. If you are on island B, on what bearing should you navigate to go to island C?

37. You are on a fishing boat that leaves its pier and heads east. After traveling for 25 miles, there is a report warning of rough seas directly south. The captain turns the boat and follows a bearing of S40°W for 13.5 miles.
   a. At this time, how far are you from the boat's pier? Round to the nearest tenth of a mile.
   b. What bearing could the boat have originally taken to arrive at this spot?

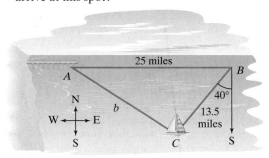

38. You are on a fishing boat that leaves its pier and heads east. After traveling for 30 miles, there is a report warning of rough seas directly south. The captain turns the boat and follows a bearing of S45°W for 12 miles.
  a. At this time, how far are you from the boat's pier? Round to the nearest tenth of a mile.
  b. What bearing could the boat have originally taken to arrive at this spot?

39. The figure shows a 400-foot tower on the side of a hill that forms 7° angle with the horizontal. Find the length of each of the two guy wires that are anchored 80 feet uphill and downhill from the tower's base and extend to the top of the tower. Round to the nearest tenth of a foot.

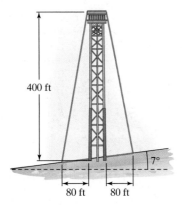

40. The figure shows a 200-foot tower on the side of a hill that forms a 5° angle with the horizontal. Find the length of each of the two guy wires that are anchored 150 feet uphill and downhill from the tower's base and extend to the top of the tower. Round to the nearest tenth of a foot.

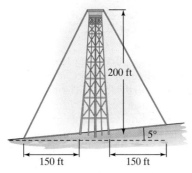

41. A Major League baseball diamond has four bases forming a square whose sides measure 90 feet each. The pitcher's mound is 60.5 feet from home plate on a line joining home plate and second base. Find the distance from the pitcher's mound to first base. Round to the nearest tenth of a foot.

42. A Little League baseball diamond has four bases forming a square whose sides measure 60 feet each. The pitcher's mound is 46 feet from home plate on a line joining home plate and second base. Find the distance from the pitcher's mound to third base. Round to the nearest tenth of a foot.

43. A commercial piece of real estate is priced at $3.50 per square foot. Find the cost, to the nearest dollar, of a triangular lot measuring 240 feet by 300 feet by 420 feet.

44. A commercial piece of real estate is priced at $4.50 per square foot. Find the cost, to the nearest dollar, of a triangular lot measuring 320 feet by 510 feet by 410 feet.

Writing in Mathematics

45. Without using symbols, state the Law of Cosines in your own words.

46. Why can't the Law of Sines be used in the first step to solve an SAS triangle?

47. Describe a strategy for solving an SAS triangle.

48. Describe a strategy for solving an SSS triangle.

49. Under what conditions would you use Heron's formula to find the area of a triangle?

50. Describe an applied problem that can be solved using the Law of Cosines, but not the Law of Sines.

51. The pitcher on your Little League team is studying angles in geometry and has a question. "Coach, suppose I'm on the pitcher's mound facing home plate. I catch a fly ball hit in my direction. If I turn to face first base and throw the ball, through how many degrees should I turn for a direct throw?" Use the information given in Exercise 42 and write an answer to your pitcher's question. Without getting too technical, describe to your pitcher how you obtained this angle.

Critical Thinking Exercises

52. The lengths of the diagonals of a parallelogram are 20 inches and 30 inches. The diagonals intersect at an angle of 35°. Find the lengths of the parallelogram's sides. (*Hint*: Diagonals of a parallelogram bisect one another.)

53. The vertices of a triangle are $A(4, -3)$, $B(2, 1)$, and $C(-2, 4)$. Find the triangle's largest angle to the nearest tenth of a degree.

54. The minute hand and the hour hand of a clock have lengths m inches and h inches, respectively. Determine the distance between the tips of the hands at 10:00 in terms of m and h.

Group Exercise

55. The group should design five original problems that can be solved using the Laws of Sines and Cosines. At least two problems should be solved using the Law of Sines and at least two problems should be solved using the Law of Cosines. At least one problem should be an application problem using the Law of Sines and at least one problem should involve an application using the Law of Cosines. The group should turn in both the problems and their solutions.

SECTION 6.3 *Polar Coordinates*

Objectives

1. Plot points in the polar coordinate system.
2. Find multiple sets of polar coordinates for a given point.
3. Convert a point from polar to rectangular coordinates.
4. Convert a point from rectangular to polar coordinates.
5. Convert an equation from rectangular to polar coordinates.
6. Convert an equation from polar to rectangular coordinates.

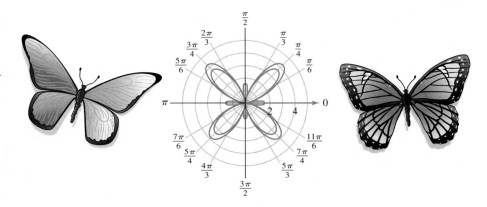

Butterflies are among the most celebrated of all insects. It's hard not to notice their beautiful colors and graceful flight. Their symmetry can be explored with trigonometric functions and a system for plotting points called the **polar coordinate system.** In many cases, polar coordinates are simpler and easier to use than rectangular coordinates.

Plotting Points in the Polar Coordinate System

1	Plot points in the polar coordinate system.

The foundation of the polar coordinate system is a horizontal ray that extends to the right. The ray is called the **polar axis** and is shown in Figure 6.18. The endpoint of the ray is called the **pole.**

A point P in the polar coordinate system is represented by an ordered pair of numbers (r, θ). Figure 6.19 shows $P = (r, \theta)$ in the polar coordinate system.

Figure 6.18

Figure 6.19 Representing a point in the polar coordinate system

- r is the directed distance from P to the pole. (We shall see that r can be positive, negative or zero.)
- θ is an angle from the polar axis to line segment OP. This angle can be measured in degrees or radians. Positive angles are measured counterclockwise from the polar axis. Negative angles are measured clockwise from the polar axis.

We refer to the ordered pair (r, θ) as the **polar coordinates** of P.

Let's look at a specific example. Suppose that the polar coordinates of a point P are $\left(3, \dfrac{\pi}{4}\right)$. Because θ is positive, we locate this point by drawing $\theta = \dfrac{\pi}{4}$ counterclockwise from the polar axis. Then we count out a distance of three units along the terminal side of the angle to reach the point P.

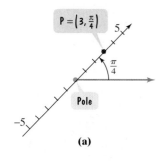

(a)

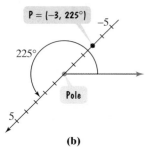

(b)

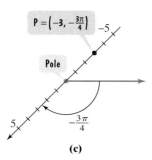

(c)

Figure 6.20 Locating points in polar coordinates

Figure 6.20(a) shows that $(r, \theta) = \left(3, \dfrac{\pi}{4}\right)$ lies three units from the pole on the terminal side of the angle $\theta = \dfrac{\pi}{4}$.

Figure 6.20 illustrates that in a polar coordinate system, a point can be represented in more than one way. In Figure 6.20(b), r is negative and θ is positive: $r = -3$ and $\theta = 225°$. Because θ is positive, we draw a 225° angle counterclockwise from the polar axis. Notice that the point P is not located on the terminal side of θ. Instead, it lies on the ray *opposite the terminal side of* θ at a distance of $|-3|$, or 3, units from the pole. In general, **when r in (r, θ) is negative, a point is located $|r|$ units along the ray opposite the terminal side of θ.**

In Figure 6.20(c), r and θ are both negative: $r = -3$ and $\theta = -\dfrac{3\pi}{4}$. Because θ is negative, we draw a $-\dfrac{3\pi}{4}$ (or $-135°$) angle clockwise from the polar axis.

Because r is negative, the point is located on the ray opposite the terminal side of θ.

Our observations indicate the importance of the sign of r in locating $P = (r, \theta)$ in polar coordinates.

The Sign of r and a Point's Location in Polar Coordinates

The point $P = (r, \theta)$ is located $|r|$ units from the pole. If $r > 0$, the point lies on the terminal side of θ. If $r < 0$, the point lies along the ray opposite the terminal side of θ. If $r = 0$, the point lies at the pole, regardless of the value of θ.

EXAMPLE 1 Plotting Points in a Polar Coordinate System

Plot the points with the following polar coordinates:

a. $(2, 135°)$ **b.** $\left(-3, \dfrac{3\pi}{2}\right)$ **c.** $\left(-1, -\dfrac{\pi}{4}\right)$.

Solution

a. To plot the point $(r, \theta) = (2, 135°)$, begin with the 135° angle. Because 135° is a positive angle, draw $\theta = 135°$ counterclockwise from the polar axis. Now consider $r = 2$. Because $r > 0$, plot the point by going out two units on the terminal side of θ. Figure 6.21(a) shows the point.

b. To plot the point $(r, \theta) = \left(-3, \dfrac{3\pi}{2}\right)$, begin with the $\dfrac{3\pi}{2}$ angle. Because $\dfrac{3\pi}{2}$ is a positive angle, we draw $\theta = \dfrac{3\pi}{2}$ counterclockwise from the polar axis. Now consider $r = -3$. Because $r < 0$, plot the point by going out three units along the ray *opposite* the terminal side of θ. Figure 6.21(b) shows the point.

c. To plot the point $(r, \theta) = \left(-1, -\dfrac{\pi}{4}\right)$, begin with the $-\dfrac{\pi}{4}$ angle. Because $-\dfrac{\pi}{4}$ is a negative angle, draw $\theta = -\dfrac{\pi}{4}$ clockwise from the polar axis.

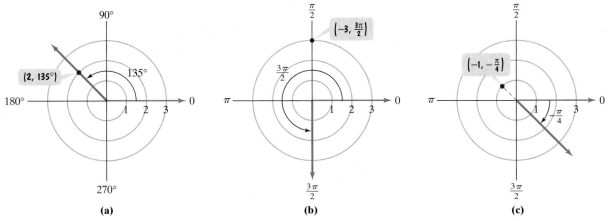

Figure 6.21 Plotting points

Now consider $r = -1$. Because $r < 0$, plot the point by going out one unit along the ray *opposite* the terminal side of θ. Figure 6.21(c) shows the point.

> **Check Point 1** Plot the points with the following polar coordinates:
>
> **a.** $(3, 315°)$ **b.** $(-2, \pi)$ **c.** $\left(-1, -\dfrac{\pi}{2}\right)$.

2 Find multiple sets of polar coordinates for a given point.

Multiple Representation of Points in the Polar Coordinate System

In rectangular coordinates, each point (x, y) has exactly one representation. By contrast, any point in polar coordinates can be represented in infinitely many ways. For example,

$$(r, \theta) = (r, \theta + 2\pi) \qquad \text{and} \qquad (r, \theta) = (-r, \theta + \pi).$$

Adding 1 revolution, or 2π radians, to the angle does not change the point's location.

Adding $\frac{1}{2}$ revolution, or π radians, to the angle and replacing r with $-r$ does not change the point's location.

Discovery

Illustrate the statements in the voice balloons by plotting:

a. $\left(1, \dfrac{\pi}{2}\right)$ and $\left(1, \dfrac{5\pi}{2}\right)$.

b. $\left(3, \dfrac{\pi}{4}\right)$ and $\left(-3, \dfrac{5\pi}{4}\right)$.

Thus, to find two other representations for the point (r, θ),

- Add 2π to the angle and do not change r.
- Add π to the angle and replace r with $-r$.

Continually adding or subtracting 2π in either of these representations does not change the point's location.

> **Multiple Representation of Points**
> If n is any integer, the point (r, θ) can be represented as
> $$(r, \theta) = (r, \theta + 2n\pi) \quad \text{or} \quad (r, \theta) = (-r, \theta + \pi + 2n\pi).$$

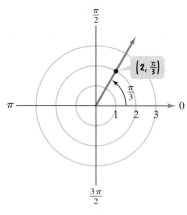

Figure 6.22 Finding other representations of a given point

EXAMPLE 2 Finding Other Polar Coordinates for a Given Point

The point $\left(2, \dfrac{\pi}{3}\right)$ is plotted in Figure 6.22. Find another representation of this point in which:

a. r is positive and $2\pi < \theta < 4\pi$.
b. r is negative and $0 < \theta < 2\pi$.
c. r is positive and $-2\pi < \theta < 0$.

Solution

a. Add 2π to the angle and do not change r.

$$\left(2, \frac{\pi}{3}\right) = \left(2, \frac{\pi}{3} + 2\pi\right) = \left(2, \frac{\pi}{3} + \frac{6\pi}{3}\right) = \left(2, \frac{7\pi}{3}\right)$$

b. Add π to the angle and replace r with $-r$.

$$\left(2, \frac{\pi}{3}\right) = \left(-2, \frac{\pi}{3} + \pi\right) = \left(-2, \frac{\pi}{3} + \frac{3\pi}{3}\right) = \left(-2, \frac{4\pi}{3}\right)$$

c. Subtract 2π from the angle and do not change r.

$$\left(2, \frac{\pi}{3}\right) = \left(2, \frac{\pi}{3} - 2\pi\right) = \left(2, \frac{\pi}{3} - \frac{6\pi}{3}\right) = \left(2, -\frac{5\pi}{3}\right)$$

Check Point 2 Find another representation of $\left(5, \dfrac{\pi}{4}\right)$ in which:

a. r is positive and $2\pi < \theta < 4\pi$.
b. r is negative and $0 < \theta < 2\pi$.
c. r is positive and $-2\pi < \theta < 0$.

Relations between Polar and Rectangular Coordinates

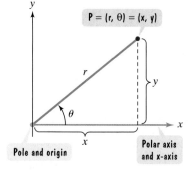

Figure 6.23 Polar and rectangular coordinate systems

We now consider both polar and rectangular coordinates simultaneously. Figure 6.23 shows the two coordinate systems. The polar axis coincides with the positive x-axis and the pole coincides with the origin. A point P, other than the origin, has rectangular coordinates (x, y) and polar coordinates (r, θ), as indicated in the figure. We wish to find equations relating the two sets of coordinates. From the figure, we see that

$$x^2 + y^2 = r^2$$

$$\sin \theta = \frac{y}{r} \qquad \cos \theta = \frac{x}{r} \qquad \tan \theta = \frac{y}{x}.$$

These relationships hold when P is in any quadrant and when $r > 0$ or $r < 0$.

Relations between Polar and Rectangular Coordinates

$$x = r \cos \theta$$
$$y = r \sin \theta$$
$$x^2 + y^2 = r^2$$

$$\tan \theta = \frac{y}{x}$$

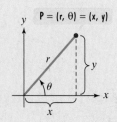

3 Convert a point from polar to rectangular coordinates.

Point Conversion from Polar to Rectangular Coordinates

To convert a point from polar coordinates (r, θ) to rectangular coordinates (x, y), use the formulas $x = r \cos \theta$ and $y = r \sin \theta$.

EXAMPLE 3   Polar-to-Rectangular Point Conversion

Find the rectangular coordinates of the points with the following polar coordinates:

a. $\left(2, \dfrac{3\pi}{2} \right)$ **b.** $\left(-8, \dfrac{\pi}{3} \right).$

Solution   We find (x, y) by substituting the given values for r and θ into $x = r \cos \theta$ and $y = r \sin \theta$.

a. We begin with the rectangular coordinates of the point $(r, \theta) = \left(2, \dfrac{3\pi}{2} \right).$

$$x = r \cos \theta = 2 \cos \frac{3\pi}{2} = 2 \cdot 0 = 0$$

$$y = r \sin \theta = 2 \sin \frac{3\pi}{2} = 2(-1) = -2$$

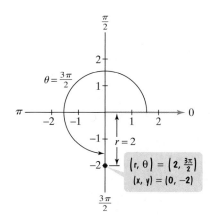

Figure 6.24 Converting $\left(2, \dfrac{3\pi}{2} \right)$ to rectangular coordinates

The rectangular coordinates of $\left(2, \dfrac{3\pi}{2} \right)$ are $(0, -2)$. See Figure 6.24.

b. We now find the rectangular coordinates of the point $(r, \theta) = \left(-8, \dfrac{\pi}{3} \right).$

$$x = r \cos \theta = -8 \cos \frac{\pi}{3} = -8 \left(\frac{1}{2} \right) = -4$$

$$y = r \sin \theta = -8 \sin \frac{\pi}{3} = -8 \left(\frac{\sqrt{3}}{2} \right) = -4\sqrt{3}$$

The rectangular coordinates of $\left(-8, \dfrac{\pi}{3} \right)$ are $(-4, -4\sqrt{3}).$

Technology

Some graphing utilities can convert a point from polar coordinates to rectangular coordinates. Consult your manual. The screen on the right verifies the polar-rectangular conversion in Example 3(a). It shows that the rectangular coordinates of $(r, \theta) = \left(2, \dfrac{3\pi}{2} \right)$ are $(0, -2)$.

Notice that the x- and y-coordinates are displayed separately.

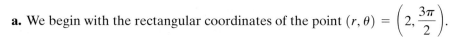

```
P▶Rx(2,3π/2)
              0
P▶Ry(2,3π/2)
             -2
```

Check Point 3 Find the rectangular coordinates of the points with the following polar coordinates:

a. $(3, \pi)$ **b.** $\left(-10, \dfrac{\pi}{6}\right)$.

4 Convert a point from rectangular to polar coordinates.

Point Conversion from Rectangular to Polar Coordinates

Conversion from rectangular coordinates (x, y) to polar coordinates (r, θ) is a bit more complicated. Keep in mind that there are infinitely many representations for a point in polar coordinates. If the point (x, y) lies in one of the four quadrants, we will use a representation in which

- r is positive, and
- θ is the smallest positive angle that lies in the same quadrant as (x, y).

These conventions provide the following procedure:

> **Converting a Point from Rectangular to Polar Coordinates**
> $(r > 0 \text{ and } 0 \le \theta < 2\pi)$
>
> **1.** Plot the point (x, y).
>
> **2.** Find r by computing the distance from the origin to (x, y): $r = \sqrt{x^2 + y^2}$.
>
> **3.** Find θ using $\tan\theta = \dfrac{y}{x}$ with θ lying in the same quadrant as (x, y).

EXAMPLE 4 Rectangular-to-Polar Point Conversion

Find polar coordinates of a point whose rectangular coordinates are $(-1, \sqrt{3})$.

Solution We begin with $(x, y) = (-1, \sqrt{3})$ and use our three-step procedure to find a set of polar coordinates (r, θ).

Step 1   Plot the point (x, y). The point $(-1, \sqrt{3})$ is plotted in quadrant II in Figure 6.25.

Step 2   Find r by computing the distance from the origin to (x, y).

$$r = \sqrt{x^2 + y^2} = \sqrt{(-1)^2 + \left(\sqrt{3}\right)^2} = \sqrt{1 + 3} = \sqrt{4} = 2$$

Step 3   Find θ using $\tan\theta = \dfrac{y}{x}$ with θ lying in the same quadrant as (x, y).

$$\tan\theta = \dfrac{y}{x} = \dfrac{\sqrt{3}}{-1} = -\sqrt{3}$$

We know that $\tan\dfrac{\pi}{3} = \sqrt{3}$. Because θ lies in quadrant II,

$$\theta = \pi - \dfrac{\pi}{3} = \dfrac{3\pi}{3} - \dfrac{\pi}{3} = \dfrac{2\pi}{3}.$$

Polar coordinates of $(-1, \sqrt{3})$ are $(r, \theta) = \left(2, \dfrac{2\pi}{3}\right)$.

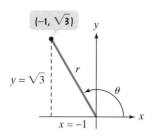

Figure 6.25 Converting $(-1, \sqrt{3})$ to polar coordinates

Technology

The screen shows the rectangular-polar conversion for $(-1, \sqrt{3})$ on a graphing utility. In Example 4, we showed that the polar coordinates of

$$(x, y) = (-1, \sqrt{3}) \text{ are } (r, \theta) = \left(2, \frac{2\pi}{3}\right). \text{ Using}$$

$\frac{2\pi}{3} \approx 2.09439510239$ verifies that our conversion is correct. Notice that the r- and (approximate) θ-coordinates are displayed separately.

```
R▸Pr(-1,√(3))
                      2
R▸Pθ(-1,√(3))
             2.094395102
```

Check Point 4 Find polar coordinates of a point whose rectangular coordinates are $(1, -\sqrt{3})$.

If a point (x, y) lies on a positive or negative axis, we use a representation in which

- r is positive, and
- θ is the smallest quadrantal angle that lies on the same positive or negative axis as (x, y).

In these cases, you can find r and θ by plotting (x, y) and inspecting the figure. Let's see how this is done.

EXAMPLE 5 Rectangular-to-Polar Point Conversion

Find polar coordinates of a point whose rectangular coordinates are $(-2, 0)$.

Solution   We begin with $(x, y) = (-2, 0)$ and find a set of polar coordinates (r, θ).

Step 1   Plot the point (x, y).   The point $(-2, 0)$ is plotted in Figure 6.26.

Step 2   Find r, the distance from the origin to (x, y).   Can you tell by looking at Figure 6.26 that this distance is 2?

$$r = \sqrt{x^2 + y^2} = \sqrt{(-2)^2 + 0^2} = \sqrt{4} = 2$$

Step 3   Find θ with θ lying on the same positive or negative axis as (x, y).   The point $(-2, 0)$ is on the negative x-axis. Thus, θ lies on the negative x-axis and $\theta = \pi$. Polar coordinates of $(-2, 0)$ are $(2, \pi)$.

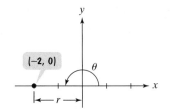

Figure 6.26 Converting $(-2, 0)$ to polar coordinates

Check Point 5 Find polar coordinates of a point whose rectangular coordinates are $(0, -4)$.

5 Convert an equation from rectangular to polar coordinates.

Equation Conversion from Rectangular to Polar Coordinates

A **polar equation** is an equation whose variables are r and θ. Two examples of polar equations are

$$r = \frac{5}{\cos\theta + \sin\theta} \qquad \text{and} \qquad r = \csc\theta.$$

To convert a rectangular equation in x and y to a polar equation in r and θ, replace x with $r \cos \theta$ and y with $r \sin \theta$.

EXAMPLE 6   Converting an Equation from Rectangular to Polar Coordinates

Convert $x + y = 5$ to a polar equation.

Solution   Our goal is to obtain an equation in which the variables are r and θ rather than x and y. We use $x = r \cos \theta$ and $y = r \sin \theta$.

$$x + y = 5 \quad \text{This is the given equation in rectangular coordinates.}$$
$$r \cos \theta + r \sin \theta = 5 \quad \text{Replace } x \text{ with } r \cos \theta \text{ and } y \text{ with } r \sin \theta.$$

Thus, the polar equation for $x + y = 5$ is $r \cos \theta + r \sin \theta = 5$. We can express this polar equation in a number of equivalent ways, including an equation that gives r in terms of θ.

$$r(\cos \theta + \sin \theta) = 5 \qquad \text{Factor out } r.$$
$$r = \frac{5}{\cos \theta + \sin \theta} \quad \text{Divide both sides of the equation by } \cos \theta + \sin \theta.$$

> **Check Point 6**   Convert $3x - y = 6$ to a polar equation. Express the polar equation with r in terms of θ.

6 Convert an equation from polar to rectangular coordinates.

Equation Conversion from Polar to Rectangular Coordinates

When we convert an equation from polar to rectangular coordinates, our goal is to obtain an equation in which the variables are x and y rather than r and θ. We use one or more of the following equations:

$$r^2 = x^2 + y^2 \qquad r \cos \theta = x \qquad r \sin \theta = y \qquad \tan \theta = \frac{y}{x}.$$

To use these equations, it is sometimes necessary to do something to the given polar equation. This could include squaring both sides, using an identity, taking the tangent of both sides, or multiplying both sides by r.

EXAMPLE 7   Converting Equations from Polar to Rectangular Form

Convert each polar equation to a rectangular equation in x and y:

a. $r = 3$ **b.** $\theta = \dfrac{\pi}{4}$ **c.** $r = \csc \theta$.

Solution

a. We use $r^2 = x^2 + y^2$ to convert the polar equation $r = 3$ to a rectangular equation.

$$r = 3 \quad \text{This is the given polar equation.}$$

$$r^2 = 9 \quad \text{Square both sides.}$$

$$x^2 + y^2 = 9 \quad \text{Use } r^2 = x^2 + y^2 \text{ on the left side.}$$

The rectangular equation for $r = 3$ is $x^2 + y^2 = 9$.

b. We use $\tan \theta = \dfrac{y}{x}$ to convert the polar equation $\theta = \dfrac{\pi}{4}$ to a rectangular equation in x and y.

$$\theta = \frac{\pi}{4} \qquad \text{This is the given polar equation.}$$

$$\tan \theta = \tan \frac{\pi}{4} \qquad \text{Take the tangent of both sides.}$$

$$\tan \theta = 1 \qquad \tan \frac{\pi}{4} = 1$$

$$\frac{y}{x} = 1 \qquad \text{Use } \tan \theta = \frac{y}{x} \text{ on the left side.}$$

$$y = x \qquad \text{Multiply both sides by } x.$$

The rectangular equation for $\theta = \dfrac{\pi}{4}$ is $y = x$.

c. We use $r \sin \theta = y$ to convert the polar equation $r = \csc \theta$ to a rectangular equation. To do this, we express the cosecant in terms of the sine.

$$r = \csc \theta \qquad \text{This is the given polar equation.}$$

$$r = \frac{1}{\sin \theta} \qquad \csc \theta = \frac{1}{\sin \theta}$$

$$r \sin \theta = 1 \qquad \text{Multiply both sides by } \sin \theta.$$
$$y = 1 \qquad \text{Use } r \sin \theta = y \text{ on the left side.}$$

The rectangular equation for $r = \csc \theta$ is $y = 1$.

Converting a polar equation to a rectangular equation may be a useful way to develop or check a graph. For example, the graph of the polar equation $r = 3$ consists of all points that are three units from the pole. Thus, the graph is a circle centered at the pole with radius $= 3$. The rectangular equation for $r = 3$, namely $x^2 + y^2 = 9$, has precisely the same graph (see Figure 6.27). We will discuss graphs of polar equations in the next section.

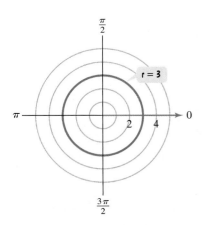

Figure 6.27 The equations $r = 3$ and $x^2 + y^2 = 9$ have the same graph.

Check Point 7 Convert each polar equation to a rectangular equation in x and y:

a. $r = 4$ **b.** $\theta = \dfrac{3\pi}{4}$ **c.** $r = \sec \theta$.

EXERCISE SET 6.3

Practice Exercises

In Exercises 1–10, indicate if the point with the given polar coordinates is represented by A, B, C, or D on the graph.

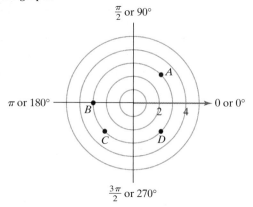

$\frac{\pi}{2}$ or 90°

π or 180°

0 or 0°

$\frac{3\pi}{2}$ or 270°

1. $(3, 225°)$

2. $(3, 315°)$

3. $\left(-3, \frac{5\pi}{4}\right)$

4. $\left(-3, \frac{\pi}{4}\right)$

5. $(3, \pi)$

6. $(-3, 0)$

7. $(3, -135°)$

8. $(3, -315°)$

9. $\left(-3, -\frac{3\pi}{4}\right)$

10. $\left(-3, -\frac{5\pi}{4}\right)$

In Exercises 11–20, use a polar coordinate system like the one shown for Exercises 1–10 to plot each point with the given polar coordinates.

11. $(2, 45°)$

12. $(1, 45°)$

13. $(3, 90°)$

14. $(2, 270°)$

15. $\left(3, \frac{4\pi}{3}\right)$

16. $\left(3, \frac{7\pi}{6}\right)$

17. $(-1, \pi)$

18. $\left(-1, \frac{3\pi}{2}\right)$

19. $\left(-2, -\frac{\pi}{2}\right)$

20. $(-3, -\pi)$

In Exercises 21–26, use a polar coordinate system like the one shown for Exercises 1–10 to plot each point with the given polar coordinates. Then find another representation (r, θ) of this point in which:

a. $r > 0, \quad 2\pi < \theta < 4\pi$

b. $r < 0, \quad 0 < \theta < 2\pi$

c. $r > 0, \quad -2\pi < \theta < 0$

21. $\left(5, \frac{\pi}{6}\right)$

22. $\left(8, \frac{\pi}{6}\right)$

23. $\left(10, \frac{3\pi}{4}\right)$

24. $\left(12, \frac{2\pi}{3}\right)$

25. $\left(4, \frac{\pi}{2}\right)$

26. $(6, \pi)$

In Exercises 27–32, select the representations that do not change the location of the given point.

27. $(7, 140°)$

 a. $(-7, 320°)$ **b.** $(-7, -40°)$

 c. $(-7, 220°)$ **d.** $(7, -220°)$

28. $(4, 120°)$

 a. $(-4, 300°)$ **b.** $(-4, -240°)$

 c. $(4, -240°)$ **d.** $(4, 480°)$

29. $\left(2, -\frac{3\pi}{4}\right)$

 a. $\left(2, -\frac{7\pi}{4}\right)$ **b.** $\left(2, \frac{5\pi}{4}\right)$

 c. $\left(-2, -\frac{\pi}{4}\right)$ **d.** $\left(-2, -\frac{7\pi}{4}\right)$

30. $\left(-2, \frac{7\pi}{6}\right)$

 a. $\left(-2, -\frac{5\pi}{6}\right)$ **b.** $\left(-2, -\frac{\pi}{6}\right)$

 c. $\left(2, -\frac{\pi}{6}\right)$ **d.** $\left(2, \frac{\pi}{6}\right)$

31. $\left(-5, -\frac{\pi}{4}\right)$

 a. $\left(-5, \frac{7\pi}{4}\right)$ **b.** $\left(5, -\frac{5\pi}{4}\right)$

 c. $\left(-5, \frac{11\pi}{4}\right)$ **d.** $\left(5, \frac{\pi}{4}\right)$

32. $(-6, 3\pi)$

 a. $(6, 2\pi)$ **b.** $(6, -\pi)$

 c. $(-6, \pi)$ **d.** $(-6, -2\pi)$

In Exercises 33–40, polar coordinates of a point are given. Find the rectangular coordinates of each point.

33. $(4, 90°)$

34. $(6, 180°)$

35. $\left(2, \frac{\pi}{3}\right)$

36. $\left(2, \frac{\pi}{6}\right)$

37. $\left(-4, \frac{\pi}{2}\right)$

38. $\left(-6, \frac{3\pi}{2}\right)$

39. $(7.4, 2.5)$

40. $(8.3, 4.6)$

In Exercises 41–48, the rectangular coordinates of a point are given. Find polar coordinates of each point.

41. $(-2, 2)$

42. $(2, -2)$

43. $(2, -2\sqrt{3})$

44. $(-2\sqrt{3}, 2)$

45. $(-\sqrt{3}, -1)$

46. $(-1, -\sqrt{3})$

47. $(5, 0)$

48. $(0, -6)$

In Exercises 49–58, convert each rectangular equation to a polar equation.

49. $3x + y = 7$ (Express r in terms of θ.)

50. $x + 5y = 8$ (Express r in terms of θ.)

51. $x = 7$ **52.** $y = 3$

53. $x^2 + y^2 = 9$ **54.** $x^2 + y^2 = 16$

55. $x^2 + y^2 = 4x$ **56.** $x^2 + y^2 = 6x$

57. $y^2 = 6x$ **58.** $x^2 = 6y$

In Exercises 59–72, convert each polar equation to a rectangular equation.

59. $r = 8$ **60.** $r = 10$

61. $\theta = \dfrac{\pi}{2}$ **62.** $\theta = \dfrac{\pi}{3}$

63. $r \sin \theta = 3$ **64.** $r \cos \theta = 7$

65. $r = 4 \csc \theta$ **66.** $r = 6 \sec \theta$

67. $r = \sin \theta$ **68.** $r = \cos \theta$

69. $r = 6 \cos \theta + 4 \sin \theta$ **70.** $r = 8 \cos \theta + 2 \sin \theta$

71. $r^2 \sin 2\theta = 2$ **72.** $r^2 \cos 2\theta = 2$

Application Exercises

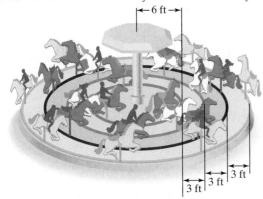

Use the figure of the merry-go-round to solve Exercises 73–74. There are four circles of horses. Each circle is three feet from the next circle. The radius of the inner circle is 6 feet.

73. If a horse in the outer circle is $\frac{2}{3}$ of the way around the merry-go-round, give its polar coordinates.

74. If a horse in the inner circle is $\frac{5}{6}$ of the way around the merry-go-round, give its polar coordinates.

The wind is blowing at 10 knots. Sailboat racers look for a sailing angle to the 10-knot wind that produces maximum sailing speed. In this application, (r, θ) describes the sailing speed, r, in knots, at an angle θ to the 10-knot wind. Use this information to solve Exercises 75–77.

75. Interpret the polar coordinates: $(6.3, 50°)$.

76. Interpret the polar coordinates: $(7.4, 85°)$.

77. Four points in this 10-knot-wind situation are $(6.3, 50°)$, $(7.4, 85°)$, $(7.5, 105°)$, $(7.3, 135°)$. Based on these points,

which sailing angle to the 10-knot wind would you recommend to a serious sailboat racer? What sailing speed is achieved at this angle?

Writing in Mathematics

78. Explain how to plot (r, θ) if $r > 0$ and $\theta > 0$.

79. Explain how to plot (r, θ) if $r < 0$ and $\theta > 0$.

80. If you are given polar coordinates of a point, explain how to find two additional sets of polar coordinates for the point.

81. Explain how to convert a point from polar to rectangular coordinates. Provide an example with your explanation.

82. Explain how to convert a point from rectangular to polar coordinates. Provide an example with your explanation.

83. Explain how to convert from a rectangular equation to a polar equation.

84. In converting $r = 5$ from a polar equation to a rectangular equation, describe what should be done to both sides of the equation and why this should be done.

85. In converting $r = \sin \theta$ from a polar equation to a rectangular equation, describe what should be done to both sides of the equation and why this should be done.

86. Suppose that (r, θ) describes the sailing speed, r, in knots, at an angle θ to a wind blowing at 20 knots. You have a list of all ordered pairs (r, θ) for integral angles from $\theta = 0°$ to $\theta = 180°$. Describe a way to present this information so that a serious sailboat racer can visualize sailing speeds at different sailing angles to the wind.

Technology Exercises

In Exercises 87–89, polar coordinates of a point are given. Use a graphing utility to find the rectangular coordinates of each point to three decimal places.

87. $\left(4, \dfrac{2\pi}{3}\right)$ **88.** $(5.2, 1.7)$

89. $(-4, 1.088)$

In Exercises 90–92, the rectangular coordinates of a point are given. Use a graphing utility to find polar coordinates of each point to three decimal places.

90. $(-5, 2)$ **91.** $(\sqrt{5}, 2)$

92. $(-4.308, -7.529)$

Critical Thinking Exercises

93. Prove that the distance, d, between two points with polar coordinates (r_1, θ_1) and (r_2, θ_2) is
$$d = \sqrt{r_1^2 + r_2^2 - 2r_1 r_2 \cos(\theta_2 - \theta_1)}.$$

94. Use the formula in Exercise 93 to find the distance between $\left(2, \dfrac{5\pi}{6}\right)$ and $\left(4, \dfrac{\pi}{6}\right)$. Express the answer in simplified radical form.

95. Convert $r = 4 \cos \theta$ from a polar equation to a rectangular equation. Use the rectangular equation to give the center and the radius.

SECTION 6.4 *Graphs of Polar Equations*

Objectives

1. Graph polar equations.
2. Use symmetry to graph polar equations.

The America's Cup is the supreme event in ocean sailing. Competition is fierce and the costs are huge. Competitors look to mathematics to provide the critical innovation that can make the difference between winning and losing. In this section's exercise set, you will see how graphs of polar equations play a role in sailing faster using mathematics.

Using Polar Grids to Graph Polar Equations

Recall that a **polar equation** is an equation whose variables are r and θ. The **graph of a polar equation** is the set of all points whose polar coordinates satisfy the equation. We use **polar grids** like the one shown in Figure 6.28 to graph polar equations. The grid consists of circles with centers at the pole. This polar grid shows five such circles. A polar grid also shows lines passing through the pole. In this grid, each line represents an angle for which we know the exact values of the trigonometric functions.

Many polar coordinate grids show more circles and more lines through the pole than in Figure 6.28. See if your campus bookstore has paper with polar grids and use the polar graph paper throughout this section.

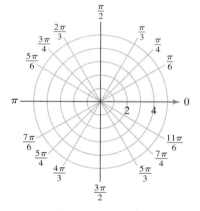

Figure 6.28 A polar coordinate grid

1 Graph polar equations.

Graphing a Polar Equation by Point Plotting

One method for graphing a polar equation such as $r = 4 \cos \theta$ is the **point-plotting method.** First, we make a table of values that satisfy the equation. Next, we plot these ordered pairs as points in the polar coordinate system. Finally, we connect the points with a smooth curve. This often gives us a picture of all ordered pairs (r, θ) that satisfy the equation.

EXAMPLE 1 **Graphing an Equation Using the Point-Plotting Method**

Graph the polar equation $r = 4 \cos \theta$ with θ in radians.

Solution We construct a partial table of coordinates using multiples of $\dfrac{\pi}{6}$. Then we plot the points and join them with a smooth curve, as shown in Figure 6.29.

θ	$r = 4\cos\theta$	(r, θ)
0	$4\cos 0 = 4\cdot 1 = 4$	$(4, 0)$
$\dfrac{\pi}{6}$	$4\cos\dfrac{\pi}{6} = 4\cdot\dfrac{\sqrt{3}}{2} = 2\sqrt{3} \approx 3.5$	$\left(3.5, \dfrac{\pi}{6}\right)$
$\dfrac{\pi}{3}$	$4\cos\dfrac{\pi}{3} = 4\cdot\dfrac{1}{2} = 2$	$\left(2, \dfrac{\pi}{3}\right)$
$\dfrac{\pi}{2}$	$4\cos\dfrac{\pi}{2} = 4\cdot 0 = 0$	$\left(0, \dfrac{\pi}{2}\right)$
$\dfrac{2\pi}{3}$	$4\cos\dfrac{2\pi}{3} = 4\left(-\dfrac{1}{2}\right) = -2$	$\left(-2, \dfrac{2\pi}{3}\right)$
$\dfrac{5\pi}{6}$	$4\cos\dfrac{5\pi}{6} = 4\left(-\dfrac{\sqrt{3}}{2}\right) = -2\sqrt{3} \approx -3.5$	$\left(-3.5, \dfrac{5\pi}{6}\right)$
π	$4\cos\pi = 4(-1) = -4$	$(-4, \pi)$
Values of r repeat.		

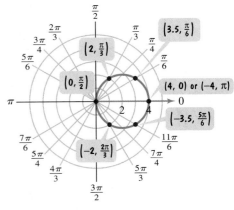

Figure 6.29 The graph of $r = 4\cos\theta$

Technology

A graphing utility can be used to obtain the graph of a polar equation. Use the polar mode with angle measure in radians. You must enter the minimum and maximum values for θ and an increment setting for θ, called θ step. θ step determines the number of points that the graphing utility will plot. Make θ step relatively small so that a significant number of points are plotted.

Shown is the graph of $r = 4\cos\theta$ in a $[-7.5, 7.5, 1]$ by $[-5, 5, 1]$ viewing rectangle with

$$\theta\min = 0$$
$$\theta\max = 2\pi$$
$$\theta\text{step} = \dfrac{\pi}{48}.$$

A square setting was used.

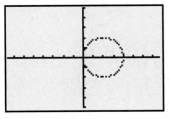

The graph of $r = 4\cos\theta$ in Figure 6.29 looks like a circle of radius 2 whose center is at the point $(x, y) = (2, 0)$. We can verify this observation by changing the polar equation to a rectangular equation.

$$r = 4\cos\theta \qquad \text{This is the given polar equation.}$$

$$r^2 = 4r\cos\theta \qquad \text{Multiply both sides by } r.$$

$$x^2 + y^2 = 4x \qquad \text{Convert to rectangular coordinates:}$$
$$r^2 = x^2 + y^2 \text{ and } r\cos\theta = x.$$

$$x^2 - 4x + y^2 = 0 \qquad \text{Subtract 4x from both sides.}$$

$$x^2 - 4x + 4 + y^2 = 4 \qquad \text{Complete the square on x:}$$
$$\tfrac{1}{2}(-4) = -2 \text{ and } (-2)^2 = 4.$$
$$\text{Add 4 to both sides.}$$

$$(x - 2)^2 + y^2 = 2^2 \qquad \text{Factor.}$$

This last equation is the standard form of the equation of a circle, $(x - h)^2 + (y - k)^2 = r^2$, with radius r and center at (h, k). Thus, the radius is 2 and the center is at $(h, k) = (2, 0)$.

In general, circles have simpler equations in polar form than in rectangular form.

Circles in Polar Coordinates

The graphs of

$$r = a \cos \theta \qquad \text{and} \qquad r = a \sin \theta$$

are circles.

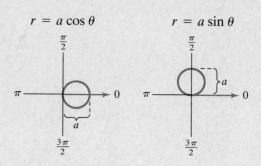

Check Point 1   Graph the equation $r = 4 \sin \theta$ with θ in radians. Use multiples of $\dfrac{\pi}{6}$ from 0 to π to generate coordinates for points (r, θ).

2 Use symmetry to graph polar equations.

Graphing a Polar Equation Using Symmetry

If the graph of a polar equation exhibits symmetry, you may be able to graph it more quickly. Three types of symmetry can be helpful.

Tests for Symmetry in Polar Coordinates

Symmetry with Respect to the Polar Axis (x-Axis)

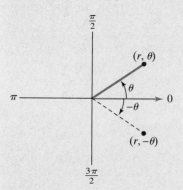

Replace θ with $-\theta$. If an equivalent equation results, the graph is symmetric with respect to the polar axis.

Symmetry with Respect to the Line $\theta = \dfrac{\pi}{2}$ (y-Axis)

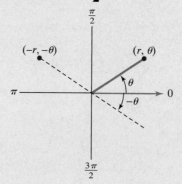

Replace (r, θ) with $(-r, -\theta)$. If an equivalent equation results, the graph is symmetric with respect to $\theta = \dfrac{\pi}{2}$.

Symmetry with Respect to the Pole (Origin)

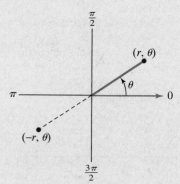

Replace r with $-r$. If an equivalent equation results, the graph is symmetric with respect to the pole.

If a polar equation passes a symmetry test, then its graph exhibits that symmetry. By contrast, if a polar equation fails a symmetry test, then its graph *may or may not* have that kind of symmetry. Thus, the graph of a polar equation may have a symmetry even if it fails a test for that particular symmetry. Nevertheless, the symmetry tests are useful. If we detect symmetry, we can obtain a graph of the equation by plotting fewer points.

EXAMPLE 2   Graphing a Polar Equation Using Symmetry

Check for symmetry and then graph the polar equation:

$$r = 1 - \cos \theta.$$

Solution   We apply each of the tests for symmetry.

Polar Axis: Replace θ with $-\theta$ in $r = 1 - \cos \theta$:

$$r = 1 - \cos(-\theta) \qquad \text{Replace } \theta \text{ with } -\theta \text{ in } r = 1 - \cos \theta.$$

$$r = 1 - \cos \theta \qquad \text{The cosine function is even:}$$
$$\cos(-\theta) = \cos \theta.$$

Because the polar equation does not change when θ is replaced with $-\theta$, the graph is symmetric with respect to the polar axis.

The Line $\theta = \dfrac{\pi}{2}$: Replace (r, θ) with $(-r, -\theta)$ in $r = 1 - \cos \theta$:

$$-r = 1 - \cos(-\theta) \qquad \text{Replace } r \text{ with } -r \text{ and } \theta \text{ with } -\theta \text{ in } r = 1 - \cos \theta.$$

$$-r = 1 - \cos \theta \qquad \cos(-\theta) = \cos \theta.$$

$$r = \cos \theta - 1 \qquad \text{Multiply both sides by } -1.$$

Because the polar equation $r = 1 - \cos \theta$ changes to $r = \cos \theta - 1$ when (r, θ) is replaced with $(-r, -\theta)$, the equation fails this symmetry test. The graph may or may not be symmetric with respect to the line $\theta = \dfrac{\pi}{2}$.

The Pole: Replace r with $-r$ in $r = 1 - \cos \theta$:

$$-r = 1 - \cos \theta \qquad \text{Replace } r \text{ with } -r.$$

$$r = \cos \theta - 1 \qquad \text{Multiply both sides by } -1.$$

Because the polar equation $r = 1 - \cos \theta$ changes to $r = \cos \theta - 1$ when r is replaced with $-r$, the equation fails this symmetry test. The graph may or may not be symmetric with respect to the pole.

Now we are ready to graph $r = 1 - \cos \theta$. Because the period of the cosine function is 2π, we need not consider values of θ beyond 2π. Recall that we discovered the graph of the equation $r = 1 - \cos \theta$ has symmetry with respect to the polar axis. Because the graph has this symmetry, we can obtain a complete graph by plotting fewer points. Let's start by finding the values of r for values of θ from 0 to π.

Technology

The graph of

$$r = 1 - \cos\theta$$

was obtained using a $[-2, 2, 1]$ by $[-2, 2, 1]$ viewing rectangle and

$$\theta\text{min} = 0, \quad \theta\text{max} = 2\pi,$$

$$\theta\text{step} = \frac{\pi}{48}.$$

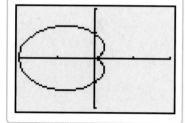

The values for r and θ are in the table above Figure 6.30. The TABLE feature on some graphing utilities is the most efficient way to create these values. The points in the table are plotted in Figure 6.30(a). Examine the graph. Keep in mind that the graph must be symmetric with respect to the polar axis. Thus, if we reflect the graph in Figure 6.30(a) about the polar axis, we will obtain a complete graph of $r = 1 - \cos\theta$. This graph is shown in Figure 6.30(b).

θ	0	$\dfrac{\pi}{6}$	$\dfrac{\pi}{3}$	$\dfrac{\pi}{2}$	$\dfrac{2\pi}{3}$	$\dfrac{5\pi}{6}$	π
r	0	0.13	0.5	1	1.5	1.87	2

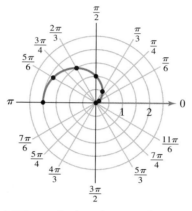

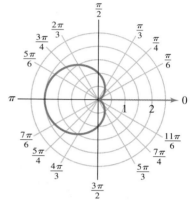

(a) The graph of $r = 1 - \cos\theta$ for $0 \le \theta \le \pi$

(b) A complete graph of $r = 1 - \cos\theta$

Figure 6.30 Graphing $r = 1 - \cos\theta$

Check Point 2   Check for symmetry and then graph the polar equation:

$$r = 1 + \cos\theta.$$

EXAMPLE 3   Graphing a Polar Equation

Graph the polar equation: $r = 1 + 2\sin\theta$.

Solution   We first check for symmetry.

$$r = 1 + 2\sin\theta$$

Polar Axis	**The Line** $\boldsymbol{\theta = \dfrac{\pi}{2}}$	**The Pole**
Replace θ with $-\theta$.	Replace (r, θ) with $(-r, -\theta)$.	Replace r with $-r$.
$r = 1 + 2\sin(-\theta)$	$-r = 1 + 2\sin(-\theta)$	$-r = 1 + 2\sin\theta$
$r = 1 + 2(-\sin\theta)$	$-r = 1 - 2\sin\theta$	$r = -1 - 2\sin\theta$
$r = 1 - 2\sin\theta$	$r = -1 + 2\sin\theta$	

None of these equations are equivalent to $r = 1 + 2\sin\theta$. Thus, the graph may or may not have each of these kinds of symmetry.

Now we are ready to graph $r = 1 + 2\sin\theta$. Because the period of the sine function is 2π, we need not consider values of θ beyond 2π. We identify points on the graph of $r = 1 + 2 \sin \theta$ by assigning values to θ and calculating the corresponding values of r. The values for r and θ are in the tables above Figures 6.31(a), (b), and (c). The complete graph of $r = 1 + 2 \sin \theta$ is shown in Figure 6.31(c). The inner loop indicates that the graph passes through the pole twice.

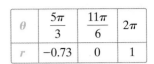

θ	0	$\dfrac{\pi}{6}$	$\dfrac{\pi}{3}$	$\dfrac{\pi}{2}$	$\dfrac{2\pi}{3}$	$\dfrac{5\pi}{6}$	π
r	1	2	2.73	3	2.73	2	1

θ	$\dfrac{7\pi}{6}$	$\dfrac{4\pi}{3}$	$\dfrac{3\pi}{2}$
r	0	-0.73	-1

θ	$\dfrac{5\pi}{3}$	$\dfrac{11\pi}{6}$	2π
r	-0.73	0	1

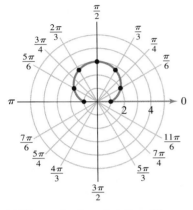

(a) The graph of $r = 1 + 2 \cos \theta$ for $0 \le \theta \le \pi$

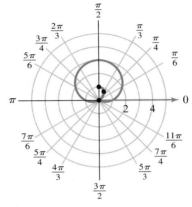

(b) The graph of $r = 1 + 2 \sin \theta$ for $0 \le \theta \le \dfrac{3\pi}{2}$

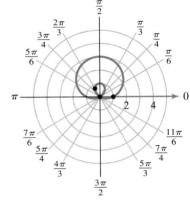

(c) The complete graph of $r = 1 + 2 \sin \theta$ for $0 \le \theta \le 2\pi$

Figure 6.31 Graphing $r = 1 + 2 \sin \theta$

We're not quite sure if the polar graph in Figure 6.31(c) looks like a snail. However, the graph is called a *limaçon*, which is a French word for snail. Limaçons come with and without inner loops.

Limaçons

The graphs of

$$r = a + b \sin \theta, \quad r = a - b \sin \theta,$$
$$r = a + b \cos \theta, \quad r = a - b \cos \theta, \quad a > 0, b > 0$$

are called **limaçons.** The ratio $\dfrac{a}{b}$ determines a limaçon's shape.

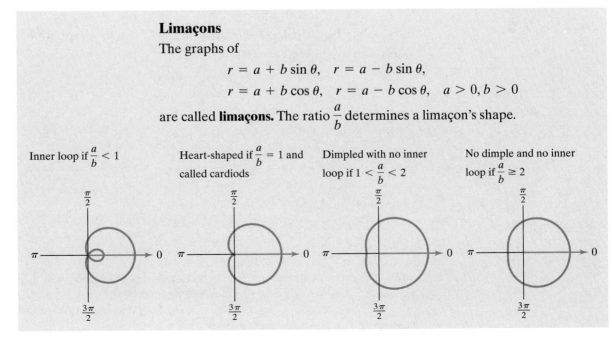

Inner loop if $\dfrac{a}{b} < 1$

Heart-shaped if $\dfrac{a}{b} = 1$ and called cardiods

Dimpled with no inner loop if $1 < \dfrac{a}{b} < 2$

No dimple and no inner loop if $\dfrac{a}{b} \ge 2$

Check Point 3 Graph the polar equation: $r = 1 - 2 \sin \theta$.

EXAMPLE 4   Graphing a Polar Equation

Graph the polar equation: $r = 4 \sin 2\theta$.

Solution   We first check for symmetry.

$$r = 4 \sin 2\theta$$

Polar Axis	**The Line $\theta = \dfrac{\pi}{2}$**	**The Pole**
Replace θ with $-\theta$.	Replace (r, θ) with $(-r, -\theta)$.	Replace r with $-r$.
$r = 4 \sin 2(-\theta)$	$-r = 4 \sin 2(-\theta)$	$-r = 4 \sin 2\theta$
$r = 4 \sin (-2\theta)$	$-r = 4 \sin (-2\theta)$	$r = -4 \sin 2\theta$
$r = -4 \sin 2\theta$	$-r = -4 \sin 2\theta$	
	$r = 4 \sin 2\theta$	
Equation changes and fails this symmetry test.	Equation does not change.	Equation changes and fails this symmetry test.

Thus, we can be sure that the graph is symmetric with respect to $\theta = \dfrac{\pi}{2}$. The graph may or may not be symmetric with respect to the polar axis or the pole.

Now we are ready to graph $r = 4 \sin 2\theta$. In Figure 6.32(a), we plot points on the graph of $r = 4 \sin 2\theta$ using values of θ from 0 to $\dfrac{\pi}{2}$ and the corresponding values of r. These coordinates are shown in the table in the margin. Because the graph is symmetric with respect to $\theta = \dfrac{\pi}{2}$, we can reflect the graph in Figure 6.32(a) about $\theta = \dfrac{\pi}{2}$ and obtain the graph from 0 to π. This graph is shown in Figure 6.32(b).

θ	0	$\dfrac{\pi}{6}$	$\dfrac{\pi}{4}$	$\dfrac{\pi}{3}$	$\dfrac{\pi}{2}$
r	0	3.46	4	3.46	0

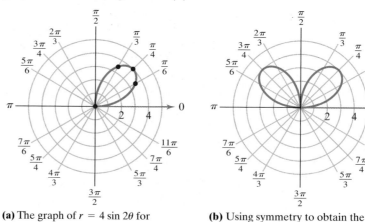

(a) The graph of $r = 4 \sin 2\theta$ for $0 \le \theta \le \dfrac{\pi}{2}$

(b) Using symmetry to obtain the graph of $r = 4 \sin 2\theta$ for $0 \le \theta \le \pi$

Figure 6.32 Partial graphs of $r = 4 \sin 2\theta$

Now we can complete the graph of $r = 4 \sin 2\theta$. The values for r and θ above the graph in Figure 6.33(a) give us the graph for $0 \le \theta \le \frac{3\pi}{2}$. Because the graph is symmetric with respect to $\theta = \frac{\pi}{2}$, we can reflect the quadrant III portion of the graph in Figure 6.33(a) about $\theta = \frac{\pi}{2}$ and obtain the complete graph from 0 to 2π. This graph is shown in Figure 6.33(b).

θ	π	$\dfrac{7\pi}{6}$	$\dfrac{5\pi}{4}$	$\dfrac{4\pi}{3}$	$\dfrac{3\pi}{2}$
r	0	3.46	4	3.46	0

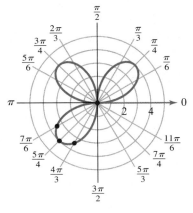

(a) The graph of $r = 4 \sin 2\theta$ for
$$0 \le \theta \le \frac{3\pi}{2}$$

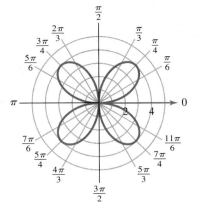

(b) Using symmetry to obtain the graph of $r = 4 \sin 2\theta$ for $0 \le \theta \le 2\pi$

Figure 6.33 Completing the graph of $r = 4 \sin 2\theta$

The curve in Figure 6.33(b) is called a **rose with four petals.**

Technology

The graph of
$$r = 4 \sin 2\theta$$
was obtained using a $[-4, 4, 1]$ by $[-4, 4, 1]$ viewing rectangle and
$$\theta \min = 0, \quad \theta \max = 2\pi,$$
$$\theta \text{ step} = \frac{\pi}{48}.$$

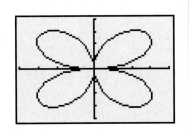

Rose Curves

The graphs of

$$r = a \sin n\theta \quad \text{and} \quad r = a \cos n\theta, \quad a \ne 0,$$

are called **rose curves.** If n is even, the rose has $2n$ petals. If n is odd, the rose has n petals.

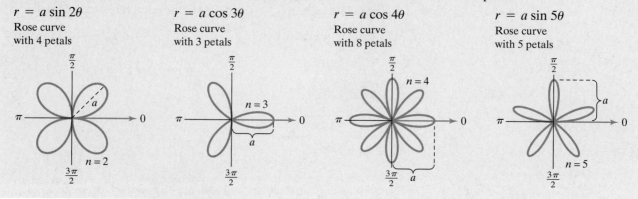

$r = a \sin 2\theta$
Rose curve
with 4 petals

$r = a \cos 3\theta$
Rose curve
with 3 petals

$r = a \cos 4\theta$
Rose curve
with 8 petals

$r = a \sin 5\theta$
Rose curve
with 5 petals

Check Point 4   Graph the polar equation: $r = 3 \cos 2\theta$.

EXAMPLE 5   Graphing a Polar Equation

Graph the polar equation: $r^2 = 4 \sin 2\theta$.

Solution   We first check for symmetry.

$$r^2 = 4 \sin 2\theta$$

Polar Axis	**The Line $\theta = \dfrac{\pi}{2}$**	**The Pole**
Replace θ with $-\theta$.	Replace (r, θ) with $(-r, -\theta)$.	Replace r with $-r$.
$r^2 = 4 \sin 2(-\theta)$	$(-r)^2 = 4 \sin 2(-\theta)$	$(-r)^2 = 4 \sin 2\theta$
$r^2 = 4 \sin (-2\theta)$	$r^2 = 4 \sin (-2\theta)$	$r^2 = 4 \sin 2\theta$
$r^2 = -4 \sin 2\theta$	$r^2 = -4 \sin 2\theta$	
Equation changes and fails this symmetry test.	Equation changes and fails this symmetry test.	Equation does not change.

Thus, we can be sure that the graph is symmetric with respect to the pole. The graph may or may not be symmetric with respect to the polar axis or the line $\theta = \dfrac{\pi}{2}$.

Now we are ready to graph $r^2 = 4 \sin 2\theta$. In Figure 6.34(a), we plot points on the graph by using values of θ from 0 to $\dfrac{\pi}{2}$ and the corresponding values of r. These coordinates are shown in the table above Figure 6.34(a). Notice that the points in Figure 6.34(a) are shown for $r \ge 0$. Because the graph is symmetric with respect to the pole, we can reflect the graph in Figure 6.34(a) about the pole and obtain the graph in Figure 6.34(b).

θ	0	$\dfrac{\pi}{6}$	$\dfrac{\pi}{4}$	$\dfrac{\pi}{3}$	$\dfrac{\pi}{2}$
r	0	±1.9	±2	±1.9	0

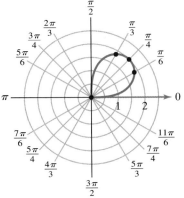

(a) The graph of $r^2 = 4 \sin 2\theta$ for $0 \le \theta \le \dfrac{\pi}{2}$

(b) Using symmetry with respect to the pole on the graph of $r^2 = 4 \sin 2\theta$

Figure 6.34 Graphing $r^2 = 4 \sin 2\theta$

Does Figure 6.34(b) show a complete graph of $r^2 = 4 \sin 2\theta$ or do we need to continue graphing in quadrants II and IV? If θ is in quadrant II, 2θ is in quadrant III or IV, where $\sin 2\theta$ is negative. Thus, $4 \sin 2\theta$ is negative. However, $r^2 = 4 \sin 2\theta$ and r^2 cannot be negative. This means that there are no points on the graph in quadrant II. The same observation applies to quadrant IV. Thus, Figure 6.34(b) shows the complete graph of $r^2 = 4 \sin 2\theta$.

The curve in Figure 6.34(b) is shaped like a propeller and is called a *lemniscate.*

Lemniscates

The graphs of

$$r^2 = a^2 \sin 2\theta \quad \text{and} \quad r^2 = a^2 \cos 2\theta, \quad a \neq 0$$

are called **lemniscates.**

$r^2 = a^2 \sin 2\theta$ is symmetric with respect to the pole.

$r^2 = a^2 \cos 2\theta$ is symmetric with respect to the polar axis, $\theta = \dfrac{\pi}{2}$, and the pole.

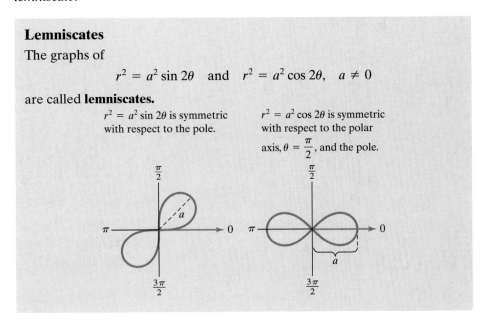

Check Point 5 Graph the polar equation: $r^2 = 4 \cos 2\theta$.

EXERCISE SET 6.4

Practice Exercises

In Exercises 1–6, the graph of a polar equation is given. Select the polar equation for each graph from the following options.

$$r = 2 \sin \theta, \quad r = 2 \cos \theta, \quad r = 1 + \sin \theta,$$

$$r = 1 - \sin \theta, \quad r = 3 \sin 2\theta, \quad r = 3 \sin 3\theta$$

1.

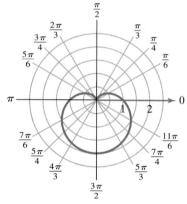

2.

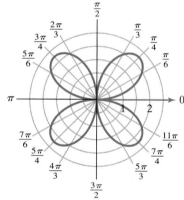

3.

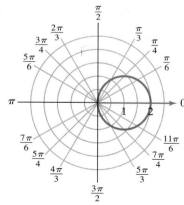

4.

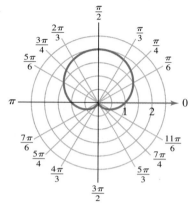

5.

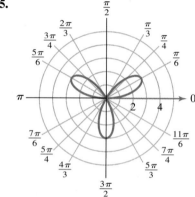

6.

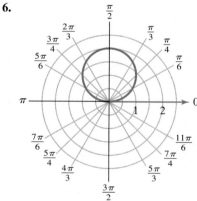

In Exercises 7–12, test for symmetry with respect to:

    a. *the polar axis* **b.** *the line* $\theta = \dfrac{\pi}{2}$ **c.** *the pole.*

7. $r = \sin \theta$ **8.** $r = \cos \theta$

9. $r = 4 + 3 \cos \theta$ **10.** $r = 2 \cos 2\theta$

11. $r^2 = 16 \cos 2\theta$ **12.** $r^2 = 16 \sin 2\theta$

In Exercises 13–34, test for symmetry and then graph each polar equation.

13. $r = 2 \cos \theta$

14. $r = 2 \sin \theta$

15. $r = 1 - \sin \theta$

16. $r = 1 + \sin \theta$

17. $r = 2 + 2 \cos \theta$

18. $r = 2 - 2 \cos \theta$

19. $r = 2 + \cos \theta$

20. $r = 2 - \sin \theta$

21. $r = 1 + 2 \cos \theta$

22. $r = 1 - 2 \cos \theta$

23. $r = 2 - 3 \sin \theta$

24. $r = 2 + 4 \sin \theta$

25. $r = 2 \cos 2\theta$

26. $r = 2 \sin 2\theta$

27. $r = 4 \sin 3\theta$

28. $r = 4 \cos 3\theta$

29. $r^2 = 9 \cos 2\theta$

30. $r^2 = 9 \sin 2\theta$

31. $r = 1 - 3 \sin \theta$

32. $r = 3 + \sin \theta$

33. $r \cos \theta = -3$

34. $r \sin \theta = 2$

Application Exercises

In Exercise Set 6.3, we considered an application in which sailboat racers look for a sailing angle to a 10-knot wind that produces maximum sailing speed. This situation is now represented by the polar graph in the figure shown. Each point (r, θ) on the graph gives the sailing speed, r, in knots, at an angle θ to the 10-knot wind. Use this information to solve Exercises 35–39.

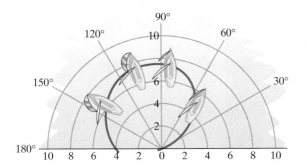

35. What is the speed, to the nearest knot, of the sailboat sailing at 60° angle to the wind?

36. What is the speed, to the nearest knot, of the sailboat sailing at a 120° angle to the wind?

37. What is the speed, to the nearest knot, of the sailboat sailing at a 90° angle to the wind?

38. What is the speed, to the nearest knot, of the sailboat sailing at a 180° angle to the wind?

39. What angle to the wind produces the maximum sailing speed? What is the speed? Round the angle to the nearest five degrees and the speed to the nearest half knot.

Writing in Mathematics

40. What is a polar equation?

41. What is the graph of a polar equation?

42. Describe how to graph a polar equation.

43. Describe the test for symmetry with respect to the polar axis.

44. Describe the test for symmetry with respect to the line $\theta = \dfrac{\pi}{2}$.

45. Describe the test for symmetry with respect to the pole.

46. If an equation fails the test for symmetry with respect to the polar axis, what can you conclude?

Technology Exercises

Use the polar mode of a graphing utility with angle measure in radians to solve Exercises 47–78. Unless otherwise indicated, use θ min $= 0$, θ max $= 2\pi$, and θ step $= \dfrac{\pi}{48}$. If you are not pleased with the quality of the graph, experiment with smaller values for θ step. However, if θ step is extremely small, it can take your graphing utility a long period of time to complete the graph.

47. Use a graphing utility to verify any six of your hand-drawn graphs in Exercises 13–34.

In Exercises 48–65, use a graphing utility to graph the polar equation.

48. $r = 4 \cos 5\theta$

49. $r = 4 \sin 5\theta$

50. $r = 4 \cos 6\theta$

51. $r = 4 \sin 6\theta$

52. $r = 2 + 2 \cos \theta$

53. $r = 2 + 2 \sin \theta$

54. $r = 4 + 2 \cos \theta$

55. $r = 4 + 2 \sin \theta$

56. $r = 2 + 4 \cos \theta$

57. $r = 2 + 4 \sin \theta$

58. $r = \dfrac{3}{\sin \theta}$

59. $r = \dfrac{3}{\cos \theta}$

60. $r = \cos \dfrac{3}{2} \theta$

61. $r = \cos \dfrac{5}{2} \theta$

62. $r = 3 \sin \left(\theta + \dfrac{\pi}{4} \right)$

63. $r = 2 \cos \left(\theta - \dfrac{\pi}{4} \right)$

64. $r = \dfrac{1}{1 - \sin \theta}$

65. $r = \dfrac{1}{3 - 2 \sin \theta}$

In Exercises 66–68, find the smallest interval for θ starting with θ min $= 0$ so that your graphing utility graphs the given polar equation exactly once without retracing any portion of it.

66. $r = 4 \sin \theta$ **67.** $r = 4 \sin 2\theta$ **68.** $r^2 = 4 \sin 2\theta$

In Exercises 69–72, use a graphing utility to graph each butterfly curve. Experiment with the range setting, particularly θ step, to produce a butterfly of the best possible quality.

69. $r = \cos^2 5\theta + \sin 3\theta + 0.3$

70. $r = \sin^4 4\theta + \cos 3\theta$

71. $r = \sin^5 \theta + 8 \sin \theta \cos^3 \theta$

72. $r = 1.5^{\sin \theta} - 2.5 \cos 4\theta + \sin^7 \dfrac{\theta}{15}$

(Use θ min $= 0$ and θ max $= 20\pi$.)

73. Use a graphing utility to graph $r = \sin n\theta$ for $n = 1, 2, 3,$ 4, 5, and 6. Use a separate viewing screen for each of the six graphs. What is the pattern for the number of loops that occur corresponding to each value of n? What is happening to the shape of the graphs as n increases? For each graph, what is the smallest interval for θ so that the graph is traced only once?

74. Repeat Exercise 73 for $r = \cos n\theta$. Are your conclusions the same as they were in Exercise 73?

75. Use a graphing utility to graph $r = 1 + 2 \sin n\theta$ for $n = 1,$ 2, 3, 4, 5, and 6. Use a separate viewing screen for each of the six graphs. What is the pattern for the number of large and small petals that occur corresponding to each value of n? How are the large and small petals related when n is odd and when n is even?

76. Repeat Exercise 75 for $r = 1 + 2 \cos n\theta$. Are your conclusions the same as they were in Exercise 75?

77. Graph the spiral $r = \theta$. Use a $[-30, 30, 1]$ by $[-30, 30, 1]$ viewing rectangle. Let θ min $= 0$ and θ max $= 2\pi$, then θ min $= 0$ and θ max $= 4\pi$, and finally θ min $= 0$ and θ max $= 8\pi$.

78. Graph the spiral $r = \dfrac{1}{\theta}$. Use a $[-1, 1, 1]$ by $[-1, 1, 1]$ viewing rectangle. Let θ min $= 0$ and θ max $= 2\pi$, then θ min $= 0$ and θ max $= 4\pi$, and finally θ min $= 0$ and θ max $= 8\pi$.

Critical Thinking Exercises

79. Describe a test for symmetry with respect to the line $\theta = \dfrac{\pi}{2}$ in which r is not replaced.

In Exercises 80–81, graph each polar equation without using a graphing utility.

80. $r = \dfrac{1}{3 - 2\cos\theta}$

81. $r = \dfrac{4}{1 + \sin\theta}$

SECTION 6.5 *Complex Numbers in Polar Form; DeMoivre's Theorem*

Objectives

1. Plot complex numbers in the complex plane.
2. Find the absolute value of a complex number.
3. Write complex numbers in polar form.
4. Convert a complex number from polar to rectangular form.
5. Find products of complex numbers in polar form.
6. Find quotients of complex numbers in polar form.
7. Find powers of complex numbers in polar form.
8. Find roots of complex numbers in polar form.

A magnification of the Mandelbrot set

Study Tip

Refer to Section 2.1 if you need to review the basics of complex numbers.

One of the new frontiers of mathematics suggests that there is an underlying order in things that appear to be random, such as the hiss and crackle of background noises as you tune a radio. Irregularities in the heartbeat, some of them severe enough to cause a heart attack, or irregularities in our sleeping patterns, such as insomnia, are examples of chaotic behavior. Chaos in the mathematical sense does not mean a complete lack of form or arrangement. In mathematics, chaos is used to describe something that appears to be random but is not actually random. The patterns of chaos appear in images like the one shown above, called the Mandelbrot set. Magnified portions of this image yield repetitions of the original structure, as well as new and unexpected patterns. The

Mandelbrot set transforms the hidden structure of chaotic events into a source of wonder and inspiration.

The Mandelbrot set is made possible by opening up graphing to include complex numbers in the form $a + bi$, where $i = \sqrt{-1}$. In this section, you will learn how to graph complex numbers and write them in terms of trigonometric functions.

1 Plot complex numbers in the complex plane.

The Complex Plane

We know that a real number can be represented as a point on a number line. By contrast, a complex number $z = a + bi$ is represented as a point (a, b) in a coordinate plane, as shown in Figure 6.35. The horizontal axis of the coordinate plane is called the **real axis.** The vertical axis is called the **imaginary axis.** The coordinate system is called the **complex plane.** Every complex number corresponds to a point in the complex plane and every point in the complex plane corresponds to a complex number. When we represent a complex number as a point in the complex plane, we say that we are **plotting the complex number.**

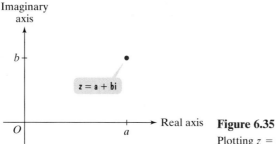

Figure 6.35
Plotting $z = a + bi$ in the complex plane

EXAMPLE 1   Plotting Complex Numbers

Plot each complex number in the complex plane:

a. $z = 3 + 4i$ **b.** $z = -1 - 2i$ **c.** $z = -3$ **d.** $z = -4i$.

Solution   See Figure 6.36.

a. We plot the complex number $z = 3 + 4i$ the same way we plot $(3, 4)$ in the rectangular coordinate system. We move three units to the right on the real axis and four units up parallel to the imaginary axis.

b. The complex number $z = -1 - 2i$ corresponds to the point $(-1, -2)$ in the rectangular coordinate system. Plot the complex number by moving one unit to the left on the real axis and two units down parallel to the imaginary axis.

c. Because $z = -3 = -3 + 0i$, this complex number corresponds to the point $(-3, 0)$. We plot -3 by moving three units to the left on the real axis.

d. Because $z = -4i = 0 - 4i$, this number corresponds to the point $(0, -4)$. We plot the complex number by moving four units down on the imaginary axis.

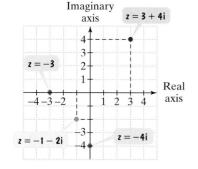

Figure 6.36 Plotting complex numbers

Check Point 1   Plot each complex number in the complex plane:

a. $z = 2 + 3i$ **b.** $z = -3 - 5i$

c. $z = -4$ **d.** $z = -i$.

2 Find the absolute value of a complex number.

Recall that the absolute value of a real number is its distance from 0 on the number line. The **absolute value of the complex number** $z = a + bi$, denoted by $|z|$, is its distance from the origin in the complex plane.

> ### The Absolute Value of a Complex Number
> The **absolute value** of the complex number $a + bi$ is
> $$|z| = |a + bi| = \sqrt{a^2 + b^2}.$$

EXAMPLE 2 Finding the Absolute Value of a Complex Number

Determine the absolute value of each of the following complex numbers:

a. $z = 3 + 4i$ **b.** $z = -1 - 2i$.

Solution

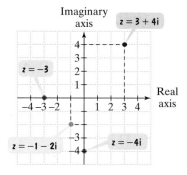

Figure 6.36, repeated

a. The absolute value of $z = 3 + 4i$ is found using $a = 3$ and $b = 4$.
$$|z| = \sqrt{3^2 + 4^2} = \sqrt{9 + 16} = \sqrt{25} = 5 \quad \text{Use } z = \sqrt{a^2 + b^2} \text{ with } a = 3 \text{ and } b = 4.$$

Thus, the distance from the origin to the point $z = 3 + 4i$, shown in quadrant I in Figure 6.36, is five units.

b. The absolute value of $z = -1 - 2i$ is found using $a = -1$ and $b = -2$.
$$|z| = \sqrt{(-1)^2 + (-2)^2} = \sqrt{1 + 4} = \sqrt{5} \quad \text{Use } z = \sqrt{a^2 + b^2} \text{ with } a = -1 \text{ and } b = -2.$$

Thus, the distance from the origin to the point $z = -1 - 2i$, shown in quadrant III in Figure 6.36, is $\sqrt{5}$ units.

Check Point 2 Determine the absolute value of each of the following complex numbers:

a. $z = 5 + 12i$ **b.** $2 - 3i$.

3 Write complex numbers in polar form.

Polar Form of a Complex Number

A complex number in the form $z = a + bi$ is said to be in **rectangular form.** Suppose that its absolute value is r. In Figure 6.37, we let θ be an angle in standard position whose terminal side passes through the point (a, b). From the figure, we see that
$$r = \sqrt{a^2 + b^2}.$$

Likewise, according to the definitions of the trigonometric functions,
$$\cos\theta = \frac{a}{r} \qquad \sin\theta = \frac{b}{r} \qquad \tan\theta = \frac{b}{a}.$$
$$a = r\cos\theta \qquad\qquad b = r\sin\theta$$

By substituting the expressions for a and b in $z = a + bi$, we write the complex number in terms of trigonometric functions.

$$z = a + bi = r\cos\theta + (r\sin\theta)i = r(\cos\theta + i\sin\theta)$$

$a = r\cos\theta$ and $b = r\sin\theta$.   Factor out r from each of the two previous terms.

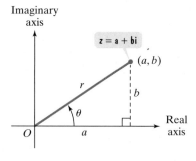

Figure 6.37

The expression $z = r(\cos\theta + i\sin\theta)$ is called the **polar form of a complex number.**

Polar Form of a Complex Number

The complex number $z = a + bi$ is written in **polar form** as

$$z = r(\cos\theta + i\sin\theta),$$

where $a = r\cos\theta, b = r\sin\theta, r = \sqrt{a^2 + b^2}$, and $\tan\theta = \dfrac{b}{a}$. The value of r is called the **modulus** (plural: moduli) of the complex number z, and the angle θ is called the **argument** of the complex number z, with $0 \le \theta < 2\pi$.

EXAMPLE 3   Writing a Complex Number in Polar Form

Plot $z = -2 - 2i$ in the complex plane. Then write z in polar form.

Solution   The complex number $z = -2 - 2i$ is in rectangular form $z = a + bi$, with $a = -2$ and $b = -2$. We plot the number by moving two units to the left on the real axis and two units down parallel to the imaginary axis, as shown in Figure 6.38.

By definition, the polar form of z is $r(\cos\theta + i\sin\theta)$. We need to determine the value for r, the modulus, and the value for θ, the argument. Figure 6.38 shows r and θ. We use $r = \sqrt{a^2 + b^2}$ with $a = -2$ and $b = -2$ to find r.

$$r = \sqrt{a^2 + b^2} = \sqrt{(-2)^2 + (-2)^2} = \sqrt{4 + 4} = \sqrt{8} = \sqrt{4 \cdot 2} = 2\sqrt{2}$$

We use $\tan\theta = \dfrac{b}{a}$ with $a = -2$ and $b = -2$ to find θ.

$$\tan\theta = \frac{b}{a} = \frac{-2}{-2} = 1$$

We know that $\tan\dfrac{\pi}{4} = 1$. Figure 6.38 shows that the argument, θ, lies in quadrant III. Thus,

$$\theta = \pi + \frac{\pi}{4} = \frac{4\pi}{4} + \frac{\pi}{4} = \frac{5\pi}{4}.$$

We use $r = 2\sqrt{2}$ and $\theta = \dfrac{5\pi}{4}$ to write the polar form. The polar form of $z = -2 - 2i$ is

$$z = r(\cos\theta + i\sin\theta) = 2\sqrt{2}\left(\cos\frac{5\pi}{4} + i\sin\frac{5\pi}{4}\right).$$

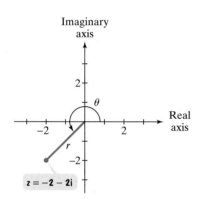

Imaginary axis

Real axis

$z = -2 - 2i$

Figure 6.38 Plotting $z = -2 - 2i$ and writing the number in polar form

Check Point 3   Plot $z = -1 - \sqrt{3}i$ in the complex plane. Then write z in polar form. Express the argument in radians.

4 Convert a complex number from polar to rectangular form.

EXAMPLE 4 Writing a Complex Number in Rectangular Form

Write $z = 2(\cos 60° + i \sin 60°)$ in rectangular form.

Solution The complex number $z = 2(\cos 60° + i \sin 60°)$ is in polar form, with $r = 2$ and $\theta = 60°$. We use exact values for $\cos 60°$ and $\sin 60°$ to write the number in rectangular form.

$$2(\cos 60° + i \sin 60°) = 2\left(\frac{1}{2} + i\frac{\sqrt{3}}{2}\right) = 1 + \sqrt{3}i$$

The rectangular form of $z = 2(\cos 60° + i \sin 60°)$ is

$$z = 1 + \sqrt{3}i.$$

Check Point 4 Write $z = 4(\cos 30° + i \sin 30°)$ in rectangular form.

5 Find products of complex numbers in polar form.

Products and Quotients in Polar Form

We can multiply and divide complex numbers fairly quickly if the numbers are expressed in polar form.

> ### Product of Two Complex Numbers in Polar Form
> Let $z_1 = r_1 (\cos \theta_1 + i \sin \theta_1)$ and $z_2 = r_2 (\cos \theta_2 + i \sin \theta_2)$ be two complex numbers in polar form. Their product, $z_1 z_2$, is
> $$z_1 z_2 = r_1 r_2 [\cos (\theta_1 + \theta_2) + i \sin (\theta_1 + \theta_2)].$$
> To multiply two complex numbers, multiply moduli and add arguments.

To prove this result, we begin by multiplying using the FOIL method. Then we simplify the product using the sum formulas for sine and cosine.

$$z_1 z_2 = \left[r_1(\cos \theta_1 + i \sin \theta_1)\right]\left[r_2(\cos \theta_2 + i \sin \theta_2)\right]$$

$$= r_1 r_2(\cos \theta_1 + i \sin \theta_1)(\cos \theta_2 + i \sin \theta_2) \qquad \text{Rearrange factors.}$$

$$= r_1 r_2(\cos \theta_1 \cos \theta_2 + i \cos \theta_1 \sin \theta_2 + i \sin \theta_1 \cos \theta_2 + i^2 \sin \theta_1 \sin \theta_2) \quad \text{Use the FOIL method.}$$

$$= r_1 r_2\left[\cos \theta_1 \cos \theta_2 + i(\cos \theta_1 \sin \theta_2 + \sin \theta_1 \cos \theta_2) + i^2 \sin \theta_1 \sin \theta_2\right] \text{ Factor } i \text{ from the second and third terms.}$$

$$= r_1 r_2\left[\cos \theta_1 \cos \theta_2 + i(\cos \theta_1 \sin \theta_2 + \sin \theta_1 \cos \theta_2) - \sin \theta_1 \sin \theta_2\right] \quad i^2 = -1$$

$$= r_1 r_2\left[(\cos \theta_1 \cos \theta_2 - \sin \theta_1 \sin \theta_2) + i(\sin \theta_1 \cos \theta_2 + \cos \theta_1 \sin \theta_2)\right] \text{ Rearrange terms.}$$

This is $\cos (\theta_1 + \theta_2)$.   This is $\sin (\theta_1 + \theta_2)$.

$$= r_1 r_2\left[\cos(\theta_1 + \theta_2) + i \sin(\theta_1 + \theta_2)\right]$$

This result gives a rule for finding the product of two complex numbers in polar form. The two parts to the rule are shown in the voice balloons below the product.

$$r_1 r_2 \left[\cos(\theta_1 + \theta_2) + i \sin(\theta_1 + \theta_2) \right]$$

Multiply moduli.   Add arguments.

EXAMPLE 5   Finding Products of Complex Numbers in Polar Form

Find the product of the complex numbers. Leave the answer in polar form.

$$z_1 = 4(\cos 50° + i \sin 50°) \qquad z_2 = 7(\cos 100° + i \sin 100°)$$

Solution

$z_1 z_2$

$= \left[4(\cos 50° + i \sin 50°) \right] \left[7(\cos 100° + i \sin 100°) \right]$ *Form the product of the given numbers.*

$= (4 \cdot 7) \left[\cos(50° + 100°) + i \sin(50° + 100°) \right]$ *Multiply moduli and add arguments.*

$= 28(\cos 150° + i \sin 150°)$ *Simplify.*

Check Point 5   Find the product of the complex numbers. Leave the answer in polar form.

$$z_1 = 6(\cos 40° + i \sin 40°) \qquad z_2 = 5(\cos 20° + i \sin 20°)$$

6 Find quotients of complex numbers in polar form.

Using algebraic methods for dividing complex numbers and the difference formulas for sine and cosine, we can obtain a rule for dividing complex numbers in polar form. The proof of this rule can be found in the appendix.

Quotient of Two Complex Numbers in Polar Form

Let $z_1 = r_1 (\cos \theta_1 + i \sin \theta_1)$ and $z_2 = r_2 (\cos \theta_2 + i \sin \theta_2)$ be two complex numbers in polar form. Their quotient, $\dfrac{z_1}{z_2}$, is

$$\frac{z_1}{z_2} = \frac{r_1}{r_2} \left[\cos(\theta_1 - \theta_2) + i \sin(\theta_1 - \theta_2) \right].$$

To divide two complex numbers, divide moduli and subtract arguments.

EXAMPLE 6   Finding Quotients of Complex Numbers in Polar Form

Find the quotient $\dfrac{z_1}{z_2}$ of the complex numbers. Leave the answer in polar form.

$$z_1 = 12 \left(\cos \frac{3\pi}{4} + i \sin \frac{3\pi}{4} \right) \qquad z_2 = 4 \left(\cos \frac{\pi}{4} + i \sin \frac{\pi}{4} \right)$$

Solution

$$\frac{z_1}{z_2} = \frac{12\left(\cos\dfrac{3\pi}{4} + i\sin\dfrac{3\pi}{4}\right)}{4\left(\cos\dfrac{\pi}{4} + i\sin\dfrac{\pi}{4}\right)}$$

Form the quotient of the given numbers.

$$= \frac{12}{4}\left[\cos\left(\frac{3\pi}{4} - \frac{\pi}{4}\right) + i\sin\left(\frac{3\pi}{4} - \frac{\pi}{4}\right)\right]$$

Divide moduli and subtract arguments.

$$= 3\left(\cos\frac{\pi}{2} + i\sin\frac{\pi}{2}\right)$$

Simplify: $\dfrac{3\pi}{4} - \dfrac{\pi}{4} = \dfrac{2\pi}{4} = \dfrac{\pi}{2}.$

Check Point 6 Find the quotient of the complex numbers. Leave the answer in polar form.

$$z_1 = 50\left(\cos\frac{4\pi}{3} + i\sin\frac{4\pi}{3}\right) \qquad z_2 = 5\left(\cos\frac{\pi}{3} + i\sin\frac{\pi}{3}\right)$$

7 Find powers of complex numbers in polar form.

Powers of Complex Numbers in Polar Form

We can use a formula to find powers of complex numbers if the complex numbers are expressed in polar form. This formula can be illustrated by repeatedly multiplying by $r(\cos\theta + i\sin\theta)$.

$z = r(\cos\theta + i\sin\theta)$	*Start with z.*
$z \cdot z = r(\cos\theta + i\sin\theta)r(\cos\theta + i\sin\theta)$	*Multiply z by z = $r(\cos\theta + i\sin\theta)$.*
$z^2 = r^2(\cos 2\theta + i\sin 2\theta)$	*Multiply moduli: $r\cdot r = r^2$. Add arguments: $\theta + \theta = 2\theta$.*
$z^2 \cdot z = r^2(\cos 2\theta + i\sin 2\theta)r(\cos\theta + i\sin\theta)$	*Multiply z^2 by z = $r(\cos\theta + i\sin\theta)$.*
$z^3 = r^3(\cos 3\theta + i\sin 3\theta)$	*Multiply moduli: $r^2\cdot r = r^3$. Add arguments: $2\theta + \theta = 3\theta$.*
$z^3 \cdot z = r^3(\cos 3\theta + i\sin 3\theta)r(\cos\theta + i\sin\theta)$	*Multiply z^3 by z = $r(\cos\theta + i\sin\theta)$.*
$z^4 = r^4(\cos 4\theta + i\sin 4\theta)$	*Multiply moduli: $r^3\cdot r = r^4$. Add arguments: $3\theta + \theta = 4\theta$.*

Do you see a pattern forming? If n is a positive integer, it appears that z^n is obtained by raising the modulus to the nth power and multiplying the argument by n. The formula for the nth power of a complex number is known as **DeMoivre's Theorem** in honor of the French mathematician Abraham DeMoivre (1667–1754).

> **DeMoivre's Theorem**
>
> Let $z = r(\cos\theta + i\sin\theta)$ be a complex number in polar form. If n is a positive integer, z to the nth power, z^n, is
>
> $$z^n = [r(\cos\theta + i\sin\theta)]^n = r^n(\cos n\theta + i\sin n\theta).$$

EXAMPLE 7   Finding the Power of a Complex Number

Find $[2(\cos 10° + i \sin 10°)]^6$. Write the answer in rectangular form, $a + bi$.

Solution   By DeMoivre's Theorem,

$[2(\cos 10° + i \sin 10°)]^6$

$= 2^6[\cos(6 \cdot 10°) + i \sin(6 \cdot 10°)]$ Raise the modulus to the 6th power and multiply the argument by 6.

$= 64(\cos 60° + i \sin 60°)$ Simplify.

$= 64\left(\dfrac{1}{2} + i\dfrac{\sqrt{3}}{2}\right)$ Write the answer in rectangular form.

$= 32 + 32\sqrt{3}i$ Multiply and express the answer in $a + bi$ form.

> **Check Point 7**   Find $[2(\cos 30° + i \sin 30°)]^5$. Write the answer in rectangular form.

EXAMPLE 8   Finding the Power of a Complex Number

Find $(1 + i)^8$ using DeMoivre's Theorem. Write the answer in rectangular form, $a + bi$.

Solution   DeMoivre's Theorem applies to complex numbers in polar form. Thus, we must first write $1 + i$ in $r(\cos \theta + i \sin \theta)$ form. Then we can use DeMoivre's Theorem. The complex number $1 + i$ is plotted in Figure 6.39. From the figure we obtain values for r and θ.

$$r = \sqrt{a^2 + b^2} = \sqrt{1^2 + 1^2} = \sqrt{2} \qquad \tan\theta = \dfrac{b}{a} = \dfrac{1}{1} = 1 \quad \text{and} \quad \theta = \dfrac{\pi}{4}$$

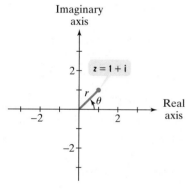

Figure 6.39 Plotting $1 + i$ and writing the number in polar form

Using these values,

$$1 + i = r(\cos\theta + i\sin\theta) = \sqrt{2}\left(\cos\dfrac{\pi}{4} + i\sin\dfrac{\pi}{4}\right).$$

Now we use DeMoivre's Theorem to raise $1 + i$ to the 8th power.

$(1 + i)^8$

$= \left[\sqrt{2}\left(\cos\dfrac{\pi}{4} + i\sin\dfrac{\pi}{4}\right)\right]^8$ Work with the polar form of $1 + i$.

$= (\sqrt{2})^8\left[\cos\left(8 \cdot \dfrac{\pi}{4}\right) + i\sin\left(8 \cdot \dfrac{\pi}{4}\right)\right]$ Apply DeMoivre's Theorem. Raise the modulus to the 8th power and multiply the argument by 8.

$= 16(\cos 2\pi + i \sin 2\pi)$ Simplify: $(\sqrt{2})^8 = (2^{1/2})^8 = 2^4 = 16$.

$= 16(1 + 0i)$ $\cos 2\pi = 1$ and $\sin 2\pi = 0$

$= 16$ Simplify.

> **Check Point 8**   Find $(1 + i)^4$ using DeMoivre's Theorem. Write the answer in rectangular form.

8 Find roots of complex numbers in polar form.

Roots of Complex Numbers in Polar Form

In Example 7, we showed that

$$[2(\cos 10° + i \sin 10°)]^6 = 64(\cos 60° + i \sin 60°).$$

We say that $2(\cos 10° + i \sin 10°)$ is a **complex sixth root** of $64(\cos 60° + i \sin 60°)$. It is one of six distinct complex sixth roots of $64(\cos 60° + i \sin 60°)$.

In general, if a complex number z satisfies the equation

$$z^n = w$$

we say that z is a **complex nth root** of w. It is one of n distinct nth complex roots that can be found using the following theorem:

> ### DeMoivre's Theorem for Finding Complex Roots
>
> Let $w = r(\cos \theta + i \sin \theta)$ be a complex number in polar form. If $w \neq 0$, w has n distinct complex nth roots given by the formula
>
> $$z_k = \sqrt[n]{r}\left[\cos\left(\frac{\theta + 2\pi k}{n}\right) + i \sin\left(\frac{\theta + 2\pi k}{n}\right)\right] \quad \text{(radians)}$$
>
> or $\quad z_k = \sqrt[n]{r}\left[\cos\left(\frac{\theta + 360°k}{n}\right) + i \sin\left(\frac{\theta + 360°k}{n}\right)\right]$ (degrees)
>
> where $k = 0, 1, 2, \ldots, n - 1$.

By raising the radian or degree formula for z_k to the nth power, you can use DeMoivre's Theorem for powers to show that $z_k^n = w$. Thus, each z_k is a complex nth root of w.

DeMoivre's Theorem for finding complex roots states that every complex number has two distinct complex square roots, three distinct complex cube roots, four distinct complex fourth roots, and so on. Each root has the same modulus, $\sqrt[n]{r}$. Successive roots have arguments that differ by the same amount, $\dfrac{2\pi}{n}$ or $\dfrac{360°}{n}$. This means that if you plot all the complex roots of any number, they will be equally spaced on a circle centered at the origin, with radius $\sqrt[n]{r}$.

EXAMPLE 9   Finding the Roots of a Complex Number

Find all the complex fourth roots of $16(\cos 120° + i \sin 120°)$. Write roots in polar form, with θ in degrees.

Solution   There are exactly four fourth roots of the given complex number. From DeMoivre's Theorem for finding complex roots, the fourth roots of $16(\cos 120° + i \sin 120°)$ are

$$z_k = \sqrt[4]{16}\left[\cos\left(\frac{120° + 360°k}{4}\right) + i \sin\left(\frac{120° + 360°k}{4}\right)\right], \quad k = 0, 1, 2, 3.$$

Use $z_k = \sqrt[n]{r}\left[\cos\left(\frac{\theta + 360°k}{n}\right) + i \sin\left(\frac{\theta + 360°k}{n}\right)\right]$.
In $16(\cos 120° + i \sin 120°)$, $r = 16$ and $\theta = 120°$.
Because we are finding fourth roots, $n = 4$.

The four fourth roots are found by substituting $0, 1, 2,$ and 3 for k in the expression for z_k above the voice balloon. Thus, the four complex fourth roots are:

$$z_0 = \sqrt[4]{16}\left[\cos\left(\frac{120° + 360° \cdot 0}{4}\right) + i\sin\left(\frac{120° + 360° \cdot 0}{4}\right)\right]$$

$$= \sqrt[4]{16}\left(\cos\frac{120°}{4} + i\sin\frac{120°}{4}\right) = 2(\cos 30° + i\sin 30°)$$

$$z_1 = \sqrt[4]{16}\left[\cos\left(\frac{120° + 360° \cdot 1}{4}\right) + i\sin\left(\frac{120° + 360° \cdot 1}{4}\right)\right]$$

$$= \sqrt[4]{16}\left(\cos\frac{480°}{4} + i\sin\frac{480°}{4}\right) = 2(\cos 120° + i\sin 120°)$$

$$z_2 = \sqrt[4]{16}\left[\cos\left(\frac{120° + 360° \cdot 2}{4}\right) + i\sin\left(\frac{120° + 360° \cdot 2}{4}\right)\right]$$

$$= \sqrt[4]{16}\left(\cos\frac{840°}{4} + i\sin\frac{840°}{4}\right) = 2(\cos 210° + i\sin 210°)$$

$$z_3 = \sqrt[4]{16}\left[\cos\left(\frac{120° + 360° \cdot 3}{4}\right) + i\sin\left(\frac{120° + 360° \cdot 3}{4}\right)\right]$$

$$= \sqrt[4]{16}\left(\cos\frac{1200°}{4} + i\sin\frac{1200°}{4}\right) = 2(\cos 300° + i\sin 300°).$$

In Figure 6.40, we have plotted each of the four fourth roots. They are equally spaced at 90° intervals on a circle with radius 2.

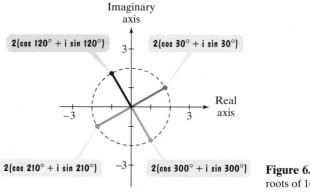

Figure 6.40 Plotting the four fourth roots of $16(\cos 120° + i\sin 120°)$

Check Point 9   Find all the complex fourth roots of $16(\cos 60° + i\sin 60°)$. Write roots in polar form, with θ in degrees.

EXAMPLE 10   Finding the Roots of a Complex Number

Find all the cube roots of 8. Write roots in rectangular form.

Solution   DeMoivre's Theorem for roots applies to complex numbers in polar form. Thus, we will first write 8 in polar form. We express θ in radians, although degrees can also be used.

$$8 = r(\cos\theta + i\sin\theta) = 8(\cos 0 + i\sin 0)$$

There are exactly three cube roots of 8. From DeMoivre's Theorem for finding complex roots, the cube roots of 8 are

$$z_k = \sqrt[3]{8}\left[\cos\left(\frac{0 + 2\pi k}{3}\right) + i\sin\left(\frac{0 + 2\pi k}{3}\right)\right], \quad k = 0, 1, 2.$$

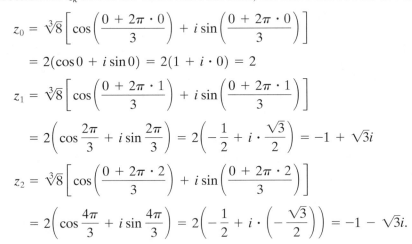

Use $z_k = \sqrt[n]{r}\left[\cos\left(\frac{\theta + 2\pi k}{n}\right) + i\sin\left(\frac{\theta + 2\pi k}{n}\right)\right]$.
In $8(\cos 0 + i\sin 0)$, $r = 8$ and $\theta = 0$.
Because we are finding cube roots, $n = 3$.

The three cube roots of 8 are found by substituting 0, 1, and 2 for k in the expression for z_k above the voice balloon. Thus, the three cube roots of 8 are

$$z_0 = \sqrt[3]{8}\left[\cos\left(\frac{0 + 2\pi \cdot 0}{3}\right) + i\sin\left(\frac{0 + 2\pi \cdot 0}{3}\right)\right]$$

$$= 2(\cos 0 + i\sin 0) = 2(1 + i \cdot 0) = 2$$

$$z_1 = \sqrt[3]{8}\left[\cos\left(\frac{0 + 2\pi \cdot 1}{3}\right) + i\sin\left(\frac{0 + 2\pi \cdot 1}{3}\right)\right]$$

$$= 2\left(\cos\frac{2\pi}{3} + i\sin\frac{2\pi}{3}\right) = 2\left(-\frac{1}{2} + i \cdot \frac{\sqrt{3}}{2}\right) = -1 + \sqrt{3}i$$

$$z_2 = \sqrt[3]{8}\left[\cos\left(\frac{0 + 2\pi \cdot 2}{3}\right) + i\sin\left(\frac{0 + 2\pi \cdot 2}{3}\right)\right]$$

$$= 2\left(\cos\frac{4\pi}{3} + i\sin\frac{4\pi}{3}\right) = 2\left(-\frac{1}{2} + i \cdot \left(-\frac{\sqrt{3}}{2}\right)\right) = -1 - \sqrt{3}i.$$

The three cube roots of 8 are plotted in Figure 6.41.

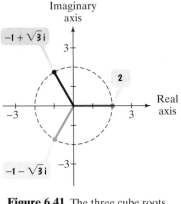

Figure 6.41 The three cube roots of 8 are equally spaced at intervals of $\frac{2\pi}{3}$ on a circle with radius 2.

Check Point 10 Find all the cube roots of 27. Write roots in rectangular form.

The Mandelbrot Set

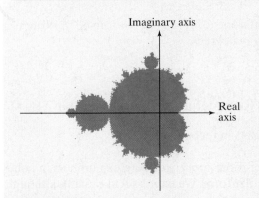

Figure 6.42

The set of all complex numbers for which the sequence

$$z, z^2 + z, (z^2 + z)^2 + z, [(z^2 + z)^2 + z]^2 + z, \dots$$

is bounded is called the **Mandelbrot set.** Plotting these complex numbers in the complex plane results in a graph that is "buglike" in shape, shown in Figure 6.42. Colors can be added to the boundary of the graph. At the boundary, color choices depend on how quickly the numbers in the boundary approach infinity when substituted into the sequence shown. The magnified boundary is shown in the introduction to this section. It includes the original buglike structure, as well as new and interesting patterns. With each level of magnification, repetition and unpredictable formations interact to create what has been called the most complicated mathematical object ever known.

EXERCISE SET 6.5

 Practice Exercises

In Exercises 1–10, plot each complex number and find its absolute value.

1. $z = 4i$ **2.** $z = 3i$

3. $z = 3$ **4.** $z = 4$

5. $z = 3 + 2i$ **6.** $z = 2 + 5i$

7. $z = 3 - i$ **8.** $z = 4 - i$

9. $z = -3 + 4i$ **10.** $z = -3 - 4i$

In Exercises 11–26, plot each complex number. Then write the complex number in polar form. You may express the argument in degrees or radians.

11. $2 + 2i$ **12.** $1 + \sqrt{3}i$ **13.** $-1 - i$

14. $2 - 2i$ **15.** $-4i$ **16.** $-3i$

17. $2\sqrt{3} - 2i$ **18.** $-2 + 2\sqrt{3}i$ **19.** -3

20. -4 **21.** $-3\sqrt{2} - 3\sqrt{3}i$ **22.** $3\sqrt{2} - 3\sqrt{2}i$

23. $-3 + 4i$ **24.** $-2 + 3i$

25. $2 - \sqrt{3}i$ **26.** $1 - \sqrt{5}i$

In Exercises 27–36, write each complex number in rectangular form. If necessary, round to the nearest tenth.

27. $6(\cos 30° + i \sin 30°)$ **28.** $12(\cos 60° + i \sin 60°)$

29. $4(\cos 240° + i \sin 240°)$ **30.** $10(\cos 210° + i \sin 210°)$

31. $8\left(\cos \dfrac{7\pi}{4} + i \sin \dfrac{7\pi}{4}\right)$ **32.** $4\left(\cos \dfrac{5\pi}{6} + i \sin \dfrac{5\pi}{6}\right)$

33. $5\left(\cos \dfrac{\pi}{2} + i \sin \dfrac{\pi}{2}\right)$ **34.** $7\left(\cos \dfrac{3\pi}{2} + i \sin \dfrac{3\pi}{2}\right)$

35. $20(\cos 205° + i \sin 205°)$ **36.** $30(\cos 2.3 + i \sin 2.3)$

In Exercises 37–44, find the product of the complex numbers. Leave answers in polar form.

37. $z_1 = 6(\cos 20° + i \sin 20°)$
$z_2 = 5(\cos 50° + i \sin 50°)$

38. $z_1 = 4(\cos 15° + i \sin 15°)$
$z_2 = 7(\cos 25° + i \sin 25°)$

39. $z_1 = 3\left(\cos \dfrac{\pi}{5} + i \sin \dfrac{\pi}{5}\right)$
$z_2 = 4\left(\cos \dfrac{\pi}{10} + i \sin \dfrac{\pi}{10}\right)$

40. $z_1 = 3\left(\cos \dfrac{5\pi}{8} + i \sin \dfrac{5\pi}{8}\right)$
$z_2 = 10\left(\cos \dfrac{\pi}{16} + i \sin \dfrac{\pi}{16}\right)$

41. $z_1 = \cos \dfrac{\pi}{4} + i \sin \dfrac{\pi}{4}$
$z_2 = \cos \dfrac{\pi}{3} + i \sin \dfrac{\pi}{3}$

42. $z_1 = \cos \dfrac{\pi}{6} + i \sin \dfrac{\pi}{6}$
$z_2 = \cos \dfrac{\pi}{4} + i \sin \dfrac{\pi}{4}$

43. $z_1 = 1 + i$ **44.** $z_1 = 1 + i$
$z_2 = -1 + i$ $z_2 = 2 + 2i$

In Exercises 45–52, find the quotient $\dfrac{z_1}{z_2}$ of the complex numbers. Leave answers in polar form. In Exercises 49–50, express the argument as an angle between 0° and 360°.

45. $z_1 = 20(\cos 75° + i \sin 75°)$
$z_2 = 4(\cos 25° + i \sin 25°)$

46. $z_1 = 50(\cos 80° + i \sin 80°)$
$z_2 = 10(\cos 20° + i \sin 20°)$

47. $z_1 = 3\left(\cos \dfrac{\pi}{5} + i \sin \dfrac{\pi}{5}\right)$

$z_2 = 4\left(\cos \dfrac{\pi}{10} + i \sin \dfrac{\pi}{10}\right)$

48. $z_1 = 3\left(\cos \dfrac{5\pi}{18} + i \sin \dfrac{5\pi}{18}\right)$

$z_2 = 10\left(\cos \dfrac{\pi}{16} + i \sin \dfrac{\pi}{16}\right)$

49. $z_1 = \cos 80° + i \sin 80°$
$z_2 = \cos 200° + i \sin 200°$

50. $z_1 = \cos 70° + i \sin 70°$
$z_2 = \cos 230° + i \sin 230°$

51. $z_1 = 2 + 2i$ **52.** $z_1 = 2 - 2i$
$z_2 = 1 + i$ $z_2 = 1 - i$

In Exercises 53–64, use DeMoivre's Theorem to find the indicated power of the complex number. Write answers in rectangular form.

53. $[4(\cos 15° + i \sin 15°)]^3$

54. $[2(\cos 10° + i \sin 10°)]^3$

55. $[2(\cos 80° + i \sin 80°)]^3$

56. $[2(\cos 40° + i \sin 40°)]^3$

57. $\left[\dfrac{1}{2}\left(\cos \dfrac{\pi}{12} + i \sin \dfrac{\pi}{12}\right)\right]^6$

58. $\left[\dfrac{1}{2}\left(\cos \dfrac{\pi}{10} + i \sin \dfrac{\pi}{10}\right)\right]^5$

59. $\left[\sqrt{2}\left(\cos \dfrac{5\pi}{6} + i \sin \dfrac{5\pi}{6}\right)\right]^4$

60. $\left[\sqrt{3}\left(\cos \dfrac{5\pi}{18} + i \sin \dfrac{5\pi}{18}\right)\right]^6$

61. $(1 + i)^5$ **62.** $(1 - i)^5$

63. $(\sqrt{3} - i)^6$ **64.** $(\sqrt{2} - i)^4$

In Exercises 65–68, find all the complex roots. Write roots in polar form with θ in degrees.

65. The complex square roots of $9(\cos 30° + i \sin 30°)$
66. The complex square roots of $25(\cos 210° + i \sin 210°)$
67. The complex cube roots of $8(\cos 210° + i \sin 210°)$
68. The complex cube roots of $27(\cos 306° + i \sin 306°)$

In Exercises 69–76, find all the complex roots. Write roots in rectangular form. If necessary, round to the nearest tenth.

69. The complex fourth roots of $81\left(\cos \dfrac{4\pi}{3} + i \sin \dfrac{4\pi}{3}\right)$

70. The complex fifth roots of $32\left(\cos \dfrac{5\pi}{3} + i \sin \dfrac{5\pi}{3}\right)$

71. The complex fifth roots of 32
72. The complex sixth roots of 64
73. The complex cube roots of 1
74. The complex cube roots of i
75. The complex fourth roots of $1 + i$
76. The complex fifth roots of $-1 + i$

 Application Exercises

In Exercises 77–78, show that the given complex number z plots as a point in the Mandelbrot set.

   a. Write the first six terms of the sequence
$$z_1, z_2, z_3, z_4, z_5, z_6, \ldots$$
   where

   $z_1 = z$: Write the given number.
   $z_2 = z^2 + z$: Square z_1 and add the given number.
   $z_3 = (z^2 + z)^2 + z$: Square z_2 and add the given number.
   $z_4 = [(z^2 + z)^2 + z]^2 + z$: Square z_3 and add the given number.

   z_5: Square z_4 and add the given number.

   z_6: Square z_5 and add the given number.

   b. If the sequence that you began writing in part (a) is bounded, the given complex number belongs to the Mandelbrot set. Show that the sequence is bounded by writing two complex numbers. One complex number should be greater in absolute value than the absolute values of the terms in the sequence. The second complex number should be less in absolute value than the absolute values of the terms in the sequence.

77. $z = i$ **78.** $z = -i$

 Writing in Mathematics

79. Explain how to plot a complex number in the complex plane. Provide an example with your explanation.
80. How do you determine the absolute value of a complex number?

81. What is the polar form of a complex number?
82. If you are given a complex number in rectangular form, how do you write it in polar form?
83. If you are given a complex number in polar form, how do you write it in rectangular form?
84. Explain how to find the product of two complex numbers in polar form.
85. Explain how to find the quotient of two complex numbers in polar form.
86. Explain how to find the power of a complex number in polar form.
87. Explain how to use DeMoivre's Theorem for finding complex roots to find the two square roots of 9.
88. Describe the graph of all complex numbers with an absolute value of 6.
89. The image of the Mandelbrot set in the section opener exhibits self-similarity: Magnified portions repeat much of the pattern of the whole structure, as well as new and unexpected patterns. Describe an object in nature that exhibits self-similarity.

 Technology Exercises

90. Use the rectangular-to-polar feature on a graphing utility to verify any four of your answers in Exercises 11–26. Be aware that you may have to adjust the angle for the correct quadrant.
91. Use the polar-to-rectangular feature on a graphing utility to verify any four of your answers in Exercises 27–36.

 Critical Thinking Exercises

92. Prove the rule for finding the quotient of two complex numbers in polar form. Begin the proof as follows, using the conjugate of the denominator:

$$\frac{r_1(\cos \theta_1 + i \sin \theta_1)}{r_2(\cos \theta_2 + i \sin \theta_2)} = \frac{r_1(\cos \theta_1 + i \sin \theta_1)}{r_2(\cos \theta_2 + i \sin \theta_2)} \cdot \frac{(\cos \theta_2 - i \sin \theta_2)}{(\cos \theta_2 - i \sin \theta_2)}.$$

Perform the indicated multiplications. Then use the difference formulas for sine and cosine.

93. Plot each of the complex fourth roots of 1.

In Exercises 94–95, use DeMoivre's Theorem for finding complex roots to find all the solutions of each equation.
94. $x^3 + 27 = 0$ **95.** $x^3 - 8i = 0$

Group Exercise

96. Group members should prepare and present a seminar on chaos. Include one or more of the following topics in your presentation: fractal images, the role of complex numbers in generating fractal images, algorithms, iterations, iteration number, and fractals in nature. Be sure to include visual images that will intrigue your audience.

SECTION 6.6 *Vectors*

Objectives

1. Use magnitude and direction to show vectors are equal.
2. Visualize scalar multiplication, vector addition, and vector subtraction as geometric vectors.
3. Represent vectors in the rectangular coordinate system.
4. Perform operations with vectors in terms of **i** and **j**.
5. Find the unit vector in the direction of **v**.
6. Write a vector in terms of its magnitude and direction.
7. Solve applied problems involving vectors.

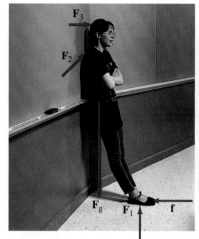

It's been a dynamic lecture, but now that it's over it's obvious that my professor is exhausted. She's slouching motionless against the board and—what's that? The forces acting against her body, including the pull of gravity, are appearing as arrows. I know that mathematics reveals the hidden patterns of the universe, but this is ridiculous. Does the arrangement of the arrows on the right have anything to do with the fact that my wiped-out professor is not sliding down the wall?

This sign shows a distance and direction for each city. Thus, the sign defines a vector for each destination.

Ours is a world of pushes and pulls. For example, suppose you are pulling a cart up a 30° incline, requiring an effort of 100 pounds. This quantity is described by giving its magnitude (a number indicating size, including a unit of measure) and also its direction. The magnitude is 100 pounds and the direction is 30° from the horizontal. Quantities that involve both a magnitude and a direction are called **vector quantities,** or **vectors** for short. Here is another example of a vector:

> You are driving due north at 50 miles per hour. The magnitude is the speed, 50 miles per hour. The direction of motion is due north.

Some quantities can be completely described by giving only their magnitude. For example, the temperature of the lecture room that you just left is 75°. This temperature has magnitude, 75°, but no direction. Quantities that involve magnitude, but no direction, are called **scalar quantities,** or **scalars** for short. Thus, a scalar has only a numerical value. Another example of a scalar is your professor's height, which you estimate to be 5.5 feet.

In the next two sections, we introduce the world of vectors, which literally surround your every move. Because vectors have both nonnegative magnitude and direction, we begin our discussion with directed line segments.

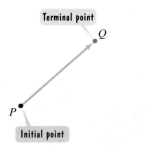

Figure 6.43 A directed line segment from P to Q

Directed Line Segments and Geometric Vectors

A line segment to which a direction has been assigned is called a **directed line segment.** Figure 6.43 shows a directed line segment from P to Q. We call P the **initial point** and Q the **terminal point.** We denote this directed line segment by

$$\overrightarrow{PQ}.$$

The **magnitude** of the directed line segment $\overrightarrow{PQ}$ is its length. We denote this by $\|\overrightarrow{PQ}\|$. Thus, $\|\overrightarrow{PQ}\|$ is the distance from point P to point Q. Because distance is nonnegative, vectors do not have negative magnitudes.

Geometrically, a **vector** is a directed line segment. Vectors are often denoted by boldface letters, such as **v**. If a vector **v** has the same magnitude and the same direction as the directed line segment $\overrightarrow{PQ}$, we write

$$\mathbf{v} = \overrightarrow{PQ}.$$

Because it is difficult to write boldface on paper, use an arrow over a single letter, such as $\vec{v}$, to denote **v**, the vector **v**.

Figure 6.44 shows four possible relationships between vectors **v** and **w**. In Figure 6.44 (a), the vectors have the same magnitude and the same direction, and are said to be *equal*. In general, vectors **v** and **w** are **equal** if they have the *same magnitude* and the *same direction.* We write this as **v** = **w**.

1 Use magnitude and direction to show vectors are equal.

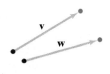   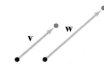

(a) v = **w** because the vectors have the same magnitude and the same direction

(b) Vectors **v** and **w** have the same magnitude, but different directions

(c) Vectors **v** and **w** have the same magnitude, but opposite directions

(d) Vectors **v** and **w** have the same direction, but different magnitudes

Figure 6.44 Relationships between vectors

EXAMPLE 1 Showing That Two Vectors Are Equal

Use Figure 6.45 to show that **u** = **v**.

Solution Equal vectors have the same magnitude and the same direction. Use the distance formula to show that **u** and **v** have the same magnitude.

Magnitude of **u**
$$\|\mathbf{u}\| = \sqrt{(x_2 - x_1)^2 + (y_2 - y_1)^2} = \sqrt{[0 - (-3)]^2 + [3 - (-3)]^2}$$
$$= \sqrt{3^2 + 6^2} = \sqrt{9 + 36} = \sqrt{45} \quad (\text{or } 3\sqrt{5})$$

Magnitude of **v**
$$\|\mathbf{v}\| = \sqrt{(x_2 - x_1)^2 + (y_2 - y_1)^2} = \sqrt{(3 - 0)^2 + (6 - 0)^2}$$
$$= \sqrt{3^2 + 6^2} = \sqrt{9 + 36} = \sqrt{45} \quad (\text{or } 3\sqrt{5})$$

Thus, **u** and **v** have the same magnitude: $\|\mathbf{u}\| = \|\mathbf{v}\|$.

One way to show that **u** and **v** have the same direction is to find the slopes of the lines on which they lie.

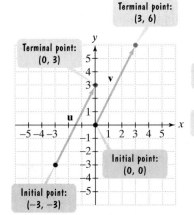

Figure 6.45

Line on which **u** lies
$$m = \frac{y_2 - y_1}{x_2 - x_1} = \frac{3 - (-3)}{0 - (-3)} = \frac{6}{3} = 2$$

Line on which **v** lies
$$m = \frac{y_2 - y_1}{x_2 - x_1} = \frac{6 - 0}{3 - 0} = \frac{6}{3} = 2$$

Because **u** and **v** are both directed toward the upper right on lines having the same slope, 2, they have the same direction.

Thus, **u** and **v** have the same magnitude and direction, and **u** = **v**.

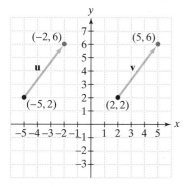

Figure 6.46

2 Visualize scalar multiplication, vector addition, and vector subtraction as geometric vectors.

Check Point 1   Use Figure 6.46 to show that $\mathbf{u} = \mathbf{v}$.

A vector can be multiplied by a real number. Figure 6.47 shows three such multiplications: $2\mathbf{v}, \frac{1}{2}\mathbf{v}$, and $-\frac{3}{2}\mathbf{v}$. **Multiplying a vector by any positive real number (except for 1) changes the magnitude of the vector, but not its direction.** This can be seen by the blue and green vectors in Figure 6.47. Compare the black and blue vectors. Can you see that $2\mathbf{v}$ has the same direction as $\mathbf{v}$ but is twice the magnitude of $\mathbf{v}$? Now, compare the black and green vectors: $\frac{1}{2}\mathbf{v}$ has the same direction as $\mathbf{v}$ but is half the magnitude of $\mathbf{v}$.

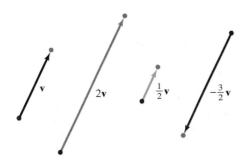

Figure 6.47 Multiplying vector $\mathbf{v}$ by real numbers

Now compare the black and red vectors in Figure 6.47. **Multiplying a vector by a negative number reverses the direction of the vector.** Notice that $-\frac{3}{2}\mathbf{v}$ has the opposite direction as $\mathbf{v}$ and is $\frac{3}{2}$ the magnitude of $\mathbf{v}$.

The multiplication of the real number, k, and the vector, $\mathbf{v}$, is called **scalar multiplication**. We write this product as $k\mathbf{v}$.

Wiped Out, But Not Sliding Down the Wall

The figure shows the sum of five vectors:

$$\mathbf{F}_1 + \mathbf{F}_2 + \mathbf{F}_3 + \mathbf{F}_g + \mathbf{f}.$$

Notice how the terminal point of each vector coincides with the initial point of the vector that's being added to it. The vector sum, from the initial point of $\mathbf{F}_1$ to the terminal point of $\mathbf{f}$, is a single point. The magnitude of a single point is zero. These forces add up to a net force of zero, allowing the professor to be motionless.

Scalar Multiplication

If k is a real number and $\mathbf{v}$ a vector, the vector $k\mathbf{v}$ is called a **scalar multiple** of the vector $\mathbf{v}$. The magnitude and direction of $k\mathbf{v}$ are given as follows:

The vector $k\mathbf{v}$ has a *magnitude* of $|k|\,\|\mathbf{v}\|$. We describe this as the absolute value of k times the magnitude of vector $\mathbf{v}$.

The vector $k\mathbf{v}$ has a *direction* that is:

- the same as the direction of $\mathbf{v}$ if $k > 0$, and
- opposite the direction of $\mathbf{v}$ if $k < 0$.

A geometric method for adding two vectors is shown in Figure 6.48. The sum of $\mathbf{u} + \mathbf{v}$ is called the **resultant vector.** Here is how we find this vector:

1. Position $\mathbf{u}$ and $\mathbf{v}$ so that the terminal point of $\mathbf{u}$ coincides with the initial point of $\mathbf{v}$.

2. The resultant vector, $\mathbf{u} + \mathbf{v}$, extends from the initial point of $\mathbf{u}$ to the terminal point of $\mathbf{v}$.

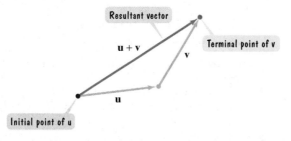

Figure 6.48 Vector addition $\mathbf{u} + \mathbf{v}$; the terminal point of $\mathbf{u}$ coincides with the initial point of $\mathbf{v}$.

The **difference of two vectors, v − u**, is defined as $\mathbf{v} - \mathbf{u} = \mathbf{v} + (-\mathbf{u})$, where $-\mathbf{u}$ is the scalar multiplication of $\mathbf{u}$ and $-1: -1\mathbf{u}$. The difference $\mathbf{v} - \mathbf{u}$ is shown geometrically in Figure 6.49.

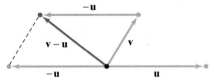

Figure 6.49 Vector subtraction $\mathbf{v} - \mathbf{u}$; the terminal point of $\mathbf{v}$ coincides with the initial point of $-\mathbf{u}$.

3 Represent vectors in the rectangular coordinate system.

Vectors in the Rectangular Coordinate System

As you saw in Example 1, vectors can be shown in the rectangular coordinate system. Now let's see how we can use the rectangular coordinate system to represent vectors. We begin with two vectors that both have a magnitude of 1. Such vectors are called **unit vectors.**

The i and j Unit Vectors

Vector **i** is the unit vector whose direction is along the positive x-axis. Vector **j** is the unit vector whose direction is along the positive y-axis.

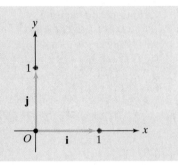

Why are the **i** and **j** unit vectors important? Vectors in the rectangular coordinate system can be represented in terms of **i** and **j**. For example, consider vector **v** with initial point at the origin, $(0, 0)$, and terminal point at $P = (a, b)$. The vector **v** is shown in Figure 6.50. We can represent **v** using **i** and **j** as $\mathbf{v} = a\mathbf{i} + b\mathbf{j}$.

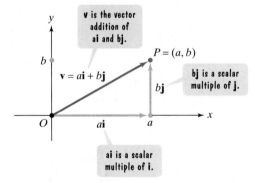

Figure 6.50 Using vector addition, vector **v** is represented as $\mathbf{v} = a\mathbf{i} + b\mathbf{j}$.

Representing Vectors in Rectangular Coordinates

Vector **v**, from $(0, 0)$ to (a, b), is represented as
$$\mathbf{v} = a\mathbf{i} + b\mathbf{j}.$$
The real numbers a and b are called the **scalar components** of **v**. Note that

- a is the **horizontal component** of **v**, and
- b is the **vertical component** of **v**.

The vector sum $a\mathbf{i} + b\mathbf{j}$ is called a **linear combination** of the vectors **i** and **j**. The magnitude of $\mathbf{v} = a\mathbf{i} + b\mathbf{j}$ is given by
$$\|\mathbf{v}\| = \sqrt{a^2 + b^2}.$$

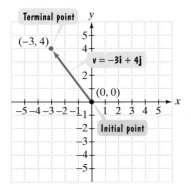

Figure 6.51 Sketching $\mathbf{v} = -3\mathbf{i} + 4\mathbf{j}$ in the rectangular coordinate system

EXAMPLE 2 **Representing a Vector in Rectangular Coordinates and Finding Its Magnitude**

Sketch the vector $\mathbf{v} = -3\mathbf{i} + 4\mathbf{j}$ and find its magnitude.

Solution For the given vector $\mathbf{v} = -3\mathbf{i} + 4\mathbf{j}$, $a = -3$ and $b = 4$. The vector can be represented with its initial point at the origin, $(0, 0)$, as shown in Figure 6.51. The vector's terminal point is then $(a, b) = (-3, 4)$. We sketch the vector by drawing an arrow from $(0, 0)$ to $(-3, 4)$. We determine the magnitude of the vector by using the distance formula. Thus, the magnitude is

$$\|\mathbf{v}\| = \sqrt{a^2 + b^2} = \sqrt{(-3)^2 + 4^2} = \sqrt{9 + 16} = \sqrt{25} = 5.$$

Check Point 2 Sketch the vector $\mathbf{v} = 3\mathbf{i} - 3\mathbf{j}$ and find its magnitude.

The vector in Example 2 was represented with its initial point at the origin. A vector whose initial point is at the origin is called a **position vector.** Any vector in rectangular coordinates whose initial point is not at the origin can be shown to be equal to a position vector. As shown in the following box, this gives us a way to represent vectors between any two points.

Representing Vectors in Rectangular Coordinates

Vector $\mathbf{v}$ with initial point $P_1 = (x_1, y_1)$ and terminal point $P_2 = (x_2, y_2)$ is equal to the position vector

$$\mathbf{v} = (x_2 - x_1)\mathbf{i} + (y_2 - y_1)\mathbf{j}.$$

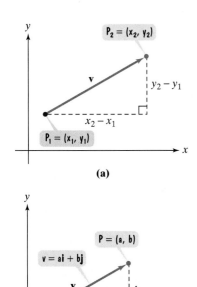

Figure 6.52

We can use congruent triangles to derive this formula. Begin with the right triangle in Figure 6.52(a). This triangle shows vector $\mathbf{v}$ from $P_1 = (x_1, y_1)$ to $P_2 = (x_2, y_2)$. In Figure 6.52(b), we move vector $\mathbf{v}$, without changing its magnitude or its direction, so that its initial point is at the origin. Using this position vector in Figure 6.52 (b), we see that

$$\mathbf{v} = a\mathbf{i} + b\mathbf{j}.$$

The equal vectors and the right angles in the right triangles in Figures 6.52(a) and (b) result in congruent triangles. The corresponding sides of these congruent triangles are equal, so that $a = x_2 - x_1$ and $b = y_2 - y_1$. This means that $\mathbf{v}$ may be expressed as

$$\mathbf{v} = a\mathbf{i} + b\mathbf{j} = (x_2 - x_1)\mathbf{i} + (y_2 - y_1)\mathbf{j}.$$

Horizontal component: x-coordinate of terminal point minus x-coordinate of initial point

Vertical component: y-coordinate of terminal point minus y-coordinate of initial point

Thus, any vector between two points in rectangular coordinates can be expressed in terms of $\mathbf{i}$ and $\mathbf{j}$. In rectangular coordinates, the term *vector* refers to the position vector expressed in terms of $\mathbf{i}$ and $\mathbf{j}$ that is equal to it.

EXAMPLE 3 Representing a Vector in Rectangular Coordinates

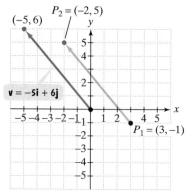

Figure 6.53 Representing the vector from $(3, -1)$ to $(-2, 5)$ as a position vector

Let **v** be the vector from initial point $P_1 = (3, -1)$ to terminal point $P_2 = (-2, 5)$. Write **v** in terms of **i** and **j**.

Solution We identify the values for the variables in the formula.

$$P_1 = (\underset{x_1}{3}, \underset{y_1}{-1}) \qquad P_2 = (\underset{x_2}{-2}, \underset{y_2}{5})$$

Using these values, we write **v** in terms of **i** and **j** as follows:

$$\mathbf{v} = (x_2 - x_1)\mathbf{i} + (y_2 - y_1)\mathbf{j} = (-2 - 3)\mathbf{i} + [5 - (-1)]\mathbf{j} = -5\mathbf{i} + 6\mathbf{j}$$

Figure 6.53 shows the vector from $P_1 = (3, -1)$ to $P_2 = (-2, 5)$ represented in terms of **i** and **j** and as a position vector.

Check Point 3 Let **v** be the vector from initial point $P_1 = (-1, 3)$ to $P_2 = (2, 7)$. Write **v** in terms of **i** and **j**.

4 Perform operations with vectors in terms of **i** and **j**.

Operations with Vectors in Terms of i and j

If vectors are expressed in terms of **i** and **j**, we can easily carry out operations such as vector addition, vector subtraction, and scalar multiplication. Recall the geometric definitions of these operations given earlier. Based on these ideas, we can add and subtract vectors using the following procedure:

Adding and Subtracting Vectors in Terms of i and j

If $\mathbf{v} = a_1\mathbf{i} + b_1\mathbf{j}$ and $\mathbf{w} = a_2\mathbf{i} + b_2\mathbf{j}$, then
$$\mathbf{v} + \mathbf{w} = (a_1 + a_2)\mathbf{i} + (b_1 + b_2)\mathbf{j}$$
$$\mathbf{v} - \mathbf{w} = (a_1 - a_2)\mathbf{i} + (b_1 - b_2)\mathbf{j}.$$

EXAMPLE 4 Adding and Subtracting Vectors

If $\mathbf{v} = 5\mathbf{i} + 4\mathbf{j}$ and $\mathbf{w} = 6\mathbf{i} - 9\mathbf{j}$, find:

    a. $\mathbf{v} + \mathbf{w}$ **b.** $\mathbf{v} - \mathbf{w}$.

Solution

a. $\mathbf{v} + \mathbf{w} = (5\mathbf{i} + 4\mathbf{j}) + (6\mathbf{i} - 9\mathbf{j})$ These are the given vectors.

$\qquad\qquad = (5 + 6)\mathbf{i} + [4 + (-9)]\mathbf{j}$ Add the horizontal components. Add the vertical components.

$\qquad\qquad = 11\mathbf{i} - 5\mathbf{j}$ Simplify.

b. $\mathbf{v} - \mathbf{w} = (5\mathbf{i} + 4\mathbf{j}) - (6\mathbf{i} - 9\mathbf{j})$ These are the given vectors.

$\qquad\qquad = (5 - 6)\mathbf{i} + [4 - (-9)]\mathbf{j}$ Subtract the horizontal components. Subtract the vertical components.

$\qquad\qquad = -\mathbf{i} + 13\mathbf{j}$ Simplify.

Check Point 4 If $\mathbf{v} = 7\mathbf{i} + 3\mathbf{j}$ and $\mathbf{w} = 4\mathbf{i} - 5\mathbf{j}$, find:
    a. $\mathbf{v} + \mathbf{w}$ **b.** $\mathbf{v} - \mathbf{w}$.

How do we perform scalar multiplication if vectors are expressed in terms of **i** and **j**? We use the following procedure to multiply the vector **v** by the scalar k:

> **Scalar Multiplication with a Vector in Terms of i and j**
> If $\mathbf{v} = a\mathbf{i} + b\mathbf{j}$ and k is a real number, then the scalar multiplication of the vector **v** and the scalar k is
> $$k\mathbf{v} = (ka)\mathbf{i} + (kb)\mathbf{j}.$$

EXAMPLE 5 Scalar Multiplication

If $\mathbf{v} = 5\mathbf{i} + 4\mathbf{j}$, find:

    a. 6**v**   **b.** −3**v**.

Solution

a. $6\mathbf{v} = 6(5\mathbf{i} + 4\mathbf{j})$	The scalar multiplication is expressed with the given vector.
$= (6 \cdot 5)\mathbf{i} + (6 \cdot 4)\mathbf{j}$	Multiply each component by 6.
$= 30\mathbf{i} + 24\mathbf{j}$	Simplify.
b. $-3\mathbf{v} = -3(5\mathbf{i} + 4\mathbf{j})$	The scalar multiplication is expressed with the given vector.
$= (-3 \cdot 5)\mathbf{i} + (-3 \cdot 4)\mathbf{j}$	Multiply each component by −3.
$= -15\mathbf{i} - 12\mathbf{j}$	Simplify.

Check Point 5 If $\mathbf{v} = 7\mathbf{i} + 10\mathbf{j}$, find:
    a. 8**v**   **b.** −5**v**.

EXAMPLE 6 Vector Operations

If $\mathbf{v} = 5\mathbf{i} + 4\mathbf{j}$ and $\mathbf{w} = 6\mathbf{i} - 9\mathbf{j}$, find $4\mathbf{v} - 2\mathbf{w}$.

Solution

$4\mathbf{v} - 2\mathbf{w} = 4(5\mathbf{i} + 4\mathbf{j}) - 2(6\mathbf{i} - 9\mathbf{j})$	Operations are expressed with the given vectors.
$= 20\mathbf{i} + 16\mathbf{j} - 12\mathbf{i} + 18\mathbf{j}$	Perform each scalar multiplication.
$= (20 - 12)\mathbf{i} + (16 + 18)\mathbf{j}$	Add horizontal and vertical components to perform the vector addition.
$= 8\mathbf{i} + 34\mathbf{j}$	Simplify.

Check Point 6 If $\mathbf{v} = 7\mathbf{i} + 3\mathbf{j}$ and $\mathbf{w} = 4\mathbf{i} - 5\mathbf{j}$, find $6\mathbf{v} - 3\mathbf{w}$.

    Properties involving vector operations resemble familiar properties of real numbers. For example, the order in which vectors are added makes no difference:

$$\mathbf{u} + \mathbf{v} = \mathbf{v} + \mathbf{u}.$$

Does this remind you of the commutative property $a + b = b + a$?

Just as 0 plays an important role in the properties of real numbers, the **zero vector 0** plays exactly the same role in the properties of vectors.

The Zero Vector

The vector whose magnitude is 0 is called the **zero vector, 0.** The zero vector is assigned no direction. It can be expressed in terms of **i** and **j** using

$$\mathbf{0} = 0\mathbf{i} + 0\mathbf{j}.$$

Properties of vector addition and scalar multiplication are given as follows:

Properties of Vector Addition and Scalar Multiplication

If **u**, **v**, and **w** are vectors, and c and d are scalars, then the following properties are true.

Vector Addition Properties

1. $\mathbf{u} + \mathbf{v} = \mathbf{v} + \mathbf{u}$	Commutative Property
2. $(\mathbf{u} + \mathbf{v}) + \mathbf{w} = \mathbf{u} + (\mathbf{v} + \mathbf{w})$	Associative Property
3. $\mathbf{u} + \mathbf{0} = \mathbf{0} + \mathbf{u} = \mathbf{u}$	Additive Identity
4. $\mathbf{u} + (-\mathbf{u}) = (-\mathbf{u}) + \mathbf{u} = \mathbf{0}$	Additive Inverse

Scalar Multiplication Properties

1. $(cd)\mathbf{u} = c(d\mathbf{u})$	Associative Property
2. $c(\mathbf{u} + \mathbf{v}) = c\mathbf{u} + c\mathbf{v}$	Distributive Property
3. $(c + d)\mathbf{u} = c\mathbf{u} + d\mathbf{u}$	Distributive Property
4. $1\mathbf{u} = \mathbf{u}$	Multiplicative Identity
5. $0\mathbf{u} = \mathbf{0}$	Multiplication Property
6. $\|c\mathbf{v}\| = |c|\,\|\mathbf{v}\|$	

5 Find the unit vector in the direction of **v**.

Unit Vectors

A unit vector is defined to be a vector whose magnitude is one. In many applications of vectors, it is helpful to find the unit vector that has the same direction as a given vector.

Discovery

To find out why the procedure in the box produces a unit vector, work Exercise 89 in Exercise Set 6.6.

Finding the Unit Vector that Has the Same Direction as a Given Nonzero Vector v

For any nonzero vector **v**, the vector

$$\frac{\mathbf{v}}{\|\mathbf{v}\|}$$

is the unit vector that has the same direction as **v**. To find this vector, divide **v** by its magnitude.

EXAMPLE 7 Finding a Unit Vector

Find the unit vector in the same direction as $\mathbf{v} = 5\mathbf{i} - 12\mathbf{j}$. Then verify that the vector has magnitude 1.

Solution We find the unit vector in the same direction as $\mathbf{v}$ by dividing $\mathbf{v}$ by its magnitude. We first find the magnitude of $\mathbf{v}$.

$$\|\mathbf{v}\| = \sqrt{a^2 + b^2} = \sqrt{5^2 + (-12)^2} = \sqrt{25 + 144} = \sqrt{169} = 13$$

The unit vector in the same direction as $\mathbf{v}$ is

$$\frac{\mathbf{v}}{\|\mathbf{v}\|} = \frac{5\mathbf{i} - 12\mathbf{j}}{13} = \frac{5}{13}\mathbf{i} - \frac{12}{13}\mathbf{j}. \quad \text{This is the scalar multiplication of } \mathbf{v} \text{ and } \tfrac{1}{13}.$$

Now we must verify that the magnitude of this vector is 1. Recall that the magnitude of $a\mathbf{i} + b\mathbf{j}$ is $\sqrt{a^2 + b^2}$. Thus, the magnitude of $\frac{5}{13}\mathbf{i} - \frac{12}{13}\mathbf{j}$ is

$$\sqrt{\left(\frac{5}{13}\right)^2 + \left(-\frac{12}{13}\right)^2} = \sqrt{\frac{25}{169} + \frac{144}{169}} = \sqrt{\frac{169}{169}} = \sqrt{1} = 1.$$

Check Point 7 Find the unit vector in the same direction as $\mathbf{v} = 4\mathbf{i} - 3\mathbf{j}$. Then verify that the vector has magnitude 1.

6 Write a vector in terms of its magnitude and direction.

Writing a Vector in Terms of Its Magnitude and Direction

Consider the vector $\mathbf{v} = a\mathbf{i} + b\mathbf{j}$. The components a and b can be expressed in terms of the magnitude of $\mathbf{v}$ and the angle θ that $\mathbf{v}$ makes with the positive x-axis. This angle is called the **direction angle** of $\mathbf{v}$ and is shown in Figure 6.54. By the definitions of sine and cosine, we have

$$\cos\theta = \frac{a}{\|\mathbf{v}\|} \qquad \text{and} \qquad \sin\theta = \frac{b}{\|\mathbf{v}\|}$$

$$a = \|\mathbf{v}\|\cos\theta \qquad\qquad b = \|\mathbf{v}\|\sin\theta.$$

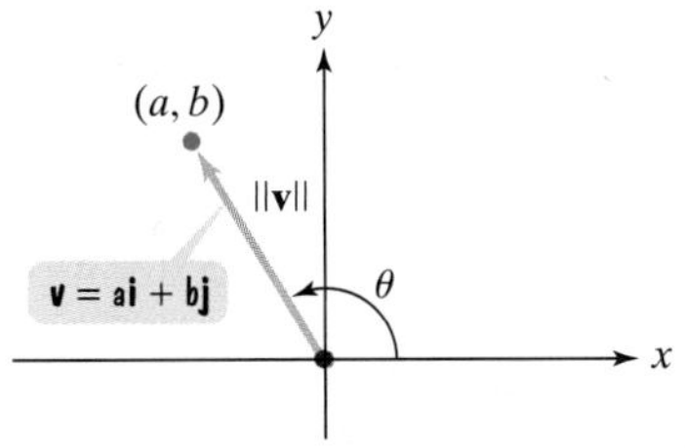

Figure 6.54 Expressing a vector in terms of its magnitude, $\|\mathbf{v}\|$, and its direction angle, θ

Thus,

$$\mathbf{v} = a\mathbf{i} + b\mathbf{j} = \|\mathbf{v}\|\cos\theta\mathbf{i} + \|\mathbf{v}\|\sin\theta\mathbf{j}.$$

Writing a Vector in Terms of Its Magnitude and Direction

Let $\mathbf{v}$ be a nonzero vector. If θ is the direction angle measured from the positive x-axis to $\mathbf{v}$, then the vector can be expressed in terms of its magnitude and direction angle as

$$\mathbf{v} = \|\mathbf{v}\|\cos\theta\mathbf{i} + \|\mathbf{v}\|\sin\theta\mathbf{j}.$$

A vector that represents the direction and speed of an object in motion is called a **velocity vector**. In Example 8, we express a wind's velocity vector in terms of the wind's magnitude and direction.

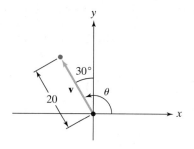

Figure 6.55 Vector **v** represents a wind blowing at 20 miles per hour in the direction N30°W.

EXAMPLE 8 Writing a Vector Whose Magnitude and Direction Are Given

The wind is blowing at 20 miles per hour in the direction N30°W. Express its velocity as a vector **v**.

Solution The vector **v** is shown in Figure 6.55. The vector's direction angle, from the positive x-axis to **v**, is

$$\theta = 90° + 30° = 120°.$$

Because the wind is blowing at 20 miles per hour, the magnitude of **v** is 20 miles per hour: $\|\mathbf{v}\| = 20$. Thus,

$$\mathbf{v} = \|\mathbf{v}\| \cos\theta\,\mathbf{i} + \|\mathbf{v}\| \sin\theta\,\mathbf{j} \qquad \text{Use the formula for a vector in terms of magnitude and direction.}$$

$$= 20\cos 120°\,\mathbf{i} + 20\sin 120°\,\mathbf{j} \qquad \|\mathbf{v}\| = 20 \text{ and } \theta = 120°.$$

$$= 20\left(-\tfrac{1}{2}\right)\mathbf{i} + 20\left(\frac{\sqrt{3}}{2}\right)\mathbf{j} \qquad \cos 120° = -\frac{1}{2} \text{ and } \sin 120° = \frac{\sqrt{3}}{2}.$$

$$= -10\mathbf{i} + 10\sqrt{3}\,\mathbf{j} \qquad \text{Simplify.}$$

The wind's velocity can be expressed in terms of **i** and **j** as $\mathbf{v} = -10\mathbf{i} + 10\sqrt{3}\,\mathbf{j}$.

Check Point 8 The jet stream is blowing at 60 miles per hour in the direction N45°E. Express its velocity as a vector **v** in terms of **i** and **j**.

7 Solve applied problems involving vectors.

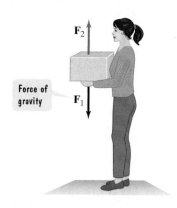

Figure 6.56 Force vectors

Applications

Many physical concepts can be represented by vectors. A vector that represents a pull or push of some type is called a **force vector.** If you are holding a 10-pound package, two force vectors are involved. The force of gravity is exerting a force of magnitude 10 pounds directly downward. This force is shown by vector $\mathbf{F}_1$ in Figure 6.56. Assuming there is no upward or downward movement of the package, you are exerting a force of magnitude 10 pounds directly upward. This force is shown by vector $\mathbf{F}_2$ in Figure 6.56. It has the same magnitude as the force exerted on your package by gravity, but it acts in the opposite direction.

If $\mathbf{F}_1$ and $\mathbf{F}_2$ are two forces acting on an object, the net effect is the same as if just the resultant force, $\mathbf{F}_1 + \mathbf{F}_2$, acted on the object. If the object is not moving, as is the case with your 10-pound package, the vector sum of all forces is the zero vector.

EXAMPLE 9 Finding the Resultant Force

Two forces, $\mathbf{F}_1$ and $\mathbf{F}_2$, of magnitude 10 and 30 pounds, respectively, act on an object. The direction of $\mathbf{F}_1$ is N20°E and the direction of $\mathbf{F}_2$ is N65°E. Find the magnitude and the direction of the resultant force. Express the direction angle to the nearest tenth of a degree.

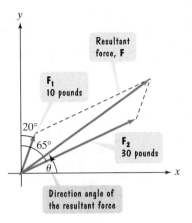

Figure 6.57

Solution The vectors $\mathbf{F}_1$ and $\mathbf{F}_2$ are shown in Figure 6.57. The direction angle for $\mathbf{F}_1$, from the positive x-axis to the vector, is 90°– 20°, or 70°. We express $\mathbf{F}_1$ using the formula for a vector in terms of its magnitude and direction.

$$\mathbf{F}_1 = \|\mathbf{F}_1\| \cos\theta\mathbf{i} + \|\mathbf{F}_1\| \sin\theta\mathbf{j}$$
$$= 10\cos 70°\mathbf{i} + 10\sin 70°\mathbf{j} \qquad \|\mathbf{F}_1\| = 10 \text{ and } \theta = 70°.$$
$$\approx 3.42\mathbf{i} + 9.40\mathbf{j} \qquad \textit{Use a calculator.}$$

Figure 6.57 illustrates that the direction angle for $\mathbf{F}_2$, from the positive x-axis to the vector, is 90°– 65°, or 25°. We express $\mathbf{F}_2$ using the formula for a vector in terms of its magnitude and direction.

$$\mathbf{F}_2 = \|\mathbf{F}_2\| \cos\theta\mathbf{i} + \|\mathbf{F}_2\| \sin\theta\mathbf{j}$$
$$= 30\cos 25°\mathbf{i} + 30\sin 25°\mathbf{j} \qquad \|\mathbf{F}_2\| = 30 \text{ and } \theta = 25°.$$
$$\approx 27.19\mathbf{i} + 12.68\mathbf{j} \qquad \textit{Use a calculator.}$$

The resultant force, $\mathbf{F}$, is $\mathbf{F}_1 + \mathbf{F}_2$. Thus,

$$\mathbf{F} = \mathbf{F}_1 + \mathbf{F}_2$$
$$\approx (3.42\mathbf{i} + 9.40\mathbf{j}) + (27.19\mathbf{i} + 12.68\mathbf{j}) \quad \textit{Use } \mathbf{F}_1 \textit{ and } \mathbf{F}_2, \textit{ found above.}$$
$$= (3.42 + 27.19)\mathbf{i} + (9.40 + 12.68)\mathbf{j} \quad \textit{Add the horizontal components. Add the vertical components.}$$
$$= 30.61\mathbf{i} + 22.08\mathbf{j}. \qquad \textit{Simplify.}$$

Now that we have the resultant force vector, $\mathbf{F}$, we can find its magnitude.

$$\|\mathbf{F}\| = \sqrt{a^2 + b^2} = \sqrt{(30.61)^2 + (22.08)^2} \approx 37.74$$

The magnitude of the resultant force is approximately 37.74 pounds.

To find θ, the direction angle of the resultant force, we can use

$$\cos\theta = \frac{a}{\|\mathbf{F}\|} \quad \text{or} \quad \sin\theta = \frac{b}{\|\mathbf{F}\|}.$$

Using the first formula, we obtain

$$\cos\theta = \frac{a}{\|\mathbf{F}\|} \approx \frac{30.61}{37.74}. \qquad \textit{Recall that the resultant force is } \mathbf{F} = 30.61\mathbf{i} + 22.08\mathbf{j}.$$

Thus,

$$\theta = \cos^{-1}\left(\frac{30.61}{37.74}\right) \approx 35.8°. \qquad \textit{Use a calculator.}$$

The direction angle of the resultant force is approximately 35.8°.

In summary, the two given forces are equivalent to a single force of approximately 37.74 pounds with a direction angle of approximately 35.8°.

Study Tip

If $\mathbf{F} = a\mathbf{i} + b\mathbf{j}$, the direction angle, θ, of $\mathbf{F}$ can also be found using

$$\tan\theta = \frac{b}{a}.$$

Check Point 9

Two forces, $\mathbf{F}_1$ and $\mathbf{F}_2$, of magnitude 30 and 60 pounds, respectively, act on an object. The direction of $\mathbf{F}_1$ is N10°E and the direction of $\mathbf{F}_2$ is N60°E. Find the magnitude, to the nearest hundredth of a pound, and the direction angle, to the nearest tenth of a degree, of the resultant force.

We have seen that velocity vectors represent the direction and speed of moving objects. Boats moving in currents and airplanes flying in winds are situations in which two velocity vectors act simultaneously. For example, suppose **v** represents the velocity of a plane in still air. Further suppose that **w** represents the velocity of the wind. The actual speed and direction of the plane is given by the vector **v** + **w**. This resultant vector describes the plane's speed and direction relative to the ground. Problems involving the resultant velocity of a boat or plane are solved using the same method that we used in Example 9 to find a single resultant force equivalent to two given forces.

EXERCISE SET 6.6

Practice Exercises

In Exercises 1–4, **u** *and* **v** *have the same direction. In each exercise:* **a.** *Find* $\|\mathbf{u}\|$. **b.** *Find* $\|\mathbf{v}\|$. **c.** *Is* **u** = **v**? *Explain.*

1.

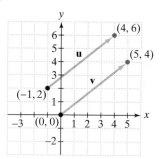

2.

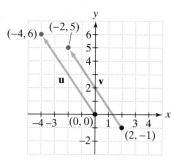

3.

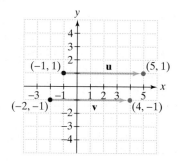

4.

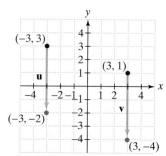

In Exercises 5–12, sketch each vector as a position vector and find its magnitude.

5. $\mathbf{v} = 3\mathbf{i} + \mathbf{j}$

6. $\mathbf{v} = 2\mathbf{i} + 3\mathbf{j}$

7. $\mathbf{v} = \mathbf{i} - \mathbf{j}$

8. $\mathbf{v} = -\mathbf{i} - \mathbf{j}$

9. $\mathbf{v} = -6\mathbf{i} - 2\mathbf{j}$

10. $\mathbf{v} = 5\mathbf{i} - 2\mathbf{j}$

11. $\mathbf{v} = -4\mathbf{i}$

12. $\mathbf{v} = -5\mathbf{j}$

In Exercises 13–20, let $\mathbf{v}$ be the vector from initial point P_1 to terminal point P_2. Write $\mathbf{v}$ in terms of $\mathbf{i}$ and $\mathbf{j}$.

13. $P_1 = (-4, -4), P_2 = (6, 2)$

14. $P_1 = (2, -5), P_2 = (-6, 6)$

15. $P_1 = (-8, 6), P_2 = (-2, 3)$

16. $P_1 = (-7, -4), P_2 = (0, -2)$

17. $P_1 = (-1, 7), P_2 = (-7, -7)$

18. $P_1 = (-1, 6), P_2 = (7, -5)$

19. $P_1 = (-3, 4), P_2 = (6, 4)$

20. $P_1 = (4, -5), P_2 = (4, 3)$

In Exercises 21–38, let

$$\mathbf{u} = 2\mathbf{i} - 5\mathbf{j}, \mathbf{v} = -3\mathbf{i} + 7\mathbf{j}, \text{ and } \mathbf{w} = -\mathbf{i} - 6\mathbf{j}.$$

Find each specified vector or scalar.

21. $\mathbf{u} + \mathbf{v}$

22. $\mathbf{v} + \mathbf{w}$

23. $\mathbf{u} - \mathbf{v}$

24. $\mathbf{v} - \mathbf{w}$

25. $\mathbf{v} - \mathbf{u}$

26. $\mathbf{w} - \mathbf{v}$

27. $5\mathbf{v}$

28. $6\mathbf{v}$

29. $-4\mathbf{w}$

30. $-7\mathbf{w}$

31. $3\mathbf{w} + 2\mathbf{v}$

32. $3\mathbf{u} + 4\mathbf{v}$

33. $3\mathbf{v} - 4\mathbf{w}$

34. $4\mathbf{w} - 3\mathbf{v}$

35. $\|2\mathbf{u}\|$

36. $\|-2\mathbf{u}\|$

37. $\|\mathbf{w} - \mathbf{u}\|$

38. $\|\mathbf{u} - \mathbf{w}\|$

In Exercises 39–46, find the unit vector that has the same direction as the vector $\mathbf{v}$.

39. $\mathbf{v} = 6\mathbf{i}$

40. $\mathbf{v} = -5\mathbf{j}$

41. $\mathbf{v} = 3\mathbf{i} - 4\mathbf{j}$

42. $\mathbf{v} = 8\mathbf{i} - 6\mathbf{j}$

43. $\mathbf{v} = 3\mathbf{i} - 2\mathbf{j}$

44. $\mathbf{v} = 4\mathbf{i} - 2\mathbf{j}$

45. $\mathbf{v} = \mathbf{i} + \mathbf{j}$

46. $\mathbf{v} = \mathbf{i} - \mathbf{j}$

In Exercises 47–52, write the vector $\mathbf{v}$ in terms of $\mathbf{i}$ and $\mathbf{j}$ whose magnitude $\|\mathbf{v}\|$ and direction angle θ are given.

47. $\|\mathbf{v}\| = 6, \quad \theta = 30°$

48. $\|\mathbf{v}\| = 8, \quad \theta = 45°$

49. $\|\mathbf{v}\| = 12, \theta = 225°$

50. $\|\mathbf{v}\| = 10, \theta = 330°$

51. $\|\mathbf{v}\| = \frac{1}{2}, \quad \theta = 113°$

52. $\|\mathbf{v}\| = \frac{1}{4}, \quad \theta = 200°$

Application Exercises

In Exercises 53–56, a vector is described. Express the vector in terms of $\mathbf{i}$ and $\mathbf{j}$. If exact values are not possible, round components to the nearest tenth.

53. A quarterback releases a football with a speed of 44 feet per second at an angle of 30° with the horizontal.

54. A child pulls a sled along level ground by exerting a force of 30 pounds on a handle that makes an angle of 45° with the ground.

55. A plane approaches a runway at 150 miles per hour at an angle of 8° with the runway.

56. A plane with an airspeed of 450 miles per hour is flying in the direction N35°W.

Vectors are used in computer graphics to determine lengths of shadows over flat surfaces. The length of the shadow for $\mathbf{v}$ in the figure shown is the absolute value of the vector's horizontal component. In Exercises 57–58, the magnitude and direction angle of $\mathbf{v}$ are given. Write $\mathbf{v}$ in terms of $\mathbf{i}$ and $\mathbf{j}$. Then find the length of the shadow to the nearest tenth of an inch.

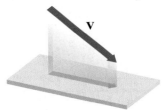

57. $\|\mathbf{v}\| = 1.5$ inches, $\quad \theta = 25°$

58. $\|\mathbf{v}\| = 1.8$ inches, $\quad \theta = 40°$

59. The magnitude and direction of two forces acting on an object are 70 pounds, S56°E, and 50 pounds, N72°E, respectively. Find the magnitude, to the nearest hundredth of a pound, and the direction angle, to the nearest tenth of a degree, of the resultant force.

60. The magnitude and direction exerted by two tugboats towing a ship are 4200 pounds, N65°E, and 3000 pounds, S58°E, respectively. Find the magnitude, to the nearest pound, and the direction angle, to the nearest tenth of a degree, of the resultant force.

The figure shows a box being pulled up a ramp inclined at 18° from the horizontal.

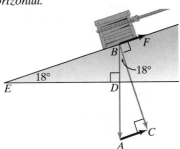

Use the following information to solve Exercises 61–62.

$$\overrightarrow{BA} = \text{force of gravity}$$
$$\|\overrightarrow{BA}\| = \text{weight of the box}$$
$$\|\overrightarrow{AC}\| = \text{magnitude of the force needed to pull the box up the ramp}$$
$$\|\overrightarrow{BC}\| = \text{magnitude of the force of the box against the ramp}$$

61. If the box weighs 100 pounds, find the magnitude of the force needed to pull it up the ramp.

62. If a force of 30 pounds is needed to pull the box up the ramp, find the weight of the box.

The forces $\mathbf{F}_1, \mathbf{F}_2, \mathbf{F}_3, \ldots, \mathbf{F}_n$ *acting on an object are in* **equilibrium** *if the resultant force is the zero vector:*

$$\mathbf{F}_1 + \mathbf{F}_2 + \mathbf{F}_3 + \cdots + \mathbf{F}_n = \mathbf{0}.$$

In Exercises 63–66, the given forces are acting on an object.
a. *Find the resultant force.*
b. *What additional force is required for the given forces to be in equilibrium?*

63. $\mathbf{F}_1 = 3\mathbf{i} - 5\mathbf{j}, \quad \mathbf{F}_2 = 6\mathbf{i} + 2\mathbf{j}$

64. $\mathbf{F}_1 = -2\mathbf{i} + 3\mathbf{j}, \quad \mathbf{F}_2 = \mathbf{i} - \mathbf{j}, \quad \mathbf{F}_3 = 5\mathbf{i} - 12\mathbf{j}$

65.

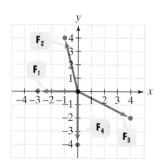

66.

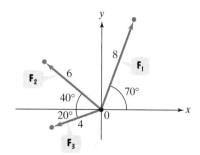

67. The figure shows a small plane flying at a speed of 180 miles per hour on a bearing of N50°E. The wind is blowing from west to east at 40 miles per hour. The figure indicates that **v** represents the velocity of the plane in still air and **w** represents the velocity of the wind.

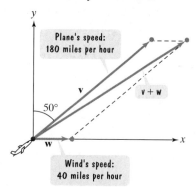

a. Express **v** and **w** in terms of their magnitudes and direction angles.
b. Find the resultant vector, **v** + **w**.
c. The magnitude of **v** + **w**, called the **ground speed** of the plane, gives its speed relative to the ground. Approximate the ground speed to the nearest mile per hour.

d. The direction angle of **v** + **w** gives the plane's true course relative to the ground. Approximate the true course to the nearest tenth of a degree. What is the plane's true bearing?

68. Use the procedure outlined in Exercise 67 to solve this exercise. A plane is flying at a speed of 400 miles per hour on a bearing of N50°W. The wind is blowing at 30 miles per hour on a bearing of N25°E.

a. Approximate the plane's ground speed to the nearest mile per hour.

b. Approximate the plane's true course to the nearest tenth of a degree. What is its true bearing?

69. A plane is flying at a speed of 320 miles per hour on a bearing of N70°E. Its ground speed is 370 miles per hour and its true course is 30°. Find the speed, to the nearest mile per hour, and the direction angle, to the nearest tenth of a degree, of the wind.

70. A plane is flying at a speed of 540 miles per hour on a bearing of S36°E. Its ground speed is 500 miles per hour and its true bearing is S44°E. Find the speed, to the nearest mile per hour, and the direction angle, to the nearest tenth of a degree, of the wind.

Writing in Mathematics

71. What is a directed line segment?

72. What are equal vectors?

73. If vector **v** is represented by an arrow, how is $-3\mathbf{v}$ represented?

74. If vectors **u** and **v** are represented by arrows, describe how the vector sum **u** + **v** is represented.

75. What is the **i** vector?

76. What is the **j** vector?

77. What is a position vector? How is a position vector represented using **i** and **j**?

78. If **v** is a vector between any two points in the rectangular coordinate system, explain how to write **v** in terms of **i** and **j**.

79. If two vectors are expressed in terms of **i** and **j**, explain how to find their sum.

80. If two vectors are expressed in terms of **i** and **j**, explain how to find their difference.

81. If a vector is expressed in terms of **i** and **j**, explain how to find the scalar multiplication of the vector and a given scalar k.

82. What is the zero vector?

83. Describe one similarity between the zero vector and the number 0.

84. Explain how to find the unit vector in the direction of any given vector **v**.

85. Explain how to write a vector in terms of its magnitude and direction.

86. You are on an airplane. The pilot announces the plane's speed over the intercom. Which speed do you think is being reported: the speed of the plane in still air or the speed after the effect of the wind has been accounted for? Explain your answer.

87. Use vectors to explain why it is difficult to hold a heavy stack of books perfectly still for a long period of time. As you become exhausted, what eventually happens? What does this mean in terms of the forces acting on the books?

Critical Thinking Exercises

88. Use the figure shown to select a true statement.

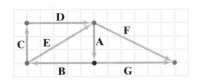

    a. $\mathbf{A} + \mathbf{B} = \mathbf{E}$
    b. $\mathbf{D} + \mathbf{A} + \mathbf{B} + \mathbf{C} = \mathbf{0}$
    c. $\mathbf{B} - \mathbf{E} = \mathbf{G} - \mathbf{F}$
    d. $\|\mathbf{A}\| \neq \|\mathbf{C}\|$

89. Let $\mathbf{v} = a\mathbf{i} + b\mathbf{j}$. Show that $\dfrac{\mathbf{v}}{\|\mathbf{v}\|}$ is a unit vector in the direction of $\mathbf{v}$.

In Exercises 90–91, refer to the navigational compass shown in the figure. The compass is marked clockwise in degrees that start at north 0°.

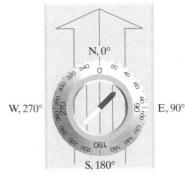

90. An airplane has an air speed of 240 miles per hour and a compass heading of 280°. A steady wind of 30 miles per hour is blowing in the direction of 265°. What is the plane's true speed relative to the ground? What is its compass heading relative to the ground?

91. Two tugboats are pulling on a large ship that has gone aground. One tug pulls with a force of 2500 pounds in a compass direction of 55°. The second tug pulls with a force of 2000 pounds in a compass direction of 95°. Find the magnitude and the compass direction of the resultant force.

92. You want to fly your small plane due north, but there is a 75 kilometer wind blowing from west to east.

    a. Find the direction angle for where you should head the plane if your speed relative to the air is 310 kilometers per hour.

    b. If you increase your air speed, should the direction angle in part (a) increase or decrease? Explain your answer.

SECTION 6.7   *The Dot Product*

Objectives

1. Find the dot product of two vectors.

2. Find the angle between two vectors.

3. Use the dot product to determine if two vectors are orthogonal.

4. Find the projection of a vector onto another vector.

5. Express a vector as the sum of two orthogonal vectors.

6. Compute work.

Talk about hard work! I can see the weightlifter's muscles quivering from the exertion of holding the barbell in a stationary position above his head. Still, I'm not sure if he's doing as much work as I, sitting at my desk with my brain quivering from studying trigonometric functions and their applications.

Would it surprise you to know that neither you nor the weightlifter are doing any work at all? The definition of work in physics and mathematics is not the same as what we mean by "work" in everyday use. To understand what is involved in real work, we turn to a new vector operation called the dot product.

1 Find the dot product of two vectors.

The Dot Product of Two Vectors

The operations of vector addition and scalar multiplication result in vectors. By contrast, the *dot product* of two vectors results in a scalar (a real number), rather than a vector.

Definition of the Dot Product

If $\mathbf{v} = a_1\mathbf{i} + b_1\mathbf{j}$ and $\mathbf{w} = a_2\mathbf{i} + b_2\mathbf{j}$ are vectors, the **dot product $\mathbf{v} \cdot \mathbf{w}$** is defined as

$$\mathbf{v} \cdot \mathbf{w} = a_1 a_2 + b_1 b_2.$$

The dot product of two vectors is the sum of the products of their horizontal and vertical components.

EXAMPLE 1 Finding Dot Products

If $\mathbf{v} = 5\mathbf{i} - 2\mathbf{j}$ and $\mathbf{w} = -3\mathbf{i} + 4\mathbf{j}$, find:

a. $\mathbf{v} \cdot \mathbf{w}$ **b.** $\mathbf{w} \cdot \mathbf{v}$ **c.** $\mathbf{v} \cdot \mathbf{v}$.

Solution To find each dot product, multiply the two horizontal components, and then multiply the two vertical components. Finally, add the two products.

a. $\mathbf{v} \cdot \mathbf{w} = 5(-3) + (-2)(4) = -15 - 8 = -23$

Multiply the horizontal components and multiply the vertical components of
$\mathbf{v} = 5\mathbf{i} - 2\mathbf{j}$ and $\mathbf{w} = -3\mathbf{i} + 4\mathbf{j}$.

b. $\mathbf{w} \cdot \mathbf{v} = -3(5) + 4(-2) = -15 - 8 = -23$

Multiply the horizontal components and multiply the vertical components of
$\mathbf{w} = -3\mathbf{i} + 4\mathbf{j}$ and $\mathbf{v} = 5\mathbf{i} - 2\mathbf{j}$.

c. $\mathbf{v} \cdot \mathbf{v} = 5(5) + (-2)(-2) = 25 + 4 = 29$

Multiply the horizontal components and multiply the vertical components of
$\mathbf{v} = 5\mathbf{i} - 2\mathbf{j}$ and $\mathbf{v} = 5\mathbf{i} - 2\mathbf{j}$.

Check Point 1 If $\mathbf{v} = 7\mathbf{i} - 4\mathbf{j}$ and $\mathbf{w} = 2\mathbf{i} - \mathbf{j}$, find:

a. $\mathbf{v} \cdot \mathbf{w}$ **b.** $\mathbf{w} \cdot \mathbf{v}$ **c.** $\mathbf{w} \cdot \mathbf{w}$.

In Example 1 and Check Point 1, did you notice that $\mathbf{v} \cdot \mathbf{w}$ and $\mathbf{w} \cdot \mathbf{v}$ produced the same scalar? The fact that $\mathbf{v} \cdot \mathbf{w} = \mathbf{w} \cdot \mathbf{v}$ follows from the definition of the dot product. Properties of the dot product are given in the box at the top of the next page. Proofs for some of these properties are given in the appendix.

Properties of the Dot Product

If **u**, **v**, and **w** are vectors, and c is a scalar, then

1. $\mathbf{u} \cdot \mathbf{v} = \mathbf{v} \cdot \mathbf{u}$
2. $\mathbf{u} \cdot (\mathbf{v} + \mathbf{w}) = \mathbf{u} \cdot \mathbf{v} + \mathbf{u} \cdot \mathbf{w}$
3. $\mathbf{0} \cdot \mathbf{v} = 0$
4. $\mathbf{v} \cdot \mathbf{v} = \|\mathbf{v}\|^2$
5. $(c\mathbf{u}) \cdot \mathbf{v} = c(\mathbf{u} \cdot \mathbf{v}) = \mathbf{u} \cdot (c\mathbf{v})$

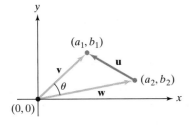

Figure 6.58

The Angle between Two Vectors

The Law of Cosines can be used to derive another formula for the dot product. This formula will give us a way to find the angle between two vectors.

Figure 6.58 shows vectors $\mathbf{v} = a_1\mathbf{i} + b_1\mathbf{j}$ and $\mathbf{w} = a_2\mathbf{i} + b_2\mathbf{j}$. By the definition of the dot product, we know that $\mathbf{v} \cdot \mathbf{w} = a_1 a_2 + b_1 b_2$. Our new formula for the dot product involves the angle between the vectors, shown as θ in the figure. Apply the Law of Cosines to the triangle shown in the figure.

$$\|\mathbf{u}\|^2 = \|\mathbf{v}\|^2 + \|\mathbf{w}\|^2 - 2\|\mathbf{v}\|\|\mathbf{w}\|\cos\theta \qquad \text{Use the Law of Cosines.}$$

$$(a_1 - a_2)^2 + (b_1 - b_2)^2 = (a_1^2 + b_1^2) + (a_2^2 + b_2^2) - 2\|\mathbf{v}\|\|\mathbf{w}\|\cos\theta \qquad \begin{array}{l}\text{Square the magnitudes of vectors} \\ \mathbf{u}, \mathbf{v}, \text{ and } \mathbf{w}.\end{array}$$

$$a_1^2 - 2a_1 a_2 + a_2^2 + b_1^2 - 2b_1 b_2 + b_2^2 = a_1^2 + b_1^2 + a_2^2 + b_2^2 - 2\|\mathbf{v}\|\|\mathbf{w}\|\cos\theta \qquad \begin{array}{l}\text{Square the binomials using} \\ (A - B)^2 = A^2 - 2AB + B^2.\end{array}$$

$$-2a_1 a_2 - 2b_1 b_2 = -2\|\mathbf{v}\|\|\mathbf{w}\|\cos\theta \qquad \begin{array}{l}\text{Subtract } a_1^2,\, a_2^2,\, b_1^2, \text{ and } b_2^2 \text{ from both} \\ \text{sides of the equation.}\end{array}$$

$$a_1 a_2 + b_1 b_2 = \|\mathbf{v}\|\|\mathbf{w}\|\cos\theta \qquad \text{Divide both sides by } -2.$$

By definition,
$\mathbf{v} \cdot \mathbf{w} = a_1 a_2 + b_1 b_2.$

$$\mathbf{v} \cdot \mathbf{w} = \|\mathbf{v}\|\|\mathbf{w}\|\cos\theta \qquad \begin{array}{l}\text{Substitute } \mathbf{v} \cdot \mathbf{w} \text{ for the expression} \\ \text{on the left side of the equation.}\end{array}$$

Alternative Formula for the Dot Product

If **v** and **w** are two nonzero vectors and θ is the smallest nonnegative angle between them, then

$$\mathbf{v} \cdot \mathbf{w} = \|\mathbf{v}\|\|\mathbf{w}\|\cos\theta.$$

2 Find the angle between two vectors.

Solving the formula in the box for $\cos\theta$ gives us a formula for finding the angle between vectors:

Formula for the Angle between Two Vectors

If **v** and **w** are two nonzero vectors and θ is the smallest nonnegative angle between **v** and **w**, then

$$\cos\theta = \frac{\mathbf{v} \cdot \mathbf{w}}{\|\mathbf{v}\|\|\mathbf{w}\|} \quad \text{and} \quad \theta = \cos^{-1}\left(\frac{\mathbf{v} \cdot \mathbf{w}}{\|\mathbf{v}\|\|\mathbf{w}\|}\right).$$

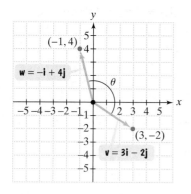

Figure 6.59 Finding the angle between two vectors

EXAMPLE 2   Finding the Angle between Two Vectors

Find the angle θ between the vectors $\mathbf{v} = 3\mathbf{i} - 2\mathbf{j}$ and $\mathbf{w} = -\mathbf{i} + 4\mathbf{j}$, shown in Figure 6.59. Round to the nearest tenth of a degree.

Solution   Use the formula for the angle between two vectors.

$$\cos \theta = \frac{\mathbf{v} \cdot \mathbf{w}}{\|\mathbf{v}\| \|\mathbf{w}\|}$$ This is the formula for the cosine of the angle between two vectors.

$$= \frac{(3\mathbf{i} - 2\mathbf{j}) \cdot (-\mathbf{i} + 4\mathbf{j})}{\sqrt{3^2 + (-2)^2}\sqrt{(-1)^2 + 4^2}}$$ Substitute the given vectors in the numerator. Find the magnitude of each vector in the denominator.

$$= \frac{3(-1) + (-2)(4)}{\sqrt{13}\sqrt{17}}$$ Find the dot product in the numerator. Simplify in the denominator.

$$= -\frac{11}{\sqrt{221}}$$ Perform the indicated operations.

The angle θ between the vectors is

$$\theta = \cos^{-1}\left(-\frac{11}{\sqrt{221}}\right) \approx 137.7°.$$ Use a calculator.

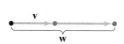

Check Point 2   Find the angle between the vectors $\mathbf{v} = 4\mathbf{i} - 3\mathbf{j}$ and $\mathbf{w} = \mathbf{i} + 2\mathbf{j}$. Round to the nearest tenth of a degree.

3 Use the dot product to determine if two vectors are orthogonal.

Parallel and Orthogonal Vectors

Two vectors are **parallel** when the angle θ between the vectors is $0°$ or $180°$. If $\theta = 0°$, the vectors point in the same direction. If $\theta = 180°$, the vectors point in opposite directions. Figure 6.60 shows parallel vectors.

$\theta = 0°$ and $\cos \theta = 1$. Vectors point in the same direction.

$\theta = 180°$ and $\cos \theta = -1$. Vectors point in opposite directions.

Figure 6.60 Parallel vectors

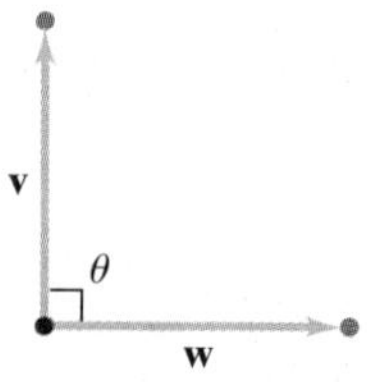

Figure 6.61 Orthogonal vectors: $\theta = 90°$ and $\cos \theta = 0$

Two vectors are **orthogonal** when the angle between the vectors is $90°$, shown in Figure 6.61. (The word "orthogonal," rather than "perpendicular," is used to describe vectors that meet at right angles.) We know that $\mathbf{v} \cdot \mathbf{w} = \|\mathbf{v}\| \|\mathbf{w}\| \cos \theta$. We also know that two vectors are orthogonal if and only if the angle between them is $90°$. Using this formula for the dot product and the meaning of orthogonal vectors gives the following result:

The Dot Product and Orthogonal Vectors

Two nonzero vectors $\mathbf{v}$ and $\mathbf{w}$ are orthogonal if and only if $\mathbf{v} \cdot \mathbf{w} = 0$. Because $\mathbf{0} \cdot \mathbf{v} = 0$, the zero vector is orthogonal to every vector $\mathbf{v}$.

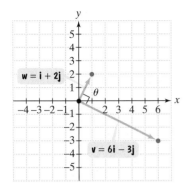

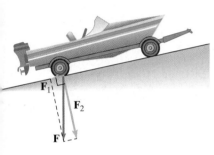

Figure 6.62 Orthogonal vectors

EXAMPLE 3 Determining Whether Vectors Are Orthogonal

Are the vectors $\mathbf{v} = 6\mathbf{i} - 3\mathbf{j}$ and $\mathbf{w} = \mathbf{i} + 2\mathbf{j}$ orthogonal?

Solution   The vectors are orthogonal if their dot product is 0. Begin by finding $\mathbf{v} \cdot \mathbf{w}$.

$$\mathbf{v} \cdot \mathbf{w} = (6\mathbf{i} - 3\mathbf{j}) \cdot (\mathbf{i} + 2\mathbf{j}) = 6(1) + (-3)(2) = 6 - 6 = 0$$

The dot product is 0. Thus, the given vectors are orthogonal. They are shown in Figure 6.62.

Check Point 3   Are the vectors $\mathbf{v} = 2\mathbf{i} + 3\mathbf{j}$ and $\mathbf{w} = 6\mathbf{i} - 4\mathbf{j}$ orthogonal?

Projection of a Vector Onto Another Vector

You know how to add two vectors to obtain a resultant vector. We now reverse this process by expressing a vector as the sum of two orthogonal vectors. By doing this, you can determine how much force is applied in a particular direction. For example, Figure 6.63 shows a boat on a tilted ramp. The force due to gravity, $\mathbf{F}$, is pulling straight down on the boat. Part of this force, $\mathbf{F}_1$, is pushing the boat down the ramp. Another part of this force, $\mathbf{F}_2$, is pressing the boat against the ramp, at a right angle to the incline. These two orthogonal vectors, $\mathbf{F}_1$ and $\mathbf{F}_2$, are called the **vector components** of $\mathbf{F}$. Notice that

$$\mathbf{F} = \mathbf{F}_1 + \mathbf{F}_2.$$

A method for finding $\mathbf{F}_1$ and $\mathbf{F}_2$ involves projecting a vector onto another vector.

Figure 6.64 shows two nonzero vectors, $\mathbf{v}$ and $\mathbf{w}$, with the same initial point. The angle between the vectors, θ, is acute in Figure 6.64(a) and obtuse in Figure 6.64(b). A third vector, called the **vector projection of v onto w**, is also shown in each figure, denoted by $\text{proj}_\mathbf{w}\,\mathbf{v}$.

How is the vector projection of $\mathbf{v}$ onto $\mathbf{w}$ formed? Draw the red line segment from the terminal point of $\mathbf{v}$ that forms a right angle with a line through $\mathbf{w}$, shown in red. The projection of $\mathbf{v}$ onto $\mathbf{w}$ lies on a line through $\mathbf{w}$, and is parallel to vector $\mathbf{w}$. This vector begins at the common initial point of $\mathbf{v}$ and $\mathbf{w}$. It ends at the point where the red line segment intersects the line through $\mathbf{w}$.

Figure 6.63

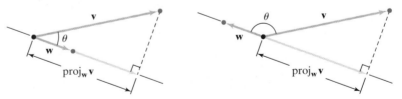

Figure 6.64(a)	**Figure 6.64(b)**

Our goal is to determine an expression for $\text{proj}_\mathbf{w}\,\mathbf{v}$. We begin with its magnitude. By the definition of the cosine function,

$$\cos \theta = \frac{\|\text{proj}_\mathbf{w}\mathbf{v}\|}{\|\mathbf{v}\|}.$$

> This is the magnitude of the vector projection of v onto w.

$$\|\mathbf{v}\| \cos \theta = \|\text{proj}_\mathbf{w}\mathbf{v}\| \qquad \text{Multiply both sides by } \|\mathbf{v}\|.$$

$$\|\text{proj}_\mathbf{w}\mathbf{v}\| = \|\mathbf{v}\| \cos \theta \qquad \text{Reverse the two sides.}$$

We can rewrite the right side of this equation and obtain another expression for the magnitude of the vector projection of $\mathbf{v}$ onto $\mathbf{w}$. To do so, use the alternate formula for the dot product, $\mathbf{v} \cdot \mathbf{w} = \|\mathbf{v}\| \|\mathbf{w}\| \cos \theta$.

Divide both sides by $\|\mathbf{w}\|$:

$$\frac{\mathbf{v} \cdot \mathbf{w}}{\|\mathbf{w}\|} = \|\mathbf{v}\| \cos \theta.$$

The expression on the right side of this equation, $\|\mathbf{v}\| \cos \theta$, is the same expression that appears in the formula for $\|\text{proj}_\mathbf{w}\mathbf{v}\|$. Thus,

$$\|\text{proj}_\mathbf{w}\mathbf{v}\| = \|\mathbf{v}\| \cos \theta = \frac{\mathbf{v} \cdot \mathbf{w}}{\|\mathbf{w}\|}.$$

$$\frac{\mathbf{v} \cdot \mathbf{w}}{\|\mathbf{w}\|} = \|\mathbf{v}\| \cos \theta$$

$$\|\text{proj}_\mathbf{w}\mathbf{v}\| = \|\mathbf{v}\| \cos \theta$$

We use the formula for the magnitude of $\text{proj}_\mathbf{w} \mathbf{v}$ to find the vector itself. This is done by finding the scalar product of the magnitude and the unit vector in the direction of $\mathbf{w}$.

$$\text{proj}_\mathbf{w}\mathbf{v} = \left(\frac{\mathbf{v} \cdot \mathbf{w}}{\|\mathbf{w}\|}\right)\left(\frac{\mathbf{w}}{\|\mathbf{w}\|}\right) = \frac{\mathbf{v} \cdot \mathbf{w}}{\|\mathbf{w}\|^2}\mathbf{w}$$

This is the magnitude of the vector projection of v onto w.

This is the unit vector in the direction of w.

4 Find the projection of a vector onto another vector.

The Vector Projection of v Onto w

If $\mathbf{v}$ and $\mathbf{w}$ are two nonzero vectors, the vector projection of $\mathbf{v}$ onto $\mathbf{w}$ is

$$\text{proj}_\mathbf{w}\mathbf{v} = \frac{\mathbf{v} \cdot \mathbf{w}}{\|\mathbf{w}\|^2}\mathbf{w}.$$

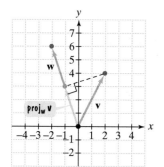

Figure 6.65 The vector projection of **v** onto **w**

EXAMPLE 4 Finding the Vector Projection of One Vector Onto Another

If $\mathbf{v} = 2\mathbf{i} + 4\mathbf{j}$ and $\mathbf{w} = -2\mathbf{i} + 6\mathbf{j}$, find the vector projection of $\mathbf{v}$ onto $\mathbf{w}$.

Solution The vector projection of $\mathbf{v}$ onto $\mathbf{w}$ is found using the formula for $\text{proj}_\mathbf{w} \mathbf{v}$.

$$\text{proj}_\mathbf{w}\mathbf{v} = \frac{\mathbf{v} \cdot \mathbf{w}}{\|\mathbf{w}\|^2}\mathbf{w} = \frac{(2\mathbf{i} + 4\mathbf{j}) \cdot (-2\mathbf{i} + 6\mathbf{j})}{(\sqrt{(-2)^2 + 6^2})^2}\mathbf{w}$$

$$= \frac{2(-2) + 4(6)}{(\sqrt{40})^2}\mathbf{w} = \frac{20}{40}\mathbf{w} = \tfrac{1}{2}(-2\mathbf{i} + 6\mathbf{j}) = -\mathbf{i} + 3\mathbf{j}$$

The three vectors $\mathbf{v}$, $\mathbf{w}$, and $\text{proj}_\mathbf{w} \mathbf{v}$, are shown in Figure 6.65.

Check Point 4 If $\mathbf{v} = 2\mathbf{i} - 5\mathbf{j}$ and $\mathbf{w} = \mathbf{i} - \mathbf{j}$, find the vector projection of $\mathbf{v}$ onto $\mathbf{w}$.

5 Express a vector as the sum of two orthogonal vectors.

We use the vector projection of $\mathbf{v}$ onto $\mathbf{w}$, $\text{proj}_\mathbf{w} \mathbf{v}$, to express $\mathbf{v}$ as the sum of two orthogonal vectors.

The Vector Components of v

Let $\mathbf{v}$ and $\mathbf{w}$ be two nonzero vectors. Vector $\mathbf{v}$ can be expressed as the sum of two orthogonal vectors, $\mathbf{v}_1$ and $\mathbf{v}_2$, where $\mathbf{v}_1$ is parallel to $\mathbf{w}$ and $\mathbf{v}_2$ is orthogonal to $\mathbf{w}$.

$$\mathbf{v}_1 = \text{proj}_\mathbf{w}\mathbf{v} = \frac{\mathbf{v} \cdot \mathbf{w}}{\|\mathbf{w}\|^2}\mathbf{w}, \quad \mathbf{v}_2 = \mathbf{v} - \mathbf{v}_1$$

Thus, $\mathbf{v} = \mathbf{v}_1 + \mathbf{v}_2$. The vectors $\mathbf{v}_1$ and $\mathbf{v}_2$ are called the **vector components** of $\mathbf{v}$. The process of expressing $\mathbf{v}$ as $\mathbf{v}_1 + \mathbf{v}_2$ is called the **decomposition** of $\mathbf{v}$ into $\mathbf{v}_1$ and $\mathbf{v}_2$.

EXAMPLE 5 Decomposing a Vector into Two Orthogonal Vectors

Let $\mathbf{v} = 2\mathbf{i} + 4\mathbf{j}$ and $\mathbf{w} = -2\mathbf{i} + 6\mathbf{j}$. Decompose $\mathbf{v}$ into two vectors, $\mathbf{v}_1$ and $\mathbf{v}_2$, where $\mathbf{v}_1$ is parallel to $\mathbf{w}$ and $\mathbf{v}_2$ is orthogonal to $\mathbf{w}$.

Solution These are the vectors we worked with in Example 4. We use the formulas in the preceding box.

$$\mathbf{v}_1 = \text{proj}_{\mathbf{w}}\mathbf{v} = -\mathbf{i} + 3\mathbf{j} \quad \text{We obtained this vector in Example 4.}$$

$$\mathbf{v}_2 = \mathbf{v} - \mathbf{v}_1 = (2\mathbf{i} + 4\mathbf{j}) - (-\mathbf{i} + 3\mathbf{j}) = 3\mathbf{i} + \mathbf{j}$$

> **Check Point 5** Let $\mathbf{v} = 2\mathbf{i} - 5\mathbf{j}$ and $\mathbf{w} = \mathbf{i} - \mathbf{j}$. (These are the vectors from Check Point 4.) Decompose $\mathbf{v}$ into two vectors, $\mathbf{v}_1$ and $\mathbf{v}_2$, where $\mathbf{v}_1$ is parallel to $\mathbf{w}$ and $\mathbf{v}_2$ is orthogonal to $\mathbf{w}$.

6 Compute work.

Work: An Application of the Dot Product

The bad news: Your car just died. The good news: It died on a level road just 200 feet from a gas station. Exerting a constant force of 90 pounds, and not necessarily whistling as you work, you manage to push the car to the gas station.

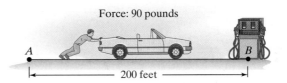

Force: 90 pounds

A B

← 200 feet →

Although you did not whistle, you certainly did work pushing the car 200 feet from point A to point B. How much work did you do? If a constant force $\mathbf{F}$ is applied to an object, moving it from point A to point B in the direction of the force, the work, W, done is

$$W = (\text{magnitude of force})(\text{distance from } A \text{ to } B).$$

You pushed with a force of 90 pounds for a distance of 200 feet. The work done by your force is

$$W = (90 \text{ pounds})(200 \text{ feet})$$

or 18,000 foot-pounds. Work is often measured in foot-pounds or in newton-meters.

The photo on the left shows an adult pulling a small child in a wagon. Work is being done. However, the situation is not quite the same as pushing your car. Pushing the car, the force you applied was along the line of motion. By contrast, the force of the adult pulling the wagon is not applied along the line of the wagon's motion. In this case, the dot product is used to determine the work done by the force.

> **Definition of Work**
> The work, W, done by a force $\mathbf{F}$ moving an object from A to B is
> $$W = \mathbf{F} \cdot \overrightarrow{AB}.$$

When computing work, it is often easier to use the alternative formula for the dot product. Thus,

$$W = \mathbf{F} \cdot \overrightarrow{AB} = \|\mathbf{F}\|\|\overrightarrow{AB}\| \cos\theta.$$

> $\|\mathbf{F}\|$ is the magnitude of the force.

> $\|\overrightarrow{AB}\|$ is the distance over which the constant force is applied.

> θ is the angle between the force and the direction of motion.

It is correct to refer to W as either the work done or the work done by the force.

EXAMPLE 6   Computing Work

A child pulls a sled along level ground by exerting a force of 30 pounds on a rope that makes an angle of 35° with the horizontal. How much work is done pulling the sled 200 feet?

Solution   The situation is illustrated in Figure 6.66. The work done is

$$W = \|\mathbf{F}\|\|\overrightarrow{AB}\| \cos\theta = (30)(200) \cos 35° \approx 4915$$

> Magnitude of the force is 30 pounds.

> Distance is 200 feet.

> The angle between the force and the sled's motion is 35°.

Thus, the work done is approximately 4915 foot-pounds.

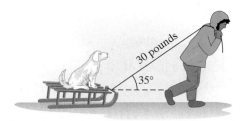

Figure 6.66 Computing work done pulling the sled 200 feet

Check Point 6   A child pulls a wagon along level ground by exerting a force of 20 pounds on a handle that makes an angle of 30° with the horizontal. How much work is done pulling the wagon 150 feet?

EXERCISE SET 6.7

Practice Exercises

In Exercises 1–8, use the given vectors to find a. $\mathbf{v} \cdot \mathbf{w}$ and b. $\mathbf{v} \cdot \mathbf{v}$.

1. $\mathbf{v} = 3\mathbf{i} + \mathbf{j}, \quad \mathbf{w} = \mathbf{i} + 3\mathbf{j}$
2. $\mathbf{v} = 3\mathbf{i} + 3\mathbf{j}, \quad \mathbf{w} = \mathbf{i} + 4\mathbf{j}$
3. $\mathbf{v} = 5\mathbf{i} - 4\mathbf{j}, \quad \mathbf{w} = -2\mathbf{i} - \mathbf{j}$
4. $\mathbf{v} = 7\mathbf{i} - 2\mathbf{j}, \quad \mathbf{w} = -3\mathbf{i} - \mathbf{j}$
5. $\mathbf{v} = -6\mathbf{i} - 5\mathbf{j}, \quad \mathbf{w} = -10\mathbf{i} - 8\mathbf{j}$
6. $\mathbf{v} = -8\mathbf{i} - 3\mathbf{j}, \quad \mathbf{w} = -10\mathbf{i} - 5\mathbf{j}$
7. $\mathbf{v} = 5\mathbf{i}, \quad \mathbf{w} = \mathbf{j}$ 8. $\mathbf{v} = \mathbf{i}, \quad \mathbf{w} = -5\mathbf{j}$

In Exercises 9–16, let

$$\mathbf{u} = 2\mathbf{i} - \mathbf{j}, \quad \mathbf{v} = 3\mathbf{i} + \mathbf{j}, \quad \text{and} \quad \mathbf{w} = \mathbf{i} + 4\mathbf{j}.$$

Find each specified scalar.

9. $\mathbf{u} \cdot (\mathbf{v} + \mathbf{w})$ 10. $\mathbf{v} \cdot (\mathbf{u} + \mathbf{w})$
11. $\mathbf{u} \cdot \mathbf{v} + \mathbf{u} \cdot \mathbf{w}$ 12. $\mathbf{v} \cdot \mathbf{u} + \mathbf{v} \cdot \mathbf{w}$
13. $(4\mathbf{u}) \cdot \mathbf{v}$ 14. $(5\mathbf{v}) \cdot \mathbf{w}$
15. $4(\mathbf{u} \cdot \mathbf{v})$ 16. $5(\mathbf{v} \cdot \mathbf{w})$

In Exercises 17–22, find the angle between $\mathbf{v}$ and $\mathbf{w}$. Round to the nearest tenth of a degree.

17. $\mathbf{v} = 2\mathbf{i} - \mathbf{j}, \quad \mathbf{w} = 3\mathbf{i} + 4\mathbf{j}$
18. $\mathbf{v} = -2\mathbf{i} + 5\mathbf{j}, \quad \mathbf{w} = 3\mathbf{i} + 6\mathbf{j}$
19. $\mathbf{v} = -3\mathbf{i} + 2\mathbf{j}, \quad \mathbf{w} = 4\mathbf{i} - \mathbf{j}$
20. $\mathbf{v} = \mathbf{i} + 2\mathbf{j}, \quad \mathbf{w} = 4\mathbf{i} - 3\mathbf{j}$
21. $\mathbf{v} = 6\mathbf{i}, \quad \mathbf{w} = 5\mathbf{i} + 4\mathbf{j}$ 22. $\mathbf{v} = 3\mathbf{j}, \quad \mathbf{w} = 4\mathbf{i} + 5\mathbf{j}$

In Exercises 23–32, use the dot product to determine whether $\mathbf{v}$ and $\mathbf{w}$ are orthogonal.

23. $\mathbf{v} = \mathbf{i} + \mathbf{j}, \quad \mathbf{w} = \mathbf{i} - \mathbf{j}$
24. $\mathbf{v} = \mathbf{i} + \mathbf{j}, \quad \mathbf{w} = -\mathbf{i} + \mathbf{j}$
25. $\mathbf{v} = 2\mathbf{i} + 8\mathbf{j}, \quad \mathbf{w} = 4\mathbf{i} - \mathbf{j}$
26. $\mathbf{v} = 8\mathbf{i} - 4\mathbf{j}, \quad \mathbf{w} = -6\mathbf{i} - 12\mathbf{j}$
27. $\mathbf{v} = 2\mathbf{i} - 2\mathbf{j}, \quad \mathbf{w} = -\mathbf{i} + \mathbf{j}$
28. $\mathbf{v} = 5\mathbf{i} - 5\mathbf{j}, \quad \mathbf{w} = \mathbf{i} - \mathbf{j}$
29. $\mathbf{v} = 3\mathbf{i}, \quad \mathbf{w} = -4\mathbf{i}$ 30. $\mathbf{v} = 5\mathbf{i}, \quad \mathbf{w} = -6\mathbf{i}$
31. $\mathbf{v} = 3\mathbf{i}, \quad \mathbf{w} = -4\mathbf{j}$ 32. $\mathbf{v} = 5\mathbf{i}, \quad \mathbf{w} = -6\mathbf{j}$

In Exercise 33–38, find $\text{proj}_{\mathbf{w}} \mathbf{v}$. Then decompose $\mathbf{v}$ into two vectors, $\mathbf{v}_1$ and $\mathbf{v}_2$, where $\mathbf{v}_1$ is parallel to $\mathbf{w}$ and $\mathbf{v}_2$ is orthogonal to $\mathbf{w}$.

33. $\mathbf{v} = 3\mathbf{i} - 2\mathbf{j}, \quad \mathbf{w} = \mathbf{i} - \mathbf{j}$
34. $\mathbf{v} = 3\mathbf{i} - 2\mathbf{j}, \quad \mathbf{w} = 2\mathbf{i} + \mathbf{j}$
35. $\mathbf{v} = \mathbf{i} + 3\mathbf{j}, \quad \mathbf{w} = -2\mathbf{i} + 5\mathbf{j}$
36. $\mathbf{v} = 2\mathbf{i} + 4\mathbf{j}, \quad \mathbf{w} = -3\mathbf{i} + 6\mathbf{j}$
37. $\mathbf{v} = \mathbf{i} + 2\mathbf{j}, \quad \mathbf{w} = 3\mathbf{i} + 6\mathbf{j}$
38. $\mathbf{v} = 2\mathbf{i} + \mathbf{j}, \quad \mathbf{w} = 6\mathbf{i} + 3\mathbf{j}$

Application Exercises

39. The components of $\mathbf{v} = 240\mathbf{i} + 300\mathbf{j}$ represent the respective number of gallons of regular and premium gas sold at a station on Monday. The components of $\mathbf{w} = 1.90\mathbf{i} + 2.07\mathbf{j}$ represent the respective prices per gallon for each kind of gas. Find $\mathbf{v} \cdot \mathbf{w}$ and describe what the answer means in practical terms.

40. The components of $\mathbf{v} = 180\mathbf{i} + 450\mathbf{j}$ represent the respective number of one-day and three-day videos rented from a video store on Monday. The components of $\mathbf{w} = 3\mathbf{i} + 2\mathbf{j}$ represent the prices to rent the one-day and three-day videos, respectively. Find $\mathbf{v} \cdot \mathbf{w}$ and describe what the answer means in practical terms.

41. Find the work done in pushing a car along a level road from point A to point B, 80 feet from A, while exerting a constant force of 95 pounds. Round to the nearest foot-pound.

42. Find the work done when a crane lifts a 6000-pound boulder through a vertical distance of 12 feet. Round to the nearest foot-pound.

43. A wagon is pulled along level ground by exerting a force of 40 pounds on a handle that makes an angle of 32° with the horizontal. How much work is done pulling the wagon 100 feet? Round to the nearest foot-pound.

44. A wagon is pulled along level ground by exerting a force of 25 pounds on a handle that makes an angle of 38° with the horizontal. How much work is done pulling the wagon 100 feet? Round to the nearest foot-pound.

45. A force of 60 pounds on a rope is used to pull a box up a ramp inclined at 12° from the horizontal. The figure shows that the rope forms an angle of 38° with the horizontal. How much work is done pulling the box 20 feet along the ramp?

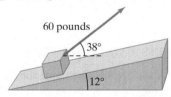

46. A force of 80 pounds on a rope is used to pull a box up a ramp inclined at 10° from the horizontal. The rope forms an angle of 33° with the horizontal. How much work is done pulling the box 25 feet along the ramp?

47. A force is given by the vector $\mathbf{F} = 3\mathbf{i} + 2\mathbf{j}$. The force moves an object along a straight line from the point $(4, 9)$ to the point $(10, 20)$. Find the work done if the distance is measured in feet and the force is measured in pounds.

48. A force is given by the vector $\mathbf{F} = 5\mathbf{i} + 7\mathbf{j}$. The force moves an object along a straight line from the point $(8, 11)$ to the point $(18, 20)$. Find the work done if the distance is measured in meters and the force is measured in newtons.

49. A force of 4 pounds acts in the direction of 50° to the horizontal. The force moves an object along a straight line from the point (3, 7) to the point (8, 10), with distance measured in feet. Find the work done by the force.

50. A force of 6 pounds acts in the direction of 40° to the horizontal. The force moves an object along a straight line from the point (5, 9) to the point (8, 20), with the distance measured in feet. Find the work done by the force.

51. Refer to Figure 6.63 on page 673. Suppose that the boat weighs 700 pounds and is on a ramp inclined at 30°. Represent the force due to gravity, **F**, using

$$\mathbf{F} = -700\mathbf{j}.$$

  a. Write a unit vector along the ramp in the upward direction.

  b. Find the vector projection of **F** onto the unit vector from part (a).

  c. What is the magnitude of the vector projection in part (b)? What does this represent?

52. Refer to Figure 6.63 on page 673. Suppose that the boat weighs 650 pounds and is on a ramp inclined at 30°. Represent the force due to gravity, **F**, using

$$\mathbf{F} = -650\mathbf{j}.$$

  a. Write a unit vector along the ramp in the upward direction.

  b. Find the vector projection of **F** onto the unit vector from part (a).

  c. What is the magnitude of the vector projection in part (b)? What does this represent?

Writing in Mathematics

53. Explain how to find the dot product of two vectors.

54. Using words and no symbols, describe how to find the dot product of two vectors with the alternative formula

$$\mathbf{v} \cdot \mathbf{w} = \|\mathbf{v}\| \|\mathbf{w}\| \cos\theta.$$

55. Describe how to find the angle between two vectors.

56. What are parallel vectors?

57. What are orthogonal vectors?

58. How do you determine if two vectors are orthogonal?

59. Draw two vectors, **v** and **w**, with the same initial point. Show the vector projection of **v** onto **w** in your diagram. Then describe how you identified this vector.

60. How do you determine the work done by a force **F** in moving an object from A to B when the direction of the force is not along the line of motion?

61. A weightlifter is holding a barbell perfectly still above his head, his body shaking from the effort. How much work is the weightlifter doing? Explain your answer.

62. Describe one way in which the everyday use of the word "work" is different from the definition of work given in this section.

 Critical Thinking Exercises

In Exercises 63–65, use the vectors

$$\mathbf{u} = a_1\mathbf{i} + b_1\mathbf{j}, \quad \mathbf{v} = a_2\mathbf{i} + b_2\mathbf{j}, \quad \text{and} \quad \mathbf{w} = a_3\mathbf{i} + b_3\mathbf{j},$$

to prove the given property.

63. $\mathbf{u} \cdot \mathbf{v} = \mathbf{v} \cdot \mathbf{u}$ **64.** $(c\mathbf{u}) \cdot \mathbf{v} = c(\mathbf{u} \cdot \mathbf{v})$

65. $\mathbf{u} \cdot (\mathbf{v} + \mathbf{w}) = \mathbf{u} \cdot \mathbf{v} + \mathbf{u} \cdot \mathbf{w}$

66. If $\mathbf{v} = -2\mathbf{i} + 5\mathbf{j}$, find a vector orthogonal to **v**.

67. Find a value of b so that $15\mathbf{i} - 3\mathbf{j}$ and $-4\mathbf{i} + b\mathbf{j}$ are orthogonal.

68. Prove that the projection of **v** onto **i** is $(\mathbf{v} \cdot \mathbf{i})\mathbf{i}$.

69. Find two vectors **v** and **w** such that the projection of **v** onto **w** is **v**.

 Group Exercise

70. Group members should research and present a report on unusual and interesting applications of vectors.

Chapter Summary, Review, and Test

Summary

DEFINITIONS AND CONCEPTS	EXAMPLES

6.1 and 6.2 The Law of Sines and the Law of Cosines

  a. The Law of Sines

$$\frac{a}{\sin A} = \frac{b}{\sin B} = \frac{c}{\sin C}$$

Ex. 1, p. 600;
Ex. 2, p. 601
Ex. 3, p. 603;

  b. The Law of Sines is used to solve SAA, ASA, and SSA (the ambiguous case) triangles. The ambiguous case may result in no triangle, one triangle, or two triangles; see the box on page 602.

Ex. 4, p. 604;
Ex. 5, p. 604

  c. The area of a triangle equals one-half the product of the lengths of two sides times the sine of their included angle.

Ex. 6, p. 606

  d. The Law of Cosines

$$a^2 = b^2 + c^2 - 2bc \cos A$$
$$b^2 = a^2 + c^2 - 2ac \cos B$$
$$c^2 = a^2 + b^2 - 2ab \cos C$$

DEFINITIONS AND CONCEPTS	EXAMPLES

e. The Law of Cosines is used to find the side opposite the given angle in an SAS triangle; see the box on page 612. The Law of Cosines is also used to find the angle opposite the longest side in an SSS triangle; see the box on page 613.

Ex. 1, p. 612;
Ex. 2, p. 614

f. Heron's Formula for the Area of a Triangle

The area of a triangle with sides a, b, and c is

$$\sqrt{s(s-a)(s-b)(s-c)},$$

where s is one-half the perimeter:

$$s = \tfrac{1}{2}(a+b+c).$$

Ex. 4, p. 616

6.3 and 6.4 Polar Coordinates and Graphs of Polar Equations

a. A point P in the polar coordinate system is represented by (r, θ), where r is the directed distance of the point from the pole and θ is the angle from the polar axis to line segment OP. The elements of the ordered pair (r, θ) are called the polar coordinates of P. See Figure 6.19 on page 619. When r in (r, θ) is negative, a point is located $|r|$ units along the ray opposite the terminal side of θ. Important information about the sign of r and the location of the point (r, θ) is found in the box on page 620.

Ex. 1, p. 620

b. Multiple Representation of Points

If n is any integer, $(r, \theta) = (r, \theta + 2n\pi)$ or $(r, \theta) = (-r, \theta + \pi + 2n\pi)$.

Ex. 2, p. 622

c. Relations between Polar and Rectangular Coordinates

$$x = r\cos\theta, \quad y = r\sin\theta, \quad x^2 + y^2 = r^2, \quad \tan\theta = \frac{y}{x}$$

d. To convert a point from polar coordinates (r, θ) to rectangular coordinates (x, y), use $x = r\cos\theta$ and $y = r\sin\theta$.

Ex. 3, p. 623

e. To convert a point from rectangular coordinates (x, y) to polar coordinates (r, θ), use the procedure in the box on page 624.

Ex. 4, p. 624;
Ex. 5, p. 625

f. To convert a rectangular equation to a polar equation, replace x with $r\cos\theta$ and y with $r\sin\theta$.

Ex. 6, p. 626

g. To convert a polar equation to a rectangular equation, use one or more of

Ex. 7, p. 626

$$r^2 = x^2 + y^2, \quad r\cos\theta = x, \quad r\sin\theta = y, \quad \text{and} \quad \tan\theta = \frac{y}{x}.$$

It is often necessary to do something to the given polar equation to use the preceding expressions.

h. A polar equation is an equation whose variables are r and θ. The graph of a polar equation is the set of all points whose polar coordinates satisfy the equation.

Ex. 1, p. 630

i. Polar equations can be graphed using point plotting and symmetry (see the box on page 632).

Ex. 2, p. 633

j. The graphs of $r = a\cos\theta$ and $r = a\sin\theta$ are circles. See the box on page 632. The graphs of $r = a \pm b\sin\theta$ and $r = a \pm b\cos\theta$ are called limaçons ($a > 0$ and $b > 0$), shown in the box on page 635. The graphs of $r = a\sin n\theta$ and $r = a\cos n\theta$, $a \neq 0$, are rose curves with $2n$ petals if n is even and n petals if n is odd. See the box on page 638. The graphs of $r^2 = a^2\sin 2\theta$ and $r^2 = a^2\cos 2\theta$, $a \neq 0$, are called lemniscates and are shown in the box on page 639.

Ex. 3, p. 634;
Ex. 4, p. 636;
Ex. 5, p. 638

6.5 Complex Numbers in Polar Form; DeMoivre's Theorem

a. The complex number $z = a + bi$ is represented as a point (a, b) in the complex plane, shown in Figure 6.35 on page 643.

Ex. 1, p. 643

b. The absolute value of $z = a + bi$ is $|z| = |a + bi| = \sqrt{a^2 + b^2}$.

Ex. 2, p. 644

c. The polar form of $z = a + bi$ is $z = r(\cos\theta + i\sin\theta)$,

where $a = r\cos\theta, b = r\sin\theta, r = \sqrt{a^2 + b^2}$, and $\tan\theta = \dfrac{b}{a}$. We call r the modulus and θ the argument of z, with $0 \le \theta < 2\pi$.

Ex. 3, p. 645;
Ex. 4, p. 646

d. Multiplying Complex Numbers in Polar Form: Multiply moduli and add arguments. See the box on page 646.

Ex. 5, p. 647

DEFINITIONS AND CONCEPTS	**EXAMPLES**

e. Dividing Complex Numbers in Polar Form: Divide moduli and subtract arguments. See the box on page 647.

Ex. 6, p. 647

f. DeMoivre's Theorem is used to find powers of complex numbers in polar form.

$$[r(\cos\theta + i\sin\theta)]^n = r^n(\cos n\theta + i\sin n\theta)$$

Ex. 7, p. 649;
Ex. 8, p. 649

g. DeMoivre's Theorem can be used to find roots of complex numbers in polar form. The n distinct nth roots of $r(\cos\theta + i\sin\theta)$ are

Ex. 9, p. 650;
Ex. 10, p. 651

$$\sqrt[n]{r}\left[\cos\left(\frac{\theta + 2\pi k}{n}\right) + i\sin\left(\frac{\theta + 2\pi k}{n}\right)\right]$$

or

$$\sqrt[n]{r}\left[\cos\left(\frac{\theta + 360°k}{n}\right) + i\sin\left(\frac{\theta + 360°k}{n}\right)\right],$$

where $k = 0, 1, 2, \ldots, n - 1$.

6.6 Vectors

a. A vector is a directed line segment.

b. Equal vectors have the same magnitude and the same direction.

Ex. 1, p. 656

c. The vector $k\mathbf{v}$, the scalar multiple of the vector $\mathbf{v}$ and the scalar k, has magnitude $|k|\|\mathbf{v}\|$. The direction of $k\mathbf{v}$ is the same as that of $\mathbf{v}$ if $k > 0$ and opposite $\mathbf{v}$ if $k < 0$.

d. The sum $\mathbf{u} + \mathbf{v}$, called the resultant vector, can be expressed geometrically. Position $\mathbf{u}$ and $\mathbf{v}$ so that the terminal point of $\mathbf{u}$ coincides with the initial point of $\mathbf{v}$. The vector $\mathbf{u} + \mathbf{v}$ extends from the initial point of $\mathbf{u}$ to the terminal point of $\mathbf{v}$.

e. The difference of two vectors, $\mathbf{u} - \mathbf{v}$, is defined as $\mathbf{u} + (-\mathbf{v})$.

f. The vector $\mathbf{i}$ is the unit vector whose direction is along the positive x-axis. The vector $\mathbf{j}$ is the unit vector whose direction is along the positive y-axis.

g. Vector $\mathbf{v}$, from $(0, 0)$ to (a, b), called a position vector, is represented as $\mathbf{v} = a\mathbf{i} + b\mathbf{j}$, where a is the horizontal component and b is the vertical component. The magnitude of $\mathbf{v}$ is given by $\|\mathbf{v}\| = \sqrt{a^2 + b^2}$.

Ex. 2, p. 659

h. Vector $\mathbf{v}$ from (x_1, y_1) to (x_2, y_2) is equal to the position vector $\mathbf{v} = (x_2 - x_1)\mathbf{i} + (y_2 - y_1)\mathbf{j}$. In rectangular coordinates, the term "vector" refers to the position vector in terms of $\mathbf{i}$ and $\mathbf{j}$ that is equal to it.

Ex. 3, p. 660

i. Operations with Vectors in Terms of $\mathbf{i}$ and $\mathbf{j}$
If $\mathbf{v} = a_1\mathbf{i} + b_1\mathbf{j}$ and $\mathbf{w} = a_2\mathbf{i} + b_2\mathbf{j}$, then

1. $\mathbf{v} + \mathbf{w} = (a_1 + a_2)\mathbf{i} + (b_1 + b_2)\mathbf{j}$

2. $\mathbf{v} - \mathbf{w} = (a_1 - a_2)\mathbf{i} + (b_1 - b_2)\mathbf{j}$

3. $k\mathbf{v} = (ka_1)\mathbf{i} + (kb_1)\mathbf{j}$

Ex. 4, p. 660;
Ex. 5, p. 661;
Ex. 6, p. 661

j. The zero vector $\mathbf{0}$ is the vector whose magnitude is 0 and is assigned no direction. Many properties of vector addition and scalar multiplication involve the zero vector. Some of these properties are listed in the box on page 662.

k. The vector $\dfrac{\mathbf{v}}{\|\mathbf{v}\|}$ is the unit vector that has the same direction as $\mathbf{v}$.

Ex. 7, p. 663

l. A vector with magnitude $\|\mathbf{v}\|$ and direction angle θ, the angle that $\mathbf{v}$ makes with the positive x-axis, can be expressed in terms of its magnitude and direction angle as

Ex. 8, p. 664;
Ex. 9, p. 664

$$\mathbf{v} = \|\mathbf{v}\|\cos\theta\mathbf{i} + \|\mathbf{v}\|\sin\theta\mathbf{j}.$$

DEFINITIONS AND CONCEPTS	EXAMPLES

6.7 The Dot Product

a. Definition of the Dot Product
If $\mathbf{v} = a_1\mathbf{i} + b_1\mathbf{j}$ and $\mathbf{w} = a_2\mathbf{i} + b_2\mathbf{j}$, the dot product of $\mathbf{v}$ and $\mathbf{w}$ is defined by $\mathbf{v} \cdot \mathbf{w} = a_1 a_2 + b_1 b_2$.

Ex. 1, p. 670

b. Alternative Formula for the Dot Product $\mathbf{v} \cdot \mathbf{w} = \|\mathbf{v}\|\|\mathbf{w}\| \cos \theta$, where θ is the smallest nonnegative angle between $\mathbf{v}$ and $\mathbf{w}$.

c. Angle between Two Vectors

Ex. 2, p. 672

$$\cos \theta = \frac{\mathbf{v} \cdot \mathbf{w}}{\|\mathbf{v}\|\|\mathbf{w}\|} \quad \text{and} \quad \theta = \cos^{-1}\left(\frac{\mathbf{v} \cdot \mathbf{w}}{\|\mathbf{v}\|\|\mathbf{w}\|}\right).$$

d. Two vectors are orthogonal when the angle between them is 90°. To show that two vectors are orthogonal, show that their dot product is zero.

Ex. 3, p. 673

e. The vector projection of $\mathbf{v}$ onto $\mathbf{w}$ is given by

Ex. 4, p. 674

$$\text{proj}_{\mathbf{w}}\mathbf{v} = \frac{\mathbf{v} \cdot \mathbf{w}}{\|\mathbf{w}\|^2}\mathbf{w}.$$

f. Expressing a vector as the sum of two orthogonal vectors, called the vector components, is shown in the box on page 674.

Ex. 5, p. 675

g. The work, W, done by a force $\mathbf{F}$ moving an object from A to B is $W = \mathbf{F} \cdot \overrightarrow{AB}$.
Thus, $W = \|\mathbf{F}\|\|\overrightarrow{AB}\| \cos \theta$, where θ is the angle between the force and the direction of motion.

Ex. 6, p. 676

Review Exercises

6.1 and 6.2

In Exercises 1–12, solve each triangle. Round lengths to the nearest tenth and angle measures to the nearest degree. If no triangle exists, state "no triangle." If two triangles exist, solve each triangle.

1. $A = 70°$, $B = 55°$, $a = 12$

2. $B = 107°$, $C = 30°$, $c = 126$

3. $B = 66°$, $a = 17$, $c = 12$

4. $a = 117$, $b = 66$, $c = 142$

5. $A = 35°$, $B = 25°$, $c = 68$

6. $A = 39°$, $a = 20$, $b = 26$

7. $C = 50°$, $a = 3$, $c = 1$

8. $A = 162°$, $b = 11.2$, $c = 48.2$

9. $a = 26.1$, $b = 40.2$, $c = 36.5$

10. $A = 40°$, $a = 6$, $b = 4$

11. $B = 37°$, $a = 12.4$, $b = 8.7$

12. $A = 23°$, $a = 54.3$, $b = 22.1$

In Exercises 13–16, find the area of the triangle having the given measurements. Round to the nearest square unit.

13. $C = 42°$, $a = 4$ feet, $b = 6$ feet

14. $A = 22°$, $b = 4$ feet, $c = 5$ feet

15. $a = 2$ meters, $b = 4$ meters, $c = 5$ meters

16. $a = 2$ meters, $b = 2$ meters, $c = 2$ meters

17. The A-frame cabin shown in the next column is 35 feet wide. The roof of the cabin makes a 60° angle with the cabin's base. Find the length of one side of the roof from its ground level to the peak. Round to the nearest tenth of a foot.

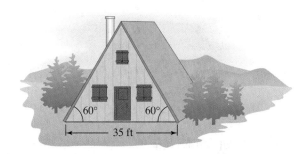

18. Two cars leave a city at the same time and travel along straight highways that differ in direction by 80°. One car averages 60 miles per hour and the other averages 50 miles per hour. How far apart will the cars be after 30 minutes? Round to the nearest tenth of a mile.

19. Two airplanes leave an airport at the same time on different runways. One flies on a bearing of N66.5°W at 325 miles per hour. The other airplane flies on a bearing of S26.5°W at 300 miles per hour. How far apart will the airplanes be after two hours?

20. The figure shows three roads that intersect to bound a triangular piece of land. Find the lengths of the other two sides of the land to the nearest foot.

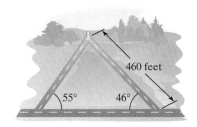

21. A commercial piece of real estate is priced at \$5.25 per square foot. Find the cost, to the nearest dollar, of a triangular lot measuring 260 feet by 320 feet by 450 feet.

6.3 and 6.4

In Exercises 22–27, plot each point in polar coordinates and find its rectangular coordinates.

22. $(4, 60°)$　　　　　　　**23.** $(3, 150°)$

24. $\left(-4, \dfrac{4\pi}{3}\right)$　　　　　**25.** $\left(-2, \dfrac{5\pi}{4}\right)$

26. $\left(-4, -\dfrac{\pi}{2}\right)$　　　　　**27.** $\left(-2, -\dfrac{\pi}{4}\right)$

In Exercises 28–30, plot each point in polar coordinates. Then find another representation (r, θ) of this point in which:

　a. $r > 0,\quad 2\pi < \theta < 4\pi$.
　b. $r < 0,\quad 0 < \theta < 2\pi$.
　c. $r > 0, -2\pi < \theta < 0$.

28. $\left(3, \dfrac{\pi}{6}\right)$　　　**29.** $\left(2, \dfrac{2\pi}{3}\right)$　　　**30.** $(3, \pi)$

In Exercises 31–36, the rectangular coordinates of a point are given. Find polar coordinates of each point.

31. $(-4, 4)$　　　　　　　**32.** $(3, -3)$
33. $(5, 12)$　　　　　　　**34.** $(-3, 4)$
35. $(0, -5)$　　　　　　　**36.** $(1, 0)$

In Exercises 37–39, convert each rectangular equation to a polar equation.

37. $2x + 3y = 8$　　　　　**38.** $x^2 + y^2 = 100$
39. $5x^2 + 5y^2 = 3y$

In Exercises 40–46, convert each polar equation to a rectangular equation.

40. $r = 3$　　　　　　　　**41.** $\theta = \dfrac{3\pi}{4}$

42. $r \cos \theta = -1$　　　　**43.** $r = 5 \sec \theta$
44. $r = 3 \cos \theta$　　　　　**45.** $5r \cos \theta + r \sin \theta = 8$
46. $r^2 \sin 2\theta = 4$

In Exercises 47–49, test for symmetry with respect to:
　a. *the polar axis*　**b.** *the line* $\theta = \dfrac{\pi}{2}$　*and*　**c.** *the pole.*

47. $r = 5 + 3 \cos \theta$　　　**48.** $r = 3 \sin \theta$
49. $r^2 = 9 \cos 2\theta$

In Exercises 50–56, graph each polar equation. Be sure to test for symmetry.

50. $r = 3 \cos \theta$　　　　　**51.** $r = 2 + 2 \sin \theta$
52. $r = \sin 2\theta$　　　　　**53.** $r = 2 + \cos \theta$
54. $r = 1 + 3 \sin \theta$　　　**55.** $r = 1 - 2 \cos \theta$
56. $r^2 = \cos 2\theta$

6.5

In Exercises 57–60, plot each complex number. Then write the complex number in polar form. You may express the argument in degrees or radians.

57. $1 - i$　　　　　　　　**58.** $-2\sqrt{3} + 2i$
59. $-3 - 4i$　　　　　　　**60.** $-5i$

In Exercises 61–64, write each complex number in rectangular form.

61. $8(\cos 60° + i \sin 60°)$　　**62.** $4(\cos 210° + i \sin 210°)$

63. $6\left(\cos \dfrac{2\pi}{3} + i \sin \dfrac{2\pi}{3}\right)$

64. $0.6(\cos 100° + i \sin 100°)$

In Exercises 65–67, find the product of the complex numbers. Leave answers in polar form.

65. $z_1 = 3(\cos 40° + i \sin 40°)$
　　$z_2 = 5(\cos 70° + i \sin 70°)$

66. $z_1 = \cos 210° + i \sin 210°$
　　$z_2 = \cos 55° + i \sin 55°$

67. $z_1 = 4\left(\cos \dfrac{3\pi}{7} + i \sin \dfrac{3\pi}{7}\right)$
　　$z_2 = 10\left(\cos \dfrac{4\pi}{7} + i \sin \dfrac{4\pi}{7}\right)$

In Exercises 68–70, find the quotient $\dfrac{z_1}{z_2}$ of the complex numbers. Leave answers in polar form.

68. $z_1 = 10(\cos 10° + i \sin 10°)$
　　$z_2 = 5(\cos 5° + i \sin 5°)$

69. $z_1 = 5\left(\cos \dfrac{4\pi}{3} + i \sin \dfrac{4\pi}{3}\right)$
　　$z_2 = 10\left(\cos \dfrac{\pi}{3} + i \sin \dfrac{\pi}{3}\right)$

70. $z_1 = 2\left(\cos \dfrac{5\pi}{3} + i \sin \dfrac{5\pi}{3}\right)$
　　$z_2 = \cos \dfrac{\pi}{2} + i \sin \dfrac{\pi}{2}$

In Exercises 71–75, use DeMoivre's Theorem to find the indicated power of the complex number. Write answers in rectangular form.

71. $[2(\cos 20° + i \sin 20°)]^3$

72. $[4(\cos 50° + i \sin 50°)]^3$

73. $\left[\dfrac{1}{2}\left(\cos \dfrac{\pi}{14} + i \sin \dfrac{\pi}{14}\right)\right]^7$

74. $(1 - \sqrt{3}i)^7$　　　　　**75.** $(-2 - 2i)^5$

In Exercises 76–77, find all the complex roots. Write roots in polar form with θ in degrees.

76. The complex square roots of $49(\cos 50° + i \sin 50°)$
77. The complex cube roots of $125(\cos 165° + i \sin 165°)$

In Exercises 78–81, find all the complex roots. Write roots in rectangular form.

78. The complex fourth roots of $16\left(\cos \dfrac{2\pi}{3} + i \sin \dfrac{2\pi}{3}\right)$

79. The complex cube roots of $8i$

80. The complex cube roots of -1

81. The complex fifth roots of $-1 - i$

6.6

In Exercises 82–84, sketch each vector as a position vector and find its magnitude.

82. $\mathbf{v} = -3\mathbf{i} - 4\mathbf{j}$ **83.** $\mathbf{v} = 5\mathbf{i} - 2\mathbf{j}$

84. $\mathbf{v} = -3\mathbf{j}$

In Exercises 85–86, let $\mathbf{v}$ be the vector from initial point P_1 to terminal point P_2. Write $\mathbf{v}$ in terms of $\mathbf{i}$ and $\mathbf{j}$.

85. $P_1 = (2,-1), \quad P_2 = (5,-3)$
86. $P_1 = (-3,0), \quad P_2 = (-2,-2)$

In Exercises 87–90, let

$$\mathbf{v} = \mathbf{i} - 5\mathbf{j} \quad \text{and} \quad \mathbf{w} = -2\mathbf{i} + 7\mathbf{j}.$$

Find each specified vector or scalar.

87. $\mathbf{v} + \mathbf{w}$ **88.** $\mathbf{w} - \mathbf{v}$
89. $6\mathbf{v} - 3\mathbf{w}$ **90.** $\|-2\mathbf{v}\|$

In Exercises 91–92, find the unit vector that has the same direction as the vector $\mathbf{v}$.

91. $\mathbf{v} = 8\mathbf{i} - 6\mathbf{j}$ **92.** $\mathbf{v} = -\mathbf{i} + 2\mathbf{j}$

93. The magnitude and direction angle of $\mathbf{v}$ are $\|\mathbf{v}\| = 12$ and $\theta = 60°$. Express $\mathbf{v}$ in terms of $\mathbf{i}$ and $\mathbf{j}$.

94. The magnitude and direction of two forces acting on an object are 100 pounds, N25°E, and 200 pounds, N80°E, respectively. Find the magnitude, to the nearest pound, and the direction angle, to the nearest tenth of a degree, of the resultant force.

95. Your boat is moving at a speed of 15 miles per hour at an angle of 25° upstream on a river flowing at 4 miles per hour. The situation is illustrated in the figure at the top of the next column.

    a. Find the vector representing your boat's velocity relative to the ground.

    b. What is the speed of your boat. to the nearest mile per hour, relative to the ground?

    c. What is the boat's direction angle, to the nearest tenth of a degree, relative to the ground?

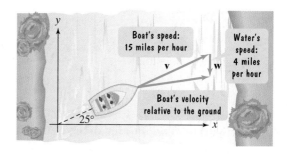

6.7

96. If $\mathbf{u} = 5\mathbf{i} + 2\mathbf{j}, \mathbf{v} = \mathbf{i} - \mathbf{j}$, and $\mathbf{w} = 3\mathbf{i} - 7\mathbf{j}$, find $\mathbf{u} \cdot (\mathbf{v} + \mathbf{w})$.

In Exercises 97–99, find the dot product $\mathbf{v} \cdot \mathbf{w}$. Then find the angle between $\mathbf{v}$ and $\mathbf{w}$ to the nearest tenth of a degree.

97. $\mathbf{v} = 2\mathbf{i} + 3\mathbf{j}, \quad \mathbf{w} = 7\mathbf{i} - 4\mathbf{j}$

98. $\mathbf{v} = 2\mathbf{i} + 4\mathbf{j}, \quad \mathbf{w} = 6\mathbf{i} - 11\mathbf{j}$

99. $\mathbf{v} = 2\mathbf{i} + \mathbf{j}, \quad \mathbf{w} = \mathbf{i} - \mathbf{j}$

In Exercises 100–101, use the dot product to determine whether $\mathbf{v}$ and $\mathbf{w}$ are orthogonal.

100. $\mathbf{v} = 12\mathbf{i} - 8\mathbf{j}, \quad \mathbf{w} = 2\mathbf{i} + 3\mathbf{j}$

101. $\mathbf{v} = \mathbf{i} + 3\mathbf{j}, \quad \mathbf{w} = -3\mathbf{i} - \mathbf{j}$

In Exercises 102–103, find $\mathrm{proj_w}\,\mathbf{v}$. Then decompose $\mathbf{v}$ into two vectors, $\mathbf{v}_1$ and $\mathbf{v}_2$, where $\mathbf{v}_1$ is parallel to $\mathbf{w}$ and $\mathbf{v}_2$ is orthogonal to $\mathbf{w}$.

102. $\mathbf{v} = -2\mathbf{i} + 5\mathbf{j}, \quad \mathbf{w} = 5\mathbf{i} + 4\mathbf{j}$

103. $\mathbf{v} = -\mathbf{i} + 2\mathbf{j}, \quad \mathbf{w} = 3\mathbf{i} - \mathbf{j}$

104. A heavy crate is dragged 50 feet along a level floor. Find the work done if a force of 30 pounds at an angle of 42° is used.

105. Explain why the weightlifter does more work in raising 300 kilograms above her head than Atlas, who is supporting the entire world.

Chapter 6 Test

1. In oblique triangle ABC, $A = 34°$, $B = 68°$, and $a = 4.8$. Find b to the nearest tenth.

2. In oblique triangle ABC, $C = 68°$, $a = 5$, and $b = 6$. Find c to the nearest tenth.

3. In oblique triangle ABC, $a = 17$ inches, $b = 45$ inches, and $c = 32$ inches. Find the area of the triangle to the nearest square inch.

4. Plot $\left(4, \dfrac{5\pi}{4}\right)$ in the polar coordinate system. Then write two other ordered pairs (r, θ) that name this point.

5. If the rectangular coordinates of a point are $(1, -1)$, find polar coordinates of the point.

6. Convert $x^2 + y^2 = 6x$ to a polar equation.

7. Convert $r = 4 \csc \theta$ to a rectangular equation.

In Exercises 8–9, graph each polar equation.

8. $r = 1 + \sin \theta$

9. $r = 1 + 3 \cos \theta$

10. Write $-\sqrt{3} + i$ in polar form.

In Exercises 11–13, perform the indicated operation. Leave answers in polar form.

11. $5(\cos 15° + i \sin 15°) \cdot 10(\cos 5° + i \sin 5°)$

12. $\dfrac{2\left(\cos \dfrac{\pi}{2} + i \sin \dfrac{\pi}{2}\right)}{4\left(\cos \dfrac{\pi}{3} + i \sin \dfrac{\pi}{3}\right)}$

13. $[2(\cos 10° + i \sin 10°)]^5$

14. Find the three cube roots of 27. Write roots in rectangular form.

15. If $P_1 = (-2, 3)$, $P_2 = (-1, 5)$, and $\mathbf{v}$ is the vector from P_1 to P_2,

   a. Write $\mathbf{v}$ in terms of $\mathbf{i}$ and $\mathbf{j}$.

   b. Find $\|\mathbf{v}\|$.

In Exercises 16–19, let
$$\mathbf{v} = -5\mathbf{i} + 2\mathbf{j} \quad \text{and} \quad \mathbf{w} = 2\mathbf{i} - 4\mathbf{j}.$$
Find the specified vector, scalar, or angle.

16. $3\mathbf{v} - 4\mathbf{w}$

17. $\mathbf{v} \cdot \mathbf{w}$.

18. the angle between $\mathbf{v}$ and $\mathbf{w}$, to the nearest degree

19. $\text{proj}_{\mathbf{w}} \mathbf{v}$

20. A small fire is sighted from ranger stations A and B. Station B is 1.6 miles due east of station A. The bearing of the fire from station A is N40°E, and the bearing of the fire from station B is N50°W. How far, to the nearest tenth of a mile, is the fire from station A?

21. The magnitude and direction of two forces acting on an object are 250 pounds, N60°E, and 150 pounds, S45°E. Find the magnitude, to the nearest pound, and the direction angle, to the nearest tenth of a degree, of the resultant force.

22. A child is pulling a wagon with a force of 40 pounds. How much work is done in moving the wagon 60 feet if the handle makes an angle of 35° with the ground? Round to the nearest foot-pound.

Cumulative Review Exercises (Chapters P–6)

Solve each equation or inequality in Exercises 1–4.

1. $x^4 - x^3 - x^2 - x - 2 = 0$

2. $2 \sin^2 \theta - 3 \sin \theta + 1 = 0$, $0 \le \theta < 2\pi$

3. $x^2 + 2x + 3 > 11$

4. $\sin \theta \cos \theta = -\frac{1}{2}$, $0 \le \theta < 2\pi$

In Exercises 5–6, graph one complete cycle.

5. $y = 3 \sin (2x - \pi)$

6. $y = -4 \cos \pi x$

In Exercises 7–8, verify each identity.

7. $\sin \theta \csc \theta - \cos^2 \theta = \sin^2 \theta$

8. $\cos\left(\theta + \dfrac{3\pi}{2}\right) = \sin \theta$

9. Find the slope and y-intercept of the line whose equation is $2x + 4y - 8 = 0$.

In Exercises 10–11, find the exact value of each expression.

10. $2 \sin \dfrac{\pi}{3} - 3 \tan \dfrac{\pi}{6}$

11. $\sin\left(\tan^{-1} \frac{1}{2}\right)$

In Exercises 12–13, find the domain of the function whose equation is given.

12. $f(x) = \sqrt{5 - x}$

13. $g(x) = \dfrac{x - 3}{x^2 - 9}$

14. A ball is thrown vertically upward from a height of 8 feet with an initial velocity of 48 feet per second. The ball's height, $s(t)$, in feet, after t seconds is given by
$$s(t) = -16t^2 + 48t + 8.$$
After how many seconds does the ball reach its maximum height? What is the maximum height?

15. An object moves in simple harmonic motion described by $d = 4 \sin 5t$, where t is measured in seconds and d in meters. Find **a.** the maximum displacement; **b.** the frequency; and **c.** the time required for one cycle.

16. Use a half-angle formula to find the exact value of $\cos 22.5°$.

17. If $\mathbf{v} = 2\mathbf{i} + 7\mathbf{j}$ and $\mathbf{w} = \mathbf{i} - 2\mathbf{j}$, find: **a.** $3\mathbf{v} - \mathbf{w}$ and **b.** $\mathbf{v} \cdot \mathbf{w}$.

18. Express as a single logarithm with a coefficient of 1: $\frac{1}{2}\log_b x - \log_b(x^2 + 1)$.

19. Write the slope-intercept form of the line passing through $(4, -1)$ and $(-8, 5)$.

20. Psychologists can measure the amount learned, L, at time t using the model $L = A(1 - e^{-kt})$. The variable A represents the total amount to be learned, and k is the learning rate. A student preparing for the SAT has 300 new vocabulary words to learn: $A = 300$. This particular student can learn 20 vocabulary words after 5 minutes: If $t = 5, L = 20$.

a. Find k, the learning rate, correct to three decimal places.

b. Approximately how many words will the student have learned after 20 minutes?

c. How long will it take for the student to learn 260 words?

Appendix
Where Did That Come From? Selected Proofs

Properties of Logarithms

The Product Rule
Let b, M, and N be positive real numbers with $b \neq 1$.
$$\log_b(MN) = \log_b M + \log_b N$$

Proof
We begin by letting $\log_b M = R$ and $\log_b N = S$.
Now we write each logarithm in exponential form.
$$\log_b M = R \quad \text{means} \quad b^R = M.$$
$$\log_b N = S \quad \text{means} \quad b^S = N.$$
By substituting and using a property of exponents, we see that
$$MN = b^R b^S = b^{R+S}.$$
Now we change $MN = b^{R+S}$ to logarithmic form.
$$MN = b^{R+S} \quad \text{means} \quad \log_b(MN) = R + S.$$
Finally, substituting $\log_b M$ for R and $\log_b N$ for S gives us
$$\log_b(MN) = \log_b M + \log_b N,$$
the property that we wanted to prove.

The quotient and power rules for logarithms are proved using similar procedures.

The Change-of-Base Property
For any logarithmic bases a and b, and any positive number M,
$$\log_b M = \frac{\log_a M}{\log_a b}.$$

Proof
To prove the change-of-base property, we let x equal the logarithm on the left side:
$$\log_b M = x.$$

Now we rewrite this logarithm in exponential form.

$$\log_b M = x \quad \text{means} \quad b^x = M.$$

Because b^x and M are equal, the logarithms with base a for each of these expressions must be equal. This means that

$$\log_a b^x = \log_a M$$

$$x \log_a b = \log_a M \qquad \text{Apply the power rule for logarithms on the left side.}$$

$$x = \frac{\log_a M}{\log_a b} \qquad \text{Solve for x by dividing both sides by } \log_a b.$$

In our first step we let x equal $\log_b M$. Replacing x on the left side by $\log_b M$ gives us

$$\log_b M = \frac{\log_a M}{\log_a b},$$

which is the change-of-base property.

SECTION 6.2 *The Law of Cosines*

Heron's Formula for the Area of a Triangle

The area of a triangle with sides a, b, and c is

$$\text{Area} = \sqrt{s(s - a)(s - b)(s - c)},$$

where s is one-half the perimeter: $s = \frac{1}{2}(a + b + c)$.

Proof

The proof of Heron's formula begins with a half-angle formula and the Law of Cosines.

$$\cos \frac{C}{2} = \sqrt{\frac{1 + \cos C}{2}} = \sqrt{\frac{1 + \frac{a^2 + b^2 - c^2}{2ab}}{2}}$$

This is the Law of Cosines: $c^2 = a^2 + b^2 - 2ab \cos C$ solved for cos C.

$$= \sqrt{\frac{a^2 + 2ab + b^2 - c^2}{4ab}} = \sqrt{\frac{(a + b)^2 - c^2}{4ab}} = \sqrt{\frac{(a + b + c)(a + b - c)}{4ab}}$$

Multiply the numerator and denominator of the radicand by $2ab$.

Factor $a^2 + 2ab + b^2$.

Factor the numerator as the differences of two squares.

We now introduce the expression for one-half the perimeter: $s = \frac{1}{2}(a + b + c)$. We replace $a + b + c$ in the numerator by $2s$. We also find an expression for $a + b - c$ as follows:

$$a + b - c = a + b + c - 2c = 2s - 2c = 2(s - c).$$

Thus,

$$\cos \frac{C}{2} = \sqrt{\frac{(a + b + c)(a + b - c)}{4ab}} = \sqrt{\frac{2s \cdot 2(s - c)}{4ab}} = \sqrt{\frac{s(s - c)}{ab}}.$$

In a similar manner, we obtain

$$\sin \frac{C}{2} = \sqrt{\frac{1 - \cos C}{2}} = \sqrt{\frac{(s - a)(s - b)}{ab}}.$$

From our work in Section 6.1, we know that the area of a triangle is one-half the product of the length of two sides times the sine of their included angle.

$$\text{Area} = \frac{1}{2} ab \sin C$$

$$= \frac{1}{2} ab \cdot 2 \sin \frac{C}{2} \cos \frac{C}{2} \qquad \sin C = \sin 2\frac{C}{2} = 2 \sin \frac{C}{2} \cos \frac{C}{2}$$

$$= ab\sqrt{\frac{(s-a)(s-b)}{ab}}\sqrt{\frac{s(s-c)}{ab}} \qquad \text{Use the expressions for } \sin\frac{C}{2} \text{ and } \cos\frac{C}{2} \text{ on page A2.}$$

$$= ab\frac{\sqrt{s(s-a)(s-b)(s-c)}}{\sqrt{a^2b^2}} \qquad \text{Multiply the radicands.}$$

$$= \sqrt{s(s-a)(s-b)(s-c)} \qquad \text{Simplify: } \frac{ab}{\sqrt{a^2b^2}} = \frac{ab}{ab} = 1.$$

SECTION 6.5 Complex Numbers in Polar Form; DeMoivre's Theorem

The Quotient of Two Complex Numbers in Polar Form

Let $z_1 = r_1(\cos\theta_1 + i\sin\theta_1)$ and $z_2 = r_2(\cos\theta_2 + i\sin\theta_2)$ be two complex numbers in polar form. Their quotient, $\frac{z_1}{z_2}$, is

$$\frac{z_1}{z_2} = \frac{r_1}{r_2}[\cos(\theta_1 - \theta_2) + i\sin(\theta_1 - \theta_2)].$$

Proof

We begin by multiplying the numerator and denominator of the quotient, $\frac{z_1}{z_2}$, by the conjugate of the denominator. Then we simplify the quotient using the difference formulas for sine and cosine.

$$\frac{z_1}{z_2} = \frac{r_1(\cos\theta_1 + i\sin\theta_1)}{r_2(\cos\theta_2 + i\sin\theta_2)} \qquad \text{This is the given quotient.}$$

$$= \frac{r_1(\cos\theta_1 + i\sin\theta_1)(\cos\theta_2 - i\sin\theta_2)}{r_2(\cos\theta_2 + i\sin\theta_2)(\cos\theta_2 - i\sin\theta_2)} \qquad \text{Multiply the numerator and denominator by the conjugate of the denominator. Recall that the conjugate of } a + bi \text{ is } a - bi.$$

$$= \frac{r_1(\cos\theta_1 + i\sin\theta_1)(\cos\theta_2 - i\sin\theta_2)}{r_2(\cos^2\theta_2 + \sin^2\theta_2)} \qquad \text{Multiply the conjugates in the denominator.}$$

$$= \frac{r_1(\cos\theta_1 + i\sin\theta_1)(\cos\theta_2 - i\sin\theta_2)}{r_2} \qquad \text{Use a Pythagorean identity: } \cos^2\theta_2 + \sin^2\theta_2 = 1.$$

$$= \frac{r_1}{r_2}(\cos\theta_1\cos\theta_2 - i\cos\theta_1\sin\theta_2 + i\sin\theta_1\cos\theta_2 - i^2\sin\theta_1\sin\theta_2) \qquad \text{Use the FOIL method.}$$

$$= \frac{r_1}{r_2}[\cos\theta_1\cos\theta_2 + i(\sin\theta_1\cos\theta_2 - \cos\theta_1\sin\theta_2) - i^2\sin\theta_1\sin\theta_2] \qquad \text{Factor } i \text{ from the second and third terms.}$$

$$= \frac{r_1}{r_2}\left[\cos\theta_1\cos\theta_2 + i(\sin\theta_1\cos\theta_2 - \cos\theta_1\sin\theta_2) - (-1)\sin\theta_1\sin\theta_2\right] \quad i^2 = -1.$$

$$= \frac{r_1}{r_2}\left[\cos\theta_1\cos\theta_2 + \sin\theta_1\sin\theta_2 + i(\sin\theta_1\cos\theta_2 - \cos\theta_1\sin\theta_2)\right] \quad \text{Rearrange terms.}$$

This is $\cos(\theta_1 - \theta_2)$. This is $\sin(\theta_1 - \theta_2)$.

$$= \frac{r_1}{r_2}\left[\cos(\theta_1 - \theta_2) + i\sin(\theta_1 - \theta_2)\right]$$

SECTION 6.7 The Dot Product

Properties of the Dot Product

If $\mathbf{u}$, $\mathbf{v}$, and $\mathbf{w}$ are vectors, and c is a scalar, then:

1. $\mathbf{u}\cdot\mathbf{v} = \mathbf{v}\cdot\mathbf{u}$
2. $\mathbf{u}\cdot(\mathbf{v}+\mathbf{w}) = \mathbf{u}\cdot\mathbf{v} + \mathbf{u}\cdot\mathbf{w}$
3. $\mathbf{0}\cdot\mathbf{v} = 0$
4. $\mathbf{v}\cdot\mathbf{v} = \|\mathbf{v}\|^2$
5. $(c\mathbf{u})\cdot\mathbf{v} = c(\mathbf{u}\cdot\mathbf{v}) = \mathbf{u}\cdot(c\mathbf{v})$

Proof To prove the second property, let

$$\mathbf{u} = u_1\mathbf{i} + u_2\mathbf{j}, \quad \mathbf{v} = v_1\mathbf{i} + v_2\mathbf{j}, \quad \text{and } \mathbf{w} = w_1\mathbf{i} + w_2\mathbf{j}.$$

Then,

$$\mathbf{u}\cdot(\mathbf{v}+\mathbf{w}) = (u_1\mathbf{i} + u_2\mathbf{j})\cdot[(v_1\mathbf{i} + v_2\mathbf{j}) + (w_1\mathbf{i} + w_2\mathbf{j})] \quad \text{These are the given vectors.}$$

$$= (u_1\mathbf{i} + u_2\mathbf{j})\cdot[(v_1 + w_1)\mathbf{i} + (v_2 + w_2)\mathbf{j}] \quad \text{Add horizontal components and add vertical components.}$$

$$= u_1(v_1 + w_1) + u_2(v_2 + w_2) \quad \text{Multiply horizontal components and multiply vertical components.}$$

$$= u_1v_1 + u_1w_1 + u_2v_2 + u_2w_2 \quad \text{Use the distributive property.}$$

$$= u_1v_1 + u_2v_2 + u_1w_1 + u_2w_2 \quad \text{Rearrange terms.}$$

This is the dot product of $\mathbf{u}$ and $\mathbf{v}$. This is the dot product of $\mathbf{u}$ and $\mathbf{w}$.

$$= \mathbf{u}\cdot\mathbf{v} + \mathbf{u}\cdot\mathbf{w}.$$

To prove the third property, let

$$\mathbf{0} = 0\mathbf{i} + 0\mathbf{j} \quad \text{and} \quad \mathbf{v} = v_1\mathbf{i} + v_2\mathbf{j}.$$

Then

$$\mathbf{0}\cdot\mathbf{v} = (0\mathbf{i} + 0\mathbf{j})\cdot(v_1\mathbf{i} + v_2\mathbf{j}) \quad \text{These are the given vectors.}$$

$$= 0\cdot v_1 + 0\cdot v_2 \quad \text{Multiply horizontal components and multiply vertical components.}$$

$$= 0 + 0$$

$$= 0.$$

To prove the first part of the fifth property, let
$$\mathbf{u} = u_1\mathbf{i} + u_2\mathbf{j} \quad \text{and} \quad \mathbf{v} = v_1\mathbf{i} + v_2\mathbf{j}.$$

Then,

$$
\begin{aligned}
(c\mathbf{u}) \cdot \mathbf{v} &= \left[c(u_1\mathbf{i} + u_2\mathbf{j})\right] \cdot (v_1\mathbf{i} + v_2\mathbf{j}) && \text{These are the given vectors.} \\
&= (cu_1\mathbf{i} + cu_2\mathbf{j}) \cdot (v_1\mathbf{i} + v_2\mathbf{j}) && \text{Multiply each component of } u_1\mathbf{i} + u_2\mathbf{j} \text{ by } c. \\
&= cu_1v_1 + cu_2v_2 && \text{Multiply horizontal components and} \\
& && \text{multiply vertical components.} \\
&= c(u_1v_1 + u_2v_2) && \text{Factor out } c \text{ from both terms.}
\end{aligned}
$$

This is the dot product of **u** and **v**.

$$= c(\mathbf{u} \cdot \mathbf{v}).$$

Answers to Selected Exercises

CHAPTER P

Section P.1

Check Point Exercises

1. a. $\sqrt{2} - 1$ **b.** $\pi - 3$ **c.** 1   **2.** 9   **3. a.** 14   **b.** 3   **4.** $38x - 19y$

Exercise Set P.1

1. a. $\sqrt{100}$ **b.** $0, \sqrt{100}$ **c.** $-9, 0, \sqrt{100}$ **d.** $-9, -\dfrac{4}{5}, 0, 0.25, 9.2, \sqrt{100}$ **e.** $\sqrt{3}$ **3. a.** $\sqrt{64}$ **b.** $0, \sqrt{64}$ **c.** $-11, 0, \sqrt{64}$

d. $-11, -\dfrac{5}{6}, 0, 0.75, \sqrt{64}$ **e.** $\sqrt{5}, \pi$ **5.** 0   **7.** Answers may vary.   **9.** true   **11.** true   **13.** true   **15.** 300   **17.** $12 - \pi$

19. $5 - \sqrt{2}$ **21.** -1 **23.** 4   **25.** 3   **27.** 7   **29.** -1 **31.** $|17 - 2|; 15$ **33.** $|5 - (-2)|; 7$ **35.** $|-4 - (-19)|; 15$

37. $|-1.4 - (-3.6)|; 2.2$ **39.** 27   **41.** -19 **43.** 25   **45.** $-\dfrac{4}{5}$ **47.** $\dfrac{5}{2}$ **49.** commutative property of addition

51. associative property of addition   **53.** commutative property of addition   **55.** distributive property of multiplication over addition
57. inverse property of multiplication   **59.** $15x + 16$ **61.** $27x - 10$ **63.** $29y - 29$ **65.** $8y - 12$ **67.** $16y - 25$ **69.** $14x$
71. $-2x + 3y + 6$ **73.** x **75.** yes   **77.** Answers may vary.   **79.** 21; In 2000, approximately 21% of American adults smoked
cigarettes.   **81. a.** $132 - 0.6a$ **b.** 120   **89.** (c) is true.   **91.** $<$ **93.** $>$

Section P.2

Check Point Exercises

1. -256 **2. a.** $16x^{12}y^{24}$ **b.** $-18x^3y^8$ **c.** $\dfrac{5y^6}{x^4}$ **d.** $\dfrac{y^8}{25x^2}$ **3. a.** 7,400,000,000   **b.** 0.000003017

4. a. 7.41×10^9 **b.** 9.2×10^{-8} **5.** $12.86   **6.** $2.5344 \times 10^3 = 2534.4$

Exercise Set P.2

1. 50   **3.** 64   **5.** -64 **7.** 1   **9.** -1 **11.** $\dfrac{1}{64}$ **13.** 32   **15.** 64   **17.** 16   **19.** $\dfrac{1}{9}$ **21.** $\dfrac{1}{16}$ **23.** $\dfrac{y}{x^2}$ **25.** y^5

27. x^{10} **29.** x^5 **31.** x^{21} **33.** x^{-15} **35.** x^7 **37.** x^{21} **39.** $64x^6$ **41.** $-\dfrac{64}{x^3}$ **43.** $9x^4y^{10}$ **45.** $6x^{11}$ **47.** $18x^9y^5$

49. $4x^{16}$ **51.** $-5a^{11}b$ **53.** $\dfrac{2}{b^7}$ **55.** $\dfrac{1}{16x^6}$ **57.** $\dfrac{3y^{14}}{4x^4}$ **59.** $\dfrac{y^2}{25x^6}$ **61.** $-\dfrac{27\,b^{15}}{a^{18}}$ **63.** 1   **65.** 4700   **67.** 4,000,000

69. 0.000786   **71.** 0.00000318   **73.** 3.6×10^3 **75.** 2.2×10^8 **77.** 2.7×10^{-2} **79.** 7.63×10^{-4} **81.** 600,000   **83.** 0.123

85. 30,000   **87.** 0.021   **89.** $\dfrac{4.8 \times 10^{11}}{1.2 \times 10^{-4}}; 4 \times 10^{15}$ **91.** $\dfrac{(7.2 \times 10^{-4})(3 \times 10^{-3})}{2.4 \times 10^{-4}}; 9 \times 10^{-3}$ **93.** $6800   **95.** 1.12×10^{12}

97. 1.06×10^{-18} gram   **107.** (b) is true.   **109.** $A = C + D$

Section P.3

Check Point Exercises

1. a. 3   **b.** $5x\sqrt{2}$ **2. a.** $\dfrac{5}{4}$ **b.** $5x\sqrt{3}$ **3. a.** $17\sqrt{13}$ **b.** $-19\sqrt{17x}$ **4. a.** $17\sqrt{3}$ **b.** $10\sqrt{2x}$ **5. a.** $\dfrac{5\sqrt{3}}{3}$ **b.** $\sqrt{3}$

6. $\dfrac{32 - 8\sqrt{5}}{11}$ **7. a.** $2\sqrt[3]{5}$ **b.** $2\sqrt[5]{2}$ **c.** $\dfrac{5}{3}$ **8.** $5\sqrt[3]{3}$ **9. a.** 9   **b.** 3   **c.** $\dfrac{1}{2}$ **10. a.** 8   **b.** $\dfrac{1}{4}$

11. a. $10x^4$ **b.** $4x^{5/2}$ **12.** $\sqrt{x}$

Exercise Set P.3

1. 6   **3.** not a real number   **5.** 13   **7.** $5\sqrt{2}$ **9.** $3|x|\sqrt{5}$ **11.** $2x\sqrt{3}$ **13.** $x\sqrt{x}$ **15.** $2x\sqrt{3x}$ **17.** $\dfrac{1}{9}$ **19.** $\dfrac{7}{4}$

21. $4x$ **23.** $5x\sqrt{2x}$ **25.** $2x^2\sqrt{5}$ **27.** $13\sqrt{3}$ **29.** $-2\sqrt{17x}$ **31.** $5\sqrt{2}$ **33.** $3\sqrt{2x}$ **35.** $34\sqrt{2}$ **37.** $20\sqrt{2} - 5\sqrt{3}$

39. $\dfrac{\sqrt{7}}{7}$ **41.** $\dfrac{\sqrt{10}}{5}$ **43.** $\dfrac{13(3 - \sqrt{11})}{-2}$ **45.** $7(\sqrt{5} + 2)$ **47.** $3(\sqrt{5} - \sqrt{3})$ **49.** 5   **51.** -2 **53.** not a real number

55. 3   **57.** −3   **59.** $-\dfrac{1}{2}$ **61.** $2\sqrt[3]{4}$ **63.** $x\sqrt[3]{x}$ **65.** $3\sqrt[3]{2}$ **67.** $2x$ **69.** $7\sqrt[3]{2}$ **71.** $13\sqrt[3]{2}$ **73.** $-y\sqrt[3]{2x}$ **75.** $\sqrt{2}+2$

77. 6   **79.** 2   **81.** 25   **83.** $\dfrac{1}{16}$ **85.** $14x^{7/12}$ **87.** $4x^{1/4}$ **89.** x^2 **91.** $5x^2|y|^3$ **93.** $27y^{2/3}$ **95.** $\sqrt{5}$ **97.** x^2

99. $\sqrt[3]{x^2}$ **101.** $\sqrt[5]{x^2y}$ **103.** $20\sqrt{2}$ mph   **105.** $\dfrac{\sqrt{5}+1}{2} \approx 1.62$ **107.** $\dfrac{7\sqrt{2\cdot2\cdot3}}{6} = \dfrac{7\sqrt{2^2\cdot3}}{6} = \dfrac{7\sqrt{2^2}\sqrt{3}}{6} = \dfrac{7\cdot2\sqrt{3}}{6} = \dfrac{7}{3}\sqrt{3}$

109. The duration of a storm whose diameter is 9 miles is 1.89 hours.   **117.** 45.00, 23.76, 15.68, 11.33, 8.59, 6.70, 5.31, 4.25, 3.41, 2.73, 2.17, 1.70, 1.30, 0.95, 0.65, 0.38; The percentage of potential employees testing positive for illegal drugs is decreasing over time.   **119.** (d) is true.   **121.** Let $\square = 25$ and $\square = 14$.   **123. a.** >   **b.** >

Section P.4

Check Point Exercises

1. a. $-x^3 + x^2 - 8x - 20$ **b.** $20x^3 - 11x^2 - 2x - 8$ **2.** $15x^3 - 31x^2 + 30x - 8$ **3.** $28x^2 - 41x + 15$
4. a. $21x^2 - 25xy + 6y^2$ **b.** $x^4 + 10x^2y + 25y^2$ **5. a.** $9x^2 + 12x + 4 - 25y^2$ **b.** $4x^2 + 4xy + y^2 + 12x + 6y + 9$

Exercise Set P.4

1. yes; $3x^2 + 2x - 5$ **3.** no   **5.** 2   **7.** 4   **9.** $11x^3 + 7x^2 - 12x - 4$; 3   **11.** $12x^3 + 4x^2 + 12x - 14$; 3   **13.** $6x^2 - 6x + 2$; 2
15. $x^3 + 1$ **17.** $2x^3 - 9x^2 + 19x - 15$ **19.** $x^2 + 10x + 21$ **21.** $x^2 - 2x - 15$ **23.** $6x^2 + 13x + 5$ **25.** $10x^2 - 9x - 9$
27. $15x^4 - 47x^2 + 28$ **29.** $8x^5 - 40x^3 + 3x^2 - 15$ **31.** $x^2 - 9$ **33.** $9x^2 - 4$ **35.** $25 - 49x^2$ **37.** $16x^4 - 25x^2$
39. $1 - y^{10}$ **41.** $x^2 + 4x + 4$ **43.** $4x^2 + 12x + 9$ **45.** $x^2 - 6x + 9$ **47.** $16x^4 - 8x^2 + 1$ **49.** $49 - 28x + 4x^2$
51. $x^3 + 3x^2 + 3x + 1$ **53.** $8x^3 + 36x^2 + 54x + 27$ **55.** $x^3 - 9x^2 + 27x - 27$ **57.** $27x^3 - 108x^2 + 144x - 64$
59. $7x^2 + 38xy + 15y^2$ **61.** $2x^2 + xy - 21y^2$ **63.** $15x^2y^2 + xy - 2$ **65.** $49x^2 + 70xy + 25y^2$ **67.** $x^4y^4 - 6x^2y^2 + 9$
69. $x^3 - y^3$ **71.** $9x^2 - 25y^2$ **73.** $x^2 + 2xy + y^2 - 9$ **75.** $9x^2 + 42x + 49 - 25y^2$ **77.** $25y^2 - 4x^2 - 12x - 9$
79. $x^2 + 2xy + y^2 + 2x + 2y + 1$ **81.** $4x^2 + 4xy + y^2 + 4x + 2y + 1$ **83.** $(A + B)^2 = A^2 + 2AB + B^2$
85. $(A + 1)^2 = A^2 + 2A + 1$ **87.** 7.567; A person earning \$40,000 feels underpaid by \$7567.   **89.** 527.53; The number of violent crimes in the United States was 527.53 per 100,000 inhabitants in 2000. The calculated value is a good approximation to the actual value, 524.7.

91. $\dfrac{2}{3}t^3 - 2t^2 + 4t$ **93.** $6x + 22$ **95.** $4x^3 - 36x^2 + 80x$ **105.** 61.2, 59.0, 56.8, 54.8, 52.8, 50.9, 49.3, 47.7, 46.4, 45.2, 44.3, 43.6, 43.1,

43.0, 43.1, 43.6, 44.4, 45.5, 47.0, 48.9, 51.2; The percentage of U.S. high school seniors who had ever used marijuana decreased from 1980, reached a low in 1993, then increased through 2000.   **107.** $V = x^3 + 7x^2 - 3x$ **109.** $6y^n - 13$

Section P.5

Check Point Exercises

1. a. $2x^2(5x - 2)$ **b.** $(x - 7)(2x + 3)$ **2.** $(x + 5)(x^2 - 2)$ **3. a.** $(x + 8)(x + 5)$ **b.** $(x - 7)(x + 2)$ **4.** $(3x - 1)(2x + 7)$
5. a. $(x + 9)(x - 9)$ **b.** $(6x + 5)(6x - 5)$ **6.** $(9x^2 + 4)(3x + 2)(3x - 2)$ **7. a.** $(x + 7)^2$ **b.** $(4x - 7)^2$
8. a. $(x + 1)(x^2 - x + 1)$ **b.** $(5x - 2)(25x^2 + 10x + 4)$ **9.** $3x(x - 5)^2$ **10.** $(x + 10 + 6a)(x + 10 - 6a)$ **11.** $\dfrac{2x - 1}{(x - 1)^{1/2}}$

Exercise Set P.5

1. $9(2x + 3)$ **3.** $3x(x + 2)$ **5.** $9x^2(x^2 - 2x + 3)$ **7.** $(x + 5)(x + 3)$ **9.** $(x - 3)(x^2 + 12)$ **11.** $(x^2 + 5)(x - 2)$
13. $(x - 1)(x^2 + 2)$ **15.** $(3x - 2)(x^2 - 2)$ **17.** $(x + 2)(x + 3)$ **19.** $(x - 5)(x + 3)$ **21.** $(x - 5)(x - 3)$
23. $(3x + 2)(x - 1)$ **25.** $(3x - 28)(x + 1)$ **27.** $(2x - 1)(3x - 4)$ **29.** $(2x + 3)(2x + 5)$ **31.** $(x + 10)(x - 10)$
33. $(6x + 7)(6x - 7)$ **35.** $(3x + 5y)(3x - 5y)$ **37.** $(x^2 + 4)(x + 2)(x - 2)$ **39.** $(4x^2 + 9)(2x + 3)(2x - 3)$ **41.** $(x + 1)^2$
43. $(x - 7)^2$ **45.** $(2x + 1)^2$ **47.** $(3x - 1)^2$ **49.** $(x + 3)(x^2 - 3x + 9)$ **51.** $(x - 4)(x^2 + 4x + 16)$
53. $(2x - 1)(4x^2 + 2x + 1)$ **55.** $(4x + 3)(16x^2 - 12x + 9)$ **57.** $3x(x + 1)(x - 1)$ **59.** $4(x + 2)(x - 3)$
61. $2(x^2 + 9)(x + 3)(x - 3)$ **63.** $(x - 3)(x + 3)(x + 2)$ **65.** $2(x - 8)(x + 7)$ **67.** $x(x - 2)(x + 2)$ **69.** prime
71. $(x - 2)(x + 2)^2$ **73.** $y(y^2 + 9)(y + 3)(y - 3)$ **75.** $5y^2(2y + 3)(2y - 3)$ **77.** $(x - 6 + 7y)(x - 6 - 7y)$
79. $(x + y)(3b + 4)(3b - 4)$ **81.** $(y - 2)(x + 4)(x - 4)$ **83.** $2x(x + 6 + 2a)(x + 6 - 2a)$ **85.** $x^{1/2}(x - 1)$ **87.** $\dfrac{4(1 + 2x)}{x^{2/3}}$
89. $-(x + 3)^{1/2}(x + 2)$ **91.** $\dfrac{x + 4}{(x + 5)^{3/2}}$ **93.** $-\dfrac{4(4x - 1)^{1/2}(x - 1)}{3}$ **95. a.** $0.36\,x$ **b.** no; It is selling at 36% of the original price.
97. $16(4 + t)(4 - t)$ **99.** $(3x + 2)(3x - 2)$ **109.** (d) is true.   **111.** $-(x + 5)(x - 1)$ **113.** $-\dfrac{10}{(x - 5)^{3/2}(x + 5)^{1/2}}$
115. $b = 0, 3, 4, -c(c + 4)$, where $c > 0$ is an integer.

Section P.6

Check Point Exercises

1. a. -5 **b.** $6, -6$ **2. a.** $x^2, x \neq -3$ **b.** $\dfrac{x-1}{x+1}, x \neq -1$ **3.** $\dfrac{x-3}{(x-2)(x+3)}, x \neq 2, x \neq -2, x \neq -3$

4. $\dfrac{3(x-1)}{x(x+2)}, x \neq 1, x \neq 0, x \neq -2$ **5.** $-2, x \neq -1$ **6.** $\dfrac{2(4x+1)}{(x+1)(x-1)}, x \neq 1, x \neq -1$ **7.** $(x-3)(x-3)(x+3)$

8. $\dfrac{-x^2+11x-20}{2(x-5)^2}, x \neq 5$ **9.** $\dfrac{2(2-3x)}{4+3x}, x \neq 0, x \neq -\dfrac{4}{3}$ **10.** $-\dfrac{1}{x(x+7)}, x \neq 0, x \neq -7$ **11.** $\dfrac{x+1}{x^{3/2}}$ **12.** $\dfrac{1}{\sqrt{x+3}+\sqrt{x}}$

Exercise Set P.6

1. 3 **3.** $5, -5$ **5.** $-1, -10$ **7.** $\dfrac{3}{x-3}, x \neq 3$ **9.** $\dfrac{y+9}{y-1}, y \neq 1, 2$ **11.** $\dfrac{x+6}{x-6}, x \neq 6, -6$ **13.** $\dfrac{(x-3)(x+3)}{x(x+4)}, x \neq 0, -4, 3$

15. $\dfrac{x-1}{x+2}, x \neq -2, -1, 2, 3$ **17.** $\dfrac{x^2+2x+4}{3x}, x \neq -2, 0, 2$ **19.** $\dfrac{(x-2)^2}{x}, x \neq 0, -2, 2$ **21.** $\dfrac{x-5}{2}, x \neq 1, -5$

23. $\dfrac{(x+2)(x+4)}{x-5}, x \neq -6, -3, -1, 3, 5$ **25.** $2, x \neq -\dfrac{5}{6}$ **27.** $\dfrac{2x-1}{x+3}, x \neq 0, -3$ **29.** $\dfrac{3}{x-3}, x \neq 3, -4$

31. $\dfrac{9x+39}{(x+4)(x+5)}, x \neq -4, -5$ **33.** $-\dfrac{3}{x(x+1)}, x \neq -1, 0$ **35.** $\dfrac{3x^2+4}{(x+2)(x-2)}, x \neq -2, 2$ **37.** $\dfrac{2x^2+50}{(x-5)(x+5)}, x \neq -5, 5$

39. $\dfrac{4x+16}{(x+3)^2}, x \neq -3$ **41.** $\dfrac{x^2-x}{(x+5)(x-2)(x+3)}, x \neq -5, 2, -3$ **43.** $\dfrac{x-1}{x+2}, x \neq -2, -1$ **45.** $\dfrac{x+1}{3x-1}, x \neq 0, \dfrac{1}{3}$

47. $\dfrac{x}{x+3}, x \neq -2, -3$ **49.** $-\dfrac{x-14}{7}, x \neq -2, 2$ **51.** $\dfrac{x-3}{x+2}, x \neq 1, 3$ **53.** $-\dfrac{2x+h}{x^2(x+h)^2}, h \neq 0$ **55.** $1 - \dfrac{1}{3x}, x > 0$

57. $-\dfrac{2}{x^2\sqrt{x^2+2}}$ **59.** $\dfrac{\sqrt{x}-\sqrt{x+h}}{h\sqrt{x}\sqrt{x+h}}, h \neq 0$ **61.** $\dfrac{1}{\sqrt{x+5}+\sqrt{x}}$ **63.** $\dfrac{1}{(x+y)(\sqrt{x}-\sqrt{y})}$ **65. a.** $86.67, 520, 1170;$
It costs \$86,670,000 to inoculate 40% of the population against this strain of flu, \$520,000,000 to inoculate 80% of the population, and \$1,170,000,000 to inoculate 90% of the population.   **b.** $x = 100$ **c.** increases rapidly; impossible to inoculate 100% of the population.

67. a. $\dfrac{100W}{L}$ **b.** round   **69.** $\dfrac{2r_1r_2}{r_1+r_2}; 24$ mph   **85.** (d) is true.   **87.** $\dfrac{2}{x+1}$

Section P.7

Check Point Exercises

1. $\{16\}$ **2.** $\{5\}$ **3.** $\{-2\}$ **4.** $\{3\}$ **5.** $\varnothing$ **6.** identity   **7.** $m = \dfrac{y-b}{x}$ **8.** $C = \dfrac{P}{1+M}$ **9.** $\{-2, 3\}$

Exercise Set P.7

1. $\{11\}$ **3.** $\{7\}$ **5.** $\{13\}$ **7.** $\{2\}$ **9.** $\{9\}$ **11.** $\{-5\}$ **13.** $\{6\}$ **15.** $\{-2\}$ **17.** $\{12\}$ **19.** $\{24\}$ **21.** $\{-15\}$

23. $\{5\}$ **25.** $\left\{\dfrac{33}{2}\right\}$ **27.** $\{-12\}$ **29.** $\left\{\dfrac{46}{5}\right\}$ **31. a.** 0 **b.** $\left\{\dfrac{1}{2}\right\}$ **33. a.** 0 **b.** $\{-2\}$ **35. a.** 0 **b.** $\{2\}$

37. a. 0 **b.** $\{4\}$ **39. a.** 1 **b.** $\{3\}$ **41. a.** -1 **b.** $\varnothing$ **43. a.** 1 **b.** $\{2\}$ **45. a.** $-2, 2$ **b.** $\varnothing$ **47. a.** $-1, 1$
b. $\{-3\}$ **49. a.** $-2, 4$ **b.** $\varnothing$ **51.** identity   **53.** inconsistent equation   **55.** conditional equation   **57.** inconsistent equation
59. $\{-7\}$ **61.** not true for any real number, $\varnothing$ **63.** $\{-4\}$ **65.** $\{8\}$ **67.** $\{-1\}$ **69.** not true for any real number, $\varnothing$

71. $\omega = \dfrac{A}{l}$ **73.** $b = \dfrac{2A}{h}$ **75.** $P = \dfrac{I}{rt}$ **77.** $m = \dfrac{E}{c^2}$ **79.** $p = \dfrac{T-D}{m}$ **81.** $a = \dfrac{2A}{h} - b$ **83.** $r = \dfrac{S-P}{Pt}$

85. $S = \dfrac{F}{B} + V$ **87.** $I = \dfrac{E}{R+r}$ **89.** $f = \dfrac{pq}{p+q}$ **91.** $\{-8, 8\}$ **93.** $\{-5, 9\}$ **95.** $\{-2, 3\}$ **97.** $\left\{-\dfrac{5}{3}, 3\right\}$ **99.** $\left\{-\dfrac{2}{5}, \dfrac{2}{5}\right\}$

101. $\varnothing$ **103.** $\left\{\dfrac{1}{2}\right\}$ **105.** approximately 41 years after 1960 in 2001   **107.** 196 lb   **125.** Answers may vary.   **127.** 2

129. $C = \dfrac{VL-SN}{L-N}$

Section P.8

Check Point Exercises

1. a. $\{0, 3\}$ **b.** $\left\{-1, \dfrac{1}{2}\right\}$ **2. a.** $\{-\sqrt{7}, \sqrt{7}\}$ **b.** $\{-5 + \sqrt{11}, -5 - \sqrt{11}\}$ **3.** $49; (x-7)^2$ **4.** $\{1 + \sqrt{3}, 1 - \sqrt{3}\}$

5. $\left\{\dfrac{-1+\sqrt{3}}{2}, \dfrac{-1-\sqrt{3}}{2}\right\}$ **6.** $-56;$ no real solutions   **7.** 1998; a good approximation   **8.** 12 in.

Exercise Set P.8

1. $\{-2, 5\}$ **3.** $\{3, 5\}$ **5.** $\left\{-\dfrac{5}{2}, \dfrac{2}{3}\right\}$ **7.** $\left\{-\dfrac{4}{3}, 2\right\}$ **9.** $\{-4, 0\}$ **11.** $\left\{0, \dfrac{1}{3}\right\}$ **13.** $\{-3, 1\}$ **15.** $\{-3, 3\}$

17. $\{-\sqrt{10}, \sqrt{10}\}$ **19.** $\{-7, 3\}$ **21.** $\left\{-\dfrac{5}{3}, \dfrac{1}{3}\right\}$ **23.** $\left\{\dfrac{1 - \sqrt{7}}{5}, \dfrac{1 + \sqrt{7}}{5}\right\}$ **25.** $\left\{\dfrac{4 - 2\sqrt{2}}{3}, \dfrac{4 + 2\sqrt{2}}{3}\right\}$

27. $36; x^2 + 12x + 36 = (x + 6)^2$ **29.** $25; x^2 - 10x + 25 = (x - 5)^2$ **31.** $\dfrac{9}{4}; x^2 + 3x + \dfrac{9}{4} = \left(x + \dfrac{3}{2}\right)^2$

33. $\dfrac{49}{4}; x^2 - 7x + \dfrac{49}{4} = \left(x - \dfrac{7}{2}\right)^2$ **35.** $\dfrac{1}{9}; x^2 - \dfrac{2}{3}x + \dfrac{1}{9} = \left(x - \dfrac{1}{3}\right)^2$ **37.** $\dfrac{1}{36}; x^2 - \dfrac{1}{3}x + \dfrac{1}{36} = \left(x - \dfrac{1}{6}\right)^2$ **39.** $\{-7, 1\}$

41. $\{1 + \sqrt{3}, 1 - \sqrt{3}\}$ **43.** $\{3 + 2\sqrt{5}, 3 - 2\sqrt{5}\}$ **45.** $\{-2 + \sqrt{3}, -2 - \sqrt{3}\}$ **47.** $\left\{\dfrac{-3 + \sqrt{13}}{2}, \dfrac{-3 - \sqrt{13}}{2}\right\}$

49. $\left\{\dfrac{1}{2}, 3\right\}$ **51.** $\left\{\dfrac{1 + \sqrt{2}}{2}, \dfrac{1 - \sqrt{2}}{2}\right\}$ **53.** $\left\{\dfrac{1 + \sqrt{7}}{3}, \dfrac{1 - \sqrt{7}}{3}\right\}$ **55.** $\{-5, -3\}$ **57.** $\left\{\dfrac{-5 + \sqrt{13}}{2}, \dfrac{-5 - \sqrt{13}}{2}\right\}$

59. $\left\{\dfrac{3 + \sqrt{57}}{6}, \dfrac{3 - \sqrt{57}}{6}\right\}$ **61.** $\left\{\dfrac{1 + \sqrt{29}}{4}, \dfrac{1 - \sqrt{29}}{4}\right\}$ **63.** 36; 2 unequal real solutions **65.** 97; 2 unequal real solutions

67. 0; 1 real solution **69.** 37; 2 unequal real solutions **71.** $\left\{-\dfrac{1}{2}, 1\right\}$ **73.** $\left\{\dfrac{1}{5}, 2\right\}$ **75.** $\{-2\sqrt{5}, 2\sqrt{5}\}$ **77.** $\{1 + \sqrt{2}, 1 - \sqrt{2}\}$

79. $\left\{\dfrac{-11 + \sqrt{33}}{4}, \dfrac{-11 - \sqrt{33}}{4}\right\}$ **81.** $\left\{0, \dfrac{8}{3}\right\}$ **83.** $\{2\}$ **85.** $\{-2, 2\}$ **87.** $\{2 + \sqrt{2}, 2 - \sqrt{2}\}$ **89.** $\left\{0, \dfrac{7}{2}\right\}$

91. $\{2 + \sqrt{10}, 2 - \sqrt{10}\}$ **93.** $\{-5, -1\}$ **95.** 19 year olds and 72 year olds; fairly well **97.** 1994 **99.** 127.28 ft **101.** 34 ft

103. width = 15 ft; length = 20 ft **105.** 10 in. **117.** $\{-3\sqrt{3}, \sqrt{3}\}$ **119.** 5 sec

Section P.9

Check Point Exercises

1. a.  **b.** **c.**

2

−4

2   6

2. a. $\{x|-2 \le x < 5\}$ **b.** $\{x|1 \le x \le 3.5\}$ **c.** $\{x|x < -1\}$ **3.** $[1, \infty)$ or $\{x|x \ge 1\}$

−2   5

1   3.5

−1

1

4. $[-1, 4)$ or $\{x|-1 \le x < 4\}$ **5.** $(-3, 7)$ or $\{x|-3 < x < 7\}$ **6.** $(-\infty, 1]$ or $[4, \infty)$ or $\{x|x \le 1 \text{ or } x \ge 4\}$

−1   4

−3   7

1   4

7. $x \le 1280$

Exercise Set P.9

1. 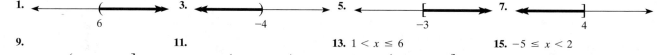 **3.** **5.** **7.**

6

−4

−3

4

9. **11.** **13.** $1 < x \le 6$ **15.** $-5 \le x < 2$

−2   5

−1   4

1   6

−5   2

17. $-3 \le x \le 1$ **19.** $x > 2$ **21.** $x \ge -3$ **23.** $x < 3$

−3   1

2

−3

3

25. $x \le 5.5$ **27.** $(-\infty, 3)$ **29.** $\left[\dfrac{20}{3}, \infty\right)$ **31.** $(-\infty, -4]$

5.5

3

$\dfrac{20}{3}$

−4

33. $\left(-\infty, -\dfrac{2}{5}\right]$ **35.** $[0, \infty)$ **37.** $(-\infty, 1)$ **39.** $[6, \infty)$

$-\dfrac{2}{5}$

0

1

6

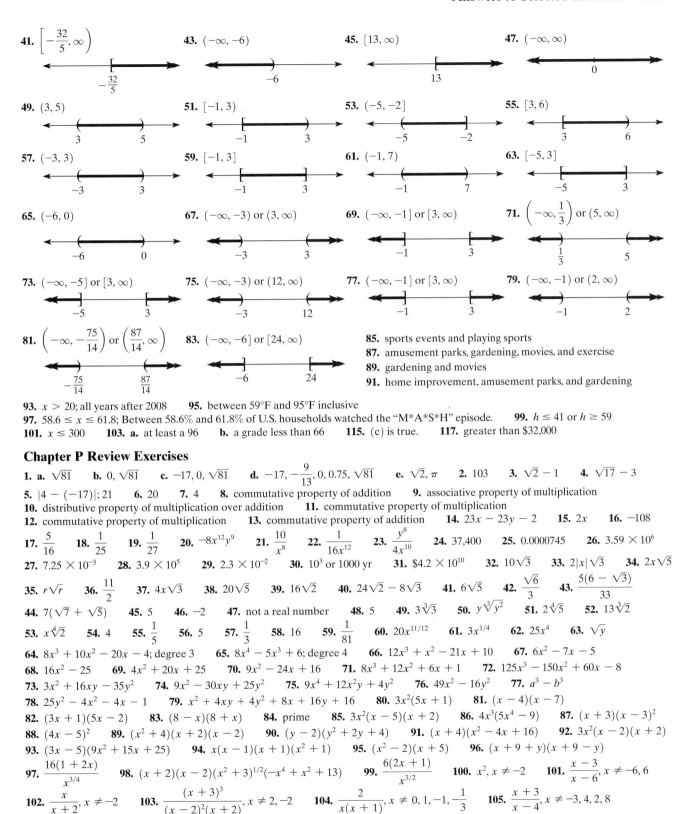

41. $\left[-\dfrac{32}{5}, \infty\right)$ **43.** $(-\infty, -6)$ **45.** $[13, \infty)$ **47.** $(-\infty, \infty)$

49. $(3, 5)$ **51.** $[-1, 3)$ **53.** $(-5, -2]$ **55.** $[3, 6)$

57. $(-3, 3)$ **59.** $[-1, 3]$ **61.** $(-1, 7)$ **63.** $[-5, 3]$

65. $(-6, 0)$ **67.** $(-\infty, -3)$ or $(3, \infty)$ **69.** $(-\infty, -1]$ or $[3, \infty)$ **71.** $\left(-\infty, \dfrac{1}{3}\right)$ or $(5, \infty)$

73. $(-\infty, -5]$ or $[3, \infty)$ **75.** $(-\infty, -3)$ or $(12, \infty)$ **77.** $(-\infty, -1]$ or $[3, \infty)$ **79.** $(-\infty, -1)$ or $(2, \infty)$

81. $\left(-\infty, -\dfrac{75}{14}\right)$ or $\left(\dfrac{87}{14}, \infty\right)$ **83.** $(-\infty, -6]$ or $[24, \infty)$

85. sports events and playing sports
87. amusement parks, gardening, movies, and exercise
89. gardening and movies
91. home improvement, amusement parks, and gardening

93. $x > 20$; all years after 2008 **95.** between 59°F and 95°F inclusive
97. $58.6 \leq x \leq 61.8$; Between 58.6% and 61.8% of U.S. households watched the "M*A*S*H" episode. **99.** $h \leq 41$ or $h \geq 59$
101. $x \leq 300$ **103. a.** at least a 96 **b.** a grade less than 66 **115.** (c) is true. **117.** greater than $32,000

Chapter P Review Exercises

1. a. $\sqrt{81}$ **b.** $0, \sqrt{81}$ **c.** $-17, 0, \sqrt{81}$ **d.** $-17, -\dfrac{9}{13}, 0, 0.75, \sqrt{81}$ **e.** $\sqrt{2}, \pi$ **2.** 103 **3.** $\sqrt{2} - 1$ **4.** $\sqrt{17} - 3$
5. $|4 - (-17)|; 21$ **6.** 20 **7.** 4 **8.** commutative property of addition **9.** associative property of multiplication
10. distributive property of multiplication over addition **11.** commutative property of multiplication
12. commutative property of multiplication **13.** commutative property of addition **14.** $23x - 23y - 2$ **15.** $2x$ **16.** -108
17. $\dfrac{5}{16}$ **18.** $\dfrac{1}{25}$ **19.** $\dfrac{1}{27}$ **20.** $-8x^{12}y^9$ **21.** $\dfrac{10}{x^8}$ **22.** $\dfrac{1}{16x^{12}}$ **23.** $\dfrac{y^8}{4x^{10}}$ **24.** 37,400 **25.** 0.0000745 **26.** 3.59×10^6
27. 7.25×10^{-3} **28.** 3.9×10^5 **29.** 2.3×10^{-2} **30.** 10^3 or 1000 yr **31.** $\$4.2 \times 10^{10}$ **32.** $10\sqrt{3}$ **33.** $2|x|\sqrt{3}$ **34.** $2x\sqrt{5}$
35. $r\sqrt{r}$ **36.** $\dfrac{11}{2}$ **37.** $4x\sqrt{3}$ **38.** $20\sqrt{5}$ **39.** $16\sqrt{2}$ **40.** $24\sqrt{2} - 8\sqrt{3}$ **41.** $6\sqrt{5}$ **42.** $\dfrac{\sqrt{6}}{3}$ **43.** $\dfrac{5(6 - \sqrt{3})}{33}$
44. $7(\sqrt{7} + \sqrt{5})$ **45.** 5 **46.** -2 **47.** not a real number **48.** 5 **49.** $3\sqrt[3]{3}$ **50.** $y\sqrt[3]{y^2}$ **51.** $2\sqrt[4]{5}$ **52.** $13\sqrt[3]{2}$
53. $x\sqrt[4]{2}$ **54.** 4 **55.** $\dfrac{1}{5}$ **56.** 5 **57.** $\dfrac{1}{3}$ **58.** 16 **59.** $\dfrac{1}{81}$ **60.** $20x^{11/12}$ **61.** $3x^{1/4}$ **62.** $25x^4$ **63.** $\sqrt{y}$
64. $8x^3 + 10x^2 - 20x - 4$; degree 3 **65.** $8x^4 - 5x^3 + 6$; degree 4 **66.** $12x^3 + x^2 - 21x + 10$ **67.** $6x^2 - 7x - 5$
68. $16x^2 - 25$ **69.** $4x^2 + 20x + 25$ **70.** $9x^2 - 24x + 16$ **71.** $8x^3 + 12x^2 + 6x + 1$ **72.** $125x^3 - 150x^2 + 60x - 8$
73. $3x^2 + 16xy - 35y^2$ **74.** $9x^2 - 30xy + 25y^2$ **75.** $9x^4 + 12x^2y + 4y^2$ **76.** $49x^2 - 16y^2$ **77.** $a^3 - b^3$
78. $25y^2 - 4x^2 - 4x - 1$ **79.** $x^2 + 4xy + 4y^2 + 8x + 16y + 16$ **80.** $3x^2(5x + 1)$ **81.** $(x - 4)(x - 7)$
82. $(3x + 1)(5x - 2)$ **83.** $(8 - x)(8 + x)$ **84.** prime **85.** $3x^2(x - 5)(x + 2)$ **86.** $4x^3(5x^4 - 9)$ **87.** $(x + 3)(x - 3)^2$
88. $(4x - 5)^2$ **89.** $(x^2 + 4)(x + 2)(x - 2)$ **90.** $(y - 2)(y^2 + 2y + 4)$ **91.** $(x + 4)(x^2 - 4x + 16)$ **92.** $3x^2(x - 2)(x + 2)$
93. $(3x - 5)(9x^2 + 15x + 25)$ **94.** $x(x - 1)(x + 1)(x^2 + 1)$ **95.** $(x^2 - 2)(x + 5)$ **96.** $(x + 9 + y)(x + 9 - y)$
97. $\dfrac{16(1 + 2x)}{x^{3/4}}$ **98.** $(x + 2)(x - 2)(x^2 + 3)^{1/2}(-x^4 + x^2 + 13)$ **99.** $\dfrac{6(2x + 1)}{x^{3/2}}$ **100.** $x^2, x \neq -2$ **101.** $\dfrac{x - 3}{x - 6}, x \neq -6, 6$
102. $\dfrac{x}{x + 2}, x \neq -2$ **103.** $\dfrac{(x + 3)^3}{(x - 2)^2(x + 2)}, x \neq 2, -2$ **104.** $\dfrac{2}{x(x + 1)}, x \neq 0, 1, -1, -\dfrac{1}{3}$ **105.** $\dfrac{x + 3}{x - 4}, x \neq -3, 4, 2, 8$

106. $\dfrac{1}{x-3}, x \neq 3, -3$ **107.** $\dfrac{4x(x-1)}{(x+2)(x-2)}, x \neq 2, -2$ **108.** $\dfrac{2x^2-3}{(x-3)(x+3)(x-2)}, x \neq 3, -3, 2$

109. $\dfrac{11x^2-x-11}{(2x-1)(x+3)(3x+2)}, x \neq \frac{1}{2}, -3, -\frac{2}{3}$ **110.** $\dfrac{3x}{x-4}, x \neq 0, 4, -4$ **111.** $\dfrac{3x+8}{3x+10}, x \neq -3, -\frac{10}{3}$ **112.** $\dfrac{25\sqrt{25-x^2}}{(5-x)^2(5+x)^2}$

113. $\{6\}$ **114.** $\{-10\}$ **115.** $\{5\}$ **116.** $\{-13\}$ **117.** $\{-3\}$ **118.** $\{-1\}$ **119.** $\{2\}$ **120.** $\{2\}$ **121.** $\left\{\dfrac{72}{11}\right\}$ **122.** $\{-12\}$

123. $\left\{\dfrac{77}{15}\right\}$ **124. a.** 0   **b.** $\{2\}$ **125. a.** 5   **b.** $\varnothing$ **126. a.** $-1, 1$ **b.** all real numbers except 1 and -1 **127. a.** $-2, 4$

b. $\{7\}$ **128.** inconsistent equation   **129.** identity   **130.** conditional equation   **131.** 2005   **132.** $h = \dfrac{3V}{B}$ **133.** $M = \dfrac{f-F}{f}$

134. $g = \dfrac{T}{r+vt}$ **135.** $\{-4, 3\}$ **136.** $\{-5, 11\}$ **137.** $\left\{-8, \dfrac{1}{2}\right\}$ **138.** $\{-4, 0\}$ **139.** $\{-8, 8\}$ **140.** $\left\{\dfrac{4+3\sqrt{2}}{3}, \dfrac{4-3\sqrt{2}}{3}\right\}$

141. $100; (x+10)^2$ **142.** $\dfrac{9}{4}; \left(x - \dfrac{3}{2}\right)^2$ **143.** $\{3, 9\}$ **144.** $\left\{2 + \dfrac{\sqrt{3}}{3}, 2 - \dfrac{\sqrt{3}}{3}\right\}$ **145.** $\{1 + \sqrt{5}, 1 - \sqrt{5}\}$

146. $\left\{\dfrac{-2+\sqrt{10}}{2}, \dfrac{-2-\sqrt{10}}{2}\right\}$ **147.** $-36; 2$ no real solutions   **148.** $81; 2$ unequal real solutions   **149.** $\left\{\dfrac{1}{2}, 5\right\}$ **150.** $\left\{-2, \dfrac{10}{3}\right\}$

151. $\left\{\dfrac{7+\sqrt{37}}{6}, \dfrac{7-\sqrt{37}}{6}\right\}$ **152.** $\{-3, 3\}$ **153.** $\{-2, 8\}$ **154.** 20 weeks   **155.** 10 years   **156.** approximately 134 m

157.

158.

159.

160. $-2 < x \leq 3$

161. $-1.5 \leq x \leq 2$

162. $x > -1$

163. $[-2, \infty)$

164. $\left[\dfrac{3}{5}, \infty\right)$

165. $\left(-\infty, -\dfrac{21}{2}\right)$

166. $(-3, \infty)$

167. $(-\infty, -2]$

168. $(2, 3]$

169. $[-9, 6]$

170. $(-\infty, -6)$ or $(0, \infty)$

171. $(-\infty, -3]$ or $[-2, \infty)$

172. Most people sleep between 5.5 and 7.5 hours.   **173.** between 50°F and 77°F inclusively

Chapter P Test

1. $-7, -\dfrac{4}{5}, 0, 0.25, \sqrt{4}, \dfrac{22}{7}$ **2.** commutative property of addition   **3.** distributive property of multiplication over addition

4. 7.6×10^{-4} **5.** $85x + 2y - 15$ **6.** $\dfrac{5y^8}{x^6}$ **7.** $3r\sqrt{2}$ **8.** $11\sqrt{2}$ **9.** $\dfrac{3(5-\sqrt{2})}{23}$ **10.** $2x\sqrt[3]{2x}$ **11.** $\dfrac{x+3}{x-2}, x \neq 2, 1$

12. $\dfrac{1}{243}$ **13.** $2x^3 - 13x^2 + 26x - 15$ **14.** $25x^2 + 30xy + 9y^2$ **15.** $(x-3)(x-6)$ **16.** $(x^2+3)(x+2)$

17. $(5x-3)(5x+3)$ **18.** $(6x-7)^2$ **19.** $(y-5)(y^2+5y+25)$ **20.** $(x+5+3y)(x+5-3y)$ **21.** $\dfrac{2x+3}{(x+3)^{3/5}}$

22. $\dfrac{2(x+3)}{x+1}, x \neq 3, -1, -4, -3$ **23.** $\dfrac{x^2+2x+15}{(x+3)(x-3)}, x \neq 3, -3$ **24.** $\dfrac{11}{(x-3)(x-4)}, x \neq 3, 4$ **25.** $\dfrac{2x}{(x+2)(x+1)}$

26. $\dfrac{10x}{\sqrt{(x^2+5)^3}}$ **27.** $\{-1\}$ **28.** $\{-6\}$ **29.** $\{5\}$ **30.** $\left\{-\dfrac{1}{2}, 2\right\}$ **31.** $\left\{\dfrac{1-5\sqrt{3}}{3}, \dfrac{1+5\sqrt{3}}{3}\right\}$ **32.** $\{1 - \sqrt{5}, 1 + \sqrt{5}\}$

33. $\{6, 12\}$ **34.** $(-\infty, 12]$

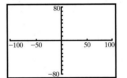

35. $\left[\dfrac{21}{8}, \infty\right)$

36. $\left[-7, \dfrac{13}{2}\right)$

37. $\left(-\infty, -\dfrac{5}{3}\right]$ or $\left[\dfrac{1}{3}, \infty\right)$

38. $h = \dfrac{3V}{lw}$ **39.** $x = x_1 + \dfrac{y - y_1}{m}$ **40.** 2004; very well   **41.** 10 ft

CHAPTER 1

Section 1.1

Check Point Exercises

1.

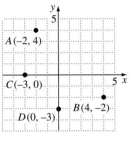

2.

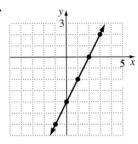

3. The minimum x-value is -100 and the maximum x-value is 100. The distance between consecutive tick marks is 50.
The minimum y-value is -80 and the maximum y-value is 80. The distance between consecutive tick marks is 10.

4. $21\dfrac{1}{2}$; 1900

Exercise Set 1.1

1.

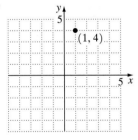

3.

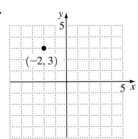

5.

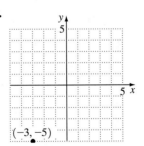

7.

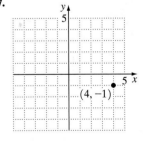

9.

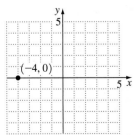

11.

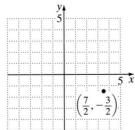

13.

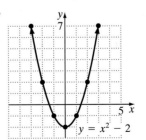

15.

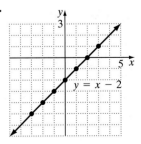

17.

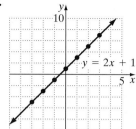

19.

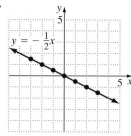

21.

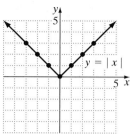

23.

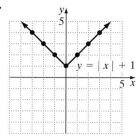

25.

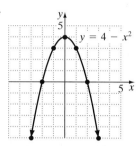

27.

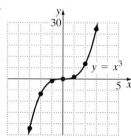

29. (c)   **31.** (b)   **33. a.** 2   **b.** -4

35. a. $1, -2$ **b.** 2   **37. a.** -1 **b.** None

39. $A(2, 7)$; The football is 7 feet high when it is 2 yards from the quarterback.

41. $C\left(6, 9\frac{1}{2}\right)$ **43.** 12 feet; 15 yards

45. 0–4 years   **47.** year 4; 8%

57. (c) gives a complete graph.

59. (c) gives a complete graph.

61. (a)   **63.** (b)

Section 1.2

Check Point Exercises

1. a. 6   **b.** $-\dfrac{7}{5}$ **2.** $y + 5 = 6(x - 2); y = 6x - 17$ **3.** $y + 1 = -5(x + 2); y = -5x - 11$

4.

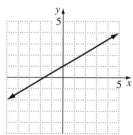

5.

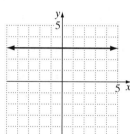

6.

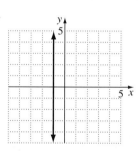

7. slope: $-\dfrac{1}{2}$; y-intercept: 2

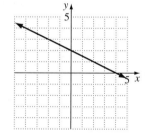

8. $y - 5 = 3(x + 2); y = 3x + 11$ **9.** 3

10. $\dfrac{2}{15} \approx 0.13$; The number of U.S. men living alone is projected to increase by 0.13 million each year.

11. $y = 2.32x + 180.1$; 319.3 million

Exercise Set 1.2

1. $\dfrac{3}{4}$; rises   **3.** $\dfrac{1}{4}$; rises   **5.** 0; horizontal   **7.** -5; falls   **9.** undefined; vertical   **11.** $y - 5 = 2(x - 3); y = 2x - 1$

13. $y - 5 = 6(x + 2); y = 6x + 17$ **15.** $y + 3 = -3(x + 2); y = -3x - 9$ **17.** $y - 0 = -4(x + 4); y = -4x - 16$

19. $y + 2 = -1\left(x + \dfrac{1}{2}\right); y = -x - \dfrac{5}{2}$ **21.** $y - 0 = \dfrac{1}{2}(x - 0); y = \dfrac{1}{2}x$ **23.** $y + 2 = -\dfrac{2}{3}(x - 6); y = -\dfrac{2}{3}x + 2$

25. using $(1, 2), y - 2 = 2(x - 1); y = 2x$ **27.** using $(-3, 0), y - 0 = 1(x + 3); y = x + 3$

29. using $(-3, -1), y + 1 = 1(x + 3); y = x + 2$ **31.** using $(-3, -2), y + 2 = \dfrac{4}{3}(x + 3); y = \dfrac{4}{3}x + 2$

33. using $(-3, -1), y + 1 = 0(x + 3); y = -1$ **35.** using $(2, 4), y - 4 = 1(x - 2); y = x + 2$

37. using $(0, 4), y - 4 = 8(x - 0); y = 8x + 4$

39. $m = 2; b = 1$

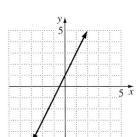

41. $m = -2; b = 1$

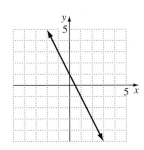

43. $m = \frac{3}{4}; b = -2$

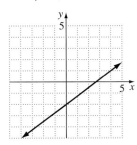

45. $m = -\frac{3}{5}; b = 7$

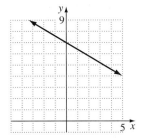

47.

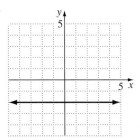

49.

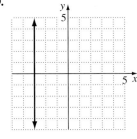

51.

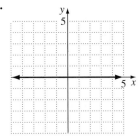

53. a. $y = -3x + 5$
b. $m = -3; b = 5$
c.

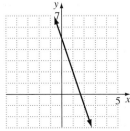

55. a. $y = -\frac{2}{3}x + 6$

b. $m = -\frac{2}{3}; b = 6$

c.

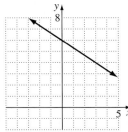

57. a. $y = 2x - 3$

b. $m = 2; b = -3$

c.

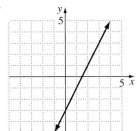

59. a. $x = 3$

b. m is undefined; no y-intercept

c.

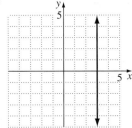

61. $y + 10 = -4(x + 8); y = -4x - 42$ **63.** $y + 3 = -5(x - 2); y = -5x + 7$ **65.** $y - 2 = \frac{2}{3}(x + 2); y = \frac{2}{3}x + \frac{10}{3}$

67. $y + 7 = -2(x - 4); y = -2x + 1$ **69.** $y = 15$ **71.** 111; The federal budget surplus is increasing $111 billion each year.
73. a. 16; In 1950, there were 16 workers per Social Security beneficiary.   **b.** −0.24; The number of workers per Social Security
beneficiary is decreasing by 0.24 worker each year.   **c.** $y = -0.24x + 16$ **d.** 1.6; 5   **75. a.** $y - 30 = 4(x - 2)$ **b.** $y = 4x + 22$
c. 74   **77.** $y = -2.3x + 255$, where x is the percentage of adult females who are literate and y is under-five mortality per thousand;
For each percent increase is adult female literacy, under-five mortality decreases by 2.3 per thousand.   **79.** $y = -0.7x + 60$; y represents
the percentage of U.S. adults who read a newspaper x years after 1995.   **81.** $y = -500x + 29,500; 4500$ shirts

93. $m = -3$

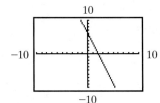

95. $m = \frac{3}{4}$

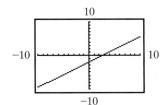

97. (c) is true.

99. a. m_1, m_3, m_2, m_4
b. b_2, b_1, b_4, b_3

Section 1.3

Check Point Exercises

1. 5   **2.** $\left(4, -\dfrac{1}{2}\right)$ **3.** $x^2 + y^2 = 16$ **4.** $(x - 5)^2 + (y + 6)^2 = 100$

5. center: $(-3, 1)$; radius: 2   **6.** $(x + 2)^2 + (y - 2)^2 = 9$

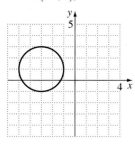   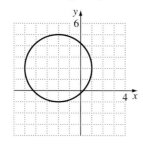

Exercise Set 1.3

1. 13   **3.** $2\sqrt{2} \approx 2.83$ **5.** 5   **7.** $\sqrt{29} \approx 5.39$ **9.** $4\sqrt{2} \approx 5.66$ **11.** $2\sqrt{5} \approx 4.47$ **13.** $2\sqrt{2} \approx 2.83$ **15.** $\sqrt{93} \approx 9.64$

17. $\sqrt{5} \approx 2.24$ **19.** $(4, 6)$ **21.** $(-4, -5)$ **23.** $\left(\dfrac{3}{2}, -6\right)$ **25.** $(-3, -2)$ **27.** $(1, 5\sqrt{5})$ **29.** $(2\sqrt{2}, 0)$ **31.** $x^2 + y^2 = 49$

33. $(x - 3)^2 + (y - 2)^2 = 25$ **35.** $(x + 1)^2 + (y - 4)^2 = 4$ **37.** $(x + 3)^2 + (y + 1)^2 = 3$ **39.** $(x + 4)^2 + (y - 0)^2 = 100$

41. center: $(0, 0)$ **43.** center: $(3, 1)$ **45.** center: $(-3, 2)$ **47.** center: $(-2, -2)$
     radius: 4   radius: 6   radius: 2   radius: 2

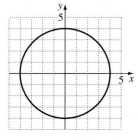   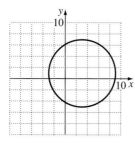   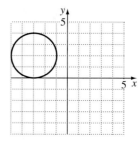   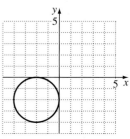

49. $(x + 3)^2 + (y + 1)^2 = 4$ **51.** $(x - 5)^2 + (y - 3)^2 = 64$ **53.** $(x + 4)^2 + (y - 1)^2 = 25$ **55.** $(x - 1)^2 + (y - 0)^2 = 16$
     center: $(-3, -1)$ center: $(5, 3)$ center: $(-4, 1)$ center: $(1, 0)$
     radius: 2   radius: 8   radius: 5   radius: 4

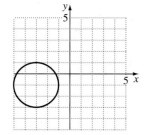   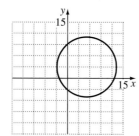   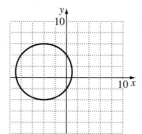   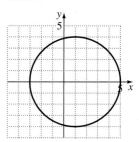

57. 0.5 hr; 31 min   **59.** $x^2 + (y - 82)^2 = 4624$ **69.**   **71.** (d) is true.

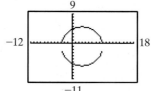

73. a. Distance between (x_1, y_1) and $\left(\dfrac{x_1 + x_2}{2}, \dfrac{y_1 + y_2}{2}\right)$

$$= \sqrt{\left(\dfrac{x_1 + x_2}{2} - x_1\right)^2 + \left(\dfrac{y_1 + y_2}{2} - y_1\right)^2}$$

$$= \sqrt{\left(\dfrac{x_1 + x_2 - 2x_1}{2}\right)^2 + \left(\dfrac{y_1 + y_2 - 2y_1}{2}\right)^2}$$

$$= \sqrt{\left(\dfrac{x_2 - x_1}{2}\right)^2 + \left(\dfrac{y_2 - y_1}{2}\right)^2}$$

$$= \sqrt{\dfrac{x_2^2 - 2x_1x_2 + x_1^2}{4} + \dfrac{y_2^2 - 2y_1y_2 + y_1^2}{4}}$$

$$= \sqrt{\dfrac{x_1^2 - 2x_1x_2 + x_2^2}{4} + \dfrac{y_1^2 - 2y_1y_2 + y_2^2}{4}}$$

$$= \sqrt{\left(\dfrac{x_1 - x_2}{2}\right)^2 + \left(\dfrac{y_1 - y_2}{2}\right)^2}$$

$$= \sqrt{\left(\dfrac{x_1 + x_2 - 2x_2}{2}\right)^2 + \left(\dfrac{y_1 + y_2 - 2y_2}{2}\right)^2}$$

$$= \sqrt{\left(\dfrac{x_1 + x_2}{2} - x_2\right)^2 + \left(\dfrac{y_1 + y_2}{2} - y_2\right)^2}$$

$$= \text{Distance between } (x_2, y_2) \text{ and } \left(\dfrac{x_1 + x_2}{2}, \dfrac{y_1 + y_2}{2}\right)$$

b. $\sqrt{\left(\dfrac{x_2 - x_1}{2}\right)^2 + \left(\dfrac{y_2 - y_1}{2}\right)^2} + \sqrt{\left(\dfrac{x_2 - x_1}{2}\right)^2 + \left(\dfrac{y_2 - y_1}{2}\right)^2}$

$$= 2\sqrt{\left(\dfrac{x_2 - x_1}{2}\right)^2 + \left(\dfrac{y_2 - y_1}{2}\right)^2}$$

$$= 2\sqrt{\dfrac{(x_2 - x_1)^2 + (y_2 - y_1)^2}{4}}$$

$$= \sqrt{(x_2 - x_1)^2 + (y_2 - y_1)^2}$$

$$= \text{Distance from } (x_1, y_1) \text{ to } (x_2, y_2)$$

75. $(x + 3)^2 + (y - 2)^2 = 16$; $x^2 + y^2 + 6x - 4y - 3 = 0$ **77.** $y + 4 = \dfrac{3}{4}(x - 3)$

Section 1.4

Check Point Exercises

1. domain: $\{5, 10, 15, 20, 25\}$; range: $\{12.8, 16.2, 18.9, 20.7, 21.8\}$ **2. a.** not a function **b.** function **3. a.** $y = 6 - 2x$; function
b. $y = \pm\sqrt{1 - x^2}$; not a function **4. a.** 42 **b.** $x^2 + 6x + 15$ **c.** $x^2 + 2x + 7$ **5. a.** $x^2 + 2xh + h^2 - 7x - 7h + 3$
b. $2x + h - 7, h \neq 0$ **6. a.** 28 **b.** 33 **7. a.** $(-\infty, \infty)$ **b.** $\{x | x \neq -7, x \neq 7\}$ **c.** $[3, \infty)$

Exercise Set 1.4

1. function; $\{1, 3, 5\}$; $\{2, 4, 5\}$ **3.** not a function; $\{3, 4\}$; $\{4, 5\}$ **5.** function; $\{-3, -2, -1, 0\}$; $\{-3, -2, -1, 0\}$
7. not a function; $\{1\}$; $\{4, 5, 6\}$ **9.** y is a function of x. **11.** y is a function of x. **13.** y is not a function of x.
15. y is not a function of x. **17.** y is a function of x. **19.** y is a function of x. **21. a.** 29 **b.** $4x + 9$ **c.** $-4x + 5$
23. a. 2 **b.** $x^2 + 12x + 38$ **c.** $x^2 - 2x + 3$ **25. a.** 13 **b.** 1 **c.** $x^4 - x^2 + 1$ **d.** $81a^4 - 9a^2 + 1$
27. a. 3 **b.** 7 **c.** $\sqrt{x} + 3$ **29. a.** $\dfrac{15}{4}$ **b.** $\dfrac{15}{4}$ **c.** $\dfrac{4x^2 - 1}{x^2}$ **31. a.** 1 **b.** -1 **c.** 1 **33.** $4, h \neq 0$ **35.** $3, h \neq 0$

37. $2x + h, h \neq 0$ **39.** $2x + h - 4, h \neq 0$ **41.** $0, h \neq 0$ **43.** $-\dfrac{1}{x(x + h)}, h \neq 0$ **45. a.** -1 **b.** 7 **c.** 19

47. a. 3 **b.** 3 **c.** 0 **49. a.** 8 **b.** 3 **c.** 6 **51.** $(-\infty, \infty)$ **53.** $\{x | x \neq 4\}$ **55.** $\{x | x \neq -4, x \neq 4\}$
57. $\{x | x \neq -3, x \neq 7\}$ **59.** $\{x | x \neq -3, x \neq -8\}$ **61.** $(-\infty, \infty)$ **63.** $[3, \infty)$ **65.** $(3, \infty)$ **67.** $[-7, \infty)$ **69.** $(-\infty, 12]$
71. $[2, 5)$ or $(5, \infty)$ **73.** $\{(1, 31), (2, 53), (3, 70), (4, 86), (5, 86)\}$; $\{1, 2, 3, 4, 5\}$; $\{31, 53, 70, 86\}$; Yes; each member of the domain
corresponds to exactly one member of the range. **75.** No; there are members of the domain that correspond to more than one member
of the range. **77.** 1713; There were 1713 gray wolves in the U.S. in 1990; Very well **79.** 19; In 1997, 19% of U.S. households were online;
Very well **81.** 5; The number of U.S. households online grew by 5% from 1998 to 2000; Okay **83.** 200; There were 200 thousand
lawyers in the United States in 1951. **85.** 1058; There were 1058 thousand or 1,058,000 lawyers in the United States in the year 2001.
87. 8873; A person earning \$40,000 owed \$8873 in taxes. **97.** $[1, \infty)$ **99.** $(-\infty, 5]$ **101.** Answers may vary.
103. $f(r_1) = 0$; r_1 is a solution to the equation $ax^2 + bx + c = 0$.

Section 1.5

Check Point Exercises

1. $(-3, 7), (-2, 2), (-1, -1),$
$(0, -2), (1, -1), (2, 2), (3, 7)$

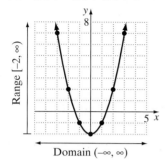

2. $f(4) = 1$; domain: $[0, 6)$; range: $(-2, 2]$
3. a. function **b.** function **c.** not a function
4. a. $f(10) \approx 16$ **b.** $x \approx 8$
5. increasing on $(-\infty, -1)$, decreasing on $(-1, 1)$, increasing on $(1, \infty)$
6. a. 1 **b.** 7 **c.** 4
7. a. 12 ft/sec **b.** 10 ft/sec **c.** 8.04 ft/sec
8. a. even **b.** odd **c.** neither

Exercise Set 1.5

1. $(-3, 11), (-2, 6), (-1, 3),$
$(0, 2), (1, 3), (2, 6), (3, 11)$

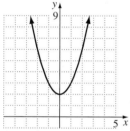

Domain: $(-\infty, \infty)$
Range: $[2, \infty)$

3. $(0, -1), (1, 0), (4, 1), (9, 2)$

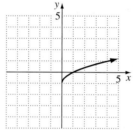

Domain: $[0, \infty)$
Range: $[-1, \infty)$

5. $(1, 0), (2, 1), (5, 2), (10, 3)$

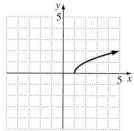

Domain: $[1, \infty)$
Range: $[0, \infty)$

7. $(-3, 2), (-2, 1), (-1, 0),$
$(0, -1), (1, 0), (2, 1), (3, 2)$

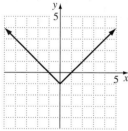

Domain: $(-\infty, \infty)$
Range: $[-1, \infty)$

9. $(-3, 4), (-2, 3), (-1, 2),$
$(0, 1), (1, 0), (2, 1), (3, 2)$

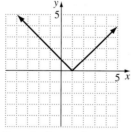

Domain: $(-\infty, \infty)$
Range: $[0, \infty)$

11. $(-3, 5), (-2, 5), (-1, 5),$
$(0, 5), (1, 5), (2, 5), (3, 5)$

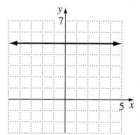

Domain: $(-\infty, \infty)$
Range: $\{5\}$

13. $(-2, -10), (-1, -3),$
$(0, -2), (1, -1), (2, 6)$

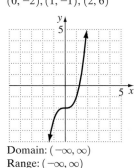

Domain: $(-\infty, \infty)$
Range: $(-\infty, \infty)$

15. a. $(-\infty, \infty)$ **b.** $[-4, \infty)$ **c.** -3 and 1 **d.** -3
17. a. $(-\infty, \infty)$ **b.** $[1, \infty)$ **c.** none **d.** 1 **e.** $f(-1) = 2$ and $f(3) = 4$
19. a. $[0, 5)$ **b.** $[-1, 5)$ **c.** 2 **d.** -1 **e.** $f(3) = 1$
21. a. $[0, \infty)$ **b.** $[1, \infty)$ **c.** none **d.** 1 **e.** $f(4) = 3$
23. a. $[-2, 6]$ **b.** $[-2, 6]$ **c.** 4 **d.** 4 **e.** $f(-1) = 5$
25. a. $(-\infty, \infty)$ **b.** $(-\infty, -2]$ **c.** none **d.** -2 **e.** $f(-4) = -5$ and $f(4) = -2$
27. a. $(-\infty, \infty)$ **b.** $(0, \infty)$ **c.** none **d.** 1
29. a. $\{-5, -2, 0, 1, 3\}$ **b.** $\{2\}$ **c.** none **d.** 2
31. function **33.** function **35.** not a function **37.** function
39. a. increasing: $(-1, \infty)$ **b.** decreasing: $(-\infty, -1)$ **c.** constant: none
41. a. increasing: $(0, \infty)$ **b.** decreasing: none **c.** constant: none
43. a. increasing: none **b.** decreasing: $(-2, 6)$ **c.** constant: none
45. a. increasing: $(-\infty, -1)$ **b.** decreasing: none **c.** constant: $(-1, \infty)$
47. a. increasing: $(-\infty, 0)$ or $(1.5, 3)$ **b.** decreasing: $(0, 1.5)$ or $(3, \infty)$ **c.** constant: none

49. a. increasing: $(-2, 4)$ **b.** decreasing: none **c.** constant: $(-\infty, -2)$ or $(4, \infty)$ **51. a.** $0; f(0) = 4$ **b.** $-3, 3; f(-3) = f(3) = 0$

53. a. $-2; f(-2) = 21$ **b.** $1; f(1) = -6$ **55.** 3 **57.** 10 **59.** $\dfrac{1}{5}$ **61. a.** 70 ft/sec **b.** 65 ft/sec **c.** 60.1 ft/sec **d.** 60.01 ft/sec

63. odd **65.** neither **67.** even **69.** even **71.** even **73.** odd **75.** even **77.** odd **79.** 1 **81.** 0 **83.** -3
85. $f(60) \approx 3.1$; In 1960, Jewish Americans made up about 3.1% of the U.S. population. **87.** $x \approx 19$ and $x \approx 64$; In 1919 and 1964, Jewish Americans made up about 3% of the U.S. population. **89.** 1940; 3.7% **91.** Each year corresponds to only one percentage.
93. increasing: $(45, 74)$; decreasing: $(16, 45)$; The number of accidents occurring per 50,000 miles driven increases with age starting at age 45, while it decreases with age starting at age 16.
95. Answers will vary; an example is 16 and 74 years old. For those ages, the number of accidents is 526.4 per 50 million miles.

97.

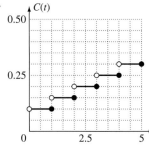

99. Answers may vary.

111.

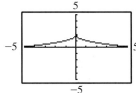

The number of doctor visits decreases during childhood and then increases as you get older. The minimum is approximately $(20.29, 3.99)$, which means that the minimum number of annual doctor visits, about 4, occurs at around age 20.

113.

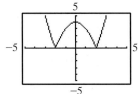

Increasing: $(-2, 0)$ or $(2, \infty)$
Decreasing: $(-\infty, -2)$ or $(0, 2)$

115.

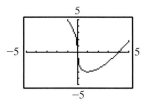

Increasing: $(1, \infty)$
Decreasing: $(-\infty, 1)$

117.

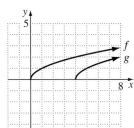

Increasing: $(-\infty, 0)$
Decreasing: $(0, \infty)$

119. (c) is true. **121.** Answers may vary.

123.

Weight at least	Cost
0 oz.	$0.37
1	0.60
2	0.83
3	1.06
4	1.29

Section 1.6

Check Point Exercises

1.

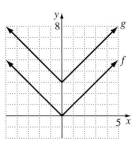

2.

3.

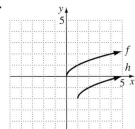

4.

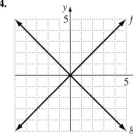

5.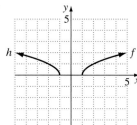

6.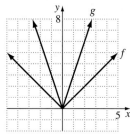

7.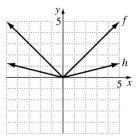

8.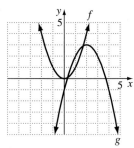

Exercise Set 1.6

1.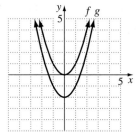

3.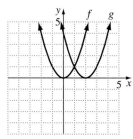

5.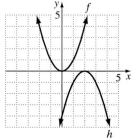

7.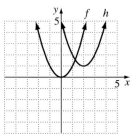

9.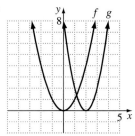

11.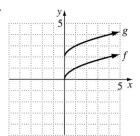

13.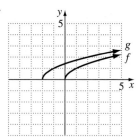

15.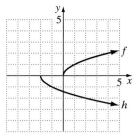

17.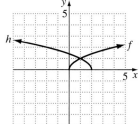

19.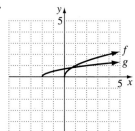

21.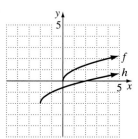

23.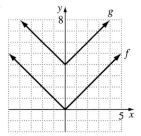

25.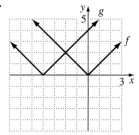

27.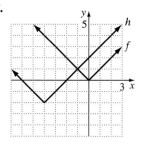

29.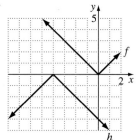

31.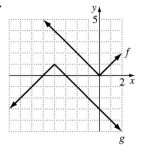

33.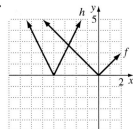

35.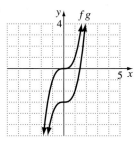

37.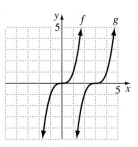

39.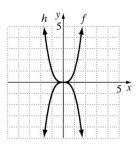

41.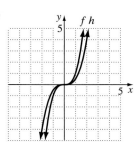

43.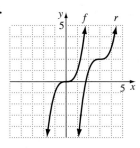

45.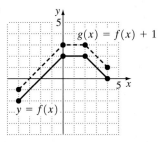

47.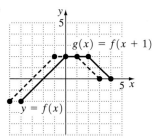

49.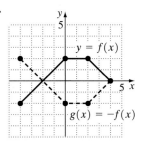

51.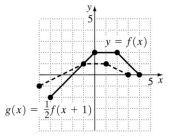

53. $f(x) = \sqrt{x-2}$ **55.** $f(x) = (x+1)^2 - 4$

57. a. First, vertically stretch the graph of $f(x) = \sqrt{x}$ by the factor 2.9; then, vertically shift the result up 20.1 units.
  b. 40.2 in.; Very well.
  c. 0.9 in. per month
  d. 0.2 in. per month; This is a much smaller rate of change. The graph is not as steep between 50 and 60 as it is between 0 and 10.

65. a. 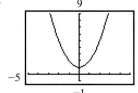   **b.** 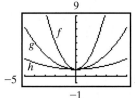   **c.** Answers may vary.   **67.** $g(x) = -(x+4)^2$
  d. Answers may vary.   **69.** $g(x) = -\sqrt{x-2} + 2$
  e. Answers may vary.   **71.** $(-a, b)$
                                73. $(a+3, b)$

Section 1.7

Check Point Exercises

1. a. $(f+g)(x) = 3x^2 + 6x + 6$ **b.** $(f+g)(4) = 78$ **2. a.** $(f+g)(x) = \sqrt{x-3} + \sqrt{x+1}$ **b.** $[3, \infty)$

3. a. $(f-g)(x) = -x^2 + x - 4$ **b.** $(fg)(x) = x^3 - 5x^2 - x + 5$ **c.** $\left(\dfrac{f}{g}\right)(x) = \dfrac{x-5}{x^2-1}, \, x \neq \pm 1$

4. a. $(f \circ g)(x) = 5x^2 + 1$ **b.** $(g \circ f)(x) = 25x^2 + 60x + 35$ **5. a.** $(f \circ g)(x) = \dfrac{4x}{2x+1}$ **b.** $\left\{ x \,\middle|\, x \neq 0 \text{ and } x \neq -\dfrac{1}{2} \right\}$

6. $f(x) = \sqrt{x}$ and $g(x) = x^2 + 5$, then $h(x) = (f \circ g)(x)$.

Exercise Set 1.7

1. a. $(f + g)(x) = 2x^2 + 3x + 2$ **b.** $(f + g)(4) = 46$ **3. a.** $(f + g)(x) = \sqrt{x - 6} + \sqrt{x + 2}$ **b.** Domain: $[6, \infty)$

5. $(f + g)(x) = 3x + 2$; Domain: $(-\infty, \infty)$; $(f - g)(x) = x + 4$; Domain: $(-\infty, \infty)$; $(fg)(x) = 2x^2 + x - 3$; Domain: $(-\infty, \infty)$;

Domain: $\left(\dfrac{f}{g}\right)(x) = \dfrac{2x + 3}{x - 1}$; Domain: $(-\infty, 1)$ or $(1, \infty)$ **7.** $(f + g)(x) = 3x^2 + x - 5$; Domain: $(-\infty, \infty)$;

$(f - g)(x) = -3x^2 + x - 5$; Domain: $(-\infty, \infty)$; $(fg)(x) = 3x^3 - 15x^2$; Domain: $(-\infty, \infty)$; $\left(\dfrac{f}{g}\right)(x) = \dfrac{x - 5}{3x^2}$;

Domain: $(-\infty, 0)$ or $(0, \infty)$ **9.** $(f + g)(x) = 2x^2 - 2$; Domain: $(-\infty, \infty)$; $(f - g)(x) = 2x^2 - 2x - 4$; Domain: $(-\infty, \infty)$;

$(fg)(x) = 2x^3 + x^2 - 4x - 3$; Domain: $(-\infty, \infty)$; $\left(\dfrac{f}{g}\right)(x) = 2x - 3$; Domain: $(-\infty, -1)$ or $(-1, \infty)$ **11.** $(f + g)(x) = \sqrt{x} + x - 4$;

Domain: $[0, \infty)$; $(f - g)(x) = \sqrt{x} - x + 4$; Domain: $[0, \infty)$; $(fg)(x) = \sqrt{x}(x - 4)$; Domain: $[0, \infty)$; $\left(\dfrac{f}{g}\right)(x) = \dfrac{\sqrt{x}}{x - 4}$;

Domain: $[0, 4)$ or $(4, \infty)$ **13.** $(f + g)(x) = \dfrac{2x + 2}{x}$; Domain: $(-\infty, 0)$ or $(0, \infty)$; $(f - g)(x) = 2$; Domain: $(-\infty, 0)$ or $(0, \infty)$;

$(fg)(x) = \dfrac{2x + 1}{x^2}$; Domain: $(-\infty, 0)$ or $(0, \infty)$; $\left(\dfrac{f}{g}\right)(x) = 2x + 1$; Domain: $(-\infty, 0)$ or $(0, \infty)$ **15.** $(f + g)(x) = \sqrt{x + 4} + \sqrt{x - 1}$;

Domain: $[1, \infty)$; $(f - g)(x) = \sqrt{x + 4} - \sqrt{x - 1}$; Domain: $[1, \infty)$; $(fg)(x) = \sqrt{x^2 + 3x - 4}$; Domain: $[1, \infty)$; $\left(\dfrac{f}{g}\right)(x) = \dfrac{\sqrt{x + 4}}{\sqrt{x - 1}}$;

Domain: $(1, \infty)$ **17. a.** $(f \circ g)(x) = 2x + 14$ **b.** $(g \circ f)(x) = 2x + 7$ **c.** $(f \circ g)(2) = 18$ **19. a.** $(f \circ g)(x) = 2x + 5$

b. $(g \circ f)(x) = 2x + 9$ **c.** $(f \circ g)(2) = 9$ **21. a.** $(f \circ g)(x) = 20x^2 - 11$ **b.** $(g \circ f)(x) = 80x^2 - 120x + 43$

c. $(f \circ g)(2) = 69$ **23. a.** $(f \circ g)(x) = x^4 - 4x^2 + 6$ **b.** $(g \circ f)(x) = x^4 + 4x^2 + 2$ **c.** $(f \circ g)(2) = 6$

25. a. $(f \circ g)(x) = \sqrt{x - 1}$ **b.** $(g \circ f)(x) = \sqrt{x} - 1$ **c.** $(f \circ g)(2) = 1$ **27. a.** $(f \circ g)(x) = x$ **b.** $(g \circ f)(x) = x$

c. $(f \circ g)(2) = 2$ **29. a.** $(f \circ g)(x) = \dfrac{2x}{1 + 3x}$ **b.** $\left\{x \middle| x \neq 0 \text{ and } x \neq -\dfrac{1}{3}\right\}$ **31. a.** $(f \circ g)(x) = \dfrac{4}{4 + x}$ **b.** $\{x | x \neq 0 \text{ and } x \neq -4\}$

33. a. $(f \circ g)(x) = \sqrt{x + 3}$ **b.** $\{x | x \geq -3\}$ **35. a.** $(f \circ g)(x) = 5 - x$ **b.** $\{x | x \leq 1\}$ **37. a.** $(f \circ g)(x) = 8 - x^2$

b. $\{x | x < -2 \text{ or } x > 2\}$ **39.** $f(x) = x^4, g(x) = 3x - 1$ **41.** $f(x) = \sqrt[3]{x}, g(x) = x^2 - 9$ **43.** $f(x) = |x|, g(x) = 2x - 5$

45. $f(x) = \dfrac{1}{x}, g(x) = 2x - 3$ **47.** 0   **49.** 10   **51.** 2   **53.** 0   **55.** 4   **57.** −6   **59.** 20; In 2000, veterinary costs in the U.S.

for dogs and cats were about $20 billion.   **61.** Domain of $D + C = \{1983, 1987, 1991, 1996, 2000\}$

63. $f + g$ represents the total world population in year x.   **65.** $(f + g)(2000) \approx 6$ billion people   **67.** $(R - C)(20,000) = -200,000$;

The company lost $200,000 since costs exceeded revenues; $(R - C)(30,000) = 0$; The company broke even since revenues equaled cost;

$(R - C)(40,000) = 200,000$; The company made a profit of $200,000.   **69. a.** f gives the price of the computer after a $400 discount.

g gives the price of the computer after a 25% discount.   **b.** $(f \circ g)(x) = 0.75x - 400$. This models the price of a computer after first a

25% discount and then a $400 discount.   **c.** $(g \circ f)(x) = 0.75(x - 400)$. This models the price of a computer after first a $400 discount

and then a 25% discount.   **d.** The function $f \circ g$ models the greater discount, since the 25% discount is taken on the regular price first.

77.

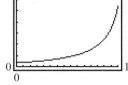

The per capita costs are
increasing over time.

79.

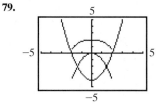

Domain of $f \circ g$ is $[-2, 2]$.

81. Assume f and g are even;
then $f(-x) = f(x)$ and $g(-x) = g(x)$.
$(fg)(-x) = f(-x) \cdot g(-x)$
$= f(x) \cdot g(x) = (fg)(x)$,
so fg is even.

83.

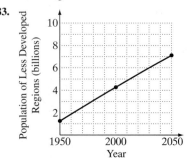

Section 1.8

Check Point Exercises

1. $f(g(x)) = x; g(f(x)) = x; f$ and g are inverses.   **2.** $f(g(x)) = x; g(f(x)) = x; f$ and g are inverses.

3. $f^{-1}(x) = \dfrac{x - 7}{2}$ **4.** $f^{-1}(x) = \sqrt[3]{\dfrac{x + 1}{4}}$ **5.** (b) and (c) have inverse functions.   **6.**

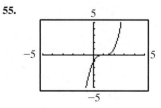

Exercise Set 1.8

1. $f(g(x)) = x; g(f(x)) = x; f$ and g are inverses.   **3.** $f(g(x)) = x; g(f(x)) = x; f$ and g are inverses.

5. $f(g(x)) = \dfrac{5x - 56}{9}; g(f(x)) = \dfrac{5x - 4}{9}; f$ and g are not inverses.   **7.** $f(g(x)) = x; g(f(x)) = x; f$ and g are inverses.

9. $f(g(x)) = x; g(f(x)) = x; f$ and g are inverses.   **11.** $f^{-1}(x) = x - 3$ **13.** $f^{-1}(x) = \dfrac{x}{2}$ **15.** $f^{-1}(x) = \dfrac{x - 3}{2}$

17. $f^{-1}(x) = \sqrt[3]{x - 2}$ **19.** $f^{-1}(x) = \sqrt[3]{x} - 2$ **21.** $f^{-1}(x) = \dfrac{1}{x}$ **23.** $f^{-1}(x) = x^2, x \geq 0$ **25.** $f^{-1}(x) = \sqrt{x - 1}; x \geq 1$

27. $f^{-1}(x) = \dfrac{3x + 1}{x - 2}; x \neq 2$ **29.** The function is not one-to-one, so it does not have an inverse function.

31. The function is not one-to-one, so it does not have an inverse function.   **33.** The function is one-to-one, so it does have an inverse function.

35.

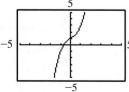

37.

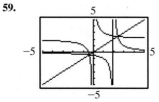

39. a. $f = \{(\text{Zambia}, -7.3), (\text{Colombia}, -4.5), (\text{Poland}, -2.8),$
$(\text{Italy}, -2.8), (\text{United States}, -1.9)\}$

b. $f^{-1} = \{(-7.3, \text{Zambia}), (-4.5, \text{Colombia}), (-2.8, \text{Poland}),$
$(-2.8, \text{Italy}), (-1.9, \text{United States})\}$; No; One member of
the domain, -2.8, corresponds to more than one member
of the range, Poland and Italy.

41. a. f is a one-to-one function.   **b.** $f^{-1}(0.25)$ is the number of people in a room for a 25% probability of two people sharing a birthday.
$f^{-1}(0.5)$ is the number of people in a room for a 50% probability of two people sharing a birthday. $f^{-1}(0.7)$ is the number of people in a
room for a 70% probability of two people sharing a birthday.

43. $f(g(x)) = \dfrac{9}{5}\left[\dfrac{5}{9}(x - 32)\right] + 32 = x$ and $g(f(x)) = \dfrac{5}{9}\left[\left(\dfrac{9}{5}x + 32\right) - 32\right] = x.$

49.

not one-to-one

51.

one-to-one

53.

not one-to-one

55.

one-to-one

57.

one-to-one

59.

f and g are inverses.

61. (d) is true.

63. $(f \circ f)(x) = x,$ so f is its own inverse.

65. 7

Section 1.9

Check Point Exercises

1. a. $f(x) = 0.08x + 15$ **b.** $g(x) = 0.12x + 3$ **c.** 300 min **2. a.** $N(x) = -100x + 18,000$ **b.** $R(x) = -100x^2 + 18,000x$

3. a. $V(x) = x(15 - 2x)(8 - 2x)$ **b.** $(0, 4)$ **4.** $x(100 - x) = 100x - x^2 \text{ ft}^2$ **5.** $A(r) = 2\pi r^2 + \dfrac{2000}{r}$

6. $I(x) = 0.07x + 0.09(25,000 - x)$ **7.** $d = \sqrt{x^6 + x^2}$

Exercise Set 1.9

1. a. $f(x) = 200 + 0.15x$ **b.** 800 mi **3. a.** $M(x) = 239.4 - 0.3x$ **b.** 2152 **5. a.** $f(x) = 1.25x$ **b.** $g(x) = 21 + 0.5x$
c. 28 times; \$35 **7. a.** $f(x) = 100 + 0.8x$ **b.** $g(x) = 40 + 0.9x$ **c.** \$600; \$580 **9. a.** $N(x) = -500x + 40,000$
b. $R(x) = -500x^2 + 40,000x$ **11. a.** $N(x) = -50x + 16,500$ **b.** $R(x) = -50x^2 + 16,500x$ **13. a.** $Y(x) = -4x + 520$
b. $T(x) = -4x^2 + 520x$ **15. a.** $V(x) = x(24 - 2x)^2$ **b.** $V(2) = 800$; If a 2-in. square is cut from each corner, the volume is
800 cubic in.; $V(3) = 972$; if a 3-in. square is cut from each corner, the volume is 972 cubic in.; $V(4) = 1024$; if a 4-in. square is cut from
each corner, the volume is 1024 cubic in.; $V(5) = 980$; if a 5-in. square is cut from each corner, the volume is 980 cubic in.; $V(6) = 864$;
if a 6-in. square is cut from each corner, the volume is 864 cubic in. Initially, as x increases, V increases. When $x = 4$, V is a maximum.
As x increases beyond 4, V decreases. **c.** $(0, 12)$ **17.** $A(x) = x(20 - 2x)$ **19.** $P(x) = x(66 - x)$ **21.** $A(x) = x(400 - x)$

23. Let x be the length of the side perpendicular to the canal. $A(x) = x(800 - 2x)$ **25.** $A(x) = \dfrac{x(1000 - 2x)}{3}$

27. $A(r) = r(440 - \pi r)$ **29.** Let x be the length of the interior wall. $C(x) = 475x + \dfrac{1,400,000}{x}$ **31.** $A(x) = \dfrac{40}{x} + x^2$

33. $V(x) = 300x^2 - 4x^3$ **35. a.** $I(x) = 0.15x + 0.07(50,000 - x)$ **b.** \$31,250 at 15%, \$18,750 at 7%
37. $I(x) = 0.12x - 0.05(8000 - x)$ **39.** $d(x) = \sqrt{x^4 - 7x^2 + 16}$ **41.** $d(x) = \sqrt{x^2 - x + 1}$ **43. a.** $A(x) = 2x\sqrt{4 - x^2}$
b. $P(x) = 4x + 2\sqrt{4 - x^2}$ **45.** $f(x) = \sqrt{x^2 + 36} + \sqrt{x^2 - 20x + 164}$ **53–57.** Answers may vary.
59. $T(x) = \dfrac{\sqrt{x^2 + 4}}{2} + \dfrac{6 - x}{5}$ **61.** $V(h) = \dfrac{\pi}{12}h^3$

Chapter 1 Review Exercises

1.

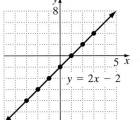

2.

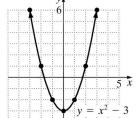

3.

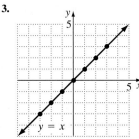

4.

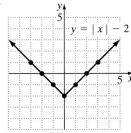

5.

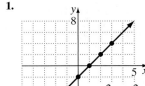

6. x-intercept: -2; y-intercept: 2 **7.** x-intercepts: 2, -2; y-intercept: -4
8. x-intercept: 5; y-intercept: none **9.** 20% **10.** 85 years
11. The percentage of Americans with Alzheimer's disease increases with age.

12. $m = -\dfrac{1}{2}$; falls **13.** $m = 1$; rises **14.** $m = 0$; horizontal **15.** $m = $ undefined; vertical

16. $y - 2 = -6(x + 3)$; $y = -6x - 16$ **17.** using $(1, 6)$, $y - 6 = 2(x - 1)$; $y = 2x + 4$

18. Slope: $\dfrac{2}{5}$; y-intercept: -1 **19.** Slope: -4; y-intercept: 5 **20.** Slope: $-\dfrac{2}{3}$; y-intercept: -2 **21.** Slope: 0; y-intercept: 4

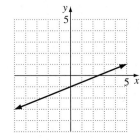

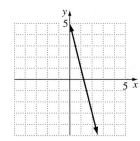

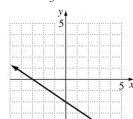

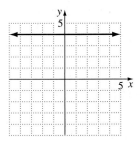

22. a. $y - 480 = 40(x - 2)$ **b.** $y = 40x + 400$ **c.** \$1200 billion.

23. Answers may vary. Possible answer:

a.

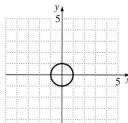

b. $y = -\dfrac{1}{40}x + 7$

c. 3.75

24. $y + 7 = -3(x - 4); y = -3x + 5$
25. $y - 6 = -3(x + 3); y = -3x - 3$
26. 13
27. $2\sqrt{2} \approx 2.83$
28. $(-5, 5)$
29. $\left(-\dfrac{11}{2}, -2\right)$
30. $x^2 + y^2 = 9$
31. $(x + 2)^2 + (y - 4)^2 = 36$

32. Center: $(0, 0)$; radius: 1

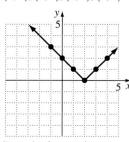

33. Center: $(-2, 3)$; radius: 3

34. Center: $(2, -1)$; radius: 3

35. Function; Domain: $\{2, 3, 5\}$; Range: $\{7\}$
36. Function; Domain: $\{1, 2, 13\}$; Range: $\{10, 500, \pi\}$
37. Not a function; Domain: $\{12, 14\}$; Range: $\{13, 15, 19\}$
38. y is a function of x.
39. y is a function of x.
40. y is not a function of x.

41. a. $f(4) = -23$ **b.** $f(x + 3) = -7x - 16$ **c.** $f(-x) = 5 + 7x$ **42. a.** $g(0) = 2$ **b.** $g(-2) = 24$
c. $g(x - 1) = 3x^2 - 11x + 10$ **d.** $g(-x) = 3x^2 + 5x + 2$ **43. a.** $g(13) = 3$ **b.** $g(0) = 4$ **c.** $g(-3) = 7$
44. a. $f(-2) = -1$ **b.** $f(1) = 1$ **c.** $f(2) = 3$ **45.** 8   **46.** $2x + h - 13$ **c.** $f(2) = 3$ **47.** $(-\infty, \infty)$
48. $(-\infty, 7)$ or $(7, \infty)$ **49.** $(-\infty, 4]$ **50.** $(-\infty, -1)$ or $(-1, 1)$ or $(1, \infty)$ **51.** $[2, 5)$ or $(5, \infty)$

52. Ordered pairs: $(-1, 9), (0, 4)$, $(1, 1), (2, 0), (3, 1), (4, 4)$.

Domain: $(-\infty, \infty)$
Range: $[0, \infty)$

53. Ordered pairs: $(-1, 3), (0, 2)$, $(1, 1), (2, 0), (3, 1), (4, 2)$.

Domain: $(-\infty, \infty)$
Range: $[0, \infty)$

54. a. Domain: $[-3, 5)$
b. Range: $[-5, 0]$
c. x-intercept: -3
d. y-intercept: -2
e. increasing: $(-2, 0)$ or $(3, 5)$
decreasing: $(-3, -2)$ or $(0, 3)$
f. $f(-2) = -3$ and $f(3) = -5$
55. a. Domain: $(-\infty, \infty)$
b. Range: $(-\infty, \infty)$
c. x-intercepts: -2 and 3
d. y-intercept: 3
e. increasing: $(-5, 0)$; decreasing: $(-\infty, -5)$ or $(0, \infty)$
f. $f(-2) = 0$ and $f(6) = -3$

56. a. Domain: $(-\infty, \infty)$ **b.** Range: $[-2, 2]$ **c.** x-intercept: 0   **d.** y-intercept: 0   **e.** increasing: $(-2, 2)$; constant: $(-\infty, -2)$ or $(2, \infty)$
f. $f(-9) = -2$ and $f(14) = 2$ **57. a.** $0; f(0) = -2$ **b.** $-2, 3; f(-2) = -3, f(3) = -5$ **58. a.** $0; f(0) = 3$ **b.** $-5; f(-5) = -6$
59. not a function   **60.** function   **61.** function   **62.** not a function   **63.** 10   **64. a.** 32 ft/sec   **b.** -32 ft/sec
c. The positive sign in part (a) means that the ball is moving up on $(0, 2)$. The negative sign in part (b) means that the ball is moving down on $(2, 4)$.   **65.** about $1167   **66.** odd; symmetric with respect to the origin   **67.** even; symmetric with respect to the y-axis
68. odd; symmetric with respect to the origin   **69. a.** yes; The graph passes the vertical line test.   **b.** Decreasing: $(3, 12)$;
The vulture descended.   **c.** Constant: $(0, 3)$ and $(12, 17)$; The vulture's height held steady during the first 3 seconds and the vulture was on the ground for 5 seconds.   **d.** Increasing: $(17, 30)$; The vulture was ascending.

70.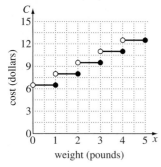

71.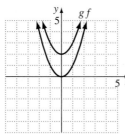

72.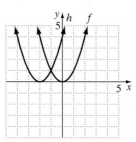

73.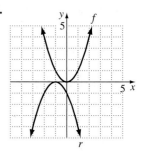

74.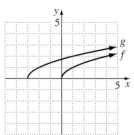

75.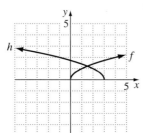

76.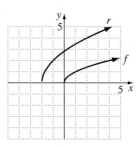

77.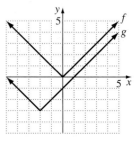

78.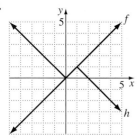

79.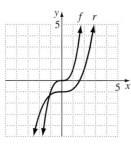

80.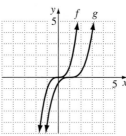

81.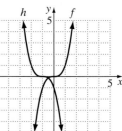

82.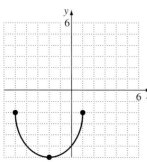

83.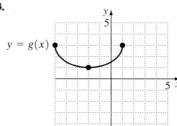

84.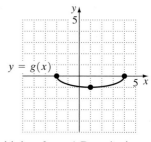

85.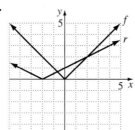

86. $(f + g)(x) = 4x - 6$; Domain: $(-\infty, \infty)$; $(f - g)(x) = 2x + 4$; Domain: $(-\infty, \infty)$;
$(fg)(x) = 3x^2 - 16x + 5$; Domain: $(-\infty, \infty)$; $\left(\dfrac{f}{g}\right)(x) = \dfrac{3x - 1}{x - 5}$; Domain: $(-\infty, 5)$ or $(5, \infty)$

87. $(f + g)(x) = 2x^2 + x$; Domain: $(-\infty, \infty)$; $(f - g)(x) = x + 2$; Domain: $(-\infty, \infty)$;
$(fg)(x) = x^4 + x^3 - x - 1$; Domain: $(-\infty, \infty)$; $\left(\dfrac{f}{g}\right)(x) = \dfrac{x^2 + x + 1}{x^2 - 1}$;
Domain: $(-\infty, -1)$ or $(-1, 1)$ or $(1, \infty)$

88. $(f + g)(x) = \sqrt{x + 7} + \sqrt{x - 2}$; Domain: $[2, \infty)$; $(f - g)(x) = \sqrt{x + 7} - \sqrt{x - 2}$;
Domain: $[2, \infty)$; $(fg)(x) = \sqrt{x^2 + 5x - 14}$; Domain: $[2, \infty)$; $\left(\dfrac{f}{g}\right)(x) = \dfrac{\sqrt{x + 7}}{\sqrt{x - 2}}$;
Domain: $(2, \infty)$

89. a. $(f \circ g)(x) = 16x^2 - 8x + 4$ **b.** $(g \circ f)(x) = 4x^2 + 11$ **c.** $(f \circ g)(3) = 124$ **90. a.** $(f \circ g)(x) = \sqrt{x + 1}$

b. $(g \circ f)(x) = \sqrt{x} + 1$ **c.** $(f \circ g)(3) = 2$ **91. a.** $\dfrac{1 + x}{1 - 2x}$ **b.** $\left\{x \middle| x \neq 0 \text{ and } x \neq \dfrac{1}{2}\right\}$ **92. a.** $\sqrt{x + 2}$ **b.** $\{x | x \geq -2\}$

93. $f(x) = x^4, g(x) = x^2 + 2x - 1$ **94.** $f(x) = \sqrt[3]{x}, g(x) = 7x + 4$ **95.** $f(g(x)) = x - \dfrac{7}{10}; g(f(x)) = x - \dfrac{7}{6};$

f and g are not inverses of each other.   **96.** $f(g(x)) = x; g(f(x)) = x; f$ and g are inverses of each other.   **97.** $f^{-1}(x) = \dfrac{x+3}{4}$

98. $f^{-1}(x) = x^2 - 2$ for $x \geq 0$ **99.** $f^{-1}(x) = \dfrac{\sqrt[3]{x-1}}{2}$ **100.** Inverse function exists.   **101.** Inverse function does not exist.

102. Inverse function exists.   **103.** Inverse function does not exist.

104.

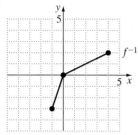

105. a. $W(x) = 15x + 567$ **b.** 2009
106. a. $f(x) = 0.05x + 15$ **b.** $g(x) = 0.07x + 5$ **c.** 500 min
107. a. $N(x) = 640 - 2x$ **b.** $R(x) = x(640 - 2x)$
108. a. $V(x) = x(16 - 2x)(24 - 2x)$ **b.** Domain: $(0, 8)$
109. $A(x) = x\left(\dfrac{400 - 3x}{2}\right)$
110. $A(x) = 2x^2 + \dfrac{32}{x}$
111. $T(x) = 0.08x + 0.12(10{,}000 - x)$

Chapter 1 Test

1.

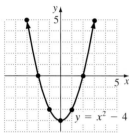

2. x-intercept: 2; y-intercept: 3   **3.** 1992; 7.8%
4. using $(2, 1)$, $y - 1 = 3(x - 2)$; $y = 3x - 5$
5. $y - 6 = 4(x + 4)$; $y = 4x + 22$ **6. a.** $y = 42x + 320$ **b.** \$866
7. Center: $(-2, 3)$; radius: 4   **8.** b, c, d
9. $f(x - 1) = x^2 - 4x + 8$
10. $g(-1) = 4; g(7) = 2$
11. Domain: $(-\infty, 4]$
12. $2x + h + 11$
13. a. $f(4) - f(-3) = 5$ **b.** Domain: $(-5, 6]$
c. Range: $[-4, 5]$ **d.** Increasing: $(-1, 2)$
e. Decreasing: $(-5, -1)$ or $(2, 6)$ **f.** 2; $f(2) = 5$
g. $-1; f(-1) = -4$ **h.** $-4, 1,$ and 5   **i.** -3
14. 48

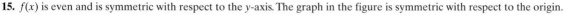

15. $f(x)$ is even and is symmetric with respect to the y-axis. The graph in the figure is symmetric with respect to the origin.
16. The graph of f is shifted 3 to the right to obtain the graph of g. Then the graph of g is stretched by a factor of 2 and reflected about the x-axis to obtain the graph of h.

17.

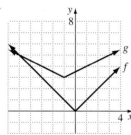

18. $(f - g)(x) = x^2 - 2x - 2$.   **19.** $\left(\dfrac{f}{g}\right)(x) = \dfrac{x^2 + 3x - 4}{5x - 2}$; Domain: $\left(-\infty, \dfrac{2}{5}\right)$ or $\left(\dfrac{2}{5}, \infty\right)$
20. $(f \circ g)(x) = 25x^2 - 5x - 6$ **21.** $(g \circ f)(x) = 5x^2 + 15x - 22$
22. $f(g(2)) = 84$ **23.** $\dfrac{7x}{2 - 4x}$; $\left\{x \mid x \neq 0 \text{ or } x \neq \dfrac{1}{2}\right\}$ **24.** $f(x) = x^7, g(x) = 2x + 13$
25. $f^{-1}(x) = x^2 + 2$ for $x \geq 0$
26. a. The graph of f passes the Horizontal Line Test.
b. $f(80) = 2000$
c. $f^{-1}(2000)$ is the income, in thousands of dollars, for those who give \$2000 to charity.

27.

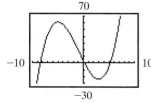

a. not one-to-one; f does not pass the Horizontal Line Test
b. neither; f is not symmetric about the origin or the y-axis
c. $(-\infty, \infty)$
d. Increasing: $(-\infty, -5)$ or $(3, \infty)$
e. Decreasing: $(-5, 3)$
f. $-5; f(-5) = \dfrac{184}{3}$
g. $3; f(3) = -24$

28. a. $T(x) = 41.78 - 0.19x$ **b.** 2004   **29. a.** $Y(x) = 50 - 1.5(x - 30)$ **b.** $T(x) = x(50 - 1.5(x - 30))$

30. $A(x) = x(300 - x)$ **31.** $A(x) = 2x^2 + \dfrac{32{,}000}{x}$

CHAPTER 2

Section 2.1

Check Point Exercises

1. a. $8 + i$ **b.** $-10 + 10i$ **2. a.** $63 + 14i$ **b.** $58 - 11i$ **3.** $\dfrac{3}{5} + \dfrac{13}{10}i$ **4. a.** $7i\sqrt{3}$ **b.** $1 - 4i\sqrt{3}$ **c.** $-7 + i\sqrt{3}$
5. $\{1 + i, 1 - i\}$

Exercise Set 2.1

1. $8 - 2i$ **3.** $-2 + 9i$ **5.** $24 + 7i$ **7.** $-14 + 17i$ **9.** $21 + 15i$ **11.** $-43 - 23i$ **13.** $-29 - 11i$ **15.** 34 **17.** 34

19. $-5 + 12i$ **21.** $\dfrac{3}{5} + \dfrac{1}{5}i$ **23.** $1 + i$ **25.** $-\dfrac{24}{25} + \dfrac{32}{25}i$ **27.** $\dfrac{7}{5} + \dfrac{4}{5}i$ **29.** $3i$ **31.** $47i$ **33.** $-8i$ **35.** $2 + 6i\sqrt{7}$

37. $-\dfrac{1}{3} + \dfrac{\sqrt{2}}{6}i$ **39.** $-\dfrac{1}{8} - \dfrac{\sqrt{3}}{24}i$ **41.** $-2\sqrt{6} - 2i\sqrt{10}$ **43.** $24\sqrt{15}$ **45.** $\{3 + i, 3 - i\}$ **47.** $\left\{-1 + \dfrac{3}{2}i, -1 - \dfrac{3}{2}i\right\}$

49. $\left\{\dfrac{4}{3} + i\dfrac{\sqrt{5}}{3}, \dfrac{4}{3} - i\dfrac{\sqrt{5}}{3}\right\}$ **59.** (d) is true.   **61.** $\dfrac{14}{25} - \dfrac{2}{25}i$ **63.** 0

Section 2.2

Check Point Exercises

1.

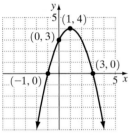

2.

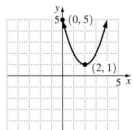

3.

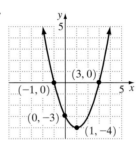

4. $45; 190$
5. $4, 4; 16$
6. 30 ft by 30 ft; 900 sq ft

Exercise Set 2.2

1. $h(x) = (x - 1)^2 + 1$ **3.** $j(x) = (x - 1)^2 - 1$ **5.** $h(x) = x^2 - 1$ **7.** $g(x) = x^2 - 2x + 1$ **9.** $(3, 1)$ **11.** $(-1, 5)$
13. $(2, -5)$ **15.** $(-1, 9)$

17. Domain: $(-\infty, \infty)$
Range: $[-1, \infty)$
axis of symmetry: $x = 4$

19. Domain: $(-\infty, \infty)$
Range: $[2, \infty)$
axis of symmetry: $x = 1$

21. Domain: $(-\infty, \infty)$
Range: $[1, \infty)$
axis of symmetry: $x = 3$

23. Domain: $(-\infty, \infty)$
Range: $[-1, \infty)$
axis of symmetry: $x = -2$

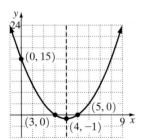

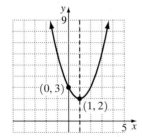

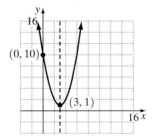

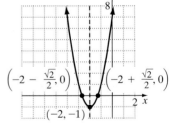

25. Domain: $(-\infty, \infty)$

Range: $(-\infty, 4]$

axis of symmetry: $x = 1$

27. Domain: $(-\infty, \infty)$

Range: $[-4, \infty)$

axis of symmetry: $x = 1$

29. Domain: $(-\infty, \infty)$

Range: $\left[-\dfrac{49}{4}, \infty\right)$

axis of symmetry: $x = -\dfrac{3}{2}$

31. Domain: $(-\infty, \infty)$

Range: $(-\infty, 4]$

axis of symmetry: $x = 1$

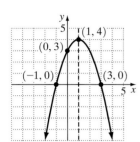

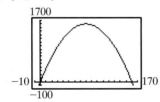

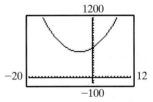

33. Domain: $(-\infty, \infty)$
Range: $(-\infty, -1]$
axis of symmetry: $x = 1$

35. minimum; $(2, -13)$ **37.** maximum; $(1, 1)$ **39.** minimum; $\left(\dfrac{1}{2}, -\dfrac{5}{4}\right)$

41. 1968; 4238 cigarettes per person; Yes   **43.** 6.25 s; 629 ft

45. The graph has the shape of a parabola.

47. 8 and 8; 64   **49.** 5 and -5; -25 **51.** 5 yd by 5 yd; 25 yd^2

53. 60 ft; 30 ft; 1800 ft^2 **55.** 150 ft by 100 ft; 15,000 ft^2 **57.** 3 in.

59. \$65; \$422,500   **61.** 25; 1250 lb   **71.** Answers may vary.

73. $(80, 1600)$ **75.** $(-4, 520)$

77.

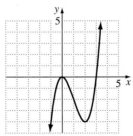

; Vertex: about $(4.41, 3.89)$; The minimum number of people in the United States holding more than one job was 3.89 million in 1974.

79. Answers may vary.
81. Answers may vary.
83. $x = 3$; $(0, 11)$
85. \$95; \$21,675

Section 2.3

Check Point Exercises

1. The graph rises to the left and to the right.   **2.** Since n is odd and the leading coefficient is negative, the function falls to the right. Since the ratio cannot be negative, the model won't be appropriate.   **3.** No; the graph should fall to the left, but doesn't appear to.

4. $\{-2, 2\}$ **5.** $\{-2, 0, 2\}$ **6.** $f(-3) = -42$; $f(-2) = 5$ **7.**

Exercise Set 2.3

1. polynomial function; degree: 3   **3.** polynomial function; degree: 5   **5.** not a polynomial function   **7.** not a polynomial function
9. not a polynomial function   **11.** possible polynomial function   **13.** not a polynomial function   **15.** (c)   **17.** (b)   **19.** (a)
21. falls to the left and rises to the right   **23.** rises to the left and to the right   **25.** falls to the left and to the right
27. $x = 5$ has multiplicity 1, the graph crosses the x-axis; $x = -4$ has multiplicity 2, the graph touches the x-axis and turns around.
29. $x = 3$ has multiplicity 1, the graph crosses the x-axis; $x = -6$ has multiplicity 3, the graph crosses the x-axis.
31. $x = 0$ has multiplicity 1, the graph crosses the x-axis; $x = 1$ has multiplicity 2, the graph touches the x-axis and turns around.
33. $x = 2$, $x = -2$ and $x = -7$ have multiplicity 1, the graph crosses the x-axis.   **35.** $f(1) = -1; f(2) = 5$
37. $f(-1) = -1; f(0) = 1$ **39.** $f(-3) = -11; f(-2) = 1$ **41.** $f(-3) = -42; f(-2) = 5$

43. a. $f(x)$ rises to the right and falls to the left.
    b. $x = -2, x = 1, x = -1$;
        $f(x)$ crosses the x-axis at each.
    c. The y-intercept is -2.
    d. neither
    e.

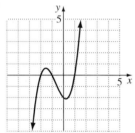

45. a. $f(x)$ rises to the left and the right.
    b. $x = 0, x = 3, x = -3$;
        $f(x)$ crosses the x-axis at -3 and 3;
        $f(x)$ touches the x-axis at 0.
    c. The y-intercept is 0.
    d. y-axis symmetry
    e.

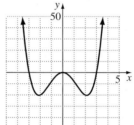

47. a. $f(x)$ falls to the left and the right.
    b. $x = 0, x = 4, x = -4$;
        $f(x)$ crosses the x-axis at -4 and 4;
        $f(x)$ touches the x-axis at 0.
    c. The y-intercept is 0.
    d. y-axis symmetry
    e.

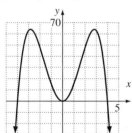

49. a. $f(x)$ rises to the left and the right.
    b. $x = 0, x = 1$;
        $f(x)$ touches the x-axis at 0 and 1.
    c. The y-intercept is 0.
    d. neither
    e.

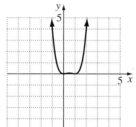

51. a. $f(x)$ falls to the left and the right.
    b. $x = 0, x = 2$;
        $f(x)$ crosses the x-axis at 0 and 2.
    c. The y-intercept is 0.
    d. neither
    e.

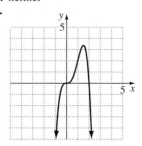

53. a. $f(x)$ rises to the left and falls to the right.
    b. $x = 0, x = \pm\sqrt{3}$;
        $f(x)$ crosses the x-axis at 0;
        $f(x)$ touches the x-axis at $\sqrt{3}$ and $-\sqrt{3}$.
    c. The y-intercept is 0.
    d. origin symmetry
    e.

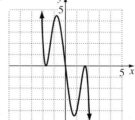

55. a. $f(x)$ rises to the left and falls to the right.
  b. $x = 0, x = 3$;
    $f(x)$ crosses the x-axis at 3;
    $f(x)$ touches the x-axis at 0.
  c. The y-intercept is 0.
  d. neither
  e.

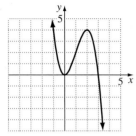

57. a. $f(x)$ falls to the left and the right.
  b. $x = 1, x = -2, x = 2$;
    $f(x)$ crosses the x-axis at -2 and 2;
    $f(x)$ touches the x-axis at 1.
  c. The y-intercept is 12.
  d. neither
  e.

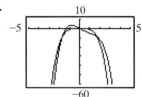

59. a. Leading Coefficient Test suggests the elk population will decline and eventually will die off.
  b.

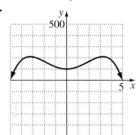

  c.

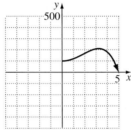

The population reaches extinction at the end of 5 years.

61. No; eventually the function would predict a negative number of thefts, which is impossible.
63. According to both the linear and the quadratic model, the number of inmates will reach 1382 thousand 14.5 years after 1985; accurate models
81. Answers may vary.
83. Answers may vary.

85.

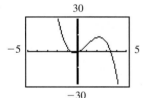

87.

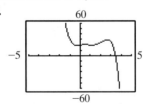

89.

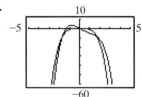

91. $f(x) = x^3 + x^2 - 12x$ **93.** Answers may vary.

Section 2.4

Check Point Exercises

1. $x + 5$ **2.** $2x^2 + 3x - 2 + \dfrac{1}{x - 3}$ **3.** $2x^2 + 7x + 14 + \dfrac{21x - 10}{x^2 - 2x}$ **4.** $x^2 - 2x - 3$ **5.** -105 **6.** $\left\{-1, -\dfrac{1}{3}, \dfrac{2}{5}\right\}$

Exercise Set 2.4

1. $x + 3$ **3.** $x^2 + 3x + 1$ **5.** $2x^2 + 3x + 5$ **7.** $4x + 3 + \dfrac{2}{3x - 2}$ **9.** $2x^2 + x + 6 - \dfrac{38}{x + 3}$

11. $4x^3 + 16x^2 + 60x + 246 + \dfrac{984}{x - 4}$ **13.** $2x + 5$ **15.** $6x^2 + 3x - 1 - \dfrac{3x - 1}{3x^2 + 1}$ **17.** $2x + 5$ **19.** $3x - 8 + \dfrac{20}{x + 5}$

21. $4x^2 + x + 4 + \dfrac{3}{x - 1}$ **23.** $6x^4 + 12x^3 + 22x^2 + 48x + 93 + \dfrac{187}{x - 2}$ **25.** $x^3 - 10x^2 + 51x - 260 + \dfrac{1300}{x + 5}$

27. $x^4 + x^3 + 2x^2 + 2x + 2$ **29.** $x^3 + 4x^2 + 16x + 64$ **31.** $2x^4 - 7x^3 + 15x^2 - 31x + 64 - \dfrac{129}{x + 2}$

33. -25 **35.** 4729 **37.** $x^2 - 5x + 6; x = -1, x = 2, x = 3$ **39.** $\left\{-\dfrac{1}{2}, 1, 2\right\}$ **41.** $\left\{-\dfrac{3}{2}, -\dfrac{1}{3}, \dfrac{1}{2}\right\}$

43. $x^3 + 5x^2 - 9x - 45$ **45. a.** 70   **b.** $80 + \dfrac{800}{x - 110}; f(30) = 70;$ yes

c. No, f is a rational function because it is a quotient of two polynomials.

57.

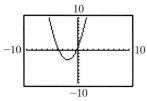

The division is correct.

59.

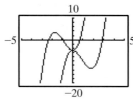

The division is not correct.
The right side should be
$3x^3 - 8x^2 - 5.$

61. $k = -12$
63. $x^{2n} - x^n + 1$

Section 2.5

Check Point Exercises

1. $\pm 1, \pm 2, \pm 3, \pm 6$ **2.** $\pm 1, \pm 3, \pm \dfrac{1}{2}, \pm \dfrac{1}{4}, \pm \dfrac{3}{2}, \pm \dfrac{3}{4}$ **3.** $\{-5, -4, 1\}$ **4.** $\{1, 2 - 3i, 2 + 3i\}$

5. a. $(x^2 - 5)(x^2 + 1)$ **b.** $(x + \sqrt{5})(x - \sqrt{5})(x^2 + 1)$ **c.** $(x + \sqrt{5})(x - \sqrt{5})(x + i)(x - i)$ **6.** $f(x) = x^3 + 3x^2 + x + 3$

7. 4, 2, or 0 positive zeros, no possible negative zeros

Exercise Set 2.5

1. $\pm 1, \pm 2, \pm 4$ **3.** $\pm 1, \pm 2, \pm 3, \pm 6, \pm \dfrac{1}{3}, \pm \dfrac{2}{3}$ **5.** $\pm 1, \pm 2, \pm 3, \pm 6, \pm \dfrac{1}{2}, \pm \dfrac{1}{4}, \pm \dfrac{3}{2}, \pm \dfrac{3}{4}$ **7.** $\pm 1, \pm 2, \pm 3, \pm 4, \pm 6, \pm 12$

9. a. $\pm 1, \pm 2, \pm 4$ **b.** 2 is a zero   **c.** $\{2, -2, -1\}$ **11. a.** $\pm 1, \pm 2, \pm 3, \pm 6, \pm \dfrac{1}{2}, \pm \dfrac{3}{2}$ **b.** 3 is a zero   **c.** $\left\{3, \dfrac{1}{2}, -2\right\}$

13. a. $\pm 1, \pm 2, \pm 4, \pm 8, \pm \dfrac{1}{3}, \pm \dfrac{2}{3}, \pm \dfrac{4}{3}, \pm \dfrac{8}{3}$ **b.** 2 is a zero   **c.** $\left\{2, -\dfrac{1}{3}, -4\right\}$ **15. a.** $\pm 1, \pm 2, \pm 3, \pm 4, \pm 6, \pm 12$ **b.** 4 is a root

c. $\{-3, 1, 4\}$ **17. a.** $\pm 1, \pm 2, \pm 3, \pm 4, \pm 6, \pm 12$ **b.** -2 is a root   **c.** $\{-2, 1 + \sqrt{7}, 1 - \sqrt{7}\}$ **19. a.** $\pm 1, \pm 5, \pm \dfrac{1}{2}, \pm \dfrac{5}{2}, \pm \dfrac{1}{3}, \pm \dfrac{5}{3}, \pm \dfrac{1}{6}, \pm \dfrac{5}{6}$

b. -5 is a root   **c.** $\left\{-5, \dfrac{1}{2}, \dfrac{1}{3}\right\}$ **21. a.** $\pm 1, \pm 2, \pm 4$ **b.** 2 is a root   **c.** $\{-2, 2, 1 + \sqrt{2}, 1 - \sqrt{2}\}$ **23. a.** $(x^2 - 5)(x^2 + 4)$

b. $(x + \sqrt{5})(x - \sqrt{5})(x^2 + 4)$ **c.** $(x + \sqrt{5})(x - \sqrt{5})(x + 2i)(x - 2i)$ **25. a.** $(x^2 - 2)(x^2 + 3)$ **b.** $(x + \sqrt{2})(x - \sqrt{2})(x^2 + 3)$

c. $(x + \sqrt{2})(x - \sqrt{2})(x + i\sqrt{3})(x - i\sqrt{3})$ **27. a.** $(x - 3)(x + 1)(x^2 + 4)$ **b.** $(x - 3)(x + 1)(x^2 + 4)$

c. $(x - 3)(x + 1)(x + 2i)(x - 2i)$ **29.** $f(x) = 2x^3 - 2x^2 + 50x - 50$ **31.** $f(x) = x^3 - 3x^2 - 15x + 125$

33. $f(x) = x^4 + 10x^2 + 9$ **35.** $f(x) = x^4 - 9x^3 + 21x^2 + 21x - 130$ **37.** no positive real roots; 3 or 1 negative real roots

39. 3 or 1 positive real roots; no negative real roots   **41.** 2 or 0 positive real roots; 2 or 0 negative real roots   **43.** $x = -2, x = 5, x = 1$

45. $\left\{-\dfrac{1}{2}, \dfrac{1 + \sqrt{17}}{2}, \dfrac{1 - \sqrt{17}}{2}\right\}$ **47.** $x = -1, x = 2 + 2i, x = 2 - 2i$ **49.** $\{-1, -2, 3 + \sqrt{13}, 3 - \sqrt{13}\}$

51. $x = -1, x = 2, x = -\dfrac{1}{3}, x = 3$ **53.** $\left\{1, -\dfrac{3}{4}, i\sqrt{2}, -i\sqrt{2}\right\}$ **55.** $\left\{-2, \dfrac{1}{2}, \sqrt{2}, -\sqrt{2}\right\}$ **57. a.** $x = 40$; about 27% of art

productivity occurs in one's 40s.   **b.** degree 2; leading coefficient: negative   **59.** $W = 3$ mm   **61.** 2 in. by 9 in. by 4 in.   **63.** 3 yr

65. Answers may vary.   **75.** $\dfrac{1}{2}, \dfrac{2}{3}, 2$ **77.** $\pm \dfrac{1}{2}$ **79.** 5, 3, or 1 positive real zeros exist

81.

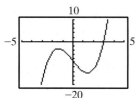

1 real zero, 2 nonreal
complex zeros

83.

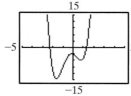

2 real zeros, 2 nonreal
complex zeros

85. (d) is true.
87. 3 in.
89. 3
91. 5
93. Answers may vary.

Section 2.6

Check Point Exercises

1. a. $\{x|x \neq 5\}$ **b.** $\{x|x \neq -5, x \neq 5\}$ **c.** all real numbers   **2. a.** $x = 1, x = -1$ **b.** $x = -1$ **c.** none
3. a. $y = 3$ **b.** $y = 0$ **c.** none
4.

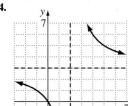

5.

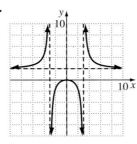

6.

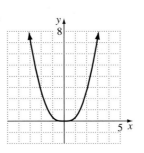

7. $y = 2x - 1$

8. a. $C(x) = 500x + 600{,}000$

b. $\overline{C}(x) = \dfrac{500x + 600{,}000}{x}$

c. $\overline{C}(1000) = 1100$, when 1000 new systems are produced, it costs \$1100 to produce each system; $\overline{C}(10{,}000) = 560$, when 10,000 new systems are produced, it costs \$560 to produce each system, $\overline{C}(100{,}000) = 506$, when 100,000 new systems are produced, it costs \$506 to produce each system.

d. $y = 500$; The cost per system approaches \$500 as more systems are produced.

9. $T(x) = \dfrac{20}{x} + \dfrac{20}{x - 10}$

Exercise Set 2.6

1. $\{x|x \neq 4\}$ **3.** $\{x|x \neq 5, x \neq -4\}$ **5.** $\{x|x \neq 7, x \neq -7\}$ **7.** All real numbers   **9.** $-\infty$ **11.** $-\infty$ **13.** 0   **15.** $+\infty$
17. $-\infty$ **19.** 1   **21.** $x = -4$ **23.** $x = 0, x = -4$ **25.** $x = -4$ **27.** no vertical asymptotes   **29.** $y = 0$ **31.** $y = 4$
33. no horizontal asymptote   **35.** $y = -\dfrac{2}{3}$

37.

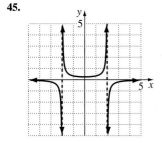

39.

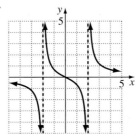

41.

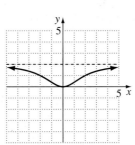

43.

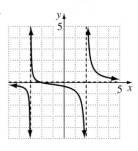

45.

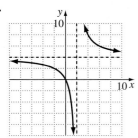

47.

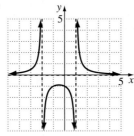

49.

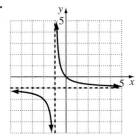

51.

53.

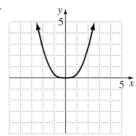

55.

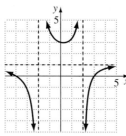

57.

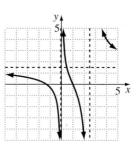

59. a. Slant asymptote: $y = x$

b.

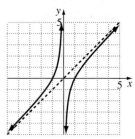

61. a. Slant asymptote: $y = x$

b.

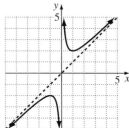

63. a. Slant asymptote: $y = x + 4$

b.

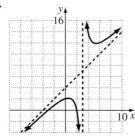

65. a. Slant asymptote: $y = x - 2$

b.

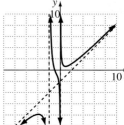

67. a. $C(x) = 100x + 100,000$

b. $\overline{C}(x) = \dfrac{100x + 100,000}{x}$

c. $\overline{C}(500) = 300$, when 500 bicycles are produced, it costs \$300 to produce each bicycle; $\overline{C}(1000) = 200$, when 1000 bicycles are produced, it costs \$200 to produce each bicycle; $\overline{C}(2000) = 150$, when 2000 bicycles are produced, it costs \$150 to produce each bicycle; $\overline{C}(4000) = 125$, when 4000 bicycles are produced, it costs \$125 to produce each bicycle.

d. $y = 100$; The cost per bicycle approaches \$100 as more bicycles are produced.

69. a. $M(x) = \dfrac{190.9x + 2413.99}{0.234x + 12.54}$

b. 355.65; $M(19) \approx 355.65$ on the graph

c. $y = \dfrac{190.9}{0.234} \approx 816$; The cost of textbooks per college student approaches \$816 as the years progress.

71. 90; An incidence ratio of 10 means 90% of the deaths are smoking related.

73. $y = 100$; The percentage of deaths cannot exceed 100% as the incidence ratio increases.

75. a. After 1 day: 35 words; after 5 days: about 12 words; after 15 days: about 7 words

b. $N(1) = 35$ words; This is the same as the estimate from the graph. $N(5) = 11$ words; This is a little less than the estimate from the graph. $N(15) = 7$ words; This is the same as the estimate from the graph.

c. The graph suggests that the students will remember 5 words over a long period of time.

d. $y = 5$; The horizontal asymptote indicates the students will remember 5 words over a long period of time.

77. $T(x) = \dfrac{10}{x} + \dfrac{5}{x}$ **79.** $A(x) = 2x + \dfrac{50}{x} + 52$ **89.** Answers may vary.

91.

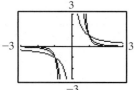

The graph approaches the horizontal asymptote faster and the vertical asymptote slower as n increases.

93.

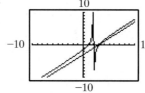

$g(x)$ is the graph of a line whereas $f(x)$ is the graph of a rational function with a slant asymptote; In $g(x)$, $x - 2$ is a factor of $x^2 - 5x + 6$.

95. (d) is true.
97. Answers may vary.
99. Answers may vary.

Section 2.7

Check Point Exercises

1. $(-\infty, -4)$ or $(5, \infty)$

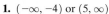

2. $(-\infty, -3]$ or $[-1, 1]$

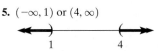

3. $(-\infty, -1)$ or $[1, \infty)$

4. Between 1 and 4 seconds

Exercise Set 2.7

1. $(-\infty, -2)$ or $(4, \infty)$

3. $[-3, 7]$

5. $(-\infty, 1)$ or $(4, \infty)$

7. $(-\infty, -4)$ or $(-1, \infty)$

9. $\left[-4, \dfrac{2}{3}\right]$

11. $(-\infty, -3)$ or $\left(\dfrac{5}{2}, \infty\right)$

13. $\left(-1, -\dfrac{3}{4}\right)$

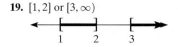

15. $\left(-\infty, -\dfrac{3}{2}\right)$ or $(0, \infty)$

17. $\varnothing$

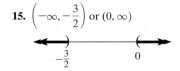

19. $[1, 2]$ or $[3, \infty)$

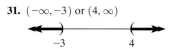

21. $[-2, -1]$ or $[1, \infty)$

23. $(-\infty, -3)$

25. $(-1, \infty)$

27. $\{0\}$ or $[9, \infty)$

29. $(-\infty, -8)$ or $(-6, 4)$ or $(6, \infty)$

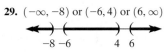

31. $(-\infty, -3)$ or $(4, \infty)$

33. $(-4, -3)$

35. $(-\infty, -5)$ or $(-3, \infty)$

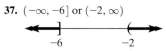

37. $(-\infty, -6]$ or $(-2, \infty)$

39. $(-3, 2)$

41. $(-\infty, -1)$ or $(1, 2)$ or $(3, \infty)$

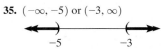

43. $[-4, 1)$ or $(1, 2]$

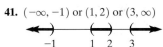

45. $\left[-6, -\dfrac{1}{2}\right]$ or $[1, \infty)$ **47.** $(-\infty, -2)$ or $[-1, 2)$

49. Between 2 and 3 seconds.

51. Between about 0.27 and about 3.73 seconds.

53. a. 200 beats per minute   **b.** Between 0 and 4 seconds after 12 seconds. The model isn't valid for the second interval. A more realistic answer is between 0 and 4 seconds. This is found by solving the inequality $H(x) > 110$.

55. Medicare spending will exceed $536.6 billion after 2007.   **57.** They must produce at least 20,000 wheelchairs.

59. Minimum side length is about 1.14 inches.   **65.** Answers may vary.   **67.** $\left[-3, \dfrac{1}{2}\right]$ **69.** $(1, 4]$

71. Answers may vary. One possible solution is $x^2 - 2x - 15 < 0$.   **73.** $(-\infty, 2)$ or $(2, \infty)$ **75.** $\varnothing$ **77. a.** $(-\infty, \infty)$ **b.** $\varnothing$

Section 2.8

Check Point Exercises

1. 137.5 lb/in^2 **2.** about 556 ft   **3.** $26 per barrel   **4.** 24 min   **5.** 96π ft^3

Exercise Set 2.8

1. 84 **3.** 25 **5.** $\dfrac{5}{6}$ **7.** 240 **9. a.** $G = kW$ **b.** $G = 0.02W$ **c.** 1.04 in. **11.** \$60 **13.** 2442 mph **15.** 607 lb

17. 0.5 hr **19.** 6.4 lb **21.** 31.78; index: about 32; not in the desirable range **23.** 11.11 foot-candles **25.** 72 ergs
27. The average number of phone calls is about 126. **29.** Yes, the wind will exert a force of 360 pounds on the window.

39. Answers may vary. **41.** The illumination is $\dfrac{1}{4}$ as much.

Chapter 2 Review Exercises

1. $-9 + 4i$ **2.** $-12 - 8i$ **3.** $29 + 11i$ **4.** $-7 - 24i$ **5.** 113 **6.** $\dfrac{15}{13} - \dfrac{3}{13}i$ **7.** $\dfrac{1}{5} + \dfrac{11}{10}i$ **8.** $i\sqrt{2}$ **9.** $-96 - 40i$

10. $2 + i\sqrt{2}$ **11.** $\{1 + i\sqrt{3}, 1 - i\sqrt{3}\}$ **12.** $\left\{\dfrac{3}{2} + \dfrac{1}{2}i, \dfrac{3}{2} - \dfrac{1}{2}i\right\}$

13.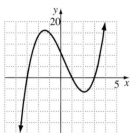

axis of symmetry: $x = 1$

14.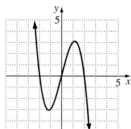

axis of symmetry: $x = -4$

15.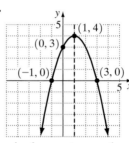

axis of symmetry: $x = 1$

16.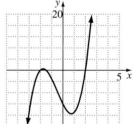

axis of symmetry: $x = 1$

17. After 2 seconds, the ball reaches a maximum height of 144 feet. **18.** $(20, 5.4)$; In 1980, the divorce rate reached a maximum of

5.4 per 1000. **19.** 250 yd by 500 yd; maximum area is 125,000 yd^2 **20.** $x = 166\dfrac{2}{3}$ ft by $y = 125$ ft **21.** 36; 5256 pounds **22.** (c)

23. (b) **24.** (a) **25.** (d) **26.** Because the degree is odd and the leading coefficient is negative, the graph falls to the right.
Therefore, the model indicates that the percentage of families below the poverty level will eventually be negative, which is impossible.
27. Since the degree is even and the leading coefficient is negative, the graph falls to the right. Therefore, the model indicates a patient will
eventually have a negative number of viral bodies, which is impossible. **28.** $x = 1$, multiplicity 1, crosses; $x = -2$, multiplicity 2, touches;
$x = -5$, multiplicity 3, crosses **29.** $x = -5$, multiplicity 1, crosses; $x = 5$, multiplicity 2, touches **30.** $f(1) = -2$ and $f(2) = 3$
31. $f(-3) = -32$ and $f(-2) = 7$

32. a. The graph falls to the
left and rises to the right.
b. no symmetry
c.

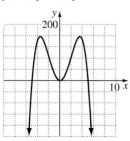

33. a. The graph rises to the
left and falls to the right.
b. origin symmetry
c.

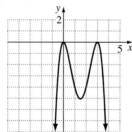

34. a. The graph falls to the
left and rises to the right.
b. no symmetry
c.

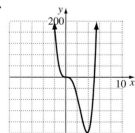

35. a. The graph falls to the
left and to the right.
b. y-axis symmetry
c.

36. a. The graph falls to the
left and to the right.
b. no symmetry
c.

37. a. The graph rises to the
left and to the right.
b. no symmetry
c.

38. $4x^2 - 7x + 5 - \dfrac{4}{x+1}$ **39.** $2x^2 - 4x + 1 - \dfrac{10}{5x-3}$ **40.** $2x^2 + 3x - 1$ **41.** $3x^3 - 4x^2 + 7$

42. $3x^3 + 6x^2 + 10x + 10 + \dfrac{20}{x-2}$ **43.** -5697 **44.** $2, \dfrac{1}{2}, -3$ **45.** $\{4, -2 \pm \sqrt{5}\}$ **46.** $\pm 1, \pm 5$

47. $\pm 1, \pm 2, \pm 4, \pm 8, \pm \dfrac{8}{3}, \pm \dfrac{4}{3}, \pm \dfrac{2}{3}, \pm \dfrac{1}{3}$ **48.** 2 or 0 positive real zeros; no negative real zeros

49. 3 or 1 positive real zeros; 2 or 0 negative real zeros **50.** No sign variations exist for either $f(x)$ or $f(-x)$, so no real roots exist.

51. a. $\pm 1, \pm 2, \pm 4$ **b.** 1 positive real zero; 2 or 0 negative real zeros **c.** 1 is a zero. **d.** $\{1, -2\}$

52. a. $\pm 1, \pm \dfrac{1}{2}, \pm \dfrac{1}{3}, \pm \dfrac{1}{6}$ **b.** 2 or 0 positive real zeros; 1 negative real zero **c.** -1 is a zero. **d.** $\left\{-1, \dfrac{1}{3}, \dfrac{1}{2}\right\}$

53. a. $\pm 1, \pm 3, \pm 5, \pm 15, \pm \dfrac{1}{2}, \pm \dfrac{1}{4}, \pm \dfrac{1}{8}, \pm \dfrac{3}{2}, \pm \dfrac{3}{4}, \pm \dfrac{3}{8}, \pm \dfrac{5}{2}, \pm \dfrac{5}{4}, \pm \dfrac{5}{8}, \pm \dfrac{15}{2}, \pm \dfrac{15}{4}, \pm \dfrac{15}{8}$

b. 3 or 1 positive real solutions; no negative real solutions **c.** $\dfrac{1}{2}$ is a zero. **d.** $\left\{\dfrac{1}{2}, \dfrac{3}{2}, \dfrac{5}{2}\right\}$

54. a. $\pm 1, \pm 2, \pm 3, \pm 6$ **b.** 2 or 0 positive real solutions; 2 or 0 negative real solutions **c.** -2 is a zero. **d.** $\{-2, -1, 1, 3\}$

55. a. $\pm 1, \pm 2, \pm \dfrac{1}{2}, \pm \dfrac{1}{4}$ **b.** 1 positive real root; 1 negative real root **c.** $\dfrac{1}{2}$ is a zero. **d.** $\left\{\dfrac{1}{2}, -\dfrac{1}{2}, i\sqrt{2}, -i\sqrt{2}\right\}$

56. a. $\pm 1, \pm 2, \pm 4, \pm \dfrac{1}{2}$ **b.** 2 or no positive zeros; 2 or no negative zeros **c.** 2 is a zero. **d.** $\left\{2, -2, \dfrac{1}{2}, -1\right\}$

57. $f(x) = x^3 - 6x^2 + 21x - 26$ **58.** $f(x) = 2x^4 + 12x^3 + 20x^2 + 12x + 18$

59. $-2, \dfrac{1}{2}, \pm i; f(x) = (x - i)(x + i)(x + 2)(2x - 1)$ **60.** $-1, 4; g(x) = (x + 1)^2(x - 4)^2$

61. 4 real zeros, one with multiplicity two **62.** 3 real zeros; 2 nonreal complex zeros

63. 2 real zeros, one with multiplicity two; 2 nonreal complex zeros **64.** 1 real zero; 4 nonreal complex zeros

65. Vertical asymptote: $x = 3$ and $x = -3$
horizontal asymptote: $y = 0$

66. Vertical asymptote: $x = -3$
horizontal asymptote: $y = 2$

67. Vertical asymptotes: $x = 3, -2$
horizontal asymptote: $y = 1$

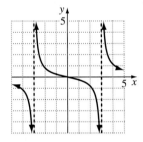

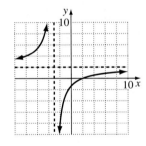

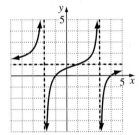

68. Vertical asymptote: $x = -2$
horizontal asymptote: $y = 1$

69. Vertical asymptote: $x = -1$
no horizontal asymptote
slant asymptote: $y = x - 1$

70. Vertical asymptote: $x = 3$
no horizontal asymptote
slant asymptote: $y = x + 5$

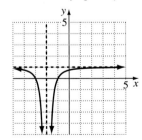

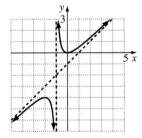

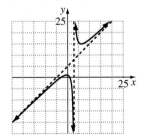

71. No vertical asymptote

no horizontal asymptote

slant asymptote: $y = -2x$

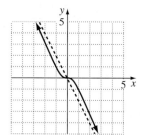

72. Vertical asymptote: $x = \dfrac{3}{2}$

no horizontal asymptote

slant asymptote: $y = 2x - 5$

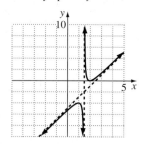

73. a. $C(x) = 25x + 50,000$

b. $\overline{C}(x) = \dfrac{25x + 50,000}{x}$

c. $\overline{C}(50) = 1025$, when 50 calculators are manufactured, it costs \$1025 to manufacture each; $\overline{C}(100) = 525$, when 100 calculators are manufactured, it costs \$525 to manufacture each; $\overline{C}(1000) = 75$, when 1000 calculators are manufactured, it costs \$75 to manufacture each; $\overline{C}(100,000) = 25.5$, when 100,000 calculators are manufactured, it costs \$25.50 to manufacture each.

d. $y = 25$; Minimum costs will approach \$25.

74. a. 1600; The difference in cost of removing 90% versus 50% of the contaminants is 16 million dollars.

b. $x = 100$; No amount of money can remove 100% of the contaminants, since $C(x)$ increases without bound as x approaches 100.

75. $y = 3000$; The number of fish in the pond approaches 3000.

76. $y = 0$; As the number of years of education increases the percentage rate of unemployment approaches zero.

77. a. $f(x) = \dfrac{1.96x + 3.14}{3.04x + 21.79}$ **b.** $y = \dfrac{49}{76}$; As the years increase, the fraction of nonviolent prisoners approaches $\dfrac{49}{76}$.

c. Answers may vary. **78.** $T(x) = \dfrac{4}{x + 3} + \dfrac{2}{x}$ **79.** $P(x) = 2x + \dfrac{2000}{x}$

80. $\left[-4, \dfrac{1}{2}\right]$

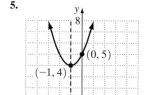

81. $\left(-\infty, \dfrac{3 - \sqrt{3}}{2}\right)$ or $\left(\dfrac{3 + \sqrt{3}}{2}, \infty\right)$

82. $(-3, 0)$ or $(1, \infty)$

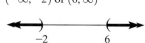

83. $(-\infty, -2)$ or $(6, \infty)$

84. $(-\infty, 4)$ or $\left[\dfrac{23}{4}, \infty\right)$

85. from 1 to 2 sec **86.** \$154

87. 1600 ft **88.** 5 hr **89.** 112 decibels

90. 16 hr **91.** 800 ft³

Chapter 2 Test

1. $47 + 16i$ **2.** $2 + i$ **3.** $38i$ **4.** $\{2 + 2i, 2 - 2i\}$

5.

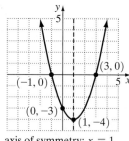

axis of symmetry: $x = -1$

6.

axis of symmetry: $x = 1$

7. maximum; $(3, 2)$

8. 23 computers; maximum daily profit $= \$16,900$

9. 7 and -7; -49

10. a. $5, 2, -2$ **b.**

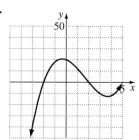

11. Since the degree of the polynomial is odd and the leading coefficient is positive, the graph of f should fall to the left and rise to the right. The x-intercepts should be 0, −1, and 1. **12. a.** 2 **b.** $\dfrac{1}{2}, \dfrac{2}{3}$ **13.** $\pm1, \pm2, \pm3, \pm6, \pm\dfrac{1}{2}, \pm\dfrac{3}{2}$

14. 3 or 1 positive real zeros; no negative real zeros. **15.** $\{-5, -3, 2\}$

16. a. $\pm1, \pm3, \pm5, \pm15, \pm\dfrac{1}{2}, \pm\dfrac{3}{2}, \pm\dfrac{5}{2}, \pm\dfrac{15}{2}$ **b.** $x = -1, x = \dfrac{3}{2}, x = \sqrt{5}, x = -\sqrt{5}$ **17.** $(x - 1)(x + 2)^2$

18. domain: $\{x | x \neq 4, x \neq -4\}$ **19.** domain: $\{x | x \neq 2\}$ **20.** domain: $\{x | x \neq -3, x \neq 1\}$ **21.** domain: all real numbers

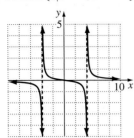   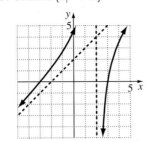   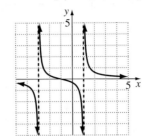   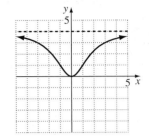

22. a. 0.9 **b.** 11 **c.** $y = 1$; as the number of learning trials increases, the proportion of correct responses approaches 1.

23. $(-3, 4)$ **24.** $(3, 10)$ **25.** 45 foot-candles

Cumulative Review Exercises (Chapters P–2)

1. $2 + \sqrt{3}$ **2.** $-3x^2 - 11x + 11$ **3.** $15\sqrt{2}$ **4.** $x^5(x - 1)(x + 1)$ **5.** $\{2, -1\}$ **6.** $\left\{ \dfrac{5 + \sqrt{13}}{6}, \dfrac{5 - \sqrt{13}}{6} \right\}$ **7.** $\left\{ \dfrac{1}{3}, -\dfrac{2}{3} \right\}$

8. $\{-3, -1, 2\}$ **9.** $(-\infty, 1)$ or $(4, \infty)$ **10.** $(-\infty, -1)$ or $\left(\dfrac{5}{3}, \infty \right)$

11. Center: $(1, -2)$; radius: 3   **12.** $t = 1 - \dfrac{V}{C}$ **13.** $(-\infty, 5]$ **14.** $x^2 - 2x - 4$ **15.** $16x^2 - 6$ **16.** -9

17. a. $x = -1, x = 1, x = 4$ **b.**

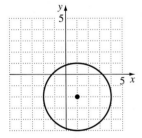   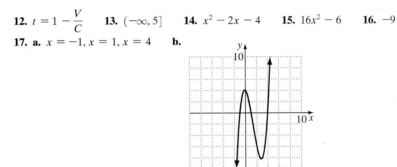

18.   **19.**   **20.**

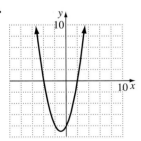   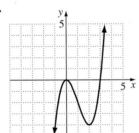   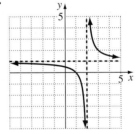

CHAPTER 3

Section 3.1

Check Point Exercises

1. 1 O-ring

2.
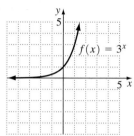

3.
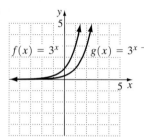

4.
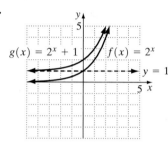

5. 11.49 billion
6. a. $14,859.47
 b. $14,918.25

Exercise Set 3.1

1. 10.556   **3.** 11.665   **5.** 0.125   **7.** 9.974   **9.** 0.387

11.
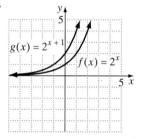

13.
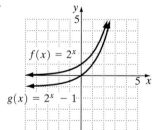

15.
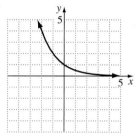

17.
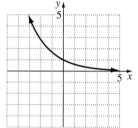

19. $H(x) = -3^{-x}$ **21.** $F(x) = -3^x$ **23.** $h(x) = 3^x - 1$

25.
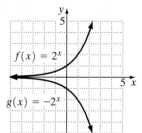

27.
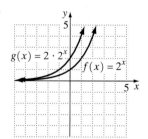

29.
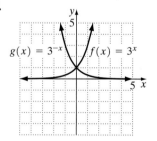

31.
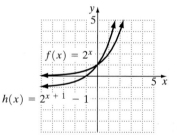

33.

35.

37.

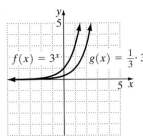

$f(x) = 3^x$ $g(x) = \frac{1}{3} \cdot 3^x$

39.

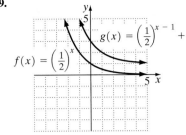

$f(x) = \left(\frac{1}{2}\right)^x$ $g(x) = \left(\frac{1}{2}\right)^{x-1} + 1$

41. a. \$13,116.51   **b.** \$13,140.67
    c. \$13,157.04   **d.** \$13,165.31
43. 7% compounded monthly
45. a. 67.38 million
    b. about 134.74 million
    c. about 269.46 million
    d. 538.85 million
    e. appears to double every 27 yr
47. $f(10) \approx 48$; 10 minutes after 8:00, 48 people have heard the rumor.

49. \$116,405.10   **51.** 3.249009585; 3.317278183; 3.321880096; 3.321995226; 3.321997068; $2^{\sqrt{3}} \approx 3.321997085$; The closer the exponent is to $\sqrt{3}$, the closer the value is to $2^{\sqrt{3}}$.   **53.** 175.6   **55. a.** 100%   **b.** $\approx 68.5\%$ **c.** $\approx 30.8\%$ **d.** 20%

57. a. 1429
    b. 24,546
    c. Growth is limited by the population; The entire population will eventually become ill.

65.

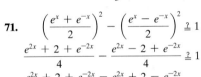

no; Nearly 4 O-rings are expected to fail.

67. a.

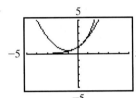

b.

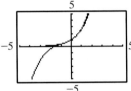

c.

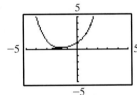

d. Answers may vary.

69. $y = 3^x$ is (d); $y = 5^x$ is (c); $y = \left(\frac{1}{3}\right)^x$ is (a); $y = \left(\frac{1}{5}\right)^x$ is (b).

71.
$$\left(\frac{e^x + e^{-x}}{2}\right)^2 - \left(\frac{e^x - e^{-x}}{2}\right)^2 \stackrel{?}{=} 1$$
$$\frac{e^{2x} + 2 + e^{-2x}}{4} - \frac{e^{2x} - 2 + e^{-2x}}{4} \stackrel{?}{=} 1$$
$$\frac{e^{2x} + 2 + e^{-2x} - e^{2x} + 2 - e^{-2x}}{4} \stackrel{?}{=} 1$$
$$\frac{4}{4} \stackrel{?}{=} 1$$
$$1 = 1$$

Section 3.2

Check Point Exercises

1. a. $7^3 = x$ **b.** $b^2 = 25$ **c.** $4^y = 26$ **2. a.** $5 = \log_2 x$ **b.** $3 = \log_b 27$ **c.** $y = \log_e 33$ **3. a.** 2   **b.** 1   **c.** $\frac{1}{2}$
4. a. 1   **b.** 0   **5. a.** 8   **b.** 17   **6.**

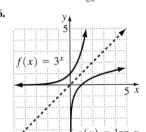

$f(x) = 3^x$

$g(x) = \log_3 x$

7. $(5, \infty)$
8. 80%
9. 4.0
10. a. $(-\infty, 4)$
    b. $(-\infty, 0)$ or $(0, \infty)$
11. a. $25x$
    b. $\sqrt{x}$
12. 4.6 ft per sec

Exercise Set 3.2

1. $2^4 = 16$ **3.** $3^2 = x$ **5.** $b^5 = 32$ **7.** $6^y = 216$ **9.** $\log_2 8 = 3$ **11.** $\log_2 \dfrac{1}{16} = -4$ **13.** $\log_8 2 = \dfrac{1}{3}$ **15.** $\log_{13} x = 2$

17. $\log_b 1000 = 3$ **19.** $\log_7 200 = y$ **21.** 2   **23.** 6   **25.** $\dfrac{1}{2}$ **27.** -3 **29.** $\dfrac{1}{2}$ **31.** 1   **33.** 0   **35.** 7   **37.** 19

39.

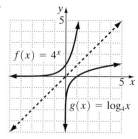

41.

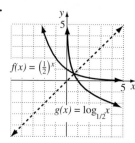

43. $H(x) = 1 - \log_3 x$
45. $h(x) = \log_3 x - 1$
47. $g(x) = \log_3(x - 1)$

49.

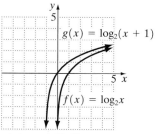

x-intercept: $(0, 0)$
vertical asymptote: $x = -1$

51.

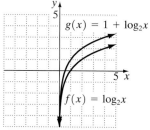

x-intercept: $(0.5, 0)$
vertical asymptote: $x = 0$

53.

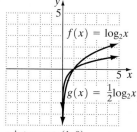

x-intercept: $(1, 0)$
vertical asymptote: $x = 0$

55. $(-4, \infty)$ **57.** $(-\infty, 2)$ **59.** $(-\infty, 2)$ or $(2, \infty)$ **61.** 2   **63.** 7   **65.** 33   **67.** 0   **69.** 6   **71.** -6 **73.** 125
75. $9x$ **77.** $5x^2$ **79.** $\sqrt{x}$ **81.** 95.4%   **83.** \$5.65 billion   **85.** ≈ 188 db; yes

87. a. 88
b. 71.5; 63.9; 58.8; 55; 52; 49.5
c.

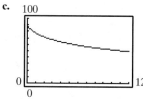

Material retention decreases
as time passes.

97.

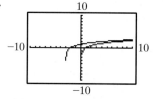

$g(x)$ is $f(x)$ shifted left 3 units left.

99.

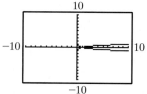

$g(x)$ is $f(x)$ reflected about the
x-axis.

101.

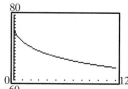

The score falls below 65 after
9 months.

103. $y = \ln x,\ y = \sqrt{x},\ y = x,$
$y = x^2,\ y = e^x,\ y = x^x$

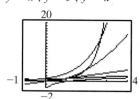

105. $\dfrac{4}{5}$

107. $\log_3 40 > \log_4 60$

Section 3.3

Check Point Exercises

1. a. $\log_6 7 + \log_6 11$ **b.** $2 + \log x$ **2. a.** $\log_8 23 - \log_8 x$ **b.** $5 - \ln 11$ **3. a.** $9 \log_6 3$ **b.** $\dfrac{1}{3} \ln x$

4. a. $4 \log_b x + \dfrac{1}{3} \log_b y$ **b.** $\dfrac{1}{2} \log_5 x - 2 - 3 \log_5 y$ **5. a.** 2   **b.** $\log \dfrac{7x + 6}{x}$

6. a. $\ln x^2 \sqrt[3]{x + 5}$ **b.** $\log \dfrac{(x - 3)^2}{x}$ **c.** $\log_b \dfrac{\sqrt[4]{x}\, y^{10}}{25}$ **7.** 4.02   **8.** 4.02

Exercise Set 3.3

1. $\log_5 7 + \log_5 3$ **3.** $1 + \log_7 x$ **5.** $3 + \log x$ **7.** $1 - \log_7 x$ **9.** $\log x - 2$ **11.** $3 - \log_4 y$ **13.** $2 - \ln 5$

15. $3 \log_b x$ **17.** $-6 \log N$ **19.** $\dfrac{1}{5} \ln x$ **21.** $2 \log_b x + \log_b y$ **23.** $\dfrac{1}{2} \log_4 x - 3$ **25.** $2 - \dfrac{1}{2} \log_6 (x + 1)$

27. $2 \log_b x + \log_b y - 2 \log_b z$ **29.** $1 + \dfrac{1}{2} \log x$ **31.** $\dfrac{1}{3} \log x - \dfrac{1}{3} \log y$ **33.** $\dfrac{1}{2} \log_b x + 3 \log_b y - 3 \log_b z$

35. $\dfrac{2}{3} \log_5 x + \dfrac{1}{3} \log_5 y - \dfrac{2}{3}$ **37.** $3 \ln x + \dfrac{1}{2} \ln(x^2 + 1) - 4 \ln(x + 1)$ **39.** $\left(1 + 2 \log x + \dfrac{1}{3} \log (1 - x)\right) - (\log 7 + 2 \log (x + 1))$

41. 1   **43.** $\ln(7x)$ **45.** 5   **47.** $\log \left(\dfrac{2x + 5}{x}\right)$ **49.** $\log (xy^3)$ **51.** $\ln(x^{1/2} y)$ or $\ln(y \sqrt{x})$ **53.** $\log_b (x^2 y^3)$ **55.** $\ln \left(\dfrac{x^5}{y^2}\right)$

57. $\ln \left(\dfrac{x^3}{y^{1/3}}\right)$ or $\ln \left(\dfrac{x^3}{\sqrt[3]{y}}\right)$ **59.** $\ln \dfrac{(x + 6)^4}{x^3}$ **61.** $\ln \left(\dfrac{x^3 y^5}{z^6}\right)$ **63.** $\log \sqrt{xy}$ **65.** $\log_5 \left(\dfrac{\sqrt{xy}}{(x + 1)^2}\right)$ **67.** $\ln \sqrt[3]{\dfrac{(x + 5)^2}{x(x^2 - 4)}}$

69. $\log \left(\dfrac{7x(x^2 - 1)}{x + 1}\right)$ or $\log (7x(x - 1))$ **71.** 1.5937   **73.** 1.6944   **75.** -1.2304 **77.** 3.6193

79.

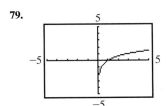

81.

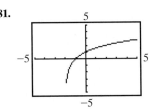

83. a. $D = 10 \log \dfrac{I}{I_0}$

b. 20 decibels louder

93. a.

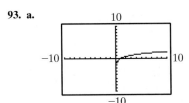

b.

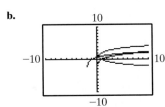

$y = 2 + \log_3 x$ shifts the graph of $y = \log_3 x$ two units upward;
$y = \log_3 (x + 2)$ shifts the graph of $y = \log_3 x$ two units left;
$y = -\log_3 x$ reflects the graph of $y = \log_3 x$ about the x-axis.

95.

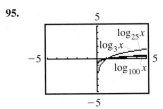

a. top graph: $y = \log_{100} x$; bottom graph: $y = \log_3 x$
b. top graph: $y = \log_3 x$; bottom graph: $y = \log_{100} x$
c. The graph of the equation with the largest b will be on the top in the interval $(0, 1)$ and on the bottom in the interval $(1, \infty)$.

97–99. Graphs may vary.
101. (d) is true.
103. $\dfrac{2A}{B}$
105. Answers may vary.

Section 3.4

Check Point Exercises

1. $\left\{\dfrac{\ln 134}{\ln 5}\right\}; \approx 3.04$ **2.** $\left\{\dfrac{\ln 9}{2}\right\}; \approx 1.10$ **3.** $\left\{\dfrac{\ln 2088 + 4 \ln 6}{3 \ln 6}\right\}; \approx 2.76$ **4.** $\{0, \ln 7\}; \ln 7 \approx 1.95$ **5.** $\{12\}$ **6.** $\{5\}$

7. $\left\{\dfrac{e^2}{3}\right\}$ **8.** 0.01   **9.** 16.2 yr   **10.** 2149

Exercise Set 3.4

1. $\left\{\dfrac{\ln 3.91}{\ln 10}\right\}; \approx 0.59$ **3.** $\{\ln 5.7\}; \approx 1.74$ **5.** $\left\{\dfrac{\ln 17}{\ln 5}\right\}; \approx 1.76$ **7.** $\left\{\ln\dfrac{23}{5}\right\}; \approx 1.53$ **9.** $\left\{\dfrac{\ln 659}{5}\right\}; \approx 1.30$

11. $\left\{\dfrac{\ln 793 - 1}{-5}\right\}; \approx -1.14$ **13.** $\left\{\dfrac{\ln 10{,}478 + 3}{5}\right\}; \approx 2.45$ **15.** $\left\{\dfrac{\ln 410}{\ln 7} - 2\right\}; \approx 1.09$ **17.** $\left\{\dfrac{\ln 813}{0.3 \ln 7}\right\}; \approx 11.48$

19. $\left\{\dfrac{3\ln 5 + \ln 3}{\ln 3 - 2\ln 5}\right\}; \approx -2.80$ **21.** $\{0, \ln 2\}; \ln 2 \approx 0.69$ **23.** $\left\{\dfrac{\ln 3}{2}\right\}; \approx 0.55$ **25.** $\{0\}$ **27.** $\{81\}$ **29.** $\{59\}$ **31.** $\left\{\dfrac{109}{27}\right\}$

33. $\left\{\dfrac{62}{3}\right\}$ **35.** $\left\{\dfrac{5}{4}\right\}$ **37.** $\{6\}$ **39.** $\{6\}$ **41.** $\{5\}$ **43.** $\{2, 12\}$ **45.** $\{e^2\}; \approx 7.39$ **47.** $\left\{\dfrac{e^4}{2}\right\}; \approx 27.30$ **49.** $\{e^{-1/2}\}; \approx 0.61$

51. $\{e^2 - 3\}; \approx 4.39$ **53.** about 0.11   **55. a.** 18.9 million   **b.** ≈ 2006 **57.** 8.2 yr   **59.** 16.8%   **61.** 8.7 yr   **63.** 15.7%
65. 1995   **67.** 2.8 days; Yes, the point (2.8, 50) appears to lie on the graph of P.   **69.** $10^{-2.4}$; 0.004 moles per liter   **75.** $\{2\}$
77. $\{4\}$ **79.** $\{2\}$ **81.** $\{-1.391606, 1.6855579\}$

83.
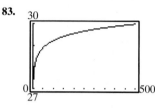
As distance from eye increases,
barometric air pressure increases,
leveling off at about 30 inches of mercury.

85.
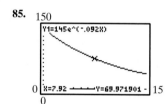
about 7.9 min

87. (c) is true.
89. $\{1, e^2\}, e^2 \approx 7.389$
91. $\{e\}, e \approx 2.718$

Section 3.5

Check Point Exercises

1. a. $A = 643\, e^{0.023t}$ **b.** 2039   **2. a.** $A = A_0 e^{-0.0248t}$ **b.** about 72 yr   **3. a.** 0.4 correct responses   **b.** 0.7 correct responses
c. 0.8 correct responses   **4. a.** $T = 30 + 70e^{-0.0673t}$ **b.** 48°C   **c.** 39 min   **5.** $y = 4e^{(\ln 7.8)x}; y = 4e^{2.054x}$

Exercise Set 3.5

1. 203 million   **3.** 2005   **5.** 2.6%   **7.** 2014   **9. a.** $A = 158{,}700e^{0.053t}$ **b.** 2007   **11.** $A = 6.04e^{0.01t}$ **13.** 8.01 g

15. 8 g; 4 g; 2 g; 1 g; 0.5 g   **17.** 15,679 years old   **19. a.** $\dfrac{A_0}{2} = A_0 e^{k(1.31)}; \dfrac{1}{2} = e^{1.31k}; \ln\dfrac{1}{2} = \ln e^{1.31k}; \ln\dfrac{1}{2} = 1.31k; k = \dfrac{\ln\frac{1}{2}}{1.31} \approx -0.52912$

b. 107 million years   **21.** $2A_0 = A_0 e^{kt}; 2 = e^{kt}; \ln 2 = \ln e^{kt}; \ln 2 = kt; t = \dfrac{\ln 2}{k}$ **23.** 63 yr   **25. a.** about 20 people

b. about 1080 people   **c.** 100,000 people   **27.** about 3.7%   **29.** about 48 years old   **31. a.** $T = 45 + 25e^{-0.0916t}$ **b.** 51°F
c. 18 min   **33.** 26 min   **35.** $y = 100e^{(\ln 4.6)x}; y = 100e^{1.526x}$ **37.** $y = 2.5e^{(\ln 0.7)x}; y = 2.5e^{-0.357x}$ **49.** $y = 1.740(1.037)^x; r \approx 0.971,$
a very good fit   **51.** $y = 0.112x + 1.547; r = 0.989;$ a very good fit   **53.** The model of best fit is the linear model; 2022
55. The logarithmic model, $y = -905{,}231.353 + 119{,}204.060 \ln x$, best fits the data. Answers for prediction may vary.
57. Answers may vary.

Chapter 3 Review Exercises

1. $g(x) = 4^{-x}$ **2.** $h(x) = -4^{-x}$ **3.** $r(x) = -4^{-x} + 3$ **4.** $f(x) = 4^x$
5.
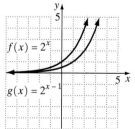

6.
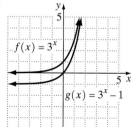

7.
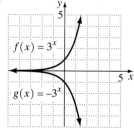

8.
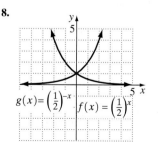

9. 5.5% compounded semiannually   **10.** 7% compounded monthly   **11. a.** 200°   **b.** 120°; 119°
c. 70°; The temperature in the room is 70°.   **12.** $49^{1/2} = 7$ **13.** $4^3 = x$ **14.** $3^y = 81$ **15.** $\log_6 216 = 3$ **16.** $\log_b 625 = 4$

17. $\log_{13} 874 = y$ **18.** 3   **19.** -2 **20.** $\varnothing$; $\log_b x$ is defined only for $x > 0$.   **21.** $\frac{1}{2}$ **22.** 1   **23.** 8

24. 5   **25.** 0   **26.**

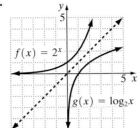

$f(x) = 2^x$ $g(x) = \log_2 x$

27.

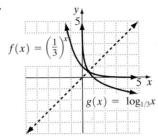

$f(x) = \left(\frac{1}{3}\right)^x$ $g(x) = \log_{1/3} x$

28. $g(x) = \log(-x)$
29. $r(x) = 1 + \log(2 - x)$
30. $h(x) = \log(2 - x)$
31. $f(x) = \log x$

32.

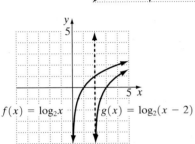

$f(x) = \log_2 x$ $g(x) = \log_2(x - 2)$

x-intercept: $(3, 0)$
vertical asymptote: $x = 2$

33.

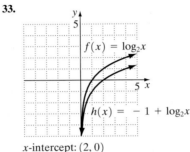

$f(x) = \log_2 x$ $h(x) = -1 + \log_2 x$

x-intercept: $(2, 0)$
vertical asymptote: $x = 0$

34.

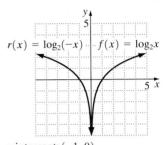

$r(x) = \log_2(-x)$ $f(x) = \log_2 x$

x-intercept: $(-1, 0)$
vertical asymptote: $x = 0$

35. $(-5, \infty)$ **36.** $(-\infty, 3)$ **37.** $(-\infty, 1) \cup (1, \infty)$ **38.** $6x$ **39.** $\sqrt{x}$ **40.** $4x^2$ **41.** 3.0

42. a. 76
b. $\approx 67, \approx 63, \approx 61, \approx 59, \approx 56$
c.

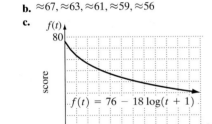

$f(t) = 76 - 18 \log(t + 1)$

time (months)
Retention decreases as time passes.

43. about 9 weeks   **44.** $2 + 3 \log_6 x$ **45.** $\frac{1}{2} \log_4 x - 3$

46. $\log_2 x + 2 \log_2 y - 6$ **47.** $\frac{1}{3} \ln x - \frac{1}{3}$ **48.** $\log_b 21$ **49.** $\log \frac{3}{x^3}$

50. $\ln(x^3 y^4)$ **51.** $\ln \frac{\sqrt{x}}{y}$ **52.** 6.2448   **53.** -0.1063

54. $\left\{\frac{\ln 12{,}143}{\ln 8}\right\}; \approx 4.523$ **55.** $\left\{\frac{1}{5} \ln 141\right\}; \approx 0.990$

56. $\left\{\frac{12 - \ln 130}{5}\right\}; \approx 1.426$ **57.** $\left\{\frac{\ln 37{,}500 - 2 \ln 5}{4 \ln 5}\right\}; \approx 1.136$

58. $\{\ln 3\}; \approx 1.099$ **59.** $\{23\}$ **60.** $\{5\}$ **61.** $\varnothing$ **62.** $\left\{\frac{1}{e}\right\}$ **63.** $\left\{\frac{e^3}{2}\right\}$

64. 2042   **65.** 2086   **66.** 2005   **67.** 7.3 yr   **68.** 14.6 yr

69. about 21.97%   **70. a.** 0.045   **b.** 55.6 million   **c.** 2012   **71.** about 15,679 years old   **72. a.** 200 people
b. about 45,411 people   **c.** 500,000 people   **73. a.** $T = 65 + 120e^{-0.144t}$ **b.** 8 min   **74.** $y = 73e^{(\ln 2.6)x}$; $y = 73e^{0.956x}$
75. $y = 6.5e^{(\ln 0.43)x}$; $y = 6.5e^{-0.844x}$ **76.** high: exponential; medium: linear; low: quadratic; Explanations will vary; negative;
The parabola opens downward.   **77.** The exponential model, $y = (3.460)(1.024)^x$, is the best fit; about 116 million

Chapter 3 Test

1.

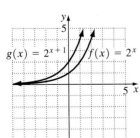

$g(x) = 2^{x+1}$ $f(x) = 2^x$

2.

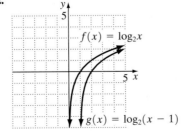

$f(x) = \log_2 x$ $g(x) = \log_2(x - 1)$

3. $5^3 = 125$ **4.** $\log_{36} 6 = \frac{1}{2}$ **5.** $(-\infty, 3)$

6. $3 + 5 \log_4 x$ **7.** $\frac{1}{3} \log_3 x - 4$ **8.** $\log(x^6 y^2)$

9. $\ln \frac{7}{x^3}$ **10.** 1.5741   **11.** $\left\{\frac{\ln 1.4}{\ln 5}\right\}$

12. $\left\{\frac{\ln 4}{0.005}\right\}$ **13.** $\{0, \ln 5\}$ **14.** $\{54.25\}$

15. $\{5\}$ **16.** $\left\{\frac{e^4}{3}\right\}$

17. 6.5% compounded semiannually; $221.15 more   **18.** 120 db   **19. a.** about 89%   **b.** decreasing; $k = -0.004 < 0$ **c.** 1995
20. $A = 509e^{0.036t}$ **21.** about 24,758 years ago   **22. a.** 14 elk   **b.** about 51 elk   **c.** 140 elk

Cumulative Review Exercises (Chapters P–3)

1. $\left\{ \dfrac{2}{3}, 2 \right\}$ **2.** $\{-1 + 2i, -1 - 2i\}$ **3.** $\{-2, -1, 1\}$ **4.** $\left\{ \dfrac{\ln 128}{5} \right\}$ **5.** $\{3\}$ **6.** $(-\infty, 4]$ **7.** $[1, 3]$

8. using $(1, 3)$, $y - 3 = -3(x - 1)$; $y = -3x + 6$ **9.** $(f \circ g)(x) = (x + 2)^2$; $(g \circ f)(x) = x^2 + 2$

10. $f^{-1}(x) = \dfrac{1}{2}x + \dfrac{7}{2}$ **11.** $x^2 + 3x - 3 + \dfrac{-4}{x + 2}$ **12.** $\pm 1, \pm \dfrac{1}{2}, \pm \dfrac{1}{4}, \pm 3, \pm \dfrac{3}{2}, \pm \dfrac{3}{4}$ **13.** 300   **14.** $f(1) = -1$ and $f(2) = 23$

15.
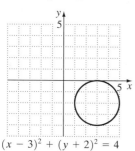
$(x - 3)^2 + (y + 2)^2 = 4$

16.
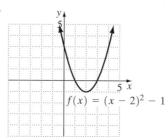
$f(x) = (x - 2)^2 - 1$

17.
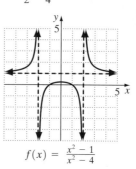
$f(x) = \dfrac{x^2 - 1}{x^2 - 4}$

18.
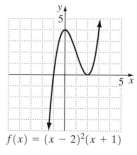
$f(x) = (x - 2)^2(x + 1)$

19. 500 yd by 500 yd; 250,000 sq yd   **20.** $\dfrac{0.5}{\ln 4} \approx 0.361$; about $\dfrac{361}{1000}$ of the people

CHAPTER 4

Section 4.1

Check Point Exercises

1. a.
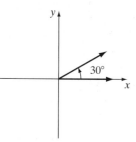

b.
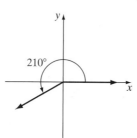

c.
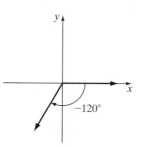

d.
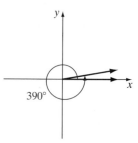

2. a. $40°$ **b.** $225°$ **3. a.** $12°$; $102°$ **b.** no complementary angle; $30°$ **4.** 3.5 radians   **5. a.** $\dfrac{\pi}{3}$ radians   **b.** $\dfrac{3\pi}{2}$ radians

c. $-\dfrac{5\pi}{3}$ radians   **6. a.** $45°$ **b.** $-240°$ **c.** $343.8°$ **7.** $\dfrac{3\pi}{2}$ in. ≈ 4.71 in.   **8.** 135π in./min ≈ 424 in./min

Exercise Set 4.1

1. quadrant II   **3.** quadrant III   **5.** quadrant I   **7.** obtuse   **9.** straight

11.
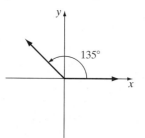

13.
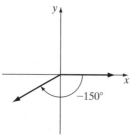

15.
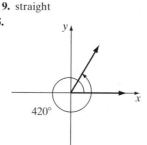

17.
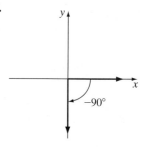

19. 35°   **21.** 210°   **23.** 315°   **25.** 38°; 128°   **27.** 52.6°; 142.6°   **29.** no complement; 69°   **31.** 4 radians   **33.** $\frac{4}{3}$ radians

35. 4 radians   **37.** $\frac{\pi}{4}$ radians   **39.** $\frac{3\pi}{4}$ radians   **41.** $\frac{5\pi}{3}$ radians   **43.** $-\frac{5\pi}{4}$ radians   **45.** 90°   **47.** 120°   **49.** 210°

51. −540°   **53.** 0.31 radians   **55.** −0.70 radians   **57.** 3.49 radians   **59.** 114.59°   **61.** 13.85°   **63.** −275.02°

65. 3π in. ≈ 9.42 in.   **67.** 10π ft ≈ 31.42 ft.   **69.** $\frac{12\pi \text{ radians}}{\text{second}}$ **71.** 60°; $\frac{\pi}{3}$ radians   **73.** $\frac{8\pi}{3}$ in. ≈ 8.38 in.

75. 12π in. ≈ 37.70 in.   **77.** 2 radians; 114.59°   **79.** 2094 mi   **81.** 1047 mph   **83.** 1508 ft/min   **97.** 30.25°   **99.** 30°25′12″

101. smaller than a right angle   **103.** 1815 mi

Section 4.2

Check Point Exercises

1. $\sin t = \frac{1}{2}$; $\cos t = \frac{\sqrt{3}}{2}$; $\tan t = \frac{\sqrt{3}}{3}$; $\csc t = 2$; $\sec t = \frac{2\sqrt{3}}{3}$; $\cot t = \sqrt{3}$ **2.** $\sin t = 0$; $\cos t = -1$; $\tan t = 0$; $\csc t = $ undefined;

$\sec t = -1$; $\cot t = $ undefined   **3.** $\sqrt{2}$; $\sqrt{2}$; 1   **4. a.** $\sqrt{2}$ **b.** $-\frac{\sqrt{2}}{2}$ **5.** $\tan t = \frac{2\sqrt{5}}{5}$; $\csc t = \frac{3}{2}$; $\sec t = \frac{3\sqrt{5}}{5}$; $\cot t = \frac{\sqrt{5}}{2}$

6. $\frac{\sqrt{3}}{2}$ **7.** 1   **8. a.** 0.7071   **b.** 1.0025

Exercise Set 4.2

1. $\sin t = \frac{8}{17}$; $\cos t = -\frac{15}{17}$; $\tan t = -\frac{8}{15}$; $\csc t = \frac{17}{8}$; $\sec t = -\frac{17}{15}$; $\cot t = -\frac{15}{8}$ **3.** $\sin t = -\frac{\sqrt{2}}{2}$; $\cos t = \frac{\sqrt{2}}{2}$; $\tan t = -1$;

$\csc t = -\sqrt{2}$; $\sec t = \sqrt{2}$; $\cot t = -1$ **5.** $\frac{1}{2}$ **7.** $-\frac{\sqrt{3}}{2}$ **9.** 0   **11.** −2   **13.** $\frac{2\sqrt{3}}{3}$ **15.** −1   **17.** undefined

19. a. $\frac{\sqrt{3}}{2}$ **b.** $\frac{\sqrt{3}}{2}$ **21. a.** $\frac{1}{2}$ **b.** $-\frac{1}{2}$ **23. a.** $-\sqrt{3}$ **b.** $\sqrt{3}$ **25.** $\tan t = \frac{8}{15}$; $\csc t = \frac{17}{8}$; $\sec t = \frac{17}{15}$; $\cot t = \frac{15}{8}$

27. $\tan t = \frac{\sqrt{2}}{4}$; $\csc t = 3$; $\sec t = \frac{3\sqrt{2}}{4}$; $\cot t = 2\sqrt{2}$ **29.** $\frac{\sqrt{13}}{7}$ **31.** $\frac{5}{8}$ **33.** 1   **35.** 1   **37.** 1   **39. a.** $\frac{\sqrt{2}}{2}$ **b.** $\frac{\sqrt{2}}{2}$

41. a. 0   **b.** 0   **43. a.** 0   **b.** 0   **45. a.** $-\frac{\sqrt{2}}{2}$ **b.** $-\frac{\sqrt{2}}{2}$ **47.** 0.7174   **49.** 0.2643   **51.** 1.1884   **53.** 0.9511

55. 3.7321   **57. a.** 12 hr   **b.** 20.3 hr   **c.** 3.7 hr   **59. a.** $1, 0, -1, 0, 1$; Answers may vary.   **b.** 28 days   **73.** 1   **75.** 16.7 ft

Section 4.3

Check Point Exercises

1. $\sin \theta = \frac{3}{5}$; $\cos \theta = \frac{4}{5}$; $\tan \theta = \frac{3}{4}$; $\csc \theta = \frac{5}{3}$; $\sec \theta = \frac{5}{4}$; $\cot \theta = \frac{4}{3}$ **2.** $\sqrt{2}$; $\sqrt{2}$; 1   **3.** $\sqrt{3}$; $\frac{\sqrt{3}}{3}$ **4. a.** $\cos 44°$ **b.** $\tan \frac{5\pi}{12}$

5. 334 yd   **6.** 54°

Exercise Set 4.3

1. 15; $\sin \theta = \frac{3}{5}$; $\cos \theta = \frac{4}{5}$; $\tan \theta = \frac{3}{4}$; $\csc \theta = \frac{5}{3}$; $\sec \theta = \frac{5}{4}$; $\cot \theta = \frac{4}{3}$ **3.** 20; $\sin \theta = \frac{20}{29}$; $\cos \theta = \frac{21}{29}$; $\tan \theta = \frac{20}{21}$; $\csc \theta = \frac{29}{20}$; $\sec \theta = \frac{29}{21}$;

$\cot \theta = \frac{21}{20}$ **5.** 24; $\sin \theta = \frac{5}{13}$; $\cos \theta = \frac{12}{13}$; $\tan \theta = \frac{5}{12}$; $\csc \theta = \frac{13}{5}$; $\sec \theta = \frac{13}{12}$; $\cot \theta = \frac{12}{5}$ **7.** 28; $\sin \theta = \frac{4}{5}$; $\cos \theta = \frac{3}{5}$; $\tan \theta = \frac{4}{3}$;

$\csc \theta = \frac{5}{4}$; $\sec \theta = \frac{5}{3}$; $\cot \theta = \frac{3}{4}$ **9.** $\frac{\sqrt{3}}{2}$ **11.** $\sqrt{2}$ **13.** $\sqrt{3}$ **15.** 0   **17.** $\frac{\sqrt{6} - 4}{4}$ **19.** $\frac{12\sqrt{3} + \sqrt{6}}{6}$ **21.** $\cos 83°$

23. $\sec 65°$ **25.** $\cot \frac{7\pi}{18}$ **27.** $\sin \frac{\pi}{10}$ **29.** 188 cm   **31.** 182 in.   **33.** 41 m   **35.** 17°   **37.** 78°   **39.** 1.147 radians

41. 0.395 radians   **43.** 529 yd   **45.** 36°   **47.** 2879 ft   **49.** 37°   **59.** 0.92106, −0.19735; 0.95534, −0.148878; 0.98007, −0.099667;

0.99500, −0.04996; 0.99995, −0.005; 0.9999995, −0.0005; 0.999999995, −0.00005; 1, −0.000005; $\frac{\cos \theta - 1}{\theta}$ approaches 0 as θ approaches 0.

61. In a right triangle, the hypotenuse is greater than either other side. Therefore, both $\frac{\text{opposite}}{\text{hypotenuse}}$ and $\frac{\text{adjacent}}{\text{hypotenuse}}$ must be less than 1 for

an acute angle in a right triangle.   **63. a.** 357 ft   **b.** 394 ft

Section 4.4

Check Point Exercises

1. $\sin \theta = -\dfrac{3}{5}$; $\cos \theta = \dfrac{4}{5}$; $\tan \theta = -\dfrac{3}{4}$; $\csc \theta = -\dfrac{5}{3}$; $\sec \theta = \dfrac{5}{4}$; $\cot \theta = -\dfrac{4}{3}$ **2. a.** 1; undefined **b.** 0; 1 **c.** −1; undefined

d. 0; −1 **3.** quadrant III **4.** $\dfrac{\sqrt{10}}{10}$; $-\dfrac{\sqrt{10}}{3}$ **5. a.** 30° **b.** $\dfrac{\pi}{4}$ **c.** 60° **d.** 0.46 **6. a.** $-\dfrac{\sqrt{3}}{2}$ **b.** 1 **c.** $\dfrac{2\sqrt{3}}{3}$

Exercise Set 4.4

1. $\sin \theta = \dfrac{3}{5}$; $\cos \theta = -\dfrac{4}{5}$; $\tan \theta = -\dfrac{3}{4}$; $\csc \theta = \dfrac{5}{3}$; $\sec \theta = -\dfrac{5}{4}$; $\cot \theta = -\dfrac{4}{3}$ **3.** $\sin \theta = \dfrac{3\sqrt{13}}{13}$; $\cos \theta = \dfrac{2\sqrt{13}}{13}$; $\tan \theta = \dfrac{3}{2}$; $\csc \theta = \dfrac{\sqrt{13}}{3}$;

$\sec \theta = \dfrac{\sqrt{13}}{2}$; $\cot \theta = \dfrac{2}{3}$ **5.** $\sin \theta = -\dfrac{\sqrt{2}}{2}$; $\cos \theta = \dfrac{\sqrt{2}}{2}$; $\tan \theta = -1$; $\csc \theta = -\sqrt{2}$; $\sec \theta = \sqrt{2}$; $\cot \theta = -1$

7. $\sin \theta = -\dfrac{5\sqrt{29}}{29}$; $\cos \theta = -\dfrac{2\sqrt{29}}{29}$; $\tan \theta = \dfrac{5}{2}$; $\csc \theta = -\dfrac{\sqrt{29}}{5}$; $\sec \theta = -\dfrac{\sqrt{29}}{2}$; $\cot \theta = \dfrac{2}{5}$ **9.** −1 **11.** −1 **13.** undefined

15. 0 **17.** quadrant I **19.** quadrant III **21.** quadrant II **23.** $\sin \theta = -\dfrac{4}{5}$; $\tan \theta = \dfrac{4}{3}$; $\csc \theta = -\dfrac{5}{4}$; $\sec \theta = -\dfrac{5}{3}$; $\cot \theta = \dfrac{3}{4}$

25. $\cos \theta = -\dfrac{12}{13}$; $\tan \theta = -\dfrac{5}{12}$; $\csc \theta = \dfrac{13}{5}$; $\sec \theta = -\dfrac{13}{12}$; $\cot \theta = -\dfrac{12}{5}$ **27.** $\sin \theta = -\dfrac{15}{17}$; $\tan \theta = -\dfrac{15}{8}$; $\csc \theta = -\dfrac{17}{15}$; $\sec \theta = \dfrac{17}{8}$;

$\cot \theta = -\dfrac{8}{15}$ **29.** $\sin \theta = \dfrac{2\sqrt{13}}{13}$; $\cos \theta = -\dfrac{3\sqrt{13}}{13}$; $\csc \theta = \dfrac{\sqrt{13}}{2}$; $\sec \theta = -\dfrac{\sqrt{13}}{3}$; $\cot \theta = -\dfrac{3}{2}$ **31.** $\sin \theta = -\dfrac{4}{5}$; $\cos \theta = -\dfrac{3}{5}$;

$\csc \theta = -\dfrac{5}{4}$; $\sec \theta = -\dfrac{5}{3}$; $\cot \theta = \dfrac{3}{4}$ **33.** $\sin \theta = -\dfrac{2\sqrt{2}}{3}$; $\cos \theta = -\dfrac{1}{3}$; $\tan \theta = 2\sqrt{2}$; $\csc \theta = -\dfrac{3\sqrt{2}}{4}$; $\cot \theta = \dfrac{\sqrt{2}}{4}$ **35.** 20°

37. 25° **39.** 5° **41.** $\dfrac{\pi}{4}$ **43.** $\dfrac{\pi}{6}$ **45.** 30° **47.** 25° **49.** 1.56 **51.** $-\dfrac{\sqrt{2}}{2}$ **53.** $\dfrac{\sqrt{3}}{3}$ **55.** $\sqrt{3}$ **57.** $\dfrac{\sqrt{3}}{2}$ **59.** −2

61. 1 **63.** $\dfrac{\sqrt{3}}{2}$ **65.** −1

Section 4.5

Check Point Exercises

1. 3 **2.** $\dfrac{1}{2}$ **3.** $2; 4\pi$ **4.** $3; \pi; \dfrac{\pi}{6}$

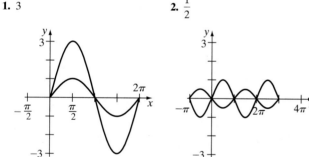

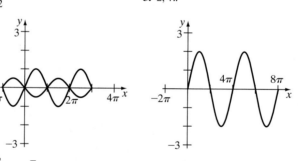

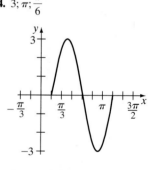

5. 4; 2 **6.** $\dfrac{3}{2}; \pi; -\dfrac{\pi}{2}$ **7.** **8.** $y = 4 \sin 4x$

9. $y = 2 \sin\left(\dfrac{\pi}{6}x - \dfrac{\pi}{2}\right) + 12$

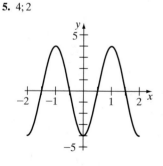

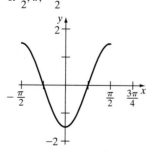

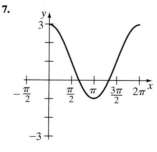

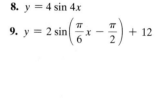

Exercise Set 4.5

1. 4

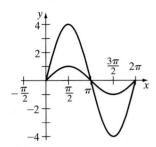

3. $\frac{1}{3}$

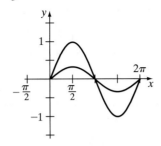

5. 3

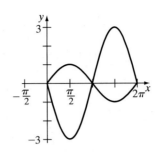

7. 1; π

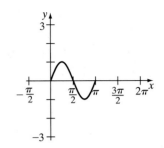

9. 3; 4π

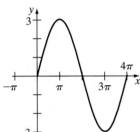

11. 4; 2

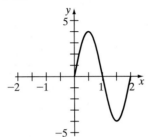

13. 3; 1

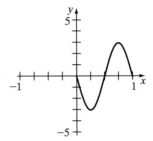

15. 1; 3π

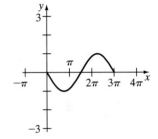

17. 1; 2π; π

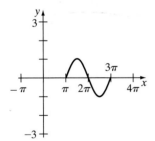

19. 1; π; $\frac{\pi}{2}$

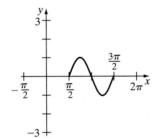

21. 3; π; $\frac{\pi}{2}$

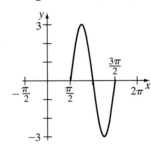

23. $\frac{1}{2}$; 2π; $-\frac{\pi}{2}$

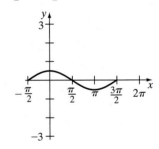

25. 2; π; $-\frac{\pi}{4}$

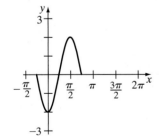

27. 3; 2; $-\frac{2}{\pi}$

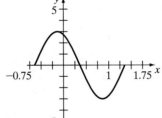

29. 2; 1; -2

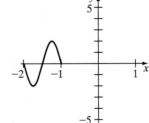

31. 2

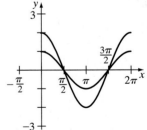

33. 2

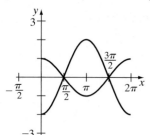

35. 1; π

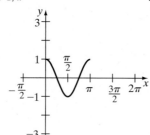

37. 4; 1

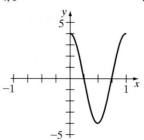

39. 4; 4π

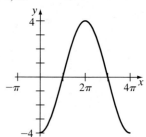

41. $\frac{1}{2}$; 6

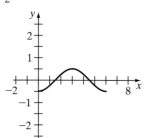

43. 3; π; $\frac{\pi}{2}$

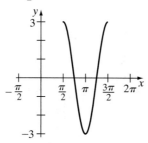

45. $\frac{1}{2}$; $\frac{2\pi}{3}$; $-\frac{\pi}{6}$

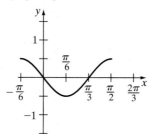

47. 3; π; $\frac{\pi}{4}$

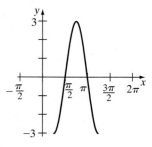

49. 2; 1; -4

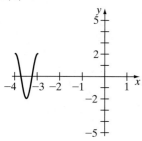

51.

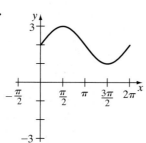

53.

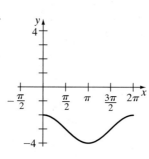

55.

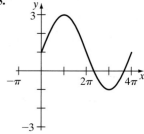

57.

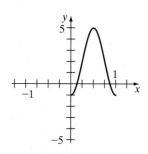

59. 33 days   **61.** 23 days   **63.** March 21   **65.** No

67.

69. a. 3   **b.** 365 days
   c. 15 hours of daylight
   d. 9 hours of daylight
   e.

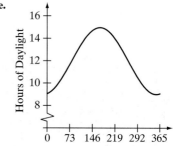

71. $y = 3 \cos \dfrac{\pi x}{6} + 9$

85.

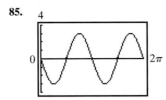

87.

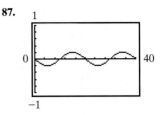

89.

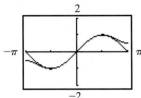

91.

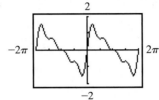

The graph is similar to $y = \sin x$, except the amplitude is greater and the curve is less smooth.

93. a.

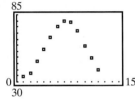

b. $y = 22.61 \sin(0.50x - 2.04) + 57.17$

c.

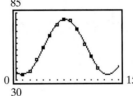

95.

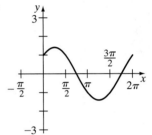

97. $y = 2 \cos(4x + \pi)$

Section 4.6

Check Point Exercises

1.

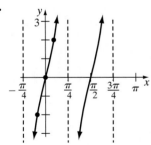

2.

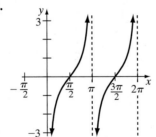

3.

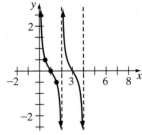

4.

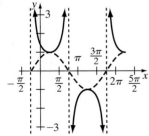

5.

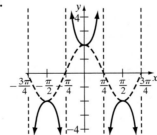

Exercise Set 4.6

1. $y = \tan(x + \pi)$ **3.** $y = -\tan\left(x - \dfrac{\pi}{2}\right)$

5.
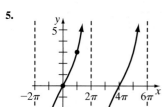

7.
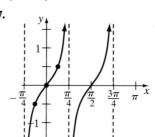

9.
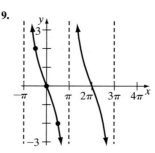

11.
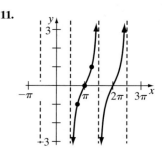

13. $y = -\cot x$ **15.** $y = \cot\left(x + \dfrac{\pi}{2}\right)$

17.
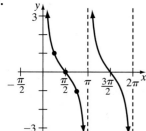

19.
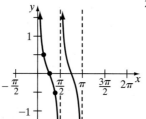

21.
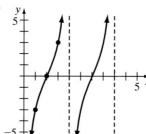

23.
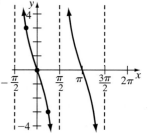

25. $y = -\dfrac{1}{2}\csc\dfrac{x}{2}$;
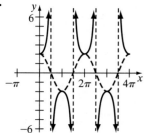

27. $y = \dfrac{1}{2}\sec 2\pi x$;

29.

31.

33.
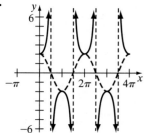

35.
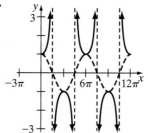

37.
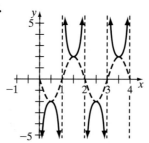

39.
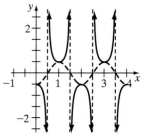

41.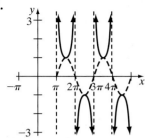

43.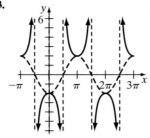

45. a.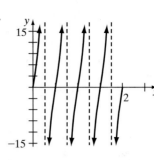

b. 0.25, 0.75, 1.25, 1.75; The beacon is shining parallel to the wall at these times.

47. $d = 10 \sec x$

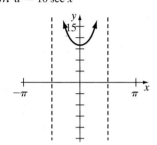

49.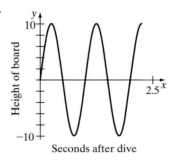

Height of board / Seconds after dive

63.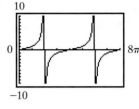

65.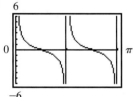

67.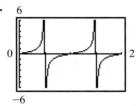

69.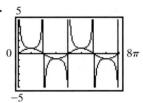

71.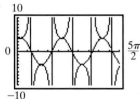

73.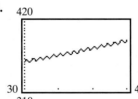

75. $y = \cot \dfrac{3}{2} x$

77. 2^{-x} decreases the amplitude as x gets larger.

Section 4.7

Check Point Exercises

1. $\dfrac{\pi}{3}$ **2.** $-\dfrac{\pi}{4}$ **3.** $\dfrac{2\pi}{3}$ **4.** $-\dfrac{\pi}{4}$ **5. a.** 1.2310 **b.** -1.5429 **6. a.** 0.7 **b.** 0 **c.** not defined **7.** $\dfrac{3}{5}$ **8.** $\dfrac{\sqrt{3}}{2}$
9. $\sqrt{x^2 + 1}$

Exercise Set 4.7

1. $\dfrac{\pi}{6}$ **3.** $\dfrac{\pi}{4}$ **5.** $-\dfrac{\pi}{6}$ **7.** $\dfrac{\pi}{6}$ **9.** $\dfrac{3\pi}{4}$ **11.** $\dfrac{\pi}{2}$ **13.** $\dfrac{\pi}{6}$ **15.** 0 **17.** $-\dfrac{\pi}{3}$ **19.** 0.30 **21.** -0.33 **23.** 1.19
25. 1.25 **27.** -1.52 **29.** -1.52 **31.** 0.9 **33.** $\dfrac{\pi}{3}$ **35.** $\dfrac{\pi}{6}$ **37.** 125 **39.** $-\dfrac{\pi}{6}$ **41.** $-\dfrac{\pi}{3}$ **43.** 0 **45.** not defined
47. $\dfrac{3}{5}$ **49.** $\dfrac{12}{5}$ **51.** $-\dfrac{3}{4}$ **53.** $\dfrac{\sqrt{2}}{2}$ **55.** $\dfrac{4\sqrt{15}}{15}$ **57.** $-2\sqrt{2}$ **59.** 2 **61.** $\dfrac{\sqrt{1-x^2}}{x}$ **63.** $\sqrt{1-4x^2}$ **65.** $\dfrac{\sqrt{x^2-1}}{x}$
67. $\dfrac{\sqrt{3}}{x}$ **69.** $\dfrac{\sqrt{x^2+4}}{2}$

71. a.

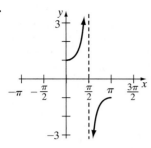

b. No horizontal line intersects the graph of $y = \sec x$ more than once, so the function is one-to-one and has an inverse function.

c.

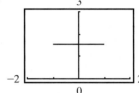

73. 0.408 radians; 0.602 radians; 0.654 radians; 0.645 radians; 0.613 radians   **75.** 1.3157 radians or 75.4°   **77.** 1.1071 sq units

91.

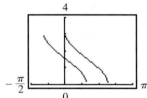

93.

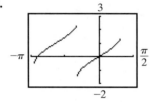

Shifted left 2 units and up 1 unit

95.

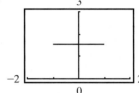

It seems
$$\sin^{-1} x + \cos^{-1} x = \frac{\pi}{2} \text{ for } -1 \le x \le 1$$

97. $x = \sin\dfrac{\pi}{8}$ **99.** $\tan \alpha = \dfrac{8}{x}$, so $\tan^{-1}\dfrac{8}{x} = \alpha$. $\tan(\alpha + \theta) = \dfrac{33}{x}$, so $\tan^{-1}\dfrac{33}{x} = \alpha + \theta$. $\theta = \alpha + \theta - \alpha = \tan^{-1}\dfrac{33}{x} - \tan^{-1}\dfrac{8}{x}$.

Section 4.8

Check Point Exercises

1. $B = 27.3°$; $b \approx 4.34$; $c \approx 9.45$ **2.** 994 ft   **3.** 29.0°   **4.** 60.3 ft   **5. a.** S 25°E   **b.** S 15°W   **6. a.** 4.2 m   **b.** S87.7°W

7. $d = -6\cos\dfrac{\pi}{2}t$ **8. a.** 12 cm   **b.** $\dfrac{1}{8}$ cm per sec   **c.** 8 sec

Exercise Set 4.8

1. $B = 66.5°$; $a \approx 4.35$; $c \approx 10.90$ **3.** $B = 37.4°$; $a \approx 42.90$; $b \approx 32.80$ **5.** $A = 73.2°$; $a \approx 101.02$; $c \approx 105.52$

7. $b \approx 39.95$; $A \approx 37.3°$; $B \approx 52.7°$ **9.** $c \approx 26.96$; $A \approx 23.6°$; $B \approx 66.4°$ **11.** $a \approx 6.71$; $B \approx 16.6°$; $A \approx 73.4°$ **13.** N 15° E

15. S 80° W   **17.** $d = -6\cos\dfrac{\pi}{2}t$ **19.** $d = 3\sin\dfrac{4\pi}{3}t$ **21. a.** 5 in.   **b.** $\dfrac{1}{4}$ in per sec   **c.** 4 sec   **23. a.** 6 in.   **b.** 1 in. per sec

c. 1 sec   **25. a.** $\dfrac{1}{2}$ in.   **b.** 0.32 in. per sec   **c.** 3.14 sec   **27. a.** 5 in.   **b.** $\dfrac{1}{3}$ in. per sec   **c.** 3 sec   **29.** 2059 ft   **31.** 695 ft

33. 1376 ft   **35.** 15.1°   **37.** 33.7 ft   **39.** 90 mi north and 120 mi east   **41.** 13.2 mi   **43.** N 53° W   **45.** N 89.5° E

47. $d = 6\sin \pi t$ **49.** $d = \sin 528 \pi t$ **59.**

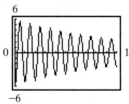

10 complete oscillations

61. 48 ft

Chapter 4 Review Exercises

1.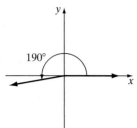

2.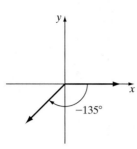

3.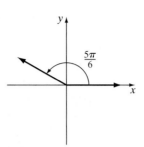

4.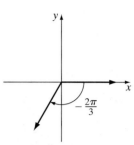

5. $40°$ **6.** $275°$ **7.** $17°; 107°$ **8.** no complement; $\dfrac{\pi}{3}$ radians **9.** 4.5 radians **10.** $\dfrac{\pi}{12}$ radians **11.** $\dfrac{2\pi}{3}$ radians

12. $\dfrac{7\pi}{4}$ radians **13.** $300°$ **14.** $252°$ **15.** $-150°$ **16.** $\dfrac{15\pi}{2}$ ft ≈ 23.56 ft **17.** 20.6π radians per min **18.** 42,412 ft per min

19. $\sin t = -\dfrac{3}{5}; \cos t = -\dfrac{4}{5}; \tan t = \dfrac{3}{4}; \csc t = -\dfrac{5}{3}; \sec t = -\dfrac{5}{4}; \cot t = \dfrac{4}{3}$ **20.** $\sin t = -\dfrac{15}{17}; \cos t = \dfrac{8}{17}; \tan t = -\dfrac{15}{8}; \csc t = -\dfrac{17}{15};$ $\sec t = \dfrac{17}{8}; \cot t = -\dfrac{8}{15}$ **21.** $-\dfrac{2\sqrt{3}}{3}$ **22.** $\sqrt{3}$ **23.** undefined **24.** undefined **25.** $\cos t = \dfrac{\sqrt{21}}{7}; \tan t = \dfrac{2\sqrt{3}}{3}; \csc t = \dfrac{\sqrt{7}}{2};$ $\sec t = \dfrac{\sqrt{21}}{3}; \cot t = \dfrac{\sqrt{3}}{2}$ **26.** 1 **27.** 1 **28.** -1 **29.** $\sin \theta = \dfrac{3}{5}, \cos \theta = \dfrac{4}{5}, \tan \theta = \dfrac{3}{4}, \csc \theta = \dfrac{5}{3}, \sec \theta = \dfrac{5}{4}, \cot \theta = \dfrac{4}{3}$ **30.** 5

31. $\dfrac{\sqrt{6}-4}{4}$ **32.** $\cos 20°$ **33.** $\sin 0$ **34.** 42 mm **35.** 23 cm **36.** 37 in. **37.** 772 ft **38.** 31 m **39.** $56°$

40. $\sin \theta = -\dfrac{5\sqrt{26}}{26}; \cos \theta = -\dfrac{\sqrt{26}}{26}; \tan \theta = 5; \csc \theta = -\dfrac{\sqrt{26}}{5}; \sec \theta = -\sqrt{26}; \cot \theta = \dfrac{1}{5}$ **41.** $\sin \theta = -1; \cos \theta = 0; \tan \theta$ is undefined;

$\csc \theta = -1; \sec \theta$ is undefined; $\cot \theta = 0$ **42.** quadrant I **43.** quadrant III **44.** $\sin \theta = -\dfrac{\sqrt{21}}{5}; \tan \theta = -\dfrac{\sqrt{21}}{2}; \csc \theta = -\dfrac{5\sqrt{21}}{21};$

$\sec \theta = \dfrac{5}{2}; \cot \theta = -\dfrac{2\sqrt{21}}{21}$ **45.** $\sin \theta = \dfrac{\sqrt{10}}{10}; \cos \theta = -\dfrac{3\sqrt{10}}{10}; \csc \theta = \sqrt{10}; \sec \theta = -\dfrac{\sqrt{10}}{3}; \cot \theta = -3$ **46.** $85°$ **47.** $\dfrac{3\pi}{8}$

48. $50°$ **49.** $-\dfrac{\sqrt{3}}{2}$ **50.** $-\sqrt{3}$ **51.** $\sqrt{2}$ **52.** $\dfrac{\sqrt{3}}{2}$ **53.** $-\sqrt{3}$ **54.** $-\dfrac{2\sqrt{3}}{3}$ **55.** $-\dfrac{\sqrt{3}}{2}$ **56.** $\dfrac{\sqrt{2}}{2}$ **57.** 1

58. $3; \dfrac{\pi}{2}$ **59.** $2; \pi$ **60.** $2; 4\pi$ **61.** $\dfrac{1}{2}; 6$

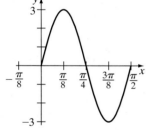

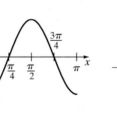

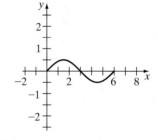

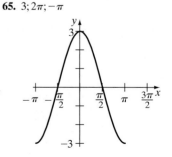

62. $1; 2$ **63.** $3; 6\pi$ **64.** $2; 2\pi; \pi$ **65.** $3; 2\pi; -\pi$

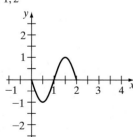

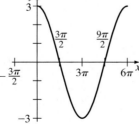

66. $\frac{3}{2}$; π; $-\frac{\pi}{8}$

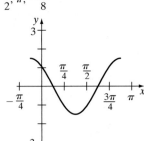

67. $\frac{5}{2}$; π; $-\frac{\pi}{4}$

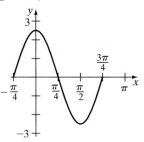

68. 3; 6; 9

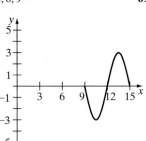

69.

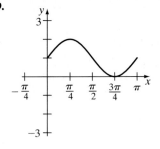

70.

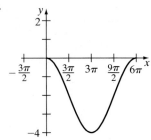

71. a. $\approx 98.52°$ **b.** 24 hr **c.** 5:00 P.M.; 98.9° **d.** 5:00 A.M.; 98.3° **e.**

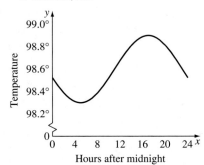

72.

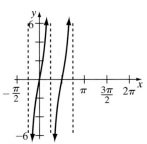

73.

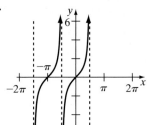

74.

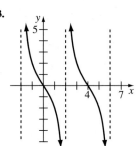

75.

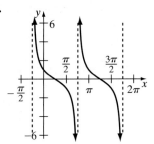

76.

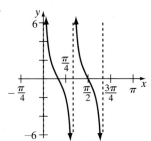

77.

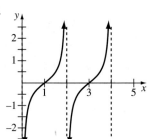

78.

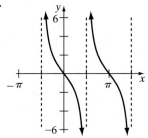

79.

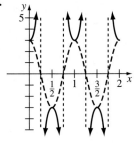

80.

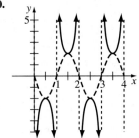

81.

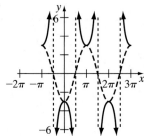

82.

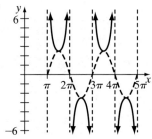

83. $\frac{\pi}{2}$ **84.** 0

85. $\frac{\pi}{4}$ **86.** $-\frac{\pi}{3}$

87. $\frac{2\pi}{3}$ **88.** $-\frac{\pi}{6}$

89. $\frac{\sqrt{2}}{2}$ **90.** 1

91. $-\frac{\sqrt{3}}{3}$ **92.** $-\frac{\sqrt{3}}{3}$

93. 2　　**94.** $\dfrac{4}{5}$　　**95.** $\dfrac{4}{5}$　　**96.** $-\dfrac{3}{4}$　　**97.** $-\dfrac{3}{4}$　　**98.** $\dfrac{\pi}{3}$　　**99.** $\dfrac{\pi}{3}$　　**100.** $-\dfrac{\pi}{6}$　　**101.** $\dfrac{2}{\sqrt{x^2+4}}$　　**102.** $\dfrac{x}{\sqrt{x^2-1}}$

103. $B \approx 67.7°; a \approx 3.79; b \approx 9.25$　　**104.** $A \approx 52.6°; a \approx 7.85; c \approx 9.88$　　**105.** $A \approx 16.6°; B \approx 73.4°; b \approx 6.71$

106. $A \approx 21.3°; B \approx 68.7°; c \approx 3.86$　　**107.** 38 ft　　**108.** 90 yd　　**109.** 21.7 ft　　**110.** N 35° E　　**111.** S 35° W

112. 24.6 mi　　**113. a.** 1282.2 mi　　**b.** S74°E　　**114. a.** 20 cm　　**b.** $\dfrac{1}{8}$ cycle per sec　　**c.** 8 sec

115. a. $\dfrac{1}{2}$ cm　　**b.** 0.64 cycle per sec　　**c.** 1.57 sec　　**116.** $d = -30 \cos \pi t$　　**117.** $d = \dfrac{1}{4} \sin \dfrac{2\pi}{5} t$

Chapter 4 Test

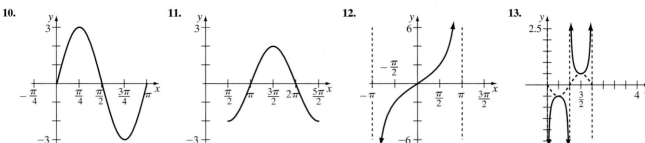

1. $\dfrac{3\pi}{4}$ radians　　**2.** $\dfrac{4\pi}{13}$　　**3.** $\dfrac{25\pi}{3}$ ft ≈ 26.18 ft　　**4.** $\sin \theta = \dfrac{5\sqrt{29}}{29}; \cos \theta = -\dfrac{2\sqrt{29}}{29}; \tan \theta = -\dfrac{5}{2}; \csc \theta = \dfrac{\sqrt{29}}{5}; \sec \theta = -\dfrac{\sqrt{29}}{2}; \cot \theta = -\dfrac{2}{5}$

5. quadrant III　　**6.** $\sin \theta = -\dfrac{2\sqrt{2}}{3}; \tan \theta = -2\sqrt{2}; \csc \theta = -\dfrac{3\sqrt{2}}{4}; \sec \theta = 3; \cot \theta = -\dfrac{\sqrt{2}}{4}$　　**7.** $\dfrac{\sqrt{3}}{6}$　　**8.** $-\sqrt{3}$　　**9.** $-\dfrac{\sqrt{2}}{2}$

10.
11.
12.
13.

14. $-\sqrt{3}$　　**15.** $B = 69°; a = 4.7; b = 12.1$　　**16.** 23 yd　　**17.** 36.1°　　**18.** N 80° W　　**19. a.** 6 in.　　**b.** $\dfrac{1}{2}$ in. per sec　　**c.** 2 sec

20. Trigonometric functions are periodic.

Cumulative Review Exercises (Chapters P–4)

1. $\{-3, 6\}$　　**2.** $\{-5, -2, 2\}$　　**3.** $\{4\}$　　**4.** $\left\{\dfrac{\ln 4500}{0.04}\right\}$　　**5.** $\{-1, 2, 3\}$　　**6.** $-3 \le x \le 8$　　**7.** $f^{-1}(x) = x^2 + 6$

8. $4x^2 - \dfrac{14}{5}x - \dfrac{17}{25} + \dfrac{284}{125x + 50}$　　**9.** $\log 1000 = 3$　　**10.** $280°$　　**11.** 3 positive real roots; 1 negative real root

12.
13.
14.

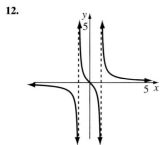
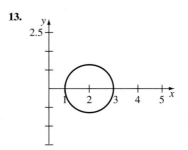
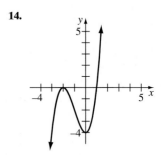

15.
16.
17. $A(x) = \dfrac{x^2}{4}$

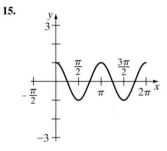
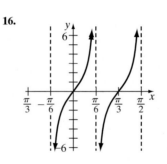
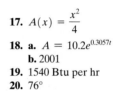

18. a. $A = 10.2e^{0.3057t}$
b. 2001
19. 1540 Btu per hr
20. 76°

CHAPTER 5

Section 5.1

Check Point Exercises

1. $\csc x \tan x = \dfrac{1}{\sin x} \cdot \dfrac{\sin x}{\cos x} = \dfrac{1}{\cos x} = \sec x$ **2.** $\cos x \cot x + \sin x = \cos x \cdot \dfrac{\cos x}{\sin x} + \sin x = \dfrac{\cos^2 x}{\sin x} + \sin x \cdot \dfrac{\sin x}{\sin x}$

$= \dfrac{\cos^2 x + \sin^2 x}{\sin x} = \dfrac{1}{\sin x} = \csc x$ **3.** $\sin x - \sin x \cos^2 x = \sin x(1 - \cos^2 x) = \sin x \cdot \sin^2 x = \sin^3 x$

4. $\dfrac{\sin x}{1 + \cos x} + \dfrac{1 + \cos x}{\sin x} = \dfrac{\sin x(\sin x)}{(1 + \cos x)(\sin x)} + \dfrac{(1 + \cos x)(1 + \cos x)}{(\sin x)(1 + \cos x)} = \dfrac{\sin^2 x + 1 + 2\cos x + \cos^2 x}{(1 + \cos x)(\sin x)}$

$= \dfrac{\sin^2 x + \cos^2 x + 1 + 2\cos x}{(1 + \cos x)(\sin x)} = \dfrac{1 + 1 + 2\cos x}{(1 + \cos x)(\sin x)} = \dfrac{2 + 2\cos x}{(1 + \cos x)(\sin x)} = \dfrac{2(1 + \cos x)}{(1 + \cos x)(\sin x)} = \dfrac{2}{\sin x} = 2\csc x$

5. $\dfrac{\cos x}{1 + \sin x} = \dfrac{\cos x(1 - \sin x)}{(1 + \sin x)(1 - \sin x)} = \dfrac{\cos x(1 - \sin x)}{1 - \sin^2 x} = \dfrac{\cos x(1 - \sin x)}{\cos^2 x} = \dfrac{1 - \sin x}{\cos x}$ **6.** $\dfrac{\sec x + \csc(-x)}{\sec x \csc x} = \dfrac{\sec x - \csc x}{\sec x \csc x}$

$= \dfrac{\dfrac{1}{\cos x} - \dfrac{1}{\sin x}}{\dfrac{1}{\cos x} \cdot \dfrac{1}{\sin x}} = \dfrac{\dfrac{\sin x}{\cos x \cdot \sin x} - \dfrac{\cos x}{\cos x \cdot \sin x}}{\dfrac{1}{\cos x \cdot \sin x}} = \dfrac{\dfrac{\sin x - \cos x}{\cos x \cdot \sin x}}{\dfrac{1}{\cos x \cdot \sin x}} = \dfrac{\sin x - \cos x}{\cos x \cdot \sin x} \cdot \dfrac{\cos x \cdot \sin x}{1} = \sin x - \cos x$

7. Left side: $\dfrac{1}{1 + \sin\theta} + \dfrac{1}{1 - \sin\theta} = \dfrac{1(1 - \sin\theta)}{(1 + \sin\theta)(1 - \sin\theta)} + \dfrac{1(1 + \sin\theta)}{(1 - \sin\theta)(1 + \sin\theta)} = \dfrac{1 - \sin\theta + 1 + \sin\theta}{(1 + \sin\theta)(1 - \sin\theta)} = \dfrac{2}{1 - \sin^2\theta}$;

Right side: $2 + 2\tan^2\theta = 2 + 2\left(\dfrac{\sin^2\theta}{\cos^2\theta}\right) = \dfrac{2\cos^2\theta}{\cos^2\theta} + \dfrac{2\sin^2\theta}{\cos^2\theta} = \dfrac{2\cos^2\theta + 2\sin^2\theta}{\cos^2\theta} = \dfrac{2(\cos^2\theta + \sin^2\theta)}{\cos^2\theta} = \dfrac{2}{\cos^2\theta} = \dfrac{2}{1 - \sin^2\theta}$

Exercise Set 5.1

For Exercises 1–59, proofs may vary.

65.

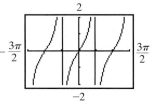

Proofs may vary

67.

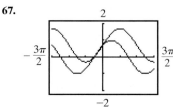

Values for x may vary.

69.

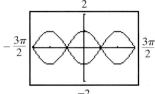

Values for x may vary.

71.

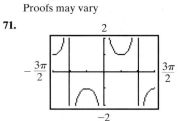

Proofs may vary.

73.

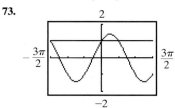

Values for x may vary.

75. Proofs may vary.
77. Answers may vary.

Section 5.2

Check Point Exercises

1. $\dfrac{\sqrt{3}}{2}$ **2.** $\dfrac{\sqrt{3}}{2}$ **3.** $\dfrac{\cos(\alpha - \beta)}{\cos\alpha\cos\beta} = \dfrac{\cos\alpha\cos\beta + \sin\alpha\sin\beta}{\cos\alpha\cos\beta} = \dfrac{\cos\alpha}{\cos\alpha} \cdot \dfrac{\cos\beta}{\cos\beta} + \dfrac{\sin\alpha}{\cos\alpha} \cdot \dfrac{\sin\beta}{\cos\beta} = 1 + \tan\alpha\tan\beta$

4. $\dfrac{\sqrt{2} + \sqrt{6}}{4}$ **5. a.** $\cos\alpha = -\dfrac{3}{5}$ **b.** $\cos\beta = \dfrac{\sqrt{3}}{2}$ **c.** $\dfrac{-3\sqrt{3} - 4}{10}$ **d.** $\dfrac{4\sqrt{3} - 3}{10}$

6. a. $y = \sin x$ **b.** $\cos\left(x + \dfrac{3\pi}{2}\right) = \cos x \cos\dfrac{3\pi}{2} - \sin x \sin\dfrac{3\pi}{2} = \cos x \cdot 0 - \sin x \cdot (-1) = \sin x$

7. $\tan(x + \pi) = \dfrac{\tan x + \tan\pi}{1 - \tan x \tan\pi} = \dfrac{\tan x + 0}{1 - \tan x \cdot 0} = \dfrac{\tan x}{1} = \tan x$

Exercise Set 5.2

1. $\dfrac{\sqrt{6} + \sqrt{2}}{4}$ **3.** $\dfrac{\sqrt{2} - \sqrt{6}}{4}$ **5. a.** $\alpha = 50°, \beta = 20°$ **b.** $\cos 30°$ **c.** $\dfrac{\sqrt{3}}{2}$ **7. a.** $\alpha = \dfrac{5\pi}{12}, \beta = \dfrac{\pi}{12}$ **b.** $\cos\left(\dfrac{\pi}{3}\right)$ **c.** $\dfrac{1}{2}$

For Exercises 9–11, proofs may vary. **13.** $\dfrac{\sqrt{6} - \sqrt{2}}{4}$ **15.** $\dfrac{\sqrt{6} + \sqrt{2}}{4}$ **17.** $\sqrt{3} + 2$ **19.** $2 - \sqrt{3}$ **21.** $-\dfrac{\sqrt{6} + \sqrt{2}}{4}$

23. $\dfrac{\sqrt{6} - \sqrt{2}}{4}$ **25.** $\sin 30°; \dfrac{1}{2}$ **27.** $\tan 45°; 1$ **29.** $\sin\dfrac{\pi}{6}; \dfrac{1}{2}$ **31.** $\tan\dfrac{\pi}{6}; \dfrac{\sqrt{3}}{3}$ For Exercises 33–55, proofs may vary.

57. a. $-\dfrac{63}{65}$ **b.** $-\dfrac{16}{65}$ **c.** $\dfrac{16}{63}$ **59. a.** $-\dfrac{4 + 6\sqrt{2}}{15}$ **b.** $\dfrac{3 - 8\sqrt{2}}{15}$ **c.** $\dfrac{54 - 25\sqrt{2}}{28}$ **61. a.** $-\dfrac{8\sqrt{3} + 15}{34}$ **b.** $\dfrac{15\sqrt{3} - 8}{34}$

c. $\dfrac{480 - 289\sqrt{3}}{33}$ **63. a.** $y = \sin x$ **b.** $\sin(\pi - x) = \sin\pi \cos x - \cos\pi \sin x = 0 \cdot \cos x - (-1)\sin x = \sin x$

65. a. $y = 2\cos x$ **b.** $\sin\left(x + \dfrac{\pi}{2}\right) + \sin\left(\dfrac{\pi}{2} - x\right) = \sin x \cos\dfrac{\pi}{2} + \cos x \sin\dfrac{\pi}{2} + \sin\dfrac{\pi}{2}\cos x - \cos\dfrac{\pi}{2}\sin x$

$= \sin x \cdot 0 + \cos x \cdot 1 + 1 \cdot \cos x - 0 \cdot \sin x = \cos x + \cos x = 2\cos x$ **67.** Proofs may vary.; amplitude is $\sqrt{13}$; period is 2π

77.

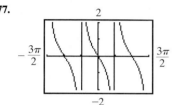

Proofs may vary.

79.

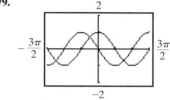

Values for x may vary.

81.

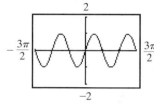

Proofs may vary.

83. Proofs may vary. **85.** $-\dfrac{24}{25}$ **87.** -1 **89.** $\dfrac{x\sqrt{1 - y^2} - y}{\sqrt{1 + x^2}}$

Section 5.3

Check Point Exercises

1. a. $-\dfrac{24}{25}$ **b.** $-\dfrac{7}{25}$ **c.** $\dfrac{24}{7}$ **2.** $\dfrac{\sqrt{3}}{2}$ **3.** $\sin 3\theta = \sin(2\theta + \theta) = \sin 2\theta \cos\theta + \cos 2\theta \sin\theta = 2\sin\theta \cos\theta \cos\theta$

$= 2\sin\theta \cos\theta \cos\theta + (2\cos^2\theta - 1)\sin\theta = 2\sin\theta \cos^2\theta + 2\sin\theta \cos^2\theta - \sin\theta = 4\sin\theta \cos^2\theta - \sin\theta = 4\sin\theta(1 - \sin^2\theta) - \sin\theta$

$= 4\sin\theta - 4\sin^3\theta - \sin\theta = 3\sin\theta - 4\sin^3\theta$

4. $\sin^4 x = (\sin^2 x)^2 = \left(\dfrac{1 - \cos 2x}{2}\right)^2 = \dfrac{1 - 2\cos 2x + \cos^2 2x}{4} = \dfrac{1}{4} - \dfrac{1}{2}\cos 2x + \dfrac{1}{4}\cos^2 2x = \dfrac{1}{4} - \dfrac{1}{2}\cos 2x + \dfrac{1}{4}\left(\dfrac{1 + \cos 2(2x)}{2}\right)$

$= \dfrac{1}{4} - \dfrac{1}{2}\cos 2x + \dfrac{1}{8} + \dfrac{1}{8}\cos 4x = \dfrac{3}{8} - \dfrac{1}{2}\cos 2x + \dfrac{1}{8}\cos 4x$

5. $-\dfrac{\sqrt{2 + \sqrt{3}}}{2}$ **6.** $\dfrac{\sin 2\theta}{1 + \cos 2\theta} = \dfrac{2\sin\theta \cos\theta}{1 + (1 - 2\sin^2\theta)} = \dfrac{2\sin\theta \cos\theta}{2 - 2\sin^2\theta} = \dfrac{2\sin\theta \cos\theta}{2(1 - \sin^2\theta)} = \dfrac{2\sin\theta \cos\theta}{2\cos^2\theta} = \dfrac{\sin\theta}{\cos\theta} = \tan\theta$

7. $\dfrac{\sec\alpha}{\sec\alpha \csc\alpha + \csc\alpha} = \dfrac{\dfrac{1}{\cos\alpha}}{\dfrac{1}{\cos\alpha}\cdot\dfrac{1}{\sin\alpha} + \dfrac{1}{\sin\alpha}} = \dfrac{\dfrac{1}{\cos\alpha}}{\dfrac{1}{\cos\alpha\sin\alpha} + \dfrac{\cos\alpha}{\cos\alpha\sin\alpha}} = \dfrac{\dfrac{1}{\cos\alpha}}{\dfrac{1 + \cos\alpha}{\cos\alpha\sin\alpha}} = \dfrac{1}{\cos\alpha}\cdot\dfrac{\cos\alpha\sin\alpha}{1 + \cos\alpha} = \dfrac{\sin\alpha}{1 + \cos\alpha} = \tan\dfrac{\alpha}{2}$

Exercise Set 5.3

1. $\dfrac{24}{25}$ **3.** $\dfrac{24}{7}$ **5.** $\dfrac{527}{625}$ **7. a.** $-\dfrac{240}{289}$ **b.** $-\dfrac{161}{289}$ **c.** $\dfrac{240}{161}$ **9. a.** $-\dfrac{336}{625}$ **b.** $\dfrac{527}{625}$ **c.** $-\dfrac{336}{527}$ **11. a.** $\dfrac{4}{5}$ **b.** $\dfrac{3}{5}$ **c.** $\dfrac{4}{3}$

13. a. $\dfrac{720}{1681}$ **b.** $\dfrac{1519}{1681}$ **c.** $\dfrac{720}{1519}$ **15.** $\dfrac{1}{2}$ **17.** $-\dfrac{\sqrt{3}}{2}$ **19.** $\dfrac{\sqrt{2}}{2}$ **21.** $\dfrac{\sqrt{3}}{3}$ For Exercises 23–33, proofs may vary.

35. $\dfrac{9}{4} - 3\cos 2x + \dfrac{3}{4}\cos 4x$ **37.** $\dfrac{1}{8} - \dfrac{1}{8}\cos 4x$ **39.** $\dfrac{\sqrt{2 - \sqrt{3}}}{2}$ **41.** $-\dfrac{\sqrt{2 + \sqrt{2}}}{2}$ **43.** $2 + \sqrt{3}$ **45.** $-\sqrt{2} + 1$

47. $\dfrac{\sqrt{10}}{10}$ **49.** $\dfrac{1}{3}$ **51.** $\dfrac{7\sqrt{2}}{10}$ **53.** $\dfrac{3}{5}$ **55. a.** $\dfrac{2\sqrt{5}}{5}$ **b.** $-\dfrac{\sqrt{5}}{5}$ **c.** -2 **57. a.** $\dfrac{3\sqrt{13}}{13}$ **b.** $\dfrac{2\sqrt{13}}{13}$ **c.** $\dfrac{3}{2}$

For Exercises 59–67, proofs may vary. **69. a.** $\dfrac{v_0^2}{32}\cdot \sin 2\theta$ **b.** $\theta = \dfrac{\pi}{4}$ **71.** $\sqrt{2 - \sqrt{2}}\cdot(2 + \sqrt{2}) \approx 2.6$

85.

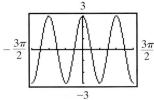

Proofs may vary.

87.

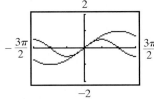

Values for x may vary.

89.

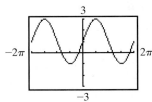

a. $y = 1 + 2 \sin x$
b. Proofs may vary.

91.

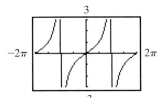

a. $y = \tan \dfrac{x}{2}$ **b.** Proofs may vary.

93. $\dfrac{5}{16} - \dfrac{7}{16} \cos 2x + \dfrac{3}{16} \cos 4x - \dfrac{1}{16} \cos 2x \cos 4x$

Section 5.4

Check Point Exercises

1. a. $\dfrac{1}{2}[\cos 3x - \cos 7x]$ **b.** $\dfrac{1}{2}[\cos 6x + \cos 8x]$ **2. a.** $2 \sin 5x \cos 2x$ **b.** $2 \cos\left(\dfrac{5x}{2}\right) \cos\left(\dfrac{x}{2}\right)$

3. $\dfrac{\cos 3x - \cos x}{\sin 3x + \sin x} = \dfrac{-2 \sin\left(\dfrac{3x + x}{2}\right) \sin\left(\dfrac{3x - x}{2}\right)}{2 \sin\left(\dfrac{3x + x}{2}\right) \cos\left(\dfrac{3x - x}{2}\right)} = \dfrac{-2 \sin\left(\dfrac{4x}{2}\right) \sin\left(\dfrac{2x}{2}\right)}{2 \sin\left(\dfrac{4x}{2}\right) \cos\left(\dfrac{2x}{2}\right)} = \dfrac{-2 \sin 2x \sin x}{2 \sin 2x \cos x} = -\dfrac{\sin x}{\cos x} = -\tan x$

Exercise Set 5.4

1. $\dfrac{1}{2}[\cos 4x - \cos 8x]$ **3.** $\dfrac{1}{2}[\cos 4x + \cos 10x]$ **5.** $\dfrac{1}{2}[\sin 3x - \sin x]$ **7.** $\dfrac{1}{2}[\sin 2x - \sin x]$ **9.** $2 \sin 4x \cos 2x$

11. $2 \sin 2x \cos 5x$ **13.** $2 \cos 3x \cos x$ **15.** $2 \sin \dfrac{3x}{2} \cos \dfrac{x}{2}$ **17.** $2 \cos x \cos \dfrac{x}{2}$ **19.** $\dfrac{\sqrt{6}}{2}$ **21.** $-\dfrac{\sqrt{2}}{2}$

For Exercises 23–29, proofs may vary.   **31. a.** $y = \cos x$ **b.** Proofs may vary.   **33. a.** $y = \tan 2x$ **b.** Proofs may vary.

35. a. $y = \sin 1704\pi t + \sin 2418\pi t$ **b.** $2 \sin 2061\pi t \cdot \cos 357\pi t$

43.

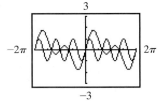

Values for x may vary.

45.

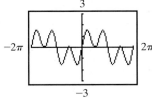

Proofs may vary.

47.

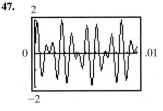

c. $\pi = 4 - \dfrac{4}{3} + \dfrac{4}{5} - \dfrac{4}{7} + \cdots$

49. a.

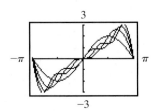

b.

For Exercises 51–55, proofs may vary.

Section 5.5

Check Point Exercises

1. $x = \dfrac{\pi}{3} + 2n\pi$ or $x = \dfrac{2\pi}{3} + 2n\pi$, where n is any integer.   **2.** $\dfrac{\pi}{6}, \dfrac{2\pi}{3}, \dfrac{7\pi}{6}, \dfrac{5\pi}{3}$ **3.** $\dfrac{\pi}{2}$ **4.** $\dfrac{\pi}{6}, \dfrac{\pi}{2}, \dfrac{5\pi}{6}$ **5.** $0, \dfrac{\pi}{4}, \pi, \dfrac{5\pi}{4}$

6. $\dfrac{\pi}{2}, \dfrac{7\pi}{6}, \dfrac{11\pi}{6}$ **7.** $\dfrac{3\pi}{4}, \dfrac{7\pi}{4}$ **8.** $\dfrac{\pi}{2}, \pi$

Exercise Set 5.5

1. Solution   **3.** Not a solution   **5.** Solution   **7.** Solution   **9.** Not a solution   **11.** $x = \dfrac{\pi}{3} + 2n\pi$ or $x = \dfrac{2\pi}{3} + 2n\pi$,

where n is any integer.   **13.** $x = \dfrac{\pi}{4} + n\pi$, where n is any integer.   **15.** $x = \dfrac{2\pi}{3} + 2n\pi$ or $x = \dfrac{4\pi}{3} + 2n\pi$, where n is any integer.

17. $x = n\pi$, where n is any integer.   **19.** $x = \dfrac{5\pi}{6} + 2n\pi$ or $x = \dfrac{7\pi}{6} + 2n\pi$, where n is any integer.

21. $\theta = \dfrac{\pi}{6} + 2n\pi$ or $\theta = \dfrac{5\pi}{6} + 2n\pi$, where n is any integer.   **23.** $\theta = \dfrac{3\pi}{2} + 2n\pi$, where n is any integer.   **25.** $\dfrac{\pi}{6}, \dfrac{\pi}{3}, \dfrac{7\pi}{6}, \dfrac{4\pi}{3}$

27. $\dfrac{5\pi}{24}, \dfrac{7\pi}{24}, \dfrac{17\pi}{24}, \dfrac{19\pi}{24}, \dfrac{29\pi}{24}, \dfrac{31\pi}{24}, \dfrac{41\pi}{24}, \dfrac{43\pi}{24}$ **29.** $\dfrac{\pi}{18}, \dfrac{7\pi}{18}, \dfrac{13\pi}{18}, \dfrac{19\pi}{18}, \dfrac{25\pi}{18}, \dfrac{31\pi}{18}$ **31.** 0   **33.** no solution   **35.** $\dfrac{4\pi}{9}, \dfrac{8\pi}{9}, \dfrac{16\pi}{9}$

37. $0, \dfrac{\pi}{3}, \pi, \dfrac{4\pi}{3}$ **39.** $\dfrac{\pi}{2}, \dfrac{7\pi}{6}, \dfrac{11\pi}{6}$ **41.** $\dfrac{2\pi}{3}, \pi, \dfrac{4\pi}{3}$ **43.** $\dfrac{3\pi}{2}$ **45.** $\dfrac{\pi}{2}, \dfrac{3\pi}{2}$ **47.** $\dfrac{\pi}{4}, \pi, \dfrac{5\pi}{4}$ **49.** $\dfrac{5\pi}{6}, \dfrac{7\pi}{6}, \dfrac{11\pi}{6}$ **51.** $\dfrac{\pi}{4}, \dfrac{5\pi}{4}$

53. $0, \dfrac{2\pi}{3}, \pi, \dfrac{4\pi}{3}$ **55.** $0, \pi$ **57.** $\dfrac{\pi}{2}, \dfrac{7\pi}{6}, \dfrac{11\pi}{6}$ **59.** π **61.** $\dfrac{\pi}{6}, \dfrac{5\pi}{6}$ **63.** $\dfrac{\pi}{6}, \dfrac{\pi}{2}, \dfrac{5\pi}{6}, \dfrac{3\pi}{2}$ **65.** $0, \dfrac{2\pi}{3}, \dfrac{4\pi}{3}$ **67.** $\dfrac{2\pi}{3}, \dfrac{4\pi}{3}$

69. $\dfrac{\pi}{8}, \dfrac{3\pi}{8}, \dfrac{9\pi}{8}, \dfrac{11\pi}{8}$ **71.** $0, \dfrac{\pi}{2}$ **73.** $\dfrac{\pi}{4}, \dfrac{3\pi}{4}$ **75.** $\dfrac{\pi}{12}, \dfrac{\pi}{4}, \dfrac{3\pi}{4}, \dfrac{11\pi}{12}, \dfrac{17\pi}{12}, \dfrac{19\pi}{12}$ **77.** 0   **79.** 0.4 sec and 2.1 sec

81. 49 days and 292 days   **83.** $t = 2 + 6n$ or $t = 4 + 6n$ where n is any nonnegative integer.   **85.** 21° or 69°.

97.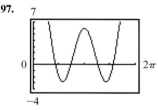

$x = 1.37, x = 2.30, x = 3.98,$
or $x = 4.91$

99.

$x = 0.37$ or $x = 2.77$

101.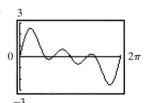

$x = 0, x = 1.57, x = 2.09, x = 3.14,$
$x = 4.19,$ or $x = 4.71$

103. no solution   **105.** $\dfrac{\pi}{3}, \dfrac{5\pi}{3}$

Chapter 5 Review Exercises

For Exercises 1–12, proofs may vary.   **13.** $\dfrac{\sqrt{6} - \sqrt{2}}{4}$ **14.** $\dfrac{\sqrt{2} - \sqrt{6}}{4}$ **15.** $2 - \sqrt{3}$ **16.** $\sqrt{3} + 2$ **17.** $\dfrac{1}{2}$ **18.** $\dfrac{1}{2}$
For Exercises 19–30, proofs may vary.

31. a. $y = \cos x$ **b.** $\sin\left(x - \dfrac{3\pi}{2}\right) = \sin x \cos \dfrac{3\pi}{2} - \cos x \sin \dfrac{3\pi}{2} = \sin x \cdot 0 - \cos x \cdot -1 = \cos x$

32. a. $y = -\sin x$ **b.** $\cos\left(x + \dfrac{\pi}{2}\right) = \cos x \cos \dfrac{\pi}{2} - \sin x \sin \dfrac{\pi}{2} = \cos x \cdot 0 - \sin x \cdot 1 = -\sin x$

33. a. $y = \tan x$ **b.** $y = \dfrac{\tan x - 1}{1 - \cot x} = \dfrac{\dfrac{\sin x}{\cos x} - 1}{1 - \dfrac{\cos x}{\sin x}} = \dfrac{\dfrac{\sin x - \cos x}{\cos x}}{\dfrac{\sin x - \cos x}{\sin x}} = \dfrac{\sin x - \cos x}{\cos x} \cdot \dfrac{\sin x}{\sin x - \cos x} = \dfrac{\sin x}{\cos x} = \tan x$

34. a. $\dfrac{33}{65}$ **b.** $\dfrac{16}{65}$ **c.** $-\dfrac{33}{56}$ **d.** $\dfrac{24}{25}$ **e.** $\dfrac{2\sqrt{13}}{13}$ **35. a.** $-\dfrac{63}{65}$ **b.** $-\dfrac{56}{65}$ **c.** $\dfrac{63}{16}$ **d.** $\dfrac{24}{25}$ **e.** $\dfrac{5\sqrt{26}}{26}$

36. a. 1   **b.** $-\dfrac{3}{5}$ **c.** Undefined   **d.** $-\dfrac{3}{5}$ **e.** $\dfrac{\sqrt{10 + 3\sqrt{10}}}{2\sqrt{5}}$ **37. a.** 1   **b.** $\dfrac{4\sqrt{2}}{9}$ **c.** Undefined   **d.** $\dfrac{4\sqrt{2}}{9}$ **e.** $-\dfrac{\sqrt{3}}{3}$

38. $\dfrac{\sqrt{3}}{2}$ **39.** $-\dfrac{\sqrt{3}}{3}$ **40.** $\dfrac{\sqrt{2-\sqrt{2}}}{2}$ **41.** $2-\sqrt{3}$ **42.** $\dfrac{1}{2}[\cos 2x - \cos 10x]$ **43.** $\dfrac{1}{2}[\sin 10x + \sin 4x]$ **44.** $-2\sin x \cos 3x$

45. $\dfrac{\sqrt{6}}{2}$ **46.** Proofs may vary.   **47.** Proofs may vary.   **48. a.** $y = \cot x$ **b.** Proofs may vary.

49. $x = \dfrac{2\pi}{3} + 2n\pi$ or $x = \dfrac{4\pi}{3} + 2n\pi$, where n is any integer.   **50.** $x = \dfrac{\pi}{4} + 2n\pi$ or $x = \dfrac{3\pi}{4} + 2n\pi$, where n is any integer.

51. $x = \dfrac{7\pi}{6} + 2n\pi$ or $x = \dfrac{11\pi}{6} + 2n\pi$, where n is any integer.   **52.** $x = \dfrac{\pi}{6} + n\pi$, where n is any integer   **53.** $\dfrac{\pi}{2}, \dfrac{3\pi}{2}$

54. $\dfrac{\pi}{6}, \dfrac{5\pi}{6}, \dfrac{9\pi}{6}$ **55.** $\dfrac{3\pi}{2}$ **56.** $0, \dfrac{\pi}{3}, \pi, \dfrac{5\pi}{3}$ **57.** π **58.** $\dfrac{\pi}{6}, \dfrac{5\pi}{6}, \dfrac{3\pi}{2}$ **59.** $\dfrac{\pi}{6}, \dfrac{5\pi}{6}, \dfrac{7\pi}{6}, \dfrac{11\pi}{6}$ **60.** $0, \pi, \dfrac{7\pi}{6}, \dfrac{11\pi}{6}$

61. $0, \dfrac{\pi}{6}, \pi, \dfrac{11\pi}{6}$ **62.** $0, \pi$ **63.** $t = \dfrac{2}{3} + 4n$ or $t = \dfrac{10}{3} + 4n$, where n is any integer.   **64.** $12°$ or $78°$

Chapter 5 Test

1. $-\dfrac{63}{65}$ **2.** $\dfrac{56}{33}$ **3.** $-\dfrac{24}{25}$ **4.** $\dfrac{3\sqrt{13}}{13}$ **5.** $\dfrac{\sqrt{6}+\sqrt{2}}{4}$ **6.** $\cos x \csc x = \cos x \cdot \dfrac{1}{\sin x} = \dfrac{\cos x}{\sin x} = \cot x$

7. $\dfrac{\sec x}{\cot x + \tan x} = \dfrac{\dfrac{1}{\cos x}}{\dfrac{\cos x}{\sin x} + \dfrac{\sin x}{\cos x}} = \dfrac{\dfrac{1}{\cos x}}{\dfrac{\cos^2 x + \sin^2 x}{\sin x \cos x}} = \dfrac{1}{\cos x} \cdot \dfrac{\sin x \cos x}{1} = \sin x$

8. $1 - \dfrac{\cos^2 x}{1 + \sin x} = 1 - \dfrac{(1 - \sin^2 x)}{1 + \sin x} = 1 - \dfrac{(1 + \sin x)(1 - \sin x)}{1 + \sin x} = 1 - (1 - \sin x) = \sin x$

9. $\cos\left(\theta + \dfrac{\pi}{2}\right) = \cos\theta \cos\dfrac{\pi}{2} - \sin\theta \sin\dfrac{\pi}{2} = \cos\theta \cdot 0 - \sin\theta \cdot 1 = -\sin\theta$

10. $\dfrac{\sin(\alpha - \beta)}{\sin\alpha \cos\beta} = \dfrac{\sin\alpha \cos\beta - \cos\alpha \sin\beta}{\sin\alpha \cos\beta} = \dfrac{\sin\alpha \cos\beta}{\sin\alpha \cos\beta} - \dfrac{\cos\alpha \sin\beta}{\sin\alpha \cos\beta} = 1 - \cot\alpha \tan\beta$

11. $\sin t \cos t(\tan t + \cot t) = \sin t \cos t\left(\dfrac{\sin t}{\cos t} + \dfrac{\cos t}{\sin t}\right) = \sin^2 t + \cos^2 t = 1$ **12.** $\dfrac{7\pi}{18}, \dfrac{11\pi}{18}, \dfrac{19\pi}{18}, \dfrac{23\pi}{18}, \dfrac{31\pi}{18},$ and $\dfrac{35\pi}{18}$

13. $\dfrac{\pi}{2}, \dfrac{7\pi}{6}, \dfrac{3\pi}{2}, \dfrac{11\pi}{6}$ **14.** $0, \dfrac{\pi}{3}, \dfrac{5\pi}{3}$ **15.** $0, \dfrac{2\pi}{3}, \dfrac{4\pi}{3}$

Cumulative Review Exercises (Chapters P–5)

1. $-3, 1 + 2i,$ and $1 - 2i$ **2.** $x = \dfrac{\log 125}{\log 11} + 1$ or $x \approx 3.01$ **3.** $(-\infty, -4] \cup [2, \infty)$ **4.** $\dfrac{\pi}{3}, \dfrac{5\pi}{3}$

5. 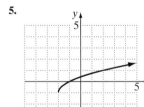   **6.** 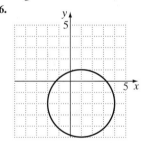   **7.** 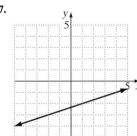   **8.**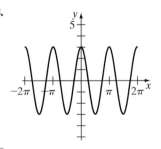

9. 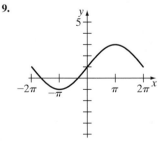   **10.**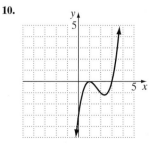

11. $2a + h + 3$ **12.** $-\dfrac{\sqrt{2}}{2}$ **13.** Proofs may vary.

14. $\dfrac{16}{9}\pi$ radians   **15.** $t \approx 19.1$ yr

16. $f^{-1}(x) = \dfrac{3x + 1}{x - 2}$ **17.** $B = 67°, b = 28.27, c = 30.71$

18. 106 mg   **19.** $h \approx 15.9$ ft   **20.** Answers may vary.

CHAPTER 6

Section 6.1

Check Point Exercises

1. $B = 34°, a \approx 13$ cm, $b \approx 8$ cm   **2.** $B = 117.5°, a \approx 9, c \approx 5$ **3.** one triangle; $C \approx 24°, b \approx 33°, b \approx 31$ **4.** no triangle
5. two triangles; $B_1 \approx 50°, C_1 \approx 95°, c_1 = 21; B_2 \approx 130°, C_2 \approx 15°, c_2 \approx 5$ **6.** approximately 34 sq m   **7.** approximately 11 mi

Exercise Set 6.1

1. $B = 42°, a \approx 8.1, b \approx 8.1$ **3.** $A = 44°, b \approx 18.6, c \approx 22.8$ **5.** $C = 95°, b \approx 81.0, c \approx 134.1$ **7.** $B = 40°, b \approx 20.9, c \approx 31.8$
9. $C = 111°, b \approx 7.3, c \approx 16.1$ **11.** $A = 80°, a \approx 39.5, c \approx 10.4$ **13.** $B = 30°, a \approx 316.0, b \approx 174.3$ **15.** $C = 50°, a \approx 7.1, b \approx 7.1$
17. one triangle; $B \approx 29°, c \approx 111°, c \approx 29.0$ **19.** one triangle; $C \approx 52°, B \approx 65°, b \approx 10.2$ **21.** one triangle; $C \approx 55°, B \approx 13°, b \approx 10.2$
23. no triangle   **25.** two triangles; $B_1 \approx 77°, C_1 \approx 43°, c_1 \approx 12.6; B_2 \approx 103°, C_2 \approx 17°, c_2 \approx 5.4$ **27.** two triangles; $B_1 \approx 54°, C_1 \approx 89°$,
$c_1 \approx 19.9; B_2 \approx 126°, C_2 \approx 17°, c_2 \approx 5.8$ **29.** two triangles; $C_1 \approx 68°, B_1 \approx 54°, b_1 \approx 21.0; C_2 \approx 112°, B_2 \approx 10°, b_2 \approx 4.5$
31. no triangle   **33.** 297 sq ft   **35.** 5 sq yd   **37.** 10 sq m   **39.** Station A is about 6 miles from the fire, station B is about 9 miles
from the fire.   **41.** The platform is about 3672 yards from one end of the beach and 3576 yards from the other.   **43.** about 184 ft
45. about 56 ft   **47.** about 30 ft   **49. a.** $a \approx 494$ ft   **b.** about 343 ft   **51.** either 9.9 mi or 2.4 mi   **63.** No   **65.** 41 ft

Section 6.2

Check Point Exercises

1. $a = 13, B \approx 28°, C \approx 32°$ **2.** $B \approx 97.9°, A \approx 52.4°, C \approx 29.7°$ **3.** approximately 917 mi apart   **4.** approximately 47 sq m

Exercise Set 6.2

1. $a \approx 6.0, B \approx 29°, C \approx 105°$ **3.** $c \approx 7.6, A \approx 52°, B \approx 32°$ **5.** $A \approx 44°, B \approx 68°, C \approx 68°$ **7.** $A \approx 117°, B \approx 36°, C \approx 27°$
9. $c \approx 4.7, A \approx 46°, B \approx 92°$ **11.** $a \approx 6.3, C \approx 28°, B \approx 50°$ **13.** $b \approx 4.7, C \approx 54°, A \approx 76°$ **15.** $b \approx 5.4, C \approx 22°, A \approx 68°$
17. $C \approx 112°, A \approx 28°, B \approx 40°$ **19.** $B \approx 100°, A \approx 19°, C \approx 61°$ **21.** $A = 60°, B = 60°, C = 60°$ **23.** no triangle
25. 4 sq ft   **27.** 22 sq m   **29.** 31 sq yd   **31.** about 61.7 mi apart   **33.** about 193 yd   **35.** N12ºE   **37. a.** about 19.3 mi
b. S58ºE   **39.** The guy wire anchored downhill is about 417.4 feet. The one anchored uphill is about 398.2 feet.   **41.** about 63.7 ft
43. \$123,454   **53.** 153.4°

Section 6.3

Check Point Exercises

1. a.

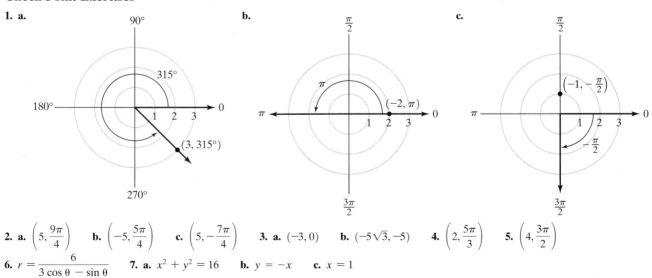

2. a. $\left(5, \dfrac{9\pi}{4}\right)$ **b.** $\left(-5, \dfrac{5\pi}{4}\right)$ **c.** $\left(5, -\dfrac{7\pi}{4}\right)$ **3. a.** $(-3, 0)$ **b.** $(-5\sqrt{3}, -5)$ **4.** $\left(2, \dfrac{5\pi}{3}\right)$ **5.** $\left(4, \dfrac{3\pi}{2}\right)$

6. $r = \dfrac{6}{3 \cos \theta - \sin \theta}$ **7. a.** $x^2 + y^2 = 16$ **b.** $y = -x$ **c.** $x = 1$

Exercise Set 6.3

1. C **3.** A **5.** B **7.** C **9.** A

11.

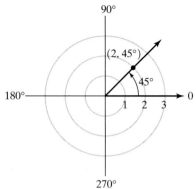

13.

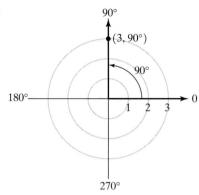

15.

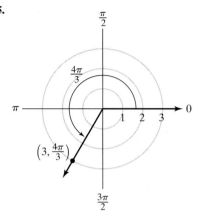

17.

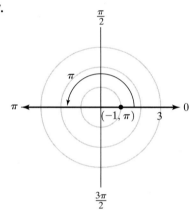

19.

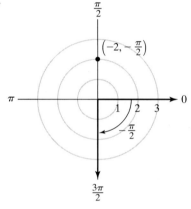

21.

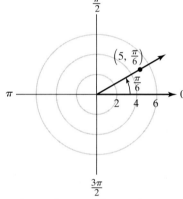

a. $\left(5, \dfrac{13\pi}{6}\right)$ **b.** $\left(-5, \dfrac{7\pi}{6}\right)$

c. $\left(5, -\dfrac{11\pi}{6}\right)$

23.

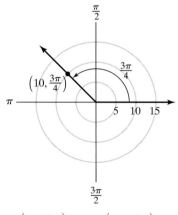

a. $\left(10, \dfrac{11\pi}{4}\right)$ **b.** $\left(-10, \dfrac{7\pi}{4}\right)$

c. $\left(10, -\dfrac{5\pi}{4}\right)$

25.

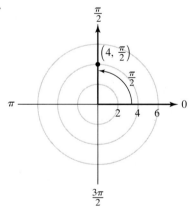

a. $\left(4, \dfrac{5\pi}{2}\right)$ **b.** $\left(-4, \dfrac{3\pi}{2}\right)$

c. $\left(4, -\dfrac{3\pi}{2}\right)$

27. a; b; d   **29.** b, d   **31.** a, b
33. $(0, 4)$ **35.** $(1, \sqrt{3})$ **37.** $(0, -4)$
39. approximately $(-5.9, 4.4)$
41. $\left(\sqrt{8}, \dfrac{3\pi}{4}\right)$ **43.** $\left(4, \dfrac{5\pi}{3}\right)$
45. $\left(2, \dfrac{7\pi}{6}\right)$ **47.** $(5, 0)$
49. $r = \dfrac{7}{3\cos\theta + \sin\theta}$ **51.** $r = \dfrac{7}{\cos\theta}$
53. $r = 3$ **55.** $r = 4\cos\theta$
57. $r = \dfrac{6\cos\theta}{\sin^2\theta}$ **59.** $x^2 + y^2 = 64$
61. $x = 0$ **63.** $y = 3$ **65.** $y = 4$
67. $x^2 + y^2 = y$ **69.** $x^2 + y^2 = 6x + 4y$

71. $y = \dfrac{1}{x}$ **73.** $\left(15, \dfrac{4\pi}{3}\right)$ **75.** 6.3 knots at an angle of 50° to the wind **77.** Answers may vary. **87.** $(-2, 3.464)$
89. $(-1.857, -3.543)$ **91.** $(3, 0.730)$ **93.** Answers may vary. **95.** $x^2 + y^2 = 4x$; center: $(2, 0)$, radius: 2

Section 6.4

Check Point Exercises

1.

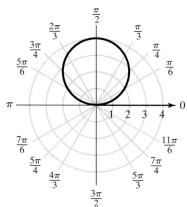

2.

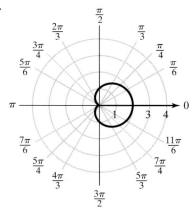

3.

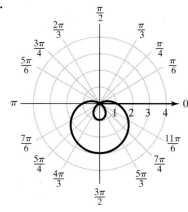

4.

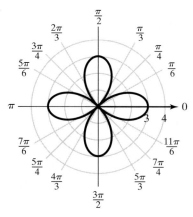

5.

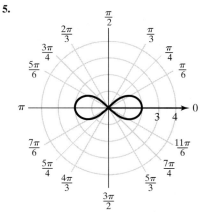

Exercise Set 6.4

1. $r = 1 - \sin\theta$ **3.** $r = 2\cos\theta$ **5.** $r = 3\sin 3\theta$

7. a. May or may not have symmetry with respect to polar axis. **b.** Has symmetry with respect to the line $\theta = \dfrac{\pi}{2}$.
c. May or may not have symmetry about the pole. **9. a.** Has symmetry with respect to polar axis.
b. May or may not have symmetry with respect to the line $\theta = \dfrac{\pi}{2}$. **c.** May or may not have symmetry about pole.
11. a. Has symmetry with respect to polar axis. **b.** Has symmetry with respect
to the line $\theta = \dfrac{\pi}{2}$. **c.** Has symmetry about the pole.

13.

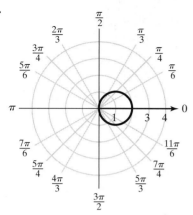

15.

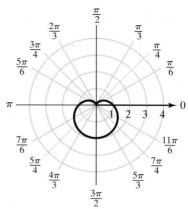

17.

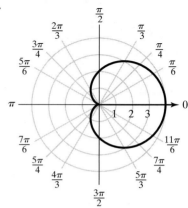

19.

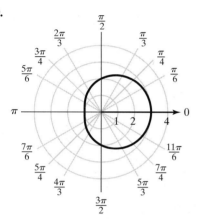

21.

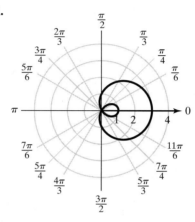

23.

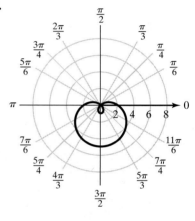

25.

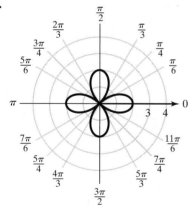

27.

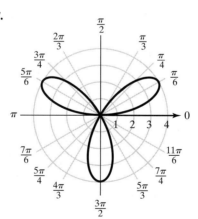

29.

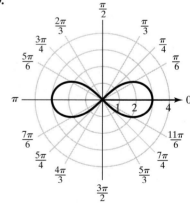

31.

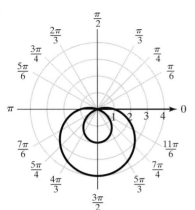

33.

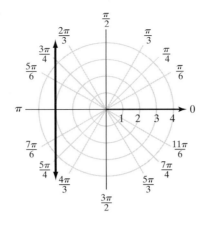

35. 6 knots   **37.** 8 knots

39. 90°; about $7\frac{1}{2}$ knots

49.

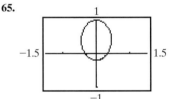

51.

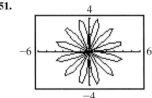

53.

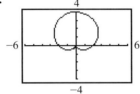

55.

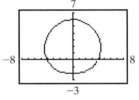

57.

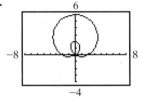

59.

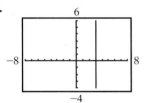

61.

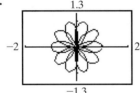

63.

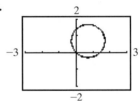

65.

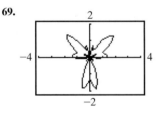

67. 2π **69.**

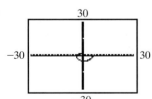

71.

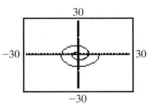

73. If n is odd, there are n loops and θmax $= \pi$ traces the graph once; while if n is even, there are $2n$ loops and θmax $= 2\pi$ traces the graph once. In each separate case, as n increases, sin $n\theta$ increases its number of loops.

75. There are n small petals and n large petals for each value of n. For odd values of n, the small petals are inside the large petals. For even n, they are between the large petals.

77.

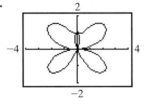

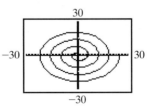

79. Answers may vary.

81.

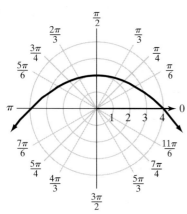

Section 6.5

Check Point Exercises

1. a.

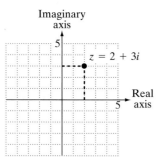

b.

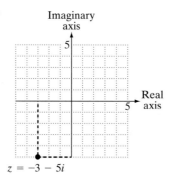

c.

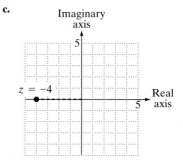

d.

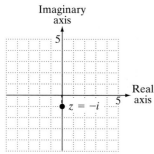

2. a. 13 **b.** $\sqrt{13}$ **3.**

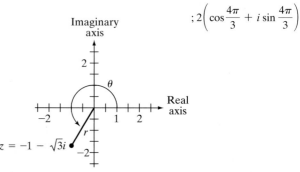

$; 2\left(\cos\dfrac{4\pi}{3} + i \sin\dfrac{4\pi}{3}\right)$

4. $z = 2\sqrt{3} + 2i$ **5.** $30(\cos 60° + i \sin 60°)$ **6.** $10(\cos \pi + i \sin \pi)$ **7.** $-16\sqrt{3} + 16i$ **8.** -4

9. $2(\cos 15° + i \sin 15°); 2(\cos 105° + i \sin 105°); 2(\cos 195° + i \sin 195°); 2(\cos 285° + i \sin 285°)$ **10.** $3; -\dfrac{3}{2} + \dfrac{3\sqrt{3}}{2}i; -\dfrac{3}{2} - \dfrac{3\sqrt{3}}{2}i$

Exercise Set 6.5

1.

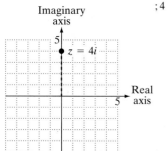

$; 4$

3.

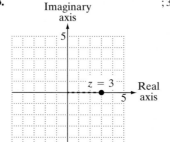

$; 3$

5.

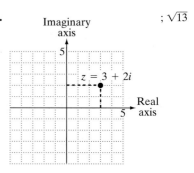

$; \sqrt{13}$

7.

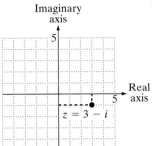

$; \sqrt{10}$

9.

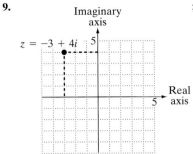

$; 5$ **11.**

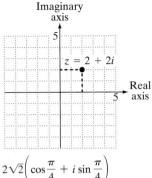

$$2\sqrt{2}\left(\cos\frac{\pi}{4} + i\sin\frac{\pi}{4}\right)$$

or $2\sqrt{2}(\cos 45° + i\sin 45°)$

13.

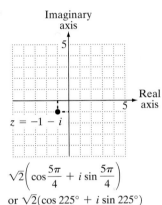

$$\sqrt{2}\left(\cos\frac{5\pi}{4} + i\sin\frac{5\pi}{4}\right)$$
or $\sqrt{2}(\cos 225° + i\sin 225°)$

15.

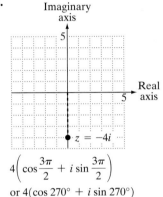

$$4\left(\cos\frac{3\pi}{2} + i\sin\frac{3\pi}{2}\right)$$
or $4(\cos 270° + i\sin 270°)$

17.

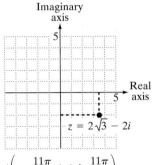

$$4\left(\cos\frac{11\pi}{6} + i\sin\frac{11\pi}{6}\right)$$
or $4(\cos 330° + i\sin 330°)$

19.

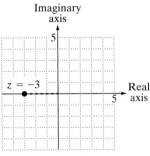

$3(\cos \pi + i\sin \pi)$
or $3(\cos 180° + i\sin 180°)$

21.

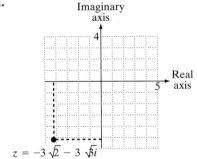

$\approx 3\sqrt{5}(\cos 230.8° + i\sin 230.8°)$

23.

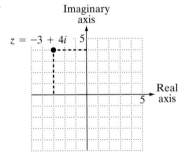

$\approx 5(\cos 126.9° + i\sin 126.9°)$

25.

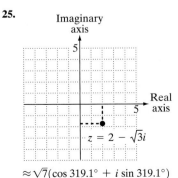

$\approx \sqrt{7}(\cos 319.1° + i\sin 319.1°)$

27. $3\sqrt{3} + 3i$ **29.** $-2 - 2\sqrt{3}i$ **31.** $4\sqrt{2} - 4\sqrt{2}i$ **33.** $5i$ **35.** $z \approx -18.1 - 8.5i$

37. $30(\cos 70° + i\sin 70°)$ **39.** $12\left(\cos\frac{3\pi}{10} + i\sin\frac{3\pi}{10}\right)$ **41.** $\cos\frac{7\pi}{12} + i\sin\frac{7\pi}{12}$

43. $2(\cos \pi + i\sin \pi)$ **45.** $5(\cos 50° + i\sin 50°)$ **47.** $\frac{3}{4}\left(\cos\frac{\pi}{10} + i\sin\frac{\pi}{10}\right)$

49. $\cos 240° + i\sin 240°$ **51.** $2(\cos 0° + i\sin 0°)$ **53.** $32\sqrt{2} + 32\sqrt{2}i$

55. $-4 - 4\sqrt{3}i$ **57.** $\frac{1}{64}i$ **59.** $-2 - 2\sqrt{3}i$ **61.** $-4 - 4i$ **63.** -64

65. $3(\cos 15° + i\sin 15°); 3(\cos 195° + i\sin 195°)$

67. $2(\cos 70° + i\sin 70°); 2(\cos 190° + i\sin 190°); 2(\cos 310° + i\sin 310°)$

69. $\dfrac{3}{2} + \dfrac{3\sqrt{3}}{2}i; \dfrac{3\sqrt{3}}{2} + \dfrac{3}{2}i; \dfrac{3}{2} - \dfrac{3\sqrt{3}}{2}i; \dfrac{3\sqrt{3}}{2} - \dfrac{3}{2}i$ **71.** $2; \approx 0.6 + 1.9i; \approx -1.6 + 1.2i; \approx -1.6 - 1.2i; \approx 0.6 - 1.9i$

73. $1; -\dfrac{1}{2} + \dfrac{\sqrt{3}}{2}i; -\dfrac{1}{2} - \dfrac{\sqrt{3}}{2}i$ **75.** $\approx 1.1 + 0.2i; \approx -0.2 + 1.1i; \approx -1.1 - 0.2i; \approx 0.2 - 1.1i$

77. a. $i; -1 + i; -i; -1 + i; -i; -1 + i$ **93.** **95.** $\sqrt{3} + i; -\sqrt{3} + i; -2i$
b. Complex numbers may vary.

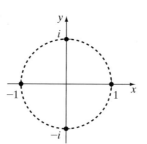

Section 6.6

Check Point Exercises

1. $\|\mathbf{u}\| = 5 = \|\mathbf{v}\|$ and $m_u = \dfrac{4}{3} = m_v$

2.

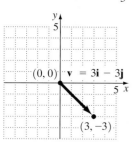

$; \|\mathbf{v}\| = 3\sqrt{2}$

3. $\mathbf{v} = 3\mathbf{i} + 4\mathbf{j}$
4. a. $11\mathbf{i} - 2\mathbf{j}$ **b.** $3\mathbf{i} + 8\mathbf{j}$
5. a. $56\mathbf{i} + 80\mathbf{j}$ **b.** $-35\mathbf{i} - 50\mathbf{j}$
6. $30\mathbf{i} + 33\mathbf{j}$

7. $\dfrac{4}{5}\mathbf{i} - \dfrac{3}{5}\mathbf{j}; \sqrt{\left(\dfrac{4}{5}\right)^2 + \left(-\dfrac{3}{5}\right)^2} = \sqrt{\dfrac{16}{25} + \dfrac{9}{25}} = \sqrt{\dfrac{25}{25}} = 1$

8. $30\sqrt{2}\mathbf{i} + 30\sqrt{2}\mathbf{j}$
9. 82.55 lb; 46.2°

Exercise Set 6.6

1. a. $\sqrt{41}$ **b.** $\sqrt{41}$ **c.** $\mathbf{u} = \mathbf{v}$ **3. a.** 6 **b.** 6 **c.** $\mathbf{u} = \mathbf{v}$

5.

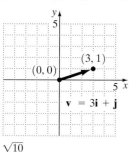

$\sqrt{10}$

7.

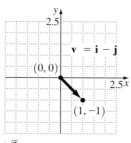

$\sqrt{2}$

9.

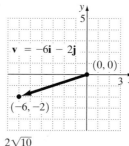

$2\sqrt{10}$

11.

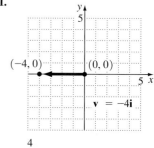

4

13. $10\mathbf{i} + 6\mathbf{j}$ **15.** $6\mathbf{i} - 3\mathbf{j}$ **17.** $-6\mathbf{i} - 14\mathbf{j}$ **19.** $9\mathbf{i}$ **21.** $-\mathbf{i} + 2\mathbf{j}$ **23.** $5\mathbf{i} - 12\mathbf{j}$ **25.** $-5\mathbf{i} + 12\mathbf{j}$ **27.** $-15\mathbf{i} + 35\mathbf{j}$

29. $4\mathbf{i} + 24\mathbf{j}$ **31.** $-9\mathbf{i} - 4\mathbf{j}$ **33.** $-5\mathbf{i} + 45\mathbf{j}$ **35.** $2\sqrt{29}$ **37.** $\sqrt{10}$ **39.** $\mathbf{i}$ **41.** $\dfrac{3}{5}\mathbf{i} - \dfrac{4}{5}\mathbf{j}$ **43.** $\dfrac{3\sqrt{13}}{13}\mathbf{i} - \dfrac{2\sqrt{13}}{3}\mathbf{j}$

45. $\dfrac{\sqrt{2}}{2}\mathbf{i} + \dfrac{\sqrt{2}}{2}\mathbf{j}$ **47.** $3\sqrt{3}\mathbf{i} + 3\mathbf{j}$ **49.** $-6\sqrt{2}\mathbf{i} - 6\sqrt{2}\mathbf{j}$ **51.** $\approx -0.20\mathbf{i} + 0.46\mathbf{j}$ **53.** $22\sqrt{3}\mathbf{i} + 22\mathbf{j}$ **55.** $148.5\mathbf{i} + 20.9\mathbf{j}$

57. $\approx 1.4\mathbf{i} + 0.6\mathbf{j}$; 1.4 in. **59.** ≈ 108.21 lbs; S 77.4° E **61.** ≈ 30.9 lbs **63. a.** $\mathbf{F} = 9\mathbf{i} - 3\mathbf{j}$ **b.** $\mathbf{F}_3 = -9\mathbf{i} + 3\mathbf{j}$

65. a. $\mathbf{F} = -2\mathbf{j}$ **b.** $\mathbf{F}_5 = 2\mathbf{j}$ **67. a.** $\mathbf{v} = 180 \cos 40°\mathbf{i} + 180 \sin 40°\mathbf{j} \approx 137.89\mathbf{i} + 115.70\mathbf{j}$ $\mathbf{w} = 40 \cos 0°\mathbf{i} + 40 \sin 0°\mathbf{j} = 40\mathbf{i}$

b. $\mathbf{v} + \mathbf{w} \approx 177.89\mathbf{i} + 115.70\mathbf{j}$ **c.** 212 mph **d.** 33.0°; N 57°E **69.** 78 mph, 75.4° **89.** Proofs may vary. **91.** $\approx 4232.1; \approx 72.7°$

Section 6.7

Check Point Exercises

1. a. 18 **b.** 18 **c.** 5 **2.** 100.3° **3.** orthogonal **4.** $\dfrac{7}{2}\mathbf{i} - \dfrac{7}{2}\mathbf{j}$ **5.** $\mathbf{v}_1 = \dfrac{7}{2}\mathbf{i} - \dfrac{7}{2}\mathbf{j}; \mathbf{v}_2 = -\dfrac{3}{2}\mathbf{i} - \dfrac{3}{2}\mathbf{j}$ **6.** approximately 2598 ft-lb

Exercise Set 6.7

1. a. 6 **b.** 10 **3. a.** −6 **b.** 41 **5. a.** 100 **b.** 61 **7. a.** 0 **b.** 25 **9.** 3 **11.** 3 **13.** 20 **15.** 20 **17.** 79.7°
19. 160.3° **21.** 38.7° **23.** orthogonal **25.** orthogonal **27.** not orthogonal **29.** not orthogonal **31.** orthogonal

33. $\mathbf{v}_1 = \text{proj}_{\mathbf{w}}\mathbf{v} = \dfrac{5}{2}\mathbf{i} - \dfrac{5}{2}\mathbf{j}; \mathbf{v}_2 = \dfrac{1}{2}\mathbf{i} + \dfrac{1}{2}\mathbf{j}$ **35.** $\mathbf{v}_1 = \text{proj}_{\mathbf{w}}\mathbf{v} = -\dfrac{26}{29}\mathbf{i} + \dfrac{65}{29}\mathbf{j}; \mathbf{v}_2 = \dfrac{55}{29}\mathbf{i} + \dfrac{22}{29}\mathbf{j}$ **37.** $\mathbf{v}_1 = \text{proj}_{\mathbf{w}}\mathbf{v} = \mathbf{i} + 2\mathbf{j}; \mathbf{v}_2 = 0$

39. 1077; $\mathbf{v} \cdot \mathbf{w} = 1077$ means $1077 in revenue was generated on Monday by the sale of 240 gallons of regular gas at $1.90 per gallon and 300 gallons of premium gas at $2.07 per gallon. **41.** 7600 foot-pounds **43.** 3392 foot-pounds **45.** 1079 foot-pounds
47. 40 foot-pounds **49.** 22.05 foot-pounds

51. a. $\dfrac{\sqrt{3}}{2}\mathbf{i} + \dfrac{1}{2}\mathbf{j}$ **b.** $-175\sqrt{3}\mathbf{i} - 175\mathbf{j}$ **c.** 350; A force of 350 pounds is required to keep the boat from rolling down the ramp.

63. $\mathbf{u} \cdot \mathbf{v} = (a_1\mathbf{i} + b_1\mathbf{j}) \cdot (a_2\mathbf{i} + b_2\mathbf{j})$
$= a_1a_2 + b_1b_2$
$= a_2a_1 + b_2b_1$
$= (a_2\mathbf{i} + b_2\mathbf{j}) \cdot (a_1\mathbf{i} + b_1\mathbf{j})$
$= \mathbf{v} \cdot \mathbf{u}$

65. $\mathbf{u} \cdot (\mathbf{v} + \mathbf{w}) = (a_1\mathbf{i} + b_1\mathbf{j}) \cdot [(a_2\mathbf{i} + b_2\mathbf{j}) + (a_3\mathbf{i} + b_3\mathbf{j})]$
$= (a_1\mathbf{i} + b_1\mathbf{j}) \cdot [(a_2 + a_3)\mathbf{i} + (b_2 + b_3)\mathbf{j}]$
$= a_1(a_2 + a_3) + b_1(b_2 + b_3)$
$= a_1a_2 + a_1a_3 + b_1b_2 + b_1b_3$
$= a_1a_2 + b_1b_2 + a_1a_3 + b_1b_3$
$= (a_1\mathbf{i} + b_1\mathbf{j}) \cdot (a_2\mathbf{i} + b_2\mathbf{j}) + (a_1\mathbf{i} + b_1\mathbf{j}) \cdot (a_3\mathbf{i} + b_3\mathbf{j})$
$= \mathbf{u} \cdot \mathbf{v} + \mathbf{u} \cdot \mathbf{w}$

67. $b = -20$ **69.** any two vectors, $\mathbf{v}$ and $\mathbf{w}$, having the same direction

Chapter 6 Review Exercises

1. $C = 55°, b \approx 10.5,$ and $c \approx 10.5$ **2.** $A = 43°, a \approx 171.9,$ and $b \approx 241.0$ **3.** $b \approx 16.3, A \approx 72°,$ and $C \approx 42°$
4. $C \approx 98°, A \approx 55°,$ and $B \approx 27°$ **5.** $C = 120°, a \approx 45.0,$ and $b \approx 33.2$ **6.** two triangles; $B_1 \approx 55°, C_1 \approx 86°,$ and $c_1 \approx 31.7$;
$B_2 \approx 125°, C_2 \approx 16°,$ and $c_2 \approx 8.8$ **7.** no triangle **8.** $a \approx 59.0, B \approx 3°,$ and $C \approx 15°$ **9.** $B \approx 78°, A \approx 39°,$ and $C \approx 63°$
10. $B \approx 25°, C \approx 115°,$ and $c \approx 8.5$ **11.** two triangles; $A_1 \approx 59°, C_1 \approx 84°, c_1 \approx 14.4; A_2 \approx 121°, C_2 \approx 22°, c_2 \approx 5.4$
12. $B \approx 9°, C \approx 148°,$ and $c \approx 73.6$ **13.** 8 sq ft **14.** 4 sq ft **15.** 4 sq m **16.** 2 sq m **17.** 35 ft **18.** 35.6 mi
19. 861 mi **20.** 404 ft; 551 ft **21.** $214,194

22. **23.** **24.**

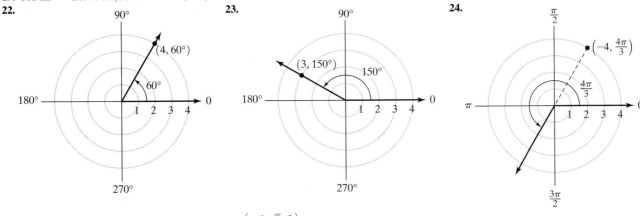

$(2, 2\sqrt{3})$ $\left(-\dfrac{3\sqrt{3}}{2}, \dfrac{3}{2}\right)$ $(2, 2\sqrt{3})$

25.

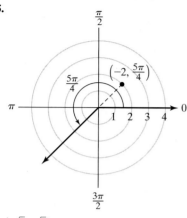

$(\sqrt{2}, \sqrt{2})$

26.

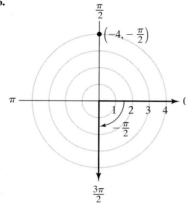

$(0, 4)$

27.

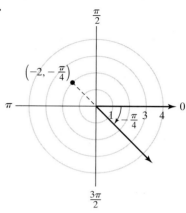

$(-\sqrt{2}, \sqrt{2})$

28.

a. $\left(3, \dfrac{13\pi}{6}\right)$ **b.** $\left(-3, \dfrac{7\pi}{6}\right)$

c. $\left(3, -\dfrac{11\pi}{6}\right)$

29.

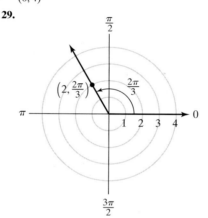

a. $\left(2, \dfrac{8\pi}{3}\right)$ **b.** $\left(-2, \dfrac{5\pi}{3}\right)$

c. $\left(2, -\dfrac{4\pi}{3}\right)$

30.

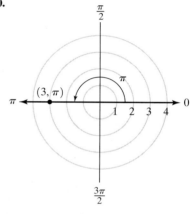

a. $(3, 3\pi)$ **b.** $(-3, 2\pi)$ **c.** $(3, -\pi)$

31. $\left(4\sqrt{2}, \dfrac{3\pi}{4}\right)$ **32.** $\left(3\sqrt{2}, \dfrac{7\pi}{4}\right)$ **33.** approximately $(13, 67°)$ **34.** approximately $(5, 127°)$ **35.** $\left(5, \dfrac{3\pi}{2}\right)$ **36.** $(1, 0)$

37. $r = \dfrac{8}{2\cos\theta + 3\sin\theta}$ **38.** $r = 10$ **39.** $5r^2 = 3r\sin\theta$ or $r = \dfrac{3}{5}\sin\theta$ **40.** $x^2 + y^2 = 9$ **41.** $y = -x$ **42.** $x = -1$

43. $x = 5$ **44.** $x^2 + y^2 = 3x$ **45.** $5x + y = 8$ **46.** $y = \dfrac{2}{x}$ **47. a.** has symmetry   **b.** may or may not have symmetry

c. may or may not have symmetry   **48. a.** may or may not have symmetry   **b.** has symmetry   **c.** may or may not have symmetry
49. a. has symmetry   **b.** has symmetry   **c.** has symmetry
50.

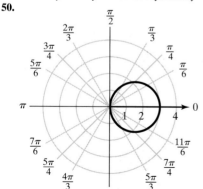

51.

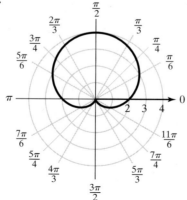

52.

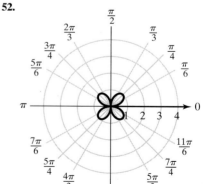

53.

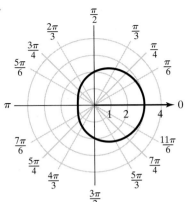

54.

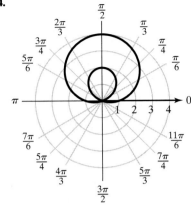

55.

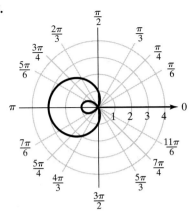

56.

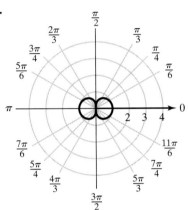

57.

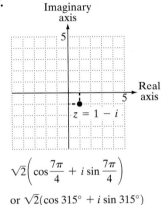

$$\sqrt{2}\left(\cos\frac{7\pi}{4} + i\sin\frac{7\pi}{4}\right)$$

or $\sqrt{2}(\cos 315° + i\sin 315°)$

58.

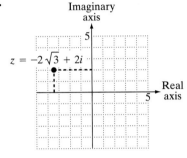

$4(\cos 150° + i\sin 150°)$ or

$$4\left(\cos\frac{5\pi}{6} + i\sin\frac{5\pi}{6}\right)$$

59.

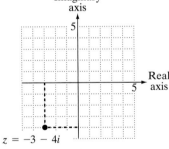

$z = -3 - 4i \approx 5(\cos 233° + i\sin 233°)$

60.

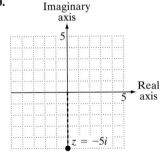

$5\left(\cos\dfrac{3\pi}{2} + i\sin\dfrac{3\pi}{2}\right)$ or $5(\cos 270° + i\sin 270°)$

61. $z = 4 + 4\sqrt{3}i$
62. $z = -2\sqrt{3} - 2i$
63. $z = -3 + 3\sqrt{3}i$
64. $z \approx -0.1 + 0.6i$
65. $15(\cos 110° + i\sin 110°)$
66. $\cos 265° + i\sin 265°$
67. $40(\cos \pi + i\sin \pi)$
68. $2(\cos 5° + i\sin 5°)$
69. $\dfrac{1}{2}(\cos \pi + i\sin \pi)$

70. $2\left(\cos\dfrac{7\pi}{6} + i\sin\dfrac{7\pi}{6}\right)$ **71.** $4 + 4\sqrt{3}i$ **72.** $-32\sqrt{3} + 32i$ **73.** $\dfrac{1}{128}i$ **74.** $64 - 64\sqrt{3}i$ **75.** $128 + 128i$

76. $7(\cos 25° + i\sin 25°); 7(\cos 205° + i\sin 205°)$ **77.** $5(\cos 55° + i\sin 55°); 5(\cos 175° + i\sin 175°); 5(\cos 295° + i\sin 295°)$

78. $\sqrt{3} + i; -1 + \sqrt{3}i; -\sqrt{3} - i; 1 - \sqrt{3}i$ **79.** $\sqrt{3} + i; -\sqrt{3} + i; -2i$

80. $\dfrac{1}{2} + \dfrac{\sqrt{3}}{2}i; -1; \dfrac{1}{2} - \dfrac{\sqrt{3}}{2}i$ **81.** $\dfrac{\sqrt[5]{8}}{2} + \dfrac{\sqrt[5]{8}}{2}i; \approx -0.49 + 0.95i; \approx -1.06 - 0.17i; \approx -0.17 - 1.06i; \approx 0.95 - 0.49i$

82.

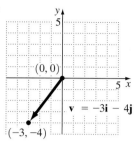

; 5

83.

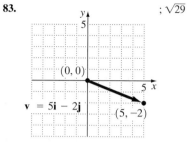

; $\sqrt{29}$

84.

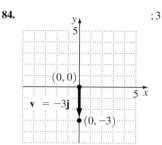

; 3

85. $3\mathbf{i} - 2\mathbf{j}$ **86.** $\mathbf{i} - 2\mathbf{j}$ **87.** $-\mathbf{i} + 2\mathbf{j}$ **88.** $-3\mathbf{i} + 12\mathbf{j}$ **89.** $12\mathbf{i} - 51\mathbf{j}$ **90.** $2\sqrt{26}$ **91.** $\dfrac{4}{5}\mathbf{i} - \dfrac{3}{5}\mathbf{j}$ **92.** $-\dfrac{1}{\sqrt{5}}\mathbf{i} + \dfrac{2}{\sqrt{5}}\mathbf{j}$

93. $6\mathbf{i} + 6\sqrt{3}\mathbf{j}$ **94.** 270 lb; 27.7°   **95. a.** $13.59\mathbf{i} + 6.34\mathbf{j}$ **b.** 14.0 mph   **c.** 13.9°   **96.** 4   **97.** 2; 86.1°   **98.** −32; 124.8°

99. 1; 71.6°   **100.** orthogonal   **101.** not orthogonal   **102.** $\mathbf{v}_1 = \text{proj}_\mathbf{w}\mathbf{v} = \dfrac{50}{41}\mathbf{i} + \dfrac{40}{41}\mathbf{j}$; $\mathbf{v}_2 = -\dfrac{132}{41}\mathbf{i} + \dfrac{165}{41}\mathbf{j}$

103. $\mathbf{v}_1 = \text{proj}_\mathbf{w}\mathbf{v} = -\dfrac{3}{2}\mathbf{i} + \dfrac{1}{2}\mathbf{j}$; $\mathbf{v}_2 = \dfrac{1}{2}\mathbf{i} + \dfrac{3}{2}\mathbf{j}$ **104.** 1115 ft-lb   **105.** Answers may vary.

Chapter 6 Test

1. 8.0   **2.** 6.2   **3.** 206 sq in.

4.

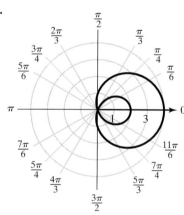

; Ordered pairs may vary.   **5.** $\left(\sqrt{2}, \dfrac{7\pi}{4}\right)$ **6.** $r^2 = 6r\cos\theta$ or $r = 6\cos\theta$ **7.** $y = 4$

8.

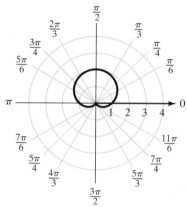

9.

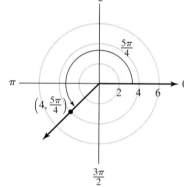

10. $2(\cos 150° + i\sin 150°)$ or $2\left(\cos\dfrac{5\pi}{6} + i\sin\dfrac{5\pi}{6}\right)$

11. $50(\cos 20° + i\sin 20°)$

12. $\dfrac{1}{2}\left(\cos\dfrac{\pi}{6} + i\sin\dfrac{\pi}{6}\right)$

13. $32(\cos 50° + i\sin 50°)$

14. $3; -\dfrac{3}{2} + \dfrac{\sqrt{3}}{2}i; -\dfrac{3}{2} - \dfrac{\sqrt{3}}{2}i$

15. a. $\mathbf{i} + 2\mathbf{j}$ **b.** $\sqrt{5}$

16. $-23\mathbf{i} + 22\mathbf{j}$ **17.** −18   **18.** 138°

19. $-\dfrac{9}{5}\mathbf{i} + \dfrac{18}{5}\mathbf{j}$ **20.** 1.0 mi

21. 323 pounds; 3.4°   **22.** 1966 ft-lb

Cumulative Review Exercises (Chapter P–6)

1. $\{-1, 2, i, -i\}$ **2.** $\dfrac{\pi}{6}, \dfrac{5\pi}{6}, \dfrac{\pi}{2}$ **3.** $\{x \mid x < -4 \text{ or } x > 2\}$ **4.** $\dfrac{3\pi}{4}, \dfrac{7\pi}{4}$

5.
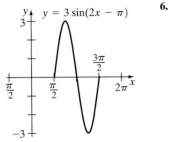

6.
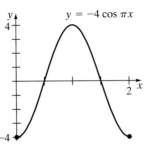

7. $\sin\theta \csc\theta - \cos^2\theta = \sin\theta\left(\dfrac{1}{\sin\theta}\right) - \cos^2\theta$

$= 1 - \cos^2\theta = \sin^2\theta$

8. $\cos\left(\theta + \dfrac{3\pi}{2}\right) = \cos\theta\cos\dfrac{3\pi}{2} - \sin\theta\sin\dfrac{3\pi}{2}$

$= \cos\theta(0) - \sin\theta(-1) = \sin\theta$

9. slope is $-\dfrac{1}{2}$; y-intercept is 2.

10. 0 **11.** $\dfrac{\sqrt{5}}{5}$ **12.** $\{x \mid x \le 5\}$

13. $\{x \mid x \ne 3, x \ne -3\}$ **14.** 1.5 sec; 44 ft **15. a.** 4 m **b.** $\dfrac{5}{2\pi}$ **c.** $\dfrac{2\pi}{5}$ sec **16.** $\dfrac{\sqrt{\sqrt{2}+2}}{2}$ **17. a.** $5\mathbf{i} + 23\mathbf{j}$ **b.** -12

18. $\log_b \dfrac{\sqrt{x}}{x^2 + 1}$ **19.** $y = -\dfrac{1}{2}x + 1$ **20. a.** 0.014 **b.** 73 words **c.** about 144 min

Subject Index

Photo Credits

Table of Contents cmcd/Getty Images, Inc., Chris Salvo/Getty Images, Inc., Tim Davis/Getty Images, Inc.-Stone, Anthony Neste, Peter Langone/ImageState/International Stock Photography Ltd., Charlie Neibergall/AP/Wide World Photos, Jim Cummins/Getty Images, Inc., TravelPix/Getty Images, Inc., AP/Wide World Photos, SuperStock, Inc., REUTERS/Colin Braley/CORBIS, Reed Saxon/AP/Wide World Photos

About the Author Robert F. Blitzer

CHAPTER P cmcd/Getty Images, Inc. **p2** SuperStock, Inc. **p13** Tom Stewart/Corbis/Stock Market **p22** William Sallaz/Duomo Photography Incorporated **p33** PEANUTS reprinted by permission of United Feature Syndicate, Inc. **p34** Purdue News Service/AP/Wide World Photos **p44** Gabe Palmer/Mug Shots/Corbis/Stock Market **p55** Telegraph Colour Library/Getty Images, Inc. **p70** A. Ramu/Stock Boston **p72** Squeak Carnwath "Equations" 1981, oil on cotton canvas 96 in. h × 72 in. w. **p84** Paramount/Getty Images, Inc.-Hulton Archive Photos **p92** CORBIS BETTMANN **p99** Stock Image/ImageState/International Stock Photography Ltd.

CHAPTER 1 Chris Salvo/Getty Images, Inc. **p116** Paramount/Shooting Star International Photo Agency **p124** David Schmidt/Masterfile Corporation **p125** Carol Simowitz/San Francisco Convention and Visitors Bureau **p142** Skip Moody/Dembinsky Photo Associates **p150** Photofest **p163** Hans Neleman/Getty Images, Inc. **p184** Douglas Kirkland/Corbis/Sygma **p189** SuperStock, Inc. **p197** Graham French/Masterfile Corporation **p209** Reuters/Fred Prouser/Getty Images, Inc.-Hulton Archive Photos **p219** Todd A. Gipstein/CORBIS

CHAPTER 2 Tim Davis/Getty Images, Inc.-Stone **p246** ©2001 Roz Chast from Cartoonbank.com. All rights reserved. **p249** Stamp from the private collection of Professor C. M. Lang, photography by Gary J. Shulfer, University of Wisconsin, Stevens Point. "Germany: #5"; Scott Standard Postage Stamp Catalogue, Scott Pub. Co., Sidney, Ohio **p253** Mary M. Steinbacher/PhotoEdit **p254** Simon Bruty/Getty Images, Inc.-Allsport; Copyright ©1981 Scott Kim, scottkim.com. All rights reserved. **p269** NIBSC/Science Photo Library/Photo Researchers, Inc. **p282** Townsend P. Dickinson/The Image Works **p283** Barbara Penoyar/Getty Images, Inc. **p292** From Hieraymi Candi, "Ni Mediolanensis Medio (Basel, 1554)," Cardano portrait, frontispiece. Smithsonian Institution Libraries. ©2001 Smithsonian Institute. **p300** Library of Congress **p305** Bob Daemmrich/Stock Boston **p326** Tom Sanders/ImageState/International Stock Photography Ltd. **p336** Stephen Simpson/Getty Images, Inc.-Taxi **p343** David Madison/Duomo Photography Incorporated **p344** Duomo Photography Incorporated **p345** UPI/CORBIS

CHAPTER 3 Anthony Neste **p356** Bruce Weaver/AP/Wide Word Photos **p367** David Weintraub/Photo Researchers, Inc. **p380** Ron Chapple/Getty Images, Inc.-Taxi **p389** KEN CHERNUS/Getty Images, Inc.-Taxi **p400** Bullit Marquez/AP/Wide World Photos **p404** Jean-Marie Chauvet/Corbis/Sygma

CHAPTER 4 Peter Langone/ImageState/International Stock Photography Ltd. **p423** San Francisco MOMA/Olivier Laude/Getty Images, Inc.-Liaison **p432** Pictor/Image State/International Stock Photography Ltd. **p437** Gary Kufner/Corbis Sharpshooters **p453** Chris Noble/Getty Images, Inc.-Stone **p461** Janet Foster/Masterfile Corporation **p463** Hugh Sitton/Getty Images, Inc.-Stone **p464** Nature Source/Raphael Macia/Photo Researchers, Inc. **p475** Jon Feingersh/Corbis/Stock Market **p497** Mehau Kulyk/Science Photo Library/Photo Researchers, Inc. **p509** Warner Bros./Picture Desk, Inc./Kobal Collection **p525** Cameramann/The Image Works **p533** Paul Sakuma/AP/Wide World Photos

CHAPTER 5 Charlie Neibergall/AP/Wide World Photos **p545** Jay Brousseau/Getty Images, Inc. **p555** David Simson/Stock Boston; Charles Gupton/Getty Images, Inc.-Stone **p565** Joe McBride/Getty Images, Inc.-Stone **p576** MPI/Getty Images, Inc.-Hulton Archive Photos; Tropical Press Agency/Getty Images, Inc.-Hulton Archive Photos **p584** Color-Pic, Inc. **p585** Doug Armand/Getty Images, Inc.–Stone

CHAPTER 6 Jim Cummins/Getty Images, Inc. **p599** Frank Clarkson/Getty Images, Inc.-Liaison **p611** Bob Daemmrich/Stock Boston **p630** Reuters/Getty Images, Inc.-Hulton Archive Photos **p642** R.F. Voss "29-Fold M-set Seahorse" computer-generated image. ©1990 R.F. Voss/IBM Research. **p655** Tom Cogill/Stephen T. Thornton; Jeff Greenberg/Photo Researchers, Inc. **p657** Tom Cogill/Stephen T. Thornton **p669** Karl Heinz Kreifelts/AP/Wide World Photos **p675** Gregory K. Scott/Photo Researchers, Inc. **p683** Robert E. Daemmrich/Getty Images, Inc.-Stone; Mimmo Jodice/CORBIS

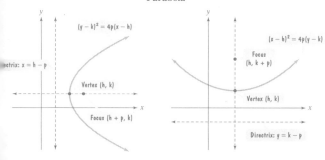

Parabola

SEQUENCES

1. Infinite Sequence:

$$\{a_n\} = a_1, a_2, a_3, \ldots, a_n, \ldots$$

2. Summation Notation:

$$\sum_{i=1}^{n} a_i = a_1 + a_2 + a_3 + \cdots + a_n$$

3. nth Term of an Arithmetic Sequence:

$$a_n = a_1 + (n - 1)d$$

4. Sum of First n Terms of an Arithmetic Sequence:

$$S_n = \frac{n}{2}(a_1 + a_n)$$

5. nth Term of a Geometric Sequence: $a_n = a_1 r^{n-1}$
6. Sum of First n Terms of a Geometric Sequence:

$$S_n = \frac{a_1(1 - r^n)}{1 - r} \ (r \neq 1)$$

7. Sum of an Infinite Geometric Series with $|r| < 1$:

$$S = \frac{a_1}{1 - r}$$

THE BINOMIAL THEOREM

1. $n! = n(n - 1)(n - 2)\cdots 3 \cdot 2 \cdot 1; 0! = 1$

2. $\displaystyle \binom{n}{r} = \frac{n!}{r!\,(n - r)!}$

3. Binomial theorem:

$$(a + b)^n = \binom{n}{0}a^n + \binom{n}{1}a^{n-1}b$$
$$+ \binom{n}{2}a^{n-2}b^2 + \cdots + \binom{n}{n}b^n$$

PERMUTATIONS, COMBINATIONS, AND PROBABILITY

1. $_nP_r$, the number of permutations of n elements taken r at a time, is given by

$$_nP_r = \frac{n!}{(n - r)!}.$$

2. $_nC_r$, the number of combinations of n elements taken r at a time, is given by

$$_nC_r = \frac{n!}{(n - r)!\,r!}.$$

3. *Probability of an Event:* $P(E) = \dfrac{n(E)}{n(S)}$, where

$n(E) = $ the number of outcomes in event E and
$n(S) = $ the number of outcomes in the sample space.

UNIT CIRCLE DEFINITIONS OF TRIGONOMETRIC FUNCTIONS

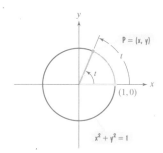

$$\sin t = y \qquad \cos t = x \qquad \tan t = \frac{y}{x}, x \neq 0$$

$$\csc t = \frac{1}{y}, y \neq 0 \quad \sec t = \frac{1}{x}, x \neq 0 \quad \cot t = \frac{x}{y}, y \neq 0.$$

RIGHT TRIANGLE DEFINITIONS OF TRIGONOMETRIC FUNCTIONS

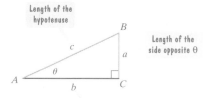

$$\sin \theta = \frac{\text{opp.}}{\text{hyp.}} \qquad \csc \theta = \frac{\text{hyp.}}{\text{opp.}}$$

$$\cos \theta = \frac{\text{adj.}}{\text{hyp.}} \qquad \sec \theta = \frac{\text{hyp.}}{\text{adj.}}$$

$$\tan \theta = \frac{\text{opp.}}{\text{adj.}} \qquad \cot \theta = \frac{\text{adj.}}{\text{opp.}}$$

TRIGONOMETRIC FUNCTIONS OF ANY ANGLE

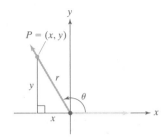

$$\sin \theta = \frac{y}{r} \qquad \csc \theta = \frac{r}{y}, \quad y \neq 0$$

$$\cos \theta = \frac{x}{r} \qquad \sec \theta = \frac{r}{x}, \quad x \neq 0$$

$$\tan \theta = \frac{y}{x}, \quad x \neq 0 \qquad \cot \theta = \frac{x}{y}, \quad y \neq 0$$

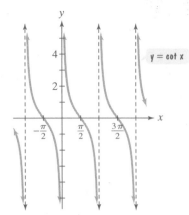

Domain: all real numbers except integral multiples of π
Range: all real numbers
Period: π

GRAPHS OF TRIGONOMETRIC FUNCTIONS

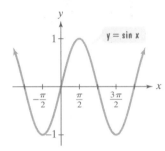

Domain: all real numbers
Range: $[-1, 1]$
Period: 2π

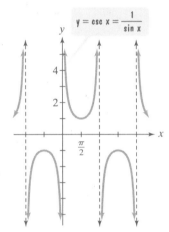

Domain: all real numbers except integral multiples of π
Range: $(-\infty, -1]$ or $[1, \infty)$
Period: 2π

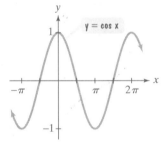

Domain: all real numbers
Range: $[-1, 1]$
Period: 2π

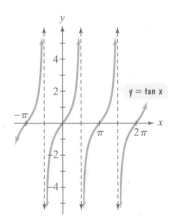

Domain: all real numbers
except odd multiples of $\dfrac{\pi}{2}$
Range: all real numbers
Period: π

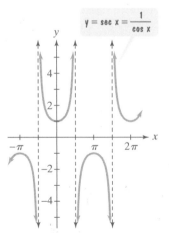

Domain: all real numbers except odd multiples of $\dfrac{\pi}{2}$
Range: $(-\infty, -1]$ or $[1, \infty)$
Period: 2π